Stereoselective Biocatalysis

edited by
Ramesh N. Patel
Bristol-Myers Squibb Pharmaceutical Research Institute
New Brunswick, New Jersey

MARCEL DEKKER, INC.

NEW YORK • BASEL

ISBN: 0-8247-8282-8

This book is printed on acid-free paper.

Headquarters
Marcel Dekker, Inc.
270 Madison Avenue, New York, NY 10016
tel: 212-696-9000; fax: 212-685-4540

Eastern Hemisphere Distribution
Marcel Dekker AG
Hutgasse 4, Postfach 812, CH-4001 Basel, Switzerland
tel: 41-61-261-8482; fax: 41-61-261-8896

World Wide Web
http://www.dekker.com

The publisher offers discounts on this book when ordered in bulk quantities. For more information, write to Special Sales/Professional Marketing at the headquarters address above.

Current printing (last digit):
10 9 8 7 6 5 4 3 2 1

PRINTED IN THE UNITED STATES OF AMERICA

Preface

Currently, research is focused on the interaction of small molecules with biological macromolecules. The search for selective enzyme inhibitors and receptor agonists/antagonists is essential for target-oriented research in the pharmaceutical, agrochemical, and food industries. Increased understanding of mechanisms of drug interactions on a molecular level has led to strong awareness of the importance of chirality as the key to the efficacy of many drug products and agrochemicals. The production of optically active chiral intermediates is a subject of increasing importance in the pharmaceutical, agricultural, and food industries. Increased regulatory pressure to market homochiral drugs by the U.S. Food and Drug Administration has led to the use of alternative approaches, including biocatalysis for the synthesis of chiral compounds.

Organic synthesis has been one of the most successful scientific disciplines and has enormous practical utility. Organic synthesis has made advances and developed practical processes in the synthesis of natural products, drugs, agricultural chemicals, polymers, and many classes of functional molecules.

One can ask the question, then, why biocatalysis? What does biocatalysis have to offer to synthetic organic chemists? Biocatalysis gives an added dimension, innovative new approaches, and enormous opportunity to prepare industrially useful chiral compounds. The advantages of biocatalysis over chemical catalysis are that enzyme-catalyzed reactions are stereoselective and regioselective and can be carried out at ambient temperature, at atmospheric pressure, and under environmentally friendly conditions. Biocatalysis minimizes the problems of isomerization, racemization, epimerization, and rearrangement of molecules that may occur during chemical processes.

Microorganisms demonstrate how complex compounds can be assembled in an efficient manner by biological systems making use of enzymes as catalysts for a variety of biotransformations. Microbial cells and enzymes can be immobilized and reused for many cycles. In addition, enzymes can be overexpressed to make biocatalytic processes economically efficient and inexpensive. Tailor-made enzymes by random and site-directed mutagenesis with modified activity, and preparation of thermostable and pH-stable enzymes will lead to the production of novel stereoselective biocatalysts. The use of enzymes in organic solvents has led to thousands of publications on enzyme-catalyzed asymmetric synthesis and resolution processes. Molecular recognition and selective catalysis are key chemical processes in life which are embodied in biocatalysts. As a consequence, the

iii

application of biocatalysts in organic synthesis has become a powerful method for conducting stereoselective and regioselective catalysis.

This book examines the use of different classes of enzymes in catalysis of different types of chemical reactions to generate chiral molecules useful in chemo-enzymatic synthesis of pharmaceutical, agricultural, and food products. Chapters are written by world-renowned scientists who are experts in their areas.

Topics include the use of hydrolytic enzymes such as lipases, esterases, proteases, dehalogenases, acylases, amidases, nitrilases, lyases, and hydantoinases in resolution of racemic compounds and in asymmetric synthesis of optically active compounds. Chapters cover the use of oxido-reductases in the synthesis of chiral alcohols, aminoalcohols, and amines; aldolase- and decarboxylase-catalyzed aldol- and acyloin-condensation reactions; oxygenases such as monooxygenases in stereoselective and regioselective hydroxylation and epoxidation reactions; dioxygenases in the chemo-enzymatic synthesis of chiral diols; stereoselective syntheses using epoxide hydrolases; microbial/enzyme-catalyzed Baeyer-Villiger reactions; enantioselective formation of chiral cyanohydrins; hydroxynitrile lyases in stereoselective synthesis; production of chiral β-hydroxy acids; and applications in organic syntheses.

Chapters are also included on yeast-mediated stereoselective biocatalysis, stereoselective synthesis of steroids, chemo-enzymatic synthesis of enantiopure arylpropionic acids, supercritical carbon dioxide as a solvent in enzyme catalysis, state-of-the-art techniques in enzyme immobilization, biocatalysis by polyethylene glycol-modified enzymes, and enzymatic deprotection techniques in organic synthesis.

All the information cited in this book provides state-of the art knowledge with more than 4,000 references and will improve the ability of the reader to use different types of enzymatic reactions in synthesis of a variety of chiral molecules. Organic chemists, chemical engineers, biochemists, microbiologists, and medicinal chemists should have knowledge of the use of enzymes in organic synthesis. This book provides such information.

It is my pleasure to acknowledge sincere appreciation to all the authors for their contribution to this book. I would like to acknowledge continual support from Joseph Stubenrauch (Production Editor) and Anita Lekhwani (Acquisitions Editor) of Marcel Dekker, Inc. My interest in biocatalysis was developed and stimulated by Drs. David Gibson, Derek Hoare, Nicholas Ornston, Allen Laskin, Ching Hou, Laszlo Szarka, Christopher Cimarusti, and John Scott and my colleagues at the University of Texas, Yale University, Exxon Research & Engineering, and Bristol-Myers Squibb. I acknowledge their support and encouragement over the years. Finally, I would like to express my sincere thanks to my wife, Lekha, and my daughter, Sapana, for their encouragement and patience while I worked on this book.

Ramesh N. Patel

Contents

Contributors

Wolf-Rainer Abraham Department of Microbiology, GBF–National Research Centre for Biotechnology, Braunschweig, Germany

Andrés-Rafael Alcántara Department of Organic and Pharmaceutical Chemistry, Universidad Complutense de Madrid, Madrid, Spain

Robert Azerad Laboratoire de Chimie et Biochimie Pharmacologiques et Toxicologiques, Université René Descartes-Paris V, Paris, France

Mark V. Baev Microbiological Biotechnology Laboratory, Department of Biology, University of Waterloo, Waterloo, Ontario, Canada

Antonio Ballesteros Department of Biocatalysis, Institute of Catalysis, Consejo Superior de Investigaciones Científicas, Madrid, Spain

Amit Banerjee Bioprocess Science, Searle, St. Louis, Missouri

Per Berglund Department of Biotechnology, Royal Institute of Technology, Stockholm, Sweden

Wilhelmus H. J. Boesten Organic Chemistry and Biotechnology Section, DSM Research, Geleen, The Netherlands

Rob C. Brown Biocatalysis Group, Chirotech Technology Ltd., Cambridge, England

Quirinus B. Broxterman Organic Chemistry and Biotechnology Section, DSM Research, Geleen, The Netherlands

Johannes Brussee Leiden Institute of Chemistry, Leiden University, Leiden, The Netherlands

Didier Combes Centre de BioIngénierie Gilbert Durand, INSA, Toulouse, France

René Csuk   Institute of Organic Chemistry, Martin Luther University Halle-Wittenberg, Halle (Saale), Germany

Paola D'Arrigo   Department of Chemistry, Politecnico di Milano, Milano, Italy

Franz Effenberger   Institute für Organische Chemie, University of Stuttgart, Stuttgart, Germany

Kurt Faber   Institute of Organic Chemistry, University of Graz, Graz, Austria

Patrizia Ferraboschi   Department of Medical Chemistry and Biochemistry, University of Milan, Milan, Italy

Miguel Ferrero   Departamento de Química Orgánica e Inorgánica, Universidad de Oviedo, Oviedo, Spain

Wolf-Dieter Fessner   Department of Organic Chemistry, Darmstadt University of Technology, Darmstadt, Germany

Fernando Formaggio   Department of Organic Chemistry, University of Padua, Padua, Italy

Iqbal Singh Gill   Biotransformations Section, Biotechnology Center of Excellence, Roche Vitamins Inc., Nutley, New Jersey

Brigitte I. Glänzer   Institute of Organic Chemistry, Martin Luther University Halle-Wittenberg, Halle (Saale), Germany

Vicente Gotor   Departamento de Química Orgánica e Inorgánica, Universidad de Oviedo, Oviedo, Spain

Thorsten Hartmann   Institute of Technical Chemistry, University of Hannover, Hannover, Germany

Junzo Hasegawa   Fine Chemical Research Laboratories, Kaneka Corporation, Hyogo, Japan

Misao Hiroto   Department of BioMedical Engineering, Toin University of Yokohama, Yokohama, Japan

Herbert L. Holland   Department of Chemistry, Brock University, St. Catharines, Ontario, Canada

Karen Elizabeth Holt   Biocatalysis Group, Chirotech Technology Ltd., Cambridge, England

Karl Hult   Department of Biotechnology, Royal Institute of Technology, Stockholm, Sweden

Yuji Inada Toin Human Science and Technology Center, Toin University of Yokohama, Yokohama, Japan

Johan Kamphuis DSM Specialty Intermediates, Sittard, The Netherlands

Bernard Kaptein Organic Chemistry and Biotechnology Section, DSM Research, Geleen, The Netherlands

P. A. Keene Biocatalysis Group, Chirotech Technology Ltd., Cambridge, England

Yoh Kodera Department of BioMedical Engineering, Toin University of Yokohama, Yokohama, Japan

U. Kragl* Institut für Biotechnologie 2, Forschungszentrum Jülich, Jülich, Germany

Wolfgang Kroutil Institute of Organic Chemistry, University of Graz, Graz, Austria

M.-R. Kula Institut für Enzymtechnologie der Heinrich-Heine-Universität Düsseldorf im Forschungszentrum Jülich, Jülich, Germany

Ayako Matsushima Department of BioMedical Engineering, Toin University of Yokohama, Yokohama, Japan

Kenji Mori Department of Chemistry, Science University of Tokyo, Tokyo, Japan

Toru Nagasawa Department of Biomolecular Science, Gifu University, Gifu, Japan

Nobuo Nagashima Fine Chemical Research Laboratories, Kaneka Corporation, Hyogo, Japan

Hiroyuki Nishimura Toin Human Science and Technology Center, Department of BioMedical Engineering, Toin University of Yokohama, Yokohama, Japan

Jun Ogawa Division of Applied Life Sciences, Graduate School of Agriculture, Kyoto University, Kyoto, Japan

Toshihisa Ohshima Department of Biological Science and Technology, University of Tokushima, Tokushima, Japan

Hiromichi Ohta Department of Chemistry, Keio University, Yokohama, Japan

Ramesh N. Patel Enzyme Technology, Process Research, Bristol-Myers Squibb Pharmaceutical Research Institute, New Brunswick, New Jersey

Tanmaya Pathak Organic Chemistry Division (Synthesis), National Chemical Laboratory, Pune, Maharasthra, India

---

Current affiliation: Lehrstuhl für Technische Chemie, Universität Rostock, Rostock, Germany.

Giuseppe Pedrocchi-Fantoni   Department of Chemistry, Politecnico di Milano, Milano, Italy

Francisco J. Plou   Department of Biocatalysis, Institute of Catalysis, Consejo Superior de Investigaciones Cientificas, Madrid, Spain

Shahrzad Reza-Elahi   Department of Medical Chemistry and Biochemistry, University of Milan, Milan, Italy

Floris P. J. T. Rutjes   Institute of Molecular Chemistry, University of Amsterdam, Amsterdam, The Netherlands

José-María Sánchez-Montero   Department of Organic and Pharmaceutical Chemistry, Universidad Complutense de Madrid, Madrid, Spain

Enzo Santaniello   Department of Medical Chemistry and Biochemistry, University of Milan, Milan, Italy

Thomas Scheper   Institute of Technical Chemistry, University of Hannover, Hannover, Germany

Hans E. Schoemaker   Organic Chemistry and Biotechnology Section, DSM Research, Geleen, The Netherlands

Eckhard Schwabe   Institute of Technical Chemistry, University of Hannover, Hannover, Germany

Nobufusa Serizawa   Biomedical Research Laboratories, Sankyo Company, Ltd., Tokyo, Japan

Stefano Servi   Department of Chemistry, Politecnico di Milano, Milano, Italy

Sakayu Shimizu   Division of Applied Life Sciences, Graduate School of Agriculture, Kyoto University, Kyoto, Japan

José-Vicente Sinisterra   Department of Organic and Pharmaceutical Chemistry, Universidad Complutense de Madrid, Madrid, Spain

Kenji Soda   Kansai University, Osaka-fu, Japan

Theo Sonke   Organic Chemistry and Biotechnology Section, DSM Research, Geleen, The Netherlands

Takeshi Sugai   Department of Chemistry, Keio University, Yokohama, Japan

Stephen J. C. Taylor   Biocatalysis Group, Chirotech Technology Ltd., Cambridge, England

Ian Nicholas Taylor   Biocatalysis Group, Chirotech Technology Ltd., Cambridge, England

Claudio Toniolo   Department of Organic Chemistry, University of Padua, Padua, Italy

Arne van der Gen   Leiden Institute of Chemistry, Leiden University, Leiden, The Netherlands

Herbert Waldmann   Max-Planck-Institute of Molecular Physiology and University of Dortmund, Dortmund, Germany

Owen P. Ward   Department of Biology, University of Waterloo, Waterloo, Ontario, Canada

Marco Wieser   Department of Biomolecular Science, Gifu University, Gifu, Japan

1

Stereoselective Synthesis Using Hydantoinases and Carbamoylases

Jun Ogawa and Sakayu Shimizu
Kyoto University, Kyoto, Japan

I. INTRODUCTION: AN OUTLINE OF HYDANTOINASE PROCESS

Optically pure D- and L-α-amino acids are of increasing interest as precursors for semi-synthetic antibiotics, new herbicides, insecticides, and physiologically active peptides, and as chiral building blocks for chemical synthesis. For the production of optically active α-amino acids, enzymatic processes have been developed other than fermentation to cover a wide range of products, not only natural L-α-amino acids but also unnatural and D-α-amino acids (Table 1).

Hydantoinase process, outlined in Fig. 1, includes two hydrolases—hydantoin-hydrolyzing enzyme (hydantoinase) and N-carbamoyl amino acid–hydrolyzing enzyme (carbamoylase)—and is one of the most efficient and versatile methods for the production of optically active α-amino acids. DL-5-Monosubstituted hydantoins, which are used as common precursors for the chemical synthesis of DL-α-amino acids [1], are the starting material of this enzymatic process. Keto-enol tautomerism is a typical feature of the hydantoin structure. Under neutral conditions, the keto form is dominant; in alkaline solution, enolization between the 4 and 5 positions can occur, as has been concluded from the fact that optically pure hydantoins readily racemize. This feature is of practical relevance for the complete conversion of racemic hydantoin derivatives to optically pure L- or D-α-amino acids without any chemical racemization step. A variety of hydantoinase and carbamoylase with different stereospecificity were found. They are D-specific hydantoinase (D-hydantoinase), L-specific hydantoinase (L-hydantoinase), none-specific hydantoinase (DL-hydantoinase), D-specific carbamoylase (D-carbamoylase), and L-specific carbamoylase (L-carbamoylase). With the combination of these enzymes, optically pure amino acids are obtained from DL-5-monosubstituted hydantoins (Fig. 2). The wide substrate range of hydantoinases and carbamoylases also gives generality to the hydantoinase process.

This chapter provides a comprehensive review of the process development and recent results of biochemical study, along with technical application of enzymatic hydantoin transformation.

Table 1 Enzymatic Process for Optically Pure Amino Acid Production

Process	Enzyme	Substrate	Enantiomer		Product example
			D	L	
Hydantoin hydrolysis	Hydantoinase with or without carbamoylase	DL-5-Monosubstituted hydantoins	+	+	D-p-Hydroxyphenylglycine
					D-Phenylglycine
					D-Valine
					L-Tryptophan
Racemate separation	Acylase	DL-N-Acetyl amino acids	+	+	L-Methionine
	Esterase	DL-Amino acid esters	+	+	D-Tryptophan
	Aminopeptidase	DL-Acetamido acid esters	+	+	D-Valine
	Amidase	DL-Amino acid amides	+	+	D- or L-Tryptophan
Lyases	Aspartase	Fumaric acid	−	+	L-Aspartate
	Tryptophan synthase	L-Serine, indole	−	+	L-Tryptophan
		L-Serine, selenol	−	+	L-Selenocysteine
	Tryptophanase	Pyruvate, indole, ammonia	−	+	L-Tryptophan
	β-Tyrosinase	Pyruvate, phenol, ammonia	−	+	L-Tyrosine
		Pyruvate, pyrocatechol, ammonia	−	+	L-DOPA
Reductive amination (NADH-dependent)	Dehydrogenases	α-Keto acids	+	+	L-Phenylalanine
					L-tert-Leucine
Transamination (PLP-dependent)	Transaminases	α-Keto acids	+	+	L-Phenylalanine
					D-p-Hydroxyphenylglycine

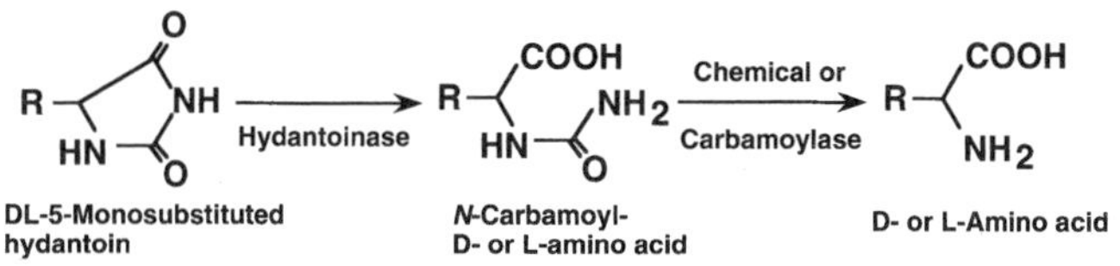

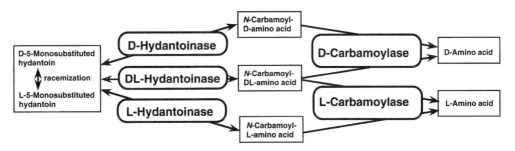

Figure 1 Outline of hydantoinase process for the production of optically pure amino acids.

II. HISTORICAL BACKGROUND OF THE PROCESS DEVELOPMENT

A biotechnological application of hydantoinase process became interesting not before the 1970s together with the growing interest of industry in producing D-α-amino acids as side chains for semisynthetic penicillins and cepharosporins. Based on the investigations of Dudley et al. [2,3] into the metabolism of N-substituted DL-5-phenylhydantoins, which were postulated to be D-stereospecific, Cecere et al. [4] found in 1975 that dihydropyrimidinase from calf liver could be used to produce several N-carbamoyl-D-amino acids from the corresponding DL-5-monosubstituted hydantoins. In 1978, Yamada and co-workers showed that microbial cells are good catalysts for stereospecific cleavage of DL-5-monosubstituted hydantoins to N-carbamoyl-D-amino acids [5–9]. Although only hydantoin-hydrolyzing activity but not N-carbamoyl amino acid–hydrolyzing activity was found in many bacteria, several bacteria that could hydrolyze N-carbamoyl-D-amino acids to D-amino acids were found thereafter [10]. With these microorganisms, enzymatic production of D-amino acids from DL-5-monosubstituted hydantoins were established.

Similar reactions transform DL-5-monosubstituted hydantoins to L-amino acids. Yokozeki et al. [11] analyzed the reaction mechanism of L-tryptophan production from DL-5-(3′-indolylmethyl)hydantoin by *Flavobacterium* sp. and found that nonstereospecific hydrolysis of hydantoin derivative and the L-specific hydrolysis of racemic N-carbamoyl amino acid were involved in this transformation. Yamashiro et al. [12] found the L-specific hydrolysis of a hydantoin derivative in L-valine production from DL-5-isopropylhydantoin by *Bacillus* sp. and that ATP was absolutely required for this amidohydrolytic reaction.

III. INDUSTRIAL APPLICATION OF HYDANTOINASE PROCESS

A. First Generation of D-*p*-Hydroxyphenylglycine Production

D-*p*-Hydroxyphenylglycine and its derivatives are important as side chain precursors for semisynthetic penicillins and cepharosporines. Yamada and co-workers found that these

Figure 2 Enzymatic reactions involved in hydantoinase process.

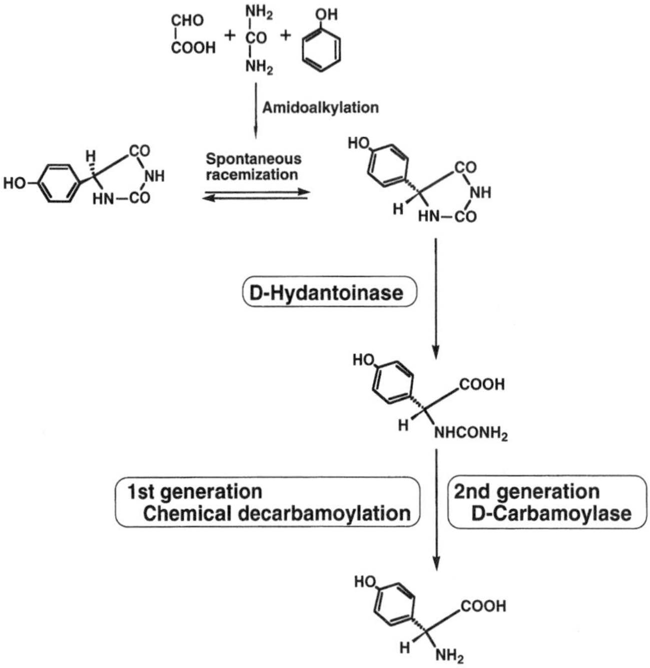

Figure 3 Industrial process for D-*p*-hydroxyphenylglycine production.

amino acids can be efficiently prepared from the corresponding 5-monosubstituted hydantoins using the microbial enzyme D-hydantoinase [5]. Interestingly, the enzyme attacked a variety of aliphatic and aromatic D-5-monosubstituted hydantoins, yielding the corresponding D form of *N*-carbamoyl-α-amino acids. Thus the enzyme can be used for the preparation of various D-amino acids.

Based on these findings, Kaneka Corporation established industrial production of D-*p*-hydroxyphenylglycine by hydantoinase process in 1979. Initially, the process involves two chemical steps and one enzymatic step [7] (Fig. 3). The substrate, DL-5-(*p*-hydroxyphenyl)hydantoin, is synthesized through an efficient chemical method involving the amidoalkylation reaction of phenol with glyoxylic acid and urea under acidic conditions. Then, the D-5-(*p*-hydroxyphenyl)hydantoin is hydrolyzed enzymatically to *N*-carbamoyl-D-*p*-hydroxyphenylglycine. This enzymatic process is carried out by a thermostable immobilized D-hydantoinase under alkaline conditions. Under these conditions, the L isomer of the remaining 5-(*p*-hydroxyphenyl)hydantoin is racemized through base catalysis. Therefore, the racemic hydantoins can be converted quantitatively into *N*-carbamoyl-D-*p*-hydroxyphenylglycine through this step. Decarbamoylation to D-*p*-hydroxyphenylglycine is performed by treating the *N*-carbamoyl-D-amino acid with equimolar nitrite under acidic conditions [8].

B. Second Generation of D-*p*-Hydroxyphenylglycine Production

Recently, the novel enzyme D-carbamoylase, which stereospecifically hydrolyzes *N*-carbamoyl-D-amino acids, was found in several bacteria [13,14]. Therefore, a sequence of two enzyme-catalyzed reactions, the D-stereospecific hydrolysis of DL-5-(*p*-hydroxyphenyl)hydantoin and subsequent hydrolysis of the D-carbamoyl derivative to D-*p*-hydroxyphenylglycine, is possible (Fig. 3). Kaneka Corporation conducted screening of the bacterium possessing D-carbamoylase and found the activity in *Agrobacterium* sp. KNK712 [15]. The D-carbamoylase gene derived from *Agrobacterium* sp. KNK712 was cloned, expressed in *Escherichia coli*, and mutagenized to obtain mutant genes coding more stable enzyme against environments in which enzymes are inactivated, such as temperature, pH, oxidation, etc. Through the selection of mutant enzymes based on their thermal stability, a mutant showing 20°C higher thermal stability than the parent enzyme was obtained [17,18]. Sequence analysis of the mutant enzyme revealed three amino acid replacements to the parent enzyme. This mutant D-carbamoylase with higher stability was immobilized and has been used in the second-generation commercial process for the production of D-*p*-hydroxyphenylglycine since 1995 [19].

IV. CHEMICAL BACKGROUND OF HYDANTOINS

A. Chemical Synthesis of 5-Monosubstituted Hydantoin Substrates

The representative methods for 5-monosubstituted hydantoin synthesis are summarized in Table 2 [20]. The most general method is Bucherer-Bergs synthesis from aldehydes, potassium cyanide, and ammonium carbonate under mild conditions. 5-Monosubstituted hydantoins are also synthesized by retrosynthesis with amino acids and potassium cyanate under acidic conditions. Introduction of C-5 substituents into the hydantoin ring with aldehyde or related compounds are also available, and results in exocyclically C-5 unsaturated hydantoins that are easily reduced to 5-monosubstituted hydantoins by common reducing agents. Amidoalkylation of phenol with glyoxylic acid and urea is useful for the synthesis of 5-(*p*-hydroxyphenyl)hydantoin [21]. According to this method, 5-(*p*-hydroxyphenyl)hydantoin can be readily prepared in high purity and good yield from relatively inexpensive starting materials.

Table 2 Major Processes of DL-5-Monosubstituted Hydantoin Synthesis

Reaction	Substrates	
Bucherer-Bergs synthesis	Aldehyde (ketone)	+ KCN + $(NH_4)_2CO_3$
Condensation with	α-Amino acids	+ KOCN
carboxyl compounds	α-Amino acid amides	+ KOCN
	α-Aminonitriles	+ KOCN
	α-Amino acids	+ Urea
	α-Hydroxy acids	+ Urea
	α-Hydroxynitriles	+ Urea
	α-Amino acid esters	+ Alkylchloroformates or $COCl_2$
	α-Amino amides	+ Alkylchloroformates or $COCl_2$
	α-Aminonitriles	+ Alkylchloroformates or $COCl_2$
	α-Aminonitriles	+ CO_2
Condensation with	Aldehyde or ketone	+ Hydantoin
hydantoin ring	(further reduction with some reducing agent)	

B.   Properties of Hydantoins

As described in the Introduction, spontaneous racemization under alkaline condition is a typical feature of hydantoin. The rate of spontaneous racemization is very much influenced by the electronic properties of the C-5 substituent. Substituents with electronegative inductive effect will stabilize the enolate structure because electron density at C-5 is lowered, thus favoring the release of the proton at C-5. Therefore, hydantoins carrying a carboxy group on an alkyl side chain such as 5-(2'-carboxyethyl)hydantoin, and those carrying arylalkylated or aryl side chain, will readily racemize often within minutes. On the other hand, it may take hours to racemize merely alkylated hydantoins. Hydantoins are more or less unstable in the presence of alkali, and the equilibrium shifts to the direction of ring opening hydrolysis.

V.   BIOCHEMICAL BACKGROUND OF HYDANTOINASE PROCESS

A.   Evaluation of Hydantoin- and Pyrimidine-transforming Activities in Microorganisms

Takahashi et al. [6] revealed that in *Pseudomonas putida* (= *P. striata*) IFO 12996 D-hydantoinase is identical with dihydropyrimidinase (EC 3.5.2.2), which catalyzes the cyclic ureide–hydrolyzing step of the reductive degradation of pyrimidine bases (Fig. 4). The same results were obtained for other hydantoinases from *Pseudomonas* sp. [22,23], *Comamonas* sp. [23], *Bacillus* sp. [9], *Arthrobacter* sp. [24], *Agrobacterium* sp. [22], and rat liver [25]. From these results, it is proposed that D-amino acid production from DL-5-monosubstituted hydantoins involves the action of the series of enzymes involved in the pyrimidine degradation pathway [24,26,27]. However, this contention has remained moot because of a lack of systematic studies on the enzymes involved in these transformations [28].

Ogawa et al. investigated the pyrimidine-transforming activity in typical hydantoin-transforming bacteria [29]. These bacteria showed dihydropyrimidinase and D-hydantoinase and/or β-ureidopropionase and D-carbamoylase activity. However, the induction profiles of pyrimidine-transforming activity and hydantoin-transforming activity for the several pyrimidine- and hydantoin-related compounds did not correspond with each other. Runser et al. reported the occurrence of D-hydantoinase without dihydropyrimidinase ac-

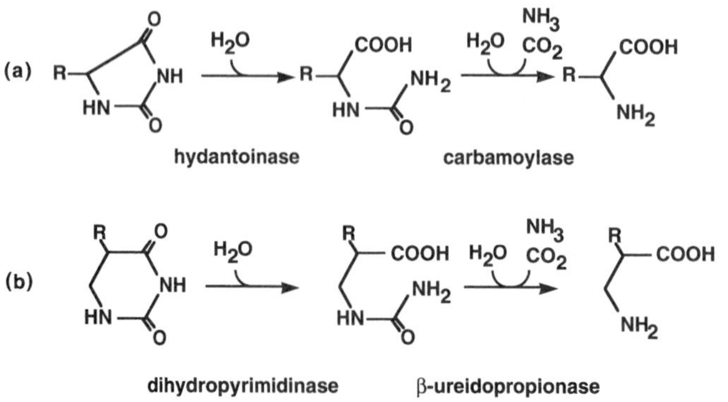

Figure 4   Hydantoin (a) and pyrimidine (b) transformation pathways.

tivity [30]. These results indicate that there are specific enzymes for hydantoin transformation not involved in pyrimidine degradation.

Although many enzymological and genetic studies have been conducted, the physiological explanation for L-specific transformation of hydantoin has not yet been established. They are specific only for L-hydantoins and *N*-carbamoyl-L-amino acids whose existence in nature is still unclear, with the exception of L-5-carboxymethylhydantoin and *N*-carbamoyl-L-aspartate.

B. Enzymes Involved in Hydantoin Hydrolysis

1. D-Hydantoinases

D-Hydantoinase was isolated from mammalian cells as well as from bacteria. The bovine liver enzyme is identical to dihydropyrimidinase and consists of four subunits [31]. It was reported to have four Zn^{2+} which are tightly bound similar to that of rat dihydropyrimidinase [32]. Bacterial D-hydantoinase was isolated from various sources such as *P. putida* (= *P. striata*) IFO 12996 [6], *Pseudomonas* sp. AJ 11220 [27], *Pseudomonas fluorescens* DSM 84 [33], *Bacillus stearothermophilus* SD1 [34], *Blastobacter* sp. A17p-4 [35], and *Arthrobacter crystallopoietes* AM2 [24]. The substrate specificities of some D-hydantoinases are summarized in Table 3. It can be seen that all bacterial D-hydantoinases other than the enzyme from *Agrobacterium* sp. IP I-671 [30] are rather similar to the mammalian dihydropyrimidinase. D-Hydantoinase from *Agrobacterium* sp. IP I-671 is specific for D-5-monosubstituted hydantoin and has no dihydropyrimidinase activity. All bacterial D-hydantoinases other than the thermostable enzyme from *B. stearothermophilus* are homotetramer with molecular masses of 190–260 kDa. The enzyme from *B. stearothermophilus* is homodimer with a molecular mass of 126 kDa. These enzymes activities require divalent cations such as Mg^{2+}, Mn^{2+}, Fe^{2+}, Co^{2+}, or Zn^{2+} for their expression of maximum activity.

2. L-Hydantoinases

The existence of L-hydantoinase might be rarer in nature than that of D-hydantoinase. It can be divided into two groups; one needs ATP for its activity and the other does not. The enzyme that needs ATP was partially purified from *Bacillus brevis* AJ-12299 [12]. The enzyme also requires Mg^{2+}, Mn^{2+}, or K^+ as cofactor for its activity. The enzyme that does not need ATP was purified from *Arthrobacter* sp. DSM 3747 [36]. The *Arthrobacter* L-hydantoinase is a homotetramer with a molecular mass of 232 kDa and showed narrow substrate specificity only toward the hydantoins that carried bulky substituents with a methylene spacer at the C-5 position. The enzyme activity was totally inhibited by the addition of EDTA and recovered by the addition of Mn^{2+} or Co^{2+} and partially by Ca^{2+}. The substrate specificities of L-hydantoinases are summarized in Table 4.

3. DL-Hydantoinases

The hydantoinases hydrolyzing both D- and L-hydantoin were also found in several bacteria. They could be divided into two groups; one needs ATP for its activity and the other does not. The ATP-requiring enzyme was purified and cloned from *Pseudomonas* sp. strain NS 671 [37,38]. The enzyme consists of two subunits with differing molecular mass of 76 and 65 kDa, and preferably hydrolyzes L-hydantoin. The enzyme that does not require ATP was purified and cloned from a moderate thermophilic bacterium *B. stearothermophilus* NS 1122A [39,40] and *Arthrobacter* sp. DSM 3745 [41]. The *Bacillus* enzyme is homotetrameric with molecular mass of 200 kDa. Although the monomer had no activity,

Table 3 Substrate Specificities of D-Hydantoinases

Substrate	Bovine liver	Pseudomonas putida IFO12996	Bacillus stearothermophilus SD1	Agrobacterium sp. IPI-671
Dihydrouracil	+	+	+	–
Hydantoin	+	+	+	+
5-Substituted hydantoins				
5-Methyl-	n.d.	+	n.d.	n.d.
5-Isopropyl-	+	+	+	+
5-Isobutyl-	+	+	n.d.	+
5-(2′-Methylthioethyl)-	+	+	n.d.	n.d.
5-Carboxymethyl-	–	n.d.	n.d.	n.d.
5-(2′-Carboxyethyl)-	–	n.d.	n.d.	n.d.
5-Phenyl-	+	+	+	n.d.
5-(p-Hydroxyphenyl)-	+	+	+	+
5-(m-Hydroxyphenyl)-	n.d.	+	n.d.	n.d.
5-(p-Methoxyphenyl)-	n.d.	+	n.d.	n.d.
5-(p-Chlorophenyl)-	n.d.	+	n.d.	n.d.
5-(m-Chlorophenyl)-	n.d.	+	n.d.	n.d.
5-(2′,4′-Dichlorophenyl)-	n.d.	+	n.d.	+
5-Benzyl-	+	n.d.	n.d.	n.d.
5-(3′-Indolylmethyl)-	–	n.d.	n.d.	n.d.
5,5-Dimethyl-	n.d.	–	n.d.	n.d.
5,5-Diphenyl-	–	–	n.d.	n.d.

+, hydrolyzed; –, not hydrolyzed; n.d., not determined.

Table 4 Substrate Specificities of L-Hydantoinases

Substrate	*Arthrobacter* sp. DSM3747	*Bacillus brevis* AJ 12299
Dihydroorotate	−	n.d.
Hydantoin	−	n.d.
5-Substituted hydantoins		
5-Methyl-	−	+
5-Isopropyl-	−	+
5-Isobutyl-	+	+
5-(2′-Methylthioethyl)-	n.d.	+
5-Carboxymethyl-	−	n.d.
5-(2′-Carboxyethyl)-	−	n.d.
5-Phenyl-	−	n.d.
5-(*p*-Hydroxyphenyl)-	−	n.d.
5-Benzyl-	+	+
5-(*p*-Hydroxybenzyl)-	+	+
5-(3′,4′-Dihydroxybenzyl)-	+	n.d.
5-Benzyloxymethyl-	+	n.d.
5-(*S*-Benzylmercaptomethyl)-	+	n.d.
5-(2′-Thienyl)	−	n.d.
5-(3′-Indolylmethyl)-	−	n.d.

+, hydrolyzed; −, not hydrolyzed; n.d., not determined.

the activity was restored by incubation with Mn^{2+} or Co^{2+}. These findings suggested that the *Bacillus* DL-hydantoinase is a metalloenzyme and the oligomeric structure supported by metal ions is required for the activity. D-Hydantoins were more effectively hydrolyzed than the L-hydantoins by the *Bacillus* DL-hydantoinase. The *Arthrobacter* DL-hydantoinase is also homotetrameric, with a molecular mass of 200 kDa, and contains 2.5 mol Zn per mol subunit, which is required both for expressing activity and retaining oligomeric structure. The stereospecificity of the *Arthrobacter* DL-hydantoinase depends on the substrate: although it is strictly L-specific for the cleavage of DL-5-(3′-indolylmethyl)hydantoin, it appears to be rather D-specific for the hydrolysis of DL-5-(2′-methylthioethyl)hydantoin. The substrate specificities of DL-hydantoinases are summarized in Table 5.

4. Other Cyclic Amide Amidohydrolases and Their Molecular Evolution

A novel enzyme imidase was found in *Blastobacter* sp. A17p-4 [42]. The imidase hydrolyzes dihydrouracil, hydantoin, and various cyclic imides very effectively (Fig. 5a). The imidase catalyzes the first reaction of the microbial cyclic imide degradation [43].

An ATP-dependent amidohydrolase *N*-methylhydantoin amidohydrolase, which catalyzes the second-step reaction in the degradation route from creatinine to glycine, via *N*-methylhydantoin, *N*-carbamoylsarcosine, and sarcosine as successive intermediates [44–52] (Fig. 5b), was found in *Pseudomonas putida* 77 [44,45]. The enzyme is inducible only with the presence of creatinine and *N*-methylhydantoin, suggesting that the role of this enzyme is in the transformation of creatinine [53]. The ATP-dependent hydrolysis of 5-monosubstituted hydantoins, e.g., 5-methylhydantoin, by the enzyme proved to be L-isomer-specific [53].

Table 5 Substrate Specificities of DL-Hydantoinases

Substrate	*Pseudomonas* sp. strain NS671	*Bacillus stearothermophilus* NS1122A	*Arthrobacter* sp. DSM 3745
Urea	n.d.	n.d.	−
Allantoin	n.d.	n.d.	−
Hydantoin	n.d.	n.d.	−
Dihydroorotate	n.d.	n.d.	−
Dihydrouracil	n.d.	n.d.	+
Dihydrothymine	n.d.	n.d.	+
5-Substituted hydantoins			
5-Methyl-	n.d.	+	+
5-Isopropyl-	+	+	+
5-Isobutyl-	+	+	n.d.
5-(1′-Methylpropyl)-	+	+	n.d.
5-(2′-Methylthioethyl)-	+	+	+
5-Carboxymethyl-	n.d.	n.d.	−
5-(2′-Carboxyethyl)-	n.d.	n.d.	−
5-Phenyl-	+	n.d.	+
5-Benzyl-	n.d.	+	+
5-(4′-Chlorobenzyl)-	n.d.	n.d.	+
5-(4′-Aminobenzyl)-	n.d.	n.d.	+
5-(4′-Fluorobenzyl)-	n.d.	n.d.	+
5-(2′-Pyridylmethyl)-	n.d.	n.d.	+
5-(4′-Nitrobenzyl)-	n.d.	n.d.	+
5-(2′-Phenylethyl)-	n.d.	n.d.	+
5-(p-Carboxybenzyl)-	n.d.	n.d.	−
5-(p-Hydroxybenzyl)-	n.d.	n.d.	−
5-(3′-Indolylmethyl)-	n.d.	n.d.	+

+, hydrolyzed; −, not hydrolyzed; n.d., not determined.

Dihydroorotase (EC 3.5.2.3) is one of the known cyclic ureide–hydrolyzing enzymes, which catalyzes the reversible cyclization of L-ureidosuccinate to dihydro-L-orotate (Fig. 5c), the third step in pyrimidine biosynthesis. Dihydroorotase from *P. putida* IFO 12996 was purified to homogeneity and characterized [54]. The enzyme only hydrolyzed dihydro-L-orotate and its methyl ester, and the reactions were reversible. Dihydroorotase is specific for the six-member cyclic ureides and shows the stereospecificity for the L-isomer.

Allantoinase (EC 3.5.2.5) is widely distributed in nature and plays an important role in the degradation of purine nucleosides (Fig. 5d). Investigation on substrate specificity of allantoinase is limited. However, it would be of interest to test allantoinase concerning the hydrolysis of various DL-5-monosubstituted hydantoins.

Recently, the complete amino acid sequences of the various cyclic amide amidohydrolases were reported and their homology search revealed that D-hydantoinase and ATP-independent L- and DL-hydantoinase are the members of superfamily of amidohydrolases related to ureases [55]. The superfamily includes dihydropyrimidinase, allantoinase, dihydroorotase, but not ATP-dependent hydantoinases. As a particular sequence, one aspartic

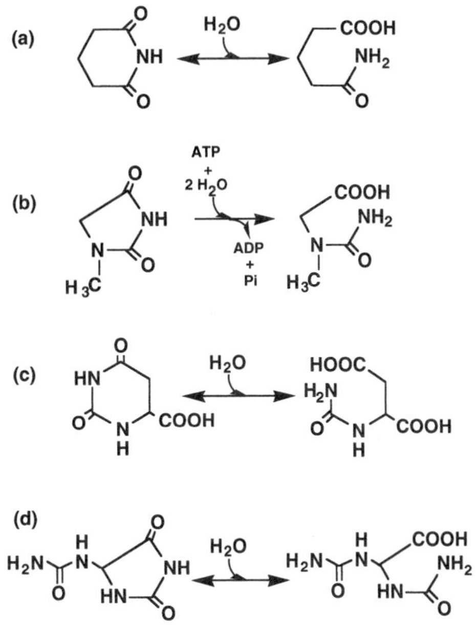

Figure 5   Reactions catalyzed by cyclic amide amidohydrolases. (a) imidase; (b) *N*-methylhydan-toin amidohydrolase; (c) dihydroorotase; (d) allantoinase.

acid and four histidine residues are found to be rigidly conserved in these amidases. These residues were found to be essential for metal binding as well as for catalysis [56].

C.   Enzymes Involved in Hydantoin Racemization

Although 5-monosubstituted hydantoins undergo spontaneous racemization under alkaline conditions, the existence of the hydantoin-racemizing enzyme hydantoin racemase was reported. Hydantoin racemase was purified from *Arthrobacter* sp. DSM 3747 [36] and *Pseudomonas* sp. strain NS671 [57], and cloned from the latter strain. Both enzymes require no special cofactor and are protected from inactivation by the addition of divalent sulfur–containing compounds, suggesting that thiol groups perform an important role in catalysis. The *Pseudomonas* hydantoin racemase is homohexameric with a molecular mass of 190 kDa. The enzyme activity was slightly stimulated by the addition of not only Mn^{2+} or Co^{2+} but also metal-chelating agents, indicating that the enzyme is not a metalloenzyme. The substrate specificities of hydantoin racemes are summarized in Table 6.

D.   Enzymes Involved in *N*-Carbamoyl Amino Acid Hydrolysis

1.   D-Carbamoylase

D-Carbamoylase was first purified homogeneously from *Comamonas* sp. E222c [13] and *Blastobacter* sp. A17p-4 [14]. The relative molecular masses of the native enzymes and those of the subunits were approximately 120,000 and 40,000, respectively. Both purified

Table 6 Substrate Specificities of Hydantoin Racemases

Substrate	*Arthrobacter* sp. DSM 3747	*Pseudomonas* sp. strain NS671
5-Substituted hydantoins		
5-Methyl-	n.d.	+
5-Isopropyl-	n.d.	+
5-Isobutyl-	+	+
5-(2′-Methylthioethyl)-	+	+
5-Benzyl-	+	+
5-(3′-Indolylmethyl)-	+	n.d.
3-N-Methyl-5-(3′-indolylmethyl)-	+	n.d.

+, hydrolyzed; n.d., not determined.

enzymes hydrolyzed various *N*-carbamoyl-D-amino acids to D-amino acids, ammonia, and carbon dioxide. *N*-Carbamoyl-D-amino acids having hydrophobic groups served as good substrates for these enzymes (Table 7). These enzymes strictly recognized the configuration of the substrate and only the D-enantiomer of the *N*-carbamoyl amino acid was hydrolyzed. This property can be applied for the optical resolution of racemic *N*-carbamoyl-α-amino acids. These enzymes did not hydrolyze β-ureidopropionate, suggesting that these enzymes are different from the enzyme involved in the pyrimidine degradation pathway, i.e., β-ureidopropionase.

Because of its instability, it was difficult to apply D-carbamoylase to industrial process, but recent genetic analysis gave the way to create a stable D-carbamoylase for the practical application. The gene coding D-carbamoylase was isolated from *Agrobacterium radiobacter* NRRL B12291 and sequenced [58]. The role of cysteine in the activity and stability toward oxidative denaturation was investigated. Among five cysteine residues, one cysteine residue (Cys 172) involved in activity and two cysteine residues (Cys 243 and Cys 279) located in proximity to external loops easily underwent oxidative denaturation; suggesting modification of these cysteine residues could enhance the enzyme stability. The gene coding D-carbamoylase was also isolated from *Agrobacterium* sp. KNK712 and mutagenized randomly to increase its thermostability [17]. From the sequence analysis of the mutant enzyme that showed improved thermostability by 5°C, the changes of the 57th amino acid histidine to tyrosine or leucine, the 203rd amino acid proline to leucine, serine, asparagine, glutamate, alanine, isoleucine, histidine, or serine, and the 263th amino acid valine to alanine, threonine, or serine were found to be effective. The mutant with amino acid arrangement of the 57th histidine to tyrosine, the 203rd proline to glutamate, and the 263th valine to alanine showed highly improved thermostability by about 20°C [18,19]. The improved *Agrobacterium* D-carbamoylase has been immobilized and used for the industrial production of D-*p*-hydroxyphenylglycine from the corresponding hydantoin in combination with the D-hydantoinase by Kaneka Corporation [19].

2. L-Carbamoylase

L-Carbamoylase activities have been found in various kinds of microorganisms [39,59–62]. The enzyme with broad substrate specificity was purified from *Alcaligenes xylosox-*

Table 7 Substrate Specificities of D-Carbamoylases

Compound	*Comamonas* sp. E222c		*Blastobacter* sp. A17p-4	
	Relative activity (%)	K_m (mM)	Relative activity (%)	K_m (mM)
Aliphatic				
N-Carbamoyl-				
D-alanine	23	12	29	4.0
D-valine	10	1.0	55	0.41
D-leucine	28	3.6	60	0.36
DL-alanine	20	n.d.	27	n.d.
DL-α-amino-*n*-butyric acid	14	n.d.	48	n.d.
DL-valine	8.4	n.d.	50	n.d.
DL-norvaline	6.3	n.d.	26	n.d.
DL-norleucine	90	4.8	92	0.79
DL-methionine	92	7.5	84	0.71
Aromatic				
N-Carbamoyl-				
D-phenylalanine	100	20	100	0.50
D-phenylglycine	24	27	170	0.88
D-*p*-hydroxyphenylglycine	47	13	130	1.7
DL-phenylalanine	59	n.d.	46	n.d.
DL-tryptophan	55	n.d.	18	n.d.
DL-phenylglycine	14	n.d.	100	n.d.
DL-*p*-hydroxyphenylglycine	39	n.d.	130	n.d.
Others				
N-Carbamoyl-				
D-serine	7.2	24	10	2.4
DL-serine	6.7	n.d.	3.0	n.d.
DL-threonine	3.0	n.d.	18	n.d.

n.d., not determined.

idans and characterized [63]. The *Alcaligenes* L-carbamoylase is a homodimeric enzyme with subunits of 65,000 and shows a broad substrate specificity not only for short chain but also for long chain and aromatic N-carbamoyl-L-amino acids (Table 8). On the other hand, the L-carbamoylase from *Arthrobacter* sp. DSM 3747 hydrolyzes only those amino acid derivatives that carry bulky substituents with a methylene spacer at the α carbon (Table 8). The gene coding L-carbamoylase was isolated from *Pseudomonas* sp. strain NS671 [37] and *B. stearothermophilus* NS 1122A [64]. All L-carbamoylases needs divalent cations such as Co^{2+}, Mn^{2+}, Ni^{2+}, or Fe^{2+} for their activity.

3. Other *N*-Carbamoyl Amino Acid Amidohydrolases

β-Ureidopropionase (EC 3.5.1.6) is the enzyme catalyzing the hydrolysis of *N*-carbamoyl-β-alanine to β-alanine and involved in pyrimidine degradation. The β-ureidopropionase from aerobic bacteria was purified from *P. putida* IFO 12996 and the characters that are

Table 8   Substrate Specificities of L-Carbamoylases

Substrate	Alcaligenes xylosoxidans	Bacillus stearothermophilus NS1122A	Arthrobacter sp. DSM 3747
N-Carbamoyl-α-amino acids			
N-Carbamoylglycine	+	+	n.d.
N-Carbamoyl-L-alanine	+	+	−
N-Carbamoyl-L-asparagine	+	+	n.d.
N-Carbamoyl-L-glutamate	−	+	−
N-Carbamoyl-L-valine	+	+	−
N-Carbamoyl-L-norvaline	+	n.d.	n.d.
N-Carbamoyl-L-leucine	+	+	−
N-Carbamoyl-L-methionine	+	+	−
N-Carbamoyl-L-isoleucine	+	+	n.d.
N-Carbamoyl-L-phenylalanine	+	+	+
N-Carbamoyl-L-serine	+	+	−
N-Carbamoyl-L-threonine	+	+	n.d.
N-Carbamoyl-L-tryptophan	−	−	+
N-Carbamoyl-L-tyrosine	−	+	+
Others			
N-Carbamoyl-β-alanine	−	n.d.	n.d.
N-Carbamoylsarcosine	−	n.d.	n.d.
N-Formyl-L-alanine	+	n.d.	n.d.
N-Formyl-L-leucine	+	n.d.	n.d.
N-Formyl-L-methionine	+	n.d.	−
N-Acetyl-L-phenylalanine	+	n.d.	−
N-Acetyl-L-norvaline	+	n.d.	n.d.

+, hydrolyzed; −, not hydrolyzed; n.d., not determined.

quite different from those of the β-ureidopropionases from other sources were revealed [65]. The *Pseudomonas* β-ureidopropionase consists of two identical polypeptide chains with relative molecular masses of 44,000. The enzyme requires a divalent metal ion, such as Co^{2+}, Ni^{2+}, or Mn^{2+}, for the activity. The enzyme showed a broad substrate specificity whereas β-ureidopropionases from mammals, protozoa, and anaerobic bacteria are specific to N-carbamoyl-β-amino acids; not only N-carbamoyl-β-amino acids, but also N-carbamoyl-γ-amino acids and several N-carbamoyl-α-amino acids such as N-carbamoylglycine, N-carbamoyl-L-alanine, N-carbamoyl-L-serine, and N-carbamoyl-L-α-amino-n-butyrate are hydrolyzed (Table 9). N-Formyl- and N-acetylalanine are also hydrolyzed by the enzyme, but the rate of hydrolysis is lower than that for N-carbamoylalanine. The hydrolysis of N-carbamoyl-α-amino acids is strictly L-stereospecific.

    N-Carbamoylsarcosine amidohydrolase (EC 3.5.1.59) is involved in the microbial degradation of creatinine and catalyzes the hydrolysis of N-carbamoylsarcosine to sarcosine, ammonia, and carbon dioxide [48]. The enzyme hydrolyzes N-carbamoyl derivatives of alanine, tryptophan, phenylalanine, phenylglycine, and p-hydroxyphenylglycine D-stereospecifically.

Table 9   Substrate Specificity of β-Ureidopropionase from *P. putida* IFO 12996

Compound	Relative activity (%)	K_m (mM)	V_{max} (μmol/min/mg)	$\dfrac{V_{max}}{K_m}$
N-Carbamoyl-β-amino acids				
N-Carbamoyl-β-alanine	100	3.7	4.1	1.1
N-Carbamoyl-DLβ-aminoisobutyrate	43	4.5	1.0	0.22
N-Carbamoyl-γ-amino acid				
N-Carbamoyl-γ-amino-*n*-butyrate	290	12	19	1.7
N-Carbamoyl-α-amino acids				
N-Carbamoylglycine	17	0.68	0.091	0.13
N-Carbamoyl-L-alanine	120	1.6	1.0	0.64
N-Carbamoyl-L-serine	34	75	3.8	0.050
N-Carbamoyl-DL-α-amino-*n*-butyrate	31	2.8	1.1	0.38
N-Carbamoyl-DL-norvaline	8.9	42	1.1	0.027
N-Carbamoyl-DL-threonine	0.97	n.d.	n.d.	n.d.
N-Carbamoyl-DL-aspartate	0.14	n.d.	n.d.	n.d.
N-Carbamoyl-L-asparagine	1.6	n.d.	n.d.	n.d.
N-Carbamoyl-L-glutamate	0.29	n.d.	n.d.	n.d.
Others				
N-Formyl-DL-alanine	75	7.7	0.84	0.11
N-Acetyl-DL-alanine	6.3	8.8	0.067	0.0077

n.d., not determined.

VI.   CONCLUSION

As described in this chapter, variety of hydantoin-hydrolyzing enzymes (Fig. 6) and *N*-carbamoyl amino acid amidohydrolases (Fig. 7) are involved in hydantoin transformation. The combinations of these enzymes provide a variety of processes for the production of optically pure α-amino acids (Fig. 8) [36,66–68]. The ability to completely convert a 100% hydantoin racemate into optically pure enantiomer renders these processes very attractive. Stereospecific D- or L-hydantoinase can produce optically pure *N*-carbamoyl-D- or L-amino acid, respectively, from DL-5-monosubstituted hydantoin with 100% yield. The use of DL-hydantoinase together with stereospecific D- or L-carbamoylase also can provide optically pure D- or L-amino acid, respectively, from DL-5-monosubstituted hydantoin with 100% yield. Construction of recombinant microorganisms carrying these enzymes and immobilization of the cells or the enzymes could enhance the efficiency of the process as already demonstrated for D-*p*-hydroxyphenylglycine production [69].

In optically pure α-amino acid production from DL-5-monosubstituted hydantoins, the wide applicability to a broad substrate range is valuable especially for the production of D-α-amino acids [70] and unnatural L-α-amino acids (Fig. 9), e.g., D-*p*-hydroxyphenylglycine [71], D-phenylglycine [71], substituted L-phenylalanine such as L-*p*-chlorophenylalanine [72] and *p*-trimethylsilylphenylalanine [73,74], L-α- and β-naphthylalanine [75], an *N*-methyl-D-aspartate receptor antagonist, (2*R*, 4*R*, 5*S*)-2-amino-4,5-(1,2-cyclohexyl)-7-

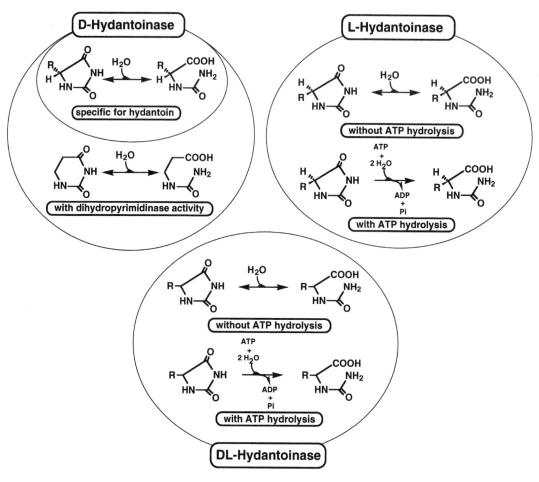

Figure 6   Reactions catalyzed by hydantoinases.

phosphonoheptanoic acid [76], and so on. Another interesting use of hydantoinase has been described by Watanabe et al. [77]. They use D-hydantoinase for the catalysis of the reverse reaction, i.e., D-selective synthesis of the hydantoin from the racemic mixture of α,α-disubstituted N-carbamoyl amino acid (Fig. 10). Optically pure 5,5-disubstituted hydantoins and α,α-disubstituted amino acids such as L-α-methyl-DOPA will gain increasing interest as a new type of antagonists of natural α-amino acids.

Other than these, N-methylhydantoin amidohydrolase and the successive enzyme in microbial creatinine transformation, i.e., N-carbamoylsarcosine amidohydrolase, are useful for the enzymatic measurement of creatinine in serum and urine, which is a marker of renal dysfunction [45]. This enzymatic method proved to be simple and precise, and has excellent sensitivity and specificity [78].

As we presented here, potential of stereoselective synthesis using hydantoinases and carbamoylases is based on the unusual diversity of the microbial enzymes and their strict specificity. Application of microbial hydantoin transformation is a good example of the applicability of enzymes from a microbially diverse world.

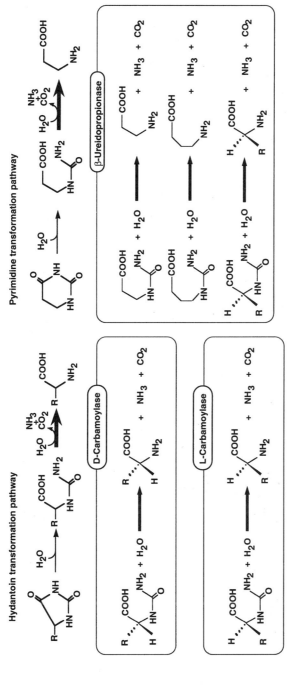

Figure 7 Reactions catalyzed by carbamoylases.

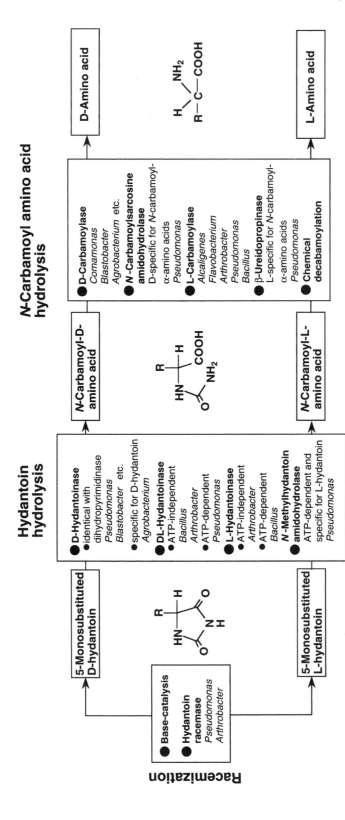

Figure 8 Optically pure α-amino acid production processes by combinations of hydantoin-transforming enzymes.

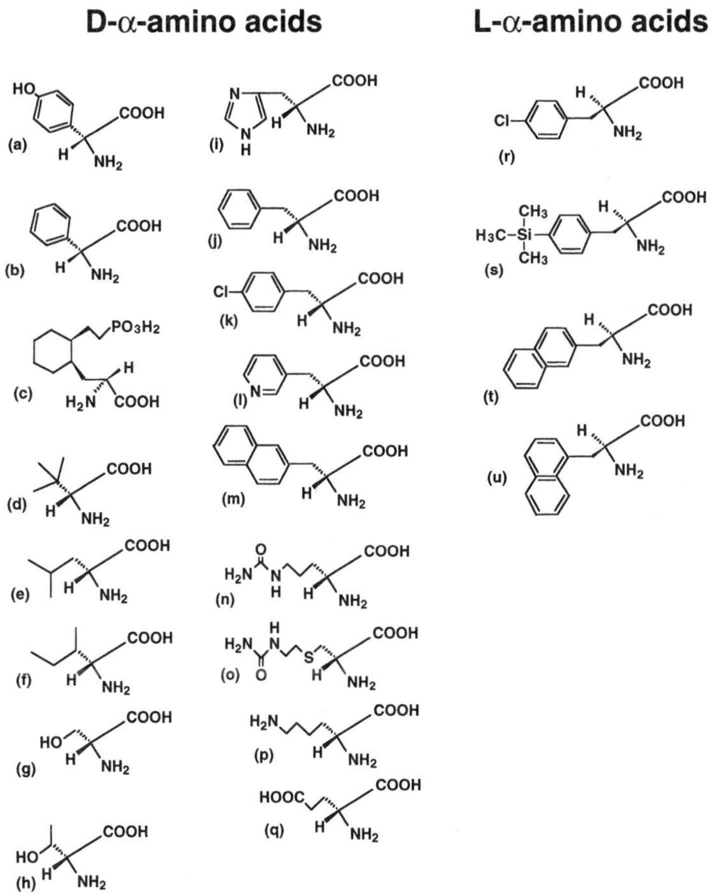

D-α-amino acids

L-α-amino acids

Figure 9 D-α-Amino acids and unnatural L-α-amino acids produced by hydantoinase process. (a) D-*p*-Hydroxyphenylglycine; (b) D-phenylglycine; (c) (2*R*,4*R*,5*S*)-2-amino-4,5-(1,2-cyclohexyl)-7-phosphonoheptanoic acid; (d) D-*tert*-leucine; (e) D-leucine; (f) D-*allo*-isoleucine; (g) D-serine; (h) D-*allo*-threonine; (i) D-histidine; (j) D-phenylalanine; (k) D-*p*-chlorophenylalanine; (l) D-3′-pyridyl-alanine; (m) D-β-naphthylalanine; (n) D-citrulline; (o) ω-ureido-D-*S*-aminoethylcysteine; (p) D-lysine; (q) D-glutamate; (r) L-*p*-chlorophenylalanine; (s) *p*-trimethylsilylphenylalanine; (t) L-β-naphthylalanine; (u) L-α-naphthylalanine.

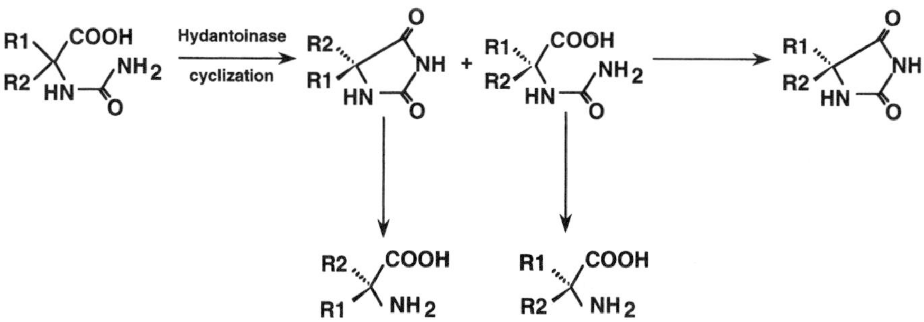

Figure 10 Stereoselective synthesis of α,α-disubstituted α-amino acid through the cyclization catalyzed by hydantoinase.

REFERENCES

1. HT Bucherer, W Steiner. J Prakt Chem 140:291–316, 1934.
2. KH Dudley, DL Bius, TC Butler. J Pharmacol Exp Ther 175:27–37, 1970.
3. KH Dudley, DL Bius. J Heterocycl Chem 10:173–180, 1973.
4. F Cecere, G Galli, F Morisi. FEBS Lett 57:192–194, 1975.
5. H Yamada, S Takahashi, Y Kii, H Kumagai. J Ferment Technol 56:484–491, 1978.
6. S Takahashi, Y Kii, H Kumagai, H Yamada. J Ferment Technol 56:492–498, 1978.
7. S Takahashi, T Ohashi, Y Kii, H Kumagai, H Yamada. J Ferment Technol 57:328–332, 1979.
8. S Shimizu, S Takahashi, H Shimada, T Ohashi, Y Tani, H Yamada. Agric Biol Chem 44:2233–2234, 1980.
9. H Yamada, S Shimizu, H Shimada, Y Tani, S Takahashi, T Ohashi. Biochimie 62:395–399, 1980.
10. R Olivier, E Fascetti, L Angelini, L Degen. Enzyme Microb Technol 1:201–204, 1979.
11. K Yokozeki, Y Hirose, K Kubota. Agric Biol Chem 51:737–746, 1978.
12. A Yamashiro, K Kubota, K Yokozeki. Agric Biol Chem 52:2857–2863, 1988.
13. J Ogawa, S Shimizu, H Yamada. Eur J Biochem 212:685–692, 1993.
14. J Ogawa, MCM Chung, S Hida, H Yamada, S Shimizu. J Biotechnol 38:11–19, 1994.
15. H Nanba, Y Ikenaka, Y Yamada, K Yajima, M Takano, S Takahashi. Biosci Biotech Biochem 62:875–881, 1998.
16. Y Ikenaka, H Nanba, Y Yamada, K Yajima, M Takano, S Takahashi. Biosci Biotech Biochem 62:882–886, 1998.
17. Y Ikenaka, H Nanba, K Yajima, Y Yamada, M Takano, S Takahashi. Biosci Biotech Biochem 62:1668–1671, 1998.
18. Y Ikenaka, H Nanba, K Yajima, Y Yamada, M Takano, S Takahashi. Biosci Biotech Biochem 62:1672–1675, 1998.
19. S Takahashi. In: Abstract of the 8th German–Japanese Workshop on Enzyme Technology, Toyama, Japan, 1994, pp 23–24.
20. E Ware. Chem Rev 46:403–470, 1950.
21. T Ohashi, S Takahashi, T Nagamachi, K Yoneda, H Yamada. Agric Biol Chem 45:831–838, 1981.
22. A Morin, W Hummel, MR Kula. Appl Microbiol Biotechnol 25:91–96, 1986.
23. G LaPointe, S Viau, D Leblanc, N Robert, A Morin. Appl Environ Microbiol 60:888–895, 1994.
24. A Möller, C Syldatk, M Schulze, F Wagner. Enzyme Microb Technol 10:618–625, 1988.
25. KH Dudley, TC Butler, DL Buis. Drug Metab Dispos 2:103–112, 1973.
26. S Runser, N Chinski, E Ohleyer. Appl Microbiol Biotechnol 33:382–388, 1990.
27. K Yokozeki, K Kubota. Agric Biol Chem 51:721–728, 1987.
28. S Kim, TP West. FEMS Microbiol Lett 77:175–180, 1991.
29. J Ogawa, T Kaimura, H Yamada, S Shimizu. FEMS Microbiol Lett 122:55–60, 1994.
30. SM Runser, PC Meyer. Eur J Biochem 213:1315–1324, 1993.
31. KP Brooks, EA Jones, BD Kim, EG Sander. Arch Biochem Biophys 226:469–483, 1983.
32. M Kikugawa, M Kaneko, S Fujimoto-Sakata, M Maeda, K Kawasaki, T Takagi, N Tamaki. Eur J Biochem 219:393–399, 1994.
33. A Morin, W Hummel, H Schütte, MR Kula. Biotechnol Appl Biochem 8:564–574, 1986.
34. SG Lee, DC Lee, SP Hong, MH Sung, HS Kim. Appl Microbiol Biotechnol 43:270–276, 1995.
35. J Ogawa, M Honda, CL Soong, S Shimizu. Biosci Biotech Biochem 59:1960–1962, 1995.
36. C Syldatk, R Müller, M Pietzsch, F Wagner. Biocatalytic Production of Amino Acids and Derivatives. München: Hanser Publishers, 1992, pp 129–176.
37. K Watabe, T Ishikawa, Y Mukohara, H Nakamura. J Bacteriol 174:962–969, 1992.
38. T Ishikawa, K Watabe, Y Mukohara, H Nakamura. Biosci Biotech Biochem 61:185–187, 1997.
39. T Ishikawa, Y Mukohara, K Watabe, S Kobayashi, H Nakamura. Biosci Biotech Biochem 58:265–270, 1994.

40. Y Mukohara, T Ishikawa, K Watabe, H Nakamura. Biosci Biotech Biochem 58:1621–1626, 1994.
41. O May, M Siemann, M Pietzsch, M Kiess, R Mattes, C Syldatk. J Biotechnol 61:1–13, 1998.
42. J Ogawa, CL Soong, M Honda, S Shimizu. Eur J Biochem 243:322–327, 1997.
43. J Ogawa, CL Soong, M Honda, S Shimizu. Appl Environ Microbiol 62:3814–3817, 1996.
44. JM Kim, S Shimizu, H Yamada. Biochem Biophys Res Commun 142:1006–1012, 1987.
45. S Shimizu, JM Kim, H Yamada. Clin Chim Acta 185:241–252, 1989.
46. H Yamada, S Shimizu, JM Kim, Y Shinmen, T Sakai. FEMS Microbiol Lett 30:337–340, 1985.
47. S Shimizu, JM Kim, Y Shinmen, H Yamada. Arch Microbiol 145:322–328, 1986.
48. JM Kim, S Shimizu, H Yamada. J Biol Chem 261:11832–11839, 1986.
49. JM Kim, S Shimizu, H Yamada. Arch Microbiol 147:58–63, 1987.
50. JM Kim, S Shimizu, H Yamada. FEBS Lett 210:77–80, 1987.
51. JM Kim, S Shimizu, H Yamada. Agric Biol Chem 50:2811–2816, 1986.
52. JM Kim, S Shimizu, H Yamada. Agric Biol Chem 51:1167–1168, 1987.
53. J Ogawa, JM Kim, W Nirdnoy, Y Amano, H Yamada, S Shimizu. Eur J Biochem 229:284–290, 1995.
54. J Ogawa, S Shimizu. Arch Microbiol 164:353–357, 1995.
55. O May, A Habenicht, R Mattes, C Syldatk, M Siemann. Biol Chem 379:743–747, 1998.
56. GJ Kim, HS Kim. Biochem J 330:295–302, 1998.
57. K Watabe, T Ishikawa, Y Mukohara, H Nakamura. J Bacteriol 174:7989–7995, 1992.
58. R Grifantini, C Pratesi, G Gall, G Gradi. J Biol Chem 271:9326–9331, 1996.
59. C Syldatk, D Cotoras, G Dombach, C Groß, H Kallwaß, F Wagner. Biotechnol Lett 9:25–30, 1987.
60. K Sano, K Yokozeki, C Eguchi, T Kagawa, I Noda, K Mitsugi. Agric Biol Chem 41:819–825, 1977.
61. T Ishikawa, K Watabe, Y Mukohara, S Kobayashi, H Nakamura. Biosci Biotech Biochem 57:982–986, 1993.
62. A Yamashiro, K Yokozeki, H Kano, K Kubota. Agric Biol Chem 52:2851–2856, 1988.
63. J Ogawa, H Miyake, S Shimizu. Appl Microbiol Biotechnol 43:1039–1043, 1995.
64. Y Mukohara, T Ishikawa, K Watabe, H Nakamura. Biosci Biotech Biochem 57:1935–1937, 1993.
65. J Ogawa, S Shimizu, Eur J Biochem 223:625–630, 1994.
66. C Syldatk, R Müller, M Siemann, K Krohn, F Wagner. Biocatalytic Production of Amino Acids and Derivatives. München: Hanser Publishers, 1992, pp 75–128.
67. C Syldatk, A Läufr, R Müller, H Höke. Adv Biochem Eng/Biotechnol 41:29–75, 1990.
68. J Ogawa, S Shimizu. J Molec Cat B: Enzymatic 2:163–176, 1997.
69. R Grifantini, G Galli, G Carpani, C Pratesi, G Frascotti, G Grandi. Microbiology 144:947–954, 1998.
70. A Bommarius, M Schwarm, K Drauz. J Molec Cat B: Enzymatic 5:1–11, 1998.
71. S Takahashi. In: K Aida, I Chibata, K Nakayama, K Takinami, H Yamada, eds. Biotechnology of Amino Acid Production. Tokyo: Kodansha, 1986, pp 269–279.
72. C Syldatk, V Lehmensiek, G Ulrichs, U Bilitewski, K Krohn, H Höke, F Wagner. Biotechnol Lett 14:99–104, 1992.
73. Y Tsuji, H Yamanaka, T Fukui, T Kawamoto, A Tanaka. Appl Microbiol Biotechnol 47:114–119, 1997.
74. H Yamanaka, T Kawamoto, A Tanaka. J Ferment Bioeng 84:181–184, 1997.
75. C Syldatk, D Völkel, U Bilitewski, K Krohn, H Höke, F Wagner. Biotechnol Lett 14:105–110, 1992.
76. D Durham, J Weber. Appl Environ Microbiol 62:739–742, 1996.
77. Kanegafuchi Kagaku Kogyo. Eur Pat 0175312 A2, 1985.
78. J Ogawa, W Nirdnoy, M Tabata, H Yamada, S Shimizu. Biosci Biotech Biochem 59:2292–2294, 1995.

2

Aminoamidase-Catalyzed Preparation and Further Transformations of Enantiopure α-Hydrogen- and α,α-Disubstituted α-Amino Acids

Theo Sonke, Bernard Kaptein, Wilhelmus H. J. Boesten, Quirinus B. Broxterman, and Hans E. Schoemaker
DSM Research, Geleen, The Netherlands

Johan Kamphuis
DSM Specialty Intermediates, Sittard, The Netherlands

Fernando Formaggio and Claudio Toniolo
University of Padua, Padua, Italy

Floris P. J. T. Rutjes
University of Amsterdam, Amsterdam, The Netherlands

I. GENERAL INTRODUCTION

Amino acids, natural as well as synthetic, are extensively used in the food, feed, agrochemical, and pharmaceutical industries. Many proteinogenic amino acids are used as infusion solutions, whereas the essential amino acids (e.g., lysine, threonine, and methionine) are used as feed additives. Sodium glutamate is widespread as a taste enhancer/seasoning agent, the commercial production of which dates back to 1908. Aspartic acid and phenylalanine methyl ester together form the low-calorie sweetener aspartame. But not only natural (proteinogenic) amino acids are applied; also synthetic amino acids are intermediates in pharmaceuticals and agrochemicals production. D-Phenylglycine (Phg) and D-*p*-hydroxyphenylglycine are produced in quantities of several thousands of tons per year for the synthesis of the semisynthetic broad-spectrum antibiotics ampicillin (**1a**), amoxicillin (**1b**), cefalexin (**2**), and others [1]. Another D-amino acid, D-valine, is used as a building block of the pyrethroid insecticide fluvalinate (**3**) [2], whereas the α-methyl-substituted amino acid (αMe)valine is a structural element in the herbicide arsenal (**4**) (marketed as a racemate by American Cyanamid although the D enantiomer is the most active) and related herbicides [3].

But also in many other fields unnatural amino acids are used, e.g., for the antihypertensive drugs Aldomet (L-α-methylDOPA) (**5**) [4], the ACE inhibitors enalapril and lisinopril (structures not shown) [5], and the recently developed HIV-protease inhibitors

Crixivan (indinavir sulfate) (**6**) [6], Invirase (saquinavir mesylate) (**7**) [7], and related compounds [8].

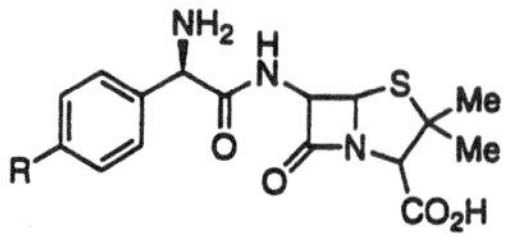

Ampicillin (R = H) **1a**
Amoxycillin (R = OH) **1b**

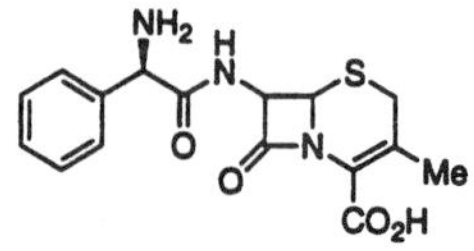

Cefalexin **2**

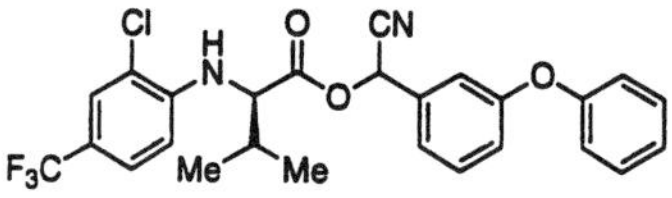

Fluvalinate **3**

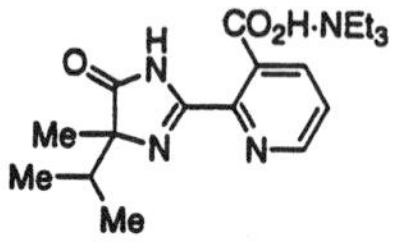

Arsenal **4**

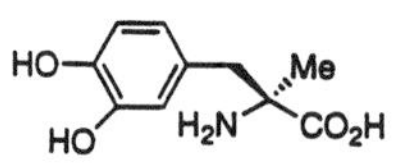

L-α-MethylDOPA **5**

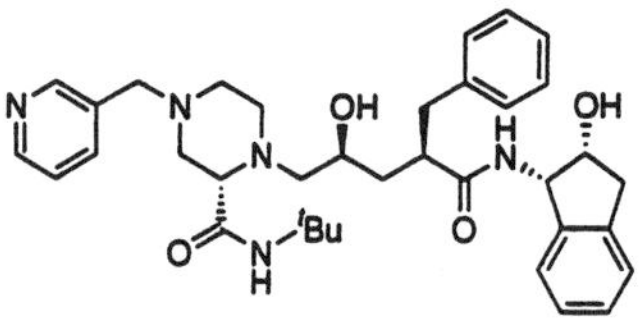

Indinavir **6**

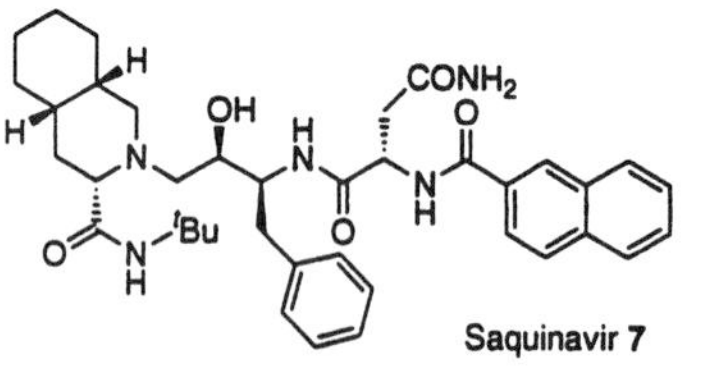

Saquinavir **7**

Although many amino acids appear in living organisms, only a few are actually isolated from nature. L-Cysteine and L-4-hydroxyproline are some of the examples in which extraction is still a commercially viable process. For most naturally occurring proteinogenic amino acids (precursor) fermentation is nowadays the preferred route for their preparation [9]. Examples include phenylalanine for the production of aspartame (**8**) (Scheme 1) [10] and glutamic acid, which are produced in amounts of multiple kilotons per year.

Only a few examples are known in which amino acids are produced by (catalytic) asymmetric synthesis. The asymmetric hydrogenation of dehydroamino acids, previously developed by Monsanto for L-phenylalanine, is today only used for the production of L-DOPA [11]. Many other asymmetric routes for the synthesis of enantiopure amino acids

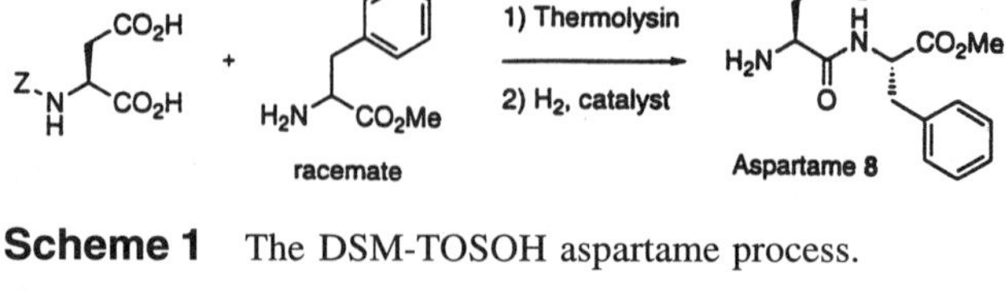

1) Thermolysin

2) H_2, catalyst

racemate

Aspartame **8**

Scheme 1 The DSM-TOSOH aspartame process.

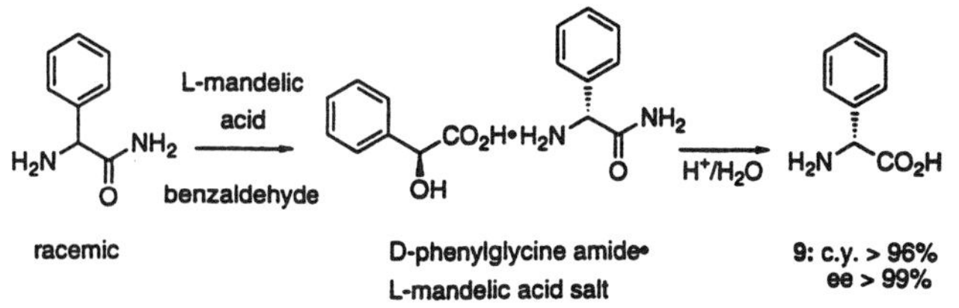

Scheme 2   Asymmetrical transformation of D-phenylglycine amide.

have been developed on a laboratory scale, but to our knowledge none of them is scaled up beyond pilot plant proportions [12]. An elegant chemical production method via an asymmetrical transformation has recently been developed at DSM. This combination of a classical resolution with an in situ racemization can be used for the production of D-phenylglycine (**9**) and related compounds (Scheme 2) [13]. Other frequently used production methods are based on enzymatic conversions with ammonia lyases or amino acid dehydrogenases [9,14].

Enzymatic resolution is another method applied for the production of L- as well as D-amino acids. The disadvantage of a resolution process, a maximal yield of 50%, can often be overcome by racemization of the unwanted enantiomer of the amino acid derivative (vide infra). Many resolution concepts have been developed and commercialized, e.g., the acylase process by Degussa and Tanabe [15], the D-hydantoinase process by Recordati and Ajinomoto [16], the aminoamidase processes of DSM, and different lipase-catalyzed processes [1,13b,17].

In this chapter we describe the DSM aminoamidase processes in more detail. Three different enzymatic resolution routes have been developed for the preparation of natural and synthetic amino acids using biocatalysts from different origin, i.e., *Pseudomonas putida, Mycobacterium neoaurum*, and *Ochrobactrum anthropi*. Scope and limitations and enzyme characterization of these amidases will be presented together with some specific examples. In addition, the use of some of these amino acids in peptide synthesis, catalytic asymmetric synthesis, and further synthetic transformations will be given.

II.   AMINOPEPTIDASE FROM *PSEUDOMONAS PUTIDA* ATCC 12633

A.   Introduction

In the mid-1970s an enzymatic process for the production of enantiopure α-hydrogen containing L- or D-amino acids (and their amides) has been developed at DSM using whole cells of *Pseudomonas putida* ATCC 12633. The process is based on the kinetic resolution of racemic amino acid amides and has been commercialized since 1988 for the production of several L- and D-amino acids [1,17b]. The scope and limitations of this process will be discussed, together with enzyme purification, characterization, and the overexpression of the gene coding for the enzyme in an *E. coli* K-12 strain.

B.   Synthesis of Racemic α-H-Amino Acid Amides

In general, the unsubstituted amides of racemic α-H-amino acids (**12**) are used as substrates that are readily available by Strecker synthesis on aldehydes. The initially formed ami-

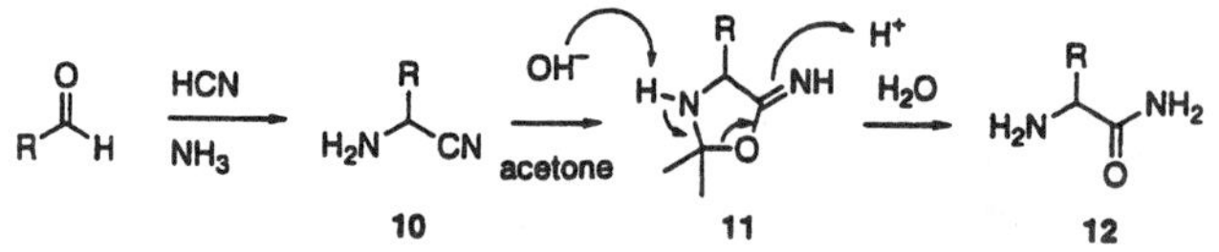

Scheme 3 Preparation of racemic α-H-amino acid amides by Strecker synthesis.

nonitriles (**10**) are hydrolyzed to the amide under mild basic conditions (room temp., pH 10–12) in the presence of an aldehyde or a ketone (Scheme 3). In general, a catalytic amount of benzaldehyde or acetone as solvent is sufficient. This hydrolysis is supposed to proceed by intramolecular addition of the acetone-hemiaminal to the nitrile leading to intermediate **11**, followed by further hydrolysis to the amide. These mild conditions prevent overhydrolysis to the racemic amino acid [18].

An alternative preparation of the amino acid amides on a laboratory scale is the alkylation of *N*-acetamidomalonate esters followed by hydrolysis and amination of the amino esters. For instance, racemic allylglycine amide, methallylglycine amide, and β-(pyridyl)alanine amides are prepared by this method [19].

C. Enzymatic Resolution

The kinetic resolution of racemic amino acid amides is performed with permeabilized whole cells of *P. putida* ATCC 12633 with a nearly 100% stereoselectivity for hydrolysis of the L-amide (enantiomeric ratio $E > 200$ [20]). Thus, both the L-acid and the D-amide can be obtained in nearly 100% e.e. at 50% conversion. The biocatalyst accepts a broad structural variety of amino acid amides, from alanine amide to, for example, β-naphthyl-glycine amide or lupinic acid amide (Scheme 4). So far more than 100 different amino acid amides have been successfully resolved.

Heteroatoms like sulfur, nitrogen, and oxygen are accepted in alkyl or (hetero)aryl substituents as well as alkenyl and alkynyl substituents (Scheme 4) [21,22]. Only for methionine amide and homomethionine amide is a slightly lower stereoselectivity observed, probably caused by (enzymatic or chemical) racemization of the L-acid under the basic reaction conditions. Cyclic amino acid amides such as proline amide and piperidine-2-carboxyamide can also be resolved. For these substrates product inhibition is observed, which leads to a decrease in enzyme activity with advancing conversion.

Substituted amides of amino acids are also hydrolyzed depending on the size of the substituent. The *N*-methyl- or *N*-methoxy amides (**13**) are hydrolyzed stereoselectively albeit with a lower activity (Scheme 5) [23], whereas, for example, the *N*-benzylamide of DL-valine is not hydrolyzed [24].

Although every resolution process is intrinsically hampered by a maximum yield of 50%, racemization of the unwanted isomer can lead to 100% yield. For the aminoamidase process this can easily be done via racemization of the benzaldehyde Schiff base of the D-amide under basic conditions (Scheme 4) [17]. Since the separation of the L-acid and the D-amide proceeds via Schiff base formation of the amide, racemization can be performed without any additional step. Evidently, if the required products are D-amino acid (amides), the L-acid can of course also be recycled via this route (after formation of the L-amide).

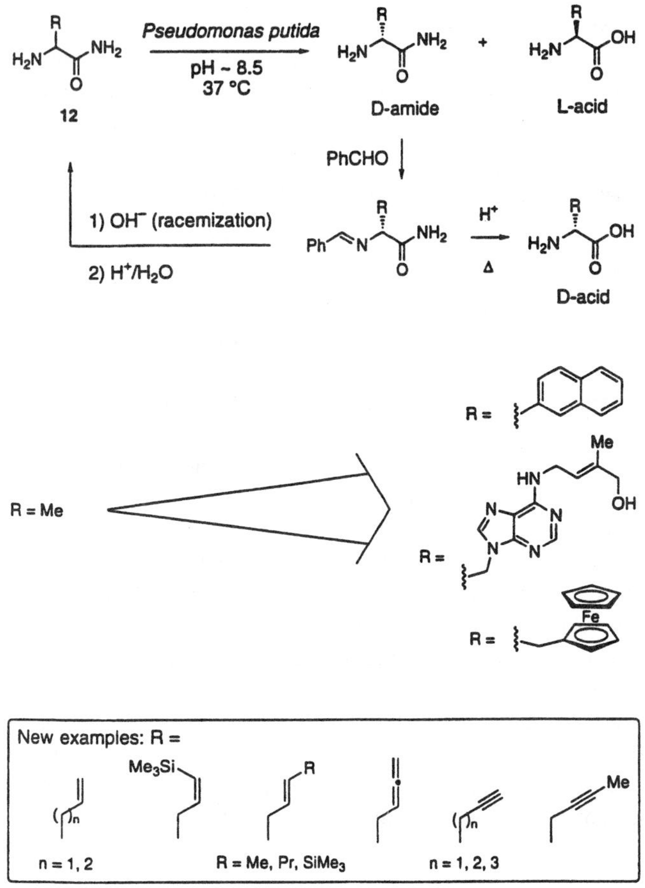

Scheme 4 Enzymatic resolution of α-H-amino acid amides by *P. putida* and substrate range.

The main advantages of the aminoamidase process can be summarized by:

1. Readily available starting material, which is a precursor of the amino acid.
2. Cheap whole-cell biocatalyst, avoiding laborious enzyme purification at production scale.
3. Both L- and D-α-H-amino acids can be prepared with high enantioselectivity.

In the resolutions described above, crude whole-cell preparations were used in the biocatalytic step. It is remarkable that so many different substrates can be resolved using this crude enzyme preparation. This raises the question of whether one enzyme is respon-

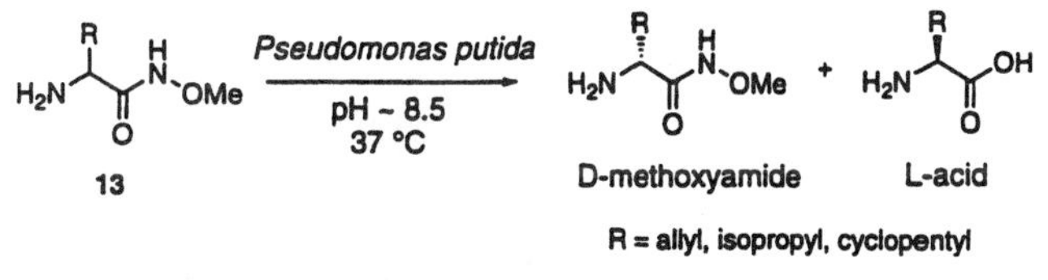

Scheme 5 Enzymatic resolution of amino acid *N*-methoxyamides by *P. putida*.

sible for the enantioselective conversions and whether the process can be improved by
cloning and expression of the genes coding for the enzyme(s) responsible for the enantioselective hydrolysis. Therefore, we embarked on a more detailed study of the aminoami-
dase(s) from this particular *P. putida* strain.

D. Purification and Characterization of an L-Aminopeptidase from *P. putida* ATCC 12633

From earlier performed transposon mutagenesis experiments, which were aimed at the
generation of an L-amidase-negative *P. putida* mutant strain, it could be concluded that
the rather broad substrate specificity of this strain is caused by more than one enzyme
[24]. However, the same experiments proved that just a single enzyme is responsible for
the L-valine amide-hydrolyzing capacity and most of the L-phenylglycine amide-hydrolyzing capacity of *P. putida* ATCC 12633.

To learn more about this amide hydrolase and to enable subsequent cloning of the
coding gene by reversed genetics, it was decided to purify and characterize this enzyme.
The basic protocol for the purification of the enzyme has been described before [13b,25].
Although this procedure led to an almost homogeneous enzyme preparation, the specific
activity of the purified enzyme was not optimal due to the limited stability of this enzyme
at pH 5.2, the value that was originally used in the Mono S cation exchange chromatography step. This enzyme inactivation can be largely prevented by performing this cation
exchange step at pH 8.0 instead of pH 5.2. The purification procedure has been further
optimized by the introduction of a Superdex 200 gel filtration column. The optimized
purification protocol is outlined in Fig. 1. The amide hydrolase can be purified by this
protocol about 300-fold with a recovery of 22% (Table 1). The specific activity of the
purified enzyme is 610 U/mg of protein, which is about twofold higher than the previous

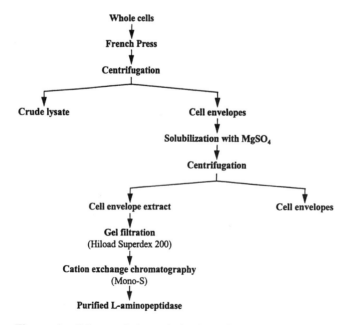

Figure 1   Scheme of the optimized purification protocol of an L-aminopeptidase from *P. putida*
ATCC 12633.

Table 1 Purification of an L-Aminopeptidase from *P. putida* ATCC 12633

Purification step	Total activity[a] (μmol min^{-1})	Total protein[b] (mg)	Sp. act. (μmol min^{-1} mg^{-1})	Recovery (%)	Purification (fold)
Disintegrated cells	9860	4870	2.0	100	1.0
Cell envelopes	8800	1600	5.5	89	2.8
Cell envelope extract	3750	26	140	38	70
Superdex-200	1820	7	260	18	130
Mono S	2200	4	610	22	305

[a]Activity was routinely assayed at 40°C and pH 9.0 with L-phenylglycine amide as substrate; 1 U has been defined as the hydrolysis of 1 μmol of substrate per minute.
[b]Protein concentrations were measured by the BCA method [55].

protocol. As demonstrated by sodium dodecyl sulfate polyacrylamide gel electrophoresis (SDS-PAGE), the purity of this enzyme preparation is at least as high as with the former purification procedure.

Characterization of the purified protein by nondenaturing gel filtration reveals a native molecular mass of approximately 400 kDa. SDS-PAGE analysis of the pure enzyme shows a single band with a molecular mass of about 52 kDa. This indicates that the L-aminopeptidase most probably is a homo-octameric enzyme. The isoelectric point of the protein is estimated as pH 10.5 using an isoelectric focusing (IEF) 3–9 Phastgel with an expanded pH range.

Divalent cations have a marked effect on the activity of this enzyme: Mg^{2+}, Co^{2+} (2- to 3-fold), and especially Mn^{2+} (12-fold) stimulate the activity (at 0.2–20 mM), whereas treatment with Cu^{2+} and Ca^{2+} (at 2 mM) cause 70% and 40% inhibition, respectively.

The L-aminopeptidase is sensitive to various classes of proteinase inhibitors. Strong inhibition of the enzyme is observed by treatment with the thiol reagents *p*-chloromercuribenzoate (pCMB) and iodoacetamide. The inhibition by pCMB can be reversed by subsequent treatment with dithiothreitol. In addition, the enzyme is inhibited by the metalchelating compounds EDTA and *o*-phenanthroline and the serine protease inhibitors phenylmethylsulfonyl fluoride and diisopropylfluorophosphate. These phenomena point to an essential serine or cysteine residue in the active site; furthermore, divalent cations seem to be involved in the catalytic mechanism and/or are important for the stability of the enzyme.

The L-aminopeptidase from *P. putida* displays clear amide-hydrolyzing activity between pH 7 and 11, with the highest activity at pH 9.0–9.5. Furthermore, the enzyme has maximum activity at 40°C (Fig. 2). The thermal stability of this enzyme in purified form is rather limited: after a 60-min incubation period at 30°C, only 65% of the initial activity remains. Addition of substrate, however, dramatically enhances the thermal stability (60 min at 30°C, 100% activity left).

The substrate specificity of the purified enzyme has been tested with a wide range of amide substrates and dipeptides (see Table 2). The enzyme is completely inactive with simple amides like acetamide, propionamide, butyramide, isobutyramide, and acrylamide. Furthermore, no activity toward nicotinamide is observed. It appears that the enzyme is only active with α-amino acid amides with a hydrogen atom at the C_α position. Highest activities are observed with L-Leu-NH$_2$ and L-Phg-NH$_2$. Also toward the α-hydroxy acid

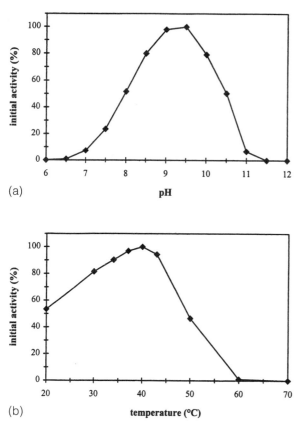

Figure 2 Effect of the pH (a) and temperature (b) on the initial activity of the purified L-aminopeptidase from *P. putida* ATCC 12633.

amide mandelic acid amide a rather low activity is observed. The enzyme is inactive with the only α-alkyl-substituted amino acid amide tested, DL-(αMe)Val-NH$_2$. Because high activities are also observed with the four dipeptides tested (L-Phe-L-Phe, L-Phe-L-Leu, L-Leu-L-Phe, and L-Leu-L-Leu), this enzyme from *P. putida* ATCC 12633 can be designated as an L-aminopeptidase.

To test the enantioselectivity of this protein, it was incubated with racemic mixtures of Leu-NH$_2$ and Phg-NH$_2$. Of both substrates, only the L enantiomer is hydrolyzed by this enzyme.

Finally, the enzyme displays normal Michaelis-Menten kinetics for the five different α-H-amino acid amides tested (Table 3). These results show that the additional methyl groups at the C$_\beta$ atom in L-Val-NH$_2$ and L-Ile-NH$_2$ lead to relatively high K_M and low V_{max} values and consequently to a low catalytic efficiency and specificity (k_{cat}/K_M). The highest k_{cat}/K_M is observed for L-Leu-NH$_2$, where this additional methyl group at the C$_\beta$ group is missing. Similarly, among the two aromatic amino acid amides tested, the lowest K_M value is obtained with L-Phe-NH$_2$, which also contains a C$_\beta$ atom without additional substituents. However, in this case the highest catalytic efficiency is observed with L-Phg-NH$_2$, pointing to a strongly positive effect of the aromatic ring directly adjacent to the C$_\alpha$ atom.

Table 2 Substrate Specificity of the Purified L-Aminopeptidase from *P. putida* ATCC 12633

Substrate	Formula	Relative activity[c] (%)
DL-Mandelic acid amide[a]	C_6H_5—CH(OH)—$CONH_2$	2
Glycine amide[b]	H—CH(NH_2)—$CONH_2$	1
L-Alanine amide[b]	CH_3—CH(NH_2)—$CONH_2$	1
L-α-Aminobutyramide[b]	CH_3—CH_2—CH(NH_2)—$CONH_2$	10
L-Valine amide[b]	$(CH_3)_2$CH—CH(NH_2)—$CONH_2$	7
L-Leucine amide[b]	$(CH_3)_2$CH—CH_2—CH(NH_2)—$CONH_2$	214
L-Isoleucine amide[b]	CH_3CH_2—CH(CH_3)—CH(NH_2)—$CONH_2$	13
L-Phenylglycine amide[a,b]	C_6H_5—CH(NH_2)—$CONH_2$	100
L-Phenylalanine amide[b]	C_6H_5—CH_2—CH(NH_2)—$CONH_2$	14
L-Methionine amide[b]	CH_3—S—CH_2—CH_2—CH(NH_2)—$CONH_2$	58
DL-Proline amide[a]	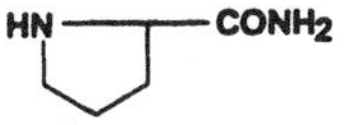	<1
L-Tryptophan amide[b]	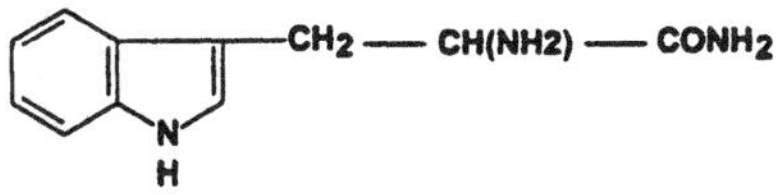	14
L-Serine amide[b]	HO—CH_2—CH(NH_2)—$CONH_2$	1
L-Glutamic acid amide[b]	HOOC—CH_2—CH_2—CH(NH_2)—$CONH_2$	<1
DL-α-Methylvaline amide[a]	$(CH_3)_2$CH—C(CH_3)(NH_2)—$CONH_2$	<1

[a]Initial activities were determined at 40°C and pH 8.0.
[b]Initial activities were determined at 40°C and pH 9.0.
[c]<1 means lower than the detection limit of the analytical procedure used.

E. Identification of the *P. putida* L-Aminopeptidase Gene (*pepA*)

To enable the construction of a much more efficient whole-cell aminopeptidase biocatalyst, it was decided to clone the L-aminopeptidase gene from *P. putida* ATCC 12633 and subsequently accomplish overexpression of it in a suitable host organism, preferably an *E. coli* K-12 strain. For the cloning of this gene from *P. putida* ATCC 12633, the reversed genetics strategy has been used. First the amino acid sequences of both the N terminus and an internal peptide of the enzyme were determined. Next, a set of degenerate poly-

Table 3 Kinetic Parameters for the Hydrolysis of Different α-H-Amino Acid Amides by Purified L-Aminopeptidase from *P. putida* ATCC 12633

Substrate	—R	K_M (mM)	V_{max} (U mg^{-1})	k_{cat} (s^{-1})	k_{cat}/K_M (s^{-1} M^{-1})
L-Phg-NH_2	—C_6H_5	65	1565	1380	20,580
L-Phe-NH_2	—CH_2—C_6H_5	15	80	68	4,525
L-Phe-$NH_2 \cdot$HCl	—CH_2—C_6H_5	15	70	65	3,910
L-Val-$NH_2 \cdot$HCl	—CH(CH_3)$_2$	130	110	95	715
L-Leu-$NH_2 \cdot$HCl	—CH_2—CH(CH_3)$_2$	15	1915	1690	99,215
L-Ile-$NH_2 \cdot$HCl	—CH(CH_3)—CH_2—CH_3	150	270	240	1,604

merase chain reaction (PCR) primers was designed based on these two L-aminopeptidase-specific amino acid sequences and the universal genetic code (Fig. 3). A PCR with these two degenerate primers and the chromosomal DNA of *P. putida* ATCC 12633 as template results in a number of different amplification products. Using each amplification product as template in a control PCR with a second set of degenerate oligonucleotides as nested primers, a 420-bps fragment can be identified as originating from the L-aminopeptidase gene. This fragment was ligated into vector pGEM-T (Promega) and sequenced. From the sequence of this fragment it can be concluded that indeed the first 420 bps of the L-aminopeptidase encoding region have been cloned. Furthermore, it became clear that this first part of the L-aminopeptidase-encoding region contains a single *Sal*I restriction site. To investigate whether this is the only *Sal*I site in the complete L-aminopeptidase-encoding region, a Southern blotting experiment was performed using the 420-bps fragment (after labeling with biotin-14-dCTP) as gene-specific probe. Two hybridizing chromosomal DNA × *Sal*I fragments of about 3 and 1.5 kb were obtained in this experiment. Because both fragments together are most likely large enough to contain the complete L-aminopeptidase gene (a protein with a subunit mass of 52 kDa should be encoded by a gene of about 1.5 kb), it can be inferred that the complete L-aminopeptidase-encoding region indeed contains only one *Sal*I site. Because of the fact that the two hybridizing chromosomal DNA × *Sal*I fragments together should contain the complete L-aminopeptidase gene from *P. putida* ATCC 12633, it was decided to clone these two fragments instead of constructing a complete genomic library of this strain in *E. coli*. Besides a strong reduction of the number of colonies to be screened, this approach would also be successful in case the intact L-aminopeptidase gene from *P. putida* is lethal for *E. coli*.

To construct both subgenomic libraries, *P. putida* ATCC 12633 chromosomal DNA was completely digested with *Sal*I. Subsequently, all fragments were separated by agarose gel electrophoresis according to size, and fragments of about 1.5 and 3 kb were isolated. These fragments were then ligated into vector pUC19, and both recombinant vector collections were transformed into *E. coli* K-12 resulting in the two desired subgenomic libraries. Next 240 colonies of both subgenomic libraries were screened by colony hybridization, using the above-mentioned biotin-labeled 420-bps fragment as probe. This resulted in the identification of one positive clone from each subgenomic library. Finally, the inserts of both positive clones (after reduction of their size by subcloning) were bidirectionally sequenced.

By combining the nucleotide sequences of these two *P. putida* chromosomal DNA fragments at the position of the overlapping *Sal*I site, the nucleotide sequence of the complete L-aminopeptidase gene (named *pep*A) including its flanking regions is obtained. This sequence has been deposited to the EMBL database and is available by accession number AJ010261. Analysis of this sequence reveals the presence of one large open reading frame (ORF) of 1491 nucleotides, which will encode a protein of 497 amino acids with a calculated molecular mass of 52,468 Da. This size agrees rather well with the

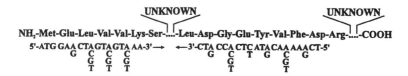

Figure 3 Amino acid sequence of the N terminus and an internal peptide of the L-aminopeptidase from *P. putida* ATCC 12633, and the degenerate PCR primers derived therefrom.

molecular mass of the L-aminopeptidase subunits as determined by SDS-PAGE, which was 52 kDa. Furthermore, the amino acid sequences of both the N terminus and the internal peptide of the purified L-aminopeptidase can be identified in the amino acid sequence, which was deduced from this ORF. Therefore, it is concluded that both cloned chromosomal DNA fragments indeed contain the complete L-aminopeptidase gene *pep*A from *P. putida* ATCC 12633.

F.   Homology of the *P. putida* L-Aminopeptidase to Other Proteins

To find protein sequences with homology to the deduced amino acid sequence of the *P. putida* L-aminopeptidase, both the SwissProt database (74,596 nonredundant sequences) and the GenBank, EMBL, DDBJ, and PDB databases (357,758 nonredundant sequences) were searched using the programs BLASTP 2.0.4 and TBLASTN 2.0.4, respectively [26]. These searches reveal a very clear homology to members of the leucine aminopeptidase (LAP) family of enzymes (EC 3.4.11.1). Some relevant data of the most homologous LAPs in these searches are given in Table 4. Among these most homologous proteins is also the bovine lens LAP, which is the only member of this family of enzymes of which

Table 4   Similarity of *P. putida* L-Aminopeptidase with Other Members of the Family of Leucine Aminopeptidases

Enzyme source	Type[a]	Accession number	Number of amino acid residues	Subunit molecular mass (Da)	Percentage identity (%)
Pseudomonas putida	B	AJ010261	497	52,468	100,0
Pseudomonas aeruginosa	B	AF054622	495	52,331	80,0
Escherichia coli (*pep*A)	B	P11648	503	54,879	53,1
Haemophilus influenzae	B	P45334	491	53,529	50,9
Mycobacterium tuberculosis	B	Q10401	343	36,152	38,8
Bacillus subtilis	B	Z99120	500	53,657	34,4
Aquifex aeolicus	B	AE000772	493	54,543	32,9
Rickettsia prowazekii	B	P27888	500	54,006	32,6
Mycobacterium tuberculosis	B	3261562	515	53,481	32,0
Synechocystis PCC 6803	B	P73971	508	53,885	32,0
Mycobacterium leprae	B	2342608	524	54,341	31,8
Chlamydia trachomatis	B	AE001279	499	54,209	31,8
Bos taurus (bovine)	M	P00727	488	53,001	31,4
Mycoplasma pneumonia	B	P75206	445	48,789	29,7
Arabidopsis thaliana	P	P30184	520	54,509	29,2
Solanum lycopersicum (mature)	P	Q42876	521	54,225	29,0
Mycoplasma salivarium	B	P47707	520	58,079	28,2
Lycopersicon esculentum (mature)	P	Q10712	518	54,419	28,2
Escherichia coli (*pep*B)	B	P37095	427	46,180	28,1
Solanum tuberosum (mature)	P	P31427	520	54,284	27,8
Schizosaccharomyces pombe	F	Q09735	513	56,195	27,0
Helicobacter pylori	B	O25294	496	54,433	26,4
Caenorhabditis elegans	N	P34629	491	52,449	24,2
Mycoplasma genitalium	B	P47631	447	49,107	23,9

[a]The following one-letter codes have been used: B, bacterium; F, fungus; N, nematode; P, plant; M, mammal.

the three-dimensional structure has been elucidated (vide infra for more details). As can be seen in the table, the overall-length similarity between the primary structure of the *P. putida* L-aminopeptidase and the most homologous LAPs is rather significant (percentage identical amino acid residues ranges from 24% to 80%). Furthermore, both the number of amino acid residues per subunit and the subunit molecular mass are typical for the LAP enzyme family. Therefore, we can conclude that the *P. putida* L-aminopeptidase also belongs to this family.

LAPs are widely distributed throughout nature (Table 4), being present in both prokaryotes (e.g., bacteria) and eukaryotes (e.g., fungi, plants, and animals), where they play a key role in the processing and regular turnover of intracellular proteins, by catalyzing the hydrolysis of amino acids from the N terminus of polypeptide chains [27]. According to these studies, catalysis by LAPs is most effective when the substrate peptides have an N-terminal leucine residue, although substantial rates of enzymatic cleavage are observed with most other amino terminal amino acids. Furthermore, LAPs also catalyze the hydrolysis of amino acid amides, alkylamides, arylamides, and hydrazides, and also have some esterase activity. So, in general, LAPs are characterized by a very broad substrate specificity. These general observations correlate well with the fact that the purified L-aminopeptidase from *P. putida* shows activity for a broad range of α-H-α-amino acid amides.

Bovine lens LAP (blLAP) is the only LAP for which structural and mechanistic data are available. It is a hexameric enzyme of total molecular mass 324 kDa, which is composed of six identical subunits of an experimentally determined molecular mass of 54 kDa [28]. For a long time it was assumed that each of these subunits had two metal ion binding sites, which are 3.0 Å apart [29]. These two metal ion binding sites are occupied by zinc ions in the native blLAP [30]. These zinc ions can be exchanged against other divalent metal cations with different exchange kinetics and different divalent metal cation affinities for both binding sites [30–32]. The substitution of the metal ions in both binding sites significantly affects the kinetic parameters (K_M and k_{cat}) of blLAP, so both metals play a role in substrate binding and activation [32]. Only recently, it has been found that the blLAP contains a third metal binding site, which is most probably also occupied with a zinc ion [33]. The distance of this third metal ion to the nearest catalytic zinc ion is approximately 12 Å. Therefore, it seems likely that this third metal ion serves a structural role (stabilization) rather than a catalytic one.

In the beginning of the 1990s, the three-dimensional structure of unliganded blLAP and its complex with the inhibitor bestatin have been determined by x-ray crystallography [34,35]. Somewhat later the x-ray structure of blLAP complexed with the inhibitors amastatin [36], L-leucine phosphonic acid [33], and L-leucinal [29], as well as an even more refined structure of the unliganded blLAP [29] were reported. Based hereon, a catalytic mechanism has been proposed for the blLAP-catalyzed hydrolysis of peptides, in which the formation of a *gem*-diolate transition state is the essential step [29,33,36]. In this mechanism, important roles have been attributed to a number of amino acid residues (see Table 5).

Because all of these important residues for the blLAP catalytic mechanism are found to be fully conserved in the *E. coli* [34] and *Rickettsia prowazekii* [37] *pep*A-encoded leucine aminopeptidases, a sequence alignment has been performed to see whether these residues are also present in the *P. putida* L-aminopeptidase. This indeed appears to be the case (see Fig. 4). Interestingly, it was observed that these residues were fully conserved in all currently known leucine aminopeptidases that are significantly homologous to the *P. putida* L-aminopeptidase. Because of these apparent similarities in (1) overall primary structure (Table 4) and (2) key active site residues among the members of the leucine

Table 5 Role of Different Amino Acid Residues of the Bovine Lens Leucine Aminopeptidase in the Two-Metal Ion Catalytic Mechanism [29,33,36]

Function	Amino acid residue	Via
Binding of zinc 488	Asp 333	O_δ and —C=O
(site 1, activation site)	Glu 335	$O_\varepsilon 2$
Binding of zinc 489	Asp 274	$O_\delta 1$
(site 2, tightly bound)	Glu 335	$O_\varepsilon 1$
	Lys 251	N_ξ
	Asp 256	$O_\delta 1$
Stabilization of *gem*-diolate	Lys 263	N_ε
transition site	Arg 337	Water molecules

aminopeptidase family, it is very likely that at least part of the conclusions drawn from structural studies of blLAP will apply for the *P. putida* L-aminopeptidase as well.

G. Construction of a Highly Efficient Whole-Cell Aminopeptidase Biocatalyst

To construct a more efficient whole-cell L-aminopeptidase biocatalyst, it was decided to clone the *P. putida* L-aminopeptidase gene in an *E. coli* K-12 host microorganism. This bacterium was chosen because of its favorable fermentation properties and the availability of a large number of specialized expression vectors.

Due to the strategy previously used for the cloning of the *P. putida pep*A gene, the intact gene was not yet available. Instead we had two recombinant clones, each containing a different part of this L-aminopeptidase gene; together both parts form the complete gene.

Because the use of PCR very often results in the introduction of unwanted mutations, it was decided to construct the whole *P. putida pep*A gene from these two clones without PCR. First, the insert fragments of both recombinant clones were isolated and fused in vector pGEM5Zf(+) (Promega). A plasmid with the two insert fragments in correct order and orientation was then identified by restriction enzyme analysis. This plasmid contains the complete *P. putida pep*A gene without any mutation.

To enable efficient heterologous expression of this gene in *E. coli*, the complete *pep*A gene was then transferred to expression plasmid pTrc99A (Pharmacia) leading to plasmid pTrcLAP. To allow maximum expression, the ATG start codon of *pep*A was fused to the *Nco*I site on this vector. In this way the distance between the ribosomal binding site on the vector and the ATG start codon of the gene is optimal. The restriction map of plasmid pTrcLAP is given in Fig. 5. With this construct, expression of the *P. putida* L-aminopeptidase gene in *E. coli* K-12 is possible. However, this expression requires the addition of the inducer isopropylthiogalactoside (IPTG) to the growth medium. Without this gratuit inducer only the basic *E. coli* L-aminopeptidase activity could be observed due to the very stringent control of expression caused by the vector-borne *lac*I^q gene. To obtain maximal L-aminopeptidase expression, IPTG has to be added in the lag or early logarithmic growth phase of the recombinant cells. Also the IPTG concentration has a strong effect on the level of L-aminopeptidase formed in the *E. coli* cells (Table 6). Although addition of 10 mM of this inducer led to the best aminopeptidase activity, 0.1 mM already results in a very good expression level.

```
MELVVKSVAA ASVKTATLVI PVGENRKLGA VAKAVDLASE GAISAVLK.R GDLAGKPGQ.  58
MEFSVKSGSP EKQRSACIVV GVFEPRRLSP IAEQLDKISD GYISALLR.R GELEGKPGQ.  58
.MLNINFVNE ESSTNQGLIV FIDEQLKLNN NLIALDQQHY ELISKTIQNK LQFSGNYGQI  59
.......... .MTKGLVLGI YSKEKEEDEP QFTSAGENFN KLVSGKLREI LNISGPSLKA  49
---------- ---------- ---------- ---------- ---------- ----G-----

.TLLLQNLQG LKAERVLLVG SGKDEALGDR TWRKLVASVA GVLKGLNGAD AVLALDDVAV 117
.TLLLHHVPN VLSERILLIG CGKERELDER QYKQVIQKTI NTLNDTGSME AVCFLTELHV 117
.TVVPSVIKS CAVKYLIIVG LGNVEKLTEA KIEELGGKIL ..QHATCAKI ATIGLKIINR 116
GKTRTFYGLH EDFPSVVVVG LGKKTAGIDE QENWHEGKEN ..IRAAVAAG CRQIQDLEIP 107
---------- --------G -G-------- ---------- ---------- ----------

NNRDAHYGKY RLLAETLLDG EYVFDRFKSQ KVEPR..ALK KVTLLADKAG QAEVERAVKH 175
KGR.NNYWKV RQAVETAKET LYSFDQLKTN KSEPRRPLRK MVFNVPTRRE LTSGERAIQH 176
INRFTSPTFT SLIASGAFLA SYRFHKYKTT LKEVEKFAVE SIEILTD..N NSEAMKLFEV 174
SVEVDPCGDA QAAAEGAVLG LYEYDDLK.. ........QK RKVVVSAKLH GSEDQEAWQR 157
---------- ---------- ---------- ---------- ---------- ----------

ASAIATGMAF TRDLGNLPPN LCHPSFLAEQ AKELGKAH.K ALKVEVLDEK KIKDLGMGAF 234
GLAIAAGIKA AKDLGNMPPN ICNAAYLASQ ARQLADSYSK NVITRVIGEQ QMKELGMHSY 236
KKLIAEAVFF TRDISNEPSN IKTPQVYAER IVEILEPL.. GVNIDVIGEH DIKNLGMGAL 232
GVLFASGQNL ARRLMETPAN EMTPTKFAEI VEENLKSASI KTDVFIRPKS WIEEQEMGSF 217
---------- -------P-N -------A-- ---------- ---------- ------M---
                         ↓        ↓           ↓                ↓
YAVGQGSDQP PRLIVLNYQG .GKKADKPFV LVGKGITFDT GGISLKPGAG MDEMKYDMCG 293
LAVGQGSQNE SLMSVIEYKG NASEDARPIV LVGKGLTFDS GGISIKPSAG MDEMKYDMCG 296
LGVGQGSQNE SKLVVMEYKG .GSRDDSTLA LVGKGVIFDT GGISLKPSSN MHLMRYDMAG 291
LSVAKGSEEP PVFLEIHYKG SPNASEPPLV FVGKGITFDS GGISIKAAAN MDLMRADMGG 277
--V--GS--- -------Y-G ---------- -VGKG--FD- GGIS-K---- M--M--DM-G
                                                         ↓ ↓ ↓
AASVFGTLRA VLELQLPVNL VCLLACAENM PSGGATRPGD IVTTMSGQTV EILNTDAEGR 353
AAAVYGVMRM VAELQLPINV IGVLAGCENM PGGRAYRPGD VLTTMSGQTV EVLNTDAEGR 356
SAAVVGTIIA LASQKVPVNV VGVVGLVENM QSGNAQRPGD VVVTMSGQTA EVLNTDAEGR 351
AATICSAIVS AAKLDLPINI VGLAPLCENM PSGKANKPGD VVRARNGKTI QVDNTDAEGR 337
-A-------- ------P-N- -------ENM --G-A--PGD ------G-T- ---NTDAEGR

LVLCDTLTYA ER.FKPQAVI DIATLTGACI VALGSHTTGL MGNNDDLVGQ LLDAGKRADD 412
LVLCDVLTYV ER.FEPEAVI DVATLTGACV IALGHHITGL MANHNPLAHE LIAASEQSGD 415
LVLADTVWYV QEKFNPKCVI DVATLTGAIT VALGSTYAGC FSNNDELADK LIKAGEAVNE 411
LILADALCYA HT.FNPKVII NAATLTGAMD IALGSGATGV FTNSSWLWNK LFEASIETGD 396
L-L-D---Y- ---F-P---I --ATLTGA-- -ALG----G- --N---L--- L--A------

RAWQLPLFDE YQ.EQLDSPF ADMGNIGGPK ..AGTITAGC FLSRFAKA.Y NWAHMDIAGT 468
RAWRLPLGDE YQ.EQLESNF ADMANIGGRP ..GGAITAGC FLSRFTRK.Y NWAHLDIAGT 471
KLWRMPLHDD YD.AMINSDI ADIANIGNVP GAAGSCTAAH FIKRFIKDGV DWAHLDIAGV 470
RVWRMPLFEH YTRQVIDCQL ADVNNIGKYR .SAGACTAAA FLKEFVTH.P KWAHLDIAGV 454
--W--PL--- Y--------- AD--NIG--- ---G--TA-- F---F----- -WAH-DIAG-

AWISGG...K DKGATGRPVP LLTQYLLDRA GA....  497     Pseudomonas putida
AWRSG....K AKGATGRPVA LLAQFLLNRA GFNGEE 503       Escherichia coli
ANSNNASALC PKGAVGYGVR LLEKFIKEYN ......  500      Rickettsia prowazekii
MTNKDEVPYL RKGMAGRPTR TLIEFLFRFS QDSA.. 488       Bovine
---------- -KG--G---- -L-------- ------
```

Figure 4 Sequence alignment of the deduced amino acid sequence of *P. putida* PepA with the primary sequences of *E. coli* aminopeptidase A, *Rickettsia prowazekii* aminopeptidase A, and bovine eye lens leucine aminopeptidase (blLAP). The bottom line is the consensus sequence, which was generated by the multiple sequence alignment algorithm Clustal W. Residues in boldface and marked with an arrow represent residues of the blLAP that have been shown to be involved in metal coordination and/or inhibitor binding.

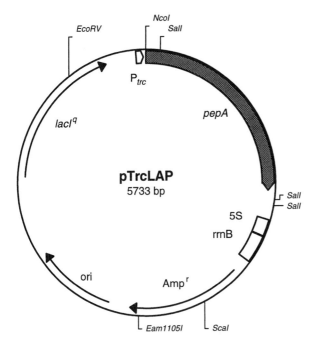

Figure 5 Restriction map of the expression vector pTrcLAP. In this vector the expression of the *P. putida* L-aminopeptidase gene (*pep*A) is controlled by the P_{trc}-*lac*I^q combination.

Although limited amounts of IPTG are required for the efficient expression of the *P. putida* L-aminopeptidase gene in *E. coli* with this construct, this system is not preferred from a commercial point of view because of the extremely high cost of this inducing compound. Therefore the *pep*A gene was transferred in the expression vectors pKK233-2 and pTrp321 too, which resulted in the constructs pKKLAP and pTrpLAP, respectively. This first construct is very similar to pTrcLAP, except the fact that pKKLAP does not contain the *lac*I^q repressor encoding gene. As a result, the expression of the *pep*A gene is nearly constitutive. In construct pTrpLAP the expression of *pep*A is driven from the promoter of the *E. coli* tryptophan operon (Fig. 6).

Table 6 Influence of the Concentration IPTG on the Expression of the *P. putida* *pep*A Gene in *E. coli* DH5α/pTrcLAP

[IPTG] (mM)	Rel. act. (%)
—	6
0.001	5
0.01	30
0.1	84
1	92
10	100

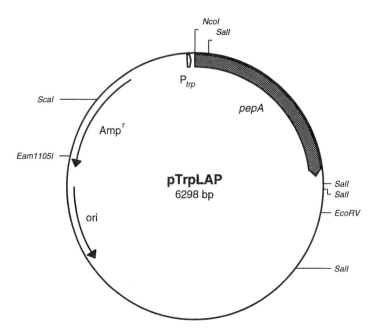

Figure 6   Restriction map of the expression vector pTrpLAP. The expression of *P. putida pep*A is driven from P$_{trp}$.

After optimization of a number of parameters, highest cellular aminopeptidase activities are obtained with *E. coli* cells containing the construct pTrpLAP. As already mentioned by its inventors [38], use of the tryptophan promoter in this construct can be further characterized by a rather simple induction strategy: the expression of the *pep*A gene automatically switches on when the tryptophan concentration in the medium drops below a certain threshold value. Because of these aspects, it was decided to further develop the fermentation of *E. coli*/pTrpLAP on lab scale. An example of a typical batch fermentation

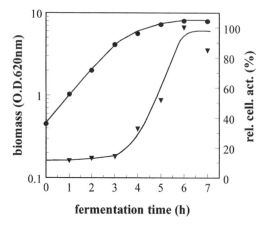

Figure 7   Fermentation of *E. coli* DH5α/pTrpLAP in a complex medium. Comparison of activity (▼) and biomass (●) profiles.

in a complex medium is given in Fig. 7. Under these circumstances the generation time is about 54 min (μ_{max} = 0.74 h^{-1}). As Fig. 7 shows, the expression of the *pep*A gene is almost completely repressed during the exponential growth phase of this fermentation, due to the high levels of free tryptophan in the growth medium. Upon reaching a status of linear growth, the expression of the enzyme from the tryptophan promoter is switched on until the maximum activity is reached upon entering the stationary growth phase. Cells obtained by this type of fermentation were harvested by centrifugation, before storage in frozen condition.

H. Application of the Improved Whole-Cell Aminopeptidase Biocatalyst

The performance of the *E. coli*/pTrpLAP whole cells, which were obtained by the fermentation described above, has been tested in a standard resolution reaction of an arbitrary α-hydrogen amino acid amide, at the optimum reaction conditions for the *P. putida* L-aminopeptidase (40°C, pH 9.5).

As the graphs in Fig. 8 show, the hydrolysis reaction with the recombinant *E. coli* cells proceeds much faster than with the *P. putida* wild-type cells. Therefore, the use of this improved whole-cell aminopeptidase biocatalyst results in significantly shorter reaction times. The reason for this better performance of the newly constructed whole-cell biocatalyst is its much higher aminopeptidase content, reflected by a 25-fold increase in specific activity compared to *P. putida*. This is caused by improved expression of the *P. putida pep*A gene, due to the use of a strong promoter in combination with a multicopy situation.

The strategy used in this case for the cloning and heterologous expression of *P. putida pep*A did not result in any mutation on protein level. Therefore, the wild-type *P. putida* enzyme will be formed in the recombinant *E. coli* strain. This means that all intrinsic properties of this L-aminopeptidase (e.g., pH optimum, substrate range, enantiospecificity) are unaltered in comparison with the properties of the enzyme found in *P. putida*.

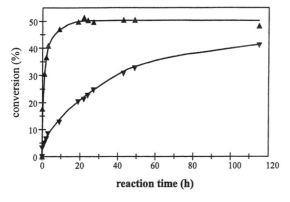

Figure 8   Resolution of an α-H-amino acid amide by *E. coli* DH5α/pTrpLAP (▲), and *P. putida* ATCC 12633 (▼). Both cells were used in the same cell-to-substrate ratio.

III. AMINO AMIDASE FROM *MYCOBACTERIUM NEOAURUM* ATCC 25795

A. Introduction

α,α-Disubstituted amino acids form an interesting class of amino acid analogs for use in pharmaceuticals. A well-proven example is the application of L-α-methyl DOPA (**5**) as an antihypertensive drug. Several routes to enantiomerically pure α,α-disubstituted amino acids have been described in the literature [39]. Methods initially described by Schöllkopf [40] and Seebach [41] are frequently used; diastereoselective alkylation of chiral lactim ethers or oxazolidinones gives access to a variety of α,α-disubstituted amino acids in e.e. > 90%. Both methods are, however, of limited use on large scale because of the great number of reaction steps. More recently, asymmetric phase transfer–catalyzed alkylations of the benzaldehyde Schiff base of alanine *tert*-butyl ester were described to give α,α-disubstituted amino acids in e.e.'s up to 82% [42].

Identically to the enzymatic resolution process for α-H-amino acid amides by *P. putida*, we searched for a new biocatalyst for the stereoselective hydrolysis of α,α-disubstituted amino acid amides (**15**). Through screening a new biocatalyst *Mycobacterium neoaurum* ATCC 25795 was obtained that fulfilled the demand for stereoselective hydrolysis [43,44].

B. Preparation of Racemic α,α-Disubstituted Amino Acid Amides

Analogous to the α-H-amino acid amides, the Strecker synthesis is the most direct way to prepare the disubstituted amino acid amides (**15**) (Scheme 6). However, in this case the basic hydrolysis of the aminonitrile (**14**) is hampered by steric interactions and is generally performed in conc. H_2SO_4 or in HCl–saturated formic acid (both containing 1 eq of water) [45].

Another route to the racemic amino acid amides is also depicted in Scheme 6, i.e., phase transfer–catalyzed alkylation of the benzaldehyde Schiff bases of amino acid amides (**16**) gives access to the desired substrates [46]. Especially alkylation with activated alkyl bromides like allylic or benzylic bromides results in high yields of the disubstituted amino acid amides.

A less common method for the preparation of disubstituted amino acid amides proceeds via cationic intermediates: reaction of *N*-acyliminium ions of amino acid amides with allylsilanes and silyl enol ethers provide unsaturated and γ-keto-disubstituted amino acid amides [47].

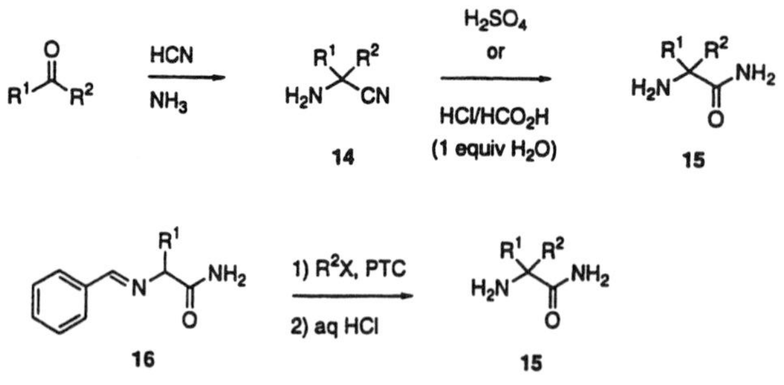

Scheme 6 Synthesis of racemic α,α-disubstituted amino acid amides.

C. Resolution of α,α-Disubstituted Amino Acid Amides Using Whole Cells of *Mycobacterium neoaurum*

Enzymatic hydrolysis of the racemic amides by the amino amidase from *M. neoaurum* affords the (S)-α,α-disubstituted amino acids and the (R)-α,α-disubstituted amino acid amides (**15**) in almost 100% e.e. at 50% conversion for most α-methyl-substituted compounds (E > 200) [43] (see Scheme 7 and Table 7). Only for glycine amides with two small substituents the enantioselectivity is decreased: for example, for isovaline amide the enantiomeric ratio E = 19 and for (α-Me)allylglycine amide E = 40 [44]. Also α-H-amino acid amides are substrates and are hydrolyzed enantioselectively; in contrast, however, dipeptides are not hydrolyzed [45]. For all α-methyl-substituted substrates the activity is high. Reactions performed at 5–10 w/w% substrate solutions in water (pH 8, 37°C) with 0.3–1.0 w/w% of freeze-dried biocatalyst are in general completed (i.e., 50% conversion) after 5–48 h. Increasing the size of the small substituent to ethyl, propyl, or allyl dramatically reduces the activity, especially if the large substituent contains no —CH$_2$— spacer at the chiral center. Due to the longer reaction times the enantioselectivity is also reduced [44].

The amino amidase from *M. neoaurum* is active over a broad pH range (vide infra) in which no natural hydrolysis of the amides is observed.

Analogous to *P. putida* which does not hydrolyze disubstituted amino acid amides, in general *M. neoaurum* is used on a preparative scale without any further purification of the fermented whole cells.

Further on in this chapter some specific examples of the use of enantiopure α,α-disubstituted amino acids will be given.

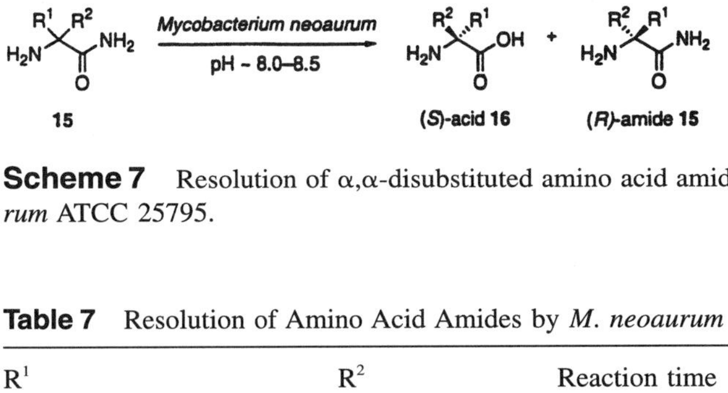

Scheme 7 Resolution of α,α-disubstituted amino acid amides (**15**) by whole cells of *M. neoaurum* ATCC 25795.

Table 7 Resolution of Amino Acid Amides by *M. neoaurum* ATCC 25795 [44]

R^1	R^2	Reaction time	Enant. ratio
CH$_3$	CH$_2$CH$_3$	5 h	19
CH$_3$	CH$_2$CH=CH$_2$	8 h	40
CH$_3$	*iso*-Propyl	24 h	>200
CH$_3$	*iso*-Butyl	8 h	>200
CH$_3$	*n*-Hexyl	24 h	>200
CH$_3$	*n*-Nonyl	24 h	>200
CH$_3$	CH$_2$Ph	24 h	>200
CH$_2$CH$_3$	CH$_2$Ph	48 h, 39% conv.	100
CH$_2$CH=CH$_2$	CH$_2$Ph	72 h, 19% conv.	15
CH$_3$	CH$_2$CH=CHPh	48 h, 33% conv.	150
CH$_3$	Ph	48 h	110
CH$_2$CH$_3$	Ph	>72 h, 13% conv.	50
CH$_2$CH=CH$_2$	Ph	>72 h, no reaction	—

D. Purification and Characterization of an L-Amino Amidase from *Mycobacterium neoaurum* ATCC 25795

In an attempt to learn more about its properties, the L-aminoamidase from *M. neoaurum* ATCC 25795, which was responsible for the enantioselective resolution of DL-α-methyl-valine amide, has been purified and characterized [1,49]. The purification procedure included ammonium sulfate fractionation, Superdex 200 gel filtration, and Mono Q anion exchange chromatography. By this procedure, the L-aminoamidase can be purified from the crude extract approximately 280-fold and with a 10% yield. The obtained enzyme preparation is homogeneous as judged by SDS-PAGE.

The molecular mass of the native enzyme is estimated at about 136 kDa. Because the subunit molecular mass is about 40 kDa, this L-aminoamidase most likely consists of three to four subunits of equal size. Isoelectric focusing showed that the isoelectric point of the purified enzyme was about 4.2.

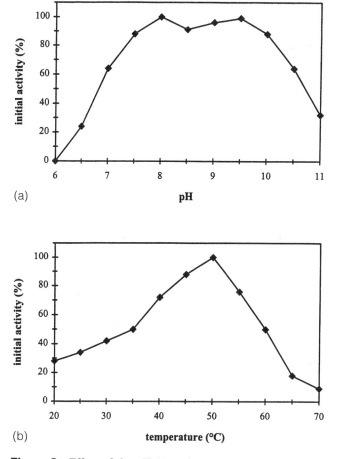

(a)

(b)

Figure 9 Effect of the pH (a) and temperature (b) on the initial activity of the purified L-amino amidase from *M. neoaurum* ATCC 25795.

Classification of this enzyme as a serine, cysteine, aspartic, or metallo-dependent enzyme [50] is somewhat problematic because occasionally inhibitors from the same class led to contradictory results. The observed effects suggest, however, that both metal ions and sulfhydryl groups may play a major role in the hydrolytic mechanism. Therefore, most probably the enzyme is a metallocysteine hydrolase.

The purified enzyme displays its highest activity at 50°C. Furthermore, the enzyme is fairly stable because more than 90% of the activity is retained after a preincubation for 30 min up to 55°C. The L-aminoamidase is active between pH values of 6.5–11.0, with a broad pH optimum at pH values of 8.0–9.5 (Fig. 9). Determination of the substrate specificity of the purified enzyme showed that this amide hydrolase was active toward a rather broad range of both α-H- and α-alkyl-substituted amino acid amides (Table 8). Within these two substrate classes, highest activity is observed for the cyclic amino acid amide DL-proline amide and the α-alkyl-substituted amino acid amides DL-isovaline amide and DL-α-allylalanine amide. Furthermore, this enzyme displays modest activity toward a number of simple amides (e.g., acetamide and propionamide). The enzyme appears to be inactive with the α-hydroxy acid amide DL-mandelic acid amide and with the dipeptide L-Phe-L-Leu. These results clearly show that this purified amide hydrolase belongs to the group of aminoamidases.

From the kinetic properties of L-aminoamidase (Table 9) it can be concluded that an additional methylene group adjacent to the C_α atom in the α-methyl-substituted substrate results in an increased affinity of the enzyme for this substrate. On the other hand, much lower affinities are observed for substrates with a bulky group directly adjacent to the C_α atom. Finally, the purified enzyme displays L-selective amino acid amide hydrolase activity toward both α-H- and α-alkyl-substituted amino acid amides. The enzyme has the lowest enantioselectivity toward alanine amide ($E \geq 25$).

Table 8 Substrate Specificity of the Purified L-Amino Amidase from *M. neoaurum* ATCC 25795

Substrate	Activity	Substrate	Activity
Acetamide	+	DL-α-Methylphenylglycine amide	+
Propionamide	+	DL-α-Methylphenylalanine amide	++
Butyramide	−	DL-α-Methylhomophenylalanine amide	+
Isobutyramide	−	DL-α-Ethylphenylglycine amide	−
Acrylamide	−	DL-α-Propylphenylglycine amide	−
Methacrylamide	−	DL-α-Allylphenylglycine amide	−
DL-Mandelic acid amide	−	DL-α-Benzylphenylglycine amide	−
DL-Valine amide	+++	DL-α-Ethylphenylalanine amide	+
DL-Phenylalanine amide	++	DL-α-Propylphenylalanine amide	+
DL-Proline amide	++++	DL-α-Allylphenylalanine amide	−
DL-α-Methylvaline amide	++	DL-α-Isopropylphenylalanine amide	−
DL-α-Methylleucine amide	+++	L-Phe-L-Leu	−
DL-Isovaline amide	++++		
DL-α-Allylalanine amide	++++		

Activities are given relative to the activity toward DL-α-methylvaline amide.
−, <5%; +, 5–50%; ++, 50–100%; +++, 100–200%; and ++++, >200% of that activity.

Table 9 Kinetic Parameters for the Hydrolysis of Different α-Alkyl-Substituted Amino Acid Amides by Purified L-Aminoamidase from *M. neoaurum* ATCC 25795

Substrate	—R^1	—R^2	K_M (mM)	V_{max} (U mg^{-1})	k_{cat} (s^{-1})	k_{cat}/K_M (s^{-1} M^{-1})
DL-α-Me-Val-NH$_2$	—CH$_3$	—CH(CH$_3$)$_2$	100	75	50	500
DL-α-Me-Leu-NH$_2$	—CH$_3$	—CH$_2$CH(CH$_3$)$_2$	10	125	85	8300
DL-Iso-Val-NH$_2$	—CH$_3$	—CH$_2$CH$_3$	30	160	105	3600
DL-α-Allyl-Ala-NH$_2$	—CH$_3$	—CH$_2$CH=CH$_2$	5	240	160	32,000
DL-α-Me-Phg-NH$_2$	—CH$_3$	—C$_6$H$_5$	335	25	15	50
DL-α-Me-Phe-NH$_2$	—CH$_3$	—CH$_2$C$_6$H$_5$	5	80	55	10,700
DL-α-Me-homo-Phe-NH$_2$	—CH$_3$	—CH$_2$CH$_2$C$_6$H$_5$	5	20	15	2700

E. Resolution of α,α-Disubstituted α-Amino Acid Esters Using PLE

Resolution of α,α-disubstituted α-amino acids (but also of α,α-disubstituted α-hydroxy acids [51]) can also be performed on their esters. We have used pig liver esterase (PLE) for resolution of a variety of α,α-disubstituted esters. Although all substrates tested are hydrolyzed, only for α-substituted phenylglycine esters (**17**, X = NH$_2$) and α-substituted mandelic esters (**17**, X = OH) reasonable enantioselectivities are observed for hydrolysis of the (*S*) enantiomer (*E* = 2–114) [44]. For these types of amino acids and hydroxy acids the PLE resolution forms a valuable extension of the *M. neoaurum* resolution technology, since the corresponding racemic amides are hydrolyzed sluggishly (Scheme 8). The PLE resolution of α-allylphenylglycine ethyl ester has been applied in the synthesis of D-α-phenylproline (vide infra).

IV. AMIDASE FROM *OCHROBACTRUM ANTHROPI* NCIMB 40321

A. Introduction

The previous parts showed that *P. putida* and *M. neoaurum* biocatalysts enable us to produce a large array of α-H- and α-alkyl-substituted amino acids in optically pure form. Nevertheless, for the kinetic resolution of very bulky α,α-dialkyl-substituted amino acid amides, α-hydroxy acid amides, and *N*-hydroxyamino acid amides we required another biocatalyst. To fulfill this desire, a classical screening program was started, aimed at the isolation of a new amidase biocatalyst [52]. Starting from various soil and sewage samples, enrichment cultures using DL-mandelic acid amide as sole nitrogen source could be readily obtained. Then 125 microorganisms were isolated and further investigated by growth experiments in media containing D- and/or L-mandelic acid amide as sole nitrogen source.

X = OH, NH$_2$
R = (CH$_2$)$_n$CH$_3$ (n = 0–3), allyl, benzyl

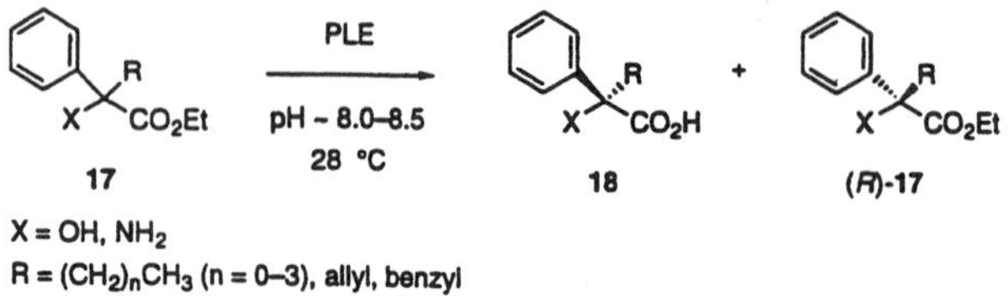

Scheme 8 Resolution of α-substituted phenylglycine esters and mandelic esters (**17**) by PLE.

By this procedure three microorganisms could be identified that hydrolyzed DL-mandelic acid amide completely L-enantiospecific. The most interesting one of these three strains, based on its high cellular activity, could be identified as an *Ochrobactrum anthropi* strain, and this strain has been deposited at the NCIMB culture collection as NCIMB 40321.

A remarkable feature of the *O. anthropi* biocatalyst is its relaxed pH profile [52]. Although the amidase displays its highest activity at pH 8.5, still 55% of this activity is retained at pH 5.0. This enables the hydrolysis of substrates, which are only very poorly soluble at weakly alkaline conditions, by just performing the hydrolysis reaction at slightly acidic conditions (e.g., see Secs. V.A and V.B). Another attractive feature of the *O. anthropi* L-amidase is its very good temperature stability. Even a preincubation at 50°C for 2 h, does not lead to a decrease in activity. Therefore, the resolution reactions with *O. anthropi* whole cells are optimally performed at 50°C and at a pH value between 5 and 8.5.

Although the amidase from *O. anthropi* has its highest hydrolytic activity for α-hydrogen-containing amino acid amides (**12**), α-(*N*-hydroxyamino)- and α-hydroxy amides (**19, 20**) and α-alkyl-substituted amino acid amides (**15**) are also hydrolyzed enantioselectively. However, other substituents at the α position did not show any reactivity as was tested for α-substituted phenacetyl amide [53] (see Table 10).

Although it is not completely clear as to which range of substituents is accepted by the enzyme, it is expected that hydrogen bond acceptors are needed at the α position since α-amido amides are not hydrolyzed and β-amino acid amides are only poor substrates. Substitution at the amide nitrogen greatly retards the hydrolytic activity of the amidase, as was proven for compound **22**. The nature of the R substituent of the amino acid amide **12** has less influence on the hydrolytic activity, although sterically more demanding substrates, like R = *tert*-butyl, decrease the reactivity [52]. The highest activity is observed for leucine amide and phenylglycine amide. For all substrates tested full (*S*) stereoselectivity (*E* > 200) is observed.

Addition of up to 50% of alcoholic solvents is possible without considerable loss of activity [53]. Typical hydrolysis reactions with *O. anthropi* are performed in water at temperatures of 40–50°C and pH between 5.0 and 8.5. Buffering of the reaction is not necessary. α,α-Disubstituted glycine amides are hydrolyzed stereoselectively by the amidase from *O. anthropi* albeit with a lower specific activity compared to the α-hydrogen analogs (see Scheme 9 and Table 11).

Table 10 Hydrolysis of α-Substituted Phenacetylamides **21** by *O. anthropi* NCIMB 40321

Substrate 21	Rel. act. (%)	Stereosel.
X = NH$_2$	100	100% S
X = NHOH	25	100% S
X = OH	5	100% S
X = H	<0.1	—
X = NH—CHO	<0.1	—
X = NH—COCH$_3$	<0.1	—
X = CONH$_2$	<0.1	—
X = CH$_3$	<0.1	—
X = SH	<0.1	—
X = Br	<0.1	—

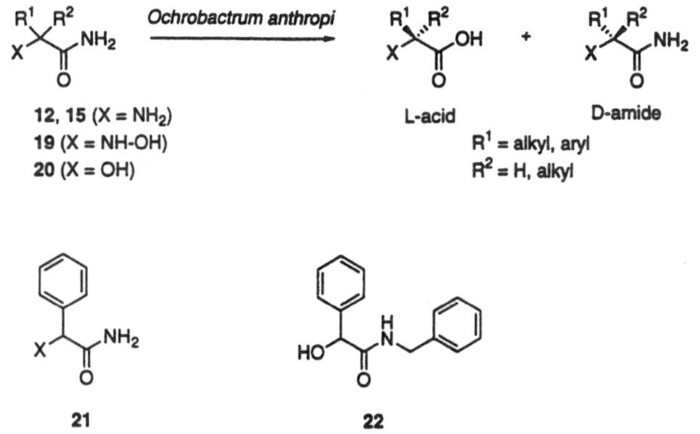

Scheme 9 Hydrolysis of α-substituted phenacetyl amides **21** by *O. anthropi* NCIMB 40321.

Most disubstituted amino acid amides are hydrolyzed by *O. anthropi* fully stereo-selectively to form the (*S*)-enantiomer (i.e., $E > 200$). Only for substrates with two small substituents is the stereoselectivity reduced. For example, for α-methylmethionine amide the enantiomeric ratio $E = 62$, while for isovaline amide $E = 6$ [54]. For the latter type of substrates the use of *M. neoaurum* is preferred over *O. anthropi*. In fact, the amidases of both species are complementary to one another since *M. neoaurum* is unable to hydro-lyze substrates with larger substituents that can still be hydrolyzed by *O. anthropi*. While the amino acid amides **15b–d** are hydrolyzed by *M. neoaurum* only with great difficulty or not at all, *O. anthropi* is sufficiently active for these substrates [44,52]. Some specific examples of the resolution with *O. anthropi* are shown in Sec. V.

Currently, the purification and biochemical characterization of an L-specific amidase from this microorganism is in progress. The results of this study will be published in the near future.

V. SPECIFIC EXAMPLES OF THE ENZYMATIC RESOLUTION PROCESS AND APPLICATION OF THE AMINO ACIDS

A. Thiamphenicol

The antibiotics thiamphenicol (**23a**) and florfenicol (**23b**) are fully synthetic analogs of chloramphenicol, active against gram-positive and gram-negative bacteria [57]. Since only

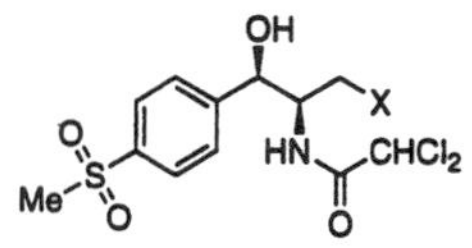

23a: X = OH; Thiamphenicol
23b: X = F; Florfenicol

the (*1R,2R*)-enantiomers are active, early synthetic routes were developed in which the enantiomerically pure intermediates were obtained by classical resolution or preferential crystallization [58]. Via these resolution methods the unwanted enantiomer remained as waste. Recently, a new method for the production of the *para*-substituted (*2S,3R*)-3-phen-ylserines **25a/b** was developed based on the enzymatic resolution of the corresponding racemic *threo*-amides **24a/b** [59]. These amino acid amides are readily prepared by aldol reaction of the corresponding para-substituted benzaldehydes with glycine amide. Reso-

Table 11 Relative Activity of *O. anthropi* NCIMB 40321 for α-Substituted Amino Acid Amides **12/15**

Substrate	R^1	R^2	Rel. act. (%)	Enant. ratio
12a	Ph	H	100	>200
15a	Ph	CH_3	2	>200
15b	Ph	CH_2CH_3	4	>200
15c	Ph	$Ch_2CH_2CH_3$	1	>200
15d	Ph	$CH_2CH=CH_2$	4	>200
15e	Ph	CH_2Ph	0	—
12b	*iso*-Pr	H	25	>200
15f	*iso*-Pr	CH_3	5	>200

lution of the amides **24a/b** using the amidase from *O. anthropi* NCIMB 40321 yields the (*2S,3R*) enantiomer of the corresponding amino acids (**25a/b**) in enantiomeric excesses of >99% and chemical yields up to 50%. Although the pH optimum of the amidase from *O. anthropi* NCIMB 40321 is at pH 8.0–8.5, the enzymatic resolution is performed at pH 5.6–6.0 to ensure a fair solubility of the otherwise unsoluble substrate **24a**. The unwanted (*2R,3S*) isomers of the amides can easily be separated from the amino acids **25a/b** after Schiff base formation with the corresponding para-substituted benzaldehyde and recycled to the racemic *threo*-amides **24a/b** under basic conditions (Scheme 10). In fact, these racemization conditions are identical to the conditions used in the preparation of the racemic *threo*-amide so that both steps can be combined. Thus, an efficient new method for the preparation of the thiamphenicol intermediates **25a/b** is developed that combines a high overall yield with a minimal number of reaction steps. Since these amino acids are already intermediates in the preparation of thiamphenicol and florfenicol, the enzymatic method can easily be implemented in the existing production process.

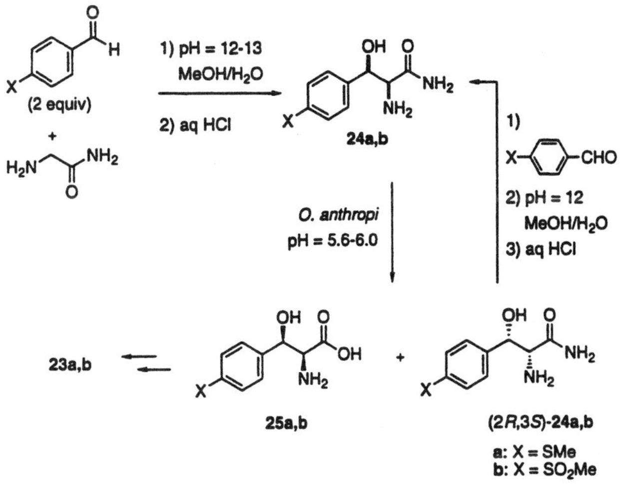

Scheme 10 Resolution of sulfur substituted *threo*-phenylserine amides using *O. anthropi* NCIMB 40321.

B. Cericlamine

β-Amino alcohols with substituents at the β position can be prepared by reduction of the corresponding amino acids and therefore form an attractive class of compounds for preparation via the above-described enzymatic resolution techniques. An example of this type of amino alcohol is the potent and selective synaptosomal 5-HT uptake inhibitor cericlamine HCl (**26**), under development by Jouvenal as an antidepressant [60].

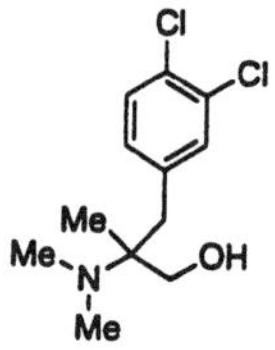

26: Cericlamine

We prepared the (*S*)-enantiomer of cericlamine via enzymatic resolution of α-methyl-3,4-dichlorophenylalanine amide (**28**) and subsequent reduction of the amino acid as described in Scheme 11 [61]. Racemic α-methyl-3,4-dichlorophenylalanine amide (**28**) is prepared by phase transfer–catalyzed benzylation of the benzaldehyde Schiff base of alanine amide (**27**) followed by acidic workup. Since this substrate is nearly insoluble in water, the enzymatic resolution using the amidase from *O. anthropi* NCIMB 40321 is performed at pH 5.3. At this pH there is a sufficient amount of the substrate in solution and still approximately 50% of the amidase activity left to allow the enzymatic hydrolysis

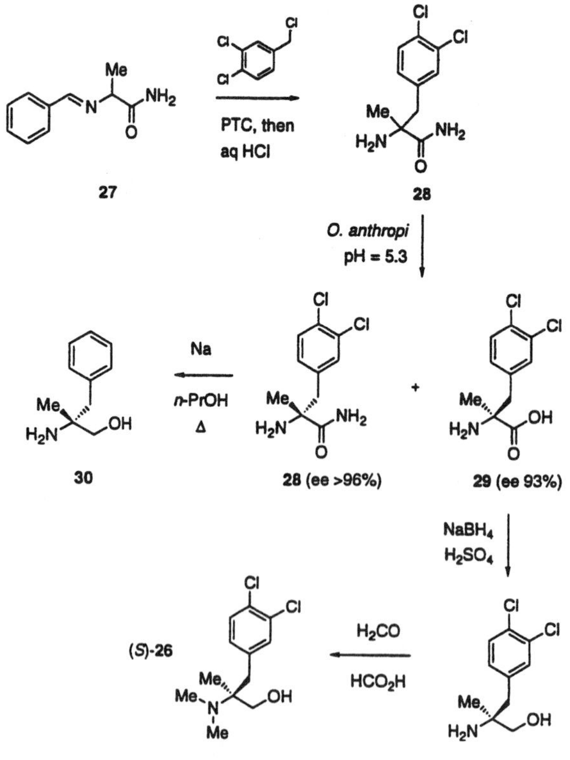

Scheme 11 Synthesis of (*S*)-cericlamine.

to proceed in a reasonable time. Depending on the conversion either the (R)-enantiomer of the substrate or the (S)-amino acid **29** can be isolated in high enantiomeric excess (enantiomeric ratio $E = 170$). The reduction of both the (R)-amide and the (S)-acid was tested with recently developed new reduction methods. Amino acids and amino acid amides can both be reduced to the corresponding amino alcohols without any further derivatization by using $NaBH_4/H_2SO_4$ [62] or sodium in boiling n-propanol [63], respectively. Although for α-H-amino acid amides the sodium/propanol reduction is accompanied by racemization, for α,α-disubstituted amino acid amides reduction proceeds in quantitative conversion without any racemization. However, the reduction of the (R)-amide **28** with sodium/propanol fails since the chloride substituents are also removed resulting in α-methylphenylalaninol (**30**). On the other hand, using $NaBH_4/H_2SO_4$ as a reducing agent, the (S)-amino acid can be transformed directly into the corresponding amino alcohol, without prior formation of an amino acid ester. Methylation of the amine under Eschweiler-Clark conditions eventually yields the (S)-enantiomer of cericlamine (**26**).

C. α-Phenyl-Substituted L-Proline

α-Alkylated prolines have been used as chiral educts in organic synthesis and as a tool in peptide research, where they can be used to influence the conformational flexibility of peptides [64]. While enantiopure α-alkylated amino acids are available via different synthetic methods, as already described in Sec. III, α-phenyl-substituted proline is not easily accessible. Recently, (R)-α-phenylproline (**31**) (having the stereochemical configuration of L-proline) was used in the synthesis of the nonpeptide substance P antagonist **32**, selective for the NK$_1$ receptor [65].

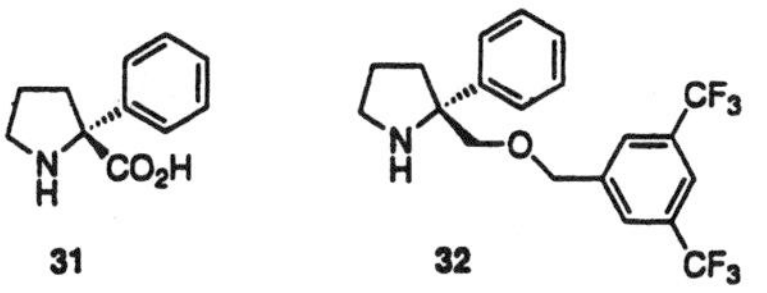

We chose to prepare (R)-α-phenylproline derivatives via enzymatic resolution of the intermediate α-allylphenylglycine ethyl ester (Scheme 12) [66]. The racemic α-allylphenylglycine ethyl ester is prepared by phase transfer–catalyzed allylation of the Schiff base of ethylphenylglycinate (**32**) and resolved with pig liver esterase (PLE) [44]. Although PLE is not fully enantioselective for this substate ($E = 23$), by running the reaction for

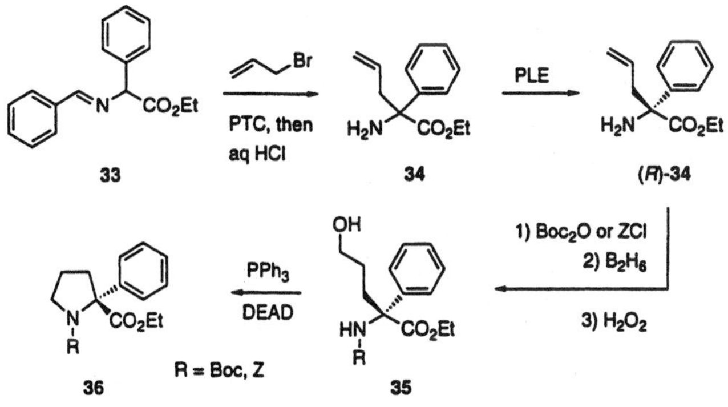

Scheme 12 Synthesis of (R)-α-phenylproline derivatives.

more than 50% conversion (i.e., 57% conversion), ethyl (*R*)-α-allylphenylglycinate (*R*-**34**) is obtained in an enantiomeric excess of 96%. Note that while the enantiopure (*S*)-α-allylphenylglycine is also available by the amidase route using *O. anthropi*, the low activity of the amidase for the amide substrate makes it difficult to obtain the remaining (*R*) enantiomer in a high enantiomeric excess. After *N*-protection of ethyl (*R*)-α-allylphenyl-glycinate with a *tert*-butyloxycarbonyl (Boc) or benzyloxycarbonyl (Z) group, hydrobor-ation, followed by oxidation, yields the α-(3-hydroxypropyl) analog. Ring closure of this compound under Mitsunobu reaction conditions yields the *N*-protected (*R*)-α-phenylproline ester **36** (see Scheme 12).

D.   Peptides Containing α-Substituted Amino Acids

α-Substituted amino acids as well as amino alcohols are found in nature in peptaibol antibiotics, like suzukacilin, alamethicin, trichotoxin, zervamicin, emerimicin, and tricho-virin [67]. Peptaibols are small peptides produced by various fungi and contain one or more α-aminoisobutyric acid (Aib, **37**) or D-isovaline (Iva, **38**) units and L-valinol, L-leucinol, or L-phenylalaninol at their carboxy terminus. Because of this, peptides contain-ing Aib residues and an amino alcohol are called "peptaibols." For example, the zervam-icins Ic, IIa, IIb, and L [48] compose a family of 16-amino-acid-residue polypeptides isolated from the fungi *Emerricellopsis salmosynnemata*. They are active against gram-positive bacteria by altering the ionic permeability of biological membranes by formation of ion channels.

Zervanicine IIa: Ac-Trp-Ile-Gln-**Aib**-Ile-Thr-**Aib**-Leu-Aib-Hyp-Gln-**Aib**-Hyp-**Aib**-Pro-**Phl**

Zervamicine IIb: Ac-Trp-Ile-Gln-**Iva**-Ile-Thr-**Aib**-Leu-**Aib**-Hyp-Gln-**Aib**-Hyp-**Aib**-Pro-**Phl**

(Ac = acetyl, Hyp = L-trans-4-hydroxyproline, Phl = phenylalaninol)

The presence of α,α-disubstituted amino acids in peptides results in some specific conformational changes compared to their α-hydrogen analog and, in addition, their con-formation tends to be more stable. The α,α-disubstituted amino acids in these peptides cause severe steric hindrance and the synthesis of related peptides is greatly complicated by this property. However, racemization in peptide coupling is not possible.

Peptides containing these residues show extremely high crystallinity, thus allowing one to perform an x-ray characterization of the conformation. The stereochemistry of peptides containing α,α-disubstituted amino acids is rather unique, since they possess significant constraints on their conformational freedom. In particular, this point is relevant to the exploitation of these compounds as precise molecular rulers or as scaffolding blocks in de novo design of protein and enzyme mimetics, but also in the three-dimensional structure–activity relationship of conformationally constrained, enzyme-resistant agonists and antagonists of bioactive peptides.

The crystal state and solution structural preferences of Aib, the prototype of the α-alkylated amino acids, has been extensively examined [69]. From the results obtained the following conclusions may be drawn: (1) Aib homopeptides, beginning at the trimer level, adopt the 3_{10}-helical structure, irrespective of the main chain length [70]. The α-helical structure is never observed. (2) Tripeptides and longer peptides containing Aib along with other amino acids are folded either in the 3_{10}- or in the α-helical structure, depending on the main-chain length, Aib content, sequence, and environmental conditions. In particular, the minimal peptide chain length for α-helix formation in the crystal state corresponds to about seven residues. Since Aib is an achiral residue, the screw sense of the helix that is formed depends on the chirality of the constituent amino acids. (3) Aib is the strongest known β-bend-forming residue [71].

Aib **37**

Iva **38**

(αMe)Val **39**

(αMe)Phe **40** (n = 0)
(αMe)Hph **41** (n = 1)

(αMe)Trp **42**

(αEt)Phe **43** (R = Et)
(α-allyl)Phe **44**
(R = $CH_2CH{=}CH_2$)

D-(αMe)Phg **45**

(αEt)Phg **46**

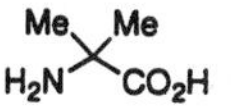

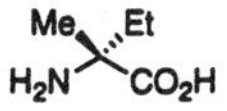

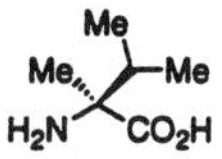

Peptides containing chiral α-alkylated amino acids have been extensively studied by Toniolo et al. [72]. The results obtained from x-ray diffraction, ^{1}H nuclear magnetic resonance (NMR), former transform infrared spectroscopy (FTIR), and circular dichroism spectroscopy (CD) may be summarized as follows: (1) In general, their tripeptides (and longer) are folded in type I(I′) and III(III′) β-bend conformations or in a 3_{10} helix, depending on the main-chain length. However, conditions known to be required for a peptide folding in an α-helical conformation (peptide main-chain length higher than seven residues, adequate presence of proteinogenic amino acids in the sequence) have not been examined so far. (2) Fully extended structures are sometimes seen in derivatives and small peptides, except in (αMe)Val (**39**) peptides. (3) β-Plated sheet structures have not been found in any of the peptides examined. (4) As for the relationship between the chirality at the α carbon and screw sense of the bend or helical structure that is formed, peptides rich in Iva with a linear side chain, (αMe)Val with a β-branched side chain, and (αMe)Hph (**41**) with a δ-branched side chain exhibit normal behavior, i.e., the same as that shown by proteinogenic amino acids (L residues give right-handed bends and helices). On the other hand, the behavior of all of the amino acids with a γ-branched side chain, i.e. (αMe)Leu, (αMe)Phe (**40**), (αMe)Trp (**42**), and (αEt)Phe (**43**), tends to be opposite.

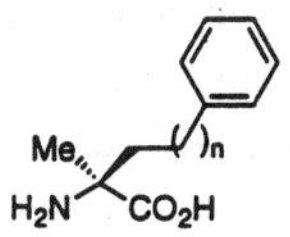

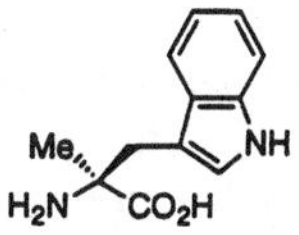

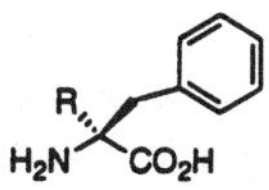

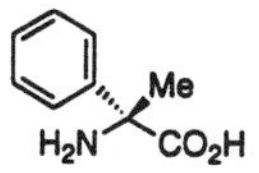

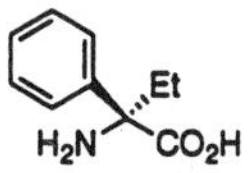

α-Alkylated amino acids are expected to become an important component in the arsenal of medicinal chemists. In this connection, bioactive analogs of the formyl methionyl tripeptide chemoattractant have been synthesized and characterized [73]. Peptides containing (αMe)Trp have been extensively studied by Horwell et al. as dipeptoid analogs of central cholestokinin (CCK-B) [74]. Another dipeptide isostere for the Phe-His portion of aspartic proteinase inhibitors is derived from L-(α-allyl)Phe [75]. Also such antibiotics as ampicillin can be prepared with α-methylated side chains, i.e., D-(αMe)Phg, **45**, and are as active as the parent antibiotic [76]. In addition, the dipeptide sweetener aspartame with (αMe)Phe incorporated is as sweet as the parent compound **8** [77]. Also fedotozine [78], containing the amino alcohols derived from (αEt)Phg (**46**), and the already mentioned (αMe)DOPA (**5**) [11] and cericlamine (**26**) [60] are α-methyl amino acid analogs with biological activity (vide supra).

E. Chiral Ligand for Asymmetrical Synthesis

Amino acids, both natural and unnatural, have always formed an outstanding source of chirality in (catalytic) asymmetrical synthesis. Of the numerous examples of the use of amino acids or oligopeptides as chiral catalysts or ligands in asymmetrical synthesis, only three examples are described here in which the original amino acids are prepared by enzymatic resolution: The use of homometphos (**47**) (prepared from D-homomethionine) in the asymmetrical Grignard cross-coupling reaction of vinyl bromide and 1-phenylethyl magnesium chloride [79], the (αMe)Phe containing diketopiperazine **48** [80] and the dipeptide Schiff base **49** [81] in asymmetrical cyanohydrin formation of benzaldehydes.

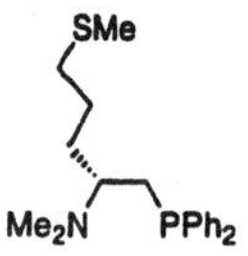

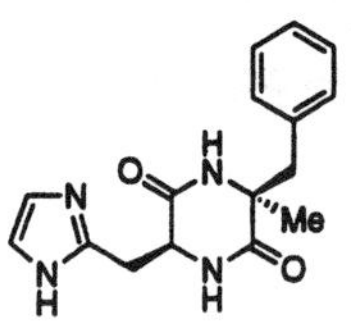

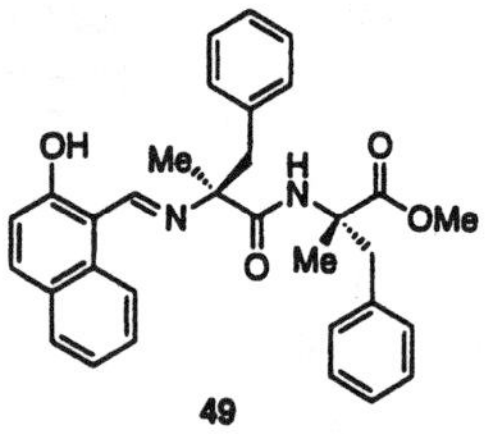

Lewis acids derived from (αMe)DOPA and menthol are also very effective catalysts in the asymmetrical aldol reaction of ketene silyl acetals with aldehydes, resulting in silylated β-hydroxy esters with e.e. values up to 99% [82].

VI. ENANTIOPURE UNSATURATED AMINO ACIDS AS BUILDING BLOCKS

Enantiopure unsaturated amino acids are ideally suited for further selective transformations. In this section some selected examples are given. Radical additions will lead to enantiopure sulfur-containing amino acids. Ring-closing olefin metathesis will give cyclic amino acids, whereas Pd-catalyzed oxy palladation, amido palladation, and cross-coupling will lead to a vast array of enantiopure heterocyclic compounds.

A. Radical Additions

Sulfur-substituted amino acid derivatives can be obtained by radical addition of mercaptans to unsaturated amino acids [83]. The addition of thiols and thioacetic acid to the double bond of allylglycine and its derivatves is performed with aza-*iso*-butyronitrile (AIBN) as

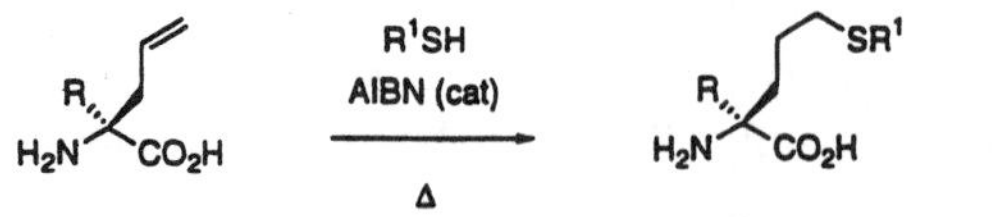

Scheme 13 Radical addition of thiols to unsaturated amino acid derivatives.

a radical chain initiator (Scheme 13). This radical addition is performed on α-H- and α-Me-substituted allylglycine (**50/51**) as well as on their N-protected- and carboxylic acid derivatives (ester, amide). In general, high yields (up to 99%) of δ-sulfur-substituted amino acid derivatives (**52**) are obtained without any racemization. Methallylglycine also adds to methylmercaptan under these conditions; however, two diastereomers are formed in a 3:1 ratio.

B. Ruthenium-Catalyzed Ring-Closing Olefin Metathesis

Olefinic amino acids are ideal starting materials for a variety of highly functionalized, conformationally restricted cyclic amino acid derivatives using a ring-closing olefin metathesis approach. A number of examples of such a process are depicted in Scheme 14. A Ru-carbene complex (**53**), recently developed by Grubbs and co-workers [84], was used. Major advantages of this catalyst are ease of handling and a high tolerance for other functionalities. Both allylglycine and homoallylglycine methyl esters with either allylic, acrylic, or homoacrylic N-substituents (i.e., compounds **54/55**) were used as substrates to form a wide variety of enantiomerically pure highly functionalized six- and seven-membered ring amino acid derivatives in good to excellent yields (Scheme 14).

C. Palladium-Catalyzed Cyclization Reactions of Acetylene Containing Amino Acids

Intramolecular oxy palladation and amido palladation reactions onto alkynes provide a convenient way for preparing concise functionalized heterocyclic structures. As nucleophiles carboxylic acids, amines, and amides have been used. Also, further functionalization of the double bond might be achieved via cross-coupling with in situ–formed organopal-

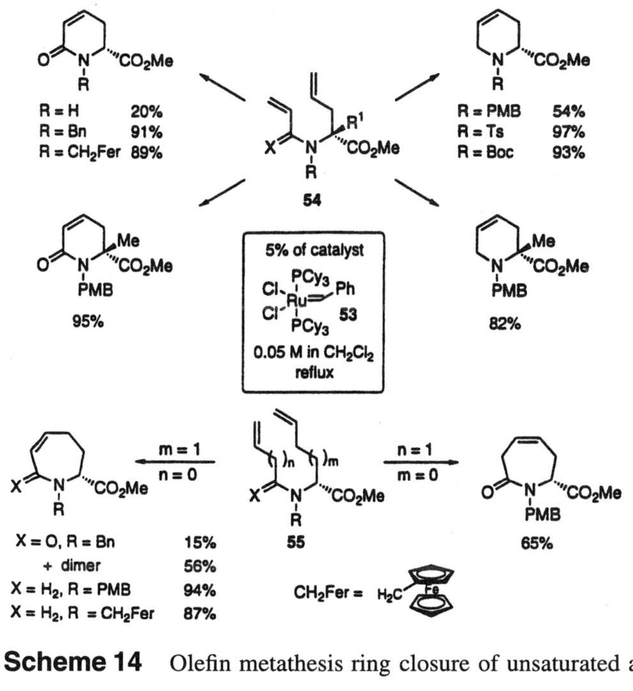

Scheme 14 Olefin metathesis ring closure of unsaturated amino acid derivatives.

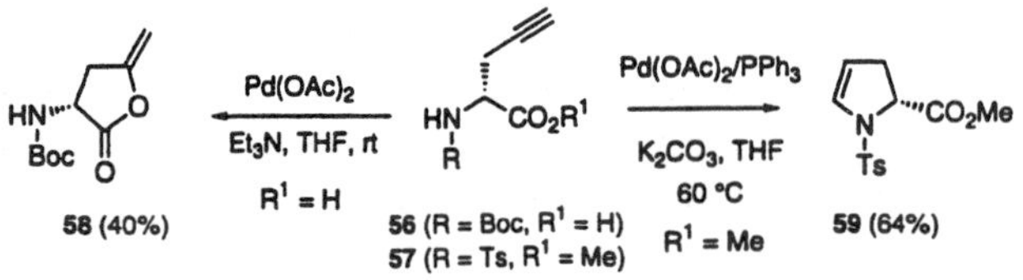

Scheme 15   Cyclization reactions of propargylglycine derivatives.

ladium(II) species. In Scheme 15 some of our results are depicted [22]. Boc-protected propargylglycine (**56**, R^1 = H, R = Boc) cyclized to form the five-membered enol lactone **58** in 40% yield in the presence of a catalytic amount of Pd(OAc)$_2$ and Et$_3$N. Starting from the *N*-tosylated propargylglycine methyl ester (**57**, R^1 = Me, R = Ts), the unsaturated proline derivative **59** was obtained in optically pure form.

Similarly, the protected homopropargyl derivative **60** was transformed in high yield into the five-membered ring enamide **61** by using Pd(PPh$_3$)$_4$ in the presence of 5 eq of K$_2$CO$_3$ (Scheme 16). Alternatively, the same starting compound **60** underwent a Pd-catalyzed cross-coupling reaction to give the substituted enamides **62** upon treatment with Pd(PPh$_3$)$_4$, K$_2$CO$_3$, TBAC, and an aryl iodide or a vinyl triflate.

VII. CONCLUSIONS

This chapter presents an overview on the use of (amino)amidases in the enzymatic resolution of amino acid amides. A wide variety of α-H-amino acids can be resolved with the aminopeptidase preparation from *P. putida* ATCC 12633. This enzyme preparation combines a high stereoselectivity with a very relaxed substrate specificity, i.e., a broad range of both L- and D-amino acid derivatives can be obtained in optically pure form, often in a single run (i.e., at 50% conversion). Although for practical reasons the crude enzyme preparation is used on a commercial scale, one of the principle enzymes involved has been cloned and expressed in an *E. coli* K-12 strain to further study its properties. The enzyme was shown to belong to the class of leucine aminopeptidases. The amino amidases from *P. putida* show an absolute requirement for a hydrogen substituent at the chiral carbon atom of the amino acid amides. In order to also resolve α,α-disubstituted amino acids, two new (amino) amidase preparations were obtained via screening, respectively from *M. neoaurum* ATCC 25795 and *O. anthropi* NCIMB 40321, which are rather complementary in their activities.

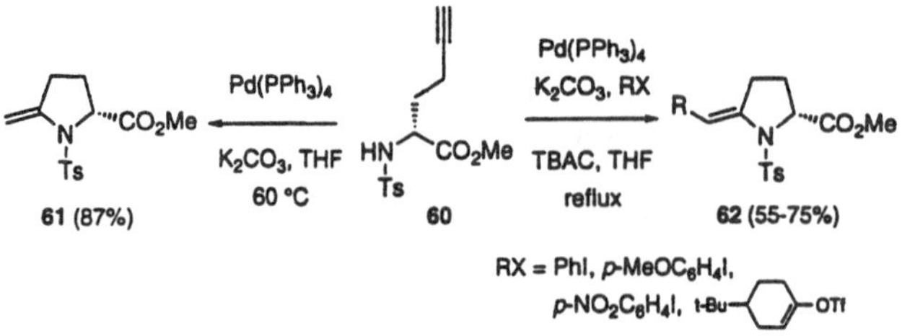

Scheme 16   Cyclization reactions of homoproparglyglycine derivatives.

M. neoaurum was shown to contain an enantioselective aminoamidase that can be used for the resolution of a broad range of α-methyl-substituted amino acid amides. Also α-ethyl, α-allyl, and α-propyl substituents are accepted by the enzyme although the reaction rate is low. Preferably the substrates should contain a —CH$_2$— linker between the chiral carbon atom and the most bulky substituent. In contrast, *O. anthropi* contains a highly stereoselective amidase, which can be used for the resolution of sterically highly demanding disubstituted amino acid amides. Interestingly, also α-hydroxy and α-*N*-hydroxyamino acid amides can be resolved using the *O. anthropi* preparation. Also, the amidase from *O. anthropi* shows a remarkably broad pH optimum, which allows resolutions to be performed under both basic and slightly acidic conditions.

Thus, a versatile toolbox of (amino)amidases is now available for the resolution of a very broad range of both α-H- and α,α-disubstituted amino acid amides. So far the applicability of this toolbox has been demonstrated for more than 100 natural and synthetic amino acids. The amino acids can be obtained in enantiomerically pure form for both the D and the L configuration.

In this chapter the versatility of this amino acid resolution technology is further illustrated by some examples of recent resolutions and further synthetic transformations. Examples include the resolution of α-methyl-substituted amino acids, e.g., α-methylvaline, and their incorporation into a number of conformationally restricted peptides that tend to adopt β-bend or 3$_{10}$-helix conformations. α-Methylphenylglycine was used to synthesize semisynthetic antibiotics. α-Methylphenylalanine was used as a building block for more stable aspartame-like sweeteners. α-Methyl-3,4-dichlorophenylalanine served as a building block for the antidepressant cericlamine, whereas α-phenylproline is a structural element of a nonpeptide substance P antagonist, selective for the NK$_1$ receptor.

The broad pH optimum of the *O. anthropi* amidase also allows for the elegant resolution of β-hydroxy-α-amino acids as precursors for thiamphenicol and florfenicol. This approach even opens up the possibility for a dynamic kinetic resolution.

Finally, the synthesis of a variety of unsaturated (alkenyl and alkynyl) amino acids is described. Those trifunctional amino acids are ideal starting materials for a series of highly functionalized, conformationally restricted cyclic amino acids using either Pd catalysis or Ru-catalyzed ring-closing metathesis. The resulting enantiopure functionalized cyclic amino acids appear to be versatile scaffolds in combinatorial approaches.

In conclusion, we have shown that aminoamidases form a class of highly versatile biocatalysts that can be used for the large-scale preparation of a wide variety of building blocks, of interest to both the pharmaceutical and the agrochemical industries.

REFERENCES

1. J Kamphuis, WHJ Boesten, B Kaptein, HFM Hermes, T Sonke, QB Broxterman, WJJ van den Tweel, HE Schoemaker. In: AN Collins, GN Sheldrake, J Crosby, eds. Chirality in Industry. Chichester, UK: J Wiley, 1992, pp 187–208.
2. (a) CA Hendrick, BA Garcia. Ger Pat 2,812,169 (1978 to Zoecon); (b) CA Hendrick. Pest Sci 11: 224, 1980.
3. (a) Eur Pat 0,123.830 (1984 to Am. Cyanamid Corp.); (b) CDS Tomlin, ed. The Pesticide Manual, 11th ed. Farnham: British Crop Protection Council, 1997, pp 694–703.
4. RP Pandey, H Meng, JC Cook, KL Reinhart. J Am Chem Soc 99:5203, 1977.
5. AA Patchett. In: D Lednicer, ed. Chronicals of Drug Discovery, Vol. 3. Washington, DC: ACS, 1993, pp 125–162.

6. (a) BD Dorsey, et al. J Med Chem 37:3443, 1994; (b) JP Vacca, BD Dorsey, et al. Proc Natl Acad Sci USA 91: 4096, 1994; (c) PE Maligres, V Upadhyay, K Rossen, SJ Ciancosi, RM Purick, KK Eng, RA Reamer, D Askin, RP Volante, PJ Reider. Tetrahedron Lett 36:2195, 1995; (d) Merck Index 12: 850, 1996.

7. (a) EB Parkes, et al. J Org Chem 59:3656, 1994; (b) NA Roberts, et al. Science 248:358, 1990; (c) Merck Index 12:1438, 1996.

8. JR Prous, ed. The Year's Drug News 1995, Barcelona: Prous Science, 1995, pp 468–475.

9. JD Rozzell, F Wagner, eds. Biocatalytic Production of Amino Acids and Derivatives. München: Hanser Publishers, 1992.

10. K Oyama. In: AN Collins, GN Sheldrake, J Crosby, eds. Chirality in Industry. Chichester, UK: J Wiley, 1992, pp 237–247.

11. WS Knowles. Acc Chem Res 16:106, 1983.

12. (a) RM Williams. Synthesis of Optically Active α-Amino Acids. Oxford: Pergamon Press, 1989; (b) MJ O'Donnell, WD Bennett, S Wu. J Am Chem Soc 111:2353–2355, 1989.

13. (a) WHJ Boesten. Dutch Pat 900,0387, 1990 (to DSM/Stamicarbon); (b) J Kamphuis, EM Meijer, WHJ Boesten, T Sonke, WJJ van den Tweel, HE Schoemaker. In: DS Clark, DA Estell, eds., Enzyme Engineering XI. New York: New York Academy of Sciences, 1992, pp 510–527.

14. K Drauz, H Waldmann. Enzyme Catalysis in Organic Synthesis, Vols. 1 and 2. Weinheim: VCH, 1995, pp 480–485 and 633–641.

15. Ref 14, pp 393–408.

16. (a) Ref 14, pp 409–431; (b) C Syldatk, R Müller, M Siemann, K Krohn, F Wagner. In Ref. 9, pp 75–128; (c) C Syldatk, R Müller, M Pietsch, F Wagner. In Ref. 9, pp 129–176.

17. (a) Ref 14, pp 377–392; (b) J Kamphuis, WHJ Boesten, QB Broxterman, JAM van Balken, EM Meijer, HE Schoemaker. Adv Biochem Eng Biotechnol 42:133–186, 1990; (c) J Kamphuis, EM Meijer, WHJ Boesten, QB Broxterman, B Kaptein, HE Schoemaker. In Ref. 9, pp 177–206.

18. (a) WHJ Boesten. Br Pat 1,548,032, 1976 (to DSM/Stamicarbon); (b) EM Meijer, WHJ Boesten, HE Schoemaker, JAM van Balken. In: J Tramper, HC van der Plas, P Linko, eds. Biocatalysts in Organic Synthesis. Amsterdam: Elsevier, 1987, pp 135–156.

19. A Kjaer, S Wagner. Acta Chem Scand 9:721, 1955.

20. In general, we consider giving $E > 200$ inappropriate due to the accuracy in the determination of the e.e. and the conversion. See C-S Chen, Y Fujimoto, G Girdaukas, CJ Sih. J Am Chem Soc 104:7294–7299, 1982.

21. FPJT Rutjes, HE Schoemaker. Tetrahedron Lett 38:677–680, 1997.

22. LB Wolf, KCMF Tjen, FPJT Rutjes, H Hiemstra, HE Schoemaker, Tetrahedron Lett 39:5081–5084, 1998

23. EC Roos, HH Mooiweer, H Hiemstra, WN Speckamp, B Kaptein, WHJ Boesten, J Kamphuis. J Org Chem 57:6769–6778, 1992.

24. Unpublished results, DSM Research, Geleen, and University of Groningen, Groningen, The Netherlands.

25. HFM Hermes, T Sonke, PJH Peters, JAM van Balken, J Kamphuis, L Dijkhuizen, EM Meijer. Appl Environ Microbiol 59:4330–4334, 1993.

26. SF Altschul, TL Madden, AA Schäffer, J Zhang, Z Zhang, W Miller, DJ Lipman. Nucleic Acids Res 25:3389–3402, 1997.

27. (a) EL Smith, RL Hill. In: PD Boyer, H Lardy, K Myrback, eds. The Enzymes. New York: Academic Press, 1960, pp 37–62; (b) RJ Delange, EL Smith. In: PD Boyer, ed. The Enzymes, Vol. 3. New York: Academic Press, 1971, pp 81–103; (c) H Hanson, M Frohne. Meth Enzymol 45:504–521, 1976; (d) H Kim, WN Lipscomb. Adv Enzymol 68:153–213, 1994.

28. SW Melbye, FH Carpenter. J Biol Chem 246:2459–2463, 1971.

29. N Sträter, WN Lipscomb. Biochemistry 34:14792–14800, 1995.

30. FH Carpenter, JM Vahl. J Biol Chem 248:294–304, 1973.

31. (a) GA Thomson, FH Carpenter. J Biol Chem 251:53–60, 1976; (b) GA Thomson, FH Carpenter. J Biol Chem 251:1618–1624, 1976.

32. MP Allen, AH Yamada, FH Carpenter. Biochemistry 22:3778–3783, 1983.
33. N Sträter, WN Lipscomb. Biochemistry 34:9200–9210, 1995.
34. SK Burley, PR David, RM Sweet, A Taylor, WN Lipscomb. J Mol Biol 224:113–140, 1992.
35. (a) SK Burley, PR David, A Taylor, WN Lipscomb. Proc Natl Acad Sci USA 87:6878–6882, 1990; (b) SK Burley, PR David, WN Lipscomb. Proc Natl Acad Sci USA 88:6916–6920, 1991; (c) H Kim, SK Burley, WN Lipscomb. J Mol Biol 230:722–724, 1993.
36. H Kim, WN Lipscomb. Biochemistry 32:8465–8478, 1993.
37. DO Wood, MJ Solomon, RR Speed. J Bacteriol 175:159–165, 1993.
38. DG Kleid, D Yansura, HL Heynek, GF Miozzari. EP 0 154 133 (1990, to Genentech).
39. (a) RM Williams. In: Synthesis of Optically Active α-Amino Acids. Oxford: Pergamon Press, 1989; (b) B Kaptein, WHJ Boesten, WJJ van den Tweel, QB Broxterman, HE Schoemaker, F Formaggio, M Crisma, C Toniolo, J Kamphuis. Chimica Oggi 14(3/4):9–12, 1996; (c) T Wirth. Angew Chem Int Ed Engl 36:225–227, 1997.
40. U Schöllkopf, R Lonsky, P Lehr. Liebigs Ann Chem 413: 1985, and references herein.
41. (a) D Seebach, A Fadel. Helv Chim Acta 68:1243, 1985; (b) M Gander-Coquoz. ibid. 71: 224, 1988.
42. (a) MJ O'Donnell, S Wu. Tetrahedron: Asymmetry 3: 591, 1992; (b) YN Belokon, KA Kochetkov, TD Churkina, NS Ikonnikov, AA Chesnokov, OV Larionov, VS Parmár, R Kumar, HB Kagan. Tetrahedron: Asymmetry 9:851–857, 1998.
43. WH Kruizinga, J Boster, RM Kellogg, J Kamphuis, WHJ Boesten, EM Meijer, HE Schoemaker. J Org Chem 53:1826–1827, 1988.
44. B Kaptein, WHJ Boesten, QB Broxterman, PJH Peters, HE Schoemaker, J Kamphuis. Tetrahedron: Asymmetry 4:1113–1116, 1993.
45. F Becke, H Flieg, P Pässler. Liebigs Ann Chem 749:198–201, 1971. See also WHJ Boesten, PJH Peters. Eur Pat 1.508.54, 1984 (Chem Abst. 104: 128238e, 1986).
46. B Kaptein, WHJ Boesten, QB Broxterman, HE Schoemaker, J Kamphuis. Tetrahedron Lett 33:6007–6010, 1992.
47. EC Roos, M Carmen Lopez, MA Brook, H Hiemstra, WN Speckamp, B Kaptein, J Kamphuis, HE Schoemaker. J Org Chem 58:3259–3268, 1993.
48. HFM Hermes. PhD dissertation, Rijksuniversiteit Groningen, 1992.
49. HFM Hermes, RF Tandler, T Sonke, L Dijkhuizen, EM Meijer. Appl Environ Microbiol 60: 153–159, 1994.
50. HM Kalisz. Adv Biochem Eng Biotechnol 36:1–65, 1988.
51. (a) H Moorlag, RM Kellogg, M Kloosterman, B Kaptein, HE Schoemaker. J Org Chem 55: 5878–5881, 1990; (b) H Moorlag, RM Kellogg. Tetrahedron: Asymmetry 2:705–720, 1991.
52. WJJ van den Tweel, TJGM van Dooren, PH de Jonge, B Kaptein, ALL Duchateau, J Kamphuis. Appl Microbiol Biotechnol 39:296–300, 1993.
53. E de Vries. PhD dissertation, Rijksuniversiteit Groningen, 1997.
54. Unpublished result, B Kaptein, DSM Research, Geleen, The Netherlands.
55. PK Smith, RI Krohn, GT Hermanson, AK Mallia, FH Gartner, MD Provenzano, EK Fujimoto, NM Goeke, BJ Olson, DC Klenk. Anal Biochem 150:76–85, 1985.
56. MM Bradford. Anal Biochem 72:248–254, 1976.
57. L Coppi, C Giordano, A Longoni, S Panossian. In: AN Collins, GN Sheldrake, J Crosby, eds. Chirality in Industry, Vol. 2. Chichester, UK: J Wiley, 1997, pp 353–362.
58. Ger pat 1.938.513, 1971 (to Sumitomo Chem Co.), US pat 2.767.213, 1956 (to Parke, Davis & Co.).
59. B Kaptein, TJGM van Dooren, WHJ Boesten, T Sonke, ALL Duchateau, QB Broxterman, J Kamphuis. Org Proc Res Dev 2:10–17, 1998.
60. (a) CJ Gouret, JG Wettstein, RD Porsolt, A Puech, JL Junien. Eur J Pharmacol. 813:1478, 1990; (b) CJ Gouret, RD Porsolt, JG Wettstein, A Puech, JL Junien. Artzniem Forsch/Drug Res 40:633, 1990.
61. B Kaptein, HM Moody, QB Broxterman, J Kamphuis. J Chem Soc Perkin Trans 2, 1495–1498, 1994.

62. A Abiko, S Masamune, Tetrahedron Lett 33:5517–5520, 1992.
63. HM Moody, B Kaptein, QB Broxterman, WHJ Boesten, J Kamphuis. Tetrahedron Lett 35: 1777–1780, 1994.
64. (a) NG Delaney, V Madison. Int J Peptide Prot Res 19:543–548, 1982; (b) VW Magaard, RM Sanchez, JW Bean. Tetrahedron Lett 34:381–384, 1993; (c) D Tourwé, J Van Betsbrugge, P Verheyden, C Hootelé. Bull Soc Chim Belg 103:201–205, 1994.
65. T Harrison, BJ Williams, CJ Swain. Bioorg Med Chem Lett 4:2733–2736, 1994.
66. J Van Betsbrugge, D Tourwé, B Kaptein, H Kierkels, QB Broxterman. Tetrahedron 53:9233–9240, 1997.
67. E Benedetti, A Bavoso, B di Blasio, V Pavone, C Pedone, C Toniolo, GM Bonora. Proc Natl Acad Sci USA 79:7951–7954, 1982.
68. AD Argoudelis, A Dietz, LE Johnson. J Antibiot 27:321–328, 1974.
69. C Toniolo, E Bennedetti. Macromolecules 24:4004, 1991.
70. C Toniolo, E Bennedetti. Trends Biochem Sci 16:350, 1991.
71. CM Venkatachalam. Biopolymers 6:1425, 1968.
72. (a) C Toniolo, M Crisma, F Formaggio, G Valle, G Cavicchioni, G Précigoux, A Aubry, J Kamphuis. Biopolymers 33:1061, 1993; (b) F Formaggio, M Pantano, M Crisma, GM Bonora, C Toniolo, J Kamphuis. J Chem Soc Perkin Trans 2, 1097, 1995; (c) F Formaggio, C Toniolo, M Crisma, G Valle, B Kaptein, HE Schoemaker, J Kamphuis, B Di Blasio, O Maglio, R Fattorusso, E Benedetti, A Santini. Int J Peptide Protein Res 45:70, 1995; (d) R Gratias, R Konat, H Kessler, M Crisma, G Valle, A Polese, F Formaggio, C Toniolo, QB Broxterman, J Kamphuis. J Am Chem Soc 120:4763–4770, 1998.
73. F Formaggio, M Pantano, M Crisma, C Toniolo, WHJ Boesten, HE Schoemaker, J Kamphuis. Bioorg Med Chem Lett 3:953, 1993.
74. DC Horwell, J Hughes, JC Hunter, MC Pritchard, RS Richardson, R Roberts, GN Woodruff. J Med Chem 34:404–414, 1991.
75. TM Zydowsky, JF Dellaria, HN Nellans. J Org Chem 53:5607–5616, 1988.
76. E Mossel, F Formaggio, M Crisma, C Toniolo, T Sonke, EC Roos, QB Broxterman, J Kamphuis. Lett Pept Sci 5:43–48, 1998.
77. (a) S Polinelli, QB Broxterman, HE Schoemaker, WHJ Boesten, M Crisma, G Valle, C Toniolo, J Kamphuis. Biorg Med Chem Lett 2:453, 1992; (b) NH Hooper, RJ Hesp, S Tieku, WHJ Boesten, C Toniolo, J Kamphuis. J Agric Food Chem 42:1397, 1994.
78. X Pascaud, C Honde, B Chanoine, F Roman, I Bueno, JL Junien. Pharm Pharmacol 42:546, 1990.
79. (a) BK Vriesema, RM Kellogg. Tetrahedron Lett 27:2049–2052, 1986; (b) G Cross, BK Vriesema, G Boven, RM Kellogg, F van Bolhuis. J Organomet Chem 370: 357, 1989.
80. R Hulst, QB Broxterman, J Kamphuis, F Formaggio, M Crisma, C Toniolo, RM Kellogg. Tetrahedron: Asymmetry 8:1987–1999, 1997.
81. B Kaptein, V Monaco, QB Broxterman, HE Schoemaker, J Kamphuis. Recl Trav Chim Pays-Bas 114:231–238, 1995.
82. ER Parmee, O Tempkin, S Masamune. J Am Chem Soc 113:9365–9266, 1991.
83. QB Broxterman, B Kaptein, J Kamphuis, HE Schoemaker. J Org Chem 57:6286–6294, 1992.
84. RH Grubbs, J Miller, GC Fu. Acc Chem Res 28:446, 1995.

3

Chemoenzymatic Synthesis of Pheromones, Terpenes, and Other Bioregulators

Kenji Mori
Science University of Tokyo, Tokyo, Japan

I. INTRODUCTION

The major project in our group since 1973 has been pheromone synthesis. We synthesized the pure enantiomers of pheromones and clarified the enantioselectivity of pheromone perception by insects [1,2]. The asymmetrical nature of pheromone receptors could be revealed by employing molecular probes as provided by enantioselective synthesis. We then thought that the asymmetrical life processes such as enzymatic or microbial transformations should be used advantageously in enantioselective synthesis. As shown in Fig. 1, our plan was to use enzyme as a tool to provide nonracemic and chiral building blocks starting from optically inactive materials. By using highly enantioselective biotransformations, it must be possible to prepare enantiomerically pure building blocks, which will give pheromones, terpenes, and other bioregulators of high enantiomeric purity.

The first attempt in 1978 along this line was the asymmetric hydrolysis of (±)-1-alkyl-2-alkynyl acetates [3,4]. The asymmetric hydrolysis of (±)-**1** with *Bacillus subtilis* var. *niger* was moderately successful, yielding the recovered (R)-**1** and (S)-**2** of 54–96% e.e. as shown in Fig. 2. After 20 years of endeavor, we could accumulate a number of examples of biotransformations using hydrolytic enzymes and yeasts, some of which are useful preparative methods in enantioselective synthesis. This chapter summarizes our results on this subject. Our objectives have been (1) to use readily available enzymes and microorganisms that are easily handled in organic laboratories, (2) to prepare versatile building blocks that can be converted to many different target molecules by organic transformations, and (3) to secure almost pure enantiomers. Our previous works were reviewed [5–8]. There is also a summary of our works on the application of biochemical methods in pheromone synthesis [9].

II. DESIGN OF VERSATILE BUILDING BLOCKS FOR ENANTIOSELECTIVE SYNTHESIS

To achieve enantioselective synthesis of a bioregulator, we need enantiomerically pure intermediate(s) unless we resolve the final product. We therefore have to secure the pure enantiomer of the intermediate. If that intermediate can also be employed in the synthesis

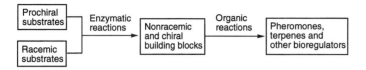

Figure 1 Roles of enzymatic and organic reactions in the synthesis of pheromones, terpenes, and other bioregulators.

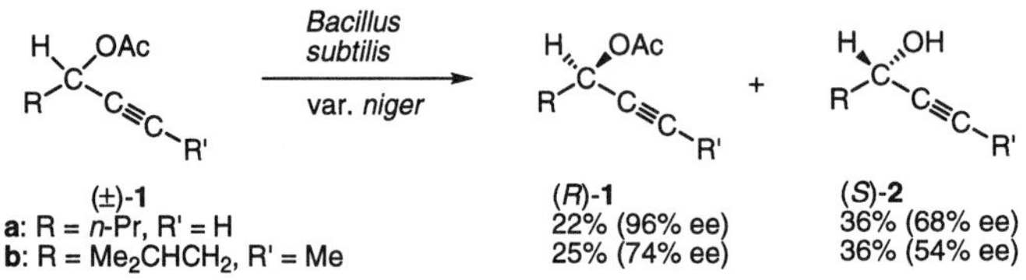

H OAc
|
R—C—C≡C—R'
(±)-**1**
a: R = *n*-Pr, R' = H
b: R = Me₂CHCH₂, R' = Me

Bacillus subtilis var. *niger* →

H OAc
|
R—C—C≡C—R'
(*R*)-**1**
22% (96% ee)
25% (74% ee)

+

H OH
|
R—C—C≡C—R'
(*S*)-**2**
36% (68% ee)
36% (54% ee)

Figure 2 Asymmetric hydrolysis of (±)-1-alkyl-2-alkynyl acetates.

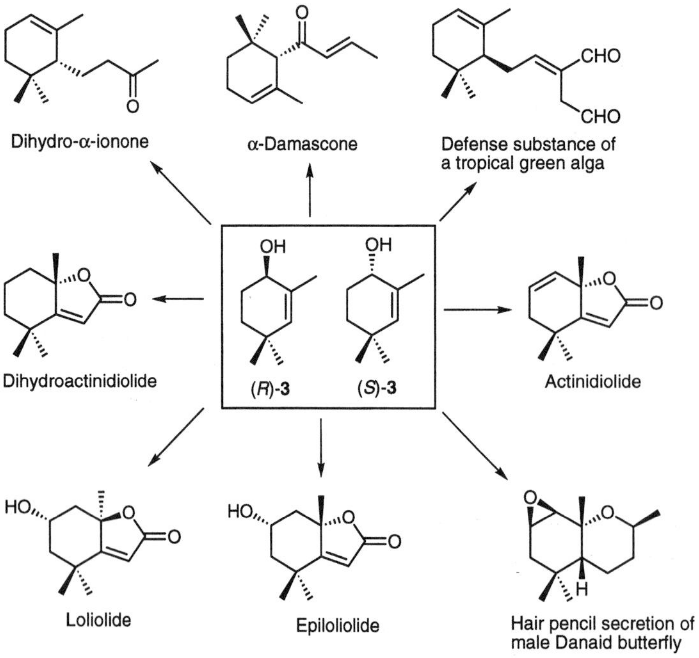

Dihydro-α-ionone

α-Damascone

Defense substance of a tropical green alga

Dihydroactinidiolide

OH   OH

(*R*)-**3**   (*S*)-**3**

Actinidiolide

Loliolide

Epiloliolide

Hair pencil secretion of male Danaid butterfly

Figure 3 Isoprenoids synthesized by starting from **3**.

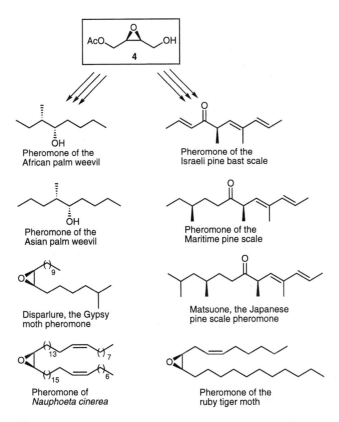

Figure 4 Pheromones synthesized by starting from **4**.

of other interesting compounds, it will be beneficial to those who work on such compounds. It is therefore worthwhile to design versatile building blocks, obtainable by biotransformation, for enantioselective synthesis. In Figs. 3–7 some versatile building blocks of enzymatic or microbial origin are shown that have been prepared and/or employed by our group.

Figure 3 shows the isoprenoids synthesized by starting from the enantiomers of **3**. They are (1) dihydro-α-ionone [10], (2) α-damascone [11], (3) a defense substance of a tropical green alga [12], (4) actinidiolide [13], (5) dihydroactinidiolide, a component of the queen recognition pheromone of the red imported fire ant (*Solenopsis invicta*) [13], (6) loliolide [13], (7) epiloliolide [13], and the major component of the hair pencil secretion of male Danaid butterfly [14].

Pheromones synthesized by starting from the optically active epoxide **4** are illustrated in Fig. 4. They are (1) the pheromone of the African palm weevil (*Rhynchophorus phoenicis*) [15], (2) the pheromone of the Asian palm weevil (*Rhynchophorus vulneratus*) [16], (3) disparlure, the pheromone of the gypsy moth (*Lymantria dispar*) [17], (4) the pheromone of the cockroach *Nauphoeta cinerea* [18], (5) the pheromone of the Israeli pine bast scale (*Matsucoccus josephi*) [19], (6) the pheromone of the Maritime pine scale (*Matsucoccus feytaudi*) [20], (7) matsuone, the pheromone of the Japanese pine scale (*Matsucoccus matsumurae*) [21], and (8) the pheromone of the ruby tiger moth (*Phragmatobia fuliginosa*) [17].

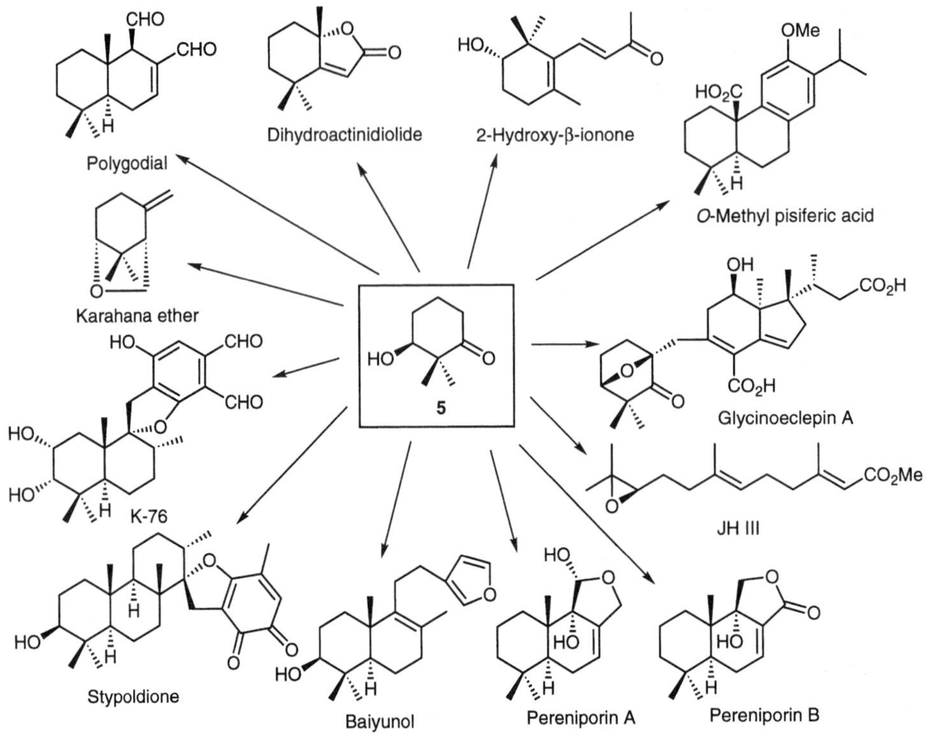

Figure 5 Isoprenoids synthesized by starting from **5**.

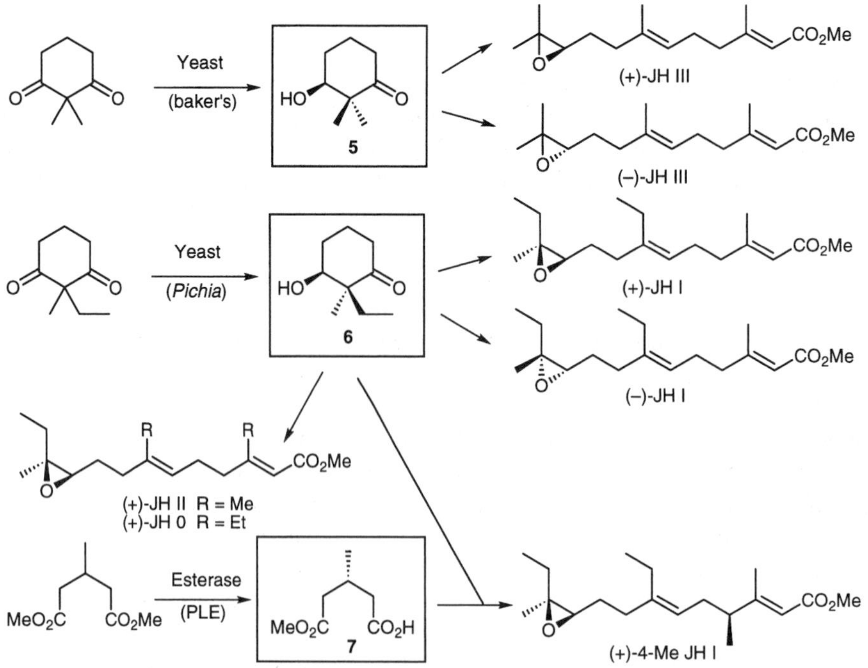

Figure 6 Synthesis of insect juvenile hormones.

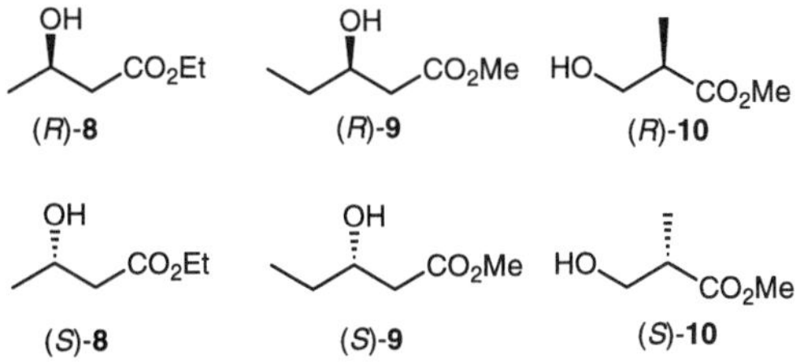

Figure 7 Commercially available building blocks.

Both of the building blocks **3** and **4** were prepared by asymmetric hydrolysis of the corresponding acetates with either pig liver esterase (PLE) [10] or pig pancreatic lipase (PPL) [17]. Building block **3** is now available by chemical asymmetric hydrogenation of the corresponding unsaturated ketone [22]. The border between biochemical and chemical processes is always moving, and therefore one must be careful to choose either an enzymatic or a chemical method for the preparation of the building blocks according to the state of the art.

Figure 5 shows the isoprenoids synthesized by starting from the hydroxy ketone **5**, which is obtainable by reduction of the corresponding β-diketone with baker's yeast [23]. The isoprenoids synthesized are (1) (S)-2-hydroxy-β-ionone [24], (2) O-methyl pisiferic acid [25], (3) glycinoeclepin A, the hatching stimulant for the eggs of soybean cyst nematode [26], (4) insect juvenile hormone III [27], (5) pereniporins A and B [28], (6) baiyunol [29], (7) stypoldione, the marine antifeedant [30], (8) the antibiotic K-76 [31], (9) karahana ether [32], polygodial, the insect antifeedant [33], and (10) dihydroactinidiolide [34].

The hydroxy ketones **5** and **6** [35] were the key building blocks in the synthesis of enantiomerically pure insect juvenile hormones [27,35] as shown in Fig. 6. In the case of the synthesis of (+)-4-methyl juvenile hormone I, the half ester **7** was also employed as the building block [36].

Figure 7 illustrates the structures of widely employed and commercially available building blocks of hydroxy ester types. Esters **8** and **9** can now be prepared by Noyori's asymmetric hydrogenation. In 1970s and in early 1980s they had to be prepared by biotransformation. These hydroxy esters will be treated in Section V.C. In this chapter, two important biotransformations were omitted, and therefore the readers should refer to the reviews describing them: (1) microbial oxidation of aromatic compounds to give cyclohexadiene-*cis*-1, 2-diols [37] and (2) enzymatic Baeyer-Villiger oxidation as catalyzed by cyclohexanone monooxygenase [38]. These two oxidation reactions provide a variety of useful building blocks.

III. AMINOACYLASE AS ENANTIOSELECTIVE CATALYST

A. α-Amino Acid as a Building Block

Aminoacylase, first obtained by Greenstein from hog kidney [39], catalyzes hydrolytic removal of the N-acetyl group of (S)-α-amino acids as shown in Fig. 8. Subsequent extensive works by Chibata and co-workers proved that the aminoacylase of *Aspergillus* is a highly enantioselective biocatalyst to effect the asymmetric hydrolysis of racemic N-acetyl-α-amino acids, giving (S)-α-amino acids and the unchanged (R)-N-acetyl-α-amino acids [40]. Aminoacylase of *Aspergillus* origin is a cheap enzyme with high enantioselec-

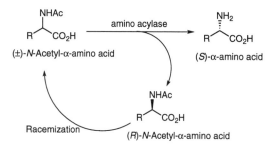

Figure 8 Asymmetric hydrolysis of (±)-*N*-acetyl-α-amino acid with aminoacylase.

tivity and broad substrate specificity tolerating alkyl substituents that do not exist among naturally occurring α-amino acids.

Gizzerosine (Fig. 9) is a toxin obtained from brown fish meal, which causes gizzard erosion or ulceration in chicks at a level of 2.2 ppm in the diet. This toxicity results in a serious disease named "black vomit," notorious in poultry production. Resolution of (±)-α-aminoadipic acid via its *N*-chloroacetyl derivative (±)-**11** by aminoacylase yielded (*S*)-**12**, which was converted to crystalline dihydrochloride of (*S*)-gizzerosine [41]. (*R*)-Gizzerosine was biologically inactive.

If pure enantiomers of α-amino acids are required, aminoacylase serves as an excellent tool for resolution.

B. α-Amino Acids as Starting Materials for α-Hydroxy Acids and Epoxides

Pure enantiomers of α-amino acids furnish pure enantiomers of α-hydroxy acids by treatment with nitrous acids followed by recrystallization. The deamination proceeds with retention of configuration. Because α-hydroxy acids readily afford epoxides, pure enantiomers of both of them are available from α-amino acids and can be used as building blocks for bioregulators.

Figure 10 shows some examples of such use of α-amino acids. Enzymatic resolution of the *N*-chloroacetyl derivative **13a** of (±)-2-aminodecanoic acid gave (*S*)-**14a**, which

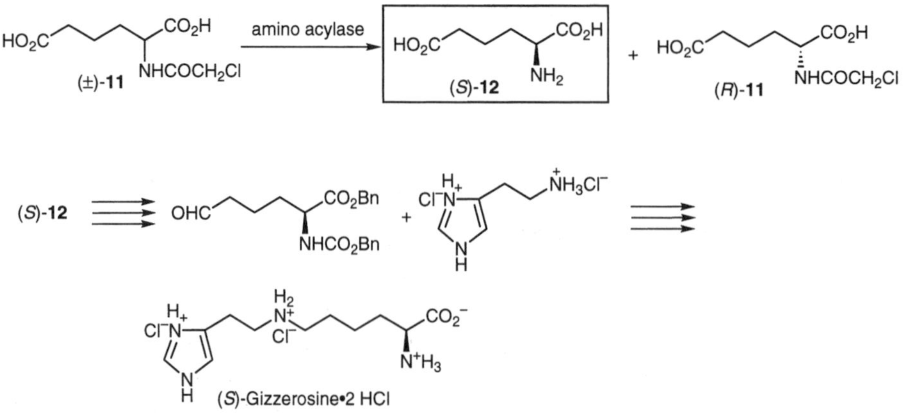

Figure 9 α-Amino acid itself as a building block. Synthesis of gizzerosine.

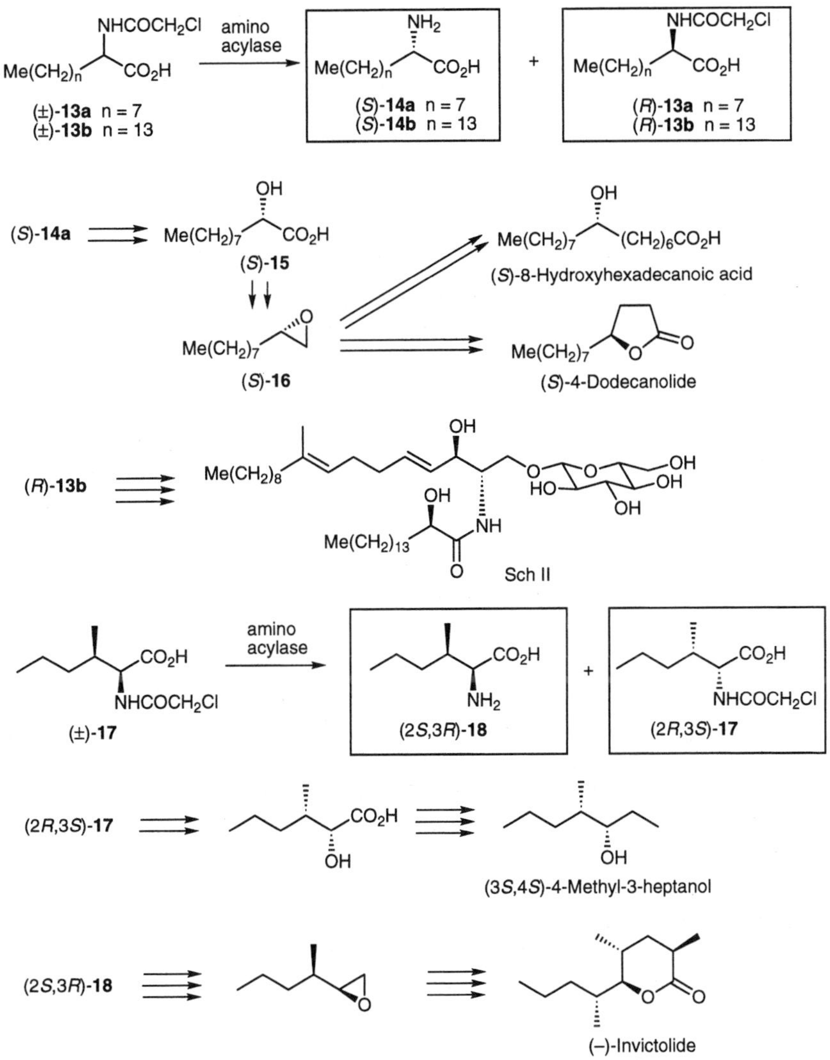

Figure 10 α-Amino acids as starting materials. Synthesis of (*S*)-8-hydroxyhexadecanoic acid, (*S*)-4-dodecanolide, fruiting-inducing cerebroside of *Schizophyllum commune* (Sch II), (*3S,4S*)-4-methyl-3-heptanol, and (−)-invictolide.

afforded (*S*)-2-hydroxydecanoic acid (**15**). This was converted to the (*S*)-epoxide **16**, from which (*S*)-8-hydroxyhexadecanoic acid, an endogenous inhibitor against spore germination of *Lygodium japonicum*, was synthesized [42]. Starting from the enantiomers of **16**, the enantiomers of 4-dodecanolide, a defense secretion of rove beetles (*Bredius mandebularis* and *B. spectabilis*), were synthesized [43]. (±)-2-*N*-Chloroacetylaminohexadecanoic acid (**13b**) was resolved, deaminated, and incorporated as a part of the cerebroside Sch II, the fruiting body inducer of *Schizophyllum commune* [44]. (*3S,4R*)-4-Methyl-3-heptanol, a component of the aggregation pheromone of the smaller European elm bark beetle (*Scolytus multistriatus*), was synthesized from (*2R,3S*)-**17** [45], while (−)-invictolide, a component of the queen recognition pheromone of the red imported fire ant (*Solenopsis invicta*), was prepared from (2S,3R)-**18** [46].

It is now more common to prepare the enantiomers of hydroxy acids by lipase-catalyzed resolution of their racemates than by deamination of the enantiomers of α-amino acids.

IV. ESTERASES AND LIPASES AS ENANTIOSELECTIVE CATALYSTS

A. Asymmetric Hydrolysis of Racemic Esters and Asymmetric Esterification or Lactonization of Racemic Alcohols

Many esterases and lipases are now commercially available and widely used in the synthesis of bioactive compounds [47]. Thanks to the rather broad substrate specificity of lipases, they are employed as resolving agents in various different cases. The enantiomers of **3** were prepared by asymmetric hydrolysis of (±)-**19** with PLE as shown in Fig. 11 [10]. Their pure 3,5-dinitrobenzoates **20** could be obtained by recrystallization, hydrolysis of which furnished the pure enantiomers of **3**. Optically active **3** is also obtainable by chemical asymmetric reduction [11] or by asymmetric hydrogenation [22] of the corresponding ketone. The enzymatic method is simpler than other methods, especially when both the enantiomers of **3** are required. As already shown in Fig. 3, **3** could be converted to various natural products.

In the above-mentioned works, **3** was converted to **21** (Fig. 12), which served as the key intermediate in the syntheses. For the preparation of an isomeric alcohol **22**, its racemate was acetylated with vinyl acetate in the presence of lipase AK (Amano) to give the acetate (R)-**23** and the remaining alcohol (S)-**22** [48]. The enantiomers of **22** served as the intermediate for the synthesis of ancistrodial, a defense substance of a termite *Ancistrotermes cavithorax* [49], (S)-γ-coronal, an ambergris odorant [48], and the enantiomer of the antibacterial terpene from *Premna oligotricha* [49].

(+)-Strigol (Fig. 13) is the germination stimulant for parasitic weeds of the genera *Striga* and *Orobanche*, which cause severe yield losses in grains and legumes. Enzymatic acetylation of (±)-**24** with vinyl acetate and lipase AK yielded (+)-**24** and (+)-**25**, the

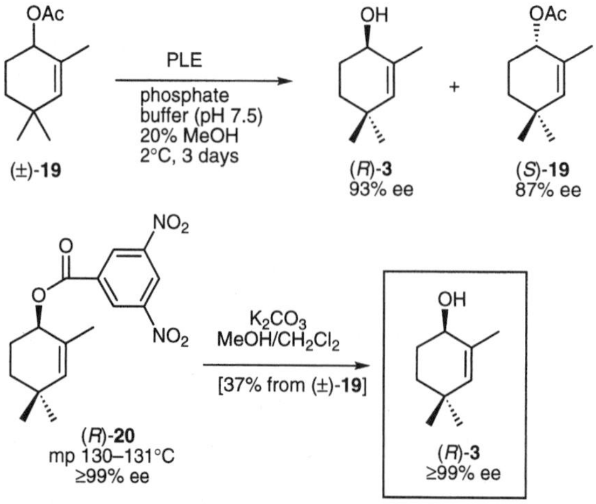

Figure 11 Preparation of the building block **3**.

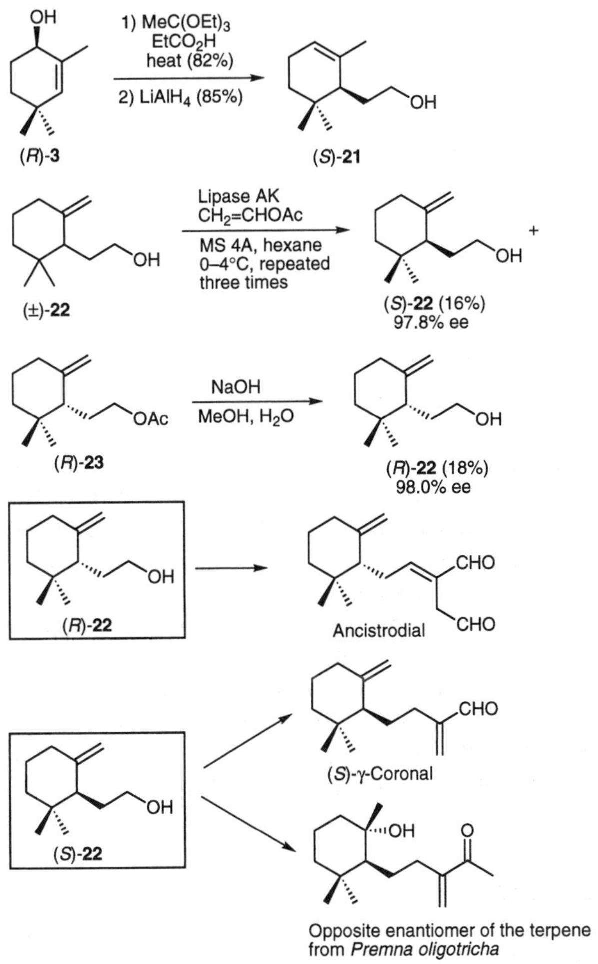

Figure 12 Preparation of the enantiomers of **22** and their conversion to terpenes.

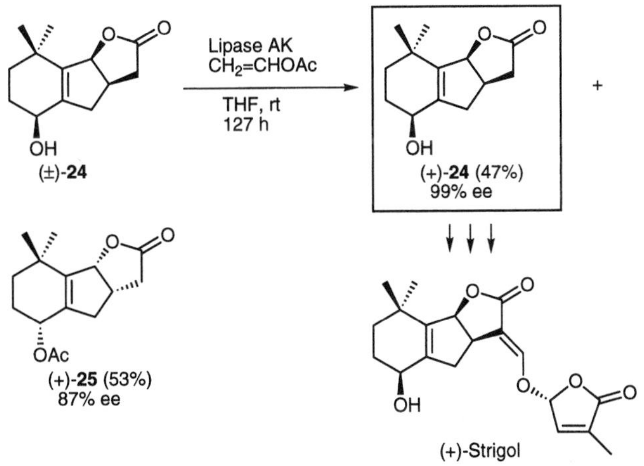

Figure 13 Preparation of (+)-**24** as the building block for (+)-strigol.

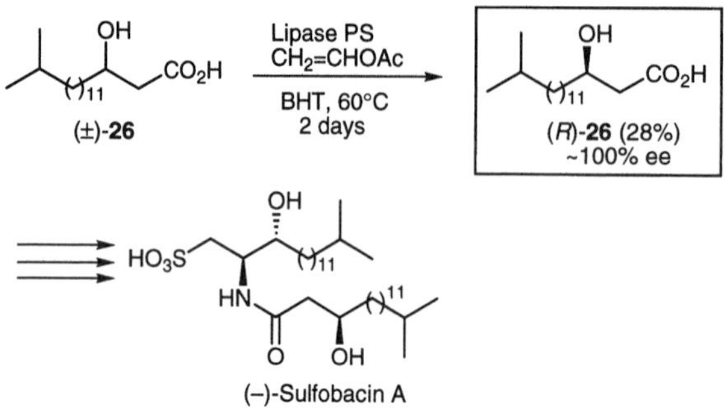

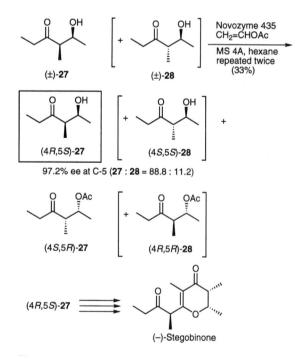

Figure 14 Preparation of (*R*)-**26** as the building block for (−)-sulfobacin A.

former of which was converted to (+)-strigol [50]. (+)-Sulfobacin A (Fig. 14) is a medicinally interesting metabolite of *Chrysobacterium* sp. Enzymatic acetylation of (±)-**26** with vinyl acetate and lipase PS (Amano) left (*R*)-**26** intact, which was converted to (+)-sulfobacin [51]. For the synthesis of (−)-stegobinone (Fig. 15), the sex pheromone of the drugstore beetle (*Stegobium paniceum*), the hydroxy ketone (*4R,5S*)-**27** was chosen as the building block to construct the dihydro-γ-pyrone ring. This ketol **27** was prepared by enzymatic acetylation of (±)-**27** with vinyl acetate and Novozyme 435 to leave (*4R,5S*)-**27** as the recovered starting material [52]. Olean (Fig. 16) is the sex pheromone of the olive fruitfly (*Bactrocera oleae*). Asymmetric hydrolysis of (±)-**29** with PLE yielded (*4S,6S*)-**30**, which was the known precursor of (*S*)-olean [53]. Lipase A (Amano, from *Aspergillus* sp.) was employed for the asymmetric hydrolysis of (±)-**31** to give (*S*)-**32**,

Figure 15 Preparation of (*4R,5S*)-**27** as the building block for (−)-stegobinone.

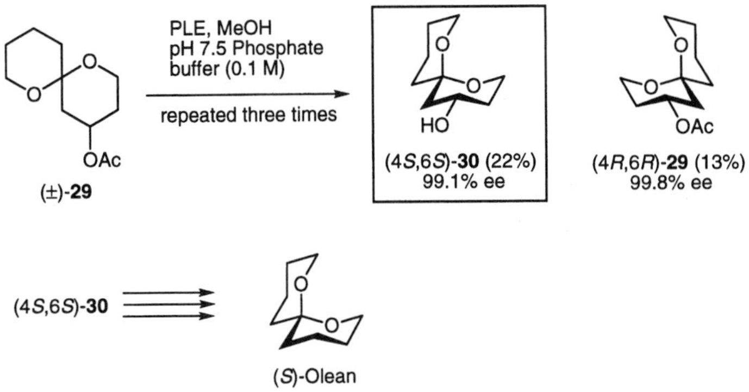

Figure 16 Preparation of (*4S,6S*)-**30** as the building block for (*S*)-olean.

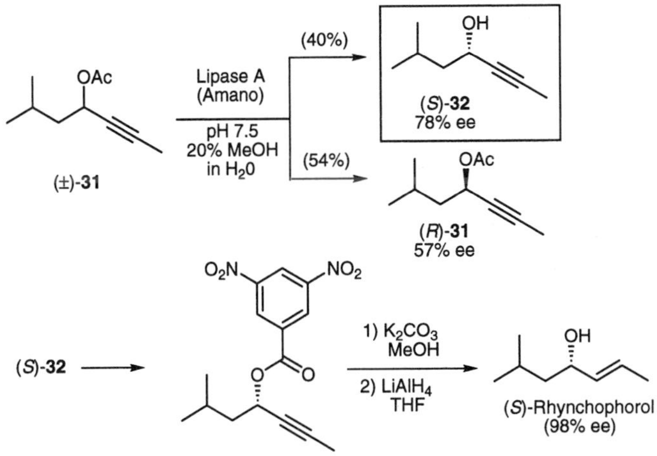

Figure 17 Preparation of (*S*)-**32** as the building block for (*S*)-rhynchophorol.

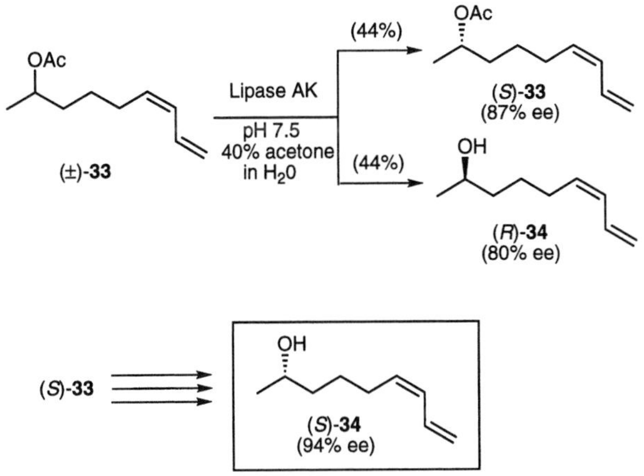

Figure 18 Synthesis of the pheromone of *Nepticula malella* (*S*)-**34**.

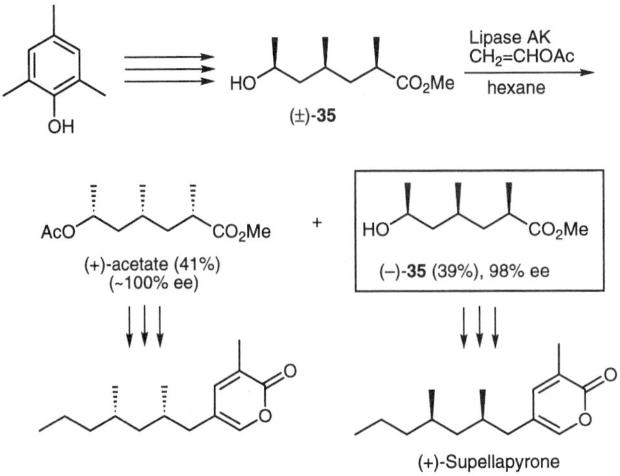

Figure 19 Preparation of (−)-**35** as the building block for (+)-supellapyrone.

which eventually gave (*S*)-rhynchophorol (Fig. 17), the pheromone of the American palm weevil (*Rhynchophorus palmarum*) [54]. Asymmetric hydrolysis of (±)-**33** with lipase AK yielded (*S*)-**33** (Fig. 18), which was converted to (*S*)-**34**, the pheromone of the leafminer *Nepticula malella* [55].

As shown in Fig. 19, asymmetric acetylation of (±)-**35** left (−)-**35** intact, which eventually afforded (2*R*,4*R*)-(+)-supellapyrone, the pheromone of the brown-banded cockroach *Supella longipalpa* [56]. Punaglandin 4 (Fig. 20) is one of the chlorinated marine prostanoids isolated from the Hawaiian octocoral *Telesto riisei*. The key building block **37** for its synthesis was prepared in 25% yield (= theoretical yield) by treating a stereoisomeric mixture of **36** with PPL [57]. In the course of the synthesis of xanthocidin (Fig. 21), an antibiotic produced by *Streptomyces xanthocidicus*, a bicyclic alcohol (±)-**38** was resolved by employing lipase AK (Amano, from *Pseudomonas* sp.) and vinyl butanoate to give (−)-**38**, which was further purified via its *p*-nitrobenzoate [58].

In 1987, Yamada and co-workers discovered that lipase can convert ω-hydroxy esters to macrolides in acceptable yield [59]. Figure 22 shows the application of Yamada's method to the synthesis of a macrolide insect pheromone [60]. Treatment of (±)-**39** with

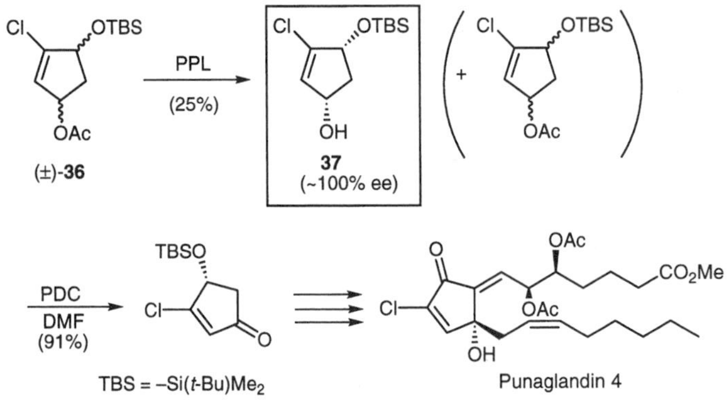

Figure 20 Preparation of **37** as the building block for punaglandin 4.

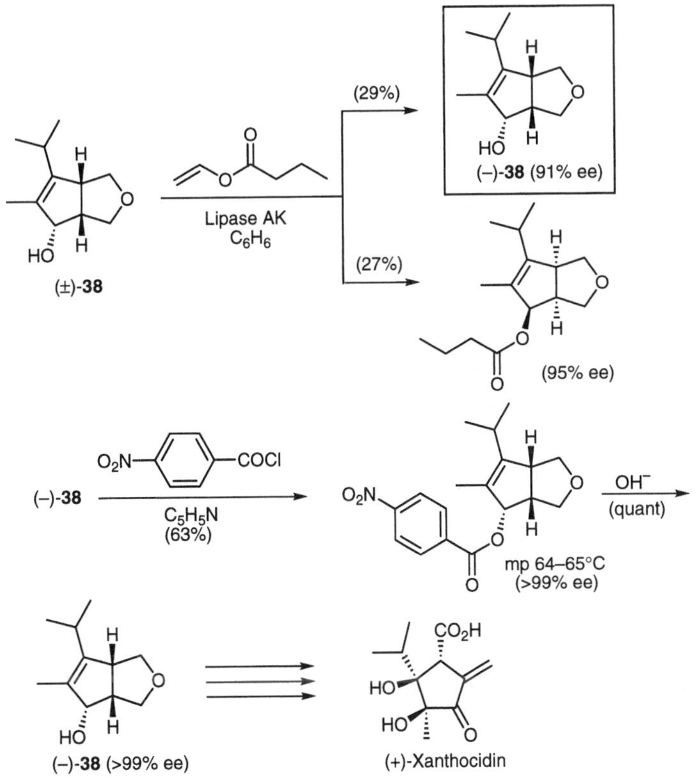

Figure 21 Preparation of (−)-**38** as the building block for xanthocidin.

lipase P (Amano, from *Pseudomonas* sp.) furnished the pheromone of the merchant grain beetle (*Oryzaephillus mercator*). Although the yield of the macrolide is modest, execution of both resolution and cyclization in a single step is the remarkable feature of lipase-catalyzed macrolactonization.

B. Asymmetric Hydrolysis of *meso*-Diesters and Asymmetric Esterification of *meso*-Diols

The ability of lipases and esterases to achieve asymmetric hydrolysis of *meso*-diesters is well known, and was reviewed by Ohno and Otsuka [61]. In theory *meso*-diesters generate the desired optically active monoesters quantitatively.

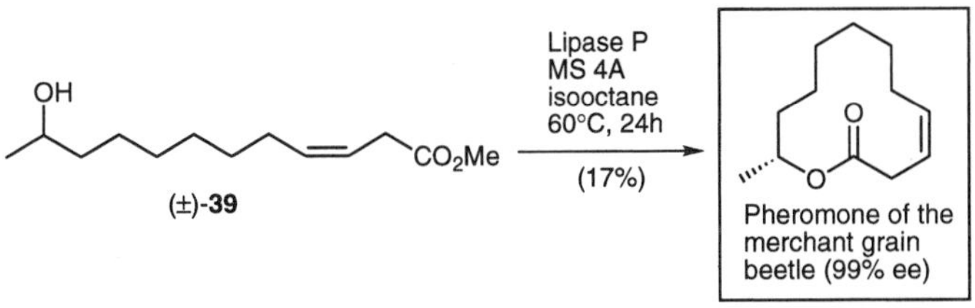

Figure 22 Synthesis of the pheromone of the merchant grain beetle.

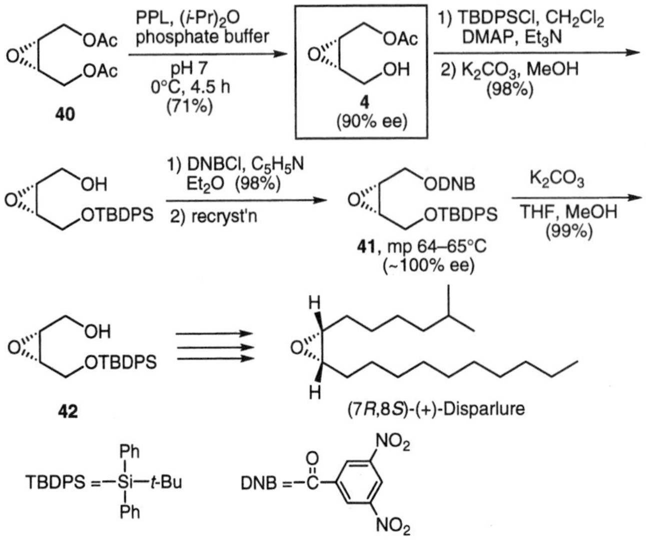

Figure 23 Preparation of the versatile building block **4** for pheromone synthesis.

Asymmetric hydrolysis of *meso*-diacetate **40** with PPL furnished monoacetate **4** (90% e.e.) as shown in Fig. 23 [17]. Further purification of **4** via crystalline **41** gave pure **42**, which was converted to (+)-disparlure and other pheromones (see Fig. 4). Asymmetric acetylation of the *meso*-diol **43** was achieved with isopropenyl acetate and lipase PS 30 (Amano) [62] to give **44** (Fig. 24), which was converted to tribolure, the pheromone of flour beetles (*Tribolium castaneum* and *T. confusum*) [62] and also to the pheromone of the pine sawfly (*Microdiprion pallipes*) [63]. The building block **44** must be useful for

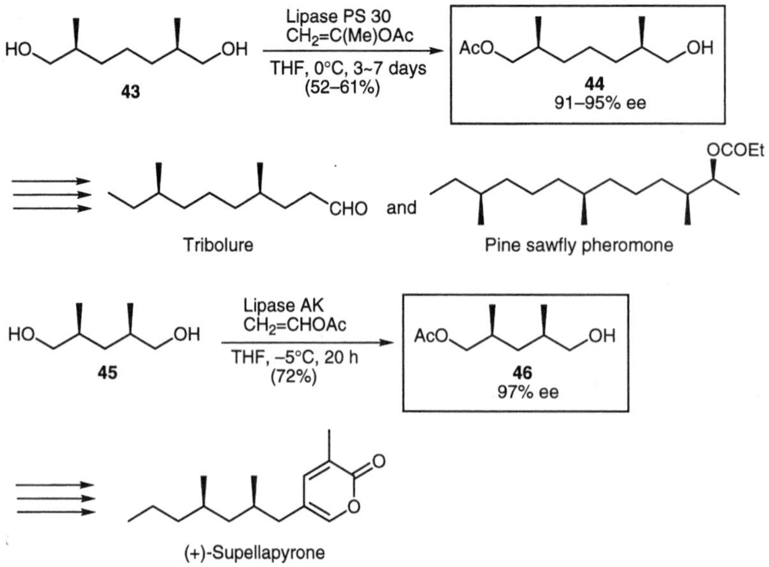

Figure 24 Preparation of the building blocks **44** and **46** for pheromone synthesis.

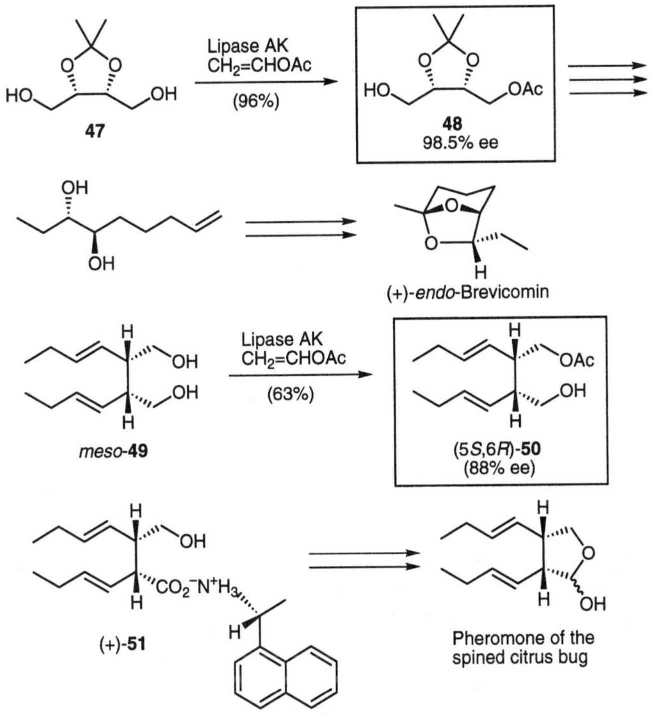

Figure 25 Preparation of the building blocks **48** and **50** for pheromone synthesis.

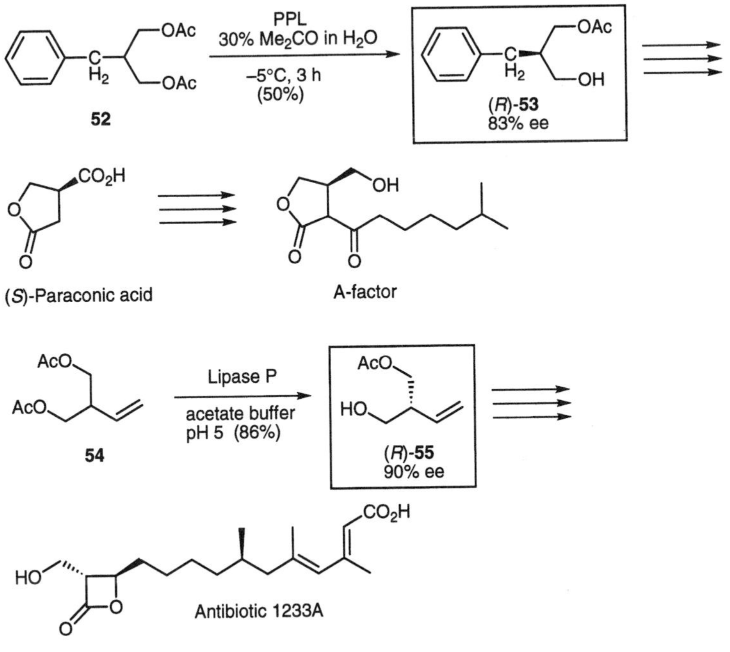

Figure 26 Preparation of the building blocks **53** and **55** for A factor and 1233A.

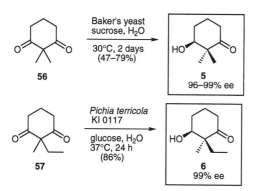

Figure 27   Reduction of monocyclic β-diketones with yeasts.

the synthesis of acyclic isoprenoids such as phytol. Another *meso*-diol **45** could be desymmetrized as shown in Fig. 24 to give **46** [64], which served as the building block for (+)-supellapyrone, the pheromone of the brown-banded cockroach (*Supella longipalpa*) [65]. (+)-*endo*-Brevicomin, the pheromone of the bark beetles (*Dendroctonus frontalis* and *Drycoetes autographus*), was synthesized by employing **48** (Fig. 25), which was prepared by asymmetric acetylation of *meso*-diol **47** with vinyl acetate and lipase AK [66]. The pheromone of the spined citrus bug (*Biprorulus bibax*) is a hemiacetal as shown in Fig. 25. Asymmetric acetylation of the *meso*-diol **49** with vinyl acetate and lipase AK gave

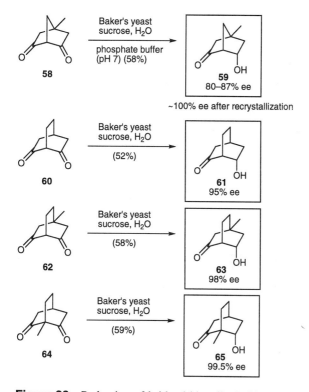

Figure 28   Reduction of bridged bicyclic β-diketones with baker's yeast.

50. This was purified by recrystallizing (+)-**51** and converted to the pheromone of ~100% e.e. [67]. Asymmetric hydrolysis of **52** with PPL yielded (*R*)-**53** (Fig. 26) [68]. This was converted to (*S*)-paraconic acid, purified as its amine salt, and eventually afforded A factor, the inducer of streptomycin biosynthesis in *Streptomyces griseus* [68]. The *meso*-diacetate **54** was hydrolyzed with lipase P to give (*R*)-**55**, from which the β-lactone part of the antibiotic 1233A was constructed (Fig. 26) [69].

V. YEASTS AS CATALYSTS FOR ENANTIOSELECTIVE REDUCTION AND OXIDATION

A. Asymmetric Reduction of β-Diketones

The ability of yeasts such as baker's yeast (*Saccharomyces cerevisiae*) was utilized extensively by chemists to reduce carbonyl compounds to alcohols [70,71]. 2,2-Dimethylcyclohexane-1,3-dione (**56**) can be reduced with fermenting baker's yeast to give the (*S*)-hydroxy ketone **5** (Fig. 27), which was employed extensively in terpene synthesis as shown in Fig. 5 [23]. Diastereo- and enantioselective reduction of **57** with *Pichia terricola* KI 0117 yielded **6** (Fig. 27) [35], which was converted to both (+)-JH I [35] and (−)-JH I [72,73] (Fig. 6). Reduction of **57** with baker's yeast was nondiastereoselective [35].

Asymmetrical reduction of some bridged bicyclic β-diketones **58**, **60**, **62**, and **64** yielded ketols **59** [26], **61** [74], **63** [74], and **65** [74], respectively, as shown in Fig. 28. The hydroxy ketones **5** and **59** were the building blocks for the synthesis of glycinoeclepin A (Fig. 29), a degraded triterpenoid with remarkable hatch-stimulating activity against the

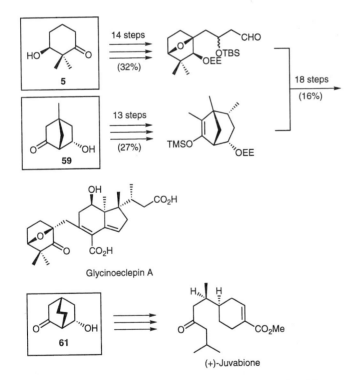

Figure 29 Synthesis of glycinoeclepin A and (+)-juvabione.

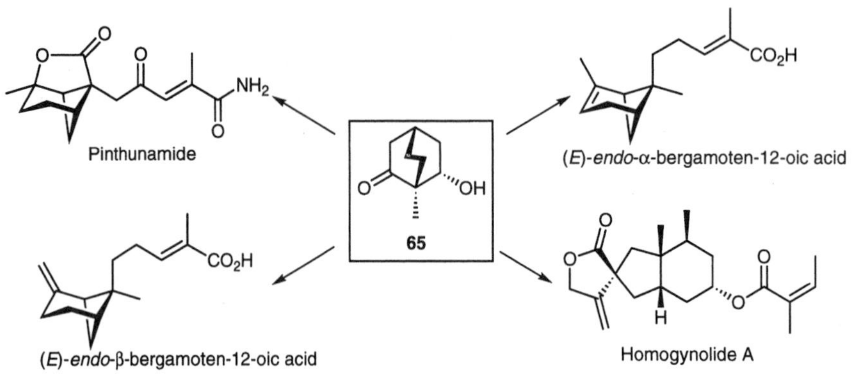

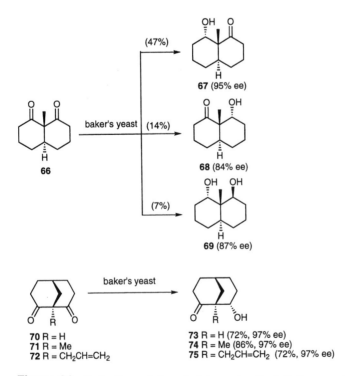

Figure 30   Terpenes synthesized from the building block **65**.

soybean cyst nematode [26]. The hydroxy ketone **61** was converted to (+)-juvabione, a terpene with juvenile hormone activity (Fig. 29) [75]. The hydroxy ketone **65** (Fig. 30) was the starting material of the following terpenes: (1) pinthunamide, a promotor of root growth of lettuce [76], (2) (*E*)-*endo*-α- and β-bergamoten-12-oic acids, sesquiterpenes isolated from wild tomato leaves as the oviposition stimulator for gravid moths [77], and (3) homogynolide A, an insect antifeedant [78].

Figure 31 shows the reduction of 9-methyl-*trans*-decalin-1,8-dione (**66**) with baker's yeast [79]. The major product **67** might be useful in isoprenoid synthesis. Reduction of

Figure 31   Reduction of 9-methyl-*trans*-decalin-1,8-dione and bicyclo[3.3.1]nonane-2,8-diones with baker's yeast.

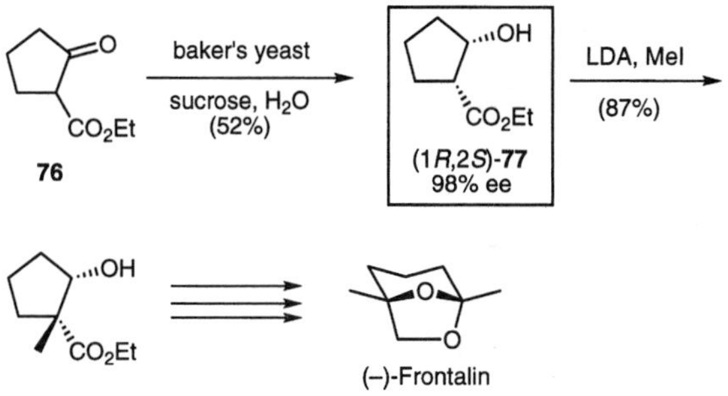

Figure 32 Preparation of **77** as the building block for frontalin.

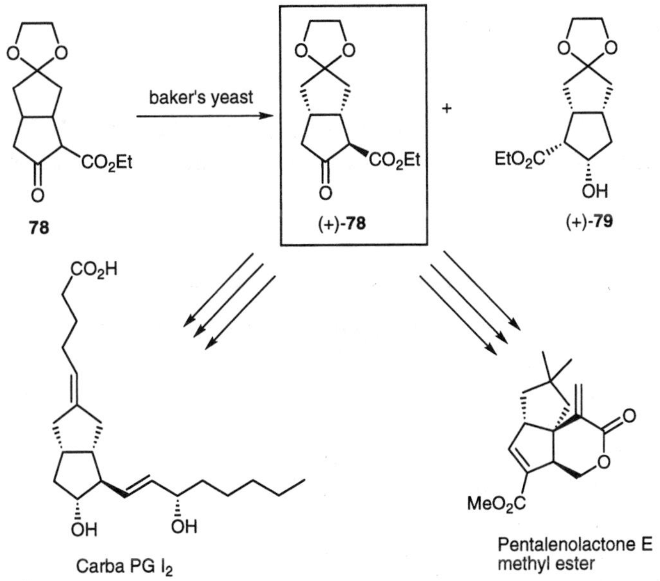

Figure 33 Preparation of the building block (+)-**78** for carbaPG I₂ and pentalenolactone E methyl ester.

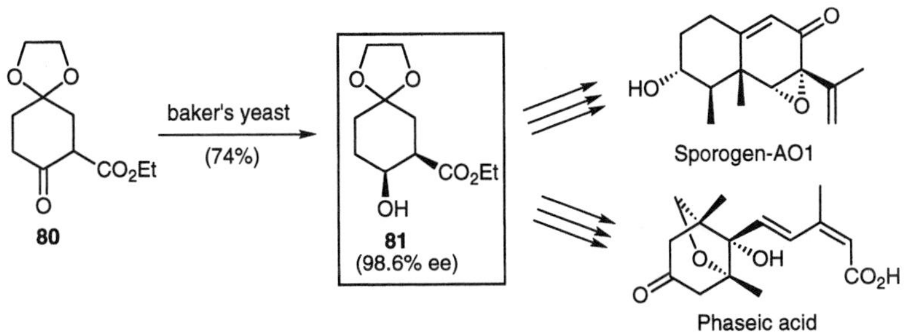

Figure 34 Preparation of the building block **81** for sporogen-AO1 and phaseic acid.

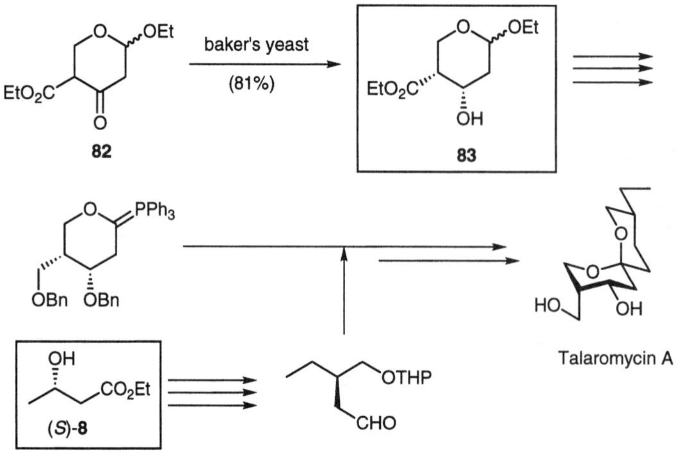

Figure 35 Preparation of the building block **83** for talaromycin A.

bicyclo[3.3.1]nonane-2,8-diones **70–72** with baker's yeast gave hydroxy ketones **73–75**, respectively (Fig. 31) [80]. These hydroxy ketones, especially **75**, might be useful in terpene synthesis.

B. Asymmetric Reduction of β-Keto Esters and a β-Keto Sulfide

Reduction of β-keto esters with baker's yeast affords optically active β-hydroxy esters. Ethyl 2-oxocyclopentane-1-carboxylate (**76**) was reduced with baker's yeast to give (*1R,2S*)-**77** (Fig. 32), which was converted to (−)-frontalin, the pheromone of the bark beetles (*Dendroctonus frontalis, D. brevicomis, D. pseudotsugae*, and *D. jeffreyi*) [81]. In the case of the reduction of (±)-**78**, only (−)-**78** was reduced to give (+)-**79**, and (+)-**78** was recovered unchanged (Fig. 33) [82]. The recovered optically active β-keto ester (+)-**78** was converted to carbaprostaglandin I$_2$ [82] and pentalenolactone E methyl ester [83]. Reduction of the β-keto ester **80** with baker's yeast afforded **81** (Fig. 34) [84], which served as the building block for sporogen-AO 1 [85], a sporogenic hormone of *Aspergillus oryzae*, and phaseic acid [86], a metabolite of abscisic acid. (−)-Talaromycin A (Fig. 35) is a fungal toxin isolated from *Talaromyces stipitatus*. It was synthesized from **83** and (*S*)-**8**, the former of which was obtained by yeast reduction of the β-keto ester **82** [87]. As

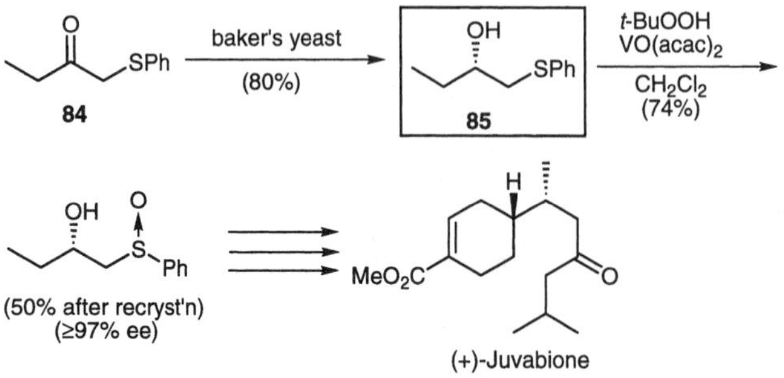

Figure 36 Preparation of the building block **85** for (+)-juvabione.

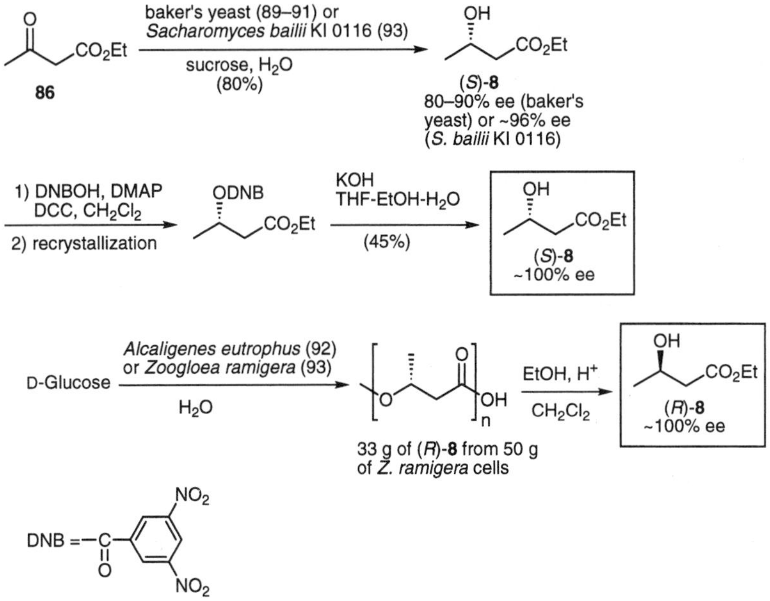

Figure 37 Biochemical preparation of the enantiomers of ethyl 3-hydroxybutanoate.

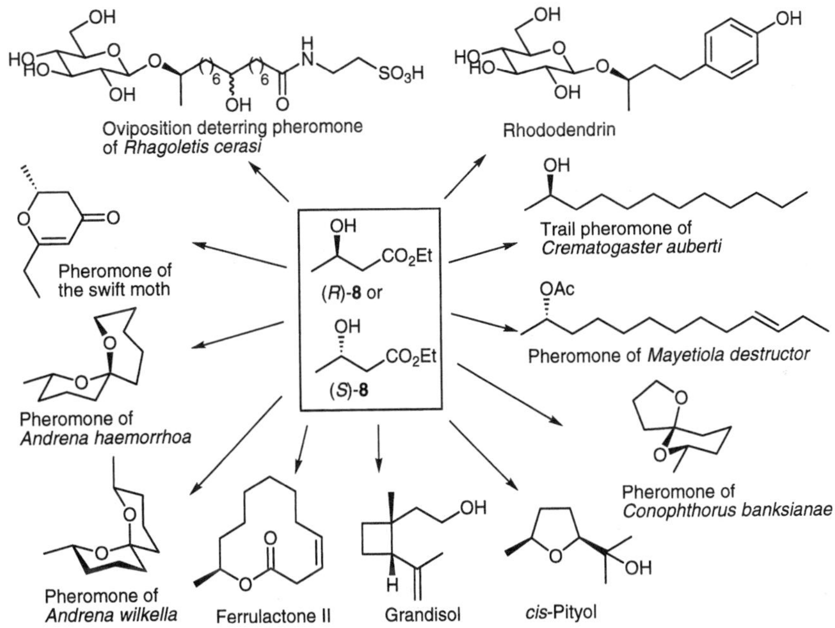

Figure 38 Compounds synthesized by starting from ethyl 3-hydroxybutanoate.

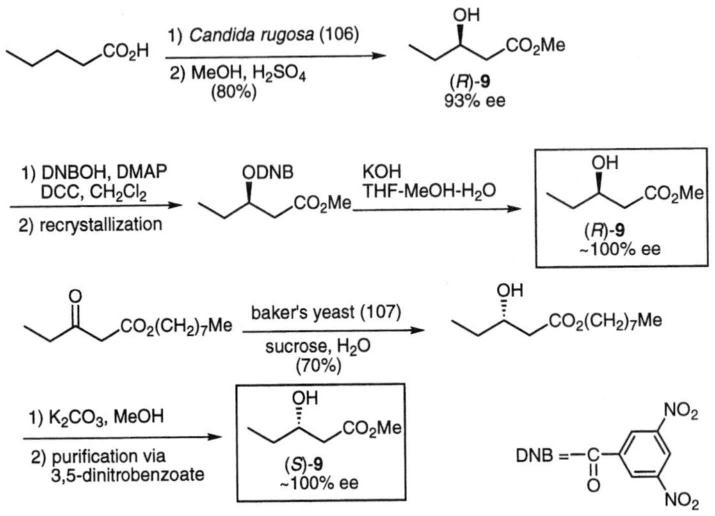

Figure 39 Biochemical preparation of the enantiomers of methyl 3-hydroxypentanoate.

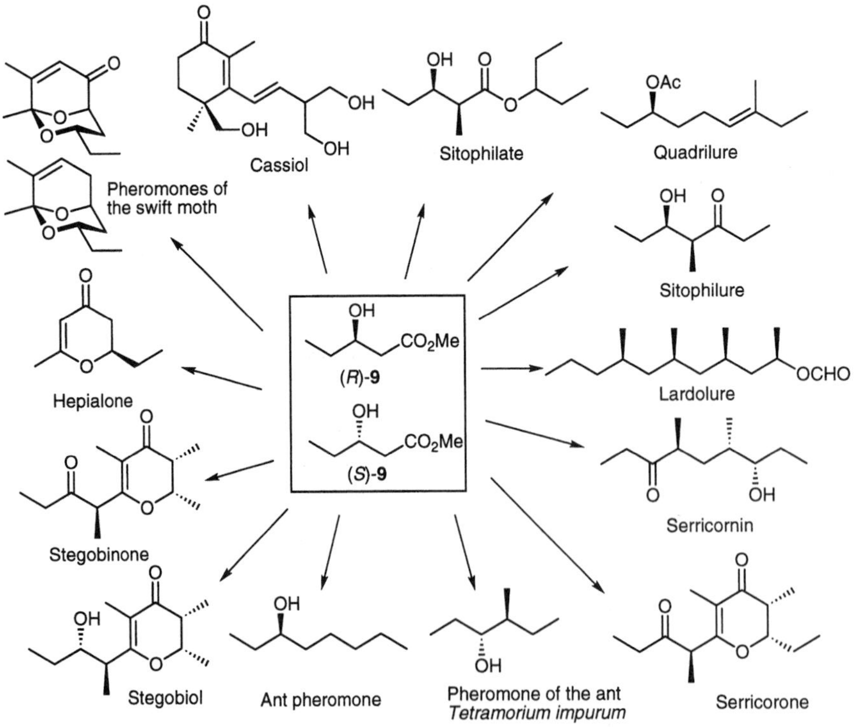

Figure 40 Compounds synthesized by starting from ethyl 3-hydroxypentanoate.

shown in Fig. 36, reduction of the β-keto sulfide **84** yielded **85**, which was converted to (+)-juvabione [88].

C. Hydroxy Esters as Building Blocks of Microbial Origin

Enantiomerically pure hydroxy esters are versatile nonracemic chiral building blocks. In the past, both enantiomers of ethyl 3-hydroxybutanoate (**8**) were prepared by biochemical methods as shown in Fig. 37 [89–93]. They are now commercially available and widely employed as building blocks. In Fig. 38 are illustrated the structures of compounds synthesized by starting from (*R*)- or (*S*)-**8**. They are (1) the oviposition-deterring pheromone of the European cherry fruit fly (*Rhagoletis cerasi*) [94], (2) rhododendrin, a glucoside isolated from *Rhododendron chrysanthum* [95], (3) the trail pheromone of worker ants of *Crematogaster auberti* [96], (4) the pheromone of the Hessian fly (*Mayetiola destructor*) [97], (5) the pheromone of the pine beetle *Conophthorus banksianae* [98], (6) *cis*-pityol, the pheromone of the elm bark beetle *Pteleobius vittatus* [99], (2) grandisol, the pheromone of the boll weevil [100], (8) ferrulactone II, the pheromone of *Cryptolestes ferrugineus* [101], (9) the pheromone of the wasp *Andrena wilkella* [102,103], (10) the pheromone of

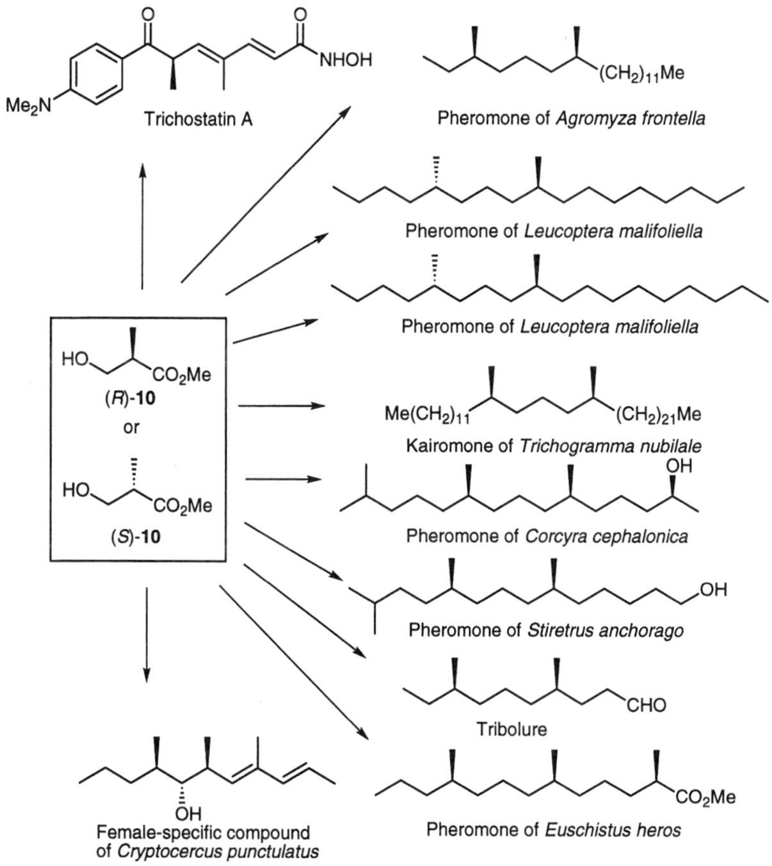

Figure 41 Compounds synthesized by starting from methyl 3-hydroxy-2-methylpropanoate.

Andrena haemorrhoa [104], and (11) the pheromone component of the swift moth *Hepialus hecta* [105].

Both enantiomers of methyl 3-hydroxypentanoate (**9**) are available by biochemical methods (Fig. 39) [106,107]. They are not commonly commercially available yet, although they can be prepared by Noyori's asymmetric hydrogenation. The methyl esters **9** could be purified via crystalline 3,5-dinitrobenzoate to give pure **9** [108]. As shown in Fig. 40, the enantiomers of **9** were converted to (1) sitophilate, the pheromone of granary weevil (*Sitophilus granarius*) [109], (2) quadrilure, the pheromone of the square-necked grain beetle (*Cathartus quadricollis*) [110], (3) lardolure, the pheromone of the acarid mite (*Lardoglyphus konoi*) [111], (4) sitophilure, the pheromone of *Sitophilus oryzae* [112], (5) serricornin, the pheromone of the cigarette beetle (*Lasioderma serricorne*) [108], (6) serricorone, the pheromone component of the cigarette beetle [113], (7) the pheromone of the ant *Tetramorium impurum* [114], (8) the pheromone of some ants [115], (9) stegobiol, the pheromone of the drugstore beetle (*Stegobium paniceum*) [116,117], (10) hepialone, the pheromone of *Hepialus californicus* [118], (11) the pheromone components of the swift moth (*Hepialus hecta*) [105], and (12) cassiol, a potent antiulcerogenic compound [119]. Both enantiomers of methyl 3-hydroxy-2-methylpropanoate (**10**) are commercially available. Hydroxylation of isobutyric acid with *Pseudomonas putida* gives (*S*)-3-hydroxy-2-methylpropanoic acid [120]. *Candida rugosa* and its mutant can oxidize isobutyric acid into either (*R*)- or (*S*)-3-hydroxy-2-methylpropanoic acid [121]. The enantiomers of **10** are popular building blocks. Figure 41 summarizes our own use of **10** as the starting material for the following compounds: (1) an antibiotic trichostatin A [122], (2) the pheromone of the alfalfa blotch leafminer (*Agromyza frontella*) [123], (3) the pheromone components of the mountain ash bentwing (*Leucoptera malifoliella*) [124], (4) the kairomone for the wasp *Trichogramma nubilale* [125], (5) the pheromone of the rice moth (*Corcyra cephalonica*) [126], (6) the pheromone of the predatory stink bug (*Stiretrus anchorago*) [127], (7) the female-specific compound of the woodroach (*Cryptocercus punctulatus*) [128], (8) tribolure, the pheromone of *Tribolium castaneum* [129], and (9) the pheromone of the stink bug *Euschistus heros* [130].

VI. CONCLUSION

Enzymes and microorganisms help synthetic chemists to prepare nonracemic chiral building blocks. At present lipases are most useful enzymes. Additional new building blocks will be designed and prepared to facilitate enantioselective synthesis. Combination of organic synthesis with biotransformation will continue to provide efficient routes for the enantioselective synthesis of bioactive natural products.

ACKNOWLEDGMENT

I thank my junior colleagues, especially Dr. H Takikawa, for their help in preparing the manuscript.

REFERENCES

1. K Mori. Chem Commun 1153–1158, 1997.
2. K Mori. Eur J Org Chem 1479–1489, 1998.

3. K Mori, H Akao. Tetrahedron Lett 4127–4130, 1978.
4. K Mori, H Akao. Tetrahedron 36: 91–96, 1980.
5. K Mori. In: Atta-ur-Rahman, ed. Studies in Natural Products Chemistry, Vol. 1, Stereoselective Synthesis (Part A). Amsterdam: Elsevier, 1988, pp 677–712.
6. K Mori. In: JR Whitaker, PE Sonnet, eds. ACS Symposium Series No. 389. Biocatalysis in Agricultural Biotechnology. Washington, DC: ACS, 1989, pp 348–358.
7. K Mori. Bull Soc Chim Belg 101: 393–405, 1992.
8. K Mori. Synlett 1097–1109, 1995.
9. K Mori. Tetrahedron 45: 3233–3298, 1989.
10. K Mori, P Puapoomchareon. Liebigs Ann Chem 1053–1056, 1991.
11. K Mori, M Amaike, M Itou. Tetrahedron 49: 1871–1878, 1993.
12. K Mori, S Aki. Acta Chem Scand 45: 625–629, 1992.
13. K Mori, V Khlebnikov. Liebigs Ann Chem 77–82, 1993.
14. K Mori, S Aki, M Kido. Liebigs Ann Chem 83–90, 1993.
15. K Mori, H Kiyota, D Rochat. Liebigs Ann Chem 865–870, 1993.
16. K Mori, H Kiyota, C Malosse, D Rochat. Liebigs Ann Chem 1201–1204, 1993.
17. JL Brevet, K Mori. Synthesis 1007–1012, 1992.
18. K Mori, NP Argade. Liebigs Ann Chem 695–700, 1994.
19. K Mori, M Amaike. J Chem Soc Perkin Trans 1, 2727–2733, 1994. (Corrigendum: J Chem Soc, Perkin Trans 1 1994; 3600).
20. K Mori, T Furuuchi, H Kiyota. Liebigs Ann Chem 971–974, 1994.
21. K Mori, T Furuuchi, K Matsuyama. Liebigs Ann Chem 2093–2099, 1995.
22. T Ohkuma, H Ikehira, T Ikariya, R Noyori. Synlett 467–468, 1997.
23. K Mori, H Mori. Org Synth Coll 8: 312–315, 1993.
24. M Yanai, T Sugai, K Mori. Agric Biol Chem 49: 2373–2377, 1985.
25. K Mori, H Mori. Tetrahedron 42: 5531–5538, 1986.
26. H Watanabe, K Mori. J Chem Soc Perkin Trans 1, 2919–2934, 1991.
27. K Mori, H Mori. Tetrahedron 43: 4097–4106, 1987.
28. K Mori, H Takaishi. Liebigs Ann Chem 939–943, 1989.
29. K Mori, M Komatsu. Tetrahedron 43: 3409–3412, 1987.
30. K Mori, Y Koga. Liebigs Ann Chem 1755–1763, 1995.
31. K Mori, M Komatau. Liebigs Ann Chem 107–119, 1988.
32. K Mori, H Mori. Tetrahedron 41: 5487–5493, 1985.
33. K Mori, H Watanabe. Tetrahedron 42: 273–281, 1986.
34. K Mori, Y Nakazono. Tetrahedron 42: 283–290, 1986.
35. K Mori, M Fujiwhara. Tetrahedron 44: 343–354, 1988.
36. K Mori, M Fujiwhara. Israel J Chem 31: 223–227, 1991.
37. T Hudlicky, JW Reed. In: A Hassner, ed. Advances in Asymmetric Synthesis, Vol. 1. Greenwich: JAI Press, 1995, pp 271–312.
38. JD Stewart. Current Org Chem 2: 195–216, 1998.
39. PJ Fodor, VE Price, JP Greenstein. J Biol Chem 178: 503–509, 1949.
40. I Chibata. In: EL Eliel, S Otsuka, ed. ACS Symposium Series No. 185. Asymmetric Reactions and Processes in Chemistry. Washington, DC: ACS, 1982, pp 195–203.
41. K Mori, T Sugai, Y Maeda, T Okazaki, T Noguchi, H Naito. Tetrahedron 41: 5307–5311, 1985.
42. Y Masaoka, M Sakakibara, K Mori. Agric Biol Chem 46:2319–2324, 1982.
43. T Sugai, K Mori. Agric Biol Chem 48: 2497–2500, 1984.
44. K Mori, Y Funaki. Tetrahedron 41: 2369–2377, 1985.
45. K Mori, H Iwasawa. Tetrahedron 36: 2209–2213, 1980.
46. S Senda, K Mori. Agric Biol Chem 51: 1379–1384, 1987.
47. F Theil. Chem Rev 95: 2203–2227, 1995.
48. S Horiuchi, H Takikawa, K Mori. Bioorg Med Chem 7: 723–726, 1999.
49. S Horiuchi, H Takikawa, K Mori. Eur J Org Chem 2851–2854, 1998.

50. K Hirayama, K Mori. Eur J Org Chem (in press).
51. H Takikawa, S Muto, D Nozawa, A Kayo, K Mori. Tetrahedron Lett 39: 6931–6934, 1998.
52. K Mori, S Sano, Y Yokoyama, M Bando, M Kido. Eur J Org Chem 1135–1142, 1998.
53. Y Yokoyama, H Takikawa, K Mori. Bioorg Med Chem 4: 409–412, 1996.
54. K Mori, K Ishigami. Liebigs Ann Chem 1195–1198, 1992.
55. K Mori, H Ogita. Liebigs Ann Chem 1065–1068, 1994.
56. K Mori, Y Takeuchi. Proc Jpn Acad Ser B 70: 143–145, 1994.
57. K Mori, T Takeuchi. Tetrahedron 44: 333–342, 1988.
58. K Mori, A Horinaka, M Kido. Liebigs Ann Chem 817–825, 1994.
59. A Makita, T Nihira, Y Yamada. Tetrahedron Lett 28: 805–808, 1987.
60. K Mori, H Tomioka. Liebigs Ann Chem 1011–1018, 1992.
61. M Ohno, M Otsuka. Org React 37: 1–55, 1989.
62. R. Chênevert, M Desjardins. J Org Chem 61: 1219–1222, 1996.
63. Y Nakamura, K Mori. Eur J Org Chem (in press).
64. K Tsuji, Y Terao, K Achiwa. Tetrahedron Lett 30: 6189–6192, 1989.
65. K Fujita, K Mori. Eur J Org Chem (in prep.).
66. K Mori, H Kiyota. Liebigs Ann Chem 989–992, 1992.
67. K Mori, M Amaike, H Watanabe. Liebigs Ann Chem 1287–1294, 1993.
68. K Mori, N Chiba. Liebigs Ann Chem 957–962, 1989.
69. K Mori, Y Takahashi. Liebigs Ann Chem 1057–1065, 1991.
70. S Servi. Synthesis 1–25, 1990.
71. R Csuk, BI Glänzer. Chem Rev 91: 49–97, 1991.
72. K Mori, M Fujiwhara. Liebigs Ann Chem 369–372, 1990.
73. K Mori. In: N Kurihara and J Miyamoto, eds. Chirality in Agrochemicals, New York: Wiley, 1998, pp 221–230.
74. K Mori, E Nagano. Biocatalysis 3: 25–36, 1990.
75. E Nagano, K Mori. Biosci Biotech Biochem 56: 1589–1591, 1992.
76. K Mori, Y Mastushima. Synthesis 406–410, 1993.
77. K Mori, Y Mastushima. Synthesis 417–421, 1994.
78. K Mori, Y Matsushima. Synthesis 845–850, 1995.
79. K Mori, S Takayama, S Yoshimura. Liebigs Ann Chem 91–95, 1993.
80. K Mori, S Takayama, M Kido. Bioorg Med Chem 2: 395–401, 1994.
81. Y Nishimura, K Mori. Eur J Org Chem 233–236, 1998.
82. K Mori, M Tsuji. Tetrahedron 42: 435–444, 1986.
83. K Mori, M Tsuji. Tetrahedron 44: 2835–2842, 1988.
84. T Kitahara, K Mori. Tetrahedron Lett 26: 451–452, 1985.
85. T Kitahara, H Kurata, K Mori. Tetrahedron 44: 4339–4349, 1988.
86. T Kitahara, K Touhara, H Watanabe, K Mori. Tetrahedron 45: 6387–6400, 1989.
87. K Mori, M Ikunaka. Tetrahedron 43: 45–58, 1987.
88. H Watanabe, H Shimizu, K Mori. Synthesis 1249–1254, 1994.
89. BS Deol, DD Ridley, GW Simpson. Aust J Chem 29: 2459–2467, 1976.
90. K Mori. Tetrahedron 37: 1341–1342, 1981.
91. E Hungerbühler, D Seebach, DS Wasmuth. Helv Chim Acta 64: 1467–1487, 1981.
92. D Seebach, MF Züger. Helv Chim Acta 65: 495–503, 1982.
93. T Sugai, M Fujita, K Mori. Nippon Kagaku Kaishi (J Chem Soc Jpn) 1315–1321, 1983
94. K Mori, Z-H Qian. Liebigs Ann Chem 291–295, 1994.
95. K Mori, Z-H Qian. Bull Soc Chim France 130: 382–387, 1993.
96. W Pinyarat, K Mori. Biosci Biotech Biochem 56: 1673, 1992.
97. Y Takeuchi, K Mori. Biosci Biotech Biochem 57: 1967–1968, 1993.
98. T Hasegawa, A Kamada, K Mori. Biosci Biotech Biochem 56: 838–839, 1992.
99. K Mori, P Puapoomchareon. Liebigs Ann Chem 1261–1262, 1989.
100. K Mori, M Miyake. Tetrahedron 43: 2229–2239, 1987.
101. T Sakai, K Mori. Agric Biol Chem 50: 177–183, 1986.

102. K Mori, K Tanida. Tetrahedron 37: 3221–3225, 1981.
103. K Mori, H Watanabe. Tetrahedron 42: 295–304, 1986.
104. K Mori, M Katsurada. Liebigs Ann Chem 157–161, 1984.
105. K Mori, H Kisida. Tetrahedron 42: 5281–5290, 1986.
106. J Hasegawa, S Hamaguchi, M Ogura, K Watanabe. J Ferm Technol 59: 257–262, 1981.
107. K Mori, H Mori, T Sugai. Tetrahedron 41: 919–925, 1985.
108. K Mori, H Watanabe. Tetrahedron 41: 3423–3428, 1985.
109. K Mori, M Ishikura. Liebigs Ann Chem 1263–1265, 1989.
110. K Mori, P Puapoomchareon. Liebigs Ann Chem 159–162, 1990.
111. K Mori, S Kuwahara. Tetrahedron 42: 5539–5544, 1986.
112. K Mori, T Ebata. Tetrahedron 42: 4421–4426, 1986.
113. T Ebata, K Mori. Agric Biol Chem 51: 2925–2928, 1987.
114. M Kato, K Mori. Agric Biol Chem 49: 3073–3075, 1985.
115. M Fujiwhara, K Mori. Agric Biol Chem 50: 2925–2927, 1986.
116. K Mori, T Ebata. Tetrahedron 42: 4685–4689, 1986.
117. K Mori, T Ebata. Tetrahedron 42: 4413–4420, 1986.
118. S Hayashi, K Mori. Agric Biol Chem 50: 3209–3210, 1986.
119. T Uno, H Watanabe, K Mori. Tetrahedron 46: 5563–5566, 1990.
120. CT Goudhue, JR Sehaeffer. Biotech Bioeng 13: 203–214, 1971.
121. J Hasegawa, M Ogura, H Kanema, H Kawaharada, K Watanabe. J Ferment Technol 61: 37–42, 1983.
122. K Mori, K Koseki. Tetrahedron 44: 6013–6020, 1988.
123. K Mori, J Wu. Liebigs Ann Chem 213–217, 1991.
124. K Mori, J Wu. Liebigs Ann Chem 439–443, 1991.
125. K Mori, J Wu. Liebigs Ann Chem 83–85, 1992.
126. K Mori, H Harada, P Zagatti, A Cork, DR Hall. Liebigs Ann Chem 259–267, 1991.
127. K Mori, J Wu. Liebigs Ann Chem 783–788, 1991.
128. K Mori, M Itou. Liebigs Ann Chem 87–93, 1992.
129. K Mori, H Takikawa. Liebigs Ann Chem 497–500, 1991.
130. K Mori, N Murata. Liebigs Ann Chem 1153–1160, 1994.

4

Stereoselective Biocatalysis for Synthesis of Some Chiral Pharmaceutical Intermediates

Ramesh N. Patel
Bristol-Myers Squibb Pharmaceutical Research Institute, New Brunswick, New Jersey

I. INTRODUCTION

Currently much attention has been focused on the interaction of small molecules with biological macromolecules. The search for selective enzyme inhibitors and receptor agonists or antagonists is one of the keys for target-oriented research in the pharmaceutical industry. Increasing understanding of the mechanism of drug interaction on a molecular level has led to the wide awareness of the importance of chirality as the key to the efficacy of many drug products. It is now known that in many cases only one stereoisomer of a drug substance is required for efficacy and the other stereoisomer either is inactive or exhibits considerably reduced acitivity. Pharmaceutical companies are aware that, where appropriate, new drugs for the clinic should be homochiral to avoid the possibility of unnecessary side effects due to an undesirable stereoisomer. The physical characteristics of enantiomers versus racemates may confer processing or formulation advantages.

Chiral drug intermediates can be prepared by different routes. One is to obtain them from naturally occurring chiral synthons mainly produced by fermentation processes. The chiral pool primarily refers to inexpensive, readily available, optically active natural products. Second is to carry out the resolution of racemic compounds. This can be achieved by preferential crystallization of stereoisomers or diastereoisomers and by kinetic resolution of racemic compounds by chemical or biocatalytic methods. Finally, chiral synthons can also be prepared by asymmetrical synthesis by either chemical or biocatalytic processes using microbial cells or enzymes derived therefrom. The advantages of microbial- or enzyme-catalyzed reactions over chemical reactions are that they are stereoselective, and can be carried out at ambient temperature and atmospheric pressure. This minimizes problems of isomerization, racemization, epimerization, and rearrangement that generally occur during chemical processes. Biocatalytic processes are generally carried out in aqueous solution. This avoids the use of environmentally harmful chemicals used in the chemical processes and solvent waste disposal. Furthermore, microbial cells or enzymes derived therefrom can be immobilized and reused in many cycles.

Recently, a number of review articles [1–14] have been published on the use of enzymes in organic synthesis. This chapter provides some specific examples of stereoselective biotransformations to prepare chiral intermediates required for the synthesis of pharmaceutical drugs.

II. ACE INHIBITORS

Captopril is designated chemically as 1-[(2S)-3-mercapto-2-methylpropionyl]-L-proline **1**.
It is used as an antihypertensive agent through suppression of the renin-angiotensin-al-
dosterone system [15–17]. Captopril and other compounds such as enalapril **2** and lisinopril **3** (Fig. 1) prevent the conversion of angiotensin I to angiotensin II by inhibition of
angiotensin-converting enzyme (ACE).

The potency of captopril **1** as an inhibitor of ACE depends critically on the configuration of the mercaptoalkanoyl moiety; the compound with the S configuration is about
100 times more active than its corresponding R enantiomer (18). The required 3-mercapto-
(2S)-methylpropionic acid moiety has been prepared from the microbially derived chiral
3-hydroxy-(2R)-methylpropionic acid, which is obtained by the hydroxylation of isobutyric
acid [19–21].

The use of extracellular lipases of microbial origin to catalyze the stereoselective
hydrolysis of 3-acylthio-2-methylpropanoic acid ester in an aqueous system has been demonstrated to produce optically active 3-acylthio-2-methylpropanoic acid [22–24]. The synthesis of chiral side chain of captopril by the lipase-catalyzed enantioselective hydrolysis
of the thioester bond of racemic 3-acetylthio-2-methylpropanoic acid **4** to yield S-(−)-**4**
has been demonstrated [25]. Among various lipases evaluated, lipase from *Rhizopus oryzae*
ATCC 24563 (heat-dried cells) and lipase PS-30 in organic solvent system (1,1,2-trichloro-
1,2,2-trifluoroethane or toluene) catalyzed the hydrolysis of thioester bond, an undesired
enantiomer of racemic **4**, to yield desired S-(−)**4**, R-(+)-3-mercapto-2-methylpropanoic
acid **5** and acetic acid **6** (Fig. 1). The reaction yield of more than 24% (theoretical maximum 50%) and enantiomeric excess (e.e.) of more than 95% were obtained for S-(−)-**4**
using each lipase in an independent experiment.

In an alternate approach to prepare the chiral side chain of captopril **1** and zofenopril
7, the lipase-catalyzed stereoselective esterification of racemic 3-benzoylthio-2-methylpro-
panoic acid **8** (Fig. 2) in an organic solvent system was demonstrated to yield R-(+)
methyl ester **9** and unreacted acid enriched in the desired S-(−) enantiomer **8** [26]. Using
lipase PS-30 with toluene as solvent and methanol as nucleophile, the desired S-(−)-**8** was
obtained in 37% reaction yield (maximum theoretical yield is 50%) and 97% e.e. Substrate
was used at 22 g/L concentration. The amount of water and concentration of methanol

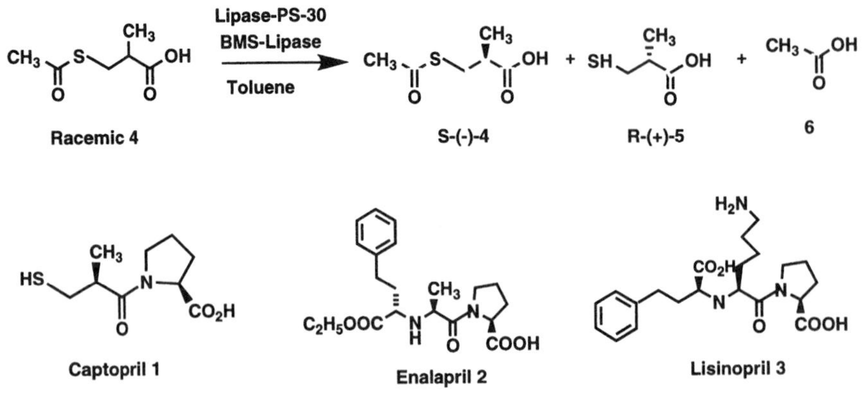

Figure 1 Synthesis of captopril side chain S-(−)-**4**: Stereoselective enzymatic hydrolysis of racemic 3-acylthio-2-methyl propanoic acid **4**.

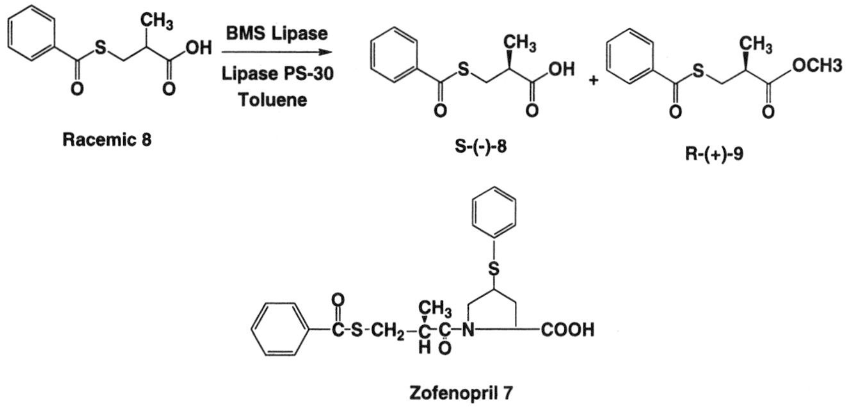

Figure 2 Synthesis of zofenopril side chain S-($-$)-**8**: Stereoselective enzymatic esterification of racemic 3-benzylthio-2-methyl propanoic acid **8**.

supplied in the reaction mixture was very critical. Water was used at 0.1% concentration in the reaction mixture. Higher than 1% water led to the aggregation of enzyme in the organic solvent with a decrease in the rate of reaction due to mass transfer limitation. The rate of esterification decreased as the methanol-to-substrate ratio was increased from 1:1 to 4:1. Higher methanol concentration probably inhibited the esterification reaction by stripping the essential water from the enzyme. Crude lipase PS-30 was immobilized on three different resins, XAD-7, XAD-2, and Accurel PP, in absorption efficiencies of about 68%, 71%, and 98.5%, respectively. These immobilized lipases were evaluated for the ability to stereoselectively esterify racemic **8**. Enzyme immobilized on Accurel PP catalyzed efficient esterification, giving 36–45% reaction yields (theoretical maximum is 50%) and more than 97.7% e.e.'s of $S(-)$-**8**. The immobilized enzyme was reused for 23 additional reaction cycles without any loss of activity and productivity. S-($-$)-**8** is a key chiral intermediate for the synthesis of captopril **1** [27] or zofenopril **7** [28]; both are antihypertensive drugs (Fig. 2).

The S-($-$)-α-[(acetylthio)methyl]benzenepropanoic acid **10** is a key chiral intermediate needed for the synthesis of neutral endopeptidase inhibitor **11** [29,30]. We have demonstrated [25] the lipase-catalyzed stereoselective hydrolysis of thioester bond of racemic α-[(acetylthio)methyl]benzenepropanoic acid **10** in organic solvent to yield R-($+$)-α-[(mercapto)methyl]benzenepropanoic acid **12** and S-($-$)-**10**. Using lipase PS-30, the S-($-$)-**10** was obtained in 40% reaction yield (theoretical maximum yield is 50%) and 98% e.e. (Fig. 3).

The S-($-$)-2-cyclohexyl-1,3-propanediol monoacetate **13** and the S-($-$)-2-phenyl-1,3-propanediol monoacetate **14** are key chiral intermediates for the chemoenzymatic synthesis of Monopril **15** (Fig. 4), a new hypertensive drug that acts as an ACE inhibitor. The asymmetrical hydrolysis of 2-cyclohexyl-1,3-propanediol diacetate **16** and 2-phenyl-1,3-propanediol diacetate **17** to the corresponding S-($-$) monoacetate **13** and S-($-$) monoacetate **14** by PPL and *Chromobacterium viscosum* lipase have been demonstrated by us [31]. In a biphasic system using 10% toluene, the reaction yield of more than 65% and e.e. of 99% were obtained for S-($-$)-**14** using each enzyme. S-($-$)-**13** was obtained in 90% reaction yield and 99.8% e.e. using *C. viscosum* lipase under similar conditions.

The stereoselective hydrolysis of dimethyl esters of symmetrical dicarboxylic acids including mesodiacids such as *cis*-1,2-cycloalkane dicarboxylic acids and diacids with

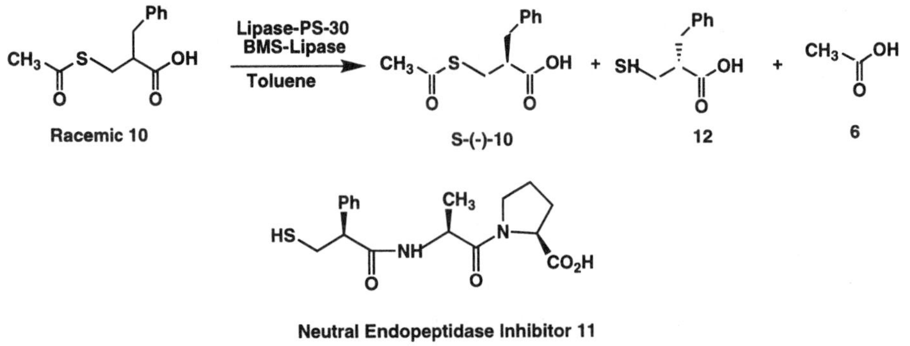

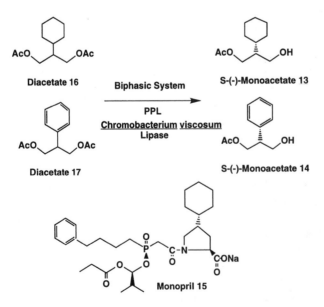

Figure 3 Preparation of chiral synthon for neutral endopeptidase inhibitor: Stereoselective enzymatic hydrolysis of racemic α-[(acetylthio)methyl]benzene propanoic acid **10**.

prochiral centers has been demonstrated by Mohr et al. [32] using pig liver esterase (PLE). The product of these stereoselective hydrolysis, chiral monoacetate of dicarboxylic acids, were obtained with e.e.'s from 10% to 90% depending on the substrate. Enantioselective hydrolysis of *cis*-1,2-diacetoxycycloalkane and 2-substituted 1,3-propanediol diacetate has been demonstrated using porcine pancreatic lipase (PPL) by Laumen and Schneider [33] and Tombo et al. [34], respectively.

Ceranopril **18** is another ACE inhibitor [35] that requires chiral intermediate carbobenzoxy(CBZ)-L-oxylysine **19** (Fig. 5). A biotransformation process was developed by Hanson et al. [36] to prepare the CBZ-L-oxylysine. *N*-α-CBZ-L-lysine **20** was first converted to the corresponding keto acid **21** by oxidative deamination using cells of *Providencia alcalifaciens* SC 9036 that contained L-amino acid oxidase and catalase. The keto

Figure 4 Preparation of chiral synthon for monopril: Asymmetrical enzymatic hydrolysis of 2-cyclohexyl- **16** and 2-phenyl-1,3-propanediol diacetate **17**.

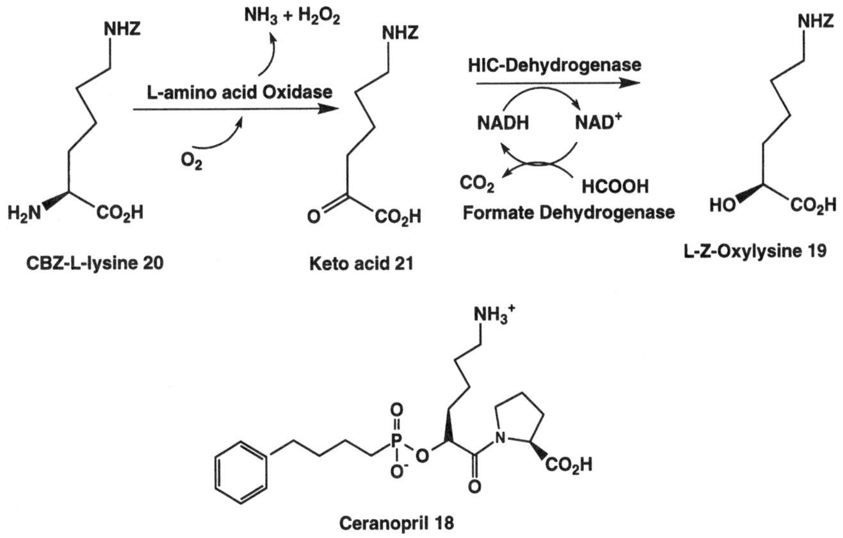

Figure 5 Synthesis of chiral synthon for ceranopril: Enzymatic conversion of CBZ-L-lysine **20** to L-Z-oxylysine **19**.

acid **21** was subsequently converted to CBZ-L-oxylysine **19** using L-2-hydroxyisocaproate dehydrogenase from *Lactobacillus confusus*. The NADH required for this reaction was regenerated using formate dehydrogenase from *Candida boidinii*. The reaction yield of 95% with 98.5% e.e. of product was obtained in the overall process.

A number of research groups have been interested in the synthesis of aminodiol **22**. This compound is believed to mimic the transition state for the renin-catalyzed hydrolysis of the peptide angiotensinogen and therefore can be useful as a potential antihypertensive drug. The enzymatic resolution of the 3-acetoxy-β-lactam **23** (Fig. 6) by lipase PS-30 has been demonstrated by Spero et al. [37]. A new approach to the BOC-protected aminodiol **24** via opening of 3,4-cis-disubstituted β-lactam has been demonstrated.

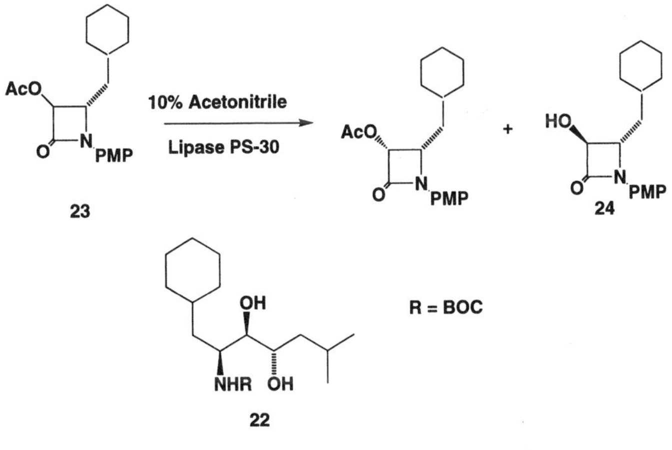

Figure 6 Lipase-catalyzed synthesis of chiral amino diol **22**.

III. PACLITAXEL SEMISYNTHESIS

Among the antimitotic agents, paclitaxel (Taxol) **25** [38,39], a complex, polycyclic diterpene, exhibits a unique mode of action on microtubule proteins responsible for the formation of the spindle during cell division. In contrast to other spindle formation inhibitors
such as vinblastine or colchicine, both of which prevent the assembly of tubuline, paclitaxel is the only compound known to inhibit the depolymerization process of microtubulin
[40]. Because of its biological activity and unusual chemical structure, paclitaxel represents
the prototype of a new series of chemotherapeutic agents. Various types of cancers have
been treated with paclitaxel and the results in treatment of ovarian cancer and metastatic
breast cancer are very promising. In collaboration with the National Cancer Institute,
Bristol-Myers Squibb developed paclitaxel for treatment of refractory ovarian cancer. Paclitaxel was originally isolated from the bark of the yew, *Taxus brevifolia* [38], and has
also been found in other *Taxus* species in relatively low yield. Taxol was initially obtained
from *T. brevifolia* bark in about 0.07% yield. It required cumbersome purification of
paclitaxel from the other related taxanes. It is estimated that about 20,000 pounds of yew
bark (equivalent of about 3000 trees) is needed to produce 1 kg of purified paclitaxel [39].
This created a concern among the environmentalists about the mass destruction of trees
to produce the required amount of paclitaxel. Alternative methods for production of paclitaxel by cell suspension cultures and by semisynthetic processes are being evaluated by
various groups [41–43]. The development of a semisynthetic process for the production
of paclitaxel from baccatin III **26** or 10-deacetylbaccatin III **27** (10-DAB) and C-13 taxol
side chain **28** (Fig. 7) was a very promising approach. Paclitaxel, related taxanes, baccatin
III, and 10-DAB can be derived from renewable resources such as extract of needles,
shoot, and young *Taxus* cultivars. The most valuable material in this mixture for semisynthesis is the taxane "nucleus" component of baccatin III **26** (paclitaxel without the C-13
side chain) and 10-DAB **27** (10-DAB, paclitaxel without the C-13 side chain and the C-
10 acetate). Conversion of taxanes to 10-DAB by cleavage of the C-10 acetate and the
C-13 paclitaxel side chain is a very attractive approach to increase the concentration of

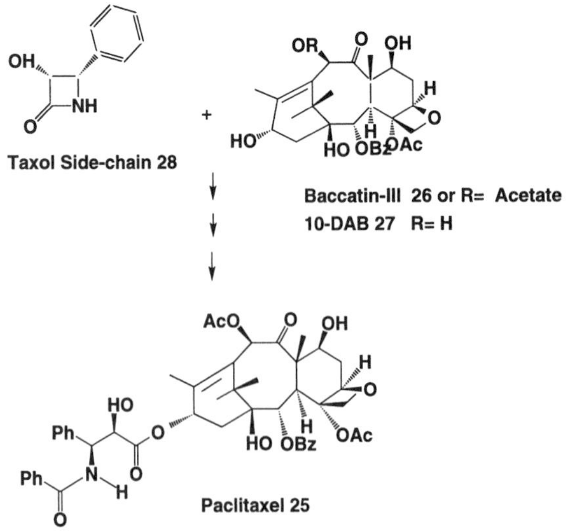

Figure 7   Semisynthesis of taxol **25**: Coupling of baccatin III **26** and taxol side chain synthon **28**.

this valuable compound in yew extracts. We reported the enzymatic conversion of complex mixture of taxanes to a single compound 10-DAB **27** by treatment of extracts prepared from a variety of yew cultivars [44–48]. Details of this process are described below.

By using selective enrichment techniques, Hanson et al. [44,47] isolated two strains of *Nocardioides* that contained novel enzymes C-13 taxolase and C-10 deacetylase. The extracellular C-13 taxolase derived from filtrate of fermentation broth of *Nocardioides albus* SC 13911 catalyzed the cleavage of C-13 side chain from paclitaxel **25** and related taxanes such as taxol C **29**, cephalomannine **30**, 7-β-xylosyltaxol **31**, 7-β-xylosyl-10-deacetyltaxol **32**, and 10-deacetyltaxol **33** (Fig. 8). The intracellular C-10 deacetylase derived from fermentation of *Nocardioides luteus* SC 13912 catalyzed the cleavage of C-10 acetate from paclitaxel **25**, related taxanes, and baccatin III to yield 10-DAB **27** (Fig. 9). Fermentation processes were developed for growth of *N. albus* SC 13911 and *N. luteus* SC 13912 to produce C-13 taxolase and C-10 deactylase, respectively [49]. A bioconversion process was demonstrated for the conversion of taxol and related taxanes in extracts of *Taxus* plant cultivars to a single compound 10-DAB using both enzymes. In the bioconversion process, ethanolic extracts of the whole young plant of five different cultivars of *Taxus* were first treated with a crude preparation of the C-13 taxolase to give complete conversion of measured taxanes to baccatin III and 10-DAB in 6 h. *Nocardioides luteus* SC 13192 whole cells were then added to the reaction mixture to give complete conversion of baccatin III to 10-DAB. The concentration of 10-DAB was increased by 5.5- to 24-fold (depending on type of *Taxus* cultivars used) in the extracts treated with the two enzymes. The bioconversion process was also applied to extracts of the bark of *T. bravifolia* to give a 12-fold increase in 10-DAB concentration. The enhancement of 10-DAB concentration in yew extracts was significant in increasing the amount and purification of this key precursor for the paclitaxel semisynthetic process using renewable resources.

Among other taxanes in bark of the specific yew and *Taxus* cultivars are 7-β-xylosyltaxanes. Using enrichment culture techniques, organisms capable of hydrolyzing 7-β-xylosyltaxanes were isolated [50]. About 125 strains with xylosidase activity (as indicated by hydrolysis of 4-methylumbelliferyl-β-D-xyloside to the fluorescent 4-methylumbelliferone) were isolated; 9 isolates produced 10-deacetyltaxol **33** and 10-DAB **27** from and 7-β-xylosyl-10-deacetyltaxol **32** and 7-β-xylosyl-10-DAB **34**, respectively (Fig. 10). The best culture that catalyzed the cleavage of xylose from 7-β-xylosyltaxol **35** and 7-β-xylosyl-10-deacetyltaxol **32** was identified as a strain of *Morexella* sp. Production of xylosidase was scaled up from a 15-L to a 500-L batch fermentation process. From 500 L of broth, 5 kg of wet cell paste was collected and used in the biotransformation process. Cell suspensions of *Moraxella* sp. in 50 mM phosphate buffer (pH 7.0) gave complete conversion of 7-β-xylosyl-10-deacetyltaxol **32** to 10-deacetyltaxol **33** and 7-β-xylosyl-10-DAB **34** to 10-DAB **27**. The C-13 side chains of 7-β-xylosyltaxol **35** and 7-β-xylosyl-10-deacetyltaxol **32** were removed by C-13 taxolase from *N. albus* SC 13911. The corresponding products were converted quantitatively to baccatin III **26** and 10-DAB **27**. Xylosidase activity in extracts of *Moraxella* sp. was found in both the soluble and particulate fractions. About 80% of the total activity was found in the pellet fraction obtained after centrifugation of soluble fraction at 101,000*g*. Various xylosyltaxanes (7-β-xylosyltaxol **35**, 7-β-xylosylcephalomannine **36**, 7-β-xylosyl-10-deacetyltaxol **32**, 7-β-xylosyl-10-deacetylcephalomannine **37**, 7-β-xylosyl-10-DAB **34**, and 7-β-xylosylbaccatin III **38** and 7-β-xylosyltaxol C **39**) were converted to 10-DAB by treatment with three enzymes: xylosidase (*Moraxella* sp.), C-13 taxolase (*N. albus*), and C-10 deacetylase (*N. luteus*) from microbial sources.

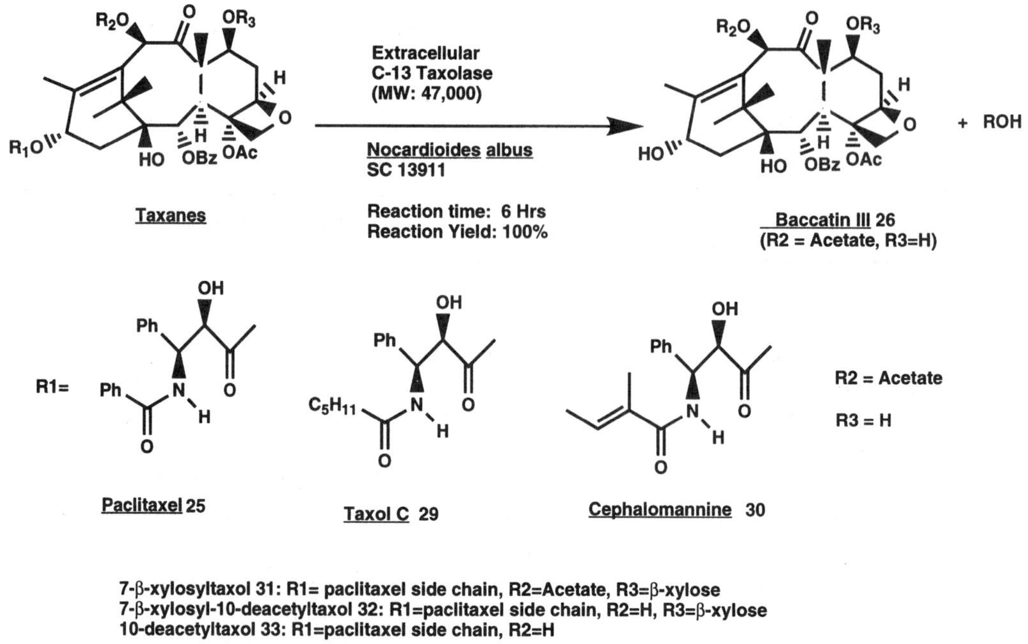

Figure 8 Hydrolysis of the C-13 side chain of taxanes by C-13 taxolase from *Nocardioides albus* SC 13911.

Another key precursor for the taxol semisynthetic process is the preparation of chiral C-13 taxol side chain. Two different stereoselective enzymatic processes were developed for the preparation of chiral C-13 taxol side chain synthon [51–54]. In one process, the stereoselective microbial reduction of 2-keto-3-(*N*-benzoylamino)-3-phenylpropionic acid ethyl ester **40** to yield (*2R,3S*)-(−)-*N*-benzoyl-3-phenylisoserine ethyl ester **41a** was dem-

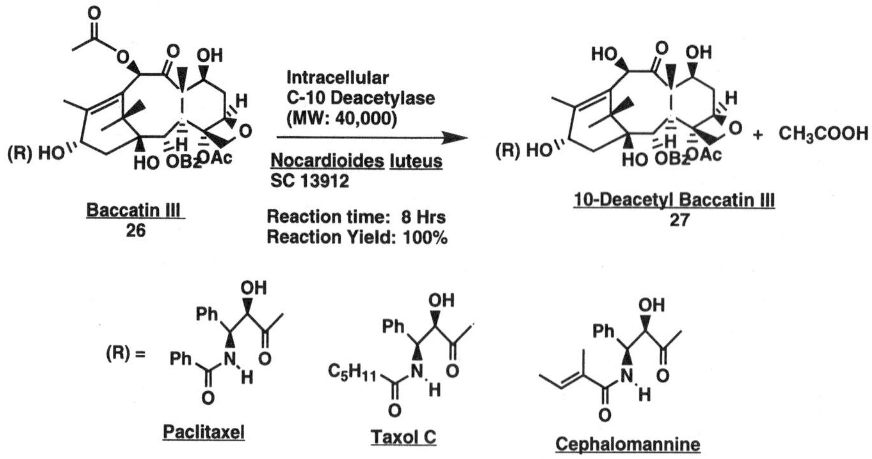

Figure 9 Hydrolysis of the C-10 acetate of taxanes by C-10 deacetylase from *Nocardioides luteus* SC 13912.

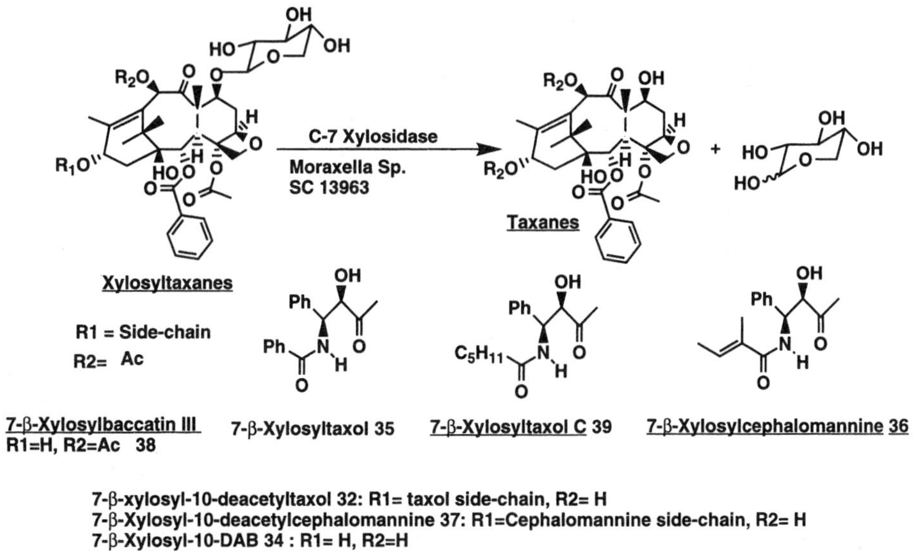

Figure 10 Hydrolysis of the C-7 xylose of xylosyl taxanes by C-7 xylosidase from *Moraxella* sp. SC 13963.

onstrated [51,52]. The reduction of compound **40** could result in the formation of four possible alcohol diastereomers (**41a–d**) (Fig. 11). Remarkably, conditions were found under which predominantly only the single (*2R,3S*) isomer was obtained by the biotransformation. After an extensive microbial screen, two strains of *Hansenula* were identified that catalyzed the stereoselective reduction of ketone **40** to the desired product **41a** in more than 80% reaction yield and more than 94% e.e. Preparative scale bioreduction of ketone **40** was demonstrated using cell suspensions of *Hansenula polymorpha* SC 13865 and *Hansenula fabianii* SC 13894 in independent experiments. In both batches, a reaction yield of more than 80% and e.e. values of more than 94% were obtained for the desired

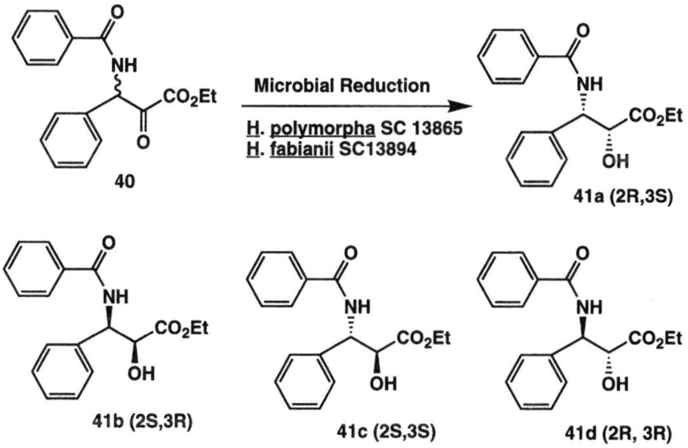

Figure 11 Synthesis of taxol side chain synthon: Stereoselective microbial reduction of 2-keto-3-(*N*-benzoylamino)-3-phenyl propionic acid, ethyl ester **40**.

alcohol isomer **41a**. A 20% yield of undesired antidiastereomers(**41c,41d**) content was obtained with *H. polymorpha* SC 13865 compared with a 10% yield with *H. fabianii* SC 13894. A 99% e.e. of desired alcohol isomer **41a** was obtained with *H. polymorpha* SC 13865 compared with a 94% e.e. with *H. fabianii* SC 13894. From one batch, 5.2 g of compound **41a** was isolated in overall 65% yield with 99.6% e.e. and 99% chemical purity. A single-stage fermentation/bioreduction process was also developed for the preparation of C-13 taxol side chain **40a**. Cells of *H. fabianii* were grown in a 15-L fermentor for 48 h; the bioreduction process was then initiated by addition of 30 g of substrate and 250 g of glucose and continued for 72 h. A reaction yield of 88% and an e.e. of 95% were obtained for the desired product **41a**.

In an alternate process for the preparation of C-13 taxol side chain, the stereoselective enzymatic hydrolysis of racemic *cis*-3-(acetyloxy)-4-phenyl-2-azetidinone **42** to the corresponding (*S*)-(−)-alcohol **43** has been demonstrated [53,54]. Lipase PS-30 from *Pseudomonas cepacia* (Amano International Enzyme Co.) and BMS lipase (extracellular lipase derived from the fermentation of *Pseudomonas* sp. SC 13856) catalyzed hydrolysis of the undesired enantiomer of racemic **42**, producing *S*-(−)-alcohol **43** and the desired *R*-(+)-acetate **44** (Fig. 12). Reaction yields of more than 48% (theoretical maximum yield is 50%) and e.e. of more than 99.5% were obtained for the desired *R*-(+)-acetate. For a very efficient enzyme source (BMS lipase), a lipase fermentation using *Pseudomonas* sp. SC 13865 was developed. In a fed-batch process using soybean oil, the fermentation resulted in 1500 U/ml of extracellular lipase activity. Crude BMS lipase (1.7 kg containing 140,000 U/g) was recovered from the filtrate by ethanol precipitation. BMS lipase and lipase PS-30 were immobilized on Accurel polypropylene (PP). These immobilized lipases were reused (10 cycles) without loss of enzyme activity, productivity, or the e.e. of the product in the resolution process. The enzymatic process for the resolution of racemic acetate **42** was scaled up to 150 L at 10 g/L substrate concentration using immobilized BMS lipase and lipase PS-30, respectively. From each reaction batch, 3-(*R*)-acetate **44** was isolated in 45 M% yield (theoretical maximum yield is 50%) and 99.5% e.e. 3-(*R*)-acetate **44** was chemically converted to 3-(*R*)-alcohol **45**. The C-13 taxol side chain (**41a** or **45**) produced

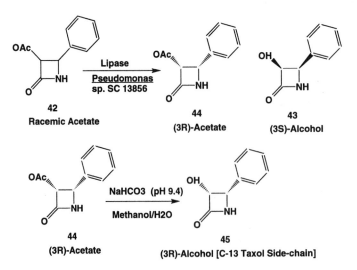

Figure 12 Synthesis of taxol side chain synthon: Stereoselective enzymatic hydrolysis of *cis*-3-(acetyloxy)-4-phenyl-2-azetidinone **42**.

either by the reductive or resolution process could be coupled to bacattin III or 10-DAB after protection and deprotection of each compound to prepare taxol by semisynthetic process.

Recently, preparation of taxol side chain precursors by the lipase-catalyzed enantio-selective esterification of methyl *trans*-β-phenylglycidate has been demonstrated [55]. Preparation of enantiomerically pure 3-hydroxy-4-phenyl β-lactam by lipase-catalyzed en-antioselective hydrolysis and transesterification of racemic esters and alcohols, respectively, have also been described by Brieva et al. [56].

IV. THROMBOXANE A2 ANTAGONIST

Thromboxane A2 (TxA2) is an exceptionally potent proaggregatory and vasoconstrictor substance produced by the metabolism of arachidonic acid in blood platelets and other tissues. Together with the potent antiaggregatory and vasodilator it is thought to play a role in the maintenance of vascular homeostasis and contribute to the pathogenesis of a variety of vascular disorders. Approaches to limiting the effect of TxA2 have focused on either inhibiting its synthesis or blocking its action at its receptor sites by means of an antagonist [57,58]. The lactol [*3aS*-(3aα,4α,7α,7aα)]-hexahydro4,7-eoxyisobenzo-furan-1-(3H)-one **46** or corresponding chiral lactone **47** (Fig. 13) is a key chiral intermediate for the total synthesis of [*1S*-[1α,2α(Z),3α,4α[[-7-[3-[[[[2-oxoheptyl)amine]acetyl]methyl]-7-oxabicyclo-[2.2.1] hept-2-yl]-5-heptanoic acid **48**, a new cardiovascular agent useful in the treatment of thrombolic disease [59,60].

Horse liver alcohol dehydrogenase (HLADH) catalyzes the oxidoreduction of a va-riety of compounds [61,62]. It has been demonstrated that HLADH catalyzes the stereo-specific oxidation of only one of the enantiopic hydroxyl groups of acyclic and monocyclic *meso*-diols [63,64]. The oxidation of *meso-exo-* and *endo*-7-oxabicyclo[2.2.1]heptane di-methanol to the corresponding enantiomerically pure γ-lactones by HLADH has been demonstrated. Nicotinamide adenine dinucleotide (NAD$^+$) and flavin adenine dinucleotide (FAD) were required for the stereoselective oxidation of substrate. Due to the high cost of enzyme and required cofactors, this process for preparing chiral lactones was econom-

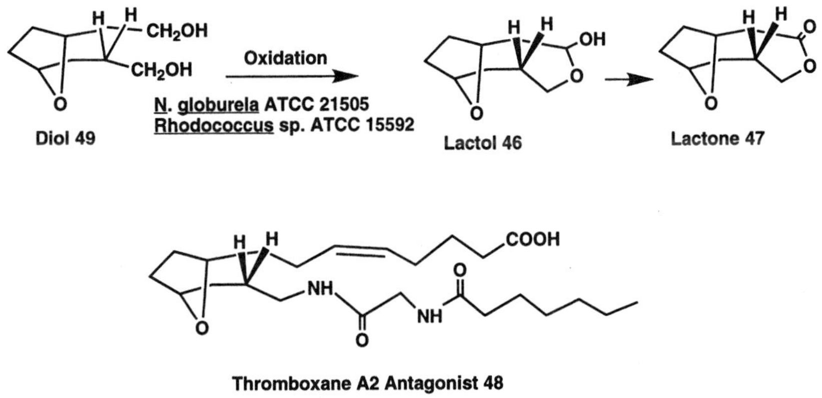

Figure 13 Synthesis of chiral synthon for thromboxane A2 antagonist: Stereoselective microbial oxidation of (*exo,exo*)-7-oxabicyclo[2.2.1]heptane-2,3-dimethanol **49** to the corresponding lactol **46** and lactone **47**.

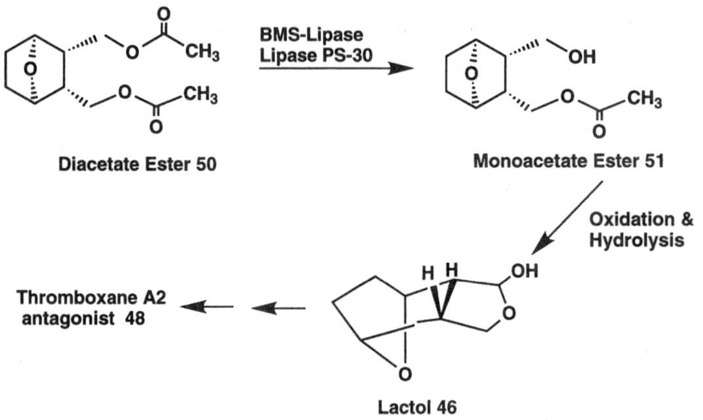

Figure 14 Synthesis of chiral synthon for thromboxane A2 antagonist: Asymmetrical enzymatic hydrolysis of (*exo,exo*)-7-oxabicyclo[2.2.1]heptane-2,3-dimethanol, diacetate ester **50**.

ically not feasible for scale-up. We have [65] developed a microbial process for the stereoselective oxidation of (*exo,exo*)-7-oxabicyclo[2.2.1]heptane-2,3-dimethanol **49** to the corresponding chiral lactol **46** and lactone **47** (Fig. 13) by cell suspensions of *Nocardia globerula* ATCC 15592 and *Rhodococcus* sp. ATCC 15592. The reaction yield of 70 M% and e.e. of 96% were for chiral lactone **47** after 96 h biotransformation process using cell suspensions of *N. globerula* ATCC 15592. An overall reaction yield of 46 M% (lactol and lactone combined) and e.e.'s of 96.7% and 98.4% were obtained for lactol **46** and lactone **47**, respectively, using cell suspensions of *Rhodococcus* sp. ATCC 15592.

The stereoselective asymmetrical hydrolysis of (*exo,exo*)-7-oxabicyclo[2.2.1]heptane-2,3-dimethanol diacetate ester **50** to the corresponding chiral monoacetate ester **51** (Fig. 14) has been demonstrated with lipases [66,67]. Lipase PS-30 from *Pseudomonas cepacia* was most effective in asymmetrical hydrolysis to obtain the desired enantiomer of monoacetate ester. The reaction yield of 75 M% and an e.e. of more than 99% was obtained when the reaction was conducted in a biphasic system containing 10% toluene. The reaction process was scaled up to 80 L (400 g of substrate) and monoacetate ester **51** was isolated in 80 M% yield with 99.3% e.e. and 99.5% chemical purity. The chiral monoacetate ester **51** was oxidized to its corresponding aldehyde and subsequently hydrolyzed to give chiral lactol **46** (Fig. 14). The chiral lactol **46** obtained by this enzymatic process was used in chemoenzymatic synthesis of thromboxane A2 antagonist **48**.

V. ANTICHOLESTEROL DRUGS

Chiral β-hydroxy esters are versatile synthons in organic synthesis specifically in the preparation of natural products [68–70]. The asymmetrical reduction of carbonyl compounds using baker's yeast has been demonstrated and reviewed [5,71,72]. In the stereoselective reduction of β-keto ester of 4-chloro- and 4-bromo-3-oxobutanoic acid, specifically 4-chloro-3-oxobutanoic acid methyl ester, Sih and Chen [73] demonstrated that the stereoselectivity of yeast-catalyzed reductions may be altered by manipulating the size of ester group using γ-chloroacetoacetate as substrate. They also indicated that the e.e. of the alcohol produced depended on the concentration of the substrate used. Nakamura et al. [74] demonstrated the reduction of β-keto ester with baker's yeast and controlled the stereoselectivity by the addition of α,β-unsaturated carbonyl compounds. The additive tended to shift the stereoselectivity of the reduction toward the production of R-hydroxy ester. The shift in stereoselectivity was accounted for on the basis of inhibition of the competitive enzyme that produced S-hydroxy ester. They also used immobilized baker's yeast to improve stereoselectivity of the reduction reactions [75]. The enantiomeric excess of alcohols produced was improved to 90% by using immobilized cells compared with 31% obtained with free cells when methyl 4-chloroacetoacetate was use as substrate.

Recently, we described the reduction of 4-chloro-3-oxobutanoic acid methyl ester **52** to S-(-)-4-chloro-3-hydroxybutanoic acid methyl ester **53** (Fig. 15) by cell suspensions of *Geotrichum candidum* SC 5469 [76]. S(−)-**53** is a key chiral intermediate in the total chemical synthesis of **54**, a cholesterol antagonist that acts by inhibiting hydroxymethyl-glutaryl CoA (HMG CoA) reductase. In the biotransformation process, a reaction yield of 95% and an e.e. of 96% were obtained for S-(−)-**53** by glucose-, acetate-, or glycerol-grown cells of *G. candidum* SC 5469. The e.e. of S-(−)-**53** was increased to 99% by heat treatment of cell suspensions (55°C for 30 min) prior to conducting the bioreduction of **52**.

Glucose-grown cells of *G. candidum* SC 5469 have also catalyzed the stereoselective reduction of ethyl, isopropyl, and tertiary butyl esters of 4-chloro-3-oxobutanoic acid and both methyl and ethyl esters of 3-bromo-3-oxobutanoic acid. A reaction yield of more than 85% and e.e.'s of more than 94% were obtained. NAD^+-dependent oxidoreductase responsible for the stereoselective reduction of β-keto esters of 4-chloro- and 4-bromo-3-oxobutanoic acid was purified 100-fold. The molecular weight of purified enzyme is 950,000. The purified oxidoreductase was immobilized on Eupergit C and used to catalyze the reduction of **52** to S-(−)-**53**. The cofactor NAD^+ required for the reduction reaction was regenerated by glucose dehydrogenase.

Nakamura et al. [77] isolated four different oxidoreductases from baker's yeast, which catalyzed the reduction of β-keto esters to β-hydroxy esters. Two oxidoreductases, namely, D-enzyme 1 (mw 25,000) and D-enzyme 2 (mw 1,600,000) catalyzed the reduction of β-keto ester stereoselectivity to D-β-hydroxy ester. In contrast, one other enzyme, L-enzyme (mw 321,000) reduced substrates to L-β-hydroxy esters. The NADP-oxidoreductases, designated the S and R enzymes, have been purified and characterized from cell extracts of *Saccharomyces cerevisiae*, which catalyze the enantioselective reductions of 3-

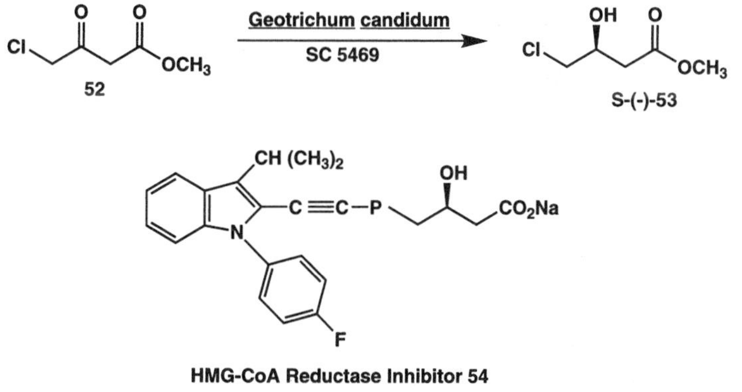

HMG-CoA Reductase Inhibitor 54

Figure 15 Synthesis of chiral synthon for anticholesterol drug: Stereoselective microbial reduction of 4-chloro-3-oxobutanoic acid methyl ester **52**.

oxo, 4-oxo, and 5-oxo esters [78]. The *S* enzyme had a molecular weight of 48,000 and reduced 3-oxo esters, 4-oxo, and 5-oxo acids and esters enantioselectively to *S*-hydroxy compounds in the presence of NADPH. This enzyme may be located in the mitochondrial fraction. The *R* enzyme, which had a molecular weight of 800,000 and contained subunits having molecular weights of 200,000 and 210,000, specifically reduced 3-oxo esters to *R*-hydroxy esters, using NADPH as coenzyme. The *R* enzyme, which occurs in the cytosol, was considered to be identical with a subunit of the fatty acid synthetase complex.

So far most microorganisms and enzymes derived therefrom have been used in the reduction β-keto or α-keto compounds involved the reduction of single keto group [79–82]. Recently, we [83] demonstrated the stereoselective reduction of a diketone 3,5-dioxo-6-(benzyloxy)hexanoic acid, ethyl ester **55** to (*3S,5R*)-dihydroxy-6-(benzyloxy)hexanoic acid, ethyl ester **56a** (Fig. 16). Compound **56a** is a key chiral intermediate required for the chemical synthesis of [4-[4α,6β(E)]]-6[4,4-bis[4-fluorophenyl)-3-(1-methyl-1H-tetra-zol-5-yl)-1,3-butadienyl]-tetrahydro-4-hydroxy-2H-pyren-2-one, compound *R*-(+)-**57**, a new anticholesterol drug acts by inhibition of HMG CoA reductase [84]. Among various microbial cultures evaluated for the stereoselective reduction of diketone **55**, cell suspensions of *Acinetobacter calcoaceticus* SC 13876 reduced **55** to **56a**. The reaction yield of 85% and an e.e. of 97% were obtained using glycerol-grown cells.

Cell extracts of *A. calcoaceticus* SC 13876 in the presence of NAD$^+$, glucose, and glucose dehydrogenase reduced **55** to the corresponding monohydroxy compounds **58** and **59** [3-hydroxy-5-oxo-6-(benzyloxy)hexanoic acid ethyl ester **58**, and 5-hydroxy-3-oxo-6-(benzyloxy)hexanoic acid ethyl ester **59**]. Both **58** and **59** were further reduced to (*3S,4R*)-dihydroxy compound **56a** by cell extracts (Fig. 16). The reaction yield of 92% and an e.e. of 99% were obtained when the reaction was carried out in a 1-L batch using cell extracts at 10 g/L substrate concentration. Product **56a** was isolated from the reaction mixture in

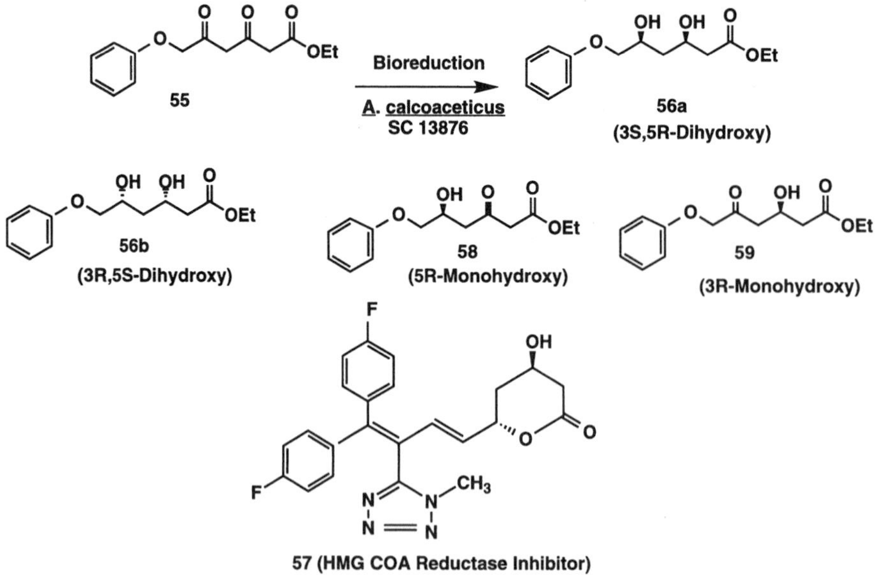

Figure 16 Synthesis of chiral synthon for anticholesterol drug: Stereoselective microbial reduction of 3,5-dioxo-6-(benzyloxy) hexanoic acid, ethyl ester **55**.

72% overall yield. Reductase that converted **55** to **56a** was purified about 200-fold from cell extracts of *A. calcoaceticus* SC 13876. The purified enzyme gave a single protein band on SDS-PAGE corresponding to 33,000 Da.

Using a resolution process, chiral alcohol *R*-(+)-**57** was also prepared by the lipase-catalyzed stereoselective acetylation of racemic **57** in organic solvent [85]. Various lipases were evaluated among which lipase PS-30 (Amano International Enzyme Co.) and BMS lipase efficiently catalyzed the acetylation of the undesired enantiomer of racemic **57** to yield *S*-(−)-acetylated product **60** and unreacted desired *R*-(+)-**57** (Fig. 17). A reaction yield of 49 M% (theoretical maximum yield is 50 M%) and an e.e. of 98.5% were obtained for *R*-(+)-**57** when the reaction was conducted in toluene as solvent in the presence of isopropenyl acetate as acyl donor. In methyl ethyl ketone at 50 g/L substrate concentration, a reaction yield of 46 M% (theoretical maximum yield is 50 M%) and optical purity of 96.4% were obtained for *R*-(+)-**57**.

Lipase PS-30 was immobilized on Accurel PP and the immobilized enzyme was reused five times without any loss of activity or productivity in the resolution process to prepare *R*-(+)-**57**. The enzymatic process was scaled up to a 640-L preparative batch using immobilized lipase PS-30 at 4 g/L racemic substrate **57** in toluene as solvent. From the reaction mixture, *R*-(+)-**57** was isolated in 40 M% overall yield (theoretical maximum yield is 50%) with 98.5% e.e. and 99.5% chemical purity. The undesired *S*-(−) acetate **60** produced by this process was enzymatically hydrolyzed by lipase PS-30 in a biphasic system to prepare the corresponding *S*-(−)-alcohol **57**. Thus both enantiomers of alcohol **57** were produced by the enzymatic process.

Pravastatin **61** and mevastatin **62** are anticholesterol drugs that act by competitively inhibiting HMG CoA reductase [86]. Pravastatin sodium is produced by two fermentation steps. The first step is the production of compound ML-236B by *Penicillium citrinum* [87–89]. Purified compound was converted to its sodium salt **63** with sodium hydroxide and in the second step was hydroxylated to pravastatin sodium **61** (Fig. 18) by *Streptomyces*

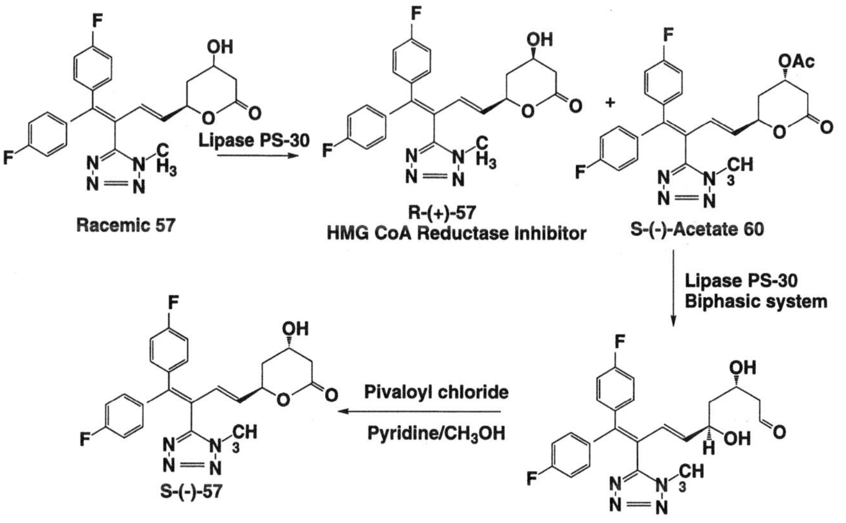

Figure 17 Synthesis of chiral anticholesterol drug *R*-(+)-**57**: Stereoselective enzymatic acetylation of racemic **57**.

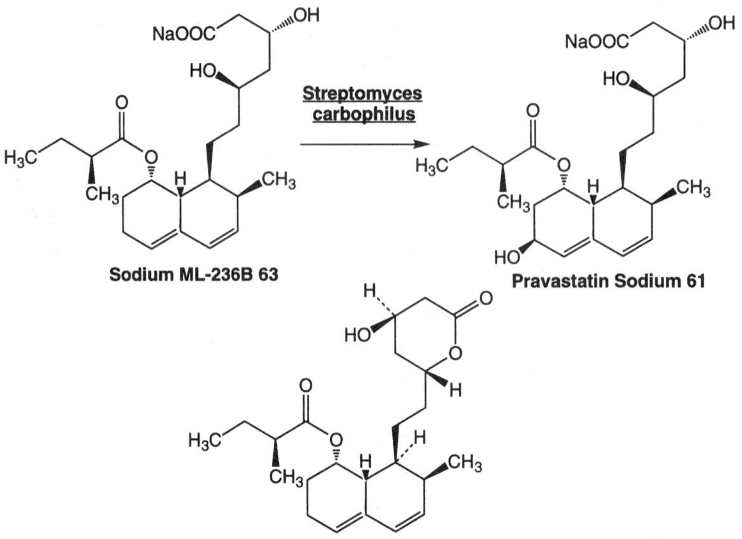

Figure 18 Stereoselective microbial hydroxylation of ML-236B to pravastatin **61**.

carbophilus [90]. A cytochrome P450–containing enzyme system that catalyzed the hydroxylation reaction has been demonstrated from *S. carbophilus* [91].

A synthetic route to the chiral lactone moiety of pravastatin and mevastatin has been prepared from *meso*-diacetate **64**. The chiral compound **65** has been prepared by the asymmetrical hydrolysis of *meso*-diacetate **64** by PLE [92]. The reaction yield of 62% and an e.e. of 94% were obtained for chiral compound **65** (Fig. 19).

The chiral intermediate 2,4-didoexyhexose derivative required for the HMG CoA reductase inhibitors has also been prepared using 2-deoxyribose-5-phosphate aldolase (DERA). This enzyme accepts a wide variety of acceptor substrates and has been useful in organic synthesis [93,94]. As shown in Fig. 20, the reactions start with a stereospecific addition of acetaldehyde **66** to a substituted acetaldehyde to form a 3-hydroxy-4-substituted butyraldehyde **67**, which reacts subsequently with another acetaldehyde to form a 2,4-dideoxyhexose derivative **68** [95]. DERA has been overexpressed in *E. coli*.

Squalene synthase is the first pathway-specific enzyme in the biosynthesis of cholesterol and catalyzes the head-to-head condensation of two molecules of farnesyl pyrophosphate (FPP) to form squalene **69**. It has been implicated the transformation of FPP

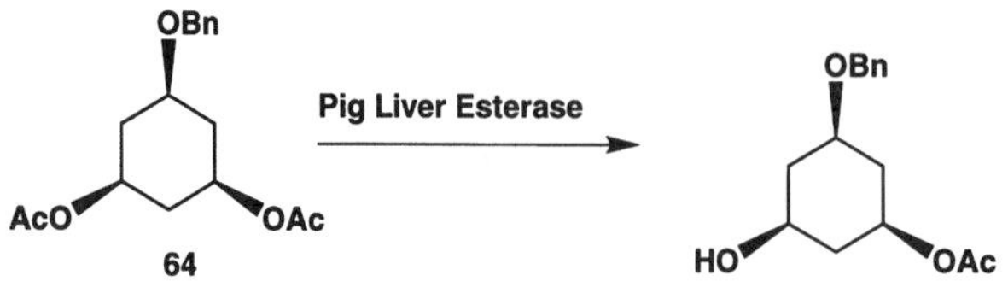

Figure 19 Enzymatic synthesis of chiral lactone moiety of pravastatin **61**.

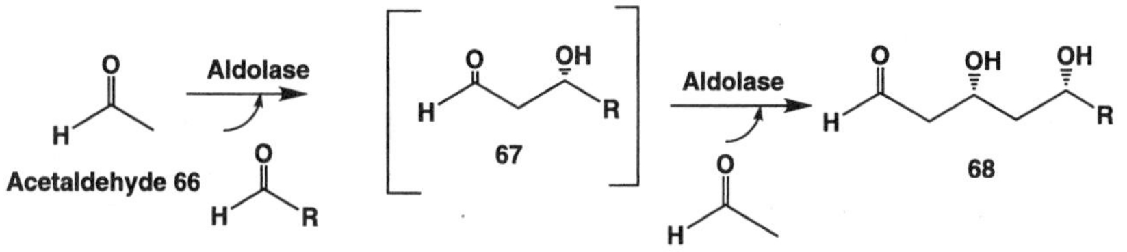

Figure 20 Preparation of 2,4-dideoxyhexose derivatives **68** by aldolase.

into presqualene pyrophosphate [96]. FPP analogs are a major class of inhibitors of squalene synthase [97,98]. However, this class of compounds lack specificity and are potential inhibitors of other FPP-consuming transferases such as geranyl geranyl pyrophosphate synthase. To increase enzyme specificity, analogs of PPP and other mechanism-based enzyme inhibitors have been synthesized [99,100]. BMS 188494 is a potent squalene synthase inhibitor effective as an anticholesterol drug [101,102]. (S)[1-(Acetoxyl)-4-(3-phenyl)butyl]phosphonic acid, diethyl ester **70** is a key chiral intermediate required for the total chemical synthesis of BMS 188494. The stereoselective acetylation of racemic [1-(hydroxy)-4-(3-phenyl)butyl]phosphonic acid, diethyl ester **71** (Fig. 21), was carried out using *Geotrichum candidum* lipase in toluene as solvent and isopropenyl acetate as acyl donor [103]. A reaction yield of 38% (theoretical maximum yield is 50%) and an e.e. of 95% were obtained for chiral **70**.

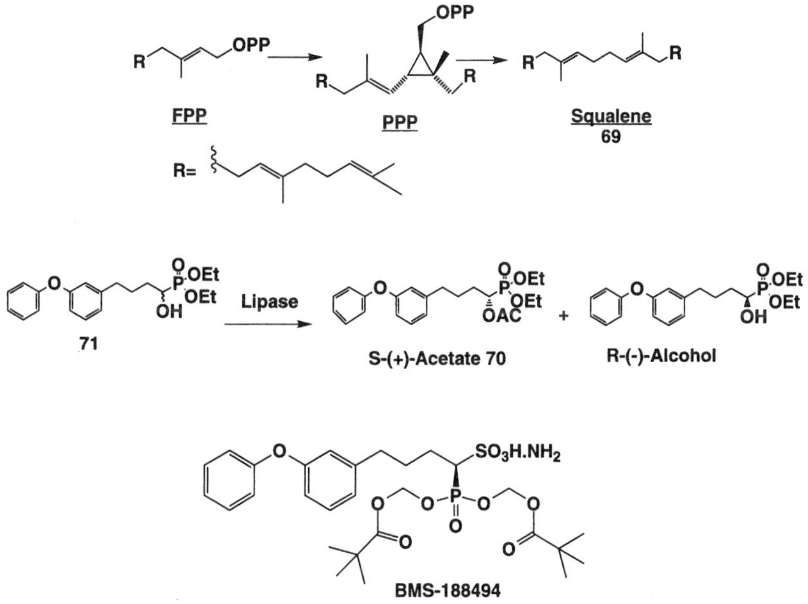

Figure 21 Enzymatic synthesis of chiral synthon for BMS 188494, a squalene synthase inhibitor: Stereoselective acetylation of racemic **71**.

VI. CALCIUM CHANNEL BLOCKING DRUGS

Diltiazem **72**, a benzothiazepinone calcium channel blocking agent that inhibits influx of extracellular calcium through L-type voltage-operated calcium channels, has been widely used clinically in the treatment of hypertension and angina [104]. Since diltiazem has a relatively short duration of action [105], recently an 8-chloro derivative was introduced in the clinic as a more potent analog [106]. Lack of extended duration of action and little information on structure–activity relationship in this class of compounds led Floyd et al. [107] and Das et al. [108,109] to prepare isosteric 1-benzazepin-2-ones and led to the identification of 6-trifluoromethyl-2-benzazepin-2-one derivative as a longer lasting and more potent antihypertensive agent. The key chiral intermediate **73** [(*3R-cis*)-1,3,4,5-tetrahydro-3-hydroxy-4-(4-methoxyphenyl)-6-(trifluromethyl)-2H-1-benzazepin-2-one] was required for the total chemical synthesis of the new calcium channel blocking agent **74** [(cis)-3-(acetoxy)-1-[2-(dimethylamino)ethyl]-1,3,4,5-tetrahydro-4-(4-methoxyphenyl)-6-trifluoromethyl)-2H-1-benzazepin-2-one]. A stereoselective microbial process (Fig. 22) has been developed for the reduction of 4,5-dihydro-4-(4-methoxyphenyl)-6-(trifluoromethyl)-1H-1-benzazepin-2,3-dione **75** to chiral **73** [110]. Compound **75** exists predominantly in the achiral enol form, which is in rapid equilibrium with the two keto form enantiomers. Reduction of **75** could give rise to formation of four possible alcohol stereoisomers. Remarkably, conditions were found under which only the single alcohol isomer **73** was obtained by microbial reduction. Among various cultures evaluated, the most effective culture, *Nocardia salmonicolor* SC 6310, catalyzed the bioconversion of **75** to **73** in 96% reaction yield with 99.9% e.e. A preparative scale fermentation process for growth of *N. salmonicolor* and a bioreduction process using cell suspensions of the organism were demonstrated.

A chiral intermediate (*2R,3S*)-3-(4-methoxyphenyl)glycidic acid methyl ester [(−)-MPGM] **76** is required for the synthesis of diltiazem **72**, a calcium channel antagonist. Matsumae et al. [111] screened over 700 microorganisms and identified lipase from *Ser-*

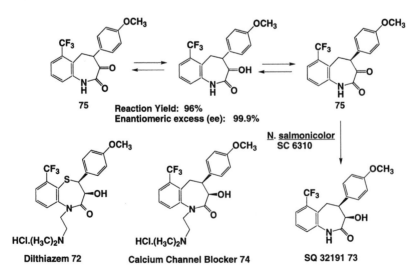

Figure 22 Synthesis of chiral synthon for calcium channel blocker: Stereoselective microbial reduction of 4,5-dihydro-4-(4-methoxyphenyl)-6-(trifluoromethyl)-1H-benzazepin-2,3-dione **75**.

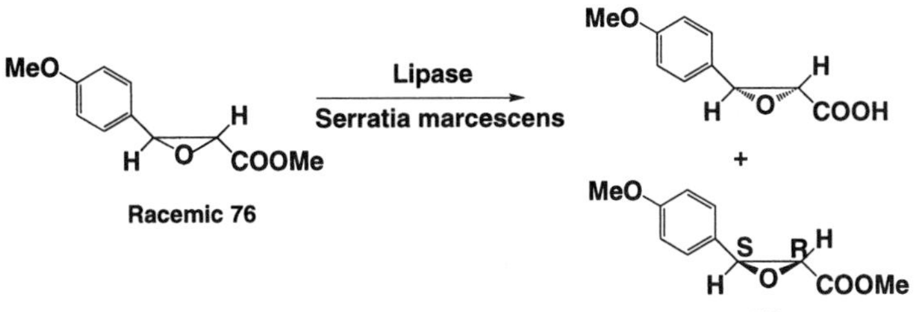

Figure 23 Synthesis of chiral synthon for calcium channel blocker, diltiazem: Enantioselective enzymatic hydrolysis of racemic **76**.

ratia marcescens, which catalyzed the enantioselective hydrolysis racemic MPGM in a biphasic system using toluene as organic phase (Fig. 23) A reaction yield of 48% (theoretical maximum yield is 50%) and an e.e. of 99.8% were obtained for [(−)-MPGM].

VII. POTASSIUM CHANNEL OPENERS

The study of potassium (K) channel biochemistry, physiology, and medicinal chemistry has fluorished, and numerous papers and reviews have been published in recent years [112–114]. It has long been known that K channels play a major role in neuronal excitability and a critical role in the basic electrical and mechanical function of a wide variety of tissues, including smooth muscle, cardiac muscle, and glands [115]. A new class of highly specific pharmacological compounds have been developed that either open or block K channels [114,116]. K-channel openers are powerful smooth muscle relaxants with in vivo hypotensive and bronchodilator activity [115]. Recently, the synthesis and antihypertensive activity of a series of novel K-channel openers [117–120] based on monosubstituted *trans*-4-amino-3,4-dihydro-2,2-dimethyl-2H-1-benzopyran-3-ol **77** have been demonstrated. Chiral epoxide **78** and diol **79** are potential intermediates for the synthesis of K-channel activators important as antihypertensive and bronchodilator agents. The stereoselective microbial oxygenation of 2,2-dimethyl-2H-1-benzopyran-6-carbonitrile **80** to the corresponding chiral epoxide **78** and chiral diol **79** (Fig. 24) has been demonstrated [121,122]. Among microbial cultures evaluated, the best culture, *Mortierella ramanniana* SC 13840, gave reaction yields of 67.5 M% and e.e. of 96% for (+)-*trans*-diol **79**.

A single-stage process (fermentation/epoxidation) for the biotransformation of **80** was developed using *M. ramanniana* SC 13840. In a 25-fermentor, the (+)-*trans*-diol **79** was obtained in the reaction yield of 60.7 M% with an e.e. of 92.5%. In the two-stage process using cell suspensions of *M. ramanniana* SC 13840 the (+)-*trans*-diol **79** was obtained in 76 M% yield with an e.e. of 96%. The reaction was carried out in a 5-L Bioflo fermentor with 2 g/L substrate and 10 g/L glucose concentration. Glucose was supplied to regenerate NADH required for this reaction. From the reaction mixture, (+)-*trans*-diol **79** was isolated in 65 M% overall yield. An e.e. of 97% and a chemical purity of 98% were obtained for the isolated (+)-*trans*-diol **79**.

In an enzymatic resolution approach, chiral (+)-*trans*-diol **79** was prepared by the stereoselective acetylation of racemic diol with lipases from *Candida cylindraceae* and *Pseudomonas cepacia*. Both enzymes catalyzed the acetylation of the undesired enantiomer of racemic diol to yield monoacetylated product and unreacted desired (+)-*trans*-diol

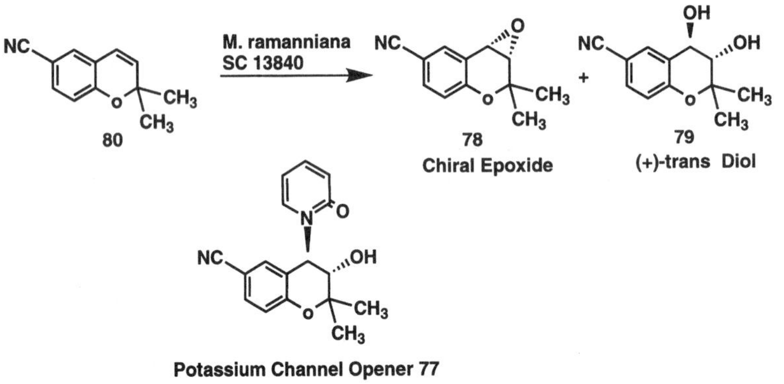

Figure 24 Oxygenation of 2,2-dimethyl-2H-1-benzopyran-6-carbonitrile **74** to the corresponding chiral epoxide **72** and (+)-*trans*-diol **73** by *Mortierella ramanniana* SC 13840.

79. A reaction yield of more than 40% (theoretical maximum yield is 50%) and an e.e. of 90% were obtained using each lipase [123].

VIII. ANTIARRHYTHMIC AGENTS

Larsen and Lish [124] reported the biological activity of a series of phenethanolamine-bearing alkyl sulfonamido groups on the benzene ring. Within this series, some compounds possessed adrenergic and antiadrenergic actions. D-(+)-Sotalol **81** is a beta blocker (125) that, unlike other beta blockers, has antiarrhythmic properties and no other peripheral actions [126]. The β-adrenergic blocking drugs such as propranolol **82** and sotalol have been separated chemically into the dextro and levo rotatory optical isomers, and it has been demonstrated that the activity of the levo isomer is 50 times that of the corresponding dextro isomer [127]. Chiral alcohol **83** is a key intermediate for the chemical synthesis of D-(+)-sotalol **81**. The stereoselective microbial reduction of *N*-[4-(2-chloroace-tyl)phenyl]methanesulfonamide **84** to the corresponding (+)-alcohol **83** (Fig. 25) has been demonstrated [128]. Among numbers of microorganisms screened for the transformation of ketone **84** to (+)-alcohol **83**, *Rhodococcus* sp. ATCC 29675, ATCC 21243, *Nocardia salmonicolor* SC 6310, and *Hansenula polymorpha* ATCC 26012 gave the desired (+)-alcohol **83** in more than 90% e.e. *H. polymorpha* ATCC 26012 catalyzed the efficient conversion of ketone **84** to (+)-alcohol **83** in 95% reaction yield and more than 99% e.e. Growth of *H. polymorpha* ATCC 26012 culture was carried out in a 380-L fermentor and cells harvested from the fermentor were used to conduct the biotransformation process. Cell suspensions (20% wet cells in 3 L of 10 mM potassium phosphate buffer, pH 7.0) were supplemented with 12 g of ketone **84** and 225 g of glucose, and the reduction reaction was carried out at 25°C, 200 rpm, pH7. Complete conversion of ketone **84** to (+)-alcohol **83** was obtained in 20 h. Using preparative high-performance liquid chromatography (HPLC), (+)-alcohol **83** were isolated from the reaction mixture in overall 68% yield with more than 99% e.e.

Both enantiomers of solketal (2,2-dimethyl-1,3-dioxolane-4-methanol) and their cor-responding aldehydes are attractive building blocks for the preparation of enantiomerically

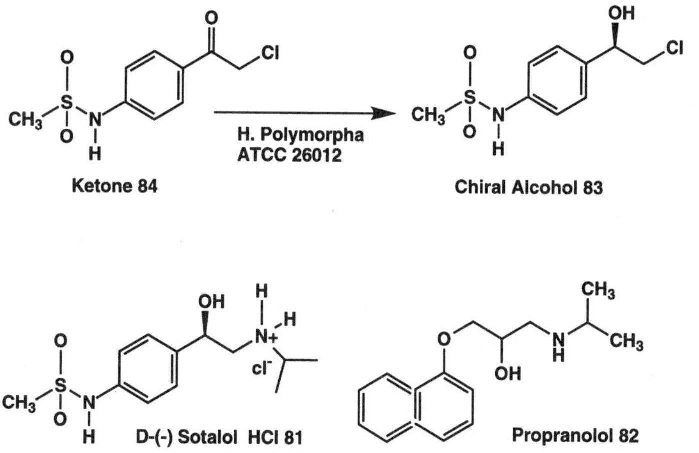

Figure 25 Synthesis of chiral synthon for D-(+)-sotalol, antiarrhythmic agent: Stereoselective microbial reduction of *N*-[4-(2-chloroacetyl)phenyl] methane sulfonamide **84**.

pure and biologically active compounds [129,130], specifically *S*-β-blocking agents. Since solketal is relatively inexpensive and commercially available, enantioselective oxidation of its alcohol function has provided an economically feasible process for the production of *R*-(−)-solketal by preferentially oxidizing the *S* enantiomer to its corresponding acid [131]. Recently, quinohemoprotein ethanol dehydrogenase *Comamonas testosteroni* has been used in the preparation of *S*-(+)-solketal [132].

The chiral compound **85** belongs to a group of 1,2,3-trisubstituted propane derivatives, which are a valuable source of chiral building blocks in the synthesis of β-adrenergic blocking agents, derivatives of 1-alkylamino-3-aryloxy-2-propanol (X = ArO, Y = HNR) **86**. (*S*-propranolol **82** (Fig. 25) has a 1-naphthyloxy for the ArO group and an isopropylamino for the HNR group whereas in (−)-timolol the ArO group is 4-mortholino-1,2,5-thiadiazol-3-yloxy and the HNR group is *tert*-butylamino. The racemic compound **87** has been prepared by Gelo and Sunjic [133] to evaluate the lipase-catalyzed enantiomeric hydrolysis. Lipase from *Pseudomonas* sp. catalyzed the enantioselective hydrolysis of racemic **87** in the presence of 10% acetone and Triton X-100 to yield chiral alcohol **85** in 40% yield (theoretical maximum yield is 50%) and 99% e.e. (Fig. 26).

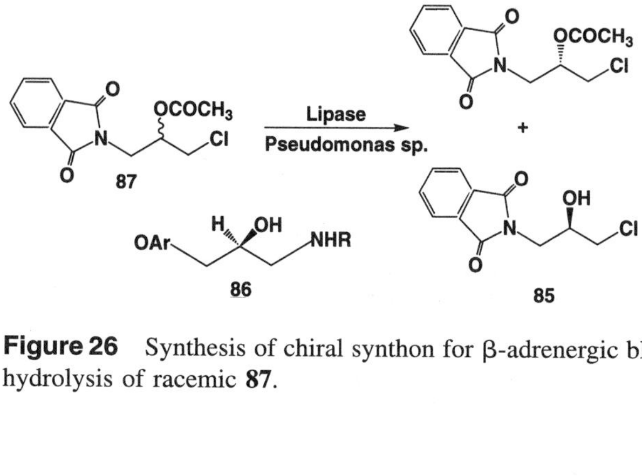

Figure 26 Synthesis of chiral synthon for β-adrenergic blocking agent: Stereoselective enzymatic hydrolysis of racemic **87**.

IX. ANTIPSYCHOTIC AGENTS

During the past few years, much effort has been directed to understanding the sigma receptor system in the brain and endocrine tissue. This effort has been motivated by the hope that the sigma site may be a target of a new class of antipsychotic drugs [134–136]. Characterization of the sigma system helped to clarify the biochemical properties of the distinct haloperidol-sensitive sigma binding site, the pharmacological effects of sigma drugs in several assay systems, and the transmitter properties of a putative endogenous ligand for the sigma site [137–140]. R-(+) compound **88** [BMY 14802] is a sigma ligand and has a high affinity for sigma binding sites and antipsychotic efficacy. The stereoselective microbial reduction of keto compound 1-(4-fluorophenyl)-4-[4-(5-fluoro-2-pyrimidinyl)-1-piperazinyl]-1-butanone **89** to yield the corresponding hydroxy compound R(+)-BMY 14802 **88** (Fig. 27) has been developed by Patel et al. [141]. Among various microorganisms evaluated for the reduction of ketone **89**, *Mortierella ramanniana* ATCC 38191 predominately reduced compound **89** to R(+)-BMY 14802 and *Pullularia pullulans* ATCC 16623 reduced compound **89** to S(−)-BMY 14802. An e.e. of more than 98% was obtained in each reaction.

In a two-stage process for reduction of compound **89**, cells of *M. ramanniana* ATCC 38191 were grown in a 380-L fermentor and cells harvested after 31 h growth were used for the reduction of ketone **89** in 15-L fermentors using 20% cell suspensions (20% w/v, wet cells). Ketone **89** was used at 2 g/L concentration and glucose was supplemented at 20 g/L concentration during the biotransformation process to generate NADH required for the reduction. After a 24-h biotransformation period, about 90% yield and 99.0% e.e. of R(+)-BMY 14802 was obtained. The R(+)-BMY 14802 was isolated from the fermentation broth in overall 70 M% yield, 99.5% e.e., and 99% chemical purity.

A single-stage fermentation/biotransformation process was demonstrated for reduction of ketone **89** to R-(+)-BMY 14802 by cells of *M. ramanniana* ATCC 38191. Cells were grown in a 20-L fermentor containing 15 L of medium. After 40 h of growth in a fermentor, the biotransformation process was initiated by addition of 30 g of ketone **89** and 300 g of glucose. The biotransformation process was completed in a 24-h period, with a reaction yield of 100% and an e.e. of 98.9% for R-(+)-BMY 14802. At the end of the biotransformation process, cells were removed by filtration and product was recovered from the filtrate in overall 80% recovery. A reductase with a molecular weight of 29,000 has been purified to homogeneity which catalyzed the conversion of ketone **89** to R-(+)-BMY 14802.

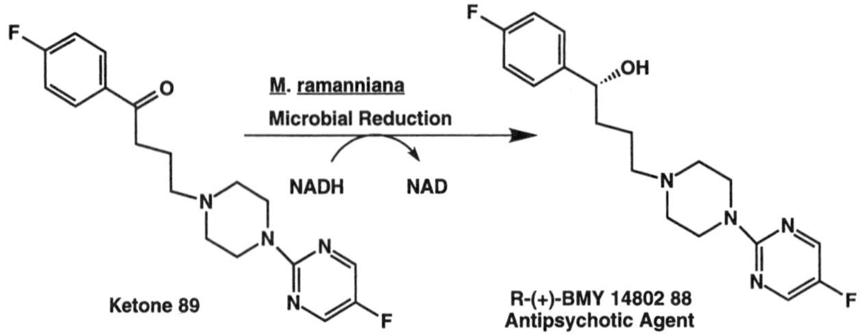

Figure 27 Preparation of R-(+)-BMY 14802, an antipsychotic agent: Stereoselective enzymatic reduction of 1-(4-fluorophenyl)-4-[4-(5-fluoro-2-pyrimidinyl)-1-piperazinyl]-1-butanone **89**.

For the production of optically active alcohols, reduction of the inexpensive prochiral ketones is a promising method. Commercially available alcohol dehydrogenases derived from horse liver and *Thermoanaerobium brockii* [2,142] have been used in the preparation of chiral alcohols. Alcohol dehydrogenase from *T. brockii* is heat-stable and has broad substrate specificity toward aliphatic ketones. Substrates with bulky side chains (such as acetophenone) are poor substrates. Alcohol dehydrogenase from yeast, horse liver, and *T. brockii* transfer the pro-*R* hydride to the reface of the carbonyl to give (*S*) alcohols, a process described by Prelog's rule [143,144]. Recently described alcohol dehydrogenases from *Pseudomonas* sp. strain PED and *Lactobacillus kefir* and *Mortierella isabellina* have been shown to catalyze the enantioselective reduction of aromatic, cyclic, and aliphatic ketones to the corresponding chiral alcohols [145,146]. Both enzymes exhibit anti-Prelog specificity transferring the pro-*R* hydride to form (*R*)-alcohols. Most oxidoreductases used for the preparation of optically active alcohols involve the use of NADH as cofactor. Simon et al. [147] have demonstrated the use of reductases from anaerobic *Clostridium* strains which catalyzed the reduction of varity of compounds to optically active alcohols using methyl or benzyl viologen as electron donor. 2-Enoate reductase and 2-oxocarboxylate reductase have been used in the stereoselective reduction of carbon-carbon– and carbon-oxygen-containing compounds.

R-(+)-BMY 14802 **88** has also been prepared by lipase-catalyzed resolution of racemic BMY 14802 acetate ester **90** [148]. Lipase from *Geotrichum candidum* (GC-20 from Amano Enzyme Co.) catalyzed the hydrolysis of acetate **90** to *R*-(+)-BMY 14802 (Fig. 28) in a biphasic solvent system in 48% reaction yield (theoretical maximum yield is 50%) and 98% e.e. The rate and enantioselectivity of the hydrolytic reaction was dependent on the organic solvent used. The enantioselectivity (*E* values) ranged from 1 in the absence of solvent to more than 100 in dichloromethane and toluene. *S*-(−)-BMY 14802 was also prepared by the chemical hydrolysis of undesired BMY 14802 acetate obtained during enzymatic resolution process.

Pipecolic acid (2-piperidine carboxylic acid) is a precursor of various bioactive compounds such as thioridazine (antipsychotic agent), pipradol (anticonvulsant agent), a potassium opioid analgesic, and immunosuppressant FK-506, all of which have been prepared from pipecolic acid [149–151]. Ng-Youn-Chem et al. [152] developed an enzymatic pro-

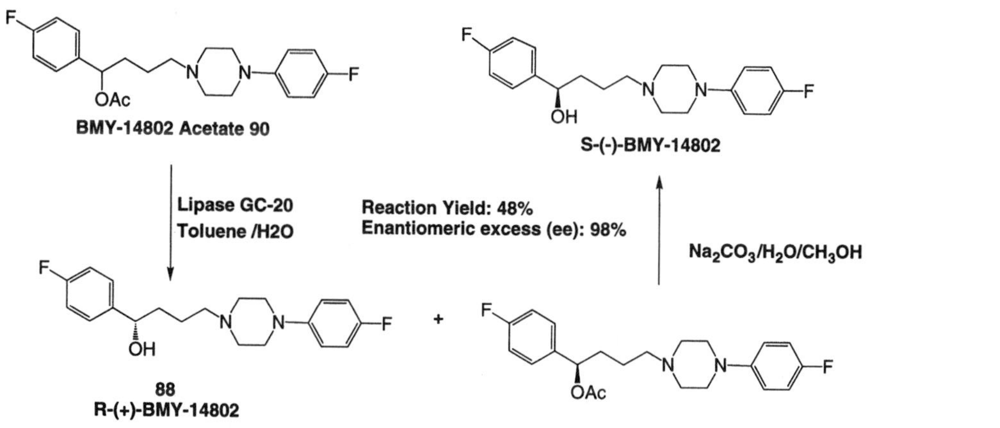

Figure 28 Preparation of *R*-(+)-BMY 14802, an antipsychotic agent: Stereoselective enzymatic hydrolysis of BMY 14802 acetate **90**.

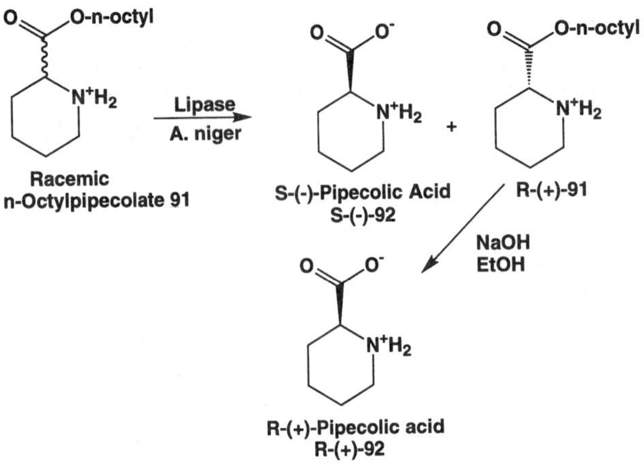

Figure 29 Enzymatic synthesis of *R*-(+)-pipecolic acid, a chiral synthon for antipsychotic and anticonvulsant agents.

cess for the kinetic resolution of racemic *n*-octyl pipecolate **91** using partially purified lipase from *Aspergillus niger*. The reaction yield of 20% (theoretical maximum yield is 50%) and the e.e. of 97% were obtained for *S*-(−)-pipecolic acid **92**. The unreacted *R*-(+)-*n*-octylpipecolic **91** was obtained in 26% reaction yield (theoretical maximum yield is 50%) and 98% e.e., which was hydrolyzed with sodium hydroxide in ethanol to yield *R*-(+)-pipecolic acid **92** in 98.5% e.e. (Fig. 29).

X. ANTI-INFECTIVE DRUGS

During the past several years synthesis of α-amino acids has been pursued intensely [153–156] because of their importance as building blocks of compounds of medicinal interest, particularly as anti-infective drugs. The asymmetrical synthesis of β-hydroxy-α-amino acids by various methods has been demonstrated [157–159] because of their utility as starting materials for the total synthesis of monobactam antibiotics. L-β-Hydroxyvaline **93** [160] is a key chiral intermediate required for the total synthesis of the orally active monobactam tigemonam **94** (Fig. 30). The resolution of CBZ-β-hydroxyvaline by chemical methods has been demonstrated [161,162]. Leucine dehydrogenase from strains of *Bacillus* [163,164] has been used for the synthesis of branched chain amino acids. Hanson et al. [165] have described the synthesis of L-β-hydroxyvaline **93** from α-keto-β-hydroxyisovalerate **95** by reductive amination using leucine dehydrogenase from *B. sphaericus* ATCC 4525 (Fig. 30). NADH required for this reaction was regenerated by either formate dehydrogenase from *Candida boidinii* or glucose dehydrogenase from *B. megaterium*. The immobilized cofactors such as polyethylene glycol-NADH and dextran-NAD were effective in the biocatalytic process. The required substrate **95** was generated either from α-keto-β-bromoisovalerate or its ethyl esters by hydrolysis with sodium hydroxide in situ. In an alternate approach, substrate **95** was also generated from methyl-2-chloro-3,3-dimethyloxiran carboxylate and the corresponding isopropyl and 1,1-dimethyl ethyl ester. These glycidic esters are converted to substrate **95** by treatment with sodium bicarbonate and sodium hydroxide. In this process, an overall reaction yield of 98% and an e.e. of 99.8% were obtained for the L-β-hydroxyvaline **93**.

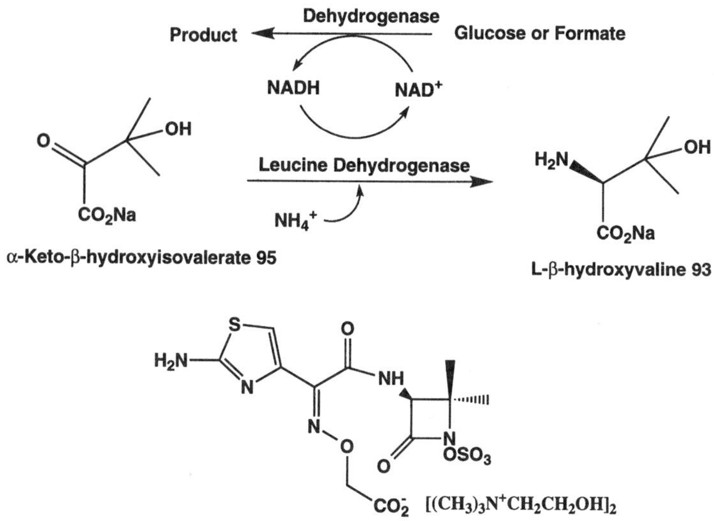

α-Keto-β-hydroxyisovalerate **95**

L-β-hydroxyvaline **93**

Tigomonam **94**

Figure 30   Enzymatic synthesis of chiral synthon for tigemonam: Stereoselective reductive amination of α-keto-β-hydroxyisovalerate **95**.

D-Phenylglycine is required for the semisynthetic antibiotic ampicillin and D-hydroxyphenylglycine (Fig. 31) is used in the production of amoxicillin **96**, cefadroxyl **97**, and caphalexin **98** [166,167]. The use of D-*p*-hydroxyphenylglycine will significantly increase because new drugs such as aspoxicillin, cefbuperazine, and cepyramide are expected to be marketed. Currently, D-amino acids are commercially produced by a chemoenzymatic route using D-hydantoinase. In this process, chemically synthesized DL-5-substituted hydantoin **99** is hydrolyzed to *N*-carbamoyl-D-amino acid **100** by microbially derived D-hydantoinase (Fig. 32). The latter compound undergoes rapid and spontaneous racemization under the reaction conditions; therefore, theoretically a 100% yield of *N*-carbamoyl-D-amino acid can be obtained, which is further chemically converted to the corresponding D-amino acid **101** [168–170]. Recently, microbial *N*-carbamoylase has been demonstrated to catalyze the conversion of *N*-carbamoyl-D-amino acid to the corresponding D-amino

Amoxicillin **96**

Cefadroxyl **97**

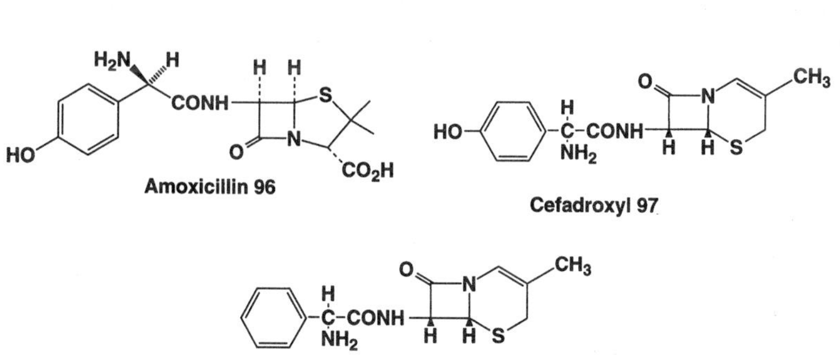

Cephalexin **98**

Figure 31   Structure of anti-infective agents ampicillin, cefadroxil, and cephalexin.

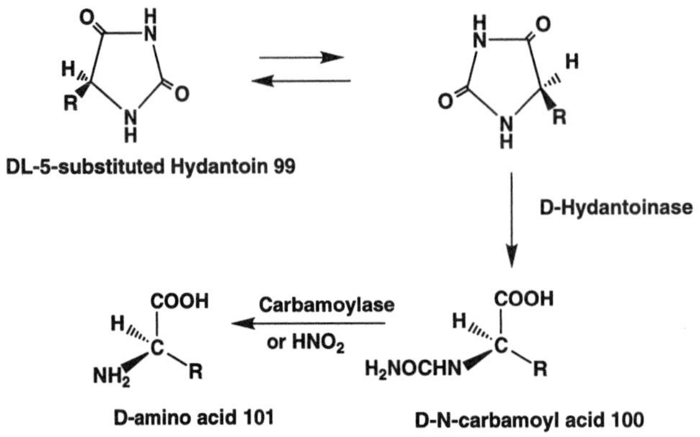

DL-5-substituted Hydantoin 99

D-Hydantoinase

D-amino acid 101   **D-N-carbamoyl acid 100**

Carbamoylase
or HNO₂

Figure 32 Synthesis of D-amino acids by D-hydantoinase and carbamoylase.

acid. Some organisms contained both D-hydantoinase and N-carbamoylase activity [171–173]. L-Hydantoinase has also been described from a microbial source that catalyzes the conversion of DL-5-substituted hydantoin to N-carbamoyl-L-amino acid. This process has been used in the production of L-amino acids [169,174,175].

D-Amino acids and L-amino acids have also been prepared by D-specific or L-specific acylases derived from microbial sources. In this process, DL-N-acetyl amino acid **102** (Fig. 33) is resolved by hydrolytic reaction to yield the D- or L-amino acid depending on D- or L-selective acylase used in the reaction [176–178].

L-α-Amino acids have been prepared by the resolution of racemic α-amino acid amide by the L-specific aminopeptidase from *Pseudomonas putida* ATCC 12633 [7]. Enzyme from *P. putida* ATCC 12633 cannot be used to resolve α-alkyl-substituted amino acid amides **103**. Aminoamidase from *Mycobacterium neoaurum* ATCC 25795 has been used in the preparation of L-α-alkyl amino acid **104** (Fig. 34) and D-amide of α-alkyl-substituted amino acids by enzymatic resolution process using racemic α-alkyl amino acid amide as a substrate [169,179]. Amidase from *Ochrobactrum anthropi* catalyzed the resolution of α,α-disubsituted amino acids, N-hydroxy amino acids, and α-hydroxy acid amides. The resolution process could lead to the production of chiral amino acids or amides in 50% yield. Recently, amino acid racemases have been used to get 100% yield of chiral amino acids [179].

Aminotransferases have been used in the production of chiral amino acids [180–182]. Aminotransferases catalyze the transfer of an amino group, a proton, and a pair of electrons from a primary amine substrate to the carbonyl group of an acceptor molecule

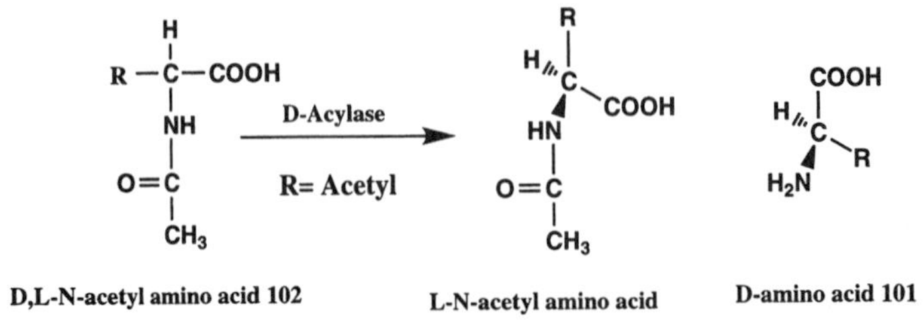

D-Acylase
R= Acetyl

D,L-N-acetyl amino acid 102   **L-N-acetyl amino acid**   **D-amino acid 101**

Figure 33 Synthesis of D-amino acids by D-acylase.

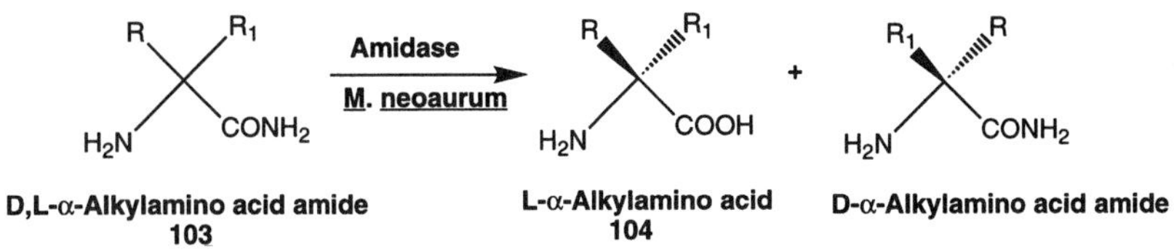

R, R₁, Amidase M. neoaurum, R R₁, +, R₁ R

H₂N CONH₂   H₂N COOH   H₂N CONH₂

D,L-α-Alkylamino acid amide   **L-α-Alkylamino acid**   **D-α-Alkylamino acid amide**
103   **104**

Figure 34 Synthesis of L-α-alkyl amino acids by amidase.

such as oxaloacetate, α-ketoglutarate, or pyruvate (Fig. 35). ω-Aminotransferases from *Pseudomonas* sp. F-126 have also been used to produce homochiral amine products [183]. The stereoselective removal of either the pro-(*S*) or the pro-(*R*) proton has been demonstrated [184,185]. ω-Aminotransferase specific for the secondary amines has also been demonstrated from *Bacillus megaterium* and *Pseudomonas aeruginosa* [186].

The bioconversion of nitriles and primary amides has been used in the production of optically active α-hydroxy or α-amino acids such as L-phenylalanine, L-lactic acid, and L-phenylglycine, which are chiral synthons in many pharmaceutical syntheses. Nitrilase has been isolated from organisms belonging to the genera *Brevibacterium, Rhodococcus,* and *Pseudomonas* [187–190].

Recently, Crich et al. [191] developed a general enzymatic asymmetrical synthesis of α-amino acids. This method describes the use of the lipase PS-30 from *Pseudomonas cepacia* to catalyze the enantioselective methanolysis of a variety of 4-substituted 2-phenyloxazolin-5-one derivatives **105** in a nonpolar organic solvent to yield optically active *N*-benzoyl-L-α-amino acid methyl esters **106** in 80–99% e.e.'s (Fig. 36). The e.e. was further improved by the protease-catalyzed kinetic resolution to yield enantiomerically pure *N*-benzoyl-L-α-amino acids **107**.

After the discovery of the antibiotic thienamycin, compounds that contain the carbapenem and penem ring systems have attracted much attention. The importance of stereochemistry of the hydroxyethyl group is demonstrated by the fact that this group must be in the (*R*) configuration for antimicrobial activity to take place. Previously, synthesis of carbapenem and penum compounds have often utilized the optically active β-lactam intermediates [192–194]. Recently, D-(−)-3-hydroxybutyric acid prepared by the microbial hydroxylation of butyric acid has been used in the preparation of β-lactam [195,196].

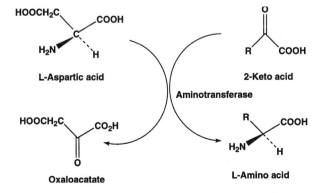

Figure 35 Synthesis of L-amino acids by aminotransferase.

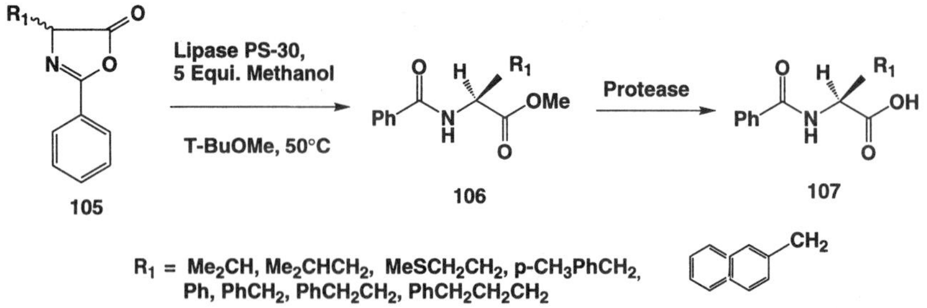

Figure 36 Enantioselective enzymatic methanolysis of 4-substituted 2-phenyloxazolin-5-one.

For the synthesis of β-lactam antibiotics, the presence of asymmetrical carbon at the 3 and 4 positions is critical to prepare optically active β-lactams [197]. Nagai et al. [198] developed enzymatic synthesis of optically active β-lactams by lipase-catalyzed kinetic resolution using the enantioselective hydrolysis of *N*-acyloxymethyl β-lactams **108** in an organic solvent (isopropyl ether saturated with water) and the transesterification of *N*-hydroxymethyl β-lactam **109** in organic solvent (methylene chloride) in the presence of vinyl acetate as acyl donor (Fig. 37). The reaction yield of 35–50% and e.e.'s of 93 to more than 99% were obtained depending on the specific substrate used in the reaction mixture. Lipase B from *Pseudomonas fragi* and lipase PS-30 from *Pseudomonas* sp. were used in the reaction mixture.

Aziridine carboxylates are chiral intermediates for the synthesis of β-lactams and amino acids [200]. The use of enantioselective ester hydrolysis in the synthesis of optically active *N*-unsubstituted and *N*-substituted aziridine carboxylate by *Candida cylindraceae* lipase has been demonstrated by Bucciareli et al. [199]. Racemic methyl aziridine-2-carboxylate and 2,3-dicarboxylate **110** were used as substrates both for enzymatic hydrolysis and for the synthesis of *N*-chloro, *N*-acyl and *N*-sulfonyl derivatives (Fig. 38). The reaction yield of 35–45% (theoretical maximum yield is 50%) and the e.e.'s of 90–98% were obtained depending on substrate used in the reaction mixture.

The synthesis of amides is critical for the development of many therapeutic activities as an amide bond is present in a large range of compounds [201]. Garcia et al. [202]

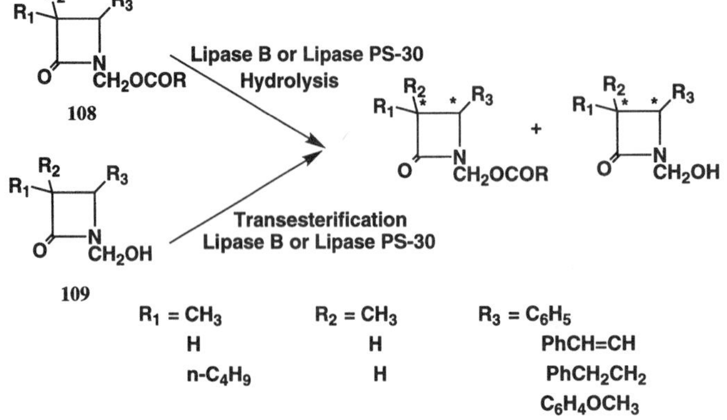

Figure 37 Enantioselective enzymatic hydrolysis of *N*-acyloxymethyl β-lactams.

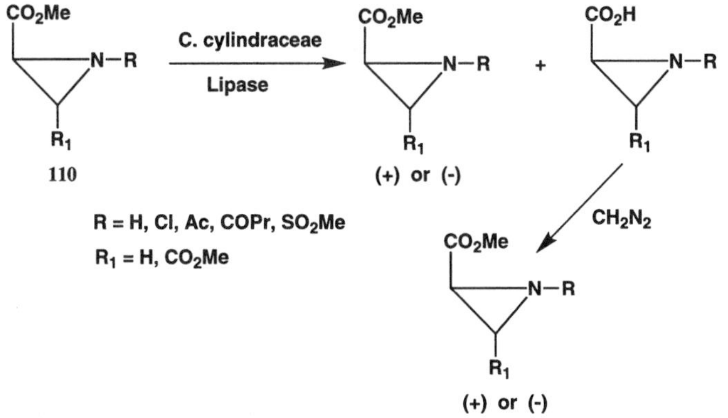

Figure 38   Enantioselective enzymatic ester hydrolysis of *N*-substituted aziridine carboxylates.

demonstrated the synthesis of optical 3-hydroxy amides **111**. Lipase from *Candida an-tarctica* efficiently catalyzed enantioselective aminolysis of various racemic 3-hydroxy esters **112** with aliphatic amines. The reaction yield of more than 54% and the e.e.'s of more than 95% were obtained depending on substrate used in the reaction mixture (Fig. 39). *Candida antarctica* lipase also catalyzed the aminolysis of racemic ethyl-3,4-epoxy-butyrate **113** to yield epoxy amide **114** in more than 92% e.e. and 27% reaction yield (Fig. 39).

Optically active 4-hydroxyalkanenitriles **115** are useful building blocks for asym-metrical synthesis as the cyano group is a functional precursor group of the amino and carbonyl groups. Lipase PS-30-catalyzed enantioselective hydrolysis of ester of 4-hydrox-yalkanenitriles **116** [Fig. 40] has been demonstrated by Takagi and Itoh [203]. The e.e.'s of 80–98% were obtained depending on the compound used in the reaction mixture.

(*S*)- and (*R*)-3-hydroxyglutaric acid monoesters are chiral precursors for the synthesis of L-carnitine and carbapenem. Monterio et al. [204] demonstrated the enantioselective hemihydrolysis of diethyl-3-hydroxyglutaric acid with esterase 30,000 to prepare (*S*)-3-hydroxyglutaric acid monoester in 90% yield and more than 98% e.e. They also demon-strated the quantitative synthesis of (*R*)-3-hydroxyglutaric acid monoester by catalyzing the hydrolysis of ethyl *t*-butyl 3-hydroxyglutaric acid by esterase 30,000. PLE-catalyzed

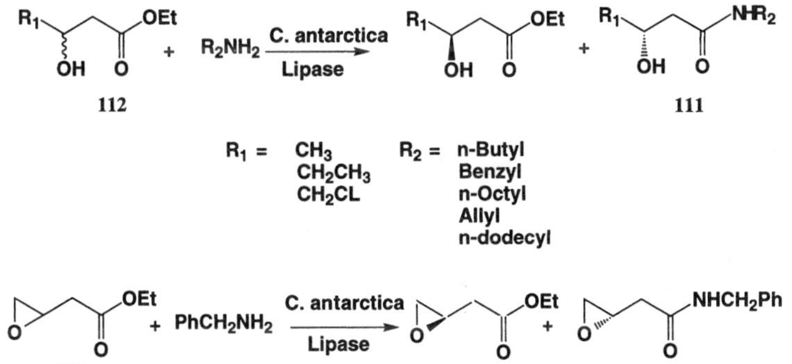

Figure 39   Enantioselective aminolysis of various racemic 3-hydroxy esters.

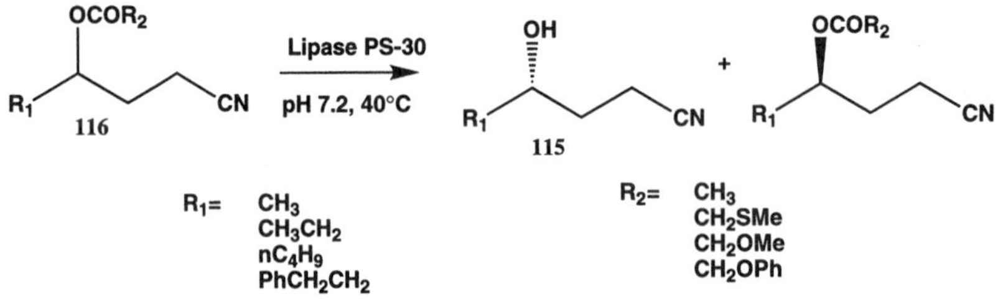

Figure 40 Enzymatic synthesis of optically active 4-hydroxyalkanenitriles.

enantioselective hydrolyses of dimethyl 3-hydroxy-, 3-methoxyethoxymethoxy-, and 3-benzyloxyglutarates have been demonstrated by Lam and Jones [205]. The enzymatic preparation of optically active 2,4-dimethylglutaric acid monomethyl esters from corresponding diester has been demonstrated by Chen et al. [206] using lipase from *Glicocledium roseum*. Asymmetrical hydrolysis of diethyl 3-hydroxy-3-methylglutarate to its corresponding monoester in high optical purity by PLE has been demonstrated [207,208].

Chiral 6-substituted 5,6-dihydro-2H-pyran-2-ones (α,β-unsaturated γ-lactones) are key structural intermediates of a variety of natural products that exhibit antifungal and antitumor activity [209]. Hasse and Schneider [210] prepared both enantiomeric series of a variety of optically pure 6-alkylated γ-lactones via an enzyme-mediated route. The key step in the process was the ring opening of enantiomerically pure alkyloxiranes **117**, accessible via the corresponding α-hydroxythioethers **118**, which were obtained enantiomerically pure by the lipase-catalyzed kinetic resolutions (Fig. 41).

Tetrahydropyran-2-methanol and tetrahydrofurfuryl alcohol are key chiral intermediates for the synthesis of polyether antibiotics. Such antibiotics are known as ionophores, which are able to transport metal ions across biological membranes. Some of their medicinal applications are in diuretics and analgesics [211]. Chiral (S)-tetrahydropyran-2-methanol **119** and **120** has been prepared from the hydrolysis of its butyrate ester **121** using porcine pancreatic lipase [212]. The unreacted (R) enantiomer of ester **122** was hydrolyzed by lipase from *Candida rugosa*. Both enantiomers were prepared in more than 97% e.e. (Fig. 42).

An antitumor antibiotic spergualin was discovered in the culture filtrate of a bacterial strain BMG162-aF2 that is related to *Bacillus laterosporus*, and its structure was determined to be (−)-(15S)-1-amino-10-guanidino-11,15-dihydroxy-4,9,12-triazanonade-

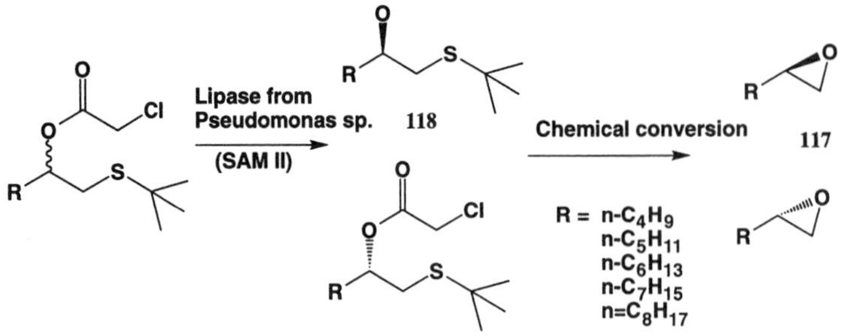

Figure 41 Preparation of chiral synthons for antifungal and antitumor agents: Lipase-catalyzed synthesis of enantiomerically pure alkyloxiranes.

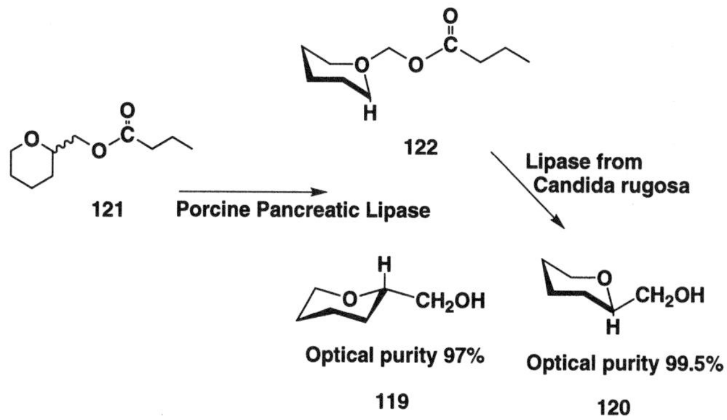

Figure 42 Preparation of chiral synthon for polyether antibiotics: Lipase-catalyzed enantioselective hydrolysis of compounds **121** and **122**.

cane-10,13-dione [213,214]. The total synthesis was accomplished by the acid-catalyzed condensation of 11-amino-1,1-dihydroxy-3,8-diazaundecane-2-one with (S)-7-guanidino-3-hydroxyheptanamide followed by the separation of the 11-epimeric mixture [215]. Antibacterial or antitumor activity of the enantiomeric mixture of spergualin was about half of that of the natural spergualin [216], indicating the importance of the configuration at C-11 for antitumor activity. Umeda et al. [217] demonstrated the optical resolution of the key intermediate of 15-deoxyspergualin **125**, racemic N-(7-guanidinoheptanoyl)-α-alkoxyglycine, by use of an exopeptidase (serine carboxypeptidase) and racemic N-(7-guanidinoheptanoyl)-α-alkoxyglycyl-L-amino acid as the substrate. Carboxypeptidase from *Penicillium janthinellum* catalyzed the hydrolysis of peptide bond of racemic N-(7-guanidinoheptanoyl)-α-methoxyglycyl-L-phenylalanine to yield (−)-N-(7-guanidinoheptanoyl)-α-methoxyglycine. They deduced that the absolute configuration of the carbon at 11 (C-11) of the bioactive (−)-enantiomer, and so that of the natural spergualin, is (S). The (−)-enantiomer of 15-deoxyspergualin was active against mouse leukemic L1210, while the (+) enantiomer was almost inactive (217).

We have demonstrated an alternate and more direct route, the lipase-catalyzed stereoselective acetylation of racemic 7-[N,N'-bis(benzyloxycarbonyl) N-(guanidinoheptanoyl)]-α-hydroxyglycine **123** to the corresponding S-(−)-acetate **124** and unreacted alcohol (+)-**123** [218] S-(−)-acetate **124** (Fig. 43) is a key intermediate for the total chemical

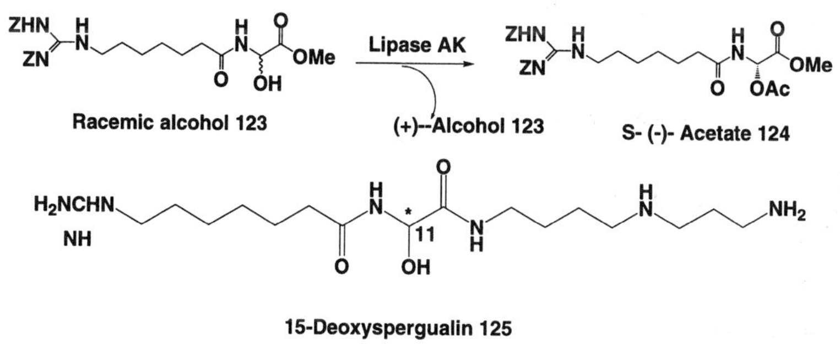

Figure 43 Preparation of chiral synthon for 15-deoxyspergaulin: Enantioselective acylation of racemic **123**.

synthesis of (−)-15-deoxyspergualin **125**, a related immunosuppressive agent and antitumor antibiotic [217,219]. The reaction was carried out in methyl ethyl ketone (MEK) using lipase from *Pseudomonas* sp. (lipase AK). Vinyl acetate was used as an acylating agent. A reaction yield of 48% (theoretical maximum yield is 50%) and an e.e. of 98% were obtained for *S*-(−)-acetate **124**. The unreacted alcohol (+)-**123** was obtained in 41% yield and 98.5% e.e.

XI. ANTI-INFLAMMATORY DRUGS

Naproxen, (*S*)-2-(6-methoxy-2-naphthyl)propanoic acid **126** is a nonsteroidal anti-inflammatory and analgesic agent first developed by Syntex [220,221]. Biologically active desired *S*-naproxen has been prepared by enantioselective hydrolysis of the methyl ester of naproxen by esterase derived from *Bacillus* subtilis Thai 1-8 [222]. The esterase was subsequently clone in *Escherichia coli* with over 800-fold increase in activity of enzyme. The resolution of racemic naproxen amide and ketoprofen amides has been demonstrated by amidases from *Rhodococcus erythropolis* MP50 and *Rhodococcus* sp. C311 (223–226). *S*-Naproxen **126** and *S*-ketoprofen **127** (Fig. 44) were obtained in 40% yields (theoretical maximum yield is 50%) and 97% e.e. Recently, the enantioselective esterification of naproxen has been demonstrated using lipase from *Candida cylindraceae* in isooctane as solvent and trimethylsilyl as alcohol. The undesired isomer of naproxen was esterified leaving desired *S* isomer unreacted [227].

Ibuprofen **128** is another well-known analgesic, anti-inflammatory drug and it is believed that it will be marketed as a single isomer drug. The kinetic enzymatic resolution of racemic ibuprofen has been reported [228]. The reaction for the resolution has been scaled up to make gram quantities of *S*-ibuprofen. This was accomplished by two enantioselective reactions each catalyzed by Novozyme 435. In the first reaction, 300 g of racemic ibuprofen was esterified with 1-dodecanol to yield the *R* ester and *S*-ibuprofen to produce 89 g of *S*-ibuprofen in 85% e.e. In the second reaction, 75 g of the 85% e.e. material was used to prepare 39 g of *S*-ibuprofen with a 97.5% e.e.

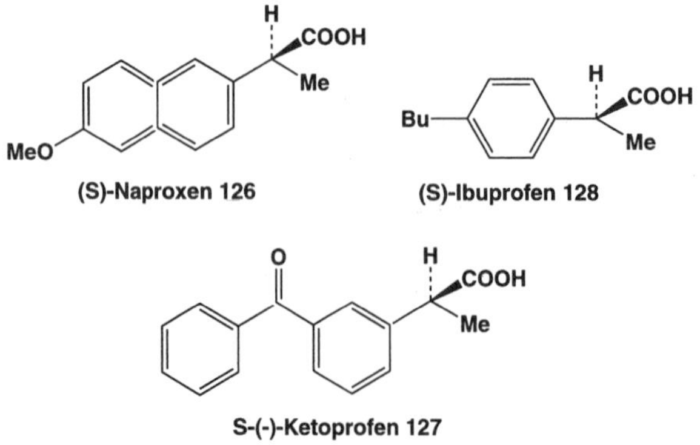

Figure 44 Structure of anti-inflammatory drugs: Naproxen **126**, ketoprofen **127**, and ibuprofen **128**.

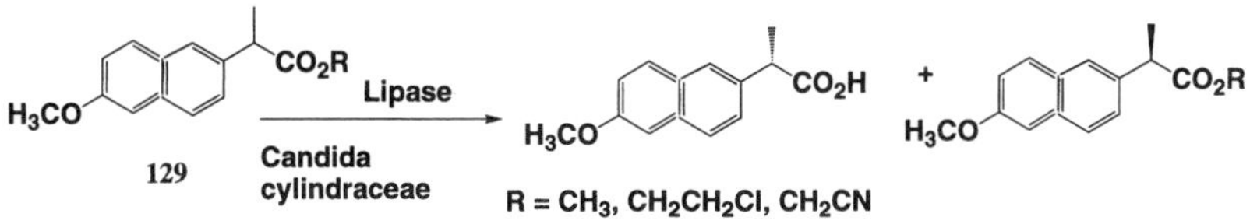

Figure 45 Preparation of chiral synthon for naproxen: Enantioselective enzymatic hydrolysis of racemic chloroethyl-2-(6-methoxy-2-naphthyl)propionate **129**.

Another approach for the enzymatic preparation of *S*-ibuprofen has been demonstrated by de Zoete et al. [229]. The enantioselective ammonolysis of ibuprofen 2-chloroethyl ester by *Candida antarctica* lipase (lipase SP435) gave the remaining ester *S*-(+) enantiomer in 44% yield and 96% e.e. The enantioselective enzymatic esterification of racemic ibuprofan has also been demonstrated using lipase from *Candida cylindraceae* [230]. The reaction was carried out in a water-in-oil microemulsion [bis(2-ethylhexyl)sulfosuccinate (AOT)/isooctane]. The lipase showed high preference for the *S*-(+) enantiomers of ibuprofen which was esterified and *R*-(−) enantiomer remained unreacted. The reaction yield of 35% was obtained using *n*-propanol in the reaction mixture as nucleophile.

Two most commonly prescribed drugs are Motrin (*p*-isobutylhydratropic acid) and Naproxen [(+)-*S*-2-(6-methoxy-2-napthyl)propionic acid]. They are widely used in the treatment of the symptoms of arthritis. Gu et al. [231] demonstrated the preparation of Naproxen via enzymatic enantioselective hydrolysis of racemic chloroethyl-2-(6-methoxy-2-naphthyl) propionate **129** by lipase from *Candida cylindraceae* (Fig. 45). The reaction yield of 40% and the e.e. of more than 98% were obtained.

S-2-Chloropropionic acid used as a chiral synthon for various nonsteroidal anti-inflammatory drugs has also been prepared by stereoselective dehalogenase reactions. *Pseudomonas putida* contained two dehalogenases, one was a low molecular weight enzyme showing 100% specificity for *S*-2-chloropropionate and the other a higher molecular weight enzyme with specificity toward *R*-2-chloropropionate [232–234]. Future use of *R*- and *S*-specific dehalogenases in enzymatic resolution processes will be very promising.

XII. ANTIVIRAL AGENTS

Purine nucleoside analogs have been used as antiviral agents [235]. Lamivudine, zidovudine, and didanosine (Fig. 46) are effective antiviral agents. Lamivudine, a highly promising drug candidate for HIV 2 and HIV 3 infection, provides a challenge to the synthetic chemist due to the presence of two acetal chiral centers, both sharing the same oxygen atom. The use of cytidine deaminase from *E. coli* [236] has been demonstrated to deaminate 2'-deoxy-3'-thiacytidine enantioselectively to prepare optically pure (*2'R-cis*)-2'-deoxy-3'-thiacytidine (3TC, lamivudine, Fig. 46).

A novel enzymatic resolution process has been developed for the preparation of chiral intermediate for lamivudine synthesis. An enzymatic resolution of α-acetoxysulfides **130** (Fig. 46) by *Pseudomonas fluorescens* lipase has been demonstrated to give chiral intermediate in more than 45% and 97% e.e. [237].

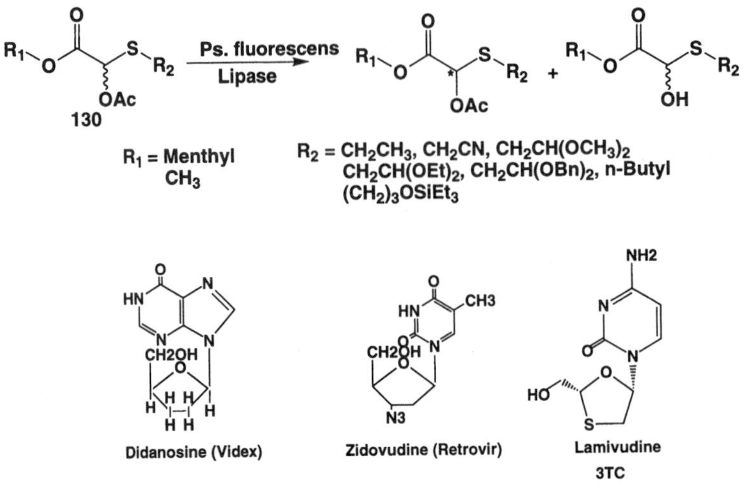

Figure 46 Preparation of chiral synthon for lamivudine syntheses: Enzymatic resolution of α-acetoxysulfides.

Biologically important carbocyclic nucleosides such as (−)-aristeromycin and (−)-neoplanocin have been prepared by the lipase-catalyzed asymmetrical hydrolysis with PLE [238]. Asymmetrical hydrolysis of the *meso*-epoxy diesters dialkyl 5,6-epoxybicyclo[2.2.1]hept-2-ene-2,3-dicarboxylate **131** with PLE quantitatively produced the optically active 6-formyl-2-(alkoxycarbonyl)bicyclo[3.1.0]hex-2-ene-1-carboxylate **132** in more than 92% e.e. (Fig. 47).

Racemic 2-azabicyclo[2.2.1]hept-5-ene-3-one has great potential as a synthetic intermediate. The bicyclic lactam is a synthon for carbocyclic sugar amines, carbonucleosides, and carbocyclic dinucleotide analogs [239]. Chiral 2-azabicyclo[2.2.1]hept-5-ene-3-one **133** is key intermediate for the synthesis of (−)-carbovir **134**, an antiviral agent effective against HIV. Nakano et al. [240] prepared chiral **133** by enantioselective trans-

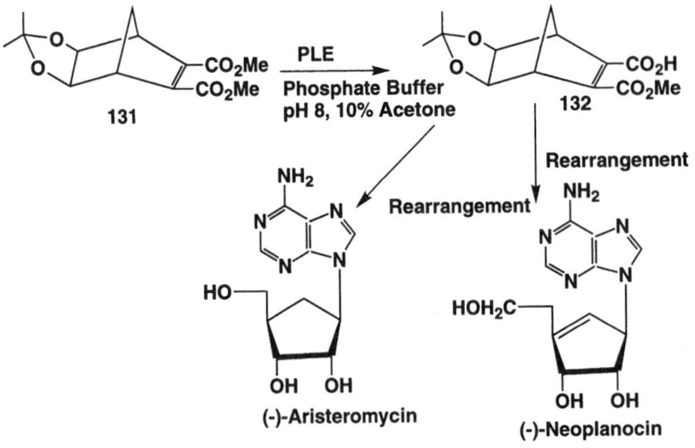

Figure 47 Preparation of chiral synthons for (−)-aristeromycin and (−)-neoplanocin: Enzymatic asymmetrical hydrolysis of *meso*-epoxy diester **131**.

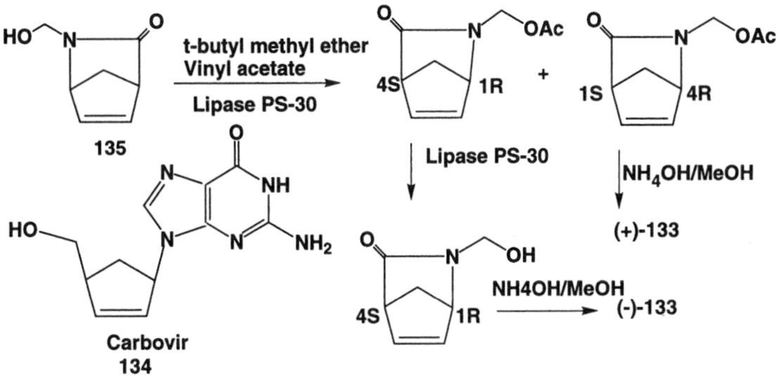

Figure 48 Synthesis of chiral intermediates for (−)-carbovir: Enantioselective transesterification of 2-hydroxymethyl-2-azabicyclo[2.2.1]hept-5-ene-3-one **135**.

esterification of 2-hydroxymethyl-2-azabicyclo[2.2.1]hept-5-ene-3-one **135** (Fig. 48). A reaction yield of 40% and e.e. of 96% were obtained.

XIII. PROSTAGLANDIN SYNTHESIS

Optically active epoxides are useful chiral synthons in the phamaceutical synthesis of prostaglandins. Microbial epoxidation of olefinic compounds was first demonstrated by van der Linden [241]. Subsequently, May et al. [242] demonstrated the epoxidation of alkenes in addition to hydroxylation of alkanes by an ω-hydroxylase system. Oxidation of alk-1-enes in the range C6–C12, α,ω-dienes from C6–C12, alkyl benzene, and allyl ethers were demonstrated using an ω-hydroxylase enzyme system from *Pseudomonas oleovorans*. *R*-Epoxy compounds in greater than 75% e.e. were produced by epoxidation reactions using the ω-hydroxylase system [243,244]. The epoxidation system from *Nocardia corallina* is very versatile, has broad substrate specificity, and reacts with unfunctionalized aliphatic as well as aromatic olefins to produce *R*-epoxides [245,246].

Chiral bicyclo[3.2.0]heptanone has been recognized as a chiral precursor for (+)-prostaglandin A2 **136** and (+)-prostaglandin-F2a **137** synthesis. The reduction of racemic bicyclo[3.2.0]hept-2-en-6-one **139** in high optical purity by cells of *Mortierella ramanniana* has been demonstrated (Fig. 49). The same organism has also been used for the stereoselective reduction of 7,7-dimethylbicyclo[3.2.0]hept-2-ene-6-one **139** to prepare the chiral synthon for (+)-leukotriene B4 synthesis [247].

Cycloalkonone oxygenase from *Pseudomonas putida* AS1 and *Acinetobacter* sp. NCIMB 9871 has been used to catalyze the regio- and stereoselective Baeyer-Villiger type of oxidation of [3.2.0]hept-2-en-6-one. The enantiomerically pure lactones prepared by this enzymatic reaction are chiral synthons for prostaglandin synthesis [248,249].

Chiral compounds **140** and **141** are key intermediates for the chemoenzymatic synthesis of some prostaglandin analogs used for the treatment of peptic ulcer disease. Babiak et al. [250] evaluated several lipases, including that from *Pseudomonas* sp., *Candida cylindraceae*, *Aspergillus niger*, and porcine pancreas to catalyze the resolution of racemic compounds **140** and **141**. It was demonstrated that all of the enzymes were selective in acylating the *R* isomer of the starting enone compound. Porcine pancreatic lipase, either free or immobilized enzyme on Amberlite XAD-8, gave 45% reaction yield (theoretical maximum yield is 50%) and 99% e.e.'s (Fig. 50).

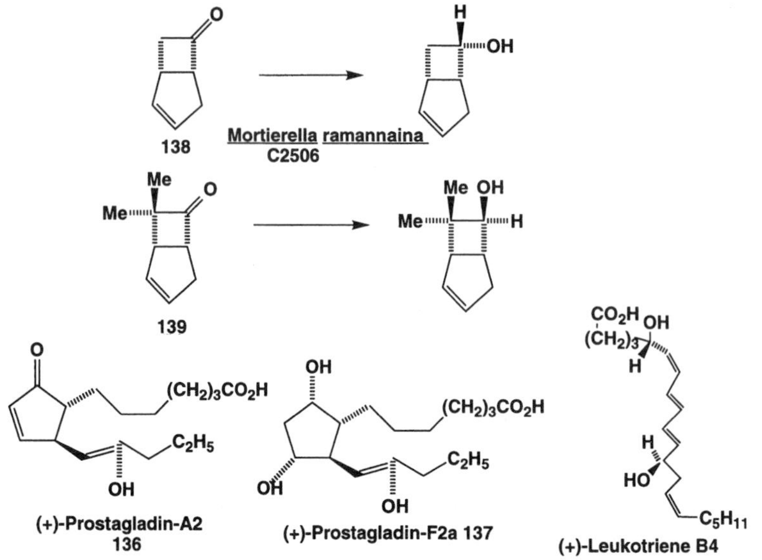

Figure 49 Synthesis of chiral synthons for prostaglandins and leukotrienes: Stereoselective reduction of bicyclo[3.2.0]hept-2-en-6-one **81** and 7,7-dimethylbicyclo[3.2.0]hept-2-ene-6-one **82**.

An efficient chemoenzymatic synthesis of both enantiomers **142** and **143** of an LTD4 antagonist have been prepared by lipase-catalyzed asymmetrical hydrolysis of prochiral and racemic dithioacetal esters **144** having up to five bonds between the prochiral/chiral center and the ester carbonyl group. The e.e. of 98% and reaction yield of 45% were obtained using lipase PS-30 (Fig. 51). LTD4 antagonists have potential for the therapeutic treatment of asthma [251].

Lipase-catalyzed transesterification of *meso*-cyclopentane diols **145–149** has been demonstrated by Theil et al. [252] with vinyl acetate as acyl donor and tetrahydrofuran/

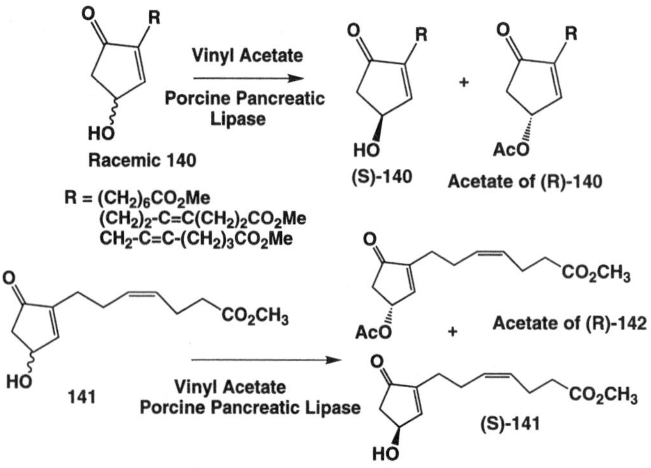

Figure 50 Enzymatic preparation of chiral synthons **140** and **141** for chemoenzymatic syntheses of prostaglandin analogs.

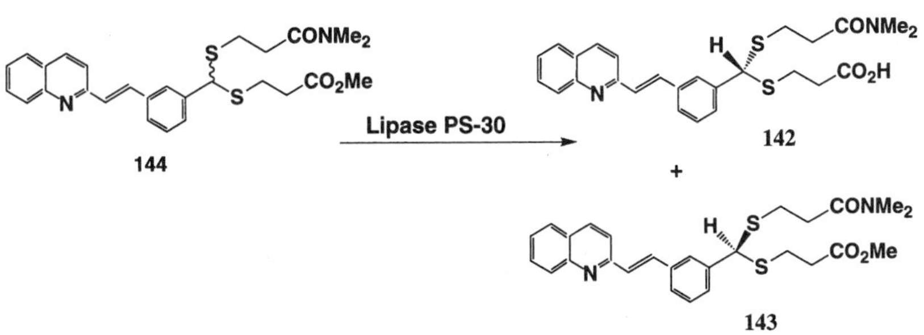

Figure 51 Chemoenzymatic synthesis of enantiomers of an LT-4 antagonist.

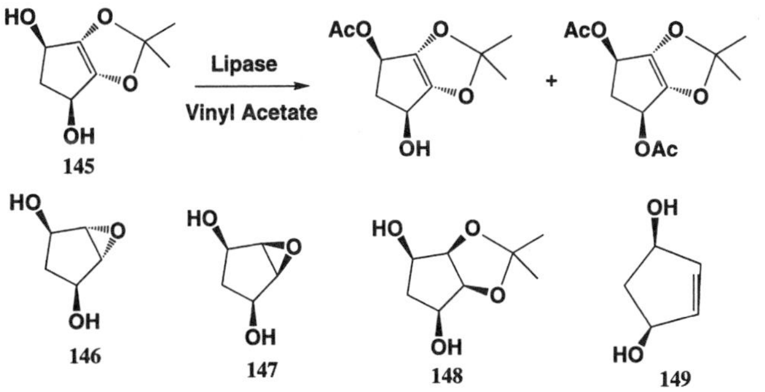

Figure 52 Lipase-catalyzed transesterification of *meso*-cyclopentane diols.

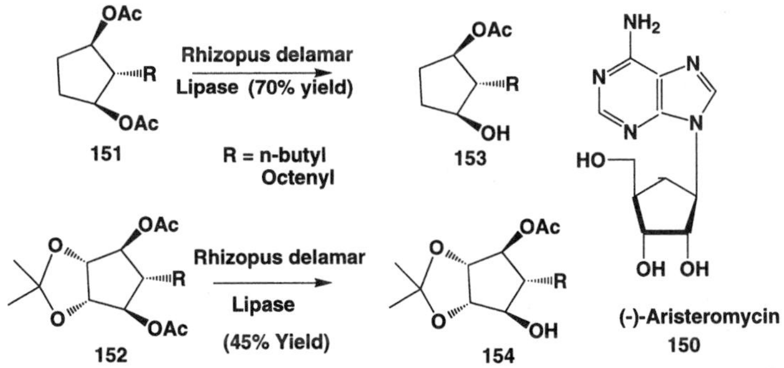

Figure 53 Enantioselective enzymatic hydrolysis of meso-1,3-bis(acetoxymethyl)-2-*trans*-alkylcyclopentanes.

triethylamine as solvent (Fig. 52). The chiral monoacetate was obtained in 65–85% reaction yield and more than 95% e.e.'s depending on the substrate used in the transesterification reaction. The chiral monoacetate served as starting material for prostaglandin synthesis.

Tanake et al. [253] demonstrated the enzymatic synthesis of the sugar moiety of carbocyclic nucleosides required for the total synthesis of (−)-aristeromycin **150**. Using lipase from *Rhizopus delamar* the enantioselective hydrolysis of *meso*-1,3-bis(acetoxymethyl)-2-*trans*-alkylcyclopentane (**151,152**) was carried out to prepare the chiral monoacetate (**153,154**) in more than 96% e.e. (Fig. 53). The chiral monoacetate was also used in the synthesis of optically active 11-deoxyprostaglandins [254].

REFERENCES

1. CJ Sih, CS Chen. Angew Chem 96:556–566, 1984.
2. JB Jones. Tetrahedron 42:3351–3403, 1986.
3. DHG Crout, S Davies, RJ Heath, CO Miles, DR Rathbone, BEP Swoboda. Biocatalysis 9: 1–30, 1994.
4. HG Davies, RH Green, DR Kelly, SM Roberts. Biotechnology 10:129–152, 1990.
5. R Csuz, BI Glanzer. Chem Rev 96:556–566, 1991.
6. J Crosby. Tetrahedron 47:4789–4846, 1991.
7. J Kamphuis, WHJ Boesten, QB Broxterman, HFM Hermes, JAM van Balken, EM Meijer, HE Schoemaker. Adv Biochem Eng Biotech 42:133–186, 1990.
8. CJ Sih, Q-M Gu, Q-M, X Holdgrun, K Harris. Chirality 4:91–97, 1992.
9. E Santaneillo, P Ferraboschi, P Grisenti, A Manzocchi. Chem Rev 92:1071–1140, 1992.
10. AL Margolin. Enzyme. Microb Technol 15:266–280, 1993.
11. DC Cole. Tetrahedron 32:9517–9582, 1994.
12. C-H Wong, GM Whitesides. Enzymes in Synthetic Organic Chemistry. Tetrahedron Organic Chemistry Series, Vol. 12. New York: Elsevier, 1994.
13. K Mori. Synlett. November, 1097–1109, 1995.
14. J-M Feng, C-H Lin, CW Bradshaw, C-H Wong. J Chem Soc Perkin Trans I:967–978, 1995.
15. MA Ondetti, DW Cushman. J Med Chem 24:355–361, 1981.
16. MA Ondetti, B Rubin, DW Cushman. Science 196:441–444, 1977.
17. DW Cushman, MA Ondetti. Biochem Pharmacol 29:1871–1875, 1980.
18. DW Cushman, MS Cheung, EF Sabo, MA Ondetti. Biochemistry 16:5484–5491, 1977.
19. CT Goodhue, JR Schaeffer. Biotechnol Bioeng 13:203–214, 1971.
20. M Schimazaki, J Hasegawa, K Kan, K Nemura, Y Nose, HKondo, AA Seymour, TN Swerdel, B Abboa-Offei. J Cardiovasc Pharmacol 17:456–465, 1991.
21. J Hasegawa, M Ogura, H Kanema, N Noda, H Kawaharada, KJ Watanabe. J Ferment Technol 60:501–508, 1982.
22. QM Gu, CS Chen, CJ Sih, C.J. Tetrahedron Lett 27:1763–1766, 1986.
23. A Sakimas, K Yuri, N Ryozo, O Hisao. Eur Pat 0172614, 1986.
24. CJ Sih. Eur Pat 87264125, Derwent International Publication #W087105328, 1987.
25. RN Patel, JM Howell, CG McNamee, KF Fortney, LJ Szarka. Biotechnol Appl Biochem 16: 34–47, 1992.
26. RN Patel, JM Howell, A Banerjee, KF Fortney, LJ Szarka. Appl Microbiol Biotechnol. 36: 29–34, 1991.
27. JL Moniot U.S. Pat App CN 88-100862, 1988.
28. MA Ondetti, A Miguel, J Krapcho. U.S. Pat 4316906,1982.
29. NG Delaney, EN Gordon, JM DeForrest, DW Cushman. Eur Pat EP361365, 1988.
30. AA Seymour, TN Swerdel, SA Fennell, SP Druckman, B Shadid, HC van der Plas, WHJ Boesten, J Kamphuis, EM Meijer, HE Schoemaker, H E. Tertrahedron 46:913–917, 1991.

31. RN Patel, RS Robison, LJ Szarka. Appl Microbiol Biotechnol 34:10–14,1990.
32. P Mohr, N Waespe-Sarcevic, C Tamm. Helv Chim Acta 66:2501–2511, 1983.
33. K Laumen, M Schneider. Tetrahedron Lett 26:2073–2076, 1985.
34. GMR Tombo, HP Schar, XF Busquets, O Ghisalba. Tetrahedron Lett 21:5707–5710, 1986.
35. DS Karenewsky, MC Badia, DW Cushman, JM DeForrest, T Dejneka, MJ Loots, MG Perri, EW Petrillo, JR Powell. J Med Chem. 31:204–212, 1988.
36. RL Hanson, KS Bembenek, RN Patel, LJ Szarka. Appl Microbiol Biotechnol 37:599–603, 1992.
37. DM Spero, S Kapadia, V Farina. Tetrahedron Lett 46:4543–4548, 1995.
38. MC Wani, HL Taylor, ME Wall, P Coggon, AT McPhail. J Am Chem Soc 93:2325–2327, 1971.
39. DGI Kingston. Pharmacol Ther 52:1–34, 1991.
40. PB Schiff, J Fant, SB Horowitz. Nature 277:665–667, 1979.
41. RA Holton, RR Juo, HB Kim, AD Williams, S Harusawa, RE Lowenthal, S Yogai. J Am Chem Soc 110:6558–6560, 1988.
42. J-N Denis, AE Greene, A Aarao Serre, M-J Luche. J Org Chem 51:46–50, 1986.
43. AA Christen, DM Gibson, J Bland. U.S. Pat 5,019504, 1991.
44. RL Hanson, JM Wasylyk, VB Nanduri, DL Cazzulino, RN Patel, LJ Szarka. J Biol Chem 269:22145–22149, 1994.
45. VB Nanduri, RL Hanson, TL LaPorte, RY Ko, RN Patel, LJ Szarka. Biotech Bioeng 48: 547–550, 1995.
46. RL Hanson, RN Patel, LJ Szarka. US Pat 5516676, 1996.
47. RL Hanson, RN Patel, LJ Szarka. US Pat 5523219, 1996.
48. RL Hanson, RN Patel, LJ Szarka. US Pat 5700669, 1997.
49. RN Patel. Annu Rev Microbiol (in press).
50. RL Hanson, JM Howell, BD Brzozowski, SA Sullivan, RN Patel, LJ Szarka. Biotechnol Appl Biochem 26:153–158, 1997.
51. RN Patel, A Banerjee, JM Howell, CG McNamee, D Brzozowski, D Mirfakhrae, JK Thottathil, LJ Szarka. Tetrahedron Asymmetry 4:2069–2084, 1993.
52. RN Patel, A Banerjee, CG McNamee, JK Thottathil, LJ Szarka. US Pat 5420337, 1995.
53. RN Patel, A Banerjee, RY Ko, JM Howell, W-S Li, FT Comezoglu, Partyka, LJ Szarka. Biotechnol Appl Biochem 20:23–33, 1994.
54. RN Patel, LJ Szarka, RA Partyka. US Pat 5567614.
55. D-M Gou, Y-C Liu, C-S Chen. J Org Chem 58:1287–1289, 1993.
56. R Brieva, JZ Crich, CJ Sih. J Org Chem 58:1068–1075, 1993.
57. M Nakane. US Pat 4,663,336, 1987.
58. AW Ford-Hutchinson. Clin Exp Allergy 21:272–276, 1991.
59. J Das, MF Haslanger, JZ Gougoutas, MF Malley. Synthesis 12:1100–1112, 1987.
60. N Hamaka, T Seko, T Miyazaki, A Kawasaki. Adv Prostaglandin, Thromboxane and Leukotriene Res 21:359–362, 1990.
61. JB Jones, JF Beck, J. F. In JB James, CJ Sih, DE Perlman, eds. Application of Biochemical Systems in Organic Synthesis. New York: John Wiley and Sons, 1987, pp 248–376.
62. H Yamada, S Shimizu. Angew Chem In Ed Engl 27:622–642, 1988.
63. KP Lok, TJ Jakovac, JB Jones. J Am Chem Soc 107:2521–2526, 1985.
64. JB Jones, CJ Francis. Can J Chem 62:2578–2584, 1984.
65. RN Patel, M Liu, A Banerjee, JK Thottathil, J Kloss, LJ Szarka. Enzyme Microb Technol 14:778–784, 1992.
66. RN Patel, M Liu, A Banerjee, LJ Szarka. Appl Microbiol Biotechnol 37:180–183, 1992.
67. RN Patel, LJ Szarka, JK Thottathil, D Kronenthal. US Pat 5266710.
68. K Mori, K Tanida. Tetrahedron Lett 40:3471–3476, 1984.
69. M Hirama, M Uei. J Am Chem Soc 104:4251–4256, 1982.
70. AS Gopalan, CJ Sih. Tetrahedron Lett 25:5235–5238, 1984
71. CJ Sih, BN Zhan, AS Gropalan, CS Chen, G Girdaukais, F vanMiddlesworth. Ann N Y Acad Sci 434:186–193, 1984.

72. OP Ward, CS Young. Enzyme Microb Technol 12:482–493, 1990.

73. CJ Sih, CS Chen. Angew Chem Int Engl 23:570–578, 1984.

74. K Nakamura, Y Kawai, S Oka, A Shue. Tetrahedron Lett 30:2245–2246, 1989.

75. K Nakamura, Y Kawai, T Miyai, A Ohno. Tetrahedron Lett 31:3631–3632, 1990.

76. RN Patel, CG McNamee, A Banerjee, JM Howell, RS Robison, LJ Szarka. Enzyme Microb Technol 14:731–738, 1992.

77. K Nakamura, Y Kawai, N Nakajima, A Ohno. J Org Chem 56:4778–4783, 1991.

78. J Heidlas, K-H Engel, R Tressl. Eur J Biochem 172:633–639, 1988.

79. E Keinan, EK Hafeli, KK Seth, RR Lamed. J Am Chem Soc 108:162–168, 1986.

80. RN Patel, CT Hou, AI Laskin, P Derelanko. J Appl Biochem 3:218–232, 1981.

81. CW Bradshaw, H Fu, GJ Shen, C-H Wong. J Org Chem 57:1526–1531, 1992.

82. M Christen, DHG Crout. J Chem Soc Chem Commun 45:264–266, 1988.

83. RN Patel, A Banerjee, CG Mc Namee, D Brzozowski, RL Hanson, LJ Szarka. Enzyme Microb Technol 15:1014–1021, 1993.

84. SY Sit, RA Parker, I Motoe, HW Balsubramanian, CD Cott, PJ Brown, WE Harte, MD Thompson, J Wright. J Med Chem 33:2982–2999, 1990.

85. RN Patel, CG McNamee, LJ Szarka. Appl Microbiol Biotechnol 38:56–60, 1992.

86. A Endo, M Kuroda, Y Tsujita. J Antibiot 29:1346–1348, 1976.

87. A Endo, M Kuroda, K Tanzawa. FEBS Lett 72:323–326, 1976.

88. M Hosobuchi, K Kurosawa, H Yoshikawa, H. Biotechnol Bioeng 42:815–820, 1993.

89. M Hosobuchi, T Shioiri, J Ohyama, M Arai, S Iwado, H Yoshikawa. Biosci Biotech Biochem 57:1414–1419, 1993.

90. N Serizawa, S Serizawa, K Nakagawa, K Furuya, T Okazaki, A Tarahara. J Antibiot 36:887–891, 1983.

91. T Matsuoka, S Miyakoshi, K Tanzawa, K Nakahara, M Hosobuchi, N Serizawa. Eur J Biochem 184:707–713, 1989.

92. H Suemume, M Takahashi, S Maeda, Z-F Xie, K Sakai. Tetrahedron: Asymmetry 1:425–429, 1990.

93. CF Barbas, Y-F Wang, C-H Wong. J Am Chem Soc 112:2013–2016, 1990.

94. L Chen, DP Dumas, C-H Wong. J Am Chem Soc 114:741–743, 1992.

95. HJ Gijsen, C-H Wong. J Am Chem Soc 116:8422–8423, 1994.

96. ARPM Valentijn, R de Hann, E de Kant, GA van der Marel, LH Cohen, JH van Boom. Recl Tran Chim Pays-Bas 114:332–336, 1995.

97. SA Biller, MJ Sofia, B DeLange, C Foster, EM Gordon, T Harrity, LC Rich, CP Ciosek. J Am Chem Soc 113:8522–8527, 1991.

98. SA Biller, C Foster, EM Gordon, T Harrity, LC Rich, J Maretta, CP Ciosek. J Med Chem Soc 34:1912–1918, 1991.

99. A Steiger, HJ Pyun, RM Coates. J Org Chem 57:3444–3450, 1992.

100. AC Oehlschlanger, SM Singh, S Sharma. J Org Chem 56:3856–3861,1991.

101. SA Biller, D Majnin. US Pat 5428028, 1995.

102. M Lawrence, SA Biller, OM Fryszman. US Pat 5447922, 1995.

103. RN Patel, A Banerjee, LJ Szarka. Tetrahedron: Asymmetry 8:1055–1059, 1997.

104. M Chaffman, RN Brogden. Drugs 29:387–390, 1985.

105. C Kawai, T Konishi, E Matsuyama, H Okazaki. Circulation 63:1035–1038, 1981.

106. T Isshiki, B Pegram, E Frohlich. Cardiovasc Drug Ther 2:539–544, 1988.

107. DM Floyd, RY Moquin, KS Atwal, SZ Ahmed, SH Spergel, JZ Gougoutas, MF Malley. J Org Chem 55:5572–5575, 1990.

108. J Das, DM Floyd, SM Kimball, RN Patel, JK Thottathil. Indian J Chem 31B, 817–820, 1992.

109. J Das, DM Floyd, SM Kimball, KJ Duff, T Vu, MW Lago, RY Moquin, VG Lee, JZ Gougoutas, MF Malley, S Moreland, RJ Brittain, SA Hedberg, G Cucinotta. J Med Chem 35: 773–778, 1992.

110. RN Patel, RS Robison, LJ Szarka, J Kloss, JK Thottathil, RH Mueller. Enzyme Microb Technol 13:906–912, 1991.

111. H Matsumae, M Furui, T Sabatani. J Ferment Bioeng 75:9397, 1993.
112. G Edwards, AH Weston. TIPS 11:417–422, 1990.
113. DW Robertson, MI Steinberg. Ann Med Chem 24:91–100, 1989.
114. DW Robertson, MI Steinberg. J Med Chem 33:1529–1533, 1990.
115. TC Hamilton, AH Weston, A. H. Gen Pharmacol 20:1–9, 1989.
116. VA Ashwood, RE Buckingham, F Cassidy, JM Evans, TC Hamilton, DJ Nash, G Stempo, KJ Willcocks. J Med Chem 29:2194–2201, 1986.
117. EN Jacobsen, W Zhang, AR Muci, JR Ecker, L Deng. J Am Chem Soc 113:7063–7067, 1991.
118. R Bergmann, V Eiermann, RJ Gericke. J Med Chem 33:2759–2761, 1990.
119. K Atwal, GJ Grover, KS Kim. US Pat 5140031, 1991.
120. JM Evans, CS Fake, TC Hamilton, RH Poyser, EA Watts. J Med Chem 26:1582–1586, 1983.
121. RN Patel, A Banerjee, B Davis, JM Howell, CG Mc Namee, D Brzozowski, J North, D Kronenthal, LJ Szarka. Bioorg Med Chem 2:535–542, 1994.
122. RN Patel, A Banerjee, CG McNamee, D Brzozowski, LJ Szarka. US Pat 5478734, 1995.
123. RN Patel, A Banerjee, CG McNamee, LJ Szarka. Tetrahedron: Asymmetry 6:123–130, 1995.
124. AA Larsen, PM Lish. Nature 203:1283–1284, 1964.
125. RH Uloth, JR Kirk, WA Gould, AA Larsen. J Med Chem 9:88–96, 1966.
126. PM Lish, JH Weikel, KW Dungan. J Pharmacol Exp Ther 149:161–173, 1965.
127. P Somani, T Bachand. Eur J Pharmacol 7:239–247, 1969.
128. RN Patel, A Banerjee, CG McNamee, LJ Szarka. Appl Microbiol Biotechnol 40:241–245, 1993.
129. G Hirth, R Barner. Helv Chim Acta 65:1059–1084, 1982.
130. U Peters, W Bankova, P Welzel. Tetrahedron 43:3803–3816, 1987.
131. MA Bertola, HS Koper, GT Phillips, AF Merx, VP Claussen. Eur Pat Appl No 0244912A1, 1987.
132. A Geerlof, J Stoorvogel, JA Jongejan, EJTM Leenen, TJGM van Dooren, WJJ Twell, JA Duine. Appl Microbiol Biotech 42:8–15, 1994.
133. M Gelo, V Sunjic. Synthesis (Sept):855–860, 1993.
134. CD Ferris, DJ Hirsch, BP Brooks, SH Snyder. J Neurochem 57:729–737, 1991.
135. JL Junien, BE Leonard. Clin Neuropharmacol 12:353–374, 1989.
136. LM Walker, WD Bower, FO Walker, RR Matsumoto, BD Costa, KC Rice. Pharm Rev 42:355–402, 1990.
137. GF Steinfels, SW Tam, L Cook. Neuropsychopharmacology 2:201–207,1989.
138. T Massamiri, SP Duckles. J Pharmacol Exp Ther 253:124–129, 1990.
139. JA Martinez, L Bueno, L. Eur J Pharmacol 202:379–383, 1991.
140. DP Taylor, MS Eison, SL Moon, FD Yocca. Adv Neuropsychol Psychopharm 1:307–315,1991.
141. RN Patel, A Banerjee, M Liu, RL Hanson, RY Ko, JM Howell, LJ Szarka. Biotechnol Appl Biochem 17:139–153, 1993.
142. JP Yevich, JS New, DW Smith. J Med Chem 29:359–369, 1986.
143. JB Jones, JF Beck. In: JB James, CJ Sih, DE Perlman, ed. Application of Biochemical Systems in Organic Synthesis. New York: John Wiley, 1986, pp 248–376.
144. E Keinan, EK Hafeli, KK Seth, RR Lamed. J Am Chem Soc 108:162–168, 1986.
145. CW Bradshaw, H Fu, GJ Shen, C-H Wong. J Org Chem 57:1526–1531, 1992.
146. CW Bradshaw, W Hummel, C-H Wong. J Org Chem 57:1532–1536, 1992.
147. H Simon, In: DA Abramowicz, ed. Biocatalysis. New York: Van Nostrand-Reinhold, 1990, pp 218–242.
148. RL Hanson, A Banerjee, FT Comezoglu, D Mirfakhrae, RN Patel, LJ Szarka. Tetrahedron: Asymmetry 5:1925–1934, 1994.
149. KS Patrick, JL Singletary. Chirality 3:208–212, 1991.
150. PS Portoghese, TL Pazdernik, WL Kuhn, G Hite, A Shafike. J Med Chem 11:12–20, 1968.
151. V Vechetti, A Giordani, G Giardina, R Colle, D Clarke. J Med Chem 34:397–402, 1991.

152. MC Ng-Youn-Chen, AN Serreqi, Q Huang, RM Kazlaukas. J Org Chem 59:417–421, 1994.
153. RM Williams. In: JE Baldwin, PD Magnus, eds. Synthesis of Optically Active α-Amino acids, Vol. 7. Oxford: Pergamon Press, 1989, pp 130–150.
154. MJ O'Donnell, DW Bennett, S Wu. J Am Chem Soc 111:2353–2355, 1989.
155. DA Evans, JA Ellman, RL Dorow. Tetrahedron Lett 28:1123–1126, 1987.
156. U Schmidt, M Respondek, A Lieberknecht, J Werner, P Fischer. Synthesis 28:256–261, 1989.
157. G Bold, RO Duthaler, M Riediker. Angew Chem Int Ed Engl 28:497–498, 1989.
158. Y Ito, M Sawamura, E Shirakawa, K Hayashizaki, T Hayashi. Tetrahedron 44:5253–5262, 1988.
159. EM Gordon, MA Ondetti, J Pluscec, CM Cimarusti, DP Bonner, RB Sykes, J Am Chem Soc 104:6053–6060, 1982.
160. RB Sykes, CM Cimarusti, DP Bonner, K Bush, DM Floyd, NH Georgopadakou, WH Koster, WC Liu, WL Parker, PA Principle, ML Rathnim, WA Slusarchyk, WH Trejo, JS Wells. Nature 291:489–491, 1981.
161. JD Godfrey, RH Mueller, DJ Van Langen. Tetrahedron Lett 27:2793–2796, 1986.
162. A Shanzer, L Somekh, D Butina. J Org Chem 44:3976–3969, 1979.
163. F Monot, Y Bemoit, J Lemal, A Honorat, D Ballerini. In: OM Neijssel, RR ven der Meer, KCAM Luyben, eds. Proc 4th Eur Cong Biotechnol, Amsterdam: Elsevier, 1987, 2:45–45.
164. H Schutte, W Hummel, H Tsai, M-R Kula. Appl Microbiol Biotechnol 22:306–317, 1985.
165. RL Hanson, J Singh, TP Kissick, RN Patel, LJ Szarka, RH Mueller. Bioorg Chem 18:116–130, 1990.
166. K Aida, I Chibata, K Nakayama, K Takinami, H Yamada, eds. Biotechnology of Amino Acid Production. Prog Ind Microbiol, Vol. 24. Amsterdam: Elsevier, 1986.
167. C Syldatk, D Cotoras, G Dombach, C Grob, H Kallwab, F Wagner. Biotechnol Lett 9:25–30, 1987.
168. H Yamada, S Takahashi, Y Kii, H Kumagai. J Ferment Technol 55:484–491, 1978.
169. J Kamphuis, HFM Hermes, JAM van Balken, HE Schoemaker, WHJ Boesten, ME Meijer, M. E. In: G Lubec, GA Rosenthal ed. Amino Acids: Chemistry, Biology and Medicine. Vienna: ESCOM Science Publ., 1190, 119–125.
170. A Morin, W Hummel, M-R Kula. Appl Microbiol Biotechnol 25:91–96, 1986.
171. R Oliveiri, E Fascetti, L Angelini, L Degen. Enzyme Microb Technol 1:201–204, 1979.
172. G-J Kim, H-S Kim. Enzyme Microb Technol 17:63–67, 1995.
173. K Yokozeki, S Nakamori, S Yamanaka, C Eguchi, K Mitsugi, F Yoshinaga. Agric Biol Chem 51:715–719, 1987.
174. Y Nishida, K Nakamichi, K Nabe, T Tosa. Enzyme Microb Technol 9:721–725, 1987.
175. R Tsugawa, S Okumura, T Ito, N Katsuya. Agric Biol Chem 30:27–34, 1966.
176. M Sugie, H Suzuki. Agric Biol Chem 44:1089–1095, 1989.
177. I Chibata, T Tosa, T Sato, T Mori. Meth Enzymol 44:746–759, 1976. (178) HK Chenault, J Dahmer, GM Whitesides. J Am Chem Soc 11:6354–6364, 1989.
178. K Sakai, K Oshima, M Moriguchi. Appl Environ Microbiol 57:2540–2543, 1991.
179. J Kamphuis, EM Meijer, WHJ Boesten, QB Broxterman, B Kaptein, HFM Hermes, HE Schoemaker, H. E. In: D Rozzell, F Wagner, eds. Production of Amino Acids and Derivatives. Münich: Hanser Publishers, 1992, pp 117–200.
180. JD Rozzell. US Pat 4,876,766, 1989.
181. SB Primrose. Eur Patent Appl 84100421.8, 1984.
182. GJ Carlton, LL Wool, MH Updike, L Lanty, JP Hamman. Biotechnology 5:317–325, 1986.
183. G Burnett, C Walch, K Yonaha, S Toyama, K Soda. J Chem Soc Chem Commun 25:826–835, 1979.
184. M Bouclier, MJ Jung, B Lippert. Eur J Biochem 98:363–367, 1979.
185. K Tanizawa, T Yoshimara, Y Asada, S Sawoda, H Misono, K Soda. Biochemistry 21:1104–1109, 1982.
186. DI Stirling. In: AN Collins, GN Sheldrake, J Crosby, eds. Chirality in Industry. New York: John Wiley, 1992, pp 209–222.

187. Y Asano, T Yasuda, Y Tani, H Yamada. Agric Biol Chem 46:1183–1189, 1982.
188. K Bui, A Arnaud, P Galzy. Enzyme Microb Technol 4:195–197, 1982.
189. M Kobayashi, T Nagasawa, H Yamada. Appl Microbiol Biotechnol 29:231–233, 1988.
190. T Nagasawa, H Nanaba, K Ryuno, K Takeuchi, H Yamada, H. Eur J Biochem 162:691–698, 1987.
191. JZ Crich, R Brieva, P Marquart, R-L Gu, S Flemming, CJ Sih. J Org Chem 58:77–82, 1993.
192. M Alpegiani, A Bedeschi, F Giudichi, E Perrone, G Franceschi. J Am Chem Soc 107:6398–6400, 1985.
193. K Fujimoto, Y Iwano, K Hirai, S Sugawara. Chem Pharm Bull 34:999–1004, 1986.
194. T Shibata, K Iino, T Tanaka, T Hashimoto, Y Kameyama, Y Sugimura. Tetrahedron Lett 26:4739–4743, 1985.
195. T Iimori, M Shibazaki. Tetrahedron Lett 26:1523–1526, 1985.
196. T Ohashi, J Hasegawa. In: AN Collins, GN Sheldrake, J Crosby, eds. Chirality in Industry. New York: John Wiley, 1992, pp 269–278.
197. R Labila, C Morin. J Antibiot 37:1103–1107.
198. H Nagai, T Shizawa, K Achiwa, Y Terao. Chem Pharm Bull 41:1933–1938.
199. M Bucciareli, A Forni, I Moretti, F Prati, G Torre. J Chem Soc Perkin Trans 3041–3045, 1993.
200. D Tanner, C Birgersson, HK Dhaliwal. Tetrahedron Lett 31:1903–1908, 1990.
201. BC Challis, JA Challis. Comprehensive Organic Chemistry, Vol. 2. New York: Pergamon Press, 1979, pp 957–985.
202. M Garcia, C del Campo, EF Llama, JM Sanchez-Montero, V Sinisterra. Tetrahedron 49:8433–8437, 1993.
203. Y Takagi, T Itoh. Bull Chem Soc Jpn 66:2949–2954, 1993.
204. J Monteiro, J Braun, F LeGoffic. Synth Commun 20:315–320, 1990.
205. LK-P Lam, JB Jones. Can J Chem 66:1422–1427, 1987.
206. C-S Chen, Y Fujimoto, CJ Sih. J Am Chem Soc 103:3580–3587.
207. F-C Haung, LFH Lee, RSD Mittal, PR Ravikumar, JA Chan, CJ Sih, E Caspi, CR Eck. J Am Chem Soc 97:4144–4149, 1975.
208. H-J Gais, KL Lukas. Chem Int Ed Engl 23:142–147, 1984.
209. MT Davies-Coleman, DEA Rivett. Fortschr Chem Org Naturst 55:1–6, 1989.
210. B Hasse, MP Schneider. Tetrahedron: Asymmetry 4:1017–1021,1993.
211. H Merz, K Stockhaus. J Med Chem 22:1475–1481, 1979.
212. EG Quartey, JA Hustad, K Faber, T Anthonsen. Enzyme Microb Technol 19:361–365, 1996.
213. T Takeuchi, H Iinuma, S Kunimoto, T Masuda, M Ishizuka, H Hamada, H Naganawa, S Kondo, H Umezawa. J Antibiot 34:1619–1621, 1981.
214. H Umezawa, S Kondo, H Iinuma, Y Kunimoto, H Iwasawa, D Ikeda, T Takeuchi. J Antibiot 34:1622–1624, 1981.
215. S Kondo, H Iwasawa, D Ikeda, Y Umeda, H Ikeda, H Inuma, H Umezawa. Antibiotics 34:1625–1627, 1981.
216. H Iwasawa, S Kondo, D Ikeda, T Takeuchi, H Umezawa. J Antibiot 35:1655–1669, 1982.
217. Y Umeda, M Moriguchi, I Katsushige, H Kuroda, T Nakamura, A Fujii, T Takeuchi, H Umezawa. J Antibiot 40:1316–1324, 1987.
218. RN Patel, A Banerjee, LJ Szarka. Tetrahedron: Asymmetry 8:1767–1771, 1997.
219. K Maeda, Y Umeda, T Saino. Ann N Y Acad Sci 685:123–125, 1993.
220. JH Fried, IT Harrison. Br Pat 1,211,134, 1967.
221. IT Harrison, B Lewis, P Nelson, W Rooks, A Roszwski, A Tomolonis, JH Fried. J Med Chem 13:203–208, 1970.
222. C Giordano, M Villa, S Panossian. In: AN Collins, GN Sheldrake, J Crosby, eds. Chirality in Industry. New York: John Wiley, 1992, pp 303–312.
223. QM Gu, RR Reddy, CJ Sih. Tetrahedron Lett 27:5203–5206, 1986.
224. E Battistel, D Bianchi, P Cesti, C Pina. Biotechnol Bioeng 38:659–664, 1991.
225. N Layh, A Stolz, J Bohme, F Effenberger, H-J Knackmuss. J Biotechnol 33:175–182, 1994.

226. K Yamamoto, Y Uneo, K Otsubo, K Kawakami, K Komatsu. Appl Environ Microbiol 56: 3125–3129, 1990.
227. S-W Tsai, H-J Wei. Enzyme Microb Technol 16:328–333, 1994.
228. M Trani, A Ducret, P Pepin, R Lortie. Biotechnol Lett 17:1095–1098, 1995.
229. MC de Zoete, CN Kock-van Dalen, F van Rantwijk, RA Sheldon. Biocatalysis 10:307–316, 1994.
230. G Hedstrom, M Backlund, JP Slotte. Biotechnol Bioeng 42:618–624, 1993.
231. QM Gu, C-H Chen, CJ Sih. Tetrahedron Lett 27:1763–1767, 1986.
232. DJ Hardman. Crit Rev Biotechnol 11:1–40, 1991.
233. PT Barth, L Bolton, JC Thomson. J Bacteriol 174:2612–2619, 1992.
234. S Fetzner, F Lingens, F. Microbiol Rev 58:641–685, 1994.
235. M Mansuri, JC Martin. Ann Rep Med Chem 22: Chap. 15, 1987.
236. M Mahmoudian, BS Baines, CS Drake, RS Hale, P Jones, JE Piercey, DS Montgomery, IJ Purvis, R Storer, MJ Dawson, GC Lawrence. Enzyme Microb Technol 15:749–754, 1993.
237. RA Milton, S Brand, MF Jones, CM Rayner. Tetrahedron Lett 36:6961–6964, 1995.
238. S Niwayama, S Kobayashi, M Ohno. J Am Chem Soc 116:3290–3295, 1994.
239. R Vince, M Hua. J Med Chem 33:17–24, 1990.
240. H Nakano, Y Okauyama, K Iwasa, H Hongo. Tetrahedron: Asymmetry 5:1155–1160.
241. AC van der Linden. Biochim Biophys Acta 77:157–162, 1963.
242. SW May, RB Schwartz, BJ Abbott, OR Zaborsky. Biochim Biophys Acta 403:245–250, 1975.
243. BJ Abbott, CT Hou. Appl Microbiol 26:86–90, 1973.
244. M-J deSmet, J Kingma, H Wynberg, B Witholt. Enzyme Microb Technol 5:352–356, 1983.
245. M Takagi, N Uemura, K Furuhashi. Ann N Y Acad Sci 613:697–702, 1990.
246. K Furuhashi. In: AN Collins, GN Sheldrake, J Crosby, eds. Chirality in Industry. New York: John Wiley and Sons, 1992, pp 167–188.
247. SM Roberts. In: MP Schneider, ed. Enzymes as Catalysts in Organic Synthesis. Boston: D Reidel, Publishing Company, 1985, pp 55–75.
248. MJ Lenn, CJ Knowles. Enzyme Microb Technol 16:964–969, 1994.
249. NF Shiptston, MJ Lenn, CJ Knowles. J Microb Meth 15:41–52, 1992.
250. KA Babaik, JS Ng, JH Dygos, CL Weyker, Y-F Wang, C-H Wong. J Org Chem 55:3377–3382, 1990.
251. DL Hughes, JJ Bergan, JS Amato, M Bhupathy, JL Leazer, JM McNamara, DR Sidler, PJ Reider, EJJ Grabowski. 1990. J Org Chem 55: 6252–6258, 1990.
252. F Theil, H Schick, G Winre, G Reck. Tetrahedron 47:7569–7574, 1991.
253. M Tanake, M Yoshioka, K Sakai. Tetrahedron: Asymmetry 4:981–985, 1993.
254. AD Borthwick, K Biggadike. Tetrahedron 48:571–575, 1992.

5
Stereoselective Hydroxylation Reactions

Herbert L. Holland
Brock University, St. Catharines, Ontario, Canada

I. INTRODUCTION

A. Hydroxylation

The conventional biocatalytic hydroxylation process involves the direct oxidation of a C-H bond to produce an alcohol (Fig. 1) and may be distinguished from other biocatalytic reactions in that no analogous single-step chemical process can compete with the biocatalytic reaction in terms of the regio- or stereoselectivity of product formation. This reaction, first applied to the production of the corticosteroids in the 1950s via the hydroxylation of progesterone at C-11α by *Rhizopus* species [1], was one of the earliest biotransformation processes to be commercially exploited. Improved versions of the original process remain in use to this day and have been supplemented during the intervening period by many other examples involving different substrates. The value of stereoselective biocatalytic hydroxylation as a method for the preparation of chiral alcohols by the direct oxidation of a specific C-H bond of the substrate has ensured continued research in this area, and this chapter will cover significant developments during the last decade in the field of hydroxylation reactions that produce stereochemically defined products from achiral substrates. During this period over 700 references dealing with biocatalytic hydroxylations have appeared in the open literature alone. Of necessity, therefore, this chapter will be selective in its coverage, but more comprehensive summaries are available: the hydroxylation reaction has been included in several general reviews of biocatalytic reactions [2–4] and has been the subject of a book dedicated to biotransformations with oxidative enzymes [5].

1. Hydroxylase Enzymes

The vast majority of enzyme-catalyzed stereoselective hydroxylations are carried out by the cytochrome P450-dependent monooxygenases [4]. The membrane-bound nature of the majority of these enzymes, together with their functional dependence on the presence of cofactors and their related electron transport proteins, has ensured that preparative biocatalytic hydroxylations are most usefully performed with whole-cell catalysts. The exceptions to this generality, the lipoxygenases and haloperoxidases, will be considered in Sec. II.B.

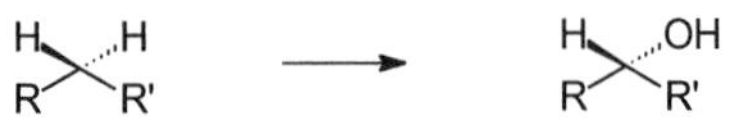

H H → H OH
R R'   R R'

Figure 1 Stereoselective hydroxylation of a methylene group.

2. Regio- and Stereoselectivity of Hydroxylation

The restriction to whole-cell biocatalysis inherent in the majority of preparative hydroxylation reactions means that, for the most part, the enzymes that perform these reactions are at best poorly characterized, and are often unidentified. This implies, in turn, that data on the substrate requirements and the selectivity of reaction of these enzymes are unavailable, a situation that often leads to the use of extensive screening for a biocatalyst suitable for the desired hydroxylation of a particular substrate. An alternative approach, the analysis of the empirical data produced by the hydroxylation of a series of related substrates by a single microorganism, has resulted in the development of predictive models for hydroxylations by several microbial biocatalysts that have recently been reviewed [6]. Hydroxylations that occur at carbons remote from any functionality of the substrate are among the most useful of such reactions and will be considered in Secs. II.A.1 to II.A.3. Microbial hydroxylations at activated carbon positions, such as those α to heteroatoms, or located adjacent to π-electron density by virtue of their allylic or benzylic positions within the substrate, will be considered in Sec. II.A.4.

B. Dihydroxylation

The process of biocatalytic dihydroxylation, the conversion of a carbon–carbon double bond to a saturated vicinal diol, is summarized in Fig. 2. This reaction, when carried out on an aromatic substrate, may be catalyzed by a monooxygenase enzyme and proceed via an arene oxide (epoxide) intermediate to give a *trans*-diol, but is exclusively the domain

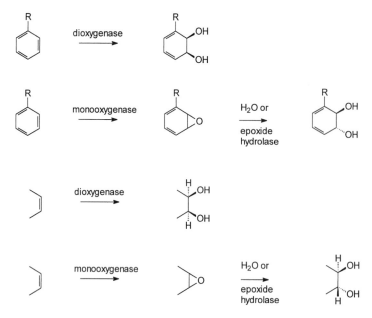

Figure 2 Stereoselective dihydroxylations.

of the arene dioxygenase enzymes when it results in the formation of a vicinal *cis*-dihydrodiol [4,7]; the latter process will be covered in Sec. III.A. The conversion of an alkene substrate to a vicinal *trans*-diol, covered in Sec. III.B, involves monooxygenase enzyme activity and proceeds via an epoxide intermediate, the hydrolysis of which to give the diol may be nonenzymic, enzyme-catalyzed, or a combination of both [4]. Like the hydroxylation reaction discussed in Sec. I.A.1, preparative biocatalytic dihydroxylations by either route are restricted for reasons of enzyme intractability and cofactor requirements to whole-cell processes.

1. Dioxygenase Enzymes

The dioxygenase enzymes such as toluene dioxygenase (TDO), naphthalene dioxygenase (NDO), and biphenyl dioxygenase (BPDO), found exclusively in prokaryotic microorganisms, possess the remarkable ability to oxidize the carbon–carbon double bond of an aromatic substrate to produce a vicinal *cis*-diol (Fig. 3). This reaction, normally part of the pathway for the metabolism of aromatic substrates by such microorganisms, may be exploited for the production of dihydroarene *cis*-diols when the diol dehydrogenase enzyme responsible for the next step of the metabolic pathway, the oxidation of the dihydroarene *cis*-diol to a catechol, is suppressed or absent. This scenario may be realized either by the use of a mutant version of the parent microorganism, or by the use of a host microorganism in which the dioxygenase enzyme is expressed in the absence of the enzymes for the remainder of the metabolic pathway. Dioxygenase-catalyzed reactions carried out by both types of biocatalyst have been recently reviewed [7].

2. Regio- and Stereoselectivity of Dihydroxylation

The regio- and stereoselectivity observed during oxidation of a wide range of substituted aromatic substrates by microorganisms expressing TDO, NDO, or BPDO have been discussed in detail [7]: in summary, TDO oxidations of monosubstituted benzenes give the product shown in Fig. 3. For di- and polysubstituted substrates, the regio- and stereose-

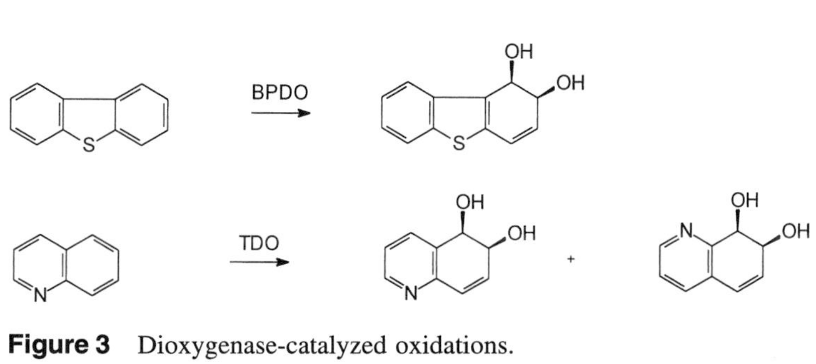

Figure 3 Dioxygenase-catalyzed oxidations.

lectivity of product formation is generally controlled by the largest substituent present on the ring. Oxidations by NDO and BPDO tend to occur at the 1,2 or equivalent position of polycyclic substrates, as shown in Fig. 3, whereas oxidations of heteroarenes show a preference for reaction in the carbocyclic over the heterocyclic ring.

The regioselectivity of alkene epoxidation leading to diol formation has not been examined in sufficient detail for predictive generalizations to be made, but for linear terpenoid substrates a preference for the terminal olefinic bond is often demonstrated, as discussed in Sec. III.C [8].

II. HYDROXYLATION REACTIONS

A. Whole-Cell-Catalyzed Hydroxylations

1. Hydroxylation at Methyl Carbon

Microbial hydroxylation of a methyl group has the potential to be stereoselective in cases where the substrate possesses enantiotopic CH_3 substituents. A classic example of such a process is the conversion of isobutyric acid to β-hydroxyisobutyric acid (Fig. 4), where the use of *Candida rugosa* IFO 0750 leads to formation of the D-(−) (*R*) isomer [9], and the L-(+) (*S*) product is obtained from oxidation using *Bullera alba* IFO 1030 [10]. Similar stereoselectivity is also observed in the oxidation of homologous acids and hydrocarbons by *Rhodococcus* species [11], and in the oxidation of cumene (**1**) to (*R*)-2-phenylpropionic acid (**2**) by *Pseudomonas oleovorans* NRRL B-3429 (Fig. 5) [12].

2. Hydroxylation at Nonactivated Methylene Carbon

Stereoselective microbial hydroxylations occurring at methylene groups of chiral substrates are frequently reported [4]. The present discussion will focus on the more unusual situation of stereoselective hydroxylations that occur at methylene groups of achiral, prochiral, or racemic substrates, resulting in the introduction of chirality to the product.

In the area of terpenoid synthesis, the production of simple chiral starting materials from achiral precursors has been examined by several groups. *Escherichia coli* containing the cloned genes for the expression of the soluble enzyme cytochrome $P450_{CAM}$ from *Pseudomonas putida* has been used for the conversion of 2,2-dimethylcyclohexanone **3** (Fig. 6) to the (*S*)-alcohol **4** in moderate yield but high enantiomeric purity [13], and 1,8-cineole (**5**) is transformed by *Bacillus cereus* to the optically pure 6-(*R*) product **6** (Fig. 7) [14]. The hydroxylation of racemic dihydro-β-campholenolactone **7** (Fig. 8) by *Fusarium culmorum* shows a lower degree of stereoselectivity, the predominant enantiomer of the product **8** being obtained in only 20% diastereomeric excess [15].

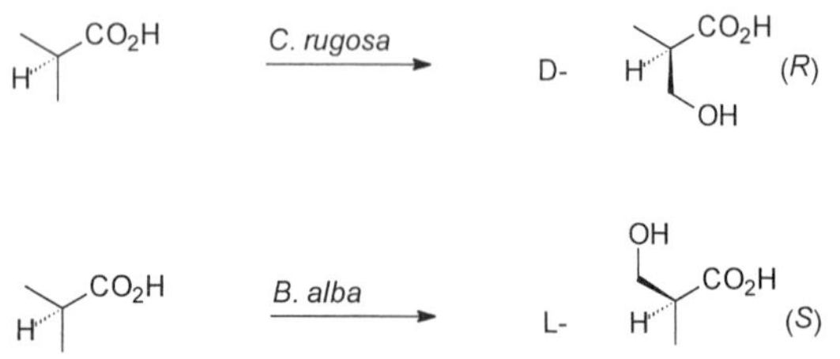

Figure 4   Stereoselective hydroxylations of isobutyric acid.

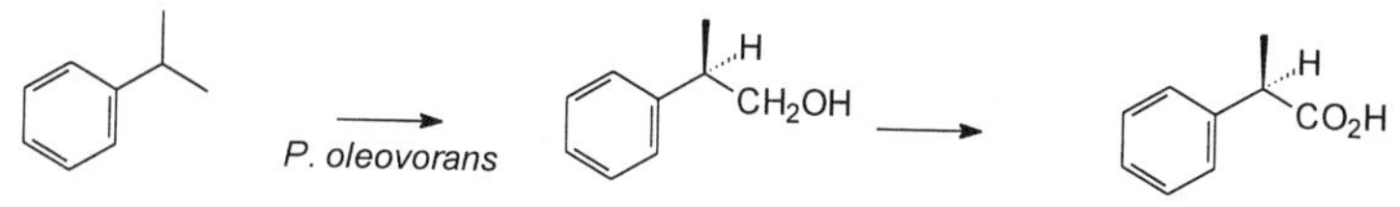

1 2

Figure 5 Stereoselective hydroxylation of cumene.

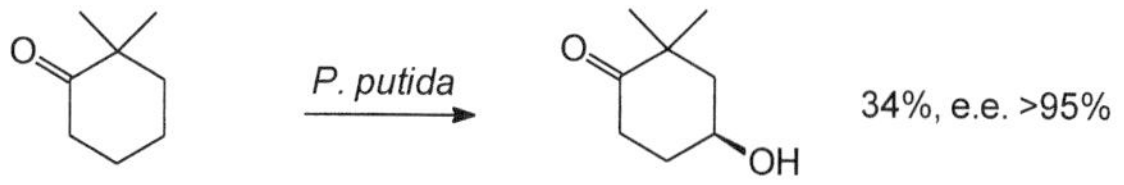

3 4

Figure 6 Stereoselective hydroxylation of 2,2-dimethylcyclohexanone.

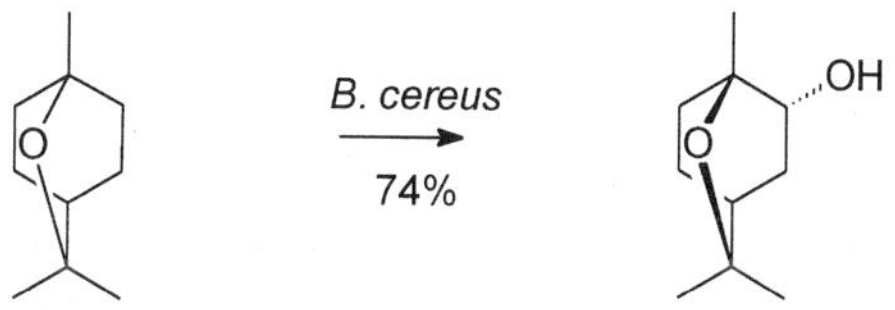

5 6

Figure 7 Hydroxylation of 1,8-cineole.

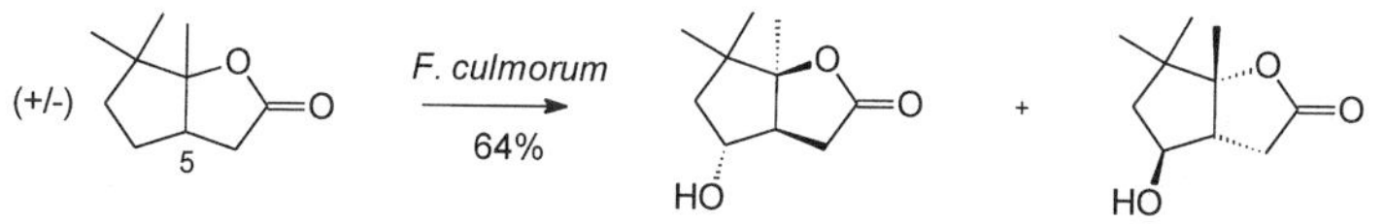

7 8

Figure 8 Hydroxylation of dihydro-β-campholenolactone.

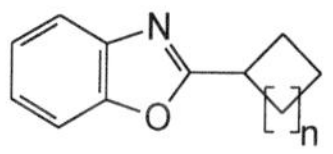

9, n = 1-5

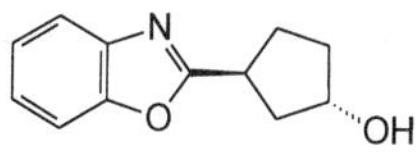

10

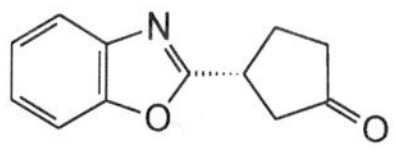

11

Figure 9 Hydroxylation of 2-cycloalkylbenzoxazoles.

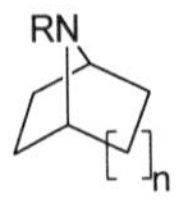

12, n = 0-2, R = sulfonamide, amide or carbamate protection

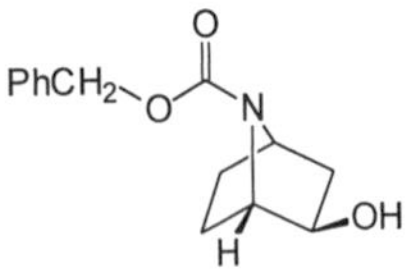

13

Figure 10 Hydroxylation of azabicycloalkanes.

The hydroxylation of a series of 2-cycloalkylbenzoxazoles (**9**) (Fig. 9) by *Cunninghamella blakesleeana* DSM 1906 and *Bacillus megaterium* DSM 32 has been examined as a potential route for the formation of chiral hydroxycycloalkylcarboxylic acids [16–19]. Products were generally obtained in moderate enantiomeric purity, but in several examples, notably the cyclopentanol **10** and its metabolic product **11**, both obtained from biotransformation of **9** (*n* = 2) with *C. blakesleeana*, enantiomeric purities of up to 95% were observed. In a search for a chiral precursor of the potent analgesic epibatidine, the hydroxylation of a series of azabicycloalkanes (**12**) (Fig. 10) has been examined using *Beauveria bassiana* ATCC 7159 [20,21], *Aspergillus ochraceus* NRRL 405 [20], *Rhizopus nigricans* UC-4285 [20], and *Rhizopus arrhizus* ATCC 11145 [20]. Although the products of methylene hydroxylation were in many cases chiral, none was obtained in higher enantiomeric purity than the 57% observed for the 1-(*S*)-*endo*-alcohol **13** obtained in low yield (8%) from *R. nigricans* [20].

A series of linear, terminally substituted alkyl derivatives **14** (Fig. 11) is converted to the corresponding (*R*)-carbinols in low isolated yield (8–19%) but high enantiomeric purity by *Helminthosporium* sp. NRRL 4671 [22].

3. Hydroxylation at Methine Carbon

Stereoselective hydroxylation at methine carbons in other than chiral substrates has rarely been reported, but the hydroxylation of the racemic lactone **7** (Fig. 8) at C-5 by *Fusarium culmorum* may be a reaction of this type. Transformation of (±)-**7** proceeds with a low degree of selectivity for the (+) 1(*S*),5(*R*) substrate, but the extent to which is this due to the formation of the major product **8** (Fig. 8), rather than hydroxylation at C-5, is not clear.

4. Hydroxylation at Activated Positions

(*a*) *Hydroxylation at Allylic Carbon.* Carbon atoms allylic or vinylogous to double bonds are often susceptible to microbial hydroxylation, and chiral products derived from such reactions have application as synthetic building blocks for the preparation of terpenes

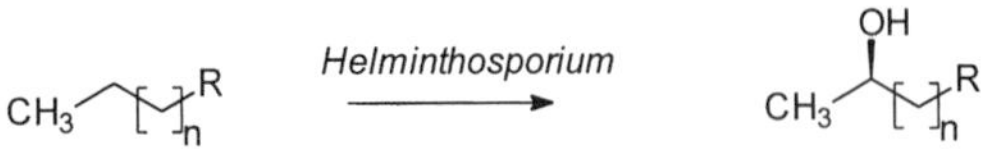

14, n = 4-8, R = CN or COCH$_3$

Figure 11 Hydroxylation of cyanoalkanes and ketones by *Helminthosporium*.

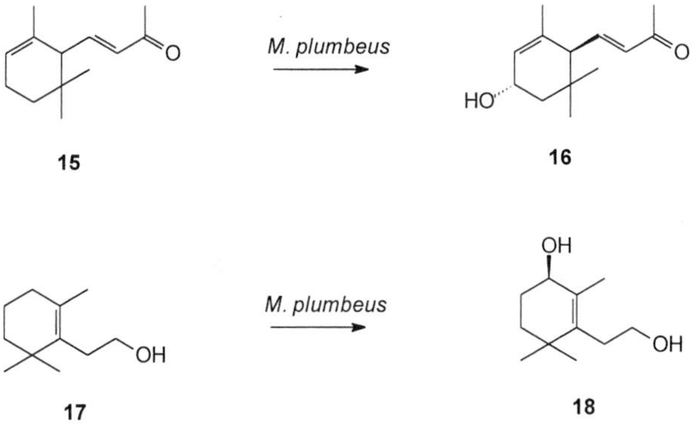

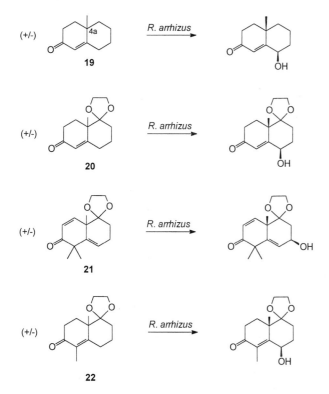

Figure 12 Formation of chiral allylic alcohols by *M. plumbeus*.

and other target molecules. The hydroxylation of (±)-α-ionone (**15**) by *Mucor plumbeus* shown in Fig. 12 to give the *trans*-allylic alcohol **16** as the major product [23], and the conversion of alcohol **17** (Fig. 12) to the allylic alcohol **18** by the same microorganism [24], are both examples of this application. In the latter case, alcohol **18** could be obtained as an enantiomerically pure product. In contrast, allylic hydroxylations of the racemic bicyclic enones **19** [25] and **20–22** [26,27] by *Rhizopus arrhizus* ATCC 11145 and that of **22** by *Aspergillus niger* ATCC 9142 [27] exhibited only a slight preference for the configuration at C-4a shown in Fig. 13, with the enantiomeric purity of the products being

Figure 13 Enantioselective hydroxylations of bicyclic enones.

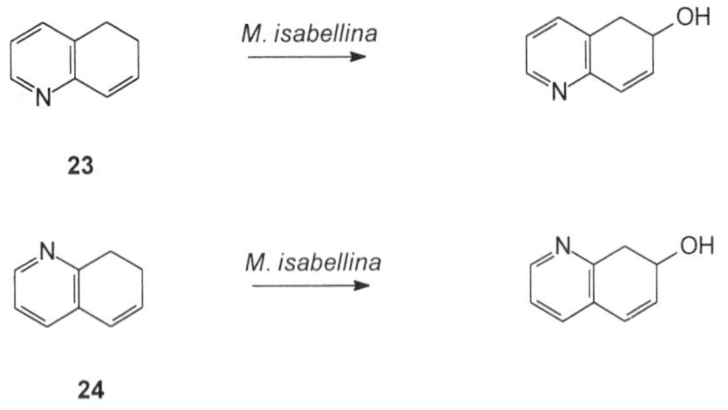

Figure 14   Allylic hydroxylations of dihydroquinolines by *M. isabellina*.

low in all cases. Allylic hydroxylation products of unknown absolute configuration were also obtained in low enantiomeric purity from the biotransformation of the dihydroquinolines **23** and **24** with *Mortierella isabellina* NRRL 1757 as shown in Fig. 14 [28].

(*b*)   *Hydroxylation α to Heteroatoms.*   The oxidative removal of alkyl groups from ethers and amines via hydroxylation α to the heteroatom and hydrolytic breakdown of the resulting hemiacetal or hemiaminal is a well-established feature of hydroxylase enzyme activity [4,5]. The reaction is frequently encountered in the context of drug metabolism, but stereoselective aspects of the process are not often expressed. Two recently reported exceptions are the enantioselective conversion of (±)-eudesmin (**25**) to (−)-pinoresinol (**26**) by O-demethylation using *Aspergillus niger* (Fig. 15) [29], and the production of (*R*)-L-carnitine in 92% enantiomeric excess and 46.5% yield via enantioselective metabolism of the racemate **27** by *Acinetobacter calcoaceticus* ATCC 39647 as shown in Fig. 16 [30].

(*c*)   *Benzylic Hydroxylation.*   Like the hydroxylations at allylic positions discussed in Sec. II.A.4.a, microbial hydroxylation at benzylic carbon is also a favored process, and often proceeds with high enantioselectivity. The chiral benzylic alcohols **29** (e.e. 97%) and **30** (e.e. 73%) (Fig. 17), which possess antidepressant and cerebral disorder–improving properties, have been prepared from **28** by biotransformations using *Streptomyces lavendulae* SANK 64687 [31] and *Cunninghamella echinulata* IFO 4443 [32], respectively, and (*R*)-4-chromanol (**31**) is produced in high (84%) enantiomeric excess by the biotransformation of chroman with *Mortierella isabellina* ATCC 42613 [33] (Fig. 18). Benzylic hydroxylation of the dihydroquinolines illustrated in Fig. 14 by *M. isabellina* also gave benzylic alcohol products, but of undetermined enantiomeric purity [28]. *M. isabellina* ATCC 42613 has also been used for the benzylic hydroxylation of a series of benzyl-substituted cycloalkanes and in one case (benzyl cyclobutane, Fig. 19, *n* = 1) gave the corresponding (*R*)-benzylic alcohol in more than 95% optical purity [34]. The hydroxylation of 2-ethylbenzoic acid by *Pseudomonas putida* ATCC 12633 or *Aspergillus niger* IFO 6661, shown in Fig. 20, was found to be enantiospecific, producing as a final product the (*S*)-methylphthalide **32** in yields of 80% and 12%, respectively [35].

Benzylic alcohols of high enantiomeric purity can also be obtained by biotransformation of suitable substrates with microorganisms expressing dioxygenase enzyme activity, illustrated in Fig. 21 for the biotransformations of the benzocycloheptene **33** [36] and the dihydronaphthalenes **34** and **35** [37] by *P. putida* UV4, the products of which were obtained as enantiopure materials with (*R*) configuration. In contrast, the hydroxylations of 1- and 2-indanone by either TDO- or NDO-expressing strains of *P. putida*, shown in Fig. 22, gave materials of lower enantiomeric purity, products from the latter substrate being formed via rearrangement of an intermediate benzylic alcohol [38]. The opposite stereo-

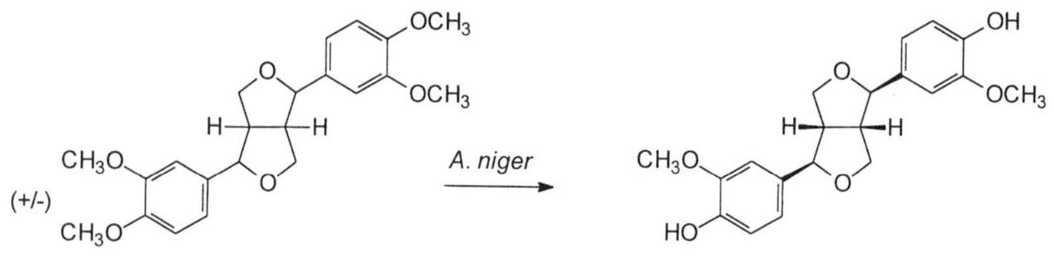

25   **26**

Figure 15   Enantioselective demethylation of eudesmin.

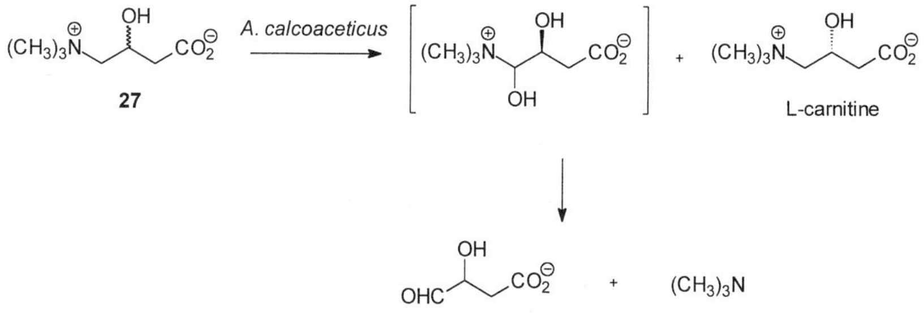

Figure 16   Enantioselective degradation of ($\pm$)-carnitine.

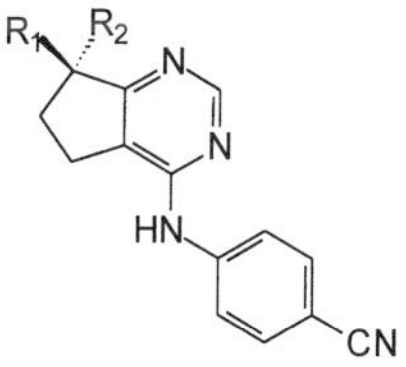

28 $R_1 = R_2 = H$

29 $R_1 = H, R_2 = OH$

30 $R_1 = OH, R_2 = H$

Figure 17   Benzylic hydroxylation of a pyrimidinoindan.

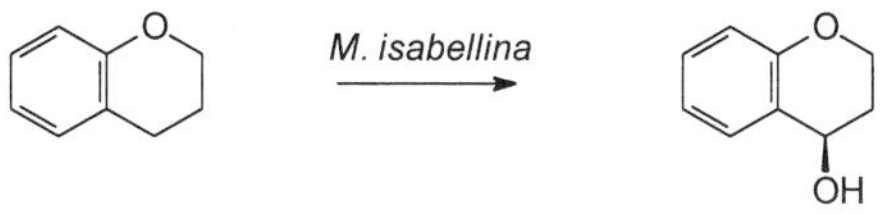

31

Figure 18   Benzylic hydroxylation of chroman by *M. isabellina*.

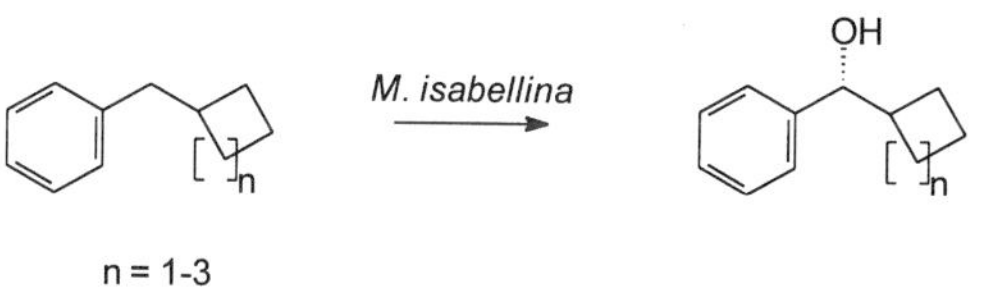

n = 1-3

Figure 19   Benzylic hydroxylations of benzylcycloalkanes by *M. isabellina*.

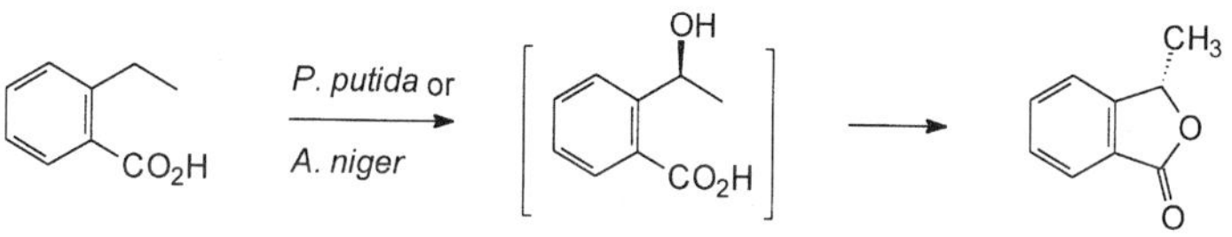

32

Figure 20   Stereoselective hydroxylation of 2-ethylbenzoic acid.

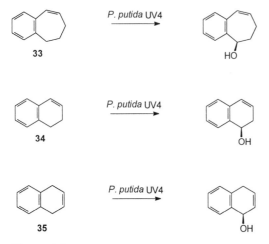

Figure 21 Stereoselective benzylic hydroxylations of benzocycloalkenes by *P. putida*.

specificity was shown in the oxidations of indan and indene by the NDO-expressing *P. putida* NCIMB 9816/11, which gave the (*S*)-alcohols shown in Fig. 23 [39], but *P. putida* UV4, which expresses TDO, has been used for the conversion of the 2-substituted indans **36** to the corresponding *cis*-1-indanols (Fig. 24) [40], and gave the products as single enantiomers with a configuration at the benzylic center consistent with that obtained for other benzylic hydroxylations involving this enzyme [36,37].

B. Hydroxylations with Isolated Enzymes

Although a significant number of hydroxylases, particularly cytochrome P450–dependent enzymes from mammalian sources, has been isolated and characterized, for the reasons discussed in Sec. I.A.1 very few such enzymes are of practical value as catalysts for stereoselective hydroxylation reactions. However, two classes of readily available enzymes do have the potential to be useful catalysts: neither the lipoxygenases or the peroxidases

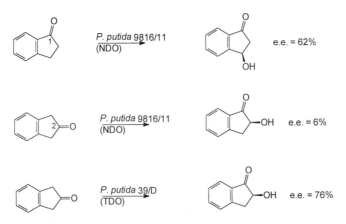

Figure 22 Hydroxylation of indanones by *P. putida*.

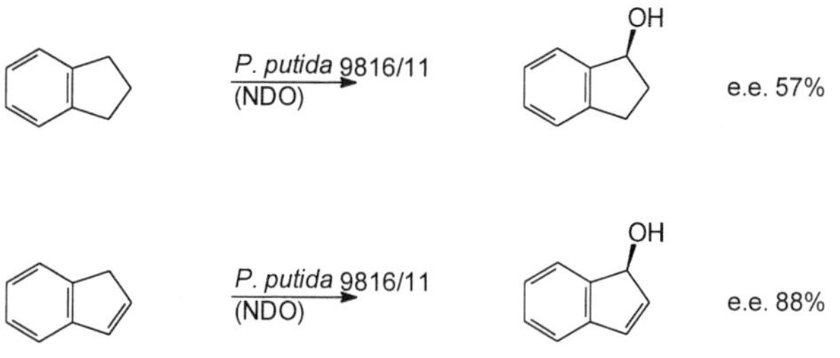

Figure 23 Benzylic hydroxylations of indan and indene by *P. putida*.

require additional cofactors for activity, and both classes of enzyme have been used for the stereoselective hydroxylation of appropriate substrates.

1. Lipoxygenases

The enzyme soybean lipoxygenase (SBLO) is a nonheme iron-containing dioxygenase that catalyzes the incorporation of molecular oxygen into polyunsaturated fatty acids according to the general equation shown in Fig. 25. Chemical reduction of the resulting hydroperoxide gives the corresponding allylic alcohol in high enantiomeric purity. The natural substrate for this enzyme is linoleic acid, but the enzyme can accept a number of different substrates provided that they contain a Z,Z-1,4-diene unit with a substituent (R, Fig. 25) of 3–10 carbons, which may also contain alcohol or keto functionality, and a carboxylic acid group appropriately spaced on the other side of the diene unit [41,42]. With these restrictions, the synthetic utility of SBLO in the production of a range of chiral alcohols has been amply demonstrated, although the regioselectivity of the reaction for the two distal vinylic carbons (a or b, Fig. 25) was found to vary significantly with both the nature of the substituent R and the length of the chain leading to the carboxylic acid group [42].

Oxidation with SBLO has been used to introduce both chiral allylic alcohol groups in a synthesis of the naturally occurring eicosanoid methyl 5(S),15(S)-diHETE **37** (Fig. 26), using a strategy of directing the regioselectivity of oxygenation by manipulation of the polar versus hydrophobic ends of the substrate chain [43]. This enzyme has also been used for oxidation of its natural substrate, linoleic acid, in a large-scale preparation of the 13(S)-hydroperoxide **38** [44] and in the preparation of (S)-2-nonen-4-olide (**39**) obtained by steam distillation of the SBLO-catalyzed reaction mixture [45] (Fig. 27). In a practical application on a larger scale, the hydrolysis of safflower oil (containing 76% linoleic acid) by a crude preparation of SBLO at substrate concentrations of up to 40 g/L can be used to produce an 80% yield of the hydroperoxide **38** (Fig. 27) [46].

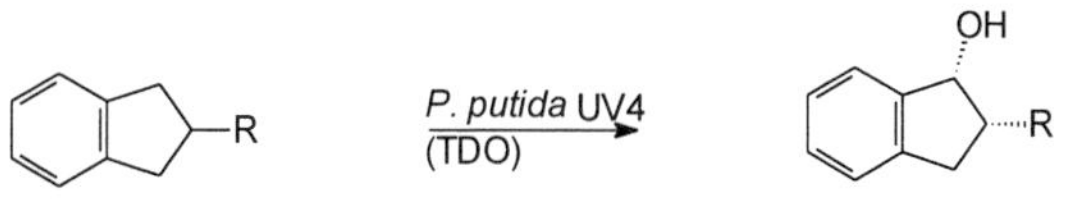

36

R = CH$_3$, N$_3$, Br, OH

Figure 24 Benzylic hydroxylations of 2-substituted indans by *P. putida*.

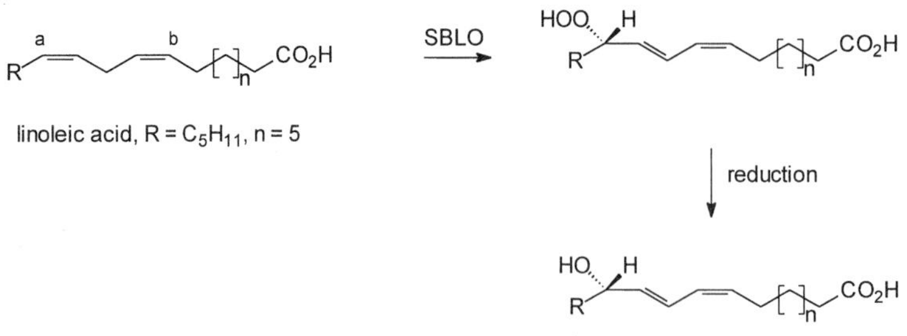

Figure 25 Soybean lipoxygenase-catalyzed oxidation of 1,4-dienes.

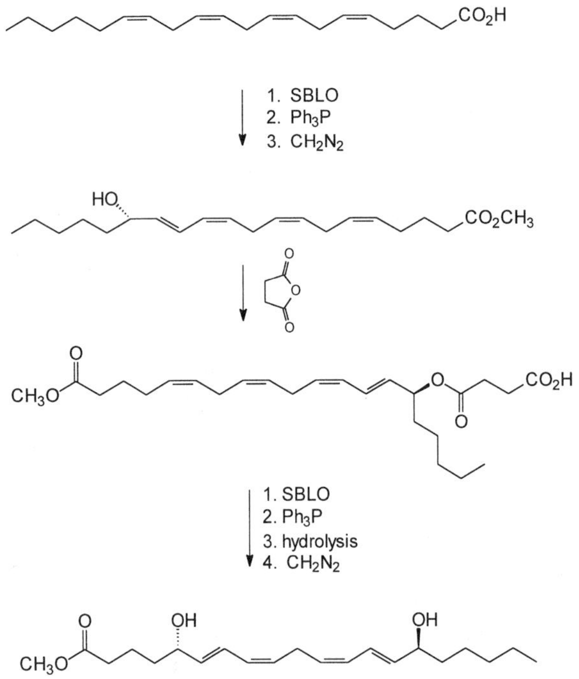

37

Figure 26 Synthesis of methyl 5(*S*),15(*S*)-HETE.

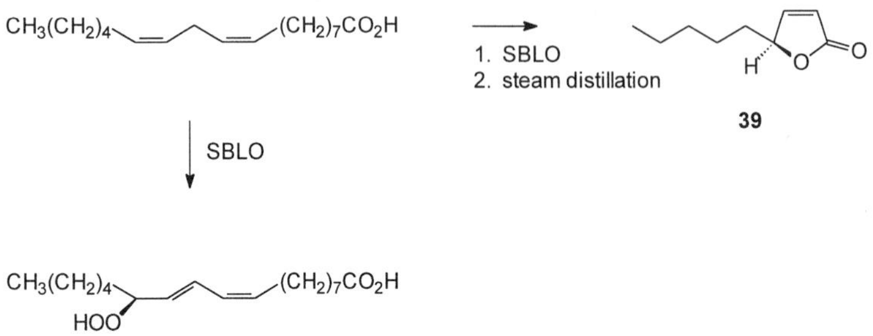

38

Figure 27 Oxidations of linoleic acid by soybean lipoxygenase.

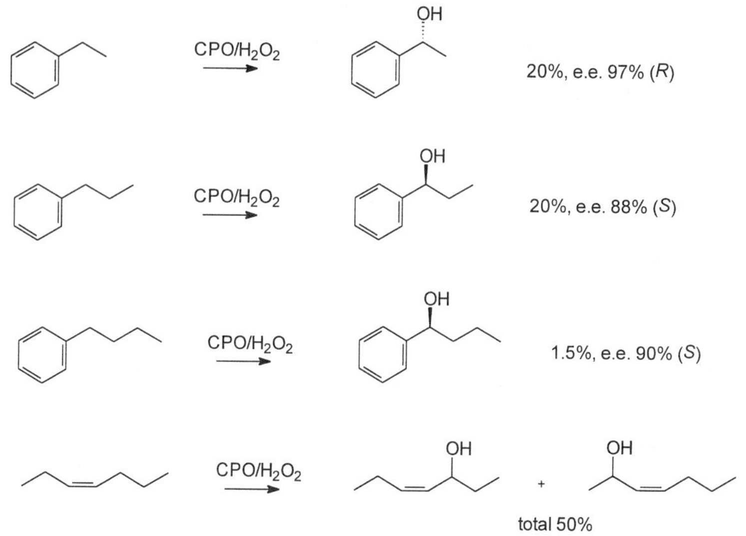

Figure 28 Benzylic and allylic hydroxylations catalyzed by chloroperoxidase.

2. Peroxidases

The heme-iron-containing peroxidases constitute a class of enzymes that oxidize organic substrates using a peroxide, usually hydrogen peroxide, as oxidant. Although several enzymes of this class are capable of the hydroxylation of suitably activated substrates [4], only one, chloroperoxidase from *Caldariomyces fumago*, currently shows significant promise as a catalyst for stereoselective hydroxylations. This enzyme, using hydrogen peroxide as oxidant, has been shown to catalyze the stereoselective benzylic hydroxylations summarized in Fig. 28 [47]. The chloroperoxidase/hydrogen peroxide combination can also perform allylic hydroxylation, notably the conversion of (*Z*)-3-heptene to the isomeric alcohols shown in Fig. 28, but with unknown stereoselectivity [47].

III. DIHYDROXYLATION REACTIONS

A. Dihydroxylation of Arene Double Bonds

The monooxygenase-catalyzed oxidation of arenes to arene oxides, resulting in the formation of *trans*-dihydrodiols via epoxide hydrolysis as outline in Fig. 2, has received much attention in view of its role in the metabolism of aromatic compounds [5] but has not been extensively developed as a means for the production of chiral dihydroarene *trans*-diols. In contrast, the dioxygenase-catalyzed oxidation of arenes illustrated in Fig. 2 has been extensively used for the preparation of chiral dihydroarene *cis*-diols. A recent review [7] tabulates over 120 different enantiopure products obtained from the oxidations of mono-, di-, and polycyclic arene substrates, together with many other examples obtained in lesser enantiomeric purity. These biotransformations are performed by microorganisms that express the TDO, NDO, or BPDO enzymes, and occur subject to the regio- and stereoselective preferences outlined in Sec. I.B.2.

The value of *cis*-dihydroarenediols as chiral building blocks for organic synthesis has led to systematic study of their large-scale production and to the investigation of methods for the production of chiral *cis*-diols from a wide range of aromatic substrates.

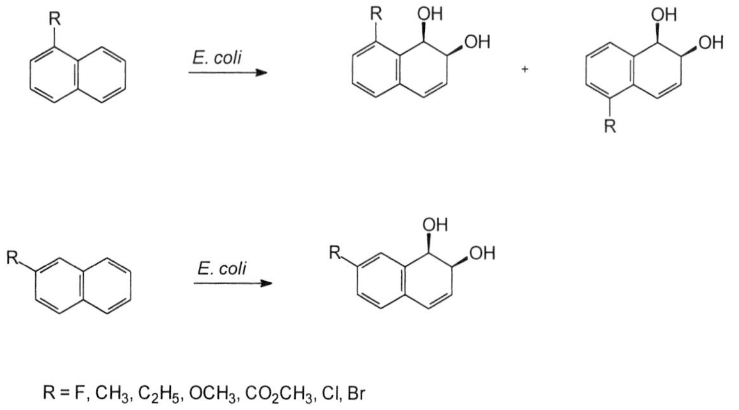

R = F, CH$_3$, C$_2$H$_5$, OCH$_3$, CO$_2$CH$_3$, Cl, Br

Figure 29   Formation of substituted naphthalene-1,2-dihydrodiols.

The conversion of toluene to the *cis*-dihydrodiol has been examined using immobilized *P. putida* UV4, and a dramatic increase in production observed under defined conditions of aeration in a fluidized-bed reactor [48]. The naphthalene dioxygenase enzyme from *Pseudomonas fluorescens* has been expressed in *E. coli*, and the resulting biocatalyst used for the conversion of a series of substituted naphthalenes to the diols shown in Fig. 29 [49], while *Pseudomonas testosteroni* AC3 converts 2-naphthoic acid to the diol **40** shown in Fig. 30 [50]. Substituted biphenyls are also substrates for TDO oxidation (Fig. 31) using *E. coli* JM109 expressing this enzyme [51].

The products from oxidation of halobenzenes with *P. putida* 39/D or the recombinant *E. coli* JM109 expressing TDO activity have been used as starting materials for many chemical syntheses, with more recent examples including the preparations of the fluoro-inositols **41** and **42** [52], 2-deoxy-2-fluoro-D-glucose (**43**) [53], and the protected 5-deoxy-5-fluoro-manno-γ-lactol (**44**) [53], shown in Fig. 32. The *cis*-diol obtained from chlorobenzene has also been used as starting materials for the syntheses of all four stereoisomers of sphingosine (**45**) (Fig. 33) [54], and *cis*-diols derived from 2-phenylethyl bromides (Fig. 34) have potential applications as chiral synthons in the preparation of morphine [55,56].

B.   Dihydroxylation of Heteroarene Double Bonds

The dioxygenase-catalyzed oxidation of heterocyclic arenes occurs predominantly in a carbocyclic ring of the substrate [7], exemplified by the conversion by *P. putida* UV4 of quinoline, isoquinoline, quinoxaline, and quinazoline to the *cis*-diol products shown in Fig. 35 [57], but a recent report of the formation of *cis*-diol **46** (Fig. 36) by oxidation of 2-chloroquinoline by *P. putida* [58] suggests the possibility that further products of this type may become available as new chiral starting materials for synthetic applications.

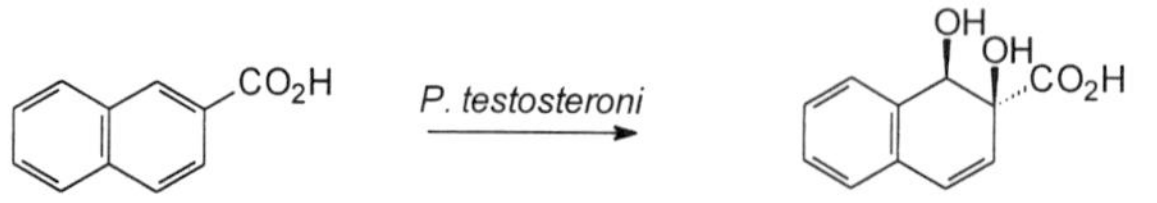

Figure 30   Dioxygenase-catalyzed oxidation of 2-naphthoic acid.

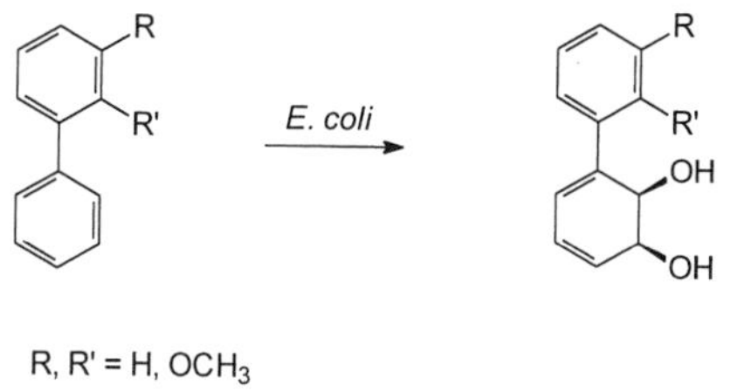

R, R' = H, OCH₃

Figure 31   Dixoygenase-catalyzed oxidation of biphenyls.

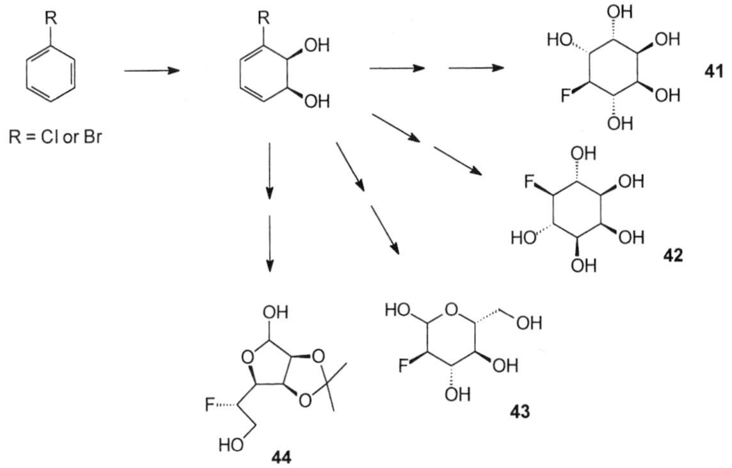

R = Cl or Br

Figure 32   Preparation of fluorinated carbohydrates from dihydrohalobenzene *cis*-diols.

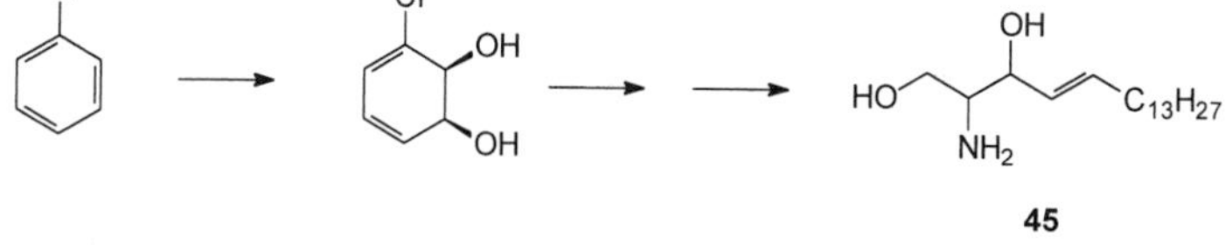

Figure 33   Preparation of sphingosine diastereomers from dihydrochlorobenzene *cis*-diol.

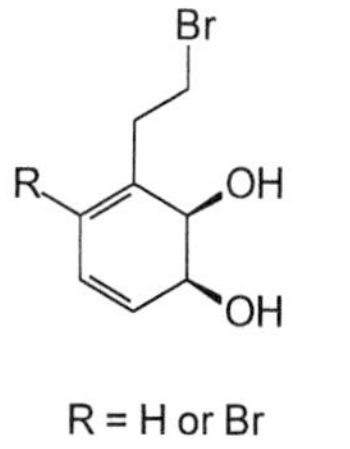

R = H or Br

Figure 34   *cis*-Dihydrodiols from 2-phenylethyl bromide.

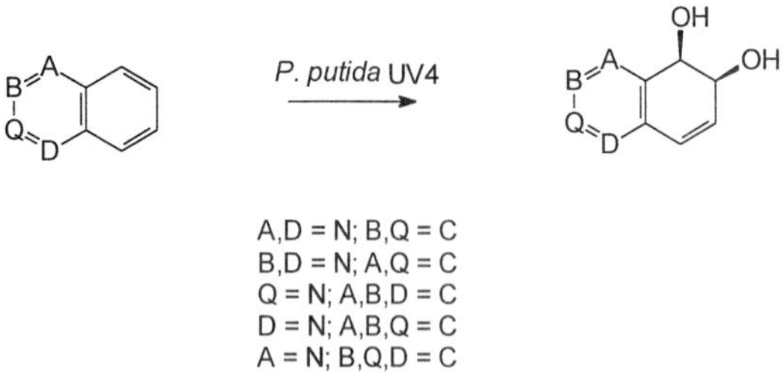

A,D = N; B,Q = C
B,D = N; A,Q = C
Q = N; A,B,D = C
D = N; A,B,Q = C
A = N; B,Q,D = C

Figure 35 Oxidations of quinoline, isoquinoline, quinoxaline, and quinazoline by *P. putida*.

C. Dihydroxylation of Nonarene Double Bonds

The conversion of alkenes into chiral *vicinal* diols by biotransformation may be conveniently divided into two groups as outlined in Fig. 2, those reactions that proceed via an epoxide intermediate to give a *trans*-diol and are presumably catalyzed by monooxygenase enzymes; and those that are catalyzed by dioxygenase enzymes in which no intermediate species is formed, and which invariably give a *cis*-diol product.

In the former area, bioconversion of substituted monoterpenes by *Aspergillus niger* LCP 521 has been used in an enantiodivergent manner to provide both optical antipodes of product. The stereoselectivity of the conversion of geranyl *N*-phenylcarbamate (**47**) to diols **48** and **49** (Fig. 37) was found to be pH-dependent, with diol **48** being produced at pH 7 as a consequence of the selective epoxide hydrolase activity of the fungus, with its enantiomer **49** being formed at pH 2 following nonenzymic acid-catalyzed hydrolysis of the intermediate epoxide [59]. Analogous biotransformations of 7-geranyloxycoumarin (**50**) by *A. niger* led to the independent formation of both enantiomers of the natural coumarin derivative marmin (**51** and **52**) (Fig. 38) [60], and the reaction has been applied independently to both enantiomers of citronellyl *N*-phenylcarbamate (**53**) to produce the four stereoisomeric diols shown in Fig. 39, each of which can be obtained as a single diastereomer [61].

The conversion of the chromene **54** to epoxide **55** (a synthetic precursor for potassium channel modulators) and diol **56** (Fig. 40) may be achieved by a number of microbial catalysts, notably *Mortierella rammaniana* SC 13840 which gives the diol **56** in 65% yield an 97% optical purity [62]. In an analogous conversion, the related microorganism *Mortierella isabellina* ATCC 42613 converted both chromenes **57** and **58** (Fig. 41) to a mixture of the corresponding *cis*- and *trans*-diols, presumably the result of regio- but nonstereoselective acid-catalyzed hydrolysis of an intermediate epoxide, with both isomeric diols being formed in identical enantiomeric purities [34].

The isolated enzyme chloroperoxidase from *Caldariomyces fumago* is reported to perform stereoselective epoxidation reactions and in one example, oxidation of α-meth-

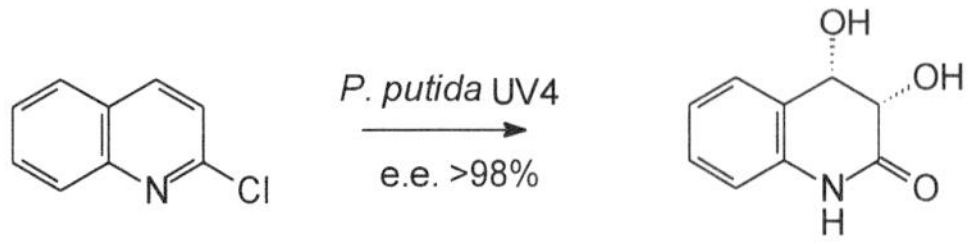

46

Figure 36 Oxidation of 2-chloroquinoline by *P. putida*.

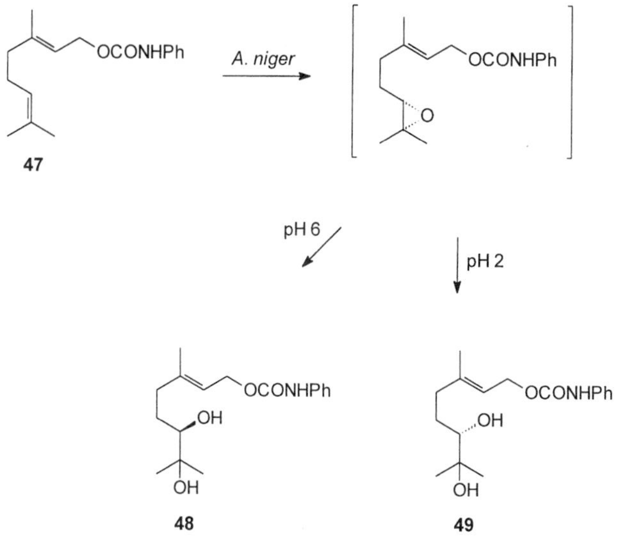

Figure 37   Enantiodivergent oxidation of geranyl *N*-phenylcarbamate by *A. niger*.

ylstyrene, the diol **59** (Fig. 42) of unspecified configuration was formed in high yield but low (25%) enantiomeric purity [47].

The oxidation of olefinic double bonds by the dioxygenase-producing bacteria results almost exclusively in the formation of *cis*-diols of high optical purity [7]. Biotransformation of the benzocycloheptene **60** [36] and of dihydronaphthalene (**61**) [37] by the TDO-producing *Pseudomonas putida* UV4 gave the corresponding diols shown in Fig. 43 in more than 98% enantiomeric purity. These products possess the opposite configuration to the *cis*-1,2-diol obtained by NDO-catalyzed biotransformation of naphthalene (cf. Figs. 3 and 29), allowing for formation of both enantiomers of the diol product **65** by catalytic hydrogenation of the naphthalene biotransformation product (Fig. 29, R = H) [37]. The

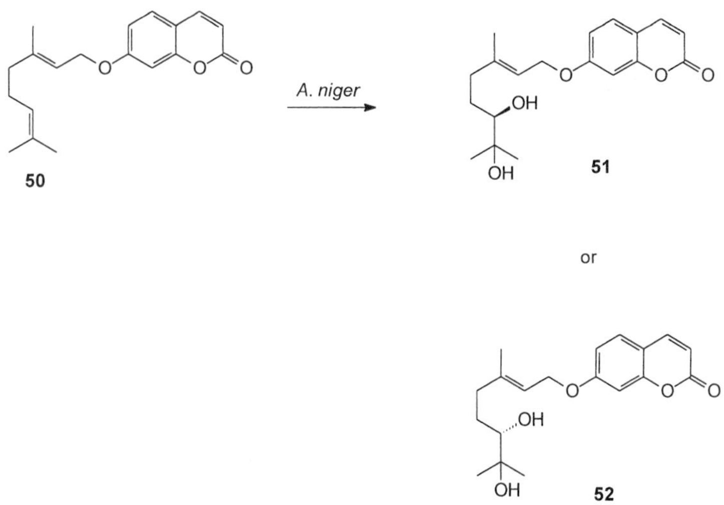

Figure 38   Enantioselective syntheses of marmin.

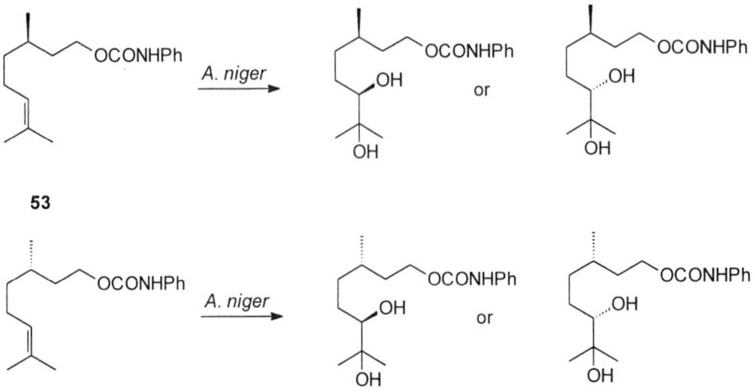

Figure 39 Preparation of diastereomeric diols from citronellyl *N*-phenylcarbamate.

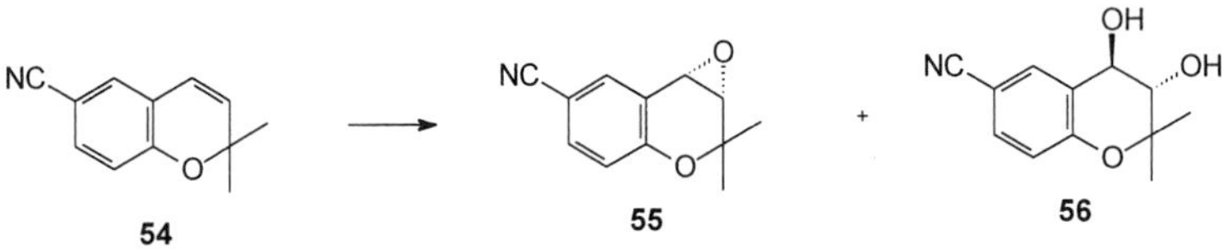

Figure 40 Microbial oxidation of 6-cyano-2,2-dimethylchromene.

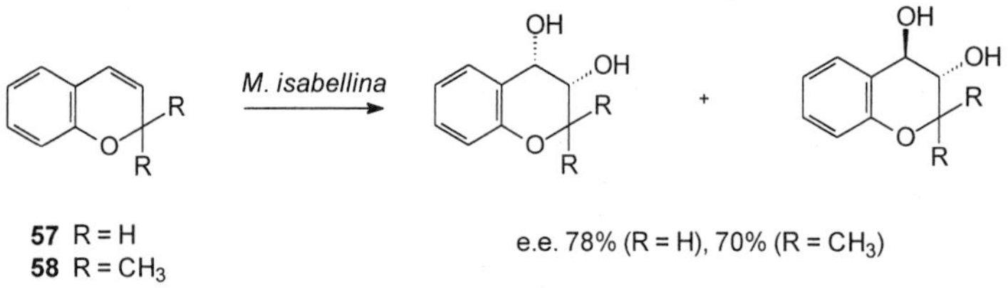

57 R = H
58 R = CH₃   e.e. 78% (R = H), 70% (R = CH₃)

Figure 41 Oxidations of chromene and dimethylchromene by *M. isabellina*.

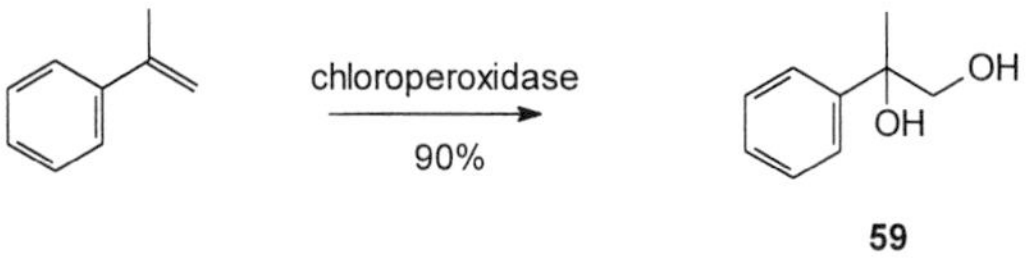

59

Figure 42 Chloroperoxidase-catalyzed oxidation of α-methylstyrene.

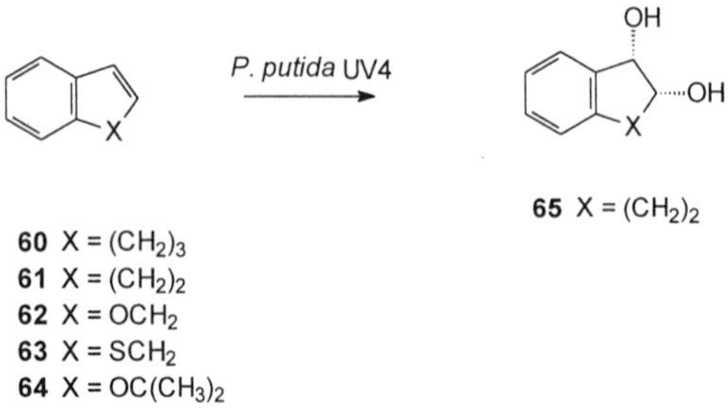

65 X = (CH₂)₂

60 X = (CH₂)₃
61 X = (CH₂)₂
62 X = OCH₂
63 X = SCH₂
64 X = OC(CH₃)₂

Figure 43 *cis*-Diol formation by oxidation of nonarene double bonds by *P. putida*.

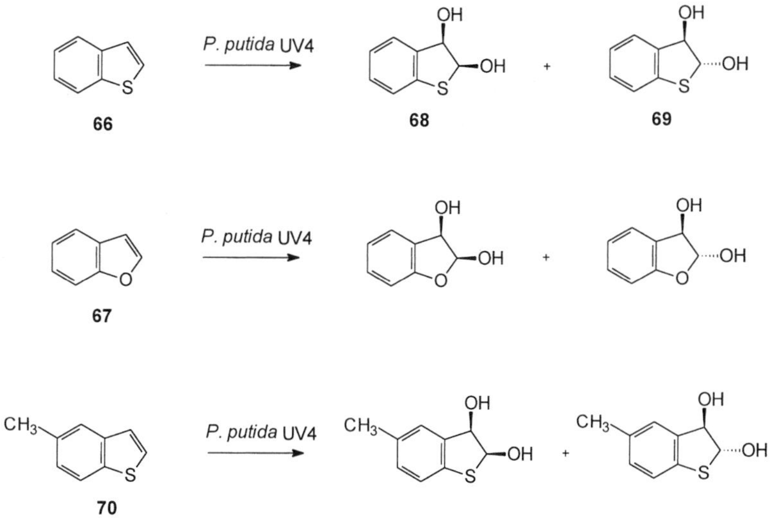

Figure 44 Oxidations of benzofuran and benzothiophenes by *P. putida*.

same biocatalyst also converted the heterocyclic substrates **62** and **63** and the dimethyl-chromene **64** to the analogous *cis*-1,2-diol products shown in Fig. 43 [63].

A more complex product profile was obtained by biotransformations of benzothiophene (**66**) and benzofuran (**67**) by *P. putida* UV4 (Fig. 44) [63,64]. The former substrate gave both *cis*- (**68**) and *trans*-diol (**69**) products with more than 98% optical purities, the latter product presumed to arise via hydrolytic opening and reclosure of the hemithioacetal **68**. Similar products are formed by oxidation of benzofuran (**67**) but in lower enantiomeric purities (55%); however, enantiopure diols are again produced from oxidation of the methylbenzothiophene **70** [64]. Analogous products could also be obtained, but in lower yields, from the 2-methyl-substituted analogs **71** and **72** (Fig. 45). However, the products from the benzofuran **71**, formed in 80% enantiomeric purity, were found to possess the opposite configuration at the benzylic carbon to those obtained from the unsubstituted analog **67**, whereas those from the benzothiophene **72** were reported to be almost racemic (e.e. 9%) [64].

IV. SUMMARY

It is apparent from the above discussion that biocatalytic stereoselective hydroxylation constitutes a powerful tool for the production of chiral alcohols of varied structure. During the past decade, considerable progress had been made in the understanding of this reaction

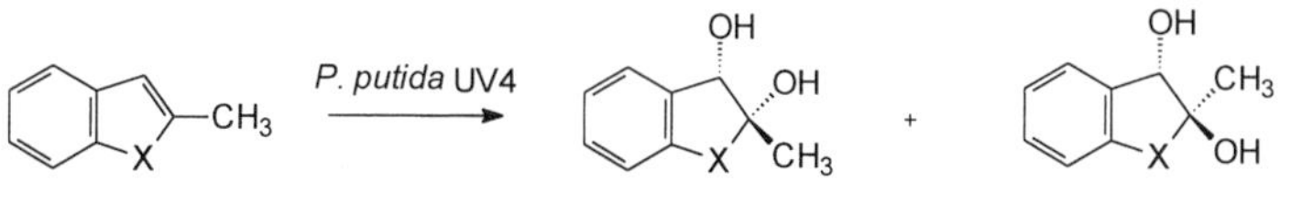

71 X = O
72 X = S

Figure 45 Oxidations of 2-methylbenzofuran and 2-methylbenzothiophene by *P. putida*.

at both the fundamental and the applied levels. Significant developments have also taken place in its practical application for the production of stereo- and regiochemically defined products to be used as starting materials for organic synthesis, as metabolic standards, and as materials that possess new or modified biological activities. The discovery and development of new biocatalysts, together with the application of existing biocatalysts to new substrate groups and a deeper understanding of the parameters that control the regio- and stereochemical selectivity of biocatalytic hydroxylation, will undoubtedly lead to an increase and improvement in the utility of biocatalytic stereochemical hydroxylation reactions.

REFERENCES

1. ERH Jones. Pure Appl Chem 33:39–52, 1973.
2. R Azerad. Bull Soc Chim Fr 132:17–51, 1995.
3. K Drauz, H Waldmann, eds. Enzyme Catalysis in Organic Synthesis. Weinheim: VCH, 1995.
4. HL Holland. In: H-J Rehm, G Reed, eds. Biotechnology, DR Kelly ed., Vol 8a, Biotransformations 1. Weinheim: Wiley-VCH, 1998, pp 475–533.
5. HL Holland. Organic Synthesis with Oxidative Enzymes. New York: VCH, 1992.
6. HL Holland. Adv Appl Microbiol 44:125–165, 1997.
7. DR Boyd, GN Sheldrake. Nat Prod Rep 15:309–324, 1998.
8. R Furstoss. In: S Servi, ed. Microbial Reagents in Organic Synthesis. Dordrecht: Kluwer, 1992, pp 333–346.
9. IY Lee, CH Kim, BK Yeon, WK Hong, ES Choi, SK Rhee, YJ Park, DH Sung, WH Baek. J Ferment Bioeng 81:79–82, 1996.
10. A Kanda, T Nakajima, H Fukuda. Jpn Pat 05,336,981, 21 Dec 1993; Chem Abs 120:321532s, 1994.
11. T Matsui. Jpn Pat 07 59,593, 7 Mar 1995; Chem Abs 122:313078z, 1995.
12. CT Hou, TA Seymour, MO Bagby. J Ind Microbiol 13:97–102, 1994.
13. Y Yamamoto, T Oritani, H Koga, T Horiuchi, K Yamashita. Agric Biol Chem 54:1915–1921, 1990.
14. WG Liu, JPN Rosazza. Tetrahedron Lett 31:2833–2836, 1990.
15. E Nobilec, M Aniol, C Wawrzenczyk. Tetrahedron 50:10339–10344, 1994.
16. A de Raadt, H Griengl, M Petch, P Plachota, N Schoo, H Weber, G Braunegg, I Kopper, M Kreiner, A Zeiser, K Kieslich. Tetrahedron: Asymmetry 7:467–472, 1996.
17. A de Raadt, H Griengl, M Petch, P Plachota, N Schoo, H Weber, G Braunegg, I Kopper, M Kreiner, A Zeiser. Tetrahedron: Asymmetry 7:473–490, 1996.
18. A de Raadt, H Griengl, M Petch, P Plachota, N Schoo, H Weber, G Braunegg, I Kopper, M Kreiner, A Zeiser. Tetrahedron: Asymmetry 7:491–496, 1996.
19. M Kreiner, G Braunegg, A de Raadt, H Griengl, I Kopper, O Sukcharoen, H Weber. Food Technol Biotechnol 35:99–106, 1997.
20. CR Davis, RA Johnson, JI Cialdella, WF Liggett, SA Mizsak, VP Marshall. J Org Chem 62:2244–2251, 1997.
21. HF Olivo, MS Hemenway, MH Gezginci. Tetrahedron Lett 39:1309–1312, 1998.
22. HL Holland, A Kohl, BG Larsen, P Andreana, J-X Gu. J Mol Cat B: Enzymatic 2:L253–L255, 1997.
23. R Azerad, A. Hammoumi. Fr Pat 2,661,190, 25 Oct 1991; Chem Abstr 116:172459m, 1992.
24. G Aranda, M Bertranne, R Azerad, M Maurs. Tetrahedron: Asymmetry 6:675–678, 1995.
25. HL Holland. Unpublished data.
26. A Hammoumi, J-P Girault, R Azerad, G Revial, J d'Angelo. Tetrahedron: Asymmetry 6:1295–1306, 1993.

27. S Arseniyadis, R Rodriquez, MM Dorado, RB Alves, J Ouazzani, G Ourisson. Tetrahedron 50:8399–8426, 1994.
28. R Agarwal, DR Boyd, ND Sharma, AS McMordie, HP Porter, B van Ommen, PJ van Bladeren. Bioorg Med Chem 2:439–446, 1994.
29. H Kasahara, M Miyazawa, H Kameoka. Phytochemistry 44:1479–1482, 1997.
30. D Ditullio, D Anderson, C-S Chen, CJ Sih. Bioorg Med Chem 2:415–420, 1994.
31. K Nakagawa, H Okazaki, H Watanabe, T Kagasaki, T Furubayashi. Jap Pat 02,174,686, 6 July 1990; Chem Abs 113:189785a, 1990.
32. K Nakagawa, H Watanabe, T Kagasaki, T Furubayashi. Jap Pat 02,174,687, 6 July 1990; Chem Abs 113:189786b, 1990.
33. HL Holland, TS Manoharan, F Schweizer. Tetrahedron: Asymmetry 2:335–338, 1991.
34. HL Holland, LJ Allen, MJ Chernishenko, M Diez, A Kohl, J Ozog, J-X Gu. J Mol Cat B: Enzymatic 3:311–324, 1997.
35. T Kitayama. Tetrahedron: Asymmetry 8:3765–3774, 1997.
36. DR Boyd, MRJ Dorrity, JF Malone, RAS McMordie, ND Sharma, H Dalton, P Williams. J Chem Soc Perkin Trans I 489–494, 1990.
37. DR Boyd, ND Sharma, NA Kerley, RAS McMordie, GN Sheldrake, P Williams, H Dalton. J Chem Soc Perkin Trans I 67–74, 1996.
38. SM Resnick, DS Torok, K Lee, JM Brand, DT Gibson. Appl Environ Microbiol 60:3323–3328, 1994.
39. DT Gibson, SM Resnick, K Lee, JM Brand, DS Torok, LP Wackett, MJ Schocken, BE Haigler. J Bacteriol 177:2615–2621, 1995.
40. DR Boyd, ND Sharma, NI Bowers, PA Goodrich, MR Groocock, AJ Blacker, DA Clarke, T Howard, H Dalton. Tetrahedron: Asymmetry 7:1559–1562, 1996.
41. P Zhang, KS Tyler. J Am Chem Soc 111:9241–9242, 1989.
42. VK Datcheva, K Kiss, L Solomon, KS Tyler. J Am Chem Soc 113:270–274, 1991.
43. D Martini, G Buono, G Iacazio. J Org Chem 61:9062–9064, 1996.
44. D Martini, G Iacazio, D Ferrand, G Buono, C Trintaphylides. Biocatalysis 11:47–63, 1994.
45. C Fuganti, R Rigoni, G Zucchi, M Barbeni, M Cisero, M Villa. Biotechnol Lett 17:301–304, 1995.
46. MBW Elshof, M Janssen, GA Veldink, JFG Vliegenthart. Rec Trav Chim Pays-Bas 115:499–504, 1996.
47. A Zaks, DR Dodds. J Am Chem Soc 117:10419–10424, 1995.
48. MG Quintana, H Dalton. Enzyme Microb Technol 22:713–720, 1998.
49. P Di Gennaro, E Galli, G Albini, F Pelizzoni, G Sello, Besetti. Res Microbiol 148:355–364, 1997.
50. B Morawski, H Griengl, DW Ribbons, DJ Williams. Tetrahedron: Asymmetry 8:845–848, 1997.
51. D Gonzalez, V Schapiro, G Seoane, T Hudlicky. Tetrahedron: Asymmetry 8:975–977, 1997.
52. BV Nguyen, C York, T Hudlicky. Tetrahedron 53:8807–8814, 1997.
53. F Yan, BV Nguyen, C York, T Hudlicky. Tetrahedron 53: 11541–11548, 1997.
54. TC Nugent, T Hudlicky. J Org Chem 63:510–520, 1998.
55. MA Endoma, G Butora, CD Claeboe, T Hudlicky, KA Abboud. Tetrahedron Lett 38:8833–8836, 1997.
56. G Butora, T Hudlicky, SP Fearnley, MR Stabile, AG Gum, D Gonzalez. Synthesis 665–681, 1998.
57. DR Boyd, ND Sharma, MRJ Dorrity, MV Hand, RAS McMordie, JF Malone, HP Porter, H Dalton, J Chima, GN Sheldrake. J Chem Soc Perkin Trans I 1065–1071, 1993.
58. DR Boyd, ND Sharma, JG Caroll, JF Malone, DG Mackerracher, CCR Allen. J Chem Soc Chem Commun 683–684, 1998.
59. XM Zhang, A Archelas, R Furstoss. J Org Chem 56:3814–3817, 1991.
60. XM Zhang, A Archelas, A Méou, R Furstoss. Tetrahedron: Asymmetry 2:247–250, 1991.

61. XM Zhang, A Archelas, R Furstoss. Tetrahedron: Asymmetry 3:1373–1376, 1992.
62. RN Patel, A Banerjee, LJ Szarka. J Am Oil Chemists Soc 72:1247–1264, 1995.
63. DR Boyd, ND Sharma, R Boyle, BT McMurray, TA Evans, JF Malone, H Dalton, J Chima, GN Sheldrake. J Chem Soc Chem Commun 49–51, 1993.
64. DR Boyd, ND Sharma, IN Brannigan, SA Haughey, JF Malone, DA Clarke, H Dalton. J Chem Soc Chem Commun 2361–2362, 1996.

6

Regio- and Stereoselective Microbial Hydroxylation of Terpenoid Compounds

Robert Azerad
Université René Descartes-Paris V, Paris, France

In the recent years there has been an increasing interest in the bioconversion of terpenes and terpenoid substrates. A number of microbial hydroxylation reactions concerning all groups from cyclic or acyclic monoterpenes to triterpenes have been reported, corresponding to a particular concern in their use as fragrant or flavor agents; their antibacterial, antiviral, or cytotoxic activities; and their increasing importance as specific pharmacological agents. For these reasons, considerable work has been systematically devoted to the search of new functionalized derivatives, more easily attained by bioconversion reactions, and susceptible to show dissociated or new activities. In addition, the functionalization of easily available and inexpensive terpenoid compounds is considered as a potential source of new chiral auxiliaries and chiral synthons for asymmetrical synthesis. The ability of microbial transformations to introduce in a one-step reaction a hydroxyl group on a saturated carbon center, with the regio- and stereoselectivity usually expected from enzymic reactions, remains unchallenged by the existing chemical procedures, even when the use of whole-cell transformation results in moderate yields.

I. ACYCLIC TERPENOIDS

The biotransformation of acyclic terpenoid hydrocarbons usually proceeds very slowly and often gives a variety of useless and difficult-to-separate metabolites. For example, when β-myrcene **1** was oxidized by *Aspergillus niger* JTS 191 for 7 days, the total content of conversion products in the neutral fraction corresponded to less than 10% of the initially added substrate [1,2]. The three main conversion products were identified as diols **2–4**, probably arising from the hydrolytic opening of the corresponding epoxides.

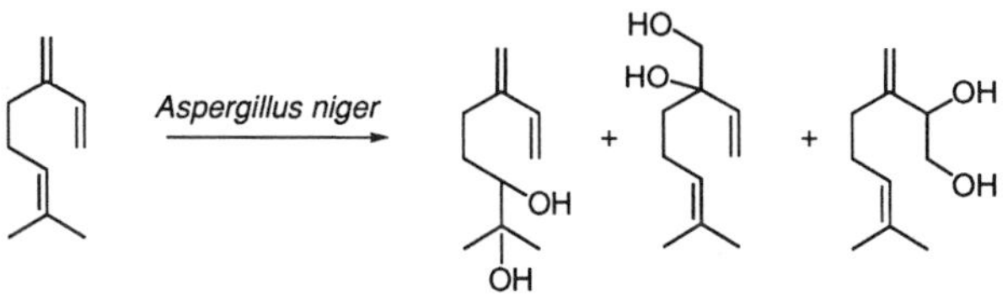

To introduce a polar function in the substrate, in order to favor its anchoring into the active site of hydroxylating enzymes, the dienyl moiety of myrcene was reacted with a dienophile such as sulfur dioxide, resulting in a polar sulfolene compound **5** that could be cleaved easily by heating. Biotransformation of the sulfolene derivative (2 g/L) by *Streptomyces albus* DSM 40763 afforded in 44% yield a 1:1 mixture of the isomeric 8-hydroxy derivatives **6–7**. A *Nocardia* sp. yielded mainly the Z-alcohol **7** (37% yield), whereas fungi mainly afforded the 8E-alcohol **6** (up to 55% yield), and sometimes an enantiomerically enriched 5(R)-hydroxy derivative **8** [3]. This method has been successfully applied to other conjugated terpenic hydrocarbons such as ocimene and farnesene, in order to produce ω-hydroxylated derivatives in good yields [4].

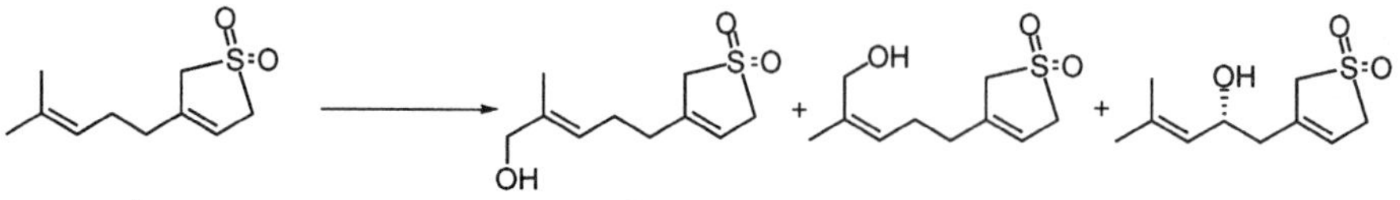

Similar results were obtained with acyclic mono- and sesquiterpenoid compounds containing an allylic alcohol group, either in a primary or a tertiary position. For example, linalool **9**, nerolidol **10**, geraniol **11**, citronellol (or their esters) are hydroxylated by *A. niger* [5–7] and many other microorganisms [8–10] in terminal position, concurrently with epoxidation (followed by hydrolysis) at the remote double bond:

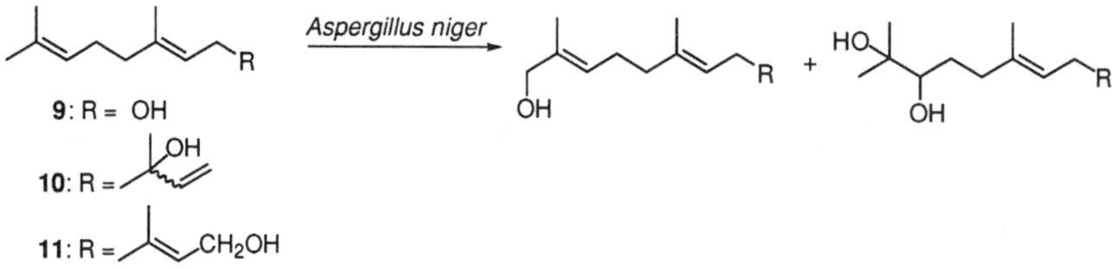

Using *Glomerella cingulata* as a biocatalyst [11–13], only the remote double bonds were regioselectively oxidized to give the corresponding diols. However, when the saturated terpenoid alcohols **12–13** were used as substrates, *G. cingulata* oxidized in fair yields the most distant isopropyl moieties, affording tertiary alcohols of high value as cosmetic and perfume ingredients [14].

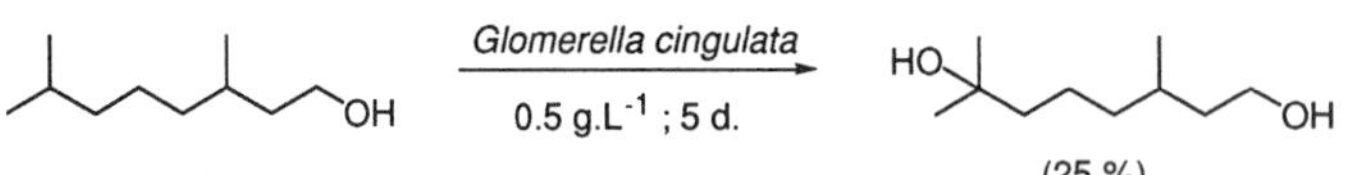

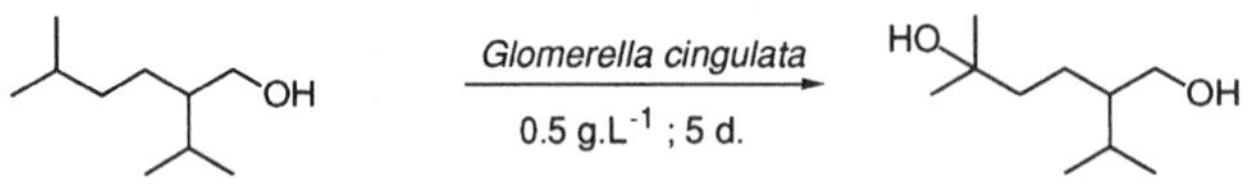

II. CYCLIC TERPENOIDS

A significant example of hydroxylation reactions of cyclic terpenic compounds is illustrated by the microbial metabolism of sclareol. Sclareol **14**, a labdane diterpene, can be easily isolated from the essential oil of *Salvia sclarea* L. (Labiatae) (clary sage oil) or from fresh tobacco leaves. It is currently used as a basic synthon for the preparation of Ambra-related fragrances and several other derivatives. Naturally occurring A-ring-oxygenated analogs of sclareol are unknown. Chemical oxidation of sclareol by the "Gif system" has been described [15] to give the 2- and 3-keto derivatives, but only in 2.5% and 0.7% yields, respectively. For such reasons, this terpene has been extensively investigated as a substrate of microbial hydroxylations. A bacterium strain isolated from soil (*Rhodococcus erythropolis* JTS 131) has been shown to selectively oxidize the C-18(α)-methyl group of sclareol and to accumulate, in the presence of added metabolic inhibitors (such as α,α-dipyridyl), the new carboxylic acid derivatives **15** and **16** (probably through primary hydroxylation to a hydroxymethyl group) [16]. *Mucor plumbeus* ATCC 4740, a fungal strain, incubated with sclareol (0.5–1 g/L) produces in high yields (Fig. 1) its 3β-hydroxylated derivative (**17**) and small amounts of the 18-hydroxymethyl (**18**) and 6α-hydroxy (**19**) derivatives [17,18]. Similar products, but in different ratios, are obtained with a *Cunninghamella* sp. (NRRL 5695) [19], *Cephalosporium aphidicola* IMI 68689 [20], *Diplodia gossypina* ATCC 10936 [21], *Bacillus sphaericus* ATCC 13805 [21], or *Bacillus cereus* UI 1477 [22]. Another *Cunninghamella* strain (*C. elegans* DSM 1908) produces, in addition, the 19-hydroxymethyl derivative **20** [21].

The same 3β-hydroxy derivative **17** is obtained in lower yield by using another fungal species, *Septomyxa affinis* ATCC 6737, because it is subsequently oxidized to the corresponding 3-keto derivative **21** [23]. This strain additionally produces smaller amounts of an unusual 2α-hydroxy derivative **22** which is also observed in incubations with *D. gossypina* (21) and *B. cereus* (22).

In a study of the metabolism of sclareol in the rat, microbial hydroxylation derivatives of sclareol were used as standards to facilitate the identification of the biliary metabolites [24]. Remarkably, on longer incubation times, the *B. cereus* strain produced significant amounts of the β-D-glucosides of each hydroxylated species [22] (Fig. 2), analogous to the derivatives usually found in mammals as detoxification and elimination metabolites. A statistical survey of the biotransformation of sclareol by microorganisms (40 bacteria and 60 fungi), undertaken to produce data allowing a better preselection of suitable strains, showed that fungi were the most active metabolizers [21]: the 3β-hydroxylation was obviously the most common reaction, followed by the 18-hydroxylation (also observed with bacteria), while 2α- and 19-hydroxylations were limited to a small number of fungi. The efficient microbial preparations of 3β-hydroxy sclareol **17** and the corresponding 3-keto derivative **21** have recently been used (Fig. 3) to provide precursors for the synthesis and stereochemical determination of a rare labdane from a Bhutanese medicinal plant, 3α-hydroxy manool **24**, a platelet aggregation inhibitor [25].

Besides sclareol, the microbial hydroxylation of a number of other naturally found and easily available terpenes of the labdane family has been investigated, particularly in view of designing chemoenzymatic routes for the hemisynthesis of forskolin **25**, an activator of adenylate cyclase used as a therapeutic agent, or for the generation of potentially active analogs. Actually, 1,9-dideoxyforskolin **26** invariably co-occurs with forskolin **25**

25

26 : R = COCH$_3$

27 : R = H

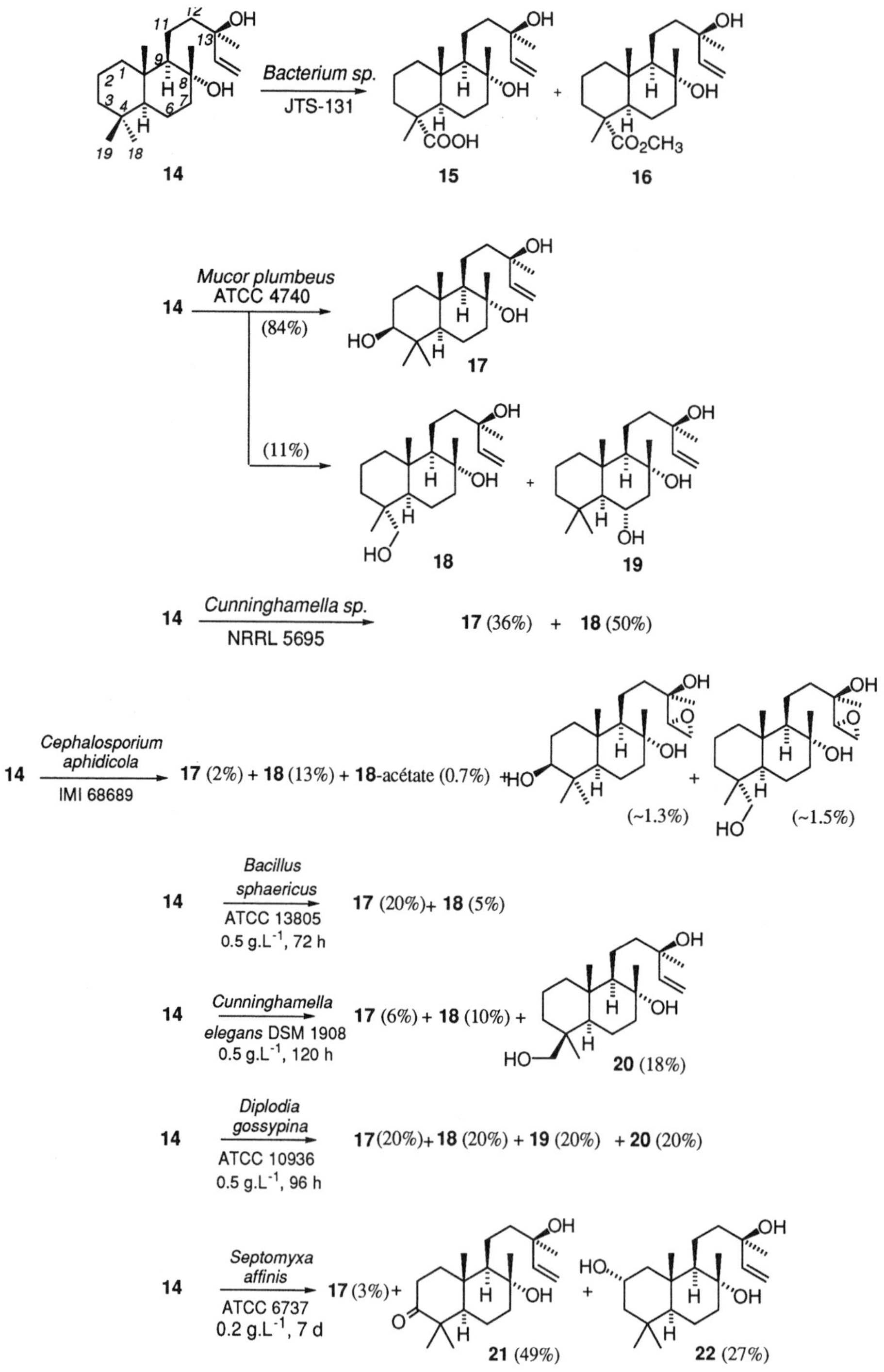

Figure 1 Hydroxylation pattern of sclareol **14** by various fungal or bacterial strains.

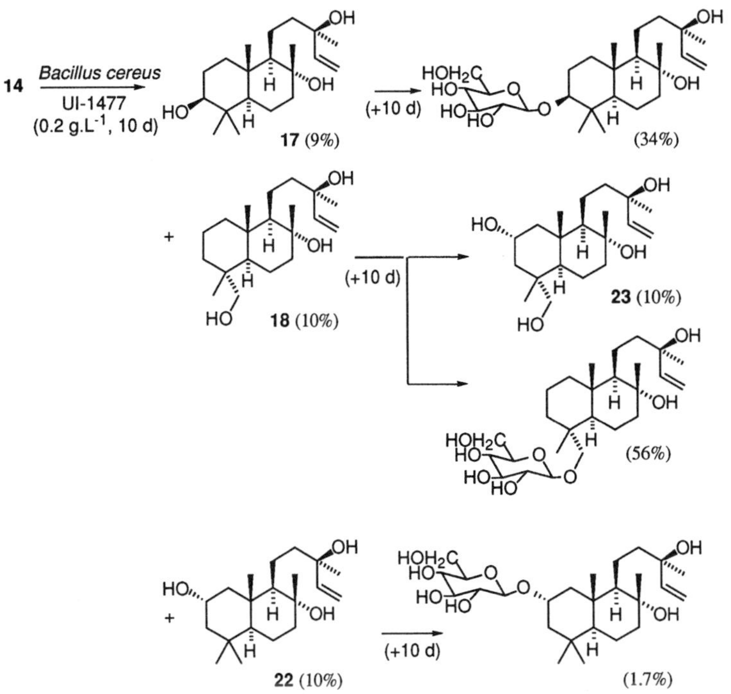

Figure 2 Hydroxylation and glucosylation pattern of sclareol **14** during shorter (10 days) or longer (20 days) incubations with *Bacillus cereus* UI-1477.

(isolated from an Indian herb, *Coleus forskolhii*) in amounts almost equal to, or sometimes even in excess of, forskolin. It was thus tempting to try to introduce the 1- and 9-hydroxy groups by microbial hydroxylation of this analog. *Mortierella isabellina* ATCC 160074 and *Neurospora crassa* ATCC 10336 afforded essentially and in moderate yields 2(α and β)-, 3-, 18- (or 19-), and 1α-hydroxylated derivatives [26,27]. On the other hand, a *Scopuloriopsis* strain (DSM 3205), selected from a screening of 263 fungal isolates from soil [28,29], was found to convert 7-deacetyl-1,9-dideoxyforskolin **27** to 7-deacetylforskolin, but in a very low yield (0.76%), besides a 2β-hydroxy derivative that was also the main product isolated in the bioconversion of 1,9-dideoxyforskolin **26**. Another fungal isolate (FF 406) produced only 3α- (20%) and 3β- (10%) hydroxy derivatives [30].

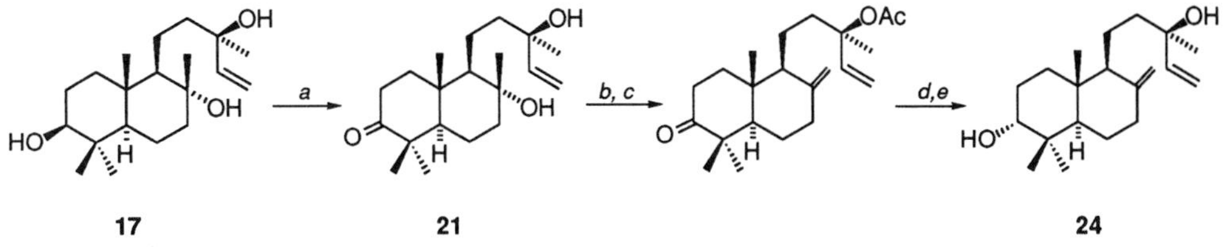

Figure 3 Hemisynthesis of 3α-hydroxymanool **24** from 3β-hydroxy sclareol **17** or 3-oxosclareol **21**, two microbial hydroxylation products. (*a*) CrO₃, pyridine; (*b*) AcCOCl, PhNMe₂; (*c*) pyrolysis; (*d*) KOH/water-MeOH; (*e*) K-Selectride.

(Z)-Abienol **28** isolated from Canada fir balsam is a predominant labdane of some fresh tobacco leaf varieties and some of its oxidation products are responsible for specific smoke aromas. In order to identify and produce some of these oxidized metabolites, (Z)-abienol metabolites created by soil microorganisms were investigated: a *Nocardia restricta* strain (JTS 162) [31,32] and a *Rhodococcus erythropolis* strain (JTS 131) [16] produced hydroxymethyl and carboxylic acid derivatives at C-17 and C-18, respectively, which accumulated in the presence of metabolic inhibitors such as α,α-dipyridyl. A recent study using plant cell cultures of *Nicotiana silvestris*, or simple incubations of (Z)-abienol with horseradish peroxidase, demonstrated the formation of 12 identified degradation metabolites [33].

Manool **29**, another common labdane, isolated from wood oil of the yellow pine *Dacrydium biforme*, was hydroxylated (but in lower yields than sclareol) by *Mucor plumbeus* at positions -3β, -2α, and/or -7α [18]. Larixol **30**, a 6-hydroxylated labdane isolated as its monoacetate from the oleoresins of European larch (*Larix europeae* D.C.) or as the free alcohol from Siberian larch (*Larix sibirica* Ledb.), and some chemically oxidized derivatives, have recently been proposed as valuable precursors for the hemisynthesis of forskolin and analogs [34]. However, incubation with an *M. plumbeus* strain led exclusively, though in good yields, to 2α-hydroxy derivatives.

Sclareolide **31**, a minor constituent of some plant extracts, can be prepared by oxidation of sclareol or manoyl oxide derivatives. The microbial oxidation of sclareolide by *M. plumbeus* ATCC 4740 was less productive than the oxidation of sclareol, but afforded, beside 3β-hydroxy and 3-keto derivatives, a small amount of 1β-hydroxy sclareolide [18]. Similar results and, in addition, the formation of 1,3-dihydroxylated derivatives have been obtained by incubation with *Curvularia lunata*, in higher conversion yields [35]. *Cephalosporium aphidicola*, a fungus producing the hydroxylated diterpenoid aphidicolin, was shown to hydroxylate sclareolide at the 3β position and, in addition, to afford a 3β,6β-dihydroxylated derivative in substantial yields [36].

Jhanol (18-hydroxymanoyl oxide **32**), a labdane diterpene isolated from *Upatorium jhanii*, a South American plant, has been recently shown to be hydroxylated in low yields (2–9%) by *Gibberella fujikuroi* at the 11α, 11β, and/or 1α positions [37]. Interestingly, feeding the strain with the 1β-hydroxy derivative resulted in oxidation to a 1-oxo compound, followed by enzymic reduction on the less hindered face to give the isomeric 1α-hydroxy derivative. Such a stereochemical inversion of a hydroxyl group, resulting from a microbial oxidation–reduction process, may be a general feature and may explain the apparent lack of stereoselectivity of some microbial hydroxylation reactions at secondary positions.

Ent-manoyl oxides **33–34**, epimeric at C-13 and deriving from abundant natural diterpenes isolated from several spanish *Sideritis* species (Labiatae), have been extensively investigated, using incubations with *Curvularia lunata* [38–40], *Rhizopus nigricans* [41–43], *Cunninghamella elegans* [43], *Fusarium moniliforme* [43], or *Gibberella fujikuroi* [44], in order to obtain highly functionalized derivatives comparable to forskolin **25**, but belonging to the *enantio* series, for biological testing. Depending on the strain and the substrate functionalization, a variety of dihydroxylated products bearing a newly introduced hydroxyl group at positions -1α, -1β, -3α, -6α, -7β, -11α, -12α, or -20 were obtained, together with trihydroxylated products, sometimes in substantial yields. Some of these derivatives exhibit significant antileishmanial activity [45].

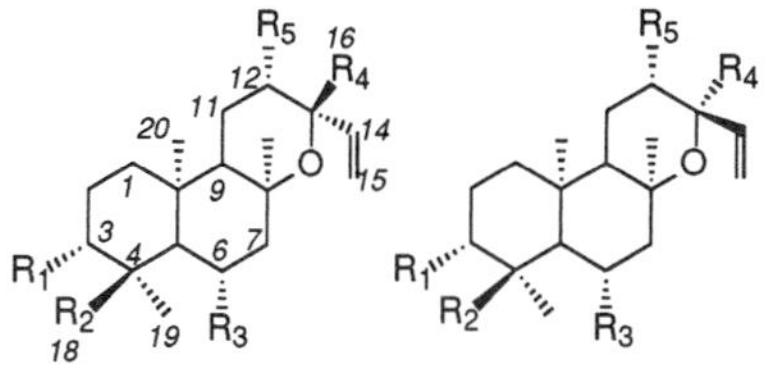

33   **34**

R_1 = H, OH or =O
R_2 = CH_3, CH_2OAc or CO_2Me
R_3 = OH or OAc
R_4 = CH_3 or CH_2OH
R_5 = H or OH

Beside labdanes, a number of other diterpene group representatives have been submitted to microbial hydroxylation in order to generate new synthetic intermediates or biologically active molecules [46–48]. Extensive studies have been carried out with the fungus *G. fujikuroi*, starting from compounds with gibbane skeletons or their biogenetic precursor, the *ent*-kaur-16-enic system, in order to elucidate the biogenetic route to giberellin formation or to produce new hydroxylated *ent*-kaur-16-ene derivatives. Incubating this strain with the chemically prepared 16β-epoxy kaurenoic acid **35** allows its conversion to the giberellin aldehyde precursor **37**, via the isolated diol **36**, the postulated C-7 hydroxylation, and the subsequent ring contraction of the C-19 oxidized derivative [49]. Other substituted derivatives, like a 15α-hydroxy, or a 15α,16α-epoxy, or a 16α,17-diol, appear to direct hydroxylation to C-11 and/or inhibit oxidation at C-19, thus precluding their transformation to a giberellin biosynthetic intermediate such as **37** [49,50].

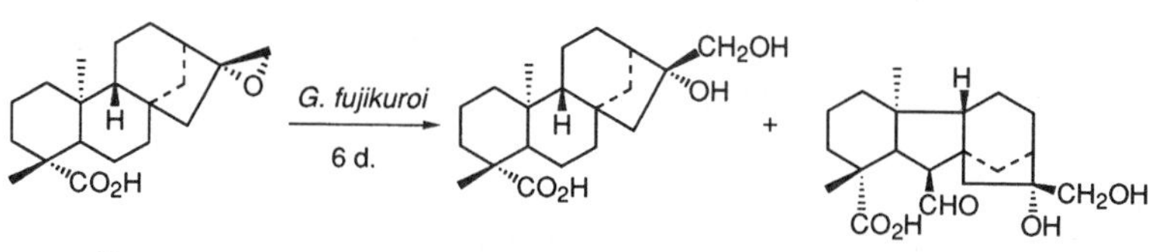

35   **36** (8 %)   **37** (15 %)

Using the same strain, candidiol **38** (*ent*-15β,18-dihydroxy-kaur-16-ene), a diterpene isolated from *Sideritis candicans*, was selectively hydroxylated in an 11β-axial position (**39**) [51]:

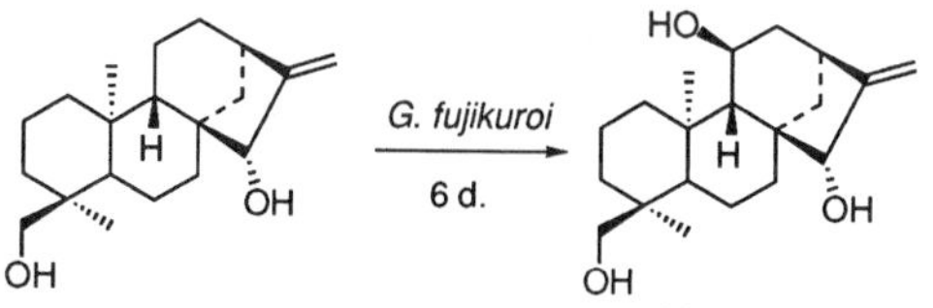

38   **39** (33 %)

Another kaurenoid derivate, steviol **40**, the aglycone of the commercially available potent sweetener stevioside, is selectively converted by *G. fujikuroi* into its 7β-axial hy-droxylated derivative **41**, while other fungal strains, such as *Rhizopus stolonifer*, simul-taneously afford a 9β-hydroxylation product [52,53]. Conversely, isosteviol **42**, obtained by acidic hydrolysis of stevioside, is hydroxylated in low yields at 7α (**43**) and 12β positions (**44**) by *Fusarium verticilloides* [54]. Because of their structural similarity to giberellin biosynthetic intermediates, the biotransformation of steviol, isosteviol, or deriv-atives by *G. fujikuroi* has been used to prepare labeled giberellins or giberellin analogs [53].

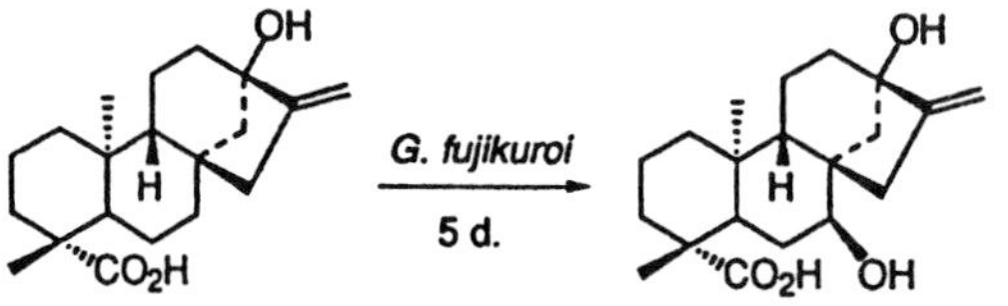

40   41 (55 %)

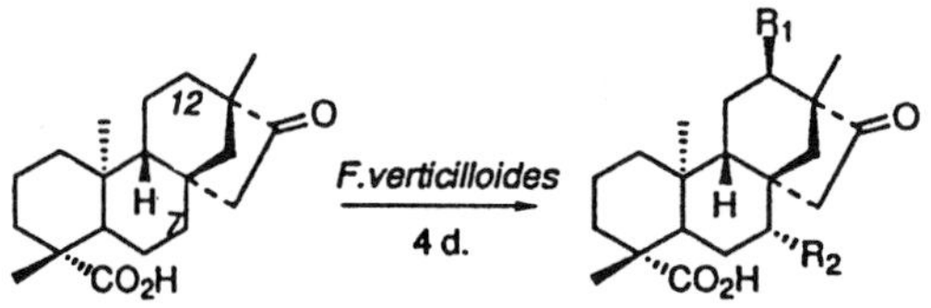

42

43 : R$_1$ = H; R$_2$ = OH (5%)
40 : R$_1$ = OH; R$_2$ = H (7%)

Several other fungal strains, such as *A. niger, Rhizopus nigricans*, and *Curvularia lunata*, have been successfully used by Garcia-Granados et al. [40,55–59] for (less regio-selective) hydroxylations of these substrates and related molecules, in relation to the cur-rent theories about steroid framework recognition, hypothetically paralleling the structure of the diterpenoid substrates [60]. While differing from the *G. fujikuroi* bioconversions, a general oxidation pattern of *ent*-kaurane-derived terpenoids [57] has been recognized, which is summarized in Fig. 4.

In another microorganism, *Cephalosporium aphidicola* IMI 68689, sequential bio-synthetic hydroxylations of the precursor aphidicolan-16β-ol **45**, first at C-18, then at C-3, and finally at C-17, give aphidicolin **46**, a diterpenoid tumor inhibitor [61,62].

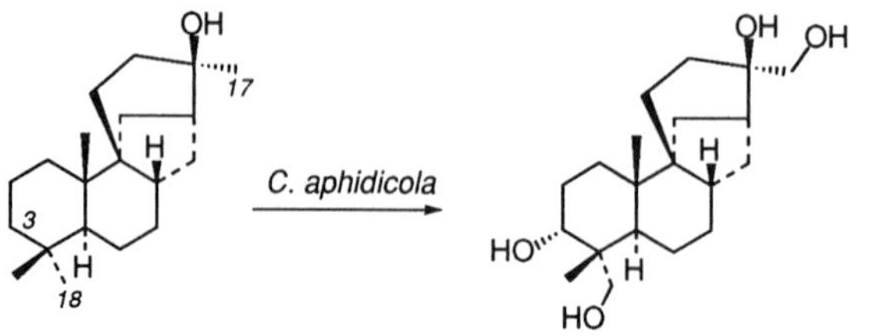

45   46

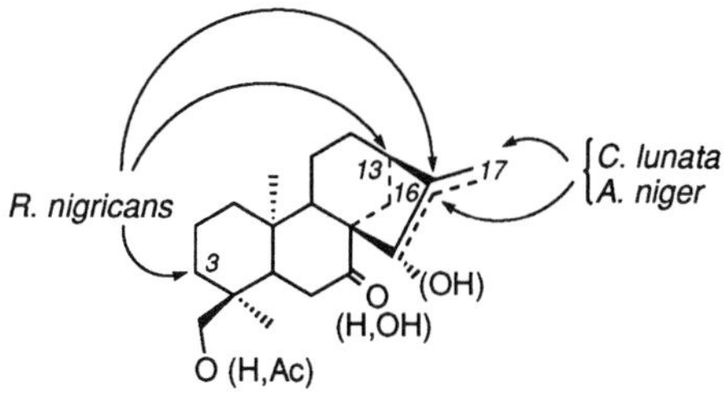

Figure 4 Favored hydroxylation sites of *ent*-kaurane derivatives by currently used fungal strains.

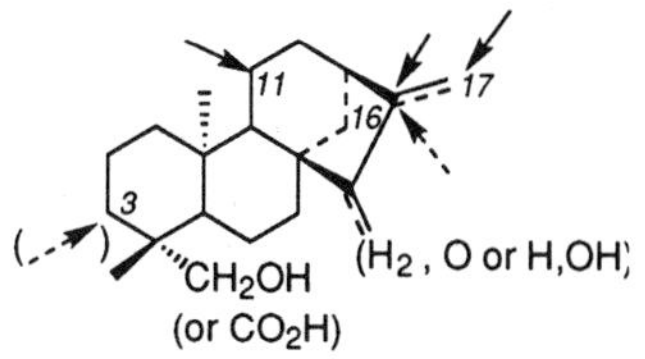

Figure 5 Main hydroxylation sites of *ent*-kaurane derivatives using *Cephalosporium aphidicola*.

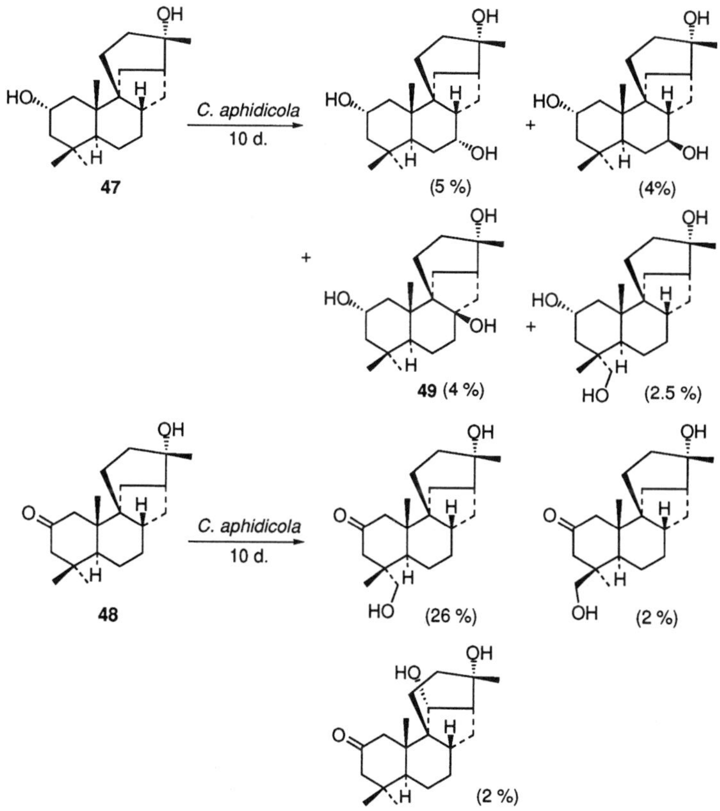

Figure 6 Hydroxylation of stemodin **47** and stemodinone **48** by *Cephalosporium aphidicola*.

Taking advantage of such biosynthetic abilities, Hanson's group has systematically investigated the biotransformation by *C. aphidicola* of some (more or less) related cyclic diterpenoids such as kaurane [63–65] or stemodane [66] derivatives. Interestingly, with most kaurane derivatives, the current hydroxylation at C-3 observed with the natural substrate was inhibited, but various other positions can be hydroxylated, including the 11β, 16α, and 17 positions (Fig. 5), the later probably after the reduction of the 16,17 double bond took place.

Various hydroxylation positions were obtained in the stemodane series, using stemodin **47**, the major diterpene isolated from *Stemodia maritima* L. (Scrophulariacae), or its 2-keto derivative, stemodinone **48**, as substrates (Fig. 6), paralleling the hydroxylation pattern already described with several other fungal microorganisms [67,68]. The 8β-hydroxylated derivative **49** exhibited high antiviral activity, and most of the hydroxylated derivatives exhibited a cytotoxic activity comparable to stemodin itself [68].

Other related diterpenes of some interest, the hydroxylation of which by microbial methods has recently been described, are isocupressic acid [69], dehydroabietic acid derivatives [70,71], grindelane derivatives [72–75], kolavenol acetate [76], norambreinolide

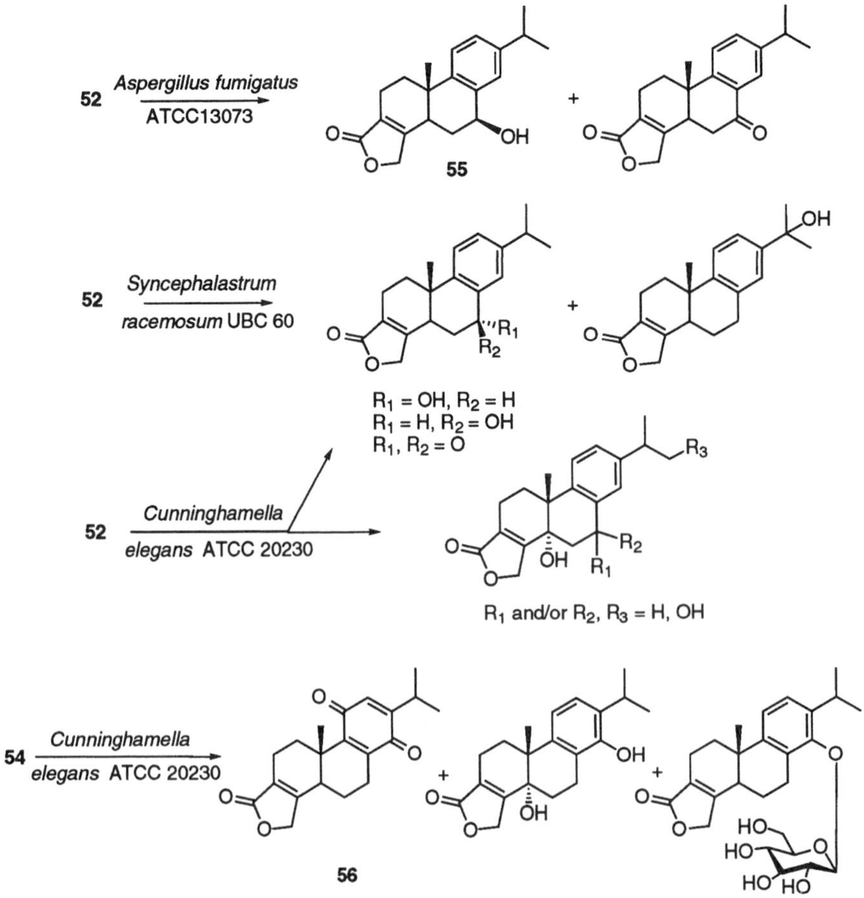

Figure 7 Oxidative metabolism of synthetic triptolide analogs (**52,54**) by selected fungi.

[18], beyerene and atiserene derivatives [59,77], and rosene derivatives [78]. In most cases, the favored hydroxylation site was again the 3β position (3α in the *ent* series) for the 4,4-dimethyl compounds, but a hydroxymethyl or a carboxylic acid group in the 4 position seems to preferentially orient the hydroxylation reaction to carbon 2.

50 : R = H
51 : R = OH

52 : $R_1 = R_2 = H$
53 : $R_1 = OH; R_2 = H$
54 : $R_1 = H; R_2 = OH$

Triepoxide diterpenes, triptolide **50** and tripdiolide **51**, isolated from the Chinese perennium herb *Tripterygium wilfordii* Hook F. (Celastraceae) have shown promising antileukemic, anti-inflammatory, and antitumor activities in vitro and in vivo. However, the toxicity of these epoxides has limited their clinical usefulness. Several analogs, deriving from synthetic abietane lactones **52–54**, have been prepared by hydroxylation with filamentous fungi [79–81], in order to introduce selectively the desired oxygenated functionalities on various positions of the ring system (Fig. 7). *Syncephalastrum racemosum* UBC 60 mainly oxidized substrate **52** at position -7, affording 7α, 7β, and 7-keto derivatives. Another metabolite was shown to be hydroxylated at C-15, into the isopropyl side chain. On the other hand, *Aspergillus fumigatus* ATCC 13073 produced nearly exclusively the 7β-hydroxy derivative **55**, in 90% yield at 0.1–0.5 g/L, in 48-h incubations. A 24-h incubation with *Cunninghamella elegans* ATCC 20230 (0.2 g/L) afforded the same hydroxylated products, new hydroxy derivatives at C-5 or C-16, and, in addition, several dihydroxylated derivatives. *Cunninghamella elegans* and *C. echinulata* ATCC 9244 were similarly used to produce oxidized metabolites from the phenolic substrates **53–54**. While glucoside conjugation was a major route of metabolism of these substrates, both strains hydroxylated substrate **54** at C-5 and, remarkably, produced the benzoquinone **56** in 35% yield [82], probably through primary hydroxylation at C-11. Quinone **56** is structurally related to the minor triptoquinones that are present in the crude plant extract and possibly responsible in part for its anti-inflammatory (and other) effects.

Paclitaxel (Taxol) **57**, a natural terpenoid product derived from the bark of the Pacific yew *Taxus brevifolia* [83–85], and its hemisynthetic analog docetaxel (Taxotere) **58**, are two recently discovered promising antitumor agents. At this moment, very few data have been published about the microbial metabolism of taxoid compounds; only site-specific hydrolyses of acyl side chains at C-13 or C-10 by extracellular and intracellular esterases of *Nocardioides albus* SC13911 and *N. lutea* SC13912, respectively, have been reported [86]. On the other hand, Hu et al. [87–89] recently described some fungal biotransformations of related more abundant natural taxane diterpenes extracted from Chinese yews or their cell cultures, in order to obtain new active substances or precursors for hemisynthesis of paclitaxel analogs.

57 : R₁ = C₆H₅CO; R₂ = CH₃CO
58 : R₁ = (CH₃)₃COCO; R₂ = H

59 : R₁ = CH₃CO
60 : R₁ = H

61 : R₁ = R₂ = H; R₃ = OH
62 : R₁ = R₃ = OH; R₂ = H
63 : R₁ = H; R₂ = R₃ = OH

64

The taxadiene **61**, a 14β-acetylated derivative lacking *inter alia* the 13α-hydroxy acyl group and the 4(20)-oxirane ring of paclitaxel, was extensively (55%) metabolized by *Cunninghamella echinulata* AS3 1990 to a 6α- (major) and a 6β- (minor) hydroxylated derivatives, with hydrolytic deacetylation at C-10 (**62,63**). In addition, epoxide **64** was formed as a minor metabolite [87]. The C-14 hydroxyl group acylation was essential for the hydroxylation at C-6 by this strain, and hydroxylation was inhibited by the presence of the epoxide ring of **64** [89]. A comparative biotransformation study with the same strain incubated with paclitaxel (**57**) or paclitaxel hemisynthesis precursors such as baccatin III (**59**) and 10-deacetyl baccatin III (**60**) showed respectively no transformation or simple deacetylation and/or epimerization at the C-7 position [89]. A similar result was obtained with other more abundant taxoid compounds found in *Taxus yunnanensis* [90]. Some fungi, such as *Taxomyces andreanae* [91,92] or *Pestalotiopsis microspora* [93], isolated from the inner bark of Pacific or Himalayan yews, respectively, were recently shown to be able to produce in culture, out of the host, small amounts of paclitaxel. This discovery raises the possibility of biosynthetic gene transfer between *Taxus* species and their corresponding endophytic fungi [93].

Sesquiterpenes are other widespread constituents of essential oils. Some of them are of considerable industrial value in the flavor and fragrance industries as well as for pharmaceutical applications. This has stimulated again the studies of their preparation from cheaper and more abundant natural precursors, frequently involving a functionalization at non-activated carbon atoms, which can be more easily attained by bioconversion methods [94].

An instructive example of the use of such microbial hydroxylation technology is found in the synthesis of odorant trace compounds present in patchouli oil (48, 95). Patchoulol **65** is the major (but odorless) constituent (35–40%) of the patchouli essential oil, obtained by steam distillation of dried leaves of *Pogostemon cablin* Benth. (Labiatae). Nor-Patchoulenol **66**, which is present in much lower amounts (0.4–0.6%), is considered to be the most powerful carrier of the olfactory properties of the essence [95].

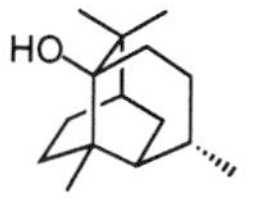

65 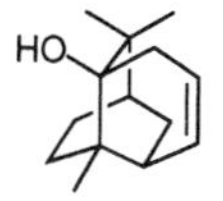   **66**

The transformation of patchoulol to nor-patchoulenol was thus considered, involving a stepwise oxidation of the 10-methyl group to a carboxylic group, followed by oxidative decarboxylation. Such a transformation is probably initiated by the hydroxylation of the 10-methyl group, which is effectively realized in the metabolism of patchoulol by rabbits, rats, and dogs (Fig. 8) [96]. Intensive screening of microorganisms was then initiated [97,98], showing that a number of fungi were able to perform the 10-methyl group hydroxylation, resulting in the required diol **67**. However, several other hydroxylation products at secondary or tertiary positions (**68–70**) were simultaneously formed (Fig. 9), depending on the strain and the incubation conditions, and had to be separated by chromatographic methods. Two *Pithomyces* species, isolated from soil in Japan, were able to produce the diol **67**, at 2–8 g/L concentration, but in apparently too low conversion yields (40–50%) for industrial use [48]. One of the metabolites, the 8-hydroxypatchoulol **70**, preferably produced by *Curvularia lunata* NRRL 2380, can be easily transformed into 7-patchoulenol **71**, a compound with good olfactory properties [95], and to patchoulione **72**, the latter being another important trace component of patchouli oil [48]. Actually, optimization of fermentation conditions and strain improvement by genetic engineering methodology could certainly increase the yield of some valuable metabolites in an industrial production.

(+)-Cedrol **73**, an abundant monohydroxylated sesquiterpene isolated from the essential oil of *Juniperus phoenicea* (Cupressaceae), has also been the subject of several biotransformation studies to prepare oxidized derivatives with potential fragrance properties, such as 3-oxocedrene **74**, which is regarded as an odorous component of cedar wood oil. While *A. niger* [99], *Beauveria sulfurescens* (= *B. bassiana*) [100], *Glomerella cingulata* [101,102], and *Cephalosporium aphidicola* [103] effectively hydroxylated (+)-cedrol predominantly at C-3 (25–45% yield), *B. cereus* afforded exclusively, though in low yield (1–2%), 2S(α)-hydroxycedrol [104]. In addition, a number of microorganisms have been reported to afford complex mixtures of derivatives mono- and dihydroxylated at positions C-2, C-4, C-9, C-10, and C-12 [105,106]. Hydroxylation of epicedrol **75** and relatives bearing hydroxyl group at C-9 and C-15 by *Cephalosporium aphidicola* also took place predominantly at C-3 [107].

73   **74**   **75**

A good example of an industrial application of a simple microbial hydroxylation reaction is the valorization of valencene **76**, a fairly cheap hydrocarbon component of

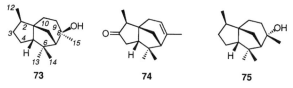

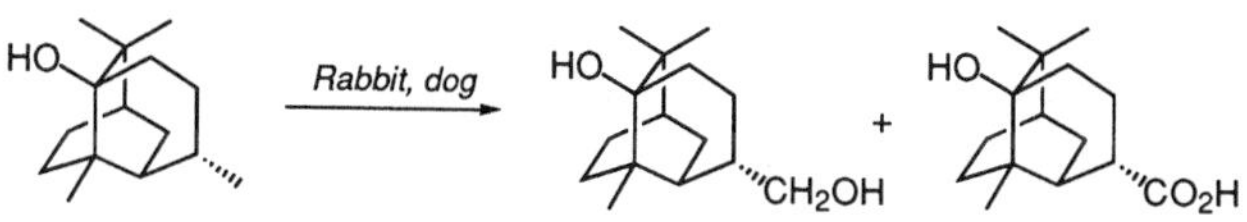

Rabbit, dog

67

Figure 8   The animal metabolism of patchoulol.

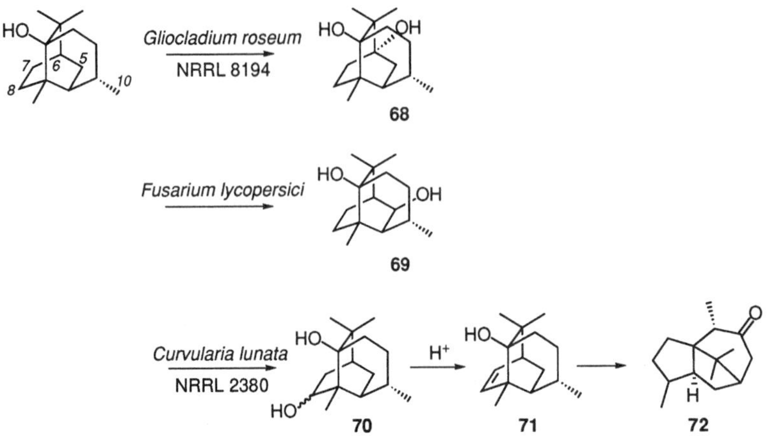

Figure 9 Production of hydroxylated metabolites of patchoulol by various fungal strains.

orange oil, which is converted by bacteria isolated from soil or infected beverages to give the more expensive sesquiterpene nootkatone **77**, an important aroma compound of grapefruit [108].

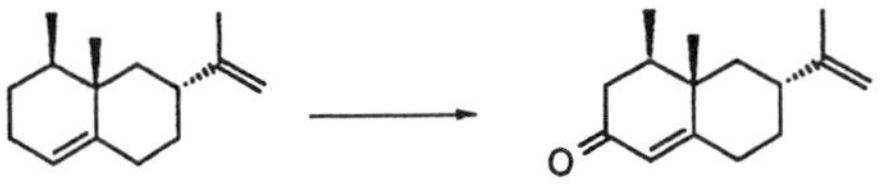

Ionones and their derivatives are other important constituents of various essential oils, with particular relevance in tobacco flavoring. Their oxidative transformation by microorganisms was used to increase their flavoring properties, due to the formation of hydroxylated or keto derivatives. For example, β-ionone **78** is mainly oxidized by A. *niger* to give in good yields and moderate-to-good optical purities 4(*R*)-hydroxy- (**79**) and 2(*S*)-hydroxy-β-ionone (**80**) [109–111]. *Cunninghamella blakesleeana* essentially affords 4-keto derivatives [110], but partial reductions of the side chain are observed (**81, 82**). Up to 10 g/L of β-ionone is rapidly converted by *Lasiodiplodia theobromae* to 2-, 3-, and 4-hydroxylated derivatives, but on prolonged incubation, extensive reduction and degradation reactions of the side chain occurred, leading ultimately to complex mixtures [112]. More recently, efficient methods for quantitative production of hydroxylated species have been described, using an immobilized A. *niger* mycelium without [113] or in the presence of [114] an organic solvent.

One enantiomer of α-ionone (**83**) is preferentially oxidized by various microorganisms, including A. *niger* and C. *blakesleeana*, to give an (allylic) 3-hydroxylated and a 3-keto derivative [110,115]. Similarly oxidized metabolites in positions -2 and/or -4 have been prepared from retinoic acid (**84**), the postulated functional form of vitamin A, using *Cunninghamella* sp. [116], and may have physiological significance as mammalian metabolites.

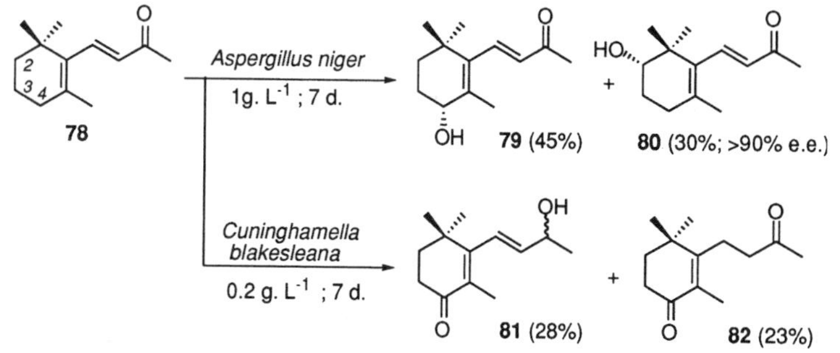

β-Damascone **85**, an isomer of β-ionone, is also majorily transformed (70–95%) into its 4-hydroxylated product by several strains of *Botrytis cinerea*. However, small amounts of 2- and 3-hydroxylated products (2–25%) were observed in these bioconversions [117]. α-Damascone **86** is more extensively metabolized, giving rise to various 3-oxidized metabolites in low yields [118].

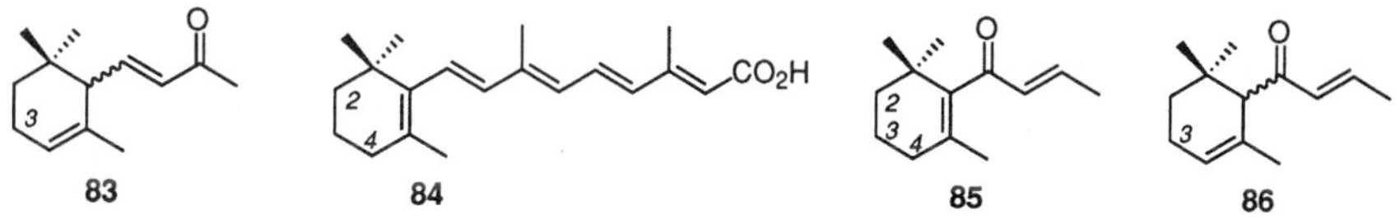

Sesquiterpene lactones with an eudesmane-6α-olide structure are abundant in nature and the biotransformations of some of them, such as α-santonin **87**, have been extensively investigated [94]. Beside reduction products, using *A. niger* MIL 5024, several monohydroxylated metabolites at positions C-8, C-11, and C-13 have recently been isolated [119], while a *Pseudomonas* strain (ATCC 43388), incubated with α-santonin in the presence of an ATPase inhibitor, accumulated a 4,5-dihydroxylated metabolite (**88**) [120].

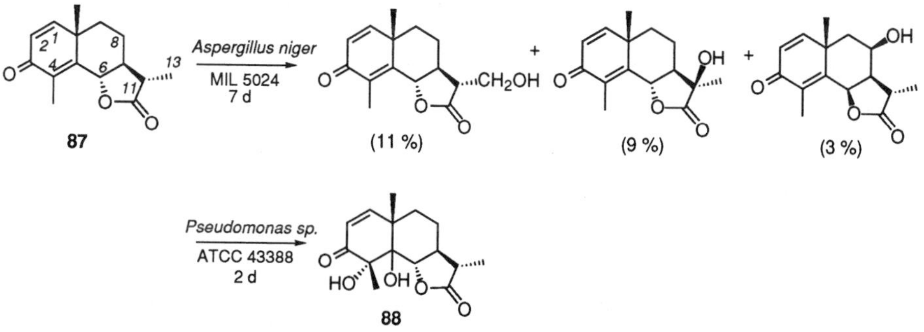

6β-Eudesmanes and eudesmanolide derivatives **89–93** isolated from or chemically obtained from isolated terpenes from *Sideritis* sp. have been used for biotransformation by *Curvularia lunata* or *Rhizopus nigricans*. Beside (sometimes predominant) reduction reactions, hydroxylations occur in substantial yields at the methyl groups of the side chain in **89** [121]; C-1, C-2, and C-4 of the A ring in **90** and **91** [122], and on various positions such as C-1, C-4, or C-8 in **92** [123] and **93** [124,125].

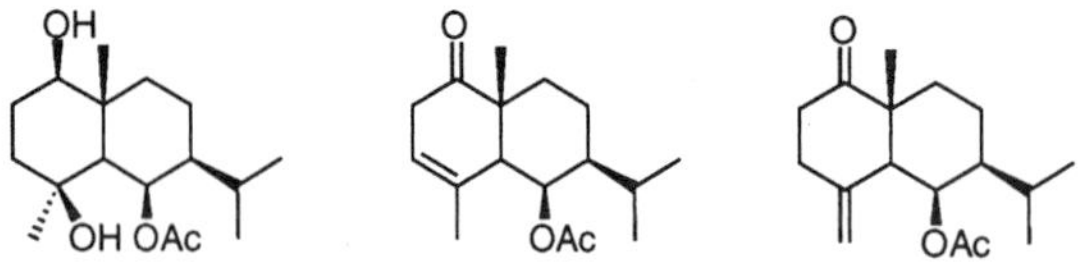

89   90   91

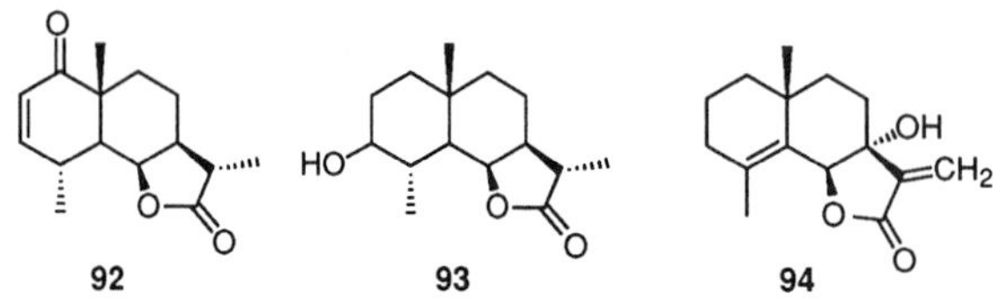

92   93   94

A related sesquiterpene lactone, 7α-hydroxyfrullanolide (**94**), abundantly present in *Sphaerantus indicus* L. (Compositae), and which was demonstrated to have in vitro cytotoxic and antitumor activity, was used for microbial transformation in order to convert it to potentially more useful metabolites. From nine fungi screened, two *Aspergillus* strains were found to reduce or acylate the side chain double bond [126], but no hydroxylation reaction was reported.

Artemisinin (**95**), an unusual endoperoxide sesquiterpenic lactone, isolated from the Chinese medicinal herb *Artemisia annua* (Asteraceae), is an active antimalarial agent used as an alternative drug (*Qinghaosu*) against strains of *Plasmodium* resistant to classical antimalarial drugs. Ethyl and methyl ether derivatives of dihydroartemisinin, arteether (**96**) and artemether (**97**), have been also used for high-risk malarial patients, including those with cerebral malaria. Extensive investigations on the animal and human metabolism of these drugs have been undertaken and mainly showed the reduction of the endoperoxidic linkage to an ether linkage, and consequently the loss of antimalarial activity. The endoperoxide linkage was more resistant in ether derivatives.

Microbial production represented the only convenient source of most of the animal metabolites (11 of them were known for arteether) and, in addition, an appropriate method for providing new analogs that may serve as prospective candidates for antimalarial evaluation or as starting materials for hemisynthesis of derivatives. Arteether (**96**) was converted by *A. niger* or *Nocardia corallina* mainly into deoxy metabolites such as **98**, the hydroxylation products **99** and **100**, and rearrangement products [127,128].

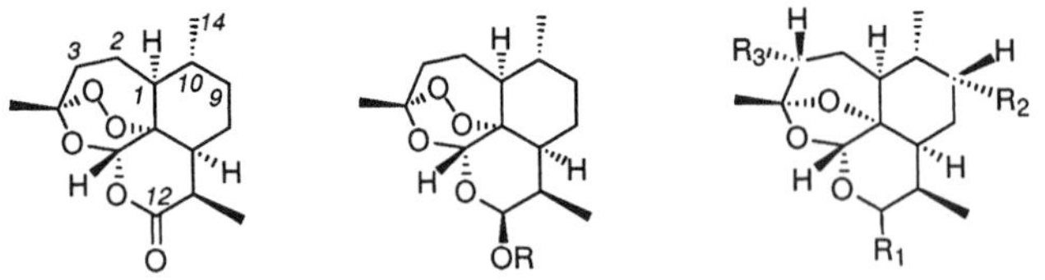

95   96 : R = Et   98 : R_1 = OH; R_2 = R_3 = H
                  97 : R = Me   99 : R_1 = OEt; R_2 = OH; R_3 = I
                                        100 : R_1 = R_3 = OH; R_2 = H

In an attempt to generate one or more hydroxylated new products retaining the endoperoxide grouping, and starting from the known hydroxylation abilities of *Beauveria bassiana* when acting on benzamido group–containing chemicals [129–132], Ziffer et al. [133,134] used as a substrate for this microorganism the *N*-phenylurethane derivative (**101**) of dihydroartemisinin, which was converted in low yield (10%) to the C-14 hydroxylated derivative **102**. Other artemisinin derivatives were also employed as substrates with *B. bassiana*. For example, arteether (**96**) was converted in fair yields into two active new

metabolites **103** and **104** containing the intact endoperoxide group [134,135]. Other metabolites of arteether (including a few retaining the endoperoxide moiety) hydroxylated in
positions -1α, -2α, -9α, -9β or -14, and corresponding to some minor metabolites found
in rat liver microsome incubations, have been recently prepared [136] using large-scale
fermentations with *Cunninghamella elegans* ATCC 9245 and *Streptomyces lavendulae* L-
105. Similar 3α-, 9α-, 9β-, and 14-hydroxy metabolites have been reported to be produced
by the same strains with artemether (**97**) as substrate.

101   102

103 (28%)   104 (10%)   105 (7%)

Anhydroartemisinin (**106**), a semisynthetic derivative with very high antimalarial
activity, was converted to the 9β-hydroxylated derivative **107** by *S. lavendulae* L-105,
whereas a *Rhizopogon* species (ATCC 36060) formed hydroxylated metabolites **108** and
109 [137]. Preparative scale incubations have led to the isolation of sufficient amounts of
these metabolites to obtain clear structural identifications (including x-ray analyses) and
to afford standard samples for HPLC/MS comparison with the 28 corresponding animal
metabolites (the 9β-hydroxy derivative **107** was found as the major metabolite in the rat)
and antimalarial testing.

106 : R = H   108   109
107 : R = OH

Much effort has been concentrated on microbial hydroxylations of dimethylcyclo-
propane-containing sesquiterpenoids, the chemical oxidation of which is particularly delicate due to the instability of the strained cyclopropane ring. Calarene **110** was used as a
substrate in the first experiments [138] but very low yields were observed in its biotransformation. By contrast, functionalized aristolenepoxide **111** [139] or the sesquiterpenoid
enone **117** [140] was more easily oxidized by a number of strains, affording in moderate
yields some new and selectively hydroxylated derivatives **112–116** or **118–120**, respectively, depending on the microorganism used.

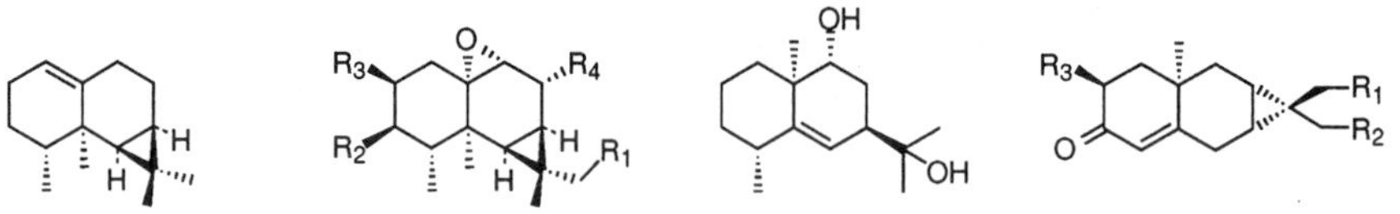

110

111 : R$_1$ = R$_2$ = R$_3$ = R$_4$ = H
112 : R$_1$ = OH; R$_2$ = R$_3$ = R$_4$ = H
113 : R$_2$ = OH; R$_1$ = R$_3$ = R$_4$ = H
114 : R$_3$ = OH; R$_1$ = R$_2$ = R$_4$ = H
115 : R$_4$ = OH; R$_1$ = R$_2$ = R$_3$ = H

116

117 : R$_1$ = R$_2$ = R$_3$ = H
118 : R$_1$ = OH; R$_2$ = R$_3$ = H
119 : R$_2$ = OH; R$_1$ = R$_3$ = H
120 : R$_3$ = OH; R$_1$ = R$_2$ = H

As usually found with terpenoid hydrocarbons, hydroxylations of aromadendrene **121** or its 8-epimer **124** are difficult: the fungal plant pathogen *Glomerella cingulata* [141] regioselectively hydroxylated one of the geminal methyl groups on the cyclopropane ring, in addition to the stereoselective dihydroxylation of the double bond (probably via hydro-lytic opening of the corresponding epoxides) to give the corresponding triols **123** or **126**; bacterial strains such as *Bacillus megaterium* DSM 32 or *Mycobacterium smegmatis* DSM 43061 have been previously shown to afford very low yields of hydroxylated derivatives **123** and **125**, respectively [138]. It is interesting to note that the direct introduction of the hydroxyl group at C-13 proceeds in both substrates from the α side, without any effect of the configuration of the ring junction; on the contrary, the stereochemistry of the dihy-droxylation of the double bond (and thus that of the intermediate epoxide) is strictly dependent on the substrate configuration at C-8.

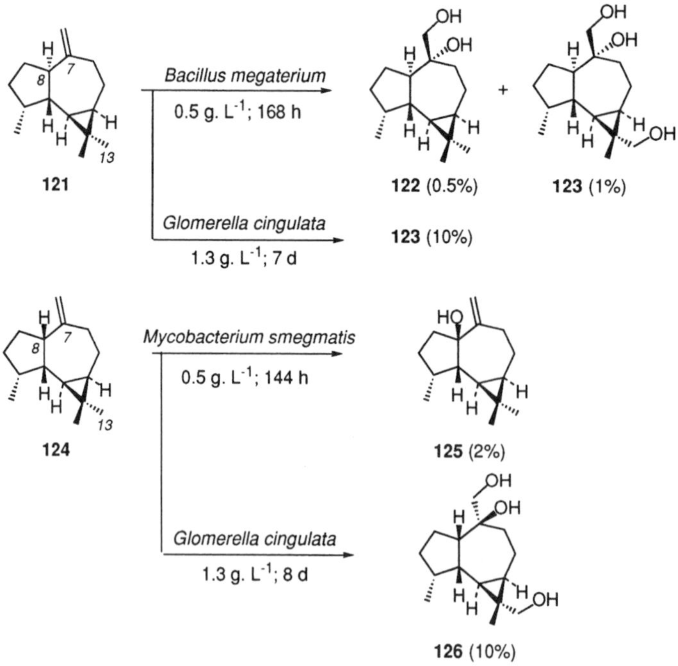

121

Bacillus megaterium
0.5 g. L^{-1}; 168 h

Glomerella cingulata
1.3 g. L^{-1}; 7 d

122 (0.5%)   +   **123** (1%)

123 (10%)

124

Mycobacterium smegmatis
0.5 g. L^{-1}; 144 h

Glomerella cingulata
1.3 g. L^{-1}; 8 d

125 (2%)

126 (10%)

Other related dimethylcyclopropane sesquiterpenoids, globulol (**127**) and 7-epiglob-ulol (**128**), are more efficiently transformed by various microorganisms [138] into similar products such as **129**, mainly hydroxylated (and sometimes further oxidized) on one (or both) geminal methyl group(s) of the cyclopropane ring. *Cephalosporium aphidicola* [142] and *G. cingulata* [143] are particularly efficient strains; with such miroorganisms, the C-7 configuration of the substrates does not appear to have any stereochemical directing effect on the regio- and stereoselectivity of the methyl group hydroxylation. Several interesting features emerge from these results. The easy opening of a cyclopropane group has been frequently used as a mechanistic probe for the formation of a free radical intermediate in enzymic (hydroxylation) reactions. However, in all examples shown above, degradation of the dimethylcyclopropane ring was only exceptionally observed. Enzymatic attack at the *exo*-methyl group generally predominates, and no stereodirecting effects of substituents at C-7 or ring junction configuration were observed, which probably reflects a constantly favored disposition of this exo group at the external and more accessible part of the substrate molecules. The importance of the microbial oxidation techniques for mild functionalization of complex molecules is particularly noticeable with such substrates.

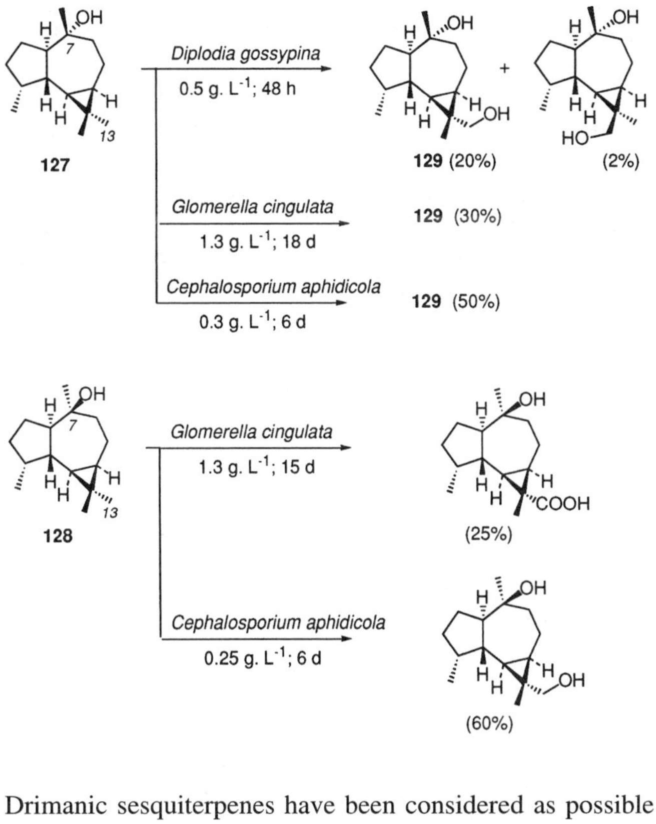

Drimanic sesquiterpenes have been considered as possible starting materials for the hemisynthesis of forskolin and several other related multifunctionalized compounds, exhibiting various biological activities. For example, drimenol **130** can be obtained from the terpenoid fraction of a South American shrub, *Drimys winteri* Forst. (Winteraceae) or by synthetic methods [144]. A key synthon, the diol **131**, can be obtained by chemical synthetic methods [145] or by oxidative degradation of labdanoid compounds [146–148], and its saturated counterpart can be easily prepared by chemical synthesis of the racemate, followed by resolution by enzymatic hydrolysis of an ester derivative [149]. As expected, the microbial hydroxylation of all of these compounds or protected derivatives, such as **132**, mainly affords 3β-hydroxy derivatives, plus minor 6α-hydroxy derivatives resulting from allylic hydroxylation of the 7,8-unsaturated substrates [145,150–152]. In no case could the desired hydroxylation at position 1(α) be observed.

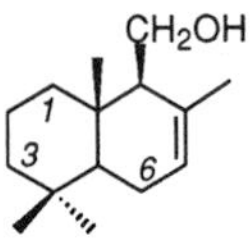

Several methods have been proposed for the transfer of the 3β-hydroxy functionalization to the 1α position, in order to prepare, say, chimeric compounds (such as **133**) containing the basic structural features of polygodial or warburganal, two known drimenic antifeedant agents for insects, and the 1α-hydroxyl group of forskolin, which are present together in deacetylscalaradial, a recently discovered marine sponge sesterterpene with cytotoxic activity [152–155]. In the simpler reaction (Fig. 10), after the classical microbial 3β-hydroxylation of a selected precursor **131** and epoxidation of the 7,8 double bond, 2,3-dehydration was obtained by a modified Mitsunobu reaction, and SeO$_2$ oxidation at the allylic position afforded the desired 1α-hydroxy derivatives, which were hydrogenated before regeneration of the 7,8 double bond [155].

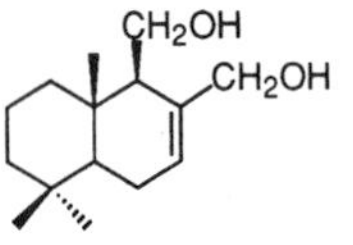

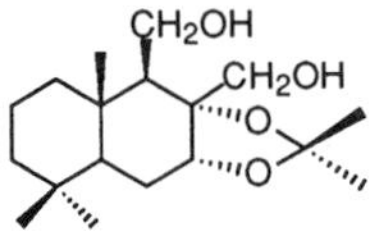

Monoterpenic cyclic hydrocarbons are readily available as inexpensive natural products that can represent useful starting materials for fragrance and flavor compounds [48,156–158]. Moreover, some monoterpenoid hydroxylated derivatives constitute a source of homochiral compounds that are frequently used as a basis for chiral auxiliaries or reagents for asymmetrical synthesis.

While R-(+)-limonene (**134**), for example, was completely and specifically converted by *Penicillium digitatum* DSM 6284 to the corresponding α-terpineol enantiomer **135** [159,160], both limonene enantiomers were oxidized by *Corynespora cassiicola* DSM

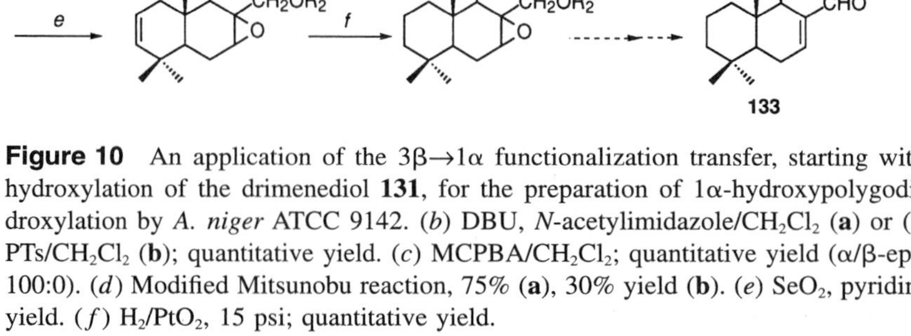

Figure 10 An application of the 3β→1α functionalization transfer, starting with microbial 3β-hydroxylation of the drimenediol **131**, for the preparation of 1α-hydroxypolygodial **133**. (*a*) Hydroxylation by *A. niger* ATCC 9142. (*b*) DBU, *N*-acetylimidazole/CH$_2$Cl$_2$ (**a**) or (CH$_3$)$_2$C(OCH$_3$)$_2$, PTs/CH$_2$Cl$_2$ (**b**); quantitative yield. (*c*) MCPBA/CH$_2$Cl$_2$; quantitative yield (α/β-epoxide: **a**, 2:8; **b**; 100:0). (*d*) Modified Mitsunobu reaction, 75% (**a**), 30% yield (**b**). (*e*) SeO$_2$, pyridine *N*-oxide; 40% yield. (*f*) H$_2$/PtO$_2$, 15 psi; quantitative yield.

62475 or *Diplodia gossypina* ATCC 10936 to afford in good yields, on a 1000-kg scale, (*1S,2S,4R*)-*p*-menth-8-ene-1,2-diol (**136**) from the (*R*)-(+) enantiomer and the corresponding (*1R,2R,4S*)-diol from the (*S*)-(−) enantiomer [159]. However, biotransformation of (*R*)- and (*S*)-limonene by another fungal microorganism, *Aspergillus cellulosae* M-77, afforded in addition several products resulting from allylic hydroxylation such as alcohols **138** and **139** and isopiperitenone **137**, as enantiomerically pure compounds retaining the initial configuration of limonene [161].

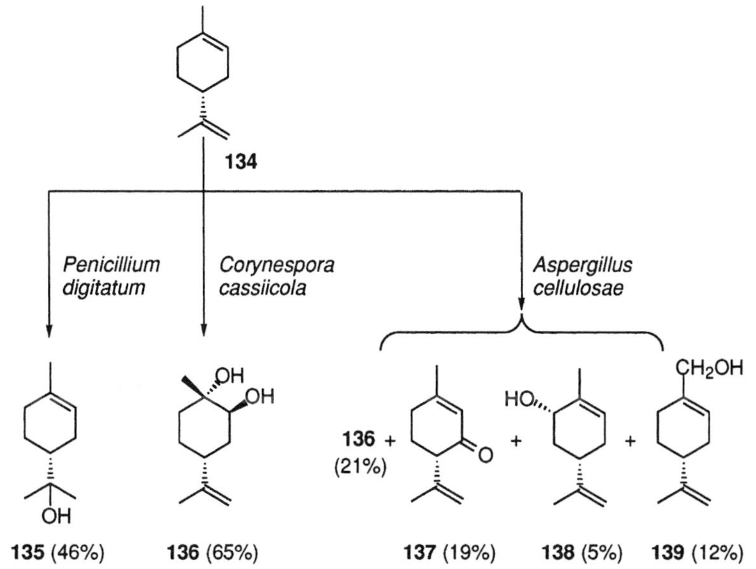

Other monoterpenic hydrocarbons, such as carenes [162] and pinenes (the most abundant terpenes in nature) [163–165], are less specifically oxidized to a number of alcohol and keto derivatives, generally in low yields [48], resulting from the high volatility of the starting material and the multiplicity of products.

The microbial hydroxylation of 1,4- and 1,8-cineoles has been more recently examined as a potential source for asymmetrical synthons or chiral auxiliaries. Both terpenes are prochiral meso substrates, susceptible to give several chiral-functionalized derivatives by regioselective and asymmetrical oxidation. The hydroxylation of 1,4-cineole (**140**) by *Bacillus cereus* UI-1477 was found to proceed on both enantiopic faces with a high regio- and stereoselectivity, affording two optically pure isomers (**141, 142**), in an endo-to-exo ratio of 7:1. A screening of a number of microorganisms showed that most of them afforded essentially the same 2-endo derivative **141** [166–170]. A different situation was found with 1,8-cineole or eucalyptol [143], the main component of *Eucalyptus* essential oils [48]. Introduction of a hydroxyl group by different microorganisms [139] was sometimes very selective, affording majorly or exclusively an optically pure 6-exo-hydroxy derivative (**145, 146**), with opposite enantiopic specificities depending on the microorganism used [171–174]; 6-endo derivatives could then be derived from the 6-exo isomers by oxidation followed by stereoselective hydride reduction [175]. The 5-exo-hydroxylated derivative (**144**) was obtained with a number of fungal strains, but generally as a racemic compound [174,176], indicating a low enantiopic discrimination in this position.

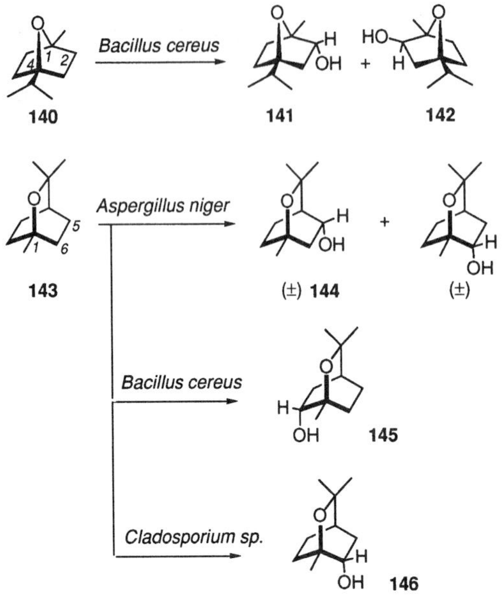

The biotransformation of monoterpenoid alcohols or ketones by microorganisms has been also extensively investigated. As a result of the presence of a polar anchoring group in such substrates, hydroxylation reactions by monooxygenases are easier and more productive, and several useful hydroxylated derivatives have been described. For example, *A. niger* biotransforms (+)- and (−)-menthol [177] and their stereoisomers [178] to give in moderate to good yields a number of nonspecifically hydroxylated derivatives, as shown with (+)-menthol **147**. Another example is illustrated by the biotransformation of isopinocampheol enantiomers **148–149** (Fig. 11), using *Mortierella isabellina* DSM 63355 [179].

Beside reduction of the carbonyl group, which often constitutes the unique biotransformation observed, some monoterpenoid ketones such as (±)-piperitone (**150**) [180], (*R*)-pulegone (**151**) [181,182], β- (**152**) [183] or α-thujone (**153**) [184] are hydroxylated in

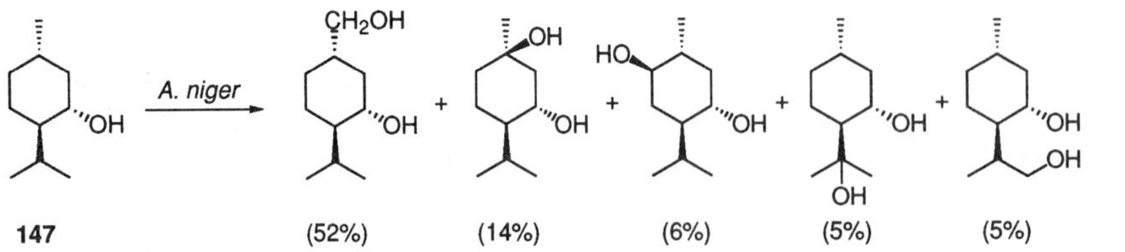

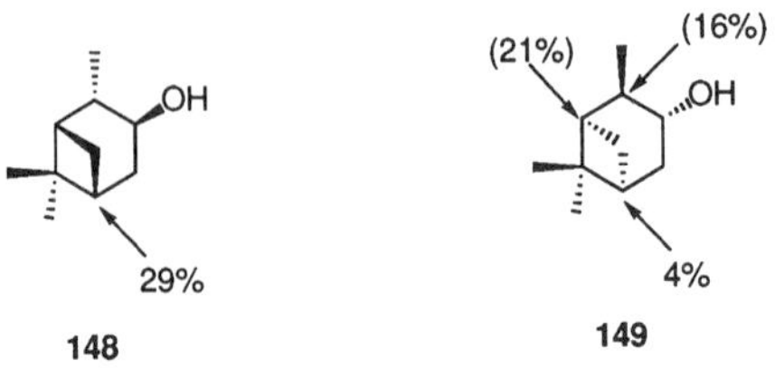

Figure 11   Hydroxylation patterns of (+)-isopinocampheol (**148**) and (−)-isopinocampheol (**149**) by *Mortierella isabellina* DSM 63355.

good yields. In most cases, as expected, hydroxylations priorly take place on allylic and tertiary positions.

III. CONCLUSION

The use of microorganisms in the field of terpenoid hydroxylations will certainly become increasingly popular. The examples shown above provide some evidence about the versatility of such reactions to produce, sometimes in high yields, a number of hydroxylated derivatives that would otherwise be impossible to access by usual chemical techniques. Despite the current multiplicity of hydroxylated products obtained, the use of whole microorganisms is still justified [185,186], owing to the actual difficulties in isolating and purifying the enzymes involved. Except in a few cases, the predictability of the regio- and stereoselectivity of such reactions remains low. However, in the future, expression of individual specific monooxygenases in various host systems would probably become feasible and help in the prediction of hydroxylation reactions.

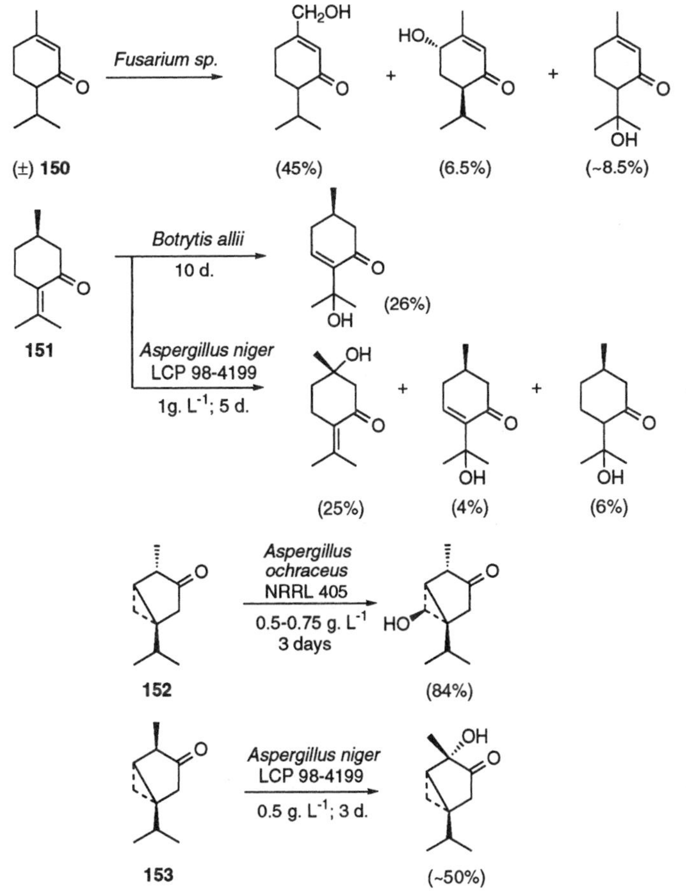

REFERENCES

1. Y Yamazaki, Y Hayashi, N Hori, Y Mikami. Agric Biol Chem 52:2921–2922, 1988.
2. D Busmann, RG Berger. J Biotechnol 37:39–43, 1994.

3. W-R Abraham, H-A Arfmann. Tetrahedron 48:6681–6688, 1992.
4. W-R Abraham, H-A Arfmann, W Giersch. Z Naturforsch 47c:851–858, 1992.
5. HA Arfmann, WR Abraham, K Kieslich. Biocatalysis 2:59–67, 1988.
6. KM Madyastha, NSR Krishna Murthy. Appl Microbiol Biotechnol 28:324–329, 1988.
7. KM Madyastha, TL Gururaja. Appl Microbiol Biotechnol 38:738–741, 1993.
8. G Bock, I Benda, P Schreier. Appl Microbiol Biotechnol 27:351, 1988.
9. DS Holmes, DM Ashworth, JA Robinson. Helv Chim Acta 73:260–271, 1990.
10. A Müller, WR Abraham, K Kieslich. Bull Soc Chim Belg 103:405–423, 1994.
11. M Miyazawa, H Nankai, H Kameoka. Phytochemistry 40:1133–1137, 1995.
12. M Miyazawa, H Nankai, H Kameoka. Phytochemistry 43:105–109, 1995.
13. M Miyazawa, H Nankai, H Kameoka. J Agric Food Chem 44:1543–1547, 1996.
14. H Nankai, M Miyazawa, H Kameoka. J Nat Prod 60:287–289, 1997.
15. DHR Barton, JC Beloeil, A Billion, J Boivin, JY Lallemand, P Lelandais, S Mergui. Helv Chim Acta 70:2187–2200, 1987.
16. T Hieda, Y Mikami, Y Obi, T Kisaki. Agric Biol Chem 47:243–250, 1983.
17. G Aranda, A Hammoumi, R Azerad, JY Lallemand. Tetrahedron Lett 32:1783–1786, 1991.
18. G Aranda, MS El Kortbi, J-Y Lallemand, A Hammoumi, I Facon, R Azerad. Tetrahedron 47:8339–8350, 1991.
19. SA Kouzi, JD McChesney. J Nat Prod 54:483–490, 1991.
20. JR Hanson, PB Hitchcock, H Nasir, A Truneh. Phytochemistry 36:903–906, 1994.
21. WR Abraham. Phytochemistry 36:1421–1424, 1994.
22. SA Kouzi, JD McChesney. Xenobiotica 21:1311–1323, 1991.
23. SA Kouzi, JD McChesney. Helv Chim Acta 73:2157–2164, 1990.
24. SA Kouzi, JD McChesney, LA Walker. Xenobiotica 23:621–632, 1993.
25. K Yasui, K Kawada, K Kagawa, K Tokura, K Kitadokoro, Y Ikenishi. Chem Pharm Bull Tokyo 41:1698–1707, 1993.
26. W Artez, D Boettger, K Sauber. Ger.Offen. DE 3,527,336; Chem Abstr 106:212559w, 1987.
27. W Aretz, D Boettger, K Sauber. Ger.Offen. DE 3,527,335; Chem Abstr 106:212558v, 1987.
28. SR Nadkarni, PM Akut, BN Ganguli, Y Khandelwal, NJ De Souza, RH Rupp. Tetrahedron Lett 27:5265–5268, 1986.
29. Y Khandelwal, PK Inamdar, NJ De Souza, RH Rupp, S Chatterjee, BN Ganguli. Tetrahedron 44:1661–1666, 1988.
30. Y Kahndelwal, NJ De Souza, S Chatterjee, BN Ganguli, RH Rupp. Tetrahedron Lett 28:4089–4092, 1987.
31. T Hieda, Y Mikami, Y Obi, T Kisaki. Agric Biol Chem 46:2249–2255, 1982.
32. T Hieda, Y Mikami, Y Obi. Agric Biol Chem 47:47–51, 1983.
33. CR Enzell, A Nasiri, I Forsblom, E Johansson, C Lundberg, J Arnap, L Björk. Acta Chem Scand 49:375–379, 1995.
34. D Herlem, J Ouazzani, F Khuong-Huu. Tetrahedron Lett 51:1241–1244, 1996.
35. Atta-Ur-Rahman, A Farooq, MI Choudhary. J Nat Prod 60:1038–1040, 1997.
36. JR Hanson, A Truneh. Phytochemistry 42:1021–1023, 1996.
37. BM Fraga, P Gonzàlez, R Guillermo, MG Hernandez. Tetrahedron 54:6159–6168, 1998.
38. JM Arias, A Garcia-Granados, MB Jimenez, A Martinez, F Rivas, ME Onorato. J Chem Res (S) 277, 1988.
39. A Garcia-Granados, A Martinez, MB Jimenez, ME Onorato, F Rivas, JM Arias. J Chem Res (S) 94–95, 1990.
40. A Garcia-Granados, MB Jimenez, A Martinez, A Parra, F Rivas, JM Arias. Phytochemistry 37:741–747, 1994.
41. A Garcia-Granados, A Martinez, JP Martinez, ME Onorato, JM Arias. J Chem Soc Perkin Trans I, 1261–1266, 1990.
42. A Garcia-Granados, E Linan, A Martinez, ME Onorato, A Parra, JM Arias. Phytochemistry 38:287–293, 1995.
43. A Garcia-Granados, E Linan, A Martinez, F Rivas, JM Arias. Phytochemistry 38:1237–1244, 1995.

44. BM Fraga, P Gonzàlez, R Guillermo, MG Hernandez, J Rovirosa. Phytochemistry 28:1851–1854, 1989.
45. A Garcia-Granados, E Linan, A Martinez, F Rivas, CM Mesa-Valle, JJ Castilla-Calvente, A Osuna. J Nat Prod 60:13–16, 1997.
46. K Kieslich. Microbial transformation of non-steroid cyclic compounds. Stuttgart: Georg Thieme Verlag, 1976.
47. OK Sebek, K Kieslich. In: D Perlman, ed. Annual Reports on Fermentation Processes. New York: Academic Press, 1977, pp 267–297.
48. V Krasnobajew. In: K Kieslich, ed. Biotechnology. Weinheim: Verlag Chemie, 1984, pp 31–77.
49. BM Fraga, P Gonzàlez, R Guillermo, JR Hanson, MG Hernandez, JA Takahashi. Phytochemistry 37:717–721, 1994.
50. BM Fraga, MG Hernandez, R Guillermo. J Nat Prod 59:952–957, 1996.
51. BM Fraga, P Gonzàlez, MG Hernandez, FG Tellado, A Perales. Phytochemistry 25:1235–1237, 1986.
52. JR Hanson, BH De Oliveira. Phytochemistry 29:3805–3807, 1990.
53. JR Hanson, BH De Oliveira. Nat Prod Rep 10:301–309, 1993.
54. BH De Oliveira, RA Strapasson. Phytochemistry 43:393–395, 1996.
55. JM Arias, A Garcia-Granados, A Martinez, ME Onorato. J Nat Prod 47:59–63, 1984.
56. A Garcia-Granados, A Martinez, ME Onorato, JM Arias. J Nat Prod 49:126–132, 1986.
57. A Garcia-Granados, A Martinez, A Ortiz, ME Onorato, JM Arias. J Nat Prod 53:441–450, 1990.
58. A Garcia-Granados, A Martinez, ME Onorato, ML Ruiz, JM Sanchez, JM Arias. Phytochemistry 29:121–126, 1990.
59. A Garcia-Granados, A Guerrero, A Martinez, A Parra, JM Arias. Phytochemistry 36:657–663, 1994.
60. BM Fraga, JCD Gomez, M Shaiq Ali, JR Hanson. Phytochemistry 34:693–696, 1993.
61. JR Hanson, AG Jarvis. Phytochemistry 36:1395–1398, 1994.
62. JR Hanson, AG Jarvis, F Laboret, J Takahashi. Phytochemistry 38:73–75, 1995.
63. MAD Boaventura, JR Hanson, PB Hitchcock, JA Takahashi. Phytochemistry 37:387–389, 1994.
64. JR Hanson, PB Hitchcock, JA Takahashi. Phytochemistry 40:797–800, 1995.
65. AB Deoliveira, JR Hanson, JA Takahashi. Phytochemistry 40:439–442, 1995.
66. JR Hanson, PB Reese, JA Takahashi, MR Wilson. Phytochemistry 36:1391–1393, 1994.
67. FA Badria, CD Hufford, Phytochemistry 30:2265–2268, 1991.
68. CD Hufford, FA Badria, M Abou-Karam, WT Shier, RD Rogers. J Nat Prod 54:1543, 1991.
69. S-J Lin, JPN Rosazza. J Nat Prod 61:922–926, 1998.
70. JP Kutney, JM Hewitt, PJ Salisbury, M Singh, JA Servizi, DW Martens, RW Gordon. Helv Chim Acta 66:2191–2197, 1983.
71. JP Kutney, LS Choi, JM Hewitt, PJ Salisbury, M Singh. Appl Environ Microbiol 49:96–100, 1985.
72. S Hiranuma, T Shimizu, H Yoshioka. Chem Pharm Bull 39:2167–2169, 1991.
73. AJ Aladesanmi, JJ Hoffmann. Phytochemistry 30:1847–1848, 1991.
74. JJ Hoffmann, LK Hutter, SJ Dentali, KH Schram, NE Mackenzie. Phytochemistry 31:3045–3049, 1992.
75. CR Carreras, J Rodriguez, HJ Siva, P Rossomando, OS Giordano, E Guerreiro. Phytochemistry 41:473–475, 1996.
76. JJ Hoffmann, BM Fraga. Phytochemistry 33:827–830, 1993.
77. A Garciagranados, A Parra, JM Arias. J Nat Prod 60:86–92, 1997.
78. M Alam, JR Hanson. Phytochemistry 29:3801–3803, 1990.
79. R Milanova, M Moore. Arch Biochem Biophys 303:165–171, 1993.
80. R Milanova, M Moore, Y Hirai. J Nat Prod-Lloydia 57:882–889, 1994.
81. R Milanova, K Han, M Moore. J Nat Prod-Lloydia 58:68–73, 1995.

82. R Milanova, N Stoynov, M Moore. Enzyme Microb Technol 19:86–93, 1996.
83. F Gueritte-Voegelein, D Guénard, P Potier. J Nat Prod 50:9–18, 1987.
84. DGI Kingston, G Samaranayake, CA Ivey. J Nat Prod 53:1–12, 1990.
85. DGI Kingston. Pharmac Ther 52:1–34, 1991.
86. RL Hanson, JM Wasylyk, VB Nanduri, DL Cazzulino, RN Patel, LJ Szarka. J Biol Chem 269:22145–22149, 1994.
87. SH Hu, XF Tian, WH Zhu, QC Fang. Tetrahedron 52:8739–8746, 1996.
88. SH Hu, XF Tian, WH Zhu, QC Fang. J Nat Prod 1006–1009, 1996.
89. S Hu, X Tian, W Zhu, Q Fang. Biocat Biotransformation 14:241–250, 1997.
90. J Zhang, L Zhang, X Wang, D Qiou, D Sun, J Gu, Q Fang. J Nat Prod 61:497–500, 1998.
91. A Stierle, G Strobel, D Stierle. Science 260:214–216, 1993.
92. A Stierle, G Strobel, D Stierle, P Grothaus, G Bignami. J Nat Prod-Lloydia 58:1315–1324, 1995.
93. G Strobel, XS Yang, J Sears, R Kramer, RS Sidhu, WM Hess. Microbiology-UK 142:435–440, 1996.
94. V Lamare, R Furstoss. Tetrahedron 46:4109–4132, 1990.
95. P Teisseire. Bull Soc Chim Fr II/66–70, 1980.
96. Luu-Bang, G Ourisson, P Teisseire. Tetrahedron Lett 26:2211, 1975.
97. E Becher, W Schüepp, PK Matzinger, PJ Teisseire, C Ehret, H Maruyama, Y Suhara, S Ito, M Ogawa, K Yokose, T Sawada, A Fujiwara, M Fujiwara, M Tazoe, Y Schlomi. Br Pat 1 586 759, 1981.
98. Y Suhara, S Ito, M Ogawa, K Yokose, T Sawada, T Sano, R Ninomiya, H Maruyama. Appl Environ Microbiol 42:187–191, 1981.
99. KC Wang, LY Ho, YS Cheng. J Chin Biochem Soc 1:53; Chem Abst 79 (1972) 113975w, 1972.
100. V Lamare, JD Fourneron, R Furstoss, C Ehret, B Corbier. Tetrahedron Lett 28:6269–6272, 1987.
101. M Miyazawa, H Nankai, H Kameoka. Chem Express 8:573–576, 1993.
102. M Miyazawa, H Nankai, H Kameoka. Phytochemistry 40:69–72, 1995.
103. JR Hanson, H Nasir. Phytochemistry 33:835–837, 1993.
104. G Maatooq, S Elsharkawy, MS Afifi, JPN Rosazza. J Nat Prod-Lloydia 56:1039–1050, 1993.
105. WR Abraham, P Washausen, K Kieslich. Z Naturforsch 42C:414–419, 1987.
106. BM Fraga, R Guillermo, JR Hanson, A Truneh. Phytochemistry 42:1583–1586, 1996.
107. E Gand, JR Hanson, H Nasir. Phytochemistry 39:1081–1084, 1995.
108. RS Dhavlikar, G Albroscheit. Dragoco Rep 20:251–258, 1973.
109. Y Mikami, E Watanabe, Y Fukunaga, T Kisaki. Agric Biol Chem 42:1075–1077, 1978.
110. DA Hartman, ME Pontones, VF Kloss, RW Curley Jr, LW Robertson. J Nat Prod 51:947–953, 1988.
111. H Kakeya, T Sugai, H Ohta. Agric Biol Chem 55:1873–1876, 1991.
112. V Krasnobajew, D Helmlinger. Helv Chim Acta 65:1590–1601, 1982.
113. C Larroche, C Creuly, J-B Gros. Appl Microbiol Biotechnol 43:222–227, 1995.
114. K Sode, I Karube, R Araki, Y Mikami. Biotechnol Bioeng 33:1191–1195, 1989.
115. Y Yamazaki, Y Hayashi, M Arita, T Hieda, Y Mikami. Appl Environ Microbiol 54:2354–2360, 1988.
116. DA Hartman, JB Basil, LW Robertson, RW Curley Jr. Pharm Res 7:270–273, 1990.
117. E Schwab, P Schreier, I Benda. Z Naturforsch C 46:395–397, 1991.
118. E Schoch, I Benda, P Schreier. Appl Environ Microbiol 57:15–18, 1991.
119. M Iida, S Wakuri, S Mineki, K Nishitani, K Yamakawa. J Ferment Bioeng 76:296–299, 1993.
120. D Colaco, I Furtado, UP Naik, S Mavinkurve, SK Paknikar. Lett Appl Microbiol 17:212–214, 1993.
121. JM Arias, A Garcia-Granados, A Martinez, ME Onorato, F Rivas. Tetrahedron Lett 29:4471–4472, 1988.
122. A Garcia-Granados, A Martinez, F Rivas, ME Onorato, JM Arias. J Nat Prod 53:436–440, 1990.

123. Y Amate, JL Breton, A Garcia-Granados, A Martinez, E Onorato, A Saenz de Buruaga. Tetrahedron 46:6939–6950, 1990.
124. Y Amate, A Garcia-Granados, A Martinez, A Saenz de Buruaga, JL Breton, E Onorato, JM Arias. Tetrahedron 47:5811–5818, 1991.
125. Y Garcia, A Garciagranados, A Martinez, A Parra, F Rivas, JM Arias. J Nat Prod-Lloydia 58:1498–1507, 1995.
126. Atta-Ur-Rahman, MI Choudhary, A Ata, M Alam, A Farooq, S Perveen, MS Shekhani, N Ahmed. J Nat Prod 57:1251–1255, 1994.
127. I-S Lee, HL ElSohly, CD Hufford. Pharm Res 7:199–203, 1990.
128. CD Hufford, I-S Lee, HL ElSohly, HT Chi, JK Baker. Pharm Res 7:923, 1990.
129. GS Fonken, RA Johnson. Chemical Oxidations with Microorganisms. New York: Marcel Dekker, 1972.
130. RA Johnson, ME Herr, HC Murray, GS Fonken. J Org Chem 33:3217–3221, 1968.
131. R Furstoss, A Archelas, B Waegell. Tetrahedron Lett 21:451–454, 1980.
132. J-D Fourneron, A Archelas, B Vigne, R Furstoss. Tetrahedron 43:2273–2284, 1987.
133. YL Hu, RJ Highet, D Marion, H Ziffer. J Chem Soc Chem Commun 1176–1177, 1991.
134. H Ziffer, Y Hu, Y Pu. In: S Servi, ed. Microbial Reagents in Organic Synthesis. Dordrecht: Kluwer Academic, 1992, pp 361–373.
135. YL Hu, H Ziffer, GY Li, HJC Yeh. Bioorg Chem 20:148–154, 1992.
136. CD Hufford, SI Khalifa, KI Orabi, FT Wiggers, R Kumar, RD Rogers, CF Campana. J Nat Prod 58:751–755, 1995.
137. SI Khalifa, JK Baker, M Jung, JD Mcchesney, CD Hufford. Pharmaceut Res 12:1493–1498, 1995.
138. WR Abraham, K Kieslich, B Stumpf, L Ernst. Phytochemistry 31:3749–3755, 1992.
139. WR Abraham, A Riep, H-P Hanssen. Bioorg Chem 24:19–28, 1996.
140. C Hebda, J Szykula, J Orpiszewski, P Fischer. Biol Chem Hoppe-Seyler 372:337–344, 1991.
141. M Miyazawa, T Uemura, H Kameoka. Phytochemistry 40:793–796, 1995.
142. JR Hanson, PB Hitchcock, R Manickavasagar. Phytochemistry 37:1023–1025, 1994.
143. M Miyazawa, T Uemura, H Kameoka. Phytochemistry 37:1027–1030, 1994.
144. H Tanimoto, T Oritani. Tetrahedron: Asymmetry 7:1695–1704, 1996.
145. DM Hollinshead, SC Howell, SV Ley, M Mahon, NM Ratcliffe, PA Worthington. J Chem Soc Perkin Trans I, 1579–1589, 1983.
146. AF Barrero, EA Manzaneda, J Altarejos, S Salido, JM Ramos. Tetrahedron Lett 35:2945–2948, 1994.
147. AF Barrero, EA Manzaneda, J Altarejos, S Salido, JM Ramos, MSJ Simmonds, WM Blaney. Tetrahedron 51:7435–7450, 1995.
148. JG Urones, IS Marcos, BG Pérez, AM Lithgow, D Diez, PM Gomez, P Basabe, NM Garrido. Tetrahedron 51:1845–1860, 1995.
149. H Akita, M Nozawa, H Shimizu. Tetrahedron: Asymmetry 9:1789–1799, 1998.
150. G Aranda, I Facon, JY Lallemand, M Leclaire, R Azerad, M Cortes, J Lopez, H Ramirez. Tetrahedron Lett 33:7845–7848, 1992.
151. HE Ramirez, M Cortes, E Agosin. J Nat Prod 56:762–764, 1993.
152. G Aranda, J-Y Lallemand, R Azerad, M Maurs, M Cortes, H Ramirez, G Vernal. Synthetic Commun 24:2525–2535, 1994.
153. T Tozyo, F Yasuda, H Nakai, H Tada. J Chim Soc Perkin Trans I, 1859–1866, 1992.
154. G Aranda, M Bertranne-Delahaye, R Azerad, M Maurs, M Cortes, H Ramirez, G Vernal, T Prangé. Synthetic Commun 27:45–60, 1997.
155. G Aranda, M Bertranne-Delahaye, J-Y Lallemand, R Azerad, M Maurs, M Cortes, J Lopez. J Mol Catalysis B Enzymatic 5:203–206, 1998.
156. FW Welsh, WD Murray, RE Williams. Crit Rev Biotechnol 9:105–169, 1989.
157. L Janssens, HL De Pooter, NM Schamp, EJ Vandamme. Proc Biochem 27:195–215, 1992.
158. S Hagedorn, B Kaphammer. Annu Rev Microbiol 48:773–800, 1994.
159. W-R Abraham, HMR Hofmann, K Kieslich, G Reng, B Stumpf. In: R Porter, S Clark, eds. CIBA Foundation Symposium 111. London: Pitman, 1985, pp 146–157.

160. Q Tan, DF Day. Appl Microbiol Biotechnol 49:96–101, 1998.
161. Y Noma, S Yamasaki, Y Asakawa. Phytochemistry 31:2725–2727, 1992.
162. B Stumpf, V Wray, K Kieslich. Appl Microbiol Biotechnol. 33:251–254, 1990.
163. PK Bhattacharyya, BR Prema, BD Kulkarni, SK Pradhan. Nature 187:689–690, 1960.
164. BR Prema, PK Bhattacharyya. Appl Microbiol 10:524–528, 1962.
165. D Busmann, RG Berger. Z Naturforsch C 49:545–552, 1994.
166. JPN Rosazza, JJ Steffens, FS Sariaslani, A Goswami, JM Beale, S Reeg, R Chapman. Appl Environ Microbiol 53:2382, 1987.
167. W-G Liu, A Goswami, RP Steffek, R Chapman, FS Sariaslani, JJ Steffens, JPN Rosazza. J Org Chem 53:5700, 1988.
168. M Miyazawa, Y Noma, K Yamamoto, H Kameoka. Chem Express 6:771–774, 1991.
169. M Miyazawa, Y Noma, K Yamamoto, H Kameoka. Chem Express 7;125–128, 1992.
170. M Miyazawa, Y Noma, K Yamamoto, H Kameoka. Chem Express 7:305–308, 1992.
171. IC MacRae, V Alberts, RM Carman, IM Shaw. Aust J Chem 32:917–922, 1979.
172. WG Liu, JPN Rosazza. Tetrahedron Lett 31:2833–2836, 1990.
173. M Miyazawa, H Nakaoka, M Hyakumachi, H Kameoka. Chem Express 6:667–670, 1991.
174. M Ismaili Alaoui, Benjilali, D Buisson, R Azerad (to be published).
175. MV DeBoggiato, CS DeHeluani, IJS DeFenik, CAN Catalàn. J Org Chem 52:1505–1511, 1987.
176. H Nishimura, Y Noma, J Mizutani. Agric Biol Chem 46:2601–2604, 1982.
177. Y Asakawa, H Takahashi, M Toyota, Y Noma. Phytochemistry 30:3981–3987, 1991.
178. H Takahashi, Y Noma, M Toyota, Y Asakawa. Phytochemistry 35:1465–1467, 1991.
179. W-R Abraham. Z. Naturforsch 49c:553–560, 1994.
180. EV Lassak, JT Pinhey, BJ Ralph, T Sheldon, JJH Simes. Aust J Chem 26:845–854, 1973.
181. M Miyazawa, H Hurono, H Kameoka. Chem Express 6:479–482, 1991.
182. M Ismaili-Alaoui, B Benjilali, D Buisson, R Azerad. Tetrahedron Lett 33:2349–2352, 1992.
183. A Ilidrissi, M Berrada, M Holeman, G Maury. Fitoterapia 61:23–29, 1990.
184. M Ismaili-Alaoui, B Benjilali, R Azerad. Nat Prod Lett 4:263–266, 1994.
185. HL Holland. Organic Synthesis with Oxidative Enzymes. New York: VCH, 1992.
186. R Azerad. In: K Faber, ed. Adv Biochem Eng Biotechnol. (Biotransformations), Vol. 63. Berlin: Springer-Verlag, 1999, pp 169–218.

7
Microbial Epoxidation: Application in Biotechnology

Wolf-Rainer Abraham
GBF–National Research Centre for Biotechnology, Braunschweig, Germany

I. INTRODUCTION

Epoxides are reactive entities due to the strain of the oxacyclopropane moiety. They can be formed from a variety of precursors but this chapter will be limited to the formation of epoxides by the oxidation of double bonds. If the double bond bears three or four different substitutes, epoxidation creates one or two chiral centers. In almost all epoxide-containing natural products these chiral centers were not created by chance and the products are nonracemic but contain an excess of one enantiomer. The chirality of the epoxides is caused by the chirality and the regioselectivity of the forming enzyme. This is a general attribute of enzymes making them ideal tools for enantiopure syntheses.

Epoxides can be opened by nucleophiles to a multitude of products and they can undergo rearrangements (Fig. 1). Due to their reactivity epoxides are formed in a number of chemical processes, used both by organisms and by chemists. Impressive examples are the mycotoxin aflatoxin B1 and benzo[a]pyrene: They are biotransformed by the cytochrome P450 monooxygenases of the human body to a number of relatively nontoxic metabolites, as well as to the ultimate toxic metabolite, aflatoxin B1-8,9-epoxide, respectively to benzo[a]pyrene epoxide [1]. These epoxides react very fast with DNA causing irreversible changes of its genetic information content that can eventually lead to cancer.

The attribute of epoxides, i.e., their ability to act as

Reactive entities
Center of asymmetry, and
Versatile intermediates in chemical syntheses

caused a broad search for microorganisms and their enzymes capable of forming epoxides. This chapter gives a short overview on the involvement of epoxides in biological processes which outlines the diversity of enzymic epoxide–forming reactions. The main part of this chapter describes the different microbial epoxidations. Here again the aim was to show the diversity of microbial epoxide formation, but not to give a complete list of this kind of reaction. Finally, epoxidations using isolated enzymes or their genes cloned into other microorganisms will give an outlook for the future research in this field.

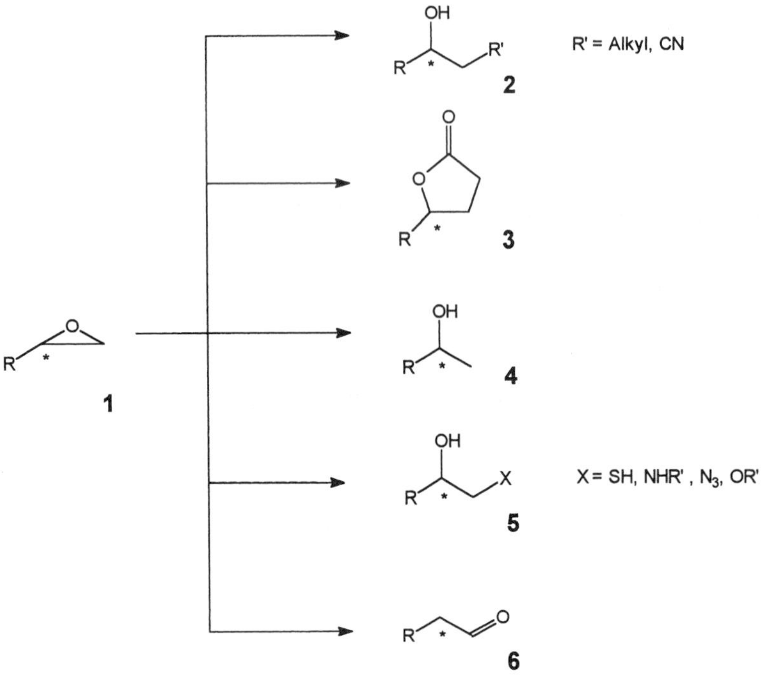

Figure 1 Epoxides are intermediates in a number of chemical and biosynthetic reactions.

II. CHIRAL EPOXIDATIONS IN SYNTHESES

A. Epoxidations by Means of Chiral Catalysts

Many attempts were undertaken to produce chiral epoxides for chemical syntheses. This can be achieved by the use of chiral catalysts. The first applicable and relatively simple procedure of chemical chiral epoxidations was described by Katsuki and Sharpless [2], later called the Katsuki-Sharpless epoxidation. In this reaction, allyl alcohols are epoxidized in the presence of tartrate esters, e.g., (−)-diethyl tartrate. This allows the production of either (R)- or (S)-epoxides depending on the selection of (R)- or (S)-tartrate ester as chiral additive. However, the reaction is limited to allylic alcohols and is somewhat sensitive to steric hindrances. In the meantime, a number of different catalysts have been developed for the epoxidation of *cis*-alkenes. The Jacobsen-Katsuki reaction allows the epoxidation of *trans*-alkenes and terminal olefins [3]. All of these approaches, however, are limited to the epoxidation of activated double bonds like allylic alcohols or require expensive catalysts, and usually the regiospecificity of these reactions is not sufficient for practical applications. Furthermore, the chiral catalysts, although usually they can be recycled, are often very expensive.

  Contrary to most chemical syntheses currently used by chemists, enzymes can epoxidize double bonds regio-, diastereo-, and enantioselectively. Microorganisms bearing these enzymes are cheap, do not produce toxic waste, and generally work under chemically mild conditions. Furthermore, one of the fundamental characteristics of the microbial world is its metabolic diversity. Coupled with an extraordinary physiological flexibility, acquired during at least 3.5 billion years of evolution, this allows microorganisms also to make use of a wide spectrum of extreme biotopes and substrates. The metabolic diversity is reflected in the species diversity and particularly in the large evolutionary distances between the main taxonomic groups of microorganisms. This biodiversity makes microorganisms the prime candidates for the search of new chiral epoxidations.

This chapter will focus on chiral epoxidations by means of microorganisms, which here include bacteria and fungi, excluding plants and animals. Enzymes producing epoxides directly will also be considered as will enzymes cloned into bacteria by genetic engineering, but these techniques will not be the prime focus of this chapter.

B. Epoxidations in Biosynthesis

Nature uses epoxides in a multitude of biosyntheses because of their high reactivity. An example is the (S)-2,3-epoxide of squalene **7**, which is cyclized to give a pentacyclic triterpene **8** of the gammacerane type [4] (Fig. 2).

The polyether antibiotics are a large group of structurally related polyketide natural products capable of transporting cations through the cell wall. One commercially important polyether antibiotic, monensin A **11**, is produced by *Streptomyces cinnamonensis* active against protozoa (coccidia, *Plasmodium falciparum*), gram-positive bacteria, mycobacteria, and fungi, and is used in the factory farming of animals. Epoxides are found as precursors for the polyether moieties in the biosyntheses of these ionophore antibiotics shown in Fig. 3 for the monensin A biosynthesis [5]. *Streptomyces cinnamonensis* was shown to epoxidize *cis*-nerolidol (Table 3) corroborating the epoxide intermediate of monensin.

Other important epoxides are postulated in the biosyntheses of pheromones derived from epoxide intermediates like the insect pheromone α-(−)-(*1S,2R,4S,5R*)-multistriatin [6]. Here again the enantioselectivity of the epoxidation is essential for the biological activity of the compounds. The fungus *Chaetomium cochlioides* rearranges enantiospecifically the epoxide **13** of a fatty acid precursor **12** to a branched aldehyde **14** (Fig. 4), an epoxide rearrangement step also reported from other pathways. This intermediate aldehyde is reduced to the alcohol **15** and gives, after another epoxidation step, the tetrahydrofuran derivative **17** found as a secondary metabolite in a number of *Chaetomium* spp. [7].

Epoxides are not only key intermediates in several biosyntheses; they are also present in a number of secondary metabolites (Fig. 5). Again, the high reactivity of the epoxy moiety contributes to the bioactivity of these metabolites. This reactivity ensures short half-lives of metabolites, e.g., in juvenile hormones, pheromones, or the toxicity of some antibiotics. The methyl ester of 10,11-epoxyfarnesenate **18** is a juvenile hormone synthesized by many insects. The pheromone disparlure **19** (see also Chapter 3 in this book) formed by the gypsy moth *Lymantria dispar* is active in only one enantiomeric form [8],

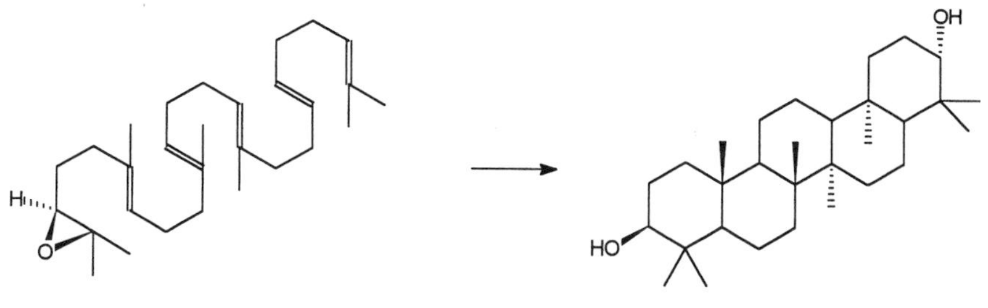

Figure 2 (S)-2,3-Squalenepoxide as key intermediate in the biosynthesis of triterpenes.

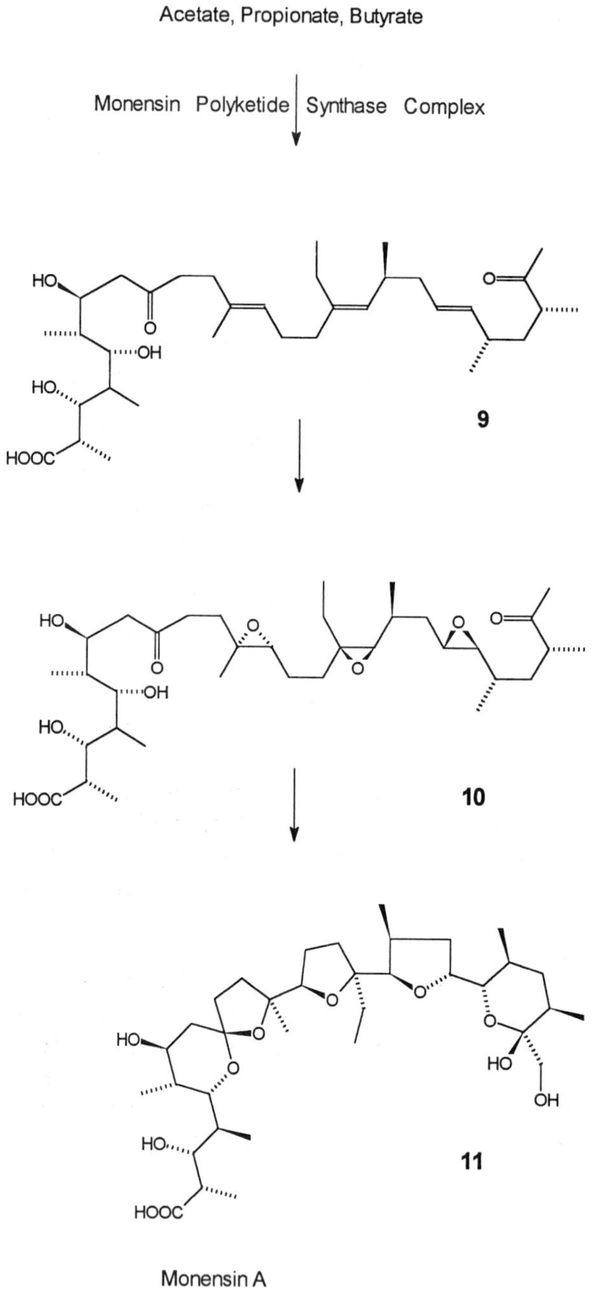

Acetate, Propionate, Butyrate

Monensin Polyketide | Synthase Complex

9

10

11

Monensin A

Figure 3   Epoxides as key intermediate in the biosynthesis of the polyether ionophore antibiotic monensin A.

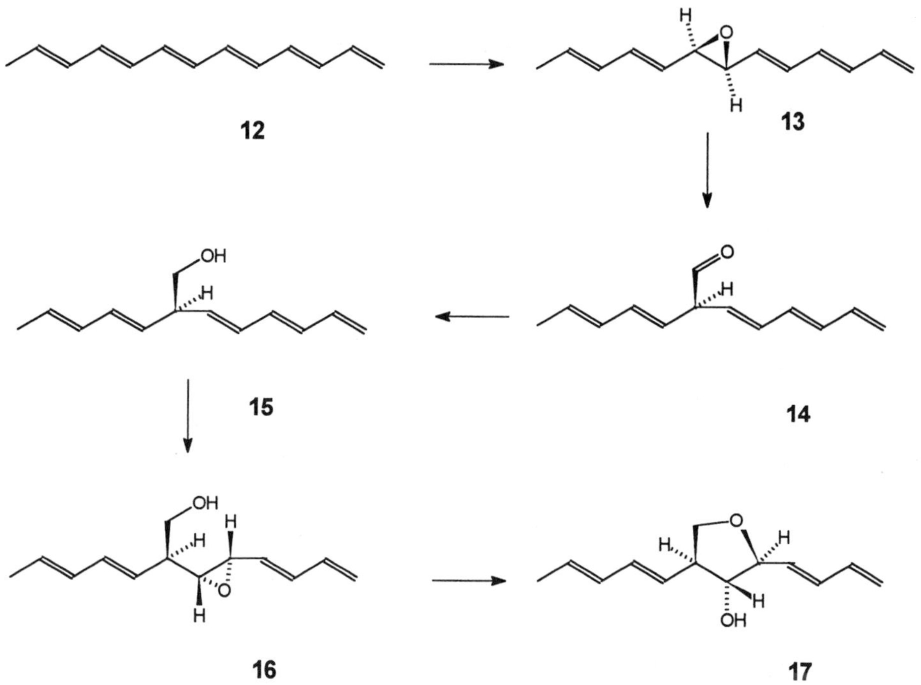

Figure 4 An epoxide is rearranged during the biosynthesis of the tetrahydrofuran metabolite of *Chaetomium cochlioides*.

which gave rise to a number of enantioselective syntheses. A number of epoxy quinonols were found in microorganisms. The smallest member of this series is epoxydon **20** found in a number of fungi [9], whereas the hemiterpene analog panepoxydon **21** was reported from *Panus* spp. [10] and *Lentinus crinitus* [11]. The sesquiterpene homologous oligosporon **22** is also known and found in *Arthrobotrys* spp. [12]. Epoxydon, panepoxydon, and oligosporon all have different configurations at the quinonol but all display cytotoxic activity, and it is assumed that the epoxide is responsible for the toxicity. The spermatozoid-releasing pheromone from the brown alga *Laminaria digitata* has been defined as an epimeric mixture of (*1*S,2*R,6S*)-lamoxirene and (*1*S,2*R,6R*)-lamoxirene **23** [13]. A number of related antibiotics and insecticides were isolated from *Streptomyces* spp. and named manumycines after the first member manumycin A1 **24** [14]. Manumycin A1 is of particular interest as a potential anticancer drug because of inhibitory activity on Ras farnesyltransferase for which it also has shown an in vivo potential [15]. The antibiotic aranorosine **25** is produced by a fungus species belonging to the genus *Pseudorachniotus* and bears two epoxides that are located adjacent to a spiro center [16]. Epoxidized quinonol result in the metabolites terreic acid **26** and terremutin **27** isolated from various strains [17]. The antibiotic LL-C10037α **28** was detected in the supernatant of a *Streptomyces* sp. [18] and its enantiomer MM 14201 **29** is produced by another *Streptomyces* sp. [19]. These are a few examples from a large number of epoxide-bearing secondary metabolites. They show the ability of enzymes to form often rather unstable compounds with high regio- and enantioselectivity. The study of the biosyntheses of these metabolites gives valuable clues for the enzymes involved in epoxide formation. Some of these enzymes

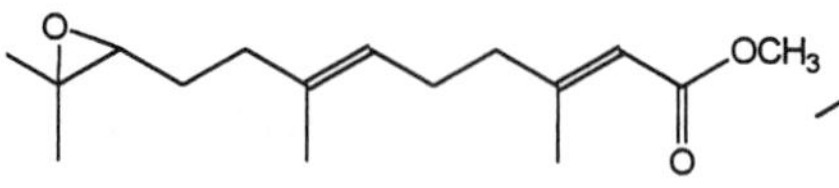

Figure 5 Some bioactive epoxides containing secondary metabolites.

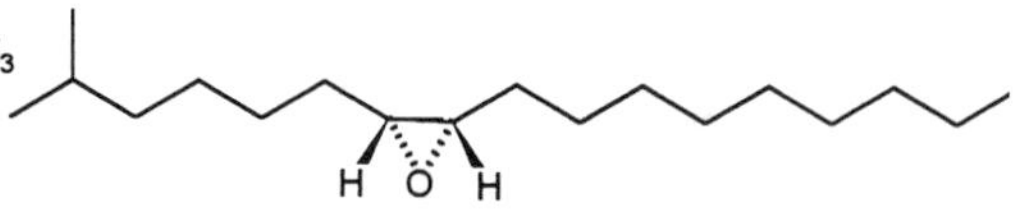

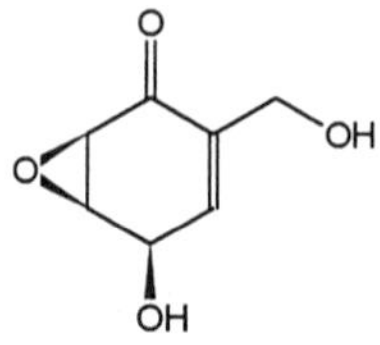

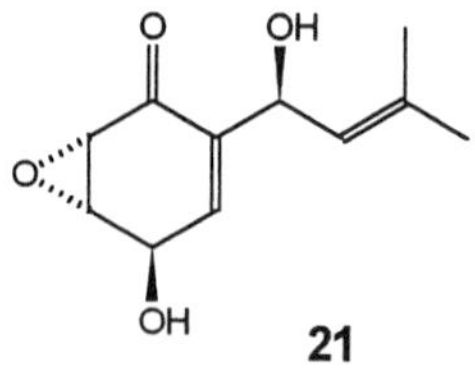

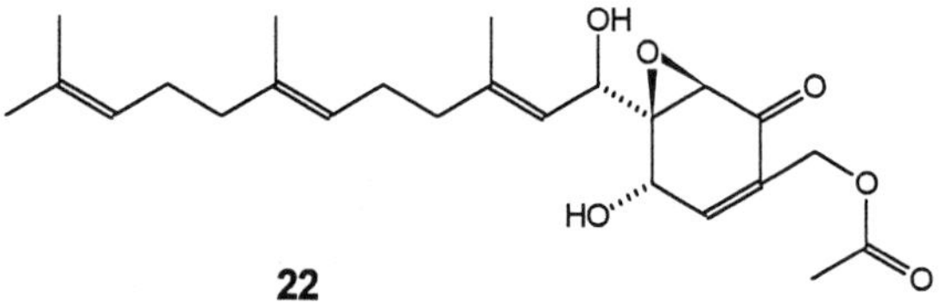

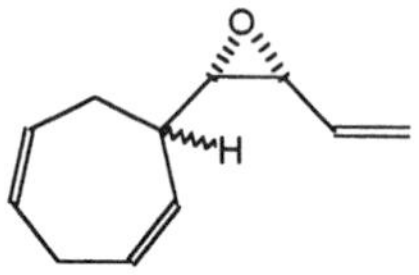

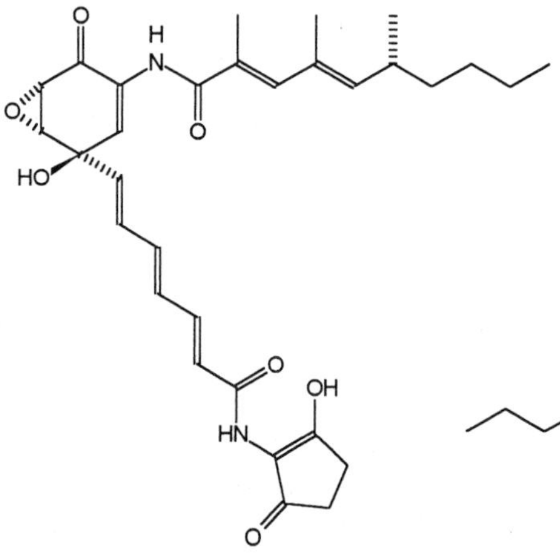

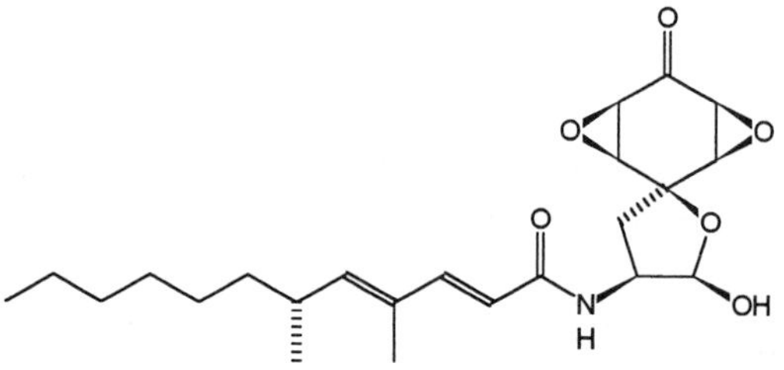

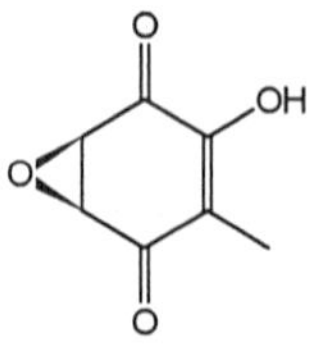

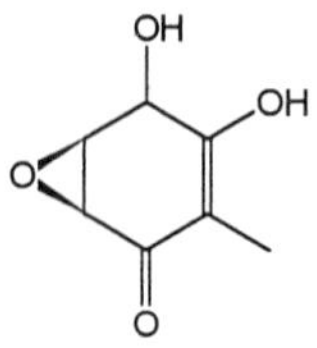

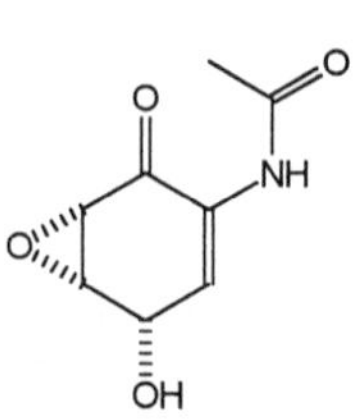

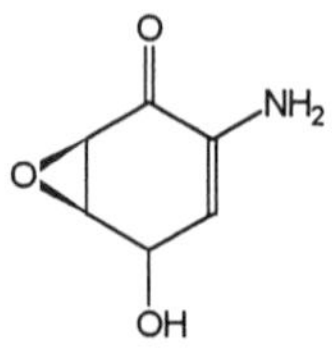

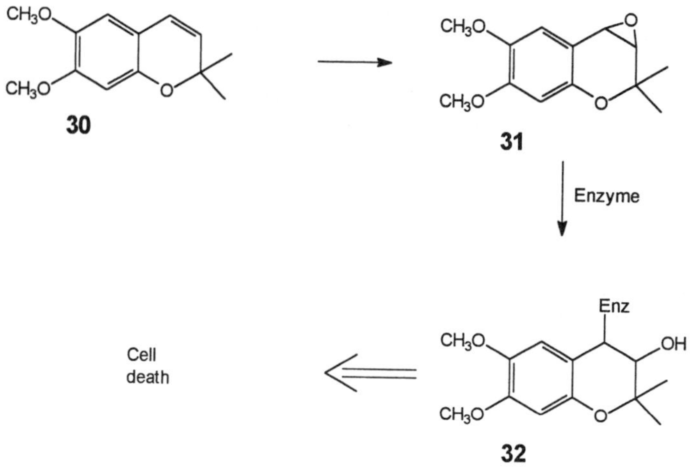

Figure 6 Conversion of the predrug precocene 2 from plants to an allatotoxin by insects.

have even been cloned and await the evaluation of their substrate spectra to be useful in biotechnological processes.

A special situation of a secondary metabolite formed by one organism and activated by epoxidation by another was found in the case of some dimethylchromenes. These compounds are formed by plants and are widespread in the family Asteraceae. Plants of the genus *Ageratum* form dimethoxychromene **30** (procecene 2), which serves as protection against insects. If an insect comes into contact with the plant it develops precocious metamorphosis, sterilization, inhibition of pheromone production, embryogenetic damage, interrupted circadian feeding rhythm, and diapause induction. All of these effects are typical for the absence of juvenile hormone. Indeed, it was found that after treatment with preococene the corpus allatum was reduced. The reason for this size reduction of the corpus allatum was found in the action of the precocene. They serve as predrugs of suicide substrates in the body of the insect because they are epoxidized at the 3,4 double bond by the enzymes normally epoxidizing farnesenate to the active juvenile hormone. The epoxide **31** of precocene, however, is so reactive that it reacts with the enzyme itself forming irreversibly an adduct—and destroying the enzyme [20] (Fig. 6). This deprives the insect of the capability to form the juvenile hormone, leading to the effects described above. The precocene of the plant is transformed into an allatotoxin by the insect itself. The mechanism is similar to the activation of benzo[a]pyrene to the toxic agent reacting with the DNA. Contrary to the interaction between pollutant and humans, however, the plant uses this epoxidation step by the insect to escape feeding.

III. EPOXIDATIONS IN BIOTRANSFORMATIONS

The idea of using microorganisms in biotransformations is to use this wealth of epoxidizing enzymes involved in biosyntheses or in the formation of secondary metabolites for the conversion of substrates to chiral epoxides. Additionally, a number of microorganisms are capable of performing epoxidations by way of cometabolism and therefore to select strains for microbial epoxidations, there is no need to screen first for epoxides in biosyntheses. Although this results in a huge number of strains principally capable of doing epoxidations, no strain performs all of the possible epoxidations and the selectivity of the enzymes becomes the limiting factor in the selection of he appropriate microorganisms. In the following a number of examples will be given to illustrate the possibilities and the limitations of microbial epoxidations. These examples were selected to show the diversity of microorganisms capable of epoxidizing double bonds and the diversity of substrates attached by them.

Detection of an epoxidation product in a biotransformation is just the beginning in the long road to optimizing this reaction to bring it to industrial application. Here the yield of the process is not everything, but the safety and the stability of the strain, the stability of the epoxide, substrate concentration, and the price of the medium are some of many factors that need to be optimized. The search for strains epoxidizing the given substrate at the given position with the correct configuration is as such often a time-consuming and therefore costly process.

A. Microbial Epoxidation of Alkenes

Chiral alkane epoxides are used as the chiral part of ferroelectric liquid crystals that are under development in some companies. Furthermore, these epoxides are the starting materials for a number of different enantiopure syntheses. Among them arylglycidyl ethers are important intermediate in the synthesis of beta blockers. Although racemic beta blockers have long been on the market, the demand for optically active pharmaceuticals is high and still rising. Styrene epoxides and related compounds are also expected to be used as intermediates in the synthesis of biologically active compounds.

In organic chemistry chiral epoxidation of 1-alkenes gives low enantiomeric yields due to the insufficient chiral activation of the terminal double bond. This promoted the search for microorganisms capable of doing the job. Screening of a large number of strains led to the identification of a number of bacteria possessing this ability (Table 1). They belong to a multitude of genera and possess a wide range of substrate specificities. The microbial epoxidation of prochiral allyl ethers by *Pseudomonas putida*, genetically engineered with the *alk* gene from *P. oleovorans*, has achieved industrial application by the Shell and Gist-Brocades companies for the production of the beta blockers Metoprolol and Atenolol [21].

One of the most versatile organisms for the epoxidations of alkenes is *Nocardia corallina* B 276 used by the Japanese company Nippon Mining. The strain epoxidizes C_6–C_{18} 1-alkenes, 2-octene, 2-methyl-1-alkenes, styrene, monochlorostyrenes, and pentafluorostyrene [22]. The monooxygenase of *Nocardia corallina* B 276 was cloned [23]. Large-scale biological production of chiral epoxides is still expensive because of the many drawbacks encountered using a conventional aqueous system caused by poorly water-soluble substrates, product inhibition, and product degradation. To overcome these limitations, a two-phase system or a reverse micelle system was proposed as a method for epoxide production [24].

B. Microbial Epoxidation of Terpenes and Terpenoids

Terpenes form a large group of highly diverse natural products. Because many terpenes are bioactive compounds and a number of them are available in large quantities at low to moderate prices, there have been many attempts to transform naturally occurring terpenes into highly valuable compounds. Obstacles in terpene chemistry are, beside many others, lack of activation of carbon–carbon bonds, instability of intermediates due to strains in

Table 1 Epoxidation of Aliphatic Alkenes by Microorganisms

Substrate	Strain	Ref.
Propene	*Methylococcus capsulatus*	46
1-Octene	*Pseudomonas* sp.	47
	Pseudomonas oleovorans	48
1-Decene	*Pseudomonas oleovorans*	48
Allylbenzene	*Pseudomonas oleovorans*	48
1-Hexadecene	*Corynebacterium equi*	49
1-Hexadecene	*Candida lipolytica*	50
1-Heptadecene	*Candida lipolytica*	50
Ethene-1-pentene	*Arthrobacter simplex*	50
	Alcaligenes eutrophus	50
	Mycobacterium sp.	50
	Nocardia neopaca	50
	Pseudomonas fluorescens	50
	Brevibacterium fuscum	50
Ethene-1-butene	*Methylobacterium organophilum*	50
Trichloroethene	*Methylosinus trichosporium*	51
1-Tetradecene	*Pseudomonas putida*(pWW0)	52
p-Methoxyethylene-phenylallylether	Recomb. *Pseudomonas putide* with *alk* gene	53
Styrene	Recomb. *E. coli* with xylene oxygenase of *P. putida*	53
	Various	54
	Nocardia corallina B 276	
Pentafluorostyrene	*Nocardia corallina* B 276	
Various allylethers	*Rhodococcus equi*	55

the molecule, and high standards concerning a very complex stereochemistry. Activation of carbon–carbon bonds and the instability of intermediates are usually better dealt with by biotransformations but the complexity of terpenes also challenges biotransformations with respect to regio- and steroselectivity (Table 2). Structure–activity comparison of a number of different terpenoids revealed that the stereochemistry of the substrate is very important for the yield and the minor products of the biotransformation [25]. Furthermore, different strains often display different regio- and enantioselectivities.

Epoxides are formed during the course of biotransformation of a number of menthene-monoterpene hydrocarbons like limonene, phellandrene, α- and γ-terpinene and terpino-lene. However, these epoxides are rapidly opened by enzymes to give the corresponding diols (see also Chapter 8 in this book). Temporarily formed epoxides of menthenols like α-terpineol are opened intramolecularly to give cyclic ethers, i.e., in the case of α-terpineol this is 2-hydroxycineol. Similar reactions were observed with acyclic terpenes like linalool, nerolidol, or farnesol. This formation of cyclic ethers was analyzed [26]. Several fungi epoxidized *trans*-nerolidol **33** at the terminal 10,11 double bond. This epoxide **34** was hydrolyzed by most strains to the vicinal diol **35**. Further oxidation of the central double bond led to an epoxide **36/38**, which was intramolecularly opened by the hydroxy group at C-10. Because the epoxidation step at the central double bond did not produce an enantiopure compound, two epimeric tetrahydrofurans (**37** and **39**) were obtained. Some fungi, like *Rhizopus arrhizus* or *Gibberella cyanea*, attacked the central double bond first and form a nonenantiopure epoxide **40**. This epoxide was immediately opened by the

Table 2 Epoxidation of Terpenes and Terpenoids by Microorganisms

Substrate	Product	Strain	Ref.
2,2-Dimethyl-6-cyano-3-chromene	-(3S,4S)-Epoxide	*Penicillium* sp.	56,57
10β,19-Dihydroxyrosa-7,15-diene	-7β,8β-epoxide	*Trichothecium roseum*	58
Deoxyvulgarin	Vulgarin	*Rhizopus nigricans*	59
		Aspergillis ochraceus	59
ent-7α,18-Diacetoxy-14β-hydroxybeyer-15-ene	ent-15α,16α-epoxide	*Rhizopus nigricans*	60
Δ⁴,⁶-3-Ketosteroid	-6β,7β-epoxide	*Rhizopus arrhizus*	61
Δ⁶-3-Ketosteroid	-6α,7α-epoxide	*Rhizopus arrhizus*	61
Linearol	-16β-17-epoxide	*Rhizopus nigricans*	62
Germacrone		*Cunninghamella blakesleeana*	63
Me(13R)-ent-16-hydroxy-8α,13-epoxylabd-14-en-18-oate		*Curvularia elegans*	64
Me(13S)-ent-16-hydroxy-3-oxo-8α,13-epoxylabd-14-en-18-oate		*Rhizopus nigricans*	65
9β,10,β-Epoxytrichodiene	12,13-epoxide	*Fusarium culmorum*	66
1-Oxo-5,6αH,4,11βH-eudesm-2-en-6,12-olide		*Rhizopus negricans*	67
ent-18-Acetoxy-3β-hydroxy-7α,15β-isopropylidenedioxykaur-16-ene		*Rhizopus nigricans*	68
Humulene	1,2-,4,5-,8,9-epoxides	*Chaetomium cochlioides*	69
Citronellol	6,7-Epoxide	*Streptomyces griseus*	70
Farnesol	10,11-epoxide	*Gibberella cyanea*	70
Geranylacetone	9,10-epoxide	*Fusarium solani*	71
		Mucor griseocyaneus	72
		Trichoderma koningii	72
Auraptene	6′,7′-epoxide	*Nocardia alba*	72
		Bacillus cereus	72
Tetramethyllimonene	8,9-epoxide	*Gibberella cyaneu*	73
trans-Nerolidol	10,11-epoxide	*Nocardia alba*	74
cis-Nerolidol	10,11-epoxide	*Streptomyces cinnamonensis*	75
Myrcene sulfone	6,7-epoxide		76
Farnesene sulfone	10,11-epoxide		77
Caryophyllene	4,5-epoxide	*Diplodia gossypina*	78
		Chaetomium cochlioides	79
α-Terpineol	1,2-epoxide	*Gibberella cyanea*	80
(S)-Sulcatol phenylurethane	5,6-epoxide	*Aspergillus niger*	81

alcohol to give two epimeric tetrahydrofurans **41** and **42**. *Rhizopus arrhizus* epoxidized these compounds further and the intermediate epoxide **43/44** was again opened by the alcohol to give two different bistetrahydrofurans **45** and **46** [27]. It is interesting to note that contrary to the central epoxide the terminal epoxide at C-10/C-11 was enantiopure in most fungi used in this study (Fig. 7). This sequence of epoxidation and intramolecular opening bears several similarities to the biosynthesis of the polyether antibiotics described above (see Fig. 3).

Biotransformation of linalool gave both tetrahydrofurans and tetrahydropyranes, but the biotransformation of *trans*-nerolidol discussed above produced only tetrahydrofurans. A set of homologous dimethylalkadienes were used as substrates to be epoxidized with the fungus *Diplodia gossypina*. It turned out that dimethylhexa-2,4-diene **53** could not form any tetrahydrofuran probably because the second double bond could not be attacked (Fig. 8). All the other hydrocarbons like **47** and **50** formed the tetrahydrofuran derivative (**49** and **52**) and sometimes also the tetrahydropyrane compound, but in much lower yields.

The sesquiterpene hydrocarbon humulene **55** is a constituent of hop. Its oxidation products are flavor impact compounds of beer, probably formed by microbial oxidation of humulene during brewing. Biotransformation of humulene with *Chaetomium cochlioides* gave a multitude of products and most of them contained epoxides. Interestingly, the first epoxidation led to a racemic 1,2-epoxide **56**. The following enzymatic steps, however, kinetically resolve this racemate leading to enantiomerically highly enriched products. The *rac*-1,2-epoxide **56** was epoxidized to the (*1S,2S,4S,5S*)- and (*1R,2R,4S,5S*)- (**58** and **62**), on the one hand, and the (*1S,2S,8S,9S*)- and (*1R,2R,8S,9S*)-diepoxides (**57** and **63**), on the other hand. In small amounts the (*1R,2R,4S,5S,8S,9S*)-triepoxide **64** was also detected (Fig. 9). Contrary to the 1,2-epoxide **56** the 4,5- and 8,9-epoxides **61** and **59** were enantiomerically pure. The ability of *C. cochlioides* for the kinetic resolution of epoxyhumulanes was further tested by using synthetic racemic 1,2-epoxy- and anti-1,2,8,9-diepoxyhumulene. These experiments corroborated the observation that this strain epoxidized the substrate exclusively with the (*4S,5S*) and (*8S,9S*) configurations. Diepoxyhumulenes contribute to the flavor product of beer [28,29] and the formation of these epoxides in the biotransformation of humulene with the fungus *C. cochlioides* stresses the view that at least some of the oxidized humulenes are formed by microbial oxidation processes.

Similar to humulene, the sesquiterpene hydrocarbon caryophyllene **65** is epoxidized by *D. gossypina* and *C. cochlioides*. Beside a number of hydroxylation products, rearranged caryophyllenes like **67** were isolated (Fig. 10). The only rearranged products discussed here are those that resulted from the rearrangement of epoxide **66** (reaction **1-6** in Fig. 1). This epoxide rearrangement is the same described above in the biosynthesis of the tetrahydrofuran metabolite formed by *C. cochlioides*. It results in a size reduction of the ring and in the transfer of the chirality of the epoxide to the newly formed quartiary carbon center implying an enantiospecific rearrangement of the epoxide that was also observed in the tetrahydrofuran metabolite **17** of this strain (see Fig. 4). This biotransformation is an example of steps in the biosynthesis of a secondary metabolites being successfully transferred to another substrate.

Nearly all of these biotransformations were done by intact microorganisms either by growing or resting cells. To identify the optimal strain for such biotransformations requires a screening of a number of strains, often hundreds. These screenings revealed that the taxonomy of the strains is reflected in its biotransformation performance. A phylogenetic approach was developed to facilitate the selection of suitable strains for different applications. The idea was to test representatives of taxonomic groups (taxa) with a set of different substrates and to group the strains according to their biotransformation behavior

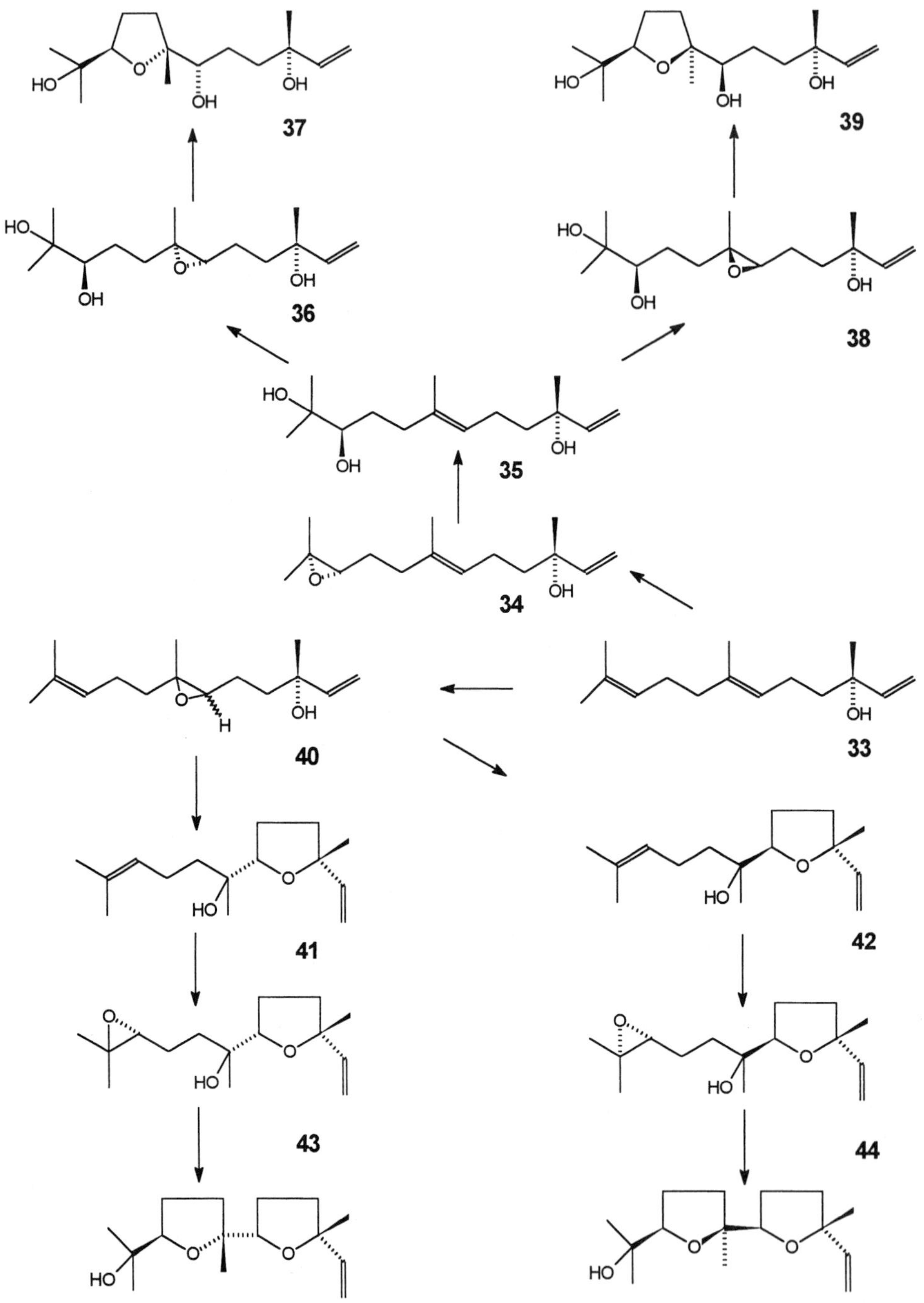

37, 39, 36, 38, 35, 34, 40, 33, 41, 42, 43, 44, 45, 46

Figure 7 Intramolecular opening of epoxides of *trans*-nerolidol by hydroxy groups results in the formation of cyclic ethers. The course of the epoxidation sequence is responsible for the formation of either mono- or bistetrahydrofuran derivatives. Note that the 10,11-epoxide is enantiomerically pure while the 6,7-epoxide had a considerable lower enantiomeric excess.

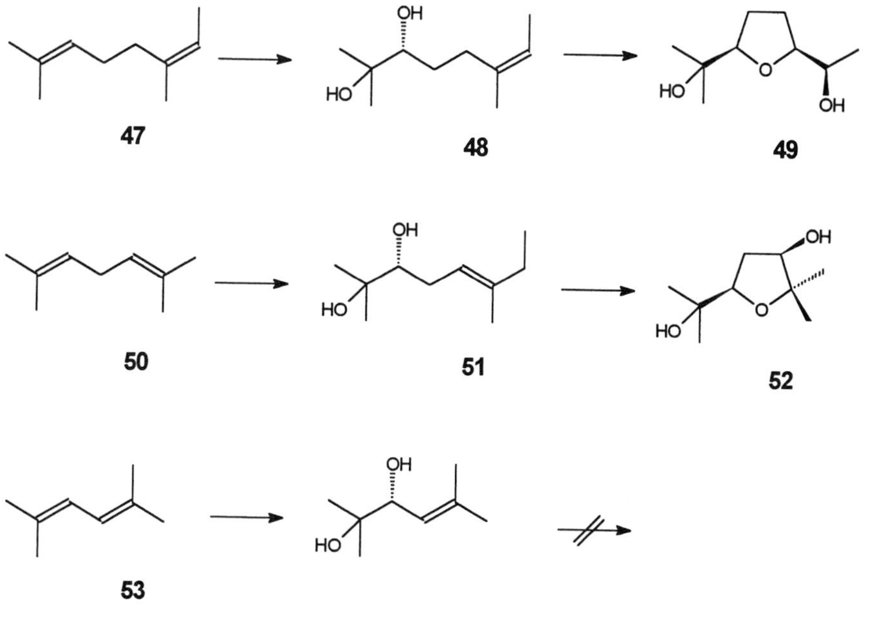

Figure 8 Ether formation by intramolecular epoxide opening depends on the chain length.

to these substrates. The aim was to form groups of microorganisms having similar enzymatic activity. With these groups in hand a "prescreen" of a relatively small number of strains from all groups could be done to determine which of these groups is most active. The strains of the main screen will then be selected only from this active group. This procedure would drastically reduce the effort for screening.

Therefore, 100 strains—60 fungi and 40 bacteria—were tested with 13 different substrates, mainly mono- and sesquiterpenes, and the products determined. Multivariate statistical analysis using the biotransformation products resulted in the identification of five groups of microorganisms corresponding almost exactly to the large phylogenetic lineages. Analysis of these groups revealed that the taxonomic position of a strain is mirrored in its ability to catalyze certain biotransformations, i.e., that the phylogeny of the strains is correlated with its biotransformation potential. In these groups, fungi and bacteria can be discerned in the course of which Basidiomycotina, Ascomycotina and Zygomycotina of the fungi formed discernable clusters. The Deuteromycotina (*Fungi imperfecti*) are only barely discernible from the Ascomycotina, which is not surprising because it is assumed that up to 80% of this group belong to the Ascomycotina. Within the bacteria gram-positive and gram-negative organisms are discernible. A further grouping could not be found with statistical significance, probably owing to the fact that the dataset contained only 40 bacterial strains, which is far too little with respect to the large diversity of these organisms [30].

At first glance the result is surprising because such a correlation is not obvious and similar attempts using secondary metabolites from microorganisms were not as unambiguous. The phylogeny used for comparison is based on the comparison of gene sequences. At the same time it is assumed that the more different the gene sequences are, the less closely organisms are related. As a gold standard for the determination of the phylogeny

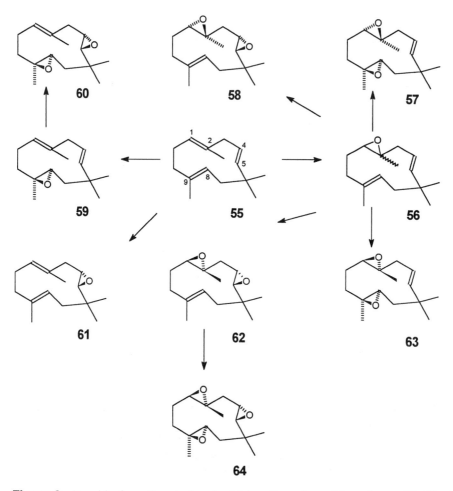

Figure 9 Epoxide formation during the biotransformation of humulene with *Chaetomium co-chlioides*. Similar processes may occur during the brewing of beer where humulene epoxides are important flavor constituents.

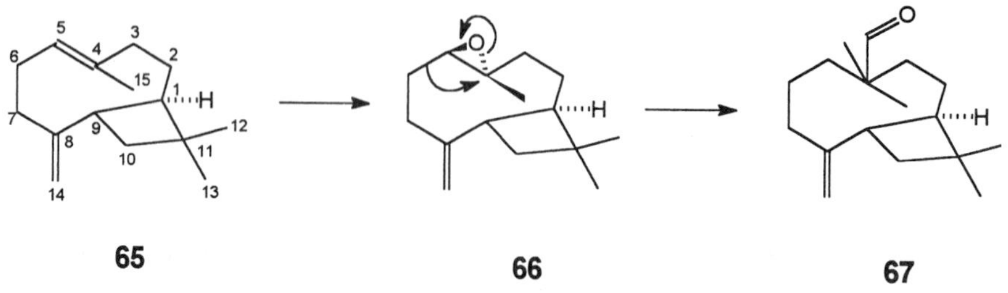

Figure 10 Epoxidation of caryophyllene and rearrangement of the 4,5-epoxide by *Chaetomium cochlioides*.

of organisms the sequences of the genes of ribosomal RNAs have been established, usually comparing 16S respectively with 18S rRNA gene sequences. Some other genes were also used to establish the phylogeny of organisms and the number is rapidly growing as more and more genome sequences become available [31]. There are some differences in the exact branching of the phylogenetic tree resulting from the comparison of the sequences of different genes, but all in all the result is the same. An excellent review concerning the molecular view of microbial diversity was recently given by Norman Pace [32]. On one hand, such a result makes one feel optimistic that we really can deduce the evolution of life. On the other hand, it also explains why a comparison of biotransformations resulted in a grouping controlled by the phylogeny of the strains. When applied to the formation of 10,11-epoxy-*trans*-nerolidol, this statistical analysis revealed that species of the Basidiomycotina are the most promising candidates because they form the epoxide but do not hydrolyze it to the vicinal diol.

The results showed that it is not necessary to screen a number of strains with a number of substrates to form groups of similar biotransformation behavior. Instead, one can use the phylogenetic position of a strain for a prediction of its general capabilities in biotransformation [33]. Hopefully this will allow a much broader use of the immense biodiversity present in the microbial world.

C.   Microbial Epoxidation of Miscellaneous Compounds

Biotransformations of drugs were also used to produce metabolites formed in the human body in larger quantities than those accessible from blood or urine. Here fungi are of special interest because like us they are eukaryotes and our metabolism is more related to theirs than to that of the bacteria. In Table 3 the biotransformation of cyproheptadine with *Cunninghamella elegans* is such an example. The formation of the antibiotic fosfomycin by epoxidation of *cis*-propenylphosphonic acid attracted much attention and a number of microorganisms capable of this epoxidation were identified. Here problems occurred concerning the toxicity of the product preventing the high substrate concentration necessary for commercial exploitation of this biotransformation.

D.   Epoxidations at Aromatic Rings

Direct epoxidations of aromatic rings are very difficult to achieve by chemical methods because of the high stability of the double bonds within such systems due to their aromatic character. For the same reason the formed epoxides are usually very unstable, undergoing a number of reactions and rearrangements. Therefore, only few thoroughly investigated examples of epoxidations at aromatic rings have been described:

Naphthalene **68** is transformed by rat liver microsomes into naphthol **70** and **71**, and it has been shown that the pathway runs via the epoxide **69** (Fig. 11). Subsequently, this epoxide is opened to the *trans*-diol which finally loses water to give naphthol [34]. Similar oxidation steps by mammalian liver enzymes were found for quinoline where the quinoline-5,6-epoxide could be detected [35].

There is strong evidence that the biosynthesis of salicylic acid **75** from benzoic acid **72** by the fungus *Phellinus tremulae* runs via an arene oxide **73** [36] (Fig. 12). The mechanism is principally the same as that for naphthalene and quinoline, but in this case the rearrangement product 2-carbomethoxyoxepin **74** was also found in the supernatant of the fungus.

Table 3 Epoxidation of Drugs and Other Metabolites by Microorganisms

Substrate	Product	Strain	Ref.
Cyproheptadine	Cyproheptadine-10,11-epoxide	*Cunninghamella elegans*	82
cis-Propenylphosphonic acid	Fosfomycin	Various	83,84
2,2-Dimethyl-2H-benzopyran-6-carbonitrile	-3S,4S-epoxide	*Corynebacterium* sp.	85
		Mortierella ramanniana	
Carbamazepine	10,11-Epoxycarbamazepine	*Streptomyces violascens*	86
N-Methyl-*N*-(4-methyleneadamant-1-yl)benzamide	-β-epoxide	*Beauveria sulfurescens*	87
Milbemycin A₄	14,15-Epoxymilbemycin A₄	*Streptomyces violascens*	88
Heptachlor		Various	89

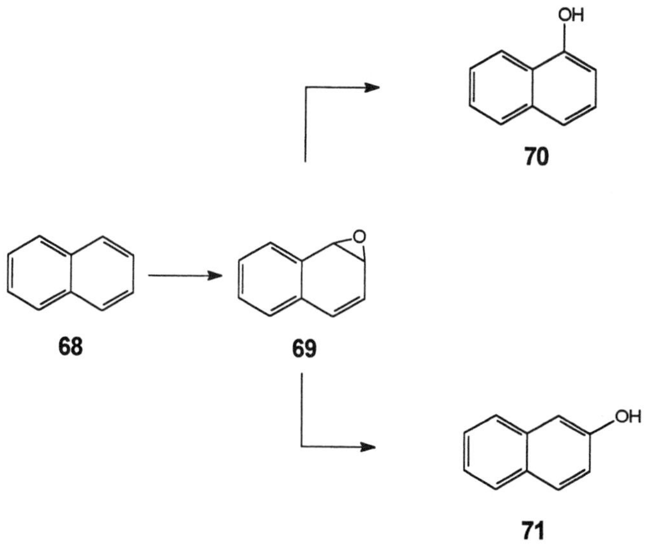

Figure 11 Biotransformation of naphthalene to naphthol via an epoxide in rats.

An interesting mechanism involving intramolecular epoxidation is used by the 4-hydroxyphenylpyruvate dioxygenase which catalyzes the oxidation of 4-hydroxyphenyl-pyruvate **76** to homogentisate **79**. This enzymes reacts molecular oxygen with the acid forming a peracid **77** which then epoxidizes the aromatic ring. Rearrangement of this intermediate **78** gives homogentisate (Fig. 13). Meanwhile the enzyme from rat liver was cloned and overexpressed in *E. coli*, making it available for biotransformation studies [37].

Antibiotic LL-C10037α **28** is formed in *Streptomyces* sp. LL-C10037 from 3-hydroxyanthranilic acid **81** and a pathway via 2,5-dihydroxyacetanilide **82** has been established. A dihydroxyacetanilide epoxidase from this strain oxidizes 2,5-dihydroxyacetanilide to 5,6-dihydro-5,6-epoxy-2-acetamidoquinone **83** utilizing molecular oxygen but without a requirement for any other cofactor. From this intermediate LL-C10037α is produced by a NADPH-dependent oxidoreductase. Using $^{18}O_2$ and $H_2^{18}O$ it has been shown that the dihydroxyacetanilide epoxidase has the characteristics of a dioxygenase [38]. This suggests that this enzyme is a dioxygenase with an epoxidation mechanism that is essentially the same as that observed for dihydrovitamin K epoxidation during the mammalian vitamin K–dependent glutamate carboxylase reaction [39].

The antibiotic MM14201 **29** is formed by *Streptomyces* sp. MPP 3051 from the same substrate but by a complementary enzyme [40], which is an impressive example of how selective epoxidizing enzymes can be and how diverse bacteria are in terms of their epoxidases (Fig. 14).

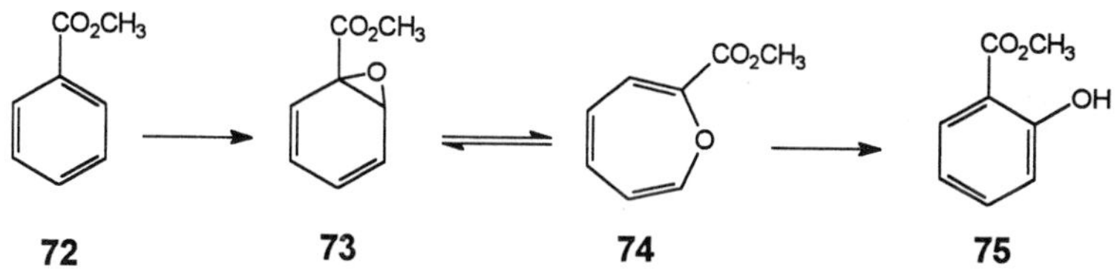

Figure 12 Biosynthesis of salicyclic acid from benzoic in *Phellinus tremulae*.

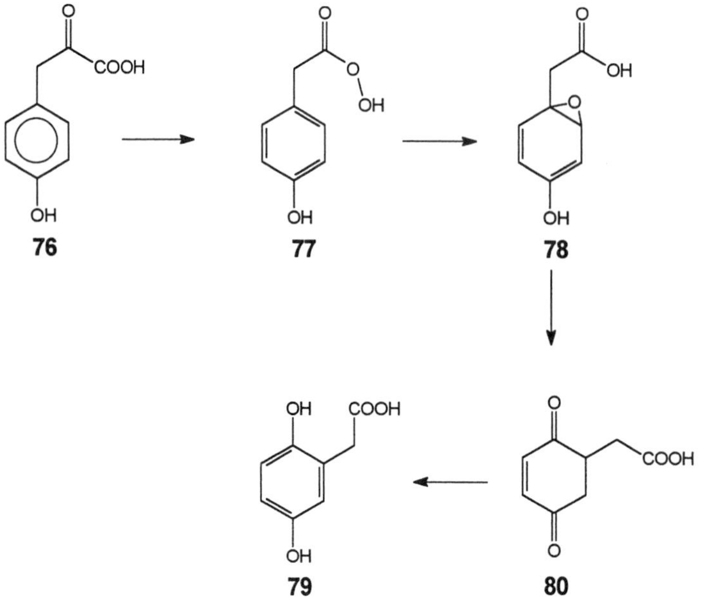

Figure 13 4-Hydroxyphenylpyruvate dioxygenase epoxidizes 4-hydroxyphenylpyruvate via an intramolecularly acting peracid.

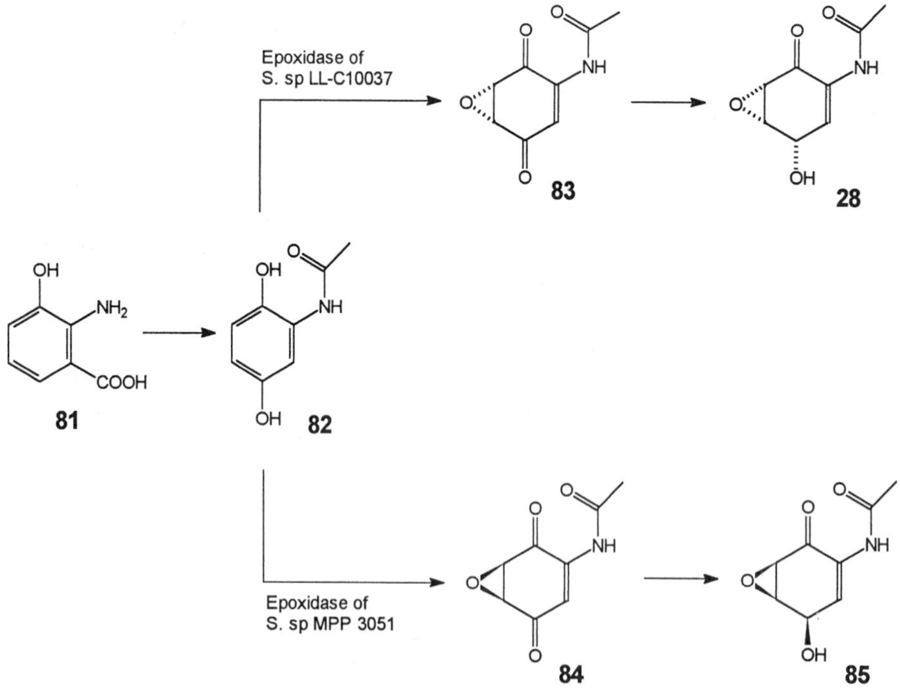

Figure 14 Epoxidation of 2,5-dihydroxyacetanilide by two different *Streptomyces* sp. yielding enantiomeric antibiotics.

E. Epoxidations Using Isolated Enzymes

Because of side reactions observed with the biotransformations of alkenes by whole cells and because of a better handling in chemical laboratories many attempts were made to use isolated enzymes for enantioselective epoxidation reactions. The use of cytochrome P450–dependent monooxygenases is hampered by the difficulty of getting most of them into solution. Often they are tightly bound to membranes and the solubilization procedure turns them into inactive proteins. However, some of them are soluble or stable in solution, and biotransformations with them have been described (Table 4). Methane monooxygenase was used for epoxidations and a solution for the problem of the regeneration of the cofactor NADH2 required for this reaction was developed [41,42]. For some of these enzymes both hydroxylation and epoxidation activities have been reported. Cloning of the gene(s) into strains that are easy to handle, such as *E. coli* or *P. putida*, may be a good choice for the use of unstable or in solution inactivated epoxidases in biotransformations (for some examples, see Table 1). Instead of using enzymes one can use catalytic antibodies generated against different transition states of the epoxidation reaction. However, the use of catalytic antibodies for enantioselective epoxidations is still in its infancy [43]. A review on the use of (halo)peroxidases were recently given [44]. However, some of these epoxidation reactions produce the corresponding allylic alcohols and/or aldehydes in quite large yields [45], severely limiting the application of this approach. Furthermore, *trans*-olefins are transformed in low yields only. Another approach is the use of lipases with hydrogen peroxide for the formation of peracids, which then do the enantioselective epoxidation.

IV. OUTLOOK

A number of very different substrates have been successfully epoxidized using a wide range of microorganisms. Until recently the best strains were identified by a random screening of easy-to-handle microorganisms. The discovery that the different lineages of bacteria and fungi show different biotransformation preferences as well can be used to speed the process of screening. Some strains produce secondary metabolites containing epoxides or are formed via epoxide intermediates. The substrate spectra of the involved enzymes can often be exploited for epoxidation reactions but usually the ranges are narrow. A successful biotransformation reaction leading to the formation of epoxides in high yields is only the first step because low enantiomeric yields can make the reaction virtually useless. The availability of chiral columns for gas chromatography or high-performance liquid chromatography allows determination of the enantiomeric excess of the products on a routine basis. Chiral alcohols as additives are an alternative to these chromatographic determinations. The optimization of the process for industrial production is expensive and time consuming but does not cause special problems.

The main problem of biotransformation to find the strain with the fastest reaction, highest yields, and highest tolerated substrate concentrations can also be overcome by the used of genetically engineered bacteria. The use of plasmids usually cannot be recommended because of the instability often observed with such systems. Integration of the gene(s) into the chromosome results in stable genetically engineered microorganisms, so-called GEMs. Such a system is used by the Shell and Gist-Brocades companies for the production of the beta blockers Metroprolol and Atenolol. A step further is the construction of new enzymes by site-directed mutagenesis to yield mutants with increased accessibility to the active site, higher turnover rates, or higher turnover numbers. Again, for such an

Table 4 Epoxidations Catalyzed by Isolated Enzymes

Substrate	Enzyme	Ref.
n-Alkenes (C$_6$–C$_{12}$)	Lipases + H$_2$O$_2$	90
Cyclic alkenes (C$_5$–C$_{12}$)	Lipases + H$_2$O$_2$	90
cis-β-Methylstyrene	Cyctochrome P450cam of *Pseudomonas putida*	91
β-Methylstyrene (*cis and trans*)	Horseradish peroxidase	92
Palmitoleic acid	Soluble enzymes from *Bacillus megaterium*	93
Styrene	Human myoglobin + H$_2$O$_2$	94
	F42Y deriv. of human myoglobin + H$_2$O$_2$	94
1,7-Octadiene	ω-Hydroxylase of *Pseudomonas oleovorans*	95
n-Alkenes (C$_2$–C$_4$)	Methane monooxygenase of *Methylosinus trichosporium*	96
	Methane monooxygenase of *Methylococcus capsulatus*	96
Propene	Halohydrin peroxidase	97
cis-2-Heptene	Chloroperoxidase from *Caldariomyces fumago*	98
cis-3-Heptene	Chloroperoxidase from *Caldariomyces fumago*	98
cis-2-Octene	Chloroperoxidase from *Caldariomyces fumago*	98
cis-3-Isoheptene	Chloroperoxidase from *Caldariomyces fumago*	98
1-Isononene	Chloroperoxidase from *Caldariomyces fumago*	98
Z-3-Methyl-3-pentene	Chloroperoxidase from *Caldariomyces fumago*	98
cis-Phenylpropene	Chloroperoxidase from *Caldariomyces fumago*	98
Dihydronaphthalene	Chloroperoxidase from *Caldariomyces fumago*	98
cis-3-Octene	Chloroperoxidase from *Caldariomyces fumago*	99
cis-2-Nonene	Chloroperoxidase from *Caldariomyces fumago*	99
cis-3-Nonene	Chloroperoxidase from *Caldariomyces fumago*	99
1,4-Pentadienol-3	Chloroperoxidase from *Caldariomyces fumago*	99
3-Methylpenta-1,4-diene	Chloroperoxidase from *Caldariomyces fumago*	99
Ethyl-3-methyl-3-butenoate	Chloroperoxidase from *Caldariomyces fumago*	100
4-Brom-2-methylbutene	Chloroperoxidase from *Caldariomyces fumago*	100
4-Methoxy-2-methylbutene	Chloroperoxidase from *Caldariomyces fumago*	100
4-Acetoxy-2-methylbutene	Chloroperoxidase from *Caldariomyces fumago*	100
α-Methylstyrene	Chloroperoxidase from *Caldariomyces fumago*	101
3-Phenylpropene	Chloroperoxidase from *Caldariomyces fumago*	101
3-Phenyl-2-methylpropene	Chloroperoxidase from *Caldariomyces fumago*	101
2-Methylallylpropionate	Chloroperoxidase from *Caldariomyces fumago*	101
Phenyl-(2-methylallyl)ether	Chloroperoxidase from *Caldariomyces fumago*	101
Isooctene	Chloroperoxidase from *Caldariomyces fumago*	101
Styrene	Horseradish peroxidase mutants	102
Linolic acid	Soybean peroxygenase	103

approach examples have been described. A consequence of this would be the application of combinatorial chemistry by using oligonucleotide primers under nonstringent conditions for the polymerase chain reaction ("dirty" PCR) or by applying phage display systems and a selection of the mutant by a highly efficient screening system. Such approaches have been summarized under the term *evolutionary biotechnology* but no application to enantioselective epoxidations has been published. The application of the various tools of genetic engineering to achieve enantioselective epoxidations will close the gap between the traditional biotransformation reactions using bacterial or fungal strains and the more chemical approach using isolated enzymes. With these new methods in hand and new ones at the horizon, more and novel epoxidation reactions will become available for industrial applications.

ACKNOWLEDGMENT

I am indebted to Peter Wolff, who kindly drew the figures. This work was supported by a grant of the German Federal Ministry for Science, Education and Research (Project No.0319433C).

REFERENCES

1. SK Yang, DW McCourt, PP Roller, HV Gelboin. Proc Natl Acad Sci USA 73:2594–2598, 1976.
2. T Katsuki, DB Sharpless. J Am Chem Soc 102:5974–5976, 1980.
3. T Katsuki. Coord Chem Rev 140:189–214, 1995.
4. I Abe, M Rohmer. J Chem Soc Perkin Trans I 783–791, 1994.
5. JA Robinson. Prog Chem Org Nat Prod 58:1–81, 1991.
6. a) WE Gore, GT Pearce, RM Silverstein. J Org Chem 40:1705–1708, 1975. b) Abs. conf: K Mori. Tetrahedron 32:1979–1981, 1976.
7. WR Abraham, HA Arfmann. Phytochemistry 31:2405–2408, 1992.
8. BA Bierl, M Beroza, CW Collier Science 170:87–89, 1970.
9. G Assante, L Camarda, L Merlini, G Nasini. Phytochemistry 20:1955–1957, 1981.
10. Z Kis, A Closse, HP Sigg, L Hruban, G Snatzke. Helv Chim Acta 53:1577–1597, 1970.
11. WR Abraham, D Abate. Z Naturforsch 50c:748–750, 1995.
12. M Stadler, O Sterner, H Anke. Z Naturforsch 48c:843–850, 1993.
13. I Maier, G Pohnert, S Pantke-Böcker, W Boland. Naturwissenschaften 83:378–379, 1996.
14. I Sattler, R Thiericke, A Zeeck. Nat Prod Rep 15:221–240, 1998.
15. M Hara, K Akasaka, S Akinaga, M Okabe, H Nakano, R Gomez, D Wood, M Uh, F Tamanoi. Proc Natl Acad Sci USA 90:2281–2285, 1993.
16. (a) K Roy, T Mukhopadhyay, GCS Reddy, KR Desikan, RH Rupp, BN Ganguli. J Antibiot 41:1780–1784, 1988; (b) HW Fehlhaber, H Kogler, T Mukhopadhyay, EKS Vijayakumar, K Roy, RH Rupp, BN Ganguli. J Antibiot 41:1785–1794, 1988; (c) AV Rama Rao, MK Gurjar, PA Sharma. Tetrahedron Lett 32:6613–6616, 1991.
17. WB Turner, DC Aldridge. Fungal Metabolites, Vol. 2. London: Academic Press, 1983.
18. MD Lee, AA Fantini, GO Morton, JC James, DB Borders, RT Testa. J Antibiot 37:1149–1152, 1984.
19. SJ Box, ML Gilpin, M Gwynn, G Hanscomb, SR Spear, AG Brown. J Antibiot 36:1631–1637, 1983.
20. WS Bowers, PH Evans, PA Marsella, DM Soderlund. Science 217:647–648, 1982.
21. SL Johnstone, GT Phillips, BW Robertson, PD Watts, MA Bertola. In: C Laane, J Tramper, MD Lily, eds. Organic Media. Amsterdam: Elsevier, pp 387–392.

22. M Takagi, N Uemura, K Furuhashi. Ann NY Acad Sci 613:697–701, 1990.
23. H Saeki, K Furuhashi. J Ferment Bioeng 78:399–406, 1994.
24. S Prichanont, DC Stuckey, DJ Leak. IChemE Res Event, Two-Day Symp (1994), Vol 1. Rugby, UK: Inst. Chem. Eng., pp 238–240.
25. WR Abraham. Zeitschr Naturforsch 49c:553–560, 1994.
26. WR Abraham, B Stumpf, HA Arfmann. J Ess Oil Res 2:251–257, 1990.
27. WR Abraham, HMR Hoffmann, K Kieslich, H Reng, B Stumpf. In: Enzymes in Organic Synthesis. Ciba Foundation Symposium No. 111. London, 1985, pp 146–160.
28. KC Lam, ML Deinzer. J Agric Food Chem 35:57–59, 1987.
29. VE Peacock, ML Deinzer. Am Soc Brew Chem J 47:4–6, 1989.
30. WR Abraham. World J Microbiol Biotechnol 10:88–92, 1994.
31. Le Maley, CR Marshall. Science 279:505–506, 1998.
32. NR Pace. Science 276:734–740, 1997.
33. WR Abraham. In: U Diederichsen, TK Lindhorst, LA Wessjohann, B Westermann, eds. New Perspectives in Bioorganic Chemistry. Weinheim: Wiley-VCH (in press).
34. DM Jerina, JW Daly, B Witkop, P Zaltzman-Nirenberg, S Udenfriend. Biochemistry 9:147–155, 1970.
35. SK Agarwal, DR Boyd, HP Porter, WB Jennings, SJ Grossman, DM Jerina. Tetrahedron Lett 27:4253–4256, 1986.
36. WA Ayer, ER Cruz. J Nat Prod 58:622–624, 1995.
37. NP Crouch, JE Baldwin, MH Lee, CH MacKinnon, ZH Zhang. Bioorg Med Chem Lett 6: 1503–1506, 1996.
38. SJ Gould, MJ Kirchmeier, RE LaFever. J Am Chem Soc 118:7663–7666, 1996.
39. S Naganathan, R Hershline, SW Ham, P Dowd. J Am Chem Soc 115:5839–5840, 1993.
40. SJ Gould, B Shen. J Am Chem Soc 113:684–86, 1991.
41. RN Patel, CT Hou, AI Laskin, A Felix. J Appl Biochem 4:175–184, 1982.
42. CT Hou, RN Patel, AI Laskin, N Barnabe. J Appl Biochem 4:379–383, 1982.
43. A Koch, JL Reymond, RA Lerner. J Am Chem Soc 116:803–804, 1994.
44. MPJ van Deurzen, F van Rantwijk, RA Sheldon. Tetrahedron 53:13183–13220, 1997.
45. AA Elfarra, RJ Duescher, CM Pasch. Arch Biochem Biophys 286:244–251, 1991.
46. RO Richards, SH Stanley, M Suzuki, H Dalton. Biocatalysis 8:253–67, 1994.
47. AC van der Linde. Biochim Biophys Acta 77:157–159, 1963.
48. MJ de Smet, J Kingma, H Wynberg, B Witholt. Enzyme Microb Technol 5:352–360, 1983.
49. H Ohta, H Tetsukawa. J Chem Soc Chem Commun 849–850, 1978.
50. K Furuhashi, Chem Econ Eng Rev 18:21–26, 1986.
51. NN Shah, ML Hanna, RT Taylor. Biotechnol Bioeng 49:161–171, 1996.
52. S Doi, K Horikoshi. (Idemitsu Kosan Co., Ltd., Japan) Eur Pat Appl EP 315949 A2 890517.
53. MG Wubbolts, S Panke, JB Beilen, B Witholt. Chimia 50:436–437, 1996.
54. C Nöthe, S Hartmans. Biocatalysis 10:219–225, 1994.
55. GT Philips, BW Robertson, MA Bertola, HS Koger, AF Marx, PD Watts. Eur Pat Appl EP 193227 A1 860903.
56. RN Patel, A Banerjee, LJ Szarka. J Am Oil Chem Soc 72:1247–1264, 1995.
57. SR Woroniecki, JT Sime, KH Baggaley, SW Elson. Biocatalysis 7:221–226, 1993.
58. M Alam, JR Hanson. Phytochemistry 29:3801–3803, 1990.
59. JM Arias, JL Breton, JA Gavin, A Garcia-Granados, A Martinez, ME Onorato. J Chem Soc Perkin Trans I 471–474, 1987.
60. A Garcia-Granados, A Martinez, ME Onorato, JM Arias. J Nat Prod 48:371–375, 1985.
61. HL Holland, PC Chenchaiah, EM Thomas, B Mader, MJ Dennis. Can J Chem 62:2740–2747, 1984.
62. A Garcia-Granados, E Onorato. A Saenz de Buruaga, JM Arias. An Quim, Ser C 78:287–289, 1982.
63. H Hikino, C Konno, T Nagashima, T Kohama, T Takemoto. Chem Pharm Bull 25:6–18, 1977.

64. A Garcia-Granados, E Linan, A Martinez, ME Onorato, JM Arias. J Nat Prod 58:1695–1701, 1995.
65. A Garcia-Grandos, E Linan, A Martinez, ME Onorato, A Parra, JM Arias. Phytochemistry 38:287–293, 1995.
66. L Gledhill, AR Hesketh, BW Bycroft, PM Dewick, J Gilbert. FEMS Microbiol Lett 81:241–245, 1991.
67. Y Amate, JL Breton, A Garcia-Granados, A Martinez, ME Onorato, A Saenz de Buruaga. Tetrahedron 46:6939–6950, 1990.
68. A Garcia-Granados, A Martinez, ME Onorato, ML Ruiz, JM Sanchez, JM Arias. Phytochemistry 29:121–126, 1990.
69. WR Abraham, B Stumpf. Z Naturforsch 42c:79–86, 1987.
70. WR Abraham, HA Arfmann, B Stumpf, P Washausen, K Kieslich. In: P. Schreier, ed. Bioflavour '87. Berlin: 1988, pp 399–414, 1988.
71. WR Abraham, HA Arfmann. Appl Microbiol Biotechnol 32:295–298, 1989.
72. A Müller, WR Abraham, K Kieslich. Bull Soc Chim Belg 103:405–423, 1994.
73. WR Abraham, B Stumpf, K Kieslich, S Reif, HMR Hoffmann. Appl Microbiol Biotechnol 24:31–34, 1986.
74. HA Arfmann, WR Abraham, K Keislich. Biocatalysis 2:59–67, 1988.
75. DS Holmes, DM Ashworth, JA Robinson. Helv Chim Acta 73:260–271, 1990.
76. WR Abraham, HA Arfmann. Tetrahedron 48:6681–6688, 1992.
77. WR Abraham, HA Arfmann, W Giersch. Z Naturforsch 47c:851–858, 1992.
78. WR Abraham, L Ernst, B Stumpf. Phytochemistry 29:115–120, 1990.
79. WR Abraham, L Ernst, HA Arfmann. Phytochemistry 29:757–763, 1990.
80. WR Abraham, B Stumpf, K Kieslich. Appl Microbiol Biotechnol 24:24–30, 1986.
81. A Archelas, R Furstoss. Tetrahedron Lett 33:5241–5242, 1992.
82. D Zhang, EBjr Hansen, J Deck, TM Heinze, A Henderson, WA Korfmacher, CE Cerniglia. Xenobiotica 27:301–315, 1997.
83. N Itoh, M Kusaka, T Hirota, A Nomura. Appl Microbiol Biotechnol 43:394–401, 1995.
84. RF White, J Birnbaum, RT Meyer, J Ten Broeke, JM Chemerda, AL Demain. Appl Microbiol 22:55–60, 1971.
85. RN Patel, A Banerjee, B Davis, J Howell, C McNamee, D Brzozowaski, J North, D Kronenthal, L Szarka. Bioorg Med Chem 2:535–542, 1994.
86. M Kittelmann, R Lattmann, O Ghisalba. Biosci Biotechnol Biochem 57:1589–1590, 1993.
87. RA Johnson, ME Herr, HC Murray, CG Chidester, F Han. J Org Chem 57:7209–7212, 1992
88. G Ramos, O Ghisalba, HP Schar, F Bruno, P Maienfisch, AC O'Sullivan. Eur Pat Appl, EP 277916 A2 880810.
89. JRW Miles, CM Tu, CR Harris. J Econ Entomol 62:1334–1338, 1969.
90. Y Miura, T Yamane. Biotechnol Lett 19:611–613, 1997.
91. PR Ortiz de Montellano, JA Fruetel, JR Collins, DL Camper, GH Loew. J Am Chem Soc 113:3195–3196, 1991.
92. S Ozaki, PR Ortiz de Montellano. J Am Chem Soc 116:4487–4488, 1994.
93. RT Ruettinger, AJ Fulco. J Biol Chem 256:5728–5734, 1981.
94. DC Levinger, JA Stevenson, LL Wong. J Chem Soc Chem Commun 22:2305–2306, 1995.
95. RT Ruettinger, GR Griffith, MJ Coon. Arch Biochem Biophys 183:528–537, 1977.
96. CT Hou, RN Patel, AI Laskin, N Barnabe. Appl Environ Microbiol 38:127–134, 1979.
97. SL Neidleman, WF Amon Jr, J Geigert (Cetus Corp). US Pat 4,284,723.
98. EJ Allain, LP Hager, L Deng, EN Jacobsen. J Am Chem Soc 115:4415–4416, 1993.
99. A Zaks, DR Dodds. J Am Chem Soc 117:10419–10424, 1995.
100. FJ Lakner, LP Hagner. J Org Chem 61:3923–3925, 1996.
101. AF Dexter, FJ Lakner. J Am Chem Soc 117:6412–6413, 1995.
102. SL Newmyer, PR Ortiz de Montellano. J Biol Chem 270:19430–19438, 1995.
103. E Blée, F Schuber. Biochem Biophys Res Commun 173:1354–1360, 1990.

8

Stereoselective Syntheses Using Microbial Epoxide Hydrolases

Wolfgang Kroutil and Kurt Faber
University of Graz, Graz, Austria

I. INTRODUCTION

Enantiopure epoxides and vicinal diols (employed as their corresponding cyclic sulfate or sulfite esters as reactive intermediates) have been shown to be highly versatile intermediates in organic synthesis [1]. The high utility of these compounds results from their ability to react with a broad variety of nucleophiles (Schemes 1 and 2) due to their electronic polarization and structural strain. Almost all of the reactions depicted in Scheme 1 involve an attack of a nucleophile at the sterically less hindered carbon atom of the oxirane ring. Other synthetically useful but lesser known reactions with epoxides are listed below (Scheme 3). For reaction R1, i.e., the nucleophilic opening of an epoxide by chloride under Ti(IV) catalysis with inversion of configuration, no loss of e.e. was claimed [2]. A nonfunctionalized chiral carbon center can be created by reduction of an oxirane using titanocene dichloride [3] or the $Et_3SiH-BH_3$ complex [4] (R2). The latter reaction represents a useful means to the creation of such chiral centers, which are difficult to achieve by other methods. Opening of the epoxide ring by Ph_3P (reaction R3) followed by a four-center elimination through a Wittig-like intermediate [5] yields an alkene of defined configuration.

II. PREPARATION OF OPTICALLY ACTIVE EPOXIDES AND VICINAL DIOLS

A. Chemical Methods

1. Enantioselective Epoxidation of Alkenes

The widely used asymmetrical epoxidation of allylic alcohols developed by Katsuki and Sharpless employs *tert*-butyl hydroperoxide (TBHP), enantiomerically pure tartaric acid esters, and isopropyl titanium(IV) (6). Similar results were obtained using TBHP and vanadyl acetyl acetonate $VO(acac)_2$ and hexacarbonyl molybdenum [7,8]. The drawback of these reactions are their restriction to allylic alcohols, which is required as anchor group for the formation of the intermediate complex.

   Nonfunctionalized, prochiral alkenes can be oxidized in an enantioselective way by employing chiral oxidants (e.g., proxycamphoric acid) or chiral catalysts, e.g., (+)-diisopropyl tartrate. Asymmetrical catalysts, like the "Jacobsen reagent," a chiral manga-

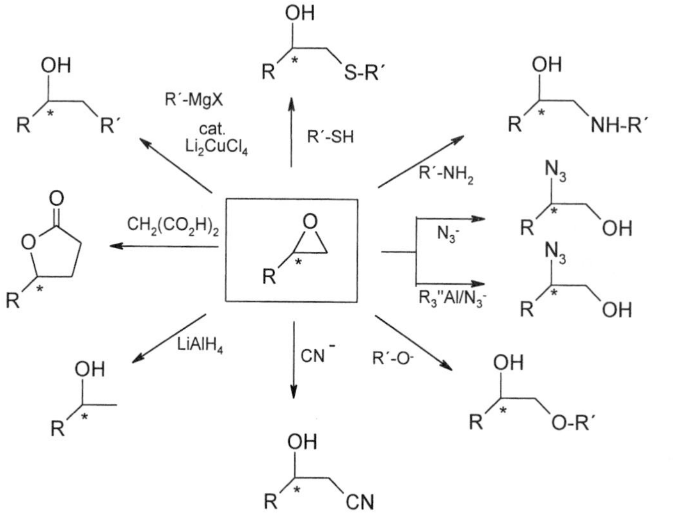

Scheme 1   Reaction of epoxides with nucleophiles. *Chirality center.

nese(III)−salen complex, have been developed for the epoxidation of nonfunctionalized olefins more recently [9–11]. Although high stereoselectivity has been achieved for the epoxidation of *cis*-alkenes, the results obtained with *trans*- and terminal olefins were less satisfactory using the latter method.

2.   Kinetic Resolution of Racemic Epoxides by Chemical Methods

Schurig and Betschinger described the rearrangement of racemic, nonfunctionalized epoxides to furnish aldehydes or ketones [12] by employing oxodiperoxomolybdenum(VI), (S)-N,N-diethyllactic amide, and a homochiral diol as selector. The selector acts as a Lewis acid, accepting preferentially only one epoxide enantiomer as ligand and catalyzing the rearrangement of only one enantiomer to the corresponding nonchiral carbonyl compound. The absolute configuration of the remaining epoxide enantiomer can be influenced by the

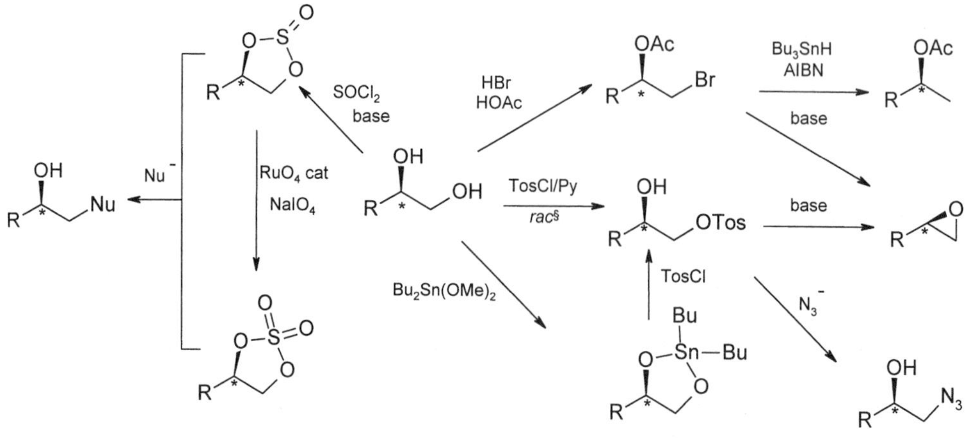

Scheme 2   Syntheses from chiral 1,2-diols. *Chirality center. §Possible racemization.

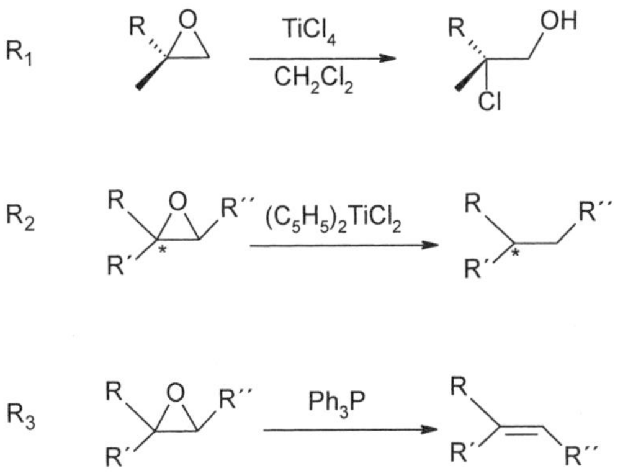

Scheme 3 Lesser known synthetically useful reactions of epoxides.

configuration of the selector. Epoxides with long chains as substituents are not transformed, most likely due to steric hindrance.

Very recently, the kinetic resolution of terminal epoxides by asymmetrical hydrolysis employing a chiral cobalt-based salen complex as catalyst in water led to the formation of 1,2-diols in high yield and with high enantiomeric enrichment [13].

Racemic epoxy alcohols or diols can be esterified with optical active acids, e.g., (−)-α-methoxy-α-trifluoromethylphenylacetic acid (MTPA, Mosher's acid) to give diastereomers that can be separated chromatographically. However, these methods suffer from expensive auxiliary reagents required in molar amounts and difficulties in the chromatographic separation of diastereoisomers, and hence are of very limited practical applicability.

3. Asymmetrical Dihydroxylation

High enantiomeric excess of diols was obtained when alkenes were cis-dihydroxylated using cat. OsO_4 in the presence of chiral quinuclidine ligands as catalyst [1,14]. The reaction is highly effective for most of the possible alkene substitution patterns. Even some mono-substituted alkenes are dihydroxylated with good enantioselectivity depending on the choice of quinuclidine ligand. *trans*-Disubstituted alkenes as well as a limited number of 1,1-di and trisubstituted alkenes are particularly good substrates, whereas *cis*-disubstituted alkenes have proven somewhat difficult to hydroxylate with high enantioselectivity [14].

4. Chiral Pool

Enantiopure organic compounds, which can easily be obtained from natural sources, e.g., D- and L-tartaric acid, D-mannitol, L-lactic acid, and L-amino acids, have served as precursors for chiral oxiranes through multistep syntheses. New chemical methods allowing for the desymmetrization of achiral epoxides have been reviewed recently [15].

B. Biocatalytic Methods

A number of biocatalytic methods provide an useful arsenal of methods as valuable alternatives to the above-mentioned techniques [16–21].

1. Hydrolytic Enzymes

(1) Prochiral or racemic synthetic precursors of epoxides, such as halohydrins, can be asymmetrized or resolved using hydrolytic enzymes [22,23]. In particular esterases and lipases have been used for such an enantioselective ester hydrolysis or esterification (Scheme 4). This methodology is well developed and high selectivities have been achieved in particular for esters of secondary alcohols, but it is impeded by the requirement of regioisomerically pure halohydrins. (2) Alternatively, α-haloacid dehalogenases catalyze the S_N2-displacement of a halogen atom at the α position of carboxylic acids with a hydroxy function. This process leads to the formation of the corresponding α-hydroxy acid with inversion of configuration [24]. However, enzymatic α-haloacid dehalogenation incurs two drawbacks, i.e., the instability of the substrates (particularly for α-bromoacids) in aqueous systems and the limited substrate tolerance, as only short chain haloacids are accepted [25].

2. Oxidoreductases

Asymmetrical biocatalytic reduction of α-keto acids [26] or α-keto alcohols (27) using D- or L-lactate dehydrogenase or glycerol dehydrogenase, respectively, provides access to chiral α-hydroxy acids or 1,2-diols (Scheme 5). The latter products can be converted to the corresponding epoxides using conventional chemical methodology. Although excellent selectivities are generally achieved, the need for the recycling of redox cofactors such as NAD(P)H has restricted the number of applications on a preparative scale. Likewise, biocatalytic asymmetrical epoxidation of alkenes catalyzed by monooxygenases cannot be performed on a preparative scale with isolated enzymes, due to their complex nature and their dependence on a redox cofactor, such as NAD(P)H. Thus, whole microbial cells are used instead. Toxic effects of the epoxide formed as well as further undesired metabolism can be minimized by employing biphasic aqueous-organic media. However, this method is not trivial and requires high bioengineering skills [28]. On the other hand, haloperoxidases are independent of nicotinamide cofactors, as they produce hypohalous acid from H_2O_2 and halide, which in turn yields a halohydrin from an alkene (Scheme 5). However, the majority of peroxidases employed for this reaction are heme-dependent, which makes them rather sensitive to the oxidant, i.e., hydrogen peroxide. As a consequence, the concentration of the latter has to be carefully monitored to prevent enzyme deactivation, e.g., by employing a peroxy-stat technique [29]. Furthermore, haloperoxidase enzymes are rare

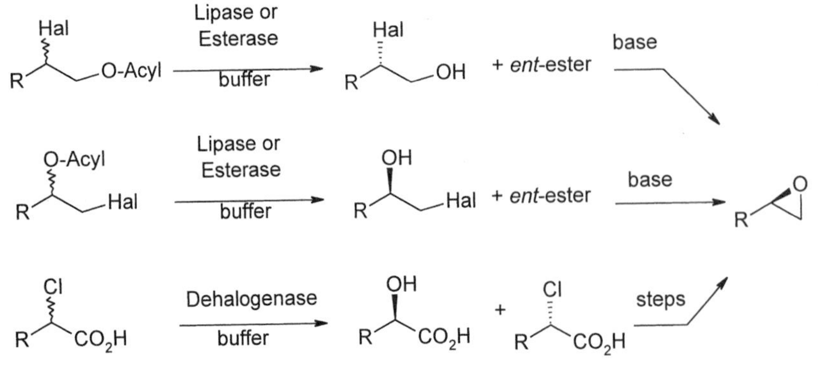

Scheme 4 Enzymatic syntheses of epoxides using hydrolytic enzymes.

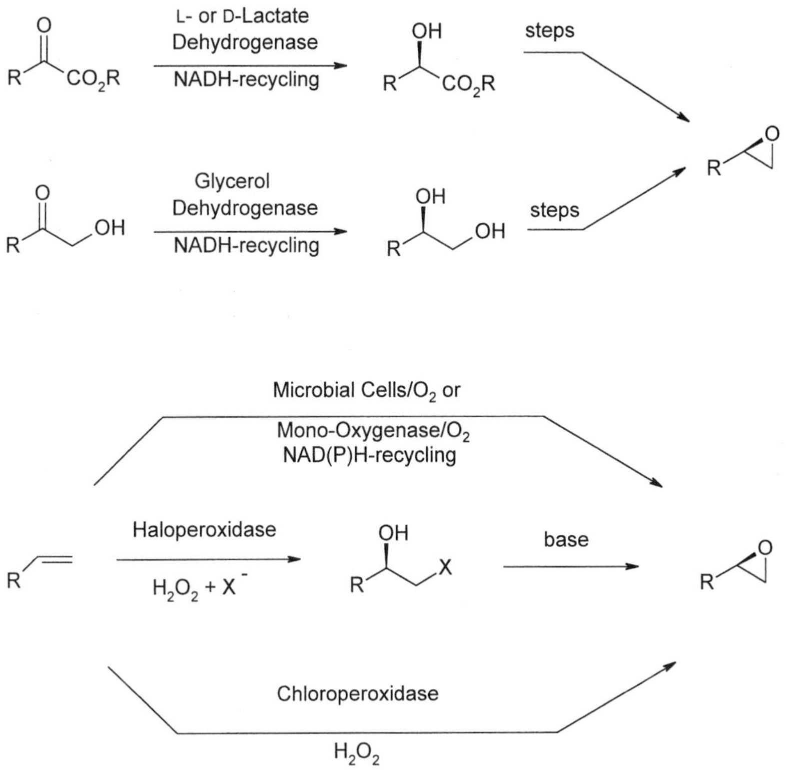

Scheme 5   Enzymatic syntheses of epoxides using dehydrogenases, monooxygenases, and peroxidases.

in nature and usually exhibit low selectivities due to the fact that the formation of halohydrins can take place not only in the active site of the enzyme but without direct enzyme catalysis [30]. Similar low selectivities have been observed with halohydrin epoxidases, which act as a "biogenic chiral base" by converting a halohydrin into the corresponding epoxide [31]. In contrast, peroxidases, such as chloroperoxidase, are cofactor-independent and thus can also be used in isolated form for the enzymatic epoxidation of alkenes. Although excellent selectivities were obtained with internal *cis*-olefins, long chain substrates and terminal alkenes were unreactive [32].

A valuable alternative to the above methods is the use of cofactor-independent epoxide hydrolases (EC 3.3.2.X) [33]. Enzymes from mammalian sources, such as rat liver tissue, have been investigated in great detail during detoxification studies [34]. However, biotransformations on a preparative scale were hampered by the limited supply of enzyme and the described transformations rarely surpass the millimolar range [35]. Furthermore, mammalian enzymes proved to be rather poor catalysts with turnover numbers within the range of about 1 s^{-1}. In contrast, during recent years, highly selective epoxide hydrolases were identified from microbial sources, which allows for an (almost) unlimited supply of these enzymes for preparative scale applications from fermentation. These interesting biocatalysts have recently received considerable attention and some of their merits have been reviewed [36–38].

III. OCCURRENCE OF EPOXIDE HYDROLASES AND THEIR BIOLOGICAL ROLE

Epoxide hydrolases have been shown to be ubiquitous in nature and were detected in various sources such as mammals [32,34,39], plants [40], insects [41–43], filamentous fungi [44], bacteria [21,45–47] as well as in red yeasts [48]. In mammals, epoxide hydrolases are assumed to play an important role in the metabolism of xenobiotics, in particular of aromatic systems, the oxides of which ("arene oxides") are thought to be a major cause for the carcinogenic nature of polycyclic aromatic systems [49]. In general, aromatics can be metabolized via two different pathways (Scheme 6). (1) Dioxetane formation via dioxygenase-catalyzed (cyclo)addition of molecular oxygen onto the C—C bond yields a (putative) dioxetane species, which is then detoxified via reductive cleavage of the O—O bond furnishing a physiologically more innocuous *cis*-1,2-diol. (2) Alternatively, formation of a highly reactive arene oxide via the introduction of a single O atom (from molecular oxygen) into the aromatic system is catalyzed by a monooxygenase. The highly reactive epoxy species is further metabolized via hydrolysis catalyzed by an epoxide hydrolase to yield a *trans*-1,2-diol.

So far five different types of epoxide hydrolases have been characterized in mammals, and they have been grouped depending on their enzymatic activity and biochemical characteristics. These groups include the soluble epoxide hydrolases (sEHs, also referred to as cytosolic epoxide hydrolases) [50], microsomal epoxide hydrolases (mEHs) [34], leukotriene A_4 hydrolase (LTA$_{44}$H) [51], cholesterol epoxide hydrolase [52], and hepoxilin hydrolase [53]. Several reviews have focused on these enzymes of medical importance [34,54].

Epoxide hydrolases have been found in such plants as soy bean [40] or in nectarines and strawberries, where they play a crucial role in the enantioselective synthesis of lactone aroma compounds [55]. Furthermore, epoxide hydrolases are involved in the biosynthesis of cutin, a polyester found in the cuticle that forms the first physical barrier against plant infection [56]. It has also been proposed that epoxide hydrolases may be involved in the production of microbial toxins [57].

In microbes, these enzymes might also have a detoxification function; however, they play a vital role in the utilization of alkenes as carbon source (Scheme 7) [58]. Epoxides

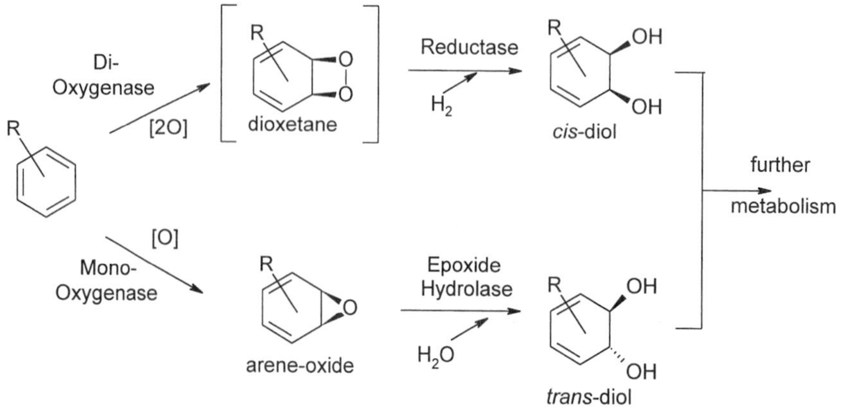

Scheme 6   Involvement of epoxide hydrolases in the biodegradation of aromatics.

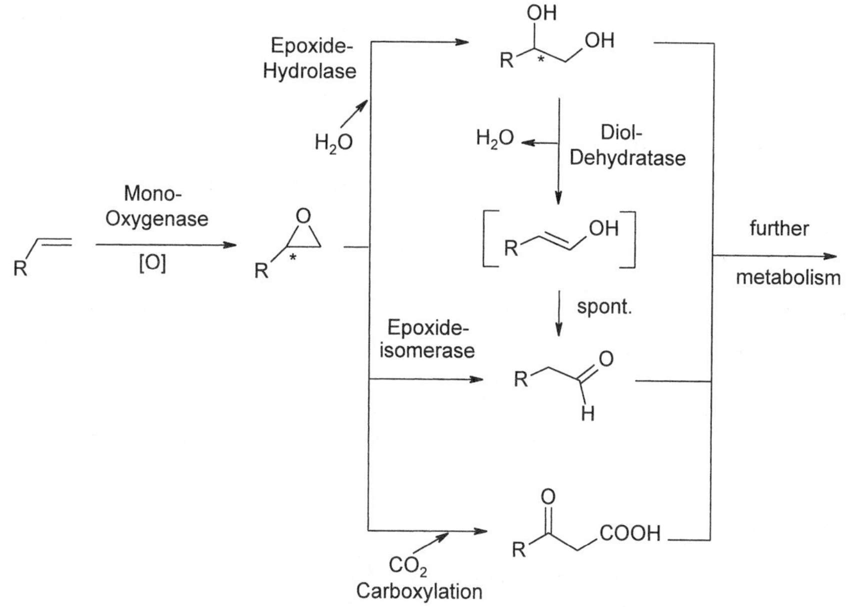

Scheme 7   Involvement of epoxide hydrolases in the biodegradation of alkenes.

may be hydrolyzed by an epoxide hydrolase to the corresponding 1,2-diol, which can be degraded either by oxidation or by elimination of water under catalysis of a diol dehydratase, yielding an aldehyde [59]. Alternatively, aldehydes are obtained via direct rearrangement of the epoxide catalyzed by an epoxide isomerase [60,61]. In a rather different pathway, propylene-grown cells of *Xanthobacter* Py2 supported the fixation of CO_2 in the degradation pathway of propylene oxide to yield a β-keto acid [62].

A.   Purification and Overexpression of Epoxide Hydrolases

Epoxide hydrolases have been purified from mammalian liver cells [63–66] but also from microbial sources such as *Bacillus megaterium* [67], *Corynebacterium* [45,46,68], *Pseudomonas* sp. [46,69], and dematiaceous fungi such as *Ulocladium atrum* and *Zopfiella karachiensis* [70]. However, some of these enzymes were only partially purified [67,68], or their enantioselectivity was not investigated [69] or was very low [45,46]. More recently, two bacterial epoxide hydrolases with high activity and high enantioselectivity were purified and characterized from *Rhodococcus* sp. NCIMB 11216 [71] and *Nocardia* TB1 [72].

Over the past few years, several purified membrane-bound and soluble epoxide hydrolases from various origin have been (at least partially) sequenced. Some of them have also been cloned and overexpressed. This is the case for the soluble epoxide hydrolase from rat liver that has been overexpressed in *E. coli* [73,74]. Other epoxide hydrolase genes that have been cloned so far are of mammalian [73,75–78], insect [42], plant [79,80], and one of bacterial [81] origin.

Regardless of the biological source, all epoxide hydrolases found to date do not possess a metal atom or a prosthetic group, and do not require any cofactors. Since the overexpression of a mammalian epoxide hydrolase did not lead to a highly active enzyme and due to the fact that the availability of these enzymes is somewhat limited, this chapter will mainly focus on epoxide hydrolases of microbial origin.

B. Structure of Epoxide Hydrolases

Despite the fact that to date no structure of an epoxide hydrolase has been fully resolved by x-ray crystallography [82], mammalian epoxide hydrolases are believed to belong to the α/β-hydrolase fold family of enzymes [83–85] since several of them show a low but significant sequence similarity with haloalkane dehalogenase from *Xanthobacter autotrophicus* GJ10 [86] of which the three-dimensional structure has been solved by x-ray crystallography [87]. Amino acid sequence similarities of a bacterial epoxide hydrolase from *Agrobacterium radiobacter* AD1 with the eukaryotic epoxide hydrolases, haloalkane dehalogenase from *X. autotrophicus* GJ10, and bromoperoxidase A2 from *Streptomyces aureofaciens* indicated that it also belonged to the α/β-hydrolase fold family.

C. Mechanism of Epoxide Hydrolases

The mechanism by which these enzymes operate was long debated. It was previously assumed that simple direct nucleophilic opening of the oxirane ring by a histidine-activated water molecule was the key step [88]. However, convincing evidence was first provided for mammalian epoxide hydrolases, showing that the reaction occurs via a covalent glycol-monoester-enzyme intermediate [89,90] and, very recently, mechanistic studies based on single-turnover experiments on *Agrobacterium radiobacter* AD1 [81] suggested the same mechanism (Scheme 8). Thus, a reaction mechanism was proposed in which the catalytic Asp[107] performs a nucleophilic attack on the primary carbon atom of the substrate epi-

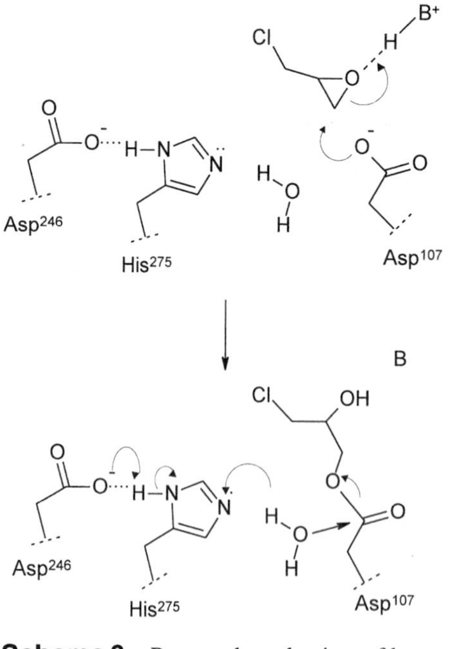

Scheme 8 Proposed mechanism of bacterial epoxide hydrolase from *Agrobacterium radiobacter* AD1.

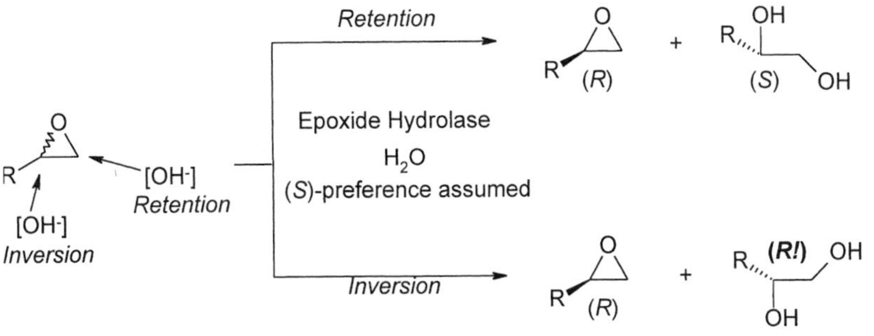

Scheme 9 Hydrolysis of epoxides proceeding with retention or inversion of configuration.

chlorohydrin, leading to the already mentioned covalently bound ester intermediate. In a next step, His[275], assisted by Asp[246], abstracts a proton from a water molecule that hydrolyzes the ester at the carbonyl function of Asp[107]. Phe[108] of epoxide hydrolase, which is located next to the nucleophile Asp[107], is probably interacting with the epoxide ring [91,92]. In addition, it has been postulated [84] that a proton-donating group (B-H) should be present in the cap domain of soluble epoxide hydrolase, which protonates the leaving group, i.e., an oxy anion that is formed upon opening of the epoxide ring.

As a consequence of this mechanism, the epoxide is generally opened in a trans-specific fashion with one oxygen from water being incorporated into the product diol [69]. For instance, (±)-*trans*-epoxysuccinate was converted to *meso*-tartrate by an epoxide hydrolase isolated from *Pseudomonas putida* [69]. In a complementary fashion, *cis*-epoxysuccinate gave D- and L-tartrate (with a *Rhodococcus* sp.) albeit in low optical purity [93]. In addition, it was shown by $^{18}OH_2$ labeling experiments that only one O atom originates from water when both mammalian or bacterial epoxide hydrolases [94] or whole fungal cells [94] were used as catalysts.

Although two cases for reactions proceeding via a formal cis-hydration process have been reported to date [96,97], they seem to be rare exceptions and—given the present knowledge for the enzyme mechanism—attempts to explain this phenomenon remain rather speculative [97].

The mechanism of enzymatic epoxide hydrolysis implies important stereochemical consequences on the outcome of the kinetic resolution of asymmetrically substituted epoxides (Scheme 9). In the majority of kinetic resolutions of esters (e.g., by ester hydrolysis and synthesis using lipases, esterases, and proteases) the absolute configuration at the stereogenic center(s) always remain the same throughout the reaction. In contrast, the enzymatic hydrolysis of epoxides may take place via attack on either carbon of the oxirane ring and it is the structure of the substrate and the nature of the enzyme involved that determine the regioselectivity of the attack, which may proceed either with retention or with inversion of configuration [94,98–100]. Consequently, the absolute configuration of *both the product and the substrate* from a kinetic resolution of a racemic epoxide has to be determined in order to elucidate the stereochemical pathway.

D. Screening for Epoxide Hydrolases

The search for novel microbial epoxide hydrolases through the screening of various fungal and bacterial sources was mainly triggered by chance observations of unexpected side reactions in microbial transformations within two research groups at about the same time.

Thus, when a terpene-type alkene was epoxidized by using whole cells of *Aspergillus niger* aiming at the preparation of the enantiomerically enriched epoxide, a certain amount of the corresponding 1,2-diol was formed as byproduct [101]. The latter fact could be attributed to the presence of a fungal epoxide hydrolase. In a related manner, the chemoselective enzymatic hydrolysis of an epoxy nitrile by a crude nitrilase preparation derived from a *Rhodococcus* sp. was expected to furnish the corresponding epoxycarboxylic acid. Instead, the dihydroxy acid product was formed through the action of the nitrilase and an unknown epoxide hydrolase being present in the crude enzyme preparation [102]. A careful investigation of the literature revealed that microbial hydrolysis of epoxides had been frequently observed as an undesired side reaction during the asymmetrical epoxidation of alkenes using whole viable microbial cells. Since it was anticipated that those strains, which were capable of epoxidizing alkenes with a high degree of enantioselectivity, would also possess the corresponding epoxide hydrolase with "matching" stereoselectivity for the further degradation of the (toxic) intermediate epoxide, the screening was mainly concentrated on strains that were selected along this principle. Fortunately, these assumptions proved to be true to a large extent and, as a consequence, an impressive number of bacteria and fungi showing epoxide hydrolase activities were identified so far, predominantly by the groups of Furstoss and Faber. It must be emphasized that the ample availability of microbial epoxide hydrolase activities from fungi and bacteria presented the key to biocatalytic transformations on a multigram scale.

IV. BIOHYDROLYSIS OF EPOXIDES

Kinetic resolution of racemic epoxides can be conveniently described by the (dimensionless) E value, if the epoxide and the diol are hydrolyzed by the same mechanism, implying that for both epoxide enantiomers the attack of the formal [OH] always occurs at the same carbon of the oxirane ring affecting *either* retention *or* inversion of configuration. As a consequence, E values can be used for the description of enantioselectivities. In all other cases, where mixed pathways (i.e., inversion *and* retention) are taking place, further information must be given for an exact description of the process and the concept of E values cannot be applied. Thus, measuring the e.e. of the diol and/or the epoxide at a certain conversion may only give a rough idea about the practical usefulness of the process, but direct comparison of selectivities between different enzymes and substrates are rather difficult. Due to the underlying kinetics, e.e./conversion datasets can be easily manipulated at will. For instance, the e.e. of the residual epoxide in a kinetic resolution at high conversion well beyond 50% will easily exceed 90%, but that does not say much about the enantioselectivity of the enzyme. Several methods for the calculation of the E value, based on the e.e. of the product and/or residual substrate and/or the degree of conversion, have been developed [103,104], which have their special merits [105]. In this review, E values are used as long as they are applicable.

Classic kinetic resolution starts from a racemate and furnishes (in the ideal case) the optically pure diol and the epoxide in each 50% yield. This so-called classic kinetic resolution pattern is often regarded as a major drawback since the theoretical chemical yield of one stereoisomer can never exceed 50% based on the racemic starting material. Subsequently, the product has to be separated from the residual starting material and often the "wrong" enantiomer is of lesser or no importance. As a consequence, methods that offer a solution to this intrinsic problem of kinetic resolution are highly desirable. On the other hand, so-called deracemization methods that allow the transformation of a racemic starting

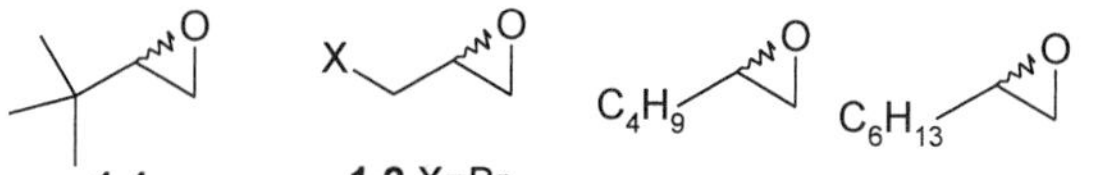

Figure 1　Monoalkyl-substituted oxiranes.

material to a single enantiomer are highly advantageous [106]. Throughout this chapter, examples for deracemization processes are given at the end of such substrate class.

A.　Mono-Substituted Oxiranes

In general, mono-substituted oxiranes (Fig. 1, Table 1) have been hydrolyzed by fungal and mammalian epoxide hydrolases with low to moderate enantioselectivity so far. It was only recently that red yeasts were identified as the biocatalysts of choice for these substrates.

In an early study, epoxide **1.1** was hydrolyzed to furnish the (R)-diol in only 50% e.e. at 14% conversion by mEH. Later on, an interesting and previously unrecorded phenomenon for this enzyme was found, i.e., upon increase of epoxide concentration, an impressive enhancement of the enantioselectivity was achieved (92% e.e. at 38% conversion). As an explanation, it was suggested that the substrate would alter the environment of the enzyme increasing its enantioselecting ability. However, this trend was not reported for other substrates.

Small mono-substituted oxiranes bearing a halogen atom in its side chain such as compounds **1.2** and **1.3** have been hydrolyzed by mammalian and yeast epoxide hydrolases with very low enantioselectivity ($E = 3$ and 5.5, respectively). More recently, three red yeasts and one bacterial strain (*Chryseomonas luteola* [107]) have been reported to trans-

Table 1　Biohydrolysis of Monoalkyl Oxiranes

Substrate	Enzyme source	e.e. (diol) (%)	e.e. (epoxide) (%)	Conversion (%)	E^a	Ref.
1.1	mEH	50 (R)	n.d.	14	3.4	108
1.1	mEH	92 (R)	n.d.	38	42^b	108
1.2	mEH	46 (R)	n.d.	17	3	109
1.3	mEH	52 (R)	n.d.	25	3.7	109
1.4	mEH	24 (R)	n.d.	16	1.7	108
1.3	*Rhodotorula glutinis*	22 (S)	>98 (R)	~10	5.5	48
1.4	*Rhodotorula glutinis*	83 (R)	>98 (S)	54	48	110
1.5	*Rhodotorula glutinis*	55 (R)	>98 (S)	64	14	110
1.5	*Rhodotorula araucariae*	87 (R)	>98 (S)	53	>200^c	111
1.5	*Rhodosporidium toruloides*	91 (R)	>98 (S)	562	>100	111
1.5	*Aspergillus niger* LCP 521	n.d.	n.d. (S)	—	3.7	112
1.5	*Aspergillus terreus* CBS 116-46	n.d.	n.d. (S)	—	6.5	112
1.5	*Nocardia* TB1	57 (R)	40 (S)	41	5.3	113

^{a}In case the E value was not reported, it was calculated from e.e. (epoxide), e.e. (diol), and/or conversion.
^{b}Increased substrate concentration (see text).
^{c}This E value was calculated from conversion and e.e. (epoxide). When e.e. (epoxide) and e.e. (diol) was used, and E value of 65 was obtained.

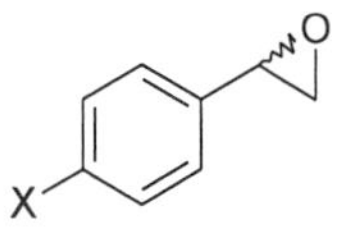

1.6 X=H	
1.7 X=F	
1.8 X=Cl	
1.9 X=NO$_2$	

Figure 2 Styrene oxide–type substrates.

form the (*R*) enantiomer of (±)-1-hexene oxide **1.4** and (±)-1-octene oxide **1.5** to the corresponding diols with good to excellent selectivity. Interestingly, it was found that *Rhodotorula glutinis* possesses an optimum of its enantioselectivity when the side chain of the oxirane ring possessed four carbons atoms **1.4** [110].

Styrene oxide–type substrates (Fig. 2) have been hydrolyzed with good enantioselectivity especially by fungal epoxide hydrolases and a recombinant epoxide hydrolase from *Agrobacterium radiobacter*. However, bacterial epoxide hydrolases, e.g., from *Nocardia* spp. and related *Rhodococcus* strains, were not useful for this substrate pattern. Similarly, enzymes from yeasts and mammals showed only low to moderate selectivity (Table 2).

It is noteworthy that all epoxide hydrolases formed the same product enantiomer, i.e., the (*R*)-diol by leaving the (*S*)-epoxide behind. However, *Beauveria bassiana* and *Beauveria densa* transform (*S*)-**1.6** with inversion into the corresponding (*R*)-diol, while all other enzymes convert (*R*)-**1.6** to the (*R*)-diol with retention. This appears to be a general rule, as long as the para substituents are Me, Cl, Br, F, and CN [117,119]. However, for the para substituent being NO$_2$, *B. bassiana* displays the same enantiopreference as *A. niger*.

1. Deracemization Technique

The matching enantiopreference and regioselectivity of *B. bassiana* and *A. niger* allowed the development of an elegant deracemization technique. Thus, combination of both strains in a single reactor led to the formation of (*R*)-phenylethane-1,2-diol as the sole product in 89% e.e. and 92% isolated yield from (±)-styrene oxide [98] (Scheme 10).

Another strategy toward an enantioconvergent process was achieved by combination of bio- and chemocatalysis (see also 2,2-di- and trisubstituted oxiranes). Thus, in a first

Table 2 Biohydrolysis of Styrene Oxide–Type Substrates

Substrate	Enzyme source	e.e. (diol) (%)	e.e. (epoxide) (%)	Ref.
1.6	mEH	45 (*R*)	76 (*S*)	115
1.6	*Rhodotorula glutinis*	48 (*R*)	>98 (*S*)	48
1.6	*Agrobacterium radiobacter*	49 (*R*)	>99 (*S*)	116
1.6	*Beauveria densa*	78 (*R*)[a]	>95 (*R*)	117
1.6	*Beauveria bassiana*	83 (*R*)[a]	98 (*R*)	118
1.6	*Aspergillus niger*	65 (*R*)	99 (*S*)	118
1.7	*Aspergillus niger*	81 (*R*)	98 (*S*)	119
1.8	*Aspergillus niger*	79 (*R*)	>98 (*S*)	119
1.9	*Aspergillus niger*	70 (*R*)	>98 (*S*)	119
1.9	*Beauveria bassiana*	49 (*R*)	20 (*S*)	119

[a]Not a kinetic resolution.

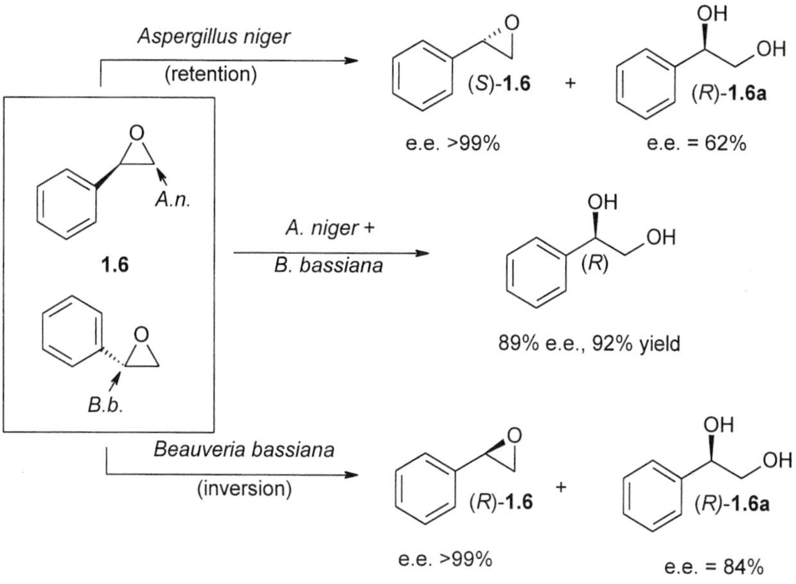

Scheme 10 Resolution and deracemization of (±)-styrene oxide by fungal cells.

step, racemic *para*-nitrostyrene oxide **1.9** was kinetically resolved by using a crude enzyme extract from *A. niger*. This biohydrolysis proceeds via attack at the less substituted C atom leading to retention of configuration at the stereogenic center with excellent regioselectivity. On the other hand, (chemical) acid-catalyzed hydrolysis of epoxides possessing a benzylic carbon atom is known to proceed preferentially with inversion at the more substituted oxirane carbon. When the biocatalytic kinetic resolution (step 1) was followed by the careful addition of acid (step 2), (*R*)-*para*-nitrophenylethane-1,2-diol **1.9a** was obtained in good yield (94%) and acceptable e.e. (80%) from the racemate. However, the process required careful tuning due to limited enantioselectivity in step 1 and due to partial racemization during step 2. A mathematical method was refined that allowed the determination of the optimum point of conversion, where the acid-catalyzed hydrolysis had to be started [120,121].

B. 2,2-Disubstituted Oxiranes

Only limited data are available on the biohydrolysis of 2,2-disubstituted oxiranes (Fig. 3, Table 3) employing mammalian and yeast epoxide hydrolases. For instance, the presence of a geminal dimethyl group in **2.1** resulted in complete enantioselectivity when mEH was used as a catalyst. Thus, epoxide **2.1** was resolved to yield the corresponding (*R*)-diol and the remaining (*S*)-epoxide **2.1** both in more than 95% e.e. at 50% conversion. On the other hand, a simple 2-methyl-2-alkyl oxirane such as **2.2** could not be resolved with high efficiency employing mEH. For this substrate pattern, bacterial epoxide hydrolases proved to be extremely useful.

Particularly substrates bearing a straight alkyl chain, such as **2.2–2.5**, even when functionalized with a terminal C=C double bond **2.6** [131] or a bromo group **2.7**, were transformed with virtually absolute enantioselectivity. As a consequence, the reactions ceased and did not proceed beyond a conversion of 50%. Interestingly, the enantio pref-

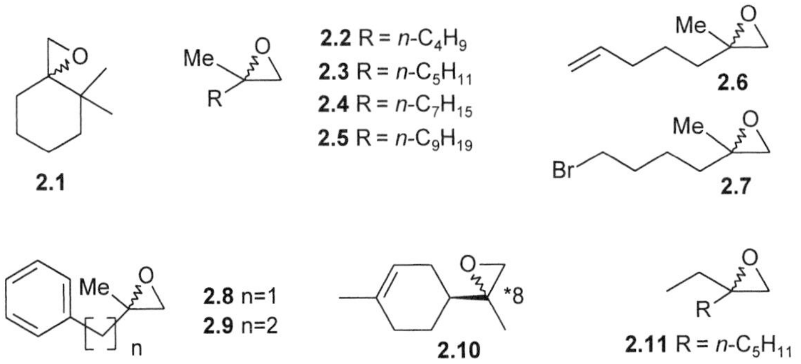

Figure 3 2,2-Disubstituted oxiranes. *Diastereomeric mixture.

erence was found to depend on the substrate structure but not on the strain used. Thus, the (R) enantiomer was preferred for mono-substituted substrates (albeit at low selectivity; see above for substrate **1.6**), but the (S) enantiomer was more quickly transformed for 2,2-disubstituted oxiranes. When the epoxide bears a synthetically useful phenyl moiety at the ω position of the alkyl chain, mimicking a masked carboxyl functionality (**2.8**), the selectivity was slightly reduced but still in a preparatively useful range (E = 123). Unexpectedly, when the carbon chain was extended by an additional CH$_2$ unit (**2.9**), the selectivity declined (E = 12). From earlier results with *Rhodococcus* sp. NCIMB 11216 it was concluded that the enantioselectivity largely depends on the relative difference in size of the two alkyl groups [21] (Scheme 11). Thus, by increasing the difference in the relative size of the two alkyl substituents, enhanced selectivities were achieved.

The fact that substrates bearing a phenyl group behaved differently might be attributed to electronic effects. High selectivity of the bacterial hydrolysis of substrate **2.10**

Table 3 Biohydrolysis of 2,2-Disubstituted Oxiranes

Substrate	Enzyme source	e.e. (diol) (%)	e.e. (epoxide) (%)	E[a]	Ref.
2.1	mEH	>95 (R)	>95 (S)	>145	122
2.2	mEH	54 (S)	40 (R)	4.9	123
2.2	*Nocardia* TB1	97 (S)	>99 (R)	>200	124
2.3	*Nocardia* TB1	>99 (S)	>99 (R)	>200	125
2.3	*Aspergillus niger* LCP 521	32 (R)	99 (S)	9	112
2.4	*Mycobacterium paraffinicum*	>99 (S)	88 (R)	>200	126
2.5	*Mycobacterium paraffinicum*	>99 (S)	52 (R)	>200	126
2.6	*Nocardia* TB1	>99 (S)	>99 (R)	>200	124
2.7	*Nocardia* TB1	96 (S)	>99 (R)	>200	127
2.8	*Nocardia* EH1	96 (S)	81 (R)	123	128
2.9	*Nocardia* EH1	80 (S)	33 (R)	12	129
2.10	*Nocardia* TB1	98 (4S,8S)	46 (4S,8R)	156	125
2.10	*Aspergillus niger*	94 (4S,8R)	98 (4S,8S)	148	130
2.11	*Nocardia* TB1	>99 (S)	28 (R)	>200	124

[a]For those cases where the E value was not given in the reference, it was calculated from e.e. (epoxide), e.e. (diol), and/or conversion.

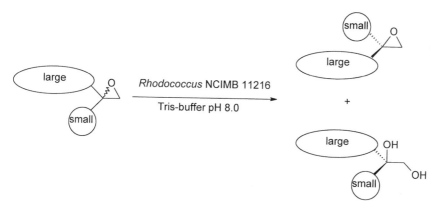

Scheme 11   Resolution of 2,2-disubstituted oxiranes by *Rhodococcus* NCIMB 11216.

showed that bulky substituents might also be present close to the oxirane ring. However, for substrate **2.10** the fungal epoxide hydrolase of *A. niger* shows opposite enantio preference as compared to the bacterial system; thus the fungus preferentially hydrolyzed the (*8R*) enantiomer, while the bacterium transformed the (*8S*) counterpart. The same result was found for substrate **2.3**, although in this case the enantioselectivity of *A. niger* was rather low (*E* = 9). More recently, enantiocomplementary bacterial strains have been identified, which show the same enantio preference as the fungi by predominantly transforming the (*R*) enantiomer [125,132]. In a complementary fashion, fungal epoxide hydrolases from *Chaetomium globosum* were found to preferentially transform the (*S*) enantiomer [112], albeit at low enantioselectivity. The substrate structure may also be varied to a certain extent, i.e., the methyl moiety in the 2 position of the oxirane ring is not strictly obligatory. When the latter was substituted by an ethyl group (**2.11**), excellent enantioselectivity was obtained by using *Nocardia* TB1 as catalyst [124].

1.   Deracemization Technique

Orru et al. showed that 2,2-disubstituted oxiranes can be hydrolyzed in an enantioconvergent fashion by making use of bio- and chemocatalysis [127]. Thus, combination of a kinetic resolution employing *Nocardia* EH1 or H8 (step 1) with a subsequent acid-catalyzed opening of the oxirane ring (step 2, Scheme 12) in a sequence yielded the corresponding (*S*)-1,2-diols in virtually enantiopure form and in high yields (more than 90%). Careful mechanistic analysis of the acid-catalyzed hydrolysis reaction using different solvent systems and mineral acids made it possible to select general experimental conditions

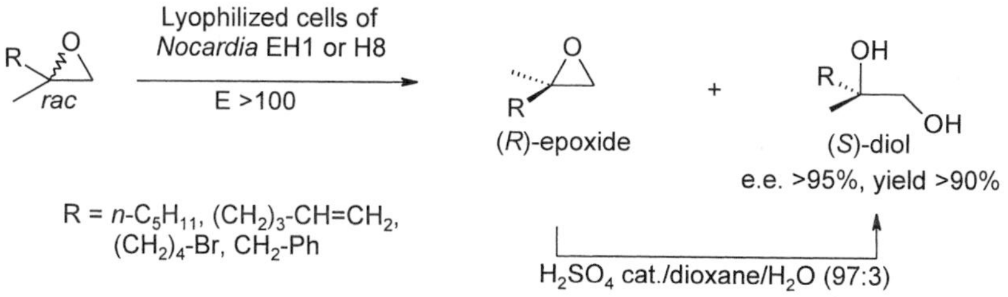

R $\begin{smallmatrix}O\\ \\ \end{smallmatrix}$
rac

⟶ Lyophilized cells of *Nocardia* EH1 or H8 / *E* >100 ⟶

$\begin{smallmatrix}O\\ \\ \end{smallmatrix}$ R
(*R*)-epoxide

+

R $\begin{smallmatrix}OH\\ \\ OH\end{smallmatrix}$
(*S*)-diol
e.e. >95%, yield >90%

R = *n*-C$_5$H$_{11}$, (CH$_2$)$_3$-CH=CH$_2$,
(CH$_2$)$_4$-Br, CH$_2$-Ph

H$_2$SO$_4$ cat./dioxane/H$_2$O (97:3)

Scheme 12   Deracemization of 2,2-disubstituted oxiranes via a resolution-inversion sequence.

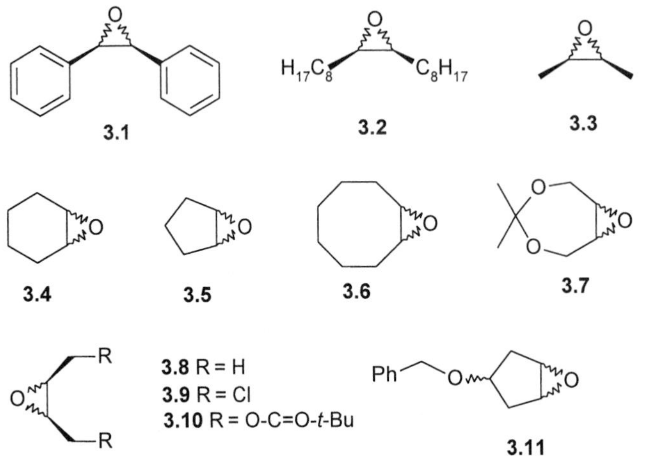

Figure 4 *meso*-Substituted oxiranes.

for the resolution-inversion procedure [127]. As a consequence, large-scale deracemization of epoxides became feasible [133].

C. 2,3-Disubstituted Oxiranes

1. *Meso*-Substrates

The asymmetrization of a *meso*-epoxide (Fig. 4, Table 4) via selective attack at one of the (enantiomeric) stereo centers of the oxirane would be an elegant application of epoxide hydrolases since it leads to a single enantiomeric *trans*-diol in 100% theoretical yield. Such asymmetrization reactions have been demonstrated with epoxide hydrolases from mammalian origin and yeast which afforded the enantiomerically enriched corresponding

Table 4 Biohydrolysis of *meso*-Epoxides

Substrate	Enzyme source	e.e. (diol) (%)	Conversion (%)	Ref.
3.1	mEH	>98 (*1R,2R*)	n.r.	137
3.2	mEH	90 (*9R,10R*)	90	138
3.3	mEH	86 (*2R,3R*)	n.r.	139
3.4	mEH	70 (*1R,2R*)	99	134
3.3	*Rhodotorula glutinis*	90 (*2R,3R*)	78[a]	48
3.4	*Rhodotorula glutinis*	90 (*1R,2R*)	~97	48
3.5	*Rhodotorula glutinis*	>98 (*1R,2R*)	~97	48
3.4	*Corynesporium cassiicola*	27[b]	n.r.	140

[a]Absolute configuration unclear (see text).
[b]The formed diol was subsequently oxidized to yield 2-hydroxy-2-butanone. However, it was not investigated as to whether the oxidation had any influence on the optical purity of the diol.
n.r. Not reported in the paper. However, in case of asymmetrization reactions, it has no consequence on the e.e. (diol).

(*R,R*)-diol of a diphenyl-substituted oxirane **3.1** as well as long and short chain dialkyl-substituted oxiranes **3.2** and **3.3**. Upon complete conversion of cyclohexene oxide **3.4** the *trans*-diol product was formed in 70% e.e. by mEH [134]. However, by switching to rabbit liver mEH, the same epoxide was hydrolyzed to form the diol in 90% e.e. [135]. Unfortunately, such highly strained bicyclic systems show a significant amount of spontaneous hydrolysis, which impedes the stereoselective transformation. It was remarked that in the absence of spontaneous hydrolysis the e.e. of the diol would be 94%. In a latter paper the e.e. (diol) of the same transformation was reported to be only 76% [136]. This discrepancy may be explained by various fractions of nonenzymatic hydrolysis, which is difficult to control.

On the contrary, only few asymmetrization reactions have been successfully verified with microbial enzymes. For instance, cyclohexene oxide **3.4** was hydrolyzed using *Corynesporium cassiicola* cells yielding *trans*-cyclohexane-1,2-diol with disappointing low e.e. (27%) [140]. It was only due to further metabolism involving an oxidation–reduction sequence by dehydrogenases present in the cells that the formed diol was transformed to furnish optically pure (*S,S*)-cyclohexane-1,2-diol. In a related study, asymmetrical hydrolysis of *cis*-epoxysuccinate using a crude enzyme preparation derived from *Rhodococcus* sp. led to D- and L-tartaric acid in almost racemic form [93]. Similar discouraging results were obtained using baker's yeast [141]. *Rhodococcus* spp. were shown to be unsuitable to hydrolyze different meso substrates (**3.4**, **3.7–3.11**) [21]. Obviously, microbial enzymes that are capable of achieving this type of reaction in a satisfactory fashion are yet to be discovered.

2. Cyclic 2,3-Disubstituted Opoxides

The (*1S,2R*) isomer of indene oxide **3.12** (Fig. 5, Table 5) is of considerable commercial interest because it is a synthetic precursor to a side chain of the human immunodeficiency virus (HIV) protease inhibitor MK 639 [142]. The yeast *Rhodotorula glutinis* and the fungus *B. sulfurescens* showed some enantioselectivity but unfortunately they displayed the wrong enantio preference, by leaving the (*1R,2S*)-epoxide unreacted. However, researchers at Merck identified several fungi that showed the desired enantio preference [143,144]. Since **3.12** is particularly prone to spontaneous hydrolysis, the reaction conditions had to be carefully optimized (0.1 M Tris, pH 7.6) to minimize the background reaction. Racemic **3.12** was resolved on a preparative scale at a concentration of 1 g/L to afford the enantiopure epoxide in 14% yield. Although these results are encouraging, further process development obviously is required to make this reaction feasible on an industrial scale. Upon extension of the ring size, **3.13** was resolved showing similar results. When bacterial epoxide hydrolases were employed, the enzymatic reaction competing with the spontaneous reaction was too slow; thus the background reaction made it impossible to obtain high e.e. values [145].

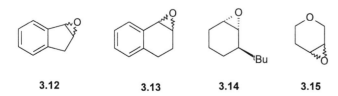

| **3.12** | **3.13** | **3.14** | **3.15** |

Figure 5   Cyclic 2,3-disubstituted oxiranes.

Table 5 Biohydrolysis of Cyclic Substituted Oxiranes

Substrate	Enzyme source	e.e. (diol) (%)	e.e. (epoxide) (%)	Ref.
3.12	*Rhodotorula glutinis*	54 (n.d.)	>98 (*1R,2S*)	48
3.12	*Beauveria bassiana*	69 (*1R,2R*)	98 (*1R,2S*)[a]	146
3.12	*Diploida gossypina*	n.d.	>99 (*1S,2R*)[b]	143
3.13	*Beauveria bassiana*	77 (*1R,2R*)	98 (*1R,2S*)[c]	146
3.14	mEH	>96 (*1R,2R,3R*)	n.d.[d]	147
3.15	mEH	>96 (*3R,4R*)	—[e]	148

[a]A 20% yield after 1 h.
[b]A 14% yield after 4.5 h.
[c]A 38% yield after 0.5 h.
[d]Conversion 50%.
[e]The process was enantioconvergent. Thus both epoxide enantiomers were transformed to yield the same product enantiomer, and no residual epoxide was obtained.
n.d. Not determined.

The enzymatic hydrolysis of (±)-*trans*-1,2-epoxy-3-*tert*-butylcyclohexane **3.14** ceased at 50% conversion yielding the corresponding (*1R,2R,3R*)-diol in more than 96% e.e. when mEH was used as a catalyst. Similar good results were obtained for the corresponding cis-configurated compound. When an oxygen atom was introduced into the ring, *rac*-3,4-epoxytetrahydrofuran **3.15** was hydrolyzed by mEH with complete stereoconvergence, and at complete conversion the sole product of the reaction was (*3R,4R*)-3,4-dihydroxytetrahydropyrane showing 96% e.e. The fact that a single enantiomeric product was obtained implies that both enantiomers were hydrolyzed with complete but opposite regioselectivity and that the attack of the formal [OH⁻] occurred at both enantiomers preferentially at the (*S*) center of the oxirane ring with inversion of configuration. Similar results were found for 2,3-dialkyl-substituted oxiranes, where the stereochemical pathways have been studied in more detail by using ^{18}O-labeling methods.

3. Noncyclic 2,3-Disubstituted Oxiranes

2,3-Dialkyl oxiranes (Fig. 6) have been efficiently hydrolyzed with high regio- and stereoselectivity by bacterial and yeast epoxide hydrolases. As an example, *cis*-2-heptene oxide **3.16** was transformed to the corresponding (*2R,3R*)-diol by whole cells of *Norcardia*

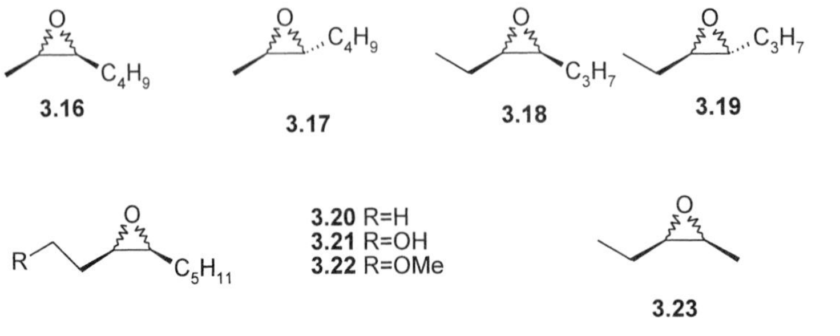

Figure 6 2,3-Disubstituted oxiranes.

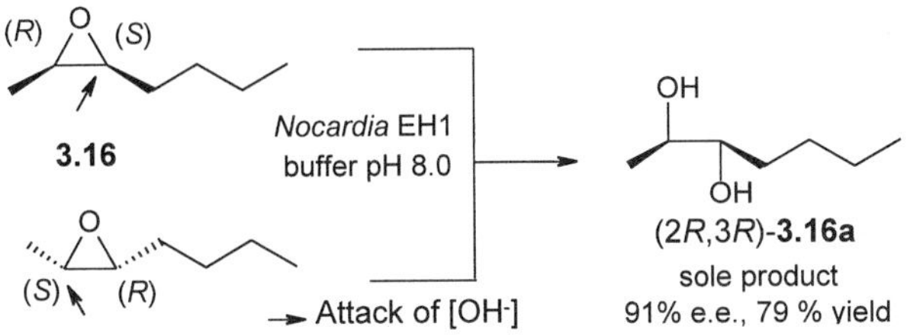

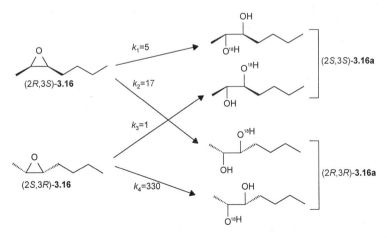

Scheme 13 Deracemization of **3.16** via enantioconvergent hydrolysis.

EH1 in 91% e.e. and 79% isolated yield (Scheme 13). This result can only be explained if the reaction proceeds through an enantioconvergent manner; thus, for the (*2R,3S*)-epoxide enantiomer, the oxirane ring is opened at C-3 (*S* configuration) and for the (*2S,3R*) enantiomer, C-2 (again *S* configuration) is attacked. The details of this intriguing process were investigated with respect to four possible enzyme-catalyzed pathways. The kinetics of this process have been solved mathematically [149] and the relative rate constants have been determined using chiral gas chromatography/mass spectroscopy (GC-MS) as analytical method (Scheme 14). It was found that the fastest reaction pathway, i.e., the attack at the (*S*)-configured C-2 (k_4), was 330 times faster than the slowest reaction, namely the attack at C-3, which is (*R*)-configured (k_3). It can be clearly seen that two phenomena are involved: (1) the preference to hydrolyze an (*S*)-configured oxirane center [(*S*) over (*R*)] and (2) the preferred attack at the less hindered carbon (C-2 over C-3). As a consequence, cis-configured substrates having a 2,3-substitution pattern can be deracemized employing one single enzyme without further bio- or chemocatalytic steps.

Similar results have been obtained with structurally related substrates and other bacterial catalysts, although the enantio- and regiospecificity was somewhat diminished (Table 6).

Wistuba et al. [139] reported that (±)-*cis*-3,4-epoxyheptane **3.18** was a very poor substrate for mEH. However, in a recent paper by Chiappe et al. [150], this compound was deracemized by mEH to give the (*3R,4R*)-diol in more than 98% e.e. at 100% conversion. An explanation for this discrepancy was not given [151]. The same good result

Scheme 14 Kinetics of the enantioconvergent hydrolysis of *rac-cis*-2-heptene oxide elucidated by ^{18}O-labeling experiments.

Table 6 Biohydrolysis of 2,3-Disubstituted Oxiranes

Substrate	Enzyme source	e.e (diol) (%)	e.e. (epoxide) (%)	Conversion	Ref.
3.16	*Nocardia* EH1	91 (2R,3R)	>99 (2R,3S)	99	149
3.17	*Nocardia* EH1	84 (2R,3S)	90 (2R,3R)	90	149
3.18	*Mycobacterium paraffinicum*	71 (3R,4R)	8 (3S,4R)	n.d.	149
3.19	*Nocardia* TB1	46 (3R,4S)	21 (3R,4R)	n.d.	149
3.16	mEH	50 (2R,3R)	n.d.	100	150
3.18	mEH	>98 (eR,4R)	n.d.	100	150
3.20	mEH	>98 (3R,4R)	n.d.	100	150
3.23	mEH	>99 (2R,3R)	>99 (2S,3R)	50	152
3.23	*Rhodotorula glutinis*	>98 (2R,3R)	>98 (2R,3S)	~50	48

n.d. Not determined.

was obtained for **3.20**. When a more polar substituent was introduced at C-1 (**3.21** and **3.22**), the selectivity decreased and at 100% conversion the (3R,4S)-diols obtained showed an e.e. of 90% and 80%, respectively. The smallest non-meso cis-substituted oxirane (*cis*-2-pantene oxide, **3.23**) was resolved by a red yeast and by mEH with high enantioselectivity. It is interesting to note that all epoxide hydrolases, independent of their origin, showed the same preference to attack the (S)-configured oxirane center with cis-substituted oxiranes being better substrates than the corresponding trans isomers [48].

4. Epoxy Fatty Acids and Derivatives

A crude enzyme preparation from the antennae of 1-day-old male gypsy moth (*Lymantria dispar*) was found to catalyze the enantioconvergent hydrolysis of (±)-disparlure **3.25** to the corresponding (7R,8R)-diol. However, for a single experiment, the antennae of 50 gypsy moths were required to provide just enough enzyme for the reaction, even at prolonged incubation times. It is obvious that this excludes these enzymes from practical applications.

A soluble epoxide hydrolase from soybeans (sbEH) hydrolyzed **3.26** to the (9R,10R)-diol in 80% e.e. at 50% conversion [153]. However, in a subsequent paper [40], the same authors report that the diol was formed in more than 99% e.e. Compound **3.26** was converted to the same (9R,10R)-diol by mEH but with significantly reduced e.e. Epoxy fatty acid **3.27** was claimed to undergo cis-hydrolysis by the fungus *Fusarium solani pisi* with concomitant retention of configuration but attempts to explain this extremely unusual phenomenon remain rather speculative [97] (Fig. 7, Table 7).

5. 2,3-Disubstituted Aryl Oxiranes

The alkyl substituent in the neighboring β position of substrates **3.28** and **3.29** had a dramatic effect on the substrate acceptance compared to styrene oxide in that neither **3.28** nor **3.29** were substrates for the fungus. *A. niger*. However, cis-configurated **3.28** was

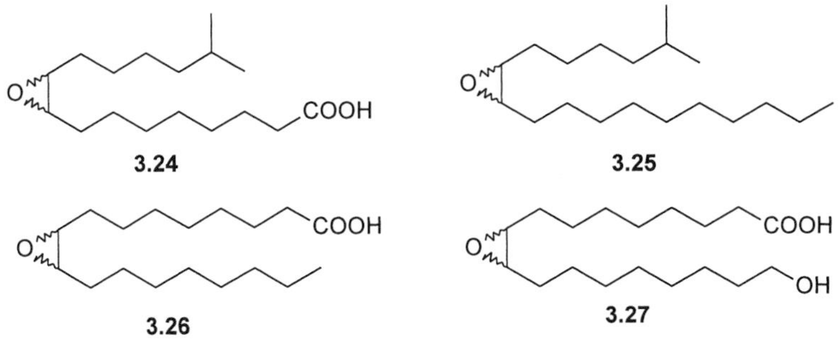

Figure 7   Epoxy fatty acids and derivatives.

hydrolyzed in a enantioconvergent manner by *Beauveria bassiana* to afford *(1R,2R)*-1-phenylpropane-1,2-diol in 98% e.e. and 85% yield. On the contrary, the corresponding trans-configurated counterpart **3.29** was kinetically resolved with high selectivity employing *B. bassiana* as catalyst (Fig. 8). The same results were obtained with the red yeast *Rhodotorula glutinis*. In contrast, mEH showed excellent enantioselectivity for the *cis*-oxirane **3.28** while the trans-configurated substrate **3.29** proved to be unreactive [155]. A series of *cis*-β-alkylstyrene oxides have been resolved by mEH by attack of the formal [OH⁻] at the *(S)*-configurated carbon atom to give the corresponding *(R,R)*-diols in more than 90% e.e. at 100% conversion [156], with the size of the alkyl moiety being in the range from ethyl to hexyl (Fig. 8, Table 8).

D.   Trisubstituted Oxiranes

Oesch et al. reported that trisubstituted oxiranes (Fig. 9) were poor substrates for mEH [158]; however, resolution of the *trans*-diastereomer *rac*-**4.1** gave both *(3R,3aS,7aR)*-**4.1** and the corresponding *(3R,3aR,7aS)*-diol **4.1a** in more than 98% e.e. The reaction was performed on a 200-mg scale but no yields were given [159]. The substituted alicyclic epoxide **4.2** (employed as diastereomeric mixture) was resolved to furnish the expected *trans*-*(1R,2R,4S)*-diol and *(1S,2R,4S)*-epoxide **4.2** by *Rhodotorula glutinis*. When the side chain of epoxide **4.2** showed *(4R)* configuration, activity and selectivity were significantly reduced. Epoxide **4.3** was used as a starting material for the synthesis of Bower's compound (see below). Although the fungi accepted the bulky substituent of substrate **4.3**, a

Table 7   Biohydrolysis of Epoxy Fatty Acids

Substrate	Enzyme source	e.e. (diol) (%)	e.e. (epoxide) (%)	Conversion (%)	Ref.
3.24	*Pseudomonas* sp.	~90 (9R,10R)	n.m. (9S,10R)	50	154
3.25	Gypsy moth antennae EH	>98 (7R,8R)	—	100	43
3.26	sbEH[a]	80 (9R,10R)	n.r.	50	153
3.26	mEH	24 (9R,10R)	n.r.	50	137
3.27	*Fusarium solani pisi*	n.r. (9S,10R)	n.r.	n.r.	96

[a]Soybean fatty acid epoxide hydrolase.
n.r. Not reported in the paper.

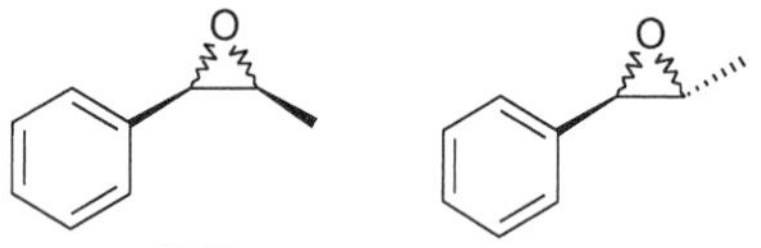

3.28   **3.29**

Figure 8 2,3-Disubstituted aryl oxiranes.

substrate possessing an aryl moiety close to the oxirane ring and two geminal methyl groups **4.4** was neither hydrolyzed by *A. niger* nor by *B. bassiana. Rhodococcus* sp. NCIMB 11216 was used for the key step in the synthesis of "linalool oxide," i.e., the hydrolysis of **4.5** (see below). A *Corynebacterium* sp. (designated C12) was isolated using cyclohexene oxide as sole carbon and energy source [160], which proved to be suitable for the hydrolysis of trisubstituted oxiranes. This strain hydrolyzed ($\pm$)-1-methyl-1,2-epoxycyclohexane **4.6** with good efficiency (Table 9).

1. Deracemization Techniques

Interestingly, the biotransformation of **4.5** by *Rhodococcus* NCIMB 11216 proved to be stereoconvergent, as both diastereomers of the substrate epoxide were transformed by the enzyme preparation to furnish the corresponding (*3R,6R*)-diol. Thus, the (*3R,6S*) isomer was hydrolyzed with inversion of configuration via attack at C-6, whereas the (*3R,6R*) diastereomer was hydrolyzed via attack at the more substituted oxirane carbon with overall retention of stereochemistry at C-7 [162]. Epoxide **4.6** was transformed into the corresponding *trans*-(*1S,2S*)-diol **4.6a** in 80% yield and more than 95% e.e. by making use of an acid-catalyzed hydrolysis of the remaining oxirane enantiomer with inversion of configuration as described above.

E. Nonnatural Nucleophiles

In reactions catalyzed by hydrolytic enzymes of the serine-hydrolase type, which form covalent acyl-enzyme intermediates during the course of the reaction, it has been shown that the "natural" nucleophile (i.e., water) can be replaced by "foreign" nucleophiles [164] such as an alcohol, amine, hydroxylamine, hydrazine, and even hydrogen peroxide. As a consequence, a wealth of synthetically useful reactions, which are usually performed in organic solvents at low water content, can be performed in a stereoselective manner. Although one requirement is fulfilled with epoxide hydrolases, i.e., a covalent enzyme—

Table 8 Biohydrolysis of 2,3-Disubstituted Aryl Oxiranes

Substrate	Enzyme source	e.e. (diol) (%)	e.e. (epoxide) (%)	Conversion (%)	Ref.
3.28	*Beauveria bassiana*	98 (*1R,2R*)	—[a]	99	157
3.29	*Beauveria bassiana*	90 (*1R,2S*)	98 (*1R,2R*)	~50	157
3.29	*Rhodotorula glutinis*	>98 (*1R,2S*)	>98 (*1R,2R*)	~50	48
3.28	mEH	>98 (*1R,2R*)	>93 (*1S,2R*)	~50	155

[a]Enantioconvergent process, no residual epoxide.

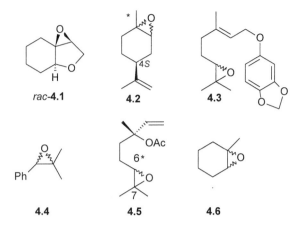

Figure 9 Trisubstituted oxiranes. *Diastereomeric mixture.

substrate intermediate is formed—the sensitivity of epoxide hydrolases to most of the water-miscible or immiscible organic solvents [71,133] poses a general problem toward the use of nonnatural nucleophiles in enzymatic epoxide hydrolysis. However, two types of transformations, i.e., aminolysis and azidolysis of an epoxide, have been reported, even when carried out in an aqueous system (Scheme 15).

When racemic arylglycidyl ethers were subjected to aminolysis in aqueous buffer catalyzed by hepatic microsomal epoxide hydrolase from rat, the corresponding (S)-configured amino alcohols were obtained in 51–88% e.e. [165]. On the other hand, when azide was employed as nucleophile for the asymmetrical opening of 2-methyl-1,2-epoxyheptane (±)-**2.3** in the presence of an immobilized crude enzyme preparation derived from *Rhodococcus* sp., the reaction revealed a complex picture [166]. The (S)-epoxide from the racemate was hydrolyzed (as in the absence of azide), and the less readily accepted (R) enantiomer was transformed into the corresponding azido alcohol (e.e. more than 60%). Although at present only speculations can be offered about the actual mechanism of both the aminolysis and azidolysis reaction, in both cases it was proven that the reaction was catalyzed by an enzyme, and that no reaction was observed in the absence of biocatalyst or by using a heat-denatured protein preparation. However, a recent related report on the aminolysis of epoxides employing crude porcine pancreative lipase [167] may likewise be explained by catalysis of a chiral protein surface rather than true lipase

Table 9 Biohydrolysis of Trisubstituted Oxiranes

Substrate	Enzyme source	e.e./d.e. (diol) (%)	e.e./e.e. (epoxide) (%)	Conversion (%)	Ref.
4.1	mEH	>98 (3R,3aR,7aS)	>98 (3R,3aS,7aR)	~50	159
4.2	*Rhodotorula glutinis*	>98 (1R,2R,4S)	>98 (1S,2R,4S)	~50	48
4.3	*Aspergillus niger*	96 (6R)	96 (6S)	~50	161
4.5	*Rhodococcus* sp.	>98 (3R,6R)	98 (3R,6R)	—	162
4.6	*Corynebacterium* C12	89 (1S,2S)	>99 (1R,2S)	n.r.	163

n.r. Not reported.

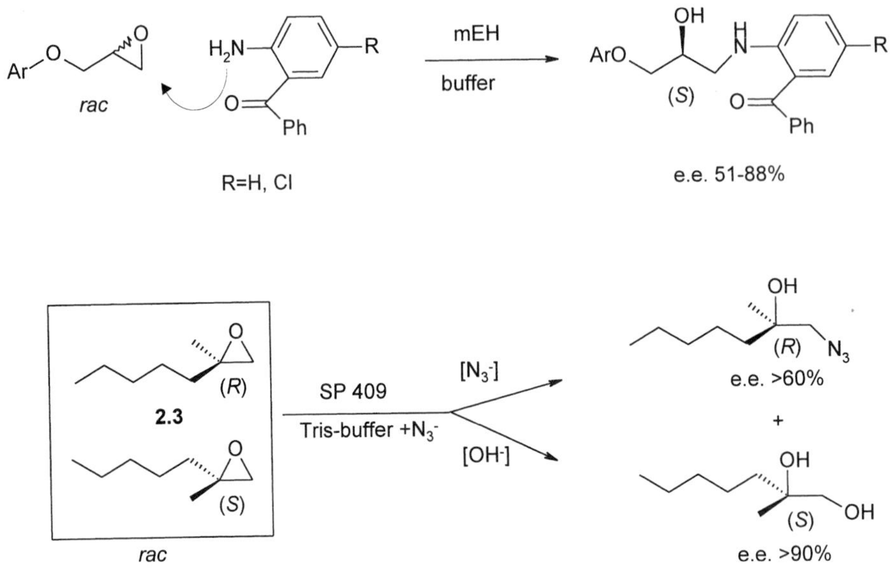

Scheme 15   Enzyme-catalyzed aminolysis and azidolysis of epoxides.

catalysis since the latter enzyme—being a serine hydrolase—is irreversibly deactivated by epoxides. In view of these facts, it remains questionable as to whether the use of nonnatural nucleophiles will be of general applicability.

V. APPLICATIONS OF EPOXIDE HYDROLASES FOR NATURAL PRODUCT SYNTHESIS

Although the use of an epoxide hydrolase was claimed for the industrial synthesis of L- and *meso*-tartaric acid already in 1969 [69], it was only recently that applications to asymmetrical synthesis appeared in the literature. This fact can be mainly attributed to the limited availability of these biocatalysts from sources such as mammals or plants. Since the production of large amounts of crude enzyme from microbial sources is now feasible, preparative scale applications are within reach of the synthetic chemist. For instance, fermentation of *Nocardia* EH1 on a 70-L scale afforded more than 700 g of lyophilized cells that can be employed as such [72].

One of the first applications of microbial hydrolysis of epoxides for the synthesis of a bioactive compound is based on the resolution of a cis-configurated 2,3-disubstituted oxirane (Scheme 16). Thus, by using an enzyme preparation derived from *Pseudomonas* sp., the (9R,10S) enantiomer was hydrolyzed in a trans-specific fashion (i.e., via inversion of configuration at C-10) yielding the (9R,10R)-*threo*-diol. The remaining (9S,10R)-epoxide was converted to (+)-disparlure, the sex pheromone of the gypsy moth in more than 95% e.e. [154].

An application of microbial epoxide hydrolases for the synthesis of α-bisabolol, one of the stereoisomers (out of four) of interest for the cosmetic industry, is illustrated in Scheme 17. This approach was based on the diastereoselective hydrolysis of a mixture of oxirane diastereoisomers obtained by chemical synthesis from (R)- or (S-limonene [130]. Thus, starting from (S)-limonene, the biohydrolysis of the mixture of (4S,8RS)-epoxides

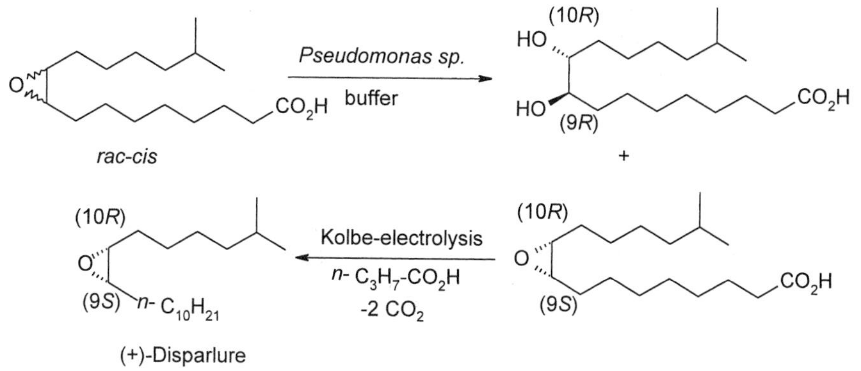

Scheme 16 Resolution of a *cis*-2,3-disubstituted epoxide and synthesis of (+)-disparlure.

led to unreacted (4S,8S)-epoxide and (4S,8R)-diol. The former was obtained in virtually diastereomerically pure form (more than 95%) and was subsequently transformed by chemical means into (4S,8R)-α-bisabolol. The formed diol from the microbiological reaction (d.e. 94%) could be cyclized back to the corresponding (4S,8R)-epoxide, thus affording access to another stereoisomer [(4S,8S)] of α-bisabolol (Scheme 17). In addition, both of the corresponding (4R) diastereoisomers of bisabolol out of the complete set of four could be prepared in an analogous manner starting from (R)-limonene.

More recently, the deracemization of (±)-**2.6** (Scheme 18) through combination of *Nocardia* EH1 epoxide hydrolase and sulfuric acid in dioxane containing a trace amount of water (vide supra) was employed for a short total synthesis of (S)-(−)-frontalin, a central aggregation pheromone of pine beetles of the *Dendroctonus* family [131]. The product [(S)-2-methyl-hept-6-ene-1,2-diol], which was obtained in 97% yield and 99% e.e. from

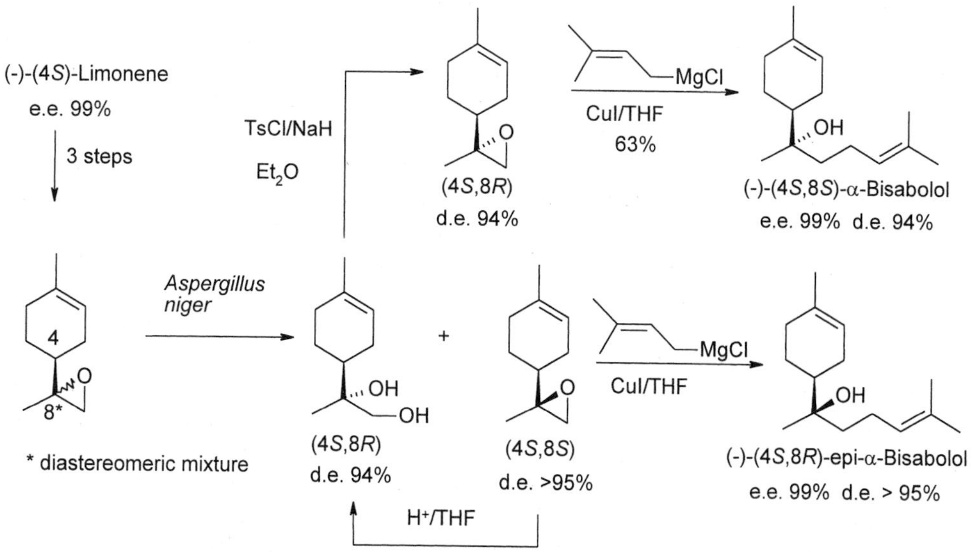

Scheme 17 Chemoenzymatic synthesis of α-bisabolol using fungal epoxide hydrolase.

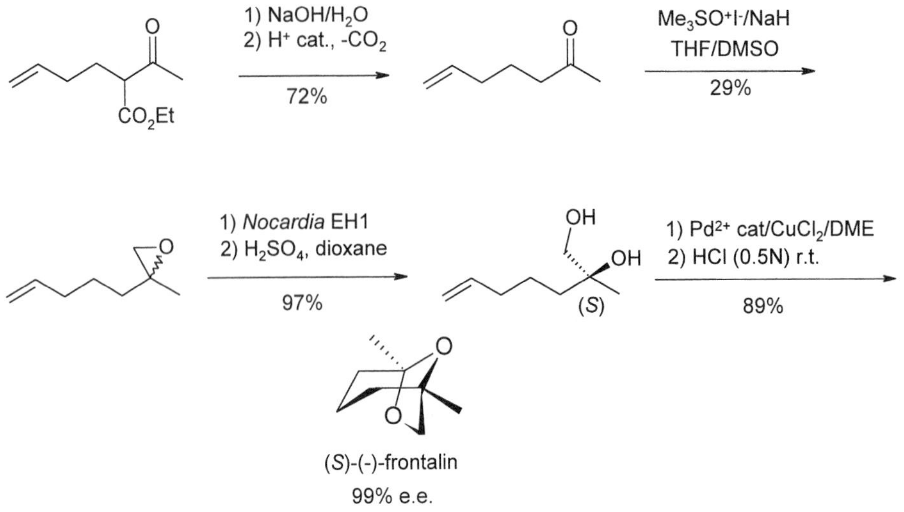

Scheme 18 showing the chemoenzymatic synthesis with reaction conditions:

Step 1: 1) NaOH/H₂O, 2) H⁺ cat., -CO₂ (72%)

Step 2: Me₃SO⁺I⁻/NaH, THF/DMSO (29%)

Step 3: 1) *Nocardia* EH1, 2) H₂SO₄, dioxane (97%)

Step 4: 1) Pd²⁺ cat/CuCl₂/DME, 2) HCl (0.5N) r.t. (89%)

(S)-(−)-frontalin
99% e.e.

Scheme 18 Chemoenzymatic synthesis of (*S*)-(−)-frontalin using bacterial epoxide hydrolase.

the deracemization reaction [127], was efficiently converted to (*S*)-(−)-frontalin via a Wacker oxidation.

Enantiomerically pure *trans*- and *cis*-linalool oxides, which are constituents of several plants and fruits, are found among the main aroma components of oolong and black tea. These compounds were prepared by employing a bacterial epoxide hydrolase as the key step from a diastereomeric mixture of 2,3-epoxylinalyl acetate **4.5** (Scheme 19) [162]. The key step consist of a separation of the diastereomeric mixture of **4.5** by employing an epoxide hydrolase preparation derived from *Rhodococcus* sp. NCIMB 11216, yielding the product diol and remaining epoxide in excellent diastereomeric excess (d.e. more than 98%). Further follow-up chemistry gave both linalool oxide isomers on a preparative scale in excellent diastereomeric and enantiomeric purities.

Both enantiomers of "Bower's compound," a potent analog of an insect juvenile hormone [161] (Scheme 20), were prepared in 96% e.e. using *A. niger* epoxide hydrolase. Interestingly, subsequent biological tests revealed that the (6*R*) antipode was about 10 times more active than the (6*S*) conterpart against the yellow mealworm *Tenebrio molitor*.

Aspergillus niger was the biocatalyst of choice for the biohydrolysis of *para*-nitrostyrene oxide. A selective kinetic resolution using a crude enzyme extract of this fungus, followed by careful acidification of the cooled crude reaction mixture, afforded the corresponding (*R*)-diol in high chemical yield (94%) and good e.e. (80%) via a deracemization protocol. This key intermediate was then transformed via a four-step sequence (Scheme 21) into enantiopure (*R*)-nifenalol, a molecule with beta blocker activity, which was obtained in 58% overall yield [120].

Finally, natural (*R*)-(−)-mevalonolactone, a key intermediate in the biosynthesis of a broad spectrum of compounds such as sterols, terpenes, and carotenoids, was synthesized via eight steps in 55% overall yield and more than 99% e.e. (Scheme 22). In the key step, the aforementioned enantioconvergent chemoenzymatic deracemization route was applied. Thus, 2-methyl-2-benzyl oxirane (±)-**2.8** was deracemized on a large scale (10 g) using lyophilized cells of *Nocardia* EH1 and sulfuric acid. The product (*S*)-diol was isolated in 94% chemical yield and 94% optical purity [133].

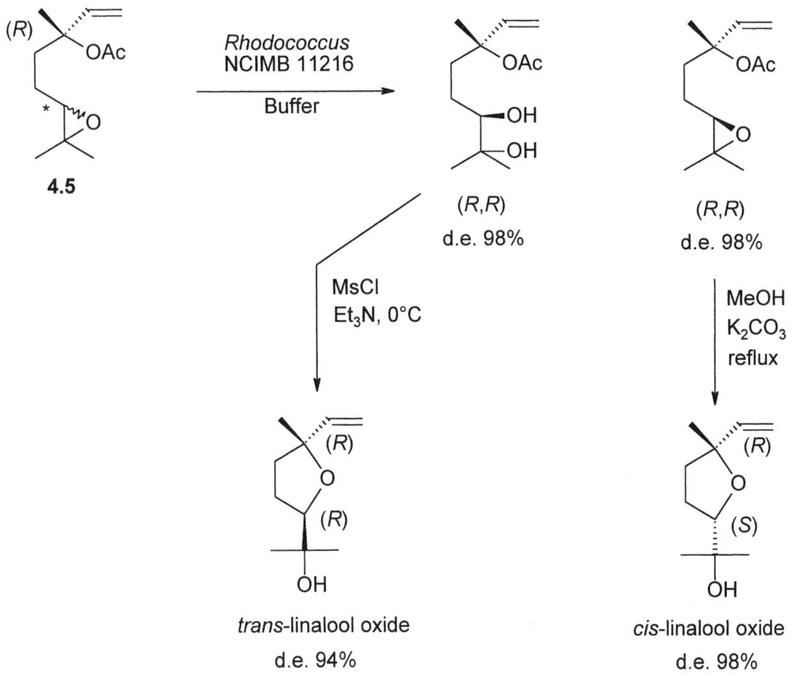

Scheme 19 Synthesis of *cis-* and *trans-*linalool oxide using bacterial epoxide hydrolase. *Diastereomeric mixture.

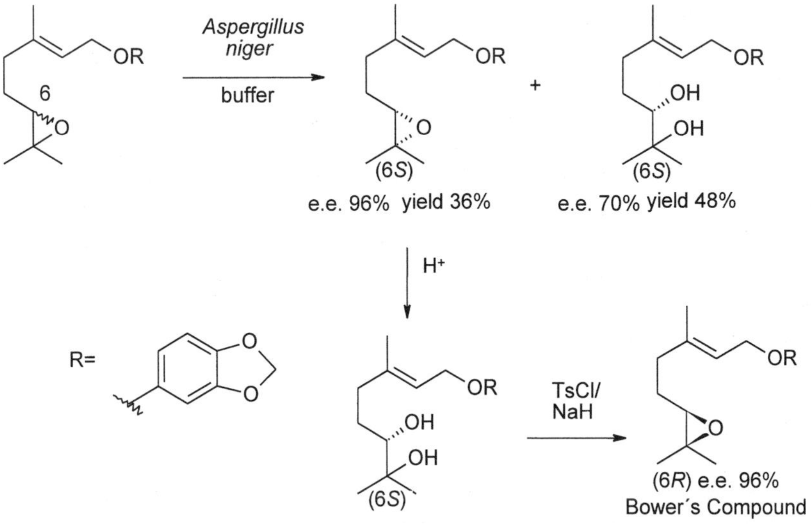

Scheme 20 Application of fungal epoxide hydrolase to the synthesis of (*R*)- and (*S*)-Bower's compound.

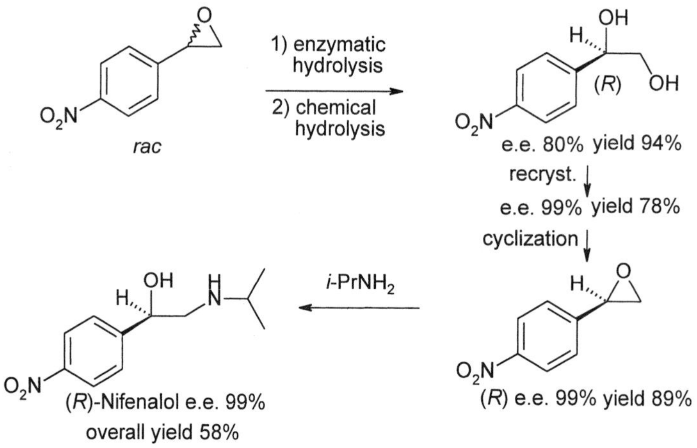

Scheme 21 Deracemization of *p*-nitrostyrene oxide and application to the synthesis of (*R*)-Nifenalol.

VI. SUMMARY AND OUTLOOK

Epoxide hydrolases from microbial sources have recently received much attention as highly versatile biocatalysts for the preparation of enantiopure epoxides and 1,2-diols, and their largely untrapped potential for asymmetrical chemoenzymatic syntheses is just being realized. The near future will certainly bring a number of useful applications of these systems to the asymmetrical synthesis of chiral bioactive compounds in enantiopure form. Microbial epoxide hydrolases are easy to use due to their independence of cofactors and their ample availability from a sufficient number of microbial strains that can be obtained from culture collections. An important additional benefit of these enzymes is the fact that, in contrast to epoxide hydrolases from mammalian systems, enzyme induction does not seem to be necessary for the majority of microbial systems. Furthermore, whole lyophilized cells, as well as crude enzymatic extracts, which can be stored in a refrigerator for several months without significant loss of activity, can be used instead of isolated enzymes. As for all enzymes, the enantioselectivity of microbial epoxide hydrolases can be low to excellent depending on the substrate structure. As this field is currently still in the development stage, more information is needed to enable predictions on the stereochemical outcome of a suitable microbial strain possessing epoxide hydrolase activity for a given substrate. This will certainly be helped when the first three-dimensional x-ray structure of such an enzyme is solved, which is expected very soon. Given the data presented above, a much more extensive use—possibly including also industrial applications—of microbial epoxide hydrolases for the preparation of enantiopure epoxides and/or vicinal diols on a preparative scale can be anticipated in the near future.

ACKNOWLEDGMENTS

We thank Romano V. A. Orru (Graz) for valuable hints and discussions. This work was financially supported by the Fonds zur Förderung der wissenschaftlichen Forschung (project F104), the Austrian Ministry of Science and Transport, and the EU contract BIO4-CT-950005.

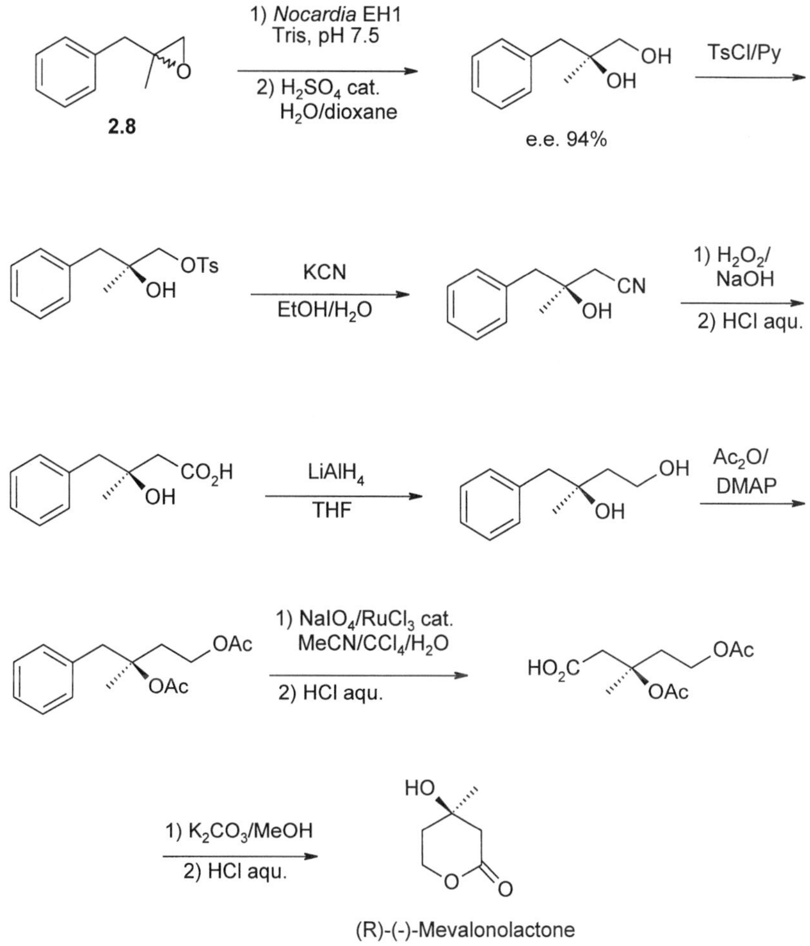

Scheme 22 Synthesis of (*R*)-(−)-mevalonolactone.

REFERENCES

1. HC Kolb, MS Van Nieuwenhze, KB Sharpless. Chem Rev 94:2483–2547, 1994.
2. LP Hager, Third International Symposium on Biocatalysis and Biotransformations, La Grande Motte, France, 1997, abstracts, p 49.
3. EE van Tamelen, JA Gladysz. J Am Chem Soc 96:5290–5291, 1974.
4. JL Fry, TJ Mraz. Tetrahedron Lett 10:849–852, 1979.
5. J March. Advanced Organic Chemistry, 4th ed. New York: John Wiley, 1992, p 1029.
6. T Katsuki, KB Sharpless. J Am Chem Soc 102:5974–5975, 1980.
7. WF Walker, WS Bowers. J Agric Food Chem 21:145–148, 1973.
8. KB Sharpless, RC Michaelson. J Am Chem Soc 95:6136–6137, 1973.
9. EN Jacobsen, W Zhang, AR Muci, JR Ecker, L Deng. J Am Chem Soc 113:7063–7064, 1991.
10. T Katsuki. Coord Chem Rev 140:189–214, 1995.
11. K Konishi, K Oda, K Nishida, T Aida, S Inoue. J Am Chem Soc 114:1313–1317, 1992.
12. V Schurig, F Betschinger. Chem Rev 92:873–888, 1992.
13. M Tokunaga, JF Larrow, F Kakiuchi, EN Jacobsen. Science 277:936–938, 1997.

14. RA Johnson, KB Sharpless. In: I Ojima, ed. Catalytic Asymmetric Synthesis. New York: VCH, 1993, pp 227–272.
15. DM Hodgson, AR Gibbs, GP Lee. Tetrahedron 52:14361–14384, 1996.
16. JAM de Bont. Tetrahedron: Asymmetry 4:1331–1340, 1993.
17. DJ Leak, PJ Aikens, M Seyed-Mahmoudian. Trends Biotechnol 10:256–261, 1992.
18. AN Onumonu, A Colocoussi, C Matthews, MP Woodland, DJ Leak. Biocatalysis 10:211–218, 1994.
19. P Besse, H Veschambre. Tetrahedron 50:8885–8927, 1994.
20. S Pedragosa-Moreau, A Archelas, R Furstoss. Bull Soc Chim Fr 132:769–800, 1995.
21. M Mischitz, W Kroutil, U Wandel, K Faber. Tetrahedron: Asymmetry 6:1261–1272, 1995.
22. M-J Kim, YK Choi. J Org Chem 57:1605–1607, 1992.
23. U Ader, MP Schneider. Tetrahedron: Asymmetry 3:205–208, 1992.
24. M Onda, K Motosugi, H Nakajima. Agric Biol Chem 54:3031–3033, 1990.
25. N Allison, AJ Skinner, RA Cooper. J Gen Microbiol 129:1283–1293, 1983.
26. G Casy, TV Lee, H Lovell. Tetrahedron Lett 33:817–820, 1992.
27. K Nakamura, T Yoneda, T Miyai, K Ushio, S Oka, A Ohno. Tetrahedron Lett 29:2453–2454, 1988.
28. K Furuhashi. Chem Econ Eng Rev 18:21–26, 1986.
29. K Seelbach, MPJ Van Deurzen, F van Rantwijk, RA Sheldon, U Kragl. Biotechnol Bioeng 55:283–288, 1997.
30. MCR Franssen. Biocatalysis 10:87–111, 1994.
31. J Geigert, SL Neidleman, T-NE Liu, SK de Witt, BM Panschar, DJ Dalietos, ER Siegel. Appl Environ Microbiol 45:1148–1149, 1983.
32. EJ Allain, LP Hager, L Deng, EN Jacobsen. J Am Chem Soc 115:4415–4416, 1993.
33. Epoxide hydrolases are occasionally also termed "epoxide hydratases" or "epoxide hydrases."
34. F Oesch. Xenobiotica 3:305–340, 1972.
35. G Berti. In: MP Schneider, ed. Enzymes as Catalysts in Organic Synthesis. Dordrecht: Reidel, NATO ASI Series C, 178, 1986, pp 349–354.
36. K Faber, M Mischitz, W Kroutil. Acta Chem Scand 50:249–258, 1996.
37. A Archelas, R Furstoss. Annu Rev Microbiol 51:491–494, 1997.
38. IVJ Archer. Tetrahedron 53:15617–15662, 1997.
39. BD Hammock, DF Grant, DH Storms. In: I Sipes, C McQueen, A Gandolfi, eds. Comprehensive Toxicology, Vol. 3. New York: Pergamon, 1996, pp 283–305.
40. E Blée, F Schuber. Eur J Biochem 230:229–234, 1995.
41. RJ Linderman, EA Walker, C Haney, RM Roe. Tetrahedron 51:10845–10856, 1995.
42. H Wojtasek, GD Prestwich. Biochem Biophys Res Commun 220:323–329, 1996.
43. GD Prestwich, SMcG Graham, W: A König, J Chem Soc Chem Commun, 575–577, 1989.
44. C Morisseau, H Nellaiah, A Archelas, R Furstoss, JC Baratti. Enzyme Microb Technol 20:446–452, 1997.
45. T Nakamura, T Nagasawa, F Yu, I Watanabe, H Yamada. Appl Environ Microbiol 60:4630–4633, 1994.
46. E Misawa, CKC Chan Kwo Chio, IV Archer, MP Woodland, N-Y Zhou, SF Carter, DA Widdowson, DJ Leak. Eur J Biochem 253:173–183, 1998.
47. MHJ Jacobs, AJ van den Wijngaard, M Pentenga, DB Janssen. Eur J Biochem 202:1217–1222, 1991.
48. CAGM Weijers. Tetrahedron: Asymmetry 8:639–647, 1997.
49. J Seidegard, JW DePierre. Biochim Biophys Acta 695:251–270, 1983.
50. DC Zeldin, S Wei, JR Falck, BD Hammock, JR Snapper, JH Capdevila. Arch Biochem Biophys 316:443–451, 1995.
51. MJ Mueller, B Samuelsson, JZ Haeggström. Biochemistry 34:3536–3543, 1995.
52. T Watabe, M Kanai, M Isobe, N Ozawa. J Biol Chem 256:2900–2907, 1981.
53. CR Pace-Asciak, WS Lee. J Biol Chem 264:9310–9314, 1989.

54. RN Wixtrom, BD Hammock. In: D Zakim, DA Vessey, eds. Biochemical Pharmacology and Toxicology: Methodological Aspects of Drug Metabolizing Enzymes, Vol. 1. New York: John Wiley, 1985, pp 1–93.

55. M Schöttler, W Boland. Helv Chim Acta 79:1488–1496, 1996.

56. E Blée, F Schuber. Plant J 4:113–123, 1993.

57. F Pinot, ED Caldas, C Schmidt, DG Gilchrist, AD Jones, CK Winter, BD Hammock. Mycopathologia 140:51–58, 1997.

58. J Swaving, JAM de Bont. Enzyme Microb Technol 22:19–26, 1998.

59. JAM de Bont, JP van Dijken, KG van Ginkel. Biochim Biophys Acta 714:465–470, 1982.

60. CAGM Weijers, H Jongejan, MCR Franssen, A de Groot, JAM de Bont. Appl Microbiol Biotechnol 42:775–781, 1995.

61. N Itoh, K Hayashi, K Okada, T Itoh, N Mizoguchi. Biosci Biotechnol Biochem 61:2058–2062, 1997.

62. FJ Small, SA Ensign. J Bacteriol 177:6170–6175, 1995.

63. P Bentley, F Oesch. FEBS Lett 59:291–295, 1975.

64. RN Wixtrom, MH Silva, BD Hammock. Anal Biochem 169:71–80, 1988.

65. L Schladt, R Hartmann, W Wörner, H Thomas, F Oesch. Eur J Biochem 176:31–37, 1988.

66. GD Prestwich, BD Hammock. Proc Natl Acad Sci USA 82:1663–1667, 1985.

67. BC Michaels, RT Ruettinger, AJ Fulco. Biochem Biophys Res Commun 92:1189–1195, 1980.

68. T Nakamura, T Nagasawa, F Yu, I Watanabe, H Yamada. J Bacteriol 174:7613–7619, 1992.

69. RH Allen, WB Jacoby. J Biol Chem 244:2078–2084, 1969.

70. G Grogan, SM Roberts, AJ Willets. FEMS Microbiol Lett 141:239–243, 1996.

71. M Mischitz, K Faber, AJ Willets. Biotechnol Lett 17:893–898, 1995.

72. W Kroutil, Y Genzel, M Pietzsch, C Syldatk, K Faber. J Biotechnol 61:143–149, 1998.

73. M Knehr, H Thomas, M Arand, T Gebel, H-D Zeller, F Oesch. J Biol Chem 268:17623–17627, 1993.

74. PA Bell, CB Kasper. J Biol Chem 268:14011–14017, 1993.

75. JK Beetham, T Tian, BD Hammock. Arch Biochem Biophys 305:197–201, 1993.

76. DF Grant, DH Storms, BD Hammock. J Biol Chem 268:17628–17633, 1993.

77. RC Skoda, A Demierre, OW McBride, FJ Gonzalez, UA Meyer. J Biol Chem 263:1549–1554, 1988.

78. CN Falany, P McQuiddy, CB Kasper. J Biol Chem 262:5924–5930, 1987.

79. A Stapleton, JK Beetham, F Pinot, JE Garbarino, DR Rockhold, M Friedman, BD Hammock, WR Belknap. Plant J 6:251–258, 1994.

80. T Kiyosue, JK Beetham, F Pinot, BD Hammock, K Yamaguchi-Shinozaki, K Shinozaki. Plant J 6:259–269, 1994.

81. R Rink, M Fennemat, M Smids, U Dehmel, DB Janssen. J Biol Chem 272:14650–14657, 1997.

82. The structure of a bacterial enzyme from *Agrobacterium radiobacter* AD1 is expected to be available in the near future.

83. GM Lacourciere, RN Armstrong. Chem Res Toxicol 7:121–124, 1994.

84. JK Beetham, D Grant, M Arand, J Garbarino, T Kiyosue, F Pinot, F Oesch, WR Belknap, K Shinozaki, BD Hammock. DNA Cell Biol 14:61–71, 1995.

85. M Arand, DF Grant, JK Beetham, T Friedberg, F Oesch, BD Hammock. FEBS Lett 338:251–256, 1994.

86. DB Janssen, F Fries, J van der Ploeg, B Kazemier, P Terpstra, B Witholt. J Bacteriol 171:6791–6799, 1989.

87. KHG Verschueren, SM Franken, HJ Rozeboom, KH Kalk, BW Dijkstra. J Mol Biol 232:856–872, 1993.

88. GC Dubois, E Apella, W Levin, AYH Lu, DM Jerina. J Biol Chem 253:2932–2939, 1978.

89. GM Lacourciere, RN Armstrong. J Am Chem Soc 115:10466–10467, 1993.

90. BD Hammock, F Pinot, JK Beetham, DF Grant, ME Arand, F Oesch. Biochem Biophys Res Commun 198:850–856, 1994.

91. KHG Verschueren, F Seljée, HJ Rozeboom, KH Kalk, BW Dijkstra. Nature 363:693–698, 1993.
92. KA Thomas, GM Smith, TB Thomas, RJ Feldmann. Proc Natl Acad Sci USA 79:4843–4847, 1982.
93. P Hechtberger, G Wirnsberger, M Mischitz, N Klempier, K Faber. Tetrahedron: Asymmetry 4:1161–1164, 1993.
94. G Bellucci, C Chiappe, A Cordoni, F Marioni. Tetrahedron Lett 45:4219–4222, 1994.
95. S Pedragosa-Moreau, A Archelas, R Furstoss. Bioorg Med Chem 2:609–616, 1994.
96. PE Kolattukudy, L Brown. Arch Biochem Biophys 166:599–607, 1975.
97. Y Suzuki, K Imai, S Marumo. J Am Chem Soc 96:3703–3705, 1974.
98. S Pedragosa-Moreau, A Archelas, R Furstoss. J Org Chem 58:5533–5536, 1993.
99. B Escoffer, J-C Prome. Bioorg Chem 17:53–63, 1989.
100. M Mischitz, C Mirtl, R Saf, K Faber. Tetrahedron: Asymmetry 7:2041–2046, 1996.
101. X M Zhang, A Archelas, R Furstoss. J Org Chem 56:3814–3817, 1991.
102. A de Raadt, N Klempier, K Faber, H Griengl. J Chem Soc Perkin Trans I, 137–140, 1992.
103. CS Chen, Y Fujimoto, G Girdaukas, CJ Sih. J Am Chem Soc 104:7294–7299, 1982.
104. JLL Rakels, AJJ Straathof, JJ Heijnen. Enzyme Microb Technol 15:1051–1056, 1993.
105. AJJ Straathof, JA Jongajan. Enzyme Microb Technol 21:559–571, 1997.
106. H Stecher, K Faber. Synthesis 1–16, 1997.
107. It must be stressed that all microorganisms reported to show preparatively useful epoxide hydrolase activity so far belong to safety class I, except *Chryseomonas luteola*, which is class II. The latter strains are known to cause ear infections in infants.
108. G Bellucci, C Chiappe, L Conti, F Marioni, G Pierini. J Org Chem 54:5978–5983, 1989.
109. D Wistuba, V Schurig. Chirality 4:178–184, 1992.
110. CAGM Weijers, AL Botes, MS van Dyk, JAM de Bont. Tetrahedron: Asymmetry 9:467–473, 1998.
111. AL Botes, CAGM Weijers, MS Dyk. Biotechnol Lett 20:421–426, 1998.
112. P Moussou, A Archelas, R Furstoss. Tetrahedron 54:1563–1572, 1998.
113. I Osprian, W Kroutil, M Mischitz, K Faber. Tetrahedron: Asymmetry 8:65–71, 1997.
114. AL Botes, JA Steenkamp, MZ Letloenyane, MS van Dyk. Biotechnol Lett 20:427–430, 1998.
115. G Bellucci, C Chiappe, A Cordoni, F Marioni, Tetrahedron: Asymmetry 4:1153–1160, 1993.
116. JHL Spelberg, R Rink, RM Kellogg, DB Janssen. Tetrahedron: Asymmetry 9:459–466, 1998.
117. G Grogan, C Rippé, A Willetts. J Mol Catal B: Enz 3:253–257, 1997.
118. S Pedragosa-Moreau, A Archelas, R Furstoss. J Org Chem 58:5533–5536, 1993.
119. S Pedragosa-Moreau, C Morisseau, J Zylber, A Archelas, J Baratti, R Furstoss. J Org Chem 61:7402–7407, 1996.
120. S Pedragosa-Moreau, C Morisseau, J Baratti, J Zylber, A Archelas, R Furstoss. Tetrahedron 53:9707–9714, 1997.
121. E Vänttinen, LT Kanerva. Tetrahedron: Asymmetry 6:1779–1786, 1995.
122. G Bellucci, C Chiappe, G Ingrosso, C Rosini. Tetrahedron: Asymmetry 6:1911–1918, 1995.
123. G Bellucci, C Chiappe, A Cordoni, F Marioni. Chirality 6:207–212, 1994.
124. W Kroutil. Biocatalytic Synthesis of Enantiopure Epoxides and Alcohols Employing Epoxide Hydrolases, Oxidases, Sulfatases and Synthetic Enzymes. PhD thesis, Graz University of Technology, 1998.
125. I Osprian. Neue prokariotische Systeme zur biokatalytischen Hydrolyse von Epoxiden. Diploma thesis, Graz University of Technology, 1996.
126. W Kroutil. Enantioselektive, bakterielle Hydrolyse von Epoxiden. Diploma thesis, Graz University of Technology, 1995.
127. RVA Orru, SF Mayer, W Kroutil, K Faber. Tetrahedron 54:859–874, 1998.
128. RVA Orru, W Kroutil, K Faber. Tetrahedron Lett 38:1753–1754, 1997.
129. I Osprian, W Kroutil, M Mischitz, K Faber. Tetrahedron: Asymmetry 8:65–71, 1997.
130. X-J Chen, A Archelas, R Furstoss. J Org Chem 58:5528–5532, 1993.
131. W Kroutil, I Osprian, M Mischitz, K Faber. Synthesis, 156–158, 1997.

132. W Krenn. Epoxidhydrolasen mit gegenläufiger Enantiopräferenz. Diploma thesis, Graz University of Technology, 1998.

133. RVA Orru, I Osprian, W Kroutil, K Faber. Synthesis (in press).

134. DM Jerina, H Ziffer, JW Daly. J Am Chem Soc 92:1056–1061, 1970.

135. G Bellucci, C Chiappe, F Marioni. J Chem Soc, Perkin Trans I, 2369–2373, 1989.

136. B Bellucci, I Capitani, C Chiappe, F Mariono. J Chem Soc Chem Commun, 1170–1173, 1989.

137. T Watabe, K Akamatsu. Biochim Biophys Acta 279:297–305, 1972.

138. G Bellucci, C Chiappe, A Cordoni, G Ingrosso. Tetrahedron Lett 37:9089–9092, 1996.

139. D Wistuba, O Träger, V Schurig. Chirality 4:185–192, 1992.

140. AJ Carnell, G Iacazio, SM Roberts, AJ Willetts. Tetrahedron Lett 35:331–334, 1994.

141. M Takeshita, N Akagi, N Akutsu, S Kuwashima, T Sato, Y Ohkubo. Tohoku Yakka Daigaku Kenyo Nempo 37:175–179, 1990. Not cited in Chem Abstr.

142. JP Vacca, BD Dorsey, WA Schleif, RB Levin, SL McDaniel, PL Darke, J Zugary, JC Quintero, OM Blahy, E Roth, VV Sardana, AJ Schlabach, PI Graham, JH Condra, L Gotlib, MK Holloway, J Lin, I-W Chen, K Vastag, D Ostovic, PS Anderson, EA Emini, JR Huff. Proc Natl Acad Sci USA 91:4096–4100, 1994.

143. J Zhang, J Reddy, C Roberge, C Senanayake, R Greasham, M Chartrain. J Ferment Bioeng 80:244–246, 1995.

144. M Chartrain, CH Senanayake, JPN Rosazza, J Zhang. Int Pat WO96/12818; 21. Oct. 1994. Chem Abstr 125:P56385w.

145. Different *Nocardia* spp. were employed under optimized conditions [143].

146. S Pedragosa-Moreau, A Archelas, R Furstoss. Tetrahedron Lett 37:3319–3322, 1996.

147. G Bellucci, G Berti, R Bianchini, P Cetera, E Mastrorilli. J Org Chem 47:3105–3112, 1982.

148. G Bellucci, G Berti, G Catelani, E Mastrorilli. J Org Chem 46:5148–5150, 1981.

149. W Kroutil, M Mischitz, K Faber. J Chem Soc, Perkin Trans I, 3629–3636, 1997.

150. C Chiappe, A Cordoni, GL Moro, CD Palese. Tetrahedron: Asymmetry 9:341–350, 1998.

151. The induction of mEH is usually performed by feeding phenobarbital to animals and is rather empirical. Depending on the procedure, different enzyme extracts showing varying activities may be obtained.

152. D Wistuba, V Schurig. Angew Chem Int Ed Engl 25:1032–1034, 1986.

153. E Blée, F Schuber. J Biol Chem 267:11881–11887, 1992.

154. PPHL Otto, F Stein, CA Van Der Willigen. Agric Ecosys Environ 21:121–123, 1988.

155. RP Hanzlik, S Heidemann, D Smith. Biochem Biophys Res Commun 82:310–315, 1978.

156. G Bellucci, C Chiappe, A Cordoni. Tetrahedron: Asymmetry 7:197–202, 1996.

157. S Pedragosa-Moreau, A Archelas, R Furstoss. Tetrahedron 52:4593–4606, 1996.

158. F Oesch, N Kaubisch, DM Jerina, JW Daly. Biochemistry 10:4858–4866, 1996.

159. PL Barili, G Berti, E Mastrorilli. Tetrahedron 49:6263–6276, 1993.

160. SF Carter, DJ Leak. Biocat Biotrans 13:111–129, 1995.

161. A Archelas, JP Delbecque, R Furstoss. Tetrahedron: Asymmetry 4:2445–2446, 1993.

162. M Mischitz, K Faber. Synlett 1996:978–980, 1996.

163. IVJ Archer, DJ Leak, DA Widdowson. Tetrahedron Lett 37:8819–8822, 1996.

164. K Faber. Biotransformations in Organic Chemistry, 3rd ed. Heidelberg: Springer-Verlag, 1997, p 270.

165. A Kamal, AB Rao, MV Rao. Tetrahedron Lett 33:4077–4080, 1992.

166. M Mischitz, K Faber. Tetrahedron Lett 35:81–84, 1994.

167. A Kamal, Y Damayanthi, MV Rao. Tetrahedron: Asymmetry 3:1361–1364, 1992.

9
Enzymatic Asymmetric Synthesis Using Aldolases

Wolf-Dieter Fessner
Darmstadt University of Technology, Darmstadt, Germany

I. INTRODUCTION

The synthesis of enantiomerically pure pharmaceuticals is of paramount importance due to the fact that the biological activity of enantiomers can differ dramatically in kind and intensity because of the chiral nature of life processes [1,2]. Supplementary to classical chemical methodology, enzymes are thus finding increasing acceptance as chiral catalysts for the in vitro synthesis of asymmetric compounds because they have been optimized by evolution for high selectivity and catalytic efficiency [3–9]. In contrast to most classical chemical operations, biocatalytic conversions can usually be performed under mild reaction conditions that are compatible with underivatized substrates, thus obviating tedious and costly protecting group manipulations [10]. Evidently, enzyme catalysis is most attractive for the synthesis and modification of biologically relevant classes of organic compounds that are typically multifunctional and water-soluble, such as amino acids [11,12] or carbohydrates [13–16], which are difficult to prepare and to handle by conventional means. Recently, enzymatic techniques have proven particularly useful in the area of oligosaccharide synthesis [17] where conventional synthetic approaches have to confront the manifold problems in control of regio- and stereoselectivity upon subunit assembly, which had long restricted the field to dedicated expert researchers. New enzymatic methodology has made an impact in facilitating the access to complex natural structures and suitable analogs to probe the medicinally relevant implications of complex oligosaccharides in central biological recognition phenomena, such as cell–cell communication for cell adhesion, immune response, inflammation, or cell differentiation during organ development and carcinogenesis [18–21].

Carbon–carbon bond formation is at the heart of organic synthesis. Catalytic C-C coupling is among the most useful synthetic methods in asymmetric synthesis because of its potential for stereoisomer generation by a convergent, "combinatorial" strategy [22]. In nature such reactions are facilitated by aldolases, which catalyze the (usually reversible) addition of carbonucleophiles to $C{=}O$ double bonds [23]. Some 30 aldolases have been identified so far [24,25], with the majority involved in carbohydrate or amino acid metabolism. Several of these enzymes have already been made commercially available for preparative applications. Accumulating protein structural information and functional knowledge has furthered major advances to a detailed understanding of catalytic events.

Besides appropriate techniques of protein handling and reaction engineering [26], for future developments of successful biocatalytic processes it will be mandatory to have ready access to an extended database that precisely defines the scope and limitations for each synthetically useful enzyme. Thus, this chapter is intended to illustrate the current status of development of the most important aldolase biocatalysts, and it will outline their potential for preparative applications toward a range of structural goals and cover recent developments of enzyme applications to general or specific problems of asymmetric synthesis. Special emphasis is placed on the synthetic power of a directed stereodivergent approach by which multiple, diastereomeric products can be derived starting from common synthetic building blocks [22,27]. Such a synthetic strategy is facilitated by the prevalence of families of related, stereocomplementary enzymes that often seem to have a very similar, broad substrate tolerance. Less common aldolases and related enzymes that have been less studied for synthetic purposes have been omitted from this coverage but have been reviewed recently in other, more comprehensive surveys [15,22,24].

II. MECHANISTIC ASPECTS

Most of the known aldolases catalyze the reversible, stereocontrolled addition of a nucleophilic ketone donor to an electrophilic aldehyde acceptor. From a synthetic perspective, the most extensively studied and synthetically most useful enzymes utilize aldol donors comprising 2-carbon or 3-carbon fragments and can be grouped into four categories depending on the structure of their nucleophilic component: (1) the pyruvate (and phosphoenolpyruvate)–dependent aldolases, (2) the dihydroxyacetone phosphate (DHAP)–dependent aldolases, (3) one acetaldehyde-dependent aldolase, and (4) the glycine-dependent aldolase (Fig. 1). Members of the first and third types generate α-methylene carbonyl compounds and thereby generate a single stereo center, whereas members of the other types form α-substituted carbonyl derivatives that contain two new vicinal chiral centers at the new C—C bond, a fact that makes them more appealing for asymmetric synthesis.

Mechanistically, the activation of the aldol donor substrates is achieved by stereospecific deprotonation along two different pathways (Fig. 2) [28]: Class I aldolases bind their substrates covalently via imine/enamine formation to an active site lysine residue to initiate bond cleavage or formation (Fig. 2a); in contrast, class II aldolases utilize transition metal ions as a Lewis acid cofactor (usually Zn^{2+}) which facilitates (Fig. 2b) deprotonation by a bidentate coordination of the donor to give the enediolate nucleophile [29]. Usually, the approach of the aldol acceptor to the enzyme-bound nucleophile occurs stereospecifically following an overall retention mechanism, while the facial differentiation of the aldehyde carbonyl is responsible for the relative stereoselectivity. In this manner, the stereochemistry of the C—C bond formation is completely controlled by the enzyme, in

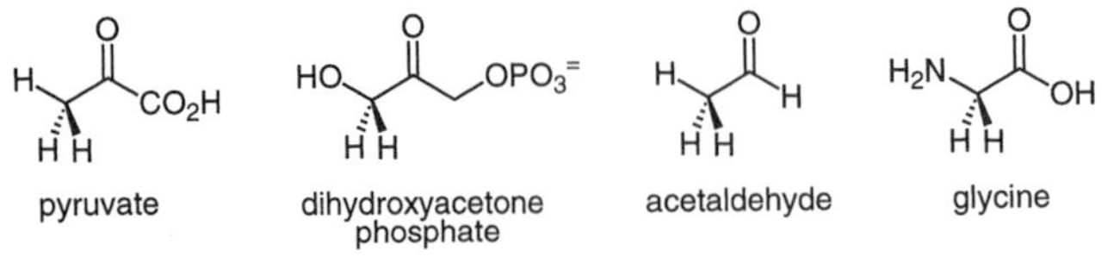

Figure 1 Substrate classes of preparatively useful aldolases.

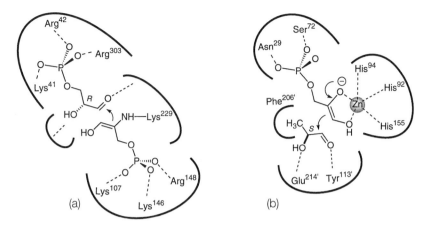

Figure 2 Schematic representation of substrate binding and C—C bond formation for the class I fructose 1,6-bisphosphate aldolase from rabbit muscle (a) and for the class II fuculose 1-phosphate aldolase from *Escherichia coli* (b).

general irrespective of the constitution or configuration of the substrate, which makes the stereochemistry of the products highly predictable. Typically, aldolases are highly specific for the nucleophilic substrate due to mechanistic requirements. This includes the necessity for a reasonably high substrate affinity as well as the general difficulty of binding and anchoring a rather small molecule in a fashion that restricts solvent access to the carbanionic site after deprotonation and shields one enantiotopical face of the nucleophile to secure correct diastereofacial discrimination (Fig. 2). On the other hand, most of the aldolases allow a broad variation of the aldehyde substrate.

Mechanistic hypotheses for both classes of aldolases have recently been substantiated by protein crystal structure analysis of liganded enzymes. Particularly, the α/β-barrel structure of the class I fructose 1,6-bisphosphate aldolase (FruA, vide infra) from rabbit muscle was the first to be uncovered by x-ray crystal structure analysis [30], followed by ones from several other species [31–34]. A complex of the aldolase with noncovalently bound substrate DHAP in the active site indicates a trajectory for the substrate traveling toward the nucleophilic Lys 229 Nᵉ [35]. Also, the three-dimensional structure for the neuraminic acid aldolase (NeuA, vide infra) from *Escherichia coli* has been determined after interception of the pyruvate Schiff base at Lys 165 Nᵉ by borohydride reduction [36], and complementary information was gained from structural investigations of the *E. coli* transaldolase [37,38] to provide a fertile ground for further modeling of substrate binding and catalysis. For class II aldolases, the zinc dependent fuculose 1-phosphate aldolase (FucA, vide infra) from *E. coli* has been solved [39,40] including the metal-chelating inhibitor phosphoglycolohydroxamate as a structural analog of the DHAP enediolate (Fig. 3) by which all key stereochemical issues could be successfully rationalized [29]. Structural details are also available for the class II FruA from *E. coli* [41,42].

Hopefully, an emerging understanding of molecular recognition in aldolase–substrate interactions during a catalytic cycle will enable a possible engineering of these catalysts toward an improved substrate tolerance and stereoselectivity for asymmetric syntheses or, in the more distant future, to rationally redesign aldolases for novel catalytic functions.

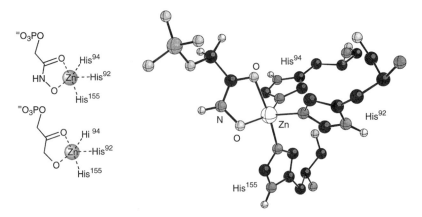

Figure 3 Coordination sphere of the zinc ion in the FucA crystal structure liganded by the chelating inhibitor phosphoglycolohydroxamate, a transition state analog for activated DHAP [29,40].

III. PYRUVATE ALDOLASES

In vivo, pyruvate lyases serve a catabolic function in the degradation of sialic acids and 2-keto-3-deoxyoctosonate (KDO), and in that of 2-keto-3-deoxyaldonic acid intermediates from hexose or pentose catabolism. Since these freely reversible aldol additions often have unfavorable equilibrium constants [43], synthetic reactions are usually driven by an excess of pyruvate to achieve a satisfactory conversion.

A few related enzymes have been identified that utilize phosphoenolpyruvate in place of pyruvate which, by release of inorganic phosphate upon C—C bond formation, renders aldol additions essentially irreversible [22,24]. Although attractive for synthetic applications, such enzymes have not been studied in close detail for preparative synthesis.

N-Acetylneuraminic acid aldolase (NeuA; EC 4.1.3.3) catalyzes the reversible addition of pyruvate to *N*-acetyl-D-mannosamine (**1**) to form the parent sialic acid (**3**) (Fig. 4). The NeuA lyases found in both bacteria and animals are type I enzymes that form an enamine intermediate with pyruvate and promote a *si*-face attack to the aldehyde carbonyl group with formation of a (*4S*) configured stereo center. Enzyme preparations from *Clostridium perfringens* and *E. coli* are commercially available, and the latter enzyme has been cloned, overexpressed [44,45], and its three-dimensional structure determined [46]. The enzyme has a broad pH optimum around 7.5 and is quite stable in solution at ambient temperature [47].

As neuraminic acid (**3**) is an important precursor to antiviral drugs, its large-scale synthesis has been developed as the prime example for an industrial aldolase bioconversion process at the multiton scale [48–54]. In an integrated approach, the expensive *N*-acetyl-glucosamine was produced by enzymatic in situ isomerization from inexpensive **1** by a combination of the NeuA with an *N*-acylglucosamine 2-epimerase (EC 5.1.3.8) catalyst in an enzyme membrane reactor [51,55]. To facilitate separation of the reaction product from excess of **2**, yeast pyruvate decarboxylase has been applied for the selective destruction of residual pyruvate into volatiles [56]. Excessive **2** may be avoided altogether by coupling of the synthesis of **3** to a thermodynamically favored process, e.g., by combination with an irreversible sialyltransferase reaction to furnish a sialosaccharide [57].

Because of the importance of sialic acids in a wide range of biological recognition events, the aldolase has become popular for the chemoenzymatic synthesis of a multitude

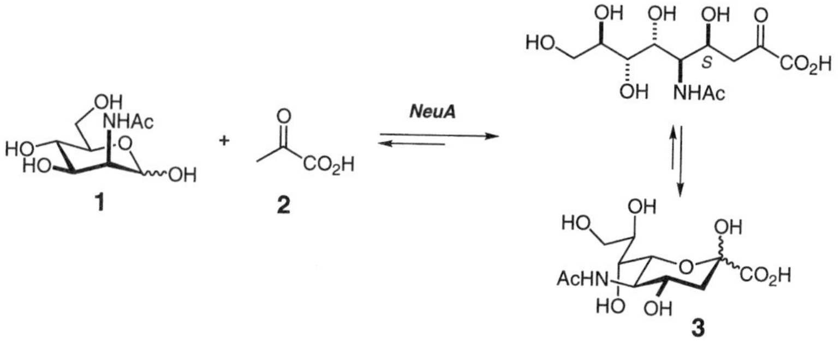

Figure 4 Natural substrates of the *N*-acetylneuraminic acid aldolase.

of derivatives and analogs of **3**, mostly for modifications at C5/C9 [16,22] such as different *N*-acylated derivatives [58,59], including amino acid conjugates [60], or 9-modified analogs [61,62] in search for new influenza inhibitors. Most notably, the *N*-acetyl group in **1** may either be omitted [63,64] or be replaced by sterically demanding substituents such as *N*-Cbz (**6**) [65,66] or even a nonpolar phenyl group [64] without destroying activity. Large acyl substituents are also tolerated at C-6 as shown by the conversion of a Boc-glycyl derivative **4** (Fig. 5) as a precursor to a fluorescent sialic acid conjugate **5** [67].

The enzyme also displays a fairly broad tolerance for stereochemically related aldehyde substrates such as a number of sugars and their derivatives larger than or equal to pentoses [47,68,69]. Several natural and unnatural sialic acid derivatives have thus been prepared by replacement of the natural D-*manno*-configurated substrate with aldose derivatives containing modifications such as epimerization, substitution, or deletion at positions C-2, -4, or -6 (Table 1) [16,22]. Epimerization at C-2, however, is restricted to small polar substituents at strongly decreased reaction rates [63,70].

In most cases, a high level of asymmetric induction for the (*4S*) configuration is retained. However, a number of carbohydrates were also found to be converted with random stereoselectivity for the C-4 configuration [56,68,71] which may be due to a contribution of thermodynamic control or to an inverse conformational preference (Fig. 6) of the different substrate classes [68]. A critical and distinctive factor seems to be recognition of the configuration by the enzymic catalyst at C-3 in the aldehydic substrate [68,69].

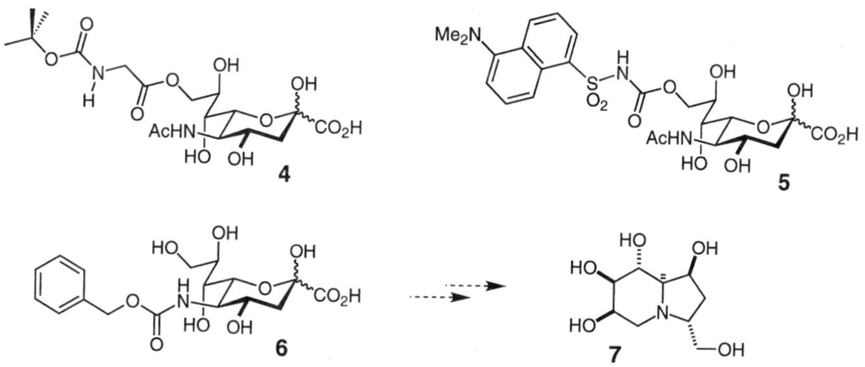

Figure 5 Neuraminic acid derivatives accessible by NeuA catalysis and related aza sugar product [66,67].

Table 1 Substrate Tolerance of Neuraminic Acid Aldolase

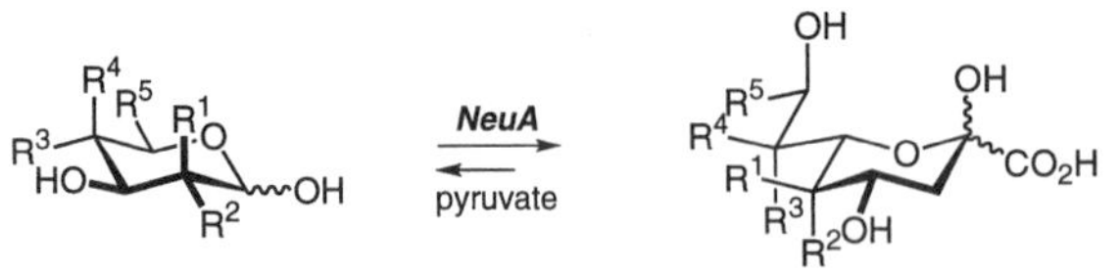

R₁	R₂	R₃	R₄	R₅	Yield (%)	Rel. rate (%)	Ref.
NHAc	H	OH	H	CH₂OH	85	100	48, 50
NHAc	H	OH	H	CH₂OAc	84	20	48, 50
NHAc	H	OH	H	CH₂OMe	59	—	48
NHAc	H	OH	H	CH₂N₃	84	60	63
NHAc	H	OH	H	CH₂OP(O)Me₂	42	—	63
NHAc	H	OH	H	CH₂O(L-lactoyl)	53	—	48, 63
NHAc	H	OH	H	CH₂O(Gly-N-Boc)	47	—	67
NHAc	H	OH	H	CH₂F	22	60	63
NHAc	H	OMe	H	CH₂OH	70	—	48
NHAc	H	H	H	CH₂OH	70	—	64
NHC(O)CH₂OH	H	OH	H	CH₂OH	61	—	48
NHCbz	H	OH	H	CH₂OH	75	—	66
OH	H	OH	H	CH₂OH	84	91	56, 64
OH	H	H	H	CH₂OH	67	35	64
OH	H	H	F	CH₂F	40	—	63
OH	H	OH	H	H	66	10	64
H	F	OH	H	CH₂OH	30	—	63
H	H	OH	H	CH₂OH	36	130	63, 64
Ph	H	OH	H	CH₂OH	76	—	64

From an application-oriented perspective, a three-point binding model has been proposed for the conversion of substrate analogs which, on the basis of conformational analysis, can predict the facial stereoselectivity of C—C bonding [22].

Products generated by the unusual *re*-face attack for (*4R*) stereochemical preference include a number of related higher ulosonic acids of biological importance such as D-KDO (**8**) [56,64,68,72] and D-2-Keto-3-deoxynonosonate (D-KDN) [47,52,73,74], or formation of the enantiomers of naturally occurring sugars (Fig. 7) such as L-NeuNAc

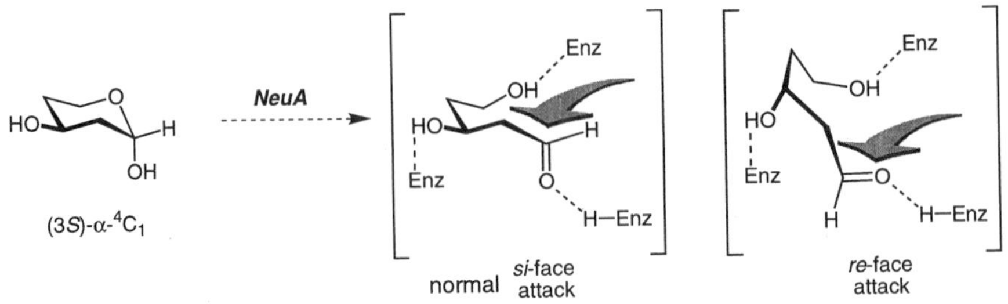

Figure 6 Three point binding model for prediction of stereoselectivity in NeuA-catalyzed reactions, based on a conformational analysis of C-3 epimeric aldopyranoses as acceptor substrates [22].

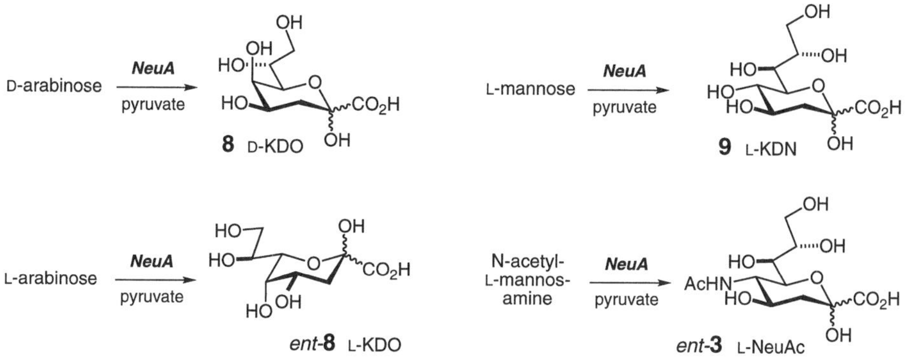

Figure 7 Natural and nonnatural sialic acids synthesized using NeuA in retention or inversion mode [56,68].

(*ent*-**3**), L-KDN (**9**), and L-KDO (*ent*-**8**) [56]. Ready access to compounds of this type may be particularly valuable for investigations on the biological activity of sialo conjugates containing nonnatural sialic acid derivatives [75].

Starting from the *N*-Cbz-protected aldolase product **6**, aza sugar **7** has been obtained stereoselectively by internal reductive amination as an analog of the bicyclic, indolizidine-type glycosidase inhibitor castanospermine [66]. Also, it had been recognized that the C12–C20 sequence of the macrolide antibiotic amphotericin B resembles the β-pyranose tautomer of **3** [76]. Thus, the branched-chain *manno*-configured substrate **10** [77] was successfully chain-extended under NeuA catalysis to yield the potential amphotericin B synthon **11** (Fig. 8) in good yield.

The 2-keto-3-deoxyoctosonate (KDO) aldolase (KdoA; EC 4.1.2.23) is an inducible enzyme in gram-negative microorganisms [78] where D-KDO **8** is a core constituent of the outer membrane lipopolysaccharide. The aldolases from *Aerobacter cloacae* and an *Aureobacterium barkerei* strain have been studied for synthetic applications [72,79]. Similar to the NeuA, the KdoA enzyme has a broad substrate specificity for aldoses in place for the natural acceptor D-arabinose (Table 2) while pyruvate was found to be irreplaceable. Preparative applications, e.g., that for the synthesis of KDO analogs **13/14** (Fig. 9), suffer

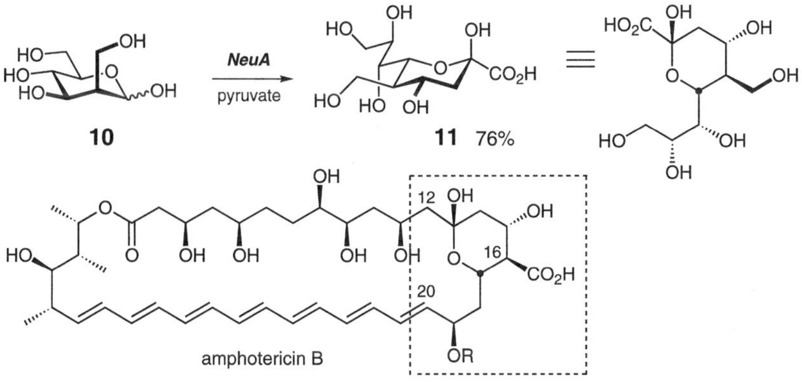

Figure 8 NeuA-catalyzed preparation of a synthetic precursor to the macrolide antibiotic amphotericin B [76].

Table 2 Substrate Tolerance of 2-Keto-3-deoxyoctosonic Acid Aldolase [72]

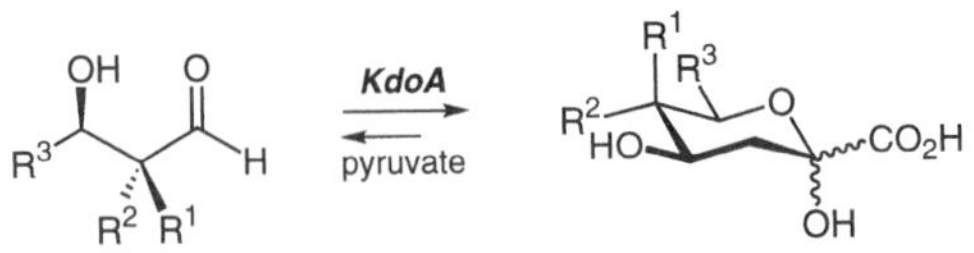

Substrate	R_1	R_2	R_3	Yield (%)	Rel. rate (%)
D-Altrose	OH	H	(RR)-$(CHOH)_2$-CH_2OH	—	25
L-Mannose	H	OH	(SS)-$(CHOH)_2$-CH_2OH	61	15
D-Arabinose	OH	H	(R)-CHOH-CH_2OH	67	100
D-Ribose	H	OH	(R)-CHOH-CH_2OH	57	72
2-Deoxy-2-fluoro-D-ribose	F	H	(R)-CHOH-CH_2OH	19	46
2-Deoxy-D-ribose	H	H	(R)-CHOH-CH_2OH	47	71
5-Azido-2,5-dideoxy-D-ribose	H	H	(R)-CHOH-CH_2N_3	—	15
D-Threose	OH	H	CH_2OH	—	128
D-Erythrose	H	OH	CH_2OH	39	93
D-Glyceraldehyde	H	OH	H	11	23
L-Glyceraldehyde	OH	H	H	—	36

from an unattractive equilibrium constant of 13 M^{-1} in the direction of synthesis [43]. The stereochemical course of aldol additions generally seems to adhere to a *re*-face attack on the aldehyde carbonyl that is complementary to that of sialic acid aldolase. On the basis of the results published so far it may be concluded that a $(3R)$ configuration is necessary (but not sufficient), and that stereochemical requirements for C-2 are less stringent [72].

The aldolase specific for cleavage of 2-keto-3-deoxy-6-phospho-D-gluconate (**16**) (KDPGlc aldolase; EC 4.1.2.14) is produced by many species of bacteria for the degradation of 6-phosphogluconate. The equilibrium constant favors synthesis (10^3 M^{-1}) [43]. The stereocomplementary aldolase (Fig. 10) acting on 2-keto-3-deoxy-6-phospho-D-galactonate (**17**) (KDPGal aldolase; EC 4.1.2.21) is less common [80].

Enzymes from *E. coli* [81–83], *Zymomonas mobilis* [84], and *Erwinia chrysanthemi* [85] have recently been cloned and can be readily isolated by two-step differential dye

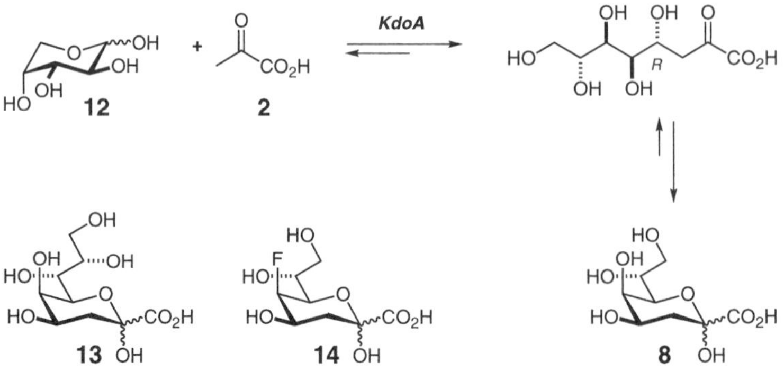

Figure 9 Natural substrates of the 2-keto-3-deoxyoctosonic acid aldolase and nonnatural sialic acids obtained by KdoA catalysis [72].

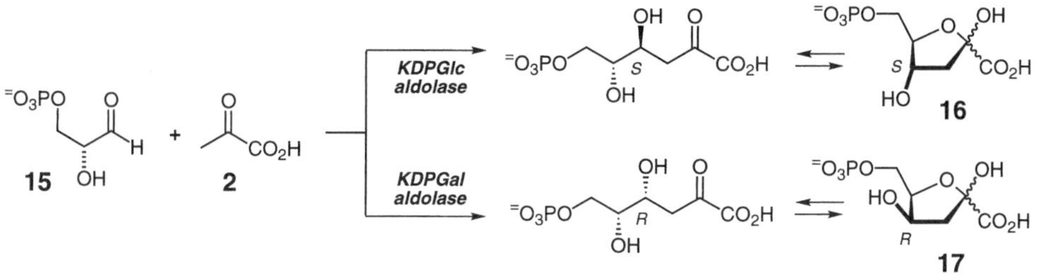

Figure 10 Aldol reactions catalyzed in vivo by the 2-keto-3-deoxy-6-phospho-D-gluconate and 2-keto-3-deoxy-6-phospho-D-galactonate aldolases.

ligand chromatography [86,87]. Comparative studies have shown that all three enzymes strongly prefer D-glyceraldehyde 3-phosphate (15) as natural acceptor [88], but at reduced reaction rates offer a rather broad substrate tolerance for polar, short chain aldehydes (Table 3) [89]. Stereoselectivity was found to be high for small polar aldehydes, which have been utilized in the preparation of compounds 20/21 and in a two-step enzymatic synthesis of 19 (Fig. 11), the amino acid portion of nikkomycin antibiotics [90].

IV. DIHYDROXYACETONE PHOSPHATE ALDOLASES

While pyruvate aldolases form only a single stereogenic center, the aldolases specific for dihydroxyacetone phosphate (DHAP, 22) as a nucleophile create two new asymmetric centers at the termini of the new C—C bond. Particularly useful for synthetic applications is the fact that nature has evolved a full set of four stereochemically unique aldolases [27] for the retroaldol cleavage of ketose 1-phosphates 23–26 (Fig. 12). In the direction of synthesis this formally allows the deliberate preparation of any one of the possible four diastereomeric aldol adducts in a building block fashion [15,22,27] by simply choosing the complementary enzyme and starting materials for full control over constitution and absolute configuration of the desired product.

Table 3 Substrate Tolerance of 2-Keto-3-deoxy-6-phospho-D-gluconate Aldolase [89]

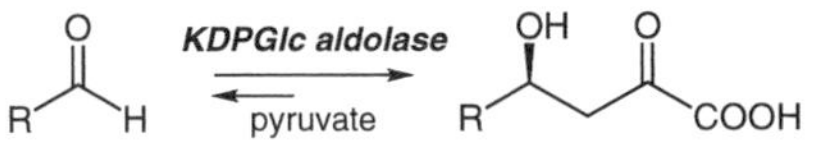

Substrate	R	Rel. rate (%)
D-Glyceraldehyde 3-phosphate	D-CH_2OH-$CH_2OPO_3^{2-}$	100.0
3-Nitropropanal	CH_2-CH_2NO_2	1.6
Chloroethanal	CH_2Cl	1.0
D-Glyceraldehyde	D-CHOH-CH_2OH	0.8
D-Lactaldehyde	D-CHOH-CH_3	0.2
D-Ribose 5-phosphate	D-*ribo*-(CHOH)$_3$-$CH_2OPO_3^{2-}$	0.04
Erythrose	*erythro*-(CHOH)$_2$-CH_2OH	0.01
Glycol aldehyde	CH_2OH	0.01

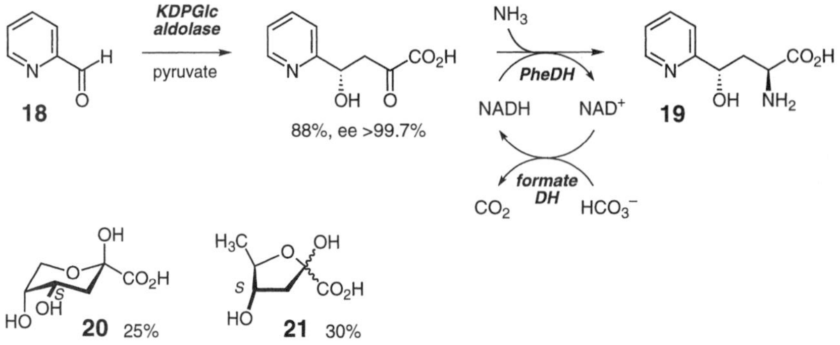

Figure 11 Stereoselective synthesis of the amino acid portion of nikkomycin antibiotics [90] and hexulosonic acids using KDPGlc aldolase [89].

The D-fructose 1,6-bisphosphate aldolase (FruA; EC 4.1.2.13) catalyzes in vivo the glycolytic equilibrium addition of **22** to D-glyceraldehyde 3-phosphate (**15**) to give D-fructose 1,6-bisphosphate (**23**). The equilibrium constant of 10^4 M^{-1} strongly favors synthesis [43]. The class I FruA isolated from rabbit muscle has been the most extensively investigated aldolase for mechanistic, structural, and preparative purposes. Its stability in solution is limiting, but the class I FruA (monomeric) from *Staphylococcus carnosus* [91] or the Zn^{2+}-dependent, type II aldolase from *E. coli* [92] are much more stable with half-lives up to several weeks or months as compared to only a few days [93]. Typically, type I FruA enzymes are tetrameric proteins composed of subunits of approximately 40 kDa, while the type II FruA are dimers of approximately 39-kDa subunits.

The D-tagatose 1,6-bisphosphate aldolase (TagA; EC 4.1.2.n) is known as a class I subtype involved in catabolism of lactose and D-galactose of different microorganisms but

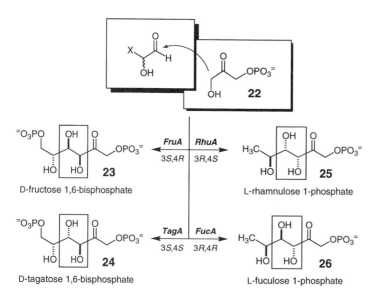

Figure 12 Aldol reactions catalyzed in vivo by the four stereocomplementary dihydroxyacetone-dependent aldolases.

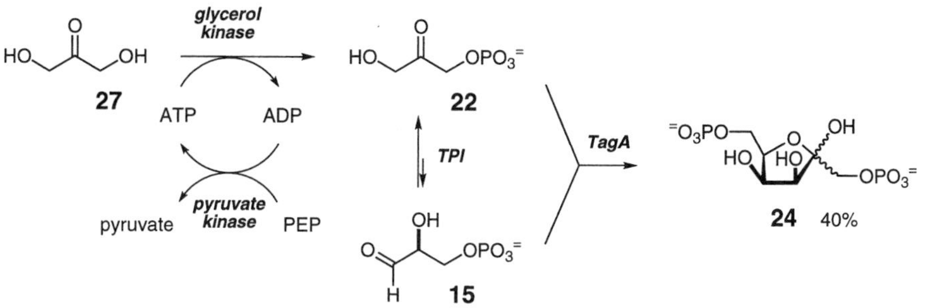

Figure 13 Enzymatic one-pot synthesis of tagatose 1,6-bisphosphate based on the stereoselective TagA from *E. coli* [95,96].

apparently has no stereochemical selectivity with regard to discrimination of **23/24** [94]. A highly stereoselective Zn^{2+}-dependent TagA of subtype II has been found in the galactitol metabolism from *E. coli* [95,96].

The L-rhamnulose 1-phosphate aldolase (RhuA; EC 4.1.2.19) and the L-fuculose 1-phosphate aldolase (FucA; EC 4.1.2.17) are found in the microbial degradation of deoxy sugars. Both enzymes from *E. coli* are homotetrameric Zn^{2+}-dependent class II aldolases with a subunit molecular weight of about 25 and 30 kDa, respectively [97]. Construction of efficient overexpression systems [97–99] has set the stage for crystallization of both proteins to enable their x-ray structure analysis [39,100]. Both metalloproteins show a very high stability in the presence of low Zn^{2+} concentrations with half-lives in the range of months at room temperature, and they even tolerate the presence of large fractions ($\geq$30%) of organic cosolvents such as dimethyl sulfoxide, dimethyl formamide, or ethanol [97].

Apparently, all DHAP aldolases are highly specific for the donor component **22** for mechanistic reasons [29]. For synthetic applications, two equivalents of **22** are conveniently generated in situ from commercial fructose 1,6-bisphosphate **23** by the combined action of FruA and triose phosphate isomerase (EC 5.3.1.1) [93,101]. The reverse, synthetic reaction can be utilized to prepare ketose bisphosphates, as has been demonstrated by an expeditious multienzymatic synthesis of the (3*S*,4*S*) all-cis-configurated D-tagatose 1,6-bisphosphate **24** (Fig. 13) from dihydroxyacetone **27**, including a cofactor-dependent phosphorylation, by employing the purified TagA from *E. coli* (Fig. 13) [95,96].

A more general technique for the clean generation of **22** in situ is based on the oxidation of L-glycerol 3-phosphate **28** (Fig. 14) catalyzed by a microbial flavin-dependent glycerol phosphate oxidase (GPO, EC 1.1.3.21; Fig. 14) [102]. Furthermore, this GPO

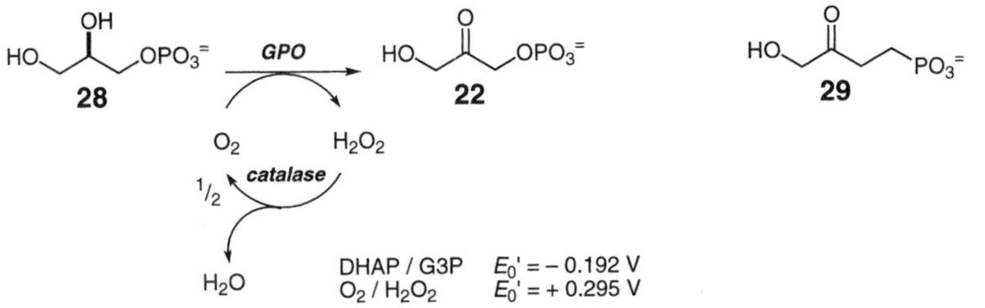

Figure 14 Enzymatic in situ generation of dihydroxyacetone phosphate (or an isosteric phosphonate analog) for stereoselective aldol reactions using DHAP aldolases [102].

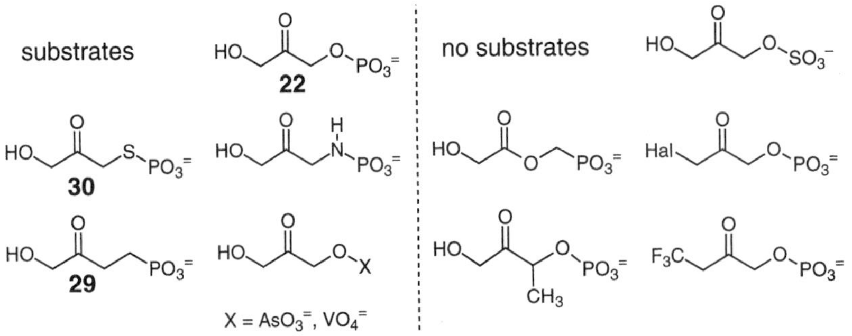

Figure 15 Substrate quality of dihydroxyacetone phosphate analogs [93,102,103].

procedure can also be used for a preparative synthesis of several DHAP analogs modified at the phosphate group (e.g., **29**, **30**) [102,103].

With the rabbit muscle FruA a few analogs modified by isosteric replacements of the ester oxygen for sulfur (**30**), nitrogen, or methylene carbon (**29**) were found to be tolerable (Fig. 15) [104–106]. The phosphonate **29** (Fig. 15) is converted by the rabbit enzyme at 10% of v_{max}, and is tolerated also by the FruA from *Staphylococcus carnosus* (class I) and by the rhamnulose 1-phosphate aldolase from *E. coli* (RhuA; EC 4.1.2.19; class II). Thus, sugar phosphonates (e.g., **31**, **32**) that mimic metabolic intermediates (Fig. 16) but lack a hydrolytically unstable phosphate ester bond can be rapidly synthesized [106]. However, several closely related enzymes surprisingly failed to react (e.g., FruA from yeast and *E. coli*; FucA from *E. coli*). The phosphorothioate analog **30** (Fig. 15) has been utilized to make terminally deoxygenated sugars accessible via a sequence of FruA, catalyzed aldolization followed by reductive desulfurization, as illustrated by the synthesis of D-olivose [107].

The DHAP aldolases accept an extensive range of differently substituted or unsubstituted aliphatic aldehydes as the acceptor component at synthetically useful rates (Tables

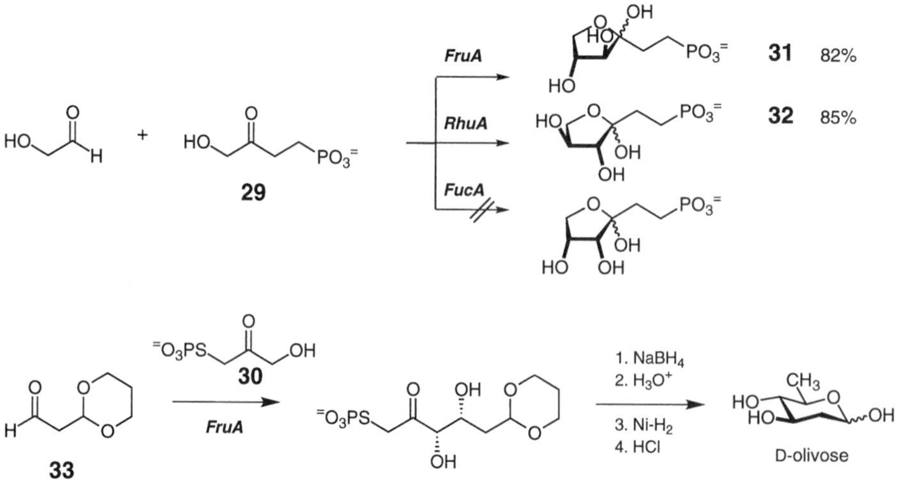

Figure 16 Stereoselective synthesis of hydrolytically stable sugar phosphonates and 2-deoxy sugars based on DHAP analogs [106,107].

Table 4   Substrate Tolerance of Fructose 1,6-Bisphosphate Aldolase

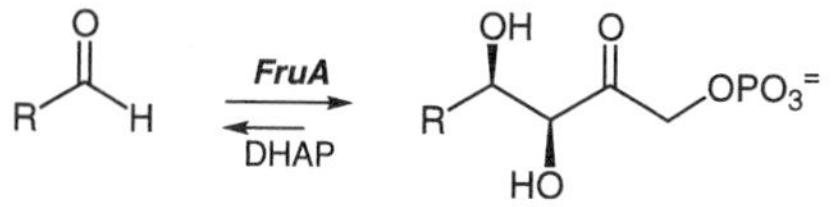

R	Rel. rate (%)	Yield (%)	Ref.
D-CHOH-CH$_2$OPO$_3^{2-}$	100	95	93, 102
H	105	—	93
CH$_3$	120	—	93
CH$_2$Cl	340	50	93, 136
CH$_2$-CH$_3$	105	73	93
CH$_2$-CH$_2$-COOH	—	81	102
CH$_2$OCH$_2$C$_6$H$_5$	25	75	93
D-CH(OCH$_3$)-CH$_2$OH	22	56	93
CH$_2$OH	33	84	93, 102
D-CHOH-CH$_3$	10	87	93, 130
L-CHOH-CH$_3$	10	80	93
DL-CHOH-C$_2$H$_5$	10	82	93
CH$_2$-CH$_2$OH	—	83	114, 130
CH$_2$-C(CH$_3$)$_2$OH	—	50	113
DL-CHOH-CH$_2$F	—	95	130
DL-CHOH-CH$_2$Cl	—	90	114
DL-CHOH-CH$_2$-CH=CH$_2$	—	85	114, 116

Table 5   Substrate Tolerance of L-Rhamnulose 1-Phosphate and L-Fuculose 1-Phosphate Aldolases [97,111]

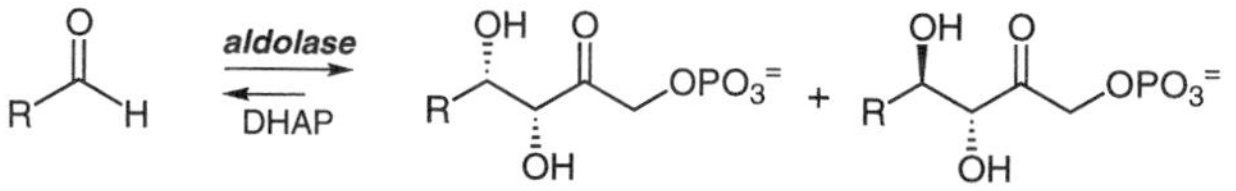

	RhuA			FucA		
R	Rel. rate (%)	Selectivity threo/erythro	Yield (%)	Rel. rate (%)	Selectivity threo/erythro	Yield (%)
L-CH$_2$OH-CH$_3$	100	>97:3	95	100	<3:97	83
CH$_2$OH	43	>97:3	82	38	<3:97	85
D-CHOH-CH$_2$OH	42	>97:3	84	28	<3:97	82
L-CHOH-CH$_2$OH	41	>97:3	91	17	<3:97	86
CH$_2$-CH$_2$OH	29	>97:3	73	11	<3:97	78
CHOH-CH$_2$OCH$_3$	—	>97:3	77	—	<3:97	83
CHOH-CH$_2$N$_3$	—	>97:3	97	—	<3:97	80
CHOH-CH$_2$F	—	>97:3	95	—	<3:97	86
H	22	—	81	44	—	73
CH$_3$	32	69:31	84	14	5:95	54
CH(CH$_3$)$_2$	22	97:3	88	20	30:70	58

4 and 5). Literally hundreds of aldehydes have so far been tested successfully by enzymatic assay and preparative experiments as a replacement for **15**, and most of the corresponding aldol products have been isolated and characterized [16,22,108]. Limitations seem to be met only with nonpolar tertiary (e.g., pivaldehyde), α,β-unsaturated, and aromatic aldehydes, which apparently are not converted for stereoelectronic reasons. Highest conversion rates, diastereoselectivities, and yields are generally achieved with 2- or 3-hydroxy aldehydes. Higher yields in the latter cases usually derive from the fact that products readily cyclize in aqueous solution to form stable furanoid or pyranoid rings. The corresponding phosphate-free compounds can be easily obtained by mild enzymatic hydrolysis using an inexpensive alkaline phosphatase (EC 3.1.3.1) at pH 7.5–9.0 [109], whereas base-labile compounds may require working at pH 4.5–6.0 using a more expensive acid phosphatase (EC 3.1.3.2) [93].

The absolute configuration of the resulting vicinal diols usually follows precisely that of the natural substrates (see Fig. 12). The configurations at C-3 are invariably conserved owing to the stereospecific deprotonation of **22** in the aldolase-active sites [29]. Kinetic stereoselectivity at C-4 is less consistent, however, with individual enzymes and with particular types of substrate analogs (Table 5) [22,97]. For example, no kinetic stereo preference is observed with the native TagA enzymes with unphosphorylated substrate analogs [95,96].

The Zn^{2+}-dependent aldolases facilitate an effective kinetic resolution of racemic 2-hydroxy aldehydes (*rac*-**34**) as substrates by an overwhelming preference (d.e. ≥95) for the L-configured enantiomers L-**34** which allows one to control three contiguous centers (e.g., **35/36**; Fig. 17) of chirality in the products [110,111]. The kinetic enantioselectivity of class I aldolases, however, with nonanionic aldehydes is rather low [93,112].

Under fully equilibrating conditions, diastereoselectivity can be influenced by thermodynamic control for the energetically favored product. Particularly strong discrimination results from utilization of 3-hydroxylated aldehydes **37** owing to the pronounced substituent effects on conformational stability of the resulting pyranoid ring products **38/39** [22,93,113–115]. Thus, in FruA-catalyzed reactions (3S)-configured hydroxyaldehydes (Fig. 18) are the preferred substrates to give a most stable all-equatorial substitution in the product **38** [93,105,113]. This technique has recently found an application in a novel

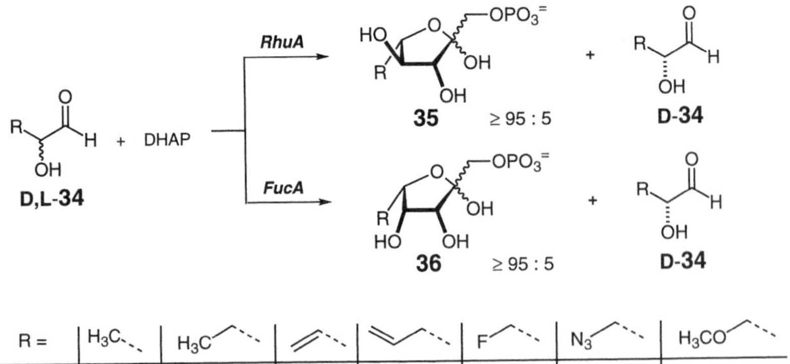

R =	H₃C	H₃C	alkene	alkene	F	N₃	H₃CO
yield [%]							
RhuA	91	86	89	84	95	97	77
FucA	85	81	84	97	86	80	83

Figure 17 Racemate resolution by class II DHAP aldolases due to their kinetic enantioselectivity for L-configured α-hydroxyaldehyde substrates [110,111].

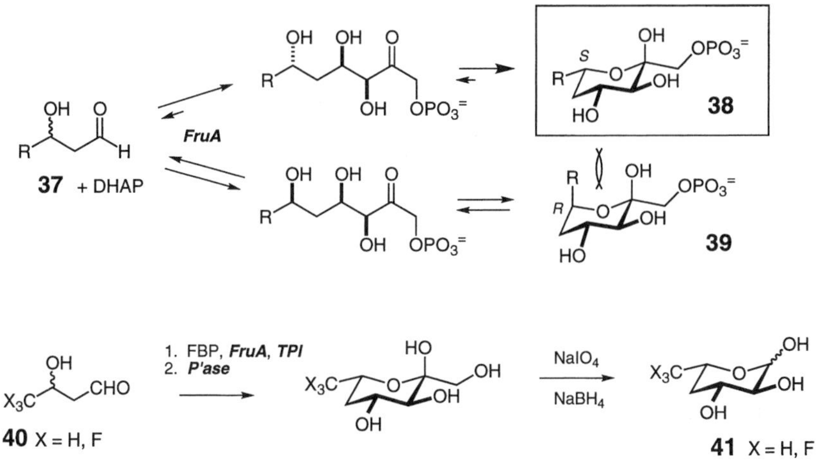

Figure 18 Diastereoselectivity in FruA-catalyzed aldol additions to 3-hydroxyaldehydes under thermodynamic control [93,113], and synthesis of L-fucose derivatives based on thermodynamic preference [116].

approach for the de novo synthesis of 4,6-dideoxy sugars such as 4-deoxy-L-fucose **41** or its trifluoromethylated analog [116].

Bidirectional chain extension of dihydroxy dialdehydes gives rise to carbon-linked disaccharide mimetics (e.g., **43**, **45**, **46**) by simple one-pot operations [115,117]. The latter may be obtained as single diastereomers in good overall yield even from racemic precursors (especially cycloolefins), if the tandem aldolization reactions are conducted under conditions of thermodynamic control (Fig. 19). Typically, the thermodynamic advantage favors the trans (**43**) and equatorial attachments (**45, 47**) of the sugar ring by far, so that the

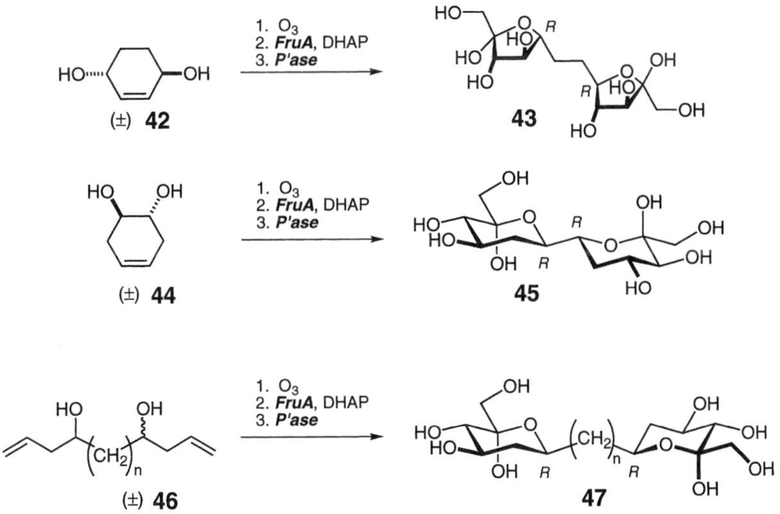

Figure 19 Applications of bidirectional chain synthesis to the generation of disaccharide mimetics using tandem enzymatic aldol additions, including racemate resolution under thermodynamic control [115,117].

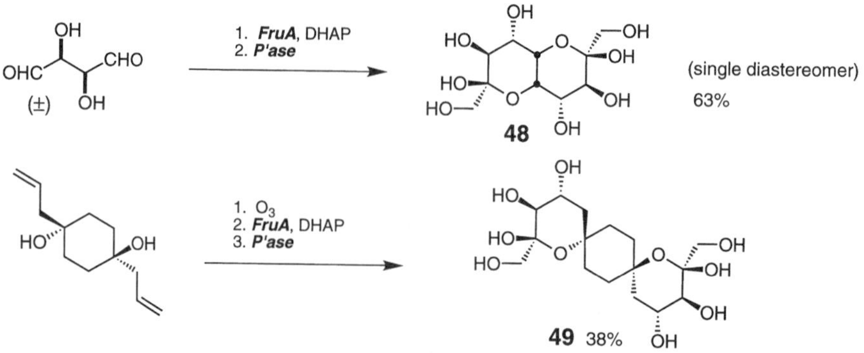

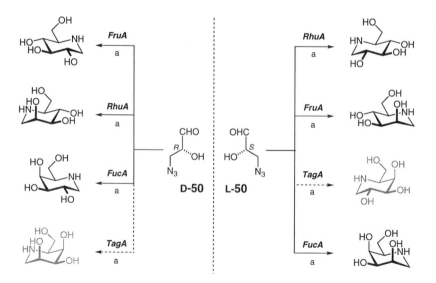

Figure 20 Bidirectional chain extension for the synthesis of anulated and spirocyclic oligosaccharide mimetics by DHAP aldolase catalysis [115].

C_2-symmetrical diastereomers can be obtained selectively, even starting from *rac/meso*-diol mixtures **46**. Similarly, highly complex structures like anulated (**48**) and spirocyclic (**49**) (Fig. 20) carbohydrate mimics may be obtained from suitably customized precursors [115].

The structural resemblance of "aza sugars" (1-deoxy sugars in which an imino group replaces the ring oxygen) to transition states or intermediates of glyco-processing enzymes has made these compounds an attractive research object because of their potential value as enzyme inhibitors for therapeutic applications. An important and flexible synthetic strategy has been developed consisting of a stereoselective enzymatic aldol addition to an azido aldehyde followed by azide hydrogenation and intramolecular reductive amination [118]. Particularly noteworthy are the stereodivergent chemoenzymatic syntheses of diastereomers of the nojirimycin type from 3-azidoglyceraldehyde **50** (Fig. 21) that have been developed independently by several groups [92,119–124].

Figure 21 Stereodivergent synthesis of 1-deoxy aza sugars of the nojirimycin type by two-step enzymatic aldolization/catalytic reductive amination [92,119–124].

Also, other 5-, 6- and 7-membered ring alkaloid analogs **51–53** (Fig. 22) can be made according to the same general procedure, but using differently substituted azido aldehydes [92,122,125,126]. The 5-membered ring compounds are powerful glycosidase inhibitors and are believed to better mimic the flattened chair transition state of glycosidic cleavage [118]. By using the tandem aldolization approach in conjunction with intramolecular reductive amination, a first example of a *C*-glycosidically linked azadisaccharide **54** as a disaccharide mimic has been prepared. Again, thermodynamic resolution of the racemic starting material could be advantageously utilized to provide the aza sugar in 20% overall yield as a single diastereomer [127].

Thio sugars are another structural variation of carbohydrates that possess interesting glycosidase-inhibiting properties. Starting from 2- and 3-thiolated aldehydes (**56, 57**), sets of stereochemically related thiosugars (e.g., **55, 58, 59**) have been prepared with DHAP aldolases (Fig. 23) having different specificities [109,128,129]. It is worth noting that an unbiased stereoselectivity indicates an equivalence of OH and SH substituents for correct substrate recognition.

The scope and synthetic usefulness of reactions catalyzed by DHAP aldolases may be further illustrated by exemplary syntheses of some interesting compounds that comprise functionalities and skeletons of a different nature. Aldol additions of **22** will typically generate ketose derivatives from which aldose isomers may be obtained by biocatalytic ketol isomerization [130]. For an alternative entry to aldoses, the "inversion strategy" [131] utilizes monoprotected dialdehydes for adolization which, after stereoselective ketone reduction, provide free aldoses from deprotection of the masked aldehyde function.

Fused bicyclic ring systems (**62**) were constructed by RhuA-catalyzed (Fig. 24) backbone extension from simple alkyl galactosides [132]; in situ oxidation of **60** was performed by the action of galactose oxidase (GalO; EC 1.1.3.9), in parallel to the generation of **22** by the GPO method. Other higher carbon sugar derivatives have also been prepared by diastereoselective chain extension of chiral pool carbohydrates or their corresponding phosphates (e.g., **63**; Fig. 25) [133]. Based on FruA-catalyzed aldol reactions, unusual spiro-anulated (**67**) and branched-chain sugars (**65, 66**) [101], a nonhydrolyzing homo-nucleoside analog **70** [134], as well as 3-deoxy-D-*arabino*-heptulosonic acid 7-phosphate

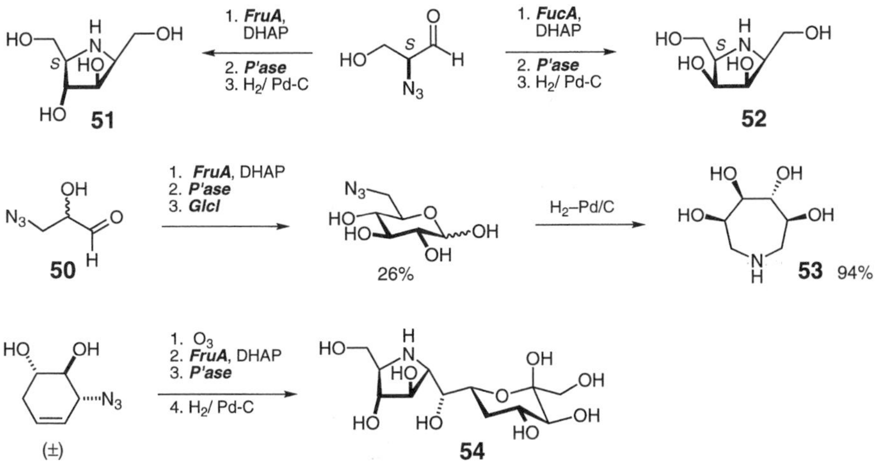

Figure 22 Stereoselective synthesis of 5- and 7-membered ring aza sugars [122,125,126] and an aza-C disaccharide as glycosidase inhibitors [127].

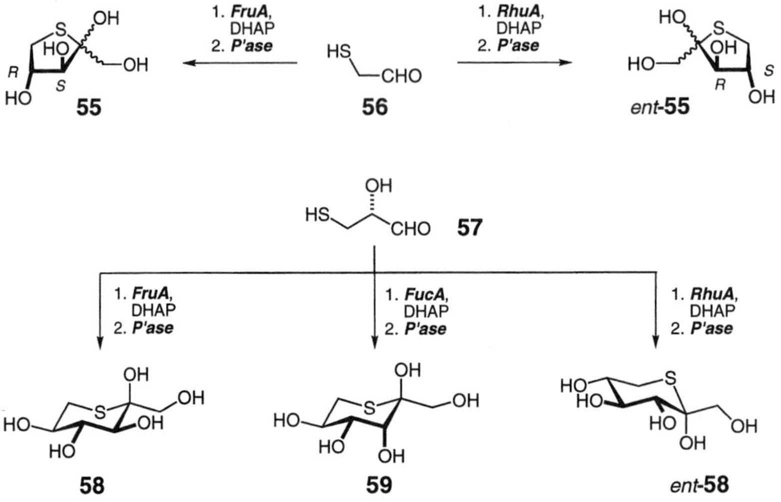

Figure 23 Stereodivergent enzymatic synthesis of thio sugars [109,128,129].

(DAHP, **64**) [135], an intermediate of the shikimic acid pathway, have been synthesized from corresponding aldehyde precursors. Several cyclitols (e.g., **71–73**) could be made from aldol products by radical or nucleophilic cyclization reactions [136–138]. An intermediate was also found to be a correctly configured precursor to the spirocyclic *Streptomyces* metabolite sphydrofuran **69** [139]. In an approach resembling the "inversion strategy" an α-C-mannoside **68** has been prepared from D-ribose 5-phosphate [140].

Applications of the aldol method are not restricted to carbohydrates or carbohydrate-derived materials, despite obvious advantages. This was impressively illustrated, e.g., by FruA-based chemoenzymatic syntheses of the insect pheromone (+)-*exo*-brevicomin (**76**; Fig. 26) [141,142]. Due to the fact that the (R,R)-configured *threo*-diol substitution pattern, which is masked in the bicyclic structure of **76** as an intramolecular acetal but comprises the only independent stereogenic centers of brevicomin, is symmetrical to chain

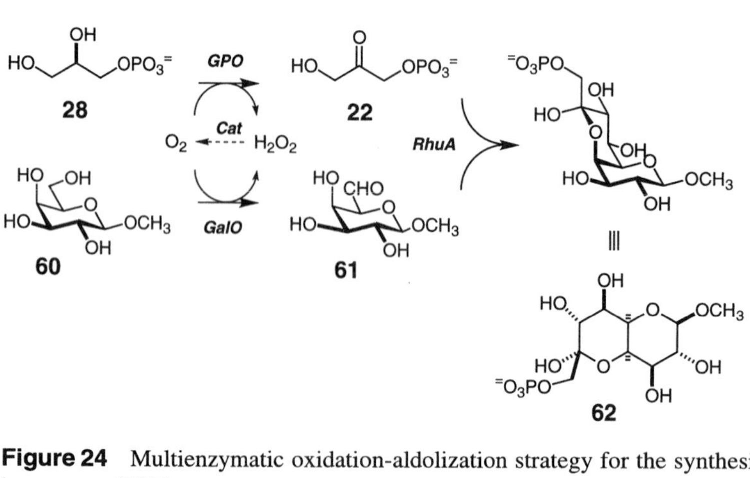

Figure 24 Multienzymatic oxidation-aldolization strategy for the synthesis of bicyclic higher carbon sugars [132].

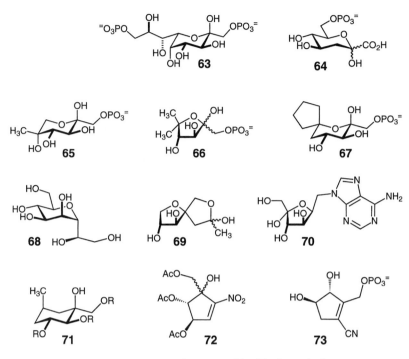

Figure 25 Exemplary compounds prepared by FruA catalysis.

reversion, suitable synthetic precursors **75** and **78** could be obtained by addition of **22** to either 5-oxohexanal (**74**) [141] or propionaldehyde (**77**) [93].

DHAP aldolases may also be advantageous in the construction of stereochemically homogeneous fragments of noncarbohydrate natural products, as was recently demonstrated by the stereoselective synthesis of **80** (Fig. 27) as a synthetic equivalent to the C3–

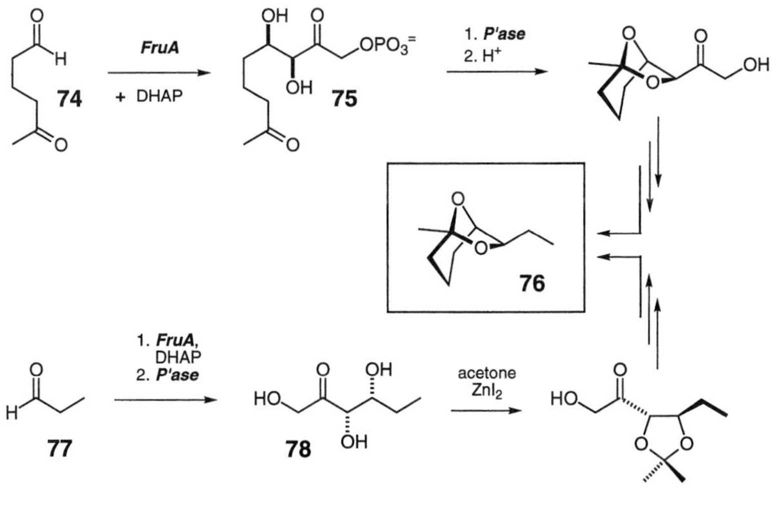

Figure 26 Inverse approaches for the FruA-based chemoenzymatic synthesis of the insect pheromone (+)-*exo*-brevicomin [141,142].

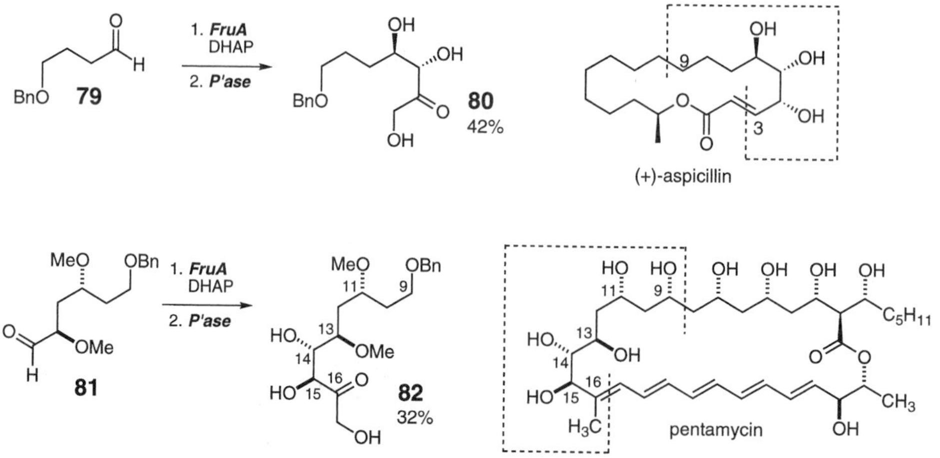

Figure 27 Stereoselective generation of synthetic precursors to the lichen macrolactone (+)-aspicillin [143] and the macrolide antibiotic pentamycin [144] using FruA catalysis.

C9 fragment of the lichen marcolactone (+)-aspicillin [143]. A FruA-mediated stereoselective addition also served as the key step in the synthesis of the "noncarbohydrate," skipped polyol C9–C16 fragment **82** of the macrolide antibiotic pentamycin [144,145].

V. 2-DEOXY-D-RIBOSE 5-PHOSPHATE ALDOLASE

The 2-deoxy-D-ribose 5-phosphate aldolase (RibA; EC 4.1.2.4) is a class I enzyme that catalyzes in vivo the addition of acetaldehyde (**85**) to D-glyceraldehyde 3-phosphate (**15**). Hence, it is a unique aldolase in that it uses two aldehydic substrates both as the aldol donor and acceptor components. RibA from *E. coli* has been cloned [146] and overexpressed. The equilibrium constant for synthesis of 2×10^{-4} M does not strongly favor synthesis [43]. Interestingly, the enzyme's relaxed acceptor specificity allows for substitution of both cosubstrates, albeit at strongly reduced ($<1\%$ of v_{max}) rates: propionaldehyde **77**, acetone **83**, or fluoroacetone **84** (Fig. 28) can replace **85** as the donor [147,148], and a number of aldehydes up to a chain length of four nonhydrogen atoms are tolerated as acceptors. However, reactions that lead to thermodynamically unfavorable products may proceed nonstereospecifically at the reaction center [149].

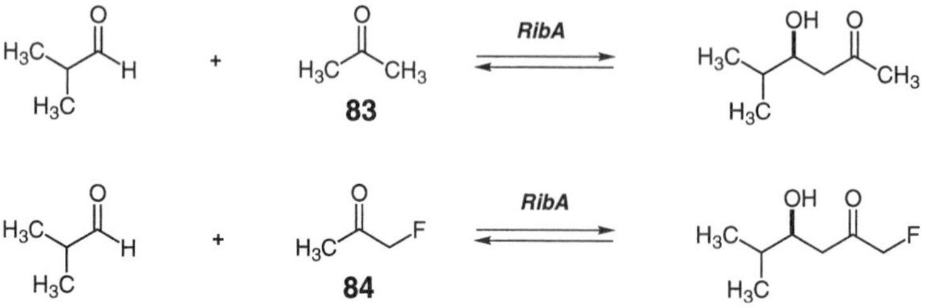

Figure 28 Aldol reactions catalyzed by RibA with nonnatural donors [147].

Table 6 Substrate Tolerance of Deoxy-D-ribose 5-Phosphate Aldolase

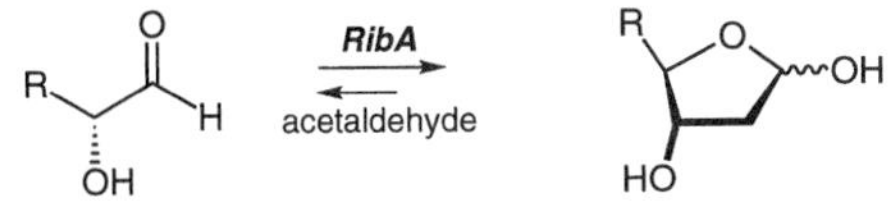

R	Yield (%)	Rel. rate (%)	Ref.
$CH_2OPO_3^{2-}$	78	100	147
CH_2OH	65[a]	0.4	147
CH_3	32	0.4	147
CH_2F	33	0.4	147
CH_2Cl	37	0.3	147
CH_2Br	30	—	151
CH_2SH	33[a]	—	151
CH_2N_3	76	0.3	147
C_2H_5	18	0.3	147
$CH=CH_2$	12	—	151
$CHOH-CH_2OH$	62[a]	0.3	147
CHN_3-CH_2OH	46	—	151
$CHOH-CH_3$	51[a]	—	151
$CHOH-CH_2-C_6H_5$	46[a]	—	151
$CH_2SCH_2-CHOH-CH_2OH$	27	—	151

[a]Equilibrates to a more stable pyranose isomer.

2-Hydroxy aldehydes are relatively good acceptors (Table 6), and the D isomers are preferred over the L isomers [147]. Starting from azido aldehydes (Fig. 29), aza sugars that are less densely functionalized (e.g., **86**) have been prepared by sequential aldolization-hydrogenation [147]. It is noteworthy that from the reaction with **77** as the donor only a single diastereomeric product (e.g., **87**) results, indicative of a stereospecific deprotonation of the donor.

The initial product from aldolization of **85** to itself can serve again as a suitable acceptor for sequential addition of a second equivalent of **85** (Fig. 30) to give (*3R,5R*)-2,4,6-trideoxyhexose **88** [150,151]. When the first acceptor is a substituted acetaldehyde, related aldol products from twofold addition of **85** could be prepared. Formation of higher

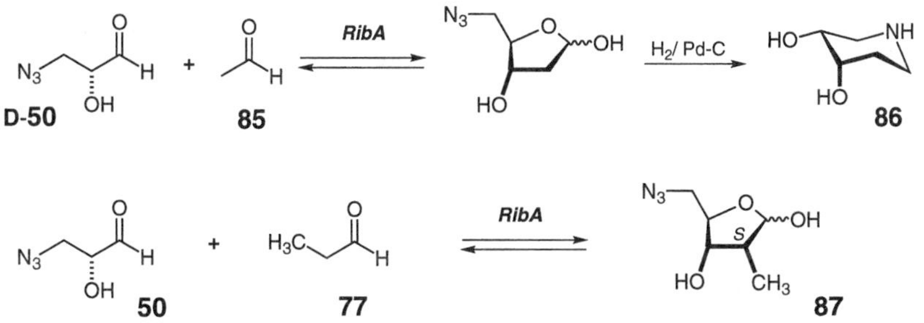

Figure 29 Aza sugar precursors prepared by RibA catalysis [147].

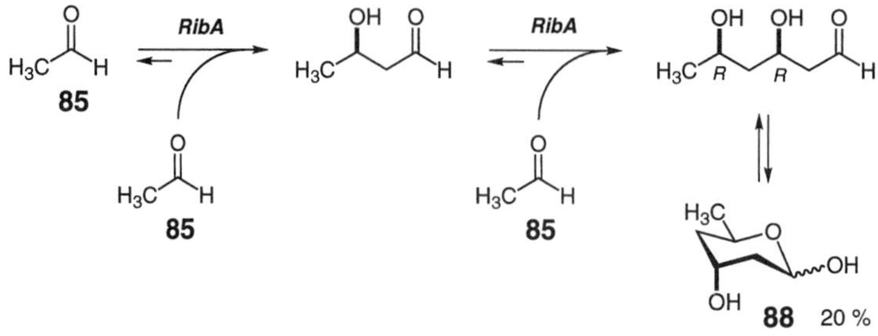

Figure 30 RibA-catalyzed sequential aldol addition [150].

order adducts is effectively precluded by rearrangement into stable cyclic hemiacetals (e.g., **88**), thus masking of the requisite free aldehyde forms.

VI. THREONINE ALDOLASE

Metabolism of β-hydroxy-α-amino acids involves pyridoxal phosphate-dependent enzymes that catalyze a reversible cleavage to aldehydes (Fig. 31) and glycine (**89**). The distinction between L-threonine aldolase (ThrA; EC 4.1.2.5), L-*allo*-threonine aldolase (EC 4.1.2.6), or serine hydroxymethyltransferase (SHMT; EC 2.1.2.1) has often been rather vague since many catalysts display only poor capacity for *erythro/threo* (i.e., **91/90**) discrimination [22]. Many enzymes display a broad substrate tolerance for the aldehyde acceptor, notably including variously substituted aliphatic as well as aromatic aldehydes (Fig. 31); however, α,β-unsaturated aldehydes are not accepted.

Aryl analogs of threonine **94** (Fig. 32) are of interest for their occurrence as constituents of a number of antibiotics (e.g., vancomycin). Because of often unfavorable equilibrium constants [43], synthetic reactions usually lead to stereoisomer mixtures due to thermodynamic control [152,153]. Resolution of diastereomeric material (**93/94**) by retroaldolization under kinetic control is more successful, as has been demonstrated by the preparation of enantiopure *p*-substituted (2R,3S)-phenylserines **94** [154] and (2R,3R)-3-hetarylserines [155] using a threonine aldolase from *Streptomyces amakusaensis* that is selective for aromatic *threo* substrates.

Rabbit liver SHMT, overexpressed in *E. coli* for industrial applications, has been employed for the highly stereoselective synthesis of L-*erythro*-2-amino-3-hydroxy-1,6-hexane dicarboxylic acid from succinic semialdehyde as a potential precursor to carbocyclic β-lactams and nucleosides [156]. In a parallel approach, the aldolase from *Candida humicola* has been applied to the synthesis of compound **95** (Fig. 33) as an intermediate en route to the immunosuppressive lipid mycestericin D [157]. Benzyl protection was found to increase the stereoselectivity, and low conversion secured a kinetic control in favor of the *erythro* product. For conversion of benzyloxyacetaldehyde, a cloned

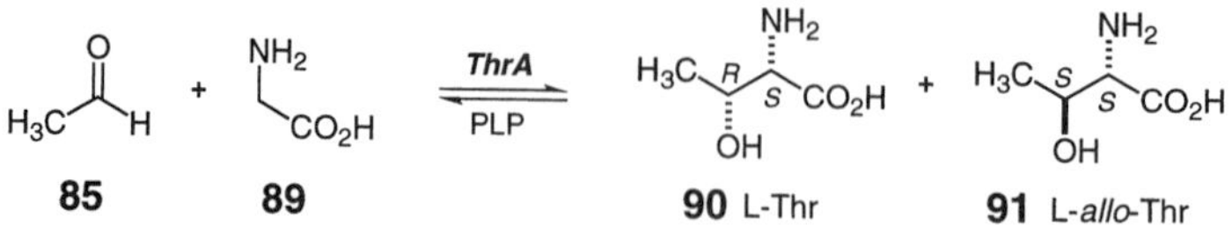

Figure 31 Aldol reactions catalyzed in vivo by L-specific threonine aldolases.

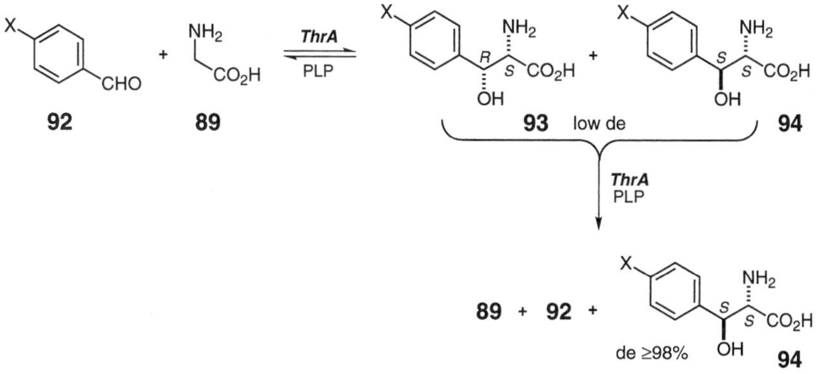

Figure 32 Preparation of enantiopure phenylserines by resolution of diastereomer mixtures using retroaldolization under kinetic control [154].

ThrA from *E. coli* showed higher diastereoselectivity than other enzymes under conditions of kinetic control, which was required in the synthesis of novel sialyl Lewis^x mimetics [158].

VII. SUMMARY

A growing number of C—C bond–forming aldolases have now been established to be of practical value as catalyst for the convergent asymmetric synthesis of interesting poly-functionalized products. The types of conversion illustrated and discussed above form a useful knowledge base to indicate the potential scope of the technique and its limitations to aid in the development of future applications. Clearly, biocatalytic C—C bond formation is particularly useful for the asymmetric synthesis of complex multifunctional molecules with important biological activity, owing to high asymmetric induction, a high level of reaction specificity, and liberation from a need for protection group schemes. It is evident that the technology is now well accepted in the chemical community and advancing a state of maturation, not the least by first processes for industrial commercialization.

With progress fueled by whole-genome sequencing and by an almost explosive growth in structural knowledge of enzymes, detailed understanding of substrate recognition and catalytic function is advancing at a rapid pace, which has set the stage for rational protein engineering toward improved catalytic or physical properties—including novel

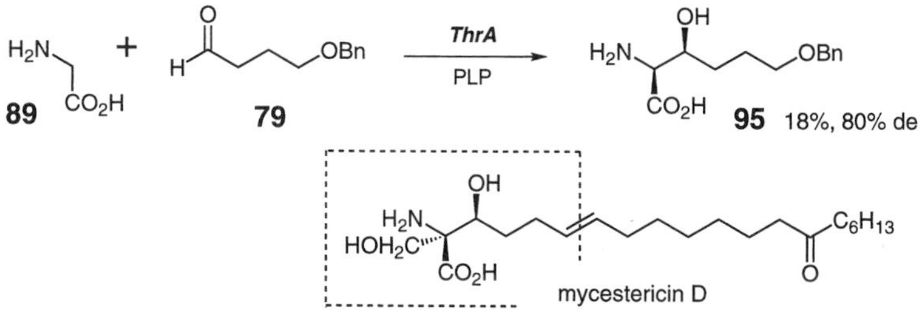

Figure 33 Synthesis of a potential precursor to the immunosuppressive lipid mycestericin D by application of ThrA catalysis under kinetic control [157].

catalytic functions that will lift the current restrictions from narrow donor specificities. Such fascinating possibilities not only open new playgrounds for creative minds but also assist each practicing scientist with effective tools for tackling the future challenges in asymmetric synthesis.

REFERENCES

1. B Testa, WF Trager. Chirality 2:129, 1990.
2. IW Wainer, DE Drayer, eds. Drug Stereochemistry. New York: Marcel Dekker, 1988.
3. HG Davies, RH Green, DR Kelly, SM Roberts. Biotransformations in Preparative Organic Chemistry. London: Academic Press. 1989.
4. W Gerhartz, ed. Enzymes in Industry: Production and Applications. Weinheim: VCH, 1991.
5. K Drauz, H Waldmann, eds. Enzyme Catalysis in Organic Synthesis. A Comprehensive Handbook. Weinheim: VCH, 1995.
6. C-H Wong, GM Whitesides. Enzymes in Synthetic Organic Chemistry. Oxford: Pergamon Press, 1994.
7. K Faber. Biotransformations in Organic Chemistry, 2nd ed. Berlin: Springer-Verlag, 1995.
8. J Halgas. Biocatalysts in Organic Synthesis. Amsterdam: Elsevier, 1992.
9. L Poppe, L Novák. Selective Biocatalysis. A Synthetic Approach. Weinheim: VCH, 1992.
10. H Waldmann, D Sebastian. Chem Rev 94:911–937, 1994.
11. D Rozzell, F Wagner, eds. Biocatalytic Production of Amino Acids and Derivatives. München: Hanser Publishers, 1992.
12. V Schellenberger, HD Jakubke. Angew Chem Int Ed Engl 30:1437–1449, 1991.
13. EJ Toone, ES Simon, MD Bednarski, GM Whitesides. Tetrahedron 45:5365–5422, 1989.
14. S David, C Augé, C Gautheron. Adv Carbohydr Chem Biochem 49:175–237, 1992.
15. S Takayama, GJ McGarvey, C-H Wong. Annu Rev Microbiol 51:285–310, 1997.
16. HJM Gijsen, L Qiao, W Fitz, C-H Wong. Chem Rev 96:443–473, 1996.
17. GM Watt, PA Lowden, SL Flitsch. Curr Opin Struct Biol 7:652–660, 1997.
18. A Varki. Proc Natl Acad Sci USA 91:7390–7397, 1994.
19. TM Carlos, JM Harlan. Blood 84:2068–2101, 1994.
20. RO Hynes, AD Lander. Cell 68:303–322, 1992.
21. CD Wegner, ed. Adhesion Molecules. London: Academic Press, 1994.
22. W-D Fessner, C Walter. Top Curr Chem 184:97–194, 1997.
23. R Kluger. Chem Rev 90:1151–1169, 1990.
24. DF Henderson, EJ Toone. Aldolases. In: BM Pinto, ed. Comprehensive Natural Product Chemistry, Vol. 3. Oxford: Pergamon Press, 1998, pp 367–440.
25. D Schomburg, M Salzmann. Enzyme Handbook. Berlin: Springer-Verlag, 1990–1995.
26. W Tischer, F Wedekind. Top Curr Chem 200:95–126, 1999.
27. W-D Fessner. In: S Servi, ed. A Building Block Strategy for Asymmetric Synthesis: The DHAP Aldolases. Dordrecht: Kluwer Academic, 1992, pp 43–55.
28. BL Horecker, O Tsolas, CY Lai. In: PD Boyer, ed. Aldolases. New York: Academic Press, 1972, pp 213–258.
29. W-D Fessner, A Schneider, H Held, G Sinerius, C Walter, M Hixon, JV Schloss. Angew Chem Int Ed Engl 35:2219–2221, 1996.
30. J Sygusch, D Beaudry, M Allaire. Proc Natl Acad Sci USA 84:7846–7850, 1987.
31. SJ Gamblin, GJ Davies, JM Grimes, RM Jackson, JA Littlechild, HC Watson. J Mol Biol 219:573–576, 1991.
32. SJ Gamblin, B Cooper, JR Millar, GJ Davies, JA Littlechild, HC Watson. FEBS Lett 262: 282–286, 1990.
33. G Hester, O Brenner-Holzach, FA Rossi, M Struck-Donatz, KH Winterhalter, JDG Smit, K Piontek. FEBS Lett 292:237–242, 1991.
34. H Kim, U Certa, H Dobeli, P Jakob, WGJ Hol. Biochemistry 37:4388–4396, 1998.

35. N Blom, J Sygusch. Nature Struct Biol 4:36–39, 1997.
36. MC Lawrence, JARG Barbosa, BJ Smith, NE Hall, PA Pilling, HC Ooi, SM Marcuccio. J Mol Biol 266:381–399, 1997.
37. J Jia, U Schörken, Y Lindqvist, GA Sprenger, G Schneider. Protein Sci 6:119–124, 1997.
38. J Jia, WJ Huang, U Schörken, H Sahm, GA Sprenger, Y Lindqvist, G Schneider. Structure 4:715–724, 1996.
39. MK Dreyer, GE Schulz. J Mol Biol 231:549–553, 1993.
40. MK Dreyer, GE Schulz. J Mol Biol 259:458–466, 1996.
41. SJ Cooper, GA Leonard, SM McSweeney, AW Thompson, JH Naismith, S Qamar, A Plater, A Berry, WN Hunter. Structure 4:1303–1315, 1996.
42. NS Blom, S Tetreault, R Coulombe, J Sygusch. Nature Struct Biol 3:856–862, 1996.
43. RN Goldberg, YB Tewari. J Phys Chem Ref Data 24:1669–1698, 1995.
44. Y Ohta, M Shimosaka, K Murata, Y Tsukada, A Kimura. Appl Microbiol Biotechnol 24:386–391, 1986.
45. K Aisaka, S Tamura, Y Arai, T Uwajima. Biotechnol Lett 9:633–637, 1987.
46. T Izard, MC Lawrence, RL Malby, GG Lilley, PM Colman. Structure 2:361–369, 1994.
47. M-J Kim, WJ Hennen, HM Sweers, C-H Wong. J Am Chem Soc 110:6481–6486, 1988.
48. C Augé, S David C Gautheron, A Malleron, B Cavayé. New J Chem 12:733–744, 1988.
49. ES Simon, MD Bednarski, GM Whitesides. J Am Chem Soc 110:7159–7163, 1988.
50. MJ Kim, WJ Hennen, HM Sweers, C-H Wong. J Am Chem Soc 110:6481–6486, 1988.
51. U Kragl, D Gygax, O Ghisalba, C Wandrey. Angew Chem Int Ed Engl 30:827–828, 1991.
52. T Sugai, A Kuboki, S Hiramatsu, H Okazaki, H Ohta. Bull Chem Soc Jpn 68:3581–3589, 1995.
53. S Blayer, JM Woodley, MD Lilly, MJ Dawson. Biotechnol Prog 12:758–763, 1996.
54. M Mahmoudian, D Noble, CS Drake, RF Middleton, DS Montgomery, JE Piercey, D Ramlakhan, M Todd, MJ Dawson. Enzyme Microb Technol 20:393–400, 1997.
55. I Maru, J Ohnishi, Y Ohta, Y Tsukada. Carbohydr Res 306:575–578, 1998.
56. CH Lin, T Sugai, RL Halcomb, Y Ichikawa, C-H Wong. J Am Chem Soc 114:10138–10145, 1992.
57. Y Ichikawa, JLC Liu, GJ Shen, C-H Wong. J Am Chem Soc 113:6300–6302, 1991.
58. A Kuboki, H Okazaki,T Sugai, H Ohta. Tetrahedron 53:2387–2400, 1997.
59. WY Wu, B Jin, DCM Kong, M Vonitzstein. Carbohydr Res 300:171–174, 1997.
60. CC Lin, CH Lin, CH Wong. Tetrahedron Lett 38:2649–2652, 1997.
61. M Murakami, K Ikeda, K Achiwa. Carbohydr Res 280:101–110, 1996.
62. DCM Kong, M Vonitzstein. Carbohydr Res 305:323–329, 1997.
63. JLC Liu, GJ Shen, Y Ichikawa, JF Rutan, G Zapata, WF Vann, C-H Wong. J Am Chem Soc 114:3901–3910, 1992.
64. C Augé, C Gautheron, S David, A Malleron, B Cavayé, B Bouxom. Tetrahedron 46;201–214, 1990.
65. MA Sparks, KW Williams, C Lukacs, A Schrell, G Priebe, A Spaltenstein, GM Whitesides. Tetrahedron 49:1–12, 1993.
66. PZ Zhou, HM Salleh, JF Honek. J Org Chem 58:264–266, 1993.
67. W Fitz, C-H Wong. J Org Chem 59:8279–8280, 1994.
68. U Kragl, A Gödde, C Wandrey, N Lubin, C Augé. J Chem Soc Perkin Trans I 119–124, 1994.
69. W Fitz JR Schwark, CH Wong. J Org Chem 60:3663–3670, 1995.
70. C Augé, B Bouxom, B Cavayé, C Gautheron. Tetrahedron Lett 30:2217–2220, 1989.
71. S David, A Malleron, B Cavaye. New J Chem 16:751–755, 1992.
72. T Sugai, GJ Shen, Y Ichikawa, C-H Wong. J Am Chem Soc 115:413–421, 1993.
73. C Augé, C Gautheron. J Chem Soc, Chem Commun 859–860, 1987.
74. C Salagnad, A Godde, B Ernst, U Kragl. Biotechnol Prog 13:810–813, 1997.
75. S Ladisch, A Hasegawa, RX Li, M Kiso. Biochemistry 34:1197–1202, 1995.
76. A Malleron, S David. New J Chem 20:153–159, 1996.

77. K Koppert, R Brossmer. Tetrahedron Lett 33:8031–8034, 1992.
78. BR Knappmann, MR Kula. Appl Microbiol Biotechnol 33:324–329, 1990.
79. MA Ghalambor, EC Heath. J Biol Chem 241:3222–3227, 1966.
80. DP Henderson, IC Cotterill, MC Shelton, EJ Toone. J Org Chem 63:906–907, 1998.
81. SE Egan, R Fliege, S Tong, A Shibata, RE Wolf, T Conway. J Bacteriol 174:4638–4646, 1992.
82. AT Carter, MB Pearson, JR Dickinson, WE Lancashire. Gene 130:155–156, 1993.
83. T Conway, KC Yi, SE Egan, REJ Wolf, DL Rowley. J Bacteriol 173:5247–5248, 1991.
84. T Conway, R Fliege, D Jones-Kilpatrick, J Liu, WO Barnell, SE Egan. Mol Microbiol 5:2901–2911, 1991.
85. N Hugouvieux-Cotte-Pattat, J Robert-Baudouy. Mol Microbiol 11:67–75, 1994.
86. RK Scopes. Anal Biochem 136:525–529, 1984.
87. MC Shelton, EJ Toone. Tetrahedron: Asymmetry 6:207–211, 1995.
88. IC Cotterill, MC Shelton, DEW Machemer, DP Henderson, EJ Toone. J Chem Soc Perkin Trans I 1335–1341, 1998.
89. MC Shelton, IC Cotterill, STA Novak, RM Poonawala, S Sudarshan, EJ Toone. J Am Chem Soc 118:2117–2125, 1996.
90. DP Henderson, MC Shelton, IC Cotterill, EJ Toone. J Org Chem 62:7910–7911, 1997.
91. HP Brockamp, MR Kula. Appl Microbiol Biotechnol 34:287–291, 1990.
92. CH von der Osten, AJ Sinskey, CF Barbas, RL Pederson, YF Wang, C-H Wong. J Am Chem Soc 111:3924–3927, 1989.
93. MD Bednarski, ES Simon, N Bischofberger, W-D Fessner, MJ Kim, W Lees, T Saito, H Waldmann, GM Whitesides. J Am Chem Soc 111:627–635, 1989.
94. DL Bissett, RL Anderson. J Biol Chem 255:8750–8755, 1980.
95. W-D Fessner, O Eyrisch. Angew Chem Int Ed Engl 31:56–58, 1992.
96. O Eyrisch, G Sinerius, W-D Fessner. Carbohydr Res 238:287–306, 1993.
97. W-D Fessner, G Sinerius, A Schneider, M Dreyer, GE Schulz, J Badia, J Aguilar. Angew Chem Int Ed Engl 30:555–558, 1991.
98. A Ozaki, EJ Toone, CH von der Osten, AJ Sinskey, GM Whitesides. J Am Chem Soc 112:4970–4971, 1990.
99. E Garcia-Junceda, GJ Shen, T Sugai, C-H Wong. Bioorg Med Chem 3:945–953, 1995.
100. MK Dreyer, GE Schulz. Acta Cryst D 52:1082–1091, 1996.
101. W-D Fessner, C Walter. Angew Chem Int Ed Engl 31:614–616, 1992.
102. W-D Fessner, G Sinerius. Angew Chem Int Ed Engl 33:209–212, 1994.
103. HL Arth, G Sinerius, WD Fessner. Liebigs Ann 2037–2042, 1995.
104. D Stribling. Biochem J 141:725–728, 1974.
105. W-D Fessner, G Sinerius. Angew Chem Int Ed Engl 33:209–212, 1994.
106. H-L Arth, W-D Fessner. Carbohydr Res 305:313–321, 1998.
107. R Duncan, DG Drueckhammer. J Org Chem 61:438–439, 1996.
108. W-D Fessner. Curr Opin Chem Biol 2:85–97, 1998.
109. W-D Fessner, G Sinerius. Bioorg Med Chem 2:639–645, 1994.
110. W-D Fessner, A Schneider, O Eyrisch, G Sinerius, J Badia. Tetrahedron: Asymmetry 4:1183–1192, 1993.
111. W-D Fessner, J Badia, O Eyrisch, A Schneider, G Sinerius. Tetrahedron Lett 33:5231–5234, 1992.
112. WJ Lees, GM Whitesides. J Org Chem 58:1887–1894, 1993.
113. JR Durrwachter, C-H Wong. J Org Chem 53:4175–4181, 1988.
114. KKC Liu, RL Pederson, C-H Wong. J Chem Soc Perkin Trans I 2669–2673, 1991.
115. M Petersen, MT Zannetti, W-D Fessner. Top Curr Chem 186:87–117, 1997.
116. G Jaeschke, C Gosse, M Petersen, W-D Fessner, unpublished.
117. O Eyrisch, W-D Fessner. Angew Chem Int Ed Engl 34:1639–1641, 1995.
118. GC Look, CH Fotsch, C-H Wong. Acc Chem Res 26:182–190, 1993.
119. RL Pederson, M-J Kim, C-H Wong. Tetrahedron Lett 29:4645–4648, 1988.

120. T Ziegler, A Straub, F Effenberger. Angew Chem 100:737–738, 1988.
121. A Straub, F Effenberger, P Fischer. J Org Chem 55:3926–3932, 1990.
122. KKC Liu, T Kajimoto, L Chen, Z Zhong, Y Ichikawa, C-H Wong. J Org Chem 56:6280–6289, 1991.
123. PZ Zhou, HM Salleh, PCM Chan, G Lajoie, JF Honek, PTC Nambiar, OP Ward. Carbohydr Res 239:155–166, 1993.
124. WJ Lees, GM Whitesides. Bioorg Chem 20:173–179, 1992.
125. RR Hung, JA Straub, GM Whitesides. J Org Chem 56:3849–3855, 1991.
126. F Moris-Varas, XH Qian, C-H Wong. J Am Chem Soc 118:7647–7652, 1996.
127. O Eyrisch. Chemoenzymatische Synthesen seltener und neuer Kohlenhydrate mit diversen Aldolasen, Isomerasen und Oxidoreduktasen. PhD dissertation, University of Freiburg, 1994.
128. F Effenberger, A Straub, V Null. Liebigs Ann Chem 1297–1301, 1992.
129. WC Chou, LH Chen, JM Fang, C-H Wong. J Am Chem Soc 116:6191–6194, 1994.
130. JR Durrwachter, DG Drueckhammer, K Nozaki, HM Sweers, C-H Wong. J Am Chem Soc 108:7812–7818, 1986.
131. CW Borysenko, A Spaltenstein, JA Straub, GM Whitesides. J Am Chem Soc 111:9275–9276, 1989.
132. O Eyrisch, M Keller, W-D Fessner. Tetrahedron Lett 35:9013–9016, 1994.
133. MD Bednarski, HM Waldmann, GM Whitesides. Tetrahedron Lett 27:5807–5810, 1986.
134. KKC Liu, C-H Wong. J Org Chem 57:4789–4791, 1992.
135. NJ Turner, GM Whitesides. J Am Chem Soc 111:624–627, 1989.
136. W Schmid, GM Whitesides. J Am Chem Soc 112:9670–9671, 1990.
137. HJM Gijsen, CH Wong. Tetrahedron Lett 36:7057–7060, 1995.
138. WC Chou, C Fotsch, CH Wong. J Org Chem 60:2916–2917, 1995.
139. BP Maliakel, W Schmid. J Carbohydr Chem 12:415–424, 1993.
140. F Nicotra, L Panza, G Russo, A Verani. Tetrahedron: Asymmetry 4:1203–1204, 1993.
141. M Schultz, H Waldmann, H Kunz, W Vogt. Liebigs Ann Chem 1019–1024, 1990.
142. DC Myles, PJI Andrulis, GM Whitesides. Tetrahedron Lett 32:4835–4838, 1991.
143. R Chenevert, M Lavoie, M Dasser. Can J Chem 75:68–73, 1997.
144. K Matsumoto, M Shimagaki, T Nakata, T Oishi. Tetrahedron Lett 34:4935–4938, 1993.
145. M Shimagaki, H Muneshima, M Kubota, T Oishi. Chem Pharm Bull 41:282–286, 1993.
146. P Valentin-Hansen, F Boetius, K Hammer-Jespersen, I Svendsen. Eur J Biochem 125:561–566, 1982.
147. L Chen, DP Dumas, C-H Wong. J Am Chem Soc 114:741–748, 1992.
148. CF Barbas, YF Wang, C-H Wong. J Am Chem Soc 112:2013–2014, 1990.
149. HJM Gijsen, C-H Wong. J Am Chem Soc 117:2947–2948, 1995.
150. HJM Gijsen, CH Wong. J Am Chem Soc 117:7585–7591, 1995.
151. C-H Wong, E Garcia-Junceda, LR Chen, O Blanco, HJM Gijsen, DH Steensma. J Am Chem Soc 117:3333–3339, 1995.
152. VP Vassilev, T Uchiyama, T Kajimoto, C-H Wong. Tetrahedron Lett 36:4081–4084, 1995.
153. T Kimura, VP Vassilev, GJ Shen, CH Wong. J Am Chem Soc 119:11734–11742, 1997.
154. RB Herbert, B Wilkinson, GJ Ellames. Can J Chem 72:114–117, 1994.
155. M Bycroft RB Herbert, GJ Ellames. J Chem Soc Perkin Trans I:2439–2442, 1996.
156. MJ Miller, SK Richardson. J Mol Catal B Enz 1:161–164, 1996.
157. K Shibata, K Shingu, VP Vassilev, K Nishide, T Fujita, M Node, T Kajimoto, C-H Wong. Tetrahedron Lett 37:2791–2794, 1996.
158. SH Wu, M Shimazaki, CC Lin, L Qiao, WJ Moree, G Weitz-Schmidt, C-H Wong. Angew Chem Int Ed Engl 35:88–90, 1996.

10

Decarboxylases in Stereoselective Catalysis

Owen P. Ward and Mark V. Baev
University of Waterloo, Waterloo, Ontario, Canada

I. INTRODUCTION

When studying phytochemical reductions that would give simple aldehydes, ketones, and diketones, in the presence of sucrose or glucose, Neuberg and co-workers [1] examined the biotransformation of benzaldehyde to benzyl alcohol using fermenting yeast. After 3–5 days no sugar or benzaldehyde remained but the amount of benzyl alcohol produced was not proportional to the amount of substrates used. The byproduct exhibited L rotation, was precipitated from Fehling's solution in the cold, and reduced alkaline copper solution in the cold. All of these properties indicated an optically active, α,β-ketone alcohol: L-phenylacetyl carbinol (*R*-1-hydroxy-1-phenylpropane-2-one) [2,3]. Neuberg called the enzyme catalyzing synthesis of this compound "carboligase."

Further studies gave evidence that the production of α-hydroxy ketones is a side reaction of pyruvate decarboxylase [4–7].

The production of (*R*)-phenylacetyl carbinol by fermenting yeast was one of the first industrial biotransformations and continues to be used as a first step in L-ephedrine synthesis.

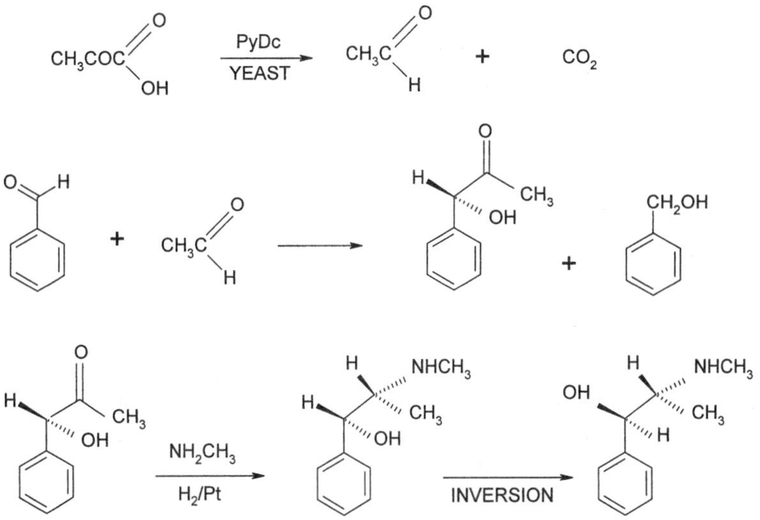

Pyruvate decarboxylase is by far the best characterized enzyme among thiamine pyrophosphate–linked α-keto acid decarboxylases. Properties of this enzyme, its structure, and its catalytic mechanism have been described in a recent review [8]. Two other enzymes belonging to the class of α-keto acid decarboxylases, benzoylformate decarboxylase and phenylpyruvate decarboxylase, are much less studied.

The objectives of this chapter are to give a short description of properties of non-oxidative α-keto acid decarboxylases and their applications in stereoselective bio-transformations.

II. PYRUVATE DECARBOXYLASE

A. Properties of Pyruvate Decarboxylase

Pyruvate decarboxylase (EC 4.1.1.1) has been characterized in different sources including yeast, bacteria, wheat, maize, sweet potato, and plants [8]. This is the first enzyme of the branch of the glycolytic pathway, which under anaerobic conditions leads to nonoxidative decarboxylation of pyruvate to reduced end-products [9]. In the case of yeast, pyruvate decarboxylase together with alcohol dehydrogenase (EC 1.1.1.1) converts pyruvate to ethanol.

All of the pyruvate decarboxylase sequences consist of subunits of about 562–610 amino acid residues, but depending on the source, the enzyme has different structures. Pyruvate decarboxylase from baker's yeast has been reported to be a tetramer, having similar subunits of about 60 kDa, whereas pyruvate decarboxylase from brewer's yeast is an $\alpha_2\beta_2$ tetramer composed of different subunits having molecular weights 59 and 61 kDa, respectively [10,11].

The pyruvate decarboxylase holoenzyme contains two to four molecules of thiamine pyrophosphate (TPP) and Mg^{2+} as cofactors. Under physiological conditions (pH $\sim$ 6.0), TPP and Mg^{2+} are bound in quasi-irreversible manner, whereas at pH values above 7.0, they are released leading to the formation of the apoenzyme [10,12,13]. The tetrameric apoenzyme subsequently dissociates into two dimers [14]. Reassociation of the apoenzyme into the holoenzyme takes place at pH values below 7.0 in the presence of TPP and Mg^{2+} [10].

Pyruvate decarboxylase is able to catalyze two different reactions: the nonoxidative decarboxylation of α-keto acids to the corresponding aldehydes [10,15–17] and a "carboxyligase" side reaction leading to the formation of hydroxy ketones [18,19]. An understanding of why the last reaction is catalyzed by pyruvate decarboxylase, the physiological role of which is to decarboxylate pyruvate to acetaldehyde, was revealed by the discovery that pyruvate decarboxylase is homologous with acetolactate synthase [20], the enzyme catalyzing an acyloin condensation in the first step of isoleucine-valine biosynthesis.

B. Decarboxylation Reaction

Although the native substrate for the enzyme is pyruvate, the enzyme is capable of decarboxylation of a number of other substrates such as hydroxy pyruvate, acetaldehyde, aliphatic α-keto acids, and p-substituted α-keto acids [11,21]. It has been reported that pyruvate decarboxylase decarboxylates higher homologues of pyruvate up to 2-oxohexanoate, also at a rate decreasing with chain length. The enzyme is activated by its substrate, pyruvate, but is practically inactive when the substrate concentration approaches zero [22].

All of the pyruvate decarboxylases investigated with respect to their kinetic properties, with the exception of the enzyme from *Zymomonas mobilis*, exhibit sigmoidal v/s plots and a lag phase during product formation. The K_m for pyruvate has been determined as $0.25-1.1$ mM. The enzyme from the yeast demonstrates specific activity in the range of $45-60$ U/mg. The pH optimum is $6.0-6.8$.

Acetaldehyde inhibits the decarboxylation of pyruvate but does not inactivate the enzyme. Activity may be restored by removal of acetaldehyde. The inhibition of decarboxylation of acetaldehyde is not competitive with pyruvate and is substrate-dependent [15]. The inhibitory effect of acetaldehyde on decarboxylation decreases with increasing chain length of the substrate. Other aldehydes also inhibit the decarboxylation of pyruvate. Benzaldehyde and acetaldehyde inhibit the enzyme by 30% and 65%, respectively. Phosphate was shown to be a competitive inhibitor for the pyruvate decarboxylase from *Saccharomyces cerevisiae* [23].

The properties of pyruvate decarboxylase have been extensively studied particularly with respect to its catalytic mechanism involving the enzyme-bound reaction intermediate 2α-hydroxyethylthiamine pyrophosphate (HETPP) [24–26]. The ability of the coenzyme (TPP) and more particularly its thiazolium ring to bind to the carbonyl group and act as an "electronic sink" makes the decarboxylation of an α-keto acid possible. This is because decarboxylation of an α-keto acid requires the buildup of negative charge on the carbonyl atom in the transition state. The reaction mechanism of pyruvate decarboxylase is described by Voet and Voet [27] and is based on isolation and identification of the HETPP intermediate.

C. Condensation Reaction

In 1961 Juni proposed the two-site theory of the mechanism of acyloin formation by pyruvate decarboxylase [15]. This theory was later confirmed by others [18,28]. According to the model, at the first site pyruvate is decarboxylated to an aldehyde–diphosphatamine complex (HETPP) called active acetaldehyde. The active acetaldehyde moiety is then irreversibly transferred to the second site, where reversible dissociation to free aldehyde takes place. The model is based on the observation that pyruvate decarboxylase not only forms free acetaldehyde as the major end-product of decarboxylation of an α-keto acid but also catalyzes formation of C-C bonds via an acyloin reaction in which free aldehyde competes with a proton for bond formation with the α carbanion of HETPP. Thus the addition of a C2 unit equivalent to acetaldehyde by means of HETPP to a carbonyl group results in an (*R*)-hydroxy ketone [29]. For instance, the production of acetoin (methylacetyl carbinol) results when acetaldehyde is allowed to accumulate or is added to the reaction mixture [28]. This phenomenon was confirmed using pyruvate decarboxylase from different sources (wheat germ, yeast, and bacteria) [15,28,30].

It was postulated that acetoin synthesis involves condensation of free acetaldehyde at the first site independently of whether or not the second site is occupied by acetaldehyde [15]. Further studies with pyruvate decarboxylase suggested that pyruvate is strongly and selectively bound at the first binding site and that acceptor aldehydes are bound at the second binding site. However, the second binding site is more flexible and accepts a wide range of substrates [18,31,32]. Acyclic and aromatic unsaturated aldehydes have been used in acyloin-type condensations [19,29,33,34].

Kren and co-workers [19] studied acyloin condensation between active acetaldehyde generated from pyruvate and a range of aromatic aldehydes. The enzyme used was yeast pyruvate decarboxylase purified to homogeneity. All of the acyloins produced were of R configuration and possessed a very high optical purity.

Bringer-Meier and Sahm [28] demonstrated that pyruvate decarboxylase from *Saccharomyces carlsbergensis* efficiently synthesizes phenylacetyl carbinol from pyruvate and benzaldehyde, whereas pyruvate decarboxylase from *Z. mobilis* is unsuitable for the biotransformations due to its low affinity for benzaldehyde and remarkable substrate inhibition.

The K_m value of the enzyme for benzaldehyde was determined as 42–50 mM [28,35].

In 1988, Fuganti and co-workers [33] demonstrated that pyruvate decarboxylase may accept other selected unnatural α-keto acids in addition to its natural substrate, pyruvate. Such an example is the decarboxylation of linear C_3, C_4, and C_5 α-keto acids in the presence of benzaldehyde by baker's yeast.

Shin and Rogers [36] investigated factors influencing phenylacetyl carbinol formation by pyruvate decarboxylase purified from *Candida utilis*. They found a significant inhibitory effect of acetaldehyde on phenylacetyl carbinol production. The inhibition constant for acetaldehyde was established as 20 mM. To overcome this negative influence, it was suggested that the biotransformation reaction be carried out at low temperature (4°C) due to reduced acetaldehyde production. Shifting the pH of the reaction mixture from 6.0 to 7.0 also appeared to be advantageous to phenylacetyl carbinol production as pH 6.0 is optimal for the acetaldehyde production.

Pyruvate decarboxylase demonstrates high resistance to denaturation by ethanol (up to a concentration of 3 M). Additions of 2.0–3.0 M of ethanol to the reaction mixture increased the biotransformation reaction rate by 30–40%. Enzyme inhibition by benzaldehyde was observed at concentrations greater than 180 mM.

While initial rate of phenylacetyl carbinol formation depends on the enzyme activity, final product concentration is not dependent on the enzyme activity to the same extent.

The efficiency of the biotransformation was found to be dependent on the pyruvate benzaldehyde molar ratio. The highest phenylacetyl carbinol concentration of 28.6 g/L was obtained using 200 mM benzaldehyde and a substrate-to-benzaldehyde ratio of 2.0.

D. Condensation Reaction Catalyzed by Yeast Cells

Though use of isolated purified enzymes is advantageous in that undesirable byproduct formation mediated by contaminating enzymes is avoided [37], in many industrial biotransformation processes for greater cost effectiveness the biocatalyst used is in the form of whole cells. For this reason baker's yeast, which is readily available, has attracted substantial attention from organic chemists as a catalyst for biotransformation processes. One of the first commercialized microbial biotransformation processes was baker's yeast–mediated production of (*R*)-phenylacetyl carbinol, where yeast pyruvate decarboxylase catalyzes acyloin formation during metabolism of sugars or pyruvate in the presence of benzaldehyde [38].

1. Production of Phenylacetyl Carbinol by Yeast Cells

Netrval and Vojtisek [39] found that a number of strains belonging to *Saccharomyces*, *Candida*, and *Hansenula* sp. are capable of L-phenylacetyl carbinol production. The highest concentration of phenylacetyl carbinol (6.3 g/L) was produced by a strain of *S. carlsbergensis* (variant "Budvar"). In other studies, typical reported yields of phenylacetyl carbinol produced during yeast fermentations were 4.5–10.4 g/L.

German patent 1,543,691 [40] claimed production of phenylacetyl carbinol in yields of 55–75% (w/w) based on benzaldehyde to give 6–8 g/L of phenylacetyl carbinol after 6–10 h. The medium contained acetaldehyde. Groger and co-workers [41] described production of phenylacetyl carbinol from benzaldehyde plus a 50% (v/v) solution of acetaldehyde/water, added incrementally in the ratio 1:1.5 to a medium containing mineral salts and molasses. They obtained maximum titers of 7–8 g/L phenylacetyl carbinol. Gupta with co-workers [42] obtained a yield of 5.24 g/L phenylacetyl carbinol using *S. cerevisiae* (CBS 1171) after an 8-h fermentation using 100 g/L yeast (wet weight, containing 70–75% moisture). Vojtisek and Netrval [43] used *S. carlsbergensis* (strain "Budvar") to produce phenylacetyl carbinol in a medium containing sodium pyruvate. They obtained 10.4 g/L phenylacetyl carbinol after 8 h when sodium pyruvate was added incrementally to the reaction mixture at hourly intervals. The first and last two additions were at rates of 1.5% (w/v), while the remaining additions were at a rate of 1% (w/v).

Seely et al. [44,45] reported that mutant strains of *S. cerevisiae* and *Candida flareri*, resistant to acetaldehyde and phenylacetyl carbinol, exhibited higher productivity than the parent strains. Nevertheless, final concentrations of phenylacetyl carbinol in the reaction mixture did not exceed 10 g/L.

Tripathi and co-workers [46] evaluated phenylacetyl carbinol biotransformation efficiency of harvested whole yeast cells, grown continuously under glucose-limited conditions at different dilution rates. They found that cells from increasing dilutions showed increasing specific rates of product formation.

Voets and co-workers [47] carried out studies on the effect of aeration on phenylacetyl carbinol production by *S. cerevisiae*. They found that although pyruvate decarboxylase participates in the fermentative catabolism of sugars, yields of phenylacetyl carbinol obtained in agitated but nonaerated batch cultures were approximately 40% lower than in aerated batches.

Negative effects of oxygen limitation on the reaction of phenylacetyl carbinol formation were also found by Agarwal with co-workers [48]. They investigated the effects of cell age, cell concentration, and dissolved oxygen on phenylacetyl carbinol productivity of yeast. The rate (mmol/h) of phenylacetyl carbinol production was found to be higher with high cell densities than when lower cell densities were used, but specific rates (mmol/g DW/h) of phenylacetyl carbinol formation were higher with lower cell densities. This finding was attributed to oxygen limitation under high-cell-density conditions. It was shown that 15- to 18-h-old cells were optimal for obtaining maximum rates of conversion (i.e., cells withdrawn from a fermenter 15–18 h after inoculation with a 5% (v/v) seed culture showed highest rates of phenylacetyl carbinol formation).

Nevertheless, recent studies carried out with *C. utilis* showed that yeast grown under conditions favorable to fermentative metabolism demonstrates much higher subsequent biotransformation productivity [35]. The respiratory quotient (RQ) was selected as the criterion of anaerobiosis. An RQ value of 1.0 for growth on glucose corresponds to respiratory-type metabolism, whereas RQ values higher than 1.0 indicated fermentative metabolism. *Candida utilis* grown in a chemostat at RQ = 1.0 resulted in a low specific rate of phenylacetyl carbinol formation. When the cultivation conditions were changed to increase RQ value, cells demonstrated significantly improved phenylacetyl carbinol production rates and yields. For the subsequent biotransformation process, RQ values of 4–5 were determined to be optimal. The maximal specific rate of phenylacetyl carbinol production was established as 0.10 g/g DW/h at RQ = 4.4.

A major problem encountered in applying enzyme or cellular catalytic systems to the biotransformation of reactive organic chemicals is the possible toxic effect of substrate and/or products on cells and their constituents with cell membranes and intracellular enzymes being the principal target of attack.

Agarwal with co-workers [48] mentioned that the rate of phenylacetyl carbinol formation from glucose and benzaldehyde in a *S. cerevisiae* catalyzing reaction decreased gradually with time. This study, determining the optimum benzaldehyde concentration, showed that beyond a concentration of 16 mM the rate of benzaldehyde to phenylacetyl carbinol conversion decreased. The process was completely inhibited at benzaldehyde concentrations beyond 20 mM. It was suggested that benzaldehyde concentration be maintained within the range 4–16 mM for maximum rates of conversion.

Tripathi et al. [46] cited the optimum concentration of benzaldehyde for phenylacetyl carbinol production as 10 mM. Benzaldehyde concentrations above 1.0 g/L caused cessation of growth of *C. utilis* [35]. A growth inhibition constant for benzaldehyde was established as 0.30 g/L. Benzaldehyde concentrations higher than 3.0 g/L caused a significant decrease in phenylacetyl carbinol productivity. The product inhibition constant for growth was estimated as 4.1 g/L phenylacetyl carbinol.

In the biotransformation experiments carried out by Long and Ward [49] increased benzaldehyde concentration also resulted in decreased yeast cell viability (determined using viable count methods using malt extract agar medium), a cessation in phenylacetyl carbinol production, and a reduced sucrose utilization rate. By recovery of yeast cells which had lost viability and reinoculation into fresh biotransformation media, it was confirmed that when cell viability is diminished the cells are no longer capable of producing significant amounts of phenylacetyl carbinol. Yeast cells could grow, albeit at a reduced rate, in the presence of 0.5 g/L benzaldehyde. At concentrations of 1–2 g/L benzaldehyde, growth was completely inhibited, but yeast viability was maintained, whereas at 3 g/L benzaldehyde viability was reduced. These results emphasized the importance of maintaining benzaldehyde concentrations at about 1 g/L in order to maintain yeast viability and prolong the biotransformation process. As benzaldehyde concentration approached zero, sucrose metabolism increased and it was concluded that supplementary substrate additions were necessary to maintain a source of acetaldehyde in extended biotransformation.

At a benzaldehyde concentration of 1.0 g/L comparison of intracellular and extracellular concentrations of substrates and product of the reaction indicated that the cell membrane maintains a permeability barrier to benzaldehyde, resulting in reduced intracellular benzaldehyde levels and possible protection of cellular proteins. At higher concentrations of benzaldehyde, this barrier appears to be damaged and intracellular enzymes are exposed to substantially higher benzaldehyde levels.

The initial and final yeast dry weight, protein, carbohydrate, and lipid determinations indicated that during the biotransformation process yeast biomass content, especially the cell protein and lipid components, was reduced. Reductions in cellular absorbance values suggested that 30–40% of the yeast cells were lysed during the 10-h biotransformation. However, pyruvate decarboxylase was found to be substantially resistant to denaturation by benzaldehyde at concentrations as high as 7 g/L. The enzyme was fairly resistant to denaturation by transformation products as well.

Thus, modification of cell permeability by benzaldehyde decreases the biotransformation ability of cells by releasing cell magnesium and thiamine pyrophosphate and reducing cofactor concentration, rather than by inactivation of pyruvate decarboxylase activity.

Vojtisek and Netrval [43] studying phenylacetyl carbinol formation from sucrose, acetaldehyde, and benzaldehyde by *S. carlsbergensis* (variant "Budvar") detected the highest initial rate of biotransformation and the highest phenylacetyl carbinol production in the cells with the lowest pyruvate decarboxylase activity. They suggested that the total amount of phenylacetyl carbinol produced would depend primarily on the actual intracellular concentration of pyruvate, i.e., biotransformation stops due to exhaustion of intracellular pyruvate, prior to inactivation of pyruvate decarboxylase. Additions of pyruvate did not influence the rate of phenylacetyl carbinol production but increased significantly the overall production of this compound.

The initial rate of the biotransformation reaction was found to be 2.74 g phenylacetyl carbinol per liter per hour in an optimized fermentation medium, which contained peptone, 6 g/L; sodium citrate, 10.5 g/L; sucrose, 40–60 g/L; yeast, 60 g DW/L; benzaldehyde, 6 g/L (increase in benzaldehyde concentration up to 8 g/L inhibited the acyloin formation almost completely); pH 4.0–5.0 [50]. The pH optimum was 4.5–5.5 when the same reaction was catalyzed by acetone powder of yeast supplemented with cofactors [51].

Therefore, when reaction times are prolonged, benzaldehyde was usually added incrementally in order to increase amount of benzaldehyde transformed and to achieve the final maximum yield [32,39,52].

A fed-batch process used in commercial production of phenylacetyl carbinol consists of two stages:

1. A yeast growth stage under partial fermentation conditions, allowing production of sufficient biomass to accumulate an intracellular pool of pyruvate and induce synthesis of pyruvate decarboxylase
2. A biotransformation stage with feeding of benzaldehyde to maintain its concentrations at a noninhibiting level

Using this strategy, Wang et al. [53] reached a level of phenylacetyl carbinol accumulation up to 22 g/L. In this study the first stage of the process (cultivation of yeast) was carried out at RQ = 4–5 and the second stage (biotransformation) conducted maintaining benzaldehyde concentration of 1–2 g/L.

2. Production of Other L-Acetyl Carbinols by Yeast Cells

Further investigations demonstrated that *S. cerevisiae* is capable of catalyzing acyloin-type condensations of acetaldehyde with other aromatic aldehydes resulting in production of a number of L-acetyl aromatic carbinols [32]:

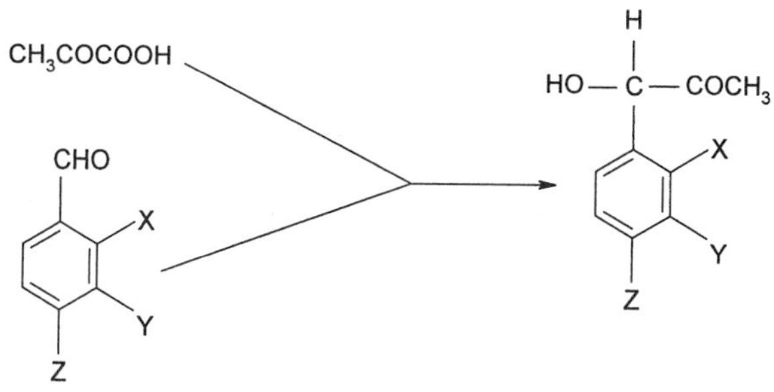

The fermentation medium used in this study contained (g/L) bacteriological peptone, 6; sodium pyruvate 61.7; citric acid adjusted to pH 4.5. Fermentations were carried out in

250-ml Erlenmeyer flasks, containing 150 ml medium, inoculated with 4.5 g fresh baker's yeast. Flasks were incubated at 30°C at 150 rpm. Following an equilibration period of 1 h, the fermentation was initiated by the addition of 2 g/L aromatic aldehyde. Another 8 g/L aromatic aldehydes was added to the reaction mixture in doses of 2 g/L at 1-h intervals. Production of L-acetyl aromatic carbinols was checked after a 6-h fermentation (Table 1).

Aldehydes in the ortho position were consistently poor substrates for carbinol production. Aldehydes with $-CH_3$, $-CF_3$ and -Cl substituents located in the para position produced higher carbinol yields than their meta counterparts. The opposite was the case with $-OCH_3$ substituent. p-Nitrobenzaldehyde, cinnamaldehyde, and salycyl aldehyde did not appear to produce carbinol products. The highest carbinol yield was observed with benzaldehyde substrate (10.1–10.2 mg/ml). The yields obtained for other aromatic aldehydes suggested that this was an efficient system for bioconversion of ring-substituted benzaldehydes to corresponding carbinols.

The initial rate of the biotransformation reaction was determined after 1-h incubations of yeast with different aromatic aldehydes in the presence of sucrose or pyruvate as sources of acetaldehyde (Table 2).

3.   Acyloin Condensation by Immobilized Yeast Cells

Immobilization of yeast cells was shown to reduce the toxic effect of benzaldehyde because of diffusional limitations and gradients of toxic compounds that are established within the immobilizing matrix [35,54,55]. Cells of *S. cerevisiae* immobilized in sodium alginate beads withstand higher concentrations of benzaldehyde (up to 6 g/L) and produce more phenylacetyl carbinol [54]. Phenylacetyl carbinol production by immobilized cells was significantly higher (1.4-, 2.5-, and 7.5-fold) than that by free cells, using initial benzaldehyde concentrations of 2, 4, and 6 g/L, respectively, in fermentation medium.

However, this increased resistance of immobilized yeast cells to high concentrations of benzaldehyde did not result in designing a prolonged semicontinuous biotransformation process with high doses of benzaldehyde added. Attempts to extend the process to three or more cycles resulted in rupturing of the cells. Continuous addition of benzaldehyde gave significantly higher yields of phenylacetyl carbinol. Slow continuous feeding decreased the inhibitory effect of benzaldehyde and much larger doses of benzaldehyde were transformed into phenylacetyl carbinol [55].

The rate and extent of biotransformation of high concentrations of benzaldehyde to phenylacetyl carbinol by immobilized cells of *S. cerevisiae* were stimulated by additions of β-cyclodextrins to fermentation medium [56]. With cyclodextrin additions of 0.5–1.5% and cumulative doses of benzaldehyde of 12 and 14 g/L, the yield of phenylacetyl carbinol obtained was about twofold higher than in the control experiment. Besides, these additions caused faster glucose consumption and benzaldehyde utilization.

However, with immobilization of yeast cells, it is not as easy to regulate metabolism with the same efficiency as can be done with free cells. In shake flasks, cells of *C. utilis*, immobilized in calcium alginate beads, exhibited enhanced resistance to benzaldehyde in comparison with free cells [35]. They also produced higher levels of phenylacetyl carbinol. But in experiments with programmed feeding of benzaldehyde in a controlled bioreactor the final phenylacetyl carbinol production by immobilized cells was 15 g/L, significantly less than the 22 g/L achieved with a free cell fed-batch system. This difference in phenylacetyl carbinol productivity was attributed to the inability to regulate yeast metabolism via RQ in immobilized cells.

Phenylacetyl carbinol production by yeast cells immobilized on carriers other than sodium alginate did not meet with great success [57]. All the six immobilized yeast systems

Table 1 Conversion of Aromatic Aldehyde to L-acetyl Aromatic Carbinol After a 6-h Fermentation

Substrate aromatic aldehyde	Product L-acetyl aromatic carbinol	Yield (mg/ml)	Location of ring substituents[a]		
			X	Y	Z
Benzaldehyde	L-Phenylacetyl carbinol	10.1–10.2	H	H	H
o-Tolualdehyde	L-2-Methylphenylacetyl carbinol	2.0–2.5	CH_3	H	H
m-Tolualdehyde	L-3-Methylphenylacetyl carbinol	5.2–6.2	H	CH_3	H
p-Tolualdehyde	L-4-Methylphenylacetyl carbinol	5.4–6.4	H	H	CH_3
2-Chlorobenzaldehyde	L-2-Chlorophenylacetyl carbinol	0.6–0.7	Cl	H	H
3-Chlorobenzaldehyde	L-3-Chlorophenylacetyl carbinol	2.1–3.2	H	Cl	H
4-Chlorobenzaldehyde	L-4-Chlorophenylacetyl carbinol	6.5–8.0	H	H	Cl
o-Anisaldehyde	L-2-Methoxyphenylacetyl carbinol	0.8–0.9	OCH_3	H	H
m-Anisaldehyde	L-3-Methoxyphenylacetyl carbinol	4.5–5.7	H	OCH_3	H
p-Anisaldehyde	L-4-Methoxyphenylacetyl carbinol	1.2–3.4	H	H	OCH_3
a,a,a-Trifluoro-o-tolualdehyde	L-2-(Trifluoromethyl)-phenylacetyl carbinol	0.2–0.3	CF_3	H	H
a,a,a-Trifluoro-m-tolualdehyde	L-3-(Trifluoromethyl)-phenylacetyl carbinol	0.3–0.4	H	CF_3	H
a,a,a-Trifluoro-p-tolualdehyde	L-4-(Trifluoromethyl)-phenylacetyl carbinol	0.5–0.8	H	H	CF_3

[a]Refer to the structure on page 273.
Source: Ref. 32.

Table 2 Rate of Formation of Aromatic Carbinol from Aromatic Aldehyde and Sucrose or Pyruvate (g/L/h)

	Sucrose	Pyruvate
Benzaldehyde	2.74	3.80
o-Tolualdehyde	0.00	0.10
m-Tolualdehyde	0.45	1.35
p-Tolualdehyde	1.71	1.86
2-Chlorobenzaldehyde	<0.05	<0.10
3-Chlorobenzaldehyde	0.43	0.10
4-Chlorobenzaldehyde	1.46	0.64

Source: Ref. 50.

investigated in that study produced phenylacetyl carbinol in amounts significantly lower than free cells.

4.   Acyloin Condensation in Organic Solvents

Conventionally yeast has been used as a biocatalyst for organic synthesis in aqueous conditions [58,59]. However, the majority of organic chemicals are poorly water-soluble but highly soluble in organic solvents, and the implementation of biotransformation systems in organic solvent-containing media offers potential to increase the solubility of poorly water-soluble substrates [60]. A wide range of reactions may be carried out in such systems [61–64]. Some of the key advantages of biocatalytic reactions carried out in nonconventional media are enhanced solubility of poorly water-soluble substrates; suppression of undesirable side reactions; reduction of substrate and product inhibition; enhanced stability of the biocatalyst; ease of product and catalyst recovery; shift of thermodynamic equilibrium in favor of synthesis; and ability to manipulate enantioselectivity in organic solvents [65].

Two types of biotransformation systems were documented in investigations using yeast enzymes in organic solvent-containing media. Where enzyme reactions have been carried out in organic solvent having a very low water content (less than 0.02%), the system is described as microaqueous [66,67]. Where organic and aqueous phases are present in excess of mutual saturation levels, the reaction medium is described as biphasic [68,69].

Only a small amount of water is required by enzymes to maintain catalytic activity [70]. Whole-cell biocatalysis tends to require more water than isolated enzymes and appears to generally require biphasic systems for biotransformations of poorly water-soluble substrates [71].

Nikolova and Ward [72,73] studied production of phenylacetyl carbinol from benzaldehyde and pyruvate by whole-cell yeast biotransformation in two-phase systems. For the biocatalyst preparation fresh pressed commercial baker's yeast (50 g) was suspended in 50 ml 0.05 M sodium citrate buffer (pH 6.0) and lyophilized. Aliquots of 300 mg of lyophilized cells were mixed with 1 g celite and the mixture was resuspended in 0.05 M sodium citrate buffer (pH 6.0). The suspension was lyophilized again and stored at 4°C. Scanning electron micrographs of the carrier celite and yeast cells lyophilized on celite are given in Fig. 1. Prior to use, organic solvents purchased in anhydrous form were saturated with 0.05 M sodium citrate buffer (pH 6.0). The same buffer was used as an aqueous component of the biphasic systems.

The effect of moisture content on production of phenylacetyl carbinol by cells immobilized on celite was investigated using hexane as organic solvent (Table 3). Maximum biotransformation activity was observed with a moisture level of 10%. The effect of the solvent type on the rate of production of phenylacetyl carbinol was investigated in two-phase systems containing 10% moisture and related to log P. The results are presented in Table 4. The highest biotransformation activities were observed with hexane and hexadecane and the lowest with chloroform and toluene.

Cell surfaces of yeast biocatalyst recovered from biphasic media containing hexane, decane, and toluene after 26 h biotransformation reaction demonstrated no apparent damage (Fig. 2). Meanwhile cell puncturing was observed after shorter biotransformation periods with hydrophilic solvents (having $P < 2.0$), chloroform, ethyl acetate, and butyl acetate (Figs. 3–5). Furthermore, cell-damaging solvents, ethyl acetate, butyl acetate, and chloroform resulted in the extraction of phospholipid from the cell membranes into the

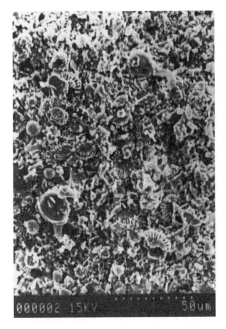 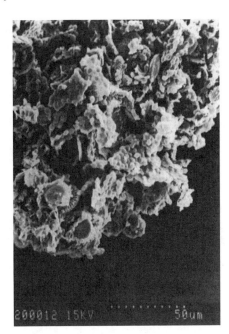

Figure 1 Scanning electron micrograph of the celite carrier (left) and yeast cells adsorbed to celite (right). (From Ref. 73.)

Table 3 Effect of Moisture Content on Phenylacetyl Carbinol Production by Whole Yeast Cells

Moisture (%)	Activity (mmol/g/h)
0.5	3.7×10^{-2}
2.0	3.5×10^{-2}
10.0	5.4×10^{-2}
20.0	4.2×10^{-2}
40.0	4.4×10^{-2}

Source: Ref. 73.

Table 4 Effect of Organic Solvents on Phenylacetyl Carbinol Production by Whole Yeast Cells

Solvent	Log p	Activity (mmol/g/h)
Ethyl acetate	0.68	2.8×10^{-2}
Butyl acetate	1.70	3.10×10^{-2}
Chloroform	2.00	1.00×10^{-2}
Toluene	2.50	0.70×10^{-2}
Hexane	3.50	6.00×10^{-2}
Dodecane	6.60	4.00×10^{-2}
Hexadecane	8.80	5.60×10^{-2}

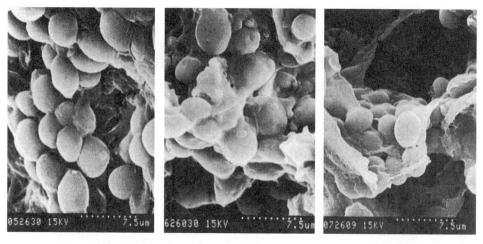

Figure 2 Scanning electron micrographs of yeast cells from toluene (left), hexane (middle), and decane (right). (From Ref. 73.)

biphasic medium, whereas little or no phospholipid was detected from cells incubated in toluene or hexane biphasic media. Phospholipids are recognized as very important components of cell membranes that influence membrane permeability and elasticity [68], and such membrane damage may affect membrane fluidity and cause cytoplasmic shrinkage and cell ultrastructural changes [74]. Biocatalytic activity was also supposed to decrease [75,76]. The results of measurements of total fatty acids and total protein released into the

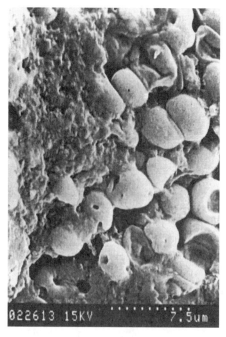

Figure 3 Scanning electron micrographs of biocatalyst recovered from biotransformation media containing ethyl acetate after 6 h (left) and 26 h (right). (From Ref. 73.)

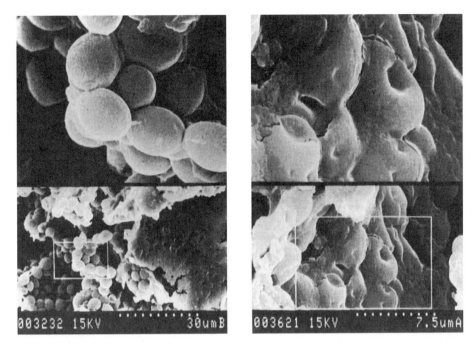

Figure 4 Scanning electron micrographs of biocatalyst recovered from two-phase systems containing butyl acetate after a biotransformation period of 2 h (left) and 6 h (right). (From Ref. 73.)

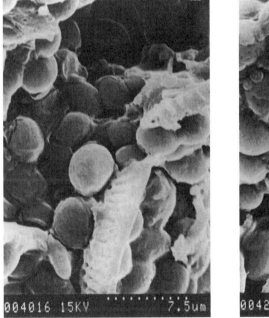
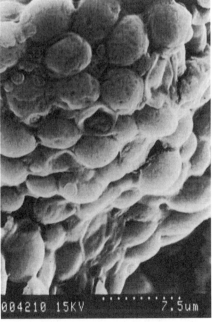

Figure 5 Scanning electron micrographs of biocatalyst recovered from two-phase systems containing chloroform at 0 time (left) and after a 2-h biotransformation (right). (From Ref. 73.)

Table 5   Comparison of Yeast Fatty Acid and Proteins Released into the
Biotransformation Medium with Observed Biocatalytic Activity

Solvent	Total fatty acid (μg/ml)	Total protein (μg/ml)	Activity (mmol/h/mg DW)
Hexane	140	45	60.0
Butyl acetate	164	50	31.0
Ethyl acetate	240	60	28.4
Chloroform	170	115	10.0
Toluene	178	80	7.0

Source: Ref. 72.

medium, done after 26 h of incubation in the biphasic biotransformants system, are consistent with the latter indicators of membrane damage (Table 5). The lowest fatty acid and protein contents were observed in hexane-containing medium, which exhibited the highest biocatalytic activity.

Monophasic systems containing water-miscible organic solvents may have potential for use in phenylacetyl carbinol production. Aqueous media containing water-miscible organic solvents (10–30% w/v of aqueous media) have been employed in combination with acetaldehyde-resistant mutant cells of *S. cerevisiae* and *C. flareri* [44,45]. A number of different aliphatic alcohols have been employed as water-miscible organic solvents including methanol, ethanol, propanol, butanol, long chain polyols (such as polyethylene glycol with MW of 200–800), and dimethyl sulfoxide (DMSO). Under these conditions it was possible to obtain up to 15 g/L of phenylacetyl carbinol in whole-cell biotransformations, while reducing benzyl alcohol formation.

5.   Byproduct Formation

A quantitative conversion of benzaldehyde to phenylacetyl carbinol was never achieved. A number of compounds were reported to be produced as byproducts during phenylacetyl carbinol biosynthesis by *S. cerevisiae*. Voets et al. [47] observed the presence of acetyl benzoyl and *trans*-cinnamaldehyde in the biotransformation mixtures. Smith and Hendlin [51] reported that benzoic acid is also produced during this biotransformation. Besides, production of the phenylacetyl carbinol isomer, 2-hydroxy-1,1-phenylpropane, was detected as a further byproduct of the reaction [77]. All of these byproducts are produced in low amounts, which do not significantly influence efficiency of the production of the main biotransformation product, phenylacetyl carbinol. The only byproduct whose formation significantly influences the efficiency of phenylacetyl carbinol production is benzyl alcohol.

Smith and Hendlin [51] suggested that there are two systems competing for benzaldehyde in the yeast cell. The first is the phenylacetyl carbinol–synthesizing system and the second is alcohol dehydrogenase, catalyzing reduction of benzaldehyde to benzyl alcohol. These researchers found that the increase in phenylacetyl carbinol production is accompanied by a decrease in the formation of benzyl alcohol and vice versa [78].

Participation of the yeast alcohol dehydrogenase in reduction of benzaldehyde and other aromatic aldehydes was confirmed in experiments in which activities of baker's yeast and purified yeast alcohol dehydrogenase toward these compounds were compared [32]. Nevertheless, no correlation between level of alcohol dehydrogenase activity in yeast cells and the capacity of the cells to produce benzyl alcohol from benzaldehyde was observed [79]. A limited supply of alcohol dehydrogenase with reduced NADH was suggested to be the limiting factor in the reaction of benzyl alcohol production. This suggestion is in keeping with the finding that the rate of aromatic alcohol production is much higher when sucrose/aromatic aldehyde–containing media are used rather than that with pyruvate [50]. In sucrose-containing media, NADH is produced in a glycolytic reaction catalyzed by glyceraldehyde-3-phosphate dehydrogenase, whereas it is not generated in the presence of pyruvate. Ose and Hironaka [80] observed that acetone powders of baker's yeast converted benzaldehyde to benzyl alcohol only when NADH was present. Thus, aromatic alcohol production may be depressed by use of pyruvate, thereby preventing generation of a supply of alcohol dehydrogenase with NADH.

A different way to suppress benzyl alcohol production was chosen by Ose and Hironaka [80]. In their experiments the proportions of benzyl alcohol to phenylacetyl carbinol formed in the presence or absence of added acetaldehyde were measured. In the absence of acetaldehyde, about 35% (w/w) of the benzaldehyde was converted to benzyl alcohol. When acetaldehyde was added, they obtained a yield of 70% (w/w) phenylacetyl carbinol and 25% (w/w) benzyl alcohol based on benzaldehyde added. Also, small amounts of acetoin were formed. Beckvarova and co-workers [52] also found that the addition of acetaldehyde reduced the amount of benzyl alcohol produced to 25% (w/w), thereby increasing the yield of phenylacetyl carbinol.

When the phenylacetyl carbinol–generating biotransformations catalyzed by yeast whole cells were carried out in monophasic systems consisting of an aqueous phase and water-miscible organic solvents, a decrease in benzyl alcohol formation was observed [44,45].

Alcohol dehydrogenase was shown to be more susceptible to benzaldehyde denaturation than pyruvate decarboxylase competing for the substrate [49].

III. BENZOYL FORMATE DECARBOXYLASE

A. Properties of Benzoylformate Decarboxylase

The second representative of the class of thiamine pyrophosphate–dependent nonoxidative α-keto acid decarboxylases is phenylglyoxylate decarboxylase (benzoylformate decarboxylase; EC 4.1.1.7). This enzyme participates in the catabolism of aromatic compounds as part of mandelate pathway in different *Pseudomonas* and *Acinetobacter* species, normally converting benzoylformate to benzaldehyde [81–83]. This pathway is induced by mandelic acid [84].

Benzoylformate decarboxylases purified from *Acinetobacter calcoaceticus* [81] and *Pseudomonas putida* [85] were shown to have similar properties. Like pyruvate decarboxylase this enzyme is thiamine pyrophosphate– and magnesium-dependent and has a tetrameric structure. The molecular weight of a subunit is 58,000 for *A. calcoaceticus* [81] and 57,000–57,500 for *P. putida* [86,87], as estimated from sodium dodecyl sulfate–polyamide gel electophoresis (SDS-PAGE).

Though benzoylformate decarboxylase was found to have significant sequence similarities with pyruvate decarboxylase [87], in immunological comparisons between these enzymes no common epitopes were detected by antibodies raised against each of them [82].

Benzoylformate decarboxylase has a higher substrate specificity than pyruvate decarboxylase. Only benzoylformate and para-substituted benzoylformates have been shown to be substrates of the enzyme [85,86,88]. The K_m for benzoylformate has been determined as 1 mM and 0.08 mM in separate studies of Hegeman [85] and Weiss and co-workers [88], respectively. The K_m for thiamine pyrophosphate has been reported as 1 μM [85]. Pyruvate, α-ketobutyrate, and α-ketoadipate are not substrates for the enzyme. The optimum pH for benzoylformate decarboxylase is 5.9 [81].

B. Acyloin Condensation Catalyzed by Benzoylformate Decarboxylase

The ability of benzoylformate decarboxylase to form an acyloin compound when incubated with benzoylformate and acetaldehyde was demonstrated for the first time with *P. putida* [89]. Cells of *P. putida* were transferred a minimum of four times at 24 h intervals in the liquid medium of Hegeman [83,85], containing 3 g/L ammonium mandelate in order to maximize induction of the synthesis of enzymes participating in mandelate catabolism. The cells were then harvested by centrifugation, washed with 50 mM sodium phosphate buffer (pH 6.0), and the pellets were stored frozen until required. Whole cells were used to catalyze acyloin formation or for preparation of cell extracts.

The biotransformation reaction mixture contained (in mM/L) sodium phosphate (pH 6.0), 200; thiamine pyrophosphate, 1.5; magnesium chloride, 2.5. Also included were *P. putida* cells (0.015 g DW/ml) or equivalent as cell extract or purified enzyme.

The product of the acyloin condensation was identified by gas chromatography as 2-hydroxypropiophenone, a tautomer of phenylacetyl carbinol. The normal reaction product of benzoylformate decarboxylase, benzaldehyde, and the acyloin product were both detected after biotransformation. This finding allowed proposal of a possible scheme of the biotransformation process.

The initial rate of benzaldehyde production was higher than the rate of acyloin production, but similar concentrations of both products were observed as the reaction progressed. It was also noted that benzaldehyde concentration declined after reaching a peak.

The process of biotransformation was more efficient with crude cell extracts rather than with whole cells. The maximal concentrations of benzaldehyde and 2-hydroxypropiophenone were 45.8 and 46.3 μmol/ml, respectively, after 2 h of crude cell extract–catalyzed biotransformation.

Accumulation of benzyl alcohol was also observed during this biotransformation reaction catalyzed by either crude cell extracts or whole cells [89,90]. This benzyl alcohol production was suggested to be the result of transformation of benzaldehyde by *P. putida* oxidoreductases. However, in this case benzaldehyde is not the substrate of the desired biotransformation reaction, but a byproduct of this reaction.

The enantiomeric excess of 2-hydroxypropiophenone was found to be 91–92% by ^{1}H NMR spectroscopy of its *O*-acetyl derivative in the presence of a chiral shift reagent. Absolute configuration of the carbinol carbon of 2-hydroxypropiophenone produced by *P. putida* was found to be *S*, and opposite to that of (*R*)-phenylacetyl carbinol prepared by using baker's yeast. The specific optical rotation for the enzymatically prepared materials was in agreement with that reported for the (*S*) enantiomeric form of 2-hydroxypropiophenone [91]. The lack of complete stereoselectivity was considered to be an intrinsic property of the enzyme reaction and is not due to slow nonenzymatic racemization of the 2-hydroxypropiophenone product.

By further manipulations of substrate concentrations, conditions for quantitative conversion of benzoylformate to 2-hydroxypropiophenone and benzaldehyde were established [90]. In the presence of 100 mM of benzoylformate and 1600 mM acetaldehyde, the biotransformation mixture after a 1-h reaction contained 61.76 mM/L of the acyloin compound and 38.2 mM/L of benzaldehyde. Under the optimized biotransformation conditions, acyloin product formation in 1 h reaction catalyzed by crude extract of 1 g DW of *P. putida* amounted to 0.617 g. The stereoselectivity of the biotransformation reaction was not improved in these experiments.

Improvement of the stereoselectivity of the reaction was achieved when instead of benzoylformate decarboxylase from *P. putida*, the enzyme from *A. calcoaceticus* was used [92,93]. *Acinetobacter calcoaceticus* NCIB 8250 was grown in the medium described by Barrowman and co-workers [81,82], containing DL-mandelic acid, 3 g/L on an orbital shaker set at 200 rpm and 30°C for 16–20 h. The culture was serially transferred as a 10% inoculum through three further passages on the same medium under the same conditions. Cells recovered from the fourth passage were used in biotransformation experiments.

The main biomass production phase was observed in the 0- to 8-h incubation period, after which the growth rate declined. Maximum acyloin-forming activity occurred after 12 h.

The optical purity (enantiomeric excess, e.e.) of the biotransformation reaction product, 2-hydroxypropiophenone, was found to be greater than 98%. The absolute configuration of the carbinol carbon of 2-hydroxypropiophenone produced by *A. calcoaceticus* (S) was the same as in the predominant enantiomer produced by *P. putida*.

Optimization of the reaction conditions resulted in formation of 8.4 g/L of product after 2 h of incubation, with 56% conversion of benzoylformate to product. An optimal yield of 6.95 g of product per liter per hour was observed. This yield corresponds to a productivity of 267 mg of acyloin product per g cell DW per hour. These results were obtained in a reaction mixture, containing whole cells, 0.09 g wet weight per ml reaction mixture; benzaldehyde, 15 mg/ml (100 mM); acetaldehyde, 70.4 mg/ml (1600 mM); sodium phosphate, 200 mM; thiamine pyrophosphate, 1.5 mM; and magnesium chloride, 2.5 mM. The reaction optimum for temperature and pH were 30°C and 6.0, respectively. Whole cells of *A. calcoaceticus* were found to be more effective as catalyst than cell free extracts [94].

Using *A. calcoaceticus*, the productivity of the process (mg of acyloin product per g cell DW per hour) was much higher than the optimal productivity observed in the less enantiospecific reaction catalyzed by *P. putida* [90]. This productivity compares very favorably with yields of the acyloin product of yeast pyruvate decarboxylase where optimal product formation in 1-h incubation periods amounted to 0.2 g per g yeast DW [50].

IV. PHENYLPYRUVATE DECARBOXYLASE

The least studied among thiamine pyrophosphate–dependent nonoxidative α-keto acid decarboxylases is phenylpyruvate decarboxylase (EC 4.1.1.43), an enzyme found in *Achromobacter auridice* [95], *Acinetobacter calcoaceticus* [81] and the denitrifying bacterium *Thauera aromatica* [96]. This enzyme participates in catabolic pathways of aromatic compounds by which phenylpyruvate is metabolized to phenylacetate through phenylacetaldehyde:

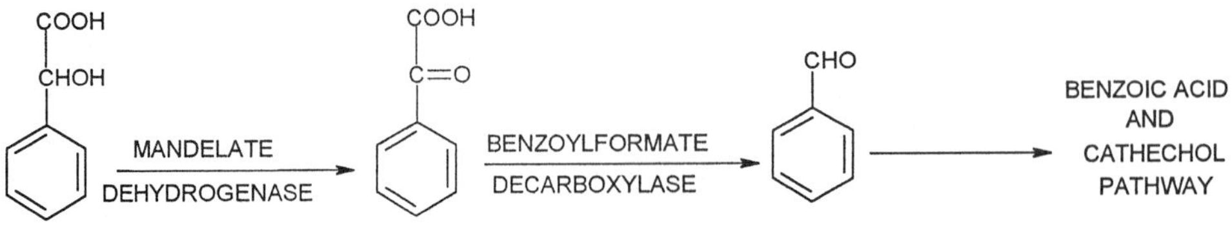

Phenylpyruvate decarboxylase is an inducible enzyme. Its activity occurred when bacteria were grown on phenylalanine or tryptophan [95] or mandelate [81]. In *T. aromatica* grown on phenylacetate, phenylglyoxylate, and 4-hydroxybenzoate, the enzyme activity reached only 4–7% of that in phenylalanine-grown cells [96]. Compounds such as phenylpyruvate, indolepyruvate, and α-keto acids with more than 6 carbon atoms in a straight chain served as substrates for the enzyme. The K_m for phenylpyruvate is 5.1×10^{-5} M. Thiamine pyrophosphate and Mg^{2+} were found to be cofactors. The pH optimum was 7.0. The native molecular weight of phenylpyruvate decarboxylase was estimated to be 235,000. It is a tetrameric enzyme consisting of identical subunits [81].

Like benzoylformate decarboxylase, phenylpyruvate decarboxylase does not demonstrate any immunological similarity to pyruvate decarboxylase [82].

No studies have been published on the ability of this enzyme to form acyloin compounds.

V. CONCLUSIONS

α-Keto acid decarboxylases can condense α-keto acids and aldehydes into optically active α,β-ketone alcohols with release of CO_2. Pyruvate decarboxylase is the best studied enzyme of this class, in terms of both its general biochemical properties and in its ability to catalyze formation of L-phenylacetyl carbinol. The latter product, produced commercially using this biotransformation, is a precursor used in synthesis of ephedrine. The acyloin-forming reaction is catalyzed by yeasts, including *Saccharomyces*, *Candida*, and *Hansenula* species. The enzyme from *Zymomonas mobilis* is unsuitable for the biotransformations because of low affinity for benzaldehyde and substrate inhibition. Pyruvate decarboxylase can decarboxylate other α-keto acids in place of pyruvate and can produce condensation products from substituted aromatic aldehydes. Benzaldehyde at concentrations of greater than 1–2 g/L reduces yeast cell viability and has a deleterious effect on the biotransformation reactions. L-Phenylacetyl carbinol is typically accumulated in concentrations of 5–10 g/L during the biotransformation, but the reaction conditions have been manipulated to produce product concentrations of up to 22 g/L. Yields of substituted acyloins produced from substituted aromatic aldehydes were lower than the corresponding yields from benzaldehyde. While some studies have shown that immobilized cells can withstand higher concentrations of benzaldehyde, the highest final phenylacetyl carbinol concentrations have been achieved in nonimmobilized cell systems. The condensation reaction has also been implemented in aqueous-organic two-phase immiscible and miscible reaction media. Generally biotransformation rates and extents were found to be lower than in aqueous media and significant damage to yeast cells has been observed in these nonconventional media.

Benzyl alcohol is a significant byproduct formed in the whole-cell biotransformation. While yeast alcohol dehydrogenase can convert benzaldehyde and substituted aromatic aldehydes to the corresponding alcohols, experiments with yeast mutants lacking one or more alcohol dehydrogenases showed that these cells still converted benzaldehyde to benzyl alcohol. Benzyl alcohol production can be reduced by using pyruvate in biotransformation media rather than sucrose as its metabolic precursor, by incorporation of acetaldehyde in the biotransformation, and by implementing the reaction in some nonconventional media.

Benzoylformate decarboxylase from *Pseudomonas* and *Acinetobacter* species, also an α-keto acid decarboxylase, has higher substrate specificity than pyruvate decarboxylase. Cells of these species grown in media inducing the mandelate pathway enzymes can convert benzoylformate and acetaldehyde to optically active 2-hydroxypropiophenone. Benzaldehyde is produced in this biotransformation reaction, as it is the normal product of benzoylformate decarboxylase. Some benzyl alcohol is also produced, in this case probably by reduction of benzaldehyde by cell oxidoreductases. In the case of *P. putida* the (S) enantiomer form of 2-hydroxypropiophenone was produced, with an e.e. of 91–92%. The same product produced by *A. calcoaceticus* had an e.e. of 98%. An optimal volumetric production of 2-hydroxypropiophenone of 6.95 g per L per h was reported.

Another α-keto acid decarboxylase, phenylpyruvate decarboxylase, has been characterized but there are no reports on condensation reactions by this enzyme.

This group of enzymes, mutated or engineered to give modified substrate specificities and improved stability, offers great potential to produce families of optically active α,β-ketone alcohols as possible bioactive molecules or their precursors.

REFERENCES

1. C Neuberg, J Hirsh. Biochem Zeitschr 115:282–310, 1921.
2. C Neuberg, L Lieberman. Biochem Zeitschr 121:322–325, 1921.
3. C Neuberg, H Ohle. Biochem Zeitschr 127:327–339, 1922.
4. W Dirscherl. Hoppe-Seyler's Z Physiol Chem 201:47–77, 1931.
5. W Langenbeck, H Wrede, W Schlockermann. Hoppe-Seyler's Z Physiol Chem 227:263, 1934.
6. TP Singer, J Pensky. Arch Biochem Biophys 31:457–459, 1951.
7. TP Singer, J Pensky. Arch Biochem Biophys 9:316–327, 1952.
8. M Pohl. Adv Biochem Eng Biotechnol 58:15–43, 1997.
9. G Hubner, M Atanasova, A Schellenberger. Biomed Biochim Acta 45:823–832, 1986.
10. G Hubner, S Konig, A Schellenberger, MHJ Koch. FEBS Lett 266:17–20, 1990.
11. D Schomburg, N Salzmann. Lyases. Pyruvate Decarboxylase. Enzyme Handbook, Vol. 1. Berlin: Springer-Verlag, 1990, pp 1–5.
12. RFW Hopman. Eur J Biochem 110:311–318, 1980.
13. H Zehender, J Ullrich. FEBS Lett 180:51–54, 1985.
14. AD Gounaris, I Turkenkopf, S Buckwalds, A Young. J Biol Chem 246:1302–1309, 1971.
15. E Juni. J Biol Chem 236:2302–2308, 1961.
16. J Ullrich, JH Wittorf, CJ Gubber. Biochim Biophys Acta 113:595–604, 1966.
17. TE Barman. Pyruvate Dehydrogenase. Enzyme Handbook, Vol. 2. New York: Springer-Verlag, 1969, p 701.
18. DHG Crout, H Dalton, DW Hutchinson, M Miyagoshi. J Chem Soc Perkin Trans, 1329–1334, 1991.
19. V Kren, DHG Crout, H Dalton, DW Hutchinson, W Konig, MM Turner, G Dean, N Thomson. J Chem Soc Chem Commun, 341–343, 1993.
20. JBA Green. FEBS Lett 246:1–5, 1989.
21. H Lehmann, G Fischer, G Hubner, KD Kohnert, A Schellenberger. Eur J Biochem 32:83–87, 1973.

22. G Hubner, A Schellenberger. Biochem Int 13:767–772, 1986.
23. A Boiteux, B Hess. FEBS Lett 9:293–296, 1973.
24. MFM Utter. In: PD Boyer, H Lardy, K Myrbak, eds. The Enzymes, Vol. 5. New York: Academic Press, 1961, p 319.
25. A Schellenberger. Angew Chem 79:1050–1061, 1961.
26. J Ullrich. Ann NY Acad Sci 378:287–305, 1982.
27. D Voet, JG Voet. Biochemistry. New York: John Wiley, 1990, pp 425–460.
28. B Bringer-Meyer, H Sahm. Biocatalysis 1:321–331, 1988.
29. R Csuk, BI Glanzer. Chem Rev 91:49–97, 1991.
30. DHG Crout, J Littlechild, SM Morrey. J Chem Soc Perkin Trans, 105–108, 1986.
31. JV Schloss, LM Ciscanik, DE Van Dyk. Nature 331:360–362, 1988.
32. A Long, P James, OP Ward. Biotech Bioeng 33:657–660, 1989.
33. C Fuganti, P Grasselli, G Poli, S Servi, A Zorella. J Chem Soc Chem Commun 1619–1621, 1988.
34. B Stumpf, K Kieslich. Appl Microbiol Biotechnol 34:598–603, 1991.
35. PL Rogers, HS Shin, B Wang. Adv Biochem Eng Biotechnol 56:33–59, 1997.
36. HS Shin, PL Rogers. Biotech Bioeng 49:52–62, 1996.
37. HGW Leuenberger. In: RH Rehm, G Reed, eds. Biotechnology, Vol. 6a. Weinheim: Verlag Chemie, 1984, pp 5–29.
38. L Kieslich. In: HJ Rehm, G Reed, eds. Biotechnology, Vol. 6a. Weinheim: Verlag Chemie, 1984, pp 1–4.
39. J Netrval, V Vojtisek. Eur J Appl Microbiol Biotechnol 16:35–38, 1982.
40. Ger Pat 1,543,691. Zwickau: Isis-chemie KG, 1966.
41. D Groger, HP Schmander, K Mothes. Zvesti Allg Mikrob 6:275–287, 1966.
42. KG Gupta, J Singh, G Sahni, S Dhawan. Biotechnol Bioeng 21:1085–1089, 1979.
43. V Vojtisek, J Netrval. Folia Microbiol 27:173–177, 1982.
44. RJ Seely, RV Hageman, MJ Yarus, SA Sullivan. US Pat WO 90/04639, 1990.
45. RJ Seely, RV Hageman, MJ Yarus, SA Sullivan. US Pat WO 90/04631, 1990.
46. CKM Tripathi, SK Basu, VC Vora, JR Mason, SJ Pirt. Biotechnol Lett 10:635–637, 1988.
47. JP Voets, EJ Vandamme, C Vlerick. Zeitschr Allg Mikrobiol 13:355–366, 1973.
48. SC Agarwal, SK Basu, VC Vora, JR Mason, SJ Pirt. Biotechnol Lett 10:635–637, 1988.
49. A Long. OP Ward. Biotechnol Bioeng 34:933–941, 1989.
50. A Long. OP Ward. J Ind Microbiol 4:49–53, 1989.
51. PF Smith, D Hendlin. J Bacteriol 65:440–445, 1953.
52. H Becvarova, O Hanc, K Macek. Folia Microbiol 8:165–169, 1963.
53. B Wang, HS Shin, PL Rogers. In: WK Teo, MGS Yap, SWK Oh, eds. Better Living Through Innovative Biochemical Engineering. Singapore: Continental Press, 1994, p 249.
54. WM Mahmoud, A-HMM El-Sayed, RW Caughlin. Biotech Bioeng 36:47–54, 1990.
55. WM Mahmoud, A-HMM El-Sayed, RW Caughlin. Biotech Bioeng 36:55–63, 1990.
56. WM Mahmoud, A-HMM El-Sayed, RW Caughlin. Biotech Bioeng 36:256–262, 1990.
57. P Nikolova, OP Ward. Biotechnol Lett 16:7–10, 1994.
58. S Servi. Synthesis, 1–25, 1990.
59. OP Ward, CS Young. Enzyme Microb Technol 12:482–493, 1990.
60. OP Ward. Bioprocessing. Milton Keynes: Open University Press, 1991, pp 160–167.
61. CS Chen, CJ Sih. Angew Chem Int Ed Engl 28:695–707, 1989.
62. JS Dordick. Enzyme Microb Technol 11:194–211, 1989.
63. JS Dordick. Curr Opin Biotechnol 2:401–407, 1991.
64. JS Dordick. In: HW Blanch, DS Clarck, eds. Applied Biocatalysis, Vol. 1, New York: Marcel Dekker, 1991, pp 1–51.
65. P Nikolova, OP Ward. J Ind Microbiol 12:627–638, 1991.
66. LV Schneider. Biotech Bioeng 37:627–638, 1991.
67. T Yamane. Biocatalysis 2:1–9, 1988.
68. A Kockova-Kratochvilova. Yeasts and Yeast-like Organisms. New York: VCH, 1990.

69. MD Lilly, JM Woodley. In: J Tramper, HC Van der Plas, P Linco, eds. Biocatalysis in Organic Synthesis. Amsterdam: Elsevier, 1985, pp 179–192.
70. AM Klibanov. Chem Technol 16:354–359, 1986.
71. MD Hochnull, MD Lilly. In: A Laane, J Tramper, MD Lilly, eds. Biocatalysis in Organic Media. Amsterdam: Elsevier, pp 393–398.
72. P Nikolova, OP Ward. In: Biocatalysis in Non-Conventional Media. Amsterdam: Elsevier, 1992, pp 675–680.
73. P Nikolova, OP Ward. J Ind Microbiol 10:169–177, 1992.
74. MD Lilly, AJ Brazier, MD Hocknull, AC Williams, JM Woodley. In: C Laane, J Tramper, MD Lilly, eds. Biocatalysis in Non-Conventional Media. Amsterdam: Elsevier, 1987, pp 3–17.
75. SJ Osborne, J Leaver, MK Turner, P Dunnhill. Enzyme Microb Technol 12:281–291, 1990.
76. LJ Bruce, AJ Daugulis. Biotechnol Prog 7:116–124, 1991.
77. P Nikolova, OP Ward. Biotech Bioeng 20:493–498, 1991.
78. PF Smith, D Hendlin. Appl Microbiol 2:294–296, 1954.
79. P Nikolova, OP Ward. Biotech Bioeng 39:870–876, 1992.
80. S Ose, J Hironaka. Studies on Production of Phenyl Acetyl Carbinol by Fermentation. Proceedings of the International Symposium on Enzyme Chemistry, Tokyo and Kyoto, 2, 1957, pp 457–460.
81. MM Barrowman, CA Fewson. Curr Microbiol 12:235–240, 1985.
82. MM Barrowman, W Harnett, AJ Scott, CA Fewson, JR Kusel. FEMS Microbiol Lett 34:57–60, 1986.
83. GD Hegeman. J Bacteriol 91:1140–1159, 1966.
84. M Mendelstam, GA Jacoby. Biochem J 94:569–577, 1965.
85. GD Hegeman. Meth Enzymol 17A:674–678, 1970.
86. LJ Reynolds, GA Garcia, JW Kozarich, GL Kenyon. Biochemistry 27:5530–5538, 1970.
87. AY Tsou, SC Ransom, GA Geret, BD Buechter, PC Babbitt, GL Kenyon. Biochemistry 29:9856–9862, 1990.
88. PM Weiss, GA Garcia, GL Kenyon, WN Clealand. Biochemistry 27:2197–2205, 1988.
89. R Wilcocks, OP Ward, S Collins, NJ Dewdney, Y Hong, W Prosen. Appl Environ Microbiol 58:1699–1704, 1992.
90. R Wilcocks, OP Ward. Biotech Bioeng 39:1058–1063, 1992.
91. FA Davis, MS Hague. J Org Chem 51:4083–4085, 1986.
92. OP Ward, R Wilcocks, E Prosen, S Collins, NJ Dewdney, Y Hong. In: S. Servi, ed. Microbial Reagents in Organic Synthesis. Amsterdam: Kluwer Academic Publishers, 1992, 67–75.
93. E Prosen, OP Ward, S Collins, NJ Dewdney, Y Hong, R Wilcocks. Biocatalysis 8:21–29, 1993.
94. E Prosen, OP Ward. J Ind Microbiol 13:287–291, 1994.
95. T Asakawa, H Wada, T Yamano. Biochim Biophys Acta 170:375–391, 1968.
96. S Schneider, ME Mohamed, G Fuchs. Arch Microb 168:310–320.

11

Biocatalysis in the Enantioselective Formation of Chiral Cyanohydrins, Valuable Building Blocks in Organic Synthesis

Johannes Brussee and Arne van der Gen
Leiden University, Leiden, The Netherlands

I. ENANTIOSELECTIVE SYNTHESIS OF *R*- AND *S*-CYANOHYDRINS USING HYDROXYNITRILE LYASE

A. Introduction

Cyanohydrins are expedient starting materials for the preparation of several important classes of compounds such as α-hydroxy acids, acyloins, α-hydroxy aldehydes, vicinal diols, ethanolamines, α- and β-amino acids, and many other biologically interesting molecules (see Sec. II of this chapter). In the last decade there has been an upsurge of interest in methods for the synthesis and application of chiral cyanohydrins [1–7]. It may be expected that these products will constitute an important broadening of the range of commercially available chiral building blocks for pharmaceuticals, pesticides, vitamins, rare amino acids, and resolving agents.

Cyanohydrins are usually prepared by the addition of hydrogen cyanide (HCN) to aldehydes and ketones or, indirectly, by the addition of trimethylsilylcyanide followed by acid hydrolysis [8]. When aldehydes or nonsymmetrical ketones are used as substrates, a new stereogenic center is created. Methods for the synthesis of chiral cyanohydrins in nonracemic form can be divided into chemical and enzymatic. In general, the chemical approaches, using chiral catalysts, have the advantage that a broad range of aldehydes can be used. Up to now major disadvantages are the moderate enantiomeric excess (e.e.) values and the problems encountered in separating the catalyst from the product.

The two most important enzymatic approaches are enzymatic kinetic resolution by lipases or esterases (whole cells or purified enzymes) [9–18] and enzymatic asymmetrical synthesis by *R*- or *S*-hydroxynitrile lyases (NHLs). Disadvantages of kinetic resolution are the need of separating the hydrolyzed product from its ester and the maximum theoretical yield of only 50%. In some cases it is possible to racemize the unwanted enantiomer (distomer) and recycle the racemate [19].

B. Discovery of Hydroxynitrile Lyase

Rosenthaler first observed the activity of HNL in 1908 [20]. It was discovered that an extract from almonds (the sweet as well as the bitter variety of *Prunus amygdalus*) is

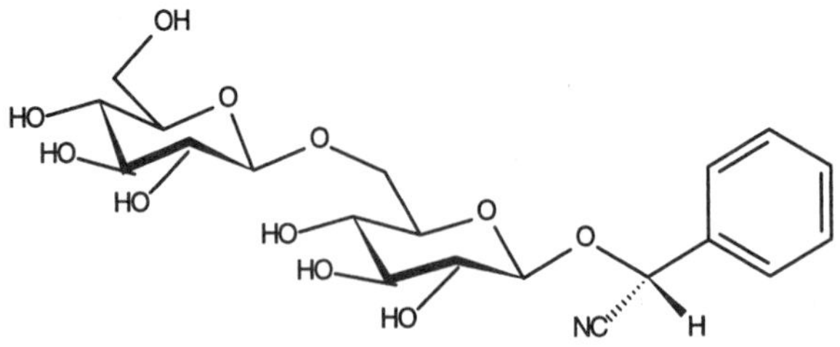

Figure 1 Amygdalin.

capable of metabolizing the cyanogenic glycoside amygdalin (Fig. 1). The extract from defatted almond meal, called emulsin, contains glucosidases (amygdalase and prunase) and HNLs, which, upon tissue damage, convert the physiological substrate amygdalin in a series of steps into glucose, benzaldehyde, and HCN.

Like all enzymes, HNL catalyzes the formation of a chemical equilibrium, in this case between α-hydroxynitriles (cyanohydrins) and their corresponding aldehydes and HCN. Conversely, in the presence of an excess of HCN, HNL efficiently catalyzes the enantioselective addition of HCN to aldehydes to form optically active cyanohydrins.

After the discovery of HNL in almonds, numerous other species were found to catalyze this type of reaction and several studies have appeared on the mechanism and kinetics of either asymmetrical synthesis or decomposition of cyanohydrins [21–34]. It was more than 50 years after the discovery by Rosenthaler that Becker and Pfeil published the first synthetically useful application in 1965 [35]. Shortly thereafter, the same authors reported a method for continuous synthesis of optically active cynaohydrins [36,37]. For that purpose, the HNL from almond (E.C. 4.1.2.10) was absorbed to ECTEOLA cellulose. After another lapse in time several research groups continued these synthetic applications from 1985 on [38–47].

Through the years, several names and abbreviations for HNLs have been used, e.g., emulsin (for extracts from defatted almond meal), oxynitrilase, mandelonitrile lyase, MDL, MNL, and acetone cyanohydrin lyase, all with different prefixes like D, L, (*R*), (*S*), (+), (−), or α. Throughout this chapter, the designation hydroxynitrile lyase, abbreviated as HNL and the prefixes *R* and *S*, will be used.

C. Distribution of Hydroxynitrile Lyases in Nature

HNL from almonds is yellow due to the presence of one flavin adenine dinucleotide (FAD) molecule in the glycoprotein (Fig. 2). Multiple forms of FAD-containing HNLs have been

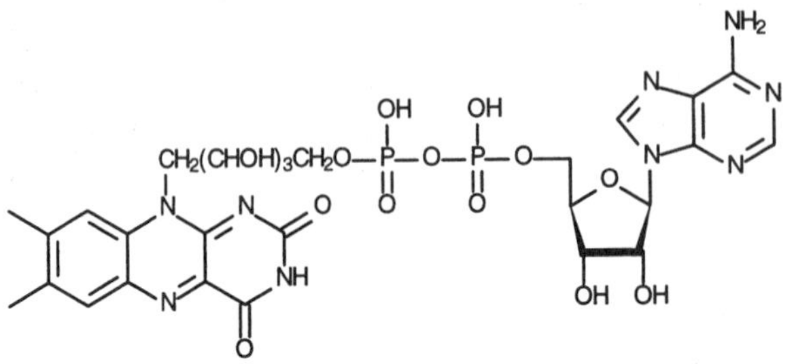

Figure 2 Flavine adenine dinucleotide (FAD).

isolated from plants, especially from the Rosaceae subfamilies Prunoideae and Maloideae, including cherry, peach, apricot, almond, plum, and apple. Enzymes from these sources were found to be quite similar with respect to their immunological behavior, molecular weight, specific activity, natural substrate (mandelonitrile), and the fact that they exist in several iso forms [26,48].

Other HNLs isolated from cyanogenic plants like flax, sorghum, the tropical rubber tree, and many other species (Table 1) basically all catalyze the same reaction, although they clearly differ in more than one property from those of the Rosaceae [4]. Striking differences are as follows: (1) There is a difference in the natural substrates for the different HNLs, e.g., 4-hydroxybenzaldehyde cyanohydrin in sorghum and acetone cyanohydrin in flax; (2) in contrast to HNLs from Rosaceae, other HNLs, including those from flax,

Table 1 Hydroxynitrile Lyase Activity in Natural Sources

Species	Ref.
Achillea millefolium	21
Aquilegia vulgaris	21
Chaenomeles japonica	26
Crataegus oxyacantha	21
Cydonia oblonga	26
Cydonia vulgaris	21
Eriobotrya japonica	21
Hevea brasiliensis (rubber tree)	34
Linum usitatissimum (flax)	32
Malus communis (apple)	26, 52, 53
Manihot esculenta (cassava)	54,55
Phlebodium aureum	55
Pirus malus	21
Pirus communis	21
Prunus amygdalus, amara (bitter almond)	21, 28, 29, 56
Prunus amygdalus, sativa (sweet almond)	21, 26, 57, 58
Prunus armeniaca (apricot)	21, 26, 52
Prunus armeniaca, dulcis	21
Prunus avium (cherry)	21, 26, 52
Prunus domestica (plum)	21, 26, 52
Prunus laurocerasus (laurel cherry)	21, 27
Prunus lyonii (california cherry)	31
Prunus padus	21
Prunus persica (peach)	21, 24, 25, 26
Prunus persica x P. armeniaca	26
Prunus serotina (black cherry)	24, 30
Prunus spinoza (plum)	21, 26
Prunus virginiana	21
Sambucus ebulus	21
Sambucus nigra (black elderberry)	21
Sorghum vulgare	29, 59
Sorghum bicolor	49, 56, 57, 58
Trifolium repens (white clover)	60
Ximenia americana	33, 57, 58

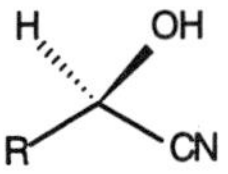

Figure 3 Cyanohydrin.

cassava, sorghum, and ximenia, contain no FAD group; (3) enzymes from *Sorghum bicolor* [45,49], *Ximenia americana* [33], *Sambucus nigra* [50], and *Hevea brasiliensis* [51] exclusively catalyze formation and decomposition of *S*-cyanohydrins, whereas the Rosaceae HNLs only catalyze the conversion of *R*-cyanohydrins with the configuration shown in Fig. 3. In most cases this corresponds to the *R* configuration. In some instances, e.g., furfural cyanohydrin, where the ring oxygen gives the substituent *R* priority over the nitrile group, the configuration is *S*.

D.   Isolation and Purification of Hydroxynitrile Lyases, Isoenzymes

In 1963 Becker and Pfeil purified the enzyme from bitter almonds for the first time [28,50]. One year later they reported to have succeeded in crystallizing the enzyme [61]. However, Pfeil communicated in 1980 that they had not been able to reproduce the experiment and at that time it was even uncertain as to whether the crystals were indeed HNL [62].

In 1991 Smitskamp-Wilms isolated four isoenzymes from sweet almonds [57]. Two iso forms dominate and constitute over 90% of the HNL content. Molecular weights of all iso forms were determined and the results revealed, in accordance with other reports [63], that the four isoenzymes all consist of a single polypeptide chain with a molecular weight of approximately 60 kDa. A few years later Effenberger isolated also four different iso forms from almonds and succeeded in obtaining three of them in crystalline form. Only preliminary x-ray diffraction results, such as unit cell parameters, were published [64].

The HNL from *Sorghum bicolor* was shown to consist of one major and two minor isoenzymes [57]. Crystallographic studies on the three isoenzymes were reported and preliminary x-ray investigations of crystallized forms were published [65].

Various species in which HNL activity has been detected are listed in Table 1. Sofar, only 11 different HNLs, belonging to six different plant families, have actually been isolated.

The amount of HNL varies largely from one species to another, from less than 1 mg (*Ximenia americana*) to 900 mg (sweet almond) per 100 g of dried or defatted material.

R-HNL from almonds (EC 4.1.2.10) and *S*-HNL from *S. bicolor* (EC 4.1.2.11) are commercially available enzymes and can be used for the laboratory synthesis of the *R* and *S* enantiomers of cyanohydrins, respectively. Larger amounts of the *S*-HNL from *Hevea brasiliensis* (EC 4.1.2.39) have been prepared by overexpression in *Pichia pastoris*. This considerably enhances the scope of substrates for *S*-NHLs [51,66,67]. The enzyme was crystallized and its three-dimensional structure was recently elucidated (see Sec. I.E.2). The very similar *S*-HNL from the leaves of *Manihot esculenta* (EC 4.1.2.37) was successfully overexpressed in *E. coli* [68].

As the available enzyme purification procedures were somewhat tedious, more rapid procedures, based on affinity chromatography, were elaborated. For this purpose a competitive inhibitor like methyl 4-(3-aminopropoxy)benzoate or methyl 4-hydroxybenzoate was bound covalently to Sepharose 4B [69] or to Eupergit C [70].

HNLs from species investigated so far have invariably been found to consist of several isoenzymes (varying from two to six) and none of these exhibit a metal requirement [26,30,57,71].

E. Catalysis Mechanism of Hydroxynitrile Lyases

Since the discovery of the synthetic capabilities of the HNL-containing emulsin system from almonds by Rosenthaler in 1908 [20,72], the mechanism underlying the formation of chiral cyanohydrins in nonracemic form has been studied intensively by several research groups. Although substantial progress has been made, it must be admitted that understanding of the catalysis mechanism is still largely incomplete. As outlined earlier, the HNLs can be classified in FAD- and non-FAD-containing enzymes. It must be assumed that fundamental differences exist between the mode of action and/or substrate binding of these two classes.

1. FAD-Containing Hydroxynitrile Lyase

A first proposal for the mechanism of action of the FAD-containing enzyme from Rosaceae was made by Albers et al. in 1934 (Scheme 1) [73]. On the basis of kinetic measurements they proposed that the enzyme contained two reactive amino groups, that reacted with two molecules of aldehyde to form imino alcohols. The imino groups were then displaced by cyanide in a reaction for which, because of its similarity to hydrolysis, the term "hydrocyanolysis" was coined.

Hustedt and Pfeil in 1961 noticed the difference in sensitivity toward the presence of protic solvents between the asymmetrical synthesis induced by cinchona cations and by the enzyme [74]. In the years thereafter Pfeil and co-workers published several important contributions to the synthetic usefulness of the enzyme. They found that the presence of the tightly but not covalently bound FAD molecule in the enzyme is essential, both for its stability and for its activity. The apoenzyme is catalytically inactive, but activity can be fully regenerated by addition of FAD in its oxidized form [28]. A reaction scheme was proposed for the asymmetrical synthesis as depicted in Scheme 2, in which the carbonyl compound first forms a complex with the chiral catalyst. This complex then reacts in a rate-determining step with a cyanide ion. In a third, fast, step the complexed cyanohydrin anion is protonated by HCN.

Important contributions to the understanding of the enzyme mechanism were made in the late 1970s and early 1980s by the group of Lothar Jaenicke at the University of Köln and the group of Marilyn Schuman Jorns, at that time at Ohio State University. Kinetic studies by Jorns [75] were consistent with an ordered uni-bi mechanisms, "uni-bi" meaning that there is one substrate and two products in the forward (decomposition)

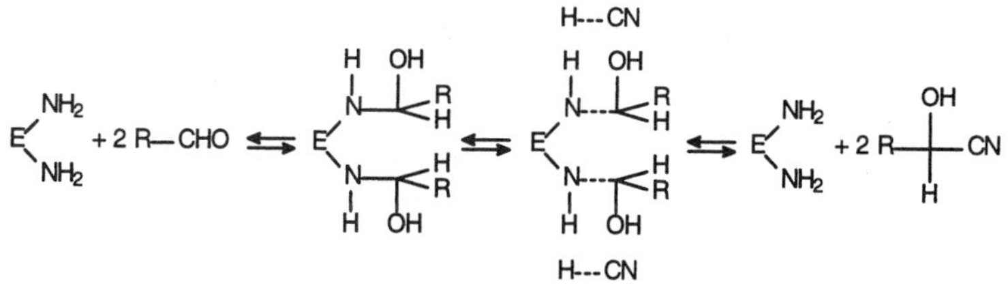

Scheme 1

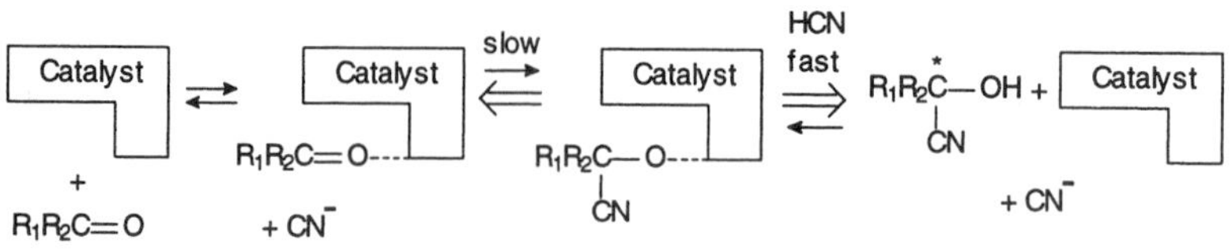

Scheme 2

reaction and "ordered" meaning that the products leave in obligatory order (in this case HCN first) [76]. Inhibitors that are reasonable analogs for cyanide ion, such as azide and thiocyanate, do not bind to the enzyme–aldehyde complex, suggesting that the binding species is HCN rather than the cyanide ion [75]. It was also found that replacement of FAD by 5-deaza-FAD leads to a large decrease in enzyme activity, indicating that binding to N-5 plays an essential role.

Surprisingly, oxidation of the 5-deaza-FAD enzyme with peracids leads to complete restoration of enzyme activity [77]. The identity of the oxidation product was subsequently determined. First, a 4α,5 epoxide is formed [78] (Scheme 3). This epoxide intermediate undergoes hydrolysis of the pyrimidine ring, followed by ring contraction [79,80]. The newly formed tricyclic compound is said not to be susceptible to nucleophilic attack, indicating that addition of a nucleophile to N-5 does not play a role in the activity of the natural enzyme.

It was observed by Bärwald and Jaenicke that no spectral change occurs upon addition of mandelonitrile, benzaldehyde, or HCN to the enzyme [81]. Apparently, binding of substrate and product does not involve the flavin molecule. It was subsequently found that one of the cysteine residues in the enzyme is essential for activity and can be selectively reacted with Michael acceptors. Upon reaction with labeled 3-oxo-3-phenylpropene, followed by reduction (to prevent a retro-Michael reaction) and extensive hydrolysis, L-2-amino-4-thia-DL-7-hydroxy-7-phenylheptanoic acid, the reduced conjugate addition product of cysteine to the inhibitor, was isolated [82] (Scheme 4).

Jaenicke and Preun then postulated a mechanism, in which the essential role of the cysteine residue and of the flavin molecule in the active site was emphasized [82]. As depicted in Scheme 5, the proposed reaction mechanism involved addition of cyanide ion to N-5 and of the carbonyl carbon to C4a, followed by an unprecedented four-center rearrangement. In view of the results reported earlier (vide supra), it is not surprising that this proposal met with severe criticism [83].

The conversion of aldehydes to cyanohydrins, which can be both acid- and base-catalyzed, does not necessitate the intermediacy of a complex transition state. Assuming that the enzymatic conversion involves base catalysis to provide the nucleophilic cyanide ion and acid catalysis to protonate the incipient oxy anion, the mechanism can in essence be presented as shown in Scheme 6. Assuming that the essential cysteine residue plays a role both in activating the carbonyl group in the first substrate and in deprotonating the HCN, a mechanism involving neither nucleophilic addition to the flavin molecule nor the presence of intermediates for which there is no evidence can be depicted [84] (Scheme 7). In view of the recently obtained structural information on the non-FAD-containing enzyme from *Hevea brasiliensis* [4,85], this proposal may have to be modified in the sense that protonation and deprotonation might involve different amino acid residues in the active site. Clarification of this point must await the elucidation of the three-dimensional structure of the almond enzyme.

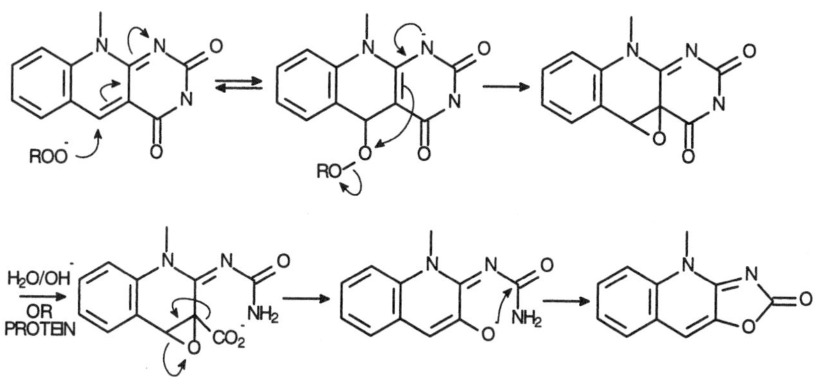

Scheme 3

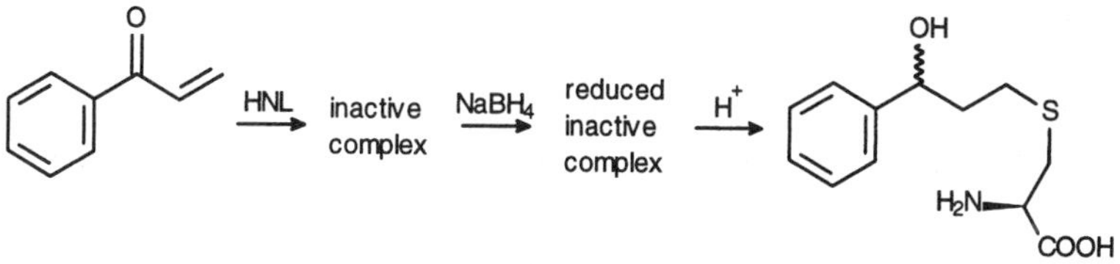

Scheme 4

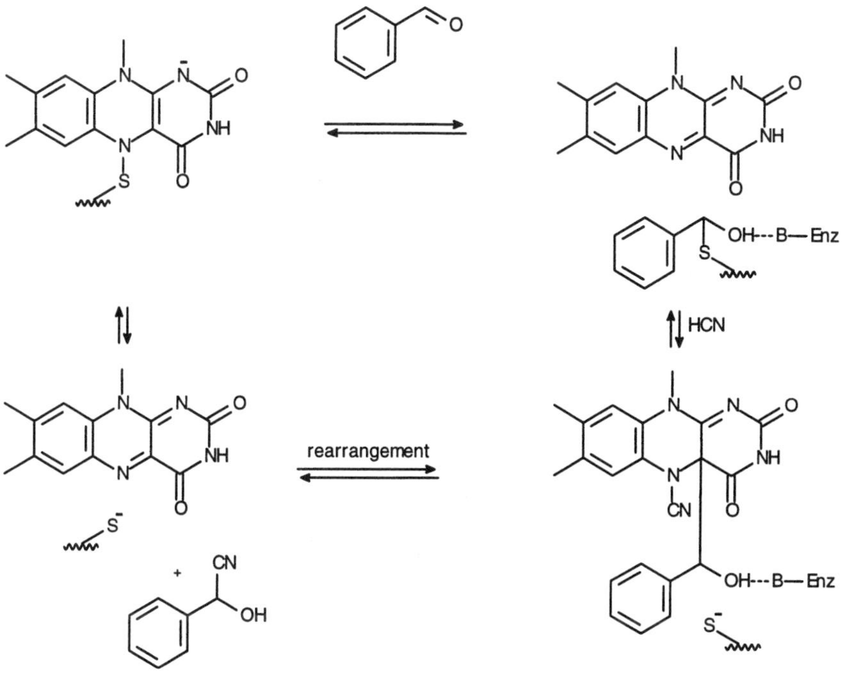

Scheme 5

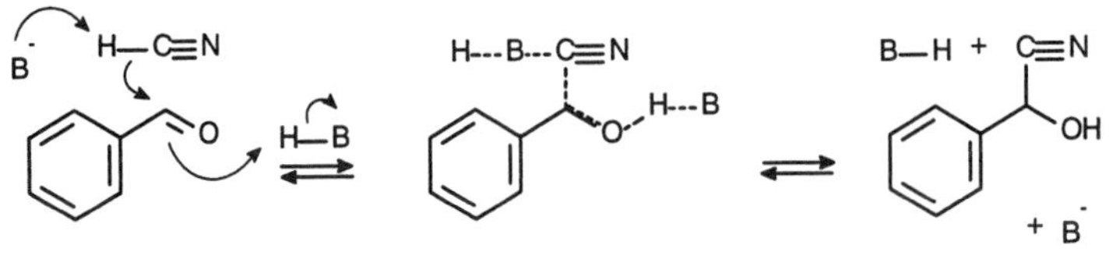

Scheme 6

2. Non-FAD-Containing Hydroxynitrile Lyases

Recently, important contributions to the understanding of the catalysis mechanism of non-FAD-containing HNLs have been made, especially by Herfried Griengl and collaborators at the Technical University of Graz. A breakthrough, with regard to both synthetic applications and determination of structure, was made by the successful overexpression of the HNL from maniok (*Manihot esculenta*) in *E. coli* and of the HNL from the rubber tree (*H. brasiliensis*) in *Pichia pastoris* [6]. Both enzymes are *S*-selective and show a wide substrate acceptance. Their molecular weight (30 kDa) and amino acid sequence were found to be highly similar. In the case of the Hb-HNL, overexpression led to extremely high levels (up to 60% of total cellular proteins) of soluble HNL, exhibiting a specific activity of 40 U/mg, which is about twice the activity found for the purified enzyme from *H. brasiliensis*. Attempts to crystallize this highly purified enzyme were successful and for the first time the three-dimensional structure of an enzyme with HNL activity could be determined by x-ray crystallography [86]. The enzyme turned out to be a new member of the large α,β-hydrolase fold family. This family comprises enzymes with a wide variety

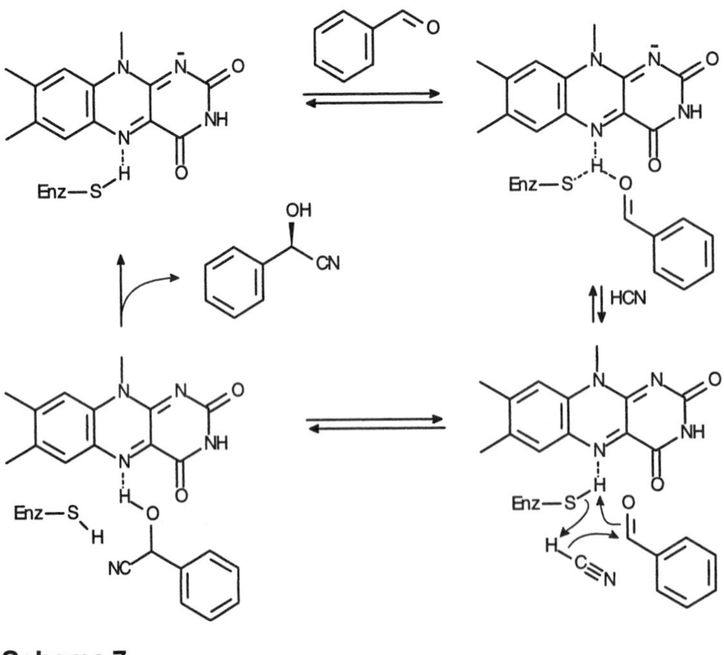

Scheme 7

of activities, including acetylcholinesterase, hydrolase, lipase, and carboxypeptidase activity, which nevertheless show large similarity in their catalytic mechanisms. On the basis of the established three-dimensional structure, site-directed mutagenisis, and inhibition experiments, Griengl and collaborators have formulated a tentative mechanism for the stereoselective formation of cyanohydrins by the Hb-HNL enzyme [4,85] (Scheme 8). Important contributing items to this proposal were the following:

> The active site is buried deep inside the protein and only accessible by a narrow channel. This is consistent with an ordered uni-bi mechanism, as predicted both for the FAD-containing [75] and for the non-FAD-containing HNLs [6, Ref. 11].
> In analogy with other members of the family of α,β-hydrolase fold enzymes, the presence of a catalytic triad in the active site was predicted and could be confirmed by site-directed mutagenesis. The triad consists of Ser[80], the oxygen atom of which is rendered nucleophilic by incipient proton abstraction by His[235], which in turn is activated by Asp[207].
> The activated Ser[80] residue is, again in analogy with other enzymes of the family, believed to act as a nucleophile and attacks the carbonyl carbon. The incipient negative charge at the carbonyl oxygen is stabilized by a set of three hydrogen bonds, stemming from the Thr[11] hydroxyl, the Cys[81] N-H, and the Cys[81] S-H, together forming the "oxy anion hole."
> The oxy anion of the resulting tetrahedral intermediate is protonated by the Cys[81] thiol group, thereby forming a hemiacetal.
> The concurrently formed thiolate anion deprotonates the HCN to give cyanide ion.
> The cyanide ion attacks the tetrahedral intermediate and displaces the Ser[80] residue in an S_N2 reaction, thereby liberating the S-cyanohydrin.

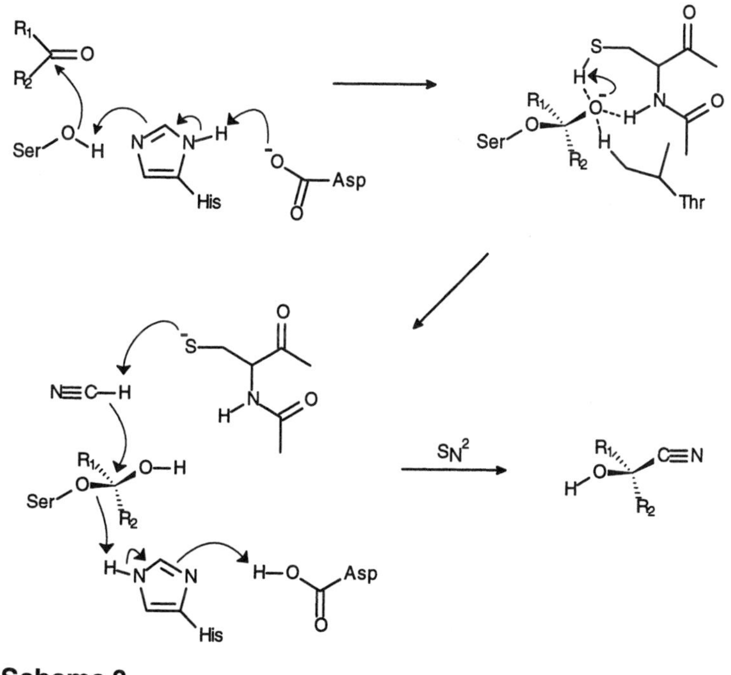

Scheme 8

F. Substrate Selectivities of Hydroxynitrile Lyases from Different Sources

The substrate specificity of the almond enzyme is relatively wide, including both saturated and unsaturated aliphatic and aromatic aldehydes, as well as aliphatic methyl ketones [87]. The chain length of aliphatic aldehydes was originally thought to be limited to C_6. Recently, it was shown that aldehydes up to C_{10} can be converted by the almond enzyme [88]. Also heteroaromatic aldehydes (thienyl, furyl) are good substrates. It appears that α-hydroxy aldehydes are not accepted by the almond enzyme [89].

The substrate specificity of the sorghum enzyme appears limited to aromatic and heteroaromatic aldehydes. The S-HNLs from *H. brasiliensis* and from *M. esculenta*, on the other hand, show a remarkably wide substrate acceptance. Both aliphatic, aromatic, and heteroaromatic aldehydes, saturated as well as α,β-unsaturated, as well as methyl ketones are accepted by these enzymes [6,68].

A survey of all substrates that have been reported to be converted to chiral cyanohydrins in at least moderate yield and useful e.e. is presented in Table 2 and Scheme 9.

G. Application of Hydroxynitrile Lyases on a Practical Scale

Interesting preparative applications in organic synthesis have been described for only four enzymes, the FAD-containing R-HNL from Rosaceae, in particular from almonds (*Prunus amygdalus*), and the non-FAD-containing S-HNLs from sorghum (*S. bicolor*), cassava (*M. esculenta*), and from the rubber tree (*H. brasiliensis*).

1. *R*-Hydroxynitrile Lyase from Almonds

(a) *Purified HNL, Immobilized on ECTEOLA Cellulose, in Aqueous Methanol.* More than 50 years after the discovery of the synthetic potential of the emulsin enzymes from almonds by Rosenthaler [21,22], the R-HNL was purified and characterized by Becker and Pfeil [28,48]. The purified enzyme combines with cellulose-based ion exchangers to form a very active and stable catalyst. (R)-(+)-Mandelonitrile with an optical purity of 94% could be obtained on a multigram scale, using aqueous ethanol or methanol as the cosolvent [36]. The small amount of racemic cyanohydrin formed originates from the spontaneous, non-enzyme-catalyzed reaction, which cannot be completely suppressed in aqueous solvents. The almond enzyme was shown [109] to consist of several isoenzymes, differing in the protein composition of the apoenzyme. For preparative purposes it is not useful to separate these isoenzymes, as they show very similar substrate specificity [57].

(b) *Crude Extract from Almond Meal in Aqueous Ethanol.* Several years later it was shown by Brussee et al. [38,46] that laborious purification of the enzyme is not necessary. Results, identical to those obtained by Becker and Pfeil, could be realized using a crude aqueous extract from defatted, ground almonds. Another improvement, in particular regarding the safety aspects, consisted of the generation of HCN in situ from a solution of KCN in an acetate buffer at pH 5.4.

(c) *Purified HNL, Supported on Cellulose, Using a Nonaqueous Solvent System.* Effenberger and co-workers showed for the first time that the enzyme-catalyzed reaction can proceed in solvents that are not miscible with water [39]. Best results were obtained using ethyl acetate. Non-enzyme-catalyzed cyanohydrin formation is almost completely suppressed under these conditions and, although longer reaction times are required, the enantiomeric purity of the cyanohydrins obtained from a variety of aldehydes was in fact

Table 2 Aldehydes as Substrates for HNLs

nr.[a]	HNL[b]	y/c[c]	e.e.[d]	Ref.[e]
1.	Pa	95	97	52
	Me	86	91	68
2.	Pa	98	95	19, 39, 42, 46, 88, 90, 91, 92, 93, 94*
	Hb	90	80	51
3.	Pa	82	99	52, 88, 92, 95*
	Me	100	91	68
4.	Pa	100	94	46, 88*
	Hb	81	96	6*, 51
5.	Pa	65	92	96
6.	Pa	98	87	88
7.	Pa	82	96	52, 97*
	Hb	35	85	51
8.	Pa	94	63	88
9.	Pa	100	95	88*, 94, 95
	Me	91	95	68
10.	Pa	95	98	52*, 88
11.	Pa	58	92	1, 39, 52, 96*
	Me	80	94	68
	Hb	90	67	51
12.	Hb	90	81	51
13.	Pa	98	96	92, 93*, 95
14.	Pa	95	96	92, 93*, 95, 96
	Me	100	92	68
	Hb	94	99	67
15.	Hb	98	94	6, 51, 66*
16.	Pa	100	99	1, 39, 42, 46, 90*, 91, 94, 98
	Me	100	92	68
	Hb	80	86	66
17.	Me	82	97	68
	Hb	46	95	66
18.	Hb	82	99	99
19.	Pa	90	95	84
20.	Pa	99	97	100
21.	Pa	100	95	100
22.	Pa	95	40	39
	Hb	44	99	67
23.	Pa	55	60	95
	Hb	88	93	67
24.	Pa	94	90	1
25.	Hb	35	80	66
26.	Pa	90	94	98
27.	Hb	88	80	6, 66*
28.	Pa	36	96	96
29.	Pa	90	99	93, 98*
30.	Pa	86	55	46
	Hb	87	99	67
31.	Pa	45	93	39*, 98, 101
	Hb	50	95	67
32.	Pa	98	96	1, 39, 93*, 96
33.	Pa	70	95	95

Table 2 Continued

nr.[a]	HNL[b]	y/c[c]	e.e.[d]	Ref.[e]
34.	Pa	81	92	102
35.	Pa	81	91	102
36.	Pa	100	89	103
37.	Pa	100	92	103
38.	Pa	97	61	103
39.	Pa	100	96	103
40.	Pa	100	73	103
41.	Pa	83	61	103
42.	Pa	81	88	103
43.	Pa	96	99	1, 39, 40, 101, 104*
	Sb	80	80	104
	Hb	55	98	67
44.	Pa	96	99	104
	Me	98	92	68
	Sb	88	87	104
	Hb	61	99	67
45.	Pa	70	99	42, 46, 90*, 91
46.	Pa	71	99	104
	Me	85	96	3
	SB	64	91	104
	Hb	52	99	67
47.	Pa	95	99	1, 104*
	Me	98	98	68
	Sb	95	98	104
	Hb	49	99	67
48.	Pa	97	82	1
49.	Pa	100	99	1, 39–42, 46, 90, 91*–94, 96, 101, 105
	Me	100	98	68
	Sb	97	97	43, 45, 49, 105*
	Hb	90	94	51
50.	Sb	80	96	43, 45*
51.	Pa	94	99	105
	Sb	88	98	44, 105
52.	Pa	65	96	96
	Hb	61	77	67
53.	Pa	85	98	18*, 39, 40
	Sb	89	93	44
	Hb	80	99	67
54.	Pa	82	98	41, 42, 46, 90, 91, 105*
	Me	82	98	68
	Sb	54	71	105
	Hb	49	95	67
55.	Pa	99	98	1*, 39, 40
	Sb	93	96	44
	Hb	98	99	67, 86*
56.	Pa	90	95	52*, 105
	Sb	30	93	105
57.	Pa	48	99	106
58.	Sb	94	92	44
59.	Pa	97	98	105

Table 2   Continued

nr.[a]	HNL[b]	y/c[c]	e.e.[d]	Ref.[e]
60.	Me	100	92	68
61.	Sb	95	91	44
62.	Pa	90	99	107
63.	Pa	90	98	99
	Sb	90	98	43, 44, 45*
64.	Pa	64	96	52*, 105
	Sb	87	99	43, 44, 45*, 105
65.	Pa	85	96	51*, 105, 106
	Sb	60	99	105
66.	Pa	82	93	18, 52, 105*
67.	Pa	98	98	42, 46, 47, 96, 105*, 108
	Sb	72	89	105
	Me	84	86	68
68.	Pa	67	98	106
69.	Pa	35	90	106
70.	Pa	68	99	106
71.	Pa	89	81	106

[a]Number refers to structures in Scheme 9.
[b]Pa for *Prunus amygdalus*, Sb for *Sorghum bicolor*, Me for *Manihot esculenta*, and Hb for *Hevea brasiliensis*.
[c]Percentage (isolated) yield or conversion.
[d]Percentage enantiomeric excess.
[e]Yield/conversion and e.e. are from reference with asterisk.

appreciably better than in the alcohol–water systems. The purified enzyme was bound to crystalline cellulose (Avicel). Other hydrophobic carriers, like celite, have also been used [110].

(*d*)   *Synthesis of Cyanohydrins Using Almond Meal in an Organic Solvent.*   A method that is particularly convenient for the preparation of cyanohydrins on a laboratory scale was published in 1991 [90,91]. Rather than first purifying the enzyme and then absorbing it on a solid support, the unpurified enzyme was used as present on its natural support, almond meal. Almond meal is commercially available (Sigma-Aldrich) and is easy to prepare in the laboratory [90,91]. By controlling the pH (3.5–5.5) and temperature (0–5°C), the non-enzyme-catalyzed conversion in the minor aqueous phase (4% v/v) is efficiently suppressed and e.e.'s of more than 99% are regularly obtained. The almond meal method has been used to great advantage in several recent investigations [52,88, 102,105,107]. Under recently described "microaqueous conditions" [101], which can be achieved by not swelling the almond meal in a buffer solution, the enzyme performs well, also at higher temperatures (up to 30°C), and can be more easily recovered.

(*e*)   *Trans-cyanation, Using Acetone Cyanohydrin as the HCN Donor.*   A different way of suppressing the nonenzymatic reaction is by keeping the concentration of HCN low at all times. This can be achieved by performing a HNL-mediated transcyanation, using acetone cyanohydrin as an HCN source [96]. The e.e. values obtained were in several cases higher than those reported previously. However, the selectivities claimed by the authors could not be reproduced [1,20]. An inherent disadvantage of this method is that the desired reaction product has to be separated from unreacted acetone cyanohydrin. This

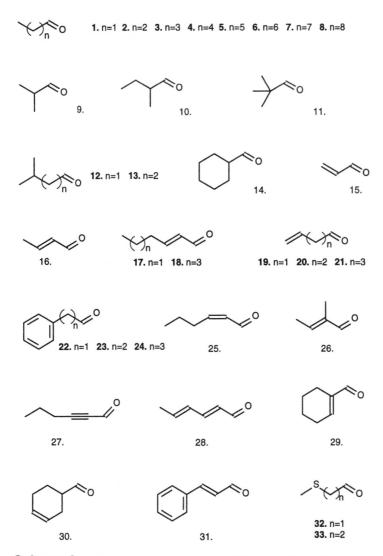

Scheme 9 Aldehydes as substrates for HNLs (see Table 2).

can be avoided by carrying out the decomposition of the transcyanating agent, either enzyme- or base-catalyzed, in a separate vessel and allow the liberated HCN to diffuse into the reaction mixture [52].

(*f*) *S-Cyanohydrins by Selective Cleavage of the R Form in a Racemate.* In principle it should be possible to obtain *S*-cyanohydrins by selective enzymatic decomposition of the *R* enantiomer in a racemic mixture. This is difficult to achieve on a practical scale because the HCN that is liberated can engage in both enzyme-catalyzed and non-enzyme-catalyzed reactions by which the *R* enantiomer is again formed. This problem has been elegantly solved by Gotor and co-workers by combining this approach with that of the transcyanation [102]. A mixture of aldehyde and racemic cyanohydrin from a methyl ketone was treated with almond meal. The *R*-cyanohydrin of the methyl ketone was selectively cleaved and the *R*-cyanohydrin of the aldehyde was selectively formed by trans-

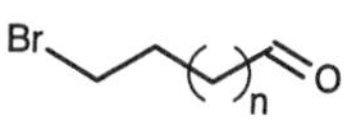

34. n=1 **35.** n=2

36. X=Br **37.** X=Cl **38.** X=AcO
39. X=MeO **40.** X=EtO **41.** BnO **42.** X= CH₂=CH₂O

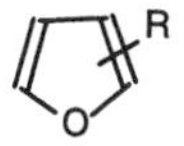

43. R= 2-CHO
44. R= 3-CHO

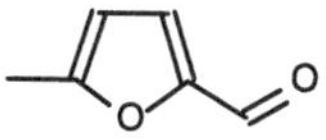

45.

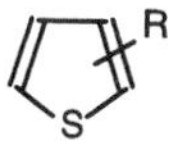

46. R= 2-CHO
47. R= 3-CHO

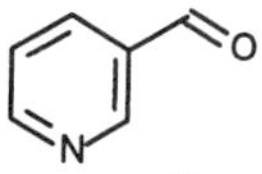

48.

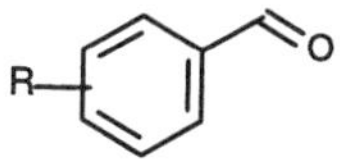

49. H   **50.** 3-Me   **51.** 4-Me
52. 2-MeO **53.** 3-MeO **54.** 4-MeO
55. 3-PhO **56.** 4-PhO
57. 4-BnO
58. 3-Br   **59.** 4-Br
60. 2-Cl **61.** 3-Cl   **62.** 4-Cl
63. 3-OH **64.** 4-OH
65. 4-AcO

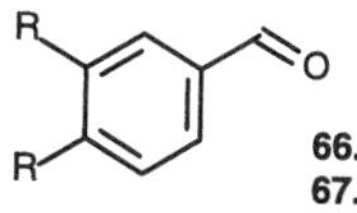

66. 3,4-diMeO
67. 3,4-methylenedioxy
68. 3-OH, 4-Me
69. 3-OH, 4-MeO

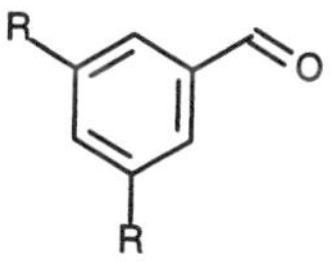

70. 3,5-di-CH₃OCH₂O
71. 3,5-di-AcO

Scheme 9 Continued

cyanation. The liberated methyl ketone does not compete with the aldehyde for the liberated HCN because its hydrocyanation proceeds at a much lower rate. After chromatographic separation, the *S*-cyanohydrins of 2-pentanone and 2-hexanone were obtained, as well as the *R*-cyanohydrins of 4-bromobutanal and 5-bromopentanal. The latter compounds are valuable starting materials for heterocyclic synthesis [111].

(*g*) *Purified Enzyme in a Membrane Reactor in a Continuous Process.* The possibility to produce cyanohydrins of high enantiomeric purity in a continuous process was studied by Kragl, Kula, and co-workers [56]. The spontaneous, non-enzyme-catalyzed reaction was suppressed by working at very low pH (3.75) and by mixing the reactants just before entering the reactor. The enantiomeric excess was further enhanced by making use of the different kinetics of the two reactions. The nonenzymatic reaction is first order in (benz)aldehyde and therefore its rate increases linearly with aldehyde concentration. The enzyme-catalyzed reaction, on the other hand, obeys typical Michaelis-Menten kinetics and reaches its V_{max} at low aldehyde concentrations. It is therefore possible to suppress the spontaneous reaction effectively by working with high enzyme and low aldehyde concentrations. The enzyme was contained by an ultrafiltration membrane. Enantiomerically pure (*R*)-mandelonitrile (e.e. > 99%) was produced with a space–time yield of up to 2400 g/L/day. The enzyme consumption was about 17,000 U/kg of product. Because of the easy accessibility of the enzyme, this was not considered to be a limiting factor for industrial application of this process. [One kilogram of almonds provides about 400,000 U, one unit being defined as the quantity that will form 1.0 μmol of benzaldehyde and HCN from (R)-mandelonitrile per min at pH 5.4 and at 25°C.]

(h) *Purified HNL in a Water/Organic Solvent Biphasic System.* A second procedure aiming at the production of chiral cyanohydrins on an industrial scale was developed by Loos et al. [112]. Purified enzyme was used in a liquid–liquid biphasic system. Applying a so-called D-optimal design [113], the process was optimized for a multitude of parameters, using piperonal and anisaldehyde as representative substrates for fast and slow cyanohydrin formation, respectively. On the basis of the results obtained, four critical parameters (pH, concentration of the two reactants, and temperature) were optimized in depth with a full factorial design (2^4-FFD) [114]. It was shown that (R)-mandelonitrile can be prepared in this two-phase system in excellent yield (98%) and e.e. (98%), using methyl *t*-butyl ether (MTBE) as the solvent and a citrate buffer as the aqueous phase at pH 5.0 at room temperature. The throughput under these conditions was 2.1 mol/L/h (corresponding to 6700 g/L/day). It was also shown that the activity of the enzyme had not measurably decreased after reusing the aqueous enzyme–containing layer in five successive experiments.

2.   *R*-Hydroxynitrile Lyases from other Rosaceae

The substrate specificity of *R*-HNL from almonds is relatively wide, including aliphatic, aromatic, heteroaromatic, and unsaturated aldehydes and aliphatic methyl ketones. As discussed in the previous paragraph, aliphatic aldehydes can be accommodated up to C_{10}. It has been shown that aromatic aldehydes with bulky substituents at positions 3 and/or 4 are not accepted by the enzyme [105]. With the aim of extending the substrate acceptance, a number of other Rosaceae *R*-HNL, namely those from apple, cherry, apricot, and plum, were investigated and found to be clearly different [52]. The meal from the seeds of these fruits was used, rather than the purified enzymes [91]. The properties of the enzyme from apple seeds turned out to be the most promising. For the more hindered substrates (3- and/or 4-substituted aromatic, long chain and branched aliphatic) this HNL gave improved results, both with regard to reaction rate and e.e., over those obtained with almond meal. The properties of the other Rosaceae HNLs were less interesting. In conclusion, only in those cases where the almond meal, which is commercially available and easy to prepare, fails to give satisfactory results, it may be worthwhile to try meal from apple seeds.

3.   *S*-Hydroxynitrile Lyase from *Sorghum Bicolor*

The first enzyme to be investigated for its potential to catalyze the preparative synthesis of *S*-cyanohydrins was the HNL isolated from the seedlings of *S. bicolor* (EC 4.1.2.11) [43,45]. The first results were rather promising. 4-Hydroxybenzaldehyde could be converted to the natural substrate, 4-hydroxymandelonitrile, on the mmol scale in high yield in optically pure form (sodium citrate buffer, pH 3.75, room temperature). Some other aromatic aldehydes also performed well. Larger amounts could be converted in a CST reactor using the enzyme, immobilized on Eupergit C. The results were confirmed and extended by Effenberger and co-workers, who used a suspension of the enzyme, immobilized at pH 5.4 on Avicel-cellulose, in an aprotic organic solvent [diisopropyl ether (DIPE)] [44].

Subsequently, Kiljunen and Kanerva showed that for preparative purposes it is not necessary to use the extensively purified enzyme. Similar results can be obtained using lyophilized and dechlorophylled *S. bicolor* shoots. Despite these promising results, it is unlikely that the *S. bicolor* enzyme will see much future application for the following reasons:

Only aromatic aldehydes are accepted by Sb-HNL. This narrow substrate range severely hampers wide application of the enzyme.

The availability of the enzyme is limited. In contrast to *Prunus amygdalus*, which contains about 4 g of HNL per kg of almonds, *Sorghum* holds only about 30 mg of HNL per kg of shoots. Attempts to produce larger amounts by genetic engineering have been frustrated by the complex posttranslational processing needed for its expression [68].

The conclusion seems justified that the *Sorghum* enzyme will not be available for the production of larger amounts of *S*-cyanohydrins in the foreseeable future.

4. *S*-Hydroxynitrile Lyases from *Manihot esculenta* and from *Hevea brasiliensis*

Much wider substrate acceptance, similar to the one shown by the *R*-HNL from almonds, is shown by the *S*-HNLs obtained from the leaves of cassava (*M. esculenta*, EC 4.1.2.37) [68] and of the rubber tree (*H. brasiliensis*, EC 4.1.2.39) [67]. Both enzymes accept aromatic, heteroaromatic, aliphatic, and α,β-unsaturated aldehydes. Also, methyl ketones have been shown to act as substrates. The enzyme from cassava leaves was the first HNL to be made available in large quantities by genetic engineering [68]. The cloned Me-HNL gene was overexpressed in *Escherichia coli* cells. The biochemical properties of recombinant Me-HNL and Me-HNL isolated from the leaves of *M. esculenta* were identical, but the specific activity of the recombinant enzyme was shown to be 25 times higher. Whereas the original experiments obtained with Me-HNL from cassava had shown unsatisfactory optical yields, the recombinant Me-HNL produced excellent results with regard to both yield and e.e. The enzyme was absorbed on nitrocellulose that had been soaked in citrate buffer (pH 3.3) and thoroughly dried. DIPE was again the solvent of choice.

The HNL from *H. brasiliensis* was first described in 1989 [34]. Its potential for the synthesis of *S*-cyanohydrins was extensively studied by Griengl and co-workers [51,66,67]. Interestingly, this enzyme, which is highly homologous to the HNL from cassava [6,54], appeared to perform best in aqueous media. The non-enzyme-catalyzed reaction was suppressed by working at low pH (citrate buffer, pH 4.0). Although transcyanation with acetone cyanohydrin can be applied to keep the HCN concentration down [51], better results were obtained by using 2 mol equivalents of potassium cyanide at pH 4.0 [67]. Striking were the excellent results obtained with 3-phenoxybenzaldehyde, the *S*-cyanohydrin of which is the alcohol component of some important pyrethroid insecticides [86,115].

A breakthrough with regard to the availability of the enzyme was achieved by its successful overexpression in several microorganisms [116]. The most efficacious expression was accomplished in the yeast *Pichea pastoris*. Yields of more than 20 g of pure HNL protein per liter of culture volume could be obtained. It was subsequently shown that the crude enzyme extract (140 IU/ml) can be used for efficient conversion of all substrates. Under standard conditions 100 IU of enzyme is used per mmol of aldehyde. Substrates that gave unsatisfactory e.e.'s under these conditions were converted with a 20-fold quantity of enzyme at pH 4.5. All reactions were carried out at 0°C.

Summarizing, it can be concluded that excellent preparative procedures have been described for the *R*-HNL from almonds as well as for the overexpressed *S*-HNL from cassava and from the rubber tree. The time seems ripe for large-scale biocatalytic production of both *R*- and *S*-cyanohydrins in enantiomerically pure form on an industrial scale. As illustrated in the second part of this chapter, this will provide the fine chemical industry with a host of valuable new chiral building blocks.

H. Safe Handling of Hydrogen Cyanide

1. Toxicity

Hydrocyanic acid is a highly toxic and easy detectable chemical. Details about the toxicity of hydrocyanic acid (CAS registry no. 74-90-8) and sodium cyanide (CAS registry no. 143-33-9) and about the hazards involved can be found in the regular handbooks [117,118]. Appropriate procedures for the safe use of HCN and related compounds, such as sodium cyanide, that generate HCN when acidified, are described in the literature [117,118]. These include the recommendation that all work with HCN be confined to well-ventilated hoods. As both liquid and dissolved HCN easily penetrates the skin, neoprene or rubber gloves should be worn at all times [119]. The TLV (threshold limit value) or MAC (maximum allowed concentration) of HCN is 10 ppm, which corresponds to 11 mg/m^3. This concentration should also be regarded as a MAC-C (ceiling) value.

2. Explosion Risks

The risk for explosion is minimalized by using only HCN solutions and by the absence of ignition sources in the fume cupboard (hood). As the explosion limits for HCN (LEL US: 5.6% v/v) are appreciably larger than the MAC value (10 ppm), the HCN alarm will go off long before explosion risk occurs. Working under the conditions described in Sec. I.G of this chapter, there is no known explosion hazard.

3. Detection

Hydrogen cyanide is a colorless, sweet-smelling gas with a characteristic odor, reminiscent of almonds. As many people cannot smell HCN at or near the MAC value, smelling does not constitute a reliable detection method. A color change in vapor detection tubes for short-term measurement allows determination of the HCN concentration in a range from 2 to 30 ppm. In conjunction with an appropriate pumping device, these tubes can also be used to measure exposure over longer periods of time. A more sophisticated but more costly way for the quantitative measurement of HCN consists of the use of a measuring and warning system for continuous monitoring of the concentration of HCN in ambient air, equipped with a measuring head and an HCN-dedicated electrochemical sensor. HCN detection equipment is commercially available from multiple sources.

4. Alternatives for the Use of Free Liquid or Gaseous HCN

Although cylinders containing hydrogen cyanide, stabilized with acid, are commercially available, their use for the laboratory preparation of chiral cyanohydrins is not recommended because of the increased risk and more complex safety precautions that have to be taken. By using sodium cyanide or a cyanohydrin as a convenient source of HCN, the handling of free liquid or gaseous HCN can be avoided.

> Sodium cyanide in an acetate buffer at pH 5.4 has been used as an in situ source of HCN in the ethanol–water system [41,46].
> A convenient procedure for the preparation of HCN solutions in an organic solvent has been described [91]. A weak acid, e.g., acetic acid, is added to a cold

solution of sodium cyanide in water, until a pH of 5.5 is reached. The HCN is then extracted with cold ethyl acetate or diisopropyl ether. The amount of HCN in the organic phase can be determined by a described procedure [120]. The organic phase can be used as such or, if the amount of water is to be minimized, it can be dried over sodium sulfate [101].

Acetone cyanohydrin has been shown to act as a practical source of HCN. It can be used as a transcyanating agent in an organic solvent in the presence of an acetate buffer at pH 5.0 [88,96]. More recently, acetone cyanohydrin has been used to generate a steady flow of HCN from a separate compartment, from which it is allowed to diffuse into the actual reaction vessel [52]. The latter procedure avoids the separation problems that are inherent to the transcyanation method.

Other cyanohydrins can also be used in the transcyanation procedure, provided that the corresponding carbonyl compound is not a good substrate for the enzyme. The successful application of the racemic cyanohydrins of 2-pentanone and 2-hexanone has been described [102,111] (see G.1.6).

5. Waste Disposal

Small amounts of HCN can be disposed of most conveniently be conversion to cyanide and oxidation to relatively innocuous cyanate with the aid of sodium hypochlorite (commercial laundry bleach) [121]. The cyanate can be further processed to ammonia and carbon dioxide by addition of sulfuric acid to pH 7, according to:

$$HCN + NaOH \rightarrow NaCN + H_2O$$

$$NaCN + NaOCl \rightarrow NaCl + NaCNO$$

$$2NaCNO + H_2SO_4 + 2H_2O \rightarrow Na_2SO_4 + 2CO_2 + 2NH_3$$

II. CYANOHYDRINS AS MULTIFUNCTIONAL SYNTHONS

Because of their multifunctional character, cyanohydrins offer a wide spectrum of opportunities for transformations (Fig. 4):

1. Reactions at the cyano group (solvolysis, Grignard reaction, hydride addition).
2. Reactions at the hydroxy group (protecting groups).
3. Conversion of the hydroxyl function to a good leaving group, followed by inversion of the stereogenic center by attack of a nucleophile.
4. Reactions at the R substituent. The presence of unsaturation in this substituent offers a whole range of reaction possibilities.

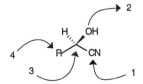

Figure 4 Transformations on cyanohydrins.

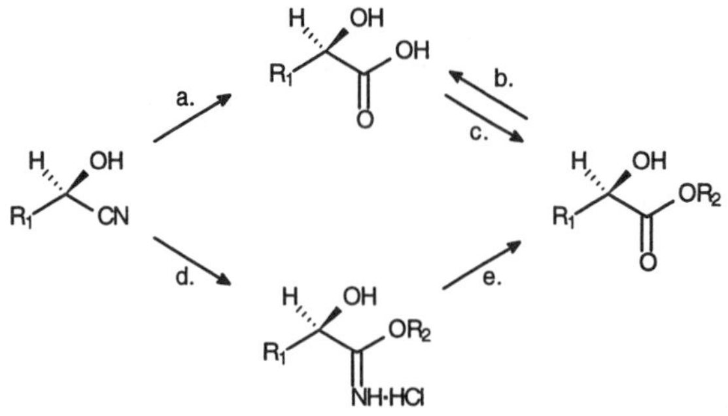

a. HCl/H$_2$O; b. NaOH/dioxane; c. ROH/H$^+$; d. HCl$_{gas}$, ROH, ether; e. H$_2$O

Scheme 10   Products from unprotected cyanohydrins.

When new stereogenic centers are created, the stereogenic center originally present can have a profound influence on the outcome of these reactions (asymmetrical induction).

A.   Reactions Under Acidic Conditions; Formation of α-Hydroxy Acids, Esters, and Lactones

Basic hydrolysis of unprotected cyanohydrins to α-hydroxy acids, as well as any other conversion of cyanohydrins under basic conditions, is attended with severe decomposition and racemization. Acid-catalyzed hydrolysis of cyanohydrins, on the other hand, is generally considered to be a simple and straightforward reaction. However, appreciable decomposition occurs when cyanohydrin and concentrated HCl are mixed and directly heated. This is hardly surprising, as most cyanohydrins are unstable at temperatures exceeding 40°C. Indeed, stirring the cyanohydrin with concentrated HCl at lower temperatures, until all of the starting material is converted to the amide, before raising the temperature in order to complete the hydrolysis, gave better results [44,47,87].

Continuous production of (R)-mandelic acid in an enzyme membrane reactor was reported by Kula et al. [122]. α-Hydroxy esters can be obtained from these acids in the usual ways. They are also accessible directly from the cyanohydrins by a modified Pinner synthesis: conversion in diethyl ether, in the presence of 2 equivalents of the alcohol and 2 equivalents of HCl gas, affords the imidate salt (Scheme 10). By dissolving this salt in water, it is rapidly and quantitatively converted to the corresponding ester [46,98]. If R-cyanohydrins of β-substituted pivaldehydes, like β-methoxy- or β-chloropivaldehyde, are hydrolized with concentrated HCl, (R)-pantolactone is obtained in good yield and high enantiomeric excess [103] (Scheme 11).

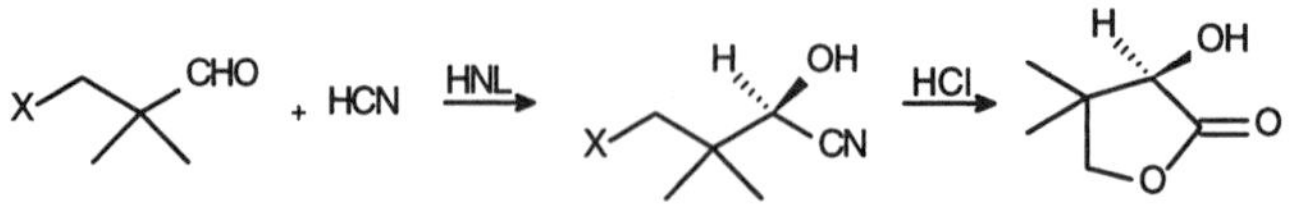

X = CH$_3$O, C$_2$H$_5$O, CH$_2$=CHCH$_2$O, C$_6$H$_5$CH$_2$O, Cl, Br, CH$_3$CO$_2$

Scheme 11   Synthesis of (R)-pantolactone.

B. Protection of the Hydroxyl Function

For application as chiral building blocks, the optically active cyanohydrins have the disadvantage of their chemical instability. In most cases purification by means of distillation, crystallization, or chromatography is accompanied by extensive degradation and racemization. Furthermore, chemical transformations using basic, reducing, or organometallic reagents often result in low yields. These problems can be overcome by introducing a protective group at oxygen (Fig. 5). Care has to be taken that this protective group is introduced under sufficiently mild conditions to avoid racemization.

Chiral cyanohydrins have been subjected to mild esterification for e.e. determination purposes [39]. Yet, because of the reactivity of the ester group, cyanohydrin esters are not suitable as starting materials in most of the subsequent reactions at the nitrile function.

Chiral cyanohydrins have been protected as tetrahydropyranyl (THP) ethers by reaction with 3,4-dihydro-2H-pyran, catalyzed by p-toluenesulfonic acid [9,13,123]. Introduction of the THP protecting group is accompanied by formation of a second chiral center and thus a mixture of diastereomers is formed. This can complicate subsequent reactions, in particular by troubling purification and analytical procedures. To avoid this problem chiral cyanohydrins have been protected as 2-methoxy-i-propyl (MIP) ethers [107,124,125,126] or 2-phenoxy-i-propyl (PIP) ethers [127]. The latter have the additional advantage of being UV-active, thereby facilitating detection of UV-inactive cyanohydrins. MIP and PIP ethers can be easily removed by mild acid hydrolysis.

Most frequently, cyanohydrins are protected as silyl ethers. Silylated cyanohydrins can be obtained, without racemization and in high yields, by reaction with a suitable silyl chloride in the presence of imidazole [41,46,128,129]. The stability of the silyl protecting group can be tuned by proper choice of the substituents at the central silicon atom [130]. The trimethylsilyl (TMS) group can be removed under mildly acidic conditions. The t-butyldimethylsilyl (TBDMS or TBS) group and the UV-active t-butyldiphenylsilyl (TBDPS) group are more acid-stable and can be removed with fluoride ions (TBAF, HF) or, in special cases, by intramolecular reductive cleavage with LiAlH$_4$ [128]. The presence of an atom bearing a proton that reacts with LiAlH$_4$, or a functional group that is reduced by LiAlH$_4$, at the carbon atom neighboring the one carrying the silyl ether is a prerequisite for this intramolecular reductive deprotection. This allows the selective cleavage of one TBS ether over another one that lacks such a structural feature [131].

C. Inversion of the Stereogenic Center

Optically active α-sulfonyloxynitriles, which can be used for inversion of the configuration of cyanohydrins, have been prepared [132]. Aliphatic α-sulfonyloxynitriles were stable at room temperature but aromatic ones already decomposed at 0°C. These α-sulfonyloxyni-

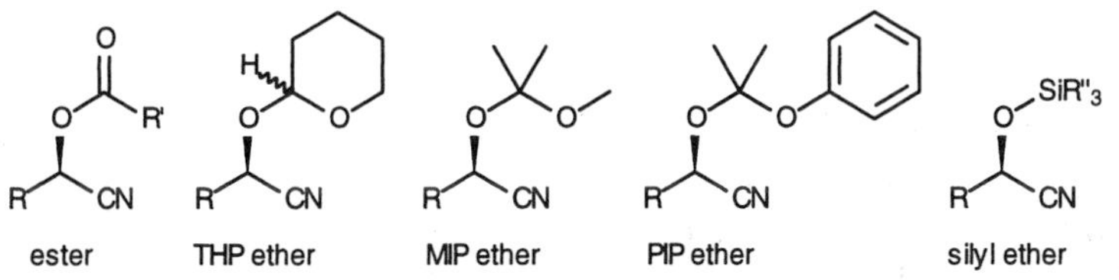

Figure 5 Protected cyanohydrins.

triles reacted with a series of reagents (Scheme 12). Reduction of aliphatic α-sulfonyloxynitriles with LiAlH$_4$ afforded aziridines in good yields and high e.e. [133].

Reactions with azide- and phthalimide nucleophiles afforded a potential entry into α-aminonitriles, which in turn are precursors for α-amino acids. α-Fluoronitriles have been prepared by fluoride substitution and, from unactivated cyanohydrins, using diethylaminosulfur trifluoride (DAST) [92]. For aliphatic substrates these reactions produced α-fluoronitriles with a high e.e. in moderate yields. With aromatic substrates considerable racemization occurred, even if the α-sulfonyloxynitrile was allowed to react in situ with the appropriate nucleophile [93].

Allylic- and benzylic cyanohydrins have been transformed into cyanohydrin esters of opposite configuration under Mitsunobu conditions (Scheme 13) [134,135]. Cyanohydrins containing strongly electron-donating substituents (e.g., 4-methoxymandelonitrile) gave extensive racemization, whereas saturated aliphatic cyanohydrins afforded esters in which the original configuration was retained. As the Mitsunobu inversion gave good results with two important categories of cyanohydrins that did not perform satisfactorily via the sulfonyloxy route, namely, aromatic cyanohydrins and α,β-unsaturated cyanohydrins, the two approaches are complementary.

Using an appropriately protected amine as a Mitsunobu nucleophile it is possible to convert aliphatic cyanohydrins to protected α-aminonitriles, which were converted to chiral amino acids [95].

D. Nucleophilic Addition to the Nitrile Function

1. DIBAL Reduction/Hydrolysis: Formation of α-Hydroxy Aldehydes

Diisobutylaluminum hydride (DIBAL) is known to be less reactive than LiAlH$_4$ and can transfer only one hydride ion per molecule. For this reason DIBAL is commonly used for partial reduction of esters and nitriles. Addition of DIBAL to a nitrile results in the formation of an imine–aluminum complex that is decomposed, during workup, to the aldehyde (Scheme 14). Cyanohydrins have to be protected during the synthesis of α-hydroxy aldehydes because dimerization and isomerization can take place. Even in the protected form special care has to be taken during synthesis (low temperatures) and workup [124,136].

2. LiAlH$_4$ or DIBAL/NaBH$_4$ Reduction: Formation of Simple Ethanolamines

In 1965 the first procedure for the asymmetrical synthesis of ethanolamines via enzyme-catalyzed addition of hydrogen cyanide to aldehydes, followed by reduction with LiAlH$_4$ was described [47,137]. Subsequently, in order to avoid decomposition and racemization, TBS-protected cyanohydrins were used [128]. Surprisingly, quantitative deprotection by an intramolecular reductive cleavage occurred and free ethanolamines were obtained in high yields [128,131]. TBS-protected ethanolamines (with one chiral center) could be obtained by DIBAL reduction at low temperature, followed by NaBH$_4$ reduction [124] (Scheme 14).

3. Grignard Reaction/Hydrolysis: Formation of Acyloins

Optically active α-hydroxy ketones (acyloins) were prepared by a Grignard reaction on TMS- or TBS-protected mandelonitriles [41] (Scheme 15). Because of the acid lability of the TMS group, in the former case the free acyloins were isolated after acidic workup. In the latter case the TBS-protected derivatives were isolated. In both cases high e.e. values

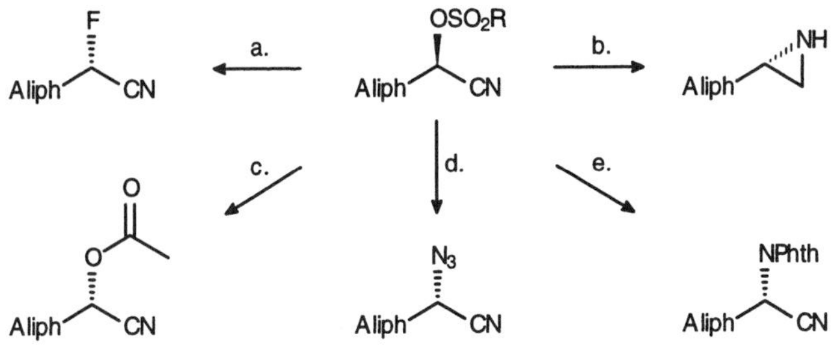

a. Amberlyst/F⁻; b. LiAlH₄; c. CH₃CO₂K/DMF; d. NaN₃; e. potassium phthalimide

Scheme 12   Substitution products from α-sulfonyloxynitriles.

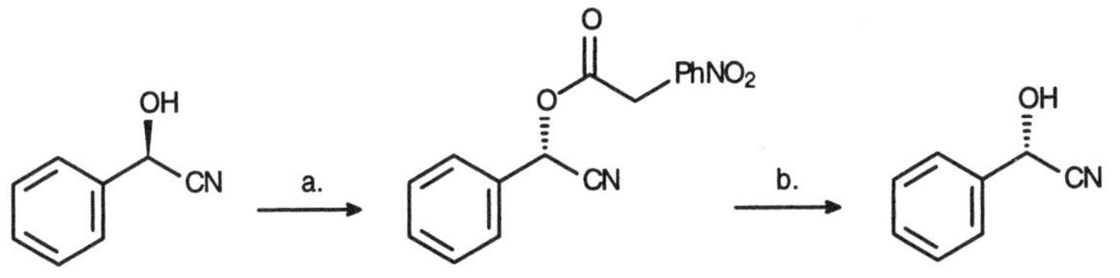

a. (Ph)₃P, DEAD, p-NO₂-phenylacetic acid; b. MsOH/MeOH

Scheme 13   Inversion under Mitsunobu conditions.

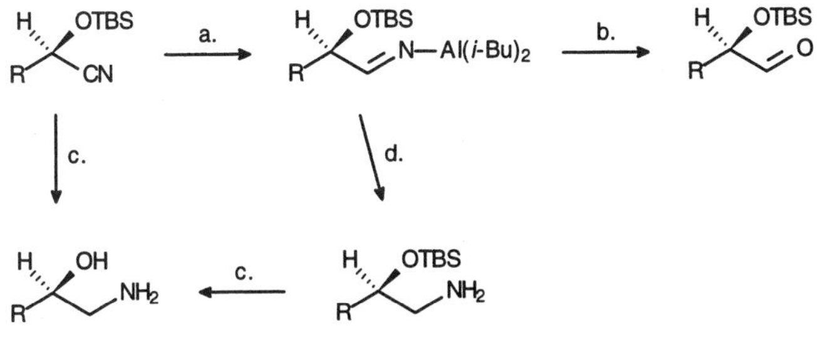

a. DIBAL; b. H₃O⁺; c. LiAlH₄; d. NaBH₄

Scheme 14   Hydride reduction of O-protected cyanohydrins.

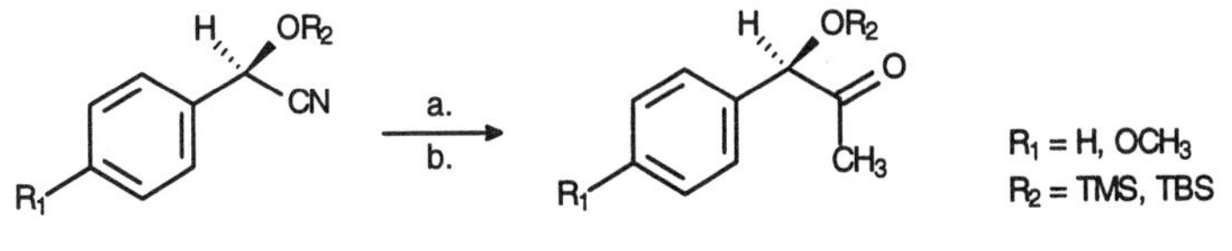

$R_1 = H, OCH_3$
$R_2 = TMS, TBS$

a. CH₃MgI; b. H₃O⁺

Scheme 15   Synthesis of O-protected acyloins.

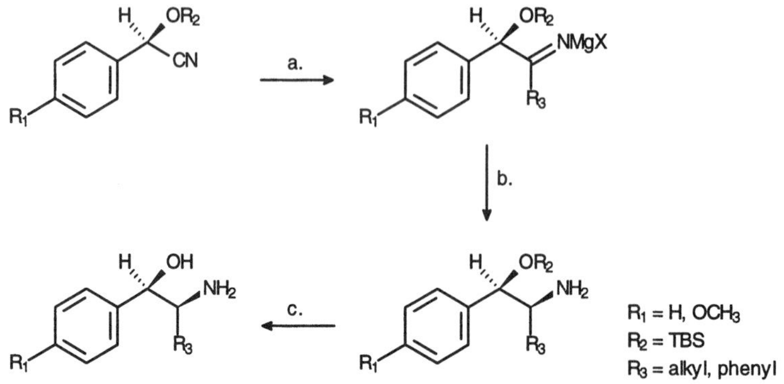

a. R$_3$MgX; b. NaBH$_4$; c. HF or LiAlH$_4$

Scheme 16   Synthesis of ethanolamines with two stereogenic centers.

were recorded. The protected acyloins are excellent starting materials for further stereo-selective transformations. For example, reductive amination in the presence of magnesium perchlorate afforded ethanolamines with high stereoselectivity [129].

4.   Grignard Reaction/Hydride Reduction: Introduction of a Second Stereogenic Center

In the previous paragraph it was shown that Grignard reagents add smoothly to silyl-protected cyanohydrins. Reduction with NaBH$_4$ of the imine intermediates, instead of acidic workup, afforded ethanolamines with two stereogenic centers (Scheme 16). Most likely, steric control during formation of the second stereogenic center is enhanced by coordination of magnesium(II) ions to both the nitrogen and the oxygen moiety. Erythro/threo ratios of the crude products varied between 85:15 and 91:9. Pure erythroethanol-amines were obtained after one crystallization [108,128].

5.   Addition of Vinyl Magnesium Bromide: Formation of 2-Amino-1,3-diols

Preparation of 2-amino-1,3-diols should be possible using functionalized organometallic reagents like Grignard compounds derived from halo ethers [138] or silylmethyl chlorides [139]. However, attempts by Effenberger et al. resulted in racemization and decomposition of the starting material. An alternative route to 2-amino-1,3-diols was found by introducing a vinyl group via Grignard addition to the nitrile function and subsequent ozonolysis, followed by reductive workup [140] (Scheme 17).

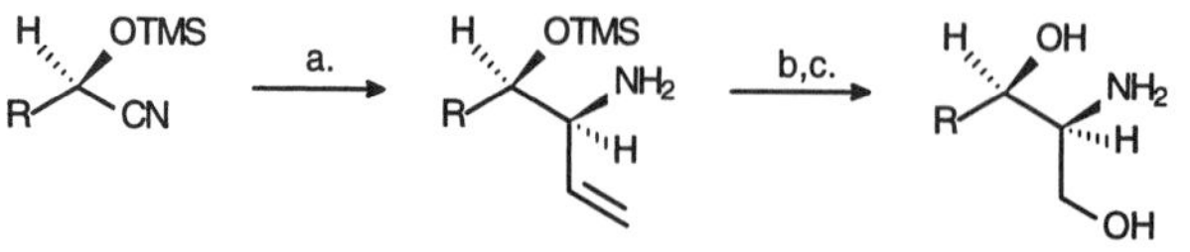

a. 1. vinylmagnesium bromide, 2. NaBH$_4$, 3. H$_3$O$^+$; b. Ac$_2$O/Py; c. 1. O$_3$, 2. NaBH$_4$

Scheme 17   Synthesis of 2-amino-1,3-diols.

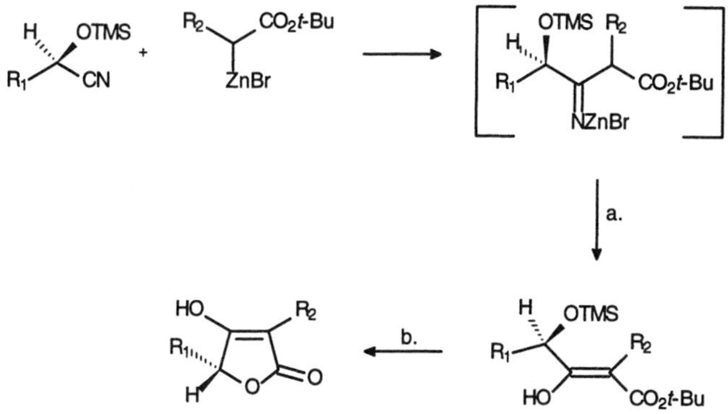

a. NH$_4$Cl/H$_2$O; b. HCl in THF or MeOH

Scheme 18 Synthesis of tetronic acids.

6. Blaise Reaction: Formation of Tetronic Acids

Tetronic acids (Scheme 18) are of great interest because their derivatives are widespread in nature. The Blaise reaction (addition of the zinc derivative of an α-bromo ester to a nitrile) has been applied to optically active O-protected cyanohydrins. Isolation of the tetronic acids was achieved, in moderate yields, by acid hydrolysis and cyclization of the intermediate imines [94]. Better results were obtained in an optimized reaction sequence [141] (Scheme 18).

E. Transimination

Transimination (interconversion of Schiff bases), also called transaldimination or trans-Schiffization, denotes the reaction of an imine with a primary amine so that the original imine is converted to a new imine and a primary amine or ammonia is liberated (Scheme 19). Chemically it is a symmetrical and reversible process.

The structures of the participating reactants strongly affect the rate of the reaction and the position of the equilibrium. The thermodynamic stability of —C=N— linkages increases with the type of amine used, in the order NH$_3$ << aliphatic amines < aromatic amines < amines with an adjacent electronegative atom bearing a free electron pair (e.g., H$_2$N—OH) [142].

Primary imines are highly reactive and water-sensitive. In most cases these compounds cannot be isolated. Most secondary imines, on the other hand, are stable enough to allow isolation. Apparently, the N-alkyl substituent exerts an appreciable stabilizing effect onto the imine double bond. Consequently, when an in situ–prepared primary imine is allowed to react with an alkyl amine under equilibrating conditions, transimination takes place and the more stable secondary imine is readily formed.

Scheme 19 Transimination.

By partial hydrogenation with DIBAL, or reaction with RMgX, of O-protected cy-anohydrins, metallated primary imines are formed initially. Upon protonation these can be subjected to transimination with primary amines to furnish, upon reduction, biologically interesting N-substituted amino alcohols [106,108,124,143], e.g., ephedrine, tembamide, aegeline, denopamine, and some analogs of salbutamol [144]. (*1R,2S*)-Amino alcohols have been converted to (*S*)-amphetamines by catalytic hydrogenation of their oxazolidi-nones [108].

1. Grignard Reaction/Transimination/Hydride Reduction: *N*-Substituted Ethanolamines

N-Substituted ethanolamines with two stereogenic centers have been prepared in a one-pot synthesis [143,145]. In this procedure the intermediate primary imine is transiminated with a primary amine followed by NaBH$_4$ reduction (Scheme 20). Even transimination with amino acid esters [123] and with ethanolamines to form diethanolamines [146] is possible.

2. DIBAL Reduction/Transimination/HCN Addition/Hydrolysis: β-Hydroxy-α-Amino Acids

MIP-protected (*R*)-mandelonitrile was reduced by DIBAL at low temperature ($-70°C$). Quenching with ammonium bromide in methanol afforded the free primary imine that was transiminated with a primary amine to introduce the *N*-alkyl group. This secondary imine was directly converted to the α-cyanoamine by addition of HCN, creating a second ste-reogenic center. During acidic workup the MIP group is removed, affording the HCl salt of the β-hydroxy-α-cyanoamine. After conversion to the oxazolidinone derivative, mild hydrolysis (solvolysis) of the nitrile group was accomplished with K$_2$CO$_3$ in ethanol. Sa-ponification of the ester and oxazolidinone moiety afforded (*2S,3R*)-N-alkyl-β-hydroxy-α-amino acids in high overall yields [126] (Scheme 21).

3. Transimination with *N*-Substituted Hydroxylamines: Chiral Aldo- and Ketonitrones

Nitrones are usually prepared by condensation of aldehydes with hydroxylamines. Cya-nohydrins can be converted to α-hydroxy aldehydes (see II.E.1), which in turn can be

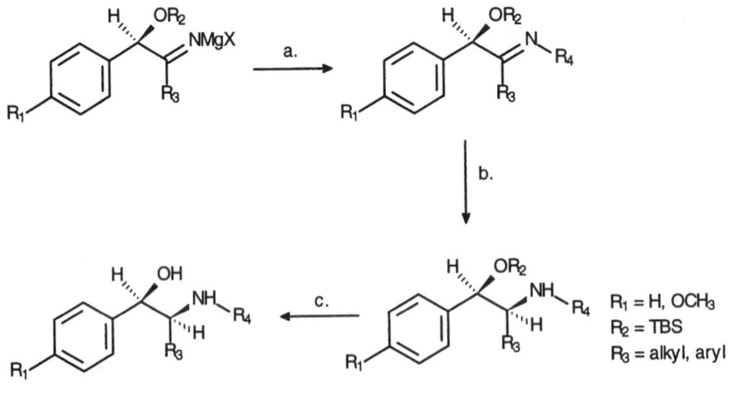

a. 1. MeOH, 2. R$_4$NH$_2$; b. NaBH$_4$; c. LiAlH$_4$

Scheme 20   Synthesis of *N*-substituted ethanolamines.

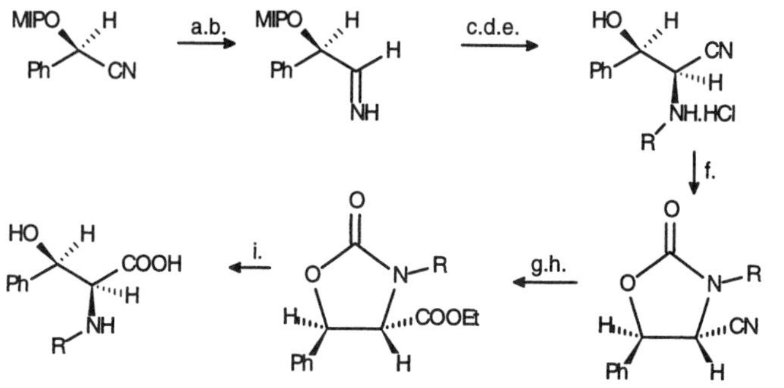

a. DIBAL; b. MeOH; c. RHN$_2$; d. HCN; e. HCl; f. CO(lm)$_2$; g. K$_2$CO$_3$, EtOH; h. HCl; i. 1. KOH, 2. H$_3$O$^+$

Scheme 21 Synthesis of β-hydroxy-α-amino acids.

used for the synthesis of nitrones. However, a particularly attractive method is the direct conversion of chiral cyanohydrins to chiral nitrones via transimination (Scheme 22). Conversion of the nitrile to a primary imine, either by DIBAL reduction or by Grignard addition, is directly followed by transimination with N-benzylhydroxylamine. Workup after 4 h provided aldo- and ketonitrones in high yields and of high enantiomeric purity [107]. These chiral nitrones in turn are valuable starting materials for intra- and intermolecular 1,3-dipolar cyclo additions [100].

F. Cyanohydrins from α,β-Unsaturated Aldehydes

Croton aldehyde (2-butenal), like many other α,β-unsaturated aldehydes, is a good substrate for almond HNL [98,125]. The unsaturated cyanohydrin (R,E)-2-hydroxy-3-pentenenitrile can be used, after protection of the hydroxyl group, as a potential chiral C$_3$-synthon in which the double bond serves as a masked oxygen functionality.

The cyano group is especially suitable for the introduction of alkyl substituents at carbon by nucleophilic addition and the formation of an amine function by reduction.

1. Ozonolysis/Reduction: Synthesis of Beta Blockers

Ozonolysis allows several modes of workup, thus providing a route for converting the olefin to an alcohol, an aldehyde, or an acid. A method was developed for the preparation of both enantiomers of 5-(hydroxymethyl)-3-isopropyloxazolidin-2-one in optically pure form [125] (Scheme 23). These compounds are known to be excellent building blocks for

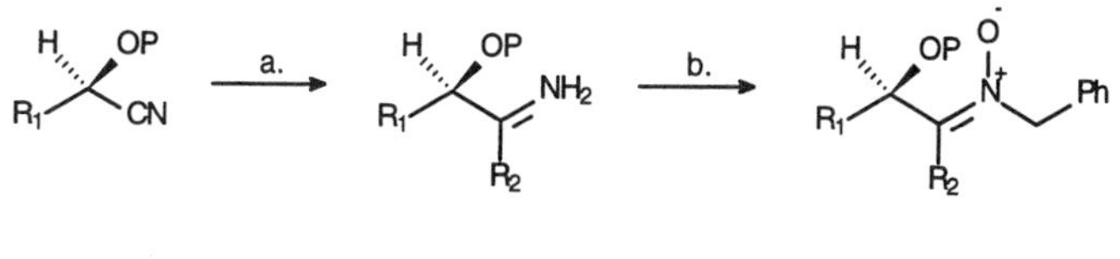

P = MIP, TBS; R$_1$, R$_2$ = alkyl, alkenyl, aryl
a. DIBAL or RMgX; b. Benzylhydroxylamine

Scheme 22   One-pot synthesis of aldo- and ketonitrones.

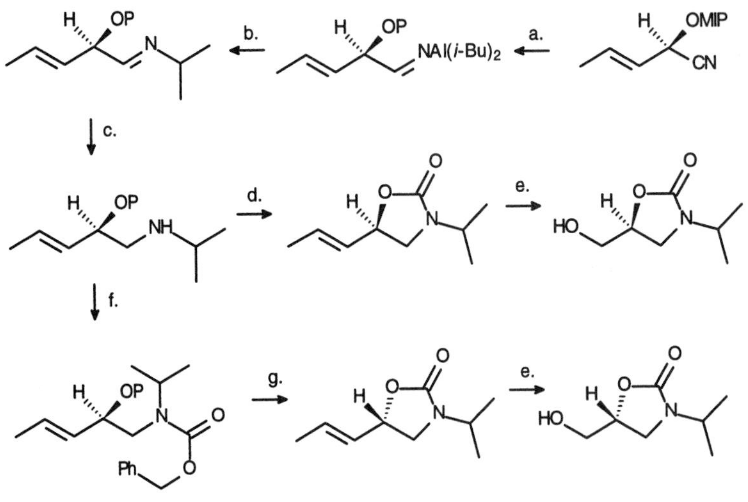

a. DIBAL; b. 1. MeOH, 2. *i*-propylamine; c. NaBH$_4$; d. CO(lm)$_2$; e. 1. O$_3$, 2. NaBH$_4$; f. benzyl chloroformate; g. SOCl$_2$

Scheme 23   Synthesis of both enantiomers of 5-hydroxymethyl-3-*i*-propyloxazolidin-2-one.

the preparation of chiral beta blockers (β-adrenoceptor antagonists), which are among the world's most applied pharmaceuticals. Most beta blockers presently marketed are racemic 1-alkylamino-3-aryloxy-2-propanols, in which the aryl group can be one of a large variety of aromatic substituents and the *N*-alkyl substituent is generally an *i*-propyl or a *t*-butyl group. The stereoselective approach depicted in Scheme 24 allows the introduction of a large variety of substituents at nitrogen, since the transimination step proceeds well with most primary amines, even with sterically hindered ones like *t*-butylamine. It was been observed earlier that transimination using *t*-butylamine was not successful when α-substi-

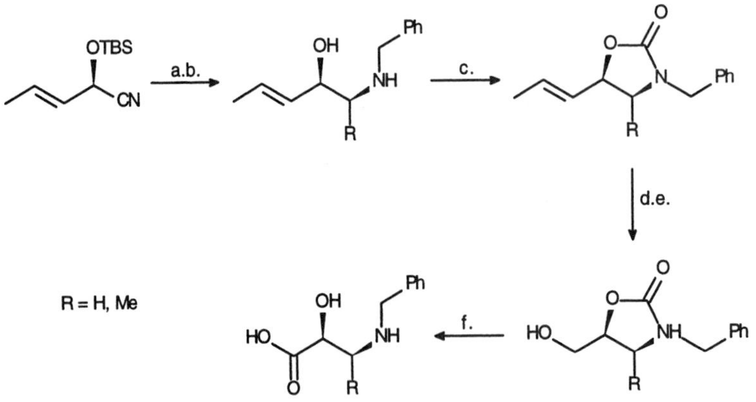

a. 1. DIBAL or CH$_3$MgI, 2. MeOH, 3. benzylamine, 4. NaBH$_4$; b. LiAlH$_4$; c. CO(lm)$_2$; d. O$_3$; e. Jones' ox.; f. 2 M KOH

Scheme 24   Synthesis of α-hydroxy-β-amino acids.

tuted imines, obtained by Grignard addition to the nitrile group, were used [146]. Inversion of the chiral center was accomplished by treating the *N*-benzyloxycarbonyl derivative with thionyl chloride [147]. Ozonolysis is carried out at a stage where both the hydroxyl and the amino substituent are protected as an oxazolidinone.

2. Ozonolysis/Oxidation: Synthesis of α-Hydroxy-β-Amino Acids

(*R,E*)-2-Hydroxy-3-pentenenitrile was also shown to be an excellent chiral starting material for the synthesis of α-hydroxy-β-amino acids of high enantiomeric and diastereomeric purity [145] (Scheme 24). In this case the first step can be a DIBAL reduction (R = H) or a Grignard reaction (R = alkyl or aryl) and the ozonolysis is directly followed by oxidative workup (Jones reagent). Hydrolysis of the oxazolidinones provided the desired *N*-protected α-hydroxy-β-amino acids.

REFERENCES

1. F Effenberger. Angew Chem Int Ed Engl 33:1555–1564, 1994.
2. LT Kanerva. Acta Chem Scand 50:234–242, 1996.
3. F Effenberger. Enantiomer 1:359–363, 1996.
4. A Hickel, M Hasslacher, H Griengl. Physiol Plant 98:891–898, 1996.
5. CG Kruse. In: An Collins, GN Sheldrake, J Crosby, eds. Chirality in Industry. New York: John Wiley, 1992, pp 279–299.
6. H Griengl, A Hickel, DV Johnson, C Kratky, M Schmidt, H Schwab. Chem Commun 1933–1940, 1997.
7. M North. Synlett 807–820, 1993
8. J March, ed. Advanced Organic Chemistry, 4th ed. New York: Wiley Interscience, 1992, p 964.
9. H Ohta, Y Kimura, Y Sugano, T Sugai. Tetrahedron 45:5469–5476, 1989.
10. H Ohta, Y Kimura, Y Sugano. Tetrahedron Lett 29:6957–6970, 1988.
11. H Ohta, Y Miyamae, G Tsuchihashi. Agric Biol Chem 53:215–222, 1989.
12. H Ohta, Y Miyamae, G Tsuchihashi. Agric Biol Chem 53:281–283, 1989.
13. Y-F Wang, S-T Chen, KK-C Liu, C-H Wong. Tetrahedron Lett 30:1917–1920, 1989.
14. N Matsuo, N Ohno. Tetrahedron Lett 26:5533–5534, 1985.
15. Y Lu, C Miet, N Kunesch, J Poisson. Tetrahedron: Asymmetry 1:707–710, 1990.
16. A Van Almsick, J Buddrus, P Hönicke-Schmidt, K Laumen, MP Schneider. J Chem Soc Chem Commun 1391–1393, 1989.
17. H Waldmann. Tetrahedron Lett 30:3057–3058, 1989.
18. I Tellitu, D Badía, E Domínguez, FJ García. Tetrahedron: Asymmetry 5:1567–1578, 1994.
19. J Roos, U Stelzer, F Effenberger. Tetrahedron: Asymmetry 9:1043–1049, 1998.
20. L Rosenthaler. Biochem Z 14:238–253, 1908.
21. L Rosenthaler. Arch Pharm 251:56–84, 1913.
22. L Rosenthaler. Biochem Z 50:486–496, 1913.
23. VK Krieble. J Am Chem Soc 34:716–735, 1912.
24. VK Krieble. J Am Chem Soc 35:1643–1647, 1913.
25. VK Krieble, WA Wieland. J Am Chem Soc 43:164, 1921.
26. E Gerstner, E Pfeil, Hoppe-Seyler's Z Physiol Chem 353:271–286, 1972.
27. E Gerstner, U Kiel, Hoppe-Seyler's Z Physiol Chem 356:1853–1857, 1975.
28. W Becker, U Benthin, E Eschenhof, E Pfeil. Biochem Z 337:156–166, 1963.
29. MK Seely, RS Criddle, EE Conn. J Biol Chem 241:4457–4462, 1966.
30. RS Yemm, JE Poulton. Arch Biochem Biophys 247:440–445, 1986.
31. L-L Xu, BK Singh, EE Conn. Arch Biochem Biophys 250:322–328, 1986.
32. L-L Xu, BK Singh, EE Conn. Arch Biochem Biophys 263:256–263, 1988.

33. GW Kuroki, EE Conn. Proc Nat Acad Sci USA 86:6978–6981, 1989.
34. D Selmar, R Lieberei, B Biehl, EE Conn. Physiologia Plantarum 75:97–101, 1989.
35. W Becker, H Freund, E Pfeil. Angew Chem Int Ed Eng 4:1079, 1965.
36. W Becker, E Pfeil. J Am Chem Soc 88:4299–4300, 1966.
37. E Pfeil, W Becker. Fed Ger Pat Appl P 36473 IV b/12 o, 1965.
38. J Brussee, ACA Jansen, A Kühn. Proc Symp Med Chem, Acta Pharm Suecica Suppl 2:479, 1985.
39. F Effenberger, T Ziegler, S Förster. Angew Chem Int Ed Engl 26:458–459, 1987.
40. F Effenberger, T Ziegler, S Förster. Eur Pat 0 276 375 A2, to Degussa A.G, 1987.
41. J Brussee, EC Roos, A van der Gen. Tetrahedron Lett 29:4485–4488, 1988.
42. J Brussee, A van der Gen. Eur Pat 322, 973 to Duphar, 1988.
43. M-R Kula, U Niedermeyer, IM Stürtz. Eur Pat Appl 0 350 908 A2, to Degussa AG, 1989.
44. F Effenberger, B Hörsch, S Förster, T Ziegler. Tetrahedron Lett 31:1249–1252, 1990.
45. U Niedermeyer, M-R Kula. Angew Chem Int Ed Engl 29:386–387, 1990.
46. J Brussee, WT Loos, CG Kruse, A van der Gen. Tetrahedron 46:979–986, 1990.
47. T Ziegler, B Hörsch, F Effenberger. Synthesis 575–578, 1990.
48. W Becker, E Pfeil. Biochem Z 346:301–321, 1966.
49. E Kiljunen, LT Kanerva. Tetrahedron: Asymmetry 5:311–314, 1994.
50. W Becker, U Benthin, E Eschenhof, E Pfeil. Angew Chem 75:93, 1963.
51. N Klempier, H Griengl, M Hayn. Tetrahedron Lett 34:4769–4772, 1993.
52. E Kiljunen, LT Kanerva. Tetrahedron: Asymmetry 8:1225–1234, 1997.
53. E Kiljunen, LT Kanerva. Tetrahedron: Asymmetry 8:1551–1577, 1997.
54. J Hughes, FJP De Carvalho, MA Hughes. Arch Biochem Biophys 311:496–502, 1994.
55. H Wajant, S Förster, H Böttinger, D Selmar, F Effenberger, K Pfizenmaier. Plant Physiol 109: 1231, 1995.
56. U Kragl, U Niedermeyer, M-R Kula, C Wandrey. Ann New York Acad Sci 167–175, 1990.
57. E Smitskamp-Wilms, J Brussee, A van der Gen, GJM van Scharrenburg, JB Sloothaak. Recl Trav Chim Pays-Bas 110:209–215, 1991.
58. GJM van Scharrenburg, JB Sloothaak, CG Kruse, E Smitskamp-Wilms, J Brussee. Ind J Chem Sect B Org Chem 32:16–19, 1993.
59. C Bové, EE Conn. J Biol Chem 236:207–210, 1961.
60. IE Coop. N.Z. J Sci Techn 71, 1940.
61. W Becker, E Pfeil. Naturwissenschaften 51:193, 1964.
62. E Pfeil. Personal communication, 1980.
63. E Gerstner, E Pfeil. Z Physiol Chem 353:271–286, 1972.
64. H Lauble, K Müller, H Schindelin, S Förster, F Effenberger. Proteins: Structure, Functions Genet 19:343–347, 1994.
65. H Lauble, S Knödler, H Schindelin, S Förster, H Wajant, F Effenberger. Acta Crystallzer D52:887–889, 1996.
66. N Klempier, U Pichler, H Griengl. Tetrahedron: Asymmetry 6:845–848, 1995.
67. M Schmidt, S Hervé, N Klempier, H Griengl. Tetrahedron 52:7833–7840, 1996.
68. S Förster, J Roos, F Effenberger, H Wajant, A Sprauer. Angew Chem Int Ed Engl 35:437–439, 1996.
69. E Hochuli. Helv Chim Acta 66:489–493, 1983.
70. R Kaul, B Mattiasson. Biotechn Appl Biochem 9:294–302, 1987.
71. EE Conn. J Agric Food Chem 17:519, 1969.
72. L Rosenthaler. Arch Pharm 246:365–366, 1908.
73. H Albers, H Albrecht. Biochem Z 269:44–62, 1934.
74. HH Hustedt, E Pfeil. Liebigs Ann Chem 640:15–28, 1961.
75. MS Jorns. Biochim Biophys Acta 613:203–209, 1980.
76. WW Cleland. Biochim Biophys Acta 67:104–137, 1963.
77. MS Jorns. J Biol Chem 254:12145–12152, 1979.
78. D Vargo, A Pokora, SW Wang, MS Jorns. J Biol Chem 256:6027–6033, 1981.

79. A Pokora, MS Jorns, D Vargo. J Am Chem Soc 104:5466–5469, 1982.
80. MS Jorns, C Ballenger, G Kinney, A Pokora, D Vargo. J Biol Chem 258;8561–8567, 1983.
81. KR Bärwald, L Jaenicke. FEBS Lett 90:255–260, 1978.
82. L Jaenicke, J Preun. Eur J Biochem 138:319–325, 1984.
83. MS Jorns. Eur J Biochem 146:481–482, 1985.
84. J Brussee, A van der Gen. Unpublished results.
85. H Griengl. Proc BioEurope, 1997.
86. U Wagner, M Hasslacher, H Griengl, H Schwab, C Kratky. Structure 4:811–822, 1996.
87. F Effenberger, B Hörsch, F Weingart, T Ziegler, S Kühner. Tetrahedron Lett 32:2605–2608, 1991.
88. TT Huuhtanen, LT Kanerva. Tetrahedron: Asymmetry 3:1223–1226, 1992.
89. F Effenberger, M Hopf, T Ziegler, J Hudelmayer. Chem Ber 124:1651–1659, 1991.
90. P Zandbergen J van der Linden, J Brussee, A van der Gen. Preparative Biotransformations. Whole cell and isolated enzymes in organic synthesis. Update 5, 4:5.1. Wiley Looseleaf Publications, Ed. SM Roberts 1995.
91. P Zandbergen, J van der Linden, J Brussee, A van der Gen. Synth Commun 21:1387–1391, 1991.
92. U Stelzer, F Effenberger. Tetrahedron: Asymmetry 4:161–164, 1993.
93. F Effenberger, U Stelzer. Chem Ber 126:779–786, 1993.
94. JJ Duffield, AC Regan. Tetrahedron: Asymmetry 7:663–666, 1996.
95. CP Decicco, P Grover. Synlett 529–530, 1997.
96. VI Ognyanov, VK Datcheva, KS Kyler. J Am Chem Soc 113:6992–6996, 1991.
97. J Syed, S Förster, F Effenberger. Tetrahedron: Asymmetry 9:805–815, 1998.
98. EGJC Warmerdam, AMCH van den Nieuwendijk, CG Kruse, J Brussee, A van der Gen. Recl Trav Chim Pays-Bas 115:20–24, 1996.
99. DV Johnson, H Griengl. Tetrahedron 53:617–624, 1997.
100. J Marcus, J Brussee, A van der Gen. Eur J Org Chem 2513–2517, 1998.
101. S Han, G Lin, Z Li, Tetrahedron: Asymmetry 9:1835–1838, 1998.
102. E Menéndez, R Brieva, F Rebolledo, V Gotor. J Chem Soc Chem Commun 989–990, 1995.
103. F Effenberger, J Eichhorn, J Roos. Tetrahedron: Asymmetry 6:271–282, 1995.
104. F Effenberger, J Eichhorn. Tetrahedron: Asymmetry 8:469–476, 1997.
105. E Kiljunen, LT Kanerva. Tetrahedron: Asymmetry 7:1105–1116, 1996.
106. F Effenberger, J Jäger. J Org Chem 62:3867–3873, 1997.
107. E Hulsbos, J Marcus, J Brussee, A van der Gen. Tetrahedron: Asymmetry 8:1061–1067, 1997.
108. F Effenberger, J Jäger. Chem Eur J 3:1370–1374, 1997.
109. HJ Aschhoff, E Pfeil. Hoppe-Seyler's Z Physiol Chem 351:818–826, 1970.
110. E Wehtje, P Adlercreutz, B Mattiasson. Biotechn Bioeng 36:39–46, 1990.
111. S Nazabadioko, R Pérez, R Brieva, V Gotor. Tetrahedron: Asymmetry 9:1597–1604, 1998.
112. WT Loos, HW Geluk, MMA Ruijken, CG Kruse, J Brussee, A van der Gen. Biocatalysis and Biotransformation 12:255–266, 1995.
113. TJ Mitchel. Technometrics 16:203–211, 1974.
114. GEP Box, JS Hunter, W Hunter. In RA Bradly, DG Kendall, JS Hunter, GS Watson, eds. Statistics for Experimenters. New York: John Wiley, 1978.
115. M Elliot, AW Farnham, NF James, PH Needham, DA Pulman. Nature 248:710–711, 1974.
116. M Hasslacher, M Schall, M Hayn, H Griengl, SD Kohlwein, H Schwab. J Biol Chem 271:5884, 1996.
117. RJ Lewis, Sr. Sax's Dangerous Properties of Industrial Materials, 8th ed. New York: Van Nostrand Reinhold, 1992.
118. National Research Council. Prudent Practices for Handling Hazardous Chemicals in Laboratories. Washington, DC: National Academy Press, 1981.
119. P Grandjean. Skin Penetration: Hazardous Chemicals at Work. London: Taylor & Francis, 1990.

120. W. Horwitz. Official Methods of Analysis of the Association of Official Analytical Chemists, 13th ed. Washington, DC: AOAC, 1980, p 318.

121. National Research Council. Prudent Practices for Disposal of Chemicals from Laboratories. Washington, DC: National Academy Press, 1983.

122. D Vasic-Racki, M Jonas, C Wandrey, W Hummel, M-R Kula. Appl Microbiol Biotechnol 31: 215–222, 1989.

123. AMCH van den Nieuwendijk, EGJC Warmerdam, J Brussee, A van der Gen. Tetrahedron: Asymmetry 6:801–806, 1995.

124. P Zandbergen, AMCH van den Nieuwendijk, J Brussee, A van der Gen, CG Kruse. Tetrahedron 48:3977–3982, 1992.

125. EGJC Warmerdam, J Brussee, CG Kruse, A van der Gen. Helv Chim Acta 77:252–256, 1994.

126. P Zandbergen, J Brussee, A van der Gen. CG Kruse. Tetrahedron: Asymmetry 3:769–774, 1992.

127. P Zandbergen, HMG Willems, GA van der Marel, J Brussee, A van der Gen. Synth Commun 22:2781–2787, 1992.

128. J Brussee, F Dofferhoff, CG Kruse, A van der Gen. Tetrahedron 46:1653–1658, 1990.

129. J Brussee, RATM van Benthem, A van der Gen. Tetrahedron. Asymmetry 1:163–166, 1990.

130. TW Greene, PG Wuts. Protective Groups in Organic Synthesis. New York: John Wiley, 1991.

131. EFJ de Vries, J Brussee, A van der Gen. J Org Chem 59:7133–7137, 1994.

132. F Effenberger, U Stelzer. Angew Chem Int Ed Engl 30:873–874, 1991.

133. F Effenberger, U Stelzer. Tetrahedron: Asymmetry 6:283–286, 1995.

134. O Mitsunobu. Synthesis 1–28, 1981.

135. EGJC Warmerdam, J Brussee, CG Kruse, A van der Gen. Tetrahedron 49:1063–1070, 1993.

136. WR Jackson, HA Jacobs, GS Jayatilake, BR Matthews, KG Watson. Aust J Chem 43:2045–2062, 1990.

137. W Becker, H Freund, E Pfeil. Angew Chem 77:1139, 1965.

138. B Castro. Bull Soc Chim France 1533–1540, 1967.

139. GJPH Boons, M Overhand, GA van der Marel, JH van Boom. Angew Chem Int Ed Engl 28: 1504–1505, 1989.

140. F Effenberger, B Gutterer, J Syed. Tetrahedron: Asymmetry 6:2933–2943, 1995.

141. F Effenberger, J Syed. Tetrahedron: Asymmetry 9:817–825, 1998.

142. MJ Mäkelä, TK Korpela. Chem Soc Rev 12:309, 1983.

143. J Brussee, A van der Gen. Recl Trav Chim Pays-Bas 110:25–26, 1991.

144. RFC Brown, AC Donohue, WR Jackson, TD McCarthy. Tetrahedron 50:13793–13752, 1994.

145. EGJC Warmerdam, RD van Rijn, J Brussee, CG Kruse, A van der Gen. Tetrahedron: Asymmetry 7:1723–1732, 1996.

146. EFJ de Vries, P Steenwinkel, J Brussee, CG Kruse, A van der Gen. J Org Chem 58:4315–4325, 1993.

147. S Kano, T Yokomatsu, H Iwasawa, Shibuya. Tetrahedron Lett 28:6331, 1987.

12

Hydroxynitrile Lyases in Stereoselective Synthesis

Franz Effenberger
University of Stuttgart, Stuttgart, Germany

I. INTRODUCTION

The release of HCN (cyanogenesis) as a defense against herbivores is widely distributed in higher plants including important food plants such as cassava or *Sorghum*. The highly toxic HCN is chemically masked in the form of cyanohydrins, which are stabilized by an *O*-β-glycosidic linkage to saccharides. During cyanogenesis (Scheme 1) first a specific β-glycosidase cleaves the cyanoglycoside into a carbohydrate and the corresponding cyanohydrin, which subsequently decomposes to the carbonyl compound and HCN. The latter step occurs spontaneously but much faster enzymatically by action of a hydroxynitrile lyase (HNL) [1]. In Scheme 1 the release of HCN and the formation of D-glucose and benzaldehyde from *prunasin* (**5**) represents an example of cyanogenesis in higher plants. It is interesting to note that an optically active cyanohydrin results by hydrolysis of the cyanoglucoside and the enzyme-catalyzed cleavage of the (*R*)-cyanohydrin requires an (*R*)-specific HNL.

As any other catalyst, enzymes also catalyze reactions in both directions. An enantioselective addition of HCN to aldehydes in the presence of a HNL, forming an optically active cyanohydrin, could therefore be deduced from the enantioselectivity of cyanogenesis.

One of the first asymmetrical syntheses and the first effected by an enzyme was the preparation of (*R*)-mandelonitrile from benzaldehyde and HCN with *emulsin* as source for an (*R*)-HNL by Rosenthaler in 1908 [2]. More than 50 years later, Pfeil et al. [3] used the isolated enzyme from bitter almond (*Prunus amygdalus*), (*R*)-PaHNL (EC 4.1.2.10), and investigated the reaction in a more general way by also using other aldehydes besides benzaldehyde. However, all efforts to improve the optical yields of the resulting (*R*)-cyanohydrins failed. Especially with slower reacting aldehydes the optical yields were very poor. As for working with enzymes customary at that time, Pfeil et al. [3] used water or water-ethanol as solvents and a pH of 5–6, the optimum of activity for (*R*)-PaHNL. Under these conditions the chemical addition of HCN to aldehydes resulting in racemic cyanohydrins prevails, especially when the enzyme-catalyzed addition is slow.

The decisive breakthrough for synthetic applications of this enzyme-catalyzed reaction was the discovery that the chemical addition of HCN to aldehydes is more or less suppressed in organic solvents not miscible with water. The enzyme-catalyzed reaction in

321

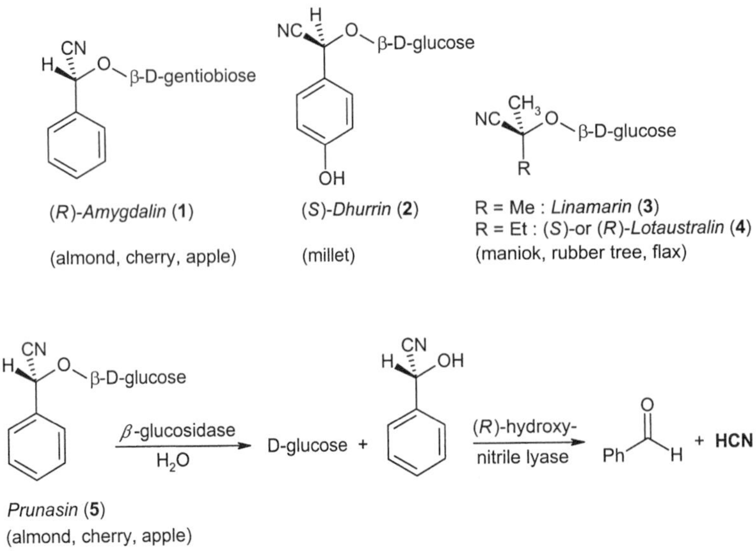

Scheme 1   Important cyanoglycosides and the cyanogenesis of prunasin.

organic solvents, however, is only slightly slower than that in water and therefore (R)-cyanohydrins of high optical purity are obtained [4].

The finding of suppressing the chemical addition of HCN to aldehydes by working in organic solvents with the result of an easy access to optically active cyanohydrins gave rise to many investigations of the preparation and follow-up reactions of (R)- as well as (S)-cyanohydrins [5–9].

II.   HYDROXYNITRILE LYASES FOR SYNTHETIC APPLICATIONS

The biosynthesis of cyanohydrins from α-amino acids is well established [1]. The majority [23] of the 28 cyanogenic glycosides are derived from five proteinaceous amino acids: L-valine, L-isoleucine, L-leucine, L-phenylalanine, and L-tyrosine. The cyanohydrins formed by biodegradation of these amino acids contain as carbonyl compounds, mainly acetone [linamarin (3)], 2-butanone [(R)-lotaustralin (4)], benzaldehyde [(R)-amygdalin (1), (R)-prunasin (5)], and p-hydroxybenzaldehyde [(S)-dhurrin (2)] (Scheme 1) [1]. In bitter almonds (R)-amygdalin (1) and (R)-prunasin (5) are the main cyanogenic glycosides. Thus, benzaldehyde cyanohydrin is the natural substrate for PaHNL-catalyzed HCN release. PaHNL accepts as substrates a surprising variety of other aldehydes, many of which are structurally very different from benzaldehyde.

Until 1987, the (R)-HNL from bitter almonds was the only HNL used as catalyst in the preparation of optically active cyanohydrins. Since the application of organic solvents enabled the preparation of many (R)-cyanohydrins in high optical yields, it was of great interest to get access to HNLs for synthetic applications, which catalyze the formation of (S)-cyanohydrins.

An enzyme, (S)-SbHNL (EC 4.1.2.11), that preferentially cleaves (S)-cyanohydrins into aldehydes and HCN was first isolated from S. bicolor [10]. Therefore reactions using this enzyme in organic solvents have been investigated. It is considerably more time consuming to isolate sufficient amounts of (S)-SbHNL from *Sorghum* than it is to isolate (R)-PaHNL from bitter almonds. Besides the problem of the more difficult accessibility the substrate range of SbHNL is much more limited. The enzyme from *Sorghum* catalyzes exclusively the addition of HCN to aromatic and heteroaromatic aldehydes, whereas it does not accept aliphatic aldehydes as substrates [11–13]. Although clones are available from both PaHNL and SbHNL, an overexpression in other organisms (*Escherichia coli, Saccharomyces cerevisiae, Pichia pastoris*) has so far not been successful.

In order to extend the substrate range and to improve accessibility, other (S)-cyanogenic glycosides have been investigated in recent years with respect to applications of the corresponding enzymes in enantioselective syntheses. The (S)-HNLs from cassava (*Manihot esculenta*) [14] and *Hevea brasiliensis* [9] proved to be highly promising. In contrast to the (S)-HNL from *Sorghum*, both enzymes accept besides aromatic and heterocyclic aldehydes, as well as aliphatic aldehydes and ketones, as substrates [9,14]. The (S)-HNLs from both *M. esculenta* [14,15] and *H. brasiliensis* [16] have been overexpressed successfully in *Escherichia coli, S. cerevisiae,* and *P. pastoris,* respectively. Hence, both recombinant (S)-HNLs are now available in amounts sufficient for synthetic and even technical applications. With the cloning and overexpression of the HNL from *Linum usitatissimum* an additional (R)-HNL for synthetic purposes has recently been developed. For some substrates (R)-LuHNL seems to have advantages in comparison with the (R)-HNL from bitter almonds [17,18].

The properties and characteristics of the five HNLs currently applied in the enzyme-catalyzed preparation of optically active cyanohydrins are listed in Table 1 [9,17,19].

Numerous investigations concerning the synthesis of optically active cyanohydrins by using chiral metal complexes, cyclic dipeptides, and lipases as catalysts have been published [5,7]. However, due to the easy access of (R)- and (S)-HNLs and the high optical and chemical yields obtained, HNL-catalyzed preparations of optically active cyanohydrins are superior to other methods.

As mentioned already, the use of organic solvents for the HNL-catalyzed addition of HCN to carbonyl compounds was decisive for many investigations concerning optically active cyanohydrins. Several variations for the practical performance of the HNL-catalyzed preparation of (R)- and (S)-cyanohydrins have been developed in recent years. Instead of pure organic solvents, a biphasic system (water/organic solvent) can be used for the reaction whereby HCN can be prepared in situ from sodium cyanide and acetic acid [20] or by transcyanation with acetone cyanohydrin [21]. It is possible to replace isolated enzymes by whole cells, e.g., by almond and apple meal instead of PaHNL or by *Sorghum* shoots instead of SbHNL [21,22].

For the reaction carried out in organic solvents it is particularly advantageous to employ the enzyme adsorbed on a suitable support, e.g., cellulose. Thus, the "support-bound" enzyme may be filtered off after completion of the reaction and reused as catalyst.

III. HNL-CATALYZED PREPARATION OF (R)- AND (S)-CYANOHYDRINS

Most of the HNL-catalyzed preparations of optically active cyanohydrins, described and discussed in the following, are performed in diisopropyl ether using the respective HNL adsorbed on a cellulose support. HCN is thereby added in free form. Thus, the differences of the HNLs in activity and selectivity can be compared.

Table 1 Properties and Characteristics of Hydroxynitrile Lyases Currently Applied in Organic Syntheses

Enzyme source	Natural substrate (cyanogenic glycosides)	R/S specificity	Molecular weight (kDa)		Optimum pH	Kinetics K_m (mM)
			Native	Subunit		
Prunus sp. (Rosaceae)	(*R*)-Mandelonitrile (amygdalin, prunasin)	*R*	55–80	55–80	5–6	0.093
Sorghum bicolor (Gramineae)	(*S*)-*p*-Hydroxymandelonitrile (dhurrin)	*S*	105	33 and 22	n.d.	0.55
Manihot esculenta (Euphorbiaceae)	Acetone cyanohydrin, (*S*)-butan-2-one cyanohydrin (linamarin, lotaustralin)	*S*	92–124	28–30	5.4	105–119
Hevea brasiliensis (Euphorbiaceae)	Acetone cyanohydrin (linamarin)	*S*	58	30	5.5–6	115
Linum usitatissimum (Linaceae)	Acetone cyanohydrin (linamarin)	*R*	82	42	5.5	2.5

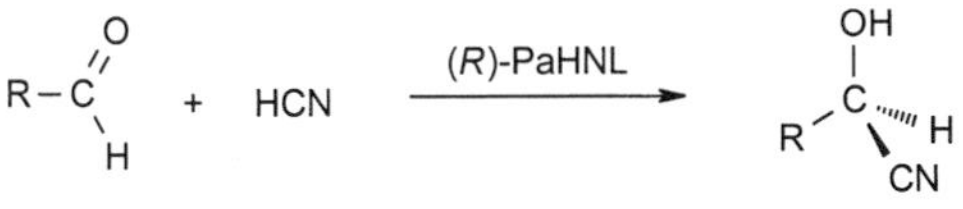

1,2 R =

1,2	R =	1,2	R =
a	Ph	g	$MeS(CH_2)_2$
b	3-PhO-C_6H_4	h	nPr
c	2-furyl	i	$Ph(CH_2)_3$
d	3-thienyl	j	Me_3C
e	3-pyridyl	k	$HOCH_2(Me_2)C$
f	MeCH=CH		

Scheme 2 Preparation of (*R*)-cyanohydrins **2** by (*R*)-PaHNL-catalyzed addition of HCN to aldehydes **1**.

A. PaHNL-Catalyzed Preparation of (*R*)-Cyanohydrins

Since (*R*)-PaHNL is readily accessible in large amounts from almond meal, the enzyme-catalyzed preparation of (*R*)-cyanohydrins with this enzyme has been investigated first and in detail (Scheme 2, Table 2) [4,5]. Of the many organic solvents investigated for PaHNL-catalyzed reactions, diisopropyl ether has proven especially advantageous with respect to the enantiomeric excess (e.e.) values obtained. PaHNL accepts in organic solvents rather high substrate concentrations (up to 2 mol/L) without decreasing e.e. values. Thus, a high space–time yield can be reached [23].

Table 2 Synthesis of (*R*)-Cyanohydrins (*R*)-**2** by PaHNL-Catalyzed Addition of HCN to Aldehydes **1**

	In H_2O/EtOH			In $i\mathrm{Pr}_2O$/Avicel[a]		
	Reaction	Cyanohydrins (*R*)-**2**		Reaction	Cyanohydrins (*R*)-**2**	
Aldehyde	time (h)	Yield (%)	e.e. (%)[b]	time (h)	Yield (%)	e.e. (%)[b]
1a	1	99	86	3	96	>99
1b	5	99	11	192[c]	99	98
1c	2	86	69	4	96	99
1d	—	—	—	6	95	>99
1e	2.5	78	7	3	97	82
1f	1.5	68	76	5.5	94	95
1g	3	87	60	16	98	96
1h	2	75	69	16	99	98
1i	—	—	—	45	94	90
1j	2.5	56	45	4.5	84	83
1k	—	—	—	1.75	84	89

[a]The enzyme was adsorbed on crystalline cellulose (Avicel).
[b]Determined by gas chromatography on β-cyclodextrin phases after either reaction with (*R*)-α-methoxy-α-trifluoromethylphenylacetoyl chloride [(*R*)-(+)-MTPA chloride] to provide the diastereomeric (*R*)-(+)-MTPA esters, or after acetylation with acetic anhydride.
[c]In ethyl acetate as solvent.

Table 3 Preparation of (R)-Cyanohydrins (R)-**2** Using Almond Meal and Acetone Cyanohydrin as HCN Source

R =	Conv. (%)	e.e. (%)	R =	Conv. (%)	e.e. (%)
n-C$_3$H$_7$	100	95	n-C$_5$H$_{11}$	100	94
i-C$_3$H$_7$	99	83	n-C$_7$H$_{15}$	98	87
n-C$_4$H$_9$	100	97	n-C$_9$H$_{19}$	94	63
i-C$_4$H$_9$	100	94	—	—	—

For comparison in Table 2 the results of (R)-PaHNL-catalyzed preparation of (R)-cyanohydrins in water-ethanol as well as in diisopropyl ether are summarized. In contrast to most other enzymes, PaHNL combines low substrate specificity with high enantioselectivity. It accepts aromatic as well as aliphatic aldehydes. Even with sterically demanding substrates such as isopentyl aldehyde relatively high e.e. values (83%) are obtained. The reaction of 3-phenoxybenzaldehyde strikingly illustrates the advantage of the organic solvent. Even after 192 h reaction time, the corresponding cyanohydrin is obtained with 98% e.e. whereas in water-ethanol after 5 h only 11% e.e. is obtained (Table 2).

Results of the PaHNL-catalyzed preparation of (R)-cyanohydrins by applying the experimental variations mentioned above are summarized in Tables 3 and 4. Table 3 shows results where the isolated (R)-PaHNL is replaced by almond meal and free HCN by acetone cyanohydrin [22a].

Results of the reaction in a biphasic system with almond meal instead of isolated PaHNL and the in situ generation of HCN from NaCN with acetic acid are summarized in Table 4 [20].

Table 4 Preparation of (R)-Cyanohydrins (R)-**2** Using Almond Meal and Generation of HCN from Sodium Cyanide with Acetic Acid

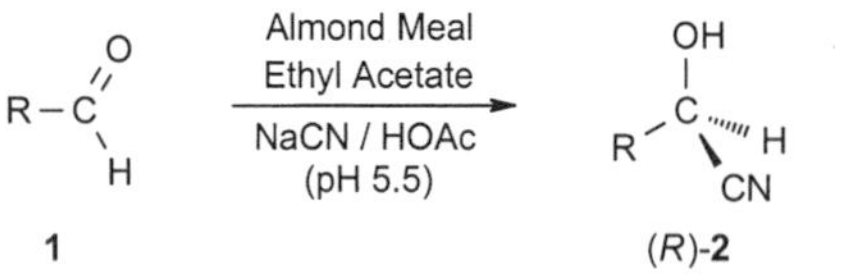

R =	Reaction condition		Cyanohydrins (R)-**2**	
	Temp. (°C)	Time (h)	Conv. (%)	e.e. (%)
C$_6$H$_5$	4	16	100	99
4-CH$_3$O-C$_6$H$_4$	20	89	47	99
2-(5-CH$_3$)furyl	4	17	70	99
n-C$_3$H$_7$	4	41	100	89
H$_3$CCH=CH	4	41	100	99

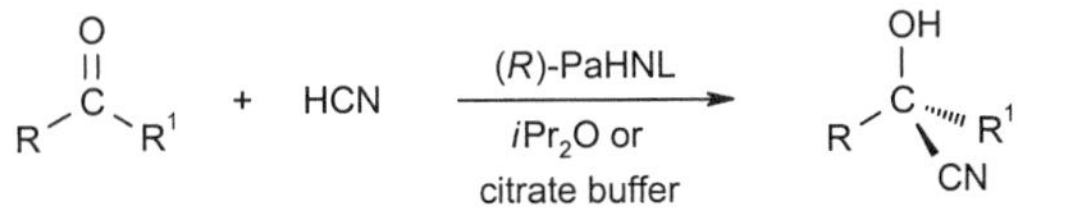

$$R \overset{O}{\underset{\|}{C}} R^1 \;+\; HCN \;\xrightarrow[\substack{iPr_2O \text{ or} \\ \text{citrate buffer}}]{(R)\text{-PaHNL}}\; R \overset{OH}{\underset{CN}{\overset{|}{C}}}\!\!\!\!\cdots R^1$$

3 (R¹ = Me)
4 (R¹ = Et)

(R)-5 (R¹ = Me)
(R)-6 (R¹ = Et)

3-6	a	b	c	d
R	C₃H₇	C₄H₉	C₅H₁₁	Me₂CHCH₂

Scheme 3 Preparation of (R)-ketone cyanohydrins **5** and **6** by (R)-PaHNL-catalyzed addition of HCN to ketones **3** and **4**.

B. (R)-HNL-Catalyzed Addition of HCN to Ketones

Only a few optically active ketone cyanohydrins are described in the literature. They have been obtained mainly by addition of HCN to ketones with a stereogenic center in the molecule through optical induction and separation of diastereoisomers [24].

Although the natural substrate of (R)-PaHNL from bitter almonds is benzaldehyde, this enzyme also catalyzes the addition of HCN to ketones to give (R)-ketone cyanohydrins (Scheme 3, Table 5) [25]. (R)-Ketone cyanohydrins derived from 2-alkanones are formed in good chemical and excellent optical yields, whereas from 3-alkanones only unsatisfactory chemical and optical yields are obtained. In contrary to the reaction behavior of aldehydes the enantioselectivity of PaHNL-catalyzed HCN addition to ketones in an organic solvent and an aqueous citrate buffer, respectively, is comparable (Table 5) [25]. The variation, working with almond meal instead of isolated (R)-PaHNL, results also in (R)-ketone cyanohydrins with surprisingly high e.e. values [26].

C. SbHNL-Catalyzed Preparation of (S)-Cyanohydrins

The (S)-hydroxynitrile lyase [EC 4.1.2.11] from S. bicolor catalyzes, as already mentioned, exclusively the addition of HCN to aromatic and hetereoaromatic aldehydes to yield the

Table 5 Preparation of (R)-Ketone Cyanohydrins (R)-**5** and (R)-**6** by PaHNL-Catalyzed Addition of HCN to Ketones **3** and **4**

	(R)-5				(R)-6	
	In iPr₂O/Avicel[a]		In citrate buffer[b]		In iPr₂O/Avicel[a]	
Ketone	Yield (%)	e.e. (%)[c]	Yield (%)	e.e. (%)[c]	Yield (%)	e.e. (%)[d]
3a, 4a	70	97	78	95	33	85
3b, 4b	90	98	94	98	21	90
3c, 4c	88	98	56	96	7	66
3d, 4d	57	98	40	98	—	—

[a]20 U enzyme per mmol ketone.
[b]50 U enzyme per mmol ketone.
[c]The e.e. values of diastereomeric (R)- or (S)-MTPA esters were determined by gas chromatography on β-cyclodextrin phases [25].
[d]The e.e. values were determined by gas chromatography on β-cyclodextrin phases after hydrolysis to the carboxylic acid and esterification with diazomethane [25].

Table 6 Preparation of (S)-Cyanohydrins (S)-**2** by SbHNL-Catalyzed Addition of HCN to
Aldehydes **1**

R =	Reaction time (h)	(S)-2 Yield (%)	(S)-2 e.e. (%)[a]	R =	Reaction time (h)	(S)-2 Yield (%)	(S)-2 e.e. (%)[a]
C_6H_5	3	91	97	$3\text{-}CH_3OC_6H_4$	20	93	89
$4\text{-}ClC_6H_4$	48	87	54	$3\text{-}PhOC_6H_4$	144	93	96
$4\text{-}H_3CC_6H_4$	32	78	87	2-Furyl	9	80	80
$3\text{-}HOC_6H_4$	24	97	91	3-Furyl	33	88	87
$3\text{-}BrC_6H_4$	18	94	92	2-Thienyl	8	64	91
$3\text{-}ClC_6H_4$	48	95	91	3-Thienyl	27	95	98
$3\text{-}F_3CC_6H_4$	20	87	52	—	—	—	—

[a]Determined by gas chromatography as (R)-MTPA esters.

respective (S)-cyanohydrins (Table 6). SbHNL does not accept aliphatic aldehydes or ketones as substrates [11,12]. The stereoselectivity of the reaction in organic solvents is
generally high and in most cases comparable with the results of the PaHNL-catalyzed
cyanohydrin formation.

D. Preparation of (S)-Cyanohydrins Catalyzed by MeHNL from *Manihot esculenta* and HbHNL from *Hevea brasiliensis*

Detailed sequence homology studies of the two enzymes, MeHNL (EC 4.1.2.37) and
HbHNL (EC 4.1.2.39), revealed that HbHNL from *H. brasiliensis* is highly homologous
to the HNL from *M. esculenta* [27]. Therefore it could be expected that these enzymes
resemble each other in substrate specificity as confirmed later by experimental investigations.

Although first experiments with the wild-type MeHNL isolated from cassava leaves
demonstrated a broad substrate range, the achieved optical yields of the (S)-cyanohydrins
obtained were unsatisfactory [15]. Not only did the overexpression of the cloned MeHNL-
gene in *E. coli* make the enzyme easier to access, but the specific activity of the recombinant MeHNL also is considerably higher than that of the wild type [14].

The (S)-cyanohydrin formation catalyzed by recombinant MeHNL adsorbed on cellulose in diisopropyl ether as solvent gave only moderate optical yields, but by using
nitrocellulose as support, the enantioselectivity was high in most cases [14]. Table 7 shows
that recombinant MeHNL exhibits a similar broad substrate range as (R)-PaHNL from
bitter almonds [14]. It catalyzes the cyanohydrin formation with aromatic and heterocyclic
as well as with aliphatic aldehydes [14].

Since the natural substrate of MeHNL in cyanogenesis is butan-2-one cyanohydrin,
it was expected that the formation of ketone cyanohydrins would be catalyzed by MeHNL.
In Table 8 the results of the MeHNL-catalyzed preparation of (S)-ketone cyanohydrins
from 2-alkanones with HCN are summarized [14].

Table 7 Preparation of (S)-Cyanohydrins (S)-**2** by MeHNL-Catalyzed Addition of HCN to Aldehydes **1** [a]

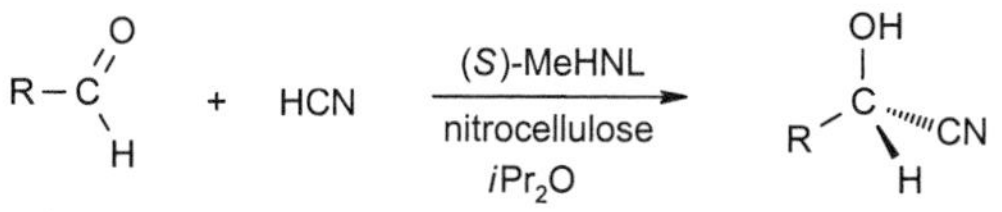

R =	Reaction time (h)	(S)-**2** Yield (%)	(S)-**2** e.e. (%)[a]	R =	Reaction time (h)	(S)-**2** Yield (%)	(S)-**2** e.e. (%)[a]
C_2H_5	4.3	86	91	C_6H_5	7.0	100	98
C_4H_9	4.0	100	91	$2\text{-}ClC_6H_4$	8.7	100	92
$(CH_3)_2CH$	6.5	91	95	$4\text{-}CH_3OC_6H_4$	9.5	82[c]	98
$(CH_3)_3C$	8.8	80	94	$3,4\text{-}CH_2O_2C_6H_3$	10.3	84[c]	86
$H_2C{=}CH$	1.0	70[b]	56	2-Thienyl	6.0	85[c]	96
$E\text{-}H_3CCH{=}CH$	1.0	100	92	3-Thienyl	4.0	98[c]	98
$E\text{-}Me(CH_2)_2CH{=}CH$	3.0	82	97	3-Furyl	6.5	98[c]	92

[a]Determined by gas chromatography on β-cyclodextrin phases after acetylation.
[b]Performed at 0°C.
[c]The yield was determined by ^1H NMR spectroscopy.
[a]Optimized procedure: To nitrocellulose [soaked in sodium citrate buffer (0.02 M, pH 3.3) for 30 min, decanted, centrifuged, and dried under high vacuum] a concentrated solution of (S)-MeHNL (900 U/ml) was dropwise added (maximal 149 U/mmol substrate). After 15 min, the enzyme-charged support was centrifuged and transferred to a flask. Diisopropyl ether, substrate (0.3–0.4 mmol), and HCN (2.6 mmol) were added, and the reaction mixture was stirred at room temperature. The support was filtered off, the filtrate dried, and solvent and unreacted substrate were distilled off to yield the pure cyanohydrin [14].

Table 8 Preparation of (S)-Ketone Cyanohydrins (S)-**5** by MeHNL-Catalyzed Addition of HCN to Ketones **3**

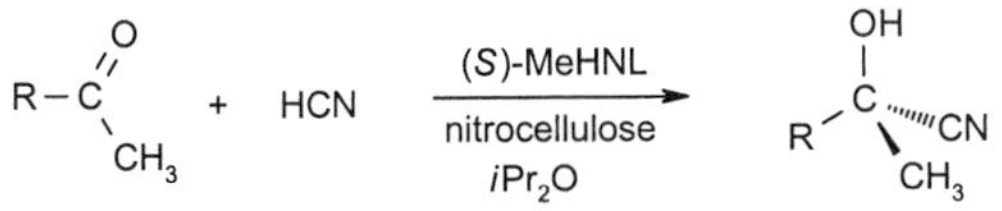

R =	Reaction time (h)	(S)-**5** Yield (%)[a]	(S)-**5** e.e. (%)[b]
C_3H_7	0.5	36	69
C_4H_9	0.5	58	80
C_5H_{11}	2.0	39	92
$(CH_3)_2CHCH_2$	0.7	69	91
$(CH_3)_3C$	0.8	81	28
C_6H_5	7.0	13[c]	78

[a]Yields of the isolated pure products.
[b]The e.e. values were determined by gas chromatography on β-cyclodextrin phases either after acetylation or as MTPA esters.
[c]The yield was determined by ^1H NMR spectroscopy.

Table 9 Preparation of (*S*)-Cyanohydrins (*S*)-**2** and (*S*)-Ketone Cyanohydrins (*S*)-**5** by HbHNL-Catalyzed Addition of HCN to Aldehydes **1** and Ketones **3** [a]

		(*S*)-**2**				(*S*)-**2** and (*S*)-**5**	
R =	R^1 =	Yield (%)	e.e. (%)	R =	R^1 =	Yield (%)	e.e. (%)
C_6H_5	H	67	99	2-Furyl	H	55	98
$C_6H_5CH_2$	H	44	99	3-Furyl	H	61	99
$C_6H_5(CH_2)_2$	H	88	93	2-Thienyl	H	52	99
E-H_5C_6CH=CH	H	50	95	3-Thienyl	H	49	99
2-$CH_3OC_6H_4$	H	61	77	$(CH_3)_2CH$	CH_3	38	88
3-$CH_3OC_6H_4$	H	80	99	C_3H_7	CH_3	51	75
4-$CH_3OC_6H_4$	H	49	95	$(CH_3)_3C$	CH_3	49	78
3-$PhOC_6H_4$	H	9	99	$C_6H_5CH_2$	CH_3	74	95
H_2C=CH	H	38	94				

[a]Optimized experimental procedure: After addition of (*S*)-HbHNL (2000 IU) to a stirred solution of substrate (1 mmol) in 1.7 ml citrate buffer (0.1 M, pH 4.5), the reaction mixture was cooled down. Subsequently, 2.5 mol equivalents of KCN, adjusted to pH 4.5 with 0.1 M citric acid, was added. After stirring for 1 h at 0–5°C, the reaction mixture was extracted three times with CH_2Cl_2, and the combined extracts were dried (Na_2SO_4). The solvent was removed and the crude cyanohydrin was chromatographed [29].

The enantioselective preparation of (*S*)-cyanohydrins catalyzed by the HNL from *H. brasiliensis* was first investigated with the wild type [28]. But only after overexpression of the cloned HbHNL-gene in *P. pastoris* did sufficient amounts of the recombinant enzyme become available [16].

In Table 9 the results of the preparation of (*S*)-cyanohydrins starting from both aldehydes and ketones with HCN and recombinant HbHNL as catalyst are summarized [9,29]. The reactions with HbHNL were performed normally in a biphasic system consisting of a concentrated aqueous enzyme solution and an organic solvent not miscible with water, e.g., *tert*-butyl methyl ether. It must be noted that the reactions and optical yields in the HbHNL-catalyzed cyanohydrin formation (Table 9) were achieved by using the 10-fold higher amount of enzyme, in comparison to MeHNL-catalyzed reactions (Tables 7 and 8) (see notes to Tables 7 and 9).

E.   LuHNL-Catalyzed Preparation of (*R*)-Cyanohydrins

As described in Sec. III.A, PaHNL from bitter almonds is an excellent biocatalyst for the preparation of (*R*)-cyanohydrins. Since the functional overexpression of PaHNL was not successful so far and some substrates, e.g., acrolein, gave only low optical yields, the catalytic suitability of the (*R*)-specific HNL from *Linum usitatissimum* [30] for the preparation of (*R*)-cyanohydrins was recently investigated [31]. The HNL from *L. usitatissimum* has been cloned and overexpressed in *P. pastoris* giving convenient access to recombinant enzyme for synthetic applications [31]. In Table 10 the LuHNL-catalyzed addition of HCN to aldehydes and ketones yielding the corresponding (*R*)-cyanohydrins is summarized [31b].

Table 10 Preparation of (*R*)-Cyanohydrins (*R*)-**2** and (*R*)-Ketone Cyanohydrins (*R*)-**5** by LuHNL-Catalyzed Addition of HCN to Aldehydes **1** and Ketones **3**

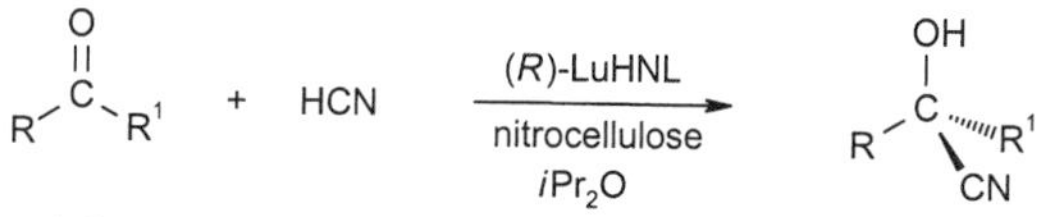

$$
\underset{\textbf{1, 3}}{\overset{O}{\underset{R}{\overset{||}{\mathbf{C}}}}\!\!_{R^1}} + HCN \xrightarrow[\substack{nitrocellulose \\ iPr_2O}]{(R)\text{-LuHNL}} \underset{(R)\text{-}\textbf{2, 5}}{\overset{OH}{\underset{R}{\overset{|}{\mathbf{C}}}}\!\!_{R^1}^{CN}}
$$

R	R^1	Reaction time (h)	(*R*)-**2** and (*R*)-**5**	
			Conv. (%)	e.e. (%)
C$_2$H$_5$	H	1.5	87	97
C$_3$H$_7$	H	2.0	91	98
(CH$_3$)$_2$CH	H	6.0	Quant.	93
(CH$_3$)$_3$C	H	4.5	Quant.	89
HOCH$_2$(CH$_3$)$_2$C	H	9.5	47	73
H$_2$C=CH	H	6.0	Quant.	74
CH$_3$CH=CH	H	9.5	21	99
H$_2$C=C(CH$_3$)	H	3.5	95	98
C$_2$H$_5$	CH$_3$	1.2	Quant.	>95
C$_3$H$_7$	CH$_3$	4.75	Quant.	93
H$_2$C=CH	CH$_3$	0.75	56	38

The results of the HNL-catalyzed syntheses of (*R*)- and (*S*)-cyanohydrins summarized in this section reveal that the development of recombinant HNLs in the last years in particular allows the preparation of almost any (*R*)- or (*S*)-cyanohydrin in high optical purity. The enzymes are available also in technical amounts; moreover, they are easy to handle due to their stability. Thus, optically active cyanohydrins are ideal chiral starting materials for the preparation of other interesting compounds with stereogenic centers.

IV. STEREOSELECTIVE TRANSFORMATION OF THE NITRILE GROUP OF (*R*)- AND (*S*)-CYANOHYDRINS

Since chiral cyanohydrins are α-substituted carboxylic acid derivatives, they have a considerable synthetic potential. Stereoselective follow-up reactions lead to several other classes of compounds with stereogenic centers. In most cases only follow-up reactions with (*R*)-cyanohydrins are described in the literature. From a few examples, however, the same reaction behavior is assured for (*S*)-cyanohydrins as well [5,11].

A. Hydrolysis of the Nitrile Group

In contrast to α-amino acids, only a few optically active α-hydroxycarboxylic acids are found in nature. Therefore general processes for the preparation of chiral α-hydroxy acids had to be developed [32].

Hydrolysis of chiral cyanohydrins offers an interesting general route to (*R*)- and (*S*)-2-hydroxycarboxylic acids. In concentrated hydrochloric acid (*R*)- and (*S*)-cyanohydrins derived from aldehydes and ketones are easily hydrolyzed to give the corresponding (*R*)-

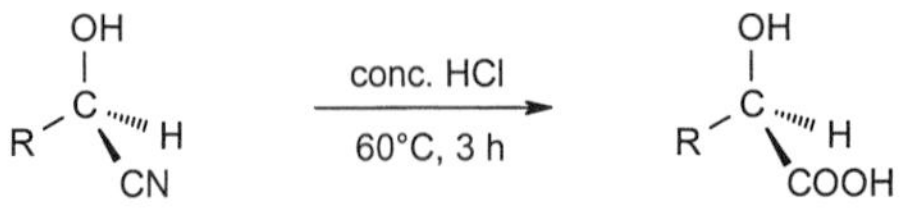

(*R*)-**2a** (R = Ph) 99%*ee* (*R*)-**7a** 99%*ee*
(*R*)-**2h** (R = Pr) 96%*ee* (*R*)-**7h** 96%*ee*

Scheme 4 Acid-catalyzed hydrolysis of (*R*)-cyanohydrins **2**.

and (*S*)-hydroxycarboxylic acids in excellent chemical yields and with complete retention of configuration (Scheme 4) [11,33]. The hydrolysis can be carried out quite simply. First the optically active cyanohydrin is prepared in the organic solvent as described before. The cellulose-adsorbed enzyme is filtered off after completion of the reaction, the solvent is removed, and the crude cyanohydrin residue is directly hydrolyzed without any further purification. From the aqueous solution the hydroxy acids are extracted with diethyl ether. A comparison of the optical purity of the crude cyanohydrins and the isolated and purified 2-hydroxycarboxylic acids **7** shows hydrolysis to proceed virtually without any racemization (Scheme 4).

The preparation of (*R*)- and (*S*)-2-hydroxycarboxylic acids, respectively, via hydrolysis of the readily accessible chiral cyanohydrins is currently the most general approach to this important class of compounds.

The preparation of (*R*)-pantolactone via the corresponding (*R*)-cyanohydrin represents an interesting application of this method. Via the cyanohydrin (*R*)-**2k**, derived from β-hydroxypivalaldehyde **1k**, optically pure (*R*)-pantolactone (*R*)-**8** is obtained in 62% yield, referred to **1k** (Scheme 5) [34].

Unsaturated chiral α-hydroxy esters are of interest due to possible reactions with the carbon double bond. They can easily be obtained from optically active cyanohydrins according to the Pinner method [35a]. Epoxidation of the chiral alcohols **9** with achiral oxidants, e.g., *m*-chloroperoxybenzoic acid (*m*-CPBA), yields a mixture of both possible epoxides **10a** and **10b**. With chiral (Sharpless titanium tartrate system) oxidants stereoselective epoxidation results. Using (+) dimethyl tartrate [(+) DMT] only the erythro isomer **10a** is obtained (Scheme 6) [35b].

A further interesting application of unsaturated chiral cyanohydrins in the synthesis of natural products is chirality transfer in cyanohydrins from the 2 to the 4 position (Scheme 7) [36].

B. Partial Reduction of the Nitrile Group

A selective hydrogenation of optically active cyanohydrins directly yielding 2-hydroxyaldehydes under complete retention of configuration is possible with Raney nickel in acidic medium but only with moderate yields [37]. O-Protected hydroxy aldehydes (*R*)-**12** are

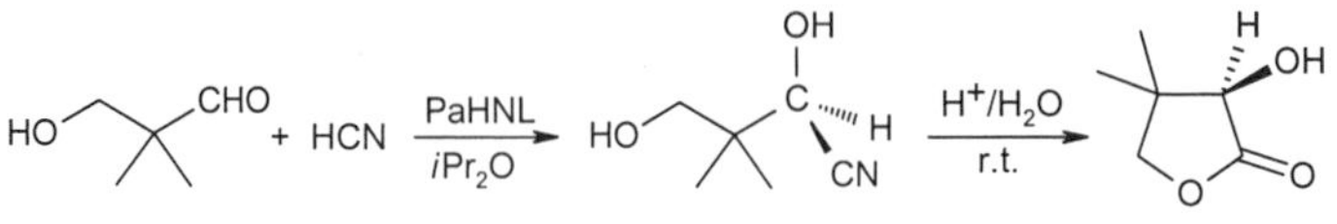

1k (*R*)-**2k** 89%*ee* (*R*)-**8** >98%*ee*

Scheme 5 Preparation of (*R*)-pantolactone (*R*)-**8**.

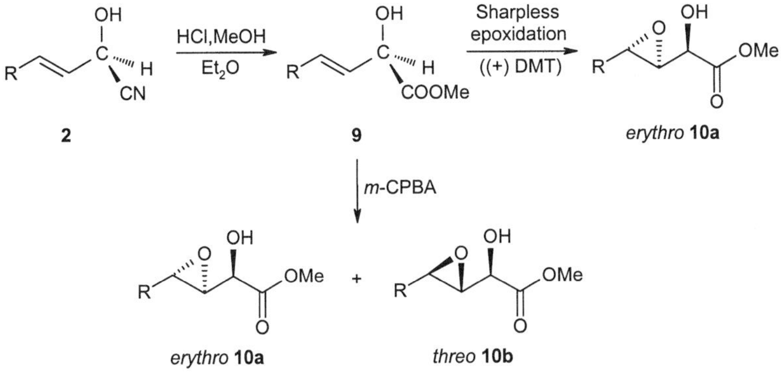

Scheme 6 Pinner reaction and subsequent epoxidation of unsaturated cyanohydrins **2**.

accessible in high yields by hydrogenation of O-protected cyanohydrins such as (*R*)-**11** with diisobutyl aluminum hydride (DIBALH) followed by mild acidic hydrolysis (Scheme 8) [38]. Besides the hydrolysis to aldehydes, the imino intermediates also allow other reactions. With primary amines NH$_3$ can be replaced by an amine (transimination) [38]. Addition of HCN to the imino intermediates opens an approach to β-hydroxy-α-amino acids **13** (Scheme 8) [38].

A partial reduction of the nitrile group is possible not only by hydrogenation with DIBALH but also by reaction of O-protected cyanohydrins with Grignard reagents to give the corresponding imino intermediates (Scheme 8). Hydrolysis of the imino compounds under mild conditions results in O-protected optically active hydroxy ketones (*R*)-**14** without any trace of racemization (Scheme 8) [38].

A partial reduction of O-protected chiral cyanohydrins is also possible by reaction with Reformatsky reagents (Blaise reaction). The primarily formed imino intermediates can be hydrolyzed under very mild conditions to give the enamines (*R*)-**15**, which yield by treatment with strong acids the tetronic acids (*R*)-**16** (Scheme 9) [39,40].

C. Preparation of Chiral 2-Amino Alcohols

2-Amino alcohols have a wide spectrum of biological activity [41]. They can be categorized as adrenaline-like compounds with one asymmetrical center at C-1 (Scheme 10) or as ephedrine-like compounds with two chiral centers at C-1 and C-2 (Scheme 11). Although it is well known that only the compounds with (*1R*) and (*1R,2S*) configuration are responsible for biological activity, so far predominantly racemates are applied as pharmaceuticals [41a]. Since in some cases (*S*) isomers seem to have noxious side effects [42], syntheses of optically pure 2-amino alcohols are gaining increasing importance.

A variety of methods for the stereoselective preparation of chiral 1,2-amino alcohols have been developed in recent years. Among these procedures the enantioselective hydro-

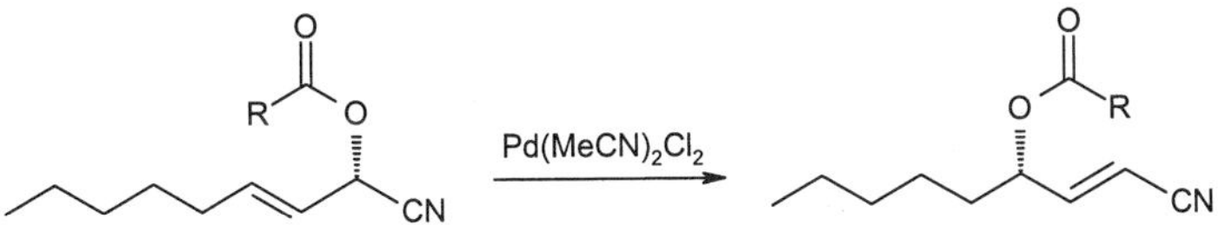

Scheme 7 Chirality transfer in unsaturated O-acyl cyanohydrins.

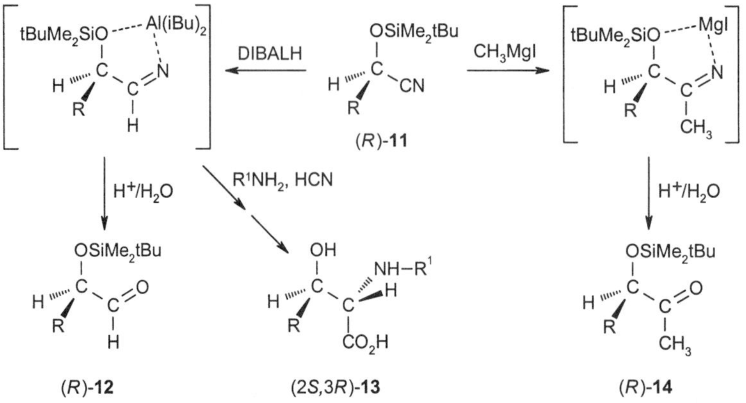

Scheme 8 Selective partial reduction of *O*-protected (*R*)-cyanohydrins (*R*)-**11**.

genation of suitable ketones [43], the diastereoselective hydrogenation of chiral α-hydroxy aldehyde hydrazones [44], and the asymmetrical amino hydroxylation of C-C double bonds [45] are most important besides some other methods [46]. In comparison to these procedures, the preparation of chiral 1,2-amino alcohols starting from optically active cyanohydrins in most cases is easier and more efficient.

Both free and O-protected optically active cyanohydrins can be hydrogenated with LiAlH$_4$ without racemization to give adrenaline-type compounds **17** (Scheme 11, Table 11) [3,33,47]. Use of the SiMe$_2$tBu-protecting group allows, as already described, partial hydrogenation with DIBALH and transimination. Subsequent hydrogenation of the imino intermediates formed gives the pharmacologically important *N*-alkyl-substituted (*1R*)-2-amino alcohols **18** (Schemes 10 and 11) [48,49].

Ephedrine-like 2-amino alcohols **19** (R^2 = H) can be prepared successfully from O-protected cyanohydrins by addition of a Grignard reagent and subsequent hydrogenation with NaBH$_4$ [50,51] (Table 12). Again NH$_3$ in the imino intermediate can be exchanged by primary amines (transimination) giving amino alcohols **19** after hydrogenation (Scheme

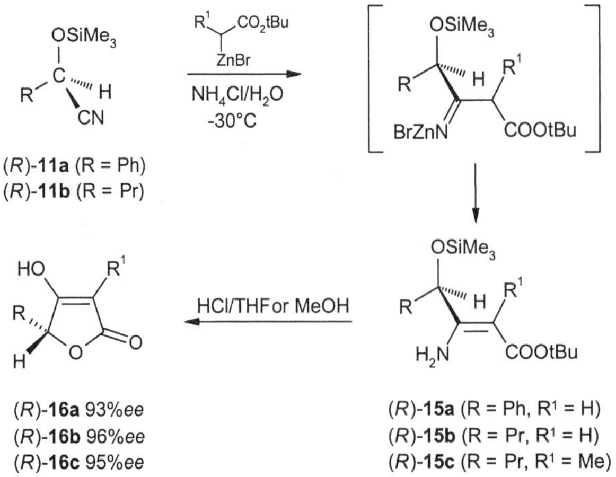

Scheme 9 Preparation of tetronic acids (*R*)-**16** via Blaise reaction.

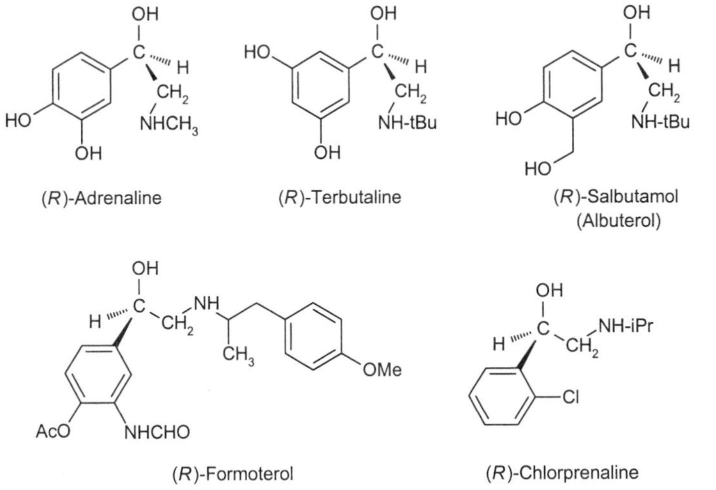

Scheme 10   Some important 2-amino alcohol pharmaceuticals.

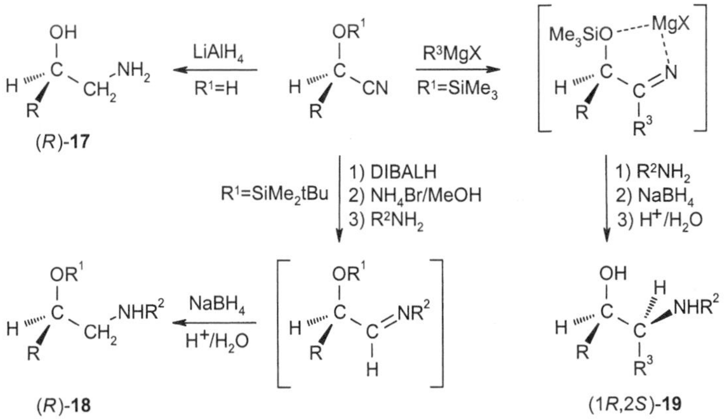

Scheme 11   Preparation of (*1R*)- and (*1R,2S*)-2-amino alcohols.

Table 11   Hydrogenation of (*R*)-Cyanohydrins (*R*)-**2** with LiAlH$_4$ to Amino Alcohols (*R*)-**17**

R =	(R)-2 e.e. (%)	(R)-17 Yield (%)	(R)-17 e.e. (%)[a]	R =	(R)-2 e.e. (%)	(R)-17 Yield (%)	(R)-17 e.e. (%)[a]
C$_6$H$_5$	99	92	99	H$_3$CS(CH$_2$)$_2$	98	31	98
3-PhOC$_6$H$_4$	98	97	95	C$_3$H$_7$	96	99	89
4-H$_3$CC$_6$H$_4$	99	100	>98	2-Furyl	98	71	98
cC$_6$H$_{11}$	91	94	80	(CH$_3$)$_3$C	93	70	>90

[a]Determined by gas chromatography after reaction with trifluoroacetic acid to the corresponding (*R*)-β-*N,O*-bis(trifluoroacetyl)amino alcohols.

Table 12 Preparation of (*1R,2S*)- or (*1S,2R*)-2-Amino Alcohols from (*R*)- and (*S*)-Cyanohydrins **2** via the Corresponding Silyl Ethers, Addition of Grignard Reagents, and Subsequent Hydrogenation with NaBH$_4$

Cyanohydrins **2**				(*1R,2S*)- and (*1S,2R*)-Amino alcohols				
R =	e.e. (%)	Config.	R^1 =	Yield (%)[a]	e.e. (%)[b]	d.e. (%)[b]	e.e. (%)[c]	d.e. (%)[c]
C$_6$H$_5$	>99	(*R*)	C$_6$H$_5$	83	99	99	100	100
C$_6$H$_5$	>99	(*R*)	C$_2$H$_5$	53	97	94	>99	>99
C$_6$H$_5$	>99	(*R*)	CH$_3$	50	99	92	99	>99
C$_3$H$_7$	95	(*R*)	C$_6$H$_5$	93	92	93	>99	>99
C$_3$H$_7$	95	(*R*)	CH$_3$	44	93	63	—	63
C$_6$H$_5$	>99	(*S*)	C$_6$H$_5$	95	99	99	100	100
C$_6$H$_5$	>99	(*S*)	CH$_3$	66	99	90	>99	>99

[a]As hydrochlorides.
[b]Determined by gas chromatography after reaction of the crude products with pivalyl chloride and filtration through silica gel.
[c]e.e. and d.e. values after one recrystallization.

11). Investigations of the stereochemistry assured for both (*R*)- and (*S*)-cyanohydrins that the Grignard addition, the transimination, and the hydrogenation proceed without any racemization at C-1. The hydrogenation of the imino intermediate at C-2 is highly stereoselective due to a chelate-controlled reaction; erythro products with (*1R,2S*) configuration are formed almost exclusively [51]. Starting from (*S*)-cyanohydrins, (*1S,2R*)-2-amino alcohols are obtained by this procedure [51,52] (Table 12). An inversion of configuration at the 1 position in 2-amino alcohols is possible by reaction with thionyl chloride [53]. Since (*1R,2S*)- and (*1S,2R*)-2-amino alcohols are accessible from (*R*)- and (*S*)-cyanohydrins, respectively, via Grignard reaction and subsequent stereoselective hydrogenation, the threo isomers with (*1S,2S*) and (*1R,2R*) configuration can be prepared stereoselectively by inversion at C-1 [51]. The preparation of L(−)-ephedrine (**19a**), which is obtained in 63% yield referring to (*R*)-**2a** (Scheme 12), illustrates an example for a

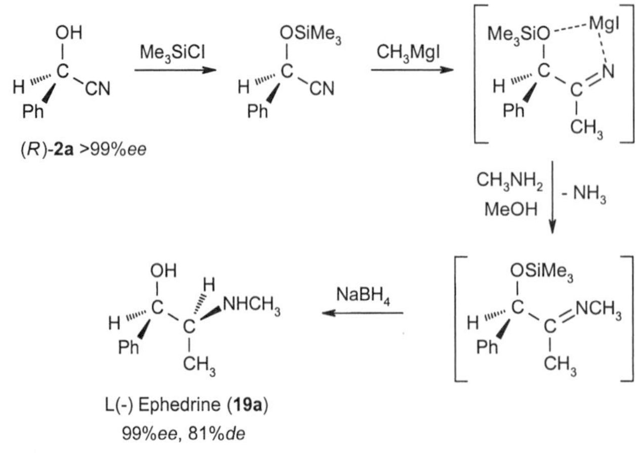

Scheme 12 Synthesis of L(−)-ephedrine (**19a**) from benzaldehyde cyanohydrin (*R*)-**2a**.

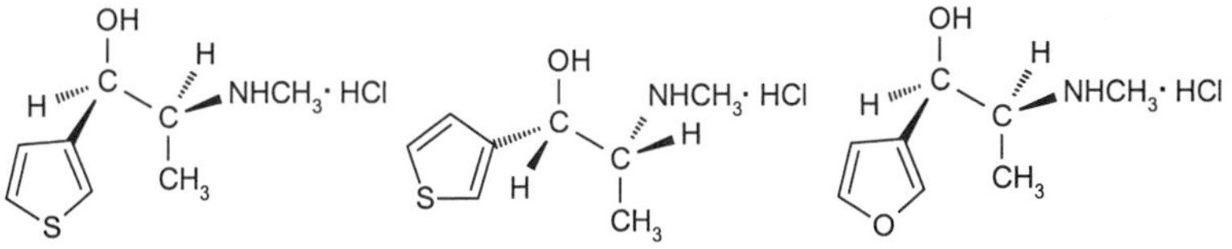

(1R,2S)-2-N-Methylamino-
1-(3-thienyl)propanol (**20a**)

(1S,2R)-2-N-Methylamino-
1-(3-thienyl)propanol (**20b**)

(1R,2S)-2-N-Methylamino-
1-(3-furyl)propanol (**21**)

Scheme 13   Heteroaromatic analogs of L(−)-ephedrine.

straightforward stereoselective synthesis of erythro 2-amino alcohols. Technically L(−)-ephedrine is produced by a fermentation process, which normally does not allow substrate variations. Starting from optically active cyanohydrins almost any structural variation is possible. In Scheme 13 two examples of heteroaromatic analogs of ephedrine, synthesized by the described procedure, are presented [52].

V.   STEREOSELECTIVE SUBSTITUTION OF THE HYDROXYL GROUP IN CHIRAL CYANOHYDRINS

The synthetic potential of optically active cyanohydrins can be extended considerably by converting the hydroxyl function to a good leaving group, which could be exchanged stereoselectively with nucleophiles.

Nucleophilic substitutions with activated α-hydroxycarboxylic acids and esters are well established [54], but little is known about the analogous reactions of cyanohydrins [55]. 2-Sulfonyloxynitriles (R)-**22**, accessible from chiral cyanohydrins (R)-**2** by sulfonylation (Scheme 14) [56], have a much higher configurational stability than the corresponding 2-halogenonitriles [55,56]. Sulfonylated cyanohydrins (R)-**22**, derived from cyanohydrins of aliphatic aldehydes (R = aliphatic), react with nucleophiles, e.g., potassium acetate (KOAc), under very mild conditions with complete inversion of configuration to give the substitution products (S)-**23** (Scheme 14) [56].

2-Sulfonyloxynitriles derived from cyanohydrins of aromatic aldehydes react with weak nucleophiles, e.g., KOAc, with partial racemization [56] (Scheme 14, R = Ph). The Mitsunobu reaction represents an alternative to the O activation of cyanohydrins combined with nucleophilic substitution [57]. Mitsunobu conditions work especially well in the exchange of allylic and benzylic hydroxyl groups in cyanohydrins [57b].

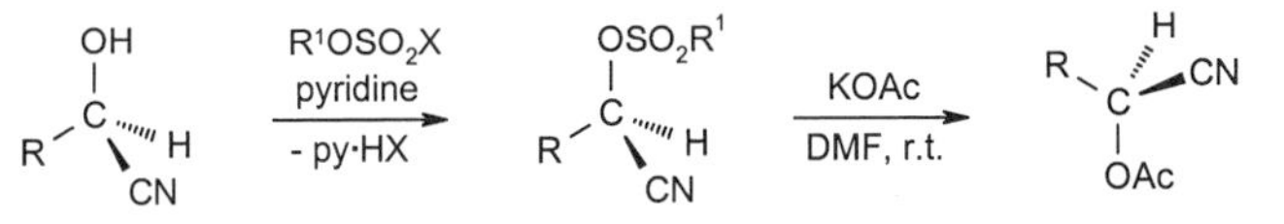

(R)-**2a** 98%ee
(R)-**2g** 96%ee
(R)-**2h** 96%ee

(R)-**22a** (R¹ = CF₃)
(R)-**22b** (R¹ = tolyl or Me)
(R)-**22c** (R¹ = tolyl)

(S)-**23a** 81%ee
(S)-**23b** 96%ee
(S)-**23c** 96%ee

	2,22,23a	2g,22,23b	2h,22,23c
R=	Ph	MeS(CH$_2$)$_2$	Pr

Scheme 14   O-Sulfonylation of (R)-cyanohydrins and reaction with O-nucleophiles.

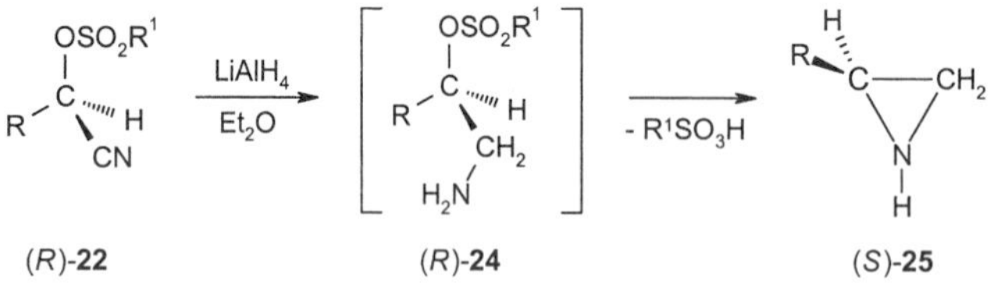

Scheme 15   Hydrogenation of *O*-sulfonylated (*R*)-cyanohydrins to (*S*)-aziridines.

A.   Reactions of 2-Sulfonyloxynitriles 22 with Nitrogen Nucleophiles

By hydrogenation of the nitrile group of α-sulfonyloxynitriles **22** the corresponding amino compounds **24** are formed primarily, which immediately react intramolecularly to give the aziridines **25** under inversion of configuration (Scheme 15) [58]. Analogous to chiral oxiranes, chiral aziridines are of interest for numerous stereoselective transformations [59].

By reaction of optically active α-sulfonyloxynitriles **22** with potassium azide the corresponding α-azidonitriles **26** are obtained under inversion of configuration. Compounds **26** can be hydrogenated selectively to give aminonitriles **29** [60] and 1,2-diamines **28**, respectively (Scheme 16) [61]. Sulfonyloxynitriles **22** react with potassium phthalimide to yield the substitution compounds **27**, which can be hydrolyzed to optically pure α-amino acids **30** (Scheme 16) [56a].

B.   Reactions of 2-Sulfonyloxynitriles with Sulfur Nucleophiles

Reactions of α-sulfonyloxynitriles with sulfur nucleophiles have not yet been described in the literature. As expected, sulfur nucleophiles react with O-activated cyanohydrins under Walden inversion to give the corresponding α-mercaptonitriles **31** and **32**, respectively (Scheme 17) [62]. As illustrated in Scheme 17, optically active 2-amino thiols **34** are accessible from substitution products **32** by hydrogenation of the nitrile function. Compounds **34** could be of interest either as complexing agents for metal ions in catalysts or as starting compounds for chiral heterocycles (Scheme 17) [62].

C.   Reductive Elimination of the Hydroxyl Group

The stereoselective synthesis of (*S*)-3,4-methylene dioxyamphetamines (*S*)-**36**, which are controversially discussed as psychoactive compounds [63], is another interesting example

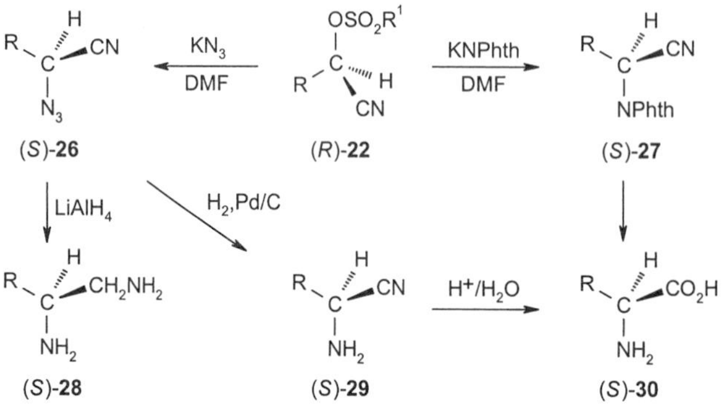

Scheme 16   Reaction of *O*-sulfonylated (*R*)-cyanohydrins with *N*-nucleophiles.

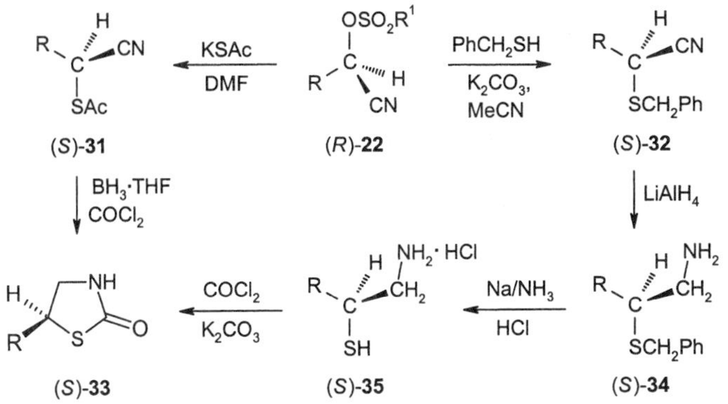

Scheme 17 Reaction of *O*-sulfonylated (*R*)-cyanohydrins with *S*-nucleophiles.

for the application of cyanohydrins as chiral starting compounds. From (*1R,2S*)-1-aryl-2-amino alcohols, prepared as described before (Scheme 11), a very efficient reductive elimination of the benzylic hydroxyl group was developed to give optically pure (*S*)-amphetamine derivatives (Scheme 18) [63]. This is an interesting case of chirality transfer from easily available chiral cyanohydrins.

Starting from (*R*)-cyanohydrins, numerous variations for the preparation of (*S*)-2-fluoronitriles have also been described that proceed in many cases with good optical yields [64].

VI. CONCLUSION

Due to the excellent accessibility, their relatively high stability and the easy handling of HNLs, (*R*)- as well as (*S*)-cyanohydrins, became very interesting and important as chiral starting compounds in stereoselective organic syntheses.

This chapter summarizes the principle reaction pathways and the most important classes of compounds that can be derived from optically active cyanohydrins. All reactions described and discussed occur without racemization and with high optical induction by

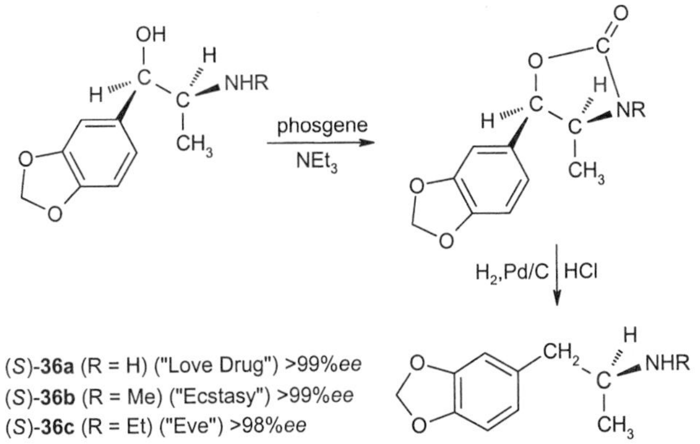

(*S*)-**36a** (R = H) ("Love Drug") >99%ee
(*S*)-**36b** (R = Me) ("Ecstasy") >99%ee
(*S*)-**36c** (R = Et) ("Eve") >98%ee

Scheme 18 Preparation of (*S*)-amphetamines (*S*)-**36** from (*1R,2S*)-2-amino alcohols.

introducing another chiral center. A variety of biologically active compounds with stereogenic centers, e.g., (1R,2S)-2-amino alcohols, are easily available in optically pure form. It can be expected that commercializing racemates and stereoisomeric mixtures of compounds as drugs will become more difficult in the future. In many cases optically active cyanohydrins could be ideal starting compounds for the selective preparation of pure stereoisomers of pharmaceuticals and plant-protecting agents.

ACKNOWLEDGMENT

The enthusiastic commitment of my co-workers, who are cited in the references, was decisive for the success of our work on optically active cyanohydrins. I would like to thank especially Drs. Siegfried Förster, Harald Wajant, and Thomas Ziegler who played a major role in the successful development of HNLs as important biocatalysts. I am strongly obliged to Dr. Angelika Baro for her commitment in preparing the manuscript.

REFERENCES

1. EE Conn. In: PK Stump, EE Conn, eds. The Biochemistry of Plants: A Comprehensive Treatise, Vol. 7. New York: Academic Press, 1981, pp 479–500.
2. L Rosenthaler. Biochem Z 14:238–253, 1908.
3. W Becker, H Freund, E Pfeil. Angew Chem Int Ed Engl 4:1079, 1965.
4. F Effenberger, T Ziegler, S Förster. Angew Chem Int Ed Engl 26:458–460, 1987.
5. F Effenberger. Angew Chem Int Ed Engl 33:1555–1564, 1994.
6. (a) CG Kruse. In: AN Collins, GN Sheldrake, J Crosby, eds. Chirality in Industry. New York: Wiley, 1992, pp 279–299; (b) WR Jackson, HA Jacobs, GS Jayatilake, BR Matthews, KG Watson. Aust J Chem 43:2045–2062, 1990.
7. M North. Synlett: 807–820, 1993.
8. LT Kanerva. Acta Chem Scand 50:234–242, 1996.
9. H Griengl, A Hickel, DV Johnson, C Kratky, M Schmidt, H Schwab. Chem Commun, 1933–1940, 1997.
10. (a) C Bové, EE Conn. J Biol Chem 236:207–210, 1961; (b) MK Seely, RS Criddle, EE Conn. J Biol Chem 241:4457–4462, 1966.
11. F Effenberger, B Hörsch, S Förster, T Ziegler. Tetrahedron Lett 31:1249–1252, 1990.
12. U Niedermeyer, MR Kula. Angew Chem Int Ed Engl 29:386–387, 1990.
13. E Smitskamp-Wilms, J Brussee, A van der Gen, GJM van Scharrenburg, JB Sloothaak. Recl Trav Chim Pays-Bas 110:209–215, 1991.
14. S Förster, J Roos, F Effenberger, H Wajant, A Sprauer. Angew Chem Int Ed Engl 35:437–439, 1996.
15. H Wajant, S Förster, H Böttinger, F Effenberger, K Pfizenmaier. Plant Sci (Limerick, Ireland) 108:1–11, 1995.
16. M Hasslacher, M Schall, M Hayn, H Griengl, SD Kohlwein, H Schwab. J Biol Chem 271: 5884–5891, 1996.
17. H Wajant, F Effenberger. Biol Chem Hoppe-Seyler 377:611–617, 1996.
18. J Albrecht, I Jansen, MR Kula. Biotechnol Appl Biochem 17:191–203, 1993.
19. (a) I Jansen, R Woker, MR Kula. Biotechnol Appl Biochem 15:90–99, 1992; (b) H Wajant, KW Mundry, K Pfizenmaier. Plant Mol Biol 26:735–746, 1994.
20. P Zandbergen, J van der Linden, J Brussee, A van der Gen. Synth Commun 21:1386–1391, 1991.
21. VI Ognyanov, VK Datcheva, KS Kyler. J Am Chem Soc 113:6992–6996, 1991.

22. (a) TT Huuhtanen, LT Kanerva. Tetrahedron: Asymmetry 3:1223–1226, 1992; (b) E Kiljunen, LT Kanerva. Tetrahedron: Asymmetry 5:311–314, 1994; (c) E Kiljunen, LT Kanerva. Tetrahedron: Asymmetry 7:1105–1116, 1996; (d) E Kiljunen, LT Kanerva. Tetrahedron: Asymmetry 8:1225–1234, 1997.

23. E Wehtje, P Adlercreutz, B Mattiasson. Biotechnol Bioeng 36:39–46, 1990.

24. (a) DA Livingston, JE Petre, CL Bergh. J Am Chem Soc 112:6449–6450, 1990; (b) MS Batra, E Brunet. Tetrahedron Lett 34:711–714, 1993.

25. (a) F Effenberger, B Hörsch, F Weingart, T Ziegler, S Kühner. Tetrahedron Lett 32:2605–2608, 1991; (b) F Effenberger, S Heid. Tetrahedron: Asymmetry 6:2945–2952, 1995.

26. E Kiljunen, LT Kanerva. Tetrahedron: Asymmetry 8:1551–1557, 1997.

27. J Hughes, FJP DeC Carvalho, MA Hughes. Arch Biochem Biophys 311:496–502, 1994.

28. (a) N Klempier, H Griengl, M Hayn. Tetrahedron Lett 34:4769–4772, 1993; (b) N Klempier, U Pichler, H Griengl. Tetrahedron: Asymmetry 6:845–848, 1995.

29. M Schmidt, S Hervé, N Klempier, H Griengl. Tetrahedron 52:7833–7840, 1996.

30. LL Xu, BK Singh, EE Conn. Arch Biochem Biophys 263:256–263, 1988.

31. (a) K Trummler, H Wajant. J Biol Chem 272:4770–4774, 1997; (b) K Trummler, J Roos, U Schwaneberg, F Effenberger, S Förster, K Pfizenmaier, H Wajant. Plant Sci 139:19–27, 1998.

32. (a) DA Evans, MM Morrissey, RL Dorow. J Am Chem Soc 107:4346–4348, 1985; (b) EJ Corey, JO Link. Tetrahedron Lett 33:3431–3434, 1992; (c) K Mikami, M Terada, T Nakai. J Am Chem Soc 112:3949–3954, 1990; (d) S Servi. Synthesis, 1–25, 1990; (d) S Tsuboi, E Nishiyama, H Furutani, M Utaka, A Takeda. J Org Chem 52:1359–1362, 1987; (e) MJ Kim, GM Whitesides. J Am Chem Soc 110:2959–2964, 1988; (f) MR Kula, C Wandrey. Methods Enzymol 136:9–21, 1987.

33. T Ziegler, B Hörsch, F Effenberger. Synthesis, 575–578, 1990.

34. F Effenberger, J Eichhorn, J Roos. Tetrahedron: Asymmetry 6:271–282, 1995.

35. (a) EGJC Warmerdam, AMCH van den Nieuwendijk, CG Kruse, J Brussee, A van der Gen. Recl Trav Chim Pays-Bas 115:20–24, 1996; (b) EGJC Warmerdam, AMCH van den Nieuwendijk, J Brussee, CG Kruse, A van der Gen. Tetrahedron: Asymmetry 7:2539–2550, 1996.

36. (a) AC Oehlschlager, P Mishra, D Dhami. Can J Chem 62:791–797, 1984; (b) DV Johnson, H Griengl. Tetrahedron 53:617–624, 1997.

37. BR Matthews, H Gountzos, WR Jackson, KG Watson. Tetrahedron Lett 30:5157–5158, 1989.

38. (a) P Zandbergen, J Brussee, A van der Gen, CG Kruse. Tetrahedron: Asymmetry 3:769–774, 1992; (b) J Brussee, EC Roos, A van der Gen. Tetrahedron Lett 29:4485–4488, 1988.

39. JJ Duffield, AC Regan. Tetrahedron: Asymmetry 7:663–666, 1996.

40. (a) J Syed, S Förster, F Effenberger. Tetrahedron: Asymmetry 9:805–815, 1998; (b) F Effenberger, J Syed. Tetrahedron: Asymmetry 9:817–825, 1998.

41. (a) A Kleemann, J Engel. Pharmazeutische Wirkstoffe, Synthesen, Patente, Anwendungen, 2nd ed. Stuttgart: Thieme, 1982 and 1987 supplement; (b) EJ Corey, JO Link. J Org Chem 56:442–444, 1991; (c) MT Reetz, MW Drewes, K Lennick, A Schmitz, X Holdgrün. Tetrahedron: Asymmetry 1:375–378, 1990; (d) T Ishizuka, S Ishibuchi, T Kunieda. Tetrahedron 49:1841–1852, 1993; (e) LJ Beeley, MA Cawthorne. Chem Br 32:31–34, 1996.

42. TJ Barberich, JW Young (Sepracor, Inc.). PCT Int. Appl. WO 91 09,596, 11.7.91 (US 5,362,755). Chem Abstr 115:142319c, 1991.

43. (a) EJ Corey, JO Link. Tetrahedron Lett 31:601–604, 1990; (b) R Hett, R Stare, P Helquist. Tetrahedron Lett 35:9375–9378, 1994; (c) PR Brodfuehrer, P Smith, JL Dillon, P Vemishetti. Org Process Res Dev 1:176–178, 1997.

44. D Enders, U Reinhold. Liebigs Ann, 11–26, 1996.

45. (a) G Li, HH Angert, KB Sharpless. Angew Chem Int Ed Engl 35:2837–2841, 1996; (b) KL Reddy, KR Dress, KB Sharpless. Tetrahedron Lett 39:3667–3670, 1998.

46. (a) O Lohse, C Spöndlin. Org Process Res Dev 1:247–249, 1997; (b) A Solladié-Cavallo, S Quazzotti, S Colonna, A Manfredi, J Fischer, A DeCian. Tetrahedron: Asymmetry 3:287–296, 1992; (c) B An, H Kim, JK Cha. J Org Chem 58:1273–1275, 1993; (d) S Vettel, C Lutz, A Diefenbach, G Haderlein, S Hammerschmidt, K Kühling, MR Mofid, T Zimmermann, P Kno-

chel. Tetrahedron: Asymmetry 8:779–800, 1997; (e) R Polt, MA Peterson. Tetrahedron Lett 31:4985–4986, 1990.

47. (a) RFC Brown, AC Donohue, WR Jackson, TD McCarthy. Tetrahedron 50:13739–13752, 1994; (b) CR Noe, M Knollmüller, P Gärtner, W Fleischhacker, E Katikarides. Monatsh Chem 126:557–564, 1995.

48. P Zandbergen, AMCH van den Nieuwendijk, J Brussee, A van der Gen. Tetrahedron 48:3977–3982, 1992.

49. F Effenberger, J Jäger. J Org Chem 62:3867–3873, 1997.

50. (a) J Brussee, F Dofferhoff, CG Kruse, A van der Gen. Tetrahedron 46:1653–1658, 1990; (b) WR Jackson, HA Jacobs, BR Matthews, GS Jayatilake, KG Watson. Tetrahedron Lett 31:1447–1450, 1990.

51. (a) F Effenberger, B Gutterer, T Ziegler. Liebigs Ann Chem, 269–273, 1991; (b) F Effenberger, B Gutterer, J Jäger. Tetrahedron: Asymmetry 8:459–467, 1997.

52. F Effenberger, J Eichhorn. Tetrahedron: Asymmetry 8:469–476, 1997.

53. (a) S Kano, T Yokomatsu, H Iwasawa, S Shibuya. Tetrahedron Lett 28:6331–6334, 1987; (b) S Kano, Y Yuasa, T Yokomatsu, S Shibuya. Heterocycles 27:1241–1248, 1988.

54. (a) U Burkard, F Effenberger. Chem Ber 119:1594–1612, 1986; (b) F Effenberger, U Burkard, J Willfahrt. Liebigs Ann Chem, 314–333, 1986; (c) U Azzena, G Delogu, G Melloni, O Piccolo. Tetrahedron Lett 30:4555–4558, 1989; (d) PR Fleming, KB Sharpless. J Org Chem 56:2869–2875, 1991.

55. (a) IA Smith. Ber Dtsch Chem Ges B71:634–643, 1938; (b) K Ichimura, M Ohta. Bull Chem Soc Jpn 43:1443–1450, 1970.

56. (a) F Effenberger, U Stelzer. Angew Chem Int Ed Engl 30:873–874, 1991; (b) F Effenberger, U Stelzer. Chem Ber 126:779–786, 1993.

57. (a) DL Hughes. Org Prep Proced Int 28:127–164, 1996; (b) EGJC Warmerdam, J Brussee, CG Kruse, A van der Gen. Tetrahedron 49:1063–1070, 1993; (c) CP Decicco, P Grover. Synlett, 529–530, 1997.

58. (a) K Ichimura, M Ohta. Bull Chem Soc Jpn 43:1443–1450, 1970; (b) F Effenberger, U Stelzer. Tetrahedron: Asymmetry 6:283–286, 1995.

59. (a) D Tanner. Angew Chem Int Ed Engl 33:599–620, 1994; (b) PF Richardson, LTJ Nelson, KB Sharpless. Tetrahedron Lett 36:9241–9244, 1995; (c) MA Poelert, RP Hof, NCMW Peper, RM Kellog. Heterocycles 37:461–475, 1994.

60. (a) MS Sigman, EN Jacobsen. J Am Chem Soc 120:5315–5316, 1998; (b) MS Iyer, KM Gigstad, ND Namdev, M Lipton. J Am Chem Soc 118:4910–4911, 1996; (c) YM Shafran, VA Bakulev, VS Mokrushin. Usp Khim 58:250–274, 1989; (d) DM Stout, LA Black, WL Matier. J Org Chem 48:5369–5373, 1983; (e) H Kunz, W Sager, D Schanzenbach, M Decker. Liebigs Ann Chem, 649–654, 1991.

61. F Effenberger, A Kremser, U Stelzer. Tetrahedron: Asymmetry 7:607–618, 1996.

62. S Gaupp. Nucleophile Substitutionen optisch aktiver α-Sulfonyloxynitrile mit Schwefel-Nucleophilen-Reaktionen α-Mercapto-substituierter Nitrile. PhD dissertation, Universität Stuttgart, 1998.

63. F Effenberger, J Jäger. Chem Eur J 3:1370–1374, 1997.

64. U Stelzer, F Effenberger. Tetrahedron: Asymmetry 4:161–164, 1993.

13

Production of Chiral β-Hydroxy Acids and Its Application in Organic Syntheses

Junzo Hasegawa and Nobuo Nagashima
Kaneka Corporation, Hyogo, Japan

I. INTRODUCTION

In recent years, a large number of reports have been published on the importance of stereochemistry for drugs and agrochemicals. These reports further clarified the distinct metabolic actions of molecules depending on their steric configurations. In the past, unexpected results were obtained on medical treatment with racemic drugs. As a result, health authorities have changed their policy, encouraging the development of single isomers as new drugs. This situation is causing the chemical industry to develop more economical and readily applicable technologies for chiral compounds.

Optically active β-hydroxycarboxylic acids are versatile chiral synthons, due to their useful dual functionality. Thus, optically active β-hydroxycarboxylic acids and their derivatives have been used as starting materials for the synthesis of optically active bioactive compounds such as vitamins, antibiotics, pheromones, and flavor compounds.

In this chapter, we review the production methods for optically active β-hydroxycarboxylic acids (esters), and chiral building blocks derived from optically active β-hydroxy acids and their use in the synthesis of optically active bioactive compounds.

II. PRODUCTION METHODS FOR OPTICALLY ACTIVE β-HYDROXYCARBOXYLIC ACIDS

In this section, we review the chemical and biochemical preparation methods that have been reported for optically active β-hydroxycarboxylic acids, especially β-hydroxybutyric acid (HBA), β-hydroxyisobutyric acid (HIBA), and β-hydroxyvaleric acid (HVA), which are most versatile for organic syntheses.

A. Chemical Methods

1. Asymmetrical Reduction of β-Keto Esters

Several groups have reported methods for the preparation of optically active β-hydroxycarboxylic acids from β-keto esters by means of asymmetrical reduction. For example, Tai et al. used a nickel catalyst modified with optically active tartrate [1], Mukaiyama et

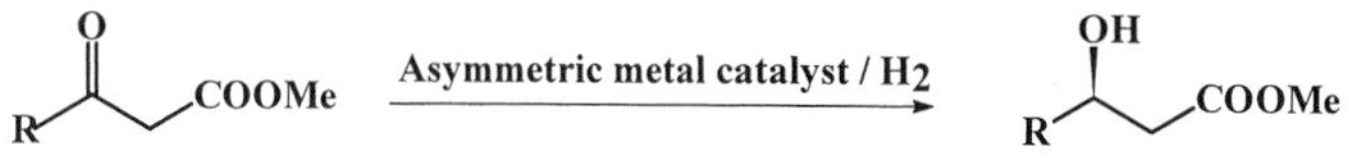

Scheme 1

al. used chiral diamine compounds and $SnCl_2$ [2], and Noyori et al. used cycloocta-1,4-dienyl-2,2′-bis(diphenylphosphosphino)-1-1′-ninaphthylruthenium, an Ru(COD)(BINAP) catalyst [3]. The Tai and Noyori methods comprise catalytic asymmetrical reduction. Among the chemical procedures. Noyori's method seems to be the most efficient for the industrial production of optically active β-hydroxycarboxylic acids from the productivity and optical quality points of view. They succeeded in the preparation of (S)- and (R)-β-hydroxybutyric acid methyl esters, (R)-β-hydroxyvaleric acid methyl ester, etc., of high optical purity (more than 99% e.e.) in high yields (Scheme 1).

2. Enantioselective Aldol Condensation

The groups of Masamune [4] and Evans [5] reported the synthesis of optically active β-hydroxycarboxylic acids by means of stereoselective chiral aldol condensation via boron enolates. (S)- and (R)-β-Hydroxyisobutyric acid of high optical purity (98% e.e.) were obtained by means of Masamune's method. However, these methods do not seem favorable for industrial production of β-hydroxycarboxylic acids because these reactions need equal amounts of the chiral starting material for the synthesis of optically active β-hydroxycarboxylic acids (Scheme 2).

3. Optical Resolution

Racemic β-hydroxycarboxylic acids can be synthesized by means of the Reformatsky reaction and the chemical reduction of β-keto acids. Lehninger et al. reported the optical resolution of racemic β-hydroxybutyric acid with quinine as a resolving agent [6] (Scheme 3).

B. Biochemical Methods

1. Oxidation of Aliphatic Glycols

It is well known that some microorganisms, e.g., *Gluconobacter*, *Hansenula*, and *Arthobacter*, can covert the alcohol groups of aliphatic glycols to the corresponding carboxylic

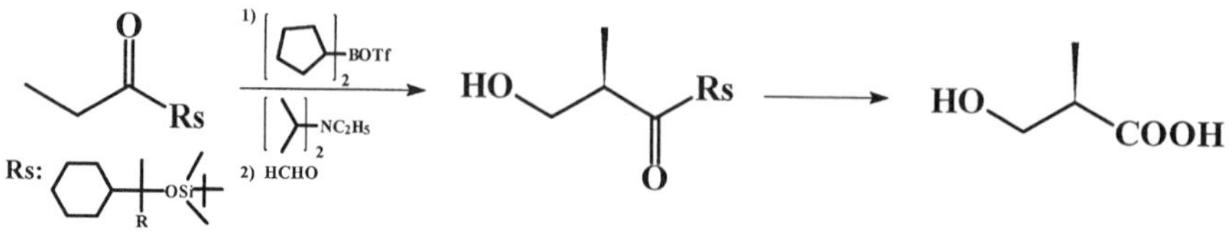

Scheme 2

Scheme 3

acids. Ohta et al. applied this oxidation for the preparation of (R)-β-hydroxyisobutyric acid from 2-methylpropane-1,3-diol using *Gluconobacter rouseus*; however, the optical purity was not very good [7] (Scheme 4).

2. Reduction of β-Keto Acids and Their Esters

This is a common biochemical transformation, which has been harnessed to produce several useful β-hydroxycarboxylic acids and esters. The first preparation of optically active β-hydroxycarboxylic acids, (S)-β-hydroxybutyric acid, (R)-β-hydroxycaproic acid, and (R)-β-hydroxycaprylic acid, through bioreduction was reported by Lemieux, who used baker's yeast [8]. Since their report, numerous studies on the bioreduction of β-keto esters with various kinds of microorganisms, especially baker's yeast, have been reported. This method was applied for the preparation of various kinds of β-hydroxycarboxylic esters, but in most cases the optical purity of the obtained β-hydroxycarboxylic acid esters was not so good because of the existence of antistereospecific reductases in the microorganisms used. Which reductase is the more dominant in the cells depends on the microorganism; therefore, the steric configuration of β-hydroxycarboxylic acid depends on the microorganism used.

At reduction of acetoacetic acid ethyl ester, baker's yeast and *Saccharomyces bailii* KI0116 afforded (S)-β-hydroxybutyric acid ethyl ester (85% e.e. and 96% e.e., respectively [9]), on the other hand, *Geotrichum candidum* CBS 233.76 afforded the (R) isomer (89.6% e.e. [10]). Kometani et al. tried to prepare (S)-β-hydroxybutyric acid ethyl ester of high optical purity using baker's yeast, and succeeded by continuously adding acetoacetic acid ethyl ester as the substrate and ethanol as the energy source to the baker's yeast solution (99.3% e.e. [11]). Mori et al. reported the preparation of optically pure (S)-β-hydroxy-butyric acid ethyl ester by use of horse liver alcohol dehydrogenase and yeast alcohol dehydrogenase coupled with NADH [9].

(S) and (R)-β-Hydroxyvaleric acid ethyl ester were obtained from β-ketovaleric acid ethyl ester by reduction with baker's yeast and *Thermoaerobium brockii*, respectively (40% e.e. [12] and 84% e.e. [13], respectively). Mori et al. reported that high optical purity of (S)-β-hydroxyvaleric acid octyl ester (97.4% e.e.) was obtained when the long chain ester of β-ketovaleric acid was used for the baker's yeast reduction [14].

Methylmalonic acid semialdehyde ethyl ester was reduced by baker's yeast and *T. brockii* to give (R)-β-hydroxyisobutyric acid ethyl ester, but the optical purity was not very good (60% e.e. and 72% e.e., respectively [15] (Scheme 5).

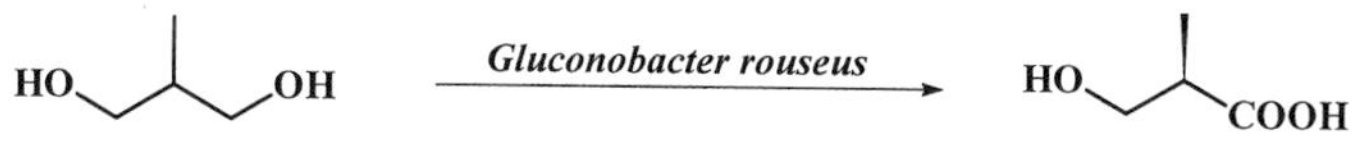

Scheme 4

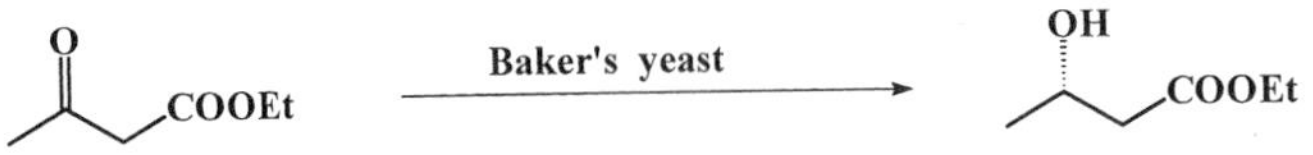

Scheme 5

3. Fermentation Process

It is well known that certain kinds of bacteria, e.g., *Azotobacter*, *Bacillus*, *Alcaligenes*, and *Zoogloea*, accumulate high amounts of poly[(R)-β-hydroxybutyrate] when they are cultured under limited nitrogen conditions. (R)-β-Hydroxybutyric acid ester of high optical purity (100% e.e.) was obtained by alcoholysis of this biopolymer [9]. When propionic acid and glucose were used as the carbon sources, *Alcaligenes eutrophus* produced a copolymer of (R)-β-hydroxybutyric acid and (R)-β-hydroxyvaleric acid. (R)-β-Hydroxy-valeric acid ester (100% e.e.) was obtained from this copolymer by a procedure similar to that described above [16] (Scheme 6).

4. β-Hydroxylation of Carboxylic Acids

This transformation has been used widely to prepare optically active β-hydroxycarboxylic acids. This process takes place in two stages: in initial dehydrogenation to the α,β-unsat-urated carboxylic acid and subsequent hydration. These steps utilize the enzymes of the β-oxidation pathway of lipid catabolism, and so the β-hydroxy acids produced are gen-erally of the "natural" (S) form. Both saturated carboxylic acids and their α,β-unsaturated counterparts have been used as raw materials. For example, β-hydroxypropionic acid (HPA) has been prepared from acrylic acid through a process mediated by *Fusarium* [17], and *Pseduomonas putida* has been used to prepare (S)-β-hydroxyisobutyric acid from isobutyric acid [18]. The preparation of C_6–C_{12} (S)-β-hydroxycarboxylic acids from the corresponding *trans*-α,β-unsaturated carboxylic acids by microbial hydration catalyzed by resting cells of *Mucor* sp. has also been reported [19] (Scheme 7).

     The authors considered microbial β-hydroxylation to be particularly suitable for fur-ther refinement. The availability and low cost of carboxylic acid precursors would favor the large-scale use of processes based on this method, but one major disadvantage is that when optically active β-hydroxycarboxylic acids were obtained, only the (S) form was observed. As noted earlier, (R)-β-hydroxycarboxylic acids exhibit potential as precursors of pharmaceuticals, and this was one of the incentives for the development of a robust, general method for the (R) forms of these intermediates.

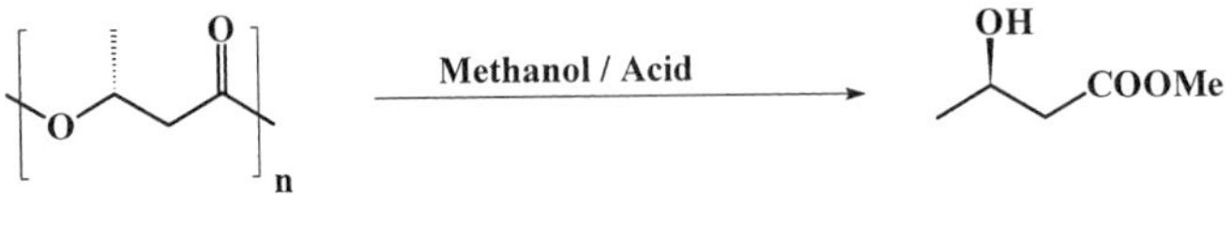

Scheme 6

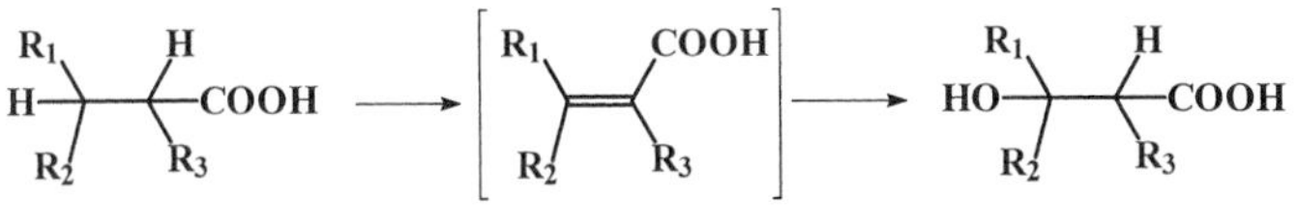

$$\begin{matrix} R_1 & H \\ H-\!\!\!\!\!-\!\!\!\!\!-COOH \\ R_2 & R_3 \end{matrix} \longrightarrow \left[\begin{matrix} R_1 & COOH \\ R_2 & R_3 \end{matrix} \right] \longrightarrow \begin{matrix} R_1 & H \\ HO-\!\!\!\!\!-\!\!\!\!\!-COOH \\ R_2 & R_3 \end{matrix}$$

Scheme 7

III. PRODUCTION METHOD FOR OPTICALLY ACTIVE β-HYDROXYCARBOXYLIC ACIDS INVOLVING MICROBIAL β-HYDROXYLATION DEVELOPED BY KANEKA [23–25]

Kaneka developed industrial production methods not only for the (S) form but also (R) form of β-hydroxycarboxylic acids involving the microbial stereoselective β-oxidation of fatty acids. And now we are producing various kinds of optically active β-hydroxycarboxylic acids by means of this method and are using them for the synthesis of the intermediates of pharmaceuticals.

A. Selection of Microorganisms that Convert Isobutyric Acid to β-Hydroxyisobutyric Acid [21]

At the beginning of this study, we planned the establishment of a new economical synthesis process for captopril, as described later. We speculated that (R)-β-hydroxyisobutyric acid was favorable as a starting material for our planned new process. Previous studies on the stereochemistry of the production of β-hydroxyisobutyric acid from isobutyric acid by *P. putida* demonstrated the production of the (S) form. Thus, the authors began this study with a search for microorganisms that could metabolize isobutyric acid to (R)-β-hydroxyisobutyric acid.

The authors screened over 700 microorganisms as to their ability to β-hydroxylate isobutyric acid. This study revealed that this capability was widely distributed in various molds, yeast, and bacteria. The conversion under the screening conditions varied from 25% to 45% and, although the majority of microorganisms produced the expected (S)-β-hydroxyisobutyric acid, a minority, notably *Candida rugosa* IFO 0750 and IFO 0591, yielded the (R) form. Some representative microorganisms that produce β-hydroxyisobutyric acid are listed in Table 1. The optical purities of the methyl ester of (R)- and (S)-β-hydroxyisobutyric acid produced by *C. rugosa* IFO 0750 and *C. rugosa* IFO 1542 (=*Saccharomycopsis lipolytica* IFO 1542) were estimated to be more than 99% e.e., and more than 99% e.e., respectively. These findings opened the door for the establishment of economical industrial production methods for both (R)- and (S)-hydroxycarboxylic acids, using inexpensive fatty acids, through microbial stereoselective β-hydroxylation.

B. Production of (R)-β-Hydroxyisobutyric Acid by a Mutant Lacking β-Hydroxyisobutyrate Dehydrogenase [23]

The use of *C. rugosa* to produce (R)-β-hydroxyisobutyric acid from isobutyric acid, although novel, was not yet efficient enough to be adopted on an industrial scale. The conversion never exceeded 50% and the byproduct, β-hydroxypropionic acid, was present at a concentration of about 10% of that of (R)-β-hydroxyisobutyric acid. It is probable that β-hydroxypropionic acid is produced from (R)-β-hydroxyisobutyric acid via propionic acid by this microorganism, through the valine catabolic pathway. It seemed reasonable,

Table 1 Stereoselectivity of β-Hydroxylation of Isobutyric Acid by Microorganisms

Microorganism	$[\alpha]_D^{25}$ (MeOH)	Configuration
P. membranaefaciens IAM 4258	15.6	(S)
T. aculeatum ATCC 22310	18.1	(S)
E. reessii CBS 179.60	17.1	(S)
G. vanryiae CBS 435.64	14.5	(S)
H. anomala IFO 1020	15.0	(S)
C. rugosa IFO 1542	17.8	(S)
C. rugosa IFO 0750	−17.6	(R)

therefore, that if a mutant unable to degrade (R)-β-hydroxyisobutyric acid could be derived from C. *rugosa* IFO 0750, a higher yield of this product could be expected (Scheme 8). This microorganism was not grown in medium containing (R)-β-hydroxyisobutyric acid as a sole carbon source; therefore, we selected mutants unable to utilize propionic acid (Table 2). A group of mutants that were unable to assimilate propionic acid was derived from the parent strain and, of them, the best mutant (NPA-104) was found to produce (R)-β-hydroxyisobutyric acid in a high yield (91%) without the byproduction of β-hydroxy-propionic acid. The mutant, NPA-104, could utilize neither propionic acid nor β-hydroxypropionic acid as a sole carbon source, and its β-hydroxyisobutyrate dehydrogenase activity as reduced to 1.2% of that of the parent strain. The degradation of (R)-β-hydroxyisobutyric acid by resting cells of strain NPA-104 was not observed at all (Table 3).

This mutant required an energy source such as glucose for high concentration production of (R)-β-hydroxyisobutyric acid. On the other hand, the parent strain did not need the addition of an energy source. These results may be due to the fact that the β-hydroxylation of isobutyric acid requires a net input of energy in the form of ATP. In the case of the parent strain, it appeared that about half of the (R)-β-hydroxyisobutyric acid produced was degraded further to provide an energy source for further β-hydroxylation. In strain NPA-104, however, the catabolic pathway for (R)-β-hydroxyisobutyric acid is blocked, and so this mutant requires an energy source for the β-hydroxylation of isobutyric acid.

Figure 1 shows a typical pattern of production of (R)-β-hydroxyisobutyric acid by a further improved mutant strain of NPA-104. The accumulation of (R)-β-hydroxyisobutyric acid reached 165 g/L on 120 h reaction with the continuous addition of isobutyric acid and glucose.

C. Production of (R)-Hydroxycarboxylic Acid [22,25]

Having demonstrated the feasibility of stereoselective production of (S)- and (R)-β-hydroxyisobutyric acid from a "natural" substrate isobutyric acid, the authors proceeded to investigate the β-hydroxylation of various other aliphatic carboxylic acids by the same microorganisms. Accordingly, the organisms that exhibited high β-hydroxylation activity toward isobutyric acid were incubated with various normal and isocarboxylic acids. However, of the normal carboxylic acids, only valeric acid gave rise to the β-hydroxy derivative

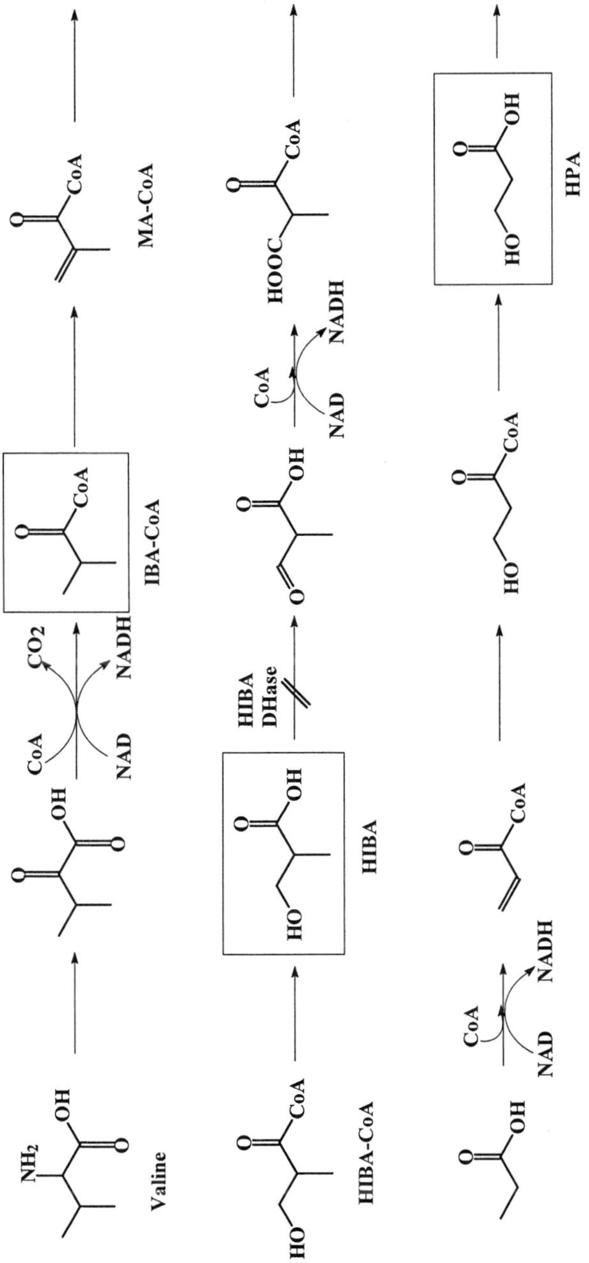

Scheme 8 Metabolic pathway of valine.

Table 2 (R)-β-Hydroxyisobutyric Acid Production by
Mutants Unable to Assimilate Propionic Acid

Strain	(R)-β-HIBA produced (mg/ml)	Yield (%)	β-HPA
NPA-101	22.4	87	−
NPA-102	22.7	96	−
NPA-103	22.7	74	+/−
NPA-104	34.5	91	−
NPA-105	29.3	84	−
Parent	13.2	33	3+

in the broth of *C. rugosa*. Although propionic acid and butyric acid were readily metabolized by the microorganisms, the corresponding β-hydroxycarboxylic acids were not accumulated, and the most normal carboxylic acids with longer chains were not metabolized at all. On the other hand, the isocarboxylic acids, isovaleric acid, isocaproic acid, and α-methylbutyric acid were converted to the corresponding β-hydroxy derivatives.

The authors' unsuccessful attempts to find culture conditions for the production of normal β-hydroxycarboxylic acids from the corresponding carboxylic acids using *C. rugosa* were described above. Although normal carboxylic acids in the range of C_1-C_7 were well metabolized by this microorganism, the corresponding β-hydroxycarboxylic acids could not be isolated, except in the case of (R)-β-hydroxyvaleric acid, which was obtained in a low yield. This latter result suggested that a general method for the production of (R)-β-hydroxycarboxylic acids with this microorganism might be possible if the further catabolism of these products could be inhibited. The next challenge, therefore, was to obtain a mutant that lacks the ability to oxidize normal β-hydroxycarboxylic acids. We selected mutants that produced β-hydroxybutyric acid from butyric acid by the similar manner to NPA-104. One mutant, NBA-2, could produce β-hydroxybutyric acid in a 92% yield from butyric acid. As shown in Table 4, the mutant grew well in a medium containing

Table 3 Properties of a Mutant NPA-104

	NPA-104	Parent
Assimilation		
(R)-β-Hydroxyisobutyric acid	−	−
Propionic acid	−	3+
β-Hydroxypropionic acid	−	3+
Glucose	5+	5+
Degradation of (R)-β-HIBA by the resting cells	0%	75.50%
Enzyme activity (U/mg protein, 10^3): (R)-β-HIBA DHase	0.8	68.6
β-HPA DHase	0	143

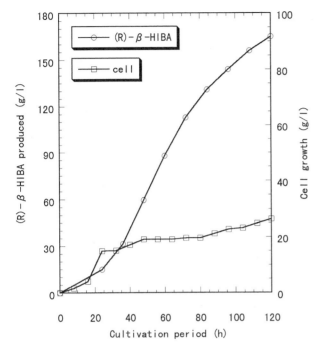

Figure 1   (*R*)-β-Hydroxyisobutyric acid production by *Candida rugosa* IFO 0750 mutant.

glucose, acetic acid, or propionic acid as the carbon source, but hardly utilized C_4–C_7 β-hydroxycarboxylic acids.

The authors proceeded to evaluate the production of (*R*)-β-hydroxycarboxylic acids from carboxylic acids by strain NBA-2. The results, shown in Table 5, were that, in addition to (*R*)-β-hydroxybutyric acid, (*R*)-β-hydroxyvaleric acid, (*R*)-β-hydroxycaproic

Table 4   Growth Response of a Mutant NBA-2 on Various
Carboxylic Acids

Carbon source	Concentration (%, v/v)	Parent	NBA-2
Glucose	1	5+	4+
Acetic acid	1	3+	2+
Propionic acid	1	3+	3+
Butyric acid	1	3+	−
(*S*)-β-Hydroxybutyric acid	1	3+	−
(*R*)-β-Hydroxybutyric acid	1	3+	−
Valeric acid	0.5	2+	−
(*R*)-β-Hydroxyvaleric acid	0.5	+	−
Caproic acid	0.5	+	−
(*R*)-β-Hydroxycaproic acid	0.5	+	−
Heptanoic acid	0.5	+	−
(*R*)-β-Hydroxyhaptanoic acid	0.5	+	−
Caprylic acid	0.5	+/−	−

Table 5 Production of (R)-β-Hydroxycarboxylic Acids by the Mutant NBA-2

			β-Hydroxycarboxylic acid			
				Yield		Optical purity
Substrate	Conc (%)			(mg/ml)	(%)	(% e.e.)
Butyric acid	1		(R)-β-HBA	10.7	95	95
Valeric acid	1		(R)-β-HVA	6.5	60	93
Caproic acid	1		(R)-β-HCA	5.2	95	96
Heptanoic acid	0.25		(R)-β-HHA	1.3	52	95

acid, and (R)-β-hydroxyheptanoic acid were all produced in good yields. The production of (R)-β-hydroxybutyric acid by an improved mutant of NBA-2 was confirmed to be at a level similar to that of (R)-β-hydroxyisobutyric acid production.

D. (S)-β-Hydroxycarboxylic Acids Production [20]

This methodology could be applied for the production of the (S) forms of β-hydroxycarboxylic acids, and (S)-β-hydroxyisobutyric acid and (S)-β-hydroxybutyric acid could be produced in high yields using mutants of *Trichosporon aculeatum* and *Endomyces tetrasperma*, respectively.

This methodology had been developed by Kaneka for use on a large scale and had furnished several (R)-β-hydroxycarboxylic acids in practical quantities for the first time. The range of β-hydroxycarboxylic acids that may be readily obtained by means of this method is given in Table 6.

Table 6 β-Hydroxycarboxylic Acids Obtained in this Study

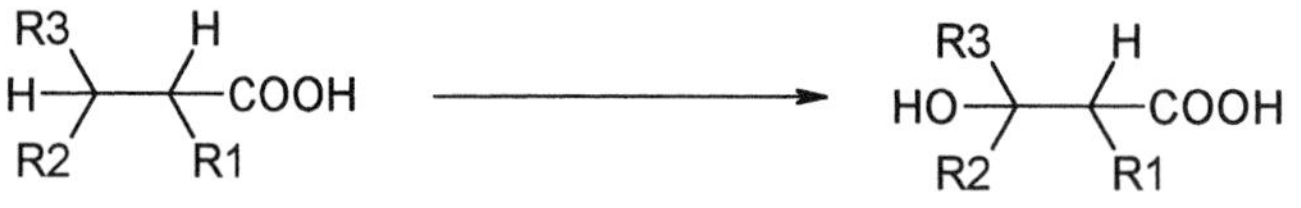

β-Hydroxycarboxylic acid	R_1	R_2	R_3	Microorganism	Optical purity (% e.e.)
β-HPA	H	H	H	*C. rugosa* IFO 0750 M*	
(R)-β-HBA	H	H	H	*C. rugosa* IFO 0750 M*	95
(S)-β-HBA	H	H	CH_3	*E. tetrasperma* CBS 765.70 M*	88
(R)-β-HVA	H	H	CH_3	*C. rugosa* IFO 0750 M*	93
(S)-β-HVA	H	H	CH_3CH_2	*E. tetrasperma* CBS 765.70 M*	82
(R)-β-HCA	H	H	CH_3CH_2	*C. rugosa* IFO 0750 M*	96
(R)-β-HHA	H	H	$CH_3CH_2CH_2$	*C. rugosa* IFO 0750 M*	95
(R)-β-HIBA	CH_3	H	H	*C. rugosa* IFO 0750 M*	99
(S)-β-HIBA	CH_3	H	H	*T. aculeatum* ATCC 22310 M*	99
(R)-α-HMBA	CH_3CH_2	H	H	*C. rugosa* IFO 0750 M*	98
(S)-α-HMBA	CH_3CH_2	H	H	*T. fermentans* CBS 2529	95
β-HIVA	H	CH_3	CH_3	*E. reessii* CBS 179.60	

*M: Mutant

IV. UTILIZATION OF OPTICALLY ACTIVE β-HYDROXY ACIDS FOR THE SYNTHESIS OF CHIRAL BIOACTIVE COMPOUNDS

In this section, chiral building blocks derived from some optically active β-hydroxy acids and their use in the synthesis of chiral bioactive compounds (including their precursors) are reviewed.

A. Useful Chiral Building Blocks Derived from Optically Active β-Hydroxy Acids

1. Chiral Building Blocks Derived from (S)- and (R)-β-Hydroxybutyric Acids

Optically active β-hydroxybutyric acid is a bifunctional molecule and is most frequently used as its ester form. In Scheme 9 are shown typical transformations of β-hydroxybutyric acid. Usually, the hydroxy group of the β-hydroxybutyrate ester is suitably protected or converted to an appropriate leaving group for further transformation. The ester group is often reduced to an alcohol or aldehyde and is occasionally hydrolyzed to a carboxylic acid. The resultant alcohol, aldehyde, or carboxylic acid is usually further transformed for various synthetic purposes.

Another type of transformation comprises substitutions at α positions of β-hydroxybutyric acid derivatives. Thus, the diastereoselective direct alkylation of β-hydroxybutyrate esters through dilithioalkoxide enolates gives α-alkylated-β-hydroxybutyrate esters, which are commonly subjected to hydroxyl protection or ester reduction, as shown in Scheme 9.

The diastereoselective indirect alkylation or amination of β-hydroxybutyric acid through 1,3-dioxanones is also known and gives α-alkylated- or α-hydrazino-β-hydroxybutyric acid derivatives, respectively. With the 1,3-dioxanone method of alkylation, it is possible to introduce two different alkyl groups diastereoselectively at the α position of β-hydroxybutyric acid. Furthermore, the diastereoselective introduction of alkyl groups at the β position of β-hydroxybutyric acid is also possible via a 1,3-dioxanone derivative.

On alkylation through a β-lactone derivative, α-alkylated-β-hydroxybutyric acids with complete inversion of the configuration at the β-carbon atom of the starting β-hydroxybutyric acid are formed.

In Figs. 2 and 3 are listed useful chiral building blocks derived from (S)- and (R)-β-hydroxybutyric acids using the above-mentioned transformations and others.

2. Chiral Building Blocks Derived from (S)- and (R)-β-Hydroxyisobutyric Acids

Optically active β-hydroxyisobutyric acid is also a bifunctional molecule and is most widely used in its ester form. Therefore, as shown in Scheme 10, the first step of a transformation is often the protection of a hydroxy group or the conversion of the hydroxy group to a suitable leaving group that is generally substituted by nucleophiles. The unchanged ester group is subsequently reduced to an alcohol in many cases, and the resultant alcohol is usually further transformed for various synthetic purposes. In a certain case, the ester group is transformed into an amine through sequential aminolysis and reduction.

Sometimes the hydroxy and carboxy groups of β-hydroxyisobutyric acid are simultaneously protected by the same protective group and the resultant protected β-hydroxyisobutyrate esters are similarly reduced to monoprotected diols.

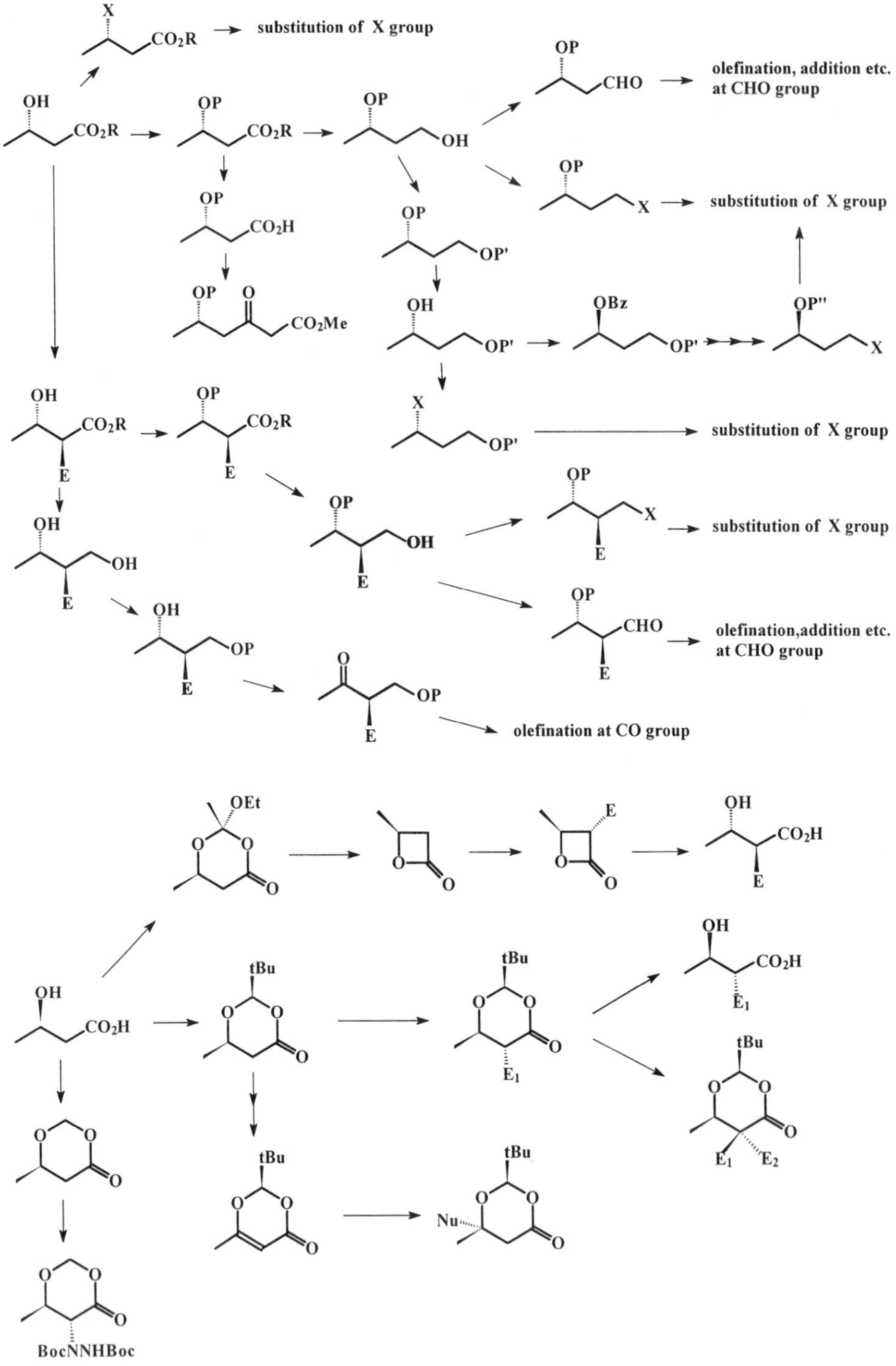

P, P', P": Protective groups
X: Leaving groups
E, E_1, E_2: Groups derived from electrophiles
Nu: Groups derived from nucleophiles
R: Residual groups of esters

Scheme 9 Typical transformations of β-hydroxybutyric acid.

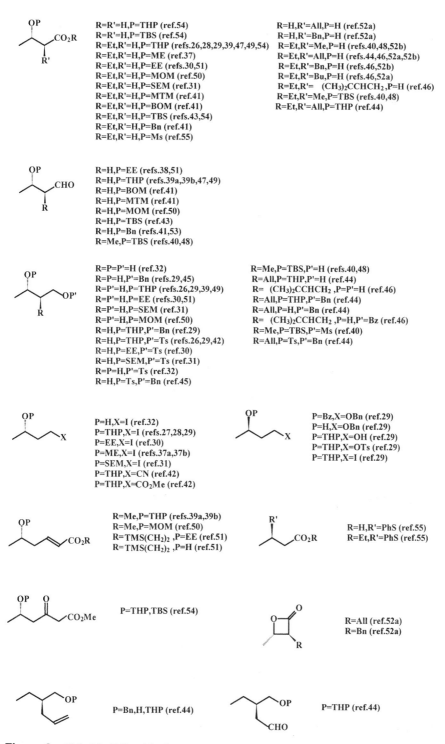

Figure 2 Chiral building blocks derived from (S)-β-hydroxybutyric acid.

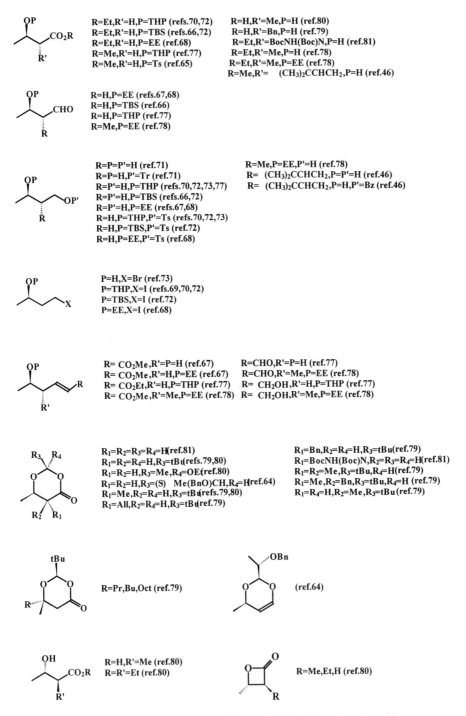

Figure 3   Chiral building blocks derived from (R)-β-hydroxybutyric acid.

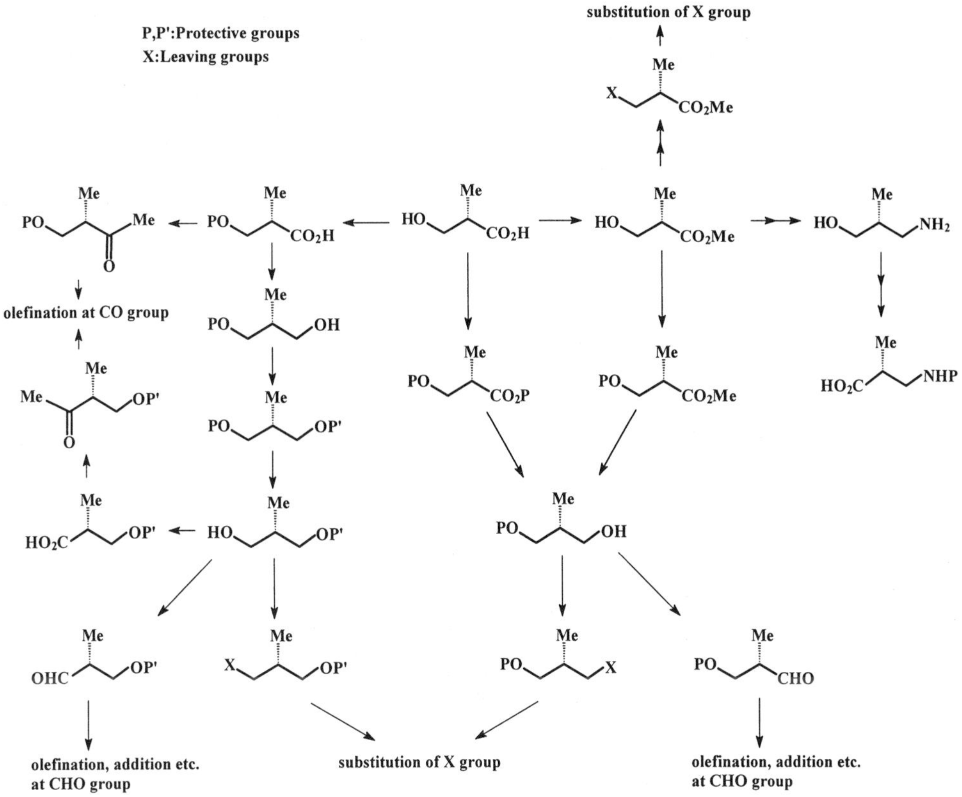

P,P':Protective groups
X:Leaving groups

substitution of X group

olefination at CO group

olefination, addition etc.
at CHO group

substitution of X group

olefination, addition etc.
at CHO group

Scheme 10 Typical transformations of β-hydroxyisobutyric acid.

It is possible to convert one enantiomer to its mirror image by means of sequential protection–deprotection and reduction–oxidation processes.

In Scheme 10 are shown the above-mentioned typical transformations of β-hydroxyisobutyric acid, and in Fig. 4 are listed useful chiral building blocks derived from (S)- and (R)-β-hydroxyisobutyric acids.

B. Synthesis of Chiral Bioactive Compounds Derived from Optically Active β-Hydroxy Acids

1. Syntheses Starting from (S)-β-Hydroxybutyric Acid and Its Derivatives

Ethyl (S)-β-hydroxybutyrate or, more rarely, methyl (S)-β-hydroxybutyrate has been used as the starting material for the synthesis of pheromones (pheromone components), antibiotics (including their intermediates), and other chiral compounds of biological origin.

Thus, starting from ethyl (S)-β-hydroxybutyrate, pheromones of diverse structural types, such as (S)-sulcatol [26], ferrulactone II [27], (2S,6R)-2-methyl-1,7-dioxaspiro[5.6]-dodecane [28], (2S,6R,8S)-2,8-dimethyl-1,7-dioxaspiro[5.5]undecane [29], (2R,5S)-2-methyl-5-hexanolide [31], and (2S,8R)-8-methyl-2-decanol propanoate [32], have been synthesized. Methyl (S)-β-hydroxybutyrate has also been used for the synthesis of (2S,3S,7S)-3,7-dimethylpentadecan-2-ol [33]. In these syntheses, the configurations at the

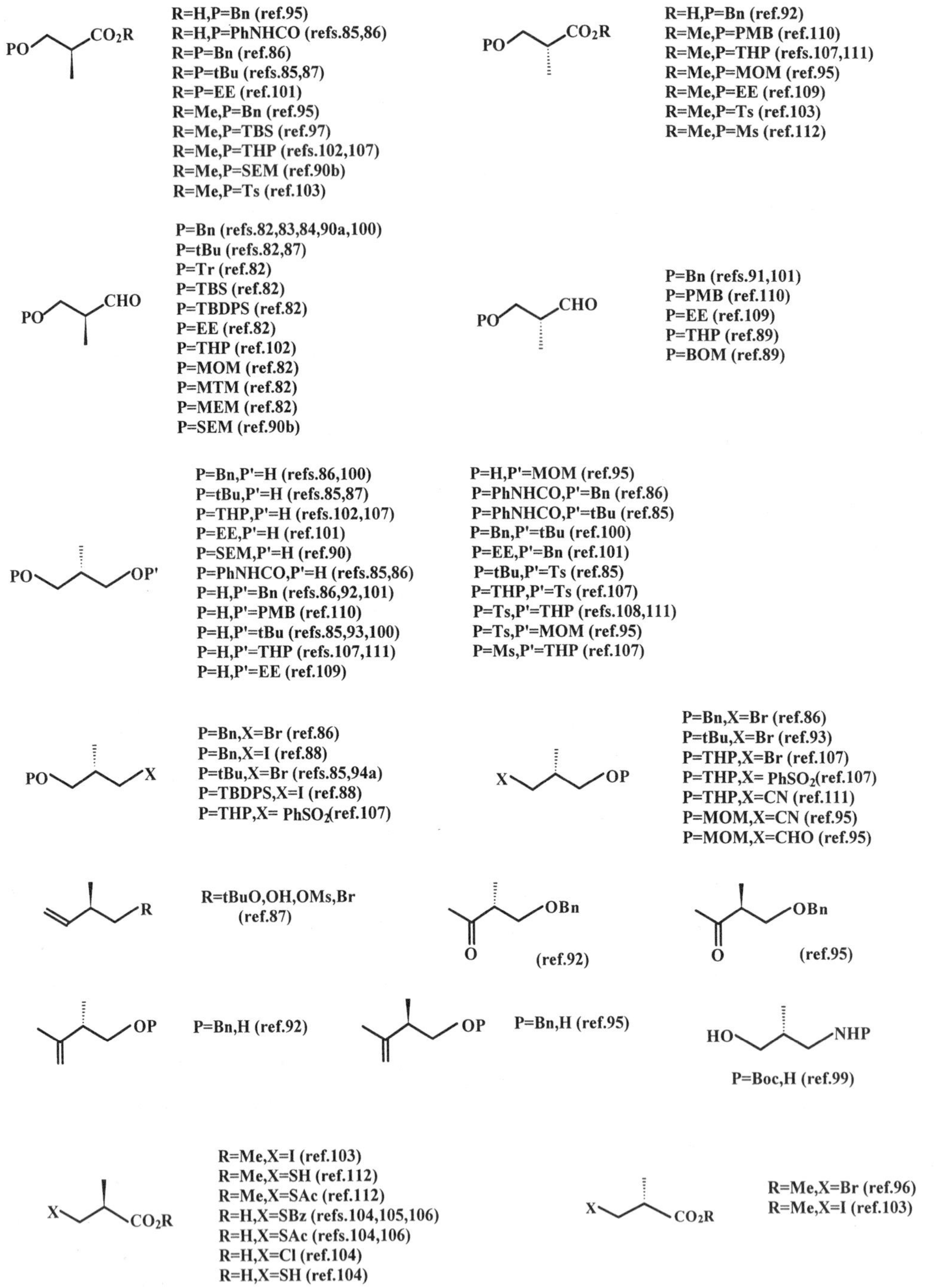

Figure 4   Chiral building blocks derived from (*S*)- and (*R*)-β-hydroxyisobutyric acids.

asymmetrical centers of the starting materials were unchanged during the series of transformations.

In the field of antibiotics, some intermediates for carbapenems have been synthesized from ethyl (S)-β-hydroxybutyrate (Fig. 5) [34–36]. In these syntheses, the construction of the hydroxy ethyl moieties of the target molecules is accompanied by inversion of the configurations at the asymmetrical centers of the starting materials. Ethyl (S)-β-hydroxybutyrate has also been employed for the synthesis of grahamimycin A1 (unnatural form) [38], colletodiol [39], and pseudomonic acid C [40], which are all other antibiotics.

For the synthesis of other chiral compounds of biological origin, we can see other examples of the use of ethyl (S)-β-hydroxybutyrate for the synthesis of rhododendrin [43] and (−)-talaromycins A and B [44].

2. Syntheses Starting from (R)-β-Hydroxybutyric Acid and Its Derivatives

(R)-β-Hydroxybutyric acid and its methyl ester have been widely used in the synthesis of various carbapenem intermediates (Fig. 6) [56–64]. Among them, (3R,4R)-4-acetoxy-3-[(R)-1-(t-butyldimethylsilyloxy)ethyl]-2-azetidinone is the most versatile intermediate for carbapenems and penems, and Kaneka developed a commercial production method [56] for this acetoxyazetidinone. Kaneka's method comprises β-lactam ring formation through the diastereoselective [2+2] cycloaddition of chlorosulfonyl isocyanate with the silyl enol ether prepared from methyl (R)-β-hydroxybutyrate as a key step. Terashima et al. also reported the synthesis of a similar acetoxyazetidinone employing the [2+2] cycloaddition reaction of chlorosulfonylisocyanate with a 4H-1,3-dioxin derivative prepared from methyl (R)-β-hydroxybutyrate [64]. All of the other synthetic approaches to carbapenem intermediates involve the condensation of lithium or boron enolates derived from (R)-β-hydroxybutyric acid with various imine derivatives [57–63].

Another example of the industrial use of methyl (R)-β-hydroxybutyrate is Zeneca's synthesis of an intermediate for MK-0507 (trade name Trusopt; Fig. 7), an antiglaucoma drug developed by Merck [65]. In this synthesis, a ketosulfone derivative is prepared from the tosylate of methyl (R)-β-hydroxybutyrate, which is asymmetrically reduced to a (4S,6S)-hydroxysulfone derivative, a key intermediate for MK-0507, by a microorganism.

Further examples of the utilization of (R)-β-hydroxybutyric acid and its esters comprise the synthesis of (R,R,R)-colletallol [68], grandisol [71], (−)-cladospolide A [72,73], mycobactin S2 [74], mycobactin S [75], and cobactin T [76].

3. Syntheses Starting from (S)-β-Hydroxyisobutyric Acid and Its Derivatives

(S)-β-Hydroxyisobutyric acid and its esters serve as bifunctional building blocks for the synthesis of a wide variety of chiral compounds of biological origin.

Thus, many chiral compounds including complex molecules, e.g., (2R,4′R,8′R)-α-tocopherol [85], (R)-muscone [86], lasalocid A [87], calcimycin [88], monensin [89], rifamycin S [90], (−)-maysine [91], (+)-phyllanthocin [92], (25S)-26-hydroxycholesterol [93], palytoxin [94], (−)-botryococcene [95], paniculidine A (unnatural form) [96], dihydrotanshinone I [97], danshexinkun A [97], aegyptinones A and B [98], and cryptophycins C and D [99], have been synthesized from (S)-β-hydroxyisobutyric acid or its derivatives.

The intermediates for 1β-methylcarbapenems have also been synthesized using methyl (S)-β-hydroxyisobutyrate as a starting material (Fig. 8) [82–84]. In these syntheses, the 1β-methyl moieties of the target molecules come from the corresponding parts of the (S)-β-hydroxyisobutyric acid derivatives.

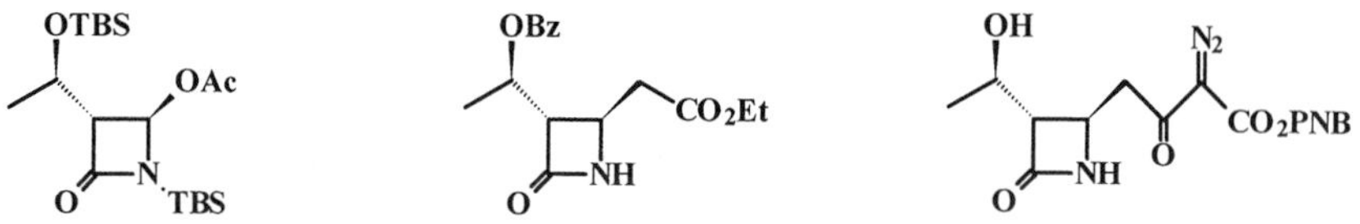

Figure 5 Carbapenem intermediates derived from ethyl (S)-β-hydroxybutyrate.

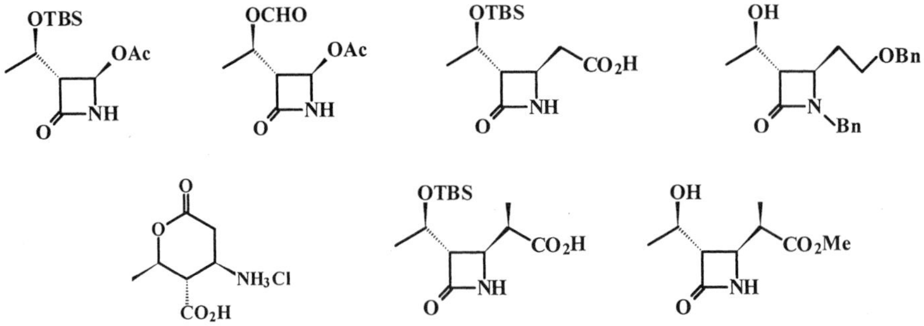

Figure 6 Carbapenem intermediates derived from methyl (R)-β-hydroxybutyrate or (R)-β-hydroxybutyric acid.

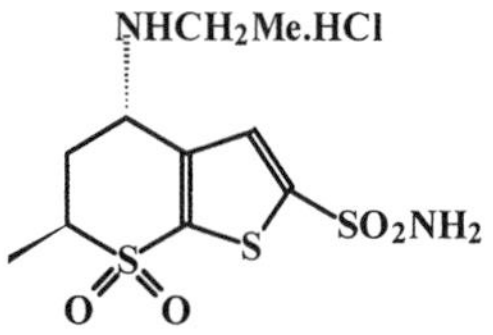

Figure 7 Structure of trusopt (antiglaucoma drug).

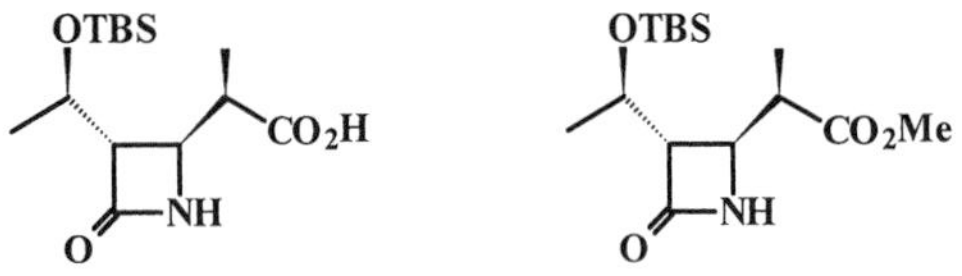

Figure 8 Carbapenem intermediates derived from methyl (S)-β-hydroxyisobutyrate.

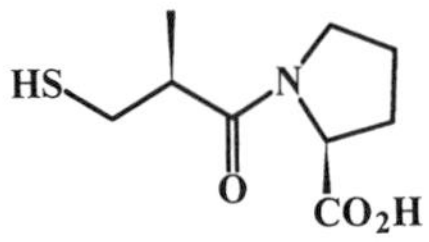

Figure 9 Structure of captopril (antihypertensive drug).

4. Syntheses Starting from (*R*)-β-Hydroxyisobutyric Acid and Its Derivatives

Compared with (*S*)-β-hydroxyisobutyric acid, the corresponding (*R*) enantiomer has been less used for the synthesis of chiral biologically active compounds. Thus, we can find several examples of the use of (*R*)-β-hydroxyisobutyric acid for the synthesis of angiotensin-converting enzyme (ACE) inhibitors [140–106], pheromones [107–109], and an immunosuppressive agent [110].

Among the ACE inhibitors, captopril (Fig. 9) is the most famous and Kaneka has developed an effective synthetic process for captopril involving (*R*)-β-hydroxyisobutyric acid [104]. In this synthesis, (*S*)-3-acetylthio-2-methylpropanoic acid, a key intermediate for captopril, is prepared from (*R*)-β-hydroxyisobutyric acid on a plant scale.

REFERENCES

1. T Harada, A Tai. Chem Lett 1195–1196, 1978.
2. T Mukaiyama, T Yura, N Iwasawa. Chem Lett 815–816, 1985.
3. R Noyori, T Ohkuma, M Kitamura. J Am Chem Soc 109:5856–5858, 1987.
4. W Choy, P Ma, S Masamune. Tetrahedron Lett 22:3555–3556, 1981.
5. D A Evans, J Bartroli, T L Shih. J Am Chem Soc 103:2127–2129, 1981.
6. A L Lehninger, G D Greville. Biochim Biophys Acta 12:188–202, 1953.
7. H Ohta, H Tetsukawa. Chem Lett 1379–1380, 1979.
8. R U Lemieux, J Giguere. Can J Chem 29:678–690, 1951.
9. T Sugai, M Fujita, K Mori. Nippon Kagaku Kashi 9:1315–1321, 1983.
10. B Wipf, E Kupfer, R Bertazzi, H G W Leuenberger. Helv Chim Acta 66:485–488, 1983.
11. T Kometani, H Yoshii, H Takeuchi, R Matsuno. J Ferment Bioeng 76:33–37, 1993.
12. G Fráter. Helv Chim Acta 62:2829–2832, 1979.
13. D Seebach, M F Züger, F Giovannini, B Sonnleitner. Angew Chem Int Ed Engl 23:151–152, 1984.
14. K Mori, H Mori, T Sugai. Tetrahedron 41:919–925, 1985.
15. M F Züger, F Giovannini, D Seebach. Angew Chem 95:1024, 1983.
16. D Seebach, M F Züger. Tetrahedron Lett 25:2747–2750, 1984.
17. T Miyoshi, T Harada. J Ferment Technol 52:388–392, 1974.
18. C T Goodhue, T R Schaeffer. Biotechnol Bioeng 13:203–214, 1971.
19. S Tahara, J Mizutani. Agric Biol Chem 42:879–883, 1978.
20. T Ohashi, J Hasegawa. Chirality in Industry. New York: John Wiley, 1992, pp 249–267.
21. J Hasegawa, M Ogura, S Hamaguchi, H Kawaharada, K Watanabe. J Ferment Technol 51:203–208, 1981.
22. J Hasegawa, S Hamaguchi, M Ogura, K Watanabe. J Ferment Technol 59:257–262, 1981.
23. J Hasegawa, M Ogura, H Kanema, N Noda, H Kawaharada, K Watanabe. J Ferment Technol 60:501–508, 1982.
24. J Hasegawa, M Ogura, H Kanema, H Kawaharada, K Watanabe. J Ferment Technol 60:591–594, 1982.
25. J Hasegawa, M Ogura, H Kanema, H Kawaharada, K Watanabe. J Ferment Technol 61:37–42, 1983.
26. K Mori. Tetrahedron 37:1341–1342, 1981.
27. T Sakai, K Mori. Agric Biol Chem 50:177–183, 1986.
28. K Mori, K Katsurada. Liebigs Ann Chem 157–161, 1984.
29. K Mori, K Tanida. Tetrahedron 37:3221–3225, 1981.
30. K Hintzer, R Weber, V Schurig. Tetrahedron Lett 22:55–58, 1981.
31. T Katsuki, M Yamaguchi. Tetrahedron Lett 28:651–654, 1987.
32. J T B Ferreira, F Simonelli. Tetrahedron 46:6311–6318, 1990.

33. A Tai, N Morimoto, M Yoshikawa, K Uehara, T Sugimura, T Kikukawa. Agric Biol Chem 54:1753–1762, 1990.
34. G Cainelli, M Contento, D Giacomini, M Panunzio. Tetrahedron Lett 26:937–940, 1985.
35. D J Hart, D-C Ha. Tetrahedron Lett 26:5493–5496, 1985.
36. T Chiba, M Nagatsuma, T Nakai. Chem Lett 1343–1346, 1985.
37. (a) D Seebach, B Seuring, H-O Kalinowski, W Lubosch, B Renger. Angew Chem 89:270–271, 1977; (b) B Seuring, D Seebach. Liebigs Ann Chem 2044–2073, 1978.
38. W Seidel, D Seebach. Tetrahedron Lett 23:159–162, 1982.
39. (a) H Tsutsui, O Mitsunobu. Tetrahedron Lett 25:2159–2162, 1984; (b) H Tsutsui, O Mitsunobu. Tetrahedron Lett 25:2163–2166, 1984.
40. G E Keck, D F Kachensky, E J Enholm. J Org Chem 49:1462–1464, 1984.
41. G E Keck, J A Murry. J Org Chem 56:6606–6611, 1991.
42. T Kitahara, K Koseki, K Mori. Agric Biol Chem 47:389–393, 1983.
43. K Mori, Z-H Qian. Bull Soc Chim Fr 130:382–387, 1993.
44. K Mori, M Ikunaka. Tetrahedron 43:45–58, 1987.
45. K Mori, T Sugai. Synthesis 752–753, 1982.
46. A Kramer, H Pfander. Helv Chim Acta 65:293–301, 1982.
47. T Sato. Heterocycles 24:2173–2174, 1986.
48. D W Brooks, R P Kellogg. Tetrahedron Lett 23:4991–4994, 1982.
49. H Redlich, B Schneider, R W Hoffmann, K-J Geueke. Liebigs Ann Chem 393–411, 1983.
50. A I Meyers, R A Amos. J Am Chem Soc 102:870–872, 1980.
51. E Hungerbühler, D Seebach, D Wasmuth. Helv Chim Acta 64:1467–1487, 1981.
52. (a) G Fráter. Helv Chim Acta 62:2825–2828, 1979; (b) G Fráter, U Müller, W Günther. Tetrahedron 40:1269–1277, 1984.
53. M T Reetz, A Jung. J Am Chem Soc 105:4833–4835, 1983.
54. L Liu, R S Tanke, M J Miller. J Org Chem 51:5332–5337, 1986.
55. S Y Dike, D H Ner, A Kumar. Synlett 433–444, 1991.
56. T Ohashi, J Hasegawa. Chirality in Industry. New York: John Wiley, 1992, pp 269–278.
57. M Hatanaka, H Nitta. Tetrahedron Lett 28:69–72, 1987.
58. M Hatanaka. Tetrahedron Lett 28:83–86, 1987.
59. T Iimori, M Shibasaki. Tetrahedron Lett 27:2149–2152, 1986.
60. T Iimori, M Shibasaki. Tetrahedron Lett 26:1523–1526, 1985.
61. T Iimori, Y Ishida, M Shibasaki. Tetrahedron Lett 27:2153–2156, 1986.
62. T Chiba, T Nakai. Tetrahedron Lett 26:4647–4648, 1985.
63. T Chiba, T Nakai. Chem Lett 651–654, 1985.
64. Y Ito, Y Kobayashi, S Terashima. Tetrahedron Lett 30:5631–5634, 1989.
65. T J Blacklock, P Sohar, J W Butcher, T Lamanec, E J J Grabowski. J Org Chem 58:1672–1679, 1993. See also, A J Blacker, R A Holt. Chirality in Industry, Vol. 2. New York: John Wiley, 1997, pp 245–261.
66. K Ohta, O Miyagawa, H Tsutsui, O Mitsunobu. Bull Chem Soc Jpn 66:523–535, 1993.
67. P Schnurrenberger, E Hungerbühler, D Seebach. Tetrahedron Lett 25:2209–2212, 1984.
68. F J Dommerholt, L Thijs, B Zwanenburg. Tetrahedron Lett 32:1495–1498, 1991.
69. K Mori, H Watanabe. Tetrahedron 42:295–304, 1986.
70. K Mori, H Watanabe. Tetrahedron 40:299–303, 1984.
71. K Mori, M Miyake. Tetrahedron 43:2229–2239, 1987.
72. K Mori, S Maemoto. Liebigs Ann Chem 863–869, 1987.
73. I Ichimoto, M Sato, M Kirihata, H Ueda. Chem Express 2:495–498, 1987.
74. P J Maurer, M J Miller. J Am Chem Soc 105:240–245, 1983.
75. J Hu, M J Miller. J Am Chem Soc 119:3462–3468, 1997.
76. J Hu, M J Miller. Tetrahedron Lett 36:6379–6382, 1995.
77. K C Nicolaou, M R Pavia, S P Seitz. J Am Chem Soc 103:1224–1226, 1981.
78. M A Sutter, D Seebach. Liebigs Ann Chem 939–949, 1983.
79. D Seebach, J Zimmermann. Helv Chim Acta 69:1147–1152, 1986.

80. A Griesbeck, D Seebach. Helv Chim Acta 70:1320–1325, 1987.
81. C Greck, L Bischoff, F Ferreira, C Pinel, E Piveteau, J P Genêt. Synlett 475–477, 1993.
82. T Kawabata, Y Kimura, Y Ito, S Terashima. Tetrahedron Lett 27:6241–6244, 1986.
83. F Shirai, T Nakai. Chem Lett 445–448, 1989.
84. F Shirai, T Nakai. Tetrahedron Lett 29:6461–6464, 1988.
85. N Cohen, W F Eichel, R J Lopresti, C Neukom, G Saucy. J Org Chem 41:3505–3511, 1976.
86. Q Branca, A Fischli. Helv Chim Acta 60:925–944, 1977.
87. T Nakata, Y Kishi. Tetrahedron Lett 2745–2748, 1978.
88. D A Evans, C E Sacks, W A Kleschick, T R Taber. J Am Chem Soc 101:6789–6791, 1979.
89. D B Collum, J H McDonald III, W C Still. J Am Chem Soc 102:2118–2120, 1980.
90. (a) H Nagaoka, Y Kishi. Tetrahedron 37:3873–3888, 1981; (b) S Masamune, B Imperiali, D S Garvey. J Am Chem Soc 104:5528–5531, 1982.
91. A I Meyers, K A Babiak, A L Campbell, D L Comins, M P Fleming, R Henning, M Heuschmann, P Hudspeth, J M Kane, P J Reider, D M Roland, K Shimizu, K Tomioka, R D Walkup. J Am Chem Soc 105:5015–5024, 1983.
92. P R McGuirk, D B Collum. J Am Chem Soc 104:4496–4497, 1982.
93. C-Y Byon, M Gut, V Toome. J Org Chem 46:3901–3903, 1981.
94. (a) L L Klein, W W McWhorter Jr, S S Ko, K-P Pfaff, Y Kishi. J Am Chem Soc 104:7362–7364, 1982; (b) S S Ko, J M Finan, M Yonaga, Y Kishi. J Am Chem Soc 104:7364–7367, 1982.
95. J D White, G N Reddy, G O Spessard. J Am Chem Soc 110:1624–1626, 1988.
96. B A Czeskis, A M Moissenkov. J Chem Soc Perkin Trans I 1353–1354, 1989.
97. R L Danheiser, D S Casebier, J L Loebach. Tetrahedron Lett 33:1149–1152, 1992.
98. R L Danheiser, D S Casebier, A H Huboux. J Org Chem 59:4844–4848, 1994.
99. R A Barrow, T Hemscheidt, J Liang, S Paik, R E Moore, M A Tius. J Am Chem Soc 117:2479–2490, 1995.
100. M R Johnson, Y Kishi. Tetrahedron Lett 4347–4350, 1979.
101. A I Meyers, J P Hudspeth. Tetrahedron Lett 22:3925–3928, 1981.
102. K R Buszek, F G Fang, C J Forsyth, S H Jung, Y Kishi, P M Scola, S K Yoon. Tetrahedron Lett 33:1553–1556, 1992.
103. M Sakurai, T Hata, Y Yabe. Tetrahedron Lett 34:5939–5942, 1993.
104. M Shimazaki, J Hasegawa, K Kan, K Nomura, Y Nose, H Kondo, T Ohashi, K Watanabe. Chem Pharm Bull 30:3139–3146, 1982.
105. D H Kim, C J Guinosso, G C Buzby Jr, D R Herbst, R J McCaully, T C Wicks, R L Wendt. J Med Chem 26:394–403, 1983.
106. J T Suh, J W Skiles, B E Williams, R D Youssefyeh, H Jones, B Loev, E S Neiss, A Schwab, W S Mann, A Khandwala, P S Wolf, I Weinryb. J Med Chem 28:57–66, 1985.
107. K Mori, S Senda. Tetrahedron 41:541–546, 1985.
108. M Kato, K Mori. Agric Biol Chem 49:2479–2480, 1985.
109. K Mori, T Ebata. Tetrahedron 42:4413–4420, 1986.
110. A B Smith III, Y Qiu, D R Jones, K Kobayashi. J Am Chem Soc 117:12011–12012, 1995.
111. K Mori, Tetrahedron 39:3107–3109, 1983.
112. C H Senanayake, R D Larsen, T J Bill, J Liu, E G Corley, P J Reider. Synlett 199–200, 1994.

14

Stereoselective Synthesis of Chiral Compounds Using Whole-Cell Biocatalysts

Paola D'Arrigo, Giuseppe Pedrocchi-Fantoni, and Stefano Servi
Politecnico di Milano, Milano, Italy

The synthesis of chiral compounds in enantiomerically pure form has become an una-voidable task in the production of chiral drugs and biologically active materials [1,2]. The demand of fine chemicals as single enantiomers has promoted a large effort in the development of effective new chiral catalysts. Successful practical applications are met with asymmetric hydrogenation catalysts [3,4] and biocatalysts [5–11], while the classical resolution approach based on diastereomeric pairs separation still remains the most widely used methodology in industry [12,13]. New technologies based on continuous quantitative chromatographic (SMB) enantiomer separation [14] or membrane separations [15] are not yet mature for large-scale applications. The interest in the use of biocatalysts is due to the possibility of effecting a chemical transformation in a reduced number of steps, avoiding the use of often aggressive costly and environmentally hazardous reagents. Typical ex-amples where biocatalysts have replaced well-established chemical processes are the use of penicillin amidase in the production of semisynthetic β-lactam antibiotics [16] and the enzymatic resolution of *p*-methoxyphenylglycidate [17–19] intermediate in the synthesis of diltiazem. Among biocatalysts the choice is between commercially available enzymes of different purity, crude enzymatic preparations obtainable from animal organs or cultures of microorganism, and whole-cell biocatalysts. Whole-cell biocatalysts can offer several advantages over the use of isolated enzymes in biocatalysis. They are a convenient and stable source of enzymes that are often synthetized by cells in response to the presence of the substrate. Once the required activity has been found, modern techniques in molecular biology allow the expression of the new enzyme in a foreign host with high activity for industrial process. This practice is often followed in industry. In cases where more enzy-matic activities are required on the same substrate, as in the case of the D-hydantoinase and carbamoylase activity in the preparation of D-amino acids from the corresponding hydantoins, the use of the whole-cell biocatalyst cannot be replaced by isolated enzymes [20]. Also reactions exploiting the oxidation and hydroxylation activity of biocatalysts are conducted with whole-cell biocatalysts due to the limited stability of isolated enzymes and the requirement and recycling of cofactor [21]. One of the features that makes whole-cell biocatalysts so attractive is that they generate cofactors together with the actual catalyst, the enzyme, and therefore they are particularly convenient in applications in which their reducing power is employed. The most common application is the reduction of carbonyl groups to secondary alcohols and of prochiral C=C double bonds to chiral tertiary-substituted carbon atoms [8–11]. The former operation is alternative to the esters hydrolysis-formation catalyzed by hydrolytic enzymes of widespread application. This group of enzymes has met with a fantastic success in the production of chiral fine chemicals, especially among pharmaceuticals. Most of the applications of hydrolytic enzymes deal with resolution processes. The costs associated with the recycling or reutilization of the *wrong* enantiomer are becoming an increasingly crucial issue, thus lessening the effectiveness of the resolution process. The asymmetrical carbonyl reduction has clear advantages over the kinetic resolution of an equivalent intermediate from the point of view of substrate utilization. In contrast, whole-cell biocatalysis usually occurs in a water medium with substrate concentration rarely exceeding 10 g/L. The ability to effectively overcome the problems of substrate availability (solubility in the medium, cell permeation), enzyme inhibition, effective space–time yield production, and product recovery is decisive in the possibility of achieving cost-effective production with this technology. Therefore whole-cell biotransformation can be competitive when applied to high-priced chiral intermediates. A relatively small number of nonpathogenic, often food grade microorganisms (yeasts, fungi, or bacteria) are used in biocatalysis for the production of chiral intermediates. These microorganisms are often the first choice for the screening of a desired activity, when a rapid methodology for the production of a limited amount of compound is required. This is often considered to be more convenient than undergoing an asymmetrical synthesis when new chiral drug candidates have to be evaluated for their biological activity in the two enantiomeric forms. It is usually possible, while conducting the screening for the required activity, to meet microorganisms showing both stereo preferences. Baker's yeast is the favorite bench reagent for the reduction of carbonyl groups and of activated C=C double bond although it displays a number of other interesting and sometimes almost unique enzymatic activities. Due to the simple approach and its application in organic chemistry laboratories, it has been the first-choice biocatalyst and a strong testimonial for the potential of biocatalysts in organic synthesis [8,9,11]. Although its performances are often surpassed by those of other organisms, industrial applications and very effective transformations of Baker's yeast are reported [22]. The stereochemistry of the resulting compound is predictable to some extents, and the substrate specificity of the enzymes produced extremely broad. Its use in biocatalysis is complemented by a number of yeasts with similar properties that are often more selective in the biotransformation. Incomplete stereochemical purity of the products can often be corrected by inhibiting or improving the synthesis of a specific enzyme by the addition of simple compounds or by controlling the substrate concentration. The use of absorbing resins in controlling substrate concentration and stereoselectivity avoids unpractical multiphase reaction mixtures and allows improved product recovery. This has emerged recently as an important tool in bringing whole-cell biocatalysis to industrial scale [23–26]. Applications of biocatalysts in organic synthesis have been thoroughly documented [5–11]. In this chapter we focus on the production of nonracemic chiral compounds, intermediates or final products, of interest as fine chemicals and how they can be obtained from the biocatalytic reduction of carbonyl groups and of activated triply substituted C=C double bonds using whole-cell microorganisms as biocatalysts. The stereochemical features of reactions leading to actual intermediates in drug manufacture will be considered.

I. PRODUCTS FROM THE REDUCTION OF AROMATIC KETONES

Substrates bearing a phenyl ring and an alkyl group on two sides of a carbonyl group are among the most extensively studied substrates for dehydrogenase enzymes. Acetophenone is an ideal substrate for assaying the stereochemical preference of a reducing biocatalyst. Due to the large difference in stereochemical requirements of the two groups flanking the carbonyl function, a simple experiment allows one to assign a microorganism to the *Prelog* specificity or to the opposite *anti-Prelog* specificity (Scheme 1). This refers to a rule developed to classify the reducing outcome of the carbonyl group catalyzed by *Curvularia lunata* simply stating that this microorganism reduces prochiral carbonyl groups giving alcohols of (*S*) absolute configuration (however, priority rules are based on dimensions) [27]. Exceptions to this rule for the same microorganism with different substrates suggest that priority in the determination of the absolute configuration will also take into account considerations other than dimensions, e.g., polarity, thus offering a less predictable answer. The rule is, however, very useful because microorganisms and enzymes usually keep the same behavior with most substrates. Many microorganisms showing (*S*) selectivity are available. Thus the observation of the selectivity in such reactions allows rapidly to evaluate the influence of the consequence of experimental conditions on the stereo preference and hence on the synthesis of enzymes in the cell. Considering the utility of having access to a set of microorganisms of known reducing capacity, researchers in industry carried out simple screening on about 200 strains of microorganisms from culture collections that had been previously known for their ability to reduce ketones. The microbial cultures were tested on phenyl- and naphthyl ketones of different structures in order to build a collection of strains of predictable absolute stereochemical requirements. In this way a list of a few best organisms was set up. Availability of such a group of biocatalysts has allowed the rapid detection of appropriate biocatalysts in a number of issues. Their idea about the economic feasibility of a whole-cell biotransformation versus an asymmetrical hydrogenation step is that if a substrate concentration of around 10 g/L can be used, a close competition between the two methodologies is reached [28]. However, this evaluation largely depends on the final product value. Whole-cell biocatalysis is often applied on a larger scale, even though such high concentration limits are often not accomplished. Results of this investigation refer to concentration of 1–2 g/L (Table 1).

Thus *Rhodotorula glutinis* reduces acetophenone and propiophenone with almost quantitative yield with *S* selectivity. Higher substrate concentrations reduce the ketone conversions. This is the rather common effect of substrate inhibition observed with most enzymatic transformations. The control of substrate concentration can be effected in various ways. 1- and 2-Acetonaphthone are also reduced efficiently by the same microorganism. Steric bulkiness around the carbonyl group reduces the conversion. More appropriate organisms for such substrates as butyrophenone, isobutyrophenone, and 2,2-dimethylpropiophenone are *Aspergillus carneus*, *Geotrichum candidum*, *Yarrowya lypolitica*, and *Y. halophila*. α-Tetralone is reduced to the secondary alcohol of (*S*) absolute configuration of 100% e.e. with *Pseudomonas methanolica*.

(*R*)-Selective enzymes are somewhat less common: *Nectria gliocladioides* has been reported to give products with (*R*) selectivity and *Clostridium kluyveri* is well known for

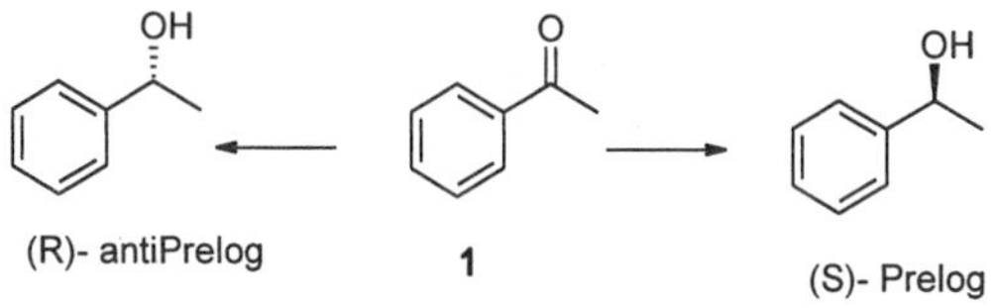

(R)- antiPrelog   **1**   (S)- Prelog

Scheme 1

Table 1 Stereo Preferences of Selected Microorganisms Toward Aromatic
Ketones [28]

Substrate	Culture	Yield (%)	e.e. (%)	Configuration
	Rhodotorula glutinis	96	100	(*S*)
	ATCC16740	20	100	(*R*)
	Geothricum candidum			
	ATCC 34614			
	Rhodotorula glutinis	82	100	(*S*)
	ATCC 16740			
	Aspergillus niveus	43	100	(*S*)
	ATCC 12276			
	Geotrichum candidum	25	100	(*S*)
	ATCC 34614			
	Yarrowia lipolytica	29	100	(+)
	ATCC 8661	55	100	(−)
	Yarrowia halophila			
	ATCC 20321			
	Hansenula subpellicosa	54	100	(+)
	ATCC 16766			
	Torulaspora hansenii	76	98	(−)
	ATCC 34022			
	Torulaspora hansenii	54	100	(−)
	ATCC 34022	44	98	(+)
	Zygosaccharomyces bailii			
	ATCC 38924			
	Rhodotorula glutinis	32	100	(*S*)
	ATCC 16740			
	Rhodotorula glutinis	79	100	(*S*)
	ATCC 16740			
	Pseudomonas methanolica	46	100	(*S*)
	ATCC 58403			

having opposite stereochemical preference with respect to most common microorganisms.
The application of this anaerobic bacteria is not as user-friendly as baker's yeast, *G.
candidum* or *B. bassiana*, and a few others yeasts and fungi that can be grown without a
particular training [29]. Alcohol dehydrogenases from *Pseudomonas* sp and *Lactobacillus
kefir* have been described that catalyze the enantioselective reduction of aromatic, cyclic,
and aliphatic ketones to their corresponding secondary alcohols with (*R*) selectivity [30–
33]. Attempts to use (*S*)-selective microorganisms in the oxidation direction on secondary
alcohols in order to obtain benzyl alcohols of (*R*) configuration gave limited yields and
enantioselectivities [34].

Although many model substrates of this class have been tested with different micro-organisms, the intermediates with the correct substitution pattern on the aromatic ring for the synthesis of actual drugs that have been submitted to biotransformation are relatively few. The extension of the outcome of the reduction of acetophenone to slightly different analogs is not so obvious. Although the stereochemical preference in carbonyl reduction is hardly altered without large modification in the proximity of the carbonyl group, substitution on the aromatic ring and the presence of other functional groups in remote parts of the molecule can profoundly influence the enantiomeric excess, product recovery, and chemical yields.

The interest in aromatic ketones reduction is due to the numerous chiral drugs bearing the 2-amino-1-phenylethanol structure. They mainly belong to a large group of adrenergic drugs with α- and β-agonist activity with pronounced difference in pharmacological activity between the two enantiomers. Due to the priority rules the absolute configuration of synthetic targets is the (R), but (S)-configurated compounds are in some case required.

The synthesis of a large group of such compounds can be effected taking as an example the preparation of (S)-isoproterenol (Scheme 2) where chirality was introduced with chiral oxazaborolidines [35]. Several authors have studied the reduction of α-halo-ketones with no substitution on the aromatic ring [36–38]. Good e.e. of the chloro alcohols are obtained when no ethanol was added to the fermenting mixture. Compounds with the (R) configuration were obtained. Halohydrines with the (S-) configuration are, however, accessible with other microorganisms. A process for the preparation of optically active 2-halo-1-phenylethanol in both configurations by microbial reduction has been reported [39]. Under suitable conditions *Candida humicola* reduces 2-chloro-1-phenylethanone to the corresponding halohydrine to the (S)-configuration at 22 g/L and 100% yield with 96% ee. This compound was used in the synthesis of levamisole [40] as outlined in Scheme 3. The selectivity shown by this microorganism could not be extended to other α-chloroke-tones with different aromatic substitution patterns showing that the effect of substituents in the aromatic ring cannot be neglected. In contrast, the strain of *Rhodotorula glutinis* var. *dairenensis* IFO 0415 could be used for the reduction of a number of chloro- and methoxy-substituted aromatic chloroketones in high e.e. and all of them with the same (R) absolute configuration.

The 3,4-dichloro phenyl analog **6** can be reduced to the corresponding hydroxy alcohols with both configurations: *Mucor mihei* affords the alcohol *R*-**7** with good selectivity, while the chloro alcohol *S*-**7** can be obtained with excellent yields and good e.e. (94%) at 2 g/L concentration with cultures of *G. candidum*. This compound is quantitatively converted into the epoxide **8** with the same absolute configuration. Partly regiose-

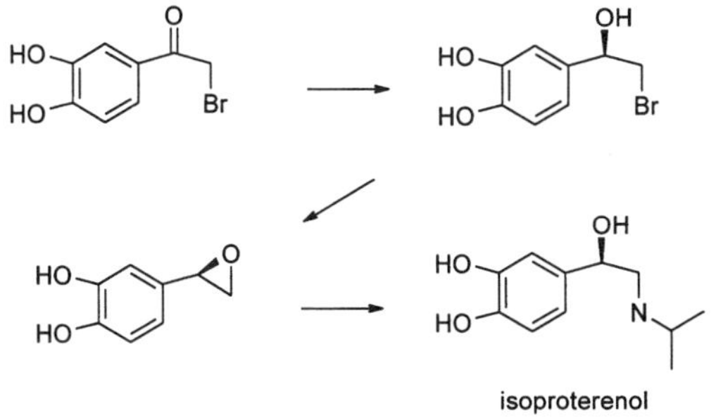

isoproterenol

Scheme 2

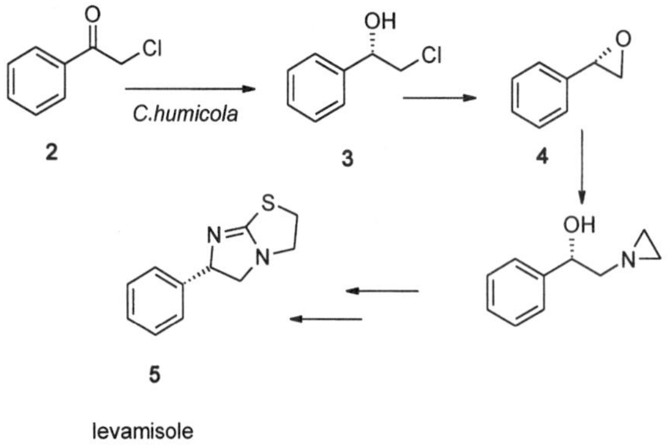

Scheme 3

lective opening with a C2 nucleophile affords the butyrolactone R-**9** intermediate in the synthesis of the pharmacologically active form of sertraline **10** [41]. It is noteworthy that the product obtained from yeast bioconversion has the expected configuration but in a considerably lower ee than the corresponding unsubstituted analog or acetophenone. The enantiomeric phenyl butyrolactone S-**9** (Scheme 4) can be obtained by direct reduction of the keto acid **11** with baker's yeast [42].

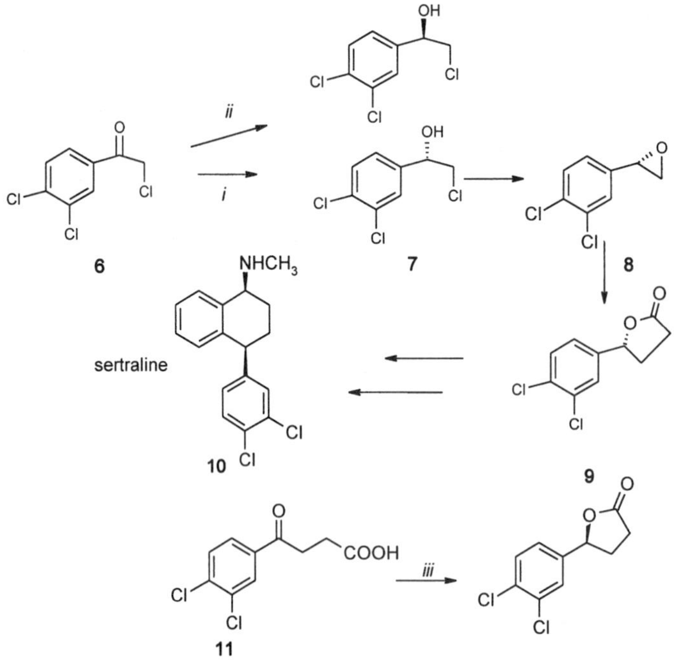

i G. candidum 96% yield, 94% ee, 2 g/l   *iii* Bakers yeast 30% yield 98% ee
ii Bakers yeast 60% yield, 60% ee
   Mucor mihei 80% yield, 85% ee

Scheme 4

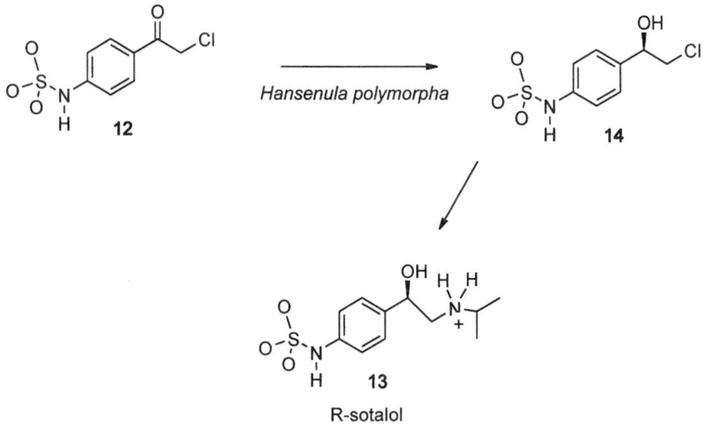

Scheme 5

In the synthesis of D-sotalol, the benzyl alcohol with opposite configuration is required. The chloroketone **12** is the substrate for *Hansenula polymorpha* in the synthesis of an intermediate in the preparation of the β-receptor antagonist **13**. Microorganisms from the genus *Nocardia*, *Rhodococcus*, and *Hansenula* reduce the substrate to the target chlorohydrin **14**. The strain of *H. polymorpha* is able to reduce the required compound to the (*R*)-chlorohydrin a possible intermediate in the synthesis of (*R*)-sotalol **13** [43]. It has been proved that the (*R*) enantiomer is 50 times more active than the epimer for its beta blocker activity, associated with antiarrhythmic activity. However, the (*S*) enantiomer is going to be developed as a type II antiarrhythmic drug with no association with beta blocker [44] (Scheme 5). When bromoacetophenone was used in the reduction, only trace amounts of product was obtained [38]. The same bromoketone was instead reduced with cultures of *Cryptococcus macerans* [45]. The enantiomeric excess of the alcohol obtained was strongly dependent on the experimental conditions. About 90% e.e. was obtained in the absence of added alcohol. With increasing fermentation time the e.e. was increased and the chemical yield decreased, suggesting a phenomenon of dynamic kinetic resolution following the reoxidation of the alcohol, or the change in oxidoreductases production with time. A strain of *G. candidum* affords the (*S*) enantiomer, a result that has precedent in the acetophenone reduction reported in Table 1.

The above-mentioned microorganism *Cryptococcus macerans* was employed in the reduction of the 2-bromo-1-indanone **15** to give the bromoindanol **16** of the configuration depicted in Scheme 6. A yield of 55% of one single enantiomer was obtained in 7 days fermentation. The fact that the yield exceeds 50% indicates that a partial racemization of the nonreduced enantiomer is occurring [46].

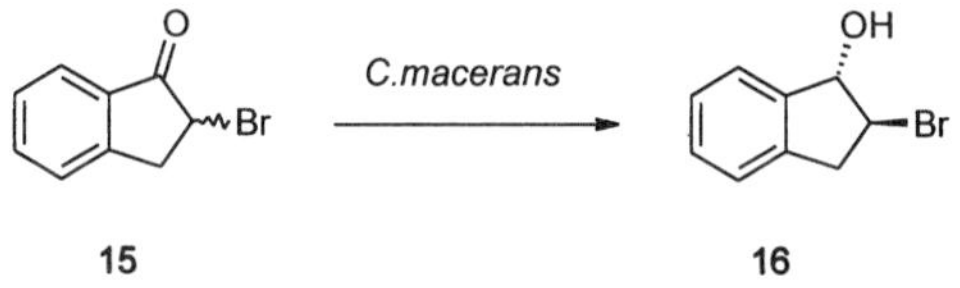

Scheme 6

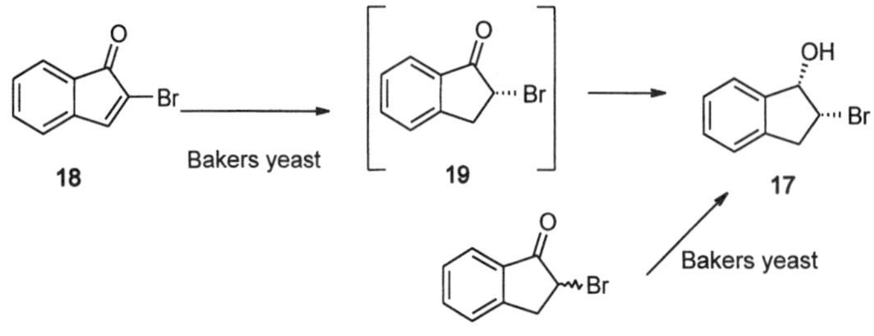

Scheme 7

The *1S,2R* stereoisomer **17** has been obtained via the Baker's yeast reduction of the unsaturated bromoketone **18** in 99% e.e. and 75% yields at 3 g/L. The reaction occurs presumably through the reduction of the initially formed saturated ketone **19**. The fact that racemic bromoketone **20** is also reduced by yeast to the same enantiomer with complete stereoselectivity is evidence that the intermediate bromoketone **19** rapidly racemizes in the reaction medium. The efficiency of the latter biotransformation makes it preferable to the corresponding yeast reduction of the unsaturated compound (Scheme 7). The preparation of such enantiomerically pure compounds found significance in the preparation of optically active 1-amino-2-indanol as ligand in the human immunodeficiency virus (HIV) protease inhibitor indinavir [47].

Baker's yeast was employed in the reduction of **21** to the optically pure allylic alcohol **22**. Pd(2) catalyzed hydride transfer of the chiral intermediate gave the indenone **23** required as an intermediate in the synthesis of a putative antihypertensive drug. At variance with the example cited above, the double bond was unaffected by the yeast [48] (Scheme 8).

The application of whole-cell biocatalysts on more advanced intermediates in the synthesis of chiral pharmaceuticals has been reported. Thus the benzyl alcohol **24** depicted in Scheme 9 was obtained with the biomass of *Mortierella ramanniana*. *R*-BMY 14802 is a sigma ligand with a high affinity for sigma binding sites [49]. The strategy of introducing the chiral center at a late stage of the synthetic path can be advantageous especially when the chiral carbon can be prone to epimerization as in the compound indicated. The biotransformation was carried out in different conditions both with resting cells, growing biomass and cell-free extract at 2 g/L. Other intermediates (Scheme 9) **26** and **27** were also obtained by bioreduction with the same organism and the same selectivity [50].

Attempts to reduce large hydrophobic substrates have usually met with limited success. However, it is often the easiest alternative for the preparation of samples of enantiopure compounds especially when the presence of reactive funtionalities in the molecules

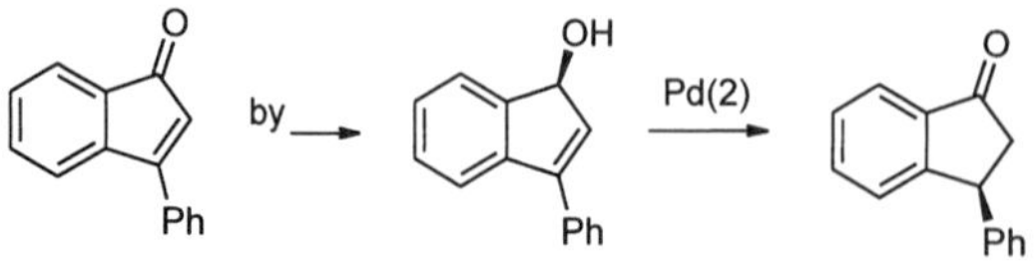

Scheme 8

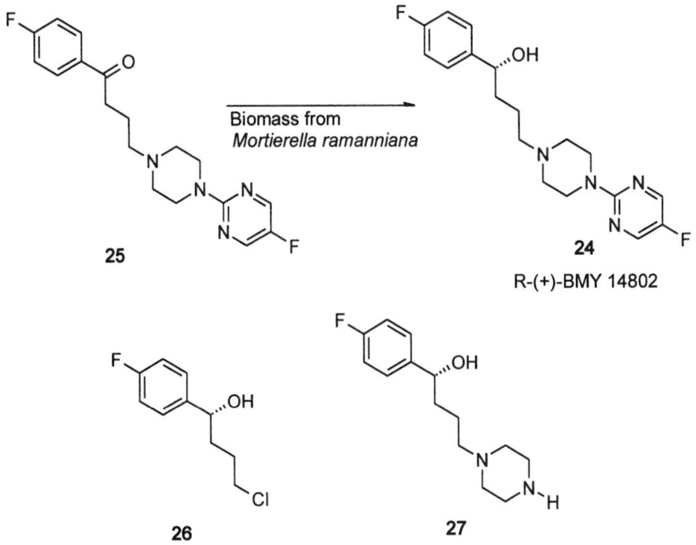

Scheme 9

prevents the use of asymmetrical hydrogenation catalysts. The reduction of ketone **28** with biomass of *Microbacterium* sp. occurs with limited space–time yields, probably due to the low water solubility of the substrate (Scheme 10). The scope of having preparative amounts of the compound for pharmacological assays is reached. The compound is a very potent cysteinyl leukotriene receptor antagonist under clinical investigations for the treatment of asthma [51]. The reduction of prochiral bisaryl ketones useful as intermediates of the synthesis of the phosphodiesterase IV inhibitor CDP 840 [52] has also been reported [53]. Aromatic ketones derived from pyridines and other heterocyclic rings have been reduced with yeast [55]. Recently, compound **30** (Scheme 11) has been reduced with the yeast *C. sorbophila* to the *S*-alcohol, a key intermediate in the synthesis of a β₃ agonist for the treatment of obesity and diabetes [56].

II. PRODUCTS FROM THE REDUCTION OF KETO ESTERS, DIKETONES, AND MULTIFUNCTIONAL CARBONYL COMPOUNDS

At least two different oxidoreductases are found in yeast able to reduce 3-oxo esters with opposite sterochemical preferences [57]. The obtainment of homochiral 3-hydroxy esters is a relevant synthetic problem due to the value of this compound as a chiral synthon [2]. A number of structurally related 3-oxo esters of (*S*) absolute configuration (L series) can be obtained with yeast and other microorganisms under different conditions with high efficiency and high e.e. values. Simple structural variations permit change in stereoselectivity of the reduction [58–66].

Since the opposite enantiomer is available either by depolymerization of natural polyhydroxybutyrate or by oxidation of butanoic acid with *Candida rugosa* [67], the application in synthesis is particularly suited. In practice both enantiomers of 3-hydroxybutyrate and valerate are available in enantiomerically pure form through biotransformation

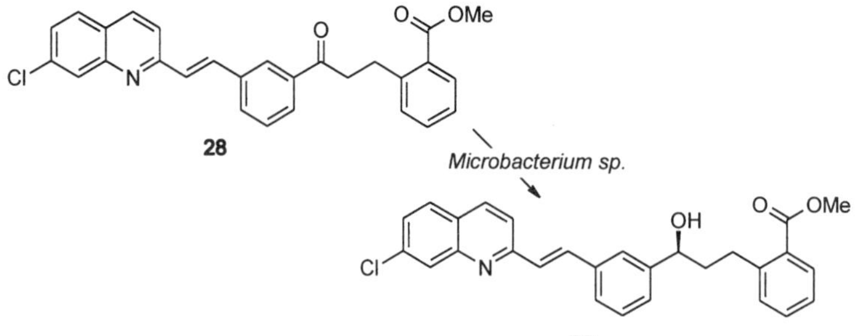

28

Microbacterium sp.

29

Scheme 10

or fermentation. Figure 1 shows some examples in which 3-hydroxy esters of high enantiomeric purity have been obtained in synthetically useful yields.

Enzymes with different stereochemical preferences for 3-oxo esters and 2-alkyl 3-
oxo esters have been isolated [57,62,63,68]. They are NADPH-dependent enzymes and
are able to catalyze the reduction of oxo esters of different type. However, they are not
available for enzyme-catalyzed reaction in substitution of the whole-cell catalyst. Synthetic
applications make use of whole-cell biocatalysts. Valuable intermediates in synthesis are
keto esters possessing additional functionality. Thus S-4-chloro-3-hydroxybutanoic acid **33**
(Scheme 12) has been obtained by reduction with suspended cells from cultures of *G.
candidum*. The compound is the intermediate in the synthesis of the cholesterol antagonist
34. In the biotransformation process a reaction yield of 95% and optical purity of 96%
were obtained at 10 g/L. The optical purity was increased to 99% by heat treatment of
cell suspensions prior to conducting the bioreduction [69].

Both enantiomers of fluoxetine and tomoxetine can be obtained from the secondary
alcohols **37** and **38** (Scheme 13) obtained by reduction of compounds **35** and **36** and
successive manipulations. The reduction of benzoylacetate with baker's yeast, *B. bassiana*,
G. candidum, gave the same S-hydroxy ester with good to high enantiomeric excess [70–
72]. A number of methods for the biocatalytic preparation of the same compounds is found
in the patent literature [73]. The pharmacologically active enantiomer of denopamine has
been obtained. Thus the biomass of *Geotrichum* sp. was used to reduce the α-keto ester
39 (Scheme 14) intermediate in the synthesis of the above compound [74].

Compound **43** (Scheme 15) belongs to the class of anticholesterol drugs acting as
hydroxymethylglutaryl CoA reductase inhibitors. The trihydroxycarboxylic acid **42** was
designed as a key intermediate in its synthesis. The reduction of the dicarbonyl compound
41 with microorganisms was attempted. A strain of *Acinetobacter calcoaceticus* was found
that was able to reduce the diketone to the dihydroxy compound with the required stereochemistry. An 85% yield and 97% optical purity were obtained with glycerol grown cells

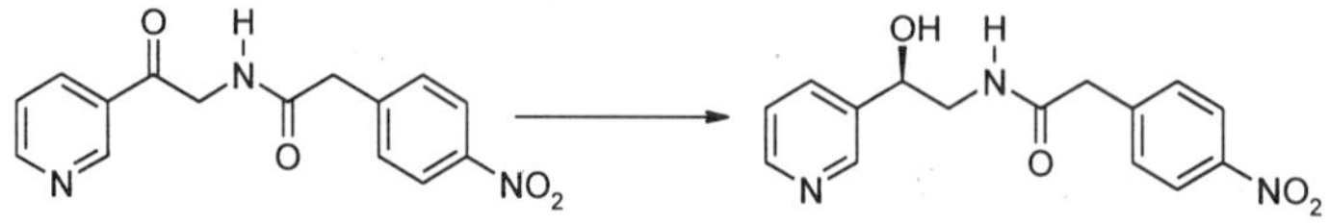

30   **31**

Scheme 11

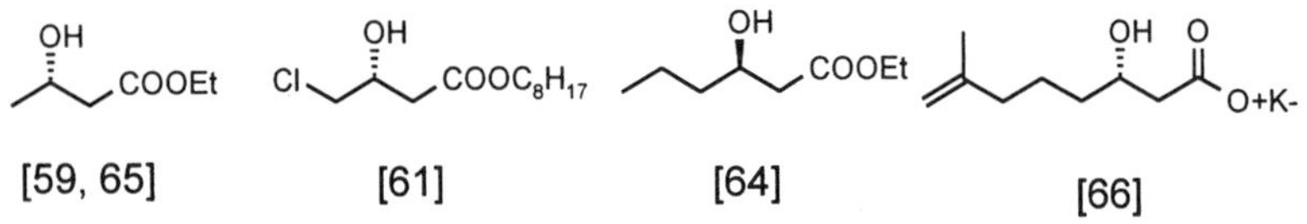

Figure 1 Stereochemistry of the reduction of some β-keto acid derivatives.

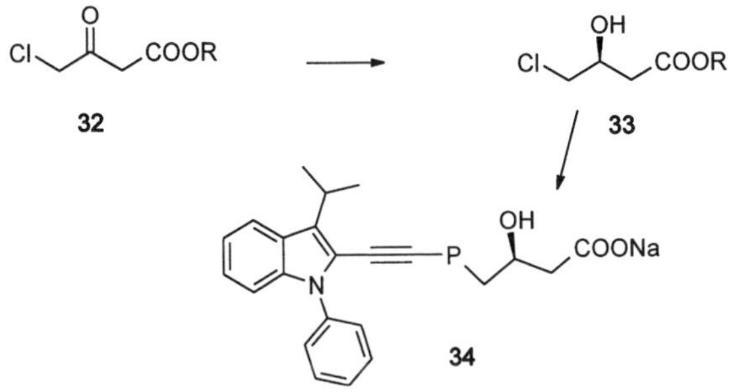

Scheme 12

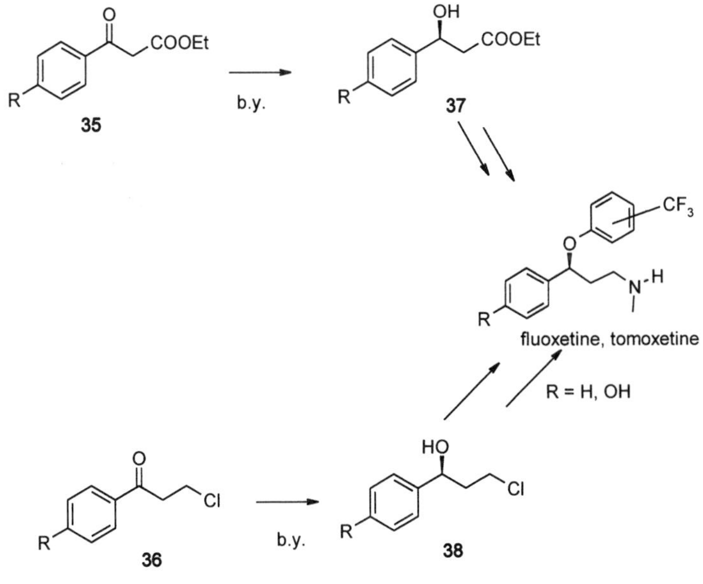

Scheme 13

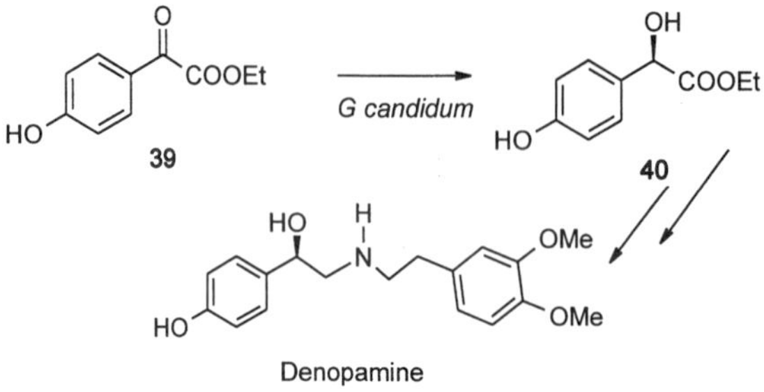

Scheme 14

at 2 g/L concentration. It was shown that the biomass initially reduced the diketone to the two isomeric hydroxy ketones which were successively further reduced to the diol with the required configuration. While the whole-cell biocatalyst also produced small amounts of isomeric anti-diols a partly purified enzyme was completely stereospecific. Although the selected microorganism is not of the kind *out of the shelf*, the result of the bioreduction is absolutely outstanding [75].

BO 2727 is a new carbapenem antibiotic with potent antipseudomonal activity. Compound **45** can be used as an intermediate in the synthesis of compound BO 2727 [76]. Obvious precursor for its preparation of the intermediate is the β-keto ester **44** (Scheme 16). Although β-keto esters are favorite substrates for yeast and many other microorganisms, only the fungal strain *Mortierella alpina* was able to reduce the target substrate out of 260 strains of different genus, among which 60 *Saccharomyces* strains evaluated. The reduction occurs with Prelog selectivity to give the required alcohol with high e.e. at 500 mg/L concentration [77].

The preparation of α-substituted β-hydroxy esters in the form of enantiomerically pure single diastereomer is not easily accomplished by reduction of the corresponding racemic ketone. Also in this case the carbonyl group is usually reduced with high enantiospecificity irrespective of the configuration of the adjacent chiral center. In fact, many examples reported always list hydroxy esters with the same relative configuration at the newly formed secondary alcohol (L series). This biotransformation has been extensively

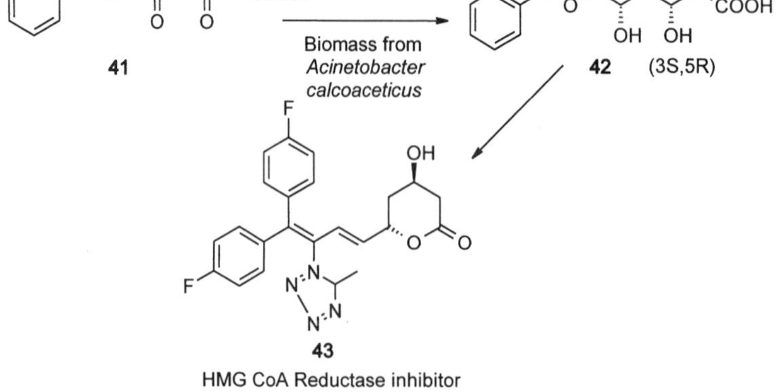

Scheme 15

Scheme 16

investigated due to the interest in the generation of highly functionalized chiral synthon with multiple applications in organic synthesis [78–93].

The diastereoselectivity is usually far from ideal giving mixtures of syn and anti products. Recently, a thorough investigation with a variety of microorganisms has been successful in preparing gram amounts of the four possible enantiomers of 2-chloro-3-hydroxy alkanoates in very high diastereoselectivity. For instance the *2R,3R*-hydroxy ester **46** (Scheme 17) has been obtained in 50% yield and 92% e.e. using cultures of a strain of *Mucor plumbeus*, while the diastereoisomer with the (*2S,3R*) configuration **47** was obtained from a strain of *S. exile* [94,95]. These chloroalcohols can be transformed into the corresponding epoxides. One of such intermediates was transformed into 2R,3S-3-phenyl isoserine [96].

Similarly, the two diastereoisomeric cyclic hydroxy esters can be obtained when baker's yeast or *R. arrhizus* is employed. The successful reduction of these racemic α-substituted β-keto esters to a single enantiomer is based on the enolization/racemization of the keto esters, which allows a dynamic kinetic resolution. If coupled with the specificity of the enzymes from different microorganisms, this can lead to preparation of compounds of diverse chirality. Thus in the series of the 5- and 6-membered ring compounds (Scheme 18) both the *1S,2R-cis-* and the *1S,2S-trans*-hydroxy ester can be obtained by switching from baker's yeast to the biomass from *R. arrhizus* [97]. Such compounds are highly versatile chiral synthons that have been currently used in the asymmetrical synthesis of biologically active compounds [98].

The results of the reduction of the β-keto acids derived from cycloalkanones of different size or from tetralones have been investigated. When larger rings are substrates for the reduction, decreased ring rigidity decreases the enantioselectivity [99]. α-Substituted cycloheptanones and cyclooctanones have also been reduced with varying and sometimes excellent diastereoselectivity to give *cis-* and *trans*-hydroxy esters according to the microorganism employed. The configuration of the secondary alcohol was (*S*) [100]. The microbial reduction of β-keto esters derived from 2-tetralone has been shown to produce good yields of 2-hydroxy-1-carboxy or 3-hydroxy-2-carboxy esters. Baker's yeast invariably affords *cis*-hydroxy ester with high enantiomeric purity, whereas fungi strains may sometimes produce exclusively *cis-* or *trans*-complementary stereochemistries. From a general survey of the baker's yeast reduction of cyclic β-keto esters, a working model for

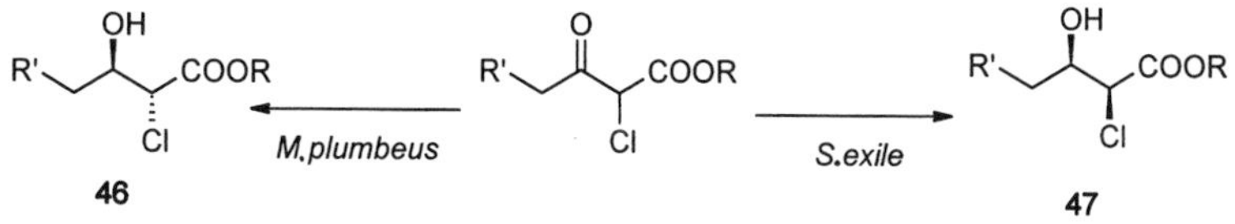

Scheme 17

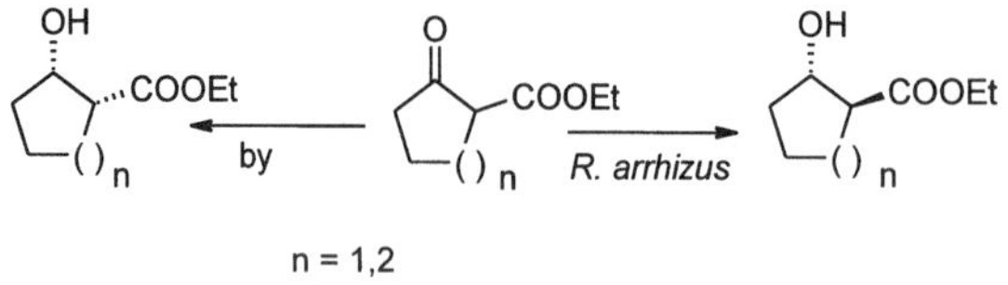

Scheme 18

predicting enantio-and diastereoselectivities of the reduction is proposed and discussed [99].

The reduction of α-keto esters is usually attributed to the action of lactate dehydrogenase (LDH) [8]. Since both L- and D-LDH may be present in microorganisms, the configuration of the hydroxy acid derivative obtained is not fully predictable. From a screening of 100 selected microorganisms, two strains with opposite stereochemical preferences were selected that would reduce **48** giving in 70–80% yields and more than 98% e.e. the α-hydroxy ester **49**, the protected form of the hydroxy amino acid **50** intermediate in the synthesis of the neurotropic drug hydroxyaniracetam (Scheme 19). The compound with the (R) configuration was obtained from cultures of C parapsilosis in kilogram amounts in 0.83 g/L/day. The enantiomer could be obtained from cultures of *Torulopsis magnoliae* [101]. R-2-hydroxy-4-phenylbutyric acid (HPBA) **53** (Scheme 20) is a precursor for the production of ACE inhibitors. Since important drugs marketed as single enantiomers like enalapril, benazepril, cilapril, and others belong to this class of compounds, many different approaches to this building block have been presented. In a thorough study, Schmidt and co-workers [102] explored several different approaches and their industrial feasibility, including homogeneous and heterogeneous catalytic asymmetric hydrogenation of **51**, the biotransformation of α-oxophenylbutyric acid with D-lactate dehydrogenase with cofactor recycling, biomass of *Proteus vulgaris* in free and immobilized form, and similar transformation on the α,β-unsaturated keto acid **52** using D-hydroxyisocaproate DH (HicDH) or free or immobilized cells of *P. mirabilis* followed by double-bond saturation. Although the e.e. values of the products were sensibly higher in the case of the enzymatic or microbial transformation, space–time yields for the catalytic hydrogenation processes were higher by one order of magnitude. However, the economics depends on several

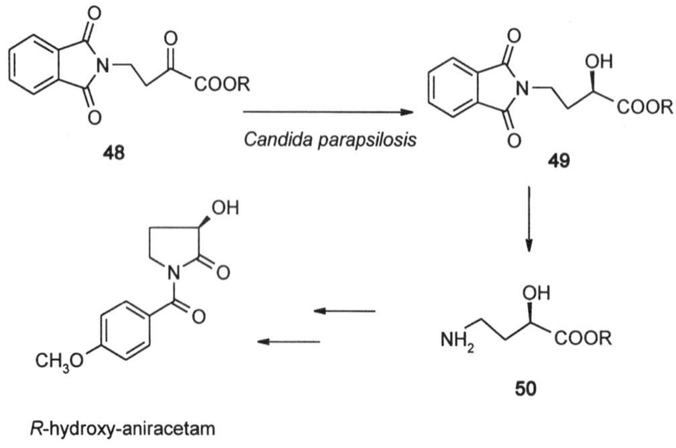

Scheme 19

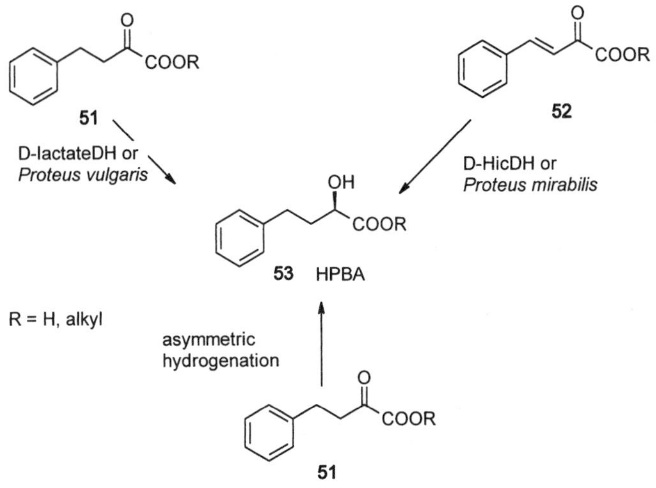

Scheme 20

factors that may change with precursor and catalyst availability, ecological costs, and other factors [102]. The reduction of **51** and **52** by yeasts and other organisms has been reported in academic and patent literature [103–108].

The yeast *Trichosporon capitatum* was identified by microbial screening as a suitable biocatalyst for the asymmetrical bioreduction of 6-bromo-β-tetralone **54** (Scheme 21) to the corresponding *S*-alcohol **55**. The outcome of the reduction was strongly influenced by the physiological state of the microbe, in particular by the culture age and the nature of the nutrients. It was also found that when ethanol was used as a cosolvent the enantiomeric excess was dependent on ethanol concentration. The presence of glucose favored the (*R*) configuration in a dose-dependent manner. High yields and e.e. values were obtained at 2 g/L in optimized conditions. Strains able to reduce to the (*R*)-configurated tetralol were

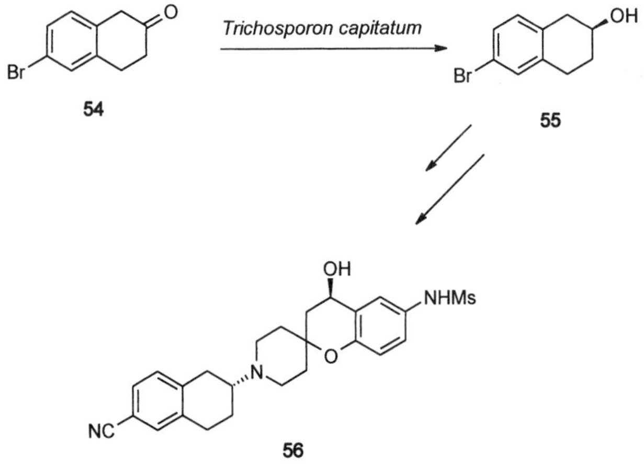

Scheme 21

also found. *Saccharomyces cerevisiae* did not show any activity on the tetralone. β-Tetralol is a precursor to the chiral drug candidate **56**, a potassium channel blocker targeted for the treatment of ventricular arrhythmia [109].

The benzazepinone **59** was designed and tested as a more potent and a longer lasting analog of diltiazem, the benzothiazepinone calcium blocker of wide use in the treatment of hypertension and angina. Synthesis of **59** (Scheme 22) required the preparation of **58**, which was prepared by microbial reduction of the dione **57**. Due to the rapid equilibration of the latter compound, the reduction to one single enantiomer with microorganisms was possible. Thus compound **58** was obtained in optimized condition at 2 g/L in 96% yields and 99.9 e.e. with cultures of *Nocardia salmonicolor* [110].

Diltiazem, one of the five best-selling synthetic chiral drugs [111], is still an important synthetic target. The approach of the direct reduction of the racemic keto amide has been successfully attempted: only the (*S*) enantiomer is reduced, while the (*R*) form is epimerized in the reaction medium. In this way high conversion of the racemate to one single enantiomerically pure diastereomer can be obtained. Baker's yeast was selected among over 500 microorganisms as giving the highest conversions (80%) and ee values (99%). Due to the extremely low solubility of the solid substrate **60** (Scheme 23), very low turnover numbers were observed. Three different reaction configurations were studied, differing in the form in which the substrate is fed to the fermenting mixture. It was found that a fed-batch operation using dimethyl formamide (DMF) solutions of the racemic starting material considerably improved the space–time yield of the biotransformation allowing transformation of 1 g of substrate in 48 h at 20 g/L concentration. In this way the efficiency of the biotransformation was improved 10-fold. In a third reaction configuration, the substrate was precipitated in a gel form from a DMF–water solution. This aggregate proved to be more soluble (100 g/L) than the crystalline material and the biotransformation proceeded smoothly. Ethanol, which was used as an energy source, was added in portion during the biotransformation at regular time [112].

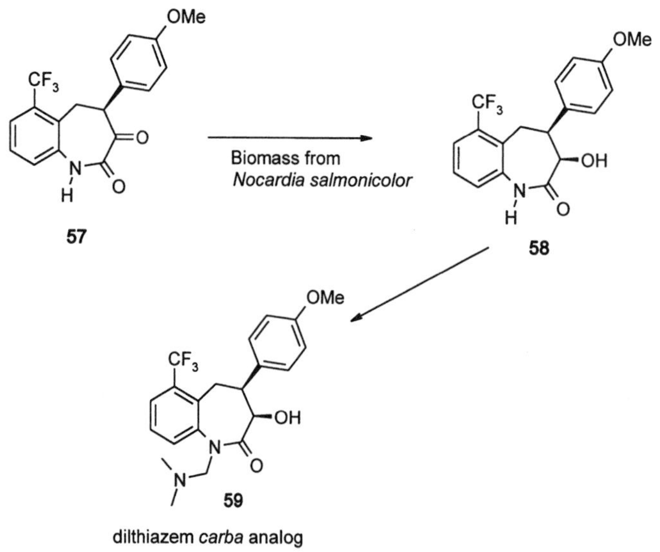

dilthiazem *carba* analog

Scheme 22

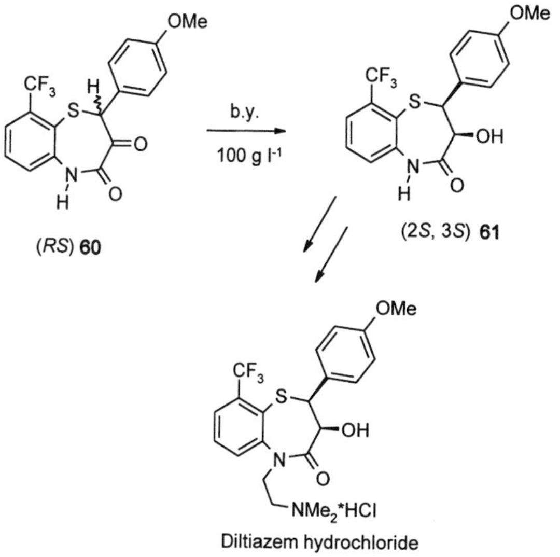

Scheme 23

The process obtained 10 g/L/day on 100-g scale [113], comparing favorably to the lipase-catalyzed kinetic resolution of p-methoxyphenylglycidate currently employed in the production of diltiazemhydrochloride [18].

III. EXTRACTIVE BIOCATALYSIS: CONTROL OF SUBSTRATE AND PRODUCT CONCENTRATION BY ABSORBING RESINS

Compound LY 300164 is a putative drug currently evaluated in clinic as a therapeutic agent for epilepsy and neurovegetative disorders. The 2,3-benzodiazepine synthesis was designed starting from optically active, suitably substituted 1-phenyl-2-propanol of (S) configuration 63 (Scheme 24). It was considered that the compound could be available through reduction with microorganisms of the corresponding phenylacetone 62. A simple screening of yeast libraries permitted selection of the yeast *Zygosaccharomyces rouxii* as providing good conversion to the desired alcohol with excellent stereo control. However, the generally favorable characteristics of Z. rouxii were compromised by its limited viability in the presence of high concentrations of organic species. While the biotransformation was complete at 2 g/L, the conversion dropped dramatically for a 10-fold increase in substrate concentration. In order to reduce both substrate and product concentration, the biotransformation was carried out in the presence of organic solvent. The presence of several different solvents proved to be toxic for the organism. A biphasic reaction design proved to be a successful alternative. A sevenfold greater substrate concentration could be reached with 95% conversion and 99.9% e.e. by simple addition of a polymeric absorbing resin to the reaction mixture. In this way by adding 80 g of substrate per liter of resin, the substrate and product concentrations in the water medium were kept around 2 g/L, thus avoiding substrate and product inhibition. Recovery of the product from the water phase was simple since most of the product remains on the resin, which can be separated from the biomass due to the more than two order of magnitude difference in the mean diameter between cells and resin beads [23].

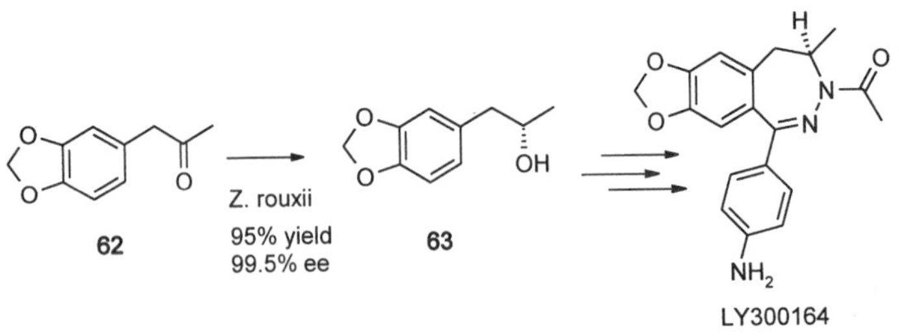

Scheme 24

Absorbing resins have been used recently for in situ product removal in fermentations. Advantages are the improved production due to the (partial) removal of the products from the solution and the cells surfaces, thus avoiding inhibition effects and cells toxicity. Together with other techniques, this approach constitutes what is defined as *extractive bioconversion* [114]. These concepts have been applied in fermentation processes in which the product is completely or largely water-soluble. In the case described in Scheme 24 the efficiency of biotransformation was dramatically improved by suppression of substrate inhibition in the biotransformation. This biotransformation could be better defined as a case of *extractive biocatalysis* [25]. The utility of this approach should be particularly effective in baker's yeast biotransformations where the problem of selectivity and limited yields due to substrate/product inhibition is often crucial. It has been proved that since the decreased substrate concentration in a multienzymatic system favors the reaction catalyzed by the enzyme with the lower K_m for the substrate, the control of substrate concentration should ultimately influence the enantiomeric excess of the product [115]. It was recently shown how the use of absorbing resins can improve the enantiomeric excess in the well-known baker's yeast reduction of 3-oxo butyrate [24]. The application of this method has also been extended to the C=C double-bond reduction of substrate **74** (Scheme 32) where the e.e. of the product was considerably improved [25,26] over conventional fermenting conditions. Thus this methodology could be of rather broad application in multienzymatic systems in controlling substrate–product inhibition, product recovery, enantioselectivity, and chemospecificity in whole-cell biotransformations. The use of low molecular weight enzyme inhibitors for stereochemical control has been reported extensively in yeast reduction [8,11].

The case history for the production of the carbonic anhydrase inhibitor MK-0507 is somewhat instructive about the possibility of a whole-cell biotransformation to enter an industrial process. The target molecule has two chiral centers: one methyl-bearing carbon atom and a secondary alcoholic function. The first chiral center is introduced using a chiral pool member: methyl-(R)-3-hydroxybutyrate obtained by depolymerization of the corresponding polyester of microbial origin. After construction of part of the structure required (**64**), the strategy of introducing the hydroxyl group with the correct stereochemistry by asymmetrical reduction of the carbonyl group was adopted. A microbial reduction was considered in order to avoid the problem of cofactor regeneration. A number of organisms—bacteria, yeasts, and fungi—were evaluated for their stereoselectivity and conversion ability. Among other organisms evaluated, *Lactobacillus plantarum* was selected for the high stereoselectivity observed in **65**. Transformation was also modest. However, it was observed that both conversion and stereoselectivity were profoundly influenced by pH. This offered the opportunity to understand that epimerization at the sulfone α-carbon

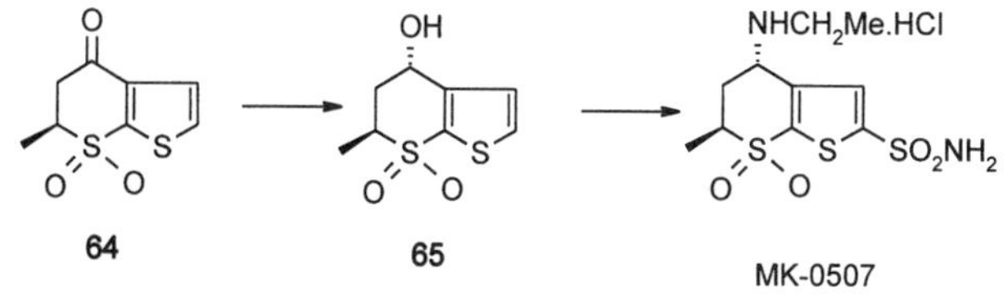

Scheme 25

was sensible at pH higher than 4.8. However, at these pH values the conversion was extremely low. The alternative biocatalyst was sought among those able to effect substantial growth and reduction at low pH values. From screening, a *Neurospora crassa* fungal strain was identified which after optimization gave 80% yield and more than 99% e.e. for the ketosulfone (Scheme 25). Attempts in isolating and cloning the gene encoding for the NADPH-dependent dehydrogenase involved in the biotransformation is currently under way [116].

The propyl analog of hydroxysulfone **65** was obtained by reduction of the corresponding ketone with the red yeast *Rhodotorula rubra* [117]. Intermediates of the same class of products obtained by microbial reduction are also present in the patent literature [118,119].

IV. PRODUCTS FROM THE BIOCATALYTIC REDUCTION OF TRIPLY SUBSTITUTED DOUBLE BONDS

The preparation of chiral compounds with a tertiary carbon atom chiral for the substitution pattern in Scheme 26 has obtained attention for the utility of those compounds as chiral synthons [120,133,151]. Products can be obtained by the asymmetrical reduction of a triply substituted C=C double bond. This reaction can be effected via asymmetrical hydrogenation with chiral catalysts or very conveniently with biocatalysts, when the double bond is activated with strongly polarizing groups. The enzymes responsible for this catalytic activity are defined as *enoate reductases* [121,122] [EC 1.3.1.31] and are not available as isolated enzymes. Their activity is found in whole-cell microorganisms, mainly yeast and microorganisms of the genus *Clostridium* [123]. While the reducing capacities of baker's yeast are quite accessible, microorganisms of the *Clostridium* type are anaerobic bacteria very sensitive to dioxygen and not as easy to grow. However, they have extraordinary reducing capacities with very high productivity numbers (10–100 times higher than the corresponding reactions with yeasts), allowing reduction of large amounts of substrate with very little biomass. They have high reducing potentials, unique among microorgan-

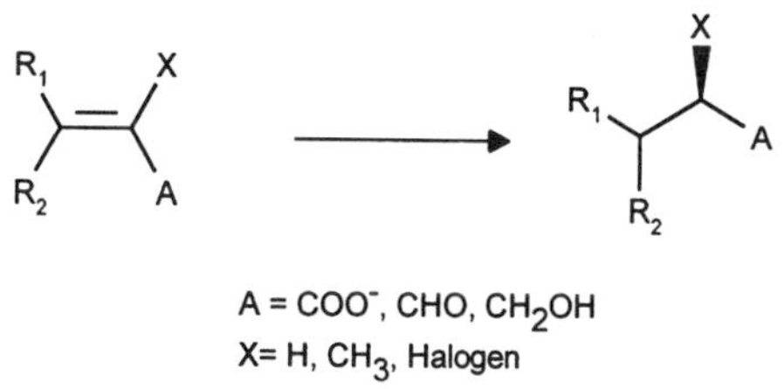

A = COO⁻, CHO, CH₂OH
X= H, CH₃, Halogen

Scheme 26

Table 2 Examples of Reductions with *Clostridium* [125,127]

Substrate	Product	Substrate	Product
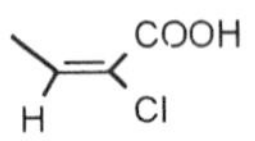	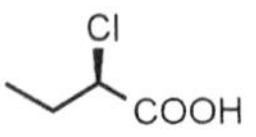	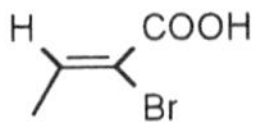	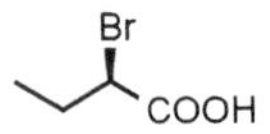
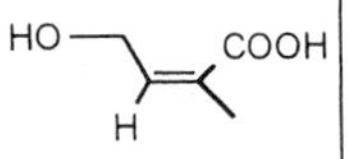	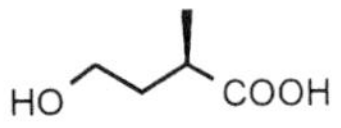	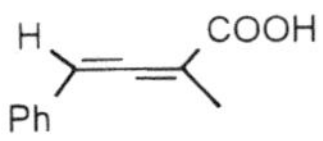	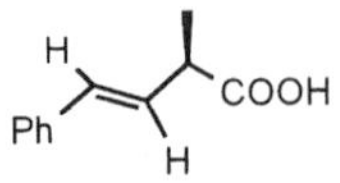

isms, allowing them to reduce carboxylic acids to primary alcohols, or, as in this case, the C=C double bond of α,β-unsaturated acids or esters. Moreover, in their use in biocatalysis as resting cells, the limited amount of cofactor present into the cells can be efficiently regenerated with hydrogen gas in the presence of small amounts of artificial mediators (viologen) or with electrochemical regeneration systems. The use of *Clostridium tyrobutyricum* DSM 1460 and *Clostridium kluyveri* DSM 555 has been fully described. These enzymatic systems have not become very popular among chemists due to the special techniques required for the cultivation of microorganisms not familiar to organic chemists. The capacities of these systems are described in detail in the literature [124].

Stereochemistry of the double-bond reduction occurring with *trans*-hydrogen delivery and the chirality of the products obtained have been studied. The stereochemical preference is opposite for *E* and *Z* double bonds. Low enantiomeric excesses are obtained if the *Z–E* conversion is favored in the reaction conditions. Substrates accepted by the enzymatic systems are restricted to those having a small X, usually —CH₃ in most examples, but also halogen or ethyl. R₁ and R₂ can be of various sizes, but one of the two is usually hydrogen. Some of the best substrates are reported in Table 2. Enoate reductase from yeast is a similar enzyme in many respects, but is cannot reduce α,β-unsaturated acids or esters, unless X is halogen (Scheme 27). Activating groups of the substrates are —CHO or NO₂. Allylic alcohols are substrates since they are usually partially transformed by yeast cells to the corresponding aldehydes by the presence of alcohol dehydrogenase.

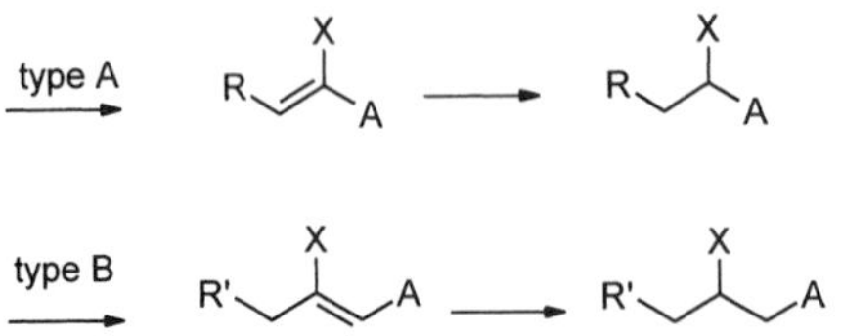

$X = H, D, CH_3, Cl, Br, CF_3$

$A = CHO, CH_2OH, CH(OCH_3)_2, COOR, NO_2$

Scheme 27

Kinetic constants and operating parameters are not as well defined as for the parent enzyme from *Clostridium*, which has been purified and used in the absence of other enzymes. All of the examples reported in the literature for yeast enoate reductase make use of whole-cell biocatalysts. The range of substrates submitted to reduction is quite large, and yields and e.e. values of products are quite remarkable. Two classes of triply substituted olefins are substrates, and they give as products methyl alkanols or corresponding compounds (Scheme 27). Examples of the two type of products obtained from baker's yeast are outlined in Table 3. Numerous are the applications of the synthesis of biologically active molecules. Compounds in entries 8, 10, and 11 in Table 3 are equivalent bifunctional chiral synthons that have been employed in the synthesis of α-tocopherol. The products in entries 11 and 17 have been transformed in enantiopure insect pheromones. The compounds shown in entries 14 and 15 are intermediates in the synthesis of zeaxanthin and of another precursor of a similar carotenoid. The products are prepared with yeast in a multikilogram scale. *Zygosaccharomyces rouxi* has been employed alternatively with a higher space–time yield than yeast. Series of compounds shown in entry 17 are obtained through the reduction of the thiophen precursor in what can be considered as a general and efficient way to 2(*S*)-2-methyl alkanols obtained after desulfurization of the thiophen ring [8].

The mechanism of the reduction of allylic alcohols (Scheme 28) has been studied in detail. The aldehyde is rapidly reduced to the allylic alcohol and its level stays low and constant during reaction. The aldehyde present at equilibrium conditions is then transformed into the saturated aldehyde and then into the final product. The reduction reaction described in Scheme 28 has been scaled to hundreds of grams for the preparation of C_5 chiral synthons [149].

The activity present in yeast is also present in the yeast enzyme concentrate type II from Sigma. Experiments carried out on different substrates showed that the enzyme is an NADPH-dependent protein. Application of this preparation is, however, much less convenient than that of whole-cell catalyst.

The yeast-mediated reductive step of conjugated endo-cyclic double bonds has been reported to occur with concomitant resolution at the chiral center removed from the actual bond reduced. The enantioselective reduction of **66** (Scheme 29) gives access to *R*-goniothalamin **67** [153]. In the case of α,β-unsaturated ketones, selectivity between the two functionalities can be observed. In compound **68** (Scheme 30) the selective reduction of the C=C double bond in the presence of a conjugated carbonyl group has been reported [154].

Compound **68a** is reduced to the allylic alcohol **71a** of (*S*) absolute configuration with no trace of the saturated species **69a** and **70a**. In contrast, the unsaturated ketone **68b** is transformed almost completely into the saturated ketone **69b** before any saturated alcohol **70b** appears in the mixture. This result is of interest in that it allows the selective production of the saturated ketone **69b**, the impact flavor of raspberries. Starting with **68b** of natural origin, the biotransformation allows one to obtain *natural* raspberry ketone **69b**. The reduction occurs using yeast enzyme concentrate (Sigma type II) in the presence of NADPH or NADH. When (*4R*)- and (*4S*)-[4-2H]-NADPH or NADH were used, the experiments showed that both cofactors were utilized but with opposite stereochemistry, thus suggesting the presence of two different enone reductases.

Reduction of compound **72** with the red yeast *Rhodotorula rubra* CBS 6469 gives access to the proposed drug **73** for the treatment of non-insulin-dependent diabetes mellitus [146]. The selection of the microorganism and the biotransformation conditions are dictated by the fact that at pH higher than 3 consistent epimerization occurs. The reduction is finally effected at pH 3, giving a product of 98% e.e. (Scheme 31).

Table 3 Examples of Products Obtained from Baker's Yeast

Entry	Substrate	Product	Notes		Yield	% e.e.	Ref.
1			R =	Et	19	44	128
				n-C$_5$H$_{11}$	28	82	
				n-C$_8$H$_7$	47	84	
2			R =	Et	23–35	>98	129
				i-Pr	16–19	>98	
				n-Bu	32–40	>98	
				CHCl$_2$	58–71	98	
				CCl$_3$	41–69	>98	
			R =	Et	23–28	47	130
				i-Pr	6–10	68	
				n-Bu	30–40	25	
				CHCl$_2$	54–65	92	
3			R$_1$ =	R$_2$ =			
			Ph	Me	50	98	
			Ph	Et	64	97[a]	
			Ph	n-Pr	23	89[a]	
			p-ClC$_6$H$_4$	Me	48	89[a]	
			p-BrC$_6$H$_4$	Me	57	94[a]	
4					30	93[c]	131
5			R =	H	80	87[c]	132
				Ac	30	>95	
					35	>95	
			R =	H	35	>90	133
				Ac	44	>92	134
				40	>95		
6					25	80	

7	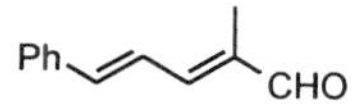	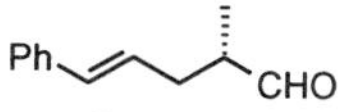		100	87	
8	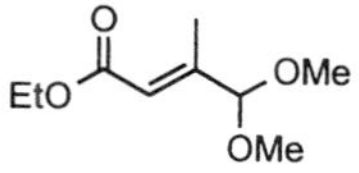	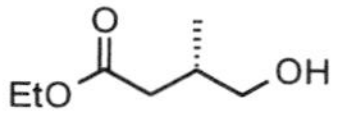		49	135	
9	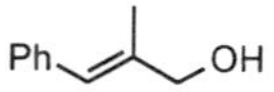	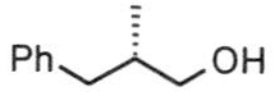		—	—	136
10	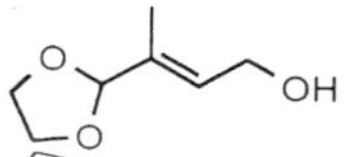	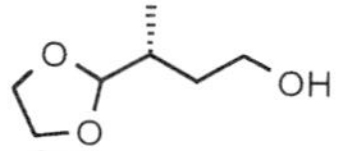	*G. candidum*			152
11	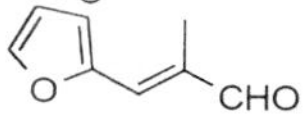	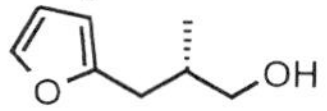		72	>99	137
12	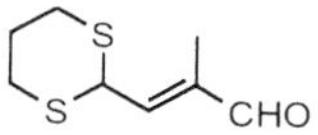	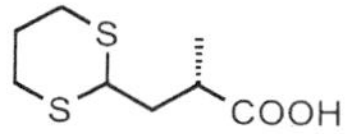		48	100	138
	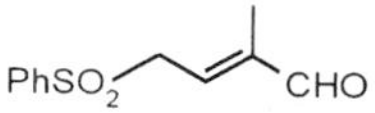	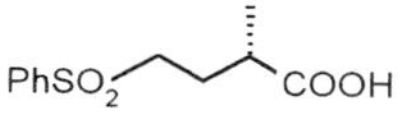		35	86	
	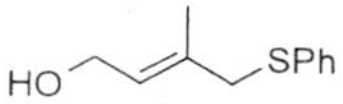	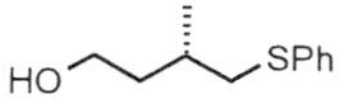		22	68	
13	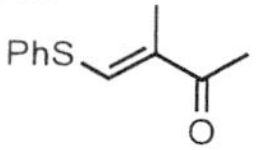	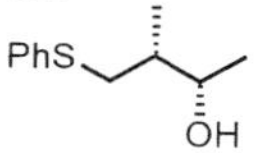		18	90	139
	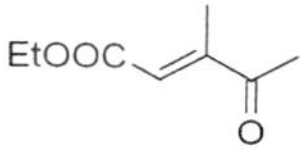	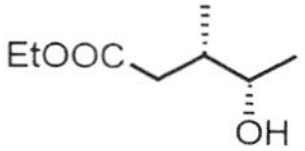			>96	
14	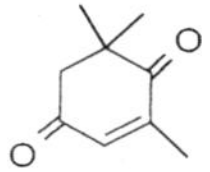	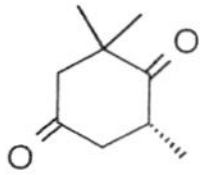		83	>97	60
			Z. rouxi	85	>98	40

Table 3 Continued

Entry	Substrate	Product	Notes		Yield	% e.e.	Ref.
15	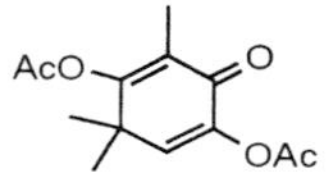 (AcO, O, OAc cyclohexenone)	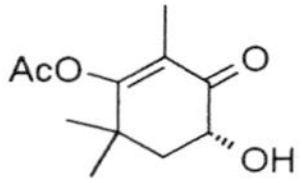 (AcO, O, OH cyclohexenone)			>98		140
16	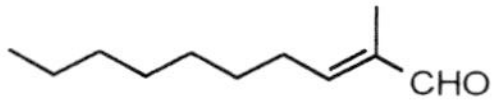 (CHO chain)	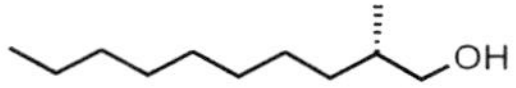 (OH chain)			34	95	137
17	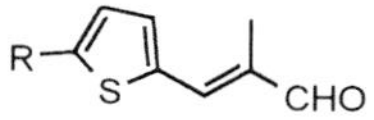 (thiophene R, CHO)	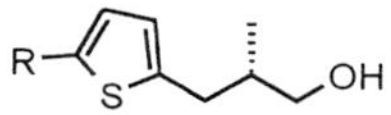 (thiophene R, OH)	R =	H	65	98	141
				Me	33	95	
				n-Pr	25	76	
	(thiophene CHO)	(thiophene OH)			74	94	151
18	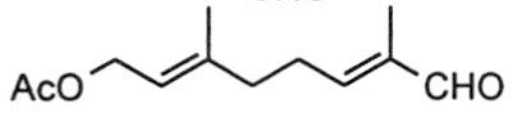 (AcO, CHO)	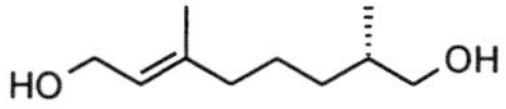 (HO, OH)			20	>97	142
19	(MeOOC, O, S-Ph, O$_n$)	(MeOOC, O, O$_n$, S-Ph)		n = 0	35	>97	143
				1	41	99	
				2	49	97	
20	(OR lactone)	(OR lactone)	R =		14	77	144
21	(R, CHO)	(R, OH)	R =	Ac	60	95	145
				PhCH$_2$	34	—	
	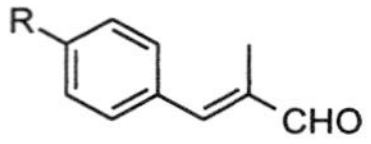	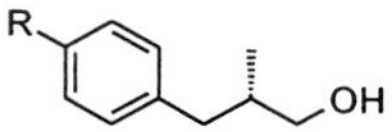		Me$_2$N	84	94[a]	
				MeO	33	78[a]	
				t-Bu	84	99	
				H	15	75[a]	
				Cl	12	78[a]	
				CN	37	92[a]	
				NO$_2$	26	88[a]	

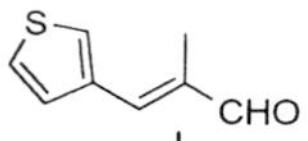

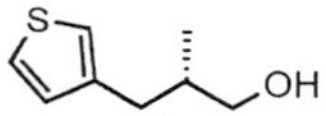

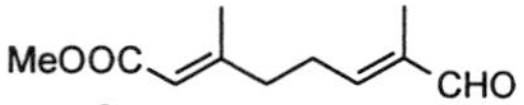

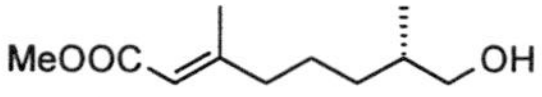

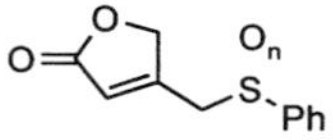

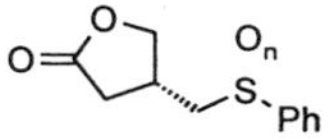

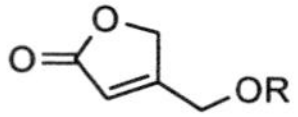

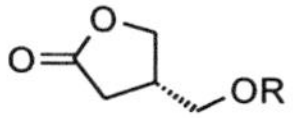

No.	Substrate	Product	Microorganism	Yield	ee	Ref.
22	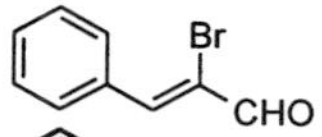	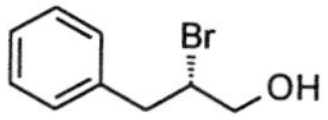		91	>98	25
23	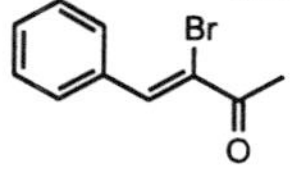	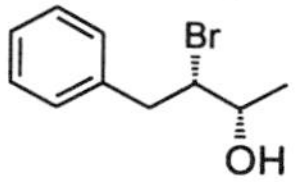		80	>99	47
24	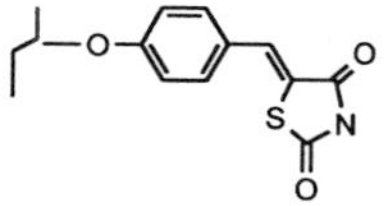	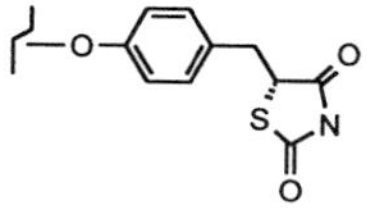	Rhodotorula rubra		98	146
25	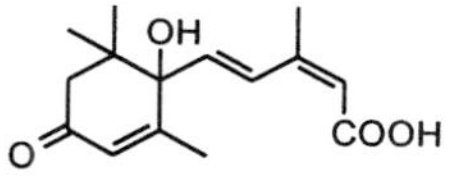	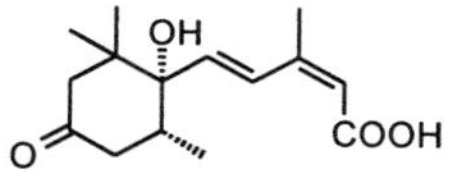	A. niger	35	95	147
26	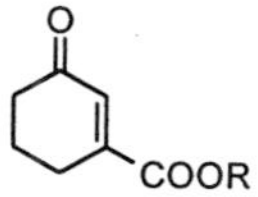	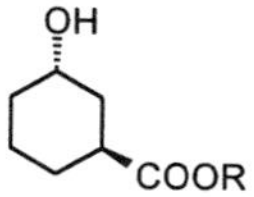	G. candidum			148

[a]Absolute configuration not directly determined.

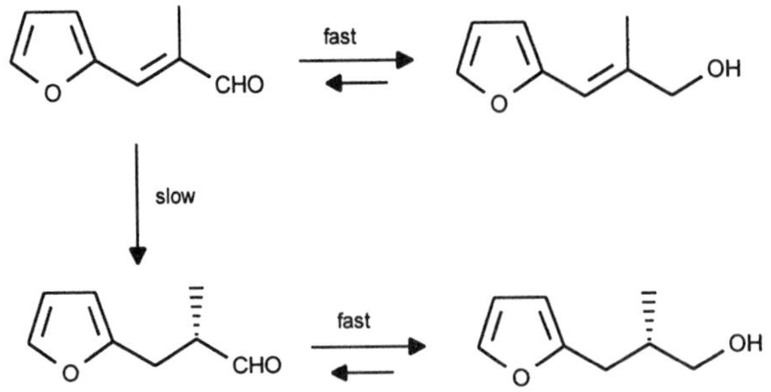

Scheme 28

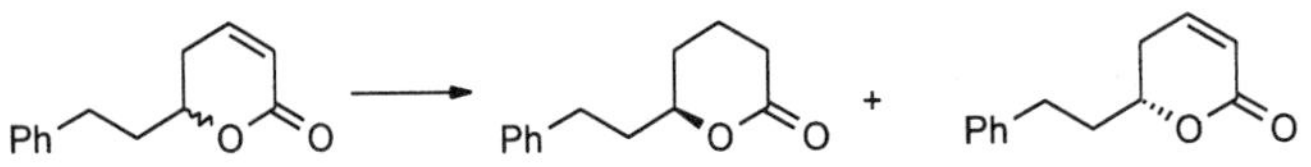

66   67

(+)-(R)-goniothalamin

Scheme 29

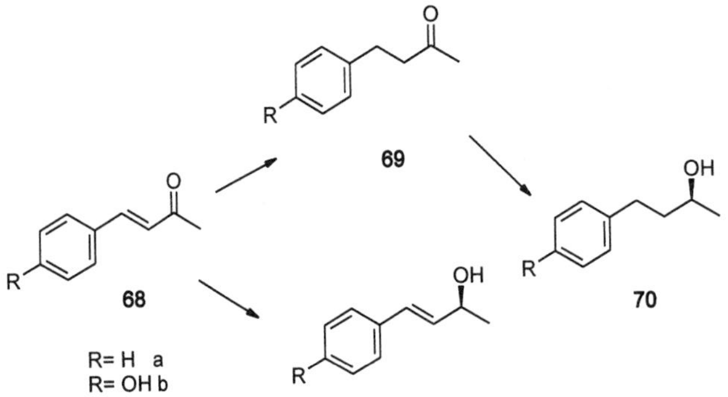

68

R= H a
R= OH b

69

70

71

Scheme 30

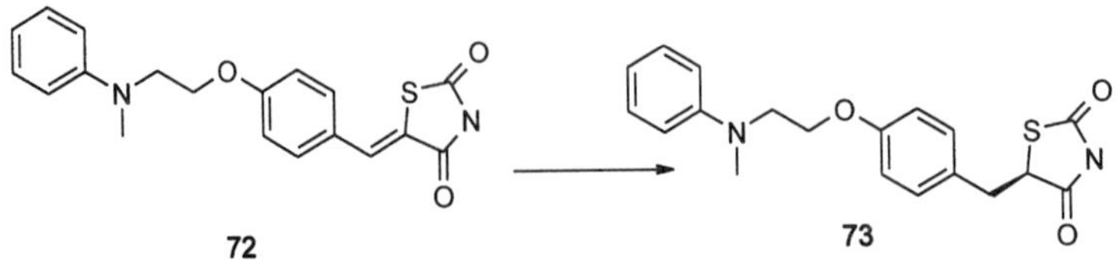

72   73

Scheme 31

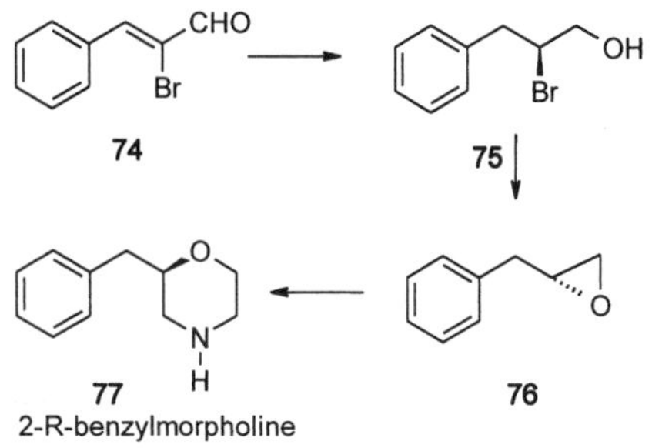

74   75

77   76
2-R-benzylmorpholine

Scheme 32

Scheme 32 outlines a high yielding approach to the enantiopure appetite suppressant drug **77**. The reduction of the unsaturated aldehyde **74** was reported to occur with modest enantioselectivity in normal fermenting conditions with baker's yeast [155]. When the biotransformation is performed at very low substrate concentration, the e.e. can be raised to more than 90%, suggesting that incomplete enantioselectivity is due to the action of enzymes operating on the same substrate with opposite stereochemical preference [115]. However, an efficient transformation can be performed if the substrate concentration is controlled with the addition of absorbing hydrophobic resins. At 5 g/L 97% recovery and 98% e.e. was obtained. The halohydrin **75** obtained was transformed into the epoxide **76** and finally into the enantiopure 2-*R*-benzylmorpholine **77**, the more active enantiomer with appetite suppressant activity [26].

Several other microorganisms are able to perform efficient activated C=C double bond reduction. The fungus *Beauveria bassiana* has enoate reductase activity toward a wide range of substrates.

The stereochemistry of the biocatalytic reductions of olefins was studied using various microorganisms, either with purified enzymes or with living cells. In particular, the reduction of various cinnamaldehydes by baker's yeast was studied by Fuganti [8], showing that this reaction proceeds in a *trans*-hydrogen addition fashion. The same *trans*-hydrogen addition was found by Kergomard et al. [156] in olefins reduced by *B. bassiana*. The development of a model of these reductions or at least the identification of a working rule for the description of the chemical course of the reaction itself may be useful to allow an effective prediction of the absolute configuration of the products [8,139]. As shown in Table 3, the enoate reductase type of activity is not confined to the few organisms mentioned above. In entry 25 of Table 3 the microorganism used was a strain of *Aspergillus niger*. Racemic abscisic acid was resolved to one single enantiomer by double-bond reduction, in 35% yield and 95% e.e. [147].

V. CONCLUSIONS

Experience from industrial and academic laboratories shows that screening for a desired reducing activity among a reasonable group of selected microorganisms is often met with success, permitting the identification of biocatalysts of both stereo preferences, in many cases. The reduction of prochiral carbonyl compounds and of triply substituted C=C double bonds with whole-cell biocatalysts finds many precedents in the literature, allowing one to make predictions regarding the stereochemical outcome of the reduction. Application of known techniques for the improvement of space–time yields and stereoselectivity will help us to meet industrially acceptable conditions.

REFERENCES

1. AN Collins, GN Sheldrake, J Crosby, eds. Chirality in Industry. Chichester: John Wiley, 1992. AN Collins, GN Sheldrake, J Crosby, eds. Chirality in Industry, Vol. 2. Chichester: John Wiley, 1997
2. RA Sheldon. Chirotechnology. New York: Marcel Dekker, 1993.
3. J-C Caille, M Mulliard, B Laboue. In AN Collins, GN Sheldrake, J Crosby, eds. Chirality in Industry, Vol. 2. Chichester: John Wiley, 1997, pp 391–401.
4. H Kumobayashi. Rec Trav Chim Pays Bas 115:201–210, 1996.
5. E Santaniello, P Ferraboschi, P Grisenti, A Manzocchi. Chem Rev 92:1071–1140, 1992.
6. K Drauz, H Waldmann. Enzyme Catalysis in Organic Synthesis. Weinheim: VCH, 1995.
7. S Servi, ed. Microbial Reagents in Organic Synthesis, NATO ASI Series, Dordrecht: Kluwer Academic, 1992. S Servi. Chim Ind 78:959–963, 1996.
8. S Servi, Yeast. In: H Rehm, ed. Biotechnology, Vol. 8a. Weinheim: VCH-Wiley, 1998, pp 363–390.
9. L Poppe, L Novak. Selective Biocatalysis. Weinheim: VCH, 1992.
10. RN Patel. Adv Appl Microbiol 43:91-140, 1997.
11. P D'Arrigo, G Pedrocchi-Fantoni, S Servi. Advances Appl Microbiol 44:81–123, 1997.
12. A Bruggink. In: AN Collins, GN Sheldrake, J Crosby, eds. Chirality in Industry, Vol. 2. Chichester: John Wiley, 1997, pp 81–98.
13. BA Astleford, LO Weigel. In: AN Collins, GN Sheldrake, J Crosby, eds. Chirality in Industry, Vol. 2. Chichester: John Wiley, 1997, pp 99–117.
14. ER Francotte. Preparative Chiral Separation by Chromatography. Proceedings of Chiral Europe '96, Strasbourg, 1996, pp 89–95.
15. JTF Keurentjes, FJM Voermans. In: AN Collins, GN Sheldrake, J Crosby, eds. Chirality in Industry, Vol. 2. Chichester: John Wiley, 1997, pp 157–180.
16. E Baldaro, C Fuganti, S Servi, A Tagliani, M Terreni. In: S Servi, ed. Microbial Reagents in Organic Synthesis, NATO ASI series Vol. C 381. Dordrecht: Kluwer Academic, 1992, pp 175–188.
17. M Furui, T Furutani, T Shibatani, Y Nakamoto, T Mori. J Ferment Bioeng 81:21, 1996.
18. H Matsumae, M Furui, T Shibatani, J Ferment Bioeng, 75:93, 1993. H Matsumae, M Furui, T Shibatani, T Tosa. J Ferment Bioeng 78:59–63, 1994.
19. A Gentile, C Giordano, C Fuganti, L Ghirotto, S Servi. J Org Chem 57:6635–6337, 1992.
20. AS Bommarius, M Kottenham, H Klenk, K Drauz. In: S Servi, ed. Microbial Reagents in Organic Synthesis, NATO ASI series Vol. C 381. Dordrecht: Kluwer Academic, 1992, pp 161–174.
21. HL Holland. Organic Synthesis with Oxidative Enzymes. New York: VCH, 1992.
22. V Crocq, C Masson, J Winter, C Richard, G Lemaitre, J Lenay, M Vivat, J Buendia, D Prat. Organic Process R&D 1:2–13, 1997.
23. BA Anderson, MM Hansen, AR Harkness, CL Henry, JT Vicenzi, MJ Zmijewski. J Am Chem Soc 117:12358–12359, 1995. BA Anderson, Proceedings Chiral USA '98. San Francisco, 1998.
24. P D'Arrigo, G Pedrocchi-Fantoni, S Servi, A Strini. Tetrahedron: Asymmetry 8:2375–2379, 1997.
25. P D'Arrigo, C Fuganti, G Pedrocchi-Fantoni, S Servi. Tetrahedron 54:15017–15026, 1998.
26. P D'Arrigo, M Lattanzio, G Pedrocchi-Fantoni, S Servi. Tetrahedron: Asymmetry 9:4021–4026, 1998.
27. V Prelog. Pure Appl. Chem. 9:119, 1964.

28. DR Dodds, DR Andrews, C Heinzelmann, R Klesse, WB Morgan, E Previte, A Sudhakar, RA Roehl, R Vail, A Zaks, A Zelazowski. Proceedings Chiral Europe '95. London, 1995, pp 55–62.

29. HG Davies, RH Green, DR Kelly, SM Roberts. Biotransformations in Preparative Organic Chemistry. London: Academic Press, 1989.

30. CW Bradshaw, W Hummel, CH Wong. J Org Chem 57:1532–1536, 1992.

31. CW Bradshaw, H Fu, GJ Shen, CH Wong. J Org Chem 57:1526–1532, 1992.

32. W Hummel, MR Kula. Eur J Biochem 184:1–13, 1989.

33. W Hummel. Biotechnol Lett 12:402–408, 1990.

34. G Fantin, M Fogagnolo, ME Guerzoni, A Medici, P Pedrini, S Poli. J Org Chem 59:924–925, 1994.

35. EJ Corey, JO Link. Tetrahedron Lett 31:601–604, 1990.

36. BS Deol, DD Ridley, GW Simpson. Aust J Chem 29:2459, 1976.

37. AEPM Sorrilha, M Marques, I Joekes, PJS Moran, JAR Rodrigues. Bioorg Med Chem Lett 2:191–196, 1992.

38. M Carvalho, MT Okamoto, PJS Moran, JAR Rodrigues. Tetrahedron 47:2073–2080, 1991.

39. H Kutsuki, I Sawa, N Mori, J Hasegawa, K Watanabe. Eur Pat Appl 0,198,440.

40. J Hasegawa, T Ohashi. Proceedings Chiral USA '93, 1993, pp 105–107.

41. GJ Quallich, TM Woodall. Tetrahedron 48:10239–10248, 1992.

42. C Barbieri, P D'Arrigo, G Pedrocchi-Fantoni, S Servi (in preparation).

43. RN Patel, A Banerjee, CG McNamee, LJ Szarka. Appl Microb Biotechnol 40:241–245, 1993.

44. A Richards. Proceedings Chiral '93 USA, 1993, pp 75–80.

45. M Imuta, KI Kawai, H Ziffer. J Org Chem 45:3352–3355, 1980.

46. M Imuta, H Ziffer. J Org Chem 43:4540–4542, 1978.

47. J Aleu, G Fronza, C Fuganti, V Perozzo, S Serra. Tetrahedron: Asymmetry 9:1589–1596, 1998.

48. W Clark, I Lantos, AP Maloney. Proceeding of Biotrans '97, La Grande Motte, 1997 p 107.

49. JM Walker, WD Bower, FO Walker, RR Matsumoto, BD Costa, KC Rice. Pharm Rev 42:355–402, 1990.

50. RN Patel, A Banerjee, M Liu, RL Hanson, RY Ko, JM Howell, LJ Szarka. Biotechnol Appl Biochem 17:139–153, 1993.

51. C Roberge, A King, V Pecore, R Greasham, M Chartrain. J Ferment Bioeng 81:532–535, 1996.

52. JE Lynch, W-B Choi, HRO Churchill, RP Volante, RA Reamer, RG Ball. J Org Chem 62:9223–9228, 1997.

53. MM Chartrain, W-B Choi, HRO Churchill, S Yamazaki. WO 0,9812,340. CA 128:256471.

54. M Takeshita, K Terada, N Akutsu, S Yoshida, T Sato. Heterocycles 26:3051–3056, 1987. D Bailey, D O'Hagan, U Dyer, RB Lamont. Tetrahedron: Asymmetry 4:1225–1228, 1993.

55. M Uchiyama, N Katoh, R Mimura, N Yokota, Y Shimogaichi, M Shimazaki, A Ohta. Tetrahedron: Asymmetry 8:3467–3474, 1997.

56. MM Chartrain, JYL Chung, C Roberge. WO 0,9803,672. CA 128:166424.

57. J Heidlas, K-H Engel, R Tressl. Eur J Biochem 172:633–639, 1988.

58. D Seebach, S Roggo, T Maetzke, H Braunschweiger, J Cercus, M Kreiger. In: SM Roberts, ed. Preparative Biotransformations. New York: John Wiley, Module 2:2.1, 1992.

59. D Seebach, MA Sutter, RH Weber, MF Züger. Org Synth 63:1–9, 1985.

60. HGW Leunberger, W Boguth, E Widmer, R Zell. Helv Chim Acta 59:1832–1849, 1976.

61. CJ Sih, B Zhou, AS Gopalan, WR Shieh, F VanMiddlesworth. In: W Bartmann, BM Trost, eds. Procesdings 14th Workshop Conference Hoechst, 1983, p 250.

62. WR Shieh, AS Gopalan, CJ Sih. J Am Chem Soc 107:2993–2994, 1985.

63. WR Shieh, CJ Sih. Tetrahedron: Asymmetry 4:1259–1269, 1993.

64. MP Dillon, MA Hayes, A Martin, TJ Simpson, JB Sweeney. Bioorg Med Chem Lett 1:223–226, 1991.

65. B Wipf, E Kupfer, R Bertazzi, HGW Leuenberger. Helv Chim Acta 66:485–488, 1983.

66. M Hirama, T Shimizu, M Iwoshita. J Chem Soc Chem Commun 599–600, 1983.
67. J Hasegawa, M Ogura, S Hamaguchi, M Shimazaki, K Watanabe. J Ferment Technol 59: 93,1981.
68. K Nakamura, Y Kawai, N Nakajima, A Ohno. J Org Chem 56:4778–4783, 1991.
69. RN Patel, M Liu, A Banerjee, JK Tottathil, J Kloss, LJ Szarka. Enzyme Microb Technol 14: 778–784, 1992.
70. G Fronza, C Fuganti, P Grasselli, A Mele. J Org Chem 56:6019–6023, 1991.
71. R Chenevert, G Fortier. Chem Lett 1603–1606, 1991.
72. R Chenevert, F Geneviève, RB Rhild. Tetrahedron 48:6769–6776, 1992.
73. I Michio, Y Kobayashi. Jpn Kokai Tokkyo Koho, 1992, CA 117;249991. I Michio, Y Kobayashi. Jpn Kokai Tokkyo Koho, 1992, CA 117;249990. RN Patel, M Liu, A Banerjee, LJ Szarka. Eur Pat Appl, 1993, CA: 119:26828.
74. G Jian-Xin, L Zu-Yi, L Guo-Qiang. Tetrahedron Lett 49:5805–5816, 1993.
75. RN Patel, CG McNamee, A Banerjee, LJ Szarka. Eur Pat Appl, 1993, CA: 120:52826. RN Patel, A Banerjee, CG McNamee, D Brzozowski, RL Hanson, LJ Szarka. Enzyme Microb Technol, 15:1014–1021, 1993.
76. S Nakagawa, T Hashizume, K Matsuda, M Sanada, O Okamoto, H Fukatsu, N Tanaka. Antimicrob Agents Chemother 37:2756–2759, 1993.
77. M Chartrain, J Armstrong, L Katz, J Keller, D Mathre, R Greasham. J Ferment Bioeng 80: 176–179, 1995.
78. BS Deol, DD Ridley, GW Simpson, Aust J Chem 29:2459–2467, 1976.
79. D Buisson, C Sanner, M Larcheveque, R Azerad. Tetrahedron Lett 28:3939–3940, 1987.
80. RW Hofmann, W Ladner, K Steinbach, W Massa, R Schmidt, G Snatzke. Chem Ber 114: 2786–2801, 1981.
81. H Akita, A Furuichi, H Koshiji, K Horikoshi, T Oishi. Chem Pharm Bull 31:4376–4383, 1983.
82. K Nakamura, T Miyai, K Nozaki, K Ushio, S Oka, A Ohno. Tetrahedron Lett 27:3155–3156, 1986.
83. D Buisson, S Henrot, M Larcheveque, R Azerad. Tetrahedron Lett 28:5033–5036, 1987.
84. D Buisson, R Azerad, C Sanner, M Larcheveque. Biocatalysis 3:85–93, 1990.
85. R Azerad, D Buisson. In: S Servi, ed. Microbial Reagents in Organic Synthesis, NATO ASI Series. Dordrecht: Kluwer Academic, 1992, pp 421–440.
86. G Frater, U Muller, W Gunther. Tetrahedron 40:1269–1277, 1984.
87. K Nakamura, T Miyai, Y Kawai, N Nakajima, A Ohno. Tetrahedron Lett 31:1159–1160, 1990.
88. K Nakamura. In: S Servi, ed. Microbial Reagents in Organic Synthesis, NATO ASI Series. Dordrecht: Kluwer Academic, 1992, pp 389–398.
89. T Sato, M Tsurumaki, T Fujisawa. Chem Lett, 1367–1370, 1986.
90. T Sakai, T Nakamura, K Fukuda, E Amano, M Utaka, A Takeda. Bull Chem Soc Jpn 59: 3185–3188, 1986.
91. RW Hoffmann, W Helbig, W Ladner. Tetrahedron Lett 23:3479–3482, 1982.
92. M Soukup, B Wipf, E Hochuli, HGW Leuenberger. Helv Chim Acta 70:232–236, 1987.
93. C Fuganti, S Lanati, S Servi, A Tagliani, A Bedeschi, G Franceschi. J Chem Soc Perkin I, 2247–2249, 1993.
94. O Cabon, D Buisson, M Larcheveque, R Azerad. Tetrahedron: Asymmetry 6:2199–2210, 1995.
95. O Cabon, D Buisson, M Larcheveque, R Azerad. Tetrahedron Lett 33:7337–7340, 1992.
96. O Cabon, D Buisson, M Larcheveque, R Azerad. Tetrahedron: Asymmetry 6:2211–2218, 1995.
97. S Danchet, D Buisson, R Azerad. J Mol Cat B: Enzymatic 5:129–132, 1998.
98. AK Gosh, WJ Thompson, PM Munson, W Liu, J Huff. Bioorg Med Chem Lett 5:83–86, 1995. J Cossy, S Ibhi, P Kahn, L Tacchini. Tetrahedron Lett 36:7877–7880, 1995. M Keppens, N De Kimpe. J Org Chem 60:3916–3918, 1995.

99. C Abalain, D Buisson, R Azerad. Tetrahedron: Asymmetry 7:2983–2996, 1996.
100. S Danchet, C Bigot, D Buisson, R Azerad. Tetrahedron: Asymmetry 8:1735–1739, 1997.
101. PK Matzinger, B Wirtz, HGW Leuenberger. Appl Microbiol Biotechnol 32:533–537, 1990.
102. E Schmidt, HU Blaser, PF Fauquex, G Sedelmeier, F Spindler. In: S Servi, ed. Microbial Reagents in Organic Synthesis. Dordrecht: Kluwer Academic, 1992, pp 377–388.
103. H Kruszewska, I Makuch, G Grynkiewicz, J Cybulski. Pol J Chem 70:1301–1307, 1996.
104. DH Dao, Y Kawai, K Hida, S Hornes, K Nakamura, A Ohno, M Okamura, T Akasaka. Bull Chem Soc Jpn 71:425–432, 1998.
105. A Matsuyama, I Takase, Y Ueda, Y Kobayashi. US Pat 5,429,935. CA 123:141884.
106. N Kawada, T Nikaido, K Kimoto. Jpn Kokai Tokkyo Koho, CA 121:203561.
107. A Matsuyama, N Teruyuki, Y Kobayashi, PCT Int Appl, CA 113:38943.
108. A Matsuyama, N Teruyuki, Y Kobayashi. PCT Int Appl, CA 112:15250.
109. J Reddy, D Tschaen, Y-J Shi, V Pecore, L Katz, R Greasham, M Chartrain. J Ferment Bioeng 81:304–309, 1996.
110. RN Patel, RS Robinson, LJ Szarka, J Kloss, JK Thottathil, RH Mueller. Enzyme Microb Technol 13:906–912, 1991.
111. RL DiCicco. Proceedings Chiral Europe '92, Manchester 1992, pp 17–19.
112. T Kometani, E Kitatsuji, R Matsuno. J Ferment Bioeng 71:197–199, 1991. T Kometani, Y Morita, H Furui, H Yoshii, R Matsuno. J Ferment Bioeng 77:13–16, 1994.
113. T Kometani, Y Sakai, H Matsumae, T Shibatani, R Matsuno. J Ferment Bioeng 84:195–199, 1997.
114. SR Roffler, HW Blanch, CR Wilke. Trends Biotech 2:129–136, 1984. A Freeman, JM Woodley, MD Lilly. Biotechnology 11:1007–1012, 1993. G Eckert, K Scugerl. Appl Microbiol Biotechnol 27:221–228, 1987. WJ Groot, KCAM Luyben. Appl Microbiol Biotechnol 25: 29–31, 1986. FA Roddick, ML Britz. J Chem Tech Biotechnol 69:383–391, 1987. E Silbiger, A Freeman. Enzyme Microb Technol 13:869–872, 1991. B Mattiasson, O Holst, eds. Extractive Bioconversion. New York: Marcel Dekker, 1991.
115. CS Chen, BN Zhou, G Girdaukas, WR Shieh, F VanMiddlesworth, AS Gopalan, CJ Sih. Bioorg Chem 12:98–117, 1984.
116. AJ Blacker, RA Holt. In: AJ Collins, GN Sheldrake, J Crosby, eds. Chirality in Industry, Vol. 2. New York: John Wiley, 1997, pp 245–261.
117. K Lorraine, S King, R Greasham, M Chartrain. Enzyme Microb Technol 19:250–255, 1996.
118. RA Holt, SR Rigby. WO 9405802. CA 120:268348.
119. Y Yasohara, Y Tari, N Ueyama, J Hasegawa, S Takahashi. EP 590549. CA 121:106675.
120. HG Leuenberger. In: S Servi, ed. Microbial Reagents in Organic Synthesis, NATO ASI series vol. C 381. Dordrecht: Kluwer Academic, 1992, pp 149–158.
121. JW Cornforth. J Lipid Res 1:3, 1959.
122. RE Dugan, LL Slakey, JW Porter. J Biol Chem 254:6312, 1970.
123. H Simon, J Bader, H Guenther, S Neumann, J Thanos. Angew Chem Int Ed Eng 24:539–553, 1985.
124. J Thanos, J Bader, H Guenther, S Neumann, F Krauss, H Simon. Meth Enzymol 136:302, 1987.
125. S Kuno, A Bacher, H Simon. Hoppe Seyler's Z Physiol Chem 366:463, 1985.
126. H Simon, J Bader, H Guenther, S Neumann, J Thanos. Ann NY Acad Sci 434:171, 1984.
127. M Buehler, H Simon. Hoppe Seyler's Z Physiol Chem 363:609, 1982.
128. M Utaka, S Konishi, A Takeda. Tetrahedron Lett 27:4737–4740, 1986.
129. M Utaka, S Konishi, A Mizuoka, T Ohkubo, T Sakai, S Tsuboi, A Takeda. J Org Chem 54: 4989–4992, 1989.
130. H Ohta, K Ozaki, G Tsuchihasi. Chem Lett 191–192, 1987. H Ohta, N Kobayashi, K Ozaki. J Org Chem 54:1802–1804, 1989.
131. P Gramatica, P Manitto, D Monti, S Speranza. Tetrahedron 44:1299–1304, 1988.
132. P Gramatica, P Manitto, L Poli. J Org Chem 50:4625–4628, 1985.
133. P Gramatica, P Manitto, BM Ranzi, A Delbianco, M Francavilla. Experientia 38:775–776, 1982.

134. C Fuganti, P Grasselli. J Chem Soc Chem Commun, 995–997, 1979.
135. HGW Leuenberger, W Boguth, R Barner, M Schmid, R Zell. Helv Chimica Acta 45:455–463, 1979.
136. C Fuganti, D Ghiringhelli, P Grasselli. J Chem Soc Chem Commun, 846–847, 1975.
137. C Fuganti, P Grasselli, HE Högberg, S Servi. J Chem Soc Perkin Trans I, 3061–3065, 1988.
138. T Sato, K Hanayama, T Fujisawa. Tetrahedron Lett 29:2197–2200, 1988.
139. S Koul, DHG Crout, W Errington, J Tax. J Chem Soc Perkin Trans I, 2969–2988, 1995.
140. R Zell, E Widmer, T Lukac, HGW Leuenberger, P Schonholzer, E Broger. Helv Chim Acta 64:2447–2462, 1981.
141. HE Högberg, E Edenström, J Fägerhag, S Servi. J Org Chem 57:2052–2059, 1992.
142. P Gramatica, G Giardina, G Speranza, P Manitto. Chem Lett, 1395–1398, 1985.
143. K Takabe, H Hiyoshi, H Sawada, M Tanaka, A Miyazaki, T Yamada, H Yoda. Tetrahedron: Asymmetry 3:1399–1400, 1992.
144. K Takabe, M Tanaka, M Sugimoto, T Yamada, H Yoda. Tetrahedron: Asymmetry 3:1385–1386, 1992.
145. V Sunjic, M Majeric, Z Hamersak. Croat Chem Acta 69:643–660, 1996.
146. BCC Cantello, DS Eggleston, D Haigh, RC Haltiwanger, CM Heath, RM Hindley, KR Jennings, JT Sime, SR Woroniecki. J Chem Soc Perkin I, 3319–3324, 1994.
147. A Arnone, R Cardillo, G Nasini, O Vajna de Pava. J Chem Soc Perkin I, 3061–3063, 1990.
148. F Trigalo, D Buisson, R Azerad. Tetrahedron Lett 29:6109–6112, 1988.
149. C Fuganti, G Pedrocchi-Fantoni, S Servi. In: SM Roberts, ed. Preparative Biotransformations. New York: John Wiley, Module 2:2.1, 1992.
150. CM Heath, RC Imrie, JJ Jones, MJ Rees, KG Robins, MS Verral. J Chem Technol Biotechnol 68:324–330, 1997.
151. HE Högberg, P Berglund, H Edlund, J Fägerhag, E Hedenström, M Lundh, O Nordin, S Servi, C Vörde. Catalysis Today 22:591–596, 1994.
152. M Schmid, R Barner. Helv Chim Acta 62:464–473, 1979.
153. C Fuganti, G Pedrocchi-Fantoni, A Sarra, S Servi. Tetrahedron: Asymmetry 5:1135–1138, 1994.
154. G Fronza, C Fuganti, M Mendozza, R Rallo, G Ottolina, D Joulain. Tetrahedron 52:4041–4052, 1996.
155. C Fuganti, P Grasselli. Chem Ind 24:983, 1977.
156. A Kergomard, MF Renard, H Veschambre. J Org Chem 50:120–123, 1985.

15

Choice of Biocatalyst in the Development of Industrial Biotransformations

Stephen J. C. Taylor, Karen Elizabeth Holt, Rob C. Brown, P. A. Keene, and Ian Nicholas Taylor
Chirotech Technology Ltd., Cambridge, England

I. INTRODUCTION

In the pharmaceutical industry a great deal of drug discovery activity is aimed at finding drugs that are potent, but are also more selective in their action and thus less toxic than the drugs of past years. Hence, pharmaceuticals are continuing to increase in structural complexity, and many contain at least one chiral carbon center. By definition, the pharmaceutical intermediates that are required are themselves becoming ever more challenging to synthesize, requiring the discovery and development of many inventive and innovative chiral synthetic methods.

One can broadly identify five main technology classes for introducing chirality: use of chiral pool materials, crystallization methods, chromatographic separations, chemocatalysis, and biocatalysis. When faced with a new chiral target to synthesize, careful evaluation of the possible routes must be undertaken to ensure that the final route is a competitive and economic one. Sometimes it is necessary to travel down two (or more) paths at the start until it becomes obvious as to which will be the chiral method of choice. Biocatalysis is very frequently one of the favored technologies, and it is the aim of this chapter to illustrate the use of biocatalytic methods for the synthesis of chiral intermediates and, in particular, highlight some of the factors that have influenced the choice of biocatalyst used.

II. COMMERCIALLY AVAILABLE ENZYMES

There are many enzymes that are commercially available, though the biotransformation literature and our own experience shows that in fact most biocatalysis is done with a relatively small subset of very frequently used enzymes. For a comprehensive review of biocatalysis in organic synthesis, the reader should refer to Drauz and Waldmann's two-volume *Enzyme Catalysis in Organic Synthesis* [1]. Many commercially available enzymes were not developed with organic synthesis in mind but are used in other industries. Examples include proteases like subtilisin, used in detergent formulations, or lipases, again used in detergents or in cheese and flavor production. There is a wide variety of ways in which enzymes are sold—most are available as fairly crude powders, others are liquids. Some are immobilized, and they may be from the original source or may have been cloned and expressed in another organism. Recently, there has been a rapid growth in companies dedicated to providing biocatalysts for organic synthesis. These include companies such as Altus with their impressive CLEC technology, and Diversa and Thermogen who specialize in thermostable enzymes. These are a valuable addition to the more established companies such as Amano or Novo who have been the traditional enzyme suppliers.

Having synthesized or otherwise acquired the appropriate substrates for biotransformation, the first key task before any enzymes are used is to develop the appropriate chiral assays for monitoring enantiomeric excesses of products and substrates. This is a field in which great advances have been made in the availability of instrumentation and chiral stationary phases over recent years, but it is outside the scope of this chapter. Such advances allow the competent analyst to develop the required assays quickly, so that very rarely should the analytical requirements be rate limiting in the route development.

Given that a biocatalytic route to the desired target is expected to give the best long-term solution, then a screen for the required biotransformation can be performed using the many commercially available enzymes of the appropriate type. It is once the data start to accumulate from these initial screens that thought must be put into which enzymes to progress further development.

There are many factors that will influence the researcher considering this choice, and some of these are discussed below. Some are obvious, like the biocatalyst cost in the biotransformation, which might be insignificant at kilo scale but could be a critical issue at manufacturing scale when faced with competitive processes. There are times when the influencing factors are more subtle. For example, if the substrate is inexpensive, a very cheap enzyme that is not particularly enantioselective and only gives a low yield of product may be preferred over a more expensive, more selective, higher yielding enzyme. However, the situation could reverse if the substrate is very expensive, and the whole synthetic picture must be considered.

In the following case studies, several biotransformations will be described and reasons for the choice of biocatalyst illustrated.

III. ISOLATED ENZYMES: BICYCLO[2.2.1]HEPT-5-EN-2-OL

Bicyclo[2.2.1]hept-5-en-2-ol (*endo*-norbornenol) is a useful starting material for a variety of compounds containing cyclopentane systems. The norbornane skeleton has been employed in the synthesis of prostaglandins, terpenes, steroids, alkaloids, and carbocyclic nucleoside analogs [2]. The potential of chiral norbornenols (and the saturated equivalent norborneols) was recognized, and a demand for multikilogram quantities of (*R*)-*endo*-norborneol suddenly arose. This is a typical situation that regularly challenges the providers of such chiral building blocks, where, to support pharmaceutical drug development programs, a process for providing the target compound must be developed and implemented quickly. In this instance, the obvious route to follow was to pick out the known chemistry from the literature and apply it. The enzymic resolution of racemic *endo*-norbornenyl acetate was known and described on multigram scale [3]. This was a useful starting point, and from this we developed a process that is shown in Scheme 1.

Provision of the biotransformation substrate is by a high-temperature pressure cycloaddition of vinyl acetate and cyclopentadiene, derived in situ from the thermolysis of dicyclopentadiene. This in itself posed the first problems on scale-up because procuring

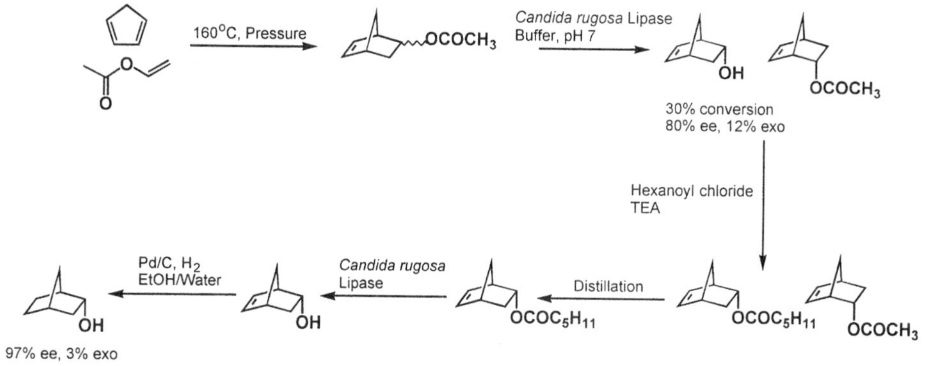

Scheme 1 Literature-based process for production of (R)-endo-norborneol.

the appropriate facility in a short time frame was difficult. The reaction also proved to be prone to the formation of byproducts via overreaction of the olefins, and of course, significant complexity was added to the process due to the unavoidable formation of a 3.6: 1 ratio of endo/exo isomers, for which there was no easy separation. Due to a relatively complex process whereby much of the cost was associated with labor, and the relatively limited requirement for the product, and indeed the uncertainty of future business, there was no driving force or time to find a better enzyme than that already reported and that was readily available. Biotransformation of the norbornenyl acetate by *Candida rugosa* lipase proceeded without difficulty but posed two more problems—(1) how to isolate the required alcohol and (2) how to raise both its enantiomeric and diastereomeric purity. A selective partition of the alcohols from the acetates was not possible due to the similarity of the two, and indeed any extraction was impossible due to the emulsions formed, caused by the high enzyme loading needed to ensure a sufficient hydrolysis. The solution, far from ideal, was to bulk-distill the whole biotransformation liquors, which allowed recovery of the esters and alcohols into a clean aqueous system from which they were readily extracted in high yield. However, direct distillation of this product mixture proved difficult, and it was much more convenient to derivatize the alcohol in situ to the hexanoate ester; separation of the lighter acetates was then easier. This left a hexanoate ester product whose enantiomeric and diastereomeric excess were (at about 80% e.e. with 12% exo isomer content) still well below the required specification. A further round of enzymic hydrolysis proved sufficient to raise the e.e. and d.e. to that required.

Due to the problems with this route, we set about finding improvements. A major breakthrough was the discovery that the initial cycloaddition reaction proceeded easily at atmospheric pressure when using a heavier vinyl ester with a boiling point in the region of 160°C [4]. This evolved into the improved route shown in Scheme 2. The hexanoate ester proved to be a better substrate for the lipase, faster and slightly more selective. An added advantage was that the crude reaction product from the cycloaddition was readily biotransformed. Again bulk distillation was used to recover the alcohol from the biotransformation, but this time the heavier esters and byproducts remained in the distillation vessel, giving a clean alcohol product. The alcohol is a crystalline solid, but unfortunately is not a conglomerate, and there was no opportunity to recrystallize it up to optical purity. This is also the case for the final saturated product. Thus another biotransformation was used to enhance the e.e. by acylation of the alcohol in organic solvent using vinyl butyrate as the acyl donor. A quick acylation screen of several enzymes known to perform well in

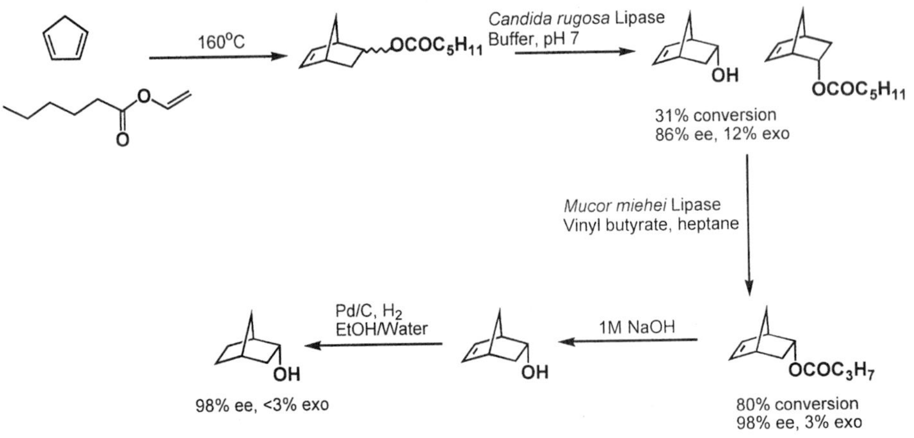

Scheme 2   Route to (R)-endo-norborneol via the hexanoate esters.

transesterifications showed that a *Mucor miehei*–derived lipase gave a selective acylation. Saponification of the product and reduction to (R)-*endo*-norborneol were then trivial.

    Subsequent to this work, further detailed process improvements have been made whereby the initial substrate synthesis has greatly improved in yield. This was achieved by using stringent feeding regimes of the components to reduce byproduct formation. Furthermore, it was found that by using an immobilized form of the enzyme the selectivity was greatly improved. Thus, the current process has been simplified to require just one bioresolution stage (Scheme 3).

IV. ISOLATED ENZYMES: C2-SYMMETRICAL 1,4-DIOLS

Enantiomerically pure C2-symmetrical diols such as 2,5-hexanediol are important intermediates in the synthesis of chiral acetals [5], 2,5-disubstituted pyrolidines [6–8], and phospholane ligands used for asymmetrical hydrogenation [9,10] (Scheme 4). Phospholane-based rhodium complexes of the DuPhos and BPE ligand families are extremely powerful catalysts for the asymmetrical hydrogenation of a wide variety of prochiral olefins, furnishing whole families of α-amino acids, amines, succinates, and others. The chiral 1,4-diols which provide the chirality for such hydrogenation catalysts have previously been obtained by electrochemical Kolbe-type coupling of chiral β-hydroxy acids [11,12], but in our hands this route was impractical at large scale, which was largely due to fouling of the electrodes. In addition, stereoselective reduction of symmetrical 2,5-hexanedione has been reported [13,14], but again the processes have not been developed into scalable methods. Enzymatic resolutions of the racemic/*meso*-hexanediol have also been investi-

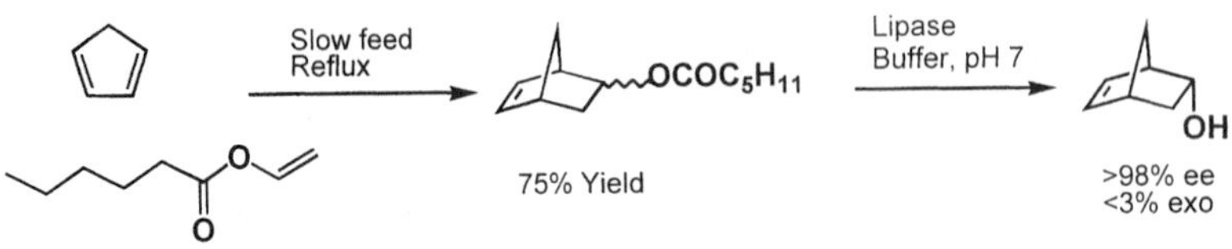

Scheme 3   Route to (R)-endo-norborneol using immobilized lipase.

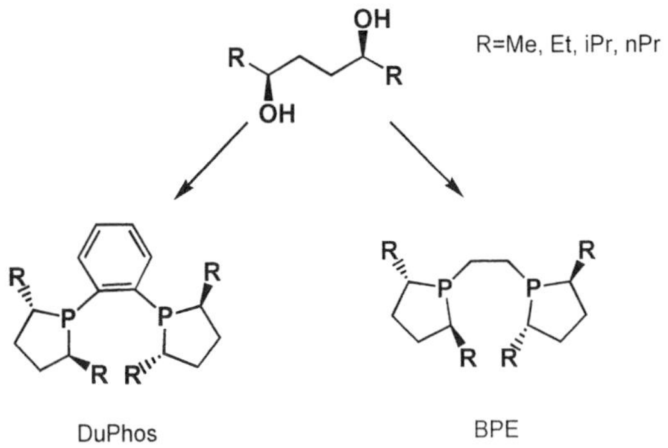

Scheme 4 Phospholane ligands for hydrogenation.

gated [15–19] but these suffered from chromatographic separation of diol/monoester/diester mixtures and the need for repeat biotransformations to increase to enantiopurity.

The commercialization of these catalysts gave rise to a requirement for tens of kilogram quantities of both (S,S)- and (R,R)-hexane-2,5-diol in a relatively short time frame. Since the starting material was readily available at small scale as a 1:1 racemic/meso mixture, the obvious approach was a solvent-based enzymic acylation, where the enzymatic transesterification ideally resolves the 1:1 racemic/meso hexanediol into a mixture of 25% (2S,5S)-diol, 50% (2R,5S)-monoester, and 25% (2R,5R)-diester. A relatively limited isolated enzyme screen was all that was required to identify a biocatalyst with the required properties to satisfy the demand for the diols. The screen of lipases with vinyl acetate in methyl t-butyl ether (MTBE) revealed good activities for Chirazyme L2 (Boehringer Mannheim), lipase PS (Amano), and ChiroCLEC-PC (Altus Biologics), but with only modest selectivities. However, by changing to vinyl butyrate, the selectivities were much improved, with Chirazyme L2 being the best (Table 1).

Analysis of the mixtures for ee/de is crucial and we were able to separate all isomers of diol, monoester, and diester on a single assay using chiral gas chromatography (GC) (Fig. 1). This allowed the easy determination of enzyme selectivity and thus the optimal conditions for this resolution. Having determined Chirazyme L2 to be the best enzyme, reaction conditions such as reaction temperature, enzyme loading, substrate concentration, and equivalents of acyl donor were investigated. Optimized conditions were determined as 1000 g/L substrate, 1.1 equivalents of vinyl butyrate 5 wt% enzyme (2.45 U/mg), 30°C, 20 h, giving the results in Table 2.

The biotransformation was developed further at the lab scale to investigate product isolation. It was discovered that a selective partition existed between the diol and the esters

Table 1 Positive Results (%) for the Enzyme-Catalyzed Transesterification of Hexanediol

Enzyme	Time (h)	Diol		Monobutyrate		Dibutyrate	
		e.e.	d.e.	e.e.	d.e.	e.e.	d.e.
Chirazyme L2	16.5	>99	>99	>99	85	97	57
Lipase PS	16.5	78	19	92	44	95	62
ChiroCLEC-PC	16.5	96	81	99	4	91	70

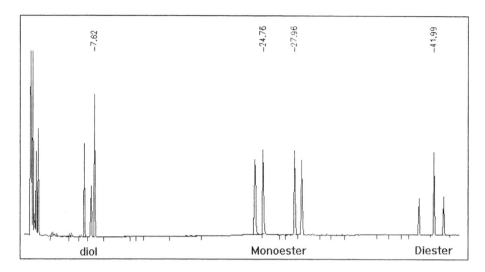

Figure 1   Example of the e.e./d.e. assay for biotransformation mixtures.

into water and heptane, respectively, thus allowing almost complete transfer of the diol to the water. The water could then simply be evaporated and the diol crystallized with MTBE. It was found that this crystallization typically increased the d.e. to 98%, thus eliminating any need for repeated biotransformation. Recrystallization of the diol from EtOAc enhanced the purity, e.e., and d.e. to more than 99% in 90% recovery as shown in Scheme 5.

To make the best possible economical use of the residual material, a method was sought whereby the (R,R) enantiomer could be recovered from the existing liquors. We believed this would be possible by converting the free hydroxyl of the monoester to a leaving group (e.g., mesylate) and inverting with an acetate nucleophile via an S_N2 displacement. With this reaction carried out in situ the resulting ester mixture could then simply be hydrolyzed to the (R,R)-diol.

After removal of the (S,S)-diol into the water washes, a heptane mixture of the (R,S)-monoester and (R,R)-diester remains. The ester mixture was treated with triethylamine and methanesulfonyl chloride in dichloromethane at low temperature. This gave a fast, clean reaction to the monomesylate in quantitative yield, leaving the dibutyrate ester untouched. We next turned our attention to the displacement reaction. Several conditions were investigated revealing cesium acetate in dimethyl formamide (DMF) to give the fastest reaction with the least racemization. However, the conditions most economic for scale-up were potassium acetate (a lot cheaper than cesium acetate) in DMF, which although slower gave the same e.e./d.e. product. Using DMF, the reaction proceeded smoothly at 90°C. After a

Table 2   Optimized Chirazyme L2 Resolution with Vinyl Butyrate (%)

Enzyme	Time (h)	Diol		Monobutyrate		Dibutyrate	
		e.e.	d.e.	e.e.	d.e.	e.e.	d.e.
Chirazyme L2	16.5	>99	88	>99	96	>99	88

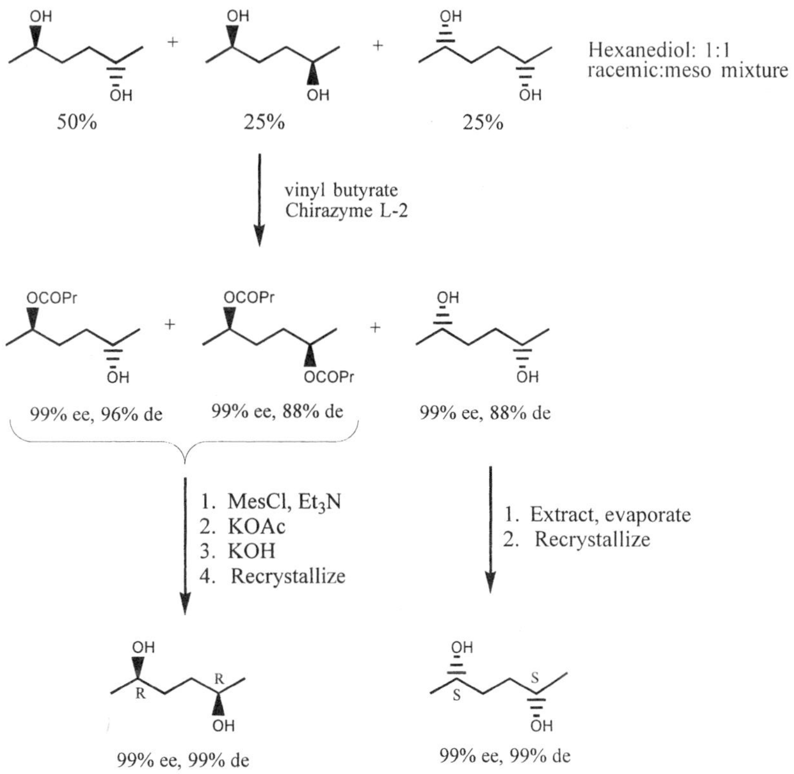

Scheme 5   Route to (R,R) and (S,S)-hexane-2,5-diol.

standard MTBE/aqueous workup, the diester mixture was isolated in 85–90% crude yield. The final step in the sequence was to hydrolyze the diester mixture to the (R,R)-diol. This was achieved in a volume-efficient reaction using KOH in methanol at room temperature. After completion the reaction was neutralized with acidic resin Amberlite IR 120. The methanol was then evaporated and the diol crystallized from MTBE to give (2R,5R)-hexanediol of 99% e.e. and 95% d.e. Recrystallization of the diol from EtOAc or MTBE enhanced the purity, e.e., and d.e. to more than 99% in 80–90% recovery.

The next challenge was to scale this chemistry up to pilot plant size. The racemic/meso hexanediol was not as readily available at large scale as might be expected. However it was possible to source 2,5-hexynediol at tonne scale and low prices. Subsequently, an efficient hydrogenation was developed to produce the saturated diol (Scheme 6). Using Raney nickel in MeOH, racemic/meso hexanediol was prepared in high yield without the

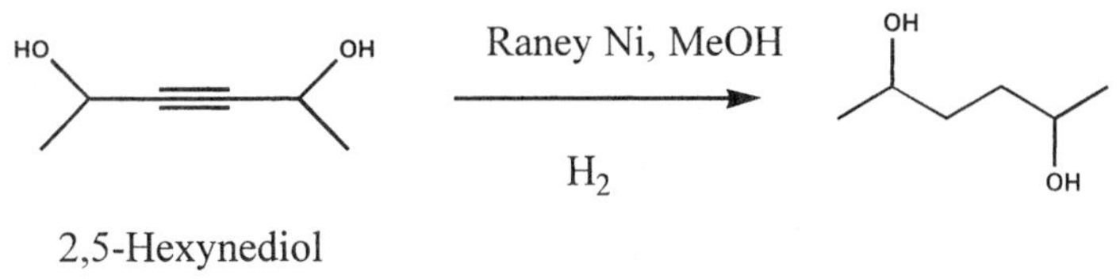

2,5-Hexynediol

Scheme 6   Large-scale source of hexane-2,5-diol.

need for an aqueous workup. This procedure was scaled up to produce multikilogram quantities of racemic/meso hexanediol, which was purified by distillation. Thus we had access to a relatively cheap source of 2,5-hexanediol (racemic/meso 3:4).

In order for the process to be more practical at large scale a number of improvements were made. As described previously, vinyl butyrate was found to be the most selective acyl donor. However, in consideration of scale-up costs, vinyl butyrate was much more expensive than vinyl propionate. Furthermore, the smell associated with butyrates was an added process consideration. Since vinyl propionate was only slightly less selective than vinyl butyrate (around 5% d.e.), on balance it was decided to scale up the biotransformation with vinyl propionate. In addition, MTBE was replaced with heptane as the reaction solvent. This prevented the solvent swap required for the diol/ester partition. The change in solvent gave no loss in selectivity. This was thus a very volume-efficient resolution that could be performed in standard chemical vessels without the need for any specialized equipment. After the reaction was complete the workup was performed as outlined previously. This process was scaled up to 40-kg input batches of racemic/meso diol without complication.

The inversion chemistry was scaled up to 40-kg input propionate ester mixture with minimal changes to the process outlined above. An alternative to using Amberlyte IR 120 to work up the final hydrolysis reaction was found. This involved extracting the (2R,5R)-hexanediol into butanone. The diol was then purified to more than 99% e.e./d.e. by partition between MTBE and water evaporation, followed by a recrystallization from ethyl acetate.

Thus, a practical, scalable synthesis of optically pure (S,S)- and (R,R)-hexane-2,5-diol was developed, providing multikilogram quantities of diols. Here the enzyme performance was perfectly adequate for the job and is unlikely to be an issue in this process if scaled up further. These procedures can also be used to synthesize enantiomerically pure (S,S)- and (R,R)-octane-3,6-diol and 2,5-dimethyloctane-3,6-diol.

V. ISOLATED ENZYMES: L-AMINOACYLASE

α-Amino acids are an extremely important class of chiral building blocks for pharmaceutical synthesis, and this is reflected in the diverse range of technologies that exist for their synthesis. These include both chemocatalytic and biocatalytic methods. Two of the most attractive methods for the synthesis of unnatural amino acids are asymmetrical hydrogenation and L-aminoacylase bioresolution. The synthesis of L-amino acids by L-aminoacylase has been known for decades (first commercialized in the 1950s by Tanabe) and is frequently used. Asymmetrical hydrogenation is also well established as exemplified by the Monsanto DIPAMP process for the production of levodopa in the 1960s, though this catalyst is limited in selectivity for a range of amino acids. By contrast, the Rh-MethylDuPhos-catalyzed asymmetrical hydrogenation of enamides is a much more recent development but is proving to be very powerful; e.e. values are typically 98% or greater, with the obvious benefit of potentially quantitative yield (Scheme 7).

The hydrogenation technology is now well established and eminently scalable, and indeed has been used by us to provide a very diverse set of amino acids. However, there are some cases where the use of L-aminoacylase is still the method of choice for inserting the chirality by resolution. These are generally where the side chain is susceptible to further transformation by the chemocatalyst, or where it might poison the catalyst, or where the enamide synthesis is difficult. Enamide hydrogenation substrates are generally synthesized

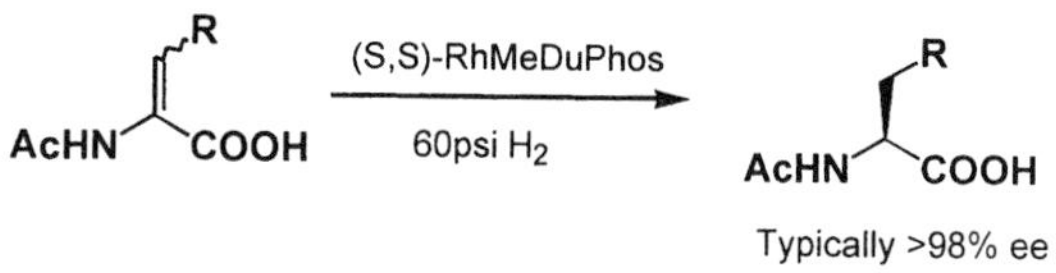

Scheme 7 α-Amino acids by asymmetrical hydrogenation.

from the appropriate aldehyde, while the L-aminoacylase substrates are conveniently derived from the alkyl halides by condensing with diethylacetamidomalonate. Thus, in some cases it is also the relative availability and cost of these key starting materials that influences the choice of route for inserting the chirality.

We recently used L-aminoacylase to provide the amino acids shown in Scheme 8. In these instances either the aldehydes were not available or they performed very poorly in the Erlenmeyer chemistry for enamide synthesis. These resolutions can be integrated very readily into an overall process, where the unwanted isomer can be racemized and recycled by way of treatment with acetic anhydride and racemization via the azalactone.

While in the process of synthesizing a large set of amino acids, primarily designed to stimulate the interest of medicinal chemists and seed future drug programs, it became obvious that acid hydrolysis of the N-acetyl amino acid hydrogenation products was not always trivial, with various examples of partial racemization or degradation, besides the requirement for desalting of the product. To alleviate these problems L-aminoacylase was used to deprotect the N-acetyl hydrogenation products. The combined selectivity of the hydrogenation catalysis and the enzymic deprotection allows for amino acids of very high enantiomeric excess, often more than 99.8% e.e., which is particularly useful for syntheses where the amino acid is entering at a late stage.

The choice of enzyme for these resolutions and deprotections was obvious, and again at multikilogram scale factors such as labor are more dominant than enzyme costs. Thus the work to date has always used the *Aspergillus* fungal enzyme.

VI. SCREENING FOR NEW BIOCATALYTIC ACTIVITIES

By conducting a screening program for biocatalysts at the microbial level, one has access to a vastly superior diversity of activities and selectivities than is available with commercially available enzymes. It is extremely useful to have on site and close to hand a collection of microbes for screening; in our case a collection of about 1400 bacterial, yeast, and fungal strains is held. All screening data are entered in a database, allowing subsequent screens to be targeted to smaller subsets of microbes that are more likely to contain the

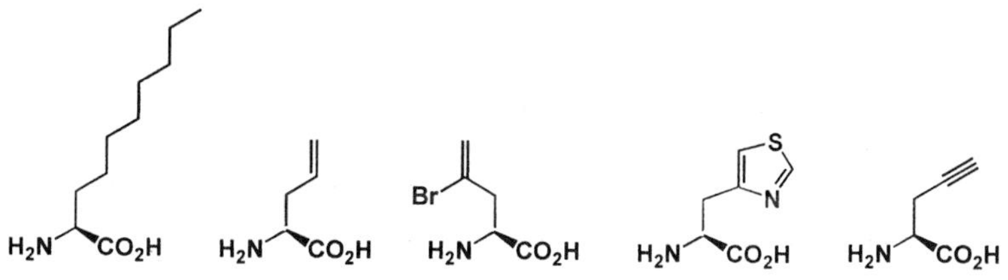

Scheme 8 Amino acids synthesized by L-aminoacylase resolution.

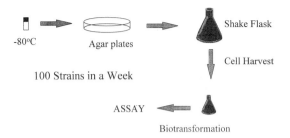

Figure 2 Screening by shake-flask culture.

desired activity. The development of techniques and apparatus for high-throughput screening has helped to greatly streamline the screening process and reduce the labor required to perform a single screen. Figure 2 shows a typical process before the omnipotent 96-well tissue culture plates were widely available and cheap.

After regeneration of the master strain on agar plates, checking for purity, individual strains had to be grown by shake-flask, harvested by centrifugation, resuspended in a biotransformation medium, the cells removed again, then assayed by the preferred method. This was very labor-intensive and space consuming, and typically about 100 strains could be assayed by a single scientist in a week. The labor and space requirements have been greatly reduced by use of 96-well plates as shown in Fig. 3. Here, once a master plate has been constructed holding about 100 strains, it can be manipulated very easily, and can be used for many replica plates to provide the working strain plates. Cells can be separated from the growth medium by centrifugation in situ, then biotransformation media applied. After centrifugation of the biotransformation media, the plates can then be assayed directly by high-performance liquid chromatography (HPLC) for activity and/or selectivity. Working volumes are also very small, reducing the need for large amounts of expensive substrates. This gives the potential for screening hundreds of strains in a few days, where the rate-limiting step is more likely to be the HPLC or GC time required to run the assays.

Having identified a promising strain, it will routinely be cloned for overexpression. Two examples, D-aminoacylase and a γ-lactamase are described below.

VII. MICROBIAL ENZYMES: D-AMINOACYLASE

It is important in drug screening and lead optimization programs to be able to study both enantiomers separately rather than in racemic form. This is because one needs to consider both the activity and the selectivity of a drug molecule, and performing a screen on each of the enantiomers generates more information than would otherwise be gained by screening with the racemate. Hence, as with other sets of chiral compounds, there is a requirement for unnatural D-amino acids. These are also often used in peptidic structures with the aim of blocking degradative enzymes.

While the hydrogenation technology can be applied with the same ease as for the L-amino acids, there is no commercially available equivalent D-aminoacylase. Here it becomes of paramount importance to be able to isolate and exploit biocatalysts from microbial strains. Specific D-aminoacylases have previously been described in strains of *Pseudomonas*, *Streptomyces*, and *Alcaligenes* [20–27]. The enzymes from these strains were isolated and characterized. In theory it should be relatively easy to use such strains

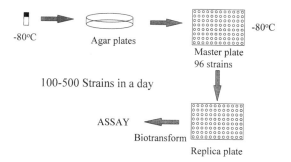

Figure 3 Screening using the 96-well format.

in whole-cell form for the resolution or deprotection of *N*-acetyl amino acids, but in practice the cells were also shown to contain L-aminoacylases, thus reducing the stereo-selectivity. In addition, the low levels of activity, even after growth on inducing media, make a purification and use of the enzyme from the whole-cell unattractive economically. A solution was foreseen by way of cloning the enzymes, and indeed this has been reported recently for the *Alcaligenes* species D-aminoacylase [28–30].

Screening of about 100 microbial strains for biotransformation of *N*-acetyl-D-phenylalanine gave several positives, and from these two were identified as strains of *Alcaligenes* species. Examination of the published DNA sequence data allowed the design of oligonucleotide probes that were used in polymerase chain reaction (PCR) experiments to see if a D-aminoacylase fragment could be cloned from a genomic digest of a strain identified as *Alcaligenes* CMC3352. A 1.4-kb fragment was isolated and shown to have D-aminoacylase activity by HPLC assay with *N*-acetyl-D-tryptophan as substrate. Sequencing the fragment showed it to have considerable homology to the published D-aminoacylase, though the newly cloned enzyme differs by its ability to hydrolyze *N*-acetyl-D-tryptophan rapidly, where the published enzyme cannot utilize this substrate.

Having cloned the D-aminoacylase, its expression and fermentation were optimized. This provided whole cells containing the recombinant D-aminoacylase for use in biotransformation of *N*-acetyl-D-amino acids. The substrate range of this enzyme is relatively unexplored, but to date several unnatural substrates have been deprotected at up to kilogram scale, and the range is increasing rapidly. The enzyme can also be used for resolution in the same manner as the L-aminoacylase (Scheme 9). Hence by making the D-aminoacylase available in large quantities at low cost, all of the technology that has been developed for the L-aminoacylase can be applied with equal ease for the production of D-amino acids.

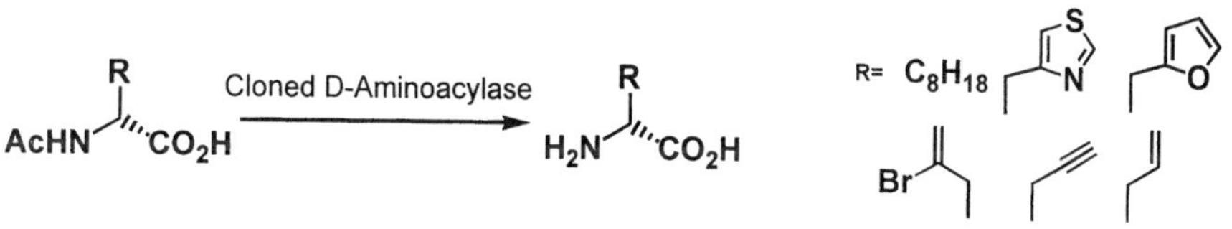

Scheme 9 Deprotections using cloned D-aminoacylase.

VIII. MICROBIAL ENZYMES: γ-LACTAMASE

2-Azabicyclo[2.2.1]hept-5-en-3-one is now well established as a very versatile synthon for the production of carbocyclic nucleosides [31] (Scheme 10). A recent example is its use in the synthesis of the potent reverse transcriptase inhibitor abacavir. Such compounds, where the ribose oxygen of the nucleoside has been replaced by a methylene, have become very valuable as chemotherapeutic agents in the fight against viral infections such as HIV or herpes.

A program of research to identify a biocatalytic route to these optically pure lactams commenced in 1990. Not surprisingly, there were no commercially available enzymes that showed lactamase activity, so one had to be found from an environmental microbial source by screening selected microbial strains. Thus we reported a biocatalytic resolution of 2-azabicyclo[2.2.1]hept-5-en-3-one several years ago [32]. By screening soil and sewage samples, using *N*-acetyl-L-phenylalanine as a sole source of carbon, many microbes with the ability to perform amide bond hydrolysis were isolated. Subsequent rescreening of these for hydrolysis of the γ-lactam revealed a strain of *Pseudomonas cepacia* that was highly selective for the hydrolysis of the (+)-lactam, leaving the desired (−) enantiomer untouched. The whole cells gave an enantiomeric ratio (*E* value) of 94 (Scheme 10).

Several attempts were made to isolate and purify the lactamase, but it proved too unstable for the manipulations required for isolation and purification. As a result, a biotransformation process was developed using the whole-cell biocatalyst. While this process allowed the development of a tonne scale (−)-lactam process, as the volumes required grew further, it became more difficult to satisfy demand. The process had several drawbacks. The use of whole cells, and the subsequent lysis of these during the biotransformation (and lysis from frozen storage), complicated the isolation of the lactam. Direct solvent extraction was impossible and a complex carbon adsorption and elution cycle was needed. While the volume efficiency of the biotransformation was tolerable, large volumes were encountered during workup. Also, the cells proved to be somewhat sensitive to the quality of racemic lactam used, which varied depending on the source. Furthermore, the quantity of cells required (1 kg of cells yielded 1 kg of (−)-lactam) contributed a significant proportion of the manufacturing cost. Thus there was a considerable driving force to invest in the research needed to find a better biocatalyst that was more efficient, selective, and—importantly—cheap. This will most commonly emerge from a focused program of screening for the microbe, enzyme isolation and characterization, followed by cloning of the enzyme for overexpression. Here we describe such a program for the production of optically pure 2-azabicyclo[2.2.1]hept-5-en-3-one [(−)-γ-lactam] using a lactamase, high-

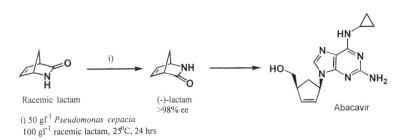

Racemic lactam   (-)-lactam
                                  >98% ee
                                                              Abacavir

i) 50 gl^{-1} *Pseudomonas cepacia*
100 gl^{-1} racemic lactam, 25°C, 24 hrs

Scheme 10   Carbocyclic nucleosides from 2-azabicyclo[2.2.1]hept-5-en-3-one.

lighting this as an example of a novel method used to identify clones, and to show the benefits of a cloned enzyme on the overall process.

A further screen for lactamases was performed, again using N-acetyl-L-phenylalanine as the sole source of carbon. From the several positives isolated, a temperature stability study was performed on the lactamase in crude cell lysates. This indicated one to be considerably more stable than the *Pseudomonas cepacia* strain [33], and indeed retained half of its activity after incubation at 60°C for 4 h. The strain was later identified as *Comamonas acidovorans*. The γ-lactamase was purified to homogeneity by a sequence of ammonium sulfate precipitation, HIC (butyl Sepharose), anion exchange chromatography (Q Sepharoses) and gel filtration (Sephacryl S200), giving a single band by sodium dodecyl sulfate–polyacrylamide gel electrophoresis (SDS-PAGE). This indicated a molecular weight of 53–55 kDa. The N-terminal sequence was then determined, as shown in Fig. 4.

A random gene library of the *C. acidovorans* was prepared by standard techniques. Genomic DNA was partially digested by *Sau*3A I, then DNA fragments from 2–6 kb isolated after size separation by electrophoresis. Fragments were ligated into the dephosphorylated *Bam*H1 site of pUC19, and the vector transformed into *E. coli* DH5α.

Detection of an expressing clone was attempted by the use of a radioactive oligonucleotide probe based on the N-terminal amino acid sequence. The probes used are shown in Fig. 4. However, this approach was not successful due to the degeneracy of the sequence. This approach may also identify nonexpressing forms of the lactamase such as truncated versions of the gene, and a direct screen for activity is more desirable and, in this case, necessary.

Conventional biocatalyst screening methods such as emulsified substrate overlay (where enzyme activity can be directly visualized in the form of a clearing zone around the biocatalyst) were not appropriate for the lactam substrate due to its high solubility (more than 500 g/L in water). Furthermore, other staining methods such as ninhydrin or glutaraldehyde did not work in an agar overlay with a soluble substrate because diffusion quickly diluted any visible effects. A different approach was needed, and so a very simple screen was developed, combining the classical techniques of replica plating and ninhydrin staining. A Whatman filter paper was dipped in a methanolic solution of (+)-lactam and allowed to dry. This impregnated the paper with (+)-lactam substrate, which was then placed directly onto the agar plate where it picked up moisture and most of the cells in each colony. The physical transfer of cells gave sufficient biomass to allow a biotransformation to occur, but the aqueous content of the paper was limited and prevented excessive diffusion of product away from the cells. The paper was incubated for several hours, dried, and stained with ninhydrin. A positive clone was clearly visualized as a colony with a

NH3 - T D T L I K V D L N R P P T D N E R V H

The underlined amino acids were selected to synthesise N-terminal degenerate oligonucleotide probes.

NH3-Asp-Thr-Leu-Ile-Lys-Val-Asp

Probe 1   5'-GAY-ACN-TTR-ATH-AAR-GTN-GA
Probe 2   5'-GAY-ACN-CTN-ATH-AAR-GTN-GA
N=A+C+T+G   R=A+G   Y=C+T

Figure 4 Derived N-terminal amino acid sequence of purified *Comamonas acidovorans* γ-lactamase.

brown halo on the light background. The paper colony pattern was then matched with the regrown cells on the agar plate, and the desired colony isolated.

DNA sequence analysis of the insert showed the fragment to incorporate an open reading frame (ORF) of 1.7 kb, shown in Fig. 5. This was driven by the upstream lac promoter of pUC19, and translates to a protein of 575 residues (61 kDa). The deduced amino acid sequence of the translated ORF showed more than 65% homology to an acetamidase from *Mycobacterium smegmatis* which has been shown to hydrolyze short chain fatty acylamides [34,35].

From the cloned gene sequence, the *N*-terminal sequence was identified. The lactamase was found to have an extra 31 amino acids upstream of the *N* terminus, which arose by a fortuitous ligation of a short piece of DNA during the cloning. However, experiments showed that enzyme expression was good, and subsequent manipulation of the gene to

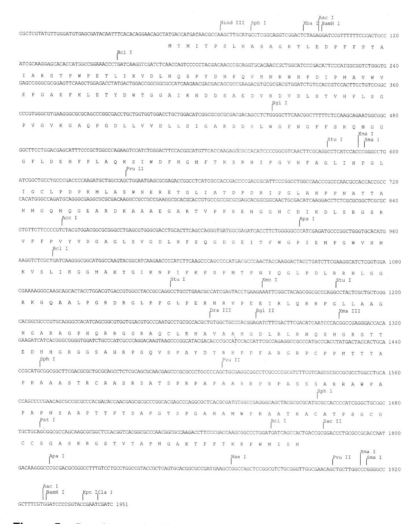

Figure 5 Complete nucleotide sequence of the γ-lactamase gene and deduced amino acid sequence from *Comamonas acidovorans* strain CMC 4093. The 1.9-kb *Sau*3AI fragment was ligated in-frame with the *Bam*HI restriction sites of pUC19 backbone.

remove these gave no added benefit. Final modification of the recombinant plasmid was the insertion of the *cer* element responsible for multimeric resolution and stable inheritance of the wild-type *E. coli* plasmid *ColEI*. The *cer* element was transferred from construct pKS4926. The final construct and expression vector were designated pPET1 (Fig. 6). The vector was transformed into *E. coli* MC1061, the designated recombinant lactamase host. The level of recombinant protein was significant enough to be visualized on Coomassie SDS-PAGE under induced growth.

Fermentation of the recombinant *E. coli* was based on a complex media/feed with glycerol as the carbon source and peptone/casein as a source of amino acids, peptides, and nitrogen. Initial growth was by a batch mode after which a continuous feed containing glycerol was used. Regulation of enzyme expression was not particularly well controlled and the fermentation did not require induction. A 500-L, 4-day fermentation yielded approximately 100 g/L (wet weight) of cell paste and 3000 U/g cell paste, sufficient to resolve about 5 tonnes of racemic substrate.

Figure 7 shows the greatly improved enzyme yield attainable from a single fermentation when the enzyme is cloned. Having optimized the fermentation, we focused on recovering the lactamase in a form suitable for use in the biotransformation. The goal was a simple process that gave sufficiently pure lactamase such that the protein content did not cause emulsions during isolation of the lactam. After lysis of the cells with a combination of lysozyme and Triton X-100, the bulk of nucleic acids and cell debris were removed by polyethyleneimine precipitation and centrifugation. The lactamase was precipitated from the filtrate by standard ammonium sulfate precipitation, before recovery by centrifugation. The pellet was finally redissolved in Tris-HCl buffer, then sterile-filtered ready for use. This was achieved in an overall yield of about 70%.

With a readily available supply of semipurified recombinant lactamase available, a much improved process was quickly developed. The biotransformation was optimized with respect to substrate concentration, enzyme loading, pH, buffer, and temperature. This highlighted 500 g/L substrate, 100 mM Tris-HCl at pH 7.5, 25°C to be optimal.

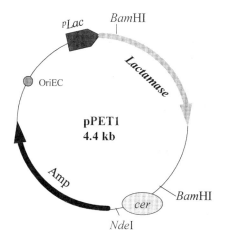

Figure 6 Schematic diagram of plasmid pPET1. *E. coli* plasmid pPET1 was derived from pUC19, which harbors a 1.9-kb *Sau*3AI genomic fragment from *Comamonas acidovorans* CMC 4093, ligated into the *Bam*HI restriction site. The *cer* stability element of the wild-type plasmid *ColEI* was inserted at the 3′ to the lactamase fragment via a *Bam*HI (partial) and *Nde*I restriction.

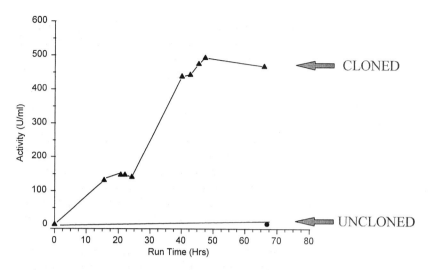

Figure 7   Comparison of wild-type and cloned enzyme expression.

      Thus the new (−)-lactam process has been much improved (Scheme 11). Volume efficiency is much higher where the enzyme tolerates 500 g/L input of lactam, a fivefold increase. Since a clean concentrated enzyme is used, the isolation is greatly simplified where direct dichloromethane solvent extraction of the lactam can be achieved, avoiding the carbon adsorption step and associated problems of low volume efficiency, handling, and safety. The final step is a recrystallization of the product from the solvent concentrate. Thus by screening for a better enzyme and cloning it, the process and its economics were greatly improved. This clearly demonstrates the importance of having the capability to perform such biocatalysis development in a fine chemicals environment.

IX. CONCLUSION

Industrial biotransformations require both isolated enzymes and microbial enzymes. Many biotransformation processes begin with the use of a commercially available enzyme, allowing process parameters to be defined and problems to be identified at an early stage, while giving access to up to multikilo quantities of the chiral target. However, as the process matures and cost considerations become important, or if there is simply not an isolated enzyme available, the ability to screen for and identify a microbial source of the enzyme is vital. Further development by cloning an enzyme has the obvious benefit of reducing the cost of the biocatalyst through overexpression. However, what can be equally

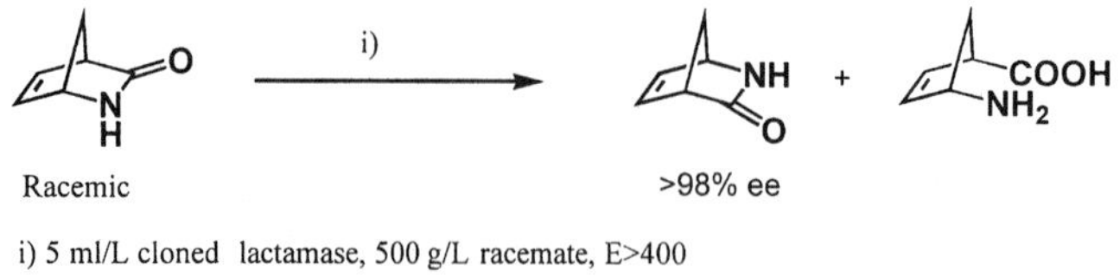

Racemic   >98% ee

i) 5 ml/L cloned lactamase, 500 g/L racemate, E>400

Scheme 11   Resolution of 2-azabicyclo[2.2.1]hept-5-en-3-one using the cloned lactamase.

important is the dramatic impact that use of a cloned enzyme can have on the overall design of a process, where product recovery in particular becomes much easier.

REFERENCES

1. K Drauz, W Waldmann. Enzyme Catalysis in Organic Synthesis: A Comprehensive Handbook. New York: VCH, 1995.
2. T Oberhauser, M Bodenteich, K Faber, G Penn, H Greingl. Tetrahedron 43(17):3931–3944, 1987.
3. G Eichberger, G Penn, K Faber, H Greingl. Tetrahedron Lett 27(5):2843–2844, 1986.
4. R McCague, SJC Taylor. Patent WO97/05093.
5. A Alexakis, P Mangeney. Tetrahedron: Asymmetry 1:477–511, 1990.
6. M Pichon, B Figadere. Tetrahedron: Asymmetry 7:927–964, 1996.
7. JK Whitesell. Chem Rev 89:1581–1590, 1989.
8. JM Chong, IS Clarke, I Koch, PC Olbach, NJ Taylor. Tetrahedron: Asymmetry 6:409–418, 1995.
9. MJ Burk, JE Feaster, RL Harlow. Tetrahedron: Asymmetry 2:569–592, 1991.
10. MJ Burk, TGP Harper, CS Kalberg. J Am Chem Soc 117:4423–4424, 1995.
11. MJ Burk, JE Feaster, RL Harlow. Organometallics 10:2653–2655, 1990.
12. K Serck-Hanssen, S Stallberg-Stenhagen, E Stenhagen. Ark Chem 18:203, 1963.
13. GJ Quallich, KN Keavy, TM Woodall. Tetrahedron Lett 36:4729–4732, 1995.
14. J Bach, R Berenguer, J Garcia, T Loscertales, J Manzanal, J Vilarrasa. Tetrahedron Lett 38: 1091–1094, 1997.
15. H Nagai, T Morimoto, K Achiwa. Synlett 1994, 289–290.
16. RJ Kazlaukas. J Am Chem Soc 111:4953–4959, 1989.
17. G Caron, RJ Kazlaukas. J Org Chem 56:7251–7256, 1991.
18. G Caron, RJ Kazlaukas. Tetrahedron: Asymmetry 5:657–664, 1994.
19. M-J Kim, IS Lee, N Jeong, YK Choi. J Org Chem 58:6483–6485, 1993.
20. M Sugie, H Suzuki. Agric Biol Chem 44:1089–1095, 1980.
21. Daicel Chemical Industries, JP 64-5488, 1989.
22. M Moriguchi, K Ideta. Appl Environ Microbiol 54:2767–2770, 1988.
23. K Sakai, K Imamura, M Goto, I Hirashiki, M Moriguchi. Agric Biol Chem 54:841–844, 1990.
24. K Sakai, T Obata, K Ideta, M Moriguchi. J Ferm Bioeng 71:79–82, 1991.
25. K Sakai, K Oshima, M Moriguchi. Appl Environ Microbiol 57:2540–2543, 1991.
26. Y-B Yang, C-S Lin, C-P Tseng, Y-J Wang, Y-C Tsai. Appl Environ Microbiol 57:1259–1260, 1991.
27. Y Kameda, E Toyoura, Y Kimura, H Yamazoe. Nature 169:1016, 1952.
28. M Moriguchi, K Sakai, Y Miyamoto, M Wakayama. Biosci Biotech Biochem 57(7):1149–1152, 1993.
29. M Wakayama, Y Katsuno, S-I Hayashi, Y Miyamoto, K Sakai, M Moriguchi. Biosci Biotech Biochem 59(11):2115–2119, 1995.
30. M Wakayama, S-I Hayashi, Y Yatsuda, Y Katsuno, K Sakai, M Moriguchi. Prot Express Purif 7:395–399, 1996.
31. S Daluge, S and RJ Vince. J Org Chem 43:2311–2320, 1978.
32. SJC Taylor, R McCague, R Wisdom, C Lee, K Dickson, G Ruecroft, F O'Brien, J Littlechild, J Bevan, SM Roberts, CT Evans. Tetrahedron: Asymmetry 4(6):1117–1128, 1993.
33. AB Brabban, J Littlechild, R Wisdom. J Ind Microbiol 16:8–14, 1996.
34. P Draper. J Gen Microbiol 46:111–123, 1967.
35. E Mahenthiralingham, P Draper, EO Davies, MJ Colston. J Gen Microbiol 139:575–583, 1993.

16

Chiral Synthons by Enzymatic Acylation and Esterification Reactions

Enzo Santaniello, Shahrzad Reza-Elahi, and Patrizia Ferraboschi
University of Milan, Milan, Italy

I. INTRODUCTION

The asymmetrical synthesis of compounds containing stereogenic centers can be achieved by a great variety of chemical and biochemical methods. Biocatalysts such as purified enzymes have been used in recent years for this purpose [1] and, among these, hydrolytic enzymes (hydrolases) have gained a special success in organic synthesis because they do not require cofactors and may catalyze the hydrolysis of substrates of different structures. From a practical point of view, the low solubility of organic compounds in aqueous media can constitute a major problem [2] that can be overcome by the discovery that many enzymes can be used in organic solvents, as shown for the first time by Klibanov in 1984 [3]. Since this report, the chemical literature has witnessed an explosion of papers dealing with applications of enzymes in nonconventional phases [4–6]. Hydrolases, and lipases in particular, seem most suitable for this purpose and reactions virtually impossible in aqueous media such as esterifications, transesterifications, and interesterifications can be catalyzed in nonconventional media such as organic solvents (Scheme 1). These reactions are especially useful in organic synthesis [7] since very often the above enzymatic reactions can proceed on a large variety of organic compounds with a high degree of stereoselectivity, and this constitutes a viable access to optically pure compounds. This biocatalytic approach can often compete with the chemical synthesis of enantiomerically pure compounds [8]. The production of fine chemicals can take advantage of biocatalysis in organic solvents [9]. Several subtle problems are related to the activity of enzymes in organic solvents with low water content and the parameters that affect such behavior have been actively investigated [10]. The effect of water on the enzyme action in organic solvents was first recognized by Zaks and Klibanov [11] and has been investigated thoroughly. Recent measurements of water molecules in apparently anhydrous enzymes have suggested the significance of a delicate balance between surface-bound and buried water essential to enzyme catalysis in organic media [12]. Studies on the protein conformation in organic solvents have shown structural integrity of the enzyme in these nonconventional conditions [13], although change in the hydrogen bond network [14] and different binding of an enzyme inhibitor [15] have also been observed. The effect of solvent characteristics and structure on the activity and enantioselectivity of enzymes has been the subject of several specific researches. However, in spite of several attempts to explain the solvent effect on

$$R_1COOH + R_2OH \rightleftharpoons R_1COOR_2 + H_2O$$

$$R_1COOH + R_2COOR_3 \rightleftharpoons R_1COOR_3 + R_2COOH$$

$$R_1COOR_2 + R_3OH \rightleftharpoons R_1COOR_3 + R_2OH$$

Scheme 1   Enzyme-catalyzed reactions in nonconventional media.

enantioselectivity of the process [16], a possible rationale of general validity is still lacking [17,18].

II. LIPASES: STRUCTURES AND PREDICTIVE MODELS

X-ray diffraction analysis of protein crystals has made three-dimensional structure available for structure–activity studies and for several lipases such data have already confirmed a catalytic triad of amino acids (Ser-His-Asp) in the active site of the enzymes [19]. It is worth mentioning that an enzyme crystal structure in neat organic solvent (acetonitrile) also is available [20]. The first structures of lipases appeared in 1990 [21,22] and at present many structural studies of lipases used in organic synthesis have confirmed that in the presence of organic solvents or lipid-like compounds conformational changes in the active site region expose the catalytic area to the solvent and to substrates [23–26]. Among the above structural studies, the three-dimensional structures of *Candida rugosa* [27–29] or *Pseudomonas cepacia* [30,31] lipases are of particular interest from an applications point of view. However, empirical rules to explain the stereochemical demand of lipases of synthetic utility are still of practical interest and a few of them are currently available [32–35]. For the *P. cepacia* lipase, a model of the active site based on molecular mechanics calculations also has appeared [36–38].

The volume of structural information and the high level of crystallographic studies [39] are becoming increasingly relevant to the rationalization of results, and predictive models try to be as accurate as possible. However, the general approach to the choice of an enzyme for practical purposes and the experimental protocols to be followed still rely on practice often based on structural analogies. In this chapter the names of enzymes will be reported as cited in the original papers, as it sometimes happens that different names correspond to the same enzyme.

III. LIPASES IN ORGANIC SOLVENTS: MODULATION OF ENANTIOSELECTIVITY

The enantioselectivity of an enzymatic reaction is usually expressed by the enantiomeric excess (e.e.) and several group studies focus on the experimental conditions that are finalized to reach the maximum e.e. In these cases, we will select the results that show the highest enantioselectivity of the process. The enantiomeric ratio E is a parameter introduced to express quantitatively the enantiopreference of an enzyme by the analysis of the kinetic data [40–42], and recent refinements of the calculation of E are based on new computer programs [43,44]. The enantioselectivity of an enzymatic reaction in organic solvents can be improved in different ways. The enzyme can be biologically modified via a site-directed mutagenesis but obviously this method requires techniques of molecular biology. The activity, stability, and enantioselectivity of an enzyme can be influenced by the mode of its preparation or the physical properties of the solid state [45]. Chemical modification of an enzyme can be obtained by several techniques of immobilization [46] and, comparing the results between free and immobilized enzymes [47,48], in some cases crosslinked lipases have been recognized as excellent biocatalysts for enantioselective hydrolysis [49] or for reactions in organic solvents [50,51]. Also noncovalent enzyme modifications can in some cases enhance the enantioselectivity of lipase-catalyzed reactions [52]. In many instances, "solvent engineering," i.e., careful selection of the most suitable solvent on the light of physicochemical properties, can contribute to obtain the highest enantioselectivity [53]. Also unusual organic solvents such as supercritical carbon dioxide can be used in lipase-catalyzed esterification reactions [54–56]. Esterification or transesterification activity and enantioselectivity can be influenced by temperature [57] or by additions of additives such as polar compounds [58,59] or crown ethers [60,61]. Addition of a ligand to an enzyme prior to its lyophilization could alter the enzyme active conformation, and the elimination of the ligand by washing does not remove the "ligand memory" from the enzyme active conformation in organic solvents [62]. This molecular bioimprinting has been applied to lipases [63,64] and to other hydrolases [65].

IV. LIPASE-CATALYZED INTERESTERIFICATIONS

The reaction of interesterification consists of an exchange of the carboxylate moiety of an ester and can be realized by reaction of an ester with a free acid (acidolysis) or reaction of two esters (Scheme 2). The interesterification process typically occurs among triglycerides and is often catalyzed by lipases [66,67]. As an example of a useful lipase-catalyzed interesterification, it has been reported (Scheme 3) that the exchange of butanoic acid in tributyrin with a polyunsaturated fatty acid may be carried out without solvent in the presence of *Candida antarctica* lipase [68].

Similar reactions can be realized in the presence of acetone powder from germinating rapeseed [69] or with immobilized lipase in supercritical carbon dioxide [70]. The application of interesterification reactions to enantioselective resolutions has found scarce examples. For instance, it has been reported that when racemic 2-(*p*-chlorophenoxy)propionic acid is used to exchange with acetic acid of the racemic acetates of a bicycloheptanol and 2-methoxycyclohexanol, optically active *p*-chlorophenoxy esters are obtained (Scheme 4). The enzymatic process is diastereoselective and doubly enantioselective for the acid and alcohols [71]. Another example of enantioselective interesterification in organic solvents leading to a nearly optically pure octanoate (Scheme 5) proceeds via acid exchange of the formate of a pyranyl alcohol with methyl octanoate [72].

$$R_1COOR_2 + R_3COOH \rightleftharpoons R_3COOR_2 + R_1COOH$$

$$R_1COOR_2 + R_3COOR_4 \rightleftharpoons R_1COOR_4 + R_3COOR_2$$

Scheme 2 Reactions of interesterification.

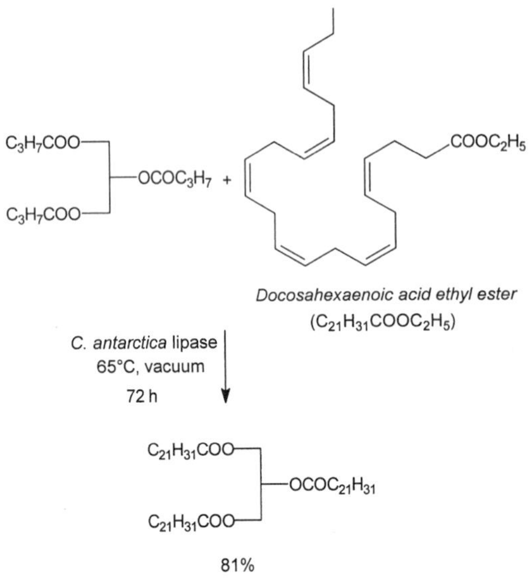

Scheme 3 Exchange of acids in a triglyceride by interesterification.

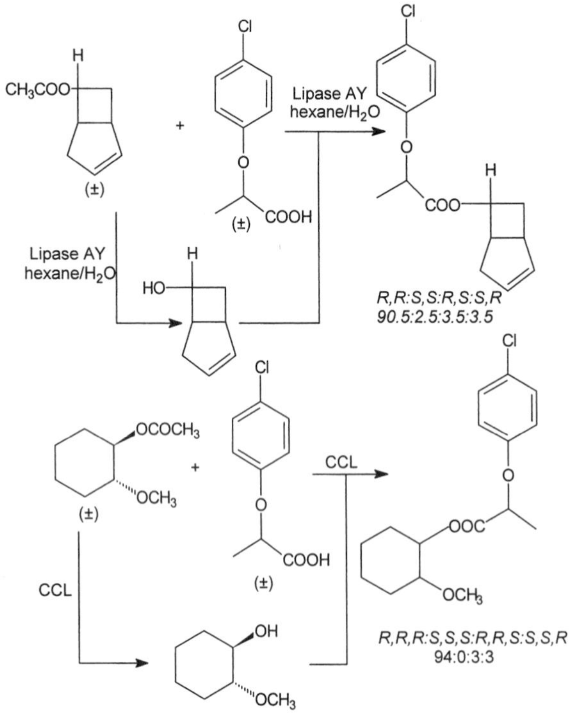

Scheme 4 Double enantioselective interesterification.

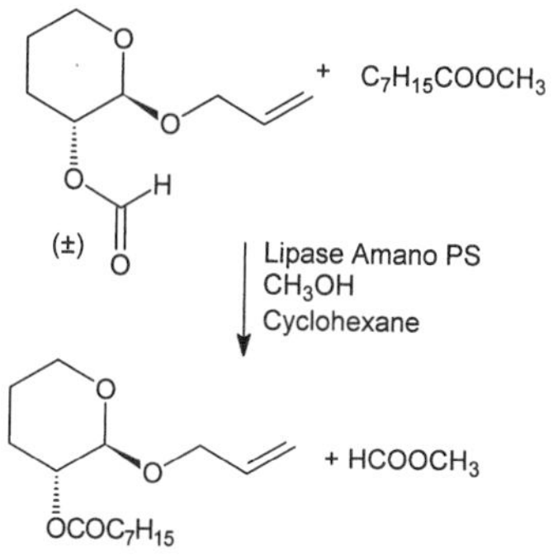

Scheme 5 An example of enantio:

V. LIPASE-CATALYZED ESTERIFICATIONS AND LACTONIZATIONS

The direct esterification reaction (Scheme 6) has found many applications in organic synthesis, although the esterification is in competition with the reverse reaction of hydrolysis [73]. The esterifications are more conveniently performed in nonaqueous media in order to displace the equilibrium in favor of the synthesis. In this case, parameters such as the amount of water and the nature of organic solvents have to be carefully considered [74]. Several 2-methyl alkanoic acids can be resolved using esterification with alcohols of various chain length, in the presence of various lipases (Scheme 7) such as immobilized *C. antarctica* lipase (more than 95% e.e. for both unreacted acid and ester) [75], *C. rugosa* lipase in cyclohexane (97.3% e.e. for the ester and 78% e.e. for the acid) [76], crosslinked *C. rugosa* lipase in heptane (99.5% e.e. for the ester and 89.4% e.e. for the acid) [77]. In some cases, the alcohol that reacts with the acid is the solvent of the reaction [76], but the esterification process may be carried out in organic solvents [75,77].

Highly enantioselective esterification of (*R,S*)-2-hydroxynonanoic acid with 1-butanol with *C. rugosa* lipase (CRL) in toluene in the presence of molecular sieves furnished the optically pure acid (Scheme 8) that has been used as a chiral synthon for the synthesis of the antimicrobial agent (*4E,7S*)-7-methoxytetradec-4-enoic acid [78].

The esterification of a racemic alcohol with lauric acid proceeds with high enantioselectivity only after imprinting with the optically active substrate and coating of the lipase with synthetic glycolipids [79]. In Scheme 9 are shown other examples of the same reaction that successfully applies to cyclic alcohols [80] and diols as well [81].

Esterification of racemic 2-phenoxypropanoic acids with arylalkanols can be doubly enantioselective for both compounds with *Candida cylindracea* lipase (CCL) in hexane [82]. The enantiomeric ratio *E* of acids and alcohols depend on the relative structures and, for instance, as shown in Scheme 10, a maximum value (*E* = 108) is reached for the *p*-chlorophenoxy acid when the alcohol counterpart is 1-phenyl-1-ethanol (*E* = 5.2).

The classical chemical reaction of acylation of alcohols with carboxylic acid anhydrides can also be carried out enantioselectively in the presence of enzymes [83]. This can be regarded as an example of esterification reaction displaced toward the products

$$RCOOH + R_1OH \rightleftharpoons RCOOR_1 + H_2O$$

Scheme 6 Esterification reaction.

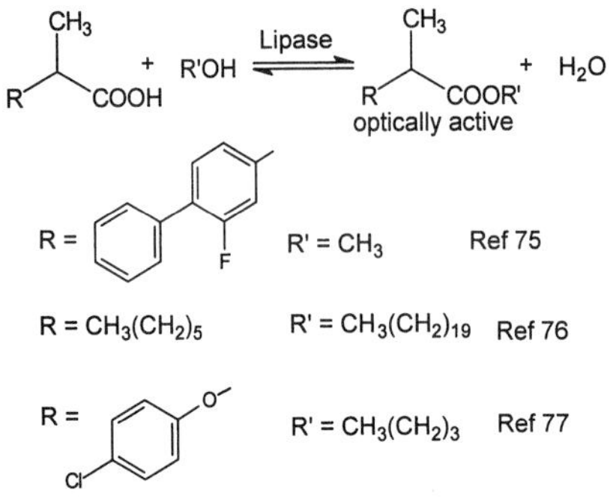

Scheme 7 Resolution of racemic 2-methylalkanoic and -arylalkanoic acids by esterification.

(Scheme 11). The reaction can be performed in organic solvents and with various acid anhydrides, as shown by a study on the factors influencing the enantioselectivity of the esterification of racemic menthol [84]. Several examples of enantioselective resolutions catalyzed by lipases with acetic anhydride [85,86] and other acid anhydrides [87,88] in organic solvents have been reported (Scheme 12).

When an anhydride such as succinic anhydride is reacted with a racemic alcohol in organic solvent with a lipase, an enantioselective resolution can be achieved [89]. The enantioselective opening of racemic or meso cyclic anhydrides can constitute a good method for the preparation of nearly optically pure esters [90–93]. Examples of these reactions are depicted in Scheme 13.

The enzymatic lactonization of hydroxy acids has been used mainly as a method for the synthesis of macrocyclic lactones [94,95] and the synthetic application of enzymatic lactonization for the purpose of enantioselective preparations refers almost exclusively to the cyclization of hydroxy esters.

VI. LIPASE-CATALYZED TRANSESTERIFICATIONS

Transesterification has been carried out in many laboratory and industrial applications involving synthesis of esters and is by far the most useful and exploited reaction used with enzymes in organic solvents [96]. Lipases are especially well suited to catalyze trans-esterifications in organic solvents [97], although other hydrolases have been studied for

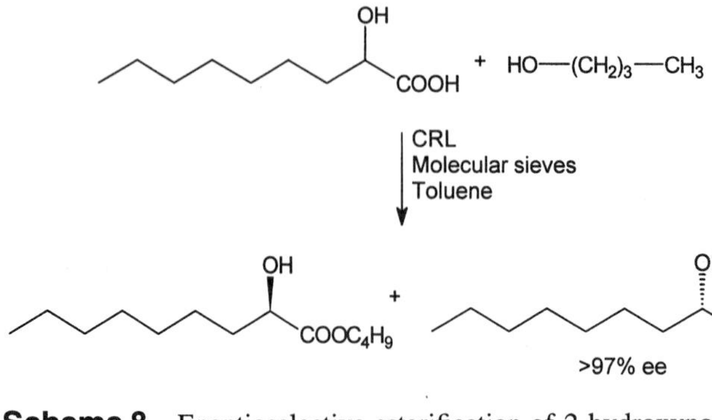

Scheme 8 Enantioselective esterification of 2-hydroxynonanoic acid.

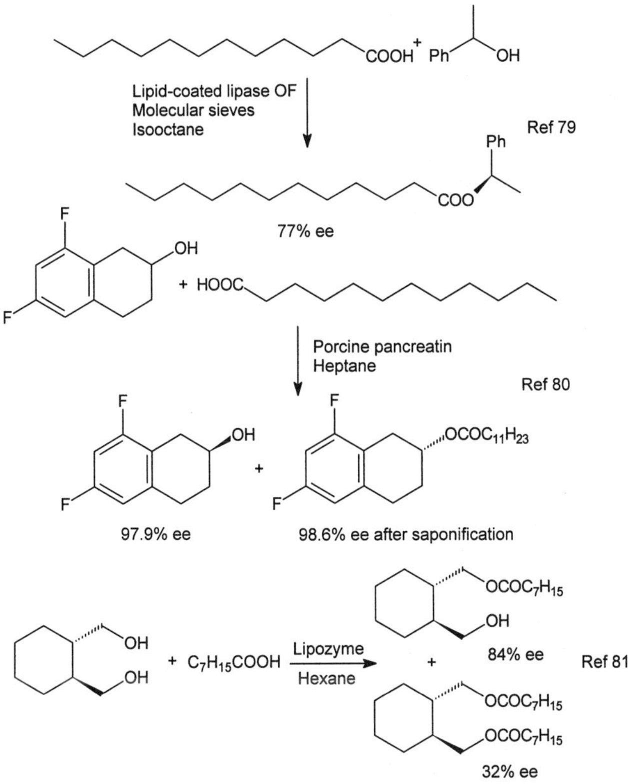

Scheme 9   Enantioselective esterifications with fatty acids.

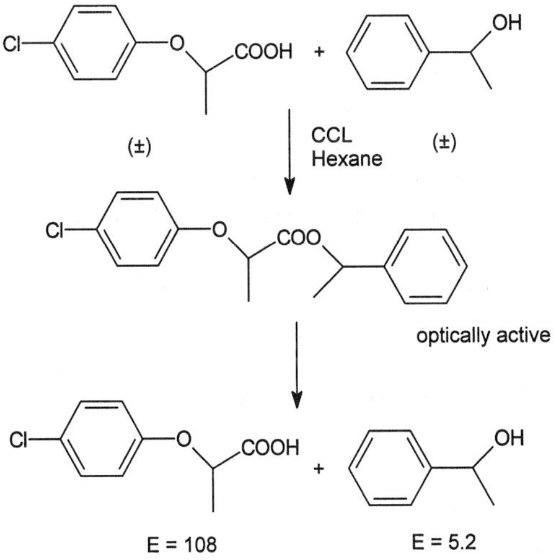

Scheme 10   Double enantioselective esterification.

$$ROH \ + \ (R_1CO)_2O \ \xrightarrow{\text{Enzyme}} \ ROCOR_1 \ + R_1COOH$$

Scheme 11   Esterification of alcohols with acid anhydrides.

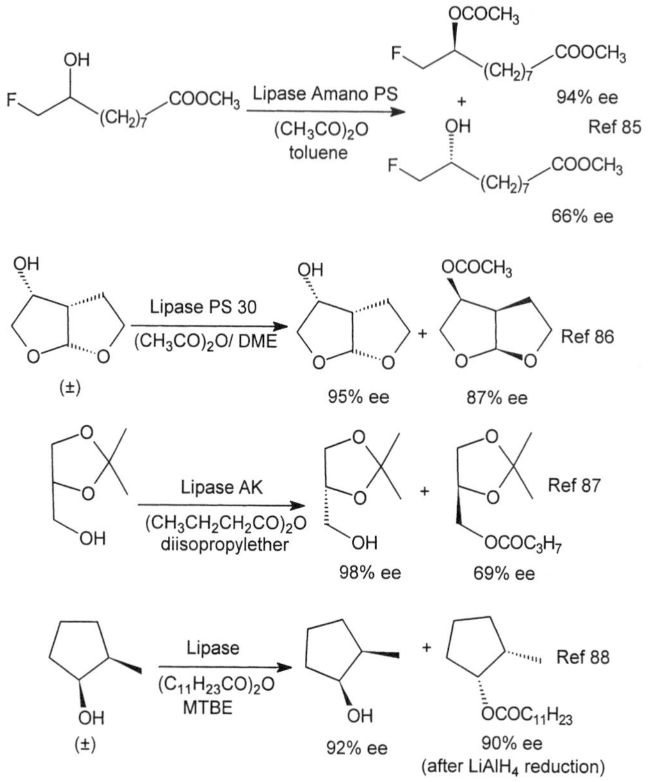

Scheme 12 Enantioselective esterification of hydroxy compounds with anhydrides.

this reaction [98,99]. The reaction of acylation of alcohols with ethyl acetate as solvent and reagent is a transesterification process that can be carried out enantioselectively in the presence of enzymes. This was one of the first methods used to carry out an enzymatic esterification in a nonconventional medium, and an interesting recent report (Scheme 14) shows the example of such applications [100].

Cyclization of hydroxy esters, i.e., lactonization, is an internal transesterification that can be carried out in the presence of lipases in organic solvents. The method, used for the synthesis of macrocyclic lactones [101], has been applied to the preparation of optically active lactones from racemic hydroxy esters [102]. This enzymatic transesterification has also been used to prepare enantiomerically pure hydroxy lactones from racemic dihydroxy esters [103–105], as shown in Scheme 15.

A typical transesterification procedure for the enzymatic enantioselective resolution of racemic esters uses an alcohol as 1-butanol in organic solvents. Depending on the lipase and the substrate, highly enantioselective preparations of optically active esters [106–108] can be achieved. The reaction applies to the resolution of ester of racemic acids, which by reaction with 1-butanol are converted to butyl esters (Scheme 16). The lipase-catalyzed alcoholysis of acetoxy groups with 1-butanol constitutes a simple, useful method for the preparation of enantiomerically pure compounds [109–116]. A few examples of this method have been selected and shown in Scheme 17. The regio- and enantioselective

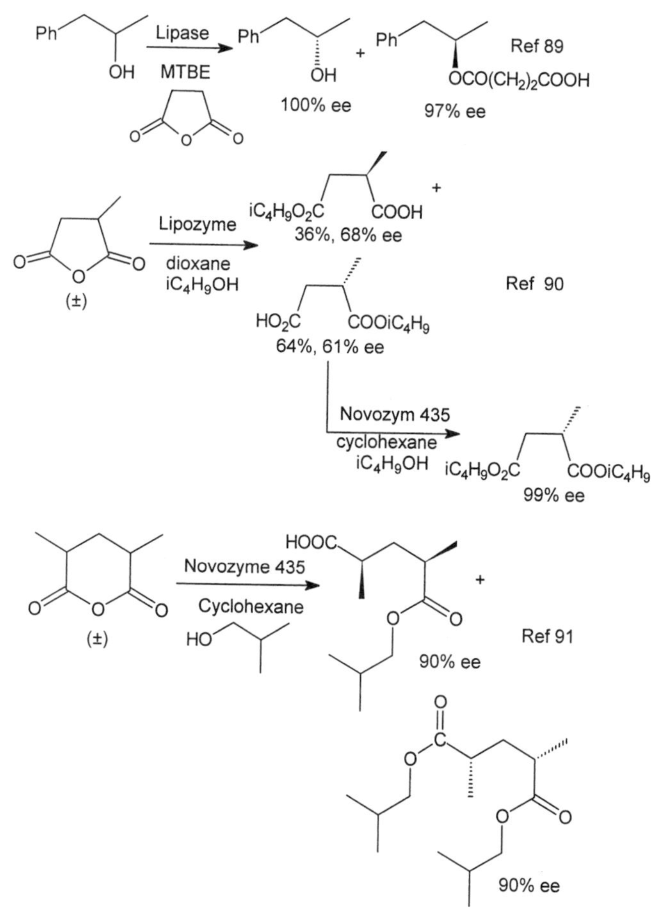

Scheme 13   Enantioselective opening of cyclic anhydrides.

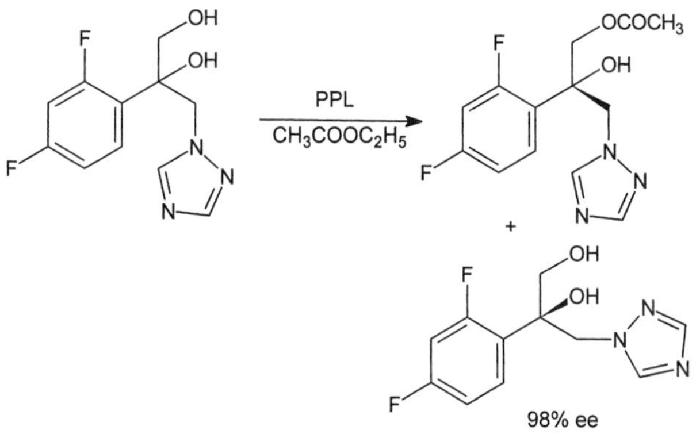

Scheme 14   Ethyl acetate as reagent and solvent of an enantioselective transesterification.

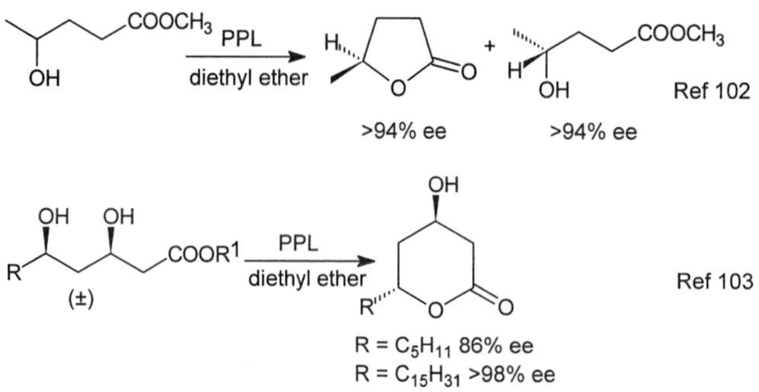

Scheme 15 Enantioselective lactonization of hydroxy esters.

opening of racemic lactones by means of alcohols in the presence of lipases with or without organic solvents is another example of transesterification (Scheme 18) that leads to esters or unreacted lactones of high optical purity [117,118].

The alcoholysis of 2-phenyloxazolin-5-one has been recently reported as an interesting enzymatic reaction for opening of lactones (Scheme 19). This compound, by reaction with N-ethoxycarbonyl 2-amino-1-butanol in the presence of a lipase in organic solvent, yields an optically active product with moderate enantioselectivity [119].

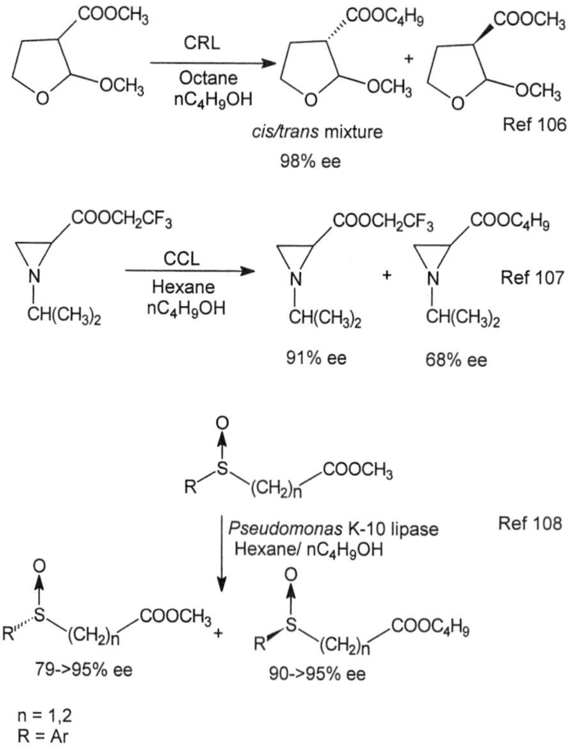

Scheme 16 Enantioselective transesterification of esters with 1-butanol.

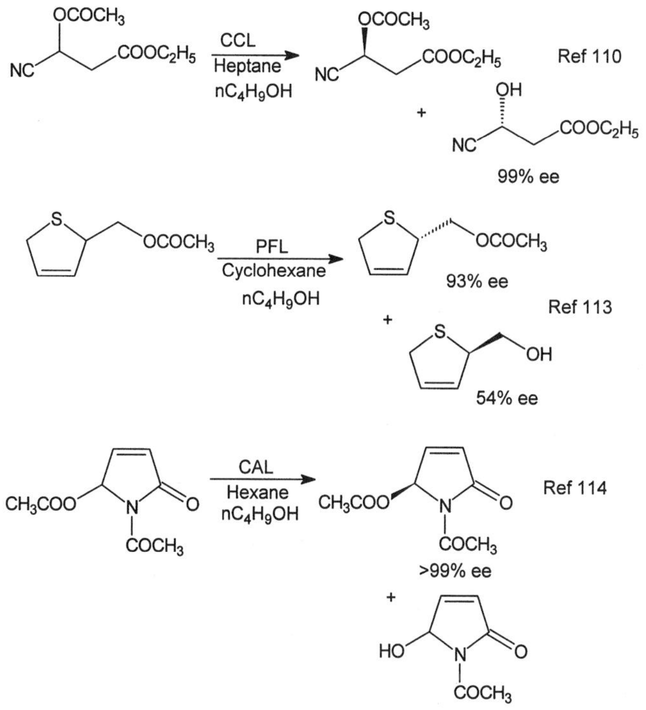

Scheme 17 Enantioselective alcoholysis of acetoxy groups.

A. Lipase-Catalyzed Irreversible Transesterifications

The lipase-catalyzed transesterifications described earlier are in general equilibrium reactions and a few methods have been proposed to make the reaction irreversible or to slow down the back reaction [120,121]. One approach consists of the use of trifluoroethyl esters as acylating reagents (Scheme 20). In this way trifluoroethanol is produced and the reverse reaction is slow, as the fluorinated alcohol is less nucleophile than ethanol [122].

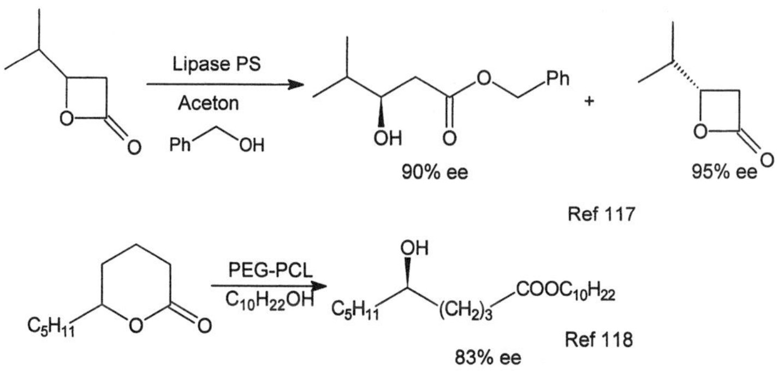

Scheme 18 Enantioselective opening of racemic lactones with alcohols.

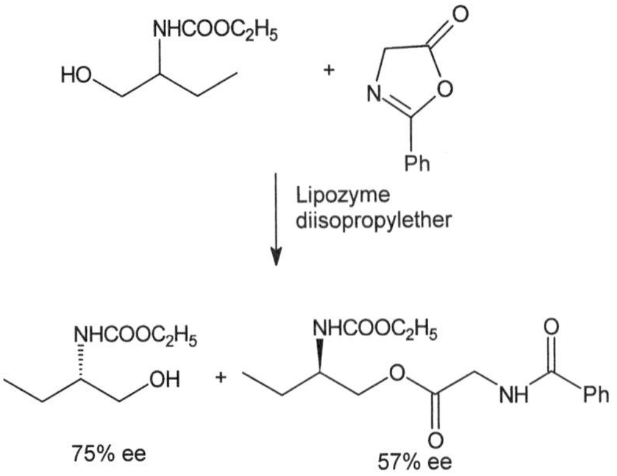

Scheme 19 Enantioselective alcoholysis of 2-phenyloxazolin-5-one.

Removal of products may be an experimental procedure adopted to shift the equilibrium of the transesterification to the forward reaction [123–125]. A similar approach consists of transesterification of hydroxy groups with ethyl thiooctanoate in the presence of *C. antarctica* lipase (Scheme 21). By this method the enantioselective resolution of alcohols can be achieved [126,127].

The lipase-catalyzed resolution of chiral acids can be accomplished via the reaction of their mixed carboxylic-carbonic anhydrides [128]. This irreversible acyl transfer reaction may proceed in a highly enantioselective manner in some cases [129], as shown in Scheme 22.

B. Lipase-Catalyzed Irreversible Transesterifications by Means of Vinyl Esters and Similar Reagents

The most satisfactory method to carry out an irreversible transesterification is the reaction of acylation of an alcohol with vinyl acylates [130,131]. In this reaction the back reaction is prevented by the irreversible tautomerization of vinyl alcohol to acetaldehyde. This latest product could cause the inhibition of the enzyme that has been immobilized to overcome this complication [132]. In some studies, however, a few cycles of reactions could be performed without affecting the enantioselectivity of the reaction [133]. Also oxime esters have been proposed as acyl transfer agents [134] for irreversible enzymatic transesterifications (Scheme 23).

Recently, 1-ethoxyvinyl acetate has been proposed as a novel and reliable acyl donor for the enzymatic resolution of alcohols (Scheme 24) in a lipase-catalyzed irreversible transesterification procedure [135]. A similar protocol has been utilized for the irreversible acetoacetylation of alcohols with diketene in the presence of lipases in organic solvents [136,137] and the reaction has been reported to proceed with high enantioselectivity [136], as shown in Scheme 25.

$$R_1\!-\!COOCH_2CF_3 \; + \; R_2OH \; \rightleftharpoons \; R\!-\!COOR_2 \; + \; CF_3CH_2OH$$

Scheme 20 Trifluoroethyl esters as acyl transfer reagents.

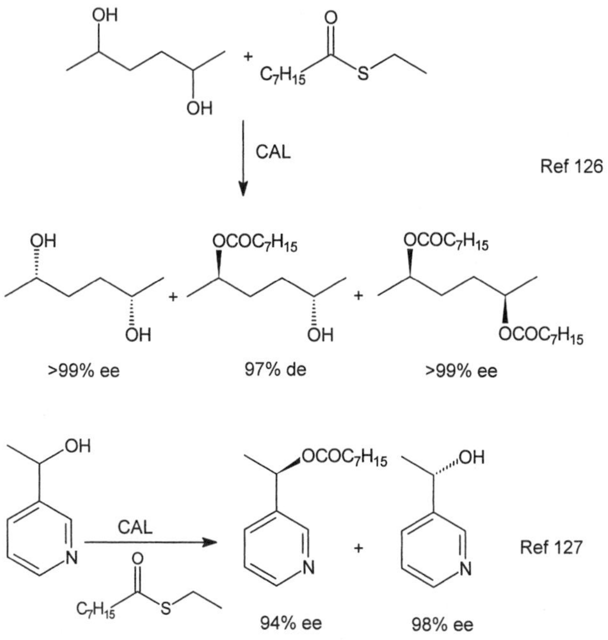

Scheme 21 Irreversible transesterifications with ethyl thiooctanoate.

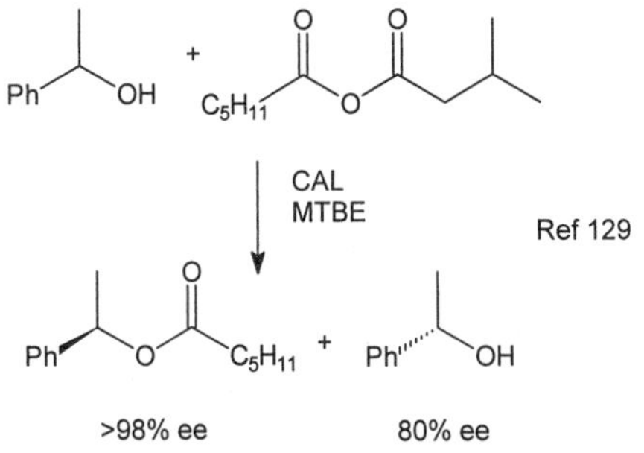

Scheme 22 Irreversible acyl transfer reaction by mixed anhydrides.

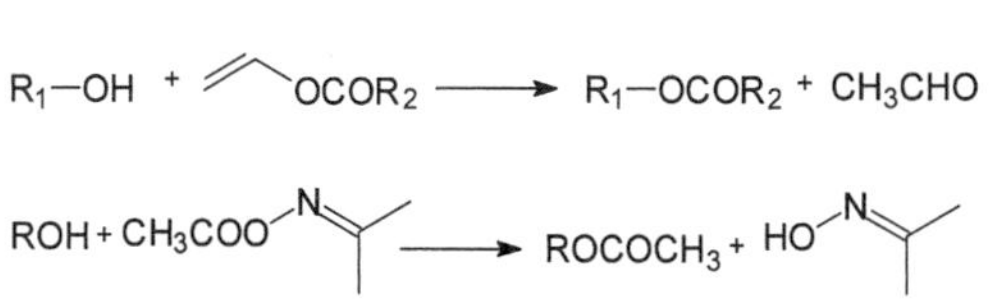

Scheme 23 Vinyl and oxime esters as reagents for irreversible transesterifications.

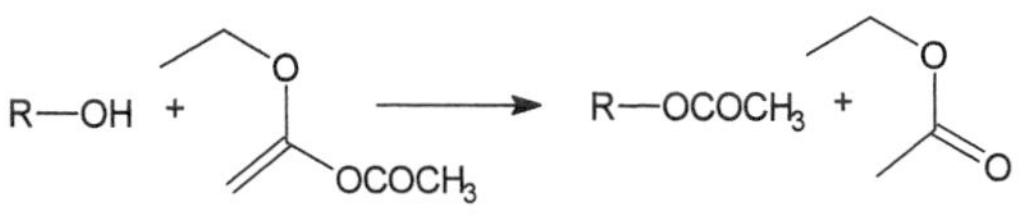

Scheme 24 1-Ethoxyvinyl acetate, a new acyl donor for irreversible transesterifications.

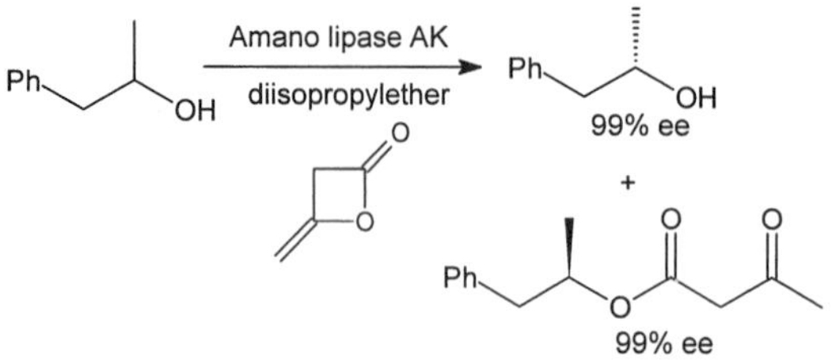
Amano lipase AK
diisopropylether
Ph
OH
O
O
99% ee
+
99% ee

Scheme 25 Enantioselective acetoacetylation of alcohols with diketene.

In conclusion, although various methods may be adopted for the enzymatic reactions, mainly vinyl acetate and a few other vinyl esters have been used as acyl transfer reagents and the applications are so numerous as to deserve a detailed report.

1. Lipase-Catalyzed Irreversible Transesterification of Primary Alcohols

2-Methyl-1-alkanols are excellent substrates for the lipase-catalyzed enantioselective transesterification with vinyl esters, mainly acetate or isopropenyl, and the best results have been obtained in organic solvents such as chloroform or dichloromethane [1a], as shown in Scheme 26. Many other examples of resolution of 2-methyl-1-alkanols have been reported (Scheme 27) and the results confirmed the enantioselectivity of the reaction and a constant stereochemical outcome that may lead to the unreacted (*R*)-alcohol and (*S*)-acetate [138–140].

Only a few reports on the resolution of 2-ethyl-1-alkanols have appeared in the literature [141,142], and 2-ethyl-1-hexanol is the substrate for the enantioselective enzymatic transesterification (Scheme 28). Also, when primary alcohols present benzamido or *t*-butylsulfonyl groups at position 2, the enzymatic resolution may be enantioselectively carried out in the presence of lipases [143,144], as shown in Scheme 29.

The enzymatic method can be applied to monocyclic primary alcohols, and the enantiomeric excess of the acetate and of the unreacted alcohol is generally high [145,146], as shown in Scheme 30. An interesting substrate is constituted by *cis*-1-diethylphosphonomethyl-2-hydroxymethylcyclohexane that contains a phosphonate ester in addition to the hydroxy group that is enzymatically acylated [147].

Dioxolane and dioxane methanols are interesting building blocks for asymmetrical synthesis of a great variety of chiral compounds because they are masked glycerol analogs with the structure of cyclic alcohols. The enzymatic transesterification of these compounds

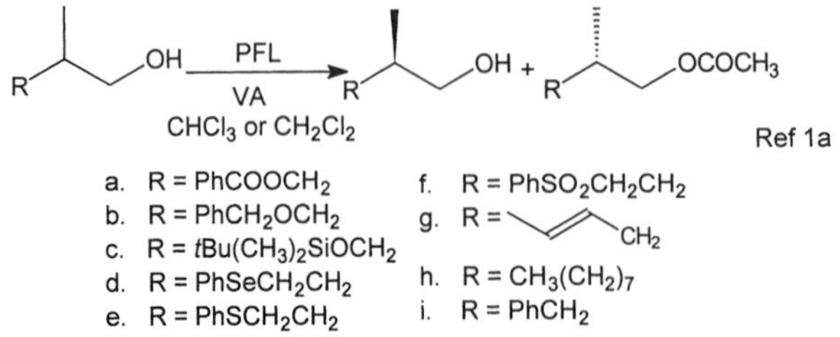
R
OH
PFL
VA
CHCl3 or CH2Cl2
OH +
OCOCH3
Ref 1a
a. R = PhCOOCH2
b. R = PhCH2OCH2
c. R = tBu(CH3)2SiOCH2
d. R = PhSeCH2CH2
e. R = PhSCH2CH2
f. R = PhSO2CH2CH2
g. R =
CH2
h. R = CH3(CH2)7
i. R = PhCH2

Scheme 26 Enantioselective resolution of 2-methyl-1-alkanols by irreversible transesterification.

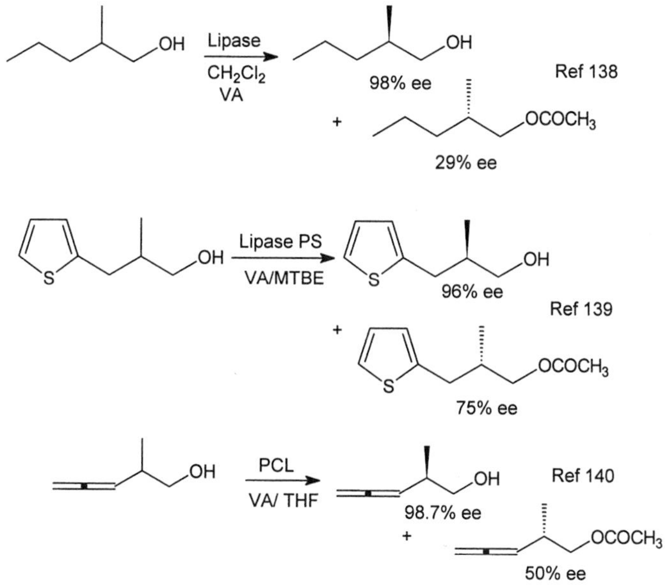

Scheme 27 Enantioselective resolution of 2-methyl-l-alkanols.

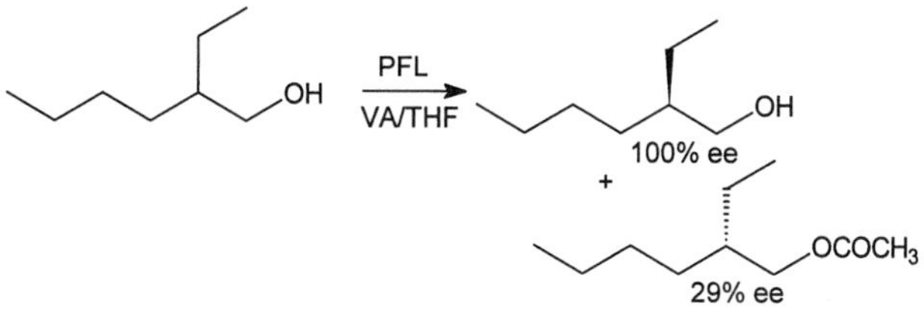

Scheme 28 Resolution of 2-ethyl-1-hexanol.

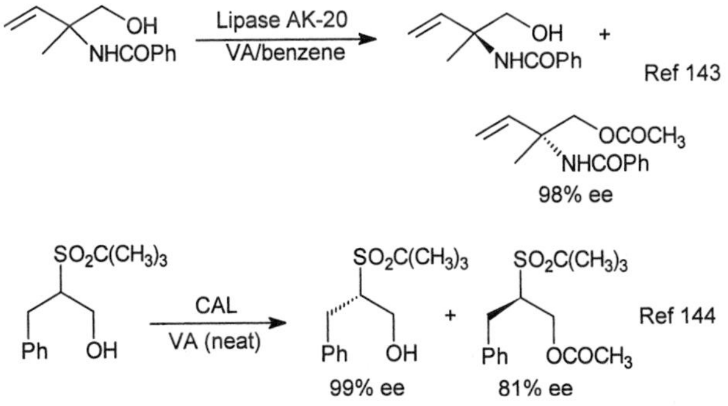

Scheme 29 Resolution of 2-substituted-1-alkanols.

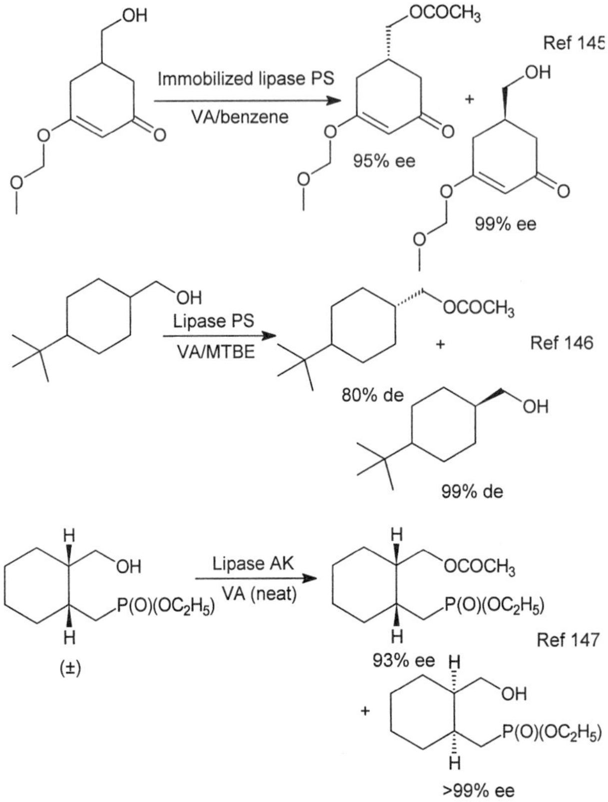

Scheme 30 Resolution of cyclic primary alcohols.

(Scheme 31) constitutes a viable method for the preparation of valuable enantiomerically pure chirons [148–153].

Also, bicyclic alcohols seem to be good substrates for the enantioselective resolution of stereogenic centers that present primary alcohols as reacting group of the enzymatic reaction [154–157]. The method also applies to cyclic hydroxymethyl aza compounds, and nearly optically pure compounds can be obtained in this way [158,159]. In Scheme 32 a few examples of enantioselective enzymatic resolutions of bicyclic compounds are shown.

The same resolution can be efficiently carried out on structurally more complex compounds that are optically active for the helical structure of the molecule, i.e., bis(hydroxymethyl) thiaetherohelicene [160]. In this way, a nearly optically pure helical molecule can be obtained (Scheme 33). This method has been successfully applied also to resolution of organometallic compounds and a few interesting results are shown in Scheme 34. Tricarbonyl [(η(6)-cycloheptatriene)chromium(0)] and ferrocenyl alcohols have been examined as substrates of the lipase-catalyzed acylation, and the reaction can be accomplished in a highly enantioselective manner [161,162].

The classic enzymatic resolution of the simplest epoxy alcohol, i.e., glycidol, reported a few years ago [163], later was extensively studied using pig pancreas lipase (PPL) and *Pseudomonas fluorescens* lipase (PFL) as biocatalyst and different solvents [164]. A

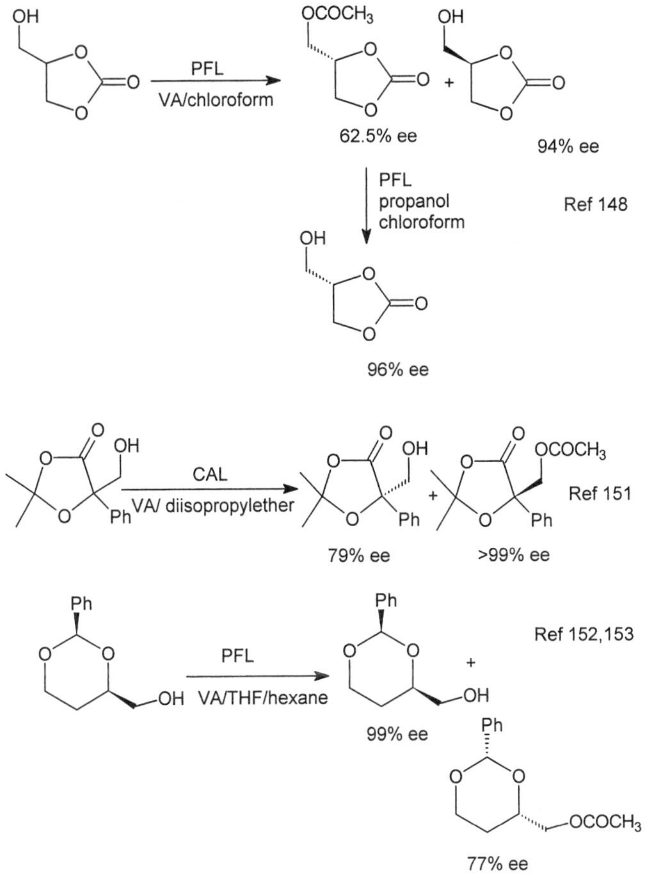

Scheme 31 Resolution of cyclic masked glycerols.

few other oxirane methanols were enantioselectively resolved by transesterification in organic solvents [165–167] or in aqueous medium as well [168], as depicted in Scheme 35.

Successful applications of the transesterification process have also been reported for primary alcohols bearing the stereogenic center at the position 3 [169–171], as shown in Scheme 36. Note that a modest enantioselectivity was observed for the enzymatic hydrolysis of similar compounds [172].

An interesting example of enzymatic resolution via the transesterification process has been reported for a racemic sulfoxide bearing the stereogenic center three carbons distant from the hydroxy group that is supposed to react at the enzymatic catalytic center [173]. The enantioselectivity of the process is especially high for the unreacted substrate (Scheme 37).

2. Lipase-Catalyzed Irreversible Transesterification of Secondary Alcohols

The stereochemical outcome of the lipase-catalyzed resolution of primary alcohols is somehow less predictable than the same enzymatic process carried out on secondary alcohols. For these substrates, thanks to the great number of results that have been reported, a more

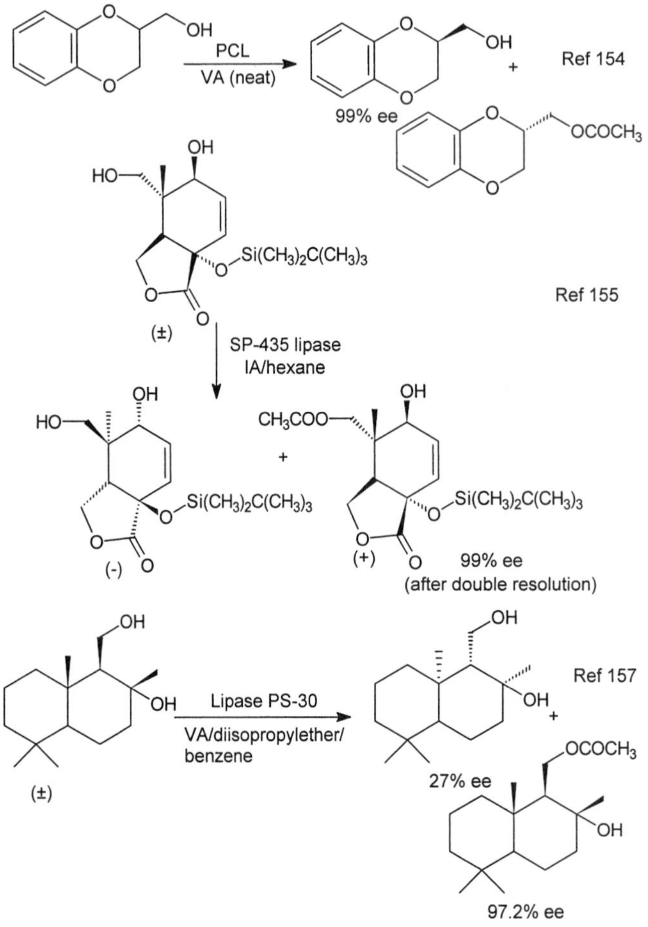

Scheme 32 Resolution of bicyclic primary alcohols.

general rule has been established for the prediction of the configuration of the resulting esters and of the unreacted alcohol from a lipase-catalyzed process [174]. The configuration of the alcohol that is preferentially acylated is shown in Scheme 38.

For the above reasons, secondary alcohols are frequently used as model substrates for studies devoted to the optimization of the enantioselectivity of the enzymatic process when immobilized lipases are used as biocatalysts [50]. This applies also to the choice of the most suitable solvent [175], or to the activity of new lipases [176], or to variations on the irreversible transesterification methodology [135,177]. The highly enantioselective resolution of racemic secondary alcohols has been reported for aliphatic and aromatic substrates [178–181] and few results are reported in Table 1 and Scheme 39.

The enantioselective esterification of unsaturated secondary alcohols has been extensively studied [182] and has found several successful additional applications with substrates containing double bonds [183–188]. The results have been collected in Table 2 and Scheme 40, whereas Scheme 41 shows interesting examples of the resolution of hydroxy compounds that contain a triple bond in the molecule [189,190].

Fluorine-containing alkanols are efficiently resolved by means of the lipase-catalyzed transesterification, fluorine being present as trifluoromethyl [191–193] or difluoro [194]

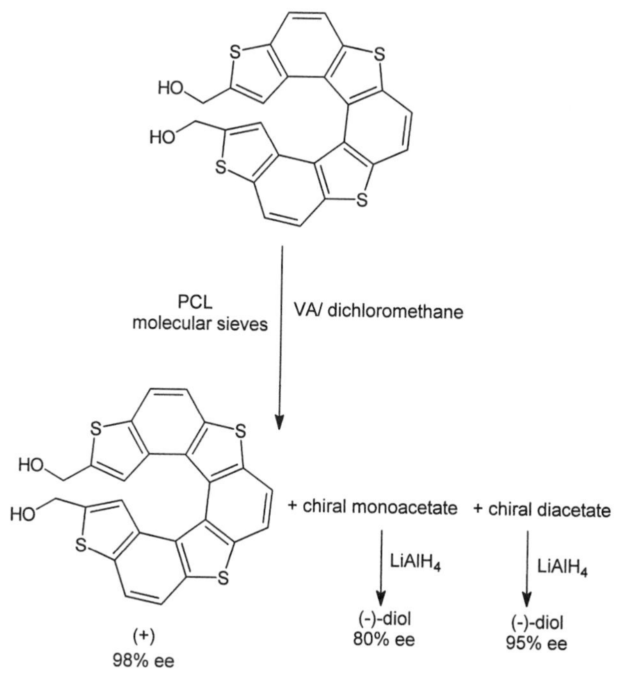

Scheme 33 Resolution of a helical compound.

moieties. As shown in Scheme 42, fluoroaromatic compounds [196] are also good substrates for the enzymatic process.

Many other hydroxy compounds bearing a secondary hydroxy group can be enantioselectively acylated. In the case of a few cyanohydrins, optically active acetates are prepared in a one-pot procedure from the aldehyde and acetone cyanohydrin, followed by a lipase-catalyzed transesterification [196]. Also, 2-hydroxy acids and esters [197–201] can be enzymatically resolved, as shown in Scheme 43.

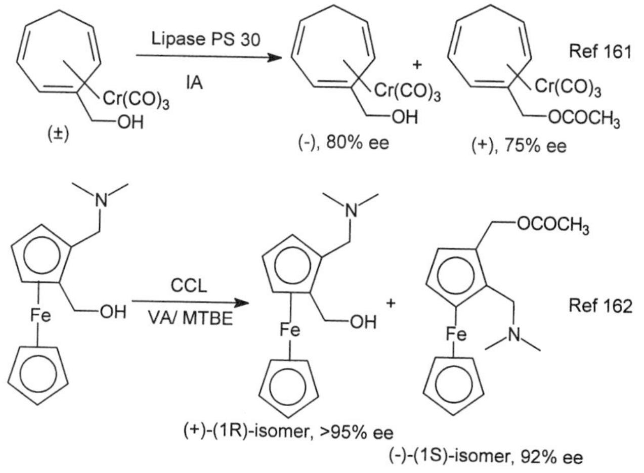

Scheme 34 Resolution of organometallic primary alcohols.

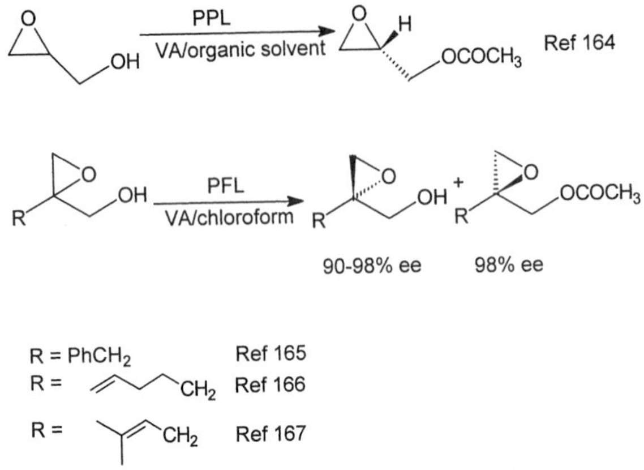

R = PhCH$_2$ Ref 165
R = ⟍⟋⟍CH$_2$ Ref 166

R = ⟍⟋⟍CH$_2$ Ref 167

Scheme 35   Resolution of oxirane methanols.

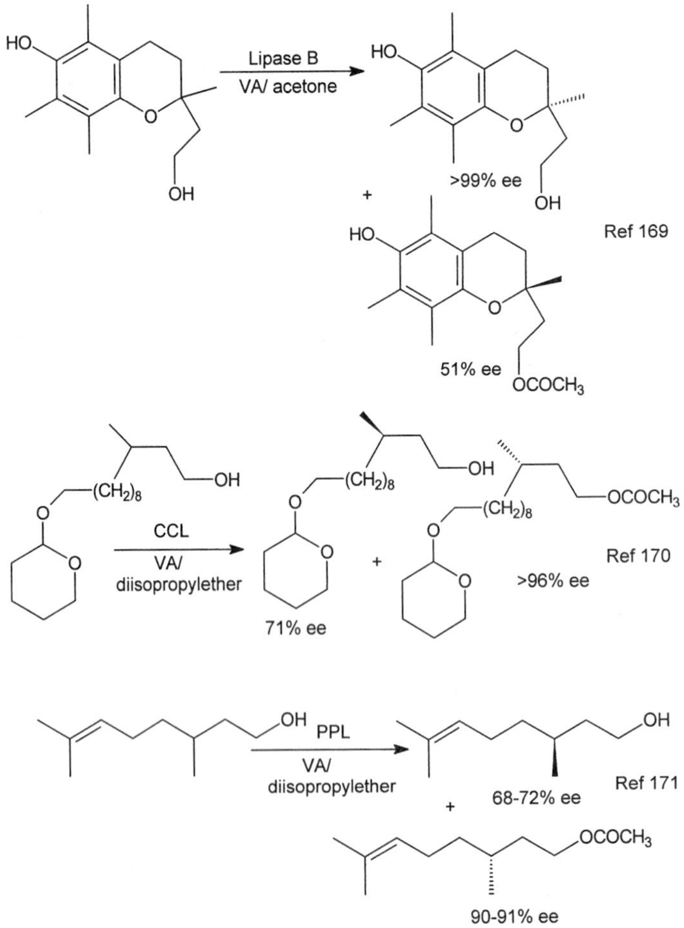

Scheme 36   Resolution of 3-substituted primary alcohols.

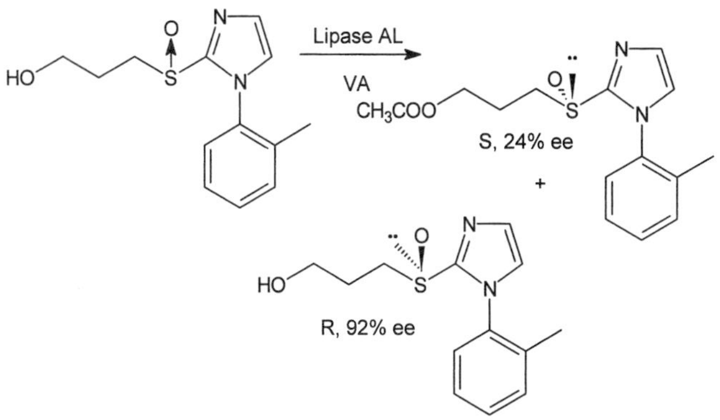

Scheme 37 Resolution of a racemic sulfoxide via acylation of a primary alcohol moiety.

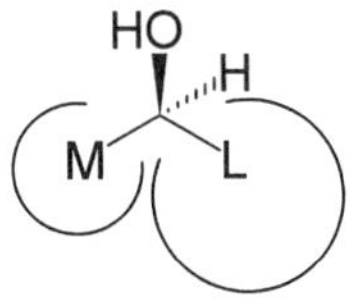

Scheme 38 Configuration of the enantiomer preferentially acylated by lipases (M, medium size substituent, L, large substituent).

Table 1 Lipase-Catalyzed Resolution of Secondary Alcohols

			Products		
			Alcohol	Acetate	
Substrate	Enzyme	Reaction conditions	(config. % e.e.)	(config. % e.e.)	Ref.
1	PFL	VA/CH$_2$Cl$_2$	S, 76	R, 74	175
2	PPL	VA/diisopropyl ether	S, 91.8	R, 81	178
3	CAL	VA/diisopropyl ether	S, 93	R, 97	179
4	Lipozyme	VA/carbon tetrachloride	S, 99	R, 87	180
5	Lipase PS	IA/heptane	S, 99	R, 99	181

Note: Refer to section "Abbreviations" preceding the references.

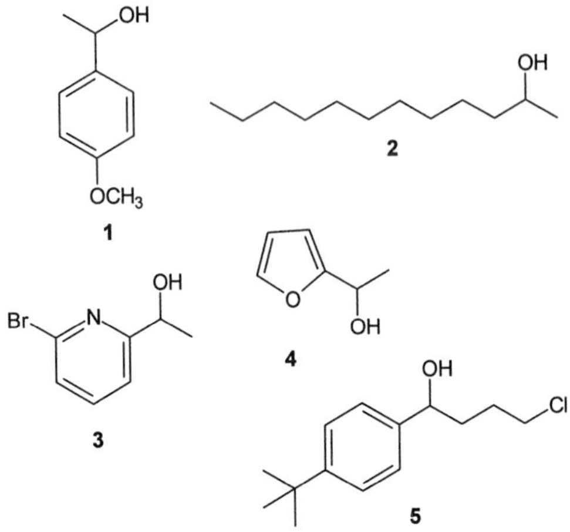

Scheme 39   Secondary alcohols as substrates for irreversible transesterifications.

Table 2   Resolution of Unsaturated Secondary Alcohols

| | | | Products | | |
| | | | Alcohol | Acetate | |
Substrate	Enzyme	Reaction conditions	(config. % e.e.)	(config. % e.e.)	Ref.
1	Lipase PS	VA/hexane	*S*, 99.9	*R*, 52	183
2	Lipase P	VA/pentane	*S*, 98	*R*, 98	184
3	*Ps.* lipase AK	VA/pentane	*S*, >98	*R*, >98	185
4	PPL	VD/toluene	*S*	*R*	186
5	Lipase PS	VA/diisopropyl ether	*S*, 98	*R*, >99	187
6	Lipase PS	VA	*S*, 96	*R*, 85	188

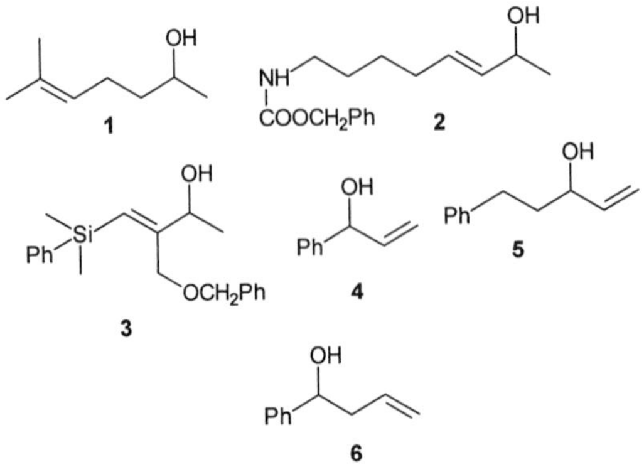

Scheme 40   Resolution of unsaturated secondary alcohols.

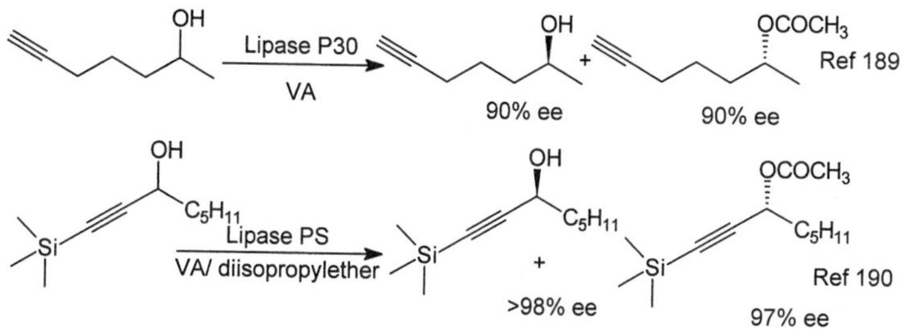

Scheme 41 Resolution of triple-bond-containing hydroxy compounds.

The same successful procedure applies to other hydroxy compounds such as hydroxy lactones [202], hydroxy amides [203], and hydroxy phosphonic acids [204,205]. The results of these enantioselective resolutions are presented in Scheme 44.

3. Lipase-Catalyzed Irreversible Transesterification of Cyclic Secondary Alcohols

Numerous resolutions of cyclic alcohols by means of the irreversible enzymatic transesterification procedure have been reported in the recent literature. The cyclic substrates

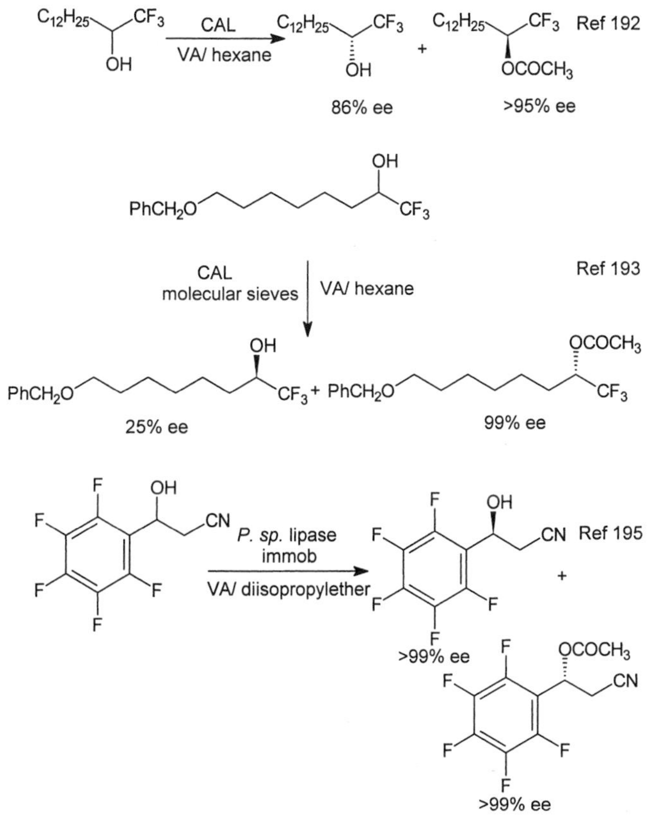

Scheme 42 Resolution of fluorine-containing secondary alcohols.

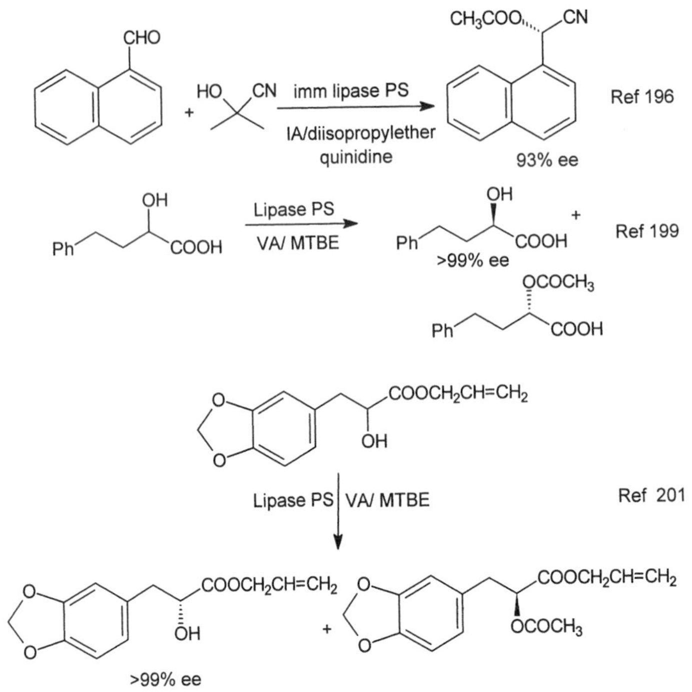

Scheme 43 Resolution of hydroxy compounds.

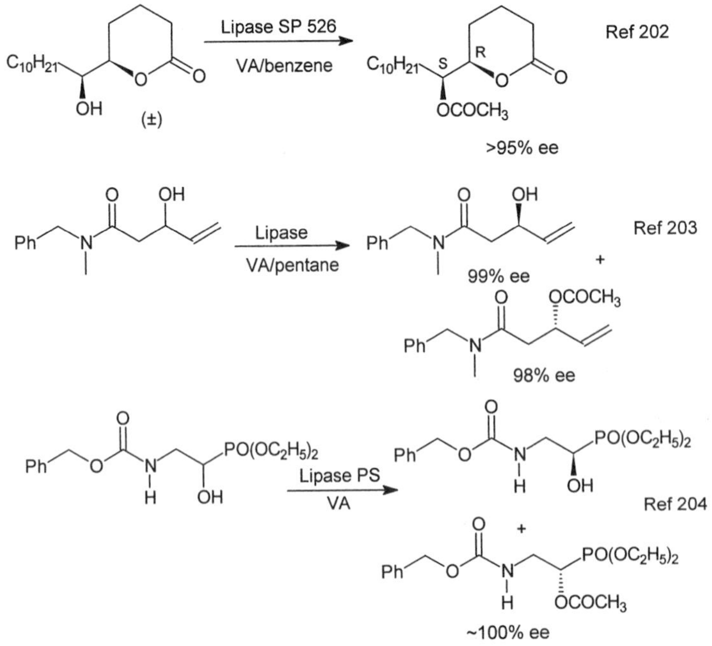

Scheme 44 Resolution of hydroxy lactones, hydroxy amides, and hydroxy phosphonamides.

include cyclopentyl alcohols that contain various groups in addition to the secondary hydroxy functionality [206–213]. Several examples of these enantioselective procedures are shown in Schemes 45 and 46.

A special group of cyclic hydroxy compounds is constituted by γ-hydroxy lactones [214–217] and cyclic semiacetals [218]. For these sensitive and relatively unstable compounds, the experimental conditions of the enzymatic process seem especially attractive and well suited to obtain efficient and enantioselective resolution of racemic substrates (Scheme 47).

The list of cyclic alcohols that are substrates for enantioselective lipase-catalyzed transesterification is very long. Among these are many cyclohexane derivatives [219–226]. Scheme 48 presents only selected examples of this class of substrate. Note that the ethynyl cyclohexenol that is enzymatically [222] resolved is a chiral intermediate for the total synthesis of vitamin D.

Other cyclic compounds have been resolved by the enzymatic process. For instance, the racemic 5-piperidein-3-ol that resolved in this way produced the optically pure (S)-alcohol, an intermediate for the synthesis of an indole alkaloid [227]. The same procedure successfully applies to a seven-member triol acetonide that is resolved with very high enantioselectivity [228]. The two results are shown in Scheme 49.

The efficient enzymatic resolution of racemic cyclic secondary alcohols can be extended to several bicyclic alcohols of different structures. These can be tetralols, indanols,

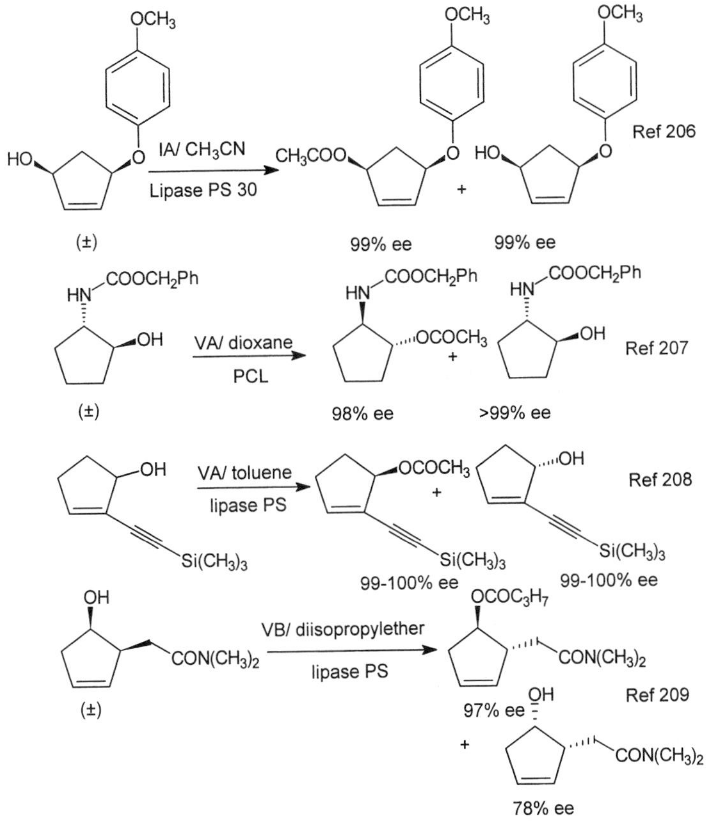

Scheme 45 Resolution of cyclopentyl secondary alcohols.

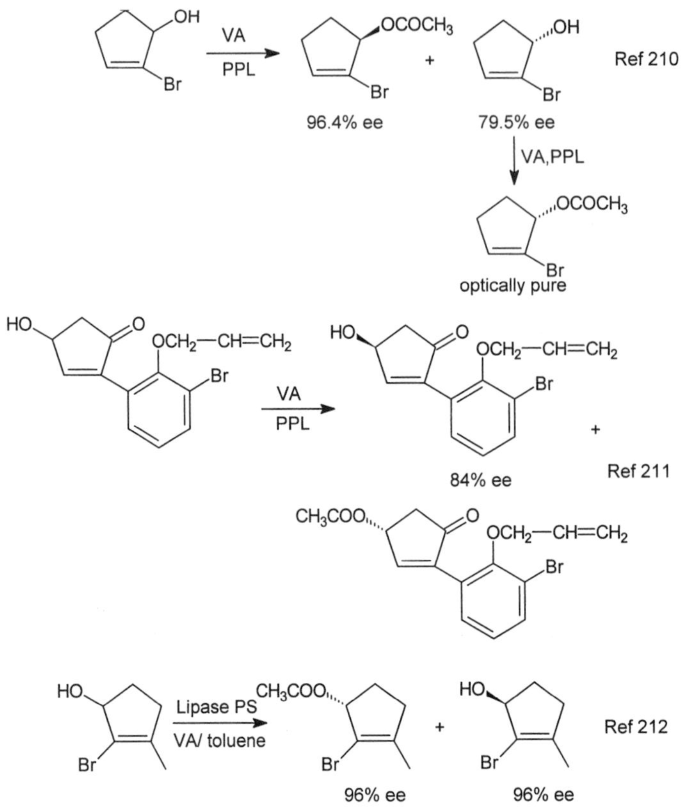

Scheme 46 Resolution of cyclopentyl secondary alcohols.

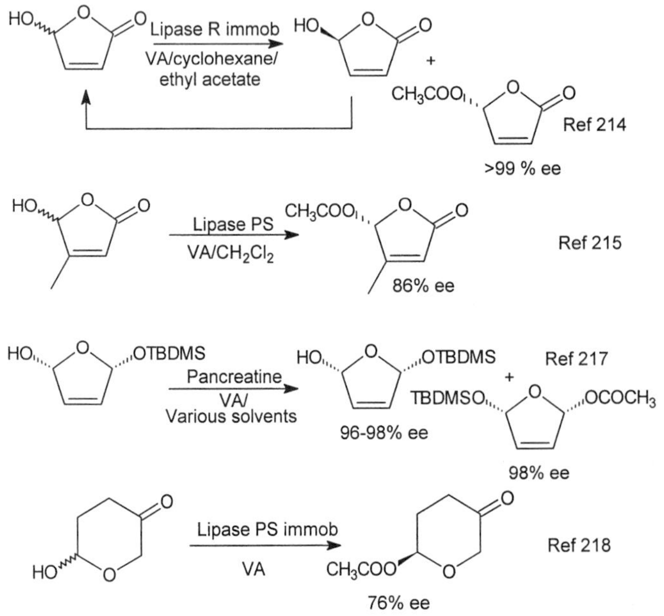

Scheme 47 Resolution of hydroxy lactones and semiacetals.

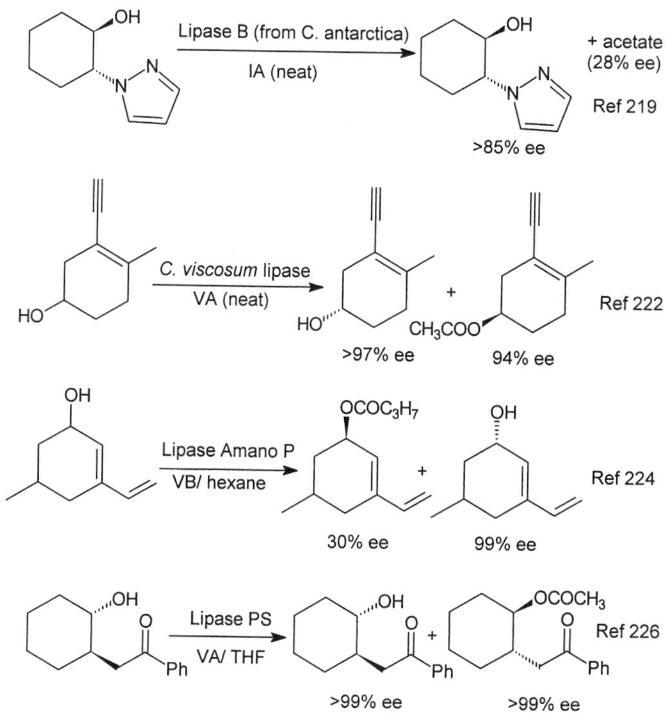

Scheme 48 Resolution of cyclohexyl secondary alcohols.

or chromanols [229–233] that afford enantiomerically pure products by procedures as shown in Scheme 50.

As for other bicyclic compounds, the resolution of a few bicyclo[3.3.0]octanols [234,235] proceeds with high stereoselectivity. By an efficient enzymatic process, a similar bicyclic structure can be obtained optically pure as a chiral intermediate for the synthesis of *d*-biotin [236]. The results are as shown in Scheme 51.

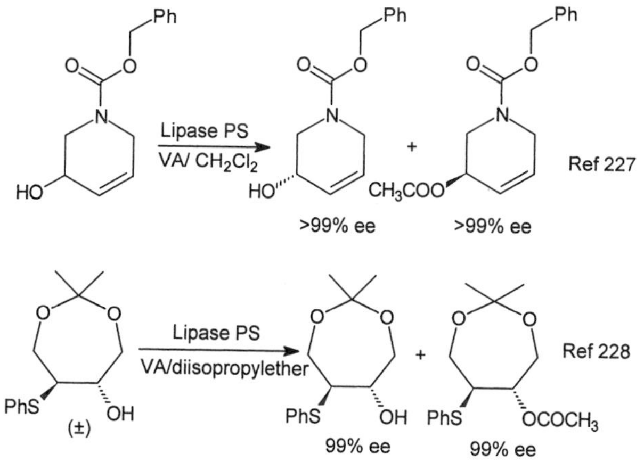

Scheme 49 Resolution of heterocyclic secondary alcohols.

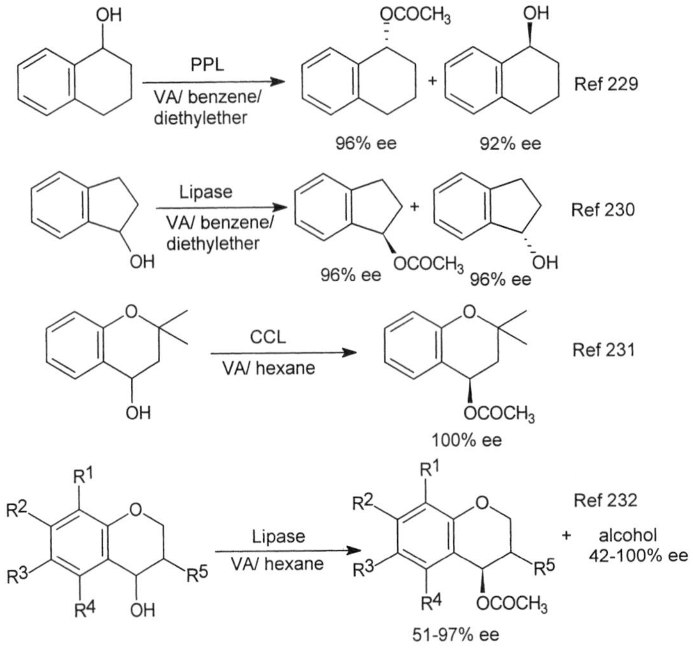

Scheme 50   Resolution of bicyclic secondary alcohols.

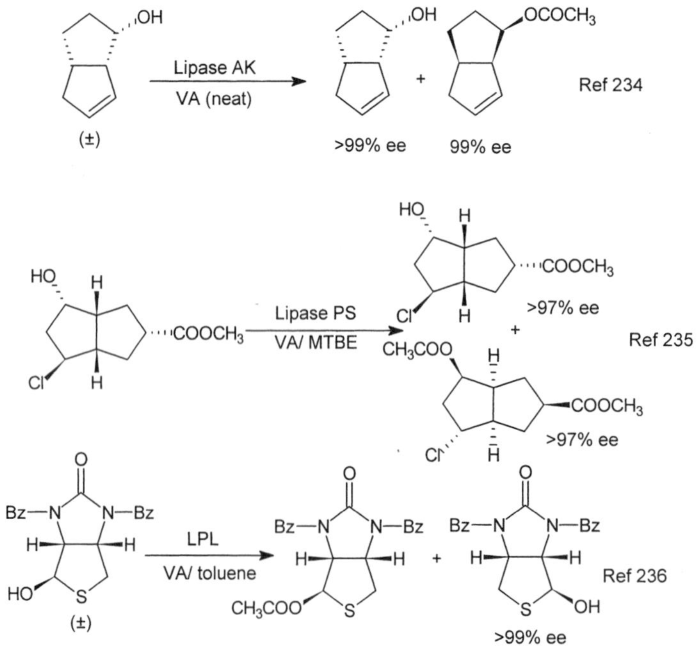

Scheme 51   Resolution of bicyclo[3.3.0]octanols and related compounds.

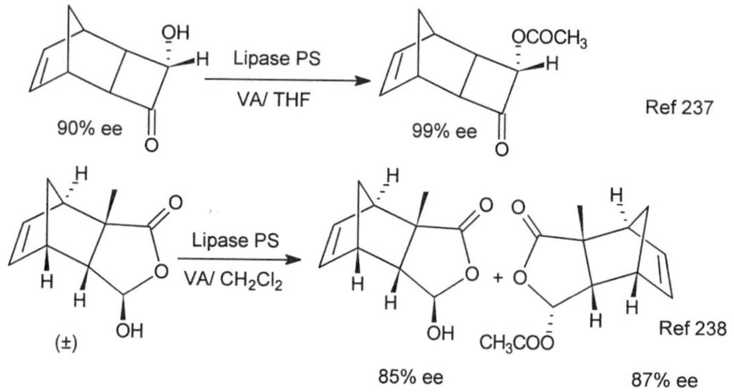

Scheme 52 Resolution of tricyclic hydroxy compounds.

Finally, the more complicated structures of tricyclic compounds that present the hydroxy ketone and hydroxy lactone moieties can be substrates for the enantioselective resolution process and some examples are presented in Scheme 52 [237,238].

(*a*) *Lipase-Catalyzed Irreversible Transesterification of Racemic Diols.* A great number of dihydroxy compounds has been examined as substrates for the enzyme-catalyzed resolution by the irreversible transesterification procedure. As shown in Scheme 53, in the case of 1,2-diols the chemoselective acylation of the primary alcohol is not enantioselective and the 1-acetate initially formed is later acylated and resolved by the lipase-catalyzed transesterification procedure to afford the optically active 1,2-diacetate [239,240].

The same method has been applied to the resolution of other 1,2-diols [241,242] as depicted in Scheme 54. For instance, the resolution of 1-benzylglycerol [241], when achieved, should lead to an important chiral synthon of general interest in asymmetrical synthesis and the above-described sequential resolution procedure can afford the enantiomerically pure monoacetate.

2-Methyl-1,2-diols do not pose the problem of the selective monoacylation of the previous 1,2-diols, since in that structure both a primary group and a less accessible tertiary hydroxy group are present. In face, as shown in Scheme 55, a few 2-methyl-1,2-diols are successfully resolved by the enzymatic transesterification process with high enantioselectivity [243].

Other sterically more crowded and functionalized 2-substituted-1,2-diols are enzymatically acylated with high enantioselectivity as well (Scheme 56). In these cases selective acylation also occurs at the primary alcohol [244–246].

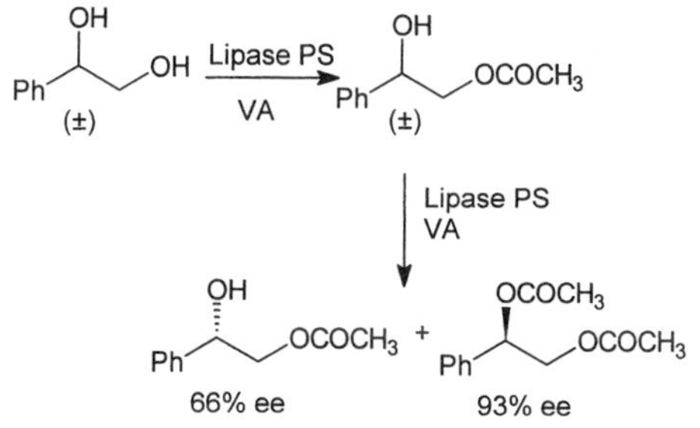

Scheme 53 Sequential resolution of 1,2-diols.

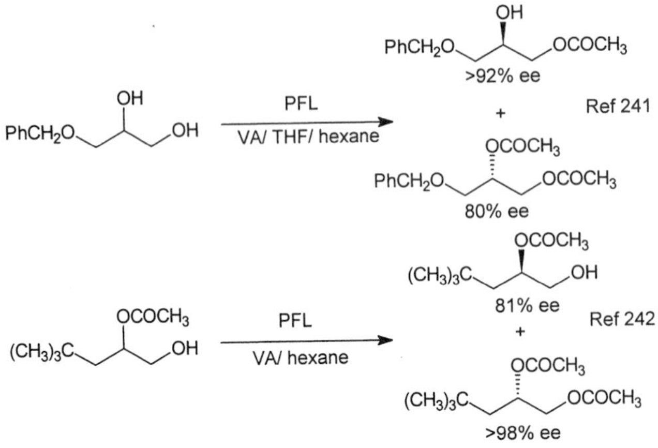

Scheme 54   Sequential resolution of 1,2-diols.

Vicinal diols in which both hydroxy groups are secondary are efficiently resolved by the enzymatic procedure [247,248] described in Scheme 57. The diastereo- and enantioselection of the process gives excellent results and in some cases enantiomerically pure products are obtained.

The enzymatic method could also be successfully applied to other less proximal diols and the diastereo- and enantioselectivity can be virtually complete [249–253], as described in Scheme 58. Depending on the substrate, the starting diol or the monoacetate or the diacetate can also be recovered as optically pure.

(b)   *Lipase-Catalyzed Irreversible Transesterification of Meso and Prochiral Diols.*   The enzymatic procedure that relies on the irreversible transesterification method applies also to *meso-* and prochiral diols, which are efficiently asymmetrized by the lipase-catalyzed irreversible transesterification in organic solvents, the process proceeding with high enantioselectivity. Among the tested compounds, acyclic *meso*-diols (Scheme 59) are stereoselectively acetylated [254–256].

Several cyclic diols have proven to be good substrates for the enzymatic process in the presence of various lipases. The method applies efficiently to various cyclic diols that present in the structure two primary (Scheme 60) and/or two secondary hydroxy groups (Scheme 61) [257–268].

As for cyclic alcohols, bi- or tricyclic diols of different structures are excellent substrates and furnish excellent results in terms of stereoselectivity [263–267]. Scheme 62 presents a few selected examples of the successful asymmetrization of bi- and tricyclic substrates.

2-Substituted-1,3-propanediols are a special class of prochiral compounds that can be efficiently asymmetrized by the lipase-catalyzed reactions. Using this procedure it is

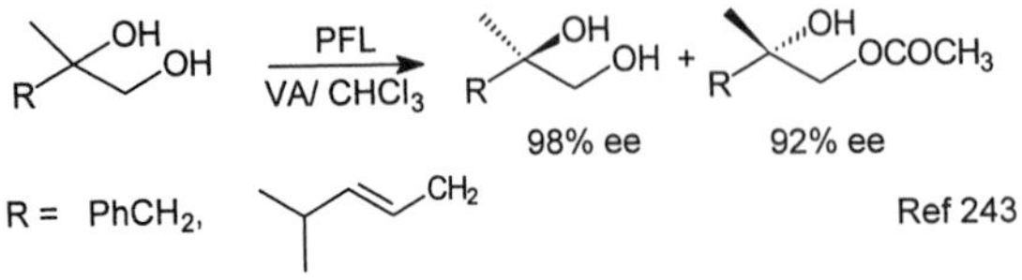

Scheme 55   Resolution of 2-methyl-1,2-diols.

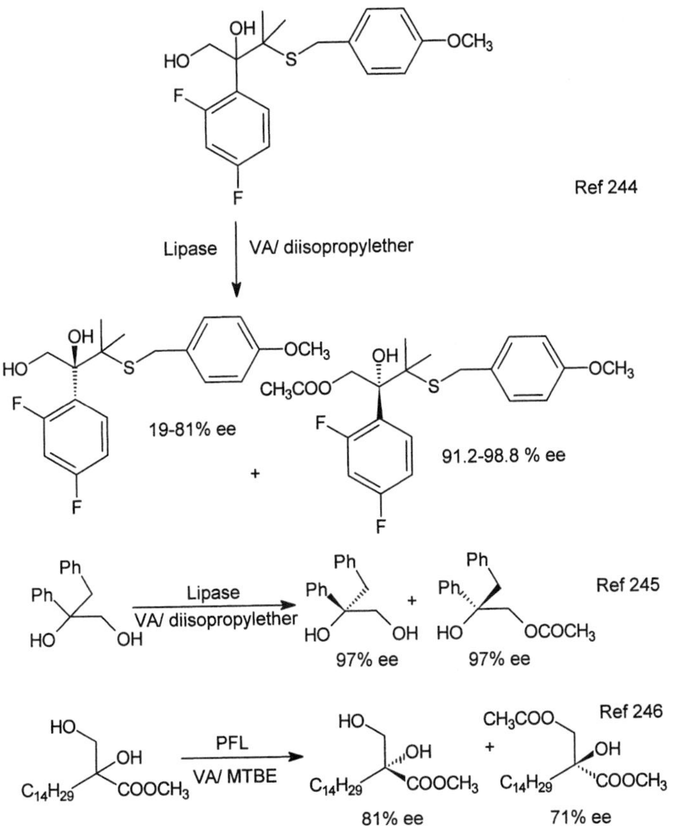

Scheme 56 Resolution of 2-substituted-1,2-diols.

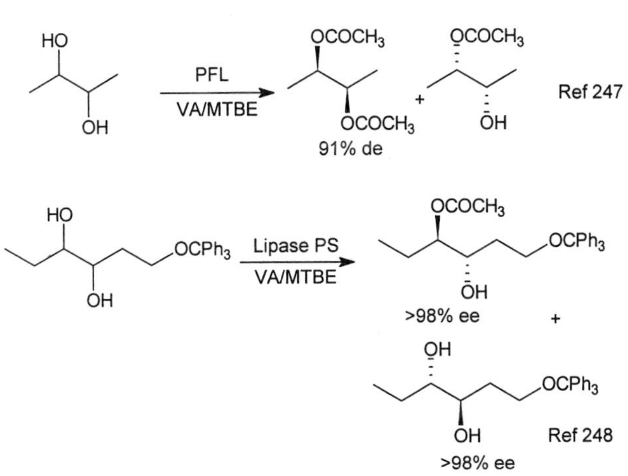

Scheme 57 Resolution of vicinal diols.

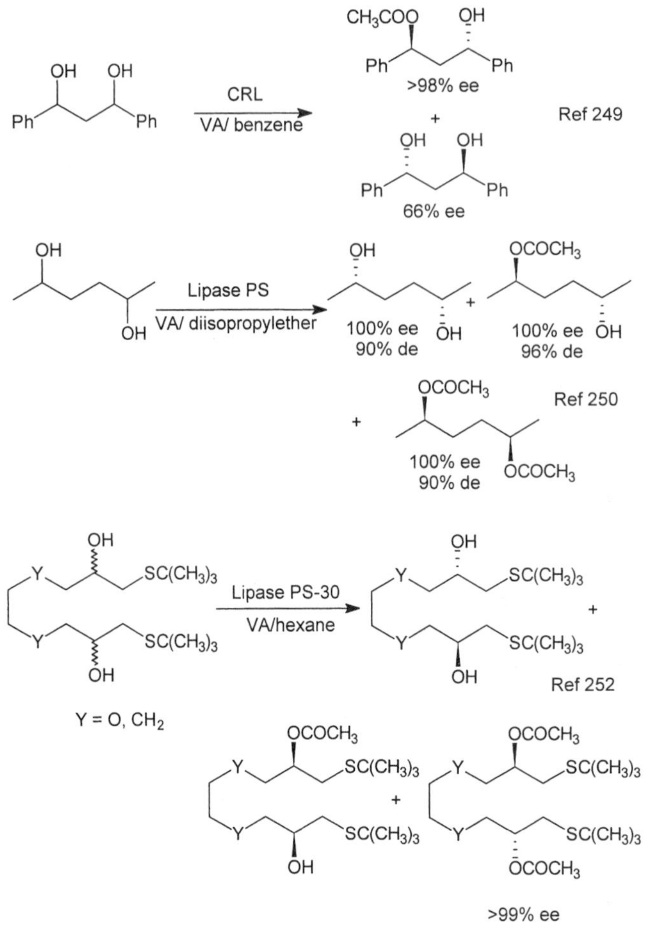

Scheme 58 Resolution of acyclic diols.

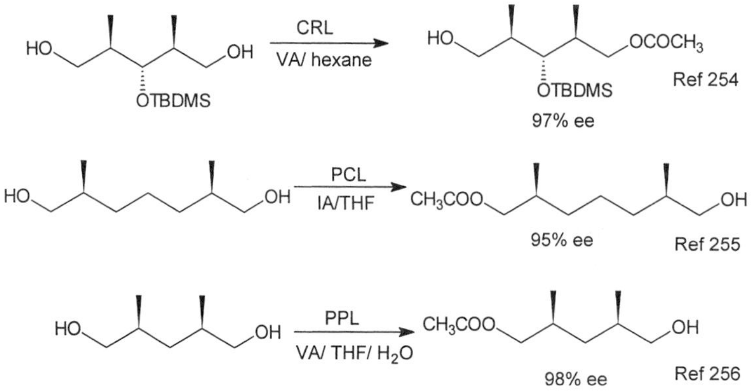

Scheme 59 Asymmetrization of acyclic *meso*-diols.

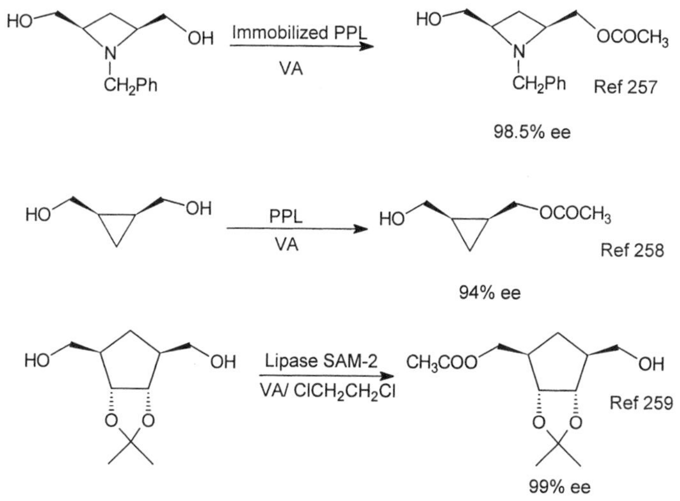

Scheme 60 Asymmetrization of cyclic *meso*-diols containing two primary alcohol moieties.

possible to prepare many important and flexible chiral synthons. Several examples of this highly enantioselective reaction are reported in the recent literature (Scheme 63), and the results are listed in Table 3 [268–275].

VII. EXPERIMENTAL CONDITIONS AND PRACTICAL CONSIDERATIONS

There are many reasons for the successful use of enzymes in organic synthesis and, as far as the enzymatic acylation and esterification reactions are concerned, lipases are by far

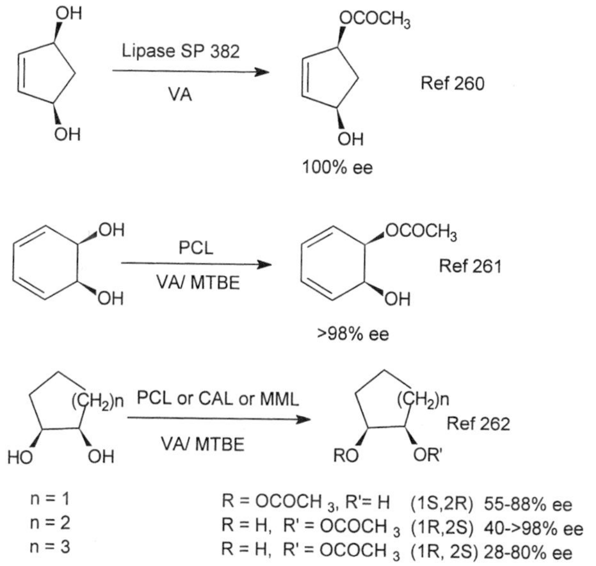

Scheme 61 Asymmetrization of cyclic *meso*-diols containing two secondary alcohol moieties.

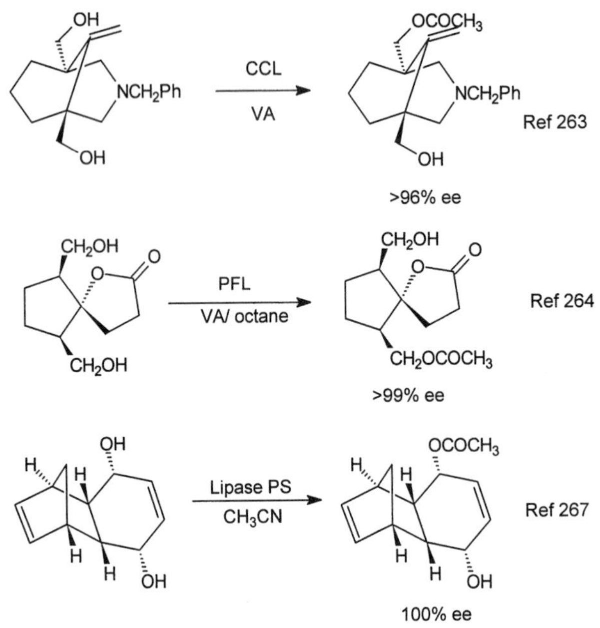

Scheme 62 Asymmetrization of bicyclic *meso*-diols.

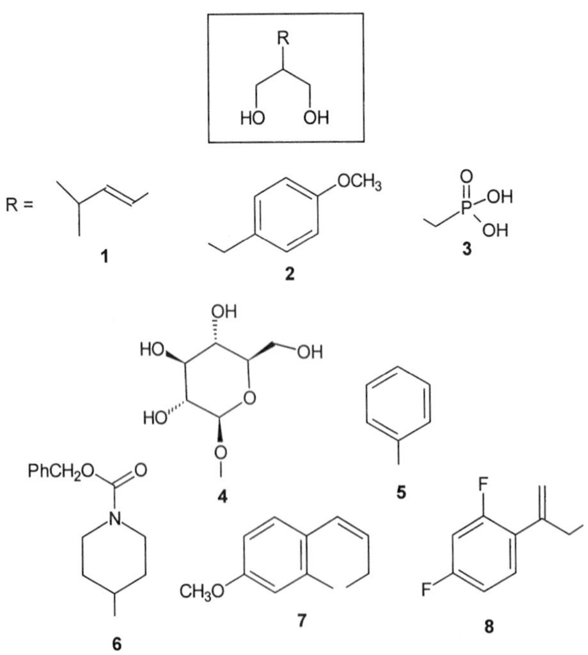

Scheme 63 Asymmetrization of prochiral diols.

Table 3 Asymmetrization of 2-Substituted 1,3-Propenediols by Irreversible Transesterification with Vinyl Acetate

Substrate	Enzyme	Solvent	Monoester (config. % e.e.)	Ref.
1	Imm. PPL	Diisopropyl ether	R, >98	268
2	Lipase PS	Diisopropyl ether	R, >98	269
3	Lipase PS	THF	R, 98	270
4	Lipase PS	Py	S, 80	271
	CAL	Py	R, 64	
5	PPL	THF[a]	R, 96.8	272
6	PPL	—[b]	R, 98	273
7	Lipase PS	MTBE	R, 94	274
8	Novo SP435	Acetonitrile	S, >98	275

[a]Vinyl butanoate as acylating agent.
[b]Vinyl acetate as solvent.

the more convenient biocatalysts. Furthermore, the above reactions can be carried out in organic solvents, and lipases are especially well suited to this purpose because they are stable at a wide range of temperatures and in a variety of organic solvents. The use of lipases is also very convenient because they are commercially available and relatively cheap. The experimental procedure is very simple since in most instances the reaction consists of the preparation of a heterogeneous mixture of reagents and enzyme in the proper organic solvent. At the end of the reaction, filtration of the enzyme (which can sometimes be reused), evaporation of the solvent, and chromatographic purification lead to purified products. Perhaps the most difficult aspect of the experimental protocol lies in the choice of the most suitable enzyme, since it is virtually impossible to have at present a general model that allows precise prediction on the structure (of the substrate)–activity (of the enzyme) relationship. Therefore, the choice of the most useful enzyme still relies on a screening of the known enzymes, and comparison with similar results appeared in the literature [276]. Sometimes, two different enzymes can be compared to obtain the same result, as in the interesting case of the resolution of hydroperoxyvinylsilanes (Scheme 64) that can be carried out either with lipases or with horseradish peroxidase [277].

A practical aspect of the use of enzymes for the preparation of chiral synthons is the fact that one can use enzymes such as lipases either in hydrolytic conditions or in organic solvents, and the most suitable method is a compromise between the solubility of the compound, the efficiency of the process, and the maximum enantioselection that can be achieved. In Scheme 65 two of these cases are presented [278,244] where esterification and hydrolysis can be compared. Interestingly, an opposite configuration of products is obtained when enzymatic hydrolysis of the ester or esterification of the alcohol is carried out.

Another problem with the procedure lies in the optimization of the enzyme/substrate ratio. This is due to the fact that the activity of an enzyme such as a lipase is referred to its hydrolytic activity on a suitable triglyceride and there are no standardized methods for determination of the activity of an enzyme in organic solvents, i.e., in a heterogeneous medium. Furthermore, the nonconventional conditions of reactions may lead to different kinetic data that strongly depend on enzyme conformation in organic media, chemico-physical properties of solvents, and the nature and properties of substrates. Therefore, often

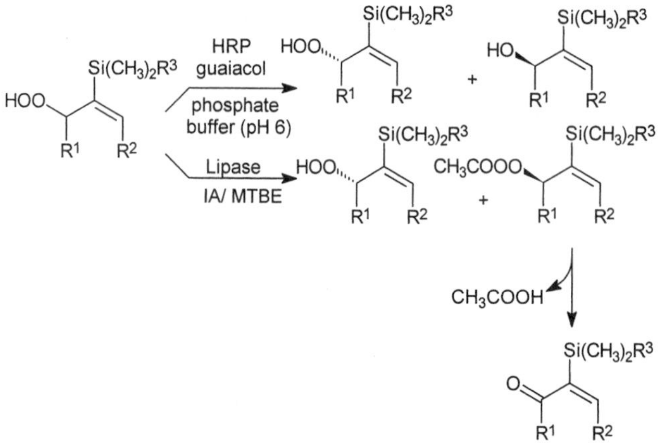

Scheme 64 Resolution of hydroperoxyvinylsilanes by different enzymes.

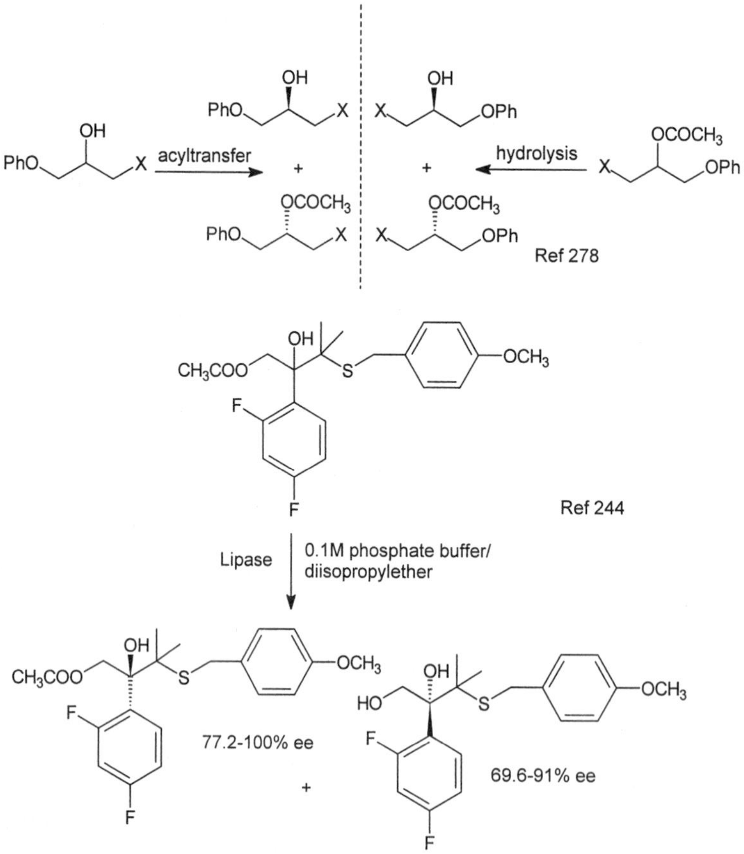

Scheme 65 Comparison of enzymatic hydrolysis and transesterification (cf. Scheme 56 for transesterification conditions).

the relative amount of the biocatalyst is a compromise between activity in a nonconventional medium and its capability to catalyze enantioselective biotransformations with the purpose of reaching the highest enantioselection. In spite of innumerable studies on the characteristics and properties of organic solvents, a rationale to guide in the choice of the most suitable organic solvent for a given reaction is still not well understood. A typical experimental procedure may be found in papers dealing with microscale laboratory projects of bioorganic chemistry for students [279].

Commercially available purified enzymes are generally used for synthetic applications. Often such preparations contain variable amounts of impurities or the enzyme itself is a mixture of isoenzymes. Unless purification of the enzyme is relatively simple [280], common laboratory practice suggests that such a commercial sample of the enzyme should not be used without further purification.

VIII. SYNTHETIC APPLICATIONS

Most of the racemic substrates reviewed in this chapter are enantioselectively resolved by enzymatic acylation and esterification reactions into products that constitute chiral synthons either of general synthetic interest or specifically prepared to build up more complicated compounds that are the targets of asymmetrical synthesis. Many papers reviewed in this work contain not only the biocatalytic preparation of the chiral building block but also their utilization, describing the overall synthesis. For other products of the enzymatic reaction the work is more concerned about the biocatalytic process, since the chiron is either of general interest or specifically required in a particular step of a previously described asymmetrical synthesis. We report here the examples of a few applications of the enzymatically prepared chiral synthons, showing the overall synthetic procedure and the step where the biocatalytic approach allows the enantioselective preparation of the proper building block. In Scheme 66 the synthesis of betaxolol is presented [281]. Betaxolol is a beta blocker that possesses the general structure of 2-amino-1-phenylethanol. Many biocatalytic ways to the synthesis of such compounds have been recently reviewed [282].

Another interesting synthetic application is the lipase-catalyzed preparation of enantiomerically pure epoxy esters as intermediates of the calcium channel blocker diltiazem [283] (Scheme 67).

IX. CONCLUSIONS

The biocatalytic approach for the preparation of enantiomerically pure compounds has gained a firm reputation among the multitude of procedures available to organic chemists. We have presented the most recent examples of resolution of racemic substrates by means of esterification and acylation procedures that are mainly catalyzed by lipases. As a general perspective, starting from the success of the enzymatic methodology, most future work should be dedicated to finding the basis of a more rational use of the biocatalysts. In order to reach this goal, one should start with information that might become available from the protein structural studies. In the meantime, applications of the well-established hydrolytic enzymes are growing extremely fast and less familiar enzymes should be investigated for biocatalytic applications. Only a few other enzymes have been studied for new applications of esterification reactions in organic solvents. Among these enzymes, one can mention phospholipases [284–286] or acylases [287]. A classical esterase such as PLE, which has

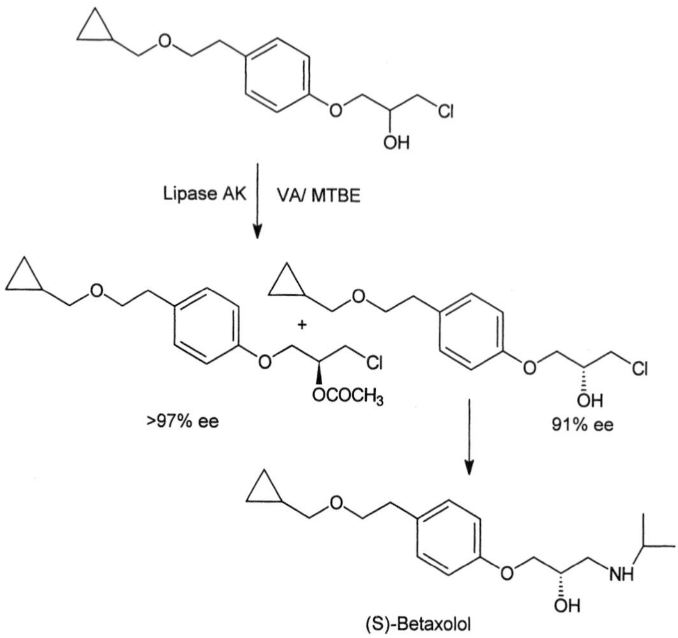

Scheme 66   Chemoenzymatic synthesis of (*S*)-betaxolol.

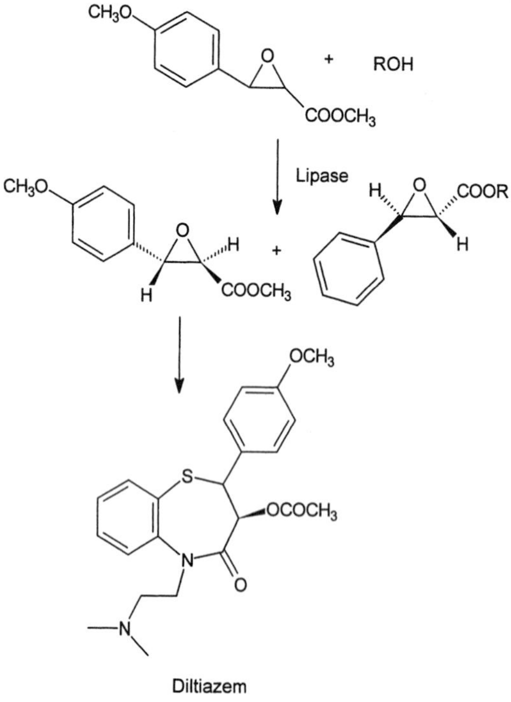

Scheme 67   Chemoenzymatic synthesis of diltiazem.

been used to hydrolyze more than 200 diesters and diacylated diols [288], has been only recently used for acylations in organic solvents [289]. However, for this biocatalytic application in nonconventional media, a polyethylene glycol–modified PLE is the biocatalyst of choice. Although the enantioselectivity is not excellent, it is interesting to note that the unmodified enzyme is not active in organic solvents. A serine protease like α-chymotrypsin can be used in a transesterification process in organic solvents in the presence of a crown ether [290].

Special attention should be devoted to less conventional applications of the enzymatic transesterification methodology such as resolution of unstable substrates as racemic secondary hydroperoxides [291]. The development of new reactions in the presence of enzymes should be pursued, as, for example, the simultaneous formation of a hemithioacetal and the irreversible transesterification in the presence of a lipase [292]. Also, for synthetic applications, the combination of enzymatic and chemical asymmetrical methods could lead to interesting results, such as the one-pot lipase-catalyzed acylation and the Mitsonobu inversion of the configuration of the unreacted alcohol, which should lead to only one enantiomeric ester [293].

ABBREVIATIONS

Enzymes

CAL	*Candida antarctica* lipase
CCL	*Candida cylindracea* lipase
CRL	*Candida rugosa* lipase
HRP	Horseradish peroxidase
LPL	Lipoprotein lipase
MML	*Mucor miehei* lipase
PCL	*Pseudomonas cepacia* lipase
PEG-PCL	*Pseudomonas cepacia* bound to polyethylene glycol (PEG)
PFL	*Pseudomonas fluorescens* lipase
PPL	Porcine pancreas lipase

Solvents and Reagents

DME	Dimethoxyethane
MTBE	Methyl *t*-butyl ether
Py	Pyridine
THF	Tetrahydrofuran
IA	Isopropenyl acetate
VA	Vinyl acetate
VB	Vinyl butyrate
VD	Vinyl dodecanoaic

REFERENCES AND NOTES

1. For a summary of the impressive amount of literature devoted to different aspects of the applications of enzyme to organic synthesis, see one of the most recent reviews on the topic: E Schoffers, A Golebiowski, CR Johnson. Tetrahedron 52:3769–3826, 1996. See also (a) E Santaniello, P Ferraboschi. In: A Hassner, ed. Advances in Asymmetric Synthesis, Vol. 2. JAI

Press, 1997, pp 237–283. (b) SM Roberts, NM Williamson. Curr Org Chem1:1–20, 1997. (c) SM Roberts. J Chem Soc Perkin Trans I:157–169, 1998.
2. W Boland, C Frössl, M Lorenz. Synthesis 1049–1072, 1991.
3. A Zaks, AM Klibanov. Science 224:1249–1251, 1984.
4. AMP Koskinen, AM Klibanov. Enzymatic Reactions in Organic Media. London: Chapman and Hall, 1996.
5. L Kvittingen. Tetrahedron 50:8253–8274, 1994.
6. A Ballesteros, U Bornscheuer, A Capewell, D Combes, JS Condoret, K Koenig, FN Kolisis, A Marty, U Menge, T Scheper, H Stamatis, A Xenakis. Biocatal Biotransform 13:1–42, 1995.
7. NJ Turner. Nat Prod Rep 11:1–15, 1994.
8. RA Sheldon, Chimia 50:418–419, 1996
9. JO Rich, P Wang, BD Martin, N Patil, MV Segueeva, JS Dordick. Chimia 50:428–429, 1996.
10. G Bell, PJ Halling, BD Moore, J Partridge, DG Rees. Trends Biotechnol 13:468–473, 1996.
11. A Zaks, AM Klibanov. J Biol Chem 263:8017–8021, 1988.
12. (a) LAS Gorman, JS Dordick. Biotechnol Bioeng 39:392–397, 1992; (b) M Dolman, PJ Halling, BD Moore, S Waldron. Biopolymers 41:313–321, 1997.
13. PA Burke, SO Smith, WW Bachovchin, AM Klibanov. J Am Chem Soc 111:8290–8291, 1989.
14. DS Hartsough, KM Merz Jr. J Am Chem Soc 115:6529–6537, 1993.
15. HP Yennawar, NH Yennawar, GK Farber. J Am Chem Soc 117:577–585, 1995.
16. T Ke, CR Wescott, AM Klibanov. J Am Chem Soc 118:3366–3374, 1996.
17. F Secundo, S Riva, G Carrea. Tetrahedron: Asymmetry. 3:267–280, 1992.
18. G Ottolina, F Gianinetti, S Riva, G Carrea. J Chem Soc Chem Commun 535–536, 1994.
19. ZS Derewenda. Adv Protein Chem 45:1–52, 1994.
20. PA Fitzpatrick, ACU Steinmetz, D Ringe, AM Klibanov. Proc Natl Acad Sci USA 90:8653–8657, 1993.
21. L Brady, AM Brzozowski, ZS Derewenda, E Dodson, G Dodson, S Tolley, JP Turkenburg, L Christiansen, B Huge-Jensen, L Norskov, L Thim, U Menge. Nature 343:767–770, 1990.
22. FK Winkler, A D'Arcy, W Hunziker. Nature 343:771–774, 1990.
23. AM Brzozowski, U Derewenda, ZS Derewenda, GG Dodson, DM Lawson, JP Turkenburg, F Bjorkling, B Huge-Jensen, SA Patkar, L Thim. Nature 351:491–494, 1991.
24. U Derewenda, AM Brzozowski, DM Lawson, ZS Derewenda. Biochemistry 31:1532–1541, 1992.
25. H van Tilbeurgh, MP Egloff, C Martinez, N Rugani, R Verger, C Cambillau. Nature 362:814–820, 1993.
26. MP Egloff, F Marguet, G Buono, R Verger, C Cambillau, H van Tilbeurgh. Biochemistry 34:2751–2762, 1995.
27. P Grochulski, Y Li, JD Schrag, F Bouthillier, P Smith, D Harrison, B Rubin, M Cygler. J Biol Chem 268:12843–12847, 1993.
28. P Grochulski, Y Li, JD Schrag, M Cygler. Protein Sci 3:82–91, 1994.
29. M Cygler, P Grochulski, RJ Kazlauskas, JD Schrag, F Bouthillier, B Rubin, AN Serreqi, AK Gupta. J Am Chem Soc 116:3180–3186. 1994.
30. KK Kim, HK Song, DH Shin, KY Hwang, SW Suh. Structure 5:173–185, 1997.
31. JD Schrag, YG Li, DM Lang, T Burgdorf, HJ Hecht, R Schmid, D Schomburg, TJ Rydel, Oliver, LC Strickland, CM Dunaway, SB Larson, J Day, A Mcpherson. Structure 5:187–202, 1997.
32. ANE Weissfloch, RJ Kazlauskas. J Org Chem 60:6959–6969, 1995.
33. M Cygler, P Grochulski, JD Schrag. Can J Microbiol 41:289–296, 1995.
34. LE Janes, RJ Kazlauskas. Tetrahedron: Asymmetry 8:3719–3733, 1997.
35. T Ohtani, H Nakatsukasa, M Kamezawa, H Tachibana, Y Naoshima. J Mol Catal B: Enzymatics 4:53–60, 1998.
36. X Grabuleda, C Jaime, A Guerrero. Tetrahedron: Asymmetry 8:3675–3683, 1997.
37. F Theil, K Lemke, S Ballschuh, A Kunath, H Schick. Tetrahedron: Asymmetry 6:1323–1344, 1995.

38. K Lemke, M Lemke, F Theil. J Org Chem 62:6268–6273, 1997.
39. DA Lang, MM Mannesse, GH DeHaas, HM Verheij, BW Dijkstra. Eur J Biochem 254:333–340, 1998.
40. CS Chen, Y Fujimoto, G Girdaukas, CJ Sih. J Am Chem Soc 104:7294–7299, 1982.
41. CJ Sih, SH Wu. Top Stereochem 19:63–125, 1989.
42. AJJ Straathof, JA Jongejan. Enzyme Microb Technol 21:559–571, 1997.
43. JLL Rakels, AJJ Straathof, JJ Heijnen. Enzyme Microb Technol 15:1051–1056, 1993.
44. (a) HW Anthonsen, BH Hoff, T Anthonsen. Tetrahedron: Asymmetry 6:3015–3022, 1995. (b) HW Anthonsen, GH Hoff, T Anthonsen. Tetrahedron: Asymmetry 7:2633–2638, 1996.
45. G Ottolina, G Carrea, S Riva, L Sartore, FM Veronese. Biotechnol Lett 14:947–952, 1992.
46. H Akita. Biocatal Biotransform 13:151–156, 1996.
47. EM Sanchez, JF Bello, MG Roig, FJ Burguillo, JM Moreno, JV Sinisterra. Enzyme Microb Technol 18:468–476, 1996.
48. G Pencreach, M Leullier, JC Baratti. Biotechnol Bioeng 56:181–189, 1997.
49. JJ Lalonde, C Govardhan, N Khalaf, AG Martinez, K Visuri, AL Margolin. J Am Chem Soc 117:6845–6852, 1995.
50. N Khalaf, CP Govardhan, JJ Lalonde, RA Persichetti, YF Wang, AL Margolin. J Am Chem Soc 118:5494–5495, 1996.
51. MT Lopez-Belmonte, AR Alcantara, JV Sinisterra. J Org Chem 62:1831–1840, 1997.
52. SH Wu, ZU Guo, CJ Sih. J Am Chem Soc 112:1990–1995, 1990.
53. CR Wescott, AM Klibanov. Biochim Biophys Acta Protein Struct Mol Enzymol 1206:1–9, 1994.
54. A Marty, W Chulalaksananukul, RM Willemot, JS Condoret. Biotechnol Bioeng 39:273–280, 1992.
55. M Rantakyla, O Aaltonen. Biotechnol Lett 16:825–830, 1994.
56. E Cernia, C Palocci, F Gasparrini, D Misiti, N Fagnano, J Mol Catal 89:11–18, 1994.
57. Y Yasufuku, S Ueji. Biotechnol Lett 17:1311–1316, 1995.
58. Y Yamamoto, H Kise. Chem Lett 1821–1824, 1993.
59. G Duan, JY Chen. Biotechnol Lett 16:1065–1068, 1994.
60. T Itoh, Y Hiyama, A Betchaku, H Tsukube. Tetrahedron Lett 34:2617–2620, 1993.
61. T Itoh, Y Takagi, T Murakami, Y Hiyama, H Tsukube. J Org Chem 61:2158–2163, 1996.
62. AJ Russell, AM Klibanov. J Biol Chem 263:11624–11626, 1988.
63. F Monot, E Paccard, R Borzeix, M Bardin, JP Vandecasteele. Appl Microbiol Biotechnol 39:483–486, 1993.
64. I Mingarro, C Abad, L Braco. Proc Natl Acad Sci USA 92:3308–3312, 1995.
65. M Stahl, MO Mansson, K Mossbach. Biotechnol Lett 12:161–166, 1990.
66. AR Macrae. J Am Oil Chem Soc 60:291–294, 1983.
67. LH Posorske, GK LeFebvre, CA Miller, TT Hansen, GL Glenvig. J Am Oil Chem Soc 65:922–926, 1988.
68. GG Haraldsson, BO Gudmundsson, O Almarsson. Tetrahedron 51:941–952, 1995.
69. I Jachmanian, KD Mukherjee. J Am Oil Chem Soc 73:1527–1532, 1996.
70. SH Yoon, H Nakaya, O Ito, O Miyawaki, KH Park, K Nakamura. Biosci Biotechnol Biochem 62:170–172, 1998.
71. ELA Macfarlane, SM Roberts, VGR Steukers, PL Taylor. J Chem Soc Perkin Trans I: 2287–2290, 1993.
72. T Sugai, H Ikeda, H Ohta. Tetrahedron 52:8123–8134, 1996.
73. R Lortie. Biotechnol Adv 15:1–15, 1997.
74. N Kamiya, M Goto. Biotechnol Prog 13:488–492, 1997.
75. R Morrone, G Nicolosi, A Patti, M Piattelli. Tetrahedron: Asymmetry 6:1773–1778, 1995.
76. EH Berglund, JM Hedenstrom, HE Hogberg. Acta Chem Scand 5:666–671, 1996.
77. RA Persichetti, JJ Lalonde, CP Govardhan, NK Khaaf, AL Margolin. Tetrahedron Lett 37:6507–6510, 1996.
78. S Sankaranarayanan, A Sharma, S Chattopadhyay. Tetrahedron: Asymmetry 7:2639–2643 (1996).

79. Y Okahata, A Hatano, K Ijiro. Tetrahedron: Asymmetry. 6:1311–1322, 1995.
80. R Martinez. Tetrahedron: Asymmetry 6:1491–1494, 1995.
81. SM Roberts, VGR Steukers, PL Taylor. Tetrahedron: Asymmetry 4:969–972, 1993.
82. PY Chen, SH Wu, KT Wang. Biotechnol Lett 15:181–184, 1993.
83. D Bianchi, P Cesti, E Battistel. J Org Chem 53:5531–5534, 1988.
84. WH Wu, CC Akoh, RS Phillips. Enzyme Microb Technol 18:536–539, 1996.
85. A Sattler, G Haufe. Tetrahedron: Asymmetry 6:2841–2848, 1995.
86. AK Ghosh, Y Chen. Tetrahedron Lett 36:505–508, 1995.
87. E Vanttinen, LT Kanerva, J Chem Soc Perkin Trans I:3459–3463, 1994.
88. AR Reinhold, RL Rosati. Tetrahedron: Asymmetry 5:1187–1190, 1994.
89. AL Gutman, D Brenner, A Boltanski. Tetrahedron: Asymmetry 4:839–844, 1993.
90. R Ozegowski, A Kunath, H Schick. Liebigs Ann 1699–1702, 1995.
91. R Ozegowski, A Kunath, H Schick. Liebigs Ann 215–217, 1994.
92. R Ozegowski, A Kunath, H Schick. Liebigs Ann 1443–1448, 1996.
93. R Ozegowski, A Kunath, H Schick. J Prakt Chem 336:544–546, 1994.
94. IL Gatfield. Ann N Y Acad Sci 434:569–572, 1984.
95. ZW Guo, CJ Sih. J Am Chem Soc 110;1999–2001, 1988.
96. J Otera. Chem Rev 93:1449–1470, 1993.
97. E Santaniello, P Ferraboschi, P Grisenti. Enzyme Microb Technol 5:367–381, 1993.
98. C Moresoli, E Flaschel, A Renken. Biocatalysis 5:213–231, 1992.
99. RC Lloyd, M Dickman, JB Jones, Tetrahedron: Asymmetry 9:551–561, 1998.
100. RG Lovey, AK Saksena, VM Girijavallabhan. Tetrahedron Lett 35:6047–6050, 1994.
101. A Makita, T Nihira, Y Yamada. Tetrahedron Lett 28:805–808, 1987.
102. AL Gutman, K Zuobi, A Boltansky. Tetrahedron Lett 28:3861–3864, 1987.
103. C Bonini, P Pucci, R Racioppi, L Viggiani. Tetrahedron: Asymmetry 3:29–32, 1991.
104. B Henkel, A Kunath, H Schick. Tetrahedron: Asymmetry 4:153–156, 1993.
105. B Henkel, A Kunath, H Schick. J Prakt Chem 339:434–440, 1997.
106. MCR Franssen, H Jongejan, H Kooijman, AL Spek, LFL Nuno, NLFLC Mondril, PMAC Boavida Dossantos, A Degroot. Tetrahedron: Asymmetry 7:497–510, 1996.
107. M Martres, G Gil, A Meou. Tetrahedron Lett. 35:8787–8790, 1994.
108. K Burgess, I Henderson, KK Ho. J Org Chem 57:1290–1295, 1992.
109. C Baldoli, S Majorana, G Carrea, S Riva. Tetrahedron: Asymmetry 4:767–772, 1993.
110. Y Lu, C Miet, N Kunesch, JE Poisson. Tetrahedron: Asymmetry 4:893–902, 1993.
111. H van der Deen, RP Hof, A van Oeveren, BL Feringa, RM Kellogg. Tetrahedron Lett 35: 8441–8444, 1994.
112. NV Thakkar, AA Banerji, HS Bevinakatti. Biotechnol Lett 16:1299–1302, 1994.
113. HJ Altenbach, GF Merhof, DJ Brauer. Tetrahedron: Asymmetry 7:2493–2496, 1996.
114. J Vanderdeen, AD Cuiper, RP Hof, A Vanoeveren, BL Feringa, RM Kellogg. J Am Chem Soc 118:3801–3803, 1996.
115. J Roos, U Stelzer, F Effenberger. Tetrahedron: Asymmetry 9:1043–1049, 1998.
116. A Fishman, M Zviely. Tetrahedron: Asymmetry 9:107–118, 1998.
117. Y Koichi, K Suginaka, Y Yamamoto. J Chem Soc, Perkin Trans I 1645–1646, 1995.
118. T Uemura, M Furukawa, Y Kodera, M Hiroto, A Matsushima, H Kuno, H Matsushima, K Sakurai, Y Inada. Biotechnol Lett 17:61–66, 1995.
119. HS Bevinakatti, RV Newadkar. Tetrahedron: Asymmetry 4:773–776, 1993.
120. K Faber, S Riva. Synthesis 895–910, 1992.
121. FJM Fang, CH Wong. Synlett 393–402, 1994.
122. B Morgan, AC Oehlschlager, TM Stokes. J Org Chem 57:3231–3236, 1992.
123. N Öhrner, M Martinelle, A Mattson, T Norin, K Hult. Biotechnol Lett 14:263–268, 1992.
124. G Lin, SH Liu. Org Prep Proced Int 25:463–466, 1993.
125. H Frykman, N Öhrner, T Norin, K Hult. Tetrahedron Lett 34:1367–1370, 1993.
126. A Mattson, N Öhrner, K Hult, T Norin. Tetrahedron: Asymmetry 4:925–930, 1993.
127. C Orrhenius, A Mattson, T Norin, N Öhrner, K Hult. Tetrahedron: Asymmetry 5:1363–1366, 1994.

128. E Guibé-Jampel, M Bassir. Tetrahedron Lett 35:421–422, 1994.

129. E Guibé-Jampel, Z Chalecki, M Bassir, M Gelopujic. Tetrahedron 52:4397–4402, 1996.

130. M Degueil-Castaing, B DeJeso, S Drouillard, B Maillard. Tetrahedron Lett 28:953–954, 1987.

131. YF Wang, JJ Lalonde, M Momongan, DE Bergbreiter, CH Wong. J Am Chem Soc 110:7200–7205, 1988.

132. B Berger, K Faber. J Chem Soc Chem Commun 1998–1199, 1991.

133. S Casati, M Caporali, P Ferraboschi, A Manzocchi, E Santaniello. Biotechnol Technol 11:81–83, 1997.

134. A Ghogare, GS Kumar. J Chem Soc Chem Commun 1533–1535, 1989.

135. Y Kita, Y Takebe, K Murata, Y Naka, S Akai. Tetrahedron Lett 37:7369–7372, 1996.

136. K Suginaka, Y Hayashi, Y Yamamoto. Tetrahedron: Asymmetry 7:1153–1158, 1996.

137. GE Jeromin, V Welsch. Tetrahedron Lett 36:6663–6664, 1995.

138. S Barth, F Effenberger. Tetrahedron: Asymmetry 4:823–833, 1993.

139. (a) F Bracher, T Papke. Tetrahedron: Asymmetry 5:1653–1656, 1994. (b) BV Nguyen, O Nordin, C Vörde, E Hedenström, HE Högberg. Tetrahedron: Asymmetry 8:983–986, 1997.

140. T Konegawa, Y Ohtsuka, H Ikeda, T Sugai, H Ohta. Synlett 1297–1299, 1997.

141. C Larpent, X Chasseray. New J Chem 17:851–855, 1993.

142. M Majeric, V Sunjic. Tetrahedron: Asymmetry 7:815–824, 1996.

143. DB Berkowitz, JA Pumphrey, QR Shen. Tetrahedron Lett 35:8743–8746, 1994.

144. HJ Gais, I Vonderweiden. Tetrahedron: Asymmetry 7:1253–1256, 1996.

145. H Miyaoka, S Sagawa, T Inoue, H Nagaoka, Y Yamada. Chem Pharm Bull Tokyo 42:405–407, 1994.

146. K Hiroya, J Hasegawa, T Watanabe, K Ogasawara. Synthesis 379–381, 1995.

147. T Yokomatsu, N Nakabayashi, K Matsumoto, S Shibuya. Tetrahedron: Asymmetry 6:3055–3062, 1995.

148. M Pallavicini, E Valoti, L Villa, O Piccolo. J Org Chem 59:1751–1754, 1994.

149. E Vänttinen, LT Kanerva. J Chem Soc Perkin Trans I 3459–3463, 1994.

150. E Vänttinen, LT Kanerva. Tetrahedron: Asymmetry 7:3037–3046, 1996.

151. RP Hof, RM Kellogg. J Org Chem 61:3423–3427, 1996.

152. B Herradon, S Valverde. Tetrahedron: Asymmetry 5:1479–1500, 1994.

153. B Herradon. J Org Chem 59:2891–2893, 1994.

154. D Mauleon, C Lobato, G Carganico. J Heterocycl Chem 31:57–59, 1994.

155. CR Johnson, YP Xu, KC Nicolaou, Z Yang, RK Guy, JG Dong, N Berova. Tetrahedron Lett 36:3291–3294, 1995.

156. MS Nair, AT Anilkumar. Tetrahedron: Asymmetry 7:511–514, 1996.

157. H Tanimoto, T Oritani. Tetrahedron: Asymmetry 7:1695–1704, 1996.

158. M Ors, A Morcuende, MI Jimenez Vacas, S Valverde, B Herradon. Synlett 449–451, 1996.

159. L Ling, JW Lown. J Chem Soc Chem Commun 1559–1560, 1996.

160. K Tanaka, Y Shogase, H Osuga, H Suzuki, K Nakamura. Tetrahedron Lett 36:1675–1678, 1995.

161. JH Rigby, P Sugathapala. Tetrahedron Lett 37:5293–5296, 1996.

162. G Nicolosi, A Patti, R Morrone, M Piattelli. Tetrahedron: Asymmetry 5:1275–1280, 1994.

163. WE Ladner, G Whitesides. J Am Chem Soc 106:7250–7251, 1984.

164. JBA Vantol, DE Kraayveld, JA Jongjean, JA Duine. Biocatal Biotransform 12:119–136, 1995.

165. P Ferraboschi, D Brembilla, P Grisenti, E Santaniello. J Org Chem 56:5478–5480, 1991.

166. P Ferraboschi, S Casati, P Grisenti, E Santaniello. Synlett 754–756, 1994.

167. P Ferraboschi, S Casati, P Grisenti, E Santaniello. Tetrahedron: Asymmetry 4:9–12, 1993.

168. YB Seu, YH Kho. Tetrahedron Lett 33:7015–7016, 1992.

169. E Mizuguchi, T Suzuki, K Achiwa. Synlett 929–930, 1994.

170. S Sankaranarayanan, A Sharma, BA Kulkarni, S Chattopadhyay. J Org Chem 60:4251–4154, 1995.

171. A Sharma, S Chattopadhyay. Liebigs Ann 529–531, 1996.

172. E Santaniello, R Canevotti, R Casati, L Ceriani, P Ferraboschi, P Grisenti. Gazz Chim Ital 119:55–57, 1989.

173. S Morita, J Matsubara, K Otsubo, K Kitano, T Ohtani, Y Kawano, M Uchida. Tetrahedron: Asymmetry 8:3707–3710, 1997.
174. RJ Kazlauskas, ANE Weissfloch, AT Rappoport, LA Cuccia. J Org Chem 56:2656–2665, 1991.
175. SM Brown, SG Davies, JAA Desousa. Tetrahedron: Asymmetry 4:813–822, 1993.
176. K Naemura, M Murata, R Tanaka, M Yano, K Hirose, Y Tobe. Tetrahedron: Asymmetry 7: 3285–3294, 1996.
177. M Schudok, G Kretzschmar. Tetrahedron Lett 38:387–388, 1997.
178. A Sharma, AS Pawar, S Chattopadhyay. Synth Commun 26:19–25, 1996.
179. J Uenishi, K Nishiwaki, S hata, K Nakamura. Tetrahedron Lett 35:7973–7976, 1994.
180. J Kaminska, I Gornicka, M Sikora, J Gora. Tetrahedron: Asymmetry 7:907–910, 1996.
181. SB Raju, TW Chiou, DF Tai. Tetrahedron: Asymmetry 6:1519–1520, 1995.
182. K Burgess, LD Jennings. J Am Chem Soc 113:6129–6139, 1991.
183. K Nakamura, M Kinoshita, A Ohno. Tetrahedron 51:8799–8808, 1995.
184. Y Hirai, M Nagatsu. Chem Lett 21–22, 1994.
185. NF Jain, PF Cirillo, JV Schaus, JS Panek. Tetrahedron Lett 36:8723–8726, 1995.
186. UV Mallavadhani, YR Rao. Tetrahedron: Asymmetry 5:23–26, 1994.
187. Y Takagi, R Ino, H Kihara, T Itoh, H Tsukube. Chem Lett 1247–1248, 1997.
188. F Bracher, T Litz. Bioorg Med Chem 4:877–880, 1996.
189. G Casy, G Gorins, R McCague, HF Olivo, SM Roberts. J Chem Soc Chem Commun 1085–1086, 1994.
190. P Allevi, P Ciuffreda, M Anastasia. Tetrahedron: Asymmetry 8:93–99, 1997.
191. K Kata, M Katayama, RK Gautam, S Fujii, H Kimoto. Biosci Biotechnol Biochem 58:1353–1354, 1994.
192. I Petschen, EA Malo, MP Bosch, A Guerrero. Tetrahedron: Asymmetry 7:2135–2143, 1996.
193. H Hamada, M Shiromoto, M Funahashi, T Itoh, K Nakamura. J Org Chem 61:2332–2336, 1996.
194. H Fukuda, M Tetsu, T Kitazume. Tetrahedron 52:157–164, 1996.
195. T Sakai, T Takayama, T Ohkawa, O Yoshio, T Ema, M Utaka. Tetrahedron Lett 38:1987–1990, 1997.
196. M Inagaki, A Hatanaka, M Mimura, J Hiratake, T Nishioka, J Oda. Bull Chem Soc Jpn 65: 111–120, 1992.
197. T Sugai, H Ohta. Tetrahedron Lett 32:7063–7064, 1991.
198. T Sugai, H Ritzen, CH Wong. Tetrahedron: Asymmetry 4:1051–1058, 1993.
199. A Chadha, M Manohar. Tetrahedron: Asymmetry 6:651–652, 1995.
200. S Vrielynck, M Vandewalle, AM Garcia, JL Mascarenas, A Mourino. Tetrahedron Lett 36: 9023–9026, 1995.
201. T Eicher, M Ott, A Speicher. Synthesis 755–762, 1996.
202. B Henkel, A Kunath, H Schick. Liebigs Ann 921–923, 1995.
203. H Takahata, Y Uchida, Y Ohkawa, T Momose. Tetrahedron: Asymmetry 4:1041–1042, 1993.
204. A Heisler, C Rabiller, R Douillard, N Goalou, G Hagel, F Levayer. Tetrahedron: Asymmetry 4:959–960, 1993.
205. M Drescher, F Hammerschmidt. Tetrahedron 53:4627–4636, 1997.
206. T Biadatti, JL Esker, CR Johnson. Tetrahedron: Asymmetry 7:2313–2320, 1996.
207. A Maestro, C Astorga, V Gotor. Tetrahedron: Asymmetry 8:3153–3159, 1997.
208. S Takano, M Suzuki, K Ogasawara. Tetrahedron: Asymmetry 4:1043–1046, 1993.
209. T Ema, S Maeno, Y Takaya, T Sakai, M Utaka. Tetrahedron: Asymmetry 7:625–628, 1996.
210. G Palmisano, M Santagostino, S Riva, M Sisti. Tetrahedron: Asymmetry 6:1229–1232, 1995.
211. Y Yoshida, Y Sato, S Okamoto, F Sato. J Chem Soc Chem Commun 811–812, 1995.
212. XM Zhou, PJ Declerq, J Gawronski. Tetrahedron: Asymmetry 6:1551–1552, 1995.
213. O Yamada, K Ogasawara. Synlett 427–428, 1995.
214. H Vanderdeen, AD Cuiper, RP Hof, A Vanoeveren, BL Feringa, RM Kellogg. J Am Chem Soc 118:3801–3803, 1996.

215. JWJF Thuring, AJH Klunder, GHL Nefkens, MA Wegman, B Zwanenburg. Tetrahedron Lett 37:4759–4760, 1996.
216. SS Kinderman, BL Feringa. Tetrahedron: Asymmetry 9:1215–1222, 1998.
217. TT Curran, DA Hay, CP Koegel, CJ Evans. Tetrahedron 53:1983–2004, 1997.
218. M Vandenheuvel, AD Cuiper, H Vanderdeen, RM Kellogg, BL Feringa. Tetrahedron Lett 38: 1655–1658, 1997.
219. M Barz, E Herdtweck, WR Thiel. Tetrahedron: Asymmetry 7:1717–1722, 1996.
220. P Stead, H Marley, M Mahmoudian, G Webb, D Noble, YT Ip, E Piga, T Rossi, S Roberts, MJ Dawson. Tetrahedron: Asymmetry 7:2247–2250, 1996.
221. S Takano, O Yamada, H Iida, K Ogasawara. Synthesis 592–596, 1994.
222. S Fernandez, M Ferrero, V Gotor, WH Okamura. J Org Chem 60:6057–6061, 1995.
223. C Orrenius, T Norin, K Hult, G Carrea. Tetrahedron: Asymmetry 6:3023–3030, 1995.
224. N Shimizu, H Akita, T Kawamata. Chem Pharm Bull 44:665–669, 1996.
225. T Fukazawa, Y Shimoji, T Hashimoto. Tetrahedron: Asymmetry 7:1649–1658, 1996.
226. P Crotti, V Dibussolo, L Favero, F Minutolo, M Pineschi. Tetrahedron: Asymmetry 7:1347–1356, 1996.
227. H Sakagami, K Samizu, T Kamikubo, K Ogasawara. Synlett 163–164, 1996.
228. O Yamada, K Ogasawara. Synthesis 1291–1294, 1995.
229. GL Lin, HC Liu. Tetrahedron Lett 36:6067–6068, 1995.
230. GL Lin, SH Liu. Org Prep Proced Int 25:463–466, 1993.
231. M Majeric, M Gelopujic, V Sunjic, A Levai, P Sebok, T Timar. Tetrahedron: Asymmetry 6: 937–944, 1995.
232. S Ramadas, GLD Krupadanam. Tetrahedron: Asymmetry 8:3059–3066, 1997.
233. T Izumi, S Murakami. J Heterocycl Chem 32:1125–1127, 1995.
234. T Iimori, I Azumaya, Y Hayashi, S Ikegami. Chem Pharm Bull 45:207–208, 1997.
235. T Yoshimitsu, Y Ohshiba, K Ogasawara. Synthesis 1029–1031, 1994.
236. S Tokuyama, T Yamano, I Aoki, K Takanohashi, K Nakahama. Chem Lett 741–744, 1993.
237. T Taniguchi, K Ogasawara. Tetrahedron Lett 38:6429–6432, 1997.
238. JWJF Thuring, GHLL Nefkens, MA Wegman, AJH Klunder, B Zwanenburg. J Org Chem 61:6931–6935, 1996.
239. F Theil, J Weidner, S Ballschuh, A Kunath, H Schick. J Org Chem 59:388–393, 1994.
240. F Theil. Catalysis Today 22:517–525, 1994.
241. B Herradon, S Cueto, A Morcuende, S Valverde. Tetrahedron: Asymmetry 4:845–864, 1993.
242. G Egri, E Baitzgacs, L Poppe. Tetrahedron: Asymmetry 7:1437–1448, 1996.
243. P Ferraboschi, S Casati, P Grisenti, E Santaniello. Tetrahedron: Asymmetry 5:1921–1924, 1994.
244. V Khlebnikov, K Mori, K Terashima, Y Tanaka, M Sato. Chem Pharm Bull 43:1659–1662, 1995.
245. RP Hof, RM Kellogg. Tetrahedron: Asymmetry 5:565–568, 1994.
246. O Jiménez, MP Bosch, A Guerrero. J Org Chem 62:3496–3499, 1997.
247. KS Bisht, VS Parmar, DHG Crout. Tetrahedron: Asymmetry 4:957–958, 1993.
248. MJ Kim, GB Choi, JY Kim, HJ Kim. Tetrahedron Lett 36:6253–6256, 1995.
249. F Levayer, C Rabiller, C Tellier. Tetrahedron: Asymmetry 6:1675–1682, 1995.
250. H Nagai, T Morimoto, K Achiwa. Synlett 289–290, 1994.
251. A Tanaka, H Yamamoto, T Oritani. Tetrahedron: Asymmetry 6:1273–1278, 1995.
252. TR Hoye, LS Tan. Synlett 615–616, 1996.
253. D Lee, MJ Kim. Tetrahedron Lett 39:2163–2166, 1998.
254. R Chenevert, G Courchesne. Tetrahedron: Asymmetry 6:2093–2096, 1995.
255. R Chenevert, M Desjardins. J Org Chem 61:1219–1222, 1996.
256. GQ Lin, WC Xu. Tetrahedron 52:5907–5912, 1996.
257. G Guanti, R Riva. Tetrahedron: Asymmetry 6:2921–2924, 1995.
258. S Pantkebocker, G Pohnert, I Fischerlui, W Boland, AF Peters. Tetrahedron 51:7927–7936, 1995.

259. B Mohar, A Stimac, J Kobe. Tetrahedron: Asymmetry 5:863–878, 1994.
260. (a) F Theil, H Schick, G Winter, G Reck. Tetrahedron 47:7569–7582, 1991; (b) F Theil. Tetrahedron: Asymmetry 6:1693–1698, 1995.
261. A Patti, C Sanfilippo, M Piattelli, G Nicolosi. J Org Chem 61:6458–6461, 1996.
262. G Nicolosi, A Patti, M Piattelli, C Sanfilippo. Tetrahedron: Asymmetry 6:519–524, 1995.
263. M Ihara, M Suzuki, A Hirabayashi, Y Tokunaga, K Fukomoto. Tetrahedron: Asymmetry 6: 2053–2058, 1995.
264. T Fujita, M Tanaka, Y Norimine, H Suemune, K Sakai. J Org Chem 62:3824–3830, 1997.
265. N Toyooka, A Nishino, T Momose. Tetrahedron Lett 34:4539–4540, 1993.
266. D Lambusta, G Nicolosi, A Patti, M Piattelli. Tetrahedron: Asymmetry 4:919–924, 1993.
267. S Takano, Y Higashi, T Kamikubo, M Moriya, K Ogasawara. J Chem Soc Chem Commun 788–789, 1993.
268. L Banfi, G Guanti, R Riva. Tetrahedron: Asymmetry 6:1345–1356, 1995.
269. S Shiotani, H Okada, T Yamamoto, K Nakamata, J Adachi, H Nakamoto. Heterocycles 43: 113–126, 1996.
270. T Yokomatsu, M Sato, S Shibuya. Tetrahedron: Asymmetry 7:2743–2754, 1996.
271. D Colombo, F Ronchetti, A Scala, IM Taino, FM Albini, L Toma. Tetrahedron: Asymmetry 5:1377–1384, 1994.
272. B Morgan, G Bylinsky, DR Dodds. Tetrahedron: Asymmetry 6:1765–1772, 1995.
273. G Guanti, L Banfi, S Brusco, E Narisano. Tetrahedron: Asymmetry 5:537–540, 1994.
274. A Fadel, P Arzel. Tetrahedron: Asymmetry 8:283–291, 1997.
275. B Morgan, BR Stockwell, DR Dodds, DR Andrews, AR Sudhakar, CM Nielsen, I Mergelsberg, A Zumbach. J Am Oil Chem Soc 74:1361–1370, 1997.
276. RN Patel, A Banerjee, LJ Szarka. J Am Oil Chem Soc 73:1363–1375, 1996.
277. W Adam, C Mock-Knoblauch, CR Saha-Möller. Tetrahedron: Asymmetry 8:1947–1950, 1997.
278. U Ader, MP Schneider. Tetrahedron: Asymmetry 3:205–208, 1992.
279. M Lee. J Chem Educ 75:217–219, 1998.
280. MJ Hernáiz, JM Sánchez-Montero, JV Sinisterra. Tetrahedron 50:10749–10760, 1994.
281. G Dibono, A Scilimati. Synthesis 699–702, 1995.
282. LT Kanerva. Acta Chem Scand 50:234–242, 1996.
283. LT Kanerva, O Sundholm. J Chem Soc Perkin Trans I 1385–1392, 1993.
284. G Lin, FC Wu, SH Liu. Tetrahedron Lett 34:1959–1962, 1993.
285. P D'Arrigo, L Deferra, V Piergianni, A Ricci, D Scarcelli, S Servi. J Chem Soc Chem Commun 1709–1710, 1994.
286. KS Bruzik, ZW Guan, S Riddle, MD Tsai. J Am Chem Soc 118:7679–7688, 1996.
287. M Ors, A Morcuende, MI Jiménez-Vacas, S Valverde, B Herradón. Synlett 449–451, 1996.
288. HG Gais In: K Drauz, H Waldmann, eds. Handbook of Enzyme Catalysis in Organic Synthesis. Weinheim: VCH, 1995.
289. L Heiss, HJ Gais. Tetrahedron Lett 36:3833–3836, 1995.
290. J Ross, JFJ Engbersen, IK Sakodinskaya, W Verboom, DN Reinhoudt. J Chem Soc Perkin Trans I 2899–2905, 1995.
291. E Höft, HJ Hamann, A Kunath, W Adam, U Hoch, CR Saha-Möller, P Schreier. Tetrahedron: Asymmetry 6:603–608, 1995.
292. S Brand, MF Jones, CM Rayner. Tetrahedron Lett 36:8493–8496, 1995.
293. E Vanttinen, LT Kanerva. Tetrahedron: Asymmetry 6:1779–1786, 1995.

17

Stereoselective Nitrile-Converting Enzymes

Marco Wieser and Toru Nagasawa
Gifu University, Gifu, Japan

I. INTRODUCTION

Nitriles are of synthetic importance owing to the ease with which they can be obtained by chemical synthesis [1] and their further transformation into higher value amides and acids. However, the chemical conversion of nitriles to the corresponding amides and acids usually requires harsh conditions including elevated temperature, the addition of expensive or elaborate reagents such as heavy metal catalysts or strongly acidic (6 N HCl) or basic (2 N NaOH) conditions with the need for a subsequent neutralization step generating huge amounts of salts [2–4]. The enzymatic hydrolysis of nitriles bears advantages regarding mild reaction conditions and high chemo-, regio-, and stereoselectivity leading to high-purity products and the absence of byproducts, salts, and metal wastes while realizing energy savings. The enzymatic conversion can proceed via two routes (Fig. 1): via nitrile hydratases to the corresponding amides, which are further hydrolyzed by amidases to the corresponding acids, or via the direct hydrolysis to the acids by nitrilases [5].

Nitrile-converting enzymes often exhibit a high activity and a broad substrate specificity, which are important characteristics for their development potential as industrial catalysts. The most famous example of an industrially applied nitrile-converting enzyme is the nitrile hydratase from *Rhodococcus rhodochrous* J1 [6] employed as the third-generation biocatalyst for the large-scale production (about 40,000 tons/year, 1998) of the plastic monomer acrylamide from acrylonitrile by the Mitsubishi Rayon Co. (formerly Nitto Chemical Co.) in Japan [7–9] and the biotechnological production (3000 tons/year) of the feed vitamin nicotinamide by Lonza AG in Switzerland [10]. Due to its high activity and stability, the *R. rhodochrous* J1 catalyst is also of potential interest for other nitrile conversions using either its nitrile hydratase [11–13] or its nitrilase [14]. After establishment of industrial nitrile hydratase processes, it is expected that nitrile-converting enzymes additionally have a great potential for the preparation of chiral building blocks as biologically active precursors of pharmaceuticals and agrochemicals. So far, the ability of nitrile-converting enzymes to catalyze stereoselective conversions is vastly unexploited considering the successful application of lipases and esterases in enantioselective preparations. However, recently the number of reports on stereoselective nitrile-converting enzymes has increased [15], in part driven by the release of U.S. Food and Drug Administration's new marketing guidelines demanding the use of the more effective stereoisomer in commercial products with an asymmetric center before clinical approval [16]. In this work, stereoselective enzymes related to nitrile metabolism, which encompass the directly nitrile-con-

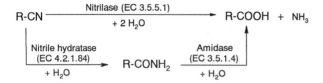

Figure 1 Nitrile-converting enzymes.

verting nitrile hydratases and nitrilases and the closely related amidases and nitrile-forming enzymes, are reviewed.

The Cahn-Mangold-Prelog (*R*,*S*) system was used here to describe the chirality of molecules, and the enantiomeric excess (e.e.) to quantify the enantiomeric purity. The e.e. values, which are usually obtained from direct concentration determinations by high-performance liquid chromatography (HPLC) or gas chromatography (GC), have replaced the operational term "optical purity" in the last decades due to a high sensitivity of optical rotations to experimental conditions [17]. However, e.e. values that were calculated in several cases from the optical rotation and the DL system of α-amino acids based on the optical rotation were also adopted from the literature. On the other hand, neither the directions in which enantiomers rotated plane-polarized light (+ or −) nor specific rotations α were stated here.

II. STEREOSELECTIVE NITRILE CONVERSIONS WITHOUT ENZYME SPECIFICATION

In a few studies of whole cell bioconversions, the type of stereoselective nitrile-converting enzymes was not specified. One of the earliest reports on stereoselective nitrile-converting enzymes describes the racemic resolution of racemic α-hydroxy nitriles (cyanohydrins) to give optically active L-α-hydroxy acids by whole cells of *Torulopsis candida* GN405 (Fig. 2). The strain was isolated from an enrichment using racemic hydroxyisovaleronitrile as the sole source of nitrogen [18]. L-α-Hydroxyisovaleric acid and L-α-hydroxyisocaproic acid were isolated as products after bioconversion of the corresponding DL-nitriles.

Another example of a nitrile conversion without enzyme specification is the biotransformation of (*R*,*S*)-3-phenyllactonitrile (phenylacetaldehyde cyanohydrin) to (*S*)-3-phenyllactic acid, a precursor for the synthesis of pharmacophores including renin inhibitors, protease inhibitors, and anti-HIV reagents, by *Pseudomonas* sp. BC-18 [19]. The enantiomeric excess of 75% after biotransformation was increased to 99.8% by repeated crystallization, and the production was enhanced by the addition of 2% (w/v) $CaCl_2$ to the reaction mixture for the precipitation of the (*S*)-acid. In a mutant strain, the specific activity and the final accumulation were enhanced 16- and 1.2-fold, respectively, compared to the parent strain. A product yield of 70% after biotransformation was obtained due to the

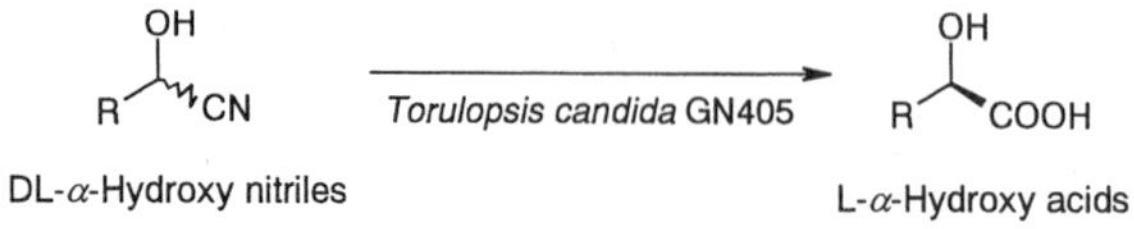

OH   OH
| |
R⟍⫲CN   ⟶ *Torulopsis candida* GN405   R⟍COOH

DL-α-Hydroxy nitriles   L-α-Hydroxy acids

Figure 2 Stereoselective conversion of racemic α-hydroxynitriles to L-α-hydroxy acids by *Torulopsis candida* GN 405.

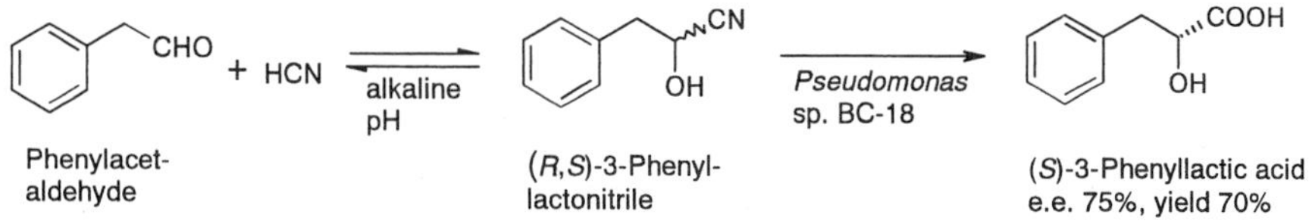

Figure 3 Enantioselective preparation of (*S*)-3-phenyllactic acid from (*R*,*S*)-3-phenyllactonitrile by *Pseudomonas* sp. BC-18.

partial dissociation of the cyanohydrin into the corresponding aldehyde and HCN and therefore the racemization of remaining (*R*)-cyanohydrin during the reaction according to the dissociation equilibrium (Fig. 3). As a result of an accelerated dissociation at alkaline pH, (*S*)-3-phenylacetic acid with the highest optical purity was obtained at pH 8.5.

III. STEREOSELECTIVE NITRILASES

Nitrilases catalyze the direct hydrolysis of nitriles to the corresponding acids. Compared to nitrile hydratases, more nitrilases bearing a high stereoselectivity have been described. Nitrilases generally comprise of one type of subunit in multiple association (α_{6-20}) with a few monomeric (α) and dimeric (α_2) exceptions. As a rough rule, they are mostly active with aromatic nitriles but exhibit high activity toward aliphatic nitriles less often [20–24].

Stereospecific nitrilases were used for the conversion of α-aminonitriles to optically active L-amino acids (Fig. 4). In an early investigation, L-alanine was formed by an L-specific nitrilase from alginate-immobilized cells of *Acinetobacter* sp. APN [25]. A decrease of the enantioselectivity with the time was supposed to be caused by a racemase forming D- from L-alanine. The stereoinversion of racemic α-aminopropionitrile led to a conversion yield above 50%. Similar L-α-amino acid preparations showed no stereoinversion and additionally accumulated the D-amide due to the presence of a nitrile hydratase/amidase system [26,27]. Additionally, a number of L-α-amino acids were synthesized by a 45-kDa monomeric nitrilase from *R. rhodochrous* PA-34 [28]. Remarkable in this case was the preferential hydrolysis of α-aminopropionitrile to D-alanine in contrast to the L-alanine formation by the *Acinetobacter* nitrilase (Fig. 4).

Rhodococcus rhodochrous NCIMB 11216 bears an enantioselective nitrilase, which is active toward a wide range of aliphatic nitriles with C-2 group substitutions [29]. The highest stereoselectivity was reached during biotransformation of (*R*,*S*)-2-methylhexani-

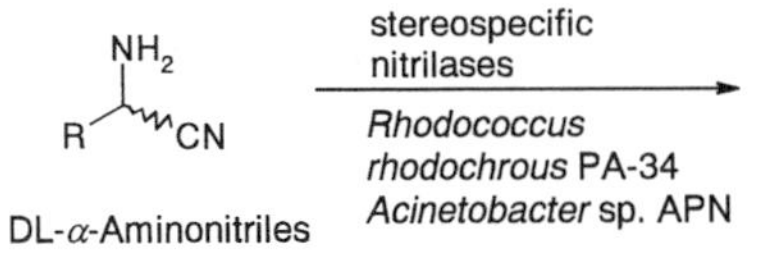

α-Amino acids	e.e. [%]
L-Leucine	90
D-Alanine	57
L-Valine	100
L-Norvaline	92
L-Norleucine	100
L-Methionine	96
Acinetobacter sp. APN: L-Alanine	74

Figure 4 Preparation of chiral α-amino acids by stereospecific nitrilases from *Rhodococcus rhodochrous* PA-34 and *Acinetobacter* sp. APN.

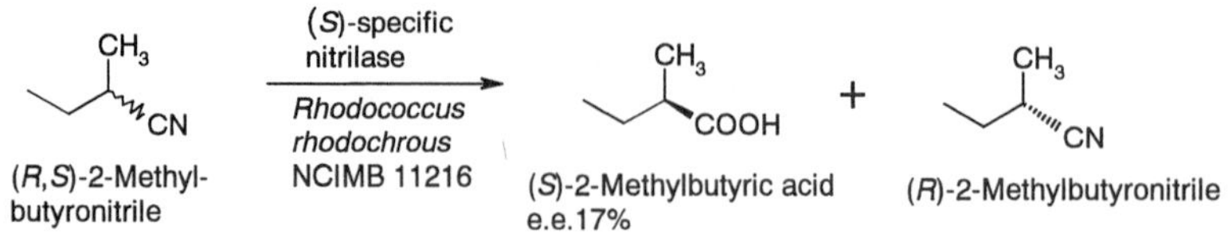

Figure 5 Racemic resolution of (R,S)-2-methylbutyronitrile by a nitrilase from *Rhodococcus rhodochrous* NCIMB 11216.

trile, where the reaction appeared enantiospecific for the (S)-nitrile with a kinetic ratio of 45:1. Evidence for a nitrilase was given by the lack of stimulation of the respiration rate of active cells by various amides and the inability to biotransform amides. However, only a moderate enantiospecificity, depending on the temperature and decreasing with time, was found. For example, during hydrolysis of (R,S)-2-methylbutyronitrile using cell-free extract, 100% e.e. was only initially reached after 11% conversion at 4°C [30]. At 30°C, the e.e. was 17% for the (S)-methylbutyric acid at the time for optimal e.e. of the remaining (R)-nitrile (Fig. 5). Therefore, the biocatalyst seemed to be more suitable for the preparation of the (R)-nitriles than the (S)-acids. With whole cells at 30°C, no enantioselectivity was found. Chiral 2-methylbutyric acid has also been prepared by a nitrile-hydrolyzing *Nocardia* sp. [31].

(R)-Mandelic acid, a widely used optical resolving agent and a precursor of semi-synthetic cephalosporins, has been obtained in optically pure form by a nitrilase-catalyzed hydrolysis of (R,S)-mandelonitrile) employing resting cells of *Alcaligenes faecalis* ATCC 8750 [32]. The process leads to (R)-mandelate in 91% yield due to a sufficient in situ racemization of the substrate facilitated by the spontaneous decomposition of mandelonitrile at slightly alkaline pH to give benzaldehyde and hydrogen cyanide (Fig. 6).

Due to the dissociation equilibrium, (R)-mandelate was also produced from benzaldehyde and HCN using whole cells. Substrate racemization has also been observed to a certain extent in related studies [19,25,33], but a similarly efficient dynamic resolution has not been achieved [34]. The strain has an additional amidase for mandelamide, but seems to have no nitrile hydratase for mandelonitrile. The stereospecific nitrilase has been purified and characterized as a typical 460-kDa homomultimer consisting of 32-kDa subunits [35].

The bioconversion of racemic mandelonitriles using stereoselective nitrilases was recently expanded for the commercial production of enantiomerically pure (R)-mandelic acid [36] and (R)-3-chloromandelic acid [37] by the Mitsubishi Rayon Company. This company has described further stereoselective nitrilases from a number of microorganisms (*Rhodococcus, Acinetobacter, Caseobacter, Aureobacterium, Alcaligenes, Pseudomonas, Nocardia, Gordona, Brevibacterium*), which were suitable for the synthesis of enantiomerically pure (R)-mandelic acid derivatives bearing 2-chloro, 4-chloro, 4-bromo, 2-fluoro, 4-methyl, 4-methoxy, 4-methylthio, and 4-nitro substituents [37].

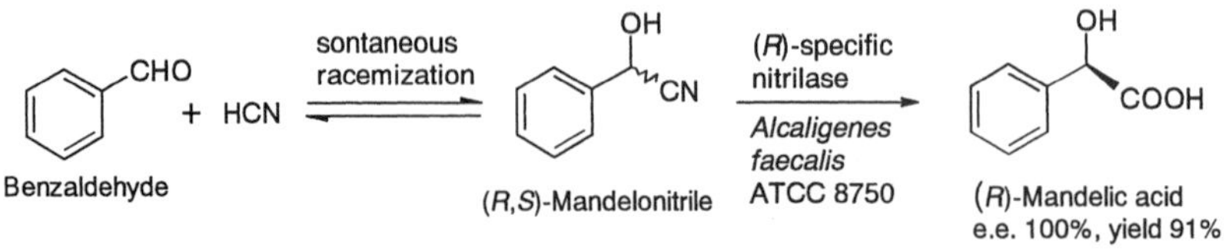

Figure 6 Synthesis of (R)-mandelic acid from (R,S)-mandelonitrile by an (R)-specific nitrilase from *Alcaligenes faecalis* ATCC 8750.

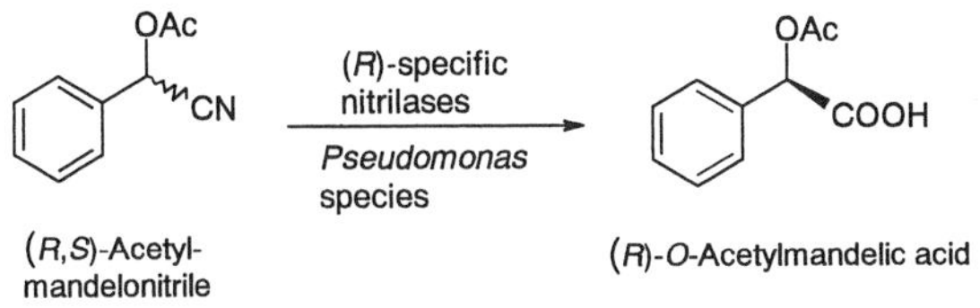

Figure 7 Synthesis of (R)-O-acetylmandelic acid by an (R)-specific nitrilase from *Pseudomonas* sp.

Nitrilases with moderate enantiospecificites toward the conversion of (R,S)-O-acetylmandelonitrile to (R)-acetylmandelic acid (Fig. 7) were found in several *Pseudomonas* species enriched with benzyl cyanide, α-methyl-, α-ethyl-, or α-methoxybenzylcyanide as the sole source of nitrogen [38]. In this reaction, mandelic acid and O-acetylmandelic acid amide were formed as byproducts, which was explained by an additional esterase activity and a nitrile hydratase side activity of the nitrilase, respectively. The chirality of the amide was not determined, but mandelic acid surprisingly had a slight excess of the (S) isomer. Instead of e.e. values, enantiomeric turnover ratios were given, varying from 10:1 to 2:1 for (R)-acetylmandelic acid and 5:1 to 1:1 for (S)-mandelic acid depending on the strain. An (R)-selective nitrilase rather than a nitrile hydratase/amidase system was supposed to be active because no amidase activity converting O-acetylmandelic acid amide could be detected. Compared with free cyanohydrins (Fig. 6), O-acetylcyanohydrins show no spontaneous racemization.

The application of the immobilized nitrilase SP 409 of *Rhodococcus* sp. from Novo Industri (Denmark), which covers a wide substrate spectrum of aliphatic, aromatic, heterocylic, and carbohydrate nitriles, proved to be synthetically viable for the mild and stereoselective transformation of base-sensitive carbohydrate nitriles [39]. Preferentially the β anomer of a diastereomeric glycosyl cyanide was hydrolyzed to the corresponding acid (Fig. 8). Using a C-7 alkoxylated glycosyl cyanide, amide intermediates were also detected, indicating the additional presence of a nitrile hydratase.

A stereoselective nitrilase from *Aspergillus furmigatus* was used to prepare (S)-α-phenylglycine in an optical purity of 80% from α-aminophenylacetonitrile (Fig. 9), which is readily available from benzaldehyde by cyanoamination [40]. However, the (R)-α-phenylglycine, which is of higher commercial interest as a raw material for cephalosporins and penicillins, could not be obtained by this biocatalyst.

2-Arylpropionitriles and -amides have been frequently tested as substrates for stereoselective nitrile-converting enzymes because the acid products, also known as profens, constitute an important class of anti-inflammatory drugs [41] with only the (S) isomer being biologically active [42]. In this class, (S)-naproxen (Figs. 11 and 18), (S)-ibuprofen

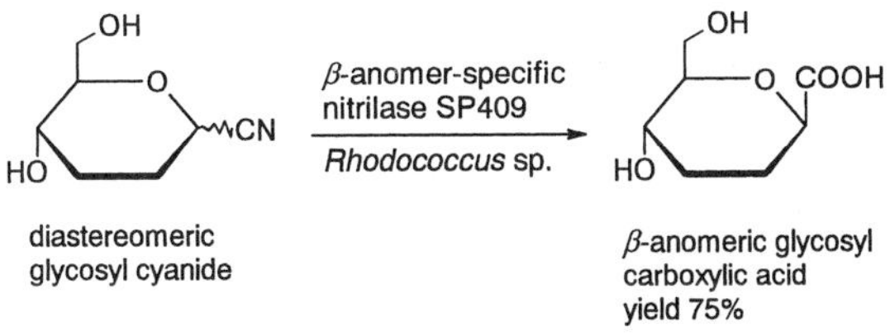

Figure 8 Stereospecific hydrolysis of a diastereomeric glycosyl cyanide by the nitrilase SP 409 from *Rhodococcus* sp.

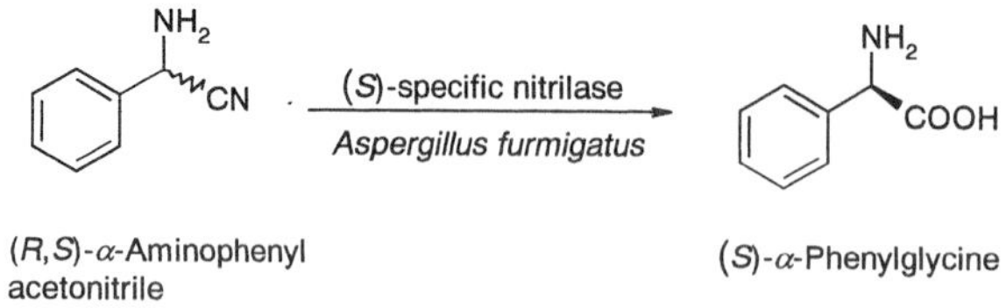

Figure 9 Synthesis of (*S*)-2-phenylglycine by a stereoselective nitrilase from *Aspergillus furmigatus*.

(Fig. 12), and (*S*)-ketoprofen (Fig. 28) are the most successfully introduced chemotherapeutic agents. Suitable enantioselective nitrile-converting enzymes have often been found, presumably in part due to the aromatic moiety of profens supporting enantiospecific recognition.

One example of an (*S*)-profen-forming enzyme is the stereoselective nitrilase from *R. rhodochrous* ATCC 21197 used for the conversion of racemic 2-aryl propionitriles to the corresponding (*S*)-acids [43,44]. The highest optical and chemical yields were obtained with 2-(4-methoxyphenyl)propionic acid (Fig. 10). In this *Rhodococcus* strain all three nitrile-converting enzymes have been found (Fig. 16, 21, and 27).

The stereospecific nitrilase of this strain was also able to convert naproxen nitrile to (*S*)-naproxen [(*S*)-2-(6-methoxy-2-napthyl)propionic acid]. However, a high enantioselectivity was only found at the expense of a low chemical yield (Fig. 11). (*S*)-Naproxen was furthermore synthesized by stereospecific nitrile hydratase/amidase enzyme systems from other *Rhodococcus* strains (Fig. 18).

(*S*)-Ibuprofen [(*S*)-2-(4-isobutylphenyl)propionic acid] has been prepared by resting cells of *Acinetobacter* sp. AK 226 [34]. Evidence for a single-step nitrilase-catalyzed reaction was obtained by the lack of an intermediate amide formation, the remaining (*R*)-nitrile (Fig. 12), and the inability to transform the racemic amide. The (*S*)-specific nitrilase was purified and characterized as a typical high molecular mass enzyme of 580 kDa [45].

IV. STEREOSELECTIVE NITRILE HYDRATASES

Nitrile hydratases catalyze the first step in the two-enzyme step conversion from nitriles to the corresponding acids, the hydration of nitriles to amides. They are metalloenzymes containing iron [46–48] or cobalt [6,49,50] in a biologically unprecedented mixed sulfur/nitrogen/oxygen coordination environment and were found in a number of gram-positive and gram-negative bacteria. Nitrile hydratases have attracted substantial interest as cata-

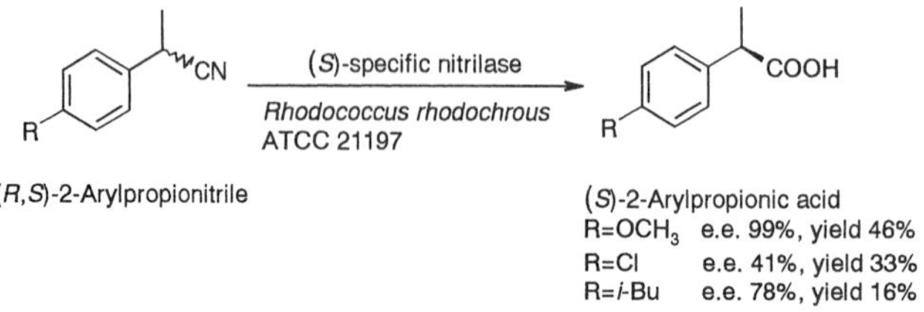

Figure 10 Synthesis of (*S*)-2-arylpropionic acids by an (*S*)-specific nitrilase from *Rhodococcus rhodochrous* ATCC 21197.

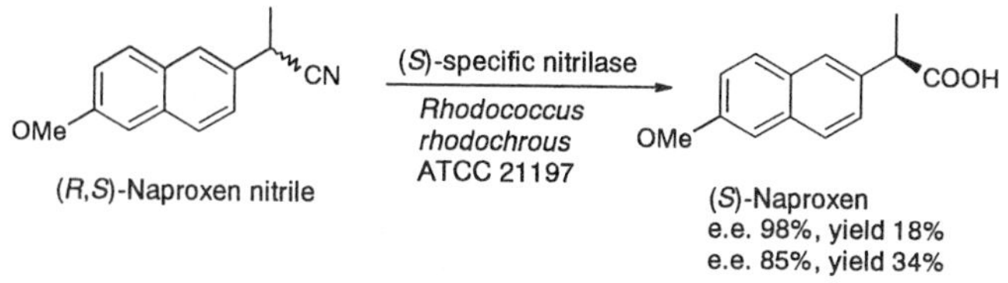

Figure 11 Preparation of (*S*)-naproxen by an (*S*)-specific nitrilase from *Rhodococcus rhodochrous* ATCC 21197.

lysts in commercial processes with the industrially applied example from *R. rhodochrous* J1 [7–10]. Usually these enzymes were found to be hetereodimers [(αβ)$_2$] with an assumed preference toward aliphatic nitrile substrates [46,48]. However, also exceptions like the high molecular mass nitrile hydratase [(αβ)$_{10}$] from *R. rhodochrous* J1 with a high activity toward aromatic nitriles have been described [6,10,13]. A number of nitrile hydratases share amino acid sequence homologies [51]. In microorganisms containing a nitrile hydratase, usually also an amidase with a corresponding substrate specificity is found. In most cases the amidase bears the higher enantioselectivity. Therefore, the enantioselectivity of a nitrile hydratase has been generally only detected with the purified enzyme, inhibition of the amidase by a specific inhibitor, or the transient or final accumulation of a chiral amide in cases of a low amidase activity. So far, only few examples of enantioselective nitrile hydratases leading to a product e.e. above 75% have been reported, as discussed in the following.

Pseudomonas putida NRRL 18668, which was recovered from soil using enrichments with (*R,S*)-2-methylglutaronitrile as sole nitrogen source, was supposed to be the first isolated gram-negative organism containing a stereospecific nitrile hydratase [52]. The strain is capable of racemic resolution of 2-(4-chlorophenyl)-3-methylbutyronitrile (2-isopropyl-4′-chlorophenylacetonitrile) to the (*S*)-amide with more than 90% e.e. at almost complete conversion [53] (Fig. 13). The corresponding (*S*)-acid was obtained by an (*S*)-specific amidase from *Pseudomonas chloraphis* B23 (Fig. 29).

Evidence for the enantioselectivity of the nitrile hydratase was given by the purified enzyme catalyzing the hydration of the (*S*)-nitrile at least 50 times faster than the hydrolysis of the (*R*)-nitrile [51]. The strain was also capable of a two-step hydrolysis of racemic ibuprofen and naproxen nitriles to the corresponding (*S*)-acids in enantiomeric purities above 90% e.e., however, with the stereoselectivity residing primarily in the amidase. In this case, the analysis of enantioselectivity was complicated due to the product inhibition of (*R*)-nitrile hydration by enzymatically formed (*S*)-amide. On the basis of the initial rate of appearance, the nitrile hydratase showed a slight preference for the (*R*) enantiomers of

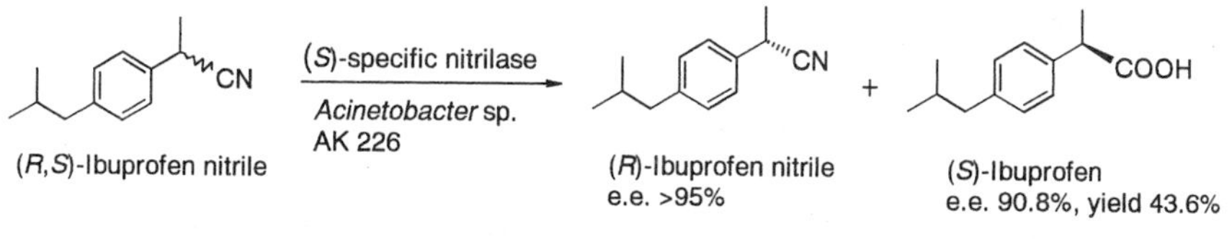

Figure 12 Synthesis of (*S*)-ibuprofen by an (*S*)-specific nitrilase from *Acinetobacter* sp. AK 226.

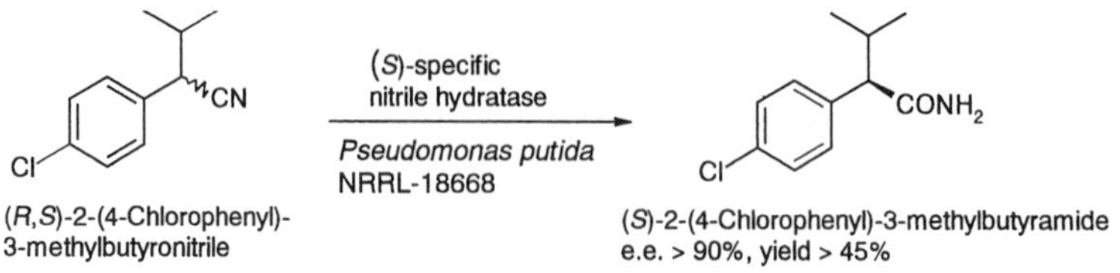

(R,S)-2-(4-Chlorophenyl)-
3-methylbutyronitrile

(S)-2-(4-Chlorophenyl)-3-methylbutyramide
e.e. > 90%, yield > 45%

Figure 13 Preparation of (S)-2-isopropyl-4'-chlorophenylacetoamide by an (S)-specific nitrile hydratase from *Pseudomonas putida* NRRL 18668.

both ibuprofen nitrile and naproxen nitrile. The enantioselectivity of the amidase was strongly influenced by the temperature with 60% e.e. at 28°C and above 90% e.e. at 50°C. As the first example of a nitrile hydratase cloning, the enzyme has been overproduced in *Escherichia coli* while retaining full stereoselectivity [54]. The *P. putida* nitrile hydratase resembles the analogous enzymes from *R. rhodochrous* J1 in its cobalt dependence [6] and highest DNA sequence homologies. Also, sequence homologies to other nitrile hydratases of 53.8–64.3% for the α subunit and 47.6–54.4% for the β subunit have been found [51].

In a number of gram-positive organisms the combination of a stereoselective nitrile hydratase and a stereoselective amidase has been described (see chapter below). However, in the case of the acetonitrile-utilizing *Rhodococcus* sp. AJ 270 no amidase activity has been detected [55]. The wide-spectrum nitrile hydratase of this microorganism was used to prepare (R)-2-phenylbutyramide from the racemic nitrile without acid as byproduct (Fig. 14).

In another rare case of the immobilized *Rhodococcus* sp. SP 361 cells from Novo Indstri, the stereoselectivity in the conversion of prochiral, protected 3-hydroxyglutaronitriles was confined to the nitrile hydratase, while the amidase was thought to be nonstereoselective [56,57]. The highest e.e. values of the resulting (S)-cyano acids were obtained from substrates bearing 3-aryl substituents (Fig. 15). In *R. rhodochrous* ATCC 21197, a bulky aryl ring in the 3-position of these substrates has been previously assumed to be essential for enantioselectivity [58]. However, such an absolute requirement could not be confirmed for *Rhodococcus* sp. SP 361 as shown by the enantioselective hydrolysis of unsubstituted 3-hydroxyglutaronitrile.

Since *Rhodococcus* sp. SP 361 lacks a nitrilase but no amide intermediates were detected, a slow, pro-(S)-selective nitrile hydratase and a fast, nonselective amidase were proposed to be active. This contrasts with most other cases bearing a fast nitrile hydratase without or with low enantioselectivity and a slow, stereospecific amidase.

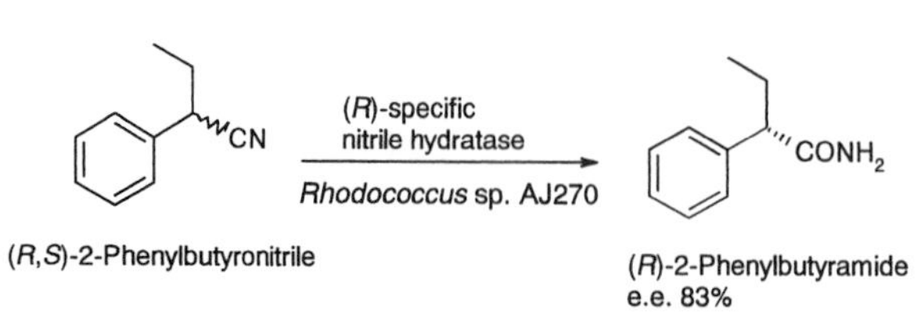

(R,S)-2-Phenylbutyronitrile

(R)-2-Phenylbutyramide
e.e. 83%

Figure 14 Synthesis of (R)-2-phenylbutyramide by an (R)-specific nitrile hydratase from *Rhodococcus* sp. AJ 270.

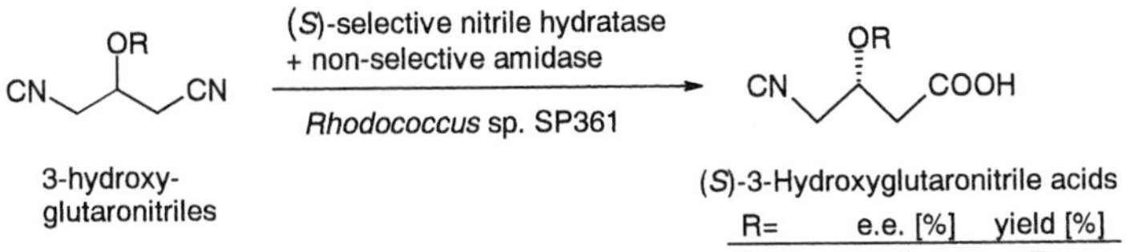

Figure 15 Asymmetric synthesis of (S)-3-hydroxyglutaronitrile acids from the prochiral dinitriles by an (S)-selective nitrile hydratase from immobilized *Rhodococcus rhodochrous* sp. SP 361.

V. STEREOSELECTIVE NITRILE HYDRATASES IN COMBINATION WITH STEREOSELECTIVE AMIDASES

Among stereoselective nitrile-converting enzymes, the combined action of a stereoselective nitrile hydratase and a stereoselective amidase has been often described, and stereospecific nitrile conversions by whole cells are frequently found in the patent literature [59,60]; however, careful analysis has usually revealed stereoselectivity in the amidase and not or only to a low extent in the hydratase [61,62]. If both enzymes were stereospecific, nitrile hydratase and amidase have been described to act either synergistically or antagonistically regarding their enantioselectivity.

Besides an (S)-specific nitrilase and an (S)-specific amidase, *Rhodococcus rhodochrous* ATCC 21197 was supposed to contain a partially (R)-selective nitrile hydratase activity toward (R,S)-2-arylpropionitriles [43] (Fig. 16).

Immobilized *Rhodococcus* sp. SP 361 is an example that shows how complex stereoselective nitrile conversions might sometimes appear [57]. In the bioconversion of (R,S)-2-alkylarylacetonitriles an unusual enantioselective diversity was observed. Whereas most racemic 2-alkylarylacetonitriles were converted to (S)-acids and (R)-amides (Fig. 17), indicating the presence of an (S)-selective amidase, ibuprofen nitrile was only converted to the (R)-acid (e.e. 32%) without any intermediate amide formation. A slow and strictly (R)-specific nitrile hydratase and a fast and less strict (S)-amidase were accounted for the (R)-specific ibuprofen synthesis. The aryl-bound isobutyl moiety of ibuprofen nitrile seemed to invert the molecule orientation at the catalytic center of the nitrile hydratase. Furthermore, in the conversion of (R,S)-2-(4-methylphenyl)propionitrile, not only the chiral (S)-acid (e.e. more than 95%, yield 41%) and (R)-amide (e.e. more than 95%, yield 18%), but also enantiomerically almost pure (R)-nitrile (e.e. more than 95%, yield 25%) was obtained. In this instance, the nitrile was hydrated with a partial (S)-selectivity. Overall, the absolute configuration of the products was rationalized according to a model assuming the presence of an (S)-selective amidase and a nitrile hydratase with (S)-, (R)-, or nonspecificity depending on the type of substrate.

Furthermore, *Agrobacterium tumefaciens* d3, which was enriched on racemic 2-phenylpropionitrile as the sole source of nitrogen, was used for the conversion of 2-arylpropionitriles to the corresponding (S)-acids [63]. The conversion of both racemic 2-phenylpropionitrile and -amide by whole cells led to the formation of the (S)-acid with an e.e. above 96% after 47.5% conversion and was caused by the combined action of an (S)-specific nitrile hydratase and a highly (S)-specific amidase (Fig. 17). The stereoselectivity of the nitrile hydratase was used for the preparation of (S)-amides by whole cells in the presence of an amidase inhibitor or with the purified nitrile hydratase [64]. The highest

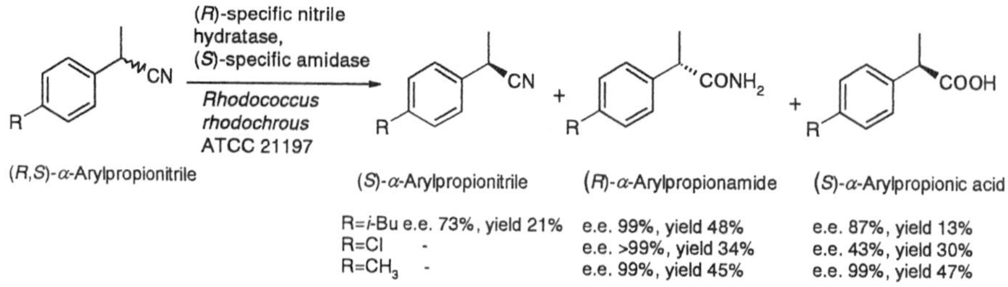

Figure 16 Conversion of (R,S)-2-arylpropionitriles by an (R)-selective nitrile hydratase and an (S)-specific amidase from *Rhodococcus rhodochrous* ATCC 21197.

enantiomeric excesses (e.e. more than 90%, yield 30%) were found in the amide products of 2-phenylpropionitrile, 2-phenylbutyronitrile, and ketoprofen nitrile using the purified enzyme. The enantioselectivity of the whole-cell hydration was increased by elevated temperatures.

Together with *R. rhodochrous* ATCC 21197 [43] and *Pseudomonas putida* NRRL 18668 [51], also *Rhodococcus* sp. C3II and *Rhodococcus erythropolis* MP 50 were used for the enantiospecific preparation of (S)-naproxen [65]. *Rhodococcus* sp. C3II lacks a nitrilase but exhibits nitrile hydratase and amidase activities, both of which are constitutive and prefer the (S)-enantiomers of naproxen derivatives. On the other hand, the enzymes from *R. erythropolis* MP 50 were induced by nitriles and its nitrile hydratase was (R)-specific [44]. Due to the presence of a strictly (S)-specific amidase, both strains finally formed (S)-naproxen with high enantioselectivity (Fig. 18). Evidence for the enantioselectivity of the nitrile hydratases of both strains was obtained by the formation of optically active amides in the presence of the amidase inhibitor diethyl phosphoramidate [63,65]. The nitrile hydratase of *Rhodococcus* sp. C3II whole cells was used for the synthesis of (S)-naproxen amide with 94% e.e. after 30% conversion in the presence of the amidase inhibitor [63]. In addition, the highly stereoselective amidases of these two strains were used to prepare (S)-ketoprofen (Fig. 28), and the amidase from *R. erythropolis* MP 50 was used to prepare (S)-2-phenylpropionic acid with more than 99% e.e. and more than 49% conversion [66,67].

Pseudomonas sp. B21C9 was isolated from soil by an enrichment culture using (R,S)-2-isopropyl-4-chlorophenylacetonitrile as the sole source of nitrogen [62]. The strain converted this nitrile to both enantiomerically pure (S)-acid and (R)-amide (Fig. 19). In order to increase the yield of the (S)-acid, which is part of a commercial pyrethroid insecticide, the (R)-amide was racemized by heating. Besides a strictly (S)-specific amidase, a poorly

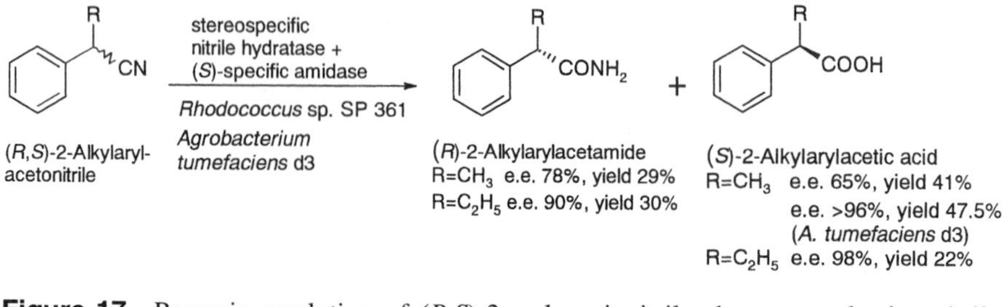

Figure 17 Racemic resolution of (R,S)-2-arylpropionitriles by stereoselective nitrile hydratase/amidase systems from *Rhodococcus* sp. SP 361 and *Agrobacterium tumefaciens* d3.

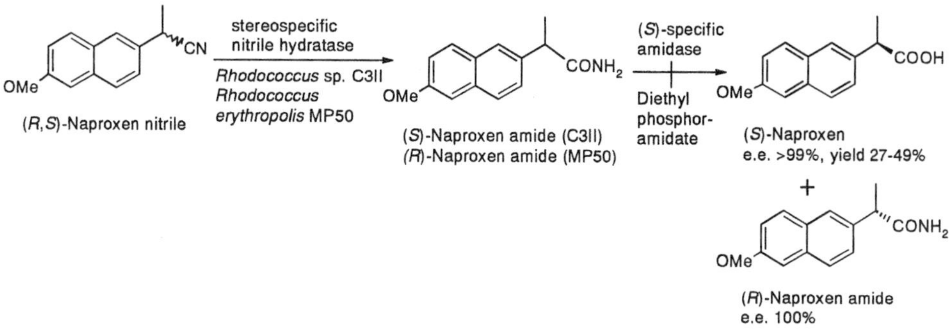

Figure 18 Stereospecific formation of the anti-inflammatory agent (S)-naproxen by the stereoselective amidase/nitrile hydratase system from two *Rhodococcus* strains.

(S)-specific nitrile hydratase was supposed to be active because in the initial stage of the reaction the amide was rich in the (S) isomer.

The stereospecific nitrile hydratase/amidase system from *Rhodococcus equi* A4 catalyzes the conversion of (R,S)-2-(4-methoxyphenyl)propionitrile and (R,S)-2-(4-chlorophenyl)propionitrile to the corresponding (S)-acids (Fig. 20). It was proposed that the electron donating effect of the methoxy group leads to a higher enantioselectivity than the electron-withdrawing chloro substituent [68]. (R,S)-2-Phenylpropionitrile itself was not enantioselectively hydrated, but the strain formed the (S)-acid from this nitrile due to its (S)-specific amidase. Partial enantioselectivity was additionally shown in the hydration of ketoprofen nitrile and the hydration plus hydrolysis of a single cyano group of aromatic nitriles. No nitrilase seemed to be present in this strain because after donation of an amidase inhibitor to a whole-cell reaction only amides but no acids were detected [69]. The (S)-specific nitrile hydratase contrasts with the analogous (R)-selective enzyme from *R. rhodochrous* ATCC 21197 (Fig. 16).

Besides the racemic resolution of 2-arylpropionitriles (Fig. 16), *R. rhodochrous* ATCC 21197 additionally catalyzes the asymmetric hydrolysis of prochiral 3-substituted glutaronitriles [56]. The highest enantioselectivity was found in the conversion of the benzoate ester of 3-hydroxyglutaronitrile to the corresponding nitrile monocarboxylic acid (e.e. more than 99%) by an enzyme system consisting of an (R)-specific nitrile hydratase and an (S)-specific amidase. The enantioselectivity depended on the number of oxygen atoms between the carbon chain and the aryl substituent (Fig. 21).

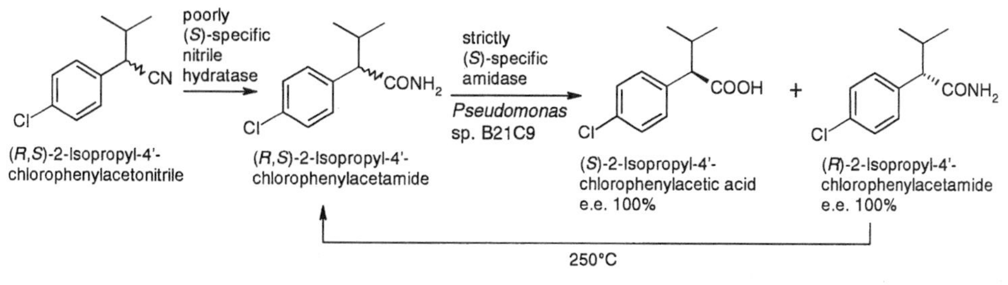

Figure 19 Preparation of (S)-2-isopropyl-4-chlorophenylacetic acid by the combined action of both stereospecific nitrile hydratase and amidase from *Pseudomonas* sp. B21C9.

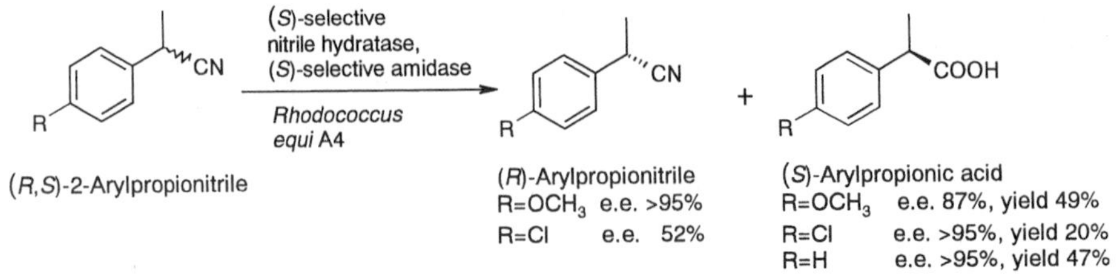

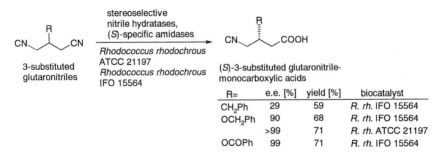

Figure 20 (*S*)-Selective conversion of (*R*,*S*)-2-(4-methoxyphenyl)propionitrile and (*R*,*S*)-2-(4-chlorophenyl)propionitrile by *Rhodococcus equi* A4.

In contrast, the nitrile hydratase of *R. rhodochrous* IFO 15564, which was isolated by a screening using 3-hydroxypropionitrile and benzonitrile as the sole sources of nitrogen, was found to be (*S*)-selective toward substituted 3-hydroxyglutaronitriles [70]. Together with an (*S*)-specific amidase, it converted 3-benzoyloxyglutarodinitrile to the sole product (*S*)-cyanocarboxylic acid in an optically pure form without leaving any intermediate (*R*)-cyano amide (Fig. 21).

VI. STEREOSELECTIVE AMIDASES

Amidases catalyze the second step in the two-enzyme step conversion of nitriles to the corresponding acids, the hydrolysis of amides. Among stereoselective nitrile-converting enzymes, the highest enantioselectivities were found in amidases. Amidases have been mostly characterized as thiol-dependent homodimers (α_2) sharing amino acid sequence homologies with other amidases. The functional relationship between nitrile hydratases and amidases is structurally documented by the close proximity of their genes [71,72]. In this chapter, stereospecific amidases alone or with nonselective nitrile hydratases are described.

Brevibacterium sp. R312, which was enriched on acetonitrile as sole nitrogen-source, contains two amidases: a nonstereoselective, wide-spectrum acylamide amidohydrolase [73] and an enantioselective α-aminoamidase. In early reports of stereoselective nitrile bioconversions, the α-aminoamidase has been used to prepare optically active L-α-amino acids from racemic α-aminonitriles [26] and α-amino amides [74]. Around 50% L-α-amino

R=	e.e. [%]	yield [%]	biocatalyst
CH_2Ph	29	59	*R. rh.* IFO 15564
OCH_2Ph	90	68	*R. rh.* IFO 15564
	>99	71	*R. rh.* ATCC 21197
OCOPh	99	71	*R. rh.* IFO 15564

Figure 21 Asymmetric synthesis of (*S*)-nitrile monocarboxylic acids from prochiral 3-substituted glutaronitriles by *Rhodococcus rhodochrous* ATCC 21197 and *Rhodococcus rhodochrous* IFO 15564.

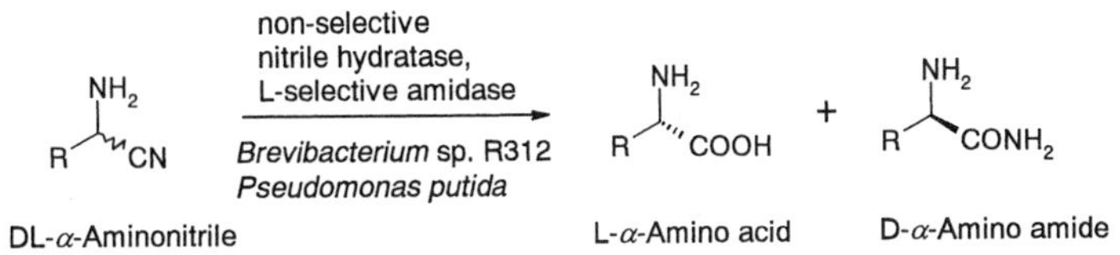

Figure 22 Formation of L-α-amino acids by L-selective amidases from *Brevibacterium* sp. R312 and a *Pseudomonas putida* strain.

acid and 50% D-α-amino amide were obtained indicating an L-selective amidase (Fig. 22). In contrast, no stereoselectivity was found in the conversion of γ-alkoxy-α-aminonitriles by this organism [75]. Using a *Pseudomonas putida* strain, the same reaction of DL-α-aminonitrile to L-amino acid and D-amide was found presumable due to a similar L-specific amidase [27].

The enantiospecific *Brevibacterium* amidase was characterized as a typical homodimer consisting of two 49-kDa subunits with sequence homologies to other amidases, and the corresponding gene was cloned and overexpressed in *E. coli* [76,77]. Its additional (*R*)-specific activity toward several 2-aryl- and 2-aryloxypropionamides was used for the racemic resolution of 2-(4-hydroxyphenoxy)propionamide (Fig. 23).

A stereospecific amidase from a *Rhodococcus* strain showed a kinetic preference in the hydrolysis of (*S*)-2-arylpropionamides resulting in the formation of the corresponding (*S*)-arylpropionic acids with high enantiomeric purity (Fig. 24). The amidase gene, which shows significant sequence homologies with other stereoselective amidases [67,70,79], was isolated and expressed in *Brevibacterium lactofermentum* [78]. The homodimeric amidase is genetically coupled to a nitrile hydratase, both of which were supposed to be translated from a polycistronic mRNA from a nitrile utilization operon. Furthermore, a possible correlation between the (*S*)-enantioselectivity toward 2-aryl propionamides and an amino acid consensus sequence was discussed [76].

Besides a highly active nitrile hydratase employed in the industrial production of acrylamide and nicotinamide [7–10] and a nitrilase [14,20–22], *R. rhodochrous* J1 contains an amidase with a high (*S*)-specificity in the hydrolysis of 2-phenylpropionamide. The corresponding gene was cloned and overexpressed in *E. coli* [71,72]. The recombinant enzyme was used for the preparation of (*S*)-2-phenylpropionic acid with high enantiomeric purity (Fig. 24) but could not recognize the configuration of 2-chloropropionitrile presumably due to the requirement of a bulky moiety for enantioselectivity.

The same (*S*)-2-phenylpropionic acid product was obtained in high enantiomeric purity, yield, and production (100 g/L) from the racemic nitrile by the combined action of a nonspecific nitrile hydratase and an (*S*)-specific amidase using whole cells of *Rho-*

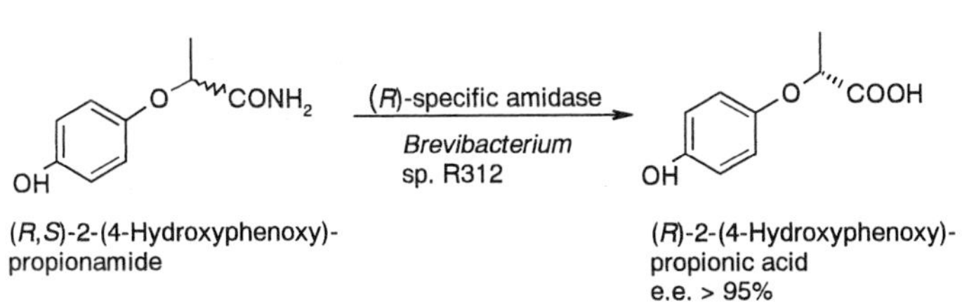

(*R,S*)-2-(4-Hydroxyphenoxy)-
propionamide

(*R*)-2-(4-Hydroxyphenoxy)-
propionic acid
e.e. > 95%

Figure 23 Formation of the (*R*)-2-(4-hydroxyphenoxy)propionic acid by a stereospecific amidase from *Brevibacterium* sp. R312.

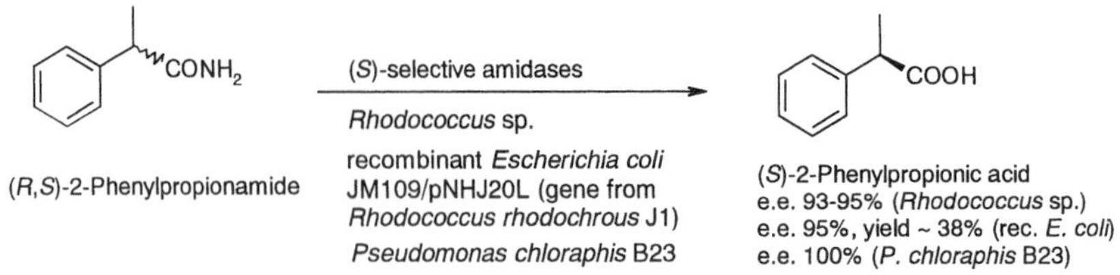

Figure 24 Preparation of (S)-2-phenylpropionic acids by an (S)-specific amidase from different strains.

dococcus equi TG 328, which was enriched on 2-methylbutyronitrile as sole source of nitrogen [61]. The enantioselectivity was increased at lowered temperatures. Also, unreacted (R)-amide was isolated with high optical purity (Fig. 25).

The same combination of a nonselective hydratase and an (S)-specific amidase in cells of R. rhodochrous IFO 15564 was employed for the racemic resolution of 2-(p-methoxyphenyl)propionitrile (Fig. 25) and 3-benzoyloxypentanamide (Fig. 26) to the corresponding (S)-acids and (R)-amides [70]. In an asymmetric synthesis using the same biocatalyst, prochiral benzylmethylmalonodinitrile was converted to the chiral amide monocarboxylic acid (Fig. 26).

Compared to a maximal bioconversion yield of 50% in racemic resolutions, asymmetric syntheses have a theoretical yield of 100%. Such an ideal yield was nearly reached with another R. rhodochrous ATCC 21197, which transformed disubstituted malononitriles such as butylmethylmalononitrile to the corresponding (R)-amide carboxylic acid with high enantiomeric excesses and yields [80]. The reaction proceeded via a fast, nonstereospecific hydration of the starting dinitrile followed by a slow, enantioselective hydrolysis of the diamide intermediate by the amidase (Fig. 27).

The (S)- and (R) forms of the anti-inflammatory agent ketoprofen [(S)-2-(3'-benzoylphenyl)propionamide] were synthesized by (S)- and (R)-selective amidases from two Rhodococcus strains [63] and Comamonas acidovorans KPO 2771-4 [81], respectively (Fig. 28). After bioconversion using C. acidovorans, which was isolated from soil using (R)-ketoprofen amide as the sole source of nitrogen, all (S)-ketoprofen amide remained in the reaction mixture. The (R)-specific monomeric amidase showed sequence homology to the amidases from Rhodococcus sp., R. rhodochrous J1, Rhodococcus N-774, Brevibabcterium sp. R312, and Pseudomonas chloraphis B23 [82].

The enantioselective amidase from R. erythropolis MP 50, which was purified and characterized as a wide-spectrum enzyme with an unusual homooctameric structure [63], was used for the stereoselective conversion of racemic 2-phenylpropionamide and na-

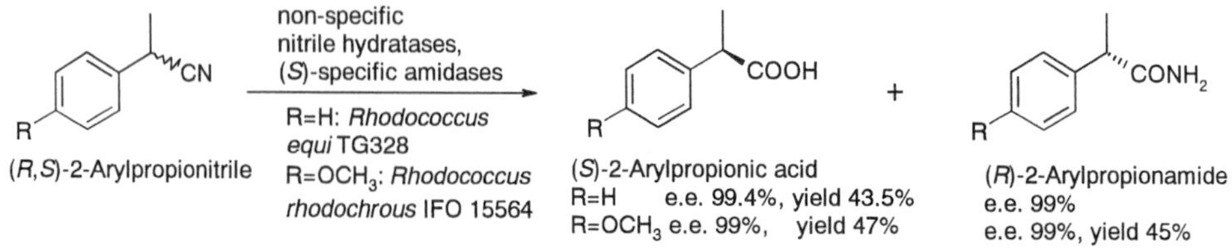

Figure 25 Preparation of (S)-2-arylpropionic acids and (R)-2-arylpropionamides from the corresponding racemic nitriles by (S)-specific amidases from a Rhodococcus equi TG 328 and Rhodococcus rhodochrous IFO 15564.

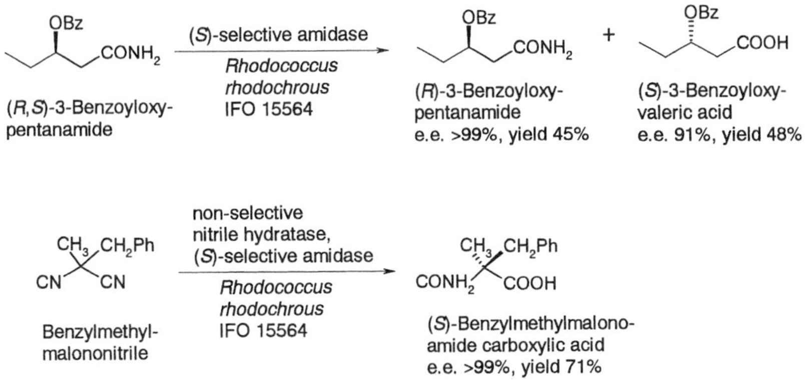

Figure 26 Racemic resolution and asymmetric synthesis catalyzed by a stereospecific amidase from *Rhodococcus rhodochrous* IFO 15564.

proxen amide to the corresponding (*S*)-acids with e.e. values of more than 99% and almost 50% conversion. The enzyme also hydrolyzed different α-amino amides but without significant enantioselectivity.

Pseudomonas chloraphis B23, the nitrile hydratase of which was used as the second-generation biocatalyst for the industrial production of acrylamide [83], contains additionally an enantioselective amidase that was purified and characterized as a typical homodimer [84]. The amidase exhibited activity against a broad range of aliphatic and aromatic amides and exhibited enantioselectivity for several aromatic amides including 2-phenylpropionamide (Fig. 24), phenylalanine amide, and 2-(4-chlorophenyl)-3-methylbutyramide (Fig. 29), but not for the amide of naproxen. The enzyme resembles in a number of characteristics other enantioselective amidases.

An (*R*)-stereospecific amidase from *B. imperiale* B222 was used for the synthesis of (*R*)-2-aryloxypropionic acids (compounds with high herbicidal activity [85]) in almost optically pure form (e.e. 86–95%). During hydrolysis of the corresponding racemic nitriles, the (*S*)-amides were recovered (e.e. 76–95%). In Fig. 30 only the e.e. values of the unsubstituted amide and acid products are shown. The racemic nitrile substrates were supposed to be quickly and nonselectively hydrated to the amides followed by a slow (*R*)-stereospecific amidase reaction [86]. This was supported by the finding of a nonspecific hydration of racemic nitriles by the partially purified nitrile hydratase and an (*R*)-specific hydrolysis of racemic amides by the partially purified amidase [87].

From a screening using racemic *N*-heterocyclic carboxamides as sole sources of nitrogen, three strains bearing stereoselective amidases were isolated and used for the

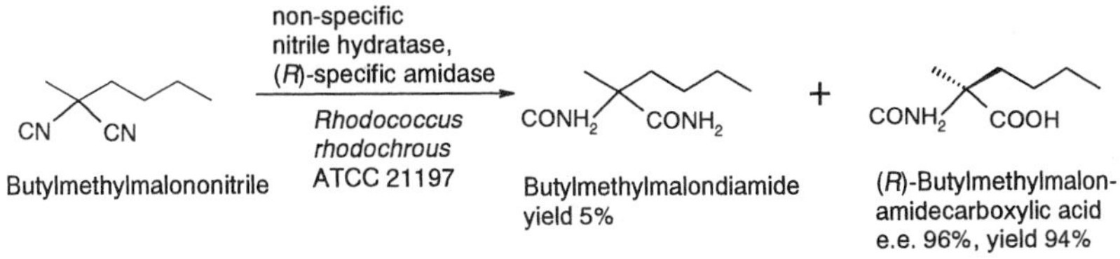

Figure 27 Asymmetric synthesis of (*R*)-butylmethylmalonamidecarboxylic acid by an (*S*)-specific amidase from *Rhodococcus rhodochrous* ATCC 21197.

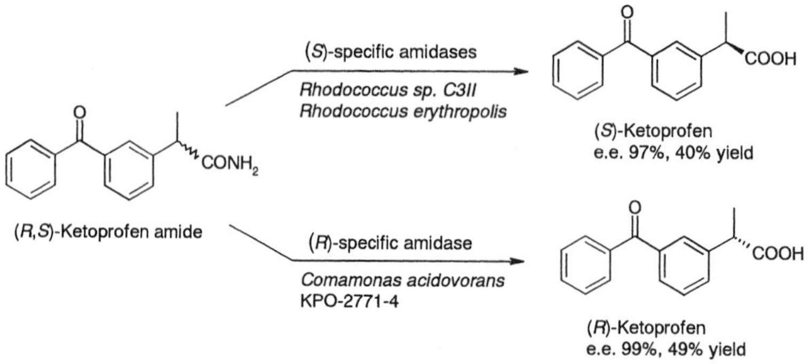

Figure 28 Racemic resolution of (R,S)-ketoprofen amide by stereospecific amidases.

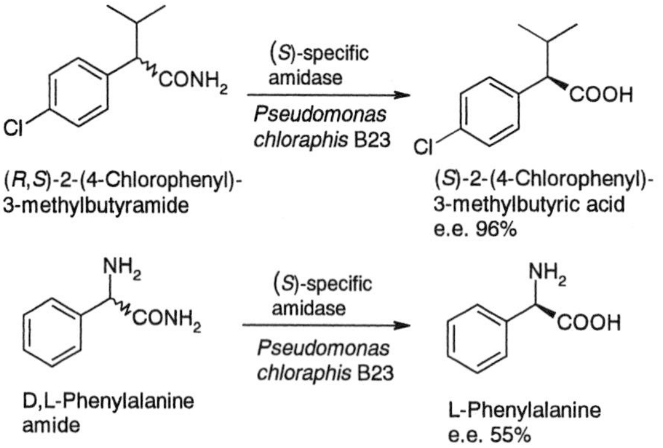

Figure 29 Racemic resolutions catalyzed by a stereospecific amidase from *Pseudomonas chloraphis* B23.

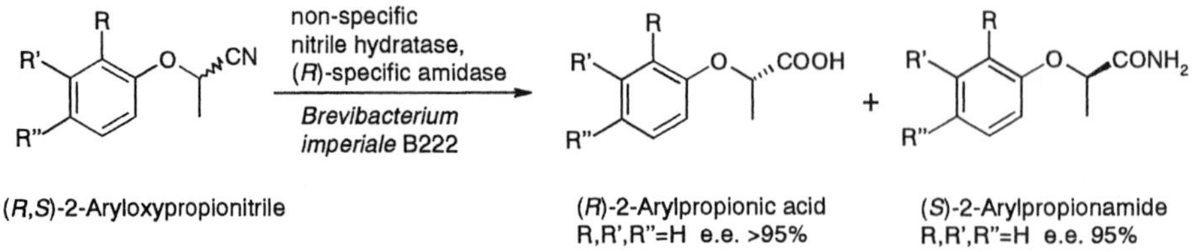

Figure 30 Synthesis of (R)-2-aryloxypropionic acids from the corresponding racemic nitriles by an (R)-specific amidase from *Brevibacterium imperiale* B222.

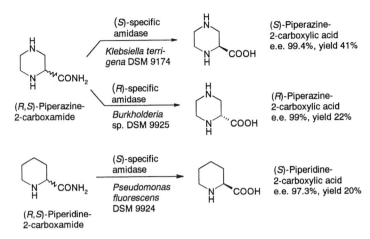

Figure 31 Racemic resolutions catalyzed by stereospecific amidases form three different microorganisms for the synthesis of chiral nonproteinogenic amino acids.

preparation of nonproteinogenic amino acids, which are precursors of numerous bioactive compounds [88]. The kinetic resolution of racemic amides to (S)- and (R)-piperazine-2-carboxylic acid and (S)-piperidine carboxylic acid in high enantiomeric purities was performed by whole cells of *Klebsiella*, *Burkholderia*, and *Pseudomonas* species, respectively (Fig. 31). As an example that shows the multiple use of nitrile-converting enzymes in one synthesis, additionally the racemic amide substrates were prepared by a nonstereospecific nitrile hydratase from the aromatic N-heterocyclic nitriles, followed by chemical reduction to the nonaromatic N heterocycles. Since (S)-piperazine-2-carboxylic acid is employed in the synthesis of an anti-HIV drug, the *Klebsiella* biotransformation is promising for commercialization by Lonza AG.

A stereospecific amidase from *Comamonas acidovorans* A18 was used for the biotechnological synthesis of (S)-dimethylcyclopropane carboxamide, which is employed as a precursor for the synthesis of the dihydropeptidase inhibitor Cilastin [89]. The (R)-acid was chemically recycled to the racemic amide (Fig. 32). The amidase gene was cloned into a faster growing *E. coli* strain leading to a constitutive enzyme expression and a shortened fermentation process.

VII. STEREOSELECTIVE NITRILE-FORMING ENZYMES

Besides the use of stereoselective nitrile-converting enzymes as described above, useful chiral building blocks have also been obtained by stereoselective nitrile-forming enzymes. The main product class of nitrile-forming enzymes are cyanohydrins (α-hydroxynitriles, 1-cyanoalkanols), which are versatile synthons in organic synthesis that are readily convertible to α-hydroxy acids [90], α-hydroxy aldehydes [91], ethanolamines [92], amino alcohols, pyrethroid insecticides [93], imidazoles, and heterocycles [94]. Examples of valuable bioactive products derived from chiral cyanohydrins are (R)-adrenaline, L-ephedrin, and (S)-amphetamines [95]. For the synthesis of chiral cyanohydrins, stereoselective enzymes from both plant and bacterial sources have been used.

Among stereoselective biotransformations based on enzymes from higher plants, mainly the hydroxynitrile lyase ("oxynitrilase")–catalyzed addition of HCN to aldehydes

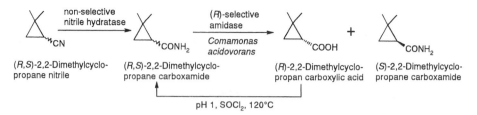

Figure 32 Synthesis of (S)-dimethylcyclopropane carboxamide by an (R)-specific amidase from *Comamonas acidovorans* A18.

has been employed for the synthesis of chiral cyanohydrins. Regarding their enantioselectivity, these enzymes can be divided into "(R)-" and "(S)-oxynitrilases" (for review, see Ref. 95).

One of the first asymmetric syntheses effected by enzymes was the preparation of (R)-mandelonitrile from benzaldehyde and HCN with emulsin as the source of the enzyme at the beginning of the twentieth century [96]. A more general procedure for the preparation of (R)-cyanohydrins using the readily accessible enzyme from bitter almonds (*Prunus amygdalus*) [97] resulted in good enantioselectivities during conversion of the natural substrate benzaldehyde [98]. In order to increase the enantioselectivity toward other aldehydes, an increasing number of these biotransformations was performed in organic solvents [99]. Recent applications of (R)-oxynitrilases for the synthesis of (R)-cyanohydrins are summarized in Fig. 33.

On the other hand, various (S)-cyanohydrins have been prepared using (S)-hydroxynitrile lyases from plants (Fig. 34). The (S)-cyanohydrins can be further converted to α-hydroxy acids by acid hydrolysis without racemization [107]. A recent example is the hydroxynitrile lyase from *Manihot esculenta*, which was cloned in *E. coli* and used as chiral catalyst for the synthesis of a broad range of optically active α-hydroxynitriles including keto-(S)-cyanohydrins using diisopropyl ether as organic solvent and HCN as cyanide source [112]. Compared to the enzymes from leaves, the overexpressed enzyme in *E. coli* showed higher enantioselectivity.

Although cyanohydrins are accessible with high enantiomeric purity via the above described oxynitrilase-catalyzed reactions, the resulting products are rather unstable with a tendency to racemization via their equilibrium with HCN and corresponding aldehydes. In contrast, enantiomerically pure cyanohydrin derivatives such as acetates are chemically quite stable also toward racemization. Therefore, bacterial ester hydrolases (esterases and lipases), which are well known for their capability of enantiomer differentiation in racemic esters and in particular commercially available lipases from *Pseudomonas* sp. (Amano, Japan), were employed for the preparation of enantiomerically pure cyanohydrin esters.

After acetylation with vinyl acetate, the Amano lipase PS was used for the (R)-specific ester cleavage of (R,S)-cyanohydrin acetate [113]. The chirality of acetylated (S)- and nonacetylated (R)-cyanohydrin products was enhanced using *R. rhodochrous* ATCC 21197 for the further conversion to optically active α-hydroxy acids (Fig. 35), which are useful starting materials in synthetic organic chemistry [90]. The *Rhodococcus* strain has been reported to contain a stereospecific nitrilase (Fig. 22), a stereospecific nitrile hydratase (Figs. 10 and 11), and a stereospecific amidase (Figs. 16 and 27).

Chiral cyanohydrins were also obtained by a lipase PS-catalyzed enantioselective esterification of racemic cyanohydrins with enol esters [114]. The lipase PS was furthermore applied for the preparation of (S)-1-acetoxy-2-arylpropionitriles by asymmetric hy-

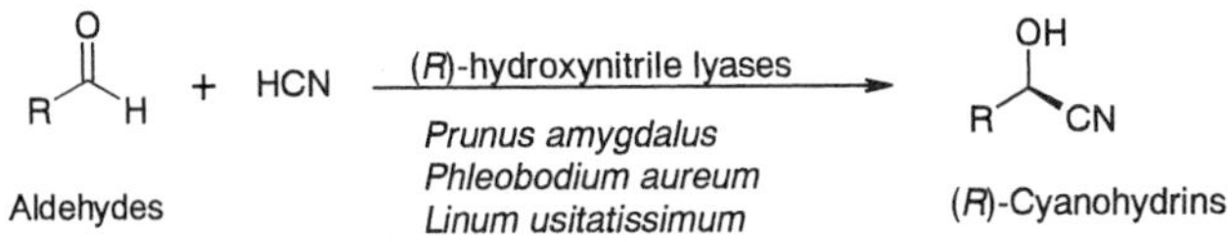

R=	e.e. [%]	yield [%]	Source of biocatalyst	Reference
Ph	94	81	Prunus amygdalus	100
	92	72	Prunus amygdalus	101
	>99	96	Prunus amygdalus	102
	99	82	Phleobodium aureum	103
	nd	nd	Linum usitatissimum	104
3,4-(CH$_2$O$_2$)Ph	93	50	Prunus amygdalus	100
	90	35	Prunus amygdalus	101
n-C$_3$H$_7$	92	100	Prunus amygdalus	100
	98	98.6	Prunus amygdalus	102
	95	100	Prunus amygdalus	105
	55	42	Phleobodium aureum	103
	nd	nd	Linum usitatissimum	104
CH$_3$CH=CH	69	99	Prunus amygdalus	100
	97	68	Prunus amygdalus	102
cC$_6$H$_{11}$	0	96	Prunus amygdalus	100
	96	72	Prunus amygdalus	101
	97	95	Prunus amygdalus	106

Figure 33 Application of (R)-hydroxynitrile lyases from plants for the synthesis of (R)-cyanohydrins.

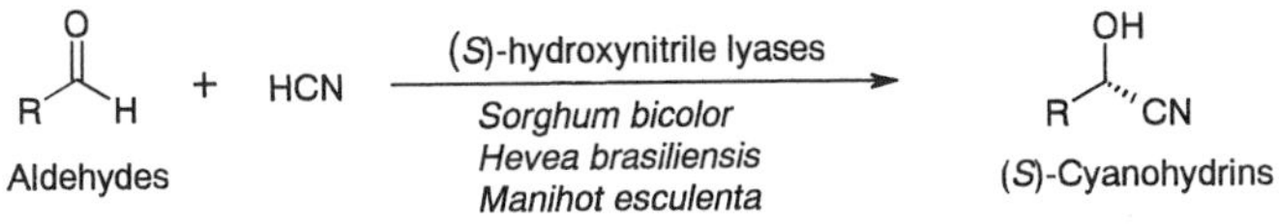

R=	e.e. [%]	Yield [%]	Source of biocatalyst	Reference
Ph	94	-	Hevea brasiliensis	108
	97	91	Sorghum bicolor	102
	90	34	Manihot esculenta	109
	98	100	Manihot esculenta	110
	>99	67	Hevea brasiliensis	111
	98	100	recombinant Escherichia coli	112
3-C$_6$H$_5$O-Ph	96	93	Sorghum bicolor	102
	99	9	Hevea brasiliensis	111
3-OH-Ph	91	97	Sorghum bicolor	102
3-thienyl	97	85	Sorghum bicolor	102
	38	90	Manihot esculenta	109
	98	98	Manihot esculenta	110
	99	49	Hevea brasiliensis	111
(CH$_3$)$_2$CH	81	-	Hevea brasiliensis	108
n-C$_3$H$_7$	95	91	Manihot esculenta	110
	80	-	Hevea brasiliensis	108
	80	70	Manihot esculenta	112
	90	92	recombinant Escherichia coli	112
i-C$_3$H$_7$	95	91	recombinant Escherichia coli	112
cC$_6$H$_{11}$	92	100	Manihot esculenta	110
	99	94	Hevea brasiliensis	111

Figure 34 Application of (S)-hydroxynitrile lyases from plants for the synthesis of (S)-cyanohydrins.

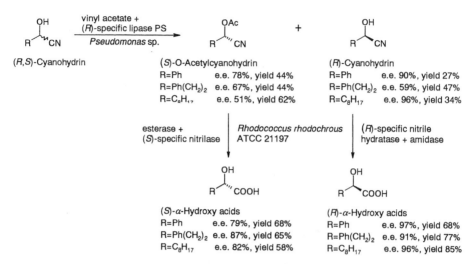

Figure 35 (*R*)-Specific conversion of cyanohydrin acetates by a *Pseudomonas* lipase PS combined with the stereospecific nitrile-converting enzymes from *Rhodococcus rhodochrous* ATCC 21197 for the synthesis of chiral α-hydroxy acids.

drolysis of the racemate (Fig. 36). The bioconversion is one of three steps in the synthesis of aryloxypropanolamines, which have hypotensive β-adrenergic blocking activity (115). A commercialized β-adrenergic blocker is (*S*)-propranolol bearing a 2-naphthyl group. Furthermore, the *Pseudomonas* lipase SAM-2 from Amano has been used for the selective hydrolysis of 47 cyanohydrin acetates of widely varying structure, which led in most cases to the remaining (*R*)-acetates in good chemical yields (30–49%) and high optical yields (e.e. more than 98%). In contrast, the actual lipase products, the cyanohydrins, were isolated with lower enantiomeric purity or in racemic form due to the equilibrium with aldehyde and HCN [116].

Other methods to introduce chirality into cyanohydrins by asymmetric hydrolysis of cyanohydrin acetates have been described with esterases of *Candida tropicalis* [117], *Bacillus coagulans* [118], and the yeast *Pichia miso* IAM 4682 [119]. *Bacillus coagulans*, for example, was used to hydrolyze asymmetricly mandelonitrile acetate esters to the chiral starting material and benzaldehydes (Fig. 37). *Meta*- and *para*-oxygen substitutions of the aromatic ring were advantageous for high enantioselectivity.

A kinetic resolution of racemic 1-cyano-1-methylalkyl and alkenyl acetates has been achieved by an esterase of *P. miso* IAM 6482, which selectively hydrolyzed the (*R*) en-

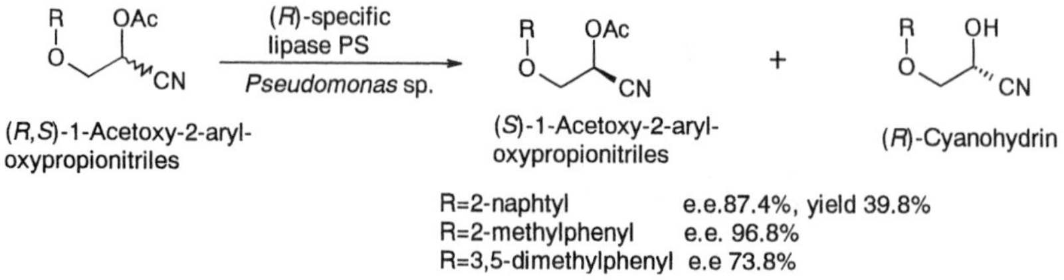

Figure 36 Racemic resolution of (*R,S*)-1-acetoxy-2-α-naphthyloxypropionitrile by the stereospecific lipase PS from *Pseudomonas* sp.

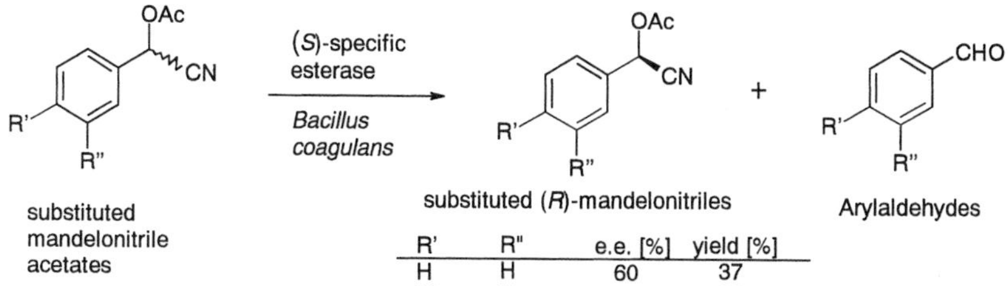

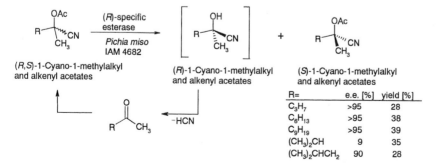

R'	R"	e.e. [%]	yield [%]
H	H	60	37
Me	H	>95	26
OMe	OMe	>95	43
OAc	OMe	>95	25
Cl	Cl	-	0

Figure 37 Application of a stereospecific esterase from *Bacillus coagulans* for the synthesis of (*R*)-mandelonitrile acetates.

antiomer, leaving behind the intact (*S*)-acetate (Fig. 38). High enantiomeric purity, determined from the optical rotation, was found in several aliphatic products [119].

Other bioprocesses for the synthesis of chiral *O*-acetylated cyanohydrins based on stereoselective esterases have been developed by the Sumitomo Chemical Co. in Japan [120], and an immobilized penicillin G acylase was used for the synthesis of (*S*)-furan-2-aldehyde cyanohydrin phenyl acetate (e.e. 72%) from the respective acylnitrile [121].

Enantioselective halohydrin hydrogen–halide lyase B from a recombinant *E. coli* bearing the corresponding gene from *Corynebacterium* sp. N-1074 was previously reported to catalyze the reversible removal of a halogen from 1,3-dichloropropanol [122] but was also found to insert a cyano group from KCN into epichlorohydrin. Using KCN as nitrile source, the enzyme was used to synthesize (*R*)-γ-chloro-β-hydroxybutyronitrile, a starting material for the synthesis of the food additive L-carnitine, from either prochiral 1,3-dichloro-2-propanol or racemic epichlorohydrin [123]. An enantiomerically almost pure product was obtained in the conversion of 1,3-dichloro-2-propanol (Fig. 39).

O-Acetylserine sulfhydrolase from *Bacillus stearothermophilus* CN3 catalyzes the substitution of an acetyl group in α-amino acids by a sulfhydryl group derived from sodium hydrogen sulfide. The presumably L-specific enzyme also acted with sodium cyanide instead of sodium hydrogen sulfide, thus incorporating a nitrile group instead of a thiol group into the acetylated amino acid [124]. As a result, β-cyano-α-amino-L-propionic acid was synthesized from *O*-acetyl-L-serine (Fig. 40). The enzyme was reported to be heat-stabile, which is of advantage for fermentation processes.

R=	e.e. [%]	yield [%]
C$_3$H$_7$	>95	28
C$_6$H$_{13}$	>95	38
C$_9$H$_{19}$	>95	39
(CH$_3$)$_2$CH	9	35
(CH$_3$)$_2$CHCH$_2$	90	28

Figure 38 Application of a stereospecific esterase from *Pichia miso* IAM 4682 for the synthesis of (*S*)-1-cyano-1-methylalkyl and -alkenyl acetates.

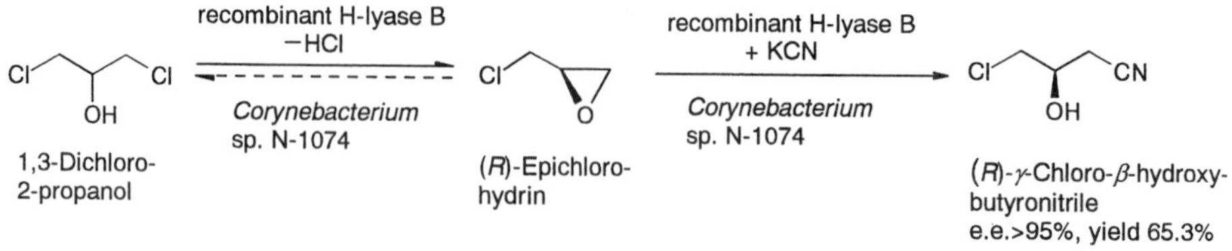

Figure 39 Synthesis of (*R*)-γ-chloro-β-hydroxybutyronitrile by halohydrin hydrogen–halide lyase B from recombinant *Escherichia coli* JM109/pST111.

The analogous reaction with the substrate bearing one more carbon (*O*-acetyl-L-homoserine) was performed by a different enzyme from the same organism named γ-cyano-α-amino-L-butyric acid synthetase [125]. The resulting optically pure γ-cyano-α-amino-L-butyric acid product can be easily further converted by nitrile-hydrolyzing enzymes to α-amino-L-glutaric acid, which is employed as a precursor for a positron emission tomography (PET) labeling and scanning substance [126].

VIII. CONCLUDING REMARKS

The capacity of organisms for nitrile conversions has been explained by the fact that nitriles, e.g., benzyl cyanide, are natural products of plants, especially in the family Cruciferae [127]. In order to find microorganisms bearing stereoselective nitrile-converting enzymes with a relevant strong and specific activity as the basis of a biotechnological process, usually enrichment cultures have been designed with racemic, chiral, or prochiral nitriles or amides that structurally resemble the desired chiral product as the sole source of nitrogen [19,28,32,53,62,63,65,88]. However, due to the usually wide substrate spectrum of nitrile-converting enzymes, there are several examples where no structural relation between the bioconversion substrate and the enrichment substrate can be found. For example, the industrial catalyst *R. rhodochrous* J1 was enriched on aliphatic acetonitrile as sole nitrogen source but is now employed in the synthesis of aromatic nicotinamide [10], and its amidase can be used for the preparation of (*S*)-2-phenylpropionic acid [72]. Some enantioselective nitrile-converting enzymes were isolated by the addition of organic solvents to the enrichment culture for the solubilization of water-insoluble enrichment substrates [64,65].

Predominantly *Rhodococcus* species have been isolated as biocatalysts bearing stereoselective nitrile-converting enzymes. It has been assumed that rhodococci generally contain a nitrile hydratase/amidase system [128], but to regard this as a rule might be

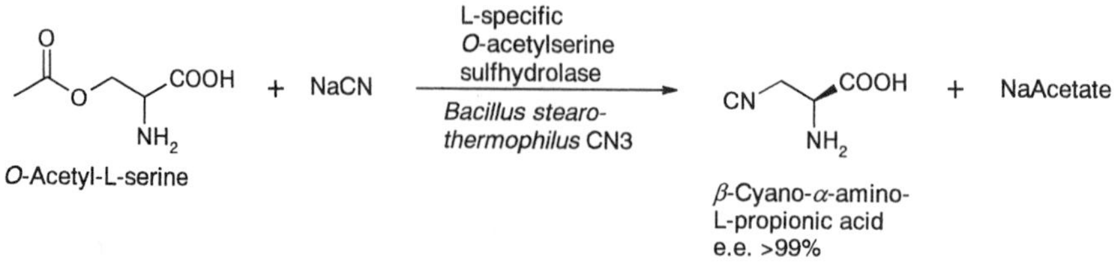

Figure 40 Preparation of β-cyano-α-amino-L-propionic acid by *O*-acetylserine sulfhydrolase from *Bacillus stearothermophilus* CN3.

premature because in a number of rhodococci only a nitrilase has been reported [28,29,32,40]. Additionally, a correlation between the type of nitrogen source and the isolated bacteria as been assumed [128]. For example, racemic 2-phenylpropionitrile as sole nitrogen source was supposed to enrich gram-negative strains, whereas other nitriles were found to enrich a majority of gram-positive strains mainly of the genus *Rhodococcus*. On the other hand, *Rhodococcus equi* TG 328 was isolated in an enrichment culture using 2-phenylpropionitrile as sole nitrogen source [61]. In conclusion, it might be difficult to predict the nature of the stereoselective nitrile-converting enzyme by the kind of nitrogen source used as enrichment substrate.

High enantioselectivity has been more often described with amidases and nitrile-forming enzymes than with the directly nitrile-converting nitrilases and nitrile hydratases. This might indicate a lower stereoselective capacity of directly nitrile-converting enzymes but is probably also in part due to the fact that the stereospecificity of these enzymes has only been the subject of intense study in the last decade.

The preferred substrates of enantioselective nitrile-converting enzymes are arenes or aliphatic compounds with long (more than three carbons) or branched carbon chains. These bulky substrate moieties seem to be a prerequisite for the stereospecific recognition by nitrile-converting enzymes.

There are two principles of biocatalytic reactions leading to chiral products: the asymmetric synthesis of meso- and prochiral compounds, and the kinetic resolution of racemates [17]. The latter is dominant by far in the number of stereospecific nitrile-converting enzymes, which might be in part due to the easier access of suitable racemates compared to prochiral substrates. In some cases racemic resolutions are convenient because both the (R) and the (S) enantiomer can be obtained [114]. However, from the commercial aspect, usually racemic resolutions are of disadvantage due to a limited theoretical yield of 50%, the subsequent, often laborious separation of product and remaining substrate, and a time-dependent decrease of the enantioselectivity due to a kinetic rather than an absolute preference for one enantiomer by the enzyme. In contrast, asymmetric syntheses are advantageous due to a theoretical yield of 100%. However, the limited yield of racemic resolutions can be partly counteracted by deracemization techniques [129], some of which have been described here (Figs. 3, 6, 19, 32, and 38).

First enantioselective nitrile conversions were recently industrialized such as the preparation of (R)-mandelic acid (Fig. 6) or are intended to be industrialized such as the synthesis of (S)-piperazine-2-carboxylic acid (Fig. 31). There are various other commercial processes in which the stereospecific conversion of nitriles is desirable. The increasing number of reports on stereoselective nitrile-converting enzymes in recent years shows that biological mechanisms hold much potential for such processes and indicates that nitrile-converting enzymes might be as useful as esterases and lipases for the synthesis of chiral building blocks.

REFERENCES

1. A de Raadt, N Klempier, K Faber, H Griengl. J Chem Soc Perkin Trans 1:137–140, 1992.
2. MA Bennet, T Yoshida. J Am Chem Soc 95:3030–3031, 1973.
3. J Chin, JH Kim. Angew Chem Int Ed Engl 29:523–525, 1990.
4. M Ravindranathan, N Kalyanam, S Sivaram. J Org Chem 47:4812–4813, 1982.
5. T Nagasawa, H Yamada. Tibtech 7:153–158, 1989.
6. T Nagasawa, K Takeuchi, H Yamada. Eur J Biochem 196:581–589, 1991.
7. M Kobayashi, T Nagasawa, H Yamada. Tibtech 10:402–408, 1992.

8. T Nagasawa, H Shimizu, H Yamada. Appl Microbiol Biotechnol 40:189–195, 1993.
9. T Nagasawa, H Yamada. Pure Appl Chem 67:1241–1256, 1995.
10. T Nagasawa, CD Mathew, J Mauger, H Yamada. Appl Environ Microbiol 54:1766–1769, 1988.
11. J Mauger, T Nagasawa, H Yamada. J Biotechnol 8:87–96, 1988.
12. J Mauger, T Nagasawa, H Yamada. Tetrahedron 45:1347–1354, 1989.
13. M Wieser, K Takeuchi, Y Wada, H Yamada, T Nagasawa. FEMS Microbiol Lett 169:17–22, 1998.
14. CD Mathew, T Nagasawa, M Kobayashi, H Yamada. Appl Environ Microbiol 54:1030–1032, 1988.
15. J Crosby, J Moilliet, JS Parratt, NJ Turner. J Chem Soc Perkin Trans 1:1679–1687, 1994.
16. Food and Drug Administration. Chirality 4:338–345, 1992.
17. D Valentine, JW Scott. Synthesis 329–356, 1978.
18. Y Fukuda, T Harada, Y Izumi. J Ferment Technol 51:393–397, 1973.
19. Y Hashimoto, E Kobayashi, T Endo, M Nishiyama, S Horinouchi. Biosci Biotech Biochem 60:1279–1283, 1996.
20. M Kobayashi, T Nagasawa, H Yamada. Eur J Biochim 182:349–356, 1989.
21. T Nagasawa, M Kobayashi, H Yamada. Arch Microbiol 105:89–94, 1988.
22. T Nagasawa, T Nakamura, H Yamada. Arch Microbiol 155:13–17, 1990.
23. T Nagasawa, J Mauger, H Yamada. Eur J Biochem 194:765–772, 1990.
24. J Mauger, T Nagasawa, H Yamada. Arch Microbiol 155:1–6, 1990.
25. AM Macadam, CJ Knowles. Biotechnol Lett 7:865–870, 1985.
26. A Arnaud, P Galzy, J Jallageas. Bull Soc Chim Fr II, 87–90, 1980.
27. J Jallageas, A Arnaud, P Galzy. Adv Biochem Eng 14:1–32, 1980.
28. TC Bhalla, A Miura, A Wakamoto, Y Ohba, K Furuhashi. Appl Microbiol Biotechnol 37: 184–190, 1992.
29. ML Gradley, CJ Knowles. Biotechnol Lett 16:41–46, 1994.
30. ML Gradley, CJF Deverson, CJ Knowles. Arch Microbiol 161:246–251, 1994.
31. Nippon Mining, Jap. Patent 172980 (91-096794/14), 1991.
32. K Yamamoto, K Oishi, I Fujimatsu, KI Komatsu. Appl Environ Microbiol 57:3028–3032, 1991.
33. MA Cohen, JS Parratt, NJ Turner. Tetrahedron: Asymmetry 3:1543–1546, 1992.
34. K Yamamoto, Y Ueno, K Otubo, K Kawakami, KI Komatsu. Appl Environ Microbiol 56: 3125–3129, 1990.
35. K Yamamoto, I Fujimatsu, KI Komatsu. J Ferm Bioeng 73:425–430, 1992.
36. T Endo, T Yamagami, K Tamura. U.S. Patent 5326702, 1994.
37. Japan Kokai Patents 4-99495, 1992; 4-99496, 1992; 4-218385, 1992; 6-95795, 1993; 6-237789, 1994; 6-284899, 1994.
38. H Layh, A Stolz, S Förster, F Effenberger, H-J Knackmuss. Arch Microbiol 158:405–411, 1992.
39. N Klempier, A de Raadt, K Faber, H Griengl. Tetrahedron Lett 32, 341–344, 1991.
40. SY Choi, YM Goo. Arch Pharm Res 9:45–47, 1986.
41. Sonawane HR, NS Bellur, JR Ahuja, DG Kulkarni. Tetrahedron: Asymmetry 3:163–192, 1992.
42. J Caldwell, A Hutt, S Fournell-Gigleux. Biochem Pharmacol 37:105–114, 1988.
43. H Kakeya, N Sakai, T Sugai, H Ohta. Tetrahedron Lett 32:1343–1346, 1991.
44. F Effenberger, J Böhme. Bioorg Med Chem 2:715–721, 1994.
45. K Yamamoto, KI Komatsu. Agric Biol Chem 55:1459–1466, 1991.
46. Y Sugiura, J Kuwahara, T Nagasawa, H Yamada. J Am Chem Soc 109:5848–5850, 1987.
47. T Nagasawa, H Nanba, K Ryuno, K Takeuchi, H Yamada. Eur J Biochem 162:691–698, 1987.
48. T Nagasawa, T Takeuchi, H Yamada. Biochem Biophys Res Commun 155:1008–1016, 1988.
49. BA Brennan, G Alms, MJ Nelson, LT Durney, RC Scarow, J Am Chem Soc 118:9194–9195, 1996.

50. Y Takashima, Y Yamaga, S Mitsuda. J Ind Microbiol Biotechnol 20:220–226, 1998.
51. S Payne, S Wu, RD Fallon, G Tudor, B Stieglitz, JM Turner, MJ Nelson. Biochemistry 36: 5447–5454, 1997.
52. D Anton, RD Fallon, W Linn, B Stieglitz, V Witterholt. International Published Application WO 92/05275 World International Property Organization, 1992.
53. RD Fallon, B Stieglitz, I Turner. Appl Microbiol Biotechnol 47:156–161, 1997.
54. S Wu, RD Fallon, MS Payne. Appl Microbiol Biotechnol 48:704–708, 1997.
55. AJ Blakey, J Colby, E Williams, C O'Reilly. FEMS Microbiol Lett 129:57–62, 1995.
56. JA Crosby, JS Parratt, NJ Turner. Tetrahedron: Asymmetry 12:1547–1550, 1992.
57. T Beard, MA Cohen, JS Parratt, NJ Turner, J Crosby, J Moillet. Tetrahedron: Asymmetry 4: 1085–1104, 1993.
58. H Kakeya, N Sakai, A Sano, M Yokoyama, T Sugai, H Ohta. Chem Lett 1823–1824, 1991.
59. JC Jallageas, A Arnaud, P Galzy. U.S.Patent 4366250, 1982.
60. K Yamamoto, K Otsubo, K Oishi. U.S. Patent 5283193, 1994.
61. T Gilligan, H Yamada, T Nagasawa. Appl Microbiol Biotechnol 39:720–725, 1993.
62. S Masutomo, A Inoue, K Kumagai, R Murai, S Mitsuda. Biosci Biotech Biochem 59:720–722, 1995.
63. R Bauer, B Hirrlinger, H Layh, A Stolz, H-J Knackmuss. Appl Microbiol Biotechnol 42:1–7, 1994.
64. R Bauer, H-J Knackmuss, A Stolz. Appl Microbiol Biotechnol 49:89–95, 1998.
65. N Layh, A Stolz, J Böhme, F Effenberger, H-J Knackmus. J Biotechnol 33:175–182, 1994.
66. N Layh, A Stolz, J Böhme, F Effenberger, H-J Knackmus. Biotechnol Lett 17:187–192, 1995.
67. B Hirrlinger, A Stolz, H-J Knackmuss. J Bacteriol 178:3501–3507, 1996.
68. L Martínková, A Stolz, H-J Knackmuss. Biotechnol Lett 18:1073–1076, 1996.
69. L Martínková, I Prepechalová, P Olsovský, V Kren. Biotechnol Lett 17:1219–1222, 1995.
70. H Ohta. Chimia 50:434–436, 1996.
71. M Kobayashi, M Nishiyama, T Nagasawa, S Horinouchi, T Beppu, H Yamada. Biochim Biophys Acta 1119:23–33, 1991.
72. M Kobayashi, H Komeda, T Nagasawa, M Nishiyama, S Horinouchi, T Beppu, H Yamada, S Shimizu. Eur J Biochem 217:327–336, 1993.
73. MA Maestracci, A Thiéry, K Bui, A Arnaud, P Galzy. Arch Microbiol 138:315–320, 1984.
74. M-P Kieny-L'Homme, A Arnaud, P Galzy. J Gen Appl Microbiol 27:307–325, 1981.
75. Y Vo-Quang, D Marais, L Vo-Quang, F LeGoffic, A Thiéry, M Maestracci, A Arnaud, P Galzy, J-C Jallageas. Tetrahedron Lett 28:4057–4060, 1987.
76. J-F Mayaux, E Cerbelaud, F Sourbrier, D Faucher, D Pétré. J Bacteriol 172:6764–6773, 1990.
77. E Cerbelaud, D Pétré. Eur Patent Appl. EP 330-529A, 1988.
78. J-F Mayaux, E Cerbelaud, F Sourbrier, P Yeh, F Blanche, D Pétré. J Bacteriol 173:6694–6704, 1991.
79. M Nishiyama, S Horinouchi, M Kobayashi, T Nagasawa, H Yamada, T Beppu. J Bacteriol 173:2465–2472, 1991.
80. M Yokoyama, T Sugai, H Ohta. Tetrahedron: Asymmetry 4:1081–1084, 1993.
81. K Yamamoto, K Otsubo, A Matsuo, T Hayashi, I Fujimatsu, KI Komatsu. Appl Environ Microbiol 62:152–155, 1996.
82. T Hayashi, Y Ueno, A Matsuo, T Hayashi, S Muramatsu, A Matsuda, KI Komatsu. J Ferm Bioeng 83:139–145, 1997.
83. T Nagasawa, H Yamada. Pure Appl Chem 72:1441–1444, 1990.
84. LM Ciskanik, JM Wilczek, RD Fallon. Appl Environ Microbiol 61:998–1003, 1995.
85. CR Worthing. The Pesticidal Manual, 6th ed. London: BCP, 1979, pp 329–332.
86. D Bianchi, A Bosetti, P Cesti, G Franzosi, S Spezia. Biotechnol Lett 13:241–244, 1991.
87. D Bianchi, E Battistel, P Cesti, P Golini, R Tassinari. Appl Microbiol Biotechnol 40:53–56, 1993.
88. E Eichhorn, JP Roduit, N Shaw, K Heinzmann, A Kiener. Tetrahedron: Asymmetry 8:1533–2536, 1997.

89. OM Birch, JM Brass, A Kiener, K Robins, D Schmidhalter, NM Shaw, T Zimmermann. Chim Oggi 13:9–13, 1995.
90. BB Corson, RA Dodge. In: H Gilman, ed. Organic Synthesis Coll. Vol. 1. New York: John Wiley, 1956, p 336.
91. JA Marshall, NH Anderson, JW Schlicher. J Org Chem 35:858–861, 1970.
92. T Satoh, S Suzuki, Y Miyaji, Z Imai. Tetrahedron Lett 4555–4558, 1969.
93. T Matsuo, T Nishioka, M Hirano, Y Suzuki, K Tsushima, N Itaya, H Yoshioka. Pest Sci 202–218, 1980.
94. DG Neilson, DAV Peters, LH Roach. J Chem Soc 2272–2273, 1962.
95. H Wajant, F Effenberger. Biol Chem 377:611–617, 1996.
96. L Rosenthaler. Biochem Z 14:238–253, 1908.
97. C Bové, EE Conn. J Biol Chem 236:207–210, 1961.
98. W Becker, H Freund, E Pfeil. Angew Chem Int Ed Engl 4:1079, 1965.
99. F Effenberger, T Ziegler, S Förster. Angew Chem Int Ed Engl 26:458–460, 1987.
100. J Brussee, WT Loos, CG Kruse, A van der Gen. Tetrahedron 46:979–986, 1990.
101. VI Ognyanov, VK Datcheva, KS Kyler. J Am Chem Soc 113:6992–6996, 1991.
102. F Effenberger. Angew Chem Int Ed Engl 33:1555–1564, 1994.
103. H Wajant, S Förster, H Böttinger, F Effenberger, K Pfizenmaier. Plant Sci 108:1–11, 1995.
104. J Albrecht, I Jansen, MR Kula. Biotechnol Appl Biochem 17:191–203, 1993.
105. TT Huuhtanen, LT Kanerva. Tetrahedron: Asymmetry 3:1223–1226, 1992.
106. F Effenberger, U Stelzer. Chem Ber 126:779–786, 1993.
107. F Effenberger, B Hörsch, S Förster, T Ziegler. Tetrahedron Lett 31:1249–1252, 1990.
108. N Klempier, H Griengl, M Hayn. Tetrahedron Lett 34:4769–4772, 1993.
109. H Wajant, S Förster, D Selmar, F Effenberger, K Pfizenmaier. Plant Physiol 109:1231–1238, 1995.
110. S Förster, J Roos, F Effenberger, H Wajant, A Sprauer. Angew Chem Int Ed Engl 35:437–439, 1996.
111. M Schmidt, S Hervé, N Klempier, H Griengl. Tetrahedron 52:7833–7840, 1996.
112. H Wajant, S Forster, A Sprauer, F Effenberger, K Pfizenmaier. Ann NY Acad Sci 799:771–776, 1996.
113. H Kakeya, N Sakai, T Sugai, H Ohta. Agric Biol Chem 55:1877–1881, 1991.
114. YF Wang, ST Chen, KKC Liu, CH Wong. Tetrahedron Lett 30:1971–1920, 1989.
115. N Matsuo, N Ohno. Tetrahedron Lett 26:5533–5534, 1985.
116. A v Almsick, J Buddrus, P Honicke-Schmidt, K Laumen, MP Schneider. J Chem Soc Chem Commun 1391–1393, 1989.
117. H Ohta, S Hiraga, K Miyamoto, G Tsuchihashi, Agric Biol Chem 52:3023–3027, 1988.
118. H Ohta, Y Miyamae, G Tsuchihashi. Agric Biol Chem 53:281–283, 1989.
119. H Ohta, V Kimura, Y Sugano, T Sugai. Tetrahedron 45:5469–5476, 1989.
120. H Hirohara, S Mitsuda, E Ando, R Komaki. In: J Tramper, HC van der Plas, P Linko, eds. Biocatalysts in Organic Synthesis. Amsterdam: Elsevier, 1985, p 119.
121. H Waldmann. Tetrahedron Lett 30:3057–3058, 1989.
122. T Nagasawa, T Nakamura, F Yu, I Watanabe, H Yamada. Appl Microbiol Biotechnol 36:478–482, 1992.
123. T Nakamura, T Nagasawa, F Yu, I Watanabe, H Yamada. Tetrahedron 50:11821–11826, 1994.
124. K Watanabe, K Omura, Y Furuya. Jap. Kokai Patent 8-56666, 1996.
125. S Shimidzu, M Ikemoto, K Omura. Jap. Kokai Patent 9-19290, 1997.
126. P Bjurling, Y Watanabe, S Oka, T Nagasawa, H Yamada, B Langström. Acta Chem Scand 44:183–188, 1990.
127. S Hashimoto, H Kameoka. J Food Sci 50:847–852, 1985.
128. N Layh, B Hirrlinger, A Stolz, H-J Knackmuss. Appl Microbiol Biotechnol 47:668–674, 1997.
129. H Stecher, K Faber. Synthesis 1:1–16, 1997.

18

Enzyme-Mediated Decarboxylation Reactions in Organic Synthesis

Hiromichi Ohta and Takeshi Sugai
Keio University, Yokohama, Japan

I. INTRODUCTION

A large number of carboxylating and decarboxylating enzymes work in biocatalytic systems, such as CO_2 fixation, biosynthesis of fatty acids, and degradation of intermediates in metabolic pathways [1]. Because of the wide variety, abundance, and recent increasing availability of those enzymes, their utilization in organic syntheses, especially as the tools for stereoselective syntheses, has been developed. In this chapter, we introduce some fundamental aspects of these enzymes and examples of biocatalysis in organic synthesis, applying to nonnatural substrates.

II. DECARBOXYLATION OF β-KETO ACIDS AND MALONIC ACIDS

Inherently, the decarboxylation of β-keto acids and malonic acids (1) proceeds very smoothly, as the resulting product bearing anion adjacent to carbonyl group stabilizes as its enolate form (2) [Eq. (1)]. Enzyme-mediated reaction sometimes utilizes this facilitated decarboxylation. Indeed, isocitric acid (3) was oxidized to the corresponding keto acid, which subsequently decarboxylated to α-ketoglutaric acid (4) by means of isocitrate dehydrogenase (EC 1.1.1.41) [Eq. (2)]. Another example is observed in the formation of acetoacetyl-CoA (5), which occupies the first step of fatty acid biosynthesis. A β-keto carboxylate 6, derived from the acetylation of malonyl-CoA with acetyl-CoA, decarboxylates to 5 by the action of 3-ketoacyl synthase [Eq. (3)].

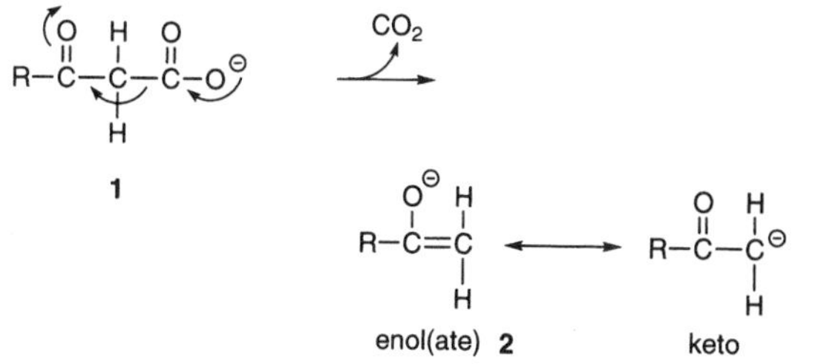

$$\text{(1)}$$

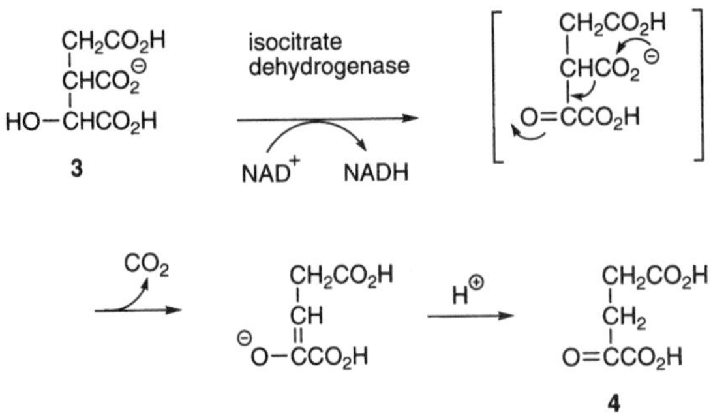

(2)

One interesting point of the decarboxylation reaction of β-keto and malonic acids should be emphasized: the formation of enol intermediates. If the protonation occurs in an enantiofacially selective manner on the enolate **7** bearing the substituents X and Y, the products become enantiomerically enriched form [**8** or *ent*-**8**, Eq. (4)].

A.   Decarboxylation of β-Keto Acids

1.   Acetolactate Decarboxylase

Acetolactate decarboxylase (EC 4.1.1.5) works on (*S*)-acetolactate **9** to give (*R*)-acetoin **10** [Eq. (5)] [2]. As stated in Eq. (4), in the course of enzyme-mediated decarboxylation, proton adds to *si* face of intermediate **11**. The enzyme itself essentially accept (*S*)-substrate; however, in the case of using crude enzyme preparation from *Klebsiella aerogenes*, both enantiomers of substrate decarboxylate to the same product, (*R*)-**10** [3] [Eq. (6)]. The reaction pathway was studied by using both enantiomers of **9** and was suggested to include the conversion of (*R*)-**9** to (*S*)-**9** via the stereospecific migration of the carboxyl group as shown in Eq. (6). Another experiment using a homolog, α-acetyl-α-hydroxybutyrate (**12**), as the substrate supported this mechanism [4]. The (*S*) enantiomer of **12** smoothly decarboxylated to (*R*)-**13** in the same manner as that of **9**. When 50% of the substrate had undergone decarboxylation, a second slower decarboxylation began accompanied by the formation of isomeric hydroxy ketone (*R*)-**14**. Compared to the normal decarboxylation product **13**, the methyl and ethyl group of **14** are replaced with each other [Eq. (7)].

In a practical sense, the substrate should be prepared under a mild and neutral condition to avoid the nonenzymatic decarboxylation. This was achieved very well by the use

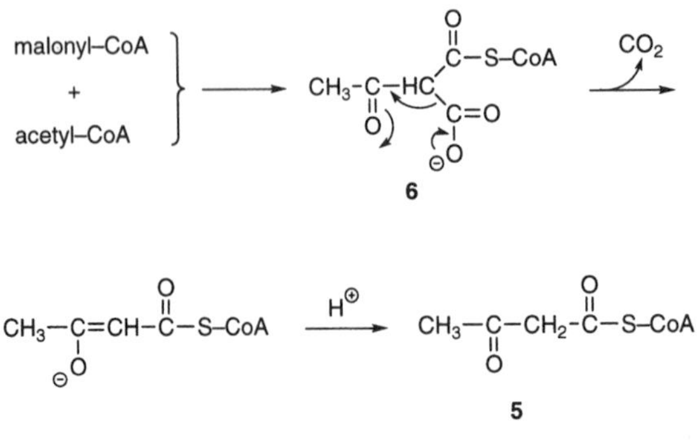

(3)

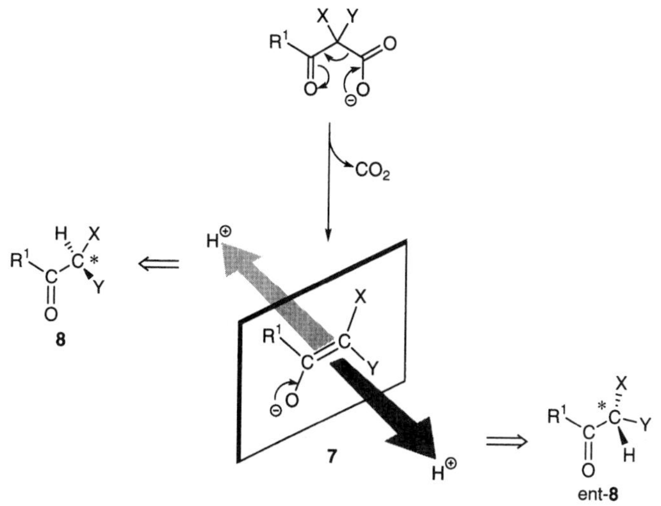

$$(4)$$

of pig liver esterase (PLE)–catalyzed hydrolysis of the precursor, and ester **15** and the e.e. of the product reached to as high as over 98% [Eq. (8)] [4].

B. Decarboxylation of Methylmalonic Acids

1. Methylmalonyl-CoA Decarboxylase

Malonyl-CoA decarboxylase (EC. 4.1.1.9) from uropygial gland only works on (*S*)-enantiomer of methylmalonyl-CoA (**16**) [5], while the (*R*) isomer remained intact. In addition, the protonation of the resulting intermediate occurred in an enantio face–selective manner, as revealed by the fact that the absolute configuration the product, 2-(^{3}H)-propionyl-CoA (**17**) was obtained in an experiment carried out in ^3H_2O [Eq. (9)], was R.

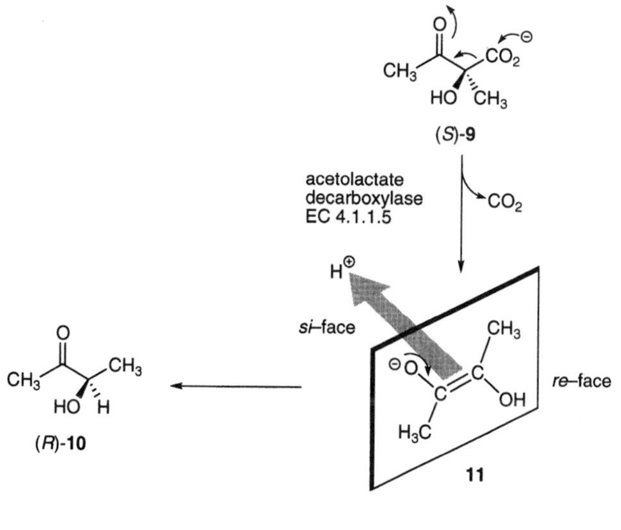

$$(5)$$

$$(6)$$

$$(7)$$

$$(8)$$

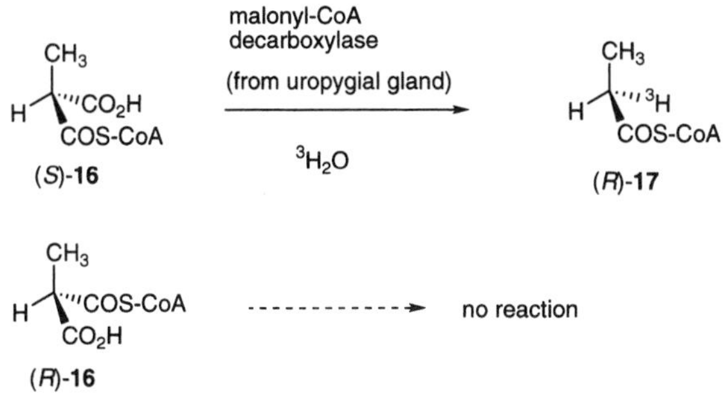

(9)

2. Serine Hydroxymethyltransferase (SHMT)

Serine hydroxymethyltransferase (SHMT) catalyzes an enantioselective decarboxylation on α-amino-α-methylmalonic acid **18**. This enzyme decarboxylates the substrate by the aid of pyridoxal-5-phosphate as shown in Eq. (10). The decarboxylation starts in an en-antiotopos-differentiating manner. Thomas and co-workers synthesized both enantiomers of the substrate **18**, containing ^{13}C in either one of two carboxy groups, and those were incubated with SHMT [6]. The fact that ^{13}C was retained in the resulting alanine (**19**) only when the starting compound was (R)-**18**, indicates that pro-(R) carboxy group of the substrate was removed as carbon dioxide. Judging from the (R)-absolute configuration of the product, the protonation of the resulted Schiff base **20** occurred in an enantiofacially

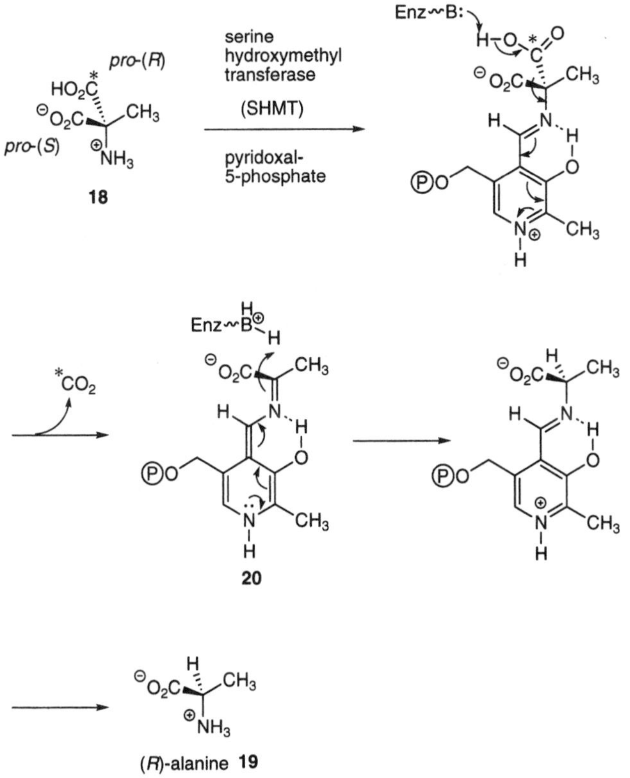

(10)

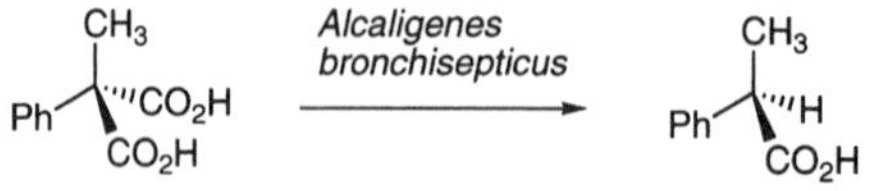

$$\text{(11)}$$

selective manner. This mechanism is also working for the case of α-aminomalonic acid [7].

C.   Decarboxylation of α-Aryl-α-methylmalonic Acid

So far we have referred to the decarboxylative enzymes that work on the naturally occurring intermediates of biochemical pathways. Due to the high affinity for the native substrates by these enzymes, however, the applicability to nonnatural synthetic substrates was somewhat limited. Thus we intended to develop a new biocatalysis, such as an enzyme which can decarboxylate α-aryl-α-methylmalonic acids to yield enantiomerically enriched α-substituted α-arylacetic acids because those products are useful compounds as anti-inflammatory agents [8–10] and the chiral derivatizing agents [11].

From an extensive screening of microorganisms, we first developed such a biocatalyst, *Alcaligenes bronchisepticus*, in 1989 [Eq. (11)] [12]. (*R*)-Phenylpropionic acid (**21a**) was obtained from α-methyl-α-phenylmalonic acid (**22a**), and the yield and e.e. of the products were extremely high. Indeed, this is a new type of biotransformation that can be performed on a preparative scale since the substrates, disubstituted malonates, are readily available via the well-established "malonate ester synthesis." To date, all of the attempts for enantioselective decarboxylation of malonates by a chemical asymmetric synthesis resulted in only low to moderate e.e. of the products. In order to understand the mechanism of this enzyme-mediated decarboxylation we embarked on the isolation of the enzyme and further study of its characteristics.

1.   Screening of Microorganism

The screening of microorganism was carried out using a medium containing phenylmalonic acid (**23**) as the sole source of carbon because the first step of the metabolic pathway of this acid would be decarboxylation to give phenylacetic acid (**24**), which would further be susceptible to a metabolism [Eq. (12)]. A wide variety of soil samples and type cultures were tested and we found a few strains able to grow on the phenylmalonic acid. Next, expecting that the introduction of a methyl group on the α position of malonate would inhibit the further degradation of compound, α-methyl-α-phenylmalonic acid was selected for microorganisms. Among them, we identified the bacterium *Alcaligenes bronchisepticus* [12,14]. This strain produces the decarboxylase as an inducible enzyme, the inducer being phenylmalonic acid.

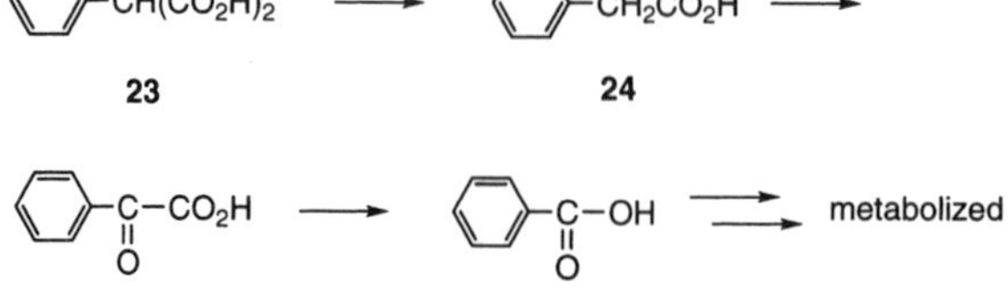

$$\text{(12)}$$

Table 1 Asymmetric Decarboxylation of α-Aryl-α-methylmalonic Acid

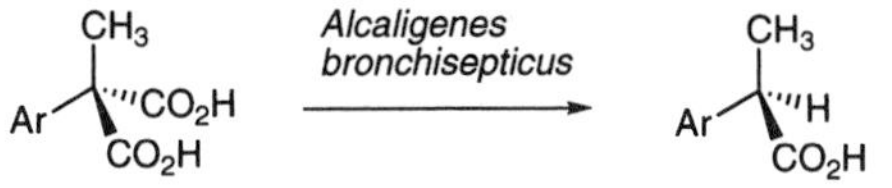

	Ar	[Sub] (%)	Yield (%)	e.e. (%)
a	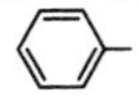 (phenyl)	0.5	80	98
b	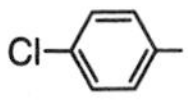 Cl–(phenyl)	0.5	95	98
c	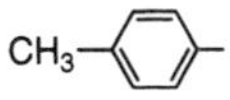 CH$_3$–(phenyl)	0.5	44	>95
d	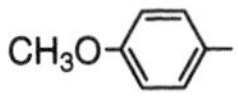 CH$_3$O–(phenyl)	0.1	48	99
e	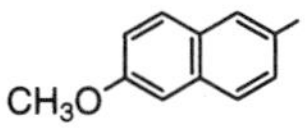 CH$_3$O–(naphthyl)	0.5	96	>95
f	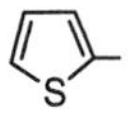 (thienyl)–S	0.3	98	95

2. Substrate Specificity

As shown in Table 1, the absolute configurations of the products were proved to be *R* and the e.e. values were determined to be over 98% in all cases. The enzyme system of *A. bronchisepticus* was also effective to other compounds with a substituent on the aromatic ring. As is clear from Table 1, an introduction of electron-withdrawing substituents promoted the reaction. The enzyme also accepted β-naphthyl ring (**22e**) and thienyl ring (**22f**) as good substrates. To the contrary, no decarboxylation was observed when the malonate has a substituent on the ortho position of the aryl ring (**22m, n**) or the alkyl group of α position is ethyl instead of methyl (Fig. 1). It is in marked contrast to the fact that **22e** is a good substrate, a compound with an α-naphthyl ring (**22p**) was completely inactive to the enzyme system. The low activity of all these compounds is due to the steric congestion rather than the electronic situation. To our surprise, the insertion of only a methylene group or a hetero atom between the α carbon and the aromatic ring made the substrate completely inactive (**25a–c** in Fig. 1). The π electrons of the aromatic ring have an essential effect to promote the reaction.

The introduction of fluorine atoms into the substrates indeed brought about a favorable effect on the reaction, as the fluorine atom is strong electron withdrawing and not bulky group (Table 2) [15]. Moreover, α-fluorinated malonic acid (**22g**) gave the corre-

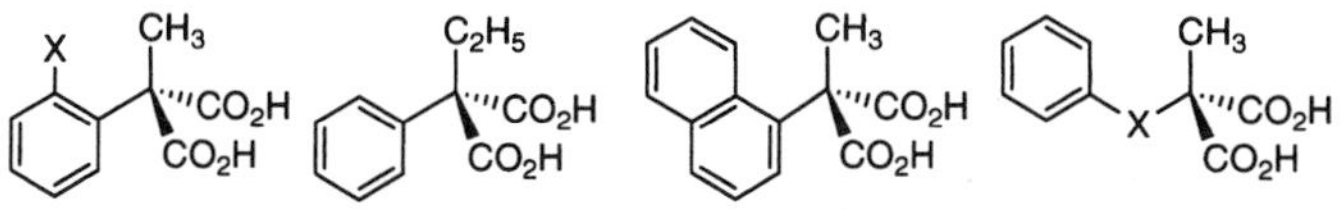

22m X: Cl	22o	22p	25a X: CH$_2$
22n X: CH$_3$			25b X: O
			25c X: S

Figure 1 Compounds that are inactive to *Alcaligenes bronchisepticus*.

Table 2   Asymmetric Decarboxylation of Fluorinated Compounds

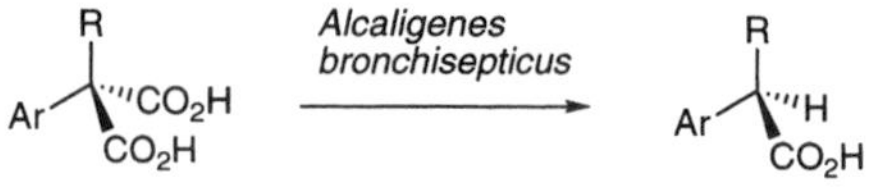

$$\underset{\textbf{22}}{\text{Ar}\overset{R}{\underset{CO_2H}{\underset{\big|}{\text{C}}}}\text{'''CO}_2\text{H}} \xrightarrow{\substack{Alcaligenes \\ bronchisepticus}} \underset{(R)\text{-}\textbf{21}}{\text{Ar}\overset{R}{\underset{CO_2H}{\underset{\big|}{\text{C}}}}\text{'''H}}$$

	Ar	R	[Sub] (%)	Yield (%)	e.e. (%)
g	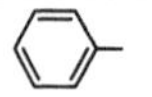 phenyl	F	0.1	64	95
h	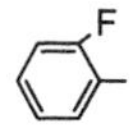 o-F-phenyl	CH$_3$	0.3	12	54
i	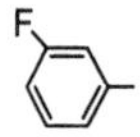 m-F-phenyl	CH$_3$	0.5	75	97
j	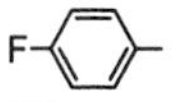 p-F-phenyl	CH$_3$	0.1	54	97
k	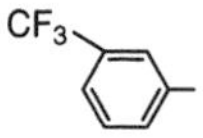 m-CF$_3$-phenyl	CH$_3$	0.5	99	>95
l	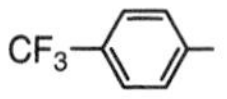 p-CF$_3$-phenyl	CH$_3$	0.3	91	95

sponding enantiomerically enriched monobasic acid in a moderate yield. The effect of substitution of ring hydrogen showed again a marked difference between ortho, and meta, or para positions. Even the steric bulkiness of as small as that of the fluorine atom has a serious effect on the reaction rate. The yield of the product after 5-day reaction of o-fluorinated compound **22h** was only 12%. The low e.e. value of the product is probably due to nonenzymatic decarboxylation. The m- and p-trifluoromethyl derivatives (**22k, l**) were good substrates, with the chemical and optical yields of the expected products being very high (Table 2).

3. Characteristics of the Enzyme and the Gene

The above results prompted us to study in detail the isolated enzyme and gene in order to elucidate the mechanism of this type of decarboxylation. The enzyme was purified from the bacterium grown in a medium as described before. The enzyme was purified to about 300-fold to 377 U/mg protein (15% yield). Sodium dodecyl sulfate–polyamide gel electrophoresis (SDS-PAGE) and high-performance liquid chromatography (HPLC) analysis showed that this enzyme was monomeric, the molecular mass being around 24 kDa. The enzyme was named as arylmalonate decarboxylase (AMDase) [16].

To a further characterization of this enzyme, the effects of some additives were examined [16]. Addition of ATP or CoA-SH to the reaction mixture did not enhance the rate, in contrast to the case of malonyl-CoA decarboxylase and other decarboxylases reported so far, where substrate forms a thiol ester with Co-A with the assistance of ATP. More surprisingly, this enzyme was not a biotin enzyme, judging from the fact that there was no influence on the rate by the addition of avidin, a potent inhibitor of the biotin–enzyme complexes [17–20]. The cofactor requirements of AMDase are entirely different from those of known analogous enzymes, such as acyl-CoA carboxylases [21], methyl-malonyl-CoA decarboxylases [17], and transcarboxylases [21,22].

A strong inhibitory effect on AMDase activity was found with sulfhydryl reagents (at 1 mM), such as $HgCl_2$ (relative activity, 0%), HgCl (8%), $AgNO_3$ (3%), iodoacetate (3%), and p-chloromercuribenzoate (PCMB) (0%). From these results, AMDase was concluded to be a thiol enzyme.

The gene of this enzyme consists of 720 base pairs (240-amino-acid residues). DNA sequence indicated that AMDase contains four cysteine residues located at 101, 148, 171, and 188 [23]. At least one of these four is estimated to play a crucial role instead of CoA-SH in the decarboxylation of disubstituted malonic acid.

4. Elucidation of the Active Site and the E-S Complex

As cysteine residue had been observed to have an essential contribution of activating the substrate in place of CoA-SH, the sulfhydryl (SH) group was replaced with a hydroxy (OH) group, considering the nucleophilicity and anion-stabilizing effect of an sulfur atom. Four mutant genes in which either one of the four codons of Cys is replaced by that of Ser were prepared and expressed in *E. coli*. Four corresponding mutant enzymes were overproduced and incubated with phenylmalonic acid (**23**) [24]. Kinetic data for the mutant enzymes as well as the wild enzyme are summarized in Table 3. Among four mutants, C188S showed a drastic decrease in the activity, due to a decrease in the catalytic turnover number (k_{cat}) rather than affinity for the substrate (K_m).

The fact that the k_{cat} value of this mutant is extremely small despite little change in conformation as seen from circular dichroic spectra and calculation of the content of secondary structure clearly indicates that SH group of Cys^{188} plays a crucial role the active site. Independent time-of-flight−mass spectroscopy (TOF-MS) and infrared (IR) studies on the inhibition by using α-bromophenylacetic acid, a substrate analog, strongly suggested the formation of thiol ester as E-S complex during the reaction process [25,26].

5. Stereochemical Course of the Reaction

This enzyme also distinguishes two enantiotopic carboxy groups, as revealed in serine hydroxymethyl transferase (SHMT) (Sec. II.B.2), based on the incubation with the labeled substrates [27] (Eq. 13). Enantioface−differentiating protonation to **26** was supposed to be responsible for the enantiomeric excess of the products. One piece of evidence supporting this enolate intermediate is an electronic effect, which was elucidated from the relationship between the logarithm of $k_{cat}(X)/k_{cat}(H)$ and Hammett σ values of the substit-

Table 3 Kinetic Parameters of Wild and Mutant Enzymes for the Decarboxylation of Phenylmalonic Acid

$$PhCH(CO_2H)_2 \xrightarrow[\text{mutant AMDase}]{\text{Wild or}} PhCH_2CO_2H$$

	k_{cat} (s^{-1})	K_m (mM)	k_{cat}/K_m
Wild	366	13.3	27.5
C101S	248	4.3	57.6
C148S	100	11.5	8.7
C171S	62	9.1	6.8
C188S	0.62	4.9	0.13

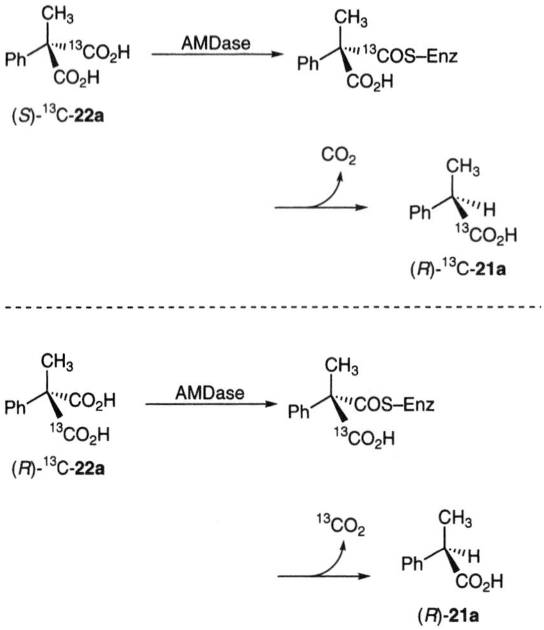

(13)

uents (p-MeO, p-Me, p-Cl, m-Cl, and H) on aromatic ring. The plus sign of the ρ value ($+1.9$) means that the transition state has some negative charge as seen in an intermediate such as **26** [Eq. (14)].

Another important conclusion is that the aromatic ring occupies the same plane as that of double bond for conjugation and this explained the low activity of ortho-substituted substrates and α-ethyl derivative as mentioned earlier (Fig. 1). If the conformation of the substrate is already restricted when it binds to the active site of the enzyme, ortho sub-stituents (X) and α substituent (CH_3) should be arranged in *syn*-periplanar, whose potential energy is higher than that of most stable, *anti*-periplanar forms [Eq. (15)]. Ab initio cal-culation indicated the difference in potential energy between *syn*- and *anti*-periplanar con-formation of α-methyl-α-(o-chlorophenyl)malonic acid (**22m**) is 5.5 kcal/mol [28], whereas that of nonmethylated compound is about 0.7 kcal/mol. In the former case, the high-energy barrier, 5.5 kcal/mol, is far beyond the gaining binding energy, 2.5–3.0 kcal/mol, estimated from the K_m values of reactive substrates (10–20 mM) to the enzyme. At present, one of the attractive forces to fix the conformation of the substrates to a rather

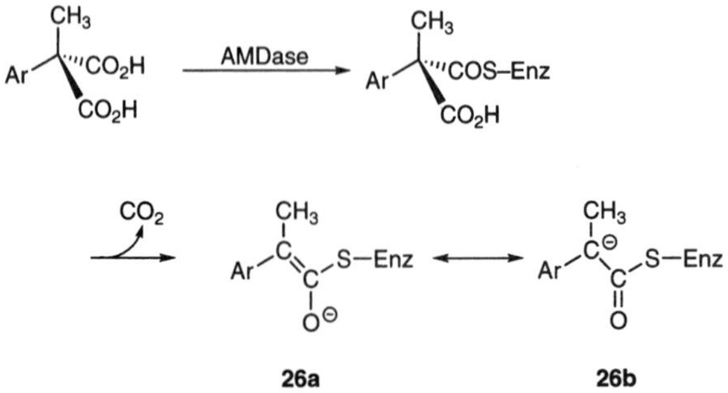

(14)

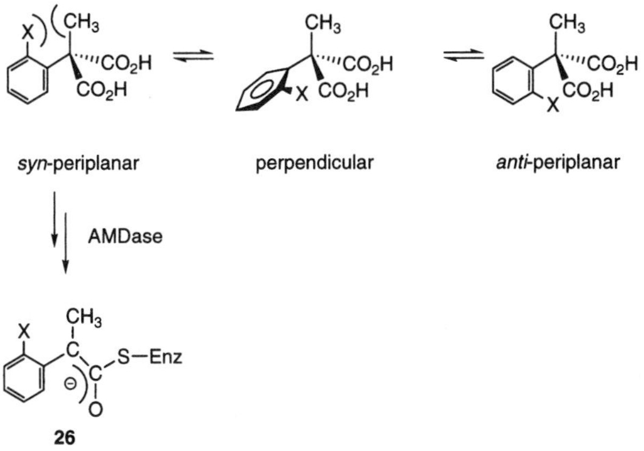

26

(15)

unfavorable *syn*-periplanar one is estimated CH$-\pi$ interaction of the aromatic rings between the amino acid components of enzyme and the substrate [29].

This supposition guided us to design a new and highly effective (low K_m) substrate, whose conformation is fixed as the *syn*-periplanar form. As expected, indane dicarboxylic acid (**27**) was effectively decarboxylated to give the corresponding monobasic acid (*R*)-**28** with high e.e. in also a high yield [Eq. (16)]. Its K_m value was 1.06 and the lowest compared with those hitherto mentioned noncyclic compounds. Indeed, the good fitness of this substrate was clarified by the calculation of $\Delta S\ddagger$ (-27.6 cal mol^{-1} K^{-1}), compared with that of phenylmalonate (-38.5) through an evaluation of the temperature effect [30].

III. DECARBOXYLATION OF α-KETO ACIDS

A. Pyruvate Decarboxylase

Pyruvate decarboxylase (PDase, EC 4.1.1.1), which catalyzes the reaction as shown in Eq. (17), has been studied since the early twentieth century as one of the key enzymes working in glycolysis [31–33]. It has been isolated from yeast [34], wheat germ [35], sweet potato [36], and a wide variety of other sources. Recently, this enzyme was revealed to also play an important role in nonmevalonate pathway for terpenoid biosynthesis [37].

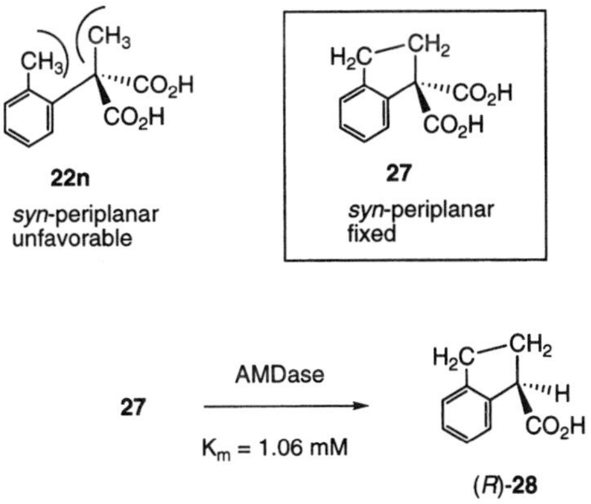

22n

syn-periplanar
unfavorable

27

syn-periplanar
fixed

(*R*)-**28**

(16)

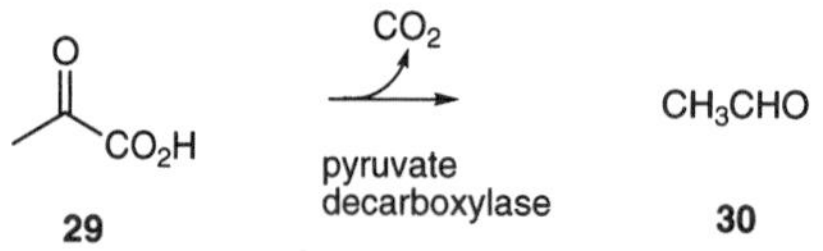

$$(17)$$

1. Mechanism of Decarboxylation Catalyzed by Pyruvate Decarboxylase

In 1937, it was found that thiamine pyrophosphate (TPP) was essential for the activity of this enzyme [38]. Through extensive mechanistic and enzymological studies it is now known that the PDase-mediated decarboxylation, as well as the related α-keto carboxylate decarboxylase, proceeds as shown in Eq. (18) [39–44]. First, C-2 of thiazole involved in TPP attacks on the carbonyl group of pyruvate (**29**) and the subsequent decarboxylation affords an enol intermediate. Protonation and hydrolysis yields acetaldehyde (**30**), accompanied with the regeneration of TPP.

2. Acetoin and 2,3-Butanediol Formation

This scheme suggests the possibility of a new C—C bond formation. The nucleophilic attack of enol intermediate (= carbanion intermediate) on another aldehyde affords ultimately a hydroxy ketone as shown in Eq. (19). The simplest product, acetoin (R = CH$_3$) (**10**), has been found for the first time in fermenting yeast [45,46]. It can be formed by the reaction between pyruvate and acetaldehyde, which itself is originated from pyruvate by the action of the same enzyme. Later, in brewers' yeast [47–51], wheat germ [52], and mammalian tissue [51], it was proved that PDase is responsible for the formation of acetoin.

Microbial acetoin formation has been extensively studied due to a special industrial interest, since this is a key step for the fermatative production of 2,3-butanediol (**31**), an important starting material in polymer industries. 2,3-Butanediol has been produced by means of *Serratia marcescens* [53], *Bacillus polymyxa* [53,54], *Klebsiella oxytoca* [55,56], and *Lactobacillus plantarum* [57]. In these cases, however, the biosynthesis of acetoin mainly proceeds through another pathway [58,59]. The key enzyme is acetolactate synthase (EC 4.1.3.18), which catalyzes the nucleophilic attack of TPP-thiazolium intermediate on another molecule of pyruvate [Eq. (20)]. The pathway from acetolactate to acetoin via the

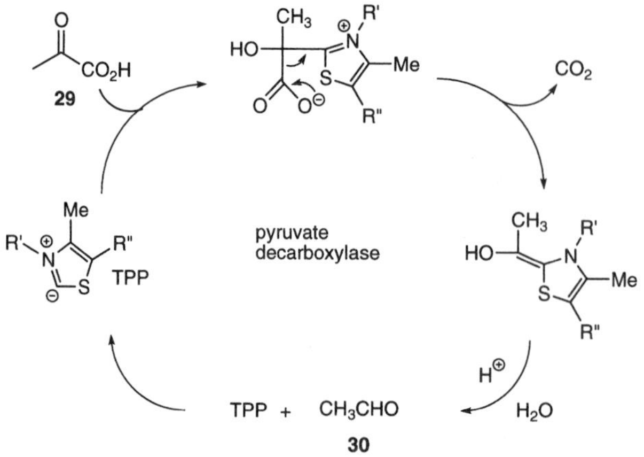

$$(18)$$

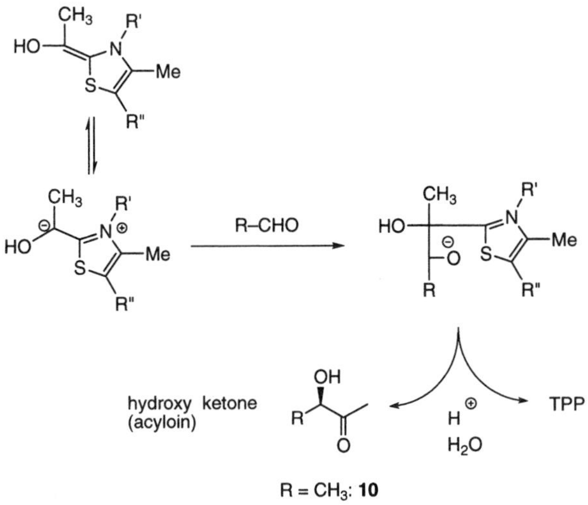

(19)

acetolactate decarboxylase–mediated decarboxylation of β-keto acid structure is described in Sec. II.A.1.

3. C—C Bond Formation: Substrate Specificity on Non-natural Aldehydes

As early as the 1910s and 1920s, acyloin (**33**) formations from some aldehydes (**32**), such as furfural [60], cinnamaldehyde [61], and benzaldehyde [62], were observed in fermenting yeast. The structure of hydroxy ketone (aceloin) from benzaldehyde was later unambigu-

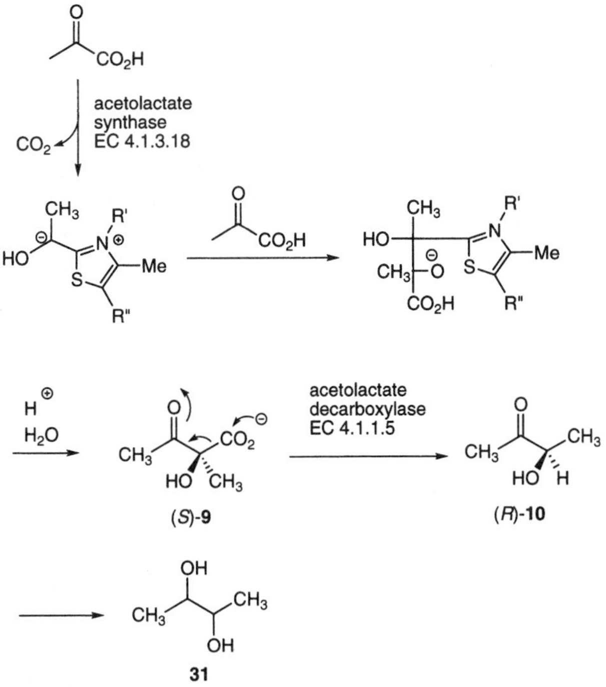

(20)

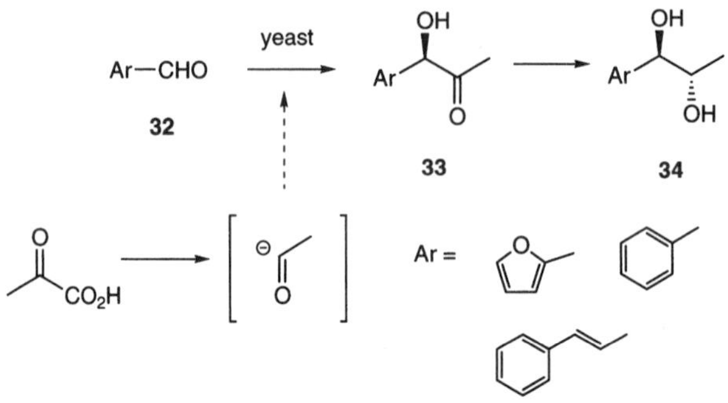

$$(21)$$

ously determined as 1-hydroxy-1-phenyl-2-propanone (phenylacetylcarbinol, PAC) [63], as shown in Eq. (21). Sometimes the concomitant reduction of carbonyl group of initially formed hydroxy ketones (**33**) occurred in whole-cell systems and the corresponding diols (**34**) were obtained as the major products. Initially the proposed enzyme "carboligase," responsible for C—C bond formation, was later revealed to be PDase by an experiment with cell-free extract [64,65], and the precursor of C_2 unit was shown to be pyruvate via the loss of CO_2.

The possibility of this new type of C—C bond formation prompted extensive studies on the substrate specificity. Fermenting yeast accepts a very wide range of aromatic and heteroaromatic aldehydes to give **33** and/or **34** (Table 4) [60–92]. Moreover, a few examples where aliphatic aldehydes were accepted for C—C bond formation have been reported [67,90,93].

Other than pyruvic acid, 2-oxobutanoic acid, 2-oxohexanoic acid, 3-methyl-2-oxo-butanoic acid, and 4-methyl-2-oxopentanoic acid have been incorporated in the resulting hydroxy ketones [48,75,86].

4. Efficiency of C—C Bond Formation Between Benzaldehyde and Pyruvic Acid

The formation of phenylacetylcarbinol (**33a**) and subsequent transformation to L-ephedrin was filed as a patent in the 1930s [94], and this procedure is the first example of the chemoenzymatic industrial synthesis of physiologically active compounds. Since then, many efforts have been devoted to improve the efficiency of the formation of **33a** from **32a** [66,68,70,73,77–78,81]. However, so far all of the procedures for acyloin formation by using fermenting microorganisms have suffered from the concomitant serious side reaction. Indeed, direct reduction of the aldehyde substrate (R-CHO, **32**) to the corresponding alcohol (R-CH$_2$OH, **35**) by the action of alcohol dehydrogenase has been the major pathway [cf. 85] [Eq. (22)]. For example, the ratio of the diol (**34a**), an ultimately reduced form of **33a**, to the undesired alcohol (**35a**) has been reported to be 1:2.7 [69,72] in the case of transformation by baker's yeast.

The remarkable improvement of the efficiency of hydroxy ketone formation was achieved by us [81]. The conditions for PDase activity in fermenting yeast was highly enhanced by tuning the concentration of the cofactors, TPP (0.4 mM) and Mg^{2+} ion (0.4 mM) [95] was applied at a high concentration of pyruvate (400 mM). As the substrate, benzaldehyde **32a** has been known to have an inhibitory effect on pyruvate decarboxylase at a high concentration [70,90], it was progressively added [77] to the reaction mixture to keep the substrate concentration low. Moreover, nitrogen gas was vigorously bubbled to

Table 4 C-2 Homologation of Aromatic or Heteroaromatic Aldehydes

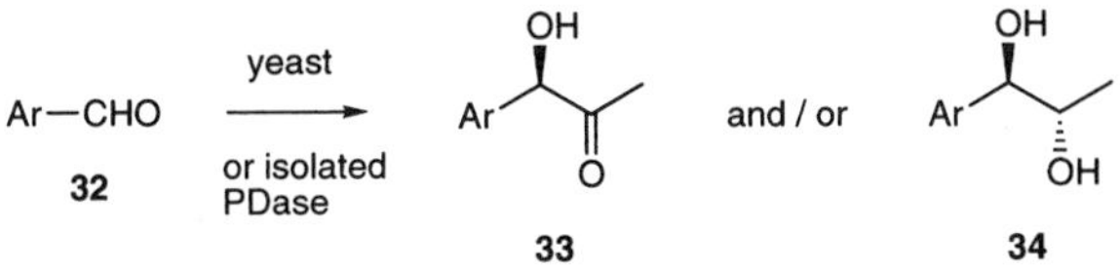

Substrate 32	Ar =	Substituent X =	Ref.
a	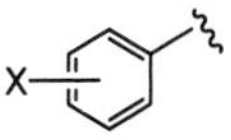	H	62–81
b		*o*-CH$_3$	72,82
c		*m*-CH$_3$	76
d		*p*-CH$_3$	67,72
e		*p*-CH(CH$_3$)$_2$	67
f		*o*-Cl	72,76,79,83
g		*m*-Cl	76,79
h		*p*-Cl	72,76,79,83
i		*o*-F	72,79
j		*m*-F	72,79
k		*p*-F	72,79
l		2,3-F$_2$	79
m		*o*-OCH$_3$	76
n		*m*-OCH$_3$	76
o		*p*-OCH$_3$	72,76
p		*o*-CF$_3$	76
q		*m*-CF$_3$	76
r		*p*-CF$_3$	76
s	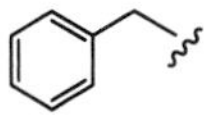		67
t	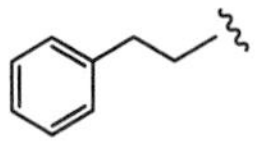		67
u	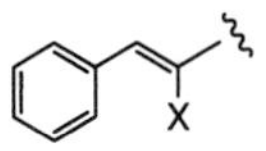	H	61,67,69,84–86
v		Br	69
w		CH$_3$	69,85,87,88
x	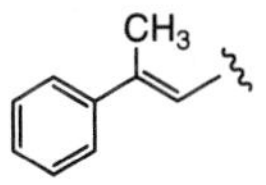		89
y	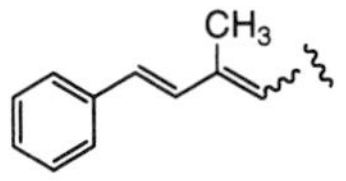		89
z	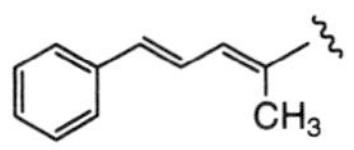		89
a1	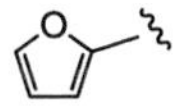		60,67,81,86,90
b1			90

Table 4 Continued

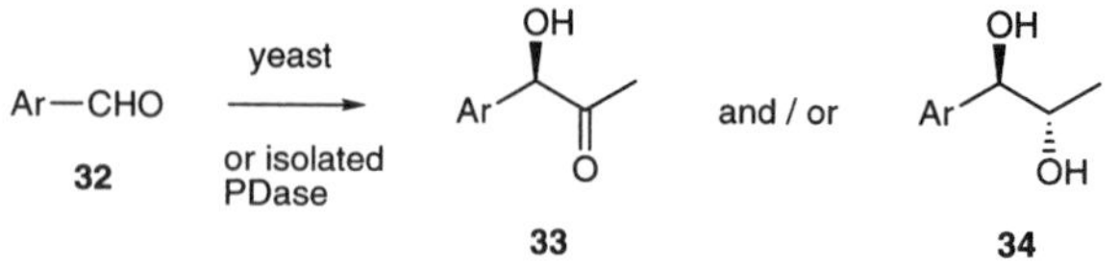

Substrate		Substituent	
32	Ar =	X =	Ref.
c1	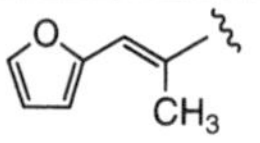		91
d1	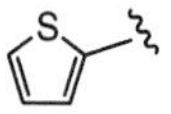		90,92
e1	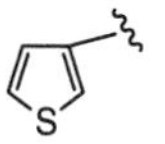		90
f1	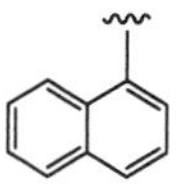		67

remove acetaldehyde, which itself reacts with the activated thaizolium intermediate and was suspected to lower the yield of the desired product **33a**. Combination of the above improvements brought about the highest accumulation of **33a** (15 g/L). The subsequent quantitative reduction of **33a** to **34a** by the addition of carbohydrates as the NADH generation source provided **34a** in a 35% overall yield from **32a**, which is also greater than those reported previously (15–26%). Moreover, as expected, the ratio of **34a** to benzyl alcohol **35a** was 2.7:1, also higher than those previously reported.

5. Pyruvate Decarboxylase-Related Hydroxy Ketones and Diols as Enantiomerically Enriched Starting Materials in Organic Synthesis

The remarkable advantage of this C—C bond formation is that the reaction proceeds in a stereoselective manner. From the screening of microorganisms [71,74,80–81,86], two types of PDase that show complementary enantiofacial selectivity have been found. PDase from yeast (*Saccharomyces*) catalyzes the attack of TPP-thiazolium intermediate on the *si* face of the aldehyde acceptor as shown in Eq. (23). In contrast, the enzyme from *Zymomonas mobilis* shows *re* face selectivity to result in the opposite enantiomer of hydroxy

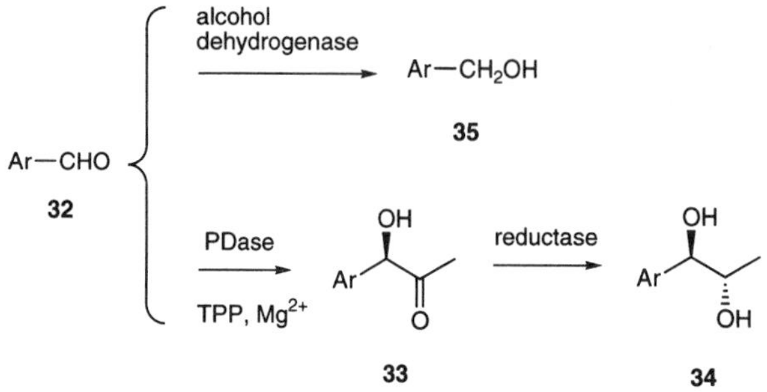

(22)

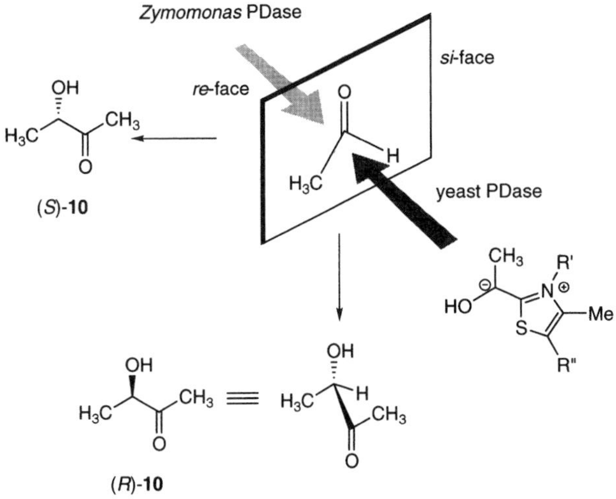

(23)

ketone to that of yeast PDase [74,80]. The enzyme from *Zymomonas*, however, has rather poor affinity to aldehyde acceptor, and accordingly the efficiency of C—C bond formation is lower than that of yeast PDase [74].

Mostly due to its availability, in situ PDase of the actively fermenting yeast [96,97] has been widely used as the biocatalysts in organic synthesis. Fuganti was the first to pay attention to the PDase-related diols **34** as the stereochemically defined and elaborately functionalized chiral synthons for organic synthesis, especially for the synthesis of natural products in enantiomerically enriched forms [98].

Needless to say that the absolute stereochemistry of each chiral center of vicinal diols **34** as well as the diastereomeric and enantiomeric excess is very important. For example, diol **34a** was a mixture of four stereoisomers, as (*1R,2S*):(*1S,2R*):(*1R,2R*):(*1S*: *2S*) = 89:0:6 (or 5):5 (or 6) [81]. The hydroxy ketone formation occurred in a highly enantioselective manner [79,80] as described in Eq. (23). Equation (24) explains the whole results: initially formed (*R*)-**33a** was reduced predominantly by an enzyme (''L-enzyme'') to afford (*1R,2S*)-**34a** (89% of the total mixture), while another minor enzyme (''D-enzyme'') seemed to be responsible for the formation of the (*1R,2R*) isomer (5–6%). The (*R*)-**33a** would be partially racemized in whole cells of yeast [79] to give the (*S*)-**33a**, and this would be further reduced in a similar manner by the two enzymes to give the (*1S,2S*) (5–6%) and (*1S,2R*) isomers, the formation of the latter being almost undetectable. In this case, single recrystallization of the corresponding dibenzoate of **34a** provided the diastereomerically and enantiomerically pure (*1R,2S*) compound [69,72,81]. In other cases also, recrystallization was very effective to obtain diastereomerically and enantiomerically pure materials [69,84,88].

The transformation of hydroxy ketones and diols to enantiomerically enriched compounds is summarized in Table 5. The diols effectively worked as the precursor of carbohydrates with *threo* configuration. L-Amicetose (**36**) [99] and D-allomuscarine (**37**) [100] were synthesized starting from **34u** [69,84] via intermediates **38a** and **38b**, respectively. The diol functionality can also be transformed into an epoxide with defined stereochemistry and further conversion to versatile functionality with concomitant C—C bond formation becomes possible. Starting from **34u**, (*S*)-4-hexanolide (**39**), the pheromone component of *Trogoderma glabrum*, was synthesized [84]. The key reaction was the reductive

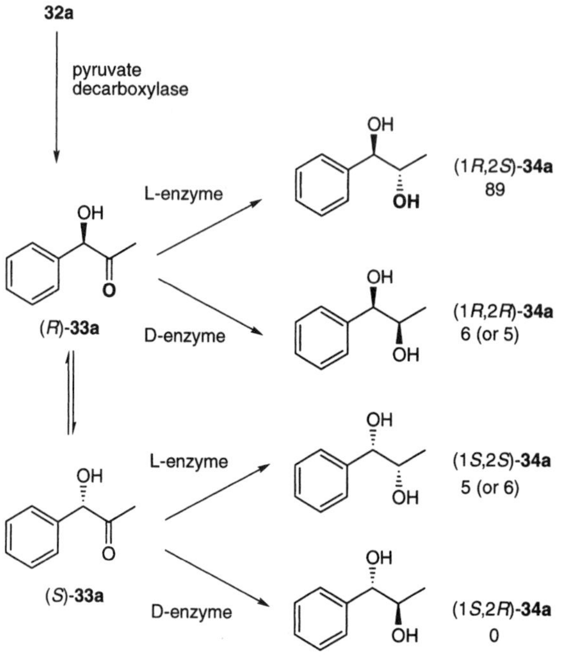

(24)

epoxide ring opening of **40**, which was derived from **34u**. Another example is (3S,4S)-4-methyl-3-heptanol (**41**), the pheromone of *Scolytus multistriatus* was synthesized [101]. The key reaction was the regioselective epoxide ring opening with dialkyl cuprate on **42**. Methyl homolog **34w** [81] is another good enantiomerically enriched starting material. α-Alkoxy (**43**) and β-alkoxy carbonyl compound (**44**) were also prepared via the rgioselective ring opening reaction of **45** [102]. In the above three cases, the regioselectively functionalized diols worked as the precursor of epoxides, the key intermediates.

Taking advantage of the allylic alcohol moiety, a new epoxy ring can be introduced via epoxidation. In the initial attempts, the expoxidation of **34w**, however, did not exhibit stereoselectivity, and accordingly two diastereomeric epoxides, which were transformed into L-mycarose **46** and L-olivomycose **47** [99], were obtained. The situation was overcome by the elaboration of the precursor **34z**, which enabled the epoxidation in a highly stereoselective manner. An improved synthesis of **47** became possible starting from the resulting epoxide [89].

Diol can be protected as acetonide, and a subsequent stereoselective introduction of nucleophiles on the adjacent (exo) carbon would be possible, as the α- and β-oxygen atoms as well as the substituents on the conformationally fixed five-membered dimethyl dioxolane ring control the direction of nucleophilic attack. For example, a 1,4 addition of ammonia to α,β-unsaturated ester **48**, derived from **34u**, proceeded stereoselectively, and the resulting amine was converted to N-trifluoroacetyl-L-acosamine **49** [103]. The inversion of the configuration at C-4 also afforded N-trifluoroacetyl-L-daunosamine **50**. In a similar manner, the key step for the synthesis of C-1 methyl analog (**52**) of **49** was the addition of ammonia to **51** [87]. A hydroxy ketone **33w**, which was produced from **32w** by only the action of PDase [85], was a good precursor of unsaturated ester **53**. From this, C-5 methyl analog (**54**) was synthesized in a similar manner [87].

Addition of a Grignard reagent on the carbonyl group can generate another chiral center with a concomitant C—C bond formation. The stereoselectivity, however, depends

Table 5 Synthesis of Enantiomerically Enriched Compounds Starting from PDase-Related Hydroxy Ketones and Diols

Starting material	Key intermediate and/or reaction	Target compound[a]	Ref.
34u	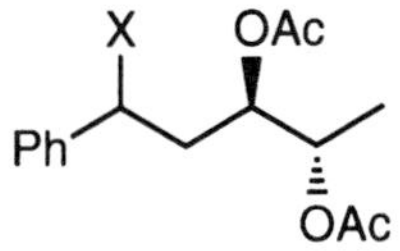 **38a** X = H	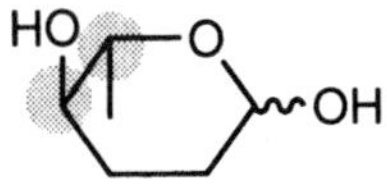 L-amicetose **36**	99
34u	**38b** X = N$_3$	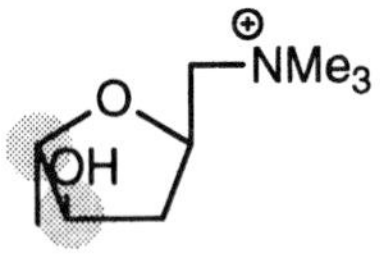 D-allomuscarine **37**	100
34u	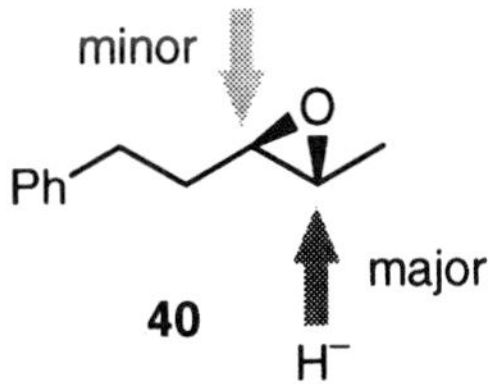 **40**	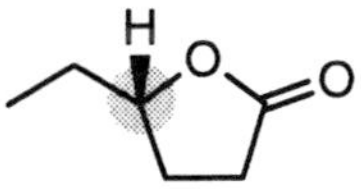 (*S*)-4-hexanolide **39**	84
34u	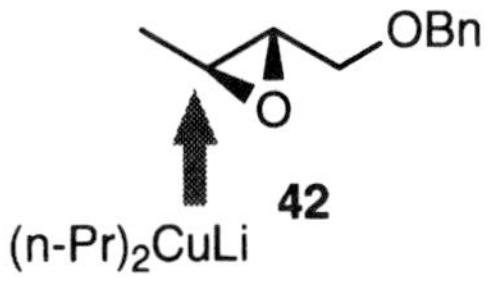 **42** (n-Pr)$_2$CuLi	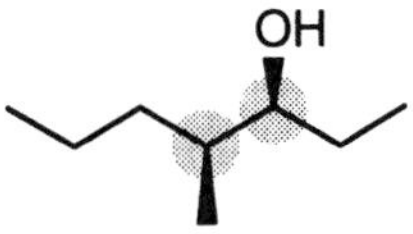 (3*S*,4*S*)-4-methyl- 3-heptanol **41**	101
34w	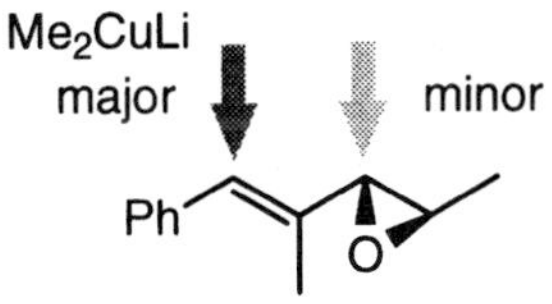 Me$_2$CuLi **45**	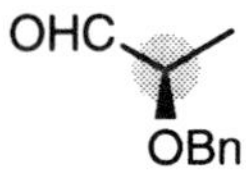 **43**	102
34w	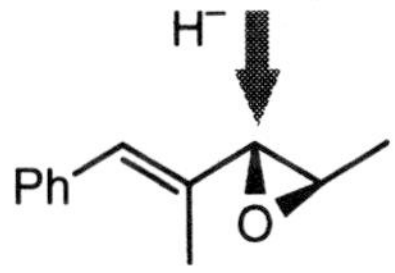 **45**	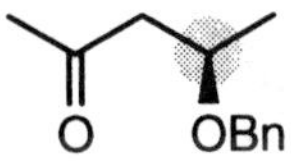 **44**	102

Table 5 Continued

Starting material	Key intermediate and/or reaction	Target compound[a]	Ref.
34w	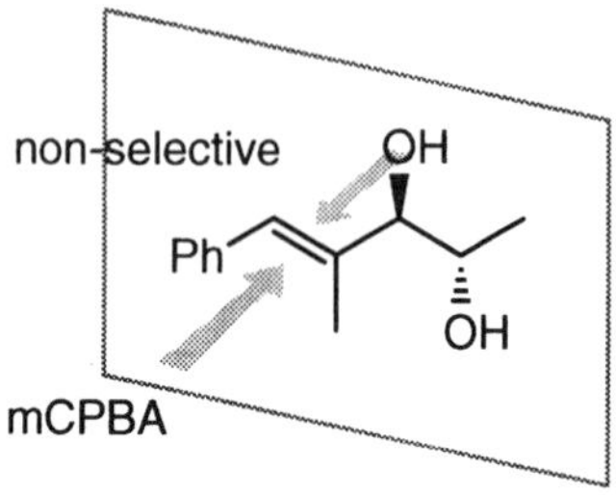 non-selective Ph OH OH mCPBA	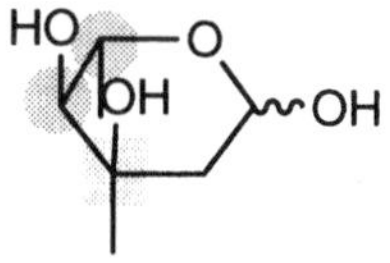 L-mycarose **46** 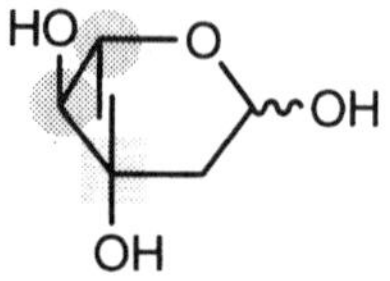 L-olivomycose **47**	99
34z	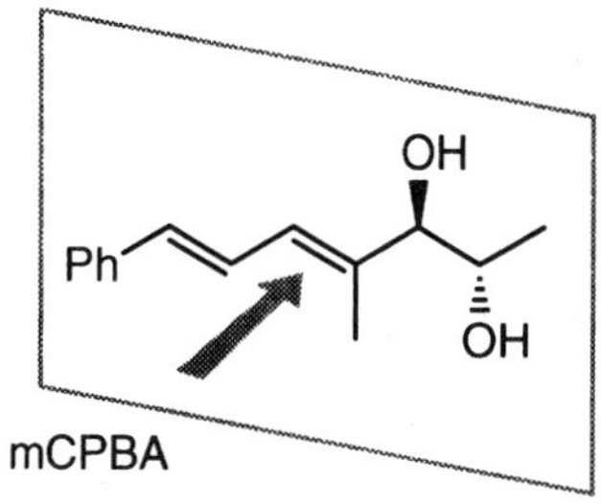 Ph OH OH mCPBA	**47**	89
34u	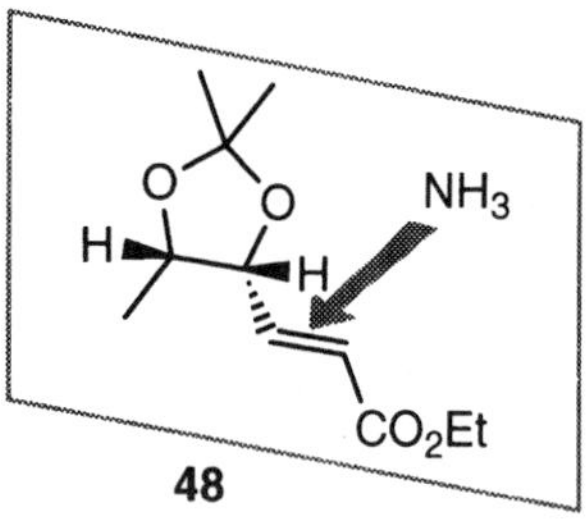 48 NH₃ CO₂Et	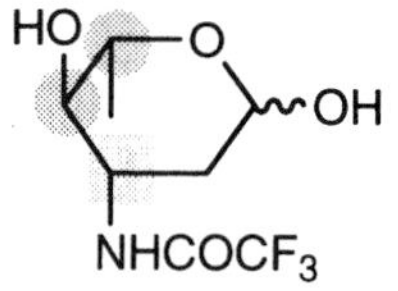 N-TFA-L-acosamine **49** 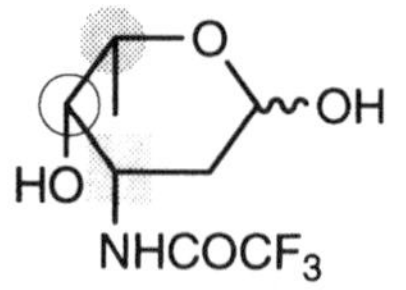 N-TFA-L-daunosamine **50**	103
34u	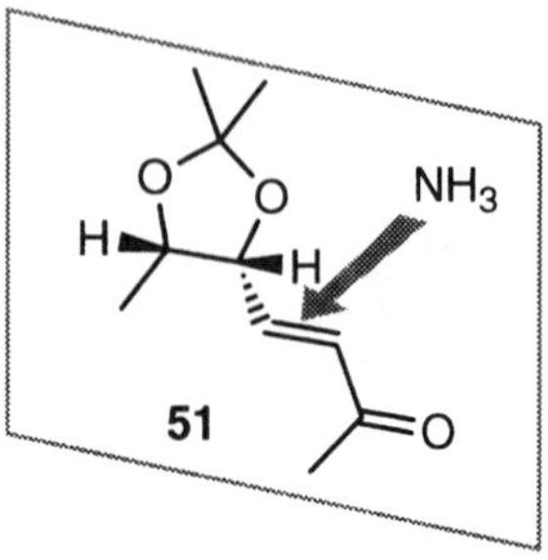 51 NH₃ O	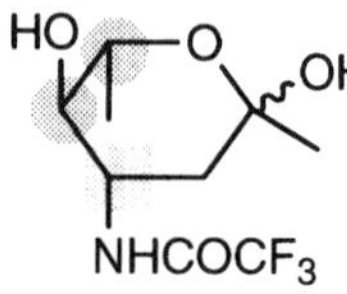 **52**	87

a

Table 5 Continued

Starting material	Key intermediate and/or reaction	Target compound[a]	Ref.
33w	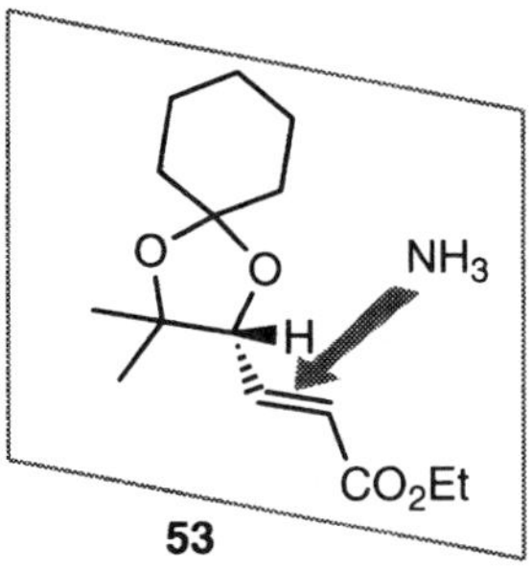 **53**	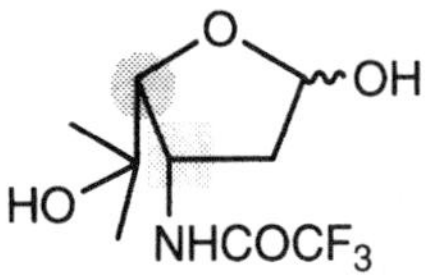 **54**	87
34u	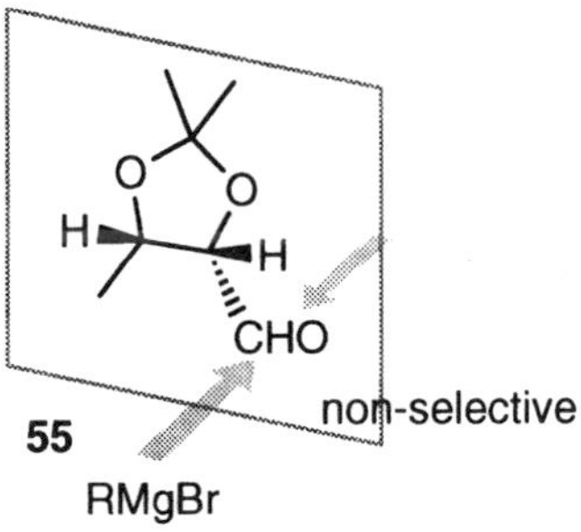 **55** RMgBr non-selective	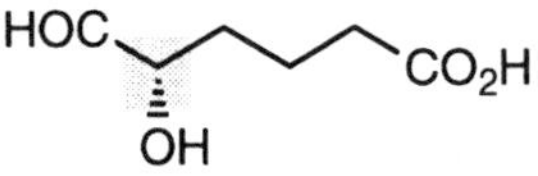 **56**	104
		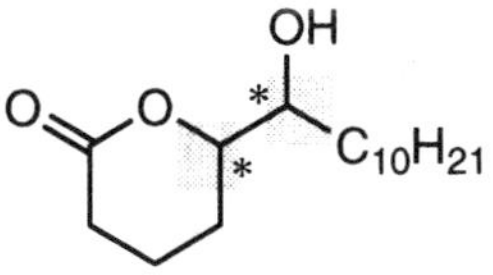 **57**	105
34c1	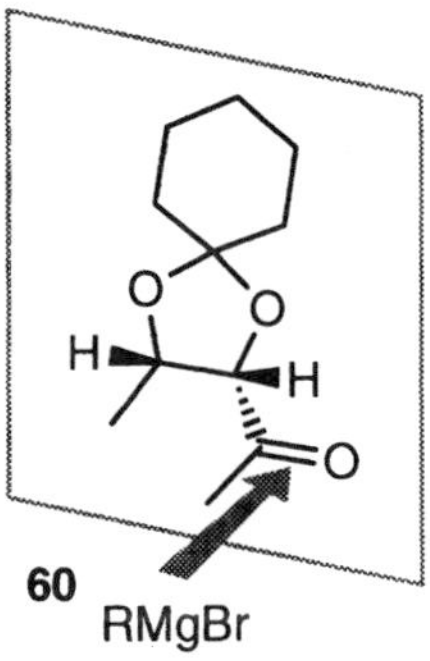 **60** RMgBr	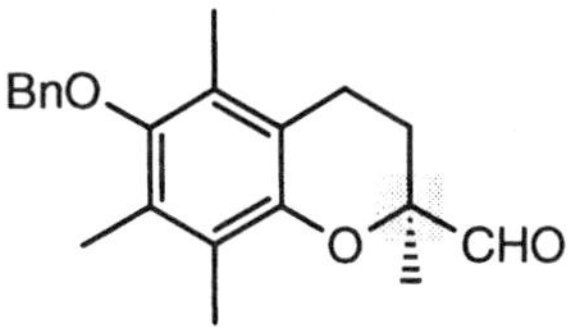 **58**	91
34w	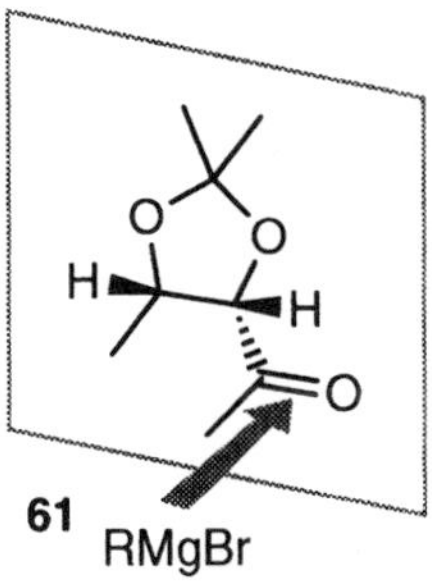 **61** RMgBr	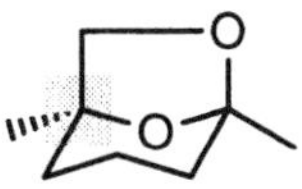 (-)-frontalin **59**	88

Table 5 Continued

Starting material	Key intermediate and/or reaction	Target compound[a]	Ref.
34w		L-arabino form **63** L-lyxo form **64**	106,107
	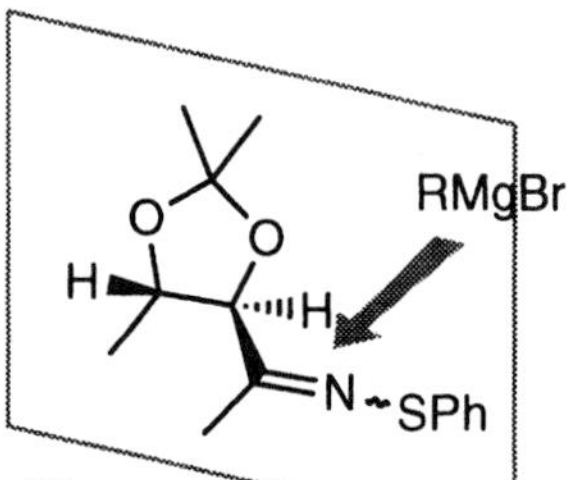	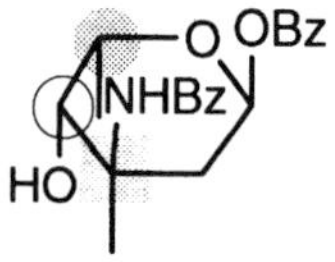 L-xylo form **66**	
34u	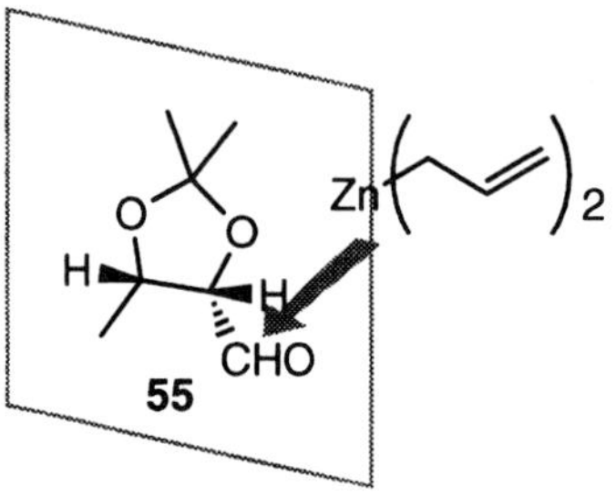	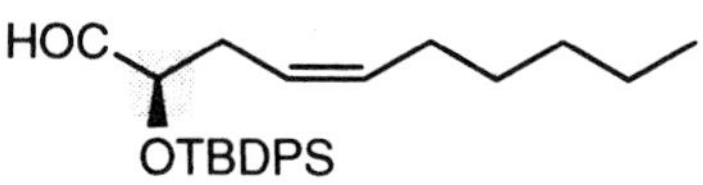 **67**	104
		(R)-**39**	108
		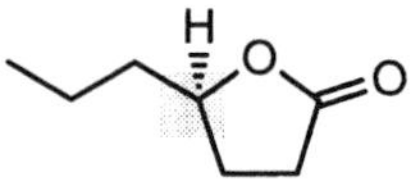 (R)-4-heptanolide **68**	108
		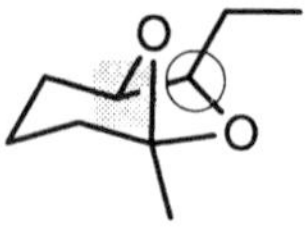 (+)-exo-brevicomin **69**	108

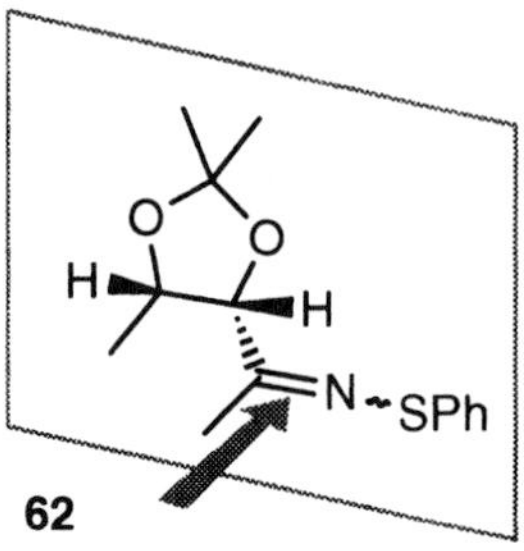

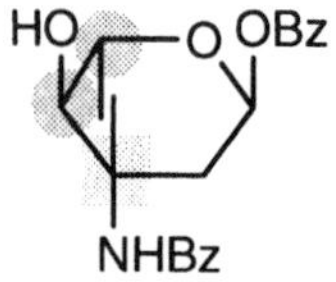

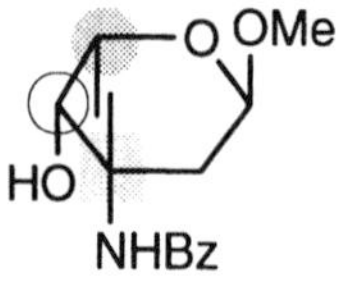

Table 5 Continued

Starting material	Key intermediate and/or reaction	Target compound[a]	Ref.

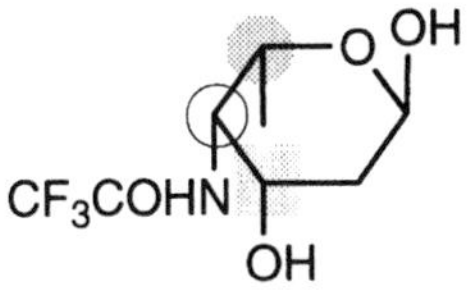

L-lyxo form **70**   L-arabino form **71**

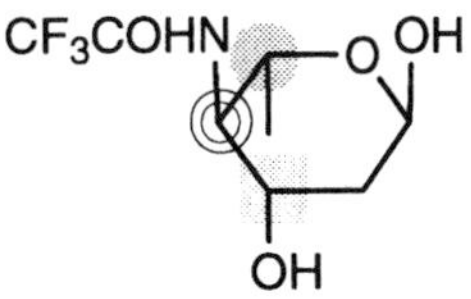

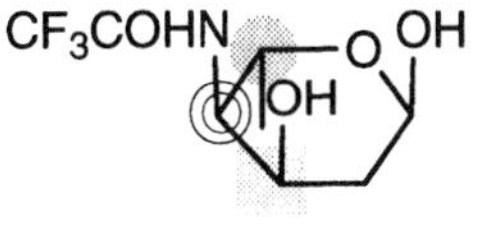

109

L-ribo form **72**

[a] ▨: Chiral Centers Originated from **33** and/or **34**; ○: Chiral Centers Originated from **33** and/or **34**, but inverted; ▮: Newly Generated Chiral Centers; ◎: Chiral Centers Originated from **33** and/or **34**, but doubly inverted.

on the structure of the substrates, the reaction conditions, as well as the nature of Grignard reagents. Indeed, the addition of 4-pentenyl or *n*-decyl magnesium bromide to aldehyde **55** yielded two diastereomers in a 6:4 ratio [104,105]. Separation of the diastereomeric products enabled the synthesis of LTB$_4$ intermediate **56**. Moreover, the second C—C bond formation also proceeds in a nondiastereomerically selective manner on the later intermediate, in total, four diastereomers of mosquito oviposition attractant pheromone **57** were obtained. Contrasting results were shown in the synthesis of tocopherol chroman **58** [91] and frontalin **59** [88] in enantiomerically enriched forms. They have been achieved by the stereoselective addition of Grignard reagent on the ketones (**60** and **61**). The stereoselective addition occurs on a similar substrate, sulfenylimine (**62**), and made the syntheses of amino sugars **63–64** possible [106,107]. It is noteworthy that in this case imine **62** was prone to epimerize at the α position of imine to give the trans isomer (**65**) under the condition of its preparation from the ketone precursor. Fortunately, the stereoselective addition also took place on **65** and L-xylo form (**66**) was synthesized from the product.

Addition of diallylzinc on **55** proceeds in highly stereoselective manner. Another LTB$_4$ intermediate **67** [104], pheromones such as (*R*)-**39**, (*R*)-4-heptanolide (**68**), (+)-*exo*-brevicomin (**69**) [108], were synthesized by utilizing newly stereochemically established chiral center. The addition worked in the synthesis of L-*lyxo* (**70**), L-*arabino* (**71**), and L-*ribo* (**72**) forms of 4-trifluoroacetamino-2,4,6-trideoxyhexose [109]. In this case, another key reaction was the subsequent regio- and stereoselective intramolecular epoxide ring opening to give oxazolidinones.

6. Pyruvate Decarboxylase in Carbohydrate Synthesis

A unique combination of pyruvate decarboxylase and aldolase-catalyzed carbohydrate syntheses was recently developed by us. In the synthesis of sialic acid starting from *N*-acetyl-D-mannosamine and pyruvic acid catalyzed by the sialic acid aldolase, the separation of the desired product, sialic acid, and the excessively used pyruvic acid has been a serious

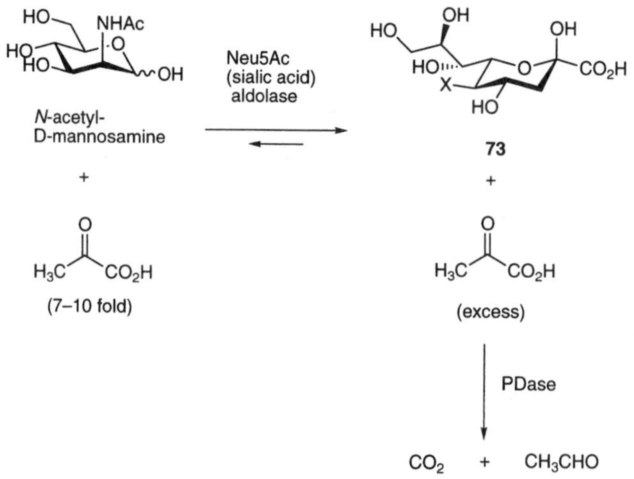

$$CO_2 \quad + \quad CH_3CHO \tag{25}$$

problem, since the pK_a values of both are nearly the same. This problem was overcome by the use of PDase, which catalyzes the decomposition of pyruvic acid into acetaldehyde and carbon dioxide, both of which are neutral and volatile materials. Indeed, by the aid of PDase as the first step of the workup procedure of sialic acid–catalyzed synthesis, the purification of the desired product was remarkably facilitated [Eq. (25)]. This protocol could be applied to the synthesis of the *N*-acetylneuraminic acid (X = AcHN, **73a**), KDN (X = HO, **73b**), *N*-glycolylneuraminic acid (X = HOCH$_2$COHN, **73c**) [95,110].

B. Transketolase

Transketolase (TKase, EC 2.2.1.1) essentially catalyzes the transfer of C-2 unit from D-xylulose-5-phosphate to ribose-5-phosphate to give D-sedoheptulose-7-phosphate, via a thiazolium intermediate as shown in Eq. (26). An important discovery was that hydroxy-pyruvate (**74**) works as the donor substrate and the reaction proceeds irreversibly via a loss of carbon dioxide [111] [Eq. (27)]. In this chapter, we put emphasis on the synthesis with hydroxypyruvate, as it is a typical TPP-mediated decarboxylation reaction of α-keto acid [112].

This enzyme is available from yeast [112,113] and spinach leaves [113–115] in quantity. Yeast enzyme is commercially available. Recently, TKase gene of *E. coli* was cloned and overproduced [116–118]. Moreover, convenient assay method has been pro-

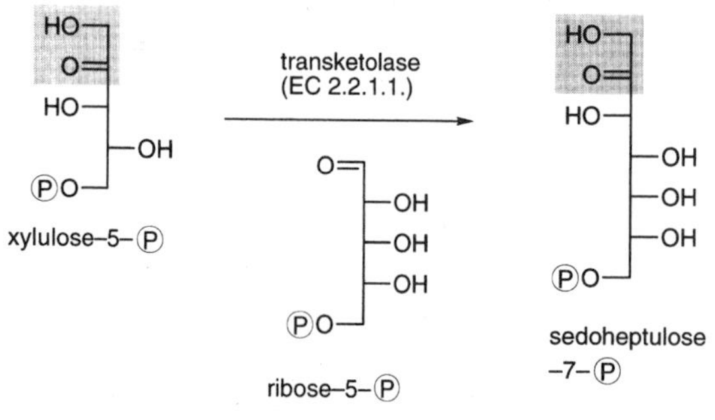

$$\tag{26}$$

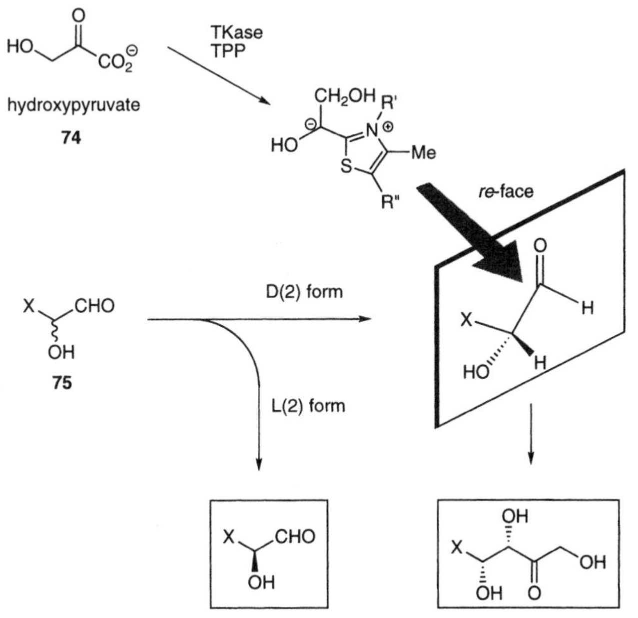

(27)

posed [119], which is very important for the evaluation of enzyme activity from any of these sources.

1. Substrate Specificity and Stereochemical Course of TKase-Catalyzed Reaction

Through extensive screening of substrates (**75a–75b3**) [112–135] it was revealed that this enzyme accepts a very wide range of substrates. In addition to phosphorylated aldoses, which are the native substrates, nonphosphorylated aldoses, simple aliphatic, aromatic, heterocyclic, and functionalized aldehydes even with an increased hydrophobicity work as the substrates (Table 6). The stereochemical course has been elucidated in Eq. (27). The hydroxyl group on the 2-position of the aldehyde is very important [132], and 2-deoxygenated aldehydes were rather weak substrates. The substrates with D-configuration at the 2-position have a stronger affinity to TKase than L forms.

2. Application of TKase-Catalyzed Reaction in Organic Syntheses

This decarboxylation reaction serves as the tool for enzyme-mediated organic synthesis [136,137]. As seen in Eq. (27), the addition of thiazolium intermediate derived from hydroxypyruvate proceeds via *re* face attack to afford the products (**76**) with stereochemically defined 2,3-*erythro* stereochemistry. The examples are summarized in Table 7. This method works very well for the synthesis of naturally occurring phosphorylated [126,134], nonphosphorylated ketoses [120,125], and deoxy sugars [115,124]. Moreover, 2,3-*erythro*-diol motif is exemplified in the chemoenzymatic synthesis of the L-series of aldoses (**77–79**) [122], aza sugars (**80,81**) [128–130] and (±)-*exo*-brevicomin (**69**), an insect pheromone [131] (Table 8). The stereochemically controlled synthesis of aldehydes with D(2) config-

(text continues on pg. 521)

Table 6 Substrate Specificity of Transketolase

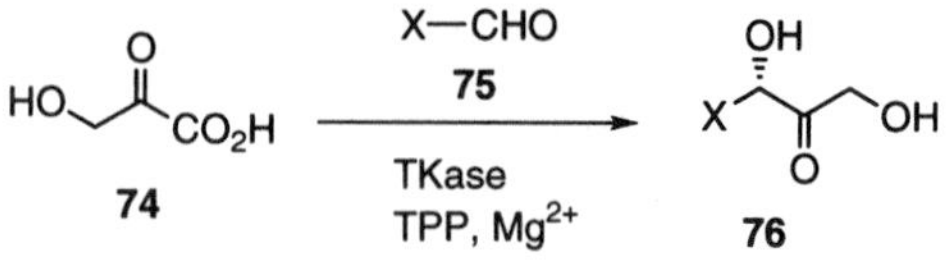

Substrate[a] 75	X =[a]	Source of enzyme[b] (relative activity)[c]	Ref.
a	 HO∕∕	Y (100)	120,121
		Y[d]	122
		S[d]	120
		S (100)	123
		E[d]	118
b	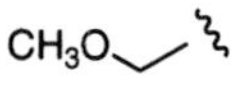 CH₃O∕∕	Y (32)	121
		S (32)	123
c	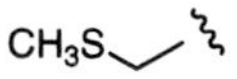 CH₃S∕∕	S (35)	123
d	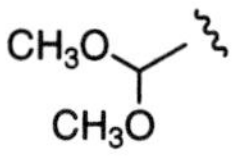 CH₃O / CH₃O	S (11)	121
e	 Cl∕∕	S (23)	123
f	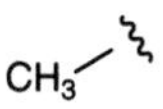 CH₃	Y (25)	121
		S (12)	123
		S[d]	124
g (DL-Glyceraldehyde)	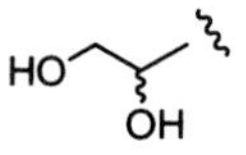 HO / OH	Y (56)	120
		Y (33)	122
		Y, S[d]	113
		Y[d]	120
		E (37)	116
		E[d]	117
h (D-)	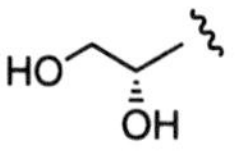 HO / OH	Y (78)	120
		Y (78)	122
		Y[d]	125
		E (66)	117
i (L-)	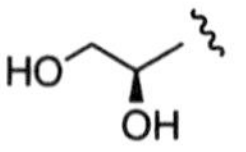 HO / OH	Y (0)	120
		E (0)	117
j (D-G–3–Ⓟ)	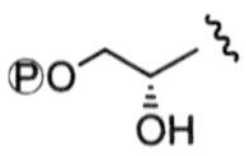 ⓅO / OH	Y[d]	112
		Y (44)	120
		Y (44)	122
		Y[d]	126
k	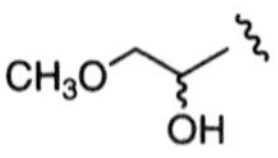 CH₃O / OH	Y (20)	121
		Y (27)	122
		Y[d]	127
l	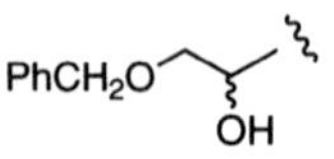 PhCH₂O / OH	Y[d]	127
		E[d]	118

Table 6 Continued

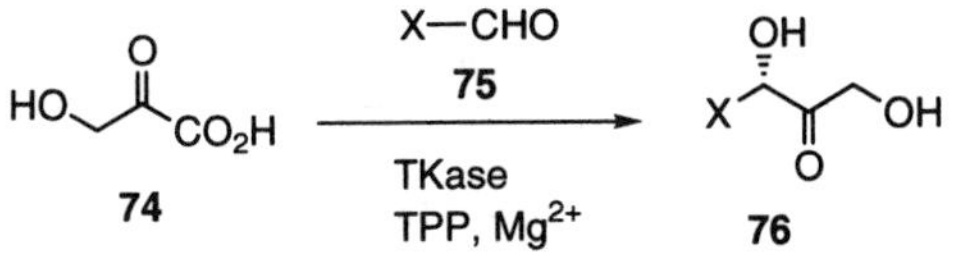

Substrate[a] 75	X =[a]	Source of enzyme[b] (relative activity)[c]	Ref.
m	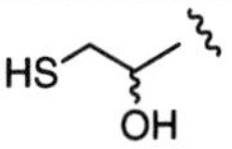 HS–CH₂–CH(OH)–	Y[d]	127
n	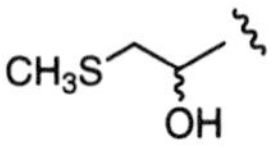 CH₃S–CH₂–CH(OH)–	Y (33)	122
o	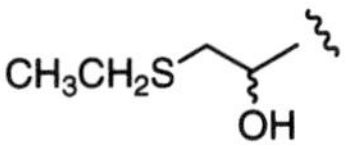 CH₃CH₂S–CH₂–CH(OH)–	Y[d]	127
p	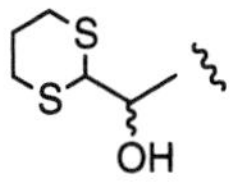 (1,3-dithian-2-yl)–CH(OH)–	S[d]	128
q	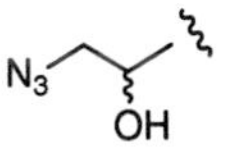 N₃–CH₂–CH(OH)–	S[d]	129
r	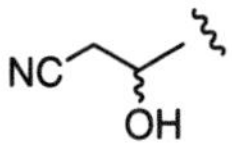 NC–CH₂–CH(OH)–	Y[d]	127,130
s	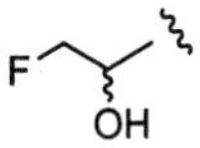 F–CH₂–CH(OH)–	Y (47) / Y[d]	121 / 127
t	CH₃–CH(OH)–	S[d] / Y (44) / Y (20)	120 / 121 / 122
u	CH₃–C(=O)–	Y (19) / E (18) / E (21)	121 / 116 / 117
v	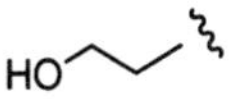 HO–CH₂CH₂–	S[d]	123
w	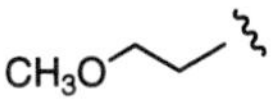 CH₃O–CH₂CH₂–	S (31)	123
x	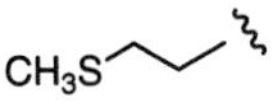 CH₃S–CH₂CH₂–	S (24)	123
y	CH₃–CH₂–	S (5) / E (20) / E (24) / E[d]	123 / 116 / 117 / 118
z (D-Erythrose)	HO–CH₂–CH(OH)–CH(OH)–	Y (56) / Y (84) / Y (56) / E (84) / E (82)	120 / 121 / 122 / 116 / 117

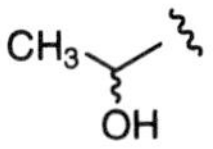

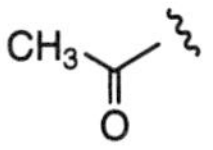

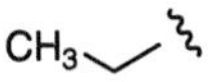

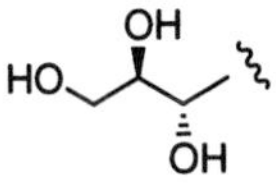

Table 6 Continued

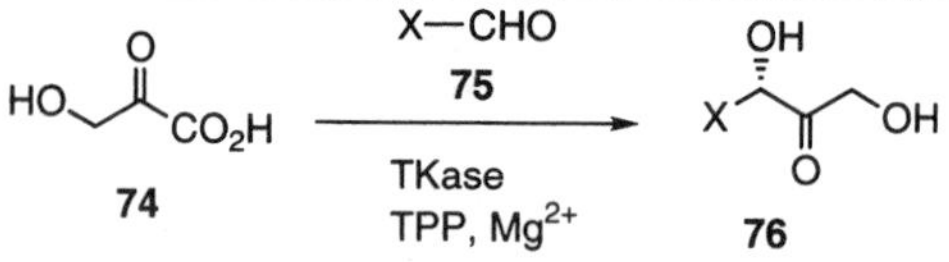

HO–CH₂–CO–CO₂H (74) + X—CHO (75) →[TKase, TPP, Mg²⁺] 76

Substrate[a] 75	X =[a]	Source of enzyme[b] (relative activity)[c]	Ref.
a1 (L-Threose)	(HO–, OH, OH structure)	Y (39)	121
		E (56)	116
		E (70)	117
b1 (D-E–4–Ⓟ)	(ⓅO–, OH, OH structure)	Y (75)	120
		Y (33)	122
c1	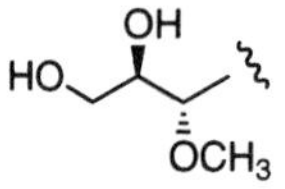 (HO–, OH, OCH₃ structure)	S (27)[e]	132
d1	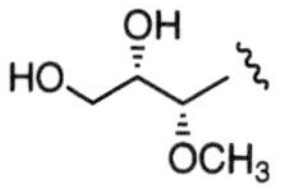 (HO–, OH, OCH₃ structure)	S (6)[e]	132
e1	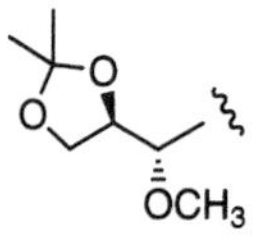 (dioxolane, OCH₃ structure)	S (11)[e]	132
f1	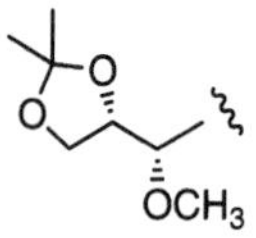 (dioxolane, OCH₃ structure)	S (10)[e]	132
g1	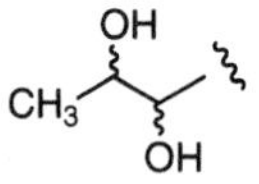 (CH₃–, OH, OH structure)	S[d]	115
h1	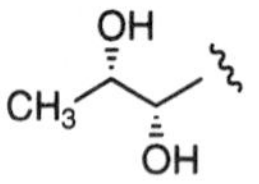 (CH₃–, OH, OH structure)	S[d]	124
i1	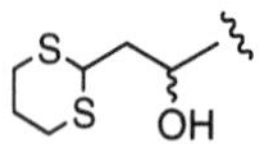 (dithiane, OH structure)	S[d]	128
j1	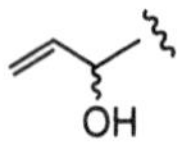 (vinyl, OH structure)	Y (56)	122
		E (30)	117
		E[d]	118
k1	(CH₃–, OH structure)	Y (33)	122
		Y[d]	127
		Y[d]	131
l1	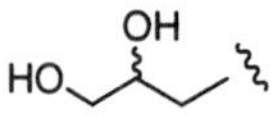 (HO–, OH structure)	Y (43)	121
m1	(HO–, OH structure)	Y (45)	121
		S (8)[e]	132

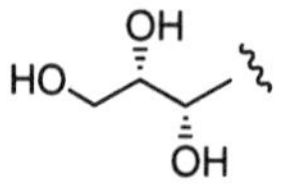

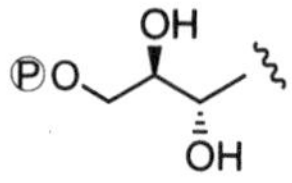

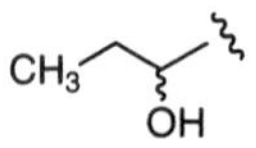

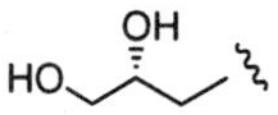

Table 6 Continued

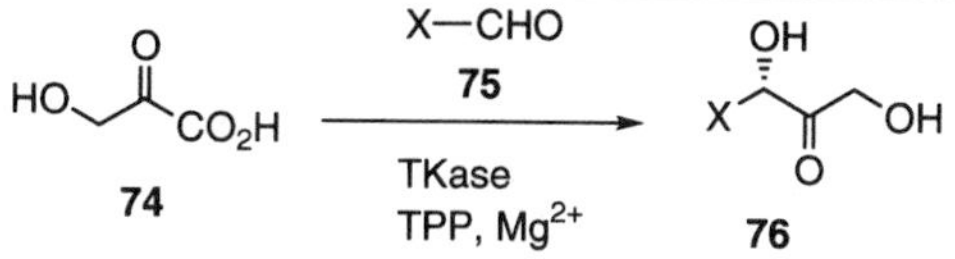

Substrate[a] 75	X =[a]	Source of enzyme[b] (relative activity)[c]	Ref.
n1	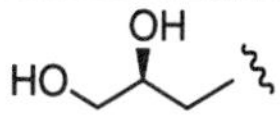	S (4)[e]	132
o1	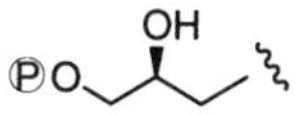	S (61)[e]	132
p1	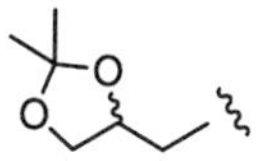	Y (11)	121
q1	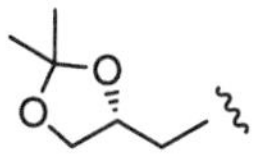	S (5)[e]	132
r1	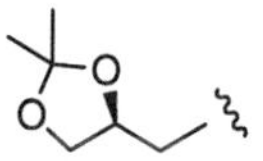	S (6)[e]	132
s1	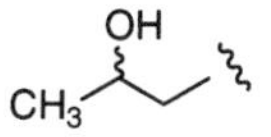	Y (29)	121
t1	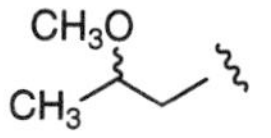	Y (30)	121
u1	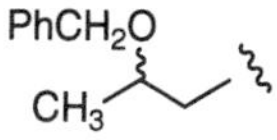	Y (14)	121
v1	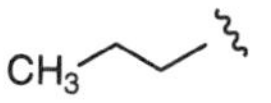	Y (11)	121
w1	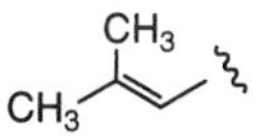	Y (11)	121
x1 (D-Ribose)	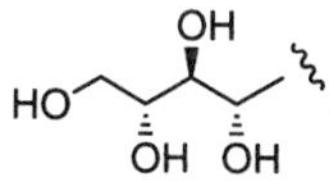	Y (30) S[d]	121 114,133
y1 (L-Lyxose)	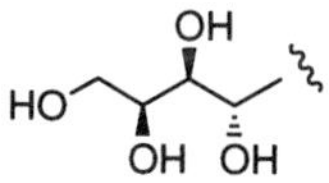	S[d]	114
z1 (D-Xylose)	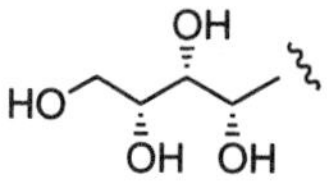	S[d]	114
a2 (L-arabinose)	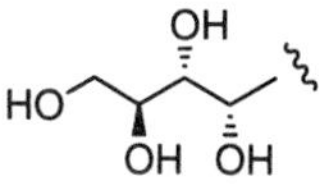	S[d]	114
b2 (D-Deoxy ribose)	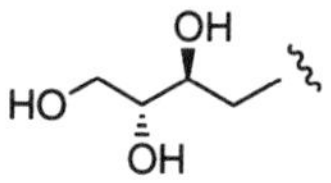	Y (16)	121

Table 6   Continued

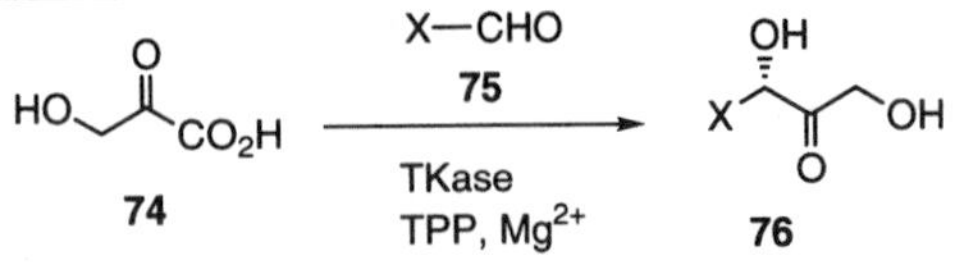

Substrate[a] 75	X =[a]	Source of enzyme[b] (relative activity)[c]	Ref.
c2 (D-Ara–5–Ⓟ)	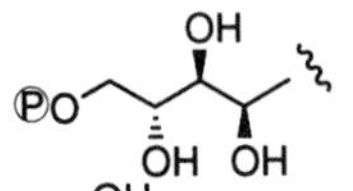	Y[d]	112
d2	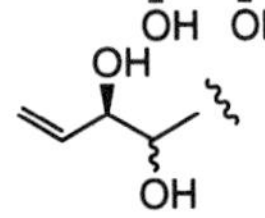	Y (36)	122
e2	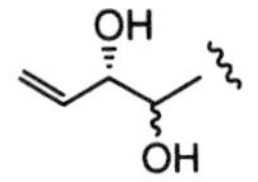	Y (32)	122
f2	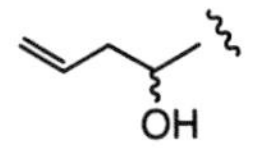	Y (28)	122
g2	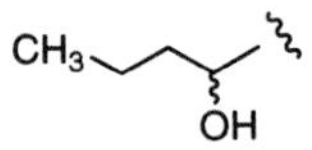	Y (22)	122
h2	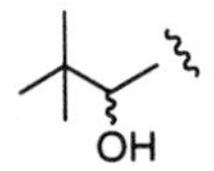	Y (11)	122
i2 (D-Glucose)	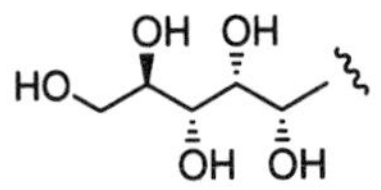	Y (4) Y (4) S (10) E (14) E (13)	121 122 120 116 117
j2 (D-Mannose)	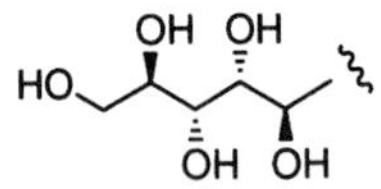	E (3) E (4)	116 117
k2 (D-G–6–Ⓟ)	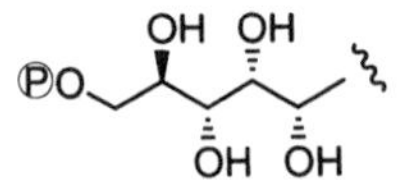	Y[d] Y (9) Y (9)	112,134 120 122
l2 (D-Allose–6–Ⓟ)	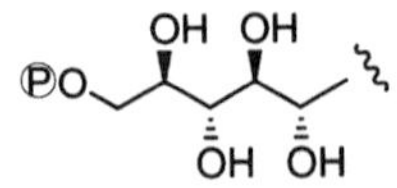	Y[d]	134
m2	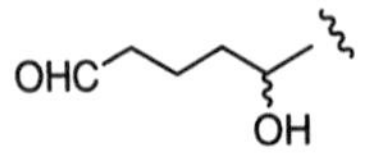	Y (27)	121
n2	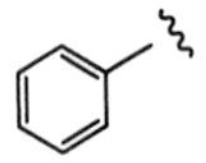	Y (10)	121
o2	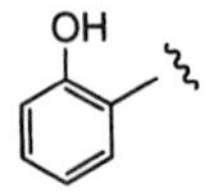	Y (28)	121

Table 6 Continued

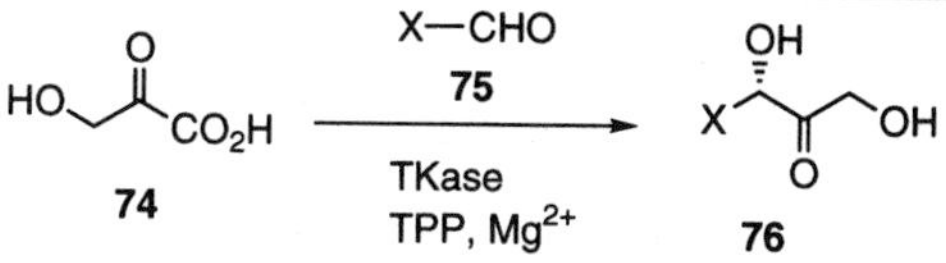

Substrate[a] 75	X =[a]	Source of enzyme[b] (relative activity)[c]	Ref.
p2	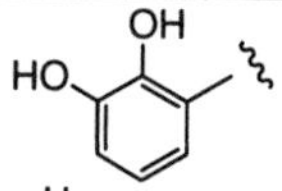	Y (32)	121
q2	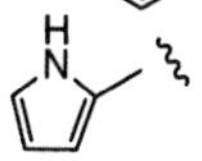	Y (21)	121
r2	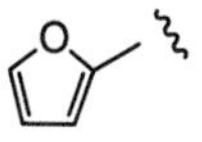	Y (11) E (3)	121 116
s2	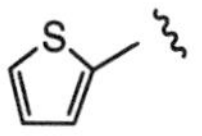	Y (32)	121
t2	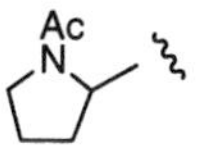	E (5)	116
u2	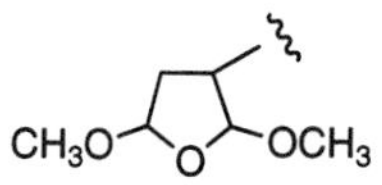	Y (13)	121
v2	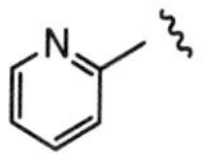	Y (13)	121
w2	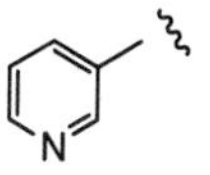	Y (13)	121
x2	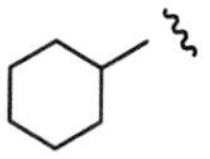	E (4)	116
y2	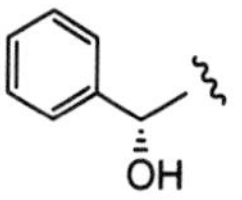	E[d]	135
z2	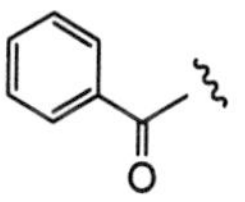	Y (11)	121
a3	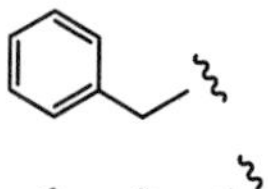	E (13) E[d]	117 118
b3	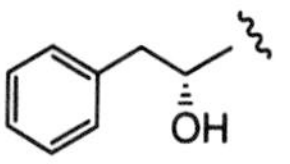	E[d]	117,135

[a] Ⓟ means phosphate.
[b] From Y, yeast; S, spinach leaves; E, *E. coli.*
[c] Glycolaldehyde (**75a**) = 100.
[d] Product was formed, but no comment on the reaction rate.
[e] Ribose-5–Ⓟ = 100.

Table 7   Transketolase-Catalyzed C—C Bond Formation

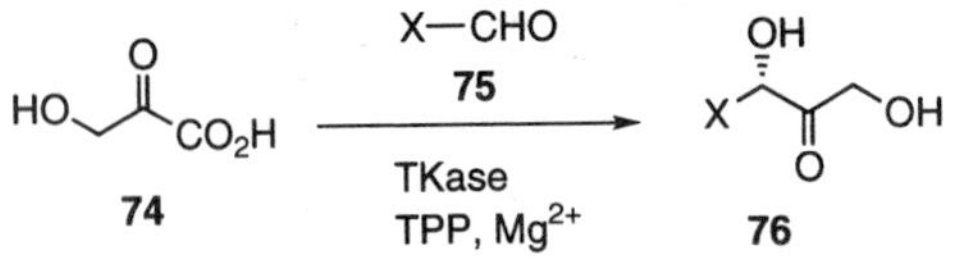

Substrate[a] 75	Product[a] 76	Ref.
a	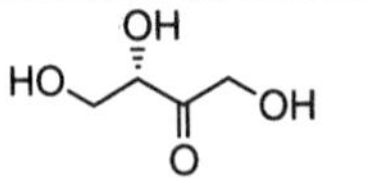 (D-erythrulose)	118,120,122,123
b	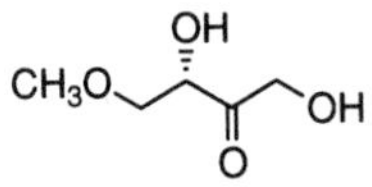	123
c	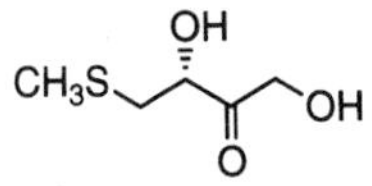	123
f	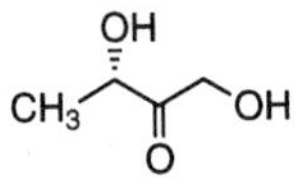	121,123,124
g ($\pm$)[b]	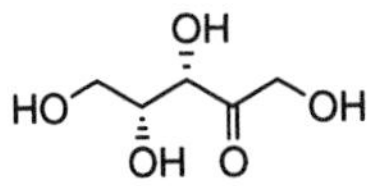 (D-xylulose)	113,120
h	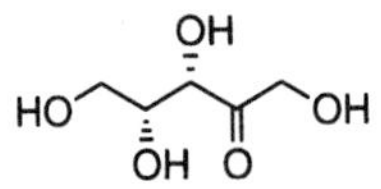 (D-xylulose)	125
j (D-G–3–(P))	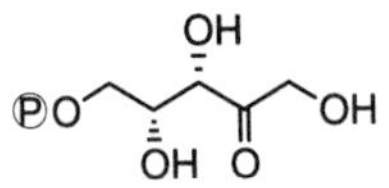 (D-xyl–5–(P))	126
k ($\pm$)[b]	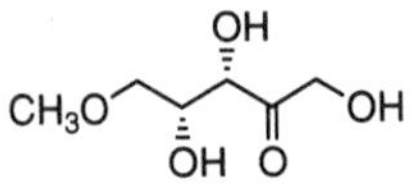	122,127
l ($\pm$)[b]	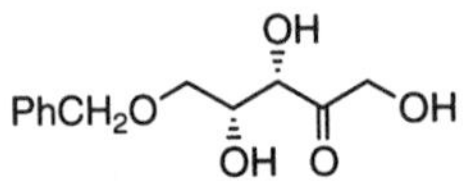	118,127
m ($\pm$)[b]	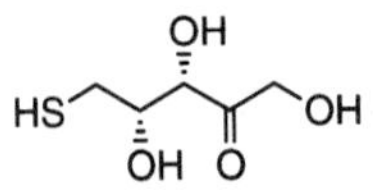	127
o ($\pm$)[b]	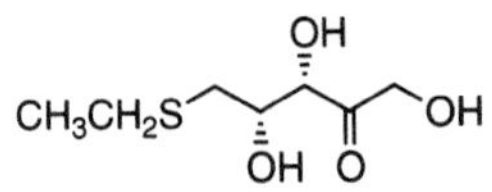	127

Table 7 Continued

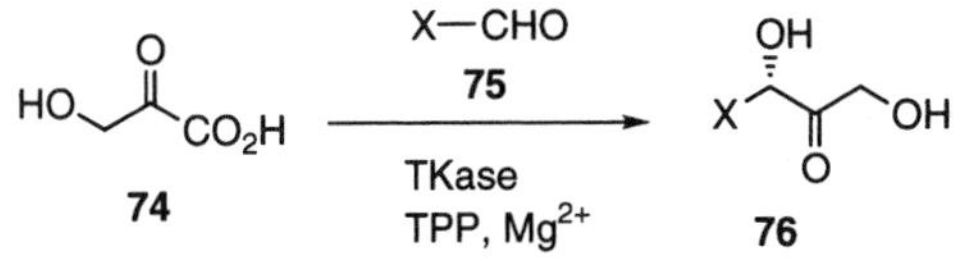

Substrate[a] 75	Product[a] 76	Ref.
p (±)[b]	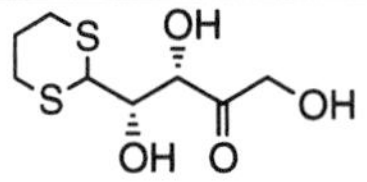	128
q (±)[b]	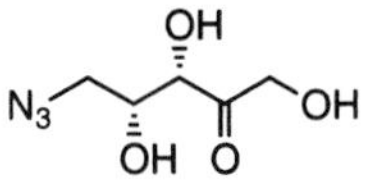	129
r (±)[b]	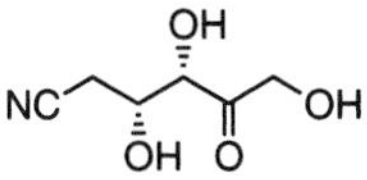	127,130
s (±)[b]	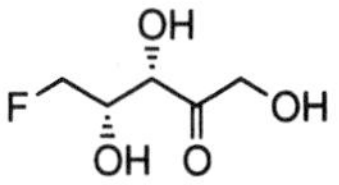	127
t (±)[b]	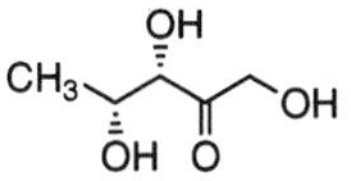	120,122
u	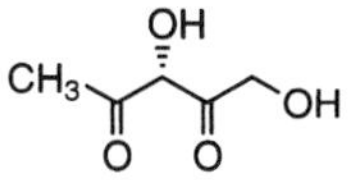	116
v	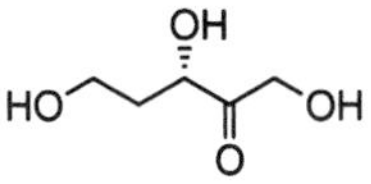	123
x	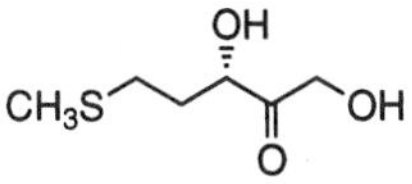	123
y	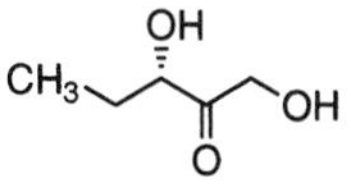	116,118
g1 (±)[b]	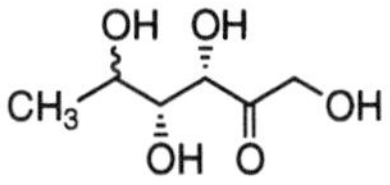	115
h1	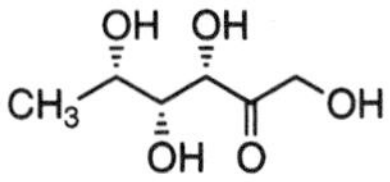	124
i1 (±)[b]	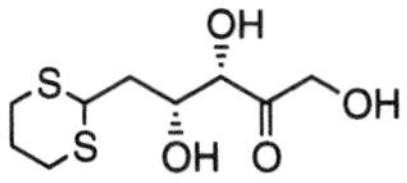	128
j1 (±)[b]	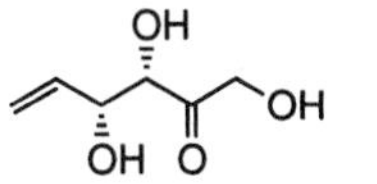	118,122

Table 7   Continued

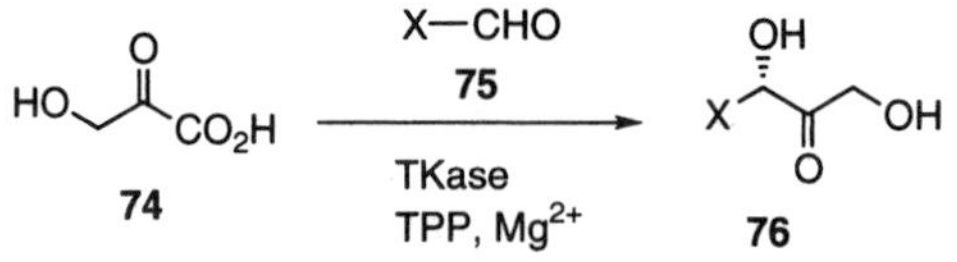

Substrate[a] 75	Product[a] 76	Ref.
k1 (±)[b]	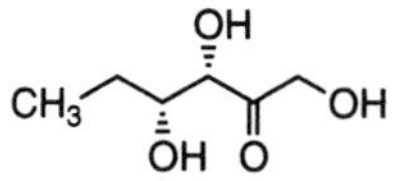	122,127
x1	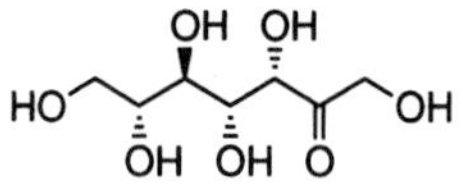 (D-sedoheptulose)	133
d2 (±)[b]	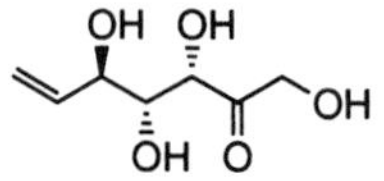	122
e2 (±)[b]	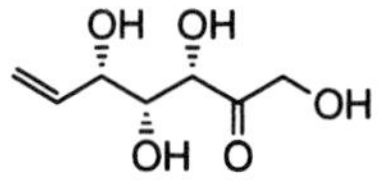	122
f2 (±)[b]	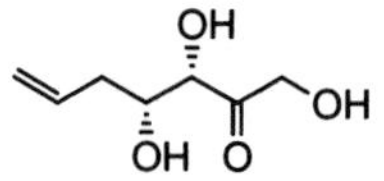	122
g2 (±)[b]	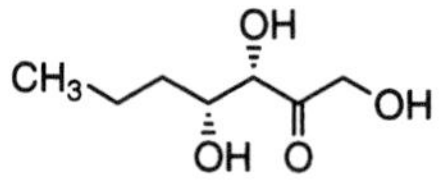	122
k2	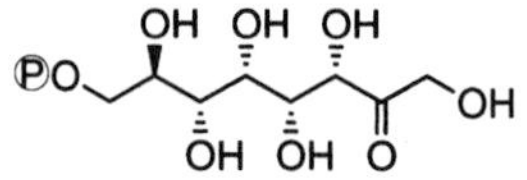	134
l2	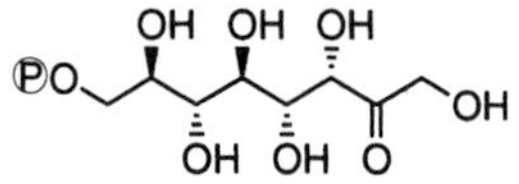	134
y2	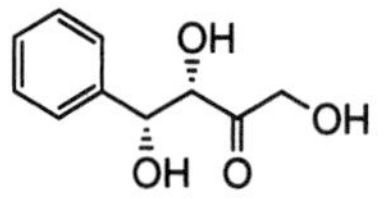	135
a3	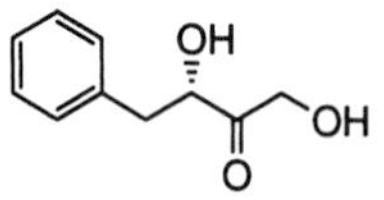	118
b3	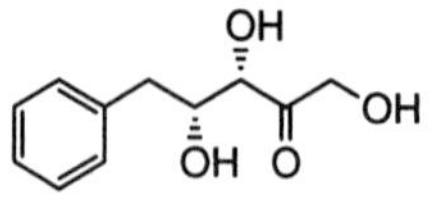	135

[a] $\textcircled{P}$ means phosphate.
[b] Only D(2) enantiomer was reacted to give hydroxy ketone.

Table 8 Utilization of TKase-Catalyzed Products and/or Kinetically Resolved Unreacted α-Hydroxy Aldehydes in Organic Syntheses

Starting material	Target compound	Ref.
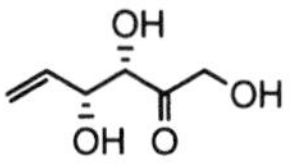 76j1	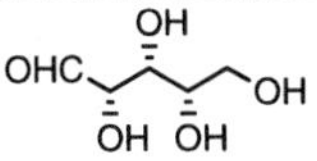 L-xylose **77**	122
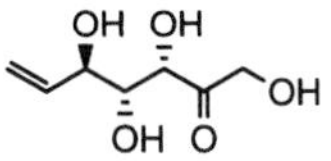 76d2	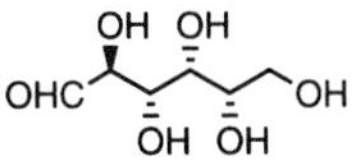 L-gulose **78**	122
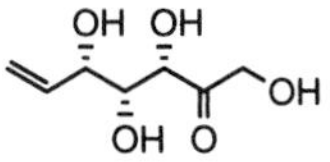 76e2	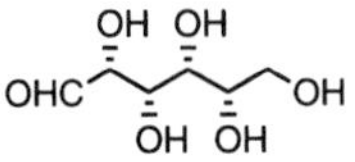 L-idose **79**	122
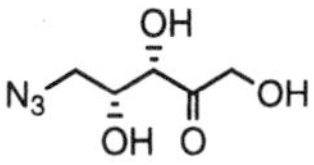 76q	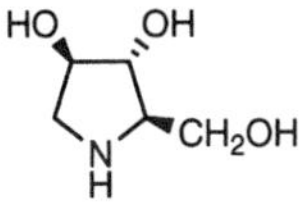 1,4-dideoxy-1,4-imino-D-arabinitol **80**	129
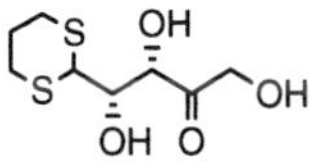 76p		128
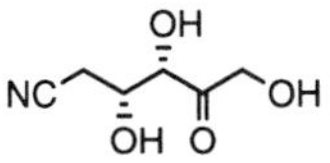 76r	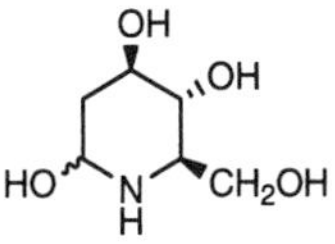 fagomin **81**	130
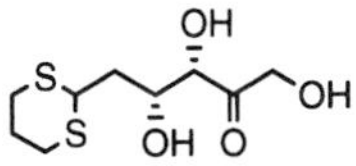 76i1		128
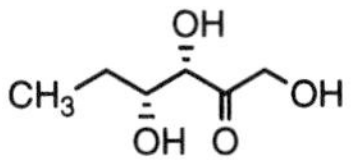 76k1	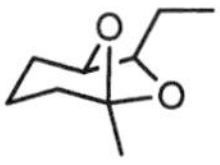 (+)-*exo*-brevicomin **69**	131

uration from naturally occurring amino acids has been proposed for the enhanced reactivity [135], although the C—C bond formation works even from the racemic substrate with a concomitant kinetic resolution.

TKase can be used as a tool for the kinetic resolution of racemic α-hydroxy aldehydes. For example, D-glycelaldehyde (**75h**) shows 66% of the relative rate compared with glycolaldehyde (**75a**), while the L form (**75i**) exhibits zero (Table 6) and, accordingly, the L isomer can be recovered after the incubation of racemic form (**75g**) with TKase and hydroxypyruvate. This protocol is indeed very good for the preparation of enantiomerically enriched forms of 2-hydroxy aldehydes with hydrophilic structures (Table 9). For this purpose, it is indispensable to provide hydroxypyruvate, a rather expensive staff, in large quantities. Elaborated procedures, such as amino acid oxidase–catalyzed preparation from

Table 9 Transketolase-Catalyzed Kinetic Resolution of 2-Hydroxy Aldehydes

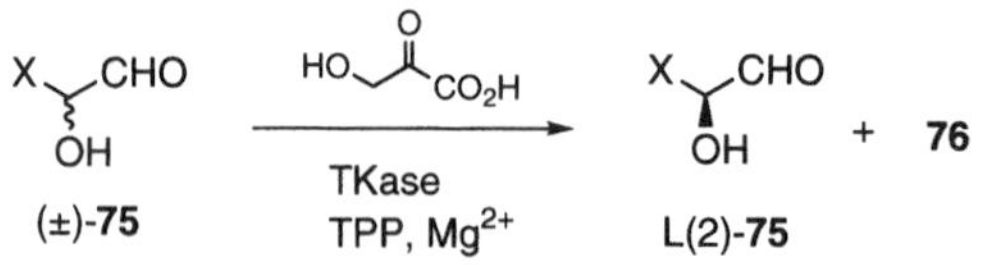

Substrate (±)-**75**	Unreacted L(2)-**75**	Ref.
k	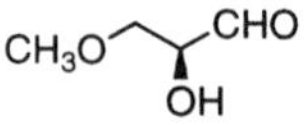 CH₃O—CHO (OH)	122,127
l	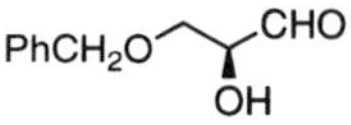 PhCH₂O—CHO (OH)	127
o	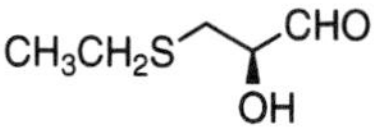 CH₃CH₂S—CHO (OH)	127
r	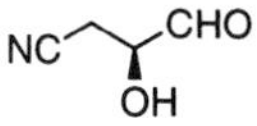 NC—CHO (OH)	127
s	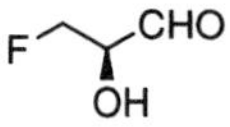 F—CHO (OH)	127
j1	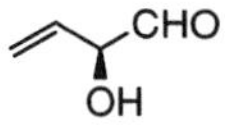 CH₂=CH—CHO (OH)	122
k1	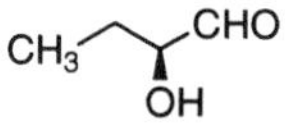 CH₃—CHO (OH)	122,127
f2	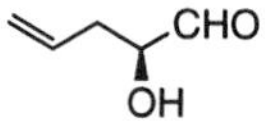 CH₂=CHCH₂—CHO (OH)	122
g2	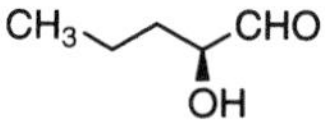 CH₃CH₂CH₂—CHO (OH)	122

D-serine [125] and serine, as well as glyoxylate aminotransferase (SGAT)–catalyzed synthesis from L-serine and glyoxylic acid [124], have been reported.

IV. ENZYMATIC CARBOXYLATION

As the reaction catalyzed by decarboxylase is essentially irreversible, the reverse carboxylation reaction is not possible in the practical sense under conventional reaction conditions. An example, however, has made a breakthrough under conditions with very high concentrations of bicarbonate ion in the solution. Pyrrole (**82**) was carboxylated to give pyrrole-2-carboxylate (**83**) by the aid of the pyrrole-2-carboxylate decarboxylase of *Bacillus megaterium* PYR 2910 in a solution containing 3 M KHCO₃ [Eq. (28)]. The conversion reached 80%, and the reaction is quite effective, considering its reaction equilibrium [138,139].

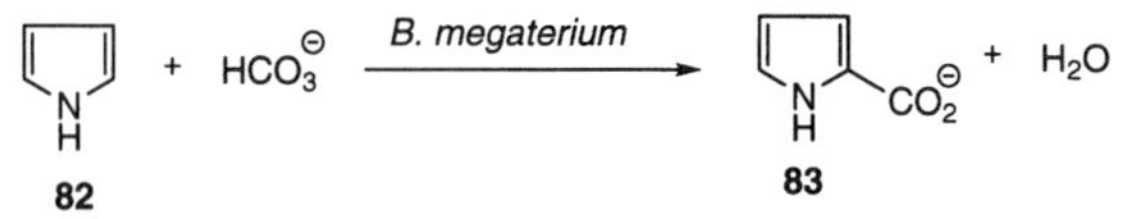

$$ (28) $$

V. CONCLUSION

As reviewed above, enzymatic decarboxylation plays an important role in organic synthe-ses, especially in the synthesis of enantiomerically enriched compounds. The choice of enzymes and/or microorganisms strongly depends on (1) the nature of the substrate, in terms of solubility in water and/or organic solvent, and volatility; (2) the ease of purifi-cation of the products from byproducts, unreacted starting materials, and reagents; and (3) the accessibility of precursors as well as that of the enzymes and microorganisms required.

ACKNOWLEDGMENT

T. S. thanks Shigeki Hiramatsu, Naoki Mochizuki, Hajime Ikeda, and Kaori Matoishi for their literature survey.

REFERENCES

1. MC Scrutton, MR Young, AW Alberts, PR Vagelos, HG Wood, MF Utter, HM Kolenbrander, MI Siegel, M Wishnick, MD Lane, BB Buchman, EA Boecker, EE Snell, I Fridovich. In PD Boyer, ed. The Enzymes, 3rd ed. Carboxylation and Decarboxylation (Nonoxidative). New York: Academic Press, 1972, pp 1–270.
2. E Juni. Meth Enzymol 1:471, Academic Press, 1955.
3. DHG Crout, J Littlechild, SM Morrey. J Chem Soc Perkin Trans 1986:105–108.
4. DHG Crout, DL Rathbone. J Chem Soc Chem Commun 1988:98–99.
5. YS Kim, PE Kolattukudy. J Biol Chem 255:686–689, 1980.
6. NR Thomas, V Schirch, D Gani. J Chem Soc Chem Commun 1990:400–402.
7. NR Thomas, JE Rose, D Gani. J Chem Soc Chem Commun 1991:908–909.
8. IT Harrison, B Lewis, P Nelson, W Rooks, A Rszkowski, A Tomalois, JH Fried. J Med Chem 13:203–205, 1970.
9. TY Shen. Angew Chem Int Engl Ed 11:460–472, 1972.
10. JP Rieu, A Boucherle, H Cousse, and G Mouzin. Tetrahedron 42:4095–4131, 1986.
11. S Hamman, M Barrelle, F Tetaz, CG Beguin. J Fluorine Chem 37:95–94, 1987.
12. K Miyamoto, H Ohta. J Am Chem Soc 112:4077–4078, 1990.
13. O Tousaint, P Capdevielle, M Maumy. Tetrahedron Lett 28:539–542, 1987.
14. K Miyamoto, H Ohta. Biocatalysis 5:49–60, 1991.
15. K Miyamoto, S Tsuchiya, H Ohta. J Fluorine Chem 59:225–232, 1992.
16. K Miyamoto, H Ohta. Eur J Biochem 210:475–481, 1992.
17. JH Galvian, SHG Allen. Arch Biochem Biophys 126:838–847.
18. A Hoffman, W Hilpert, P. Dimroth. Eur J Biochem 179:645–650, 1988.
19. NM Green. Biochem J 94:23–24, 1965.
20. NM Green, EJ Tomes. Biochem J 118:67–70, 1970.
21. PD Boyer. The Enzyme, Vol 6. New York: Academic Press, 1972, pp 37–115.
22. HG Wood, H Lochmuller, C Riepertinger, F Lynen. Biochem Z 337:247–266, 1963.
23. K Miyamoto, H Ohta. Appl Microbiol Biotechnol 38:234–238, 1992.
24. M Miyazaki, H Kakidani, S Hanzawa, H Ohta. Bull Chem Soc Jpn 70:2765–2769, 1997.

25. T Kawasaki, M Watanabe, H Ohta. Bull Chem Soc Jpn 68:2017–2020, 1995.
26. T Kawasaki, Y Fujioka, K Saito, H Ohta. Chem Lett 1996:195–196.
27. K Miyamoto, S Tsuchiya, H Ohta. J Am Chem Soc 114:6256–6257, 1992.
28. K Miyamoto, H Ohta, Y Osamura. BioMed Chem 2:469–475, 1994.
29. T Kawasaki, K Saito, H Ohta. Chem Lett 1997:351–352.
30. T Kawasaki, E Horimai, H Ohta. Bull Chem Soc Jpn 69:3591–3594, 1996.
31. C Neuberg, P Rosenthal. Biochem Z 51:128–142, 1913.
32. C Neuberg, P Rosenthal. Biochem Z 61:171–183, 1913.
33. DE Green, D Herbert, V Subrahmanyan. J Biol Chem 138:327–339, 1941.
34. J Ullrich, JH Wittorf, CJ Gubler. Biochim Biophys Acta 113:595–604, 1966.
35. TP Singer, J Pensky. J Biol Chem 190:375–388, 1952.
36. K Oba, I Uritani. Meth Enzymol 90:528–532, 1982.
37. M Rohmer, M Seemann, S Horbach, S Bringer-Meyer, H Sahm. J Am Chem Soc 118:2564–2566, 1996.
38. K Lohmann, P Schuster. Biochem Z 294:188–214, 1937.
39. E Juni. J Biol Chem 236:2302–2308, 1961.
40. A Schellenberger. Angew Chem Int Ed Engl 6:1024–1035, 1967.
41. R Kluger. Chem Rev 87:863–876, 1987.
42. F Dyda, W Furey, Swaminathan, M Sax, B Farrenkopf, F Jordan. Biochemistry 32:6165–6170, 1993.
43. M Lobell, DHG Crout. J Am Chem Soc 118:1867–1873, 1996.
44. S Sun, GS Smith, MH O'Leary, RL Schowen. J Am Chem Soc 119:1507–1515, 1997.
45. C Neuberg, E Reinfurth. Biochem Z 143:553–565, 1923.
46. WZ Dirscherl, A Schöllig. Z Physiol Chem 252:53–69, 1938.
47. NH Gross, CH Werkman. Arch Biochem 15:125–131, 1947.
48. H Suomalainen, T Linnahalme. Arch Biochem Biophys 114:502–513, 1966.
49. G Chao, F Jordan. Biochemistry 23:3576–3582, 1984.
50. J Stivers, MW WAshabaugh. Biochemistry 32:13472–13482, 1993.
51. E Juni. J Biol Chem 195:727–734, 1952.
52. TP Singer, J Pensky. Biochim Biophys Acta 9:316–327, 1952.
53. S Ui, H Masuda, H Muraki. J Ferment Technol 61:253–259, 1983.
54. C de Mas, NB Jansen, GT Tsao. Biotechnol Bioeng 31:366–377, 1988.
55. NB Jansen, MC Flickinger, GT Tsao. Biotechnol Bioeng 26:362–369, 1984.
56. PB Beronio Jr., GT Tsao. Biotechnol Bioeng 42:1263–1269, 1993.
57. TJ Montville, AH-M Hsu, ME Meyer. Appl Environ Microbiol 53:1798–1802, 1987.
58. E Juni. J Biol Chem 195:715–726, 1952.
59. S Ui, T Masuda, H Masuda, H Muraki. J Ferment Technol 64:481–486, 1986.
60. CJ Lintner, HJ von Liebig. Hoppe Seyler's Z Physiol Chem 88:109–121, 1913.
61. E Róna. Biochem Z 67:137–142, 1914.
62. C Neuberg, J Hirsch. Biochem Z 115:282–310, 1921.
63. C Neuberg, H Ohle. Biochem Z 128:610–618, 1922.
64. O Hanc, B Kanak. Naturwissenschaften 43:498, 1956.
65. PF Smith, D Hendlin. J Bacteriol 65:440–445, 1953.
66. PF Smith, D Hendlin. Appl Microbiol 2:294–296, 1954.
67. H-P Schmauder, D Gröger. Pharmazie 23:320–331, 1968.
68. JP Voets, EJ Vandamme, C Vlerick. Zwitschrift Allg Mikrobiol 13:355–366, 1975.
69. C Fuganti, P Graselli. Chem Ind 17:983, 1977.
70. KG Gupta, J Singh, G Sahni, S Dhawan. Biotechnol Bioeng 21:1085–1089, 1979.
71. J Netrval, V Vojtisek. Eur J Appl Microbiol Biotechnol 16:35–38, 1982.
72. H Ohta, K Ozaki, J Konishi, G Tsuchihashi. Agric Biol Chem 50:1261–1266, 1986.
73. SC Agarwal, SK Basu, VC Vora, JR Mason, SJ Pirt. Biotechnol Bioeng 29:783–785, 1987.
74. S Bringer-Meyer, H Sahm. Biocatalysis 1:321–331, 1988.
75. C Fuganti, P Grasselli, G Poli, S Servi, A Zorzella. J Chem Soc Chem Commun 1619–1621, 1988.

76. A Long, P James, OP Ward. Biotechnol Bioeng 34:657–660, 1989.
77. A Long, OP Ward. Biotechnol Bioeng 34:933–941, 1989.
78. A Long, OP Ward. J Ind Microbiol 4:49–53, 1989.
79. V Křen, DHG Crout, H Dalton, DW Hutchinson, W König, MM Turner, G Dean, N Thomson. J Chem Soc Chem Commun 341–343, 1993.
80. S Bornemann, DHG Crout, H Dalton, DW Hutchinson, G Dean, N Thomson, MM Turner. J Chem Soc Perkin I 309–311, 1993.
81. N Mochizuki, S Hiramatsu, T Sugai, H Ohta, H Morita, H Itokawa. Biosci Biotechnol Biochem 59:2282–2291, 1995.
82. M Behrens, NN Iwanoff. Biochem Z 196:478–481, 1926.
83. C Neuberg, L Liebermann. Biochem Z 121:311–325, 1921.
84. R Bernardi, C Fuganti, P Grasselli, G Marinoni. Synthesis 50–52, 1980.
85. G Bertolli, G Fronza, C Fuganti, P Grasselli, L Majori, F Spreafico. Tetrahedron Lett 965–968, 1981.
86. R Cardillo, S Servi, C Tinti. Appl Microbiol Biotechnol 36:300–303, 1991.
87. G Fronza, C Fuganti, P Grasselli, L Majori, G Pedrocchi-Fantoni, F Spreafico. J Org Chem 47:3289–3296, 1982.
88. C Fuganti, P Grasselli, S Servi. J Chem Soc Perkin I 241–244, 1983.
89. C Fuganti, P Grasselli. Tetrahedron Lett 1161–1164, 1979.
90. DHG Crout, H Dalton, DW Hutchinson, J Miyagoshi. J Chem Soc Perkin I 1329–1334, 1991.
91. C Fuganti, P Grasselli. J Chem Soc Chem Commun 205–206, 1982.
92. N Mochizuki, T Sugai, H Ohta, T Yokomatsu, S Shibuya. Heterocycles 45:331–338, 1997.
93. W-R Abraham, B Stumpf. Z Naturforsch 42c:559–566, 1987.
94. M Bockmühl, G Ehrhart, L Stein, D.R.P. 571229; Chem Abstr 27:2691–2692, 1933.
95. T Sugai, A Kuboki, S Hiramatsu, H Okazaki, H Ohta. Bull Chem Soc Jpn 68:3581–3589, 1995.
96. S Servi. Synthesis 1–25, 1990.
97. R Csuk, BI Glänzer. Chem Rev 91:49–97, 1991.
98. C Fuganti. Pure Appl Chem 62:1449–1452, 1990.
99. C Fuganti, P Grasselli. J Chem Soc Chem Commun 299–300, 1978.
100. G Fronza, C Fuganti, P Grasselli. Tetrahedron Lett 3941–3942, 1978.
101. C Fuganti, P Grasselli, S Servi, C Zirotti. Tetrahedron Lett 23:4269–4272, 1982.
102. C Fuganti, P Grasselli, F Spreafico, C Zirotti, P Casati. J Org Chem 49:543–546, 1984.
103. G Fronza, C Fuganti, P Grasselli. J Chem Soc Chem Commun 442–444, 1980.
104. C Fuganti, S Servi, C Zirotti. Tetrahedron Lett 24:5285–5286, 1983.
105. C Fuganti, P Grasselli, S Servi. J Chem Soc Chem Commun 1285–1286, 1982.
106. G Fronza, C Fuganti, P Grasselli. Tetrahedron Lett 5073–5076, 1981.
107. G Fronza, C Fuganti, P Grasselli, G Pedrocchi-Fantoni. J Carbohydr Chem 2:225–248, 1983.
108. C Fuganti, P Grasselli, G Pedrocchi-Fantoni, S Servi, C Zirotti. Tetrahedron Lett 24:3753–3756, 1983.
109. G Fronza, C Fuganti, P Grasselli, G Pedrocchi-Fantoni. Carbohydr Res 136:115–124, 1985.
110. A Kuboki, H Okazaki, T Sugai, H Ohta. Tetrahedron 53:2387–2400, 1997.
111. P Srere, JR Cooper, M Tabachnick, E Racker. Arch Biochem Biophys 74:295–305, 1958.
112. AS Datta, E Racker. J Biol Chem 236:617–623, 1961.
113. C Demuynck, F Fisson, I Bennani-Baiti, H Samaki, J-C Mani. Agric Biol Chem 54:3073–3078, 1990.
114. JJ Vilafranca, B Axelrod. J Biol Chem 246:3126–3131, 1971.
115. L Hecquet, J Bolte, C Demuynck. Tetrahedron 50:8677–8684, 1994.
116. GR Hobbs, MD Lilly, NJ Turner, JM Ward, AJ Willets, JM Woodley. J Chem Soc Perkin I 165–166, 1993.
117. MD Lilly, R Chauhan, C French, M Gyamerah, GR Hobbs, A Humphery, M Isupov, JA Littlechild, RK Mitra, KG Morris, M Rupprecht, NJ Turner, JM Ward, AJ Willetts, JM Woodley. Ann NY Acad Sci 783:513–525, 1996.

118. KG Morris, MEB Smith, NJ Turner, MD Lilly, RK Mitra, JM Woodley. Tetrahedron: Asymmetry 7:2185–2188, 1996.
119. L Hecquet, J Botle, C Demuynck. Biosci Biotechnol Biochem 57:2174–2176, 1993.
120. J Bolte, C Demuynck, H Samaki. Tetrahedron Lett 28:5525–5528, 1987.
121. C Demuynck, J Bolte, L Hecquet, V Balmas. Tetrahedron Lett 32:5085–5088, 1991.
122. Y Kobori, DC Myles, GM Whitesides. J Org Chem 57:5899–5907, 1992.
123. V Dalmas, C Demuynck. Tetrahedron: Asymmetry 4:2383–2388, 1993.
124. L Hecquet, J Bolte, C Demuynck. Tetrahedron 52:8223–8232, 1996.
125. C Demuynck, J Bolte, L Hecquet, H Samaki. Carbohydr Res 206:79–85, 1990.
126. A Mocali, D Aldinucci, F Paoletti. Carbohydr Res 143:288–293, 1985.
127. F Effenberger, V Null, G Ziegler. Tetrahedron Lett 33:5157–5160, 1992.
128. L Hecquet, M Lemaire, J Bolte, C Demuynck. Tetrahedron Lett 35:8791–8794, 1994.
129. T Ziegler, A Straub, F Effenberger. Angew Chem Int Ed Engl 27:716–717, 1988.
130. F Effenberger, V Null. Liebigs Ann Chem 1211–1212, 1992.
131. DC Myles, PJ Andrulis III, GM Whitesides. Tetrahedron Lett 32:4835–4838, 1991.
132. C André, C Demuynck, T Gefflaut, C Guérard, L Hecquet, M Lemairre, J Bolte. J Mol Catal B: Enz 5:113–118, 1998.
133. V Dalmas, C Demuynck. Tetrahedron: Asymmetry 4:1169–1172, 1993.
134. M Kapuscinski, FP Franke, I Flanigan, JK MacLeod, JR Williams. Carbohydr Res 140:69–79, 1985.
135. AJ Humphrey, NJ Turner, R McCague, SJC Taylor. J Chem Soc Chem Commun 2475–2476, 1995.
136. S David, C Augé, C Gautheron. Adv Carbohydr Chem Biochem 49:175–237, 1991.
137. S Takayama, GJ McGarvey, C-H Wong. Annu Rev Microbiol 51:285–310, 1997.
138. M Weiser, T Yoshida, T Nagasawa. Tetrahedron Lett 39:4309–4310, 1998.
139. H Omura, M Weiser, T Nagasawa. Eur J Biochem 253:480–484, 1998.

19

Yeast-Mediated Stereoselective Biocatalysis

René Csuk and Brigitte I. Glänzer
*Institute of Organic Chemistry, Martin Luther University Halle-Wittenberg, Halle
(Saale), Germany*

I. INTRODUCTION

Microbial transformations, and yeast-mediated conversions in particular, have been widely used since the early days of mankind for the production of dairy products, bread, and alcoholic beverages. Whereas all of these early applications used mixed cultures of microorganisms it was the merit of Pasteur in 1862 [1] to lay a scientific foundation of one of these early applications, namely, the oxidation of alcohol to acetic acid by using a pure culture of *Bacterium xylinium*. All of these early biotechnological operations have been more or less directed in the areas of agricultural and humane nutrition; the reduction of furfural to furfuryl alcohol under anaerobic conditions of fermentation, however, by means of living yeast [2,3] was the first "phytochemical reduction" of an organic molecule described in the literature.

Nowadays, numerous yeast or bacterial or enzymatic biotransformations, biodegenerations, or bioconversions are known, and, as Chaleff [4] pointed out, in the initial excess of enthusiasm [5] that invariably accompanies the birth of a new field [6], biotransformations were hailed as a panacea that would ultimately displace traditional organic chemistry [7,8]. However, bioconversions should be employed when a given reaction step is not easily accomplished by classical chemical methods [9]; thus, the role of biotransformations is one of support rather than replacement.

As a consequence, the stereoselective biotransformations have to be carried out by pure cultures of microorganisms (or with purified enzymes). Thus, they should always be considered as a way of performing selective transformations starting from a well-defined pure starting material into well-defined products [10]. Further differences between biotransformations and fermentations have already been listed in the literature [11].

Applications in the areas of environmental pollution problems as well as energy sector–relevant applications [12] are of tremendous impact; however, this chapter is confined to stereoselective biotransformations.

The general aims of yeast-mediated biotransformations focus on the introduction of chiral centers, the resolution of racemates, and the selective conversion of functional groups among several groups of similar reactivities. Several excellent reviews have been published on microbial transformations also covering yeast-mediated conversions [13–17]. To avoid excessive echoing of these reviews, background material will be limited to the minimum commensurate with both the readability of this chapter and the chemical nature of the discussion to follow.

Since the availability of certain microorganisms is often a deciding factor for an organic chemist turning to the use of biotransformations in synthesis, baker's yeast (BY), i.e., *Saccharomyces cerevisiae* as a readily available microorganism (world output is more than 600,000 tons/year) [18], seems to be an ideal "reagent" fulfilling all of the requirements for convenient application. For example, it is cheap and readily available from local breweries even in large amounts, thereby reducing the disadvantage associated with the use of whole cells for laboratory scale operations that the sterile growing of microbial cells is more or less required to guarantee clean bioconversions.

It must be pointed out that yeast-mediated biotransformations may be complicated by side reactions that interfere or even dominate the desired conversion; workup may be somewhat time consuming and messy due to the separation of the product from the huge amount of biomass. The ideal interactions between the yeast and the substrate are scarcely found in praxi. Ideal interactions between substrate yeast and product include that both substrate and product are able to pass the cell membrane, should be soluble in the fermentation medium, and must not inactivate the catalytic activity of the involved microbial enzymes. Finally, high turnover rates and high regio- and stereospecificity are also of high practical relevance. Some advice on how to deal with these basic problems often encountered in such biotransformations is provided in Fig. 1 [9,11,13,19,20].

In addition, reactions with intact yeasts can be performed by using either previously grown cells or under fermentative conditions or immobilized cells; turnover rates and stereospecificity are likely to differ in all three experimental setups. Although loss of

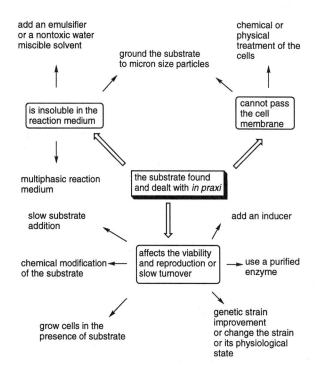

Figure 1 Procedure for managing basic problems encountered in yeast-mediated biotransformations.

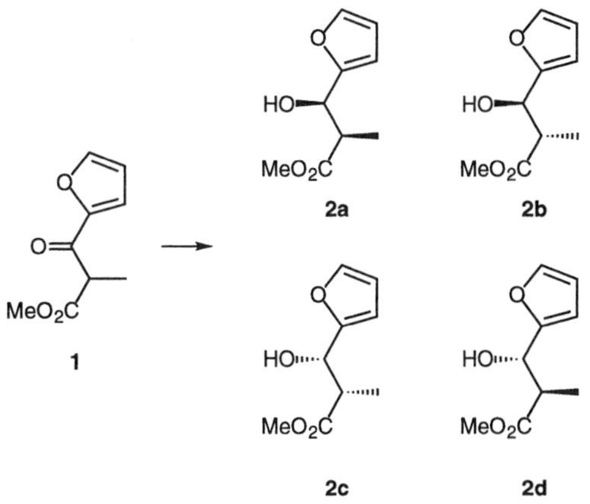

Figure 2

prepolymer	yield [%]	2a ee[%]	2b ee[%]	2c ee[%]	2d	syn/anti
none	43	90	91	--	--	39/61
polyurethane	33	56	74	--	--	41/59
photon-crosslinking	64	--	--	81	57	50/50

activity due to an additional permeability barrier [21] is observed, bioconversions with immobilized yeast cells are very attractive. Four major categories of immobilized yeasts can be recognized:

1. Entrapment in a gel or a membrane or within capsules (urethane, alginate, cellulose, agar, chitosan, collagen, κ-carrageenan, and polyamide have been used as polymerous porous networks)
2. Covalent attachment to a carrier material, e.g., carboxymethylcellulose
3. Immobilization by chemical or physical adsorption, i.e., surface adsorption to a water-insoluble solid support (ion exchange resin, cellulose derivatives, or metal oxides)
4. Cell aggregation of the microorganism by physical or chemical crosslinking

Product formation rates are usually high, the operational stability of yeast is increased, and an easier isolation of the products is provided. Some differences in yield and stereoselectivity, however, have been encountered depending on the kind of immobilization (Figs. 2 and 3) [22–31].

II. REDUCTIONS

A. Reduction of Monocarbonyl Compounds

The asymmetrical reduction of carbonyl compounds by BY constitutes one of the most widely applicable reactions. Thus, ketones with varying alkyl/aryl substituents are reduced by yeasts [32–35] and the secondary alcohols obtained were mainly of (S) configuration. Sterically hindered ketones are not reduced at all. From these findings a hydrogen transfer

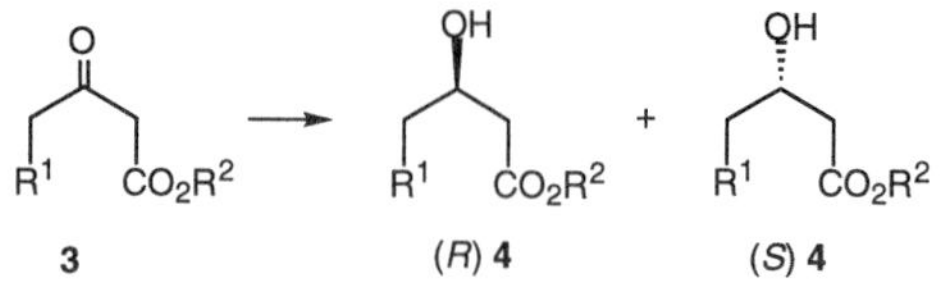

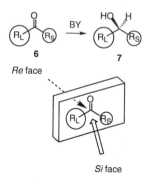

entrapment	product	ee [%]
free BY	(S) 4	42
alginate	(S) 4	16
carrageenan	(S) 4	11
polyurethane	(R) 4	82

Figure 3

to the *re* face of the prochiral ketone **6**, with R_S representing a small substituent and R_L a large substituent adjacent to the carbonyl group (Fig. 4) to yield alcohol **7**. However, one should exercise considerable caution upon applying Prelog's rule [36] to intact cell systems [37].

Whereas there are many reports on the reduction of steroids [38–43] (although some of these reductions are attributable to the action of bacteria having contaminated the yeast [44]), only a few examples for the reduction of unfunctionalized cycloalkanones [45] have been described [46–48].

Thus, **8** (Fig. 5) gave a 1:1 mixture of the *cis*- and *trans*-cyclohexanols **9** and **10** [49]. More examples can be provided for the reduction of bi- and polycyclic cycloalkanones. Thus, racemic (1α,4α,5α)-4-(benzyloxycarbonyl)-2-oxabicyclo[3.2.0]heptane-6-one (**11**) has been reduced by BY (Fig. 6) to afford 59% of a separable mixture of the corresponding diastereomeric 6-hydroxybicycloheptanols **12** and **13** in 10% and 2% isolated yields. Their separate reoxidation by pyridinium chlorochromate gave the resolved enantiomers (*1R, 4R, 5R*)-**11** and (*1S, 4S, 5S*)-**11** [50].

An analogous sequence was elaborated for the synthesis of enantiomerically pure 2-oxo-bicyclo[2.2.1]heptane-7-carboxylates [51]. (±)-Bicyclo[3.2.0]hept-2-en-6-one (**14**) (Fig. 7) gave **15** and **16** upon reduction with BY in the presence of added riboflavin and commercial yeast nutrient [52]. In order to improve the enantioselectivity achieved by commercially available BY, other different strains of yeasts were screened and marked differences established. It was shown that the rate of yeast reduction is influenced by the

Figure 4

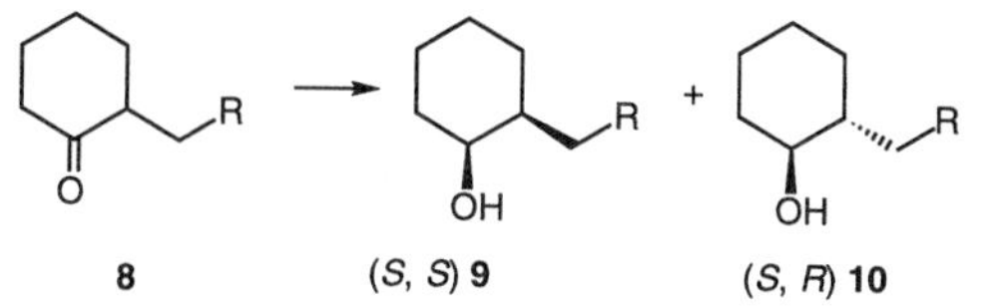

Figure 5

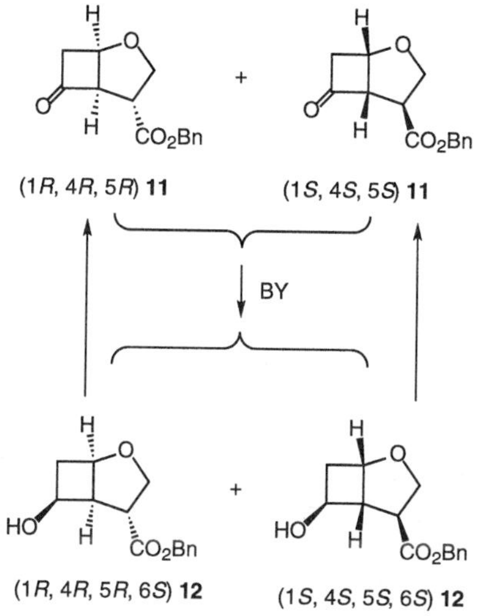

Figure 6

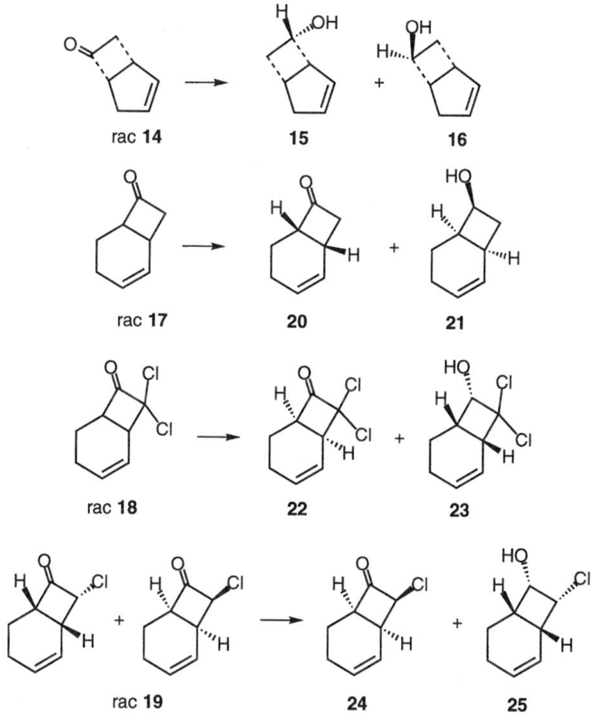

Figure 7

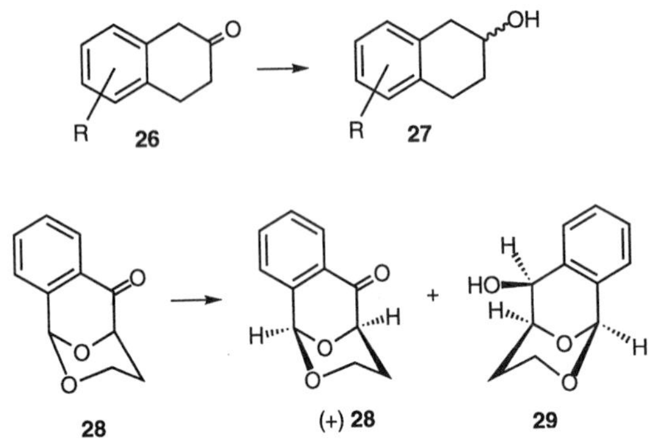

Figure 8

concentration of **14** and that the ratio **15:16** depends both on incubation time and the glucose concentration. Whereas reduction of **14** with *Mortierella ramanniana* gave rise to an endo-to-exo ratio of more than 30:1, the reduction of bicyclo[4.2.0]octenones **17**, **18**, and **19** was found to be completely diastereoselective for the reduction from the ketone's exo face in addition to being highly enantioselective [53,54]. In addition, the reduction of 4-methyl- or 1,4-dimethylbicyclo[3.2.0]hept-3-en-6-one gave a mixture of the enantio-merically pure *exo-* and *endo*-alcohols [55].

2-Tetralones **26** mono- or disubstituted with methoxy or hydroxy group in the aro-matic ring were reduced to 2-tetralols **27** in good yields (Fig. 8) and with high e.e.'s by nonfermenting BY [56]. With BY the tricyclic ketone **28** was stereoselectively reduced to afford the alcohol **29** albeit with moderate yield [57].

The asymmetrical reduction of flavone **30** with fermenting BY (Fig. 9) led to the formation of (*2S, 4S*)-*cis*-4-hydroxyflavan (**31**) and (*2R*)-flavone (**32**) [58]. From the cy-cloalkanone derivatives **33** the corresponding alcohols **34** were obtained albeit at low yields [59].

Moderate enantioselectivity has been observed for the reduction of norbornenone **35** [60] (Fig. 10) and bicyclo[2.2.2]octan-2-one (**36**) [61].

Interestingly enough, under fermenting conditions the structural analog **39** (Fig. 11) gave 60% of enantiometrically pure **40** [62].

From the twistanone **41** (Fig. 12) *exo*-**42** and *endo*-**43** were obtained [61]. Higher e.e. values and yields, however, were achieved for the reduction of the latter compound

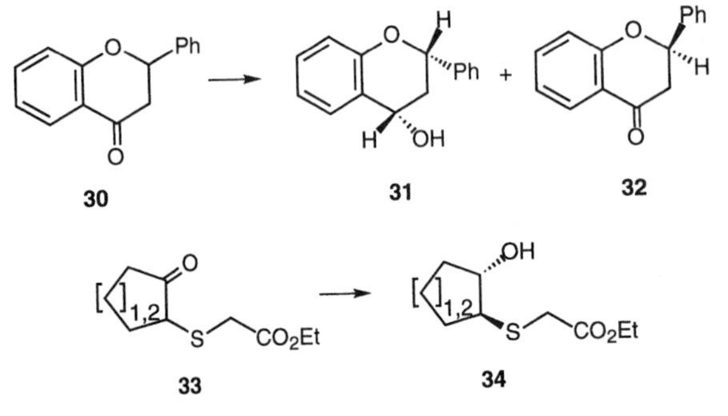

Figure 9

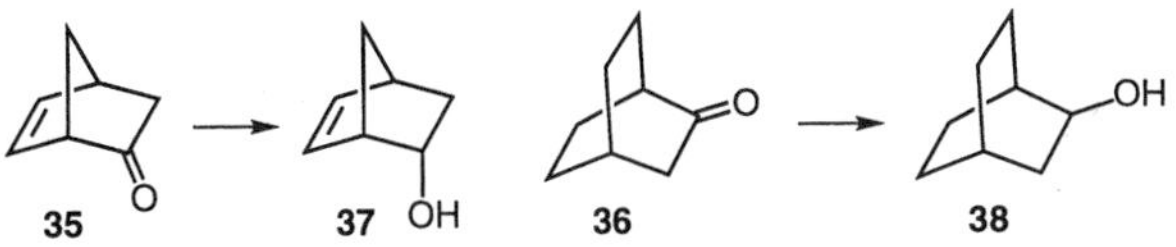

Figure 10

employing *Rhodotorula rubra* [61]. Even pentacyclic compounds could be reduced by BY as exemplified with the cage compound **44** that afforded after 5d incubation a 1:1 mixture of the alcohols **45** and **46** that were easily separated after acetylation [63].

It was observed that the incubation of a pyrrolo[2,1-*c*]-[1,4]-benzodiazepine-2,5,11-trione (**47**) with BY (Fig. 13) gave stereo- and regioselective reduction of the C-2 carbonyl group to yield **48** in good yields and with high e.e values [64].

Good results were also obtained [65] for the reduction of aliphatic ketones; this subject had already been covered in several reviews [21,47,66–68]. The stereochemical course of these reductions has been investigated by several deuteration studies [69–71]. Thus, the reduction of α-hydroxy ketones **49** gave the corresponding 1,2-diols **50** (Fig. 14) [72–78]. From 1-methylsulfonylated ketones the corresponding alcohols were obtained with e.e. values up to 87% [79].

Fermentative reduction of substituted acetophenones **51** gave (*S*)-1-arylethanols **52** (Fig. 15) in moderate yields and with e.e. values between 82% and 96% [32,33,60,80–87]. To obtain good yields and high e.e. values it seems crucial to optimize the incubation conditions very carefully [88,89].

In analogy, α-(acylamino)acetophenones afforded the (*R*)-configured 2-acylamino-1-arylethanols [90]. The BY reduction of amino- or hydroxy-substituted trifluoroacetophenone derivatives proceeded in an enantioselective way to produce (*R*)-trifluorobenzylalcohols in up to 92% e.e. [91]; these findings are in strong contrast to the results with simple trifluoroacetophenone analogs [92]. A very detailed mechanistic study was performed for the reduction of 3-chloropropiophenones [93,94].

The cerebrovascular and peripheral vascular active drug pentoxyfylline (**53**) (Fig. 16) is reduced by *S. cerevisiae* to the alcohol **54** albeit with only 8% yield within 72 h. The highest conversion rates, however, have been obtained using *Rhodotorula mucilaginosa*, *Eryptococcus macerans*, or *Curvularia falcata* [95].

The reduction of **55** gave **56** but mainly (*S*)-**57** showing an e.e. of more than 99%. From the ketoacetate **58** the diol **59** was obtained with high e.e. value [96]; high e.e. values were also reported for the reduction of 4-acetylpyridine, 3-(benzyloxy)-1-hydroxypropanone, and 1-(acyloxy)-3-azido-2-propanols [13].

The reduction of the acetates **60** (Fig. 17) afforded the (*S*)-monoacetates **61** with 72–96% e.e. while the methyl ethers **62** gave the (*R*)-monoethers **63** with 64–76% e.e. [97].

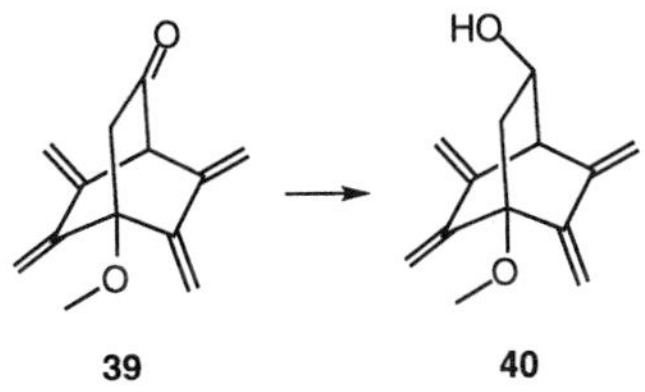

Figure 11

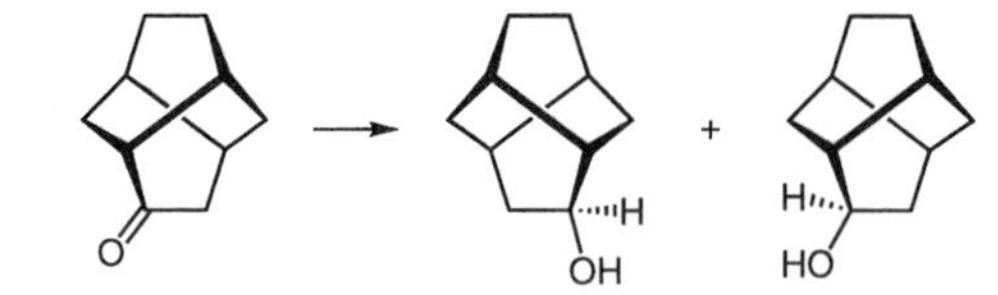

41   (–)-(4S) *exo* **42**   (+)-(4S) *endo* **43**

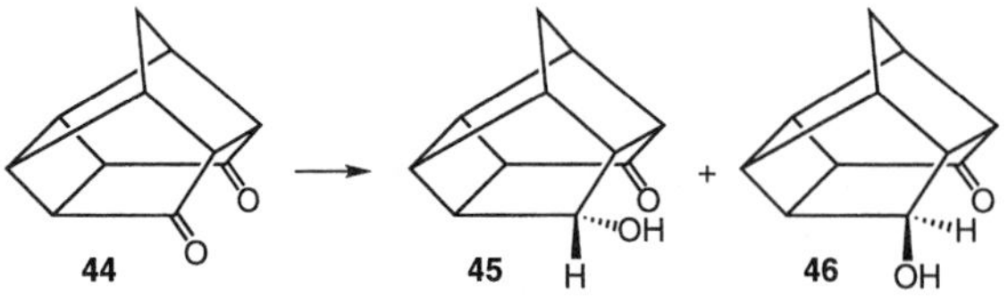

44   **45**   **46**

Figure 12

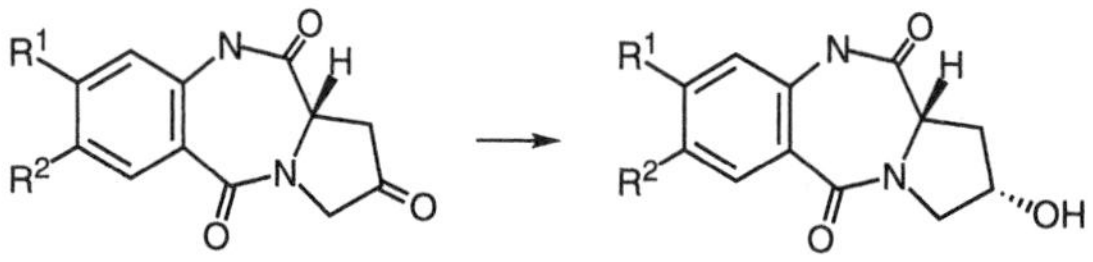

47   **48**

Figure 13

49   **50**

Figure 14

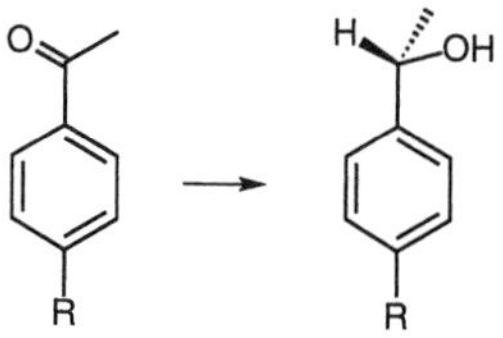

51   (S) **52**

Figure 15

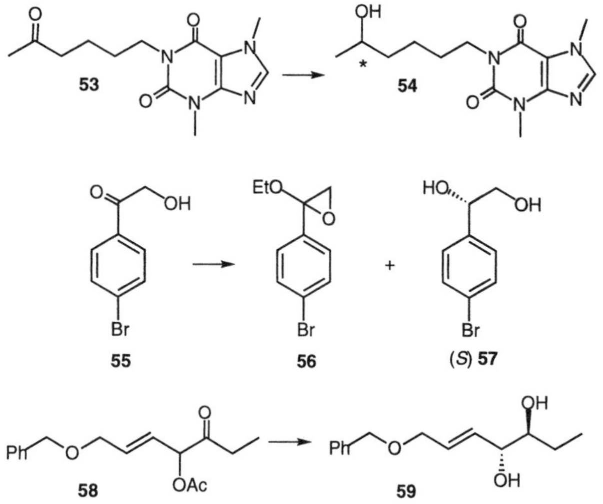

Figure 16

While α-fluoro-, α-chloro- and α-bromoacetophenones **64** (Fig. 18) gave the corresponding (−)-halohydrines **65**, the reduction of α-iodoacetophenone gave acetophenone and (−)-1-phenylethanol (**66**) [98]. In addition, several α-bromoketones furnished the corresponding bromohydrines [13].

The versatility of these yeast-mediated reactions is exemplified by the synthesis of masked 1,2- or 1,4-diols by the *Eridania*-mediated bioreduction of **67** to afford (*S*)-configurated alcohols **68** (Fig. 19) possessing e.e. values of 90–95% [99] as well as by the convenient access to the aggregation pheromone sulcatol [(*S*)-**69**] that has been obtained by the reduction of 6-methyl-hept-5-en-2-one (**70**) in 80% yield and 94% e.e. [100].

In addition, whereas the bioreduction of the keto acid **73** (Fig. 20) with immobilized BY entrapped in carrageenan beads did not have satisfactory results, use of BY entrapped in calcium alginate gave **74** in 40% yield and with an e.e. of 95% [101]. Several α-ketols have been reduced by BY and gave the corresponding diols [102–104] or lactones [105].

Finally, an elegant synthesis of brefeldin A started from **71** to afford (*S*)-**72** with 99% e.e. [106], and a kinetic resolution by BY has been used for the synthesis of *endo*-brevicomin [107,108] (Fig. 21). Thus, reduction of racemic **75** under anaerobic conditions afforded via **76** a mixture of *endo*-**77** (corresponds to a syn-selective reduction) and *exo*-**78** [107].

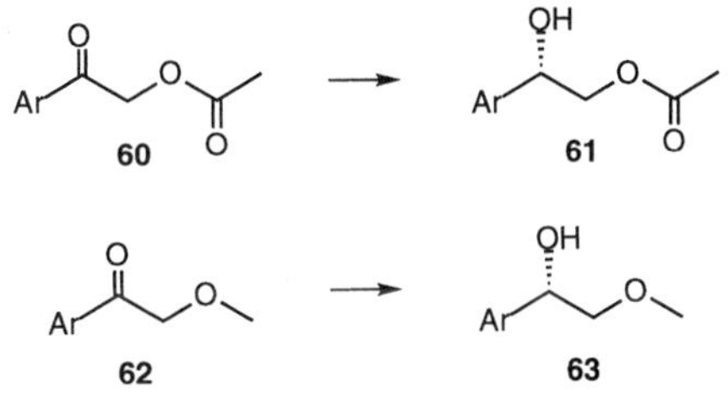

Figure 17

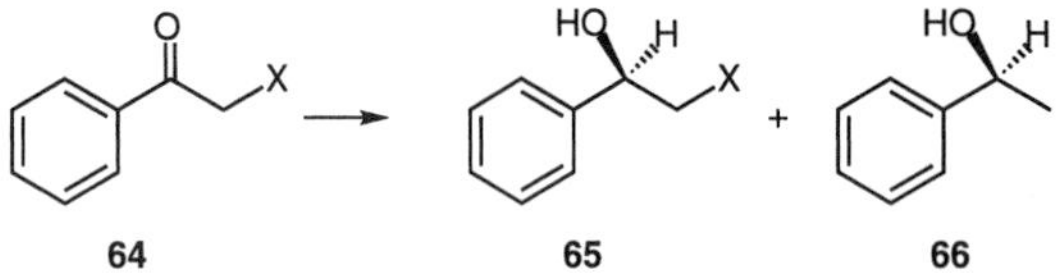

Figure 18

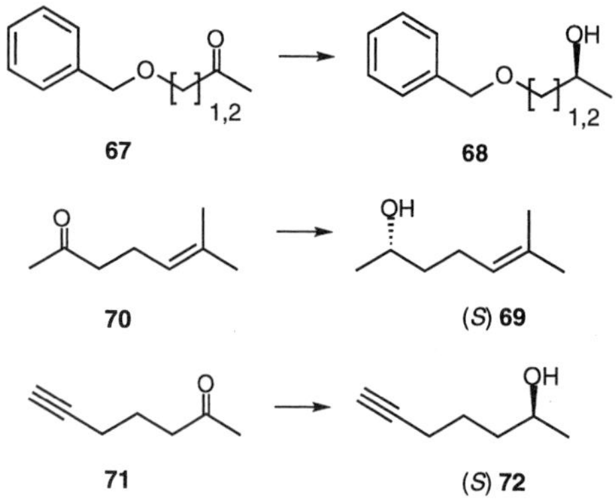

Figure 19

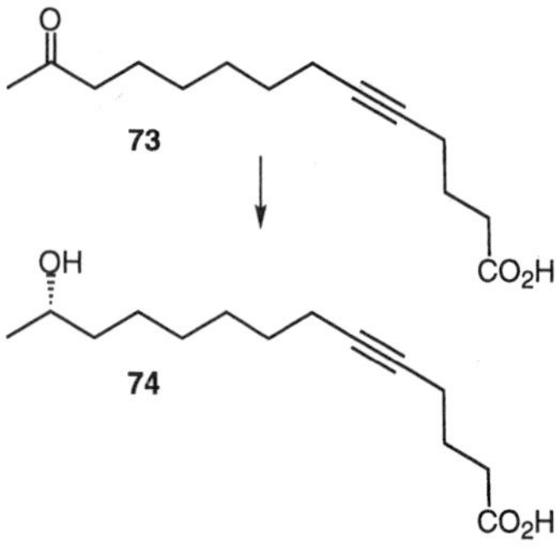

Figure 20

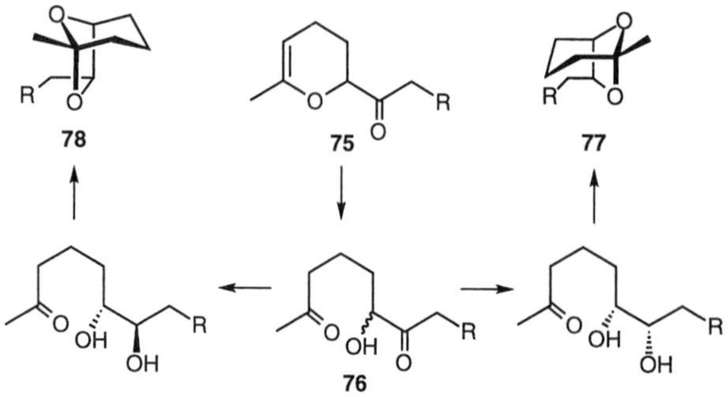

Figure 21

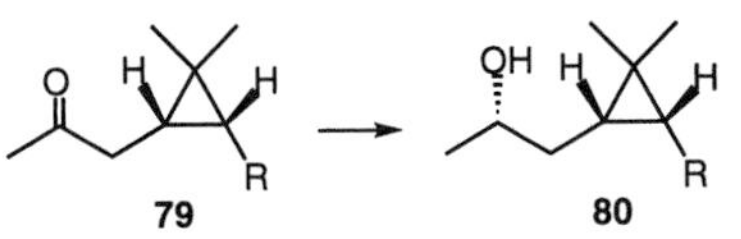

Figure 22

Several cis-3-substituted 2,2-dimethyl-1-(2-oxopropyl)cyclopropanes **79** (Fig. 22) have effectively been reduced by BY to give predominantly (*S*) isomers of the corresponding 2-hydroxypropyl derivatives **80** [109]. In this context it was shown that the reduction of the 2-oxopropyl group proceeds 3.5 times more rapidly and with higher stereoselectivity [(*S*)/(*R*) = 98:2] when the group is attached to the (*S*) carbon of the cyclopropane unit than of the keto group attached to the (*R*) carbon [(*S*)(*R*) = 88:12 [109].

Baker's yeast reduction of α-azidoketones furnished the corresponding syn- and anti-configurated azido alcohols with excellent e.e. values [110,111].

Due to their potential as drugs, fluoro-containing compounds have been the focus of interest for many years. Recently, the reduction of these compounds by microbial or enzymatic systems has gained much interest as tools for the preparation of enantiomerically pure compounds. Thus, several trifluoromethyl ketones **81–84** (Fig. 23) have been reduced in good yields but with fair e.e. to the corresponding (*R*)-carbinols **85–88**, respectively [112]. By comparative studies it has been established that trifluoromethyl ketones are reduced more quickly than the corresponding methyl ketones but more slowly than their bromomethyl analogs [84].

Perfluoroalkyl methyl ketones afforded products of high optical purity [113,114] whereas bis(perfluoroalkyl)ketones were scarcely reduced at all [114]. From the α,β-un-

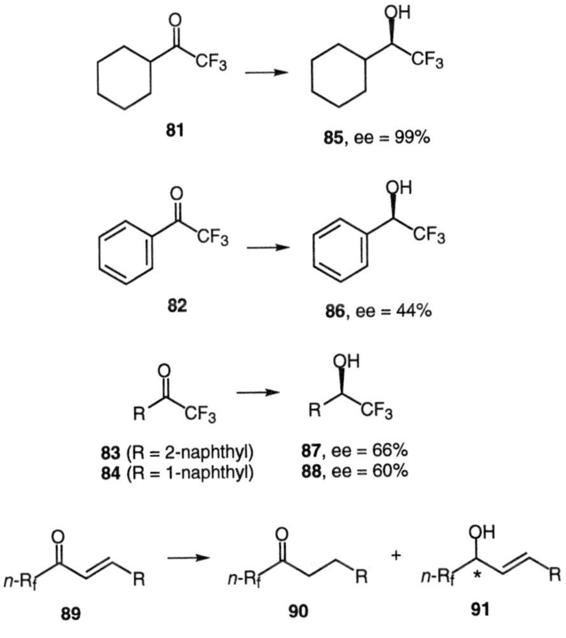

81   **85,** ee = 99%

82   **86,** ee = 44%

83 (R = 2-naphthyl)   **87,** ee = 66%
84 (R = 1-naphthyl)   **88,** ee = 60%

89   **90**   **91**

Figure 23

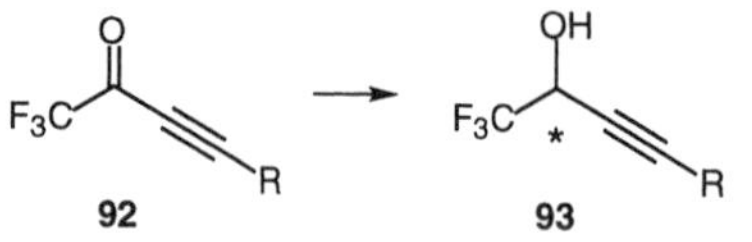

Figure 24

saturated compounds **89** the products **90** were obtained together with traces of **91** [115,116].

From the fluoroalkynones **92** (Fig. 24) the fluoroalkynoles **93** were obtained with moderate e.e. [115]. In addition, it has been established that fluorine substitution on the methyl group of acetylferrocene and acetophenone strengthened *anti*-Prelog-type stereoselectivity in the microbial reduction with six strains of yeast [117].

Recently, the yeast reduction of organometallic compounds has been investigated in more detail although the BY reduction of porphyrins and hemoglobins is a long known process [118–120]. Besides some reductions of inorganic materials [121–123] ferrocenyl-type molecules **94** (Fig. 25) gave **95** with e.e. greater than 90% [124–126] and arylketone-Cr(CO)$_3$ complexes **96** afforded the alcohols **97** [127]. From the metallocenic aldehyde **98** after a 55% conversion (*R*)-**99** was obtained with an e.e. of 66% [128].

In addition, the kinetic resolution of (2,4-hexadien-1-al)-FE(CO)$_3$ or of (2,4-hexadien-1,6-dial)-Fe(CO)$_3$ with BY proceeded with high efficiency and enantioselectivity [222]. Recently, the course of the asymmetrical reduction of acyclic (diene)-FE(CO)$_3$ complexes

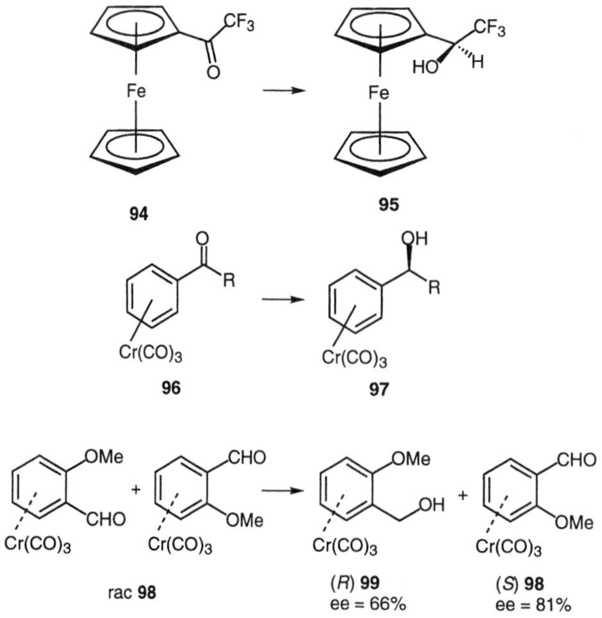

Figure 25

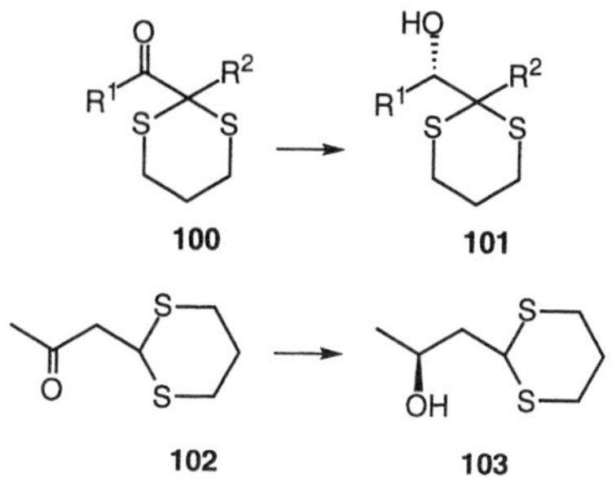

Figure 26

bearing aldehyde and hydroxymethyl substituents [129] and the reductions of (sorbalde-hyde)-Fe(CO)$_3$ complexes have been studied in detail [130].

B. Reduction of Thiocarbonyl- and Thio-Substituted Compounds

Thioaldehydes are readily reduced by yeast to afford the corresponding thiols [131–133]; diethyl thioether was reported to cleave to ethanthiol [134] whereas thioketones gave optically active thiols [135].

The reduction of carbonyl groups with adjacent sulfur substituents by actively fermenting BY is critically dependent on the substituents attached to the sulfur-containing group and to the carbonyl group; the relative ease of reduction decreases from β-keto sulfones to β-keto sulfoxides to β-keto sulfides [136] but in cases where adjacent substituents are bulky, very little reaction occurs. α-Keto thioacetals **100** (Fig. 26) were asymmetrically reduced by fermenting BY to afford optically active α-hydroxy thioacetals **101** predominantly of (S) configuration [137–142]. The β-keto thioacetal (**102**) gave (S)-1-(1,3-dithian-2-yl)-2-hydroxypropane (**103**) with 99% e.e. [137]. The 1,1-bis (p-tolylthio)-ketones **104** (Fig. 27) afforded the corresponding alcohols **105** [139,143]; β-keto dithio-esters **106** were reduced by BY to the optically pure (3S)-hydroxy thioesters **107a,b** [139,143].

Several β-keto sulfides **108** (Fig. 28) have been reduced by yeasts [136,144–147]; thus, (phenylthio)alkanones **108** gave the (S)-hydroxyalkylphenylsulfides **109** with high e.e. but moderate yield. Generally, the reduction of these compounds proceeds with relative difficulty and only at low substrate concentrations [136].

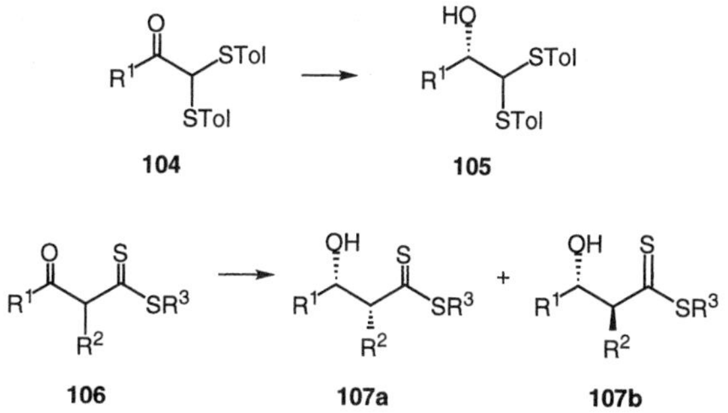

Figure 27

Figure 28

The reduction of β-keto sulfones [86,136,145,148–152] offers a synthetic approach to β-hydroxy esters of opposite configuration than those obtained from the yeast-mediated reduction of the corresponding β-keto esters. Thus, from **110** (Fig. 29) **111** was obtained (in 98% e.e.) that gave upon desulfuration (*R*)-**112**; (*S*)-**112** was obtained by BY reduction of **113** [86] (Fig. 30). The BY reduction of γ- and δ-keto sulfones has been used as a key step in the synthesis of several natural products [153].

α-Heterocyclic-substituted ketones have been reduced by yeasts. Thus, high e.e. values were obtained for 5-acetyl-2-isoxazolines [154,155] and ketoisoxazoles [156], whereas yields and enantioselectivity were low for 2-acyl thiazoles [139]. The reduction of 2,3,5-triphenyltetrazolium chloride **114** (Fig. 31) gave **115** [101,154,157,158].

The reduction of the benzoylpyridines **116** (Fig. 32) gave the optically active α-hydroxybenzylpyridines **117** with 26–86% e.e., whereas the asymmetrical reduction of the corresponding *N*-oxide **118** gave **119** with e.e. values up to 93% [159]. The double reduction of 2,6-diacetylpyridines **120** (Fig. 33) by BY in the presence of allyl alcohol led to **121** with essentially 100% enantiomeric purity [160].

Interestingly enough, no reduction of aliphatic, unactivated nitro compounds seems to occur although the reduction of aromatic nitro compounds by BY is well documented [47,161,162]; a few examples for the reduction of ketones containing an additional nitro moiety have been reported. However, these reactions gave only low yields of the desired hydroxy nitroalkanes [163,164]. Better results have been obtained using the masked amino ketones **122** (Fig. 34) to afford the (*S*)-configured alcohols **123**. α,β-Unsaturated nitroalkenes **124** were reduced with moderate to excellent e.e. to afford nitroalkanes **125**.

The treatment of (*Z*)-3-aryl- and (*Z*)-3-alkyl-2-phenyl-3-nitropropenenitriles **126** (Fig. 35) gave access to 5-amino-3-aryl/alkyl-4-phenylisoxazoles **127** in good to moderate yields [165]. BY reduction of prochiral γ-nitroketones led to the enantioselective formation of (*S*)-4-nitroalcohols. Thus, **128b** (Fig. 36) gave **130** in 74% yield with an e.e. of 99% [166]. Direduction was observed for the symmetrical diketone **128a** that gave 58% of (2*S*,8*S*)-**129** with an e.e. of 95% [167].

α-β-Epoxy ketones on treatment with BY yielded different types of products depending on their substitution pattern. Thus, small groups (H, Me) attached to the epoxy end protect that end from attack, whereas 1-acyl epoxides with H, Me, or propyl as the 2-epoxy substituent gave solely the epoxy alcohol with moderate stereoselectivity (13–

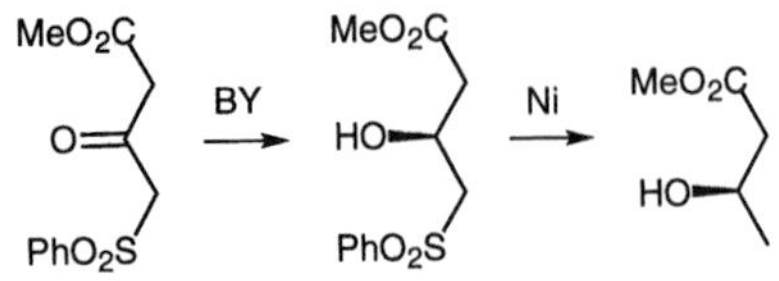

Figure 29

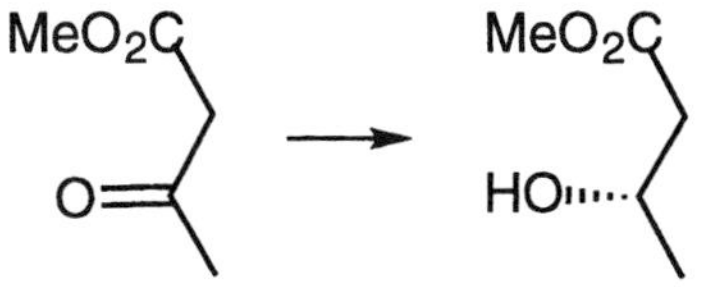

113 (*S*) 112

Figure 30

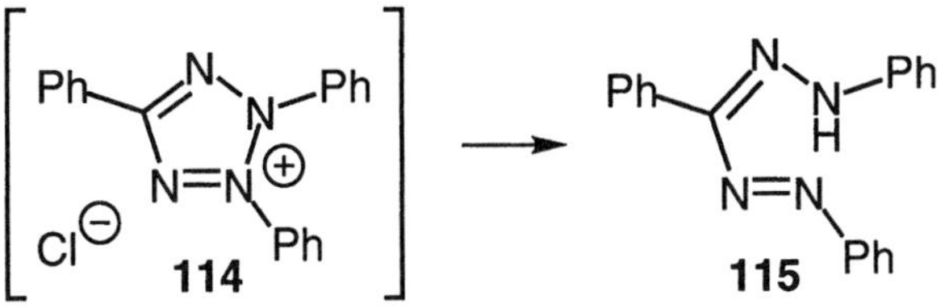

114 115

Figure 31

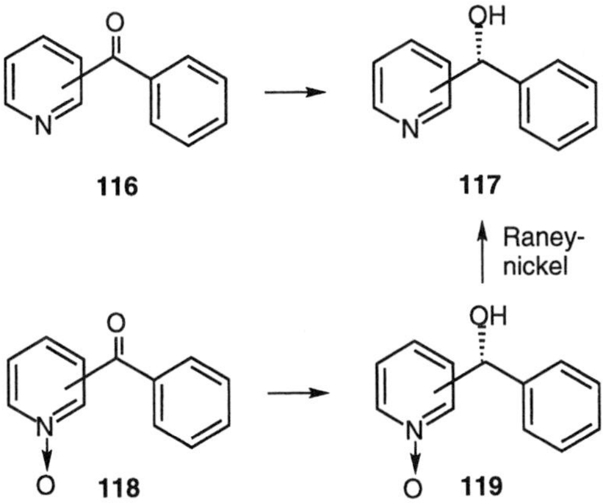

116 117 118 119

Figure 32

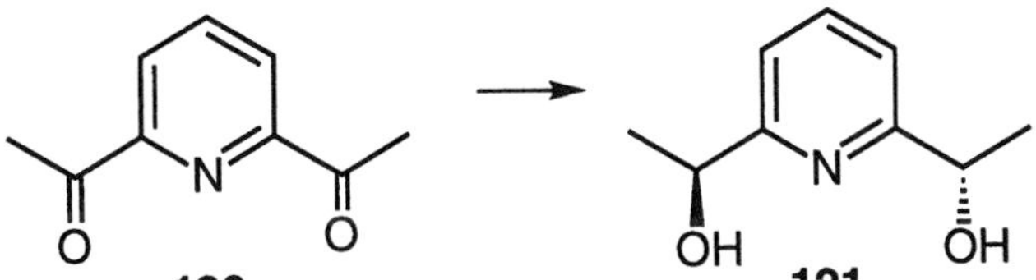

120 121

Figure 33

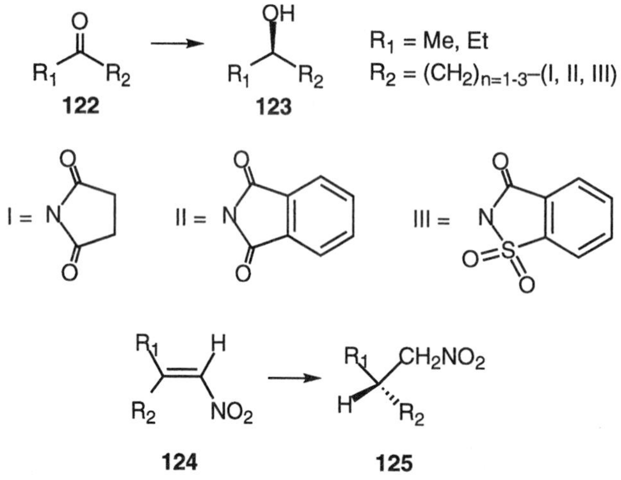

Figure 34

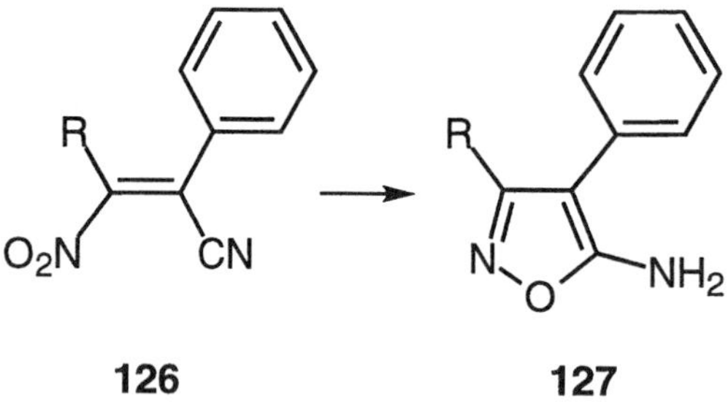

Figure 35

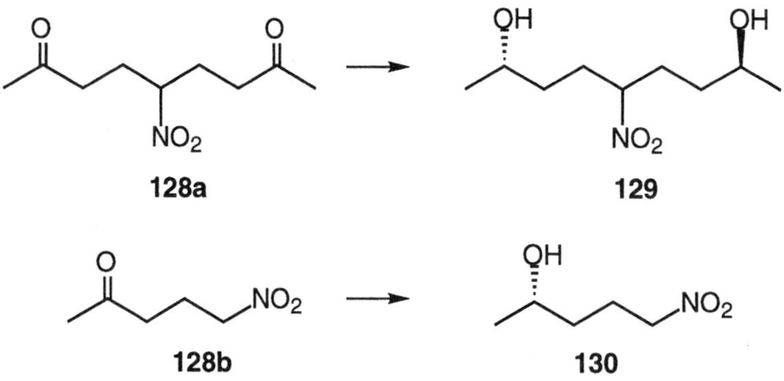

Figure 36

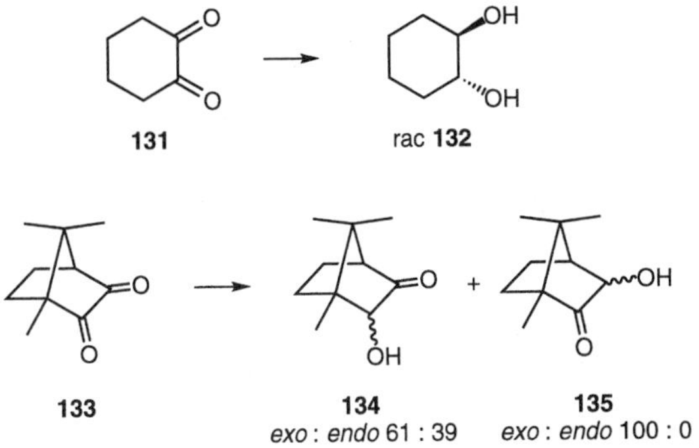

Figure 37

64% diastereomeric excess). With a 2-phenylsubstituent besides reduction of the carbonyl group [168], either the 1,2,3-triol [169,170] or the 2,3-diol is obtained [169,171].

C. Reduction of Dicarbonyl Compounds

1. Cyclic Diketones

The reduction of cyclohexane-1,2-dione (**131**) (Fig. 37) gave racemic *trans*-cyclohexane-1,2-diol (**132**) [172]. In analogy, camphor quinone **133** gave in 63% yield 3-hydroxy camphor **134** and *exo*-2-hydroxy camphor **135** [47].

The reduction of cyclic 2,2-disubstituted 1,3-diketones **136** (Fig. 38) can be regarded as an example of an enzyme-catalyzed distinction of a substrate containing two trigonal carbonyl centers with stereotopic faces and one prochiral tetrahedral carbon center where monoreduction generates two chiral centers. All of the products **137** and **138** were obtained with high e.e. [173–185]. Cyclohexanoid 1,3-diketones have been reduced with low diastereoselectivity as compared to the cyclopentanoid series [173,184,186–196]. Dimerization under participation of acetaldehyde produced by fermenting yeast has occasionally

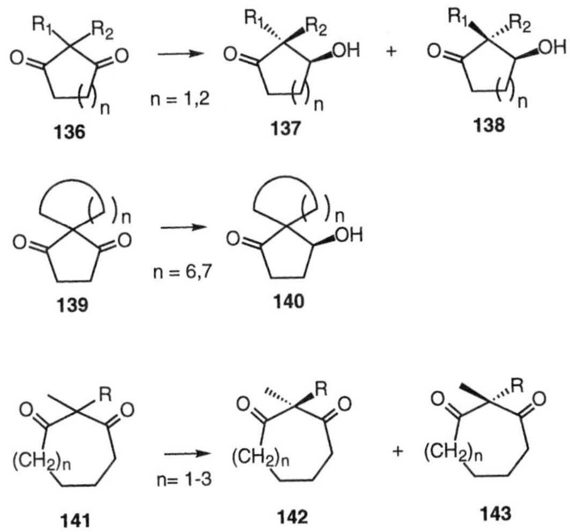

Figure 38

been observed. Even spiro-fused diones **139** were efficiently reduced to the ketols **140** showing an enantiomeric purity of more than 96% in each case [197].

Reductions of cyclic 1,3-diketones being part of a medium sized ring are not effectively achieved as in the case of the five- or six-membered rings [198]. Reductions of the bridged 1,3-diketones **144** (Fig. 39) gave enantiomerically pure (*1R,4S,6S*)-6-hydroxy-1-methyl-bicyclo[2.2.2]octan-2-ones **145** [199,200].

An analogous reduction has been carried out successfully on 4-methylbicyclo[2.21]heptane-2,6-dione in 42% yield and an e.e. of 82% [201]. The bicyclo[3.3.0]octane-2,8-dione (**146**) gave upon reduction with BY 41% of enantiomerically pure alcohol **147** [202].

As far as cyclic 1,4-diketones are concerned the reduction of 2,2,5,5-tetramethyl-1,4-cyclohexanedione (**148**) with BY (Fig. 40) was relatively inefficient; better yields of **149** were achieved employing *Curvularia lunata* [203]. Better results, in general, were achieved for 3,5,5-trimethyl-2-cyclohexene-14-dione **150** to give 83% yield of **151** [204,205]. From the 4-hydroxy isophoron analog **152**, 30% yield of **153** [206] and from the diacetate **154**, 32% yield of **155** [207] were obtained.

By reduction of the hexahydro-1,4-methanonaphthalene-5,8-dione **156** (Fig. 41) afforded the optically active 8-hydroxynaphthalene-5(*1H*)-one **157** with 67% e.e. [208].

Racemic bicyclo[3.3.1]nonane-2,6-dione (**158**) (Fig. 42) was reduced by BY, and the (*1S, 5S*)-**158** dione could be recovered. (*1R, 2S, 5R*)-**159** and (*1S, 2S, 5S*)-**160** in an 85:15 ratio and 5% yield of diol were isolated [209]. No reduction of the cyclic diketone was observed for **161** (Fig. 43) but reduction occurred at the side chain to yield **162** in 19% yield [210]. Microbial reduction of the diketone **163** afforded 40% yield of an inseparable mixture of (*8S,9S*)-**164** and (*8S,9R*)-**165** in a ratio of 77:23 [211]. Contrary to the reduction of **166**, BY cannot reduce *trans*-anellated 9-methyldecalin-1,8-dione (**167**) [212].

Early attempts at the reductions of quinones were successful though the reported yields were somewhat low [213–215].

2. Acyclic Diketones

The reduction of acyclic 1,2-diketones by BY is a long known reaction [66,172,216,217]. The reduction of butane-2,3-dione (**168**) (Fig. 44) gave 2,3-butanediol (**169**) in 60% yield [216]. The BY reduction of benzil (**170**) stopped at the monoreduction stage with benzoin (**171**) [172,218]. Employing *Saccharomyces montanus* [219] or *Rhodotorula glutinis* [219] gave the corresponding (*S,S*)-diol with 94% e.e.; the (*R,R*)-diol was obtained with 96% e.e. by the use of *Candida macerans* [220]. In contrast to these findings with **170**, [218]

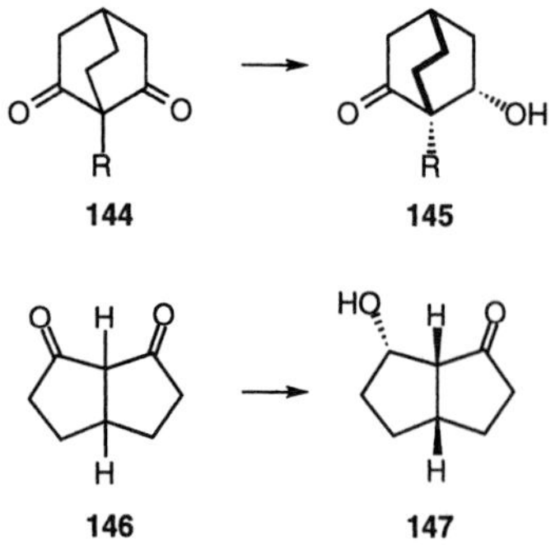

Figure 39

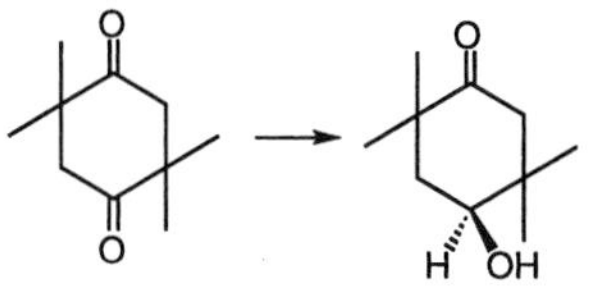

Figure 40

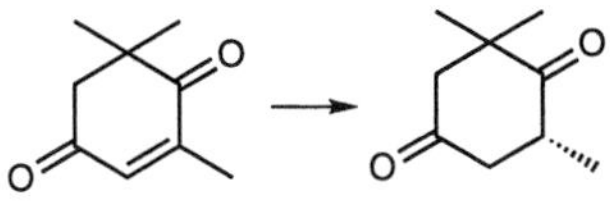

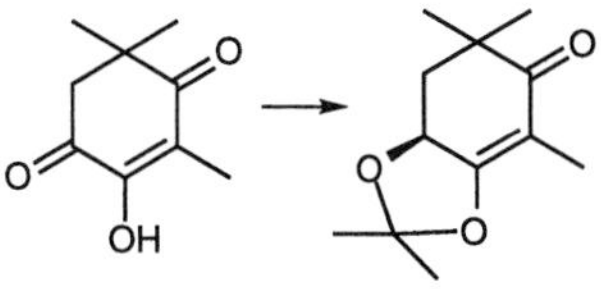

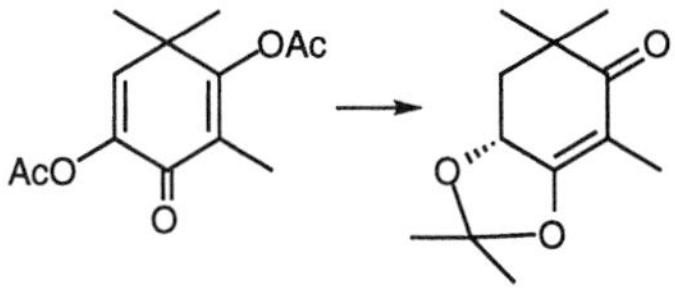

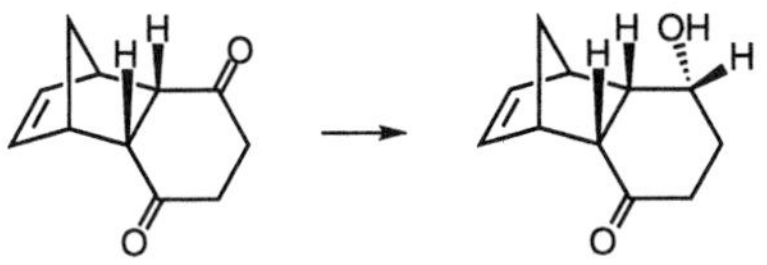

Figure 41

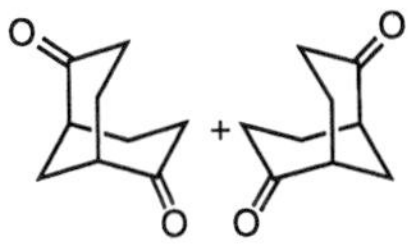

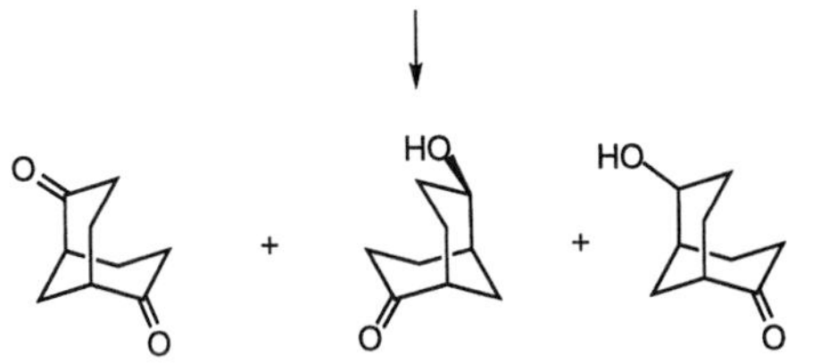

Figure 42

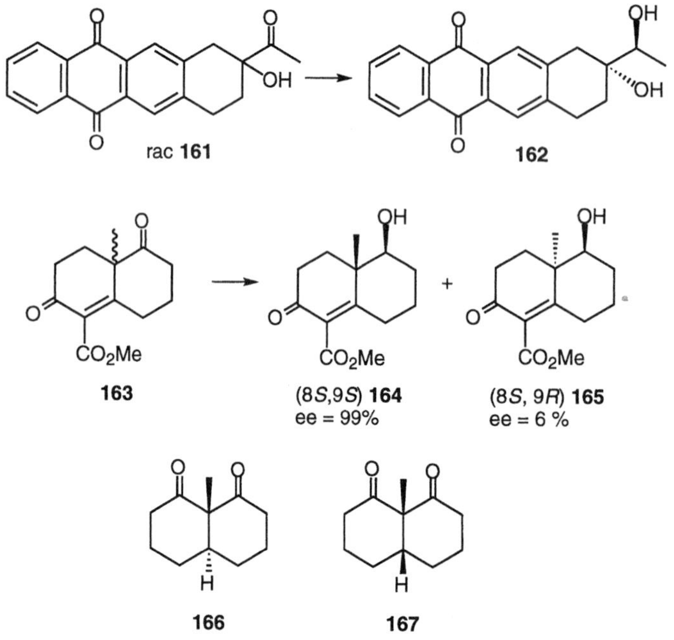

Figure 43

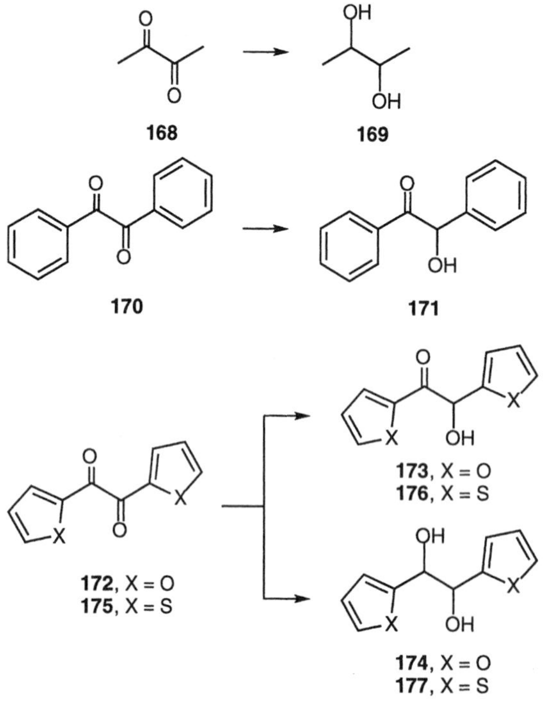

Figure 44

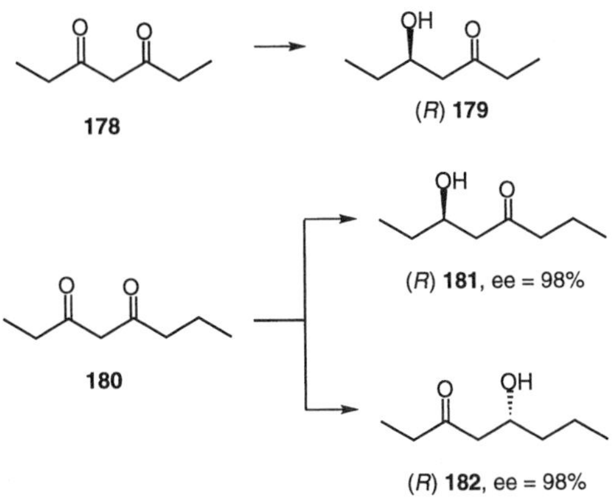

Figure 45

furil (**172**) was very quickly reduced via furoin (**173**) into hydrofuroin (**174**) [221]. Similarly, **175** and **176** afforded **177** [172,218].

In addition, reductions of acyclic 1,3-diketones often gave moderate to poor results [222–229]. Compound **178**, however, was easily reduced within 3 days by Hirondelle BY in quantitative yield but only with low e.e. (30%) to (*R*)-**179** (Fig. 45) [223]. In comparison, the reduction of the more lipophilic **180** gave after 100% conversion a 33:67 mixture of the ketones **181** and **182**, each exhibiting an e.e. of 98% [230].

The 3-methyl-branched compound **183** (Fig. 46) gave a 4:1 mixture of *syn*-**184** and *anti*-**185** [223,230]. 4-Methylheptane-3,5-dione (**186**) was reduced by *Geotrichum candidum* under aerobic conditions to give **187**; the stereoisomer **188** was also obtained by reduction under anaerobic conditions.

A detailed investigation of the reduction of fluorinated β-diketones **190** (Fig. 47) showed that the presence of additives in the BY reduction turns the stereoselectivity of the reaction toward the (*R*) or (*S*) enantiomer of the corresponding ketols. Thus, **190**

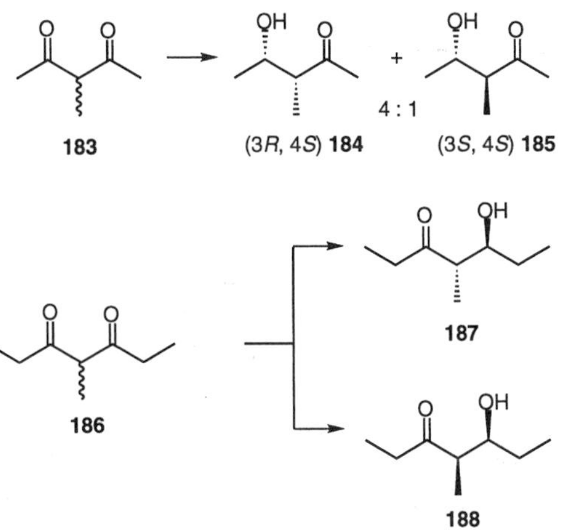

Figure 46

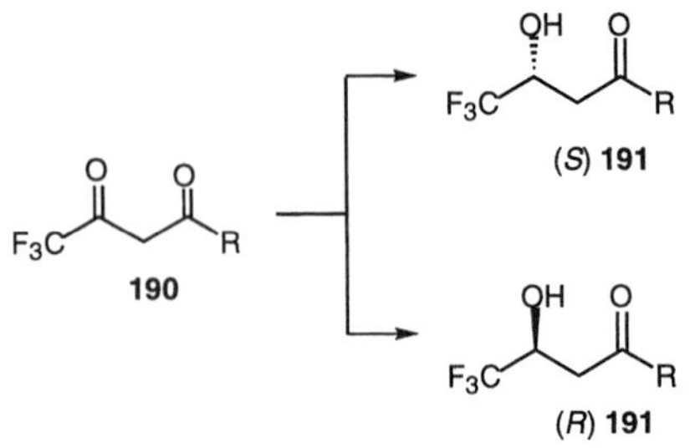

Figure 47

afforded (S)-**191** upon addition of allyl alcohol, whereas (R)-**191** was formed with allyl bromide as an additive [231].

In this context, the effect of additives onto the course of reductions has also been established by BY cell-free extracts [232]. BY reduction of 3-acyltetrahydrothiopyran-4-ones **192** (Fig. 48) gave regio- and enantioselective 3-acyl-4-hydroxy-tetrahydrothiopyranes (3R,4S)-**193** and (3S,4R)-**193** [233].

1,4-Diketones are good substrates for BY but the selectivity of these reductions is very low [234,235]. In contrast to the exclusive monoreduction of 1,3-diketones, 2,5-hexanedione (**194**) (Fig. 49) was cleanly reduced to (2S,5S)-hexane-2,5-diol (**195**).

Fluorinated 1,5-diketones gave primarily products of monoreduction remote of the fluorine substituent whereas on prolonged reaction time the corresponding diols could be obtained [236].

3.   α-Keto Esters

α-Keto esters have also been successfully reduced by BY. Thus, 2-oxo-2-arylacetic acid derivatives **189a–c** [237,238] gave the optically pure α-hydroxy acids **196a–c** (Fig. 50). Ethyl pyruvate (**197b**) is reduced to (R)-ethyl lactate (**198b**) but pyruvate is reduced by purified yeast alcohol dehydrogenase in the presence of NADH into (S)-lactate [239].

Further examples, albeit with somewhat lower e.e. values, have been reported [237,240–245]. The course of the reduction of α-keto esters **199** (Fig. 51) can be altered and controlled by using organic solvents instead of water. In addition, it has been found that α-hydroxy esters **200** produced by the yeast reduction are decomposed enzymically in water, and this asymmetrical decomposition does not take place during the reduction in an organic solvent such as benzene [246–248].

High e.e. values have been obtained for the BY reduction of 3-chloro-2-oxoalkanoates **201** leading to syn-**202** and anti-**203** in good yields [249].

The e.e. values have been obtained for the BY reduction of 3-chloro-2-oxoalkanoates **201** leading to syn-**202** and anti-**203** in good yields [249].

The synthesis of enantiomerically pure (R)-pantolactone (**204**) was achieved via enantiospecific reduction of the ketopantolactone (**205**) (Fig. 52). Among a broad variety of microorganisms tested [250], highest yields and e.e. values have been reported with *Rhodotorula minuta* [251], *Aspergillus niger* [252], *Candida parapsilosis* [252], and *Byssochlamys fulva*, whereas the reductions mediated with different strains of *Saccharomyces cerevisiae* [253] gave low yields [254] or low e.e. value [250]. The reduction of **205** in the presence of organic electron mediatos by several microorganisms has been reported [255]. In addition, the enantioselectivity in the BY reduction of **205** was improved by the addition of β-cyclodextrin to the reaction mixture [256].

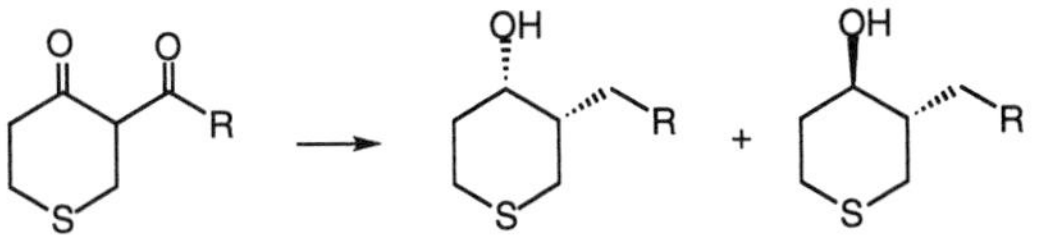

192 (3*R*, 4*S*) **193** (3*S*, 4*R*) **193**

Figure 48

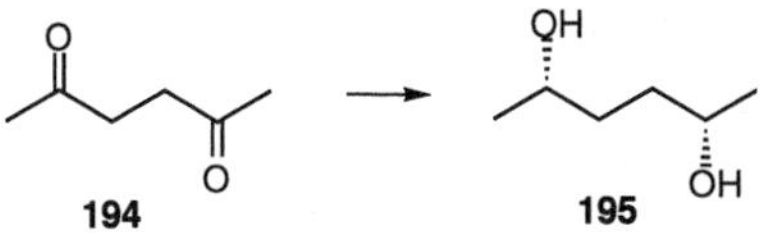

194 **195**

Figure 49

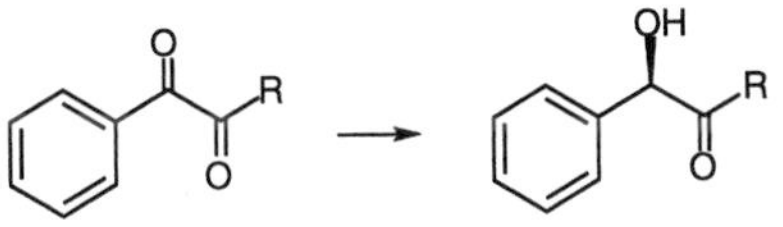

189a R = OMe **196a** 59%, ee = 100%
189b R = OEt **196b** 68%, ee = 100 %
189c R = NH_2 **196c** 70%, ee = 100 %

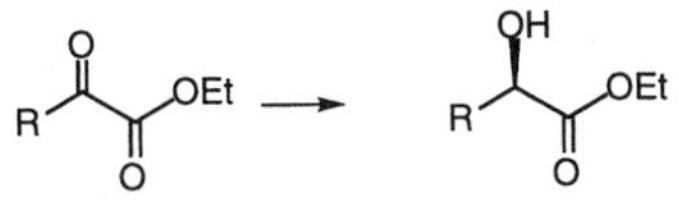

197a R = thiophenyl **198a**, ee = 100%
197b R = CH_3 **198b**, ee = 92 %

Figure 50

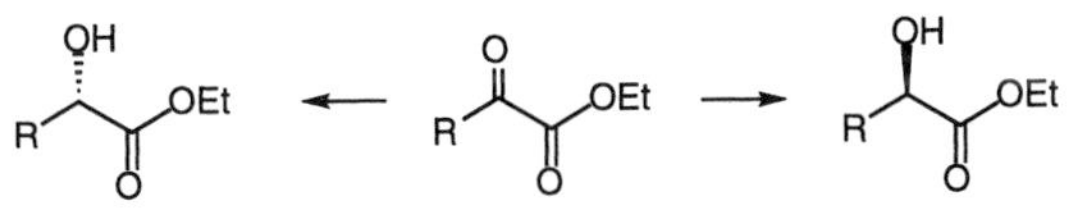

(*S*) **200** **199** (*R*) **200**

in water, 5-79%
ee= 19-93%

R = Me, Et, Prop,
But, Pent, i-Prop

in benzene, 26-90%
ee = 13-90%

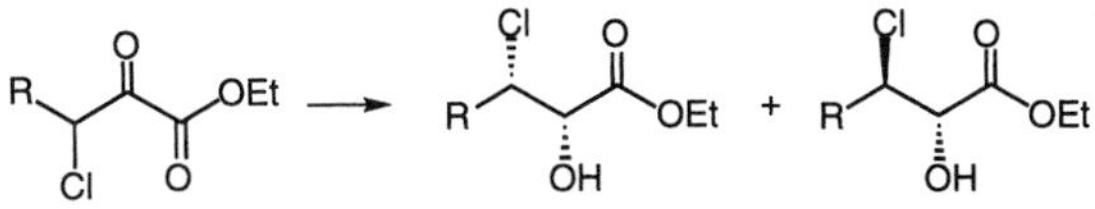

201 **202** **203**

Figure 51

205   *(R)* **204**

Figure 52

4.   Reduction of α,γ-Dioxo-Substituted Esters and Oxo-Substituted α,γ-Diesters

Diethyl 2-methyl-3-oxosuccinate (**206a**) (Fig. 53) gave a 43:57 mixture of *syn*-(*2R,3R*)-**207a** (79% e.e.) and *anti*-(*2S,3R*)-**208a** (31% e.e.). Higher e.e. values were obtained employing *Candida albicans* instead of BY [257]. The corresponding dimethyl analog **206b** afforded an inseparable mixture of the dimethyl 2-methyl malates **207b** and **208b** [258,259]. These reductions seem to depend strongly on the fermentation conditions used for growth of BY [243,260–265]. It is worthwile mentioning in this context that the BY reduction of acetoacetylated Meldrum's acid (**209**) gave 60% of enantiomerically pure **210** [266].

5.   Reduction of β-Keto Esters

Reductions of β-keto esters with yeasts have been performed quite often; it seems generally accepted that the enantioselectivity of the reduction of cyclic esters seems to be higher than those of open chain β-keto esters substituted at C-2. Recently, a neuronal network model for these reactions was developed [267,268]. The influence of organic solvents has been investigated [269]. The reduction of six-membered cyclic β-keto esters is rather widespread; thus, the reduction of the simplest compound in this context **211** (Fig. 54) gave 65–85% of ethyl(+)(*1R,2S*)-2-hydroxycyclohexane carboxylate (**212**) in 86–99% e.e. [270–272] and a diastereoselectivity of 76–99% [270]. Due to an equilibrium by enolization of the starting material followed by a kinetic resolution only one of the possible diastereomers is produced in excess.

Yeasts (*Saccharomyces cerevisiae, Saccharomyces montanus*) as well as fungi reduced structural analogous methoxy-substituted 2-cyano-1-tetralones **213** (Fig. 55) to *cis*-2-cyano-1-tetraols **214** with high enantiomeric purity [273]. Several analogs have been

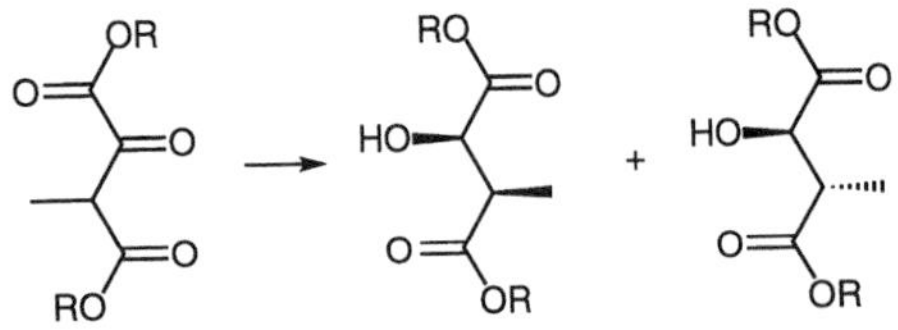

206   **207**   **208**

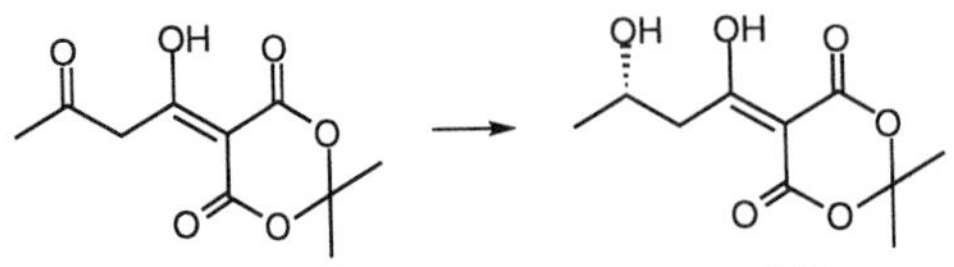

209   **210**

Figure 53

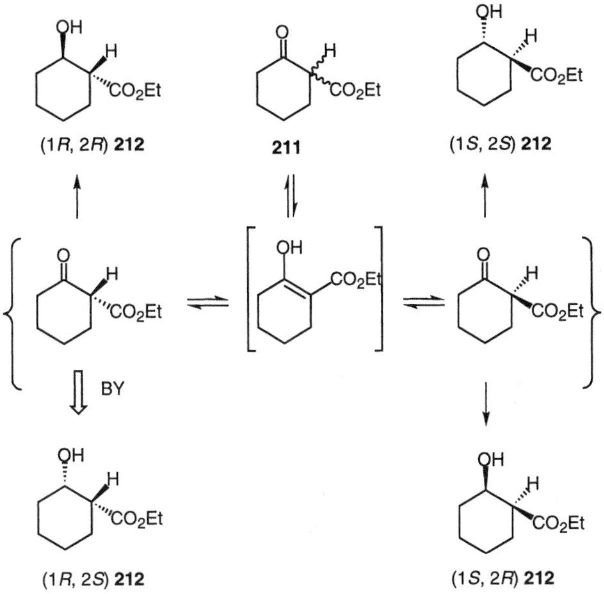

Figure 54

subjected to yeast-mediated reductions, although for several of these reactions the results were not quite convincing [271,274–280]. As far as five-membered rings are concerned, the reduction of the simplest compound **215** (Fig. 56) gave 80% of diastereomerically pure (*1R,2S*)-**216** [237,270,272,281,282].

Although for these types of compounds the reduction with BY often gives mixtures of optically pure diastereomeric hydroxy esters, predominantly of (*2S*) configuration, reductions performed with mold strains exhibit a very high enantio- and diastereoselectivity, thus leading to only one optically pure cis or trans stereoisomer [270,275,283–285].

The reduction of cyclic β-keto esters has been extended to five- or six-membered rings containing one additional heteroatom instead of carbon. Thus, reaction of methyl tetrahydro-3-oxothiophene-2-carboxylate (**217**) (Fig. 57) gave 62% yield of (*2R,3S*)-**218** [286] that upon treatment with Raney-Nickel gave (*3S*)-**219**, which cannot be obtained by BY reduction of the keto ester **220** [287]. The piperidone **221b** gave **222b** with an e.e. of 95% whereas the analogous reduction of **221a** gave in 71% yield and 98% diastereomeric pure **222a** [288]. From the reduction of β-oxoproline derivatives or oxopiperidine carboxylates **223** the *cis*-esters **224** were obtained [289–291].

The reductions of aliphatic β-keto esters by yeasts are well documented [13,21, 67,68,292,293]. The general feature for these reductions is that the absolute configuration

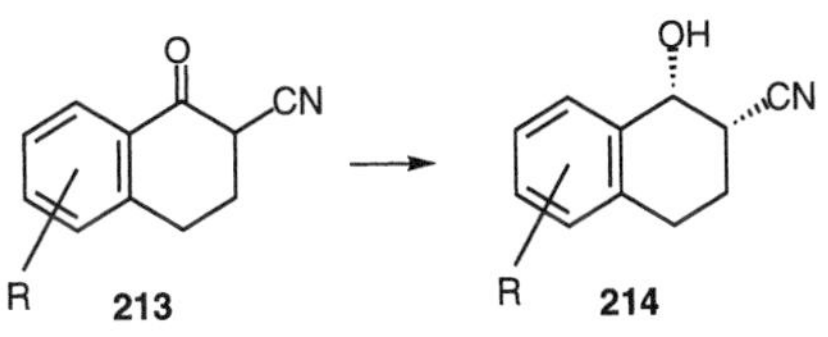

Figure 55

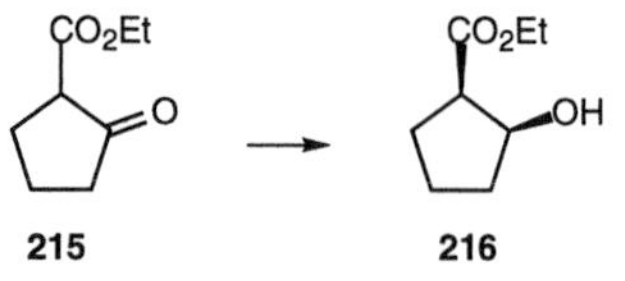

Figure 56

and the optical purity of the products depend strongly on the size and the nature of the substituents adjacent to the carbonyl group and of the ester moiety but also the pH, the glucose concentration [294], the substrate concentration [295,296], the cultivation conditions of the yeast [297], and the presence of additives [175]. In general, the absolute configuration of the products can be predicted by Prelog's rule [36]; but several exceptions to the rule are known [295,298]. It seems reasonable that low concentrations of the substrate gave better enantioselectivity.

The changes in the stereochemical course caused by different physiological states of the yeasts are due to the induction of different oxidoreductases [37,294,297,299–301]. Thus, three different oxidoreductases have been isolated from the cytosolic fraction of *S. cerevisiae* (Red Star) and characterized [13].

Detailed studies have been performed by ethyl acetoacetate (**255**) (Fig. 58) conversion to the (*S*)-hydroxybutanoate **226** [73,175,302–317] and for ethyl γ-chloroacetate

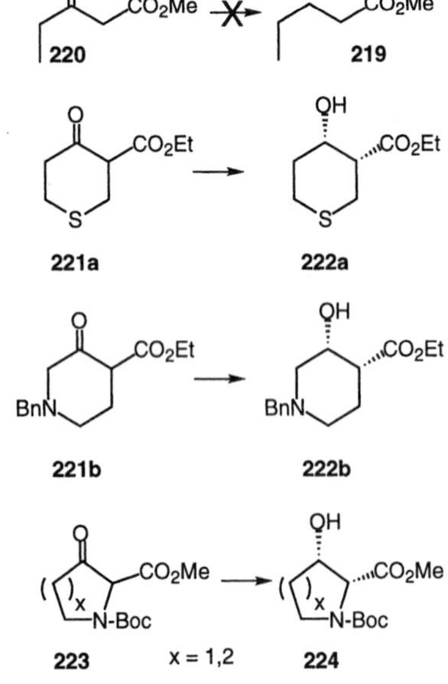

Figure 57

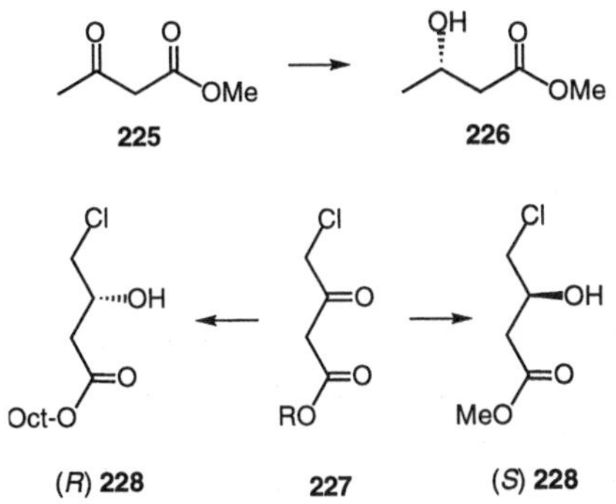

Figure 58

(**227**) conversing to (*S*)-**228** [37,318]. Many "best" conditions were claimed especially for the synthesis of (*S*)-**226** [175,302,319] and this process has already been scaled up [320]. The reduction of **225** by many other microorganisms has been investigated and the results have been compared [81,82].

Also, the bioreduction with immobilized BY in hexane in the presence of alcohols [321,296] as well as the reduction with nonfermenting yeast in hexane in the presence of a small amount of water has been reported to yield products with an excellent e.e. [321,322].

In addition, alkyl-, aryl-, or halogen-substituted β-keto esters [37,175,212,237,294–298, 303,304,315,323–340] as well as δ- or ε-heteroatomic substituted β-keto esters have been studied very intensively [175,233,294,295,298,304,341–348].

By reduction of methyl 3-oxo-4-pentynoate (**229**) (Fig. 59) gave the corresponding (*S*)-3-hydroxy ester **230** (80% e.e.); whereas its 5-trimethylsilyl derivative **231** gave the enantiomer **232** [349].

The reduction of polyfluorinated β-keto esters **233, 234**; and **235** gave the desired reduction products **236, 237**, and **238** in good yields [113,114,343].

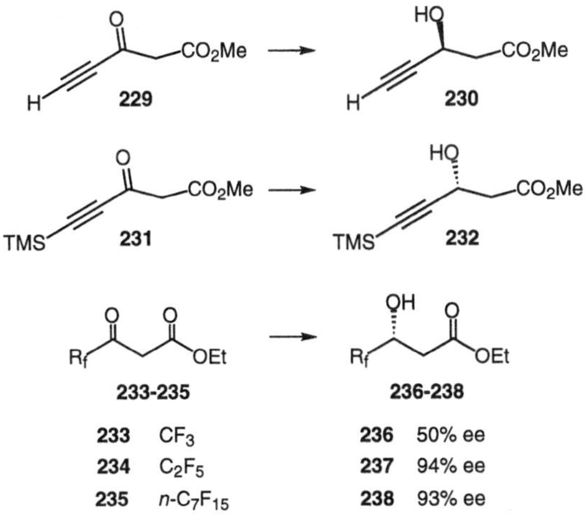

233 CF$_3$	**236** 50% ee
234 C$_2$F$_5$	**237** 94% ee
235 *n*-C$_7$F$_{15}$	**238** 93% ee

Figure 59

Figure 60

Enantioselectivity of the yeast-mediated reduction of 5-(benzyloxy)-3-oxopentanoate esters **239** (Fig. 60) [350,351] was influenced by changes in the esters' alkoxy group in a way that the enantioselectivity increases with increasing chain length for the *n*-alkyl ester.

From reduction of compounds **241** (Fig. 61) with BY predominantly (*R*)-**242** were obtained, [351,341,352,353], whereas the use of *Candida guilliermondi* for some of these β-keto esters gave the (*S*)-configured stereoisomers [341].

β-Keto esters with an additional center of chirality in the γ position [354] or in the alcohol part [355] as well as α-sulfenyl- [356] or α-hydroxy-substituted derivatives [357] have been reduced in this way and a general mode for predicting the diastereoselectivity for the latter compounds has been suggested.

In addition, various formyl derivatives **243** have been reduced to afford (*S*)-**244** [305] in 70–83% yield and e.e. values ranging from 46% to 91%. Aryl-substituted derivatives **245** were reduced to the corresponding alcohols **246** that served as valuable intermediates for the synthesis of enantiomerically pure drugs [358,359].

The stereochemical course of the yeast-mediated reduction of **247** (Fig. 62) could be altered by the addition of ethyl chloroacetate; thus, under usual conditions **248** was obtained, whereas upon addition of ethyl chloroacetate, **249** was obtained with an e.e. of 91% [360].

Recently, it has been shown that the BY-mediated reduction of α-substituted β-keto esters led to the reduction of the carbonyl group with high stereospecificity at low pH values, whereas at higher pH values cleavage of the C—C bond is observed. Thus, **250** gave at pH = 4 **251** and **252** each of 99% e.e. with a syn:anti ratio of 10:1, whereas at pH = 7 a 1:1 ratio was observed and the e.e. values were significantly lower. Finally, at pH = 8 a cleavage of the C—C bond with more than 75% formation of benzoic acid (**229**) was observed [361].

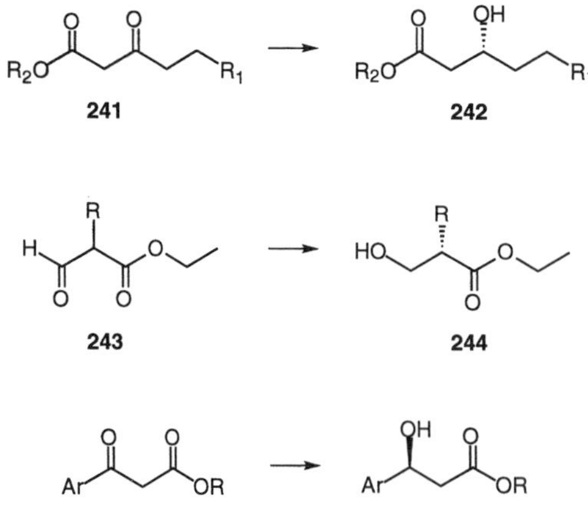

Figure 61

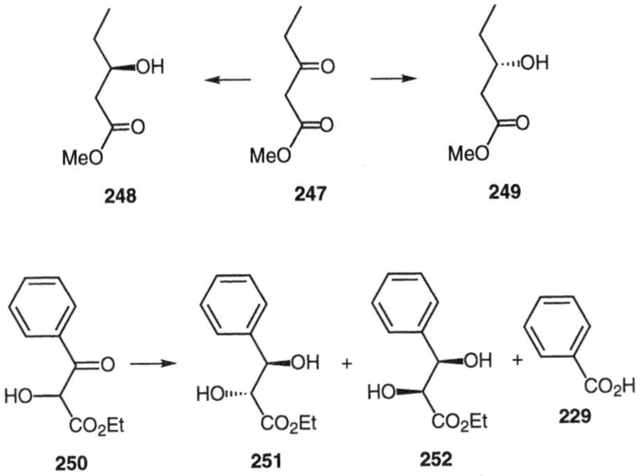

Figure 62

It is worthwhile mentioning the reduction of β-keto butanoates substituted at the α position [362–369], of 4-substituted 3-oxobutanamides [370,371], of prochiral 3-keto adipates [372], and of the heteroanalogous keto esters **253** (Fig. 63) that gave β-hydroxy-phosphonates **254** upon BY reduction [373,374].

Upon treatment with BY, 2-alkyl-3-oxobutyronitriles **255** gave optically pure 2-alkyl-3-hydroxybutyronitriles **256** and **257** [375].

BY reduction of the γ- and δ-keto acids/esters **258** or **259** (Fig. 64) afforded the corresponding γ- or δ-lactones **260** and **261**, respectively [298,376–387]. In general, δ-keto acids are more rapidly reduced than the corresponding γ-keto esters.

BY reduction of 4-oxo-3-tetrahydrothiopyranyl acetates **262** provided optically pure diastereomeric alcohols **263** and **264** [388].

III. C—C BOND-FORMING AND BOND-BREAKING REACTIONS

The longest known C—C bond-forming reaction mediated by yeast are acyloin-type condensations [389–393]. Thus, for example, substituted benzaldehydes **265** (Fig. 65) led via **266** to the products **267** showing predominantly an anticonfiguration [287,394,395]. These

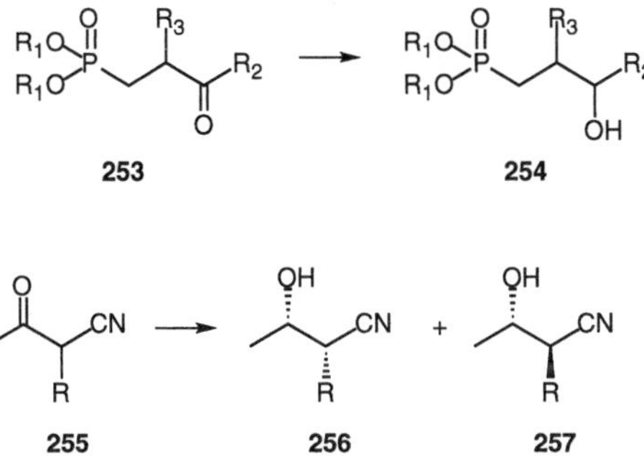

Figure 63

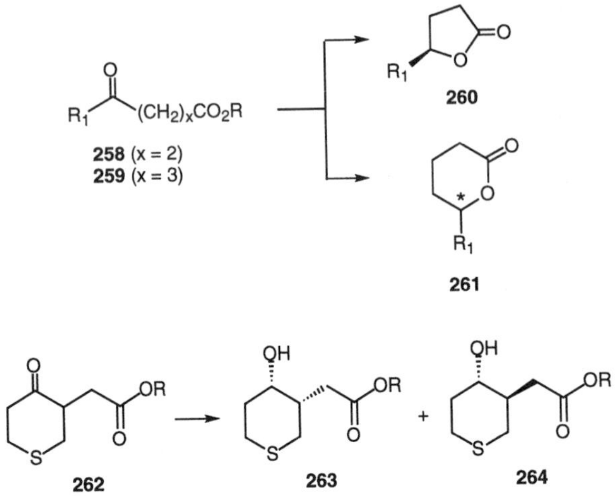

Figure 64

investigations have been extended to other aldehydes as well [390,394–399]; interestingly, salicyl aldehyde [399–401] and heptanal [399] have been shown to be the most cell-toxic aldehydes employed in this reaction. The formation of the acyloins **267** can be explained by the addition of a C$_2$ unit equivalent of acetaldehyde onto the *si* face of the carbonyl group to form (*R*)-α-hydroxy ketones [319] that are subsequently reduced on their *re* face by alcohol dehydrogenase(s). Whereas a broad structural tolerance for the first aldehyde to be used is found, only acetaldehyde is transferred by the living cells; no incorporation of added aldehydes was detected. The other limitation is found with the use of α,β-unsaturated carbonyl compounds that are predominantly biohydrogenated; the chain extension reaction is found only as a side reaction [242,319,392,393,402–409].

It seems that the condensing enzyme(s) are very specific whereas the substrate specificity for the subsequent reduction step is not so restricted. Thus, racemic **268** (Fig. 66) that were not formed by BY-mediated acyloin-type condensations were reduced to the alcohols **269** and **270** [319,410]. Additional examples have been reported [319,410–414].

The reduction of the α,β-unsaturated compound (*Z*)-3-methyl-2,4-pentadienal (**271**) (Fig. 67) gave 65% yield of **272**, 20% of (*Z*)-**273**, and 5% of (*E*)-**274**; the latter compound was subjected to a second treatment with BY to give 25% yield of enantiomerically pure **272** that was also obtained by BY treatment of **273** in 45% yield [414].

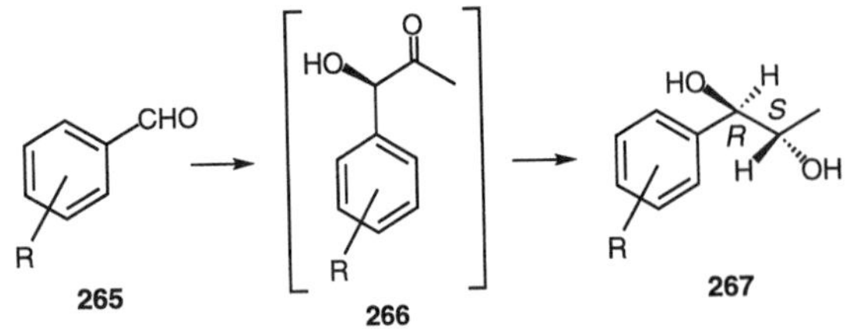

Figure 65

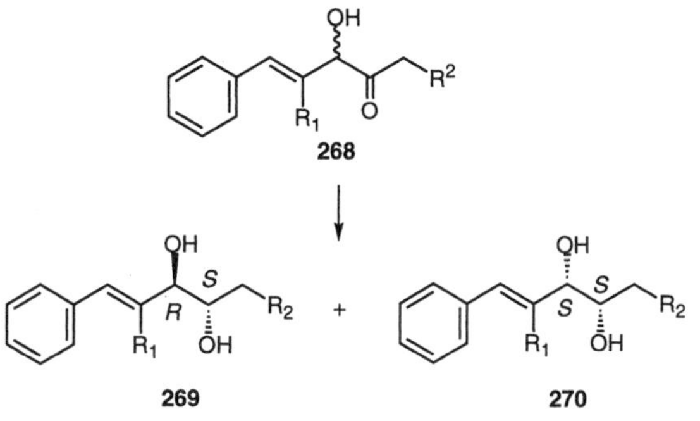

Figure 66

Complex mixtures were obtained from ethyl 4,4-dimethoxy-3-methylcrotonate [415,416] and analogs [417]. Recently, the BY-mediated bioreductions of many α,β-unsaturated ketones have been the subject of very detailed investigations [418–426]. It is worthwhile mentioning in this context the access to C_{10} synthons. Thus, geraniol, 3,7-dimethyl-2,6-octadien-1-ol (**275**) (Fig. 68), was transformed into its corresponding aldehyde **276** that by a one-pot double-hydrogenation reaction gave pure **277**, which was also accessible from **275** via reduction (via **278**). The selenium dioxide oxidation of **278** gave **279**, which was subjected to another BY-mediated reduction [427,428]. Compound **275** was also accessible from **280** where **281** gave **278** [429–431].

Several examples for the reduction of α,β-unsaturated alcohols, e.g., substituted cinnamyl alcohols [432], have been reported whereas the reduction of analogous α,β-unsaturated ketones proceeded very slowly and even failed in several cases [302,432–436]. Upon reduction of the glycoside **282** (Fig. 69) 83% yield of tetra-*O*-acetylconiferin (**283**) was obtained; interestingly, no deacetylations were found to occur during this transformation [437].

Clean reduction of the double bond occurred with **284**, **285**, and **286** (Fig. 70) after 7d incubation with fermenting yeast to yield **287**, **288**, and **289**, respectively [113].

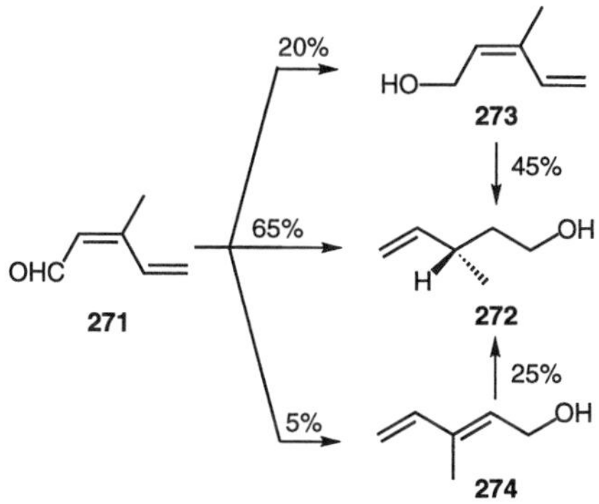

Figure 67

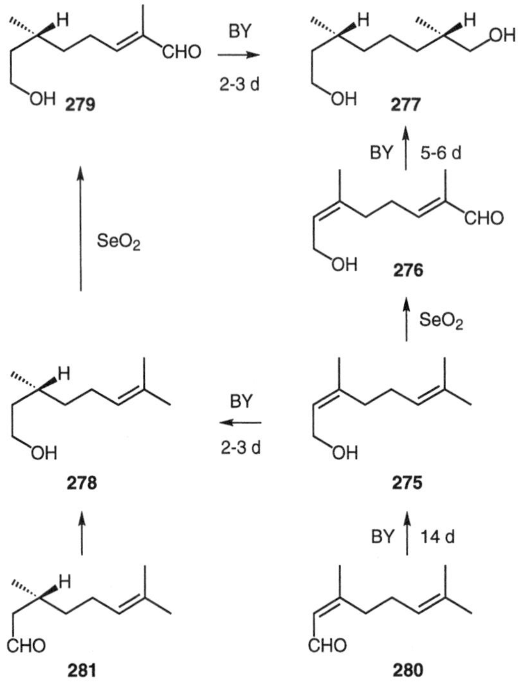

Figure 68

IV. OXIDATIONS

Oxidations by means of yeasts have rarely been performed since the oxidation capabilities of yeasts seem to be limited [438] and chemical methods are often adequate [46].

Whereas chiral sulfoxidation has been performed quite successfully with many fungi including *Rhizopus arrhizus, Aspergillus niger, Penicillium* sp., and *Rhodotorula* sp., almost no examples are available for those transformations that are mediated by yeasts. Thus, oxidation of **290** (Fig. 71) gave only 5% of **291** [138]; 9-thiastearate [301,439] and

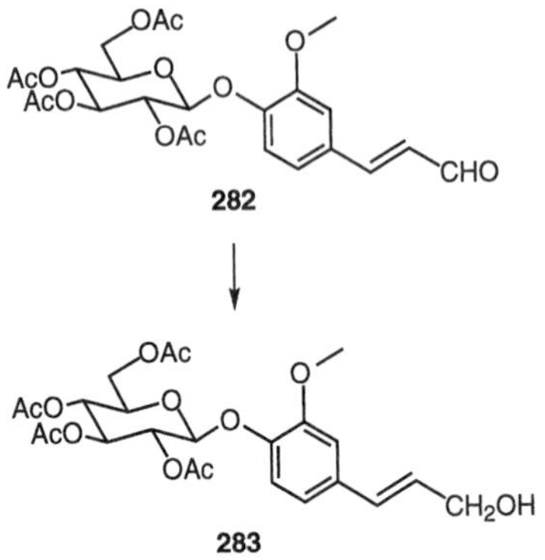

Figure 69

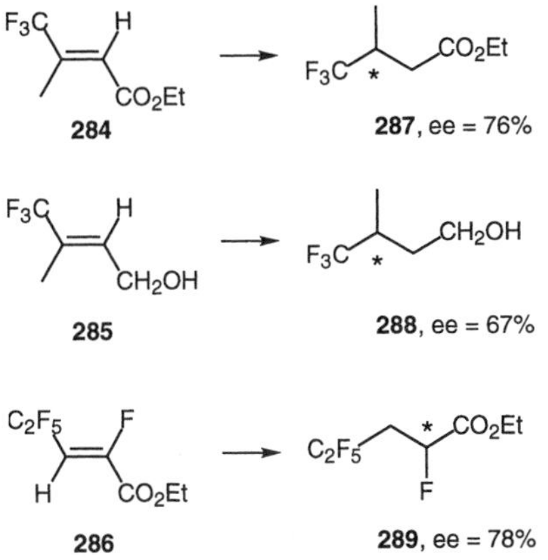

Figure 70

methyl styryl sulfide were also oxidized to the corresponding sulfoxide with yeast. Methyl *p*-tolyl sulfide (**292**) was oxidized by fermenting yeast to afford the (*R*)-configurated sulfide **293** in 60% yield with an e.e. of 92% [440,441]. The α-allenic alcohol **294** was oxidized to **295**, which isomerized to afford **296** in 26% yield [442].

The oxidation of racemic 1,2-alkane-diols under aerobic conditions furnished 1-hydroxy-2-alkanones that were produced by enantioselective oxidation of (*R*)-1,2-alkanediols [443]. Similarly, racemic 1-aryl and 1-heteroaryl alcohols **297** (Fig. 72) upon BY oxidation gave corresponding ketones **298** and the (*R*)-configurated alcohols **297** with 40–100% e.e. [444].

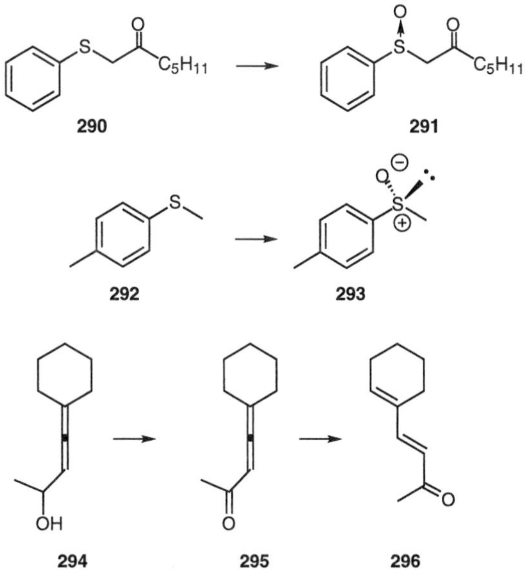

Figure 71

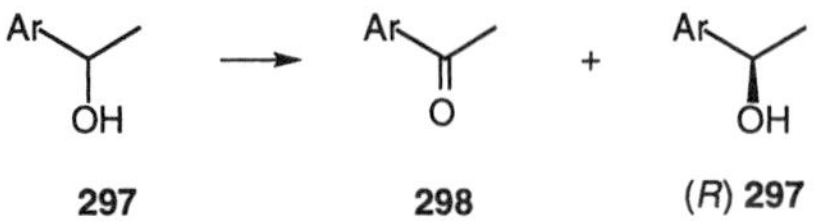

297   **298**   (*R*) **297**

Figure 72

V. DECARBOXYLATIONS

The decarboxylation of substituted cinnamic acids **299** (Fig. 73) has been shown to proceed with retention of configuration to yield **300**, hence confirming earlier findings obtained for *Bacillus pumillus* [445,146]. Similar to the decarboxylations of α,β-unsaturated carboxylic acids upon treatment with BY, there is one report describing the bioconversion of (*E*)-cinnamyl aldehyde (**301**) to styrene (**302**) by means of *S. cerevisiae* instead of its reduction to cinnamyl alcohol [242].

It is worthwhile mentioning the BY-mediated decarboxylative incorporation of linear C_3, C_4, and C_5 α-oxo acids into (*R*)-α-hydroxy ketones (**303**) upon incubation with benzaldehyde (**265**) [446].

VI. CYCLIZATION OF SQUALENOIDS

Cyclization of **304a** to **305a** (Fig. 74) by a cyclase was achieved in an enantioselective manner by treatment of **304a** with ultrasonically activated BY; similarly, from **304b** lanosterol (**305b**) was obtained in 83% yield. It was suggested that the ultrasound effect is more likely associated either with facilitating substrate diffusion by removing the outer membrane or by liberating membrane-associated sterol carrier protein factors rather than by activating the cyclase [447].

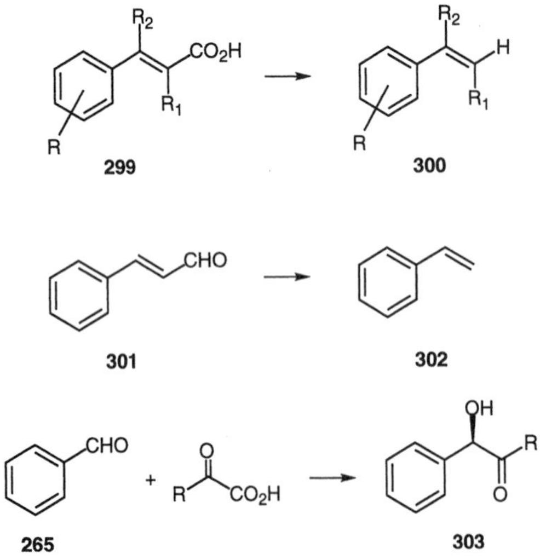

299   **300**

301   **302**

265   **303**

Figure 73

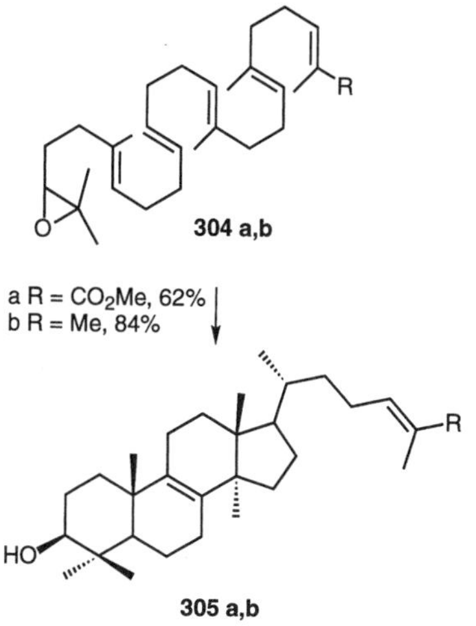

304 a,b

a R = CO$_2$Me, 62%
b R = Me, 84%

305 a,b

Figure 74

VII. MISCELLANEOUS C—C BOND-FORMING REACTIONS

α,β-Unsaturated ketones **406** (Fig. 75) upon BY-mediated reaction with 2,2,2-trifluoro-ethanol gave the product of a conjugate 1,4 addition **307** albeit in low yields (26–41%) but with high e.e. values (91–93%) [448].

Imidazo-fused quinazolines have been prepared from *N*-(allylcarbamoyl)-anthra-nilonitriles by ultrasonically treated BY [449], an approach that has also been used for the synthesis of other anellated heterocycles [450]. In addition, BY catalyzed the regioselective cycloaddition of nitrile oxides **308** (Fig. 76) to cinnamic esters **309** to afford predominantly **310** (BY9) [451]. Similar reactions have been investigated employing vinyl pyridines [451,452].

Finally, a Diels-Alder reaction between the pyradizinones **311** and diethyl azodicar-boxylate (**312**) gave the optically active cycloadducts **313** with e.e. values up to 63% [453].

VIII. HYDROLYSES

Yeast-mediated deacylations have been regarded merely as simple and more or less useless reactions. Originally observed in the steroid field [454], yeast-mediated hydrolyses and the enzymes involved in these processes have been investigated very carefully. Thus, protein-ase [455–457], esterase [458–467], phospholipase [468–470], lipase [468,471], tributy-rinase [472,473], as well as triacylglycerol lipase activities [471,474–476] have been de-

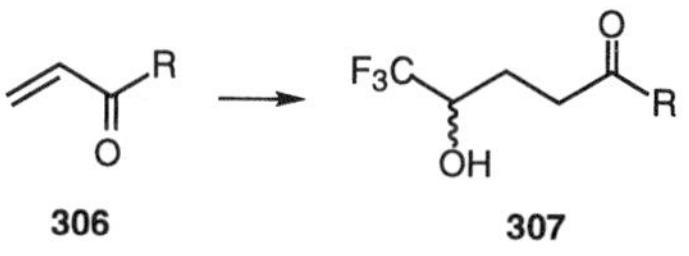

306   307

Figure 75

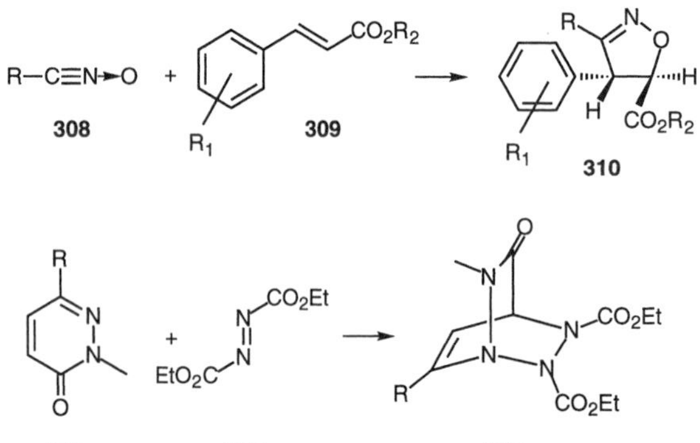

Figure 76

tected and investigated; a mathematical model of kinetic resolution of enantiomers has been proposed and several resolutions in the field of prostaglandin precursors have been published [13,477–481].

As an alternative to the enzymic de-N-acetylation of racemic N-acetyl-α-amino acids by amino acylases, the resolution of the esters of N-acetyl amino acids **314** (Fig. 77) by BY-mediated ester hydrolysis gave enantiomerically pure esters of **315** [482,483].

Generally, the unreacted D-configurated esters were isolated with 3–100% e.e., with high e.e. values for substrates containing unbranched alkyl or arylalkyl substituents, whereas γ substituents showed no effect. The introduction of additional polar groups lowered the e.e. values and β substituents inhibited the hydrolysis reactions. Variation of the alcohol part showed little or no influence on the course of hydrolysis; no reaction, however, was obtained for tert-butyl esters [482,483].

As an alternative to the use of fresh yeast, the use of acetone-dried powders [396], lyophilized yeast [484], or reverse micellar suspensions has been suggested. It is of interest to note that racemic α-substituted carboxylic esters other than amino acid esters generally gave poor results in these hydrolyses [485,486].

Regio- and enantiodifferentiation, however, can be found for the selective hydrolysis of 2-O-acyllactones. Thus, 2-O-acylpantoyllactones **316** (Fig. 78) gave (S)-**204** and (R)-**316** [317,487]. Similarly, the deacetylation of the carbohydrate-derived α-acetoxylactone **318** afforded **319**, whereas **320** was not affected by lyophilized BY [488,489].

Further applications [26] were performed on cyclopentene diacetates [490,491], cyclopentanone dicarboxylates [491], methyl acetate [492], citronellyl acetate [492], and a disaccharide [493]. In addition, both enantiomers of optically active 1-alkyn-3-ols **321** (Fig. 79) of high optical purity can be obtained by resolution of their racemic acetates **322** by use of lyophilized yeast [484].

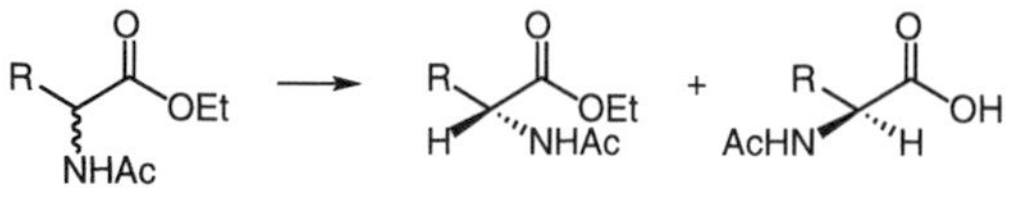

Figure 77

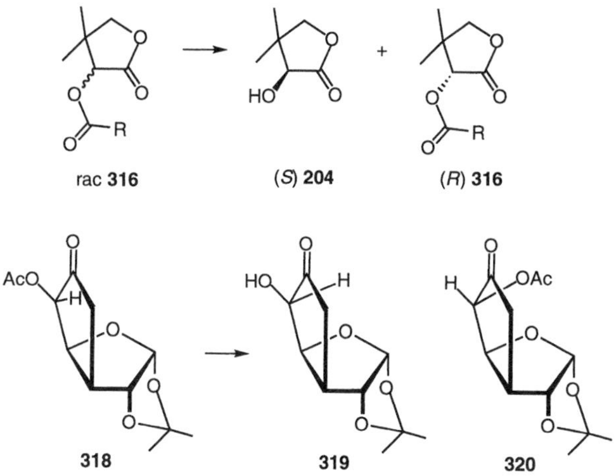

rac **316**　　　(S) **204**　　　(R) **316**

318　　　**319**　　　**320**

Figure 78

Several other hydrolyses have been performed, although only products of low enantiomeric purity were obtained [494].

IX. MISCELLANEOUS REACTIONS

Although there are many examples of yeast-mediated hydrogenations of alkenes to yield alkanes, examples for the reverse process are scarcely found. Thus, methyl-5-thiastearate (**323**) (Fig. 80) upon BY treatment gave methyl 5-thiaoleate (**324**) in 66% yield [495]. In addition, several α-allenic alcohols **325** have been reduced to the corresponding β-ethylenic alcohols **326**, whereas the β-allenic alcohol **327** gave 35% of γ-acetylenic alcohol **328** [442].

Besides this, several yeast-mediated phosphorylations [11,66,496–505], as well as cleavage of glycosidic bonds [66] and the reduction of galactose to dulcitol [506], have been reported. From phytic acid, D-myoinositol-1,2,6-trisphosphate was obtained [507].

Of importance in environmental chemistry is the transformation of DDT (**329**) (Fig. 81) into DDD (**330**) by reductive dechlorination of the DDT, whereas DDE (**331**) was not reduced to DDD (**330**) upon treatment with BY [508].

The reduction of (Z)-3-chloro-3-alken-2-ones **332** gave the optically active α-chloroketones **333** that were reduced on further treatment with BY to optically pure chlorohydrins **334** and **335**, respectively [509].

A dechlorinating rearrangement was observed upon treatment of **337** (Fig. 82) with BY to give 68% yield of **336** [510].

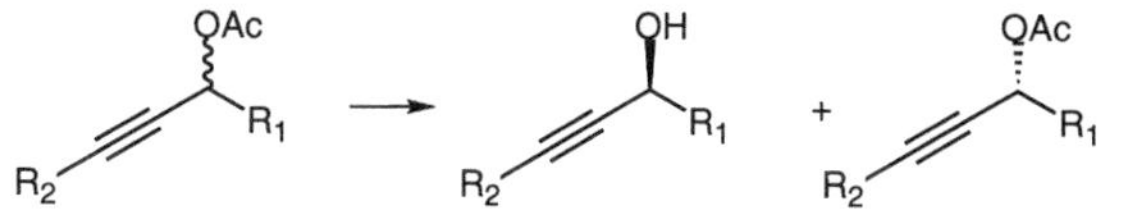

rac **322**　　　**321**　　　**322**

Figure 79

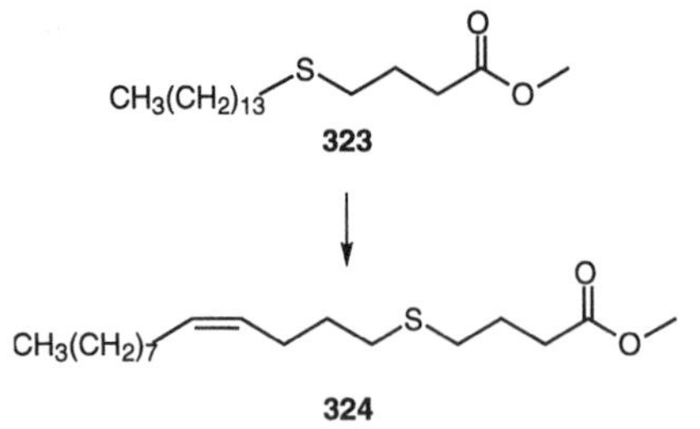

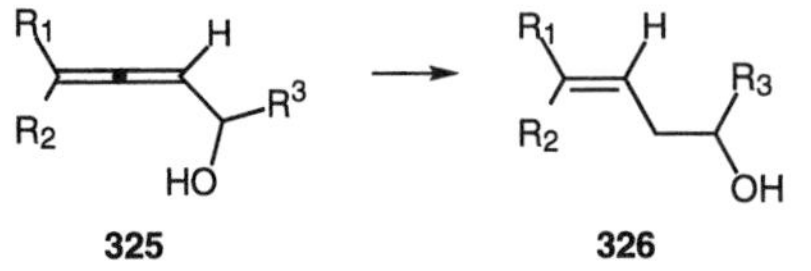

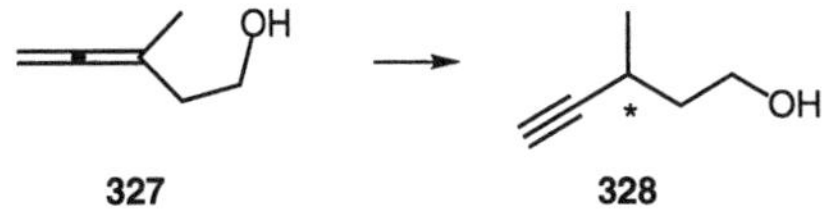

Figure 80

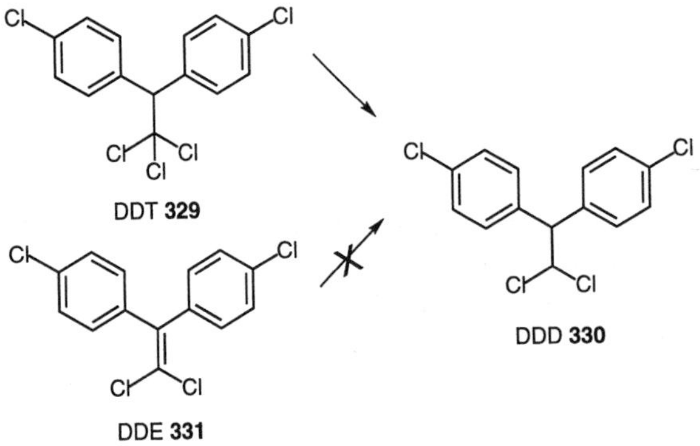

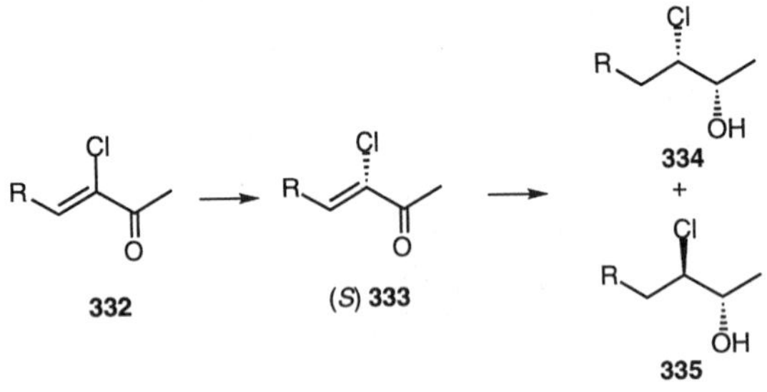

Figure 81

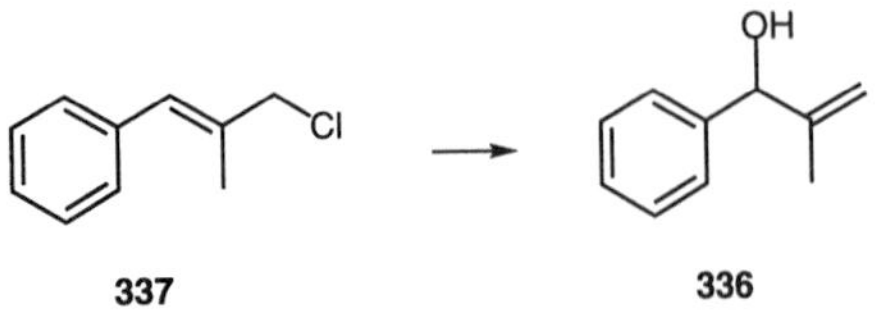

Figure 82

564

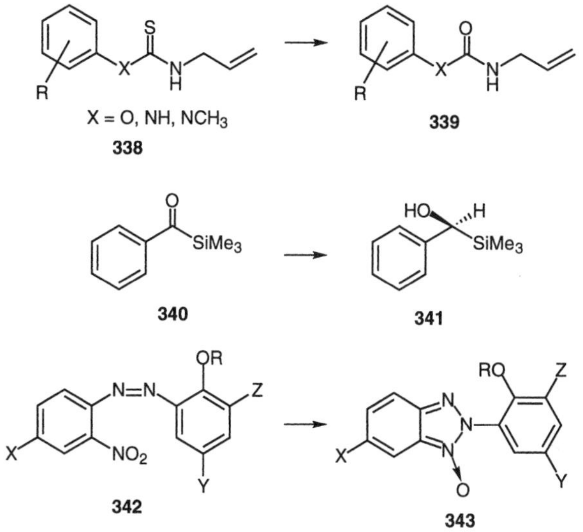

Figure 83

Desulfuration of arylalkylthiocarbamates and 1-alkyl-3-arylthioureas **338** (Fig. 83) to the corresponding carbamates and ureas **339** was obtained in good yields by BY [511].

As far as unusual reactions are concerned, phenylacylsilane **340** was reduced by BY to provide the optically active carbinol **341** in 73–90% e.e., although the yields for this reduction were rather low (10–30%) [512].

The reduction of several indolo[2,3-*a*]quinolizidines by BY gave novel products resulting from reduction of the indole double bond, cleavage of the C–D ring junction, or reduction of a lactame to a carbinol amine [371]. The known reduction of aromatic nitro compounds has been used as a key step in the synthesis of 2-aryl-2*H*-benzotriazoles. Thus, treatment of the *o*-nitrophenylazo dye **342** gave the 2-aryl-2*H*-benzotriazole-1-oxides **343** in good yields [513].

Recently, several aromatic mono- [514] and dinitro-substituted benzene derivatives [515] reported to be were reduced by BY. From the *N*-oxides **344** and **345** products **346**,

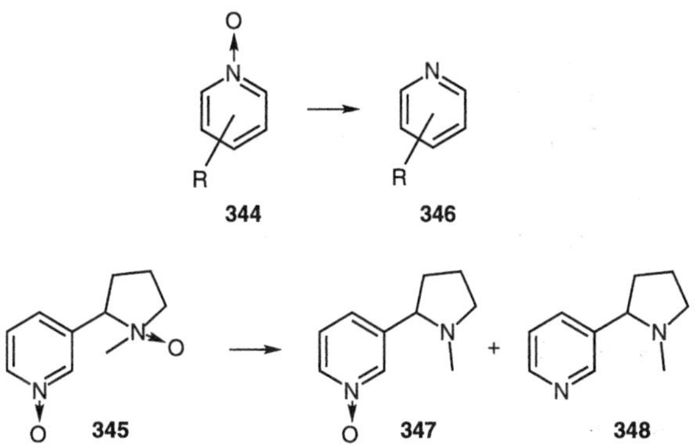

Figure 84

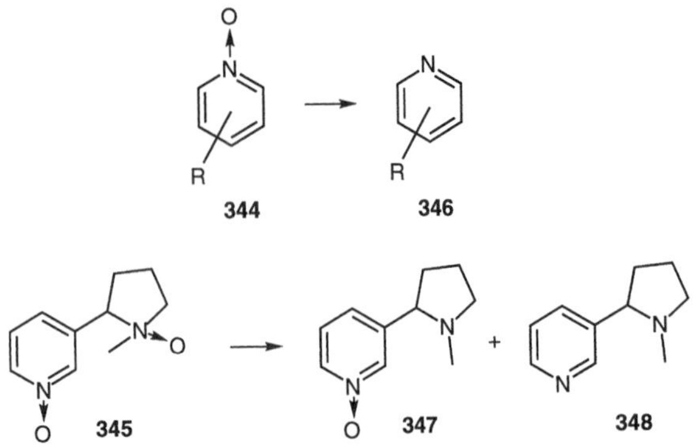

Figure 85

Figure 86

347, and **348** were obtained, respectively (Fig. 84) [516]. The deoxygenation of more complex *N*-oxides has been described [517].

Aromatic nitroso compounds **349** (Fig. 85) gave the amines **350** in 22–99% yield [518] and from the azidoarenes **351** the amines **352** were obtained [519,520].

Aromatic and heteroaromatic nitroalkenes **353** gave the nitroalkanes **354** [521] in excellent yields. From oximes [522] chiral amines were obtained [308], whereas hydrazones were hydrolyzed to the corresponding carbonyl compounds [523].

Calcium alginate–immobilized yeast cells were successfully used for the synthesis of small peptides [524]. Of special synthetic interest seems to be the biohydrogenation of 2-substituted allyl alcohols **355** (Fig. 86) to afford enantioselectively the (*R*)-2-methylalkanols **356** [525,526]. For natural product synthesis several butenolides have been reduced very successfully with BY [527–529].

REFERENCES

1. L Pasteur. Compt Rend Acad Sci 55:28–33, 1862.
2. W Windisch. Wochenschrift Brau 15:189–191, 1898.
3. CJ Lintner, HJ von Liebig. Hoppe Seylers Z Physiol Chem 88:109–113, 1913.
4. RS Chaleff. Pure Appl Chem 60:821–824, 1988.
5. D Perlman. Dev Ind Microbiol 21:15, 1980.
6. K Mori, T Sugai. J Synth Org Chem 41:1044–1053, 1983.
7. DE Eveleigh. Sci Am 245:120, 1981.
8. TH Maugh II. Science 221:351, 1983.
9. HGW Leuenberger. In: K Kieslich, ed. Biotransformations. Weinheim: Verlag Chemie, 1984, vol. 8, pp 5–29.
10. K Kieslich. In: K Kieslich, ed. Biotransformations. Weinheim: Verlag Chemie, 1984, vol. 8, pp 1–4.
11. H Yamada, S Shimizu. Angew Chem 100:640–661, 1988.

12. AM Chakrabarty. Pure Appl Chem 60:837–840, 1988.
13. R Csuk, BI Glänzer. Chem Rev 91:49–97, 1991.
14. S Servi. Synthesis 1990; 1–25.
15. S Roberts, N Williamson. Curr Org Chem 1:1–20, 1997.
16. DW Knight, EL Trown. Curr Org Chem 1:95–107, 1997.
17. LT Kanerva. Acta Chem Scand 50:234–242, 1996.
18. HK Kölbl, H Hildebrand, N Piel, T Schröder, W Zitzmann. Pure Appl Chem 60:825–831, 1988.
19. E Kondo, E Mazur. J Gen Appl Microbiol 7:113–117, 1961.
20. S Ohlson, PO Larson, K Mosbach. Eur J Appl Microbiol Biotechnol 7:103–110, 1979.
21. H Simon, J Bader, H Günther, S Neumann, J Thanos. Angew Chem 97:541–555, 1985.
22. A Margeritas, TJ Merchant. CRC Crit Rev Biotechnol 1:339–393, 1984.
23. B Mattiasson, B Hahn-Hagerdal. Eur J Appl Microbiol Biotechnol 16:52–64, 1982.
24. M Kierstan, C Bucke. Biotechnol Bioeng 19:387–394, 1977.
25. T Tosa, T Sato, T Mori, K Yamamoto, I Takata, Y Nishida, I Chibata. Biotech Bioeng 21: 1697–1709, 1979.
26. T Sakai, T Nakamura, K Fukuda, E Amano, M Utaka, A Takeda. Bull Chem Soc Jpn 59: 3185–3188, 1986.
27. T Sato, Y Nishida, T Tosa, I Chibata. Biochem Biophys Acta 570:190–198, 1979.
28. T Sato, Y Nishida, T Tosa, I Chibata. Appl Microbiol 27:878–895, 1974.
29. Y Naoshima, H Hasegawa. Chem Lett 1987; 2379–2382.
30. M Utaka, S Konishi, T Okuba, S Tsuboi, A Takeda. Tetrahedron Lett 28:1447–1450, 1987.
31. AD Argoudelis, RR Herr, DJ Mason, TR Pyhe, JF Zieserl. Biochemistry 6:165–172, 1967.
32. R MacLeod, H Prosser, L Fikentscher, J Lanyi, HS Mosher. Biochemistry 3:838–846, 1964.
33. O Cervinka, L Hub. Collect Czech Chem Commun 31:2615–2617, 1966.
34. HE Högberg, E Hedenström, J Fägerhag. J Org Chem 57:2052–2059, 1992.
35. J Sakaki, Y Sugita, M Sato, C Kaneko. Tetrahedron 47:6197–6214, 1991.
36. V Prelog. Pure Appl Chem 9:119–130, 196.
37. CJ Sih, B Zhou, AS Gopalan, WR Shieh, F VanMiddlesworth. In: W Barth, ed. Proc 14th Workshop Conference Hoechst. Weinheim: Verlag Chemie, 1983, pp 250–259.
38. L Mamoli, A Vercellone. Ber Deut Chem Ges 40:470–471, 1937.
39. A Wettstein. Ber Deutsch Chem Ges 40:250–252, 1937.
40. L Mamoli, A Vercellone. Ber Deutsch Chem Ges 70:2079–2082, 1937.
41. A Butenandt, H Dannenberg. Ber Deutsch Chem Ges 71:1681–1685, 1938.
42. B Camerino, CG Alberti, A Vercellone. Helv Chim Acta 37:1945–1948, 1953.
43. DHR Barton, T Shioiri, DA Widdowson. Chem Commun 1970; 939–940.
44. L Mamoli, A Vercellone. Ber Deutsch Chem Ges 71:1686–1687, 1938.
45. RL Crumbie, DD Ridley, GW Simpson. J Chem Soc Chem Commun 1977; 315–316.
46. A Kergomard. In: K Kieslich, ed. Biotransformations. Weinheim: Verlag Chemie, 1984, vol. 8, pp 127–205.
47. C Neuberg, Adv Carbohydr Chem Biochem 4:75–117, 1948.
48. S Akamatsu. Biochem Z 142:188, 1923.
49. Z Wimmer, M Budesinsky, T Macek, A Svatos, D Saman, S Vasickova, M Romanuk. Coll Czech Chem Commun 52:2326–2337, 1987.
50. G Lowe, S Swain. J Chem Soc Perkin Trans 1 1985; 391–398.
51. I Stibor, I Vesely, J Palecek, J Mostecky. Synthesis 1986; 640–642.
52. RF Newton, J Paton, DP Reynolds, S Young. J Chem Soc Chem Commun 1979; 908–909.
53. DJ Kertesz, AF Kluge. J Org Chem 53:4962–4968, 1988.
54. HG Davies, TCC Gartenmann, J Leaver, SM Roberts, MK Turner. Tetrahedron Lett 1986, 1093–1094.
55. G Fantin, M Fogagnolo, E Marotta, A Medici, P Pedrini, P Righi. Chem Lett 1996, 511–512.
56. P Manitto, G Speranza, D Monti, G Fontana, E Panosetti. Tetrahedron 51:11531–11546, 1995.

57. B Wunsch, H Diekmann. Heterocycles 38:709–712, 1994.
58. T Izumi, T Hino, A Kasahara. J Chem Soc Perkin Trans 1 1992; 1265–1267.
59. YD Vankar, K Shah, A Bawa, SP Singh. Tetrahedron 47:8883–8906, 1991.
60. G Eichberger, K Faber, H Griengl. Monatshefte Chem 116:1233–1236, 1985.
61. M Nakazaki, H Chikamatsu, K Naemura, M Asao. J Org Chem 45:4432–4440, 1980.
62. G Burnier, P Vogel. Helv Chim Acta 73:984–1001, 1990.
63. AP Marchand, GM Reddy. Tetrahedron Lett 31:1811–1814, 1990.
64. A Kamal. J Org Chem 56:2237–2240, 1991.
65. A Archelas, R Furstoss. Tetrahedron Lett 33:5241–5242, 1992.
66. K Kieslich. Microbial Transformations of Non-Steroid Cyclic Compounds. Weinheim: Verlag Chemie, 1976.
67. CJ Sih, C-S Chen. Angew Chem 96:556–565, 1984.
68. T Fujisawa, T Sato, T Itoh. J Synth Org Chem 44:519–531, 1986.
69. VE Althouse, DM Feigl, WA Sanderson, HS Mosher. J Am Chem Soc 88:3595–3599, 1966.
70. U Nagai, JI Kobayashi. Tetrahedron Lett 1976; 2873–2874.
71. H Günther, MA Alizade, M Kellner, F Biller, H Simon. Z Naturforsch 28C:241–246, 1973.
72. J Barry, HB Kagan. Synthesis 1981; 453–454.
73. DD Ridley, M Stralow. J Chem Soc Chem Commun 1975; 400.
74. PA Levene, A Walti. Organic Syntheses Coll 2:545–546, 1943.
75. JP Guette, N Spassky, D Boucherot. Bull Soc Chim 1972; 4217–4224.
76. E Färber, FF Nord, C Neuberg. Biochem Z 112:313–323, 1920.
77. E Schroetter, J Weidner, H Schick. Liebigs Ann Chem 1991; 397–398.
78. B Jakob, SG Voss, H Gerlach. Tetrahedron: Asymmetry 7:3255–3262, 1996.
79. AR Maguire, DG Lowney. J Chem Soc Perkin Trans 1 1997; 235–238.
80. DR Deardorff, DC Myles, KD MacFerrin. Tetrahedron Lett 1985; 5615–5618.
81. F Aragozzini, E Maconi, R Craveri. Appl Microbiol Biotechnol 24:175–177, 1986.
82. F Aragozzini, E Maconi. Ann Microbiol 35:87–92, 1985.
83. C Neuberg, FF Nord. Ber Dtsch Chem Ges 52:2237–2256, 1919.
84. M Bucciarelli, A Forni, I Moretti, G Torre. Synthesis 1983; 897–899.
85. CO Meese. Liebigs Ann Chem 1986; 2004–2007.
86. K Nakamura, K Ushio, S Oka, A Ohno, S Yasui. Tetrahedron Lett 1984; 3979–3982.
87. T Izumi, O Itou, K Kodera. J Chem Technol Biotechnol 67:89–95, 1996.
88. P Ferraboschi, P. Grisenti, A Manzocchi, E Santaniello. J Chem Soc Perkin Trans 1 1990; 2469–2474.
89. G Fantin, M Fogagnolo, ME Guerzoni, A Medici, P Pedrini, S Poli. J Org Chem 59:924–925, 1994.
90. T Izumi, K Fukaya. Bull Chem Soc Jpn 66:1216–1221, 1993.
91. T Fujisawa, T Sugimoto, M Shimizu. Tetrahedron: Asymmetry 5:1095–1098, 1994.
92. T Fujisawa, K Ichikawa, M Shimizu. Tetrahedron: Asymmetry 4:1237–1240, 1993.
93. G Fronza, C Fuganti, P Grasselli. Tetrahedron: Asymmetry 4:1909–1916, 1993.
94. G Fronza, C Fuganti, P. Grasselli, A Mele. J Org Chem 56:6019–6023, 1991.
95. PJ Davies, S-K Yang, RV Smith. Appl Environm Microbiol 48:327–331, 1984.
96. G Pedrocchi-Fantoni, S Servi. J Chem Soc, Perkin Trans 1 1991; 1764–1765.
97. P Ferraboschi, P Grisenti, A Manzocchi, E Santaniello. Tetrahedron 50:10539–10548, 1994.
98. M De Carvalho, MT Okamoto, PJ Samenho Moran, JAR Rodrigues. Tetrahedron 47:2073–2080, 1991.
99. A Manzocchi, A Fiecchi, E Santaniello. Synthesis 1987; 1007–1009.
100. A Belan, J Bolte, A Fauve, JG Gourcy, H Veschambre. J Org Chem 52:256–260, 1987.
101. Y Naoshima, A Nakamura, Y Munakata, M Kamezawa, H Tachibana. Bull Chem Soc Jpn 63:1263–1265, 1990.
102. M Kodama, H Minami, Y Mima, Y Fukuyama. Tetrahedron Lett 31:4025–4026, 1990.
103. S Ramaswamy, AC Oehlschlager. Tetrahedron 47:1145–1156, 1991.
104. T Sakai, K Wada, T Murakami, K Kohra, N Imajo, Y Ooga, S Tsuboi, A Takeda, M Utaka. Bull Chem Soc Jpn 65:631–638, 1992.

105. G Fronza, C Fuganti, P Grasselli, R Pulido-Fernandez, S Servi, A Tagliani, M Terreni. Tetrahedron 47:9247–9252, 1991.
106. C Le Drian, AE Greene. J Am Chem Soc 104:5473–5483, 1982.
107. S Ramaswamy, AC Oehlschlager. J Org Chem 54:255–257, 1989.
108. N Ibrahim, T Eggimann, EA Dixon, H Wieser. Tetrahedron 46:1503–1514, 1990.
109. AV Tkachev, AV Rukavishnikov, YV Gatilov, IY Bagryanskaya. Tetrahedron: Asymmetry 3: 1165–1187, 1992.
110. P Besse, H Veschambre, R Chenevert, M Dickman. Tetrahedron: Asymmetry 5:1727–1744, 1994.
111. P Besse, H Veschambre, M Dickman, R Chenevert. J Org Chem 59:8288–8291, 1994.
112. M Bucciarelli, A Forni, I Moretti, G Torre. J Chem Soc Chem Commun 1978; 456–457.
113. T Kitazume, N Ishikawa. Chem Lett 1983; 237–238.
114. T Kitazume, T Yamazaki, T Ishikawa. Nippon Kagaku Kaishi 1983; 1363–1368.
115. T Kitazume, T Sato. J Fluorine Chem 30:189–202, 1985.
116. T Kitazume, N Ishikawa. J Fluorine Chem 29:431–444, 1985.
117. Y Yamazaki, H Kobayashi. Tetrahedron: Asymmetry 4:1287–1294, 1993.
118. HR Gutmann, BJ Jandorf, O Bodansky. J Biol Chem 169:145–149, 1947.
119. E Stier. Z Physiol Chem 281:181–188, 1935.
120. HR Gutmann, BJ Jandorf, O Bodansky. Fed Proc 6:257–261, 1947.
121. PJG Mann, JH Quastel. Nature 158:154, 1946.
122. L Boutroux. Compt Rend Acad Sci 91:236, 1880.
123. E Minarik. Mitteilungen Rebe Wein Obst Früchteverwert 22:245–252, 1972.
124. VI Sokolov, LL Troitskaya, OA Reutov. Dokl Akad Nauk SSSR 237:1376, 1977.
125. A Ratajczak, B Misterkiewicz. Chem Ind 1976; 902.
126. VI Sokolov, LL Troitskaya, T Rozhkova. Gazz Chim Ital 117:525–527, 1987.
127. J Gillois, D Buisson, R Azerad, G Jaouen. J Chem Soc Chem Commun 1988; 1224–1225.
128. S Top, G Jaouen, J Gillois, C Baldoli, S Maiorana. J Chem Soc Chem Commun 1988; 1284–1285.
129. JAS Howell, MG Palin, G Jaouen, S Top, H El Hafa, JM Cense. Tetrahedron: Asymmetry 4:1241–1252, 1993.
130. JAS Howell, PJ O'Leary, MG Palin, G Jaouen, S Top. Tetrahedron: Asymmetry 7:307–315, 1996.
131. C Neuberg, FF Nord. Ber Dtsch Chem Ges 47:2264–2269, 1914.
132. C Neuberg, FF Nord. Biochem Z 67:46–52, 1914.
133. FF Nord. Ber Dtsch Chem Ges 52:1207–1211, 1919.
134. C Neuberg, E Schwenk. Biochem Z 71:118–125, 1915.
135. JK Nielsen, JO Madsen. Tetrahedron: Asymmetry 5:403–410, 1994.
136. RL Crumbie, BS Deol, JE Nemorin, DD Ridley. Aust J Chem 31:1965–1980, 1978.
137. T Fujisawa, E Kojima, T Itoh, T Sato. Chem Lett 1985; 1751–1754.
138. CQ Han, D DiTullio, YF Wang, CJ Sih. J Org Chem 51:1253–1258, 1986.
139. G Guanti, L Banfi, E Narisano. Tetrahedron Lett 27:3547–3550, 1986.
140. Y Takaishi, YL Yang, D DiTullio, CJ Sih. Tetrahedron Lett 23:5489–5492, 1982.
141. Y Noda, M Kikuchi. Chem Lett 1989; 1755–1756.
142. CAM Afonso, MT Barros, LS Godinho, CD Maycock. Synthesis 1991; 575–580.
143. G Guanti, L Banfi, A Guaragna, E Narisano. J Chem Soc Chem Commun 1986; 138–140.
144. T Yamazaki, M Asai, T Ohnogi, JT Lin, T Kitazume. J Fluorine Chem 35:537–553, 1987.
145. S Iruchijma, N Kojima. Agric Biol Chem 42:451–455, 1978.
146. SR Indahl, RR Scheline. Appl Microbiol 16:667–673, 1968.
147. T Fujisawa, T Itoh, M Nakai, T Sato. Tetrahedron Lett 1985; 771–774.
148. AP Kozikowski, BB Mugrage, CS Li, L Felder. Tetrahedron Lett 27:4817–4820, 1986.
149. R Tanikaga, K Hosoya, A Kaji. J Chem Soc Perkin Trans 1 1987; 1799–1803.
150. R Bernardi, P Bravo, R Cardillo, D Ghiringhelli, G Resnati. J Chem Soc Chem Commun 1988; 2831–2834.

151. T Sato, Y Okumura, J Itai, T Fujisawa. Chem Lett 1988; 1537–1540.
152. R Tanikaga, K. Hosoya, A Kaji. Chem Lett 1987; 829–832.
153. A Gopalan, R Lucero, H Jacobs, K Berryman. Synth Commun 21:1321–1329, 1991.
154. C Ticozzi, A Zanarotti. Tetrahedron Lett 29:6167–6170, 1988.
155. C Ticozzi, A Zanarotti. Liebigs Ann Chem 1989;1257–1259.
156. G Bianchi, G Comi, I Venturini. Gazz Chim Ital 114:285–286, 1984.
157. R Kuhn, D Jerschel. Ber Dtsch Chem Ges 77:591–598, 1944.
158. HJ Cotrell. Nature 159:748–749, 1947.
159. M Takeshita, S Yoshida, T Sato, N Akutsu. Heterocycles 35:879–884, 1993.
160. D Bailey, D O'Hagan, U Dyer, RB Lamont. Tetrahedron: Asymmetry 4:1255–1258, 1993.
161. W Baik, DI Kim, HJ Lee, W-J Chung, BH Kim, SW Lee. Tetrahedron Lett 38:4579–4580, 1997.
162. JA Blackie, N Turner, A Wells. Tetrahedron Lett 38:3043–3046, 1997.
163. K Nakamura, Y Inoue, J Shibahara, S Oka, A Ohno. Tetrahedron Lett 29:4769–4770, 1988.
164. C Forzato, P Nitti, G Pitacco, E Valentin. Tetrahedron: Asymmetry 8:1811–1820,1997.
165. A Navarro-Ocana, M Jimenez-Estrada, MB Gonzalez-Paredes, E Barzana. Synlett 1996; 695–696.
166. A Guarna, EG Occhiato, LM Spinetti, ME Vallecchi, D Scarpi. Tetrahedron 51:1775–1788, 1995.
167. EG Occhiato, A Guarna, F De Sarlo, D Scarpi. Tetrahedron: Asymmetry 6:2971–2976, 1995.
168. M Takeshita, N Akutsu. Tetrahedron: Asymmetry 3:1381–1384, 1992.
169. G Fouche, RM Horak, O Meth-Cohn. J Chem Soc Chem Commun 1993;119–120.
170. O Meth-Cohn, RM Horak, G Fouche. J Chem Soc Perkin Trans 1 1994; 1517–1527.
171. MR Horak, RA Learmonth, VJ Maharaj. Tetrahedron Lett 36:1541–1544, 1995.
172. R Chenevert, S Thiboutot. Chem Lett 1988; 1191–1192.
173. LM Kogan, VE Gulaya, IV Torgov. Tetrahedron Lett 1967; 4673–4676.
174. DW Brooks, PG Grothaus, WL Irwin. J Org Chem 47:2820–2821, 1982.
175. K Nakamura, K Inoue, K Ushio, S Oka, A Ohno. Chem Lett 1987; 679–682.
176. RP Lanzilotta, DG Bradley, CC Beard. Appl Microbiol 29:427–429, 1975.
177. RR Rando. Biochem Pharmacol 23:2328–2330, 1974.
178. DW Brooks, PG Grothaus, H Mazdiyasni. J Am Chem Soc 105:4472–4473, 1983.
179. Z Weishan, Z Zhiping, W Zhongqi. Scientia Sinica Series B 27:1217–1225, 1984.
180. Z-P Zhuang, W-S Zhou. Tetrahedron 41:3633–3641, 1985.
181. W-M Dai, W-S Thou. Tetrahedron 41:4475–4482, 1985.
182. DW Brooks, H Mazdiyasni, S Chakrabarti. Tetrahedron Lett 25:1241–1244, 1984.
183. DW Brooks, PG Grothaus, JT Palmer. Tetrahedron Lett 23:4187–4190, 1982.
184. DW Brooks, H Mazdiyasni, PG Grothaus. J Org Chem 52:3223–3232, 1987.
185. VE Gulaya, SN Ananchenko, IV Torgov, KA Koshcheenko, GG Bychkova. Bioorg Khim 55: 768–775, 1979.
186. K Mori, H Mori, M Yanai. Tetrahedron 42:291–294, 1986.
187. M Yanai, T Sugai, K Mori. Agric Biol Chem 49:2373–2377, 1985.
188. Y Lu, G Barth, K Kleslich, PD Strong, WL Duax, C Djerassi. J Org Chem 48:4549–4554, 1983.
189. A Murai, N Tanimoto, N Sakamoto, T Masamune. J Am Chem Soc 110:1985–1986, 1988.
190. NF Taylor, FH White, R Eisenthal. Biochem Pharmacol 21:347–353, 1972.
191. I Kubo, Y-W Lee, M Pettei, F Pilkiewicz, K Nakanishi. J Chem Soc Chem Commun 1976; 1013.
192. CS Barnes, JW Loder. Aust J Chem 15:322–328, 1962.
193. Y Naya, M Kotake. Nippon Kagaku Zasshi 91:275–278, 1970.
194. Y Naya, M Kotake. Nippon Kagaku Zasshi 89:1113–1117, 1968.
195. Y Naya, M Kotake. Tetrahedron Lett 1968; 1645–1646.
196. K Mori, H Mori. 41 41:5487–5493, 1985.
197. Y-Y Zhu, DJ Burnell. Tetrahedron: Asymmetry 7:3295–3304, 1996.

198. DW Brooks, H Mazdiyasni, P Sallay. J Org Chem 50:3411–3414, 1985.
199. K Mori, Y Matsushima. Synthesis 1993; 406–410.
200. T Kitahara, M Miyake, M Kido, K Mori. Tetrahedron: Asymmetry 1:775–782, 1990.
201. H Watanabe, K Mori. J Chem Soc Perkin Trans 1 1991; 2919–2934.
202. T Inoue, K Hosomi, M Araki, K Nishide, M Node. Tetrahedron: Asymmetry 6:31–34, 1995.
203. J d Angelo, G Revial, R Azerad, D Buisson. J Org Chem 51:40–45, 1986.
204. HGW Leuenberger, W Boguth, E Widmer, R Zell. Helv Chim Acta 59:1832–1849, 1976.
205. N Lamb, SR Abrams. Can J Chem 68:1151–1162, 1990.
206. HGW Leuenberger. Biocatalysts Org Synth 22:99–118, 1985.
207. R Zell, E Widmer, T Lukac, HGW Leuenberger, P Schönholzer, EA Broger. Helv Chim Acta 64:2447–2462, 1981.
208. AP Marchand, D Xing, Y Wang, SG Bott. Tetrahedron: Asymmetry 6:2709–2714, 1995.
209. G Hoffmann, R Wiartalla. Tetrahedron Lett 23:3887–3888, 1982.
210. S Terashima, K Tamoto. Tetrahedron Lett 23:3715–3718, 1982.
211. S Inayama, N Shimizu, T Ohkura, H Akita, T Oishi, Y Iitaka. Chem Pharm Bull 34:2660–2663, 1986.
212. K Mori, S Takayama, S Yoshimura. Liebigs Ann Chem 1993; 91–95.
213. H Lüers, J Mengele. Biochem Z 179:238–241, 1926.
214. A Vercellone. Biochem Z 279;137–149, 1935.
215. C Neuberg, E Simon. Biochem Z 171:256–263, 1926.
216. C Neuberg, FF Nord. Ber 52:2248–2250, 1919.
217. T Ishibashi, K Okada, Y Yamamoto, T Sasahara. Rept Res Lab Kirin Brewery Co Ltd 15:1–6, 1972.
218. C Neuberg, FF Nord. Ber 52:2251–2256, 1919.
219. D Buisson, S El Baba, R Azerad. Tetrahedron Lett 27:4453–4454, 1986.
220. M Imuta, H Ziffer. J Org Chem 43:3319–3323, 1978.
221. C Neuberg, H Lustig, RN Cogan. Arch Biochem 19:163–169, 1948.
222. S Veibel, E Bach, C35 (1936). Kgl Danske Videnskab Selskab Math Fys Medd 13:18–22, 1936.
223. J Bolte, JG Gourcy, H Veschambre. Tetrahedron Lett 1986; 565–568.
224. H Ohta, K Ozaki, GI Tsuchihashi. Agric Biol Chem 50:2499–2502, 1986.
225. H Ohta, K Ozaki, G-I Tsuchihashi. Chem Lett 1987; 2225–2226.
226. R Chenevert, S Thiboutot. Can J Chem 64:1599–1601, 1986.
227. M Utaka, H Ito, T Mizumoto, S Tsuboi. Tetrahedron: Asymmetry 6:685–686, 1995.
228. F Bracher, T Litz. Arch Pharm 327:591–593, 1994.
229. JN Cui, R Teraoka, T Ema, T Sakai, M Utaka. Tetrahedron Lett 38:3021–3024, 1997.
230. A Fauve, H Veschambre. J Org Chem 53:5215–5219, 1988.
231. A Forni, I Moretti, F Prati, G Torre. Tetrahedron 50:11995–12000, 1994.
232. K Ishihara, Y Higashi, S Tsuboi, M Utaka. Chem Lett 1995; 253–254.
233. M Eh, M Kalesse. Synlett 1995; 837–838.
234. T Fujisawa, E Kojima, T Sato. Chem Lett 1987; 2227–2228.
235. T Fujisawa, E Kojima, T Itoh, T Sato. Tetrahedron Lett 1985; 6089–6092.
236. T Kitazume, Y Nakayama. J Org Chem 51:2795–2799, 1986.
237. BS Deol, DD Ridley, GW Simpson. Aust J Chem 29:2459–2467, 1976.
238. S Iriuchijima, M Ogawa. Synthesis 1982; 41–42.
239. JV Eys, NO Kaplan. J Amer Chem Soc 79:2782, 1957.
240. DHR Barton, BD Brown, DD Ridley, DA Widdowson, AJ Keys, CJ Leaver. J Chem Soc Perkin Trans 1 1975; 2069–2076.
241. H Suemune, Y Mizuhara, H Akita, T Oishi, K Sakai. Chem Pharm Bull 35:3112–3118, 1987.
242. E Rona. Biochem Z 67:137–142, 1914.
243. S Tsuboi, E Nishiyama, H Furutani, M Utaka, A Takeda. J Org Chem 52:1359–1362, 1987.
244. J Kearns, MM Kayser. Tetrahedron Lett 35:2845–2848, 1994.
245. F Blaser, PF Deschenaux, T Kallimopoulos, A Jacot-Guillarmod. Helv Chim Acta 74:787–790, 1991.

246. K Nakamura, S-i Kondo, N Nakajima, A Ohno. Tetrahedron 51:687–694, 1995.
247. K Nakamura, S Kondo, Y Kawai, A Ohno. Tetrahedron Lett 32:7075–7078, 1991.
248. K Nakamura, S Kondo, Y Kawai, A Ohno. Bull Chem Soc Jpn 66:2738–2743, 1993.
249. S Tsuboi, H Furutani, MH Ansari, T Sakai, M Utaka, A Takeda. J Org Chem 58:486–492, 1993.
250. RP Lanzilotta, DG Bradley, KM McDonald. Appl Microbiol 27:130–134, 1974.
251. S Shimizu, H Yamada, H Hata, T Morishita, S Akutsu, M Kawamura. Agric Biol Chem 51:289–290, 1987.
252. S Shimizu, H Hata, Y Yamada. Agric Biol Chem 48:2285–2291, 1984.
253. HL King Jr, DR Wilken. J Biol Chem 247:4096–4105, 1972.
254. R Kuhn, T Wieland. Ber Dtsch Chem Ges 75:121–123, 1942.
255. R Eck, H Simon. Tetrahedron: Asymnmetry 5:1419–1422, 1994.
256. K Nakamura, S Kondo, Y Kawai, A Ohno. Tetrahedron: Asymmetry 4:1253–1254, 1993.
257. H Akita, H Matsukura, T Oishi. Chem Pharm Bull 34:2656–2659, 1986.
258. H Akita, A Furuichi, H Koshiji, K Horikoshi, T Oishi. Chem Pharm Bull 31:4384–4390, 1983.
259. L Poppe, L Novak, P Kolonits, A Bata, C Szantay. Tetrahedron 44:1477–1487, 1988.
260. S Tsuboi, E Nishiyama, M Utaka, A Takeda. Tetrahedron Lett 27:1915–1916, 1986.
261. SH Wu, LQ Zhang, CS Chen, G Girdaukas, CJ Sih. Tetrahedron Lett 1985; 4323–4326.
262. T Arslan, SA Benner, J Org Chem 58:2260–2264, 1993.
263. G Fronza, C Fuganti, P Grasselli, L Malpezzi, A Mele. J Org Chem 59:3487–3489, 1994.
264. PG Baraldi, S Manfredini, GP Pollini, R Romagnoli, D Simoni, V Zanirato. Tetrahedron Lett 33:2871–2874, 1992.
265. G Fogliato, G Fronza, C Fuganti, P Grasselli, S Servi. J Org Chem 60:5693–5695, 1995.
266. M Sato, J Sakaki, Y Sugita, T Nakano, C Kaneko. Tetrahedron Lett 31:7463–7466, 1990.
267. D Zakarya, L Farhaoui, S Fkih-Tetouani. Tetrahedron Lett 35:4985–4988, 1994.
268. D Zakarya, L Farhaoui, S Fkih-Tetouani. J Phys Org Chem 9:672–676, 1996.
269. O Rotthaus, D Krueger, M Demuth, K Schaffner. Tetrahedron 53:935–938, 1997.
270. D Buisson, R Azerad. Tetrahedron Lett 27:2631–2634, 1986.
271. G Frater. Helv Chim Acta 63:1383–1390, 1980.
272. V Spiliotis, D Papahatjis, N Ragoussis. Tetrahedron Lett 31:1615–1616, 1990.
273. M Mehmandoust, D Buisson, R Azerad. Tetrahedron Lett 36:6461–6462, 1995.
274. T Kitahara, K Mori. Tetrahedron Lett 1985; 451–452.
275. T Tirilly, J Kloosterman, G Sipma, JGK van den Bosch. Phytochem 22:2082–2087, 1983.
276. S Tanaka, K Wada, M Katayama, S Marumo. Agric Biol Chem 48:3189–3192, 1984.
277. K Mori, M Ikunaka. Tetrahedron 43:45–58, 1987.
278. S Tanaka, K Wada, S Marumo, H Hattari. Tetrahedron Lett 25:5907–5910, 1984.
279. JC Gilbert, RD Selliah. J Org Chem 58:6255–6265, 1993.
280. T Kitahara, H Kurata, K Mori. Tetrahedron 44:4339–4349, 1988.
281. E Didier, B Loubinoux, GM Ramos Tombo, G Rihs. Tetrahedron 47:4941–4958, 1991.
282. A Wahhab, DF Tavares, A Rauk. Can J Chem 68:1559–1563, 1990.
283. K Mori, M Tsuji. Tetrahedron 44:2835–2842, 1988.
284. K Mori, M Tsuji. Tetrahedron 42:435–444, 1986.
285. ZF Xie, K Funakoshi, H Suemune, T Oishi, H Akita, K Sakai. Chem Pharm Bull 34:3058–3060, 1986.
286. RW Hoffmann, W Helbig, W Ladner. Tetrahedron Lett 23:3479–3482, 1982.
287. H Ohta, K Ozaki, J Konishi, GI Tsuchihashi. Agric Biol Chem 50:1261–1266, 1986.
288. RW Hoffmann, W Ladner, W Helbig. Liebigs Ann Chem 1984; 1170–1179.
289. J Cooper, PT Gallagher, DW Knight. J Chem Soc Perkin Trans 1 1993; 1313–1317.
290. DW Knight, N Lewis, AC Share, D Haigh. Tetrahedron: Asymmetry 4:625–628, 1993.
291. MP Sibi, JW Christensen. Tetrahedron Lett 31:5689–5692, 1990.
292. H Ohta. J Synth Org Chem 41:1018–1030, 1983.
293. HU Reissig. Nachr Chem Tech Lab 34:782–784, 1986.

294. C Fuganti, P. Grasselli, P. Casati, M Carmeno. Tetrahedron Lett 26:101–104, 1985.
295. K Nakamura, M Higaki, K Ushio, S Oka, A Ohno. Tetrahedron Lett 26:4213–4216, 1985.
296. AS Gopalan, HK Jacobs. Tetrahedron Lett 30:5705–5708, 1989.
297. K Ushio, K Inouye, K Nakamura, S Oka, A Ohno. Tetrahedron Lett 27:2657–2660, 1986.
298. A Manzocchi, R Casati, A Fiecchi, E Santaniello. J Chem Soc Perkin Trans 1 1986; 2653–2757.
299. WR Shieh, AS Gopalan, CJ Sih. J Am Chem Soc 107:2993–2994, 1985.
300. C-S Chen, B-N Zhou, G Girdaukas, W-R Shieh, F VanMiddleswort, AS Gopalan. Bioorg Chem 12:98–117, 1984.
301. BN Zhou, AS Gopalan, F VanMiddlesworth, WR Shieh, CJ Sih. J Am Chem Soc 105:5925–5926, 1983.
302. K Mori. Tetrahedron 37:1341–1342, 1981.
303. G Frater. Helv Chim Acta 62:2829–2832, 1979.
304. M Hirama, M Shimizu, M Iwashita. J Chem Soc Chem Commun 1983; 599–600.
305. J Ehrler, F Giovannini, B Lamatsch, D Seebach. Chimia 40:172–173, 1986.
306. B Seuring, D Seebach. Helv Chim Acta 60:1175–1181, 1977.
307. K Mori, K Tanida. Tetrahedron 37:3221–3225, 1981.
308. E Hungerbühler, D Seebach, D Wasmuth. Helv Chim Acta 64:1467–1487, 1981.
309. K Mori, T Sugai. Synthesis 1982; 752–753.
310. H Tsutsui, O Mitsunobu. Tetrahedron Lett 25:2159–2162, 1984.
311. T Chiba, M Nagatsuma, T Nakai. Chem Letters 1985; 1343–1346.
312. Al Meyers, RA Amos. J Am Chem Soc 102:870–872, 1980.
313. T Katsuki, M Yamaguchi. Tetrahedron Lett 28:651–654, 1987.
314. K Hintzer, B Koppenhoefer, V Schurig. J Org Chem 47:3850–3854, 1982.
315. RU Lemieux, J Giguere. Can J Chem 29:678–690, 1951.
316. LY Jayasinghe, D Kodituwakku, AJ Smallridge, MA Trewhella. Bull Chem Soc Jpn 67:2528–2531, 1994.
317. M Hamdani, B De Jeso, H Deleuze, A Saux, B Maillard. Tetrahedron: Asymmetry 4:1233–1236, 1993.
318. GHL Nefkens, RJF Nivard. Rec Trav Chim Pays Bas 83:199–214, 1964.
319. C Fuganti, P Grasselli, S Servi, F Spreafico, C Zirotti. J Org Chem 49:4087–4089, 1984.
320. B Wipf, E Kupfer, R Bertazzi, HGW Leuenberger. Helv Chim Acta 66:485–488, 1983.
321. Y Naoshima, J Maeda, Y Munakata. J Chem Soc Perkin Trans 1 1992; 659–660.
322. M North. Tetrahedron Lett 37:1699–1702, 1996.
323. M Utaka, H Higashi, A Takeda. J Chem Soc Chem Commun 1987; 1368–1369.
324. M Hirama, M Uei. J Am Chem Soc 104:4251–4253, 1982.
325. T Sato. Can J Chem 65:2732–2733, 1987.
326. D Seebach, M Züger. Helv Chim Acta 65:495–503, 1982.
327. R Müller, F Lingens. Angew Chem 98:778, 1986.
328. P DeShong, M-T Lin, JJ Perez. Tetrahedron Lett 27:2091–2094, 1986.
329. D Seebach, MF Züger, F Giovannini, B Sonnleitner, A Fiechter. Angew Chem 96:155–156, 1984.
330. Y Kawai, M Tsujimoto, S Kondo, K Takanobe, K Nakamura, A Ohno. Bull Chem Soc Jpn 67:524–528, 1994.
331. Y Kawai, S-i Kondo, M Tsujimoto, K Nakamura, A Ohno. Bull Chem Soc Jpn 67:2244–2247, 1994.
332. WR Shieh, CJ Sih. Tetrahedron: Asymmetry 4:1259–1269, 1993.
333. H Watabu, M Ohkubo, H Matsubara, T Sakai, S Tsuboi, M Utaka. Chem Lett 1989; 2183–2184.
334. M Hamdani, B De Jeso, H Deleuze, B Maillard. Res Chem Intermed 19:681–692, 1993.
335. K Kurumaya, K Takatori, R Isii, M Kajiwara. Heterocycles 30:745–748, 1990.
336. H Akita, R Todoroski, H Endo, Y Ikari, T Oishi. Synthesis 1993; 513–516.
337. AS Gopalan, HK Jacobs. J Chem Soc Perkin Trans 1 1990; 1897–1900.

338. K Nakamura, Y Kawai, A Ohno. Tetrahedron Lett 32:2927–2928, 1991.
339. M Hamdani, B De Jeso, H Deleuze, B Maillard. Tetrahedron: Asymmetry 4:1229–1232, 1993.
340. Y Kawai, K Hida, K Nakamura, A Ohno. Tetrahedron Lett 36:591–592, 1995.
341. M Christen, DHG Crout. J Chem Soc Chem Commun 1988; 264–266.
342. E Santaniello, R Casati, F Milani. J Chem Res (S) 1984; 132–133.
343. D Seebach, P Renaud, WB Schweizer, MF Züger, MJ Brienne. Helv Chim Acta 67:1843–1853, 1984.
344. D Seebach, M Eberle. Synthesis 1986; 37–40.
345. H Chikashita, K Ohkawa, K Itoh. Bull Chem Soc Jpn 62:3513–3517, 1989.
346. S Hashiguchi, A Kawada, H Natsugari. Synthesis 1992; 403–408.
347. R Hayakawa, M Shimizu, T Fujisawa. Tetrahedron Lett 37:7533–7536, 1996.
348. S Hashiguchi, A Kawada, H Natsugari. J Chem Soc Perkin Trans 1 1991; 2435–2444.
349. MH Ansari, T Kusumoto, T Hiyama. Tetrahedron Lett 34:8271–8276, 1993.
350. DW Brooks, RP Kellogg, CS Cooper. J Org Chem 52:192–196, 1987.
351. M Hirama, T Nakamine, S Ito. Chem Lett 1986; 1381–1384.
352. K Mori, H Mori, T Sugai. Tetrahedron 41:919–925, 1985.
353. A Furuichi, H Akita, H Matsukura, T Oishi, K Horikoshi. Agric Biol Chem 51:293, 1987.
354. M Hirama, T Nakamine, S Ito. Tetrahedron Lett 27:5281–5284, 1986.
355. T Hudlicky, T Tsunoda, KG Gadamasetti, JA Murry, GE Keck. J Org Chem 56:3619–3623, 1991.
356. T Fujisawa, T Itoh, T Sato. Tetrahedron Lett 25:5083–5086, 1984.
357. T Sato, M Tsurumaki, T Fujisawa. Chem Lett 1986; 1367–1370.
358. A Kumar, DH Ner, SY Dike. Tetrahedron Lett 32:1901–1904, 1991.
359. R Chenevert, G Fortier, R Bel Rhlid. Tetrahedron 48:6769–6776, 1992.
360. K Nakamura, Y Kawai, A Ohno. Tetrahedron Lett 31:267–270, 1990.
361. NW Fadnavis, SK Vadivel, UT Bhalerao. Tetrahedron: Asymmetry 8:2355–2359, 1997.
362. K Nakamura, T Miyai, K Nozaki, K Ushio, S Oka, A Ohno. Tetrahedron Lett 27:3155–3156, 1986.
363. H Akita, A Furuichi, H Koshiji, K Horikoshi, T Oishi. Chem Pharm Bull 31:4376–4383, 1983.
364. G Frater, U Müller, W Günther. Tetrahedron 40:1269–1277, 1984.
365. K Maurer. Biochem Z 189:216–223, 1927.
366. D Buisson, S Henrot, M Larcheveque, R Azerad. Tetrahedron Lett 28:5033–5036, 1987.
367. H Akita, H Koshiji, A Furuichi, K Horikoshi, T Oishi. Tetrahedron Lett 24:2009–2010, 1983.
368. H Akita, H Matsukura, T Oishi. Tetrahedron Lett 27:5397–5400, 1986.
369. RW Hoffmann, W Ladner, K Steinbach, W Massa, R Schmidt, G Snatzke. Chem Ber 114:2786–2801, 1981.
370. C Fuganti, P Grasselli, PF Seneci, P Casati. Tetrahedron Lett 27:5275–5276, 1986.
371. T Hudlicky, G Gillman, C Andersen. Tetrahedron: Asymmetry 3:281–286, 1992.
372. K Ogura, T Iihama, S Kiuchi, T Kajiki, O Koshikawa, K Takahashi, H Iida. J Org Chem 51:700–705, 1986.
373. A Pondaven-Raphalen, G Sturtz. Bull Soc Chim Fr 1978; 215–229.
374. E Zymanczyk-Duda, B Lejczak, P Kafarski, J Grimaud, P Fischer. Tetrahedron 51:11809–11814, 1995.
375. T Itoh, T Fukuda, T Fujisawa. Bull Chem Soc Jpn 62:3851–3855, 1989.
376. M Utaka, H Watabu, A Takeda. J Org Chem 52:4363–4368, 1987.
377. M Gessner, C Günther, A Mosandl. Z Naturforsch 42C:1159–1164, 1987.
378. Y Naoshima, H Ozawa, H Kondo, S Hayashi. Agric Biol Chem 47:1431–1434, 1983.
379. G Tynenburg Muys, Bv der Ven, AP de Jonge. Nature 194:995–996, 1962.
380. JP Vigneron, R Meric, M Dhaenens. Tetrahedron Lett 21:2057–2060, 1980.
381. M Utaka, H Watabu, A Takeda. Chem Lett 1985; 1475–1476.
382. K Mori, T Otsuka. Tetrahedron 41:547–551, 1985.
383. S Tsuboi, J Sakamoto, T Sakai, M Utaka. Synlett 1991; 867–868.

384. S Tsuboi, J Sakamoto, T Kawano, M Utaka, A Takeda. J Org Chem 56:7177–7179, 1991.
385. M Aquino, S Cardani, G Fronza, C Fuganti, R Pulido Fernandez, A Tagliani. Tetrahedron 47:7887–7896, 1991.
386. P Mazur, K Nakanishi. J Org Chem 57:1047–1051, 1992.
387. B Sarmah, NC Barua. Tetrahedron 49:2253–2260, 1993.
388. T Fujisawa, BI Mobele, M Shimizu. Tetrahedron Lett 32:7055–7058, 1991.
389. C Neuberg, J Hirsch. Biochem Z 115:282–287, 1921.
390. C Neuberg, E Reinfurth. Biochem Z 143:553–558, 1923.
391. W Dirscherl. Z Physiol Chem 47:47–57, 1931.
392. C Fuganti, P Grasselli, F Spreafico, C Zirotti. J Org Chem 49:543–546, 1984.
393. C Fuganti, P Grasselli. Chem Ind 1977; 983.
394. C Neuberg, L Liebermann. Biochem Z 120:311–337, 1921.
395. M Behrens, NN Iwanoff. Biochem Z 196:478–484, 1926.
396. PF Smith, D Hendlin. J Bacteriol 65:440–445, 1953.
397. NH Gross, CH Werkman. Arch Biochem 15:125–131, 1947.
398. J Hirsch. Biochem Z 131:178–185, 1922.
399. H-P Schmauder, D Gröger. Pharmazie 23:320–331, 1968.
400. P Mayer. Biochem Z 62:459–467, 1914.
401. O Hanc, B Kakac. Naturwissenschaften 43:498–500, 1956.
402. C Fuganti, P Grasselli, G Marinoni. Tetrahedron Lett 13:1161–1164, 1979.
403. G Bertolli, G Fronza, C Fuganti, P Grasselli, L Mojori, F Spreafico. Tetrahedron Lett 22: 965–968, 1981.
404. C Fuganti, D Ghiringhelli, P Grasselli. J Chem Soc Chem Commun 1975; 846–847.
405. G Fronza, C Fuganti, P Grasselli, G Poli, aS Servi. J Org Chem 53:6154–6156, 1988.
406. C Fuganti, P Grasselli. J Chem Soc Chem Commun 1982; 205–206.
407. C Fuganti, P Grasselli. J Chem Soc Chem Commun 1979; 995–997.
408. C Fuganti, P Grasselli, S Servi, F Spreafico, C Zirotti, P Casati. J Chem Res (S) 1984; 112–113.
409. C Fuganti, P Grasselli, F Spreafico, C Zirotti, P Casati. J Chem Res (S) 1985; 22–23.
410. G Fronza, C Fuganti, G Pedrocchi-Fantoni, S Servi. J Org Chem 52:1141–1144, 1987.
411. G Fronza, C Fuganti, P Grasselli, S Servi. Tetrahedron Lett 1985; 4961–4964.
412. G Fronza, C Fuganti, P Grasselli, S Servi. J Org Chem 52:2086–2089, 1987.
413. G Fronza, C Fuganti, P Grasselli, S Servi. Tetrahedron Lett 27:4363–4366, 1986.
414. P Gramatica, P Manitto, D Monti, G Speranza. Tetrahedron 44:1299–1304, 1988.
415. P Ferraboschi, P Grisenti, R Casati, A Fiecchi, E Santaniello. J Chem Soc Perkin Trans 1 1987; 1743–1748.
416. P Ferraboschi, A Fiecchi, P Grisenti, E Santaniello. J Chem Soc Perkin Trans 1 1987; 1749–1752.
417. HGW Leuenberger, W Boguth, R Barner, M Schmid, R Zell. Helv Chim Acta 62:455–463, 1979.
418. G Fronza, C Fuganti, M Mendozza, RS Rallo, GH Ottolina, D Joulain. Tetrahedron 52:4041–4052, 1996.
419. G Fogliato, G Fronza, C Fuganti, S Lanati, R Rallo, R Rigoni, S Servi. Tetrahedron 51: 10231–10240, 1995.
420. G Fronza, G Fogliato, C Ruganti, S Lanati, R Rallo, S Servi. Tetrahedron Lett 36:123–124, 1995.
421. G Fronza, C Fuganti, P Grasselli, S Lanati, R Rallo, S Tchilibon. J Chem Soc Perkin Trans 1 1994; 2927–2930.
422. Y Kawai, K Saitou, K Hida, DH Dao, A Ohno. Bull Chem Soc Jpn 69:2633–2638, 1996.
423. T Sakai, S Matsumoto, S Hidaka, N Imajo, S Tsuboi, M Utaka. Bull Chem Soc Jpn 64:3473–3475, 1991.
424. J Sakaki, M Suzuki, S Kobayashi, M Sato, C Kaneko. Chem Lett 1990; 901–904.
425. G Fronza, C Fuganti, G Pedrocchi-Fantoni, S Servi. Chem Lett 1989; 2141–2144.

426. UT Bhalerao, Y Chandraprakash, RL Babu, NW Fadnavis. Synth Commun 23:1201–1208, 1993.
427. P Gramatica, P Manitto, D Monti, G Speranza. Tetrahedron 43:4481–4486, 1987.
428. P Gramatica, P Manitto, BM Ranzi, A Delbianco, M Francavilla. Experientia 38:775–776, 1982.
429. C Neuberg, E Kerb. Biochem Z 92:111–123, 1918.
430. P Mayer, C Neuberg. Biochem Z 71:174–179, 1915.
431. L Poppe, L Novak, J Devenyi, C Szantay. Tetrahedron Lett 32:2643–2646, 1991.
432. P Gramatica, BM Ranzi, P Manitto. Bioorg Chem 10:22–28, 1981.
433. M Ito, R Masahara, K Tsukida. Tetrahedron Lett 18:2767–2770, 1977.
434. IS Liu, TH Lee, H Yokohama, KL Simpson, CO Chichester. Phytochemistry 12:2953–2956, 1977.
435. CJ Sih, RG Salomon, P Price, R Sood, G Peruzzotti. Tetrahedron Lett 1972; 2435–2437.
436. K Mori, DMS Wheeler, JO Jilek, B Kakac, M Protiva. Coll Czech Chem Commun 30:2236–2240, 1965.
437. H Pauly, K Feuerstein. Ber 60:1031–1034, 1927.
438. CT Hou, R Patel, AI Laskin, N Barnabe, I Marczak. Appl Environ Microbiol 38:135–142, 1979.
439. PH Buist, DM Marecak. J Am Chem Soc 113:5877–5878, 1991.
440. J Tang, I Brackenridge, SM Roberts, J Beecher, AJ Willetts. Tetrahedron 51:13217–13238, 1995.
441. J Beecher, I Brackenridge, SM Roberts, J Tang, AJ Willetts. J Chem Soc, Perkin Trans 1 1995; 1641–1643.
442. G Gil, E Ferre, M Barre, J Le Petit. Tetrahedron Lett 29:3797–3798, 1988.
443. T Kometani, Y Morita, H Furui, H Yoshii, R Matsuno. Chem Lett 1993; 2123–2124.
444. G Fantin, M Fogagnolo, A Medici, P Pedrini, S Poli. Tetrahedron Lett 34:883–884, 1993.
445. P Gramatica, BM Ranzi, P Manitto. Bioorg Chem 10:14–21, 1981.
446. C Fuganti, P Grasselli, G Poli, S Servi, A Zorzella. J Chem Soc Chem Commun 1988; 1619–1621.
447. EJ Corey, SC Virgil, S Sarshar. J Am Chem Soc 113:8171–8172, 1991.
448. T Kitazume, N Ishikawa. Chem Lett 1984; 1815–1818.
449. A Kamal, MV Rao, AB Rao. J Chem Soc Perkin Trans 1 1990; 2755–2757.
450. A Kamal, MV Rao, AB Rao. Heterocycles 31:577–579, 1990.
451. CJ Easton, CMM Hughes, ERT Tiekink, GP Savage, GW Simpson. Tetrahedron Lett 36:629–632, 1995.
452. KR Rao, N Bhanumathi, PB Sattur. Tetrahedron Lett 31:3201–3204, 1990.
453. RR Kakulapati, B Nanduri, VDN Yadavalli, NS Trichinapally. Chem Lett 1992; 2059–2060.
454. L Mamoli. Ber Dtsch Chem Ges 71:2696–2703, 1938.
455. JF Lenney. J Biol Chem 221:919–930, 1956.
456. T Achstetter, O Emter, C Ehmann, DH Wolf. J Biol Chem 259:13334–13343, 1984.
457. T Achstetter, DH Wolf. Yeast 1:139–157, 1985.
458. GE Wheeler, AH Rose. J Gen Microbiol 74:189–192, 1973.
459. S Taketani, T Osumi, H Katsuki. Biochim Biophys Acta 525:87–92, 1987.
460. T Nurminen, E Oura, H Suomalainen. Biochem J 116:61–69, 1970.
461. K Wöhrmann, P Lange. J Inst Brew 86:174–177, 1980.
462. S Taketani, T Nishino, H Katsuki. J Biochem 89:1667–1673, 1981.
463. E Parkkinen, H Suomalainen. J Inst Brew 88:98–101, 1982.
464. FH Schermers, JH Duffus, AM MacLeod. J Inst Brew 82:170–174, 1976.
465. E Parkkinen, H Suomalainen. J Inst Brew 88:34–38, 1982.
466. E Parkinnen. Cell Mol Biol 26:147–154, 1980.
467. E Parkkinen, E Oura, H Suomalainen. J Inst Brew 84:5–8, 1978.
468. T Nurminen, H Suomalainen. Biochem J 118:759–763, 1970.
469. W Witt, ME Schweingruber, A Mertsching. Biochim Biophys Acta 795:108–116, 1984.

470. W Witt, A Mertsching, E König. Biochim Biophys Acta 795:117–124, 1984.
471. I Schousboe. Biochim Biophys Acta 450:165–174, 1976.
472. JA Anderson. J Bactiol 27:69–83, 1934.
473. J Bours, DAA Mossel. Arch Lebensmittelhygiene 24:197–203, 1973.
474. I Schousboe. Biochem Biophys Acta 424:366–375, 1976.
475. EJF Demant. FEBS Lett 85:109–113, 1978.
476. G Schaffner, P Matile. Biochem Physiol Pflanz 174:811–821, 1979.
477. WJ Marsheck, M Miyano. Biochem Biophys Acta 316:363–368, 1973.
478. CJ Sih, P Price, R Sood, RG Salomon, G Peruzzotti, M Casey. J Am Chem Soc 94:3643–3645, 1972.
479. WP Schneider, HC Murray. J Org Chem 38:397–403, 1973.
480. CJ Sih, RG Salomon, P Price, R Sood, G Peruzzotti. J Am Chem Soc 97:857, 1975.
481. CJ Sih, RG Salomon, P Price, G Peruzzotti, R Sood. J Chem Soc Chem Commun 1972; 240–241.
482. BI Glänzer, K Faber, H Griengl. Tetrahedron 43:771–778, 1987.
483. BI Glänzer, K Faber, H Griengl. Tetrahedron Lett 1986; 4293–4296.
484. BI Glänzer, K Faber, H Griengl. Tetrahedron 43:5791–5796, 1987.
485. JE Lynch, RP Volante, RV Wattley, I Shinkai. Tetrahedron Lett 28:1385–1388, 1987.
486. BI Glänzer, K Faber, H Griengl, M Röhr, W Wöhrer. Enzyme Microb Technol 10:744–753, 1988.
487. BI Glänzer, K Faber, H Griengl. Enzyme Microb Technol 10:689–695, 1988.
488. R Csuk, BI Glänzer. Carbohydr Res 168:C9–C11, 1987.
489. R Csuk, BI Glänzer. Zt Naturforsch 43B:1355–1357.
490. T Tanaka, S Kurozumi, T Toru, S Miura, M Kobayashi, S Ishimoto. Tetrahedron 32:1713–1718, 1976.
491. H Suemune, M Tanaka, H Obaishi, K Sakai. Chem Pharm Bull 36:15–21, 1988.
492. Y Yamaguchi, T Oritani, N Tajima, A Komatsu, T Moroe. Nippon Nogei Kagakukaishi 50: 475–480, 1976.
493. Z Szurmai, L Janossy. Carbohydr Res 296:279–284, 1996.
494. H Suemune, Y Mizuhara, H Akita, K Sakai. Chem Pharm Bull 34:3440–3444, 1986.
495. PH Buist, HG Dallmann, RT Rymerson, PM Seigel. Tetrahedron Lett 28:857–860, 1987.
496. K Lohmann. Biochem Z 233:460–466, 1955.
497. HK Barrenscheen, W Filz. Biochem Z 250:281–284, 1952.
498. AL Dounce, A Rothstein, GT Beyer, R Meier, RM Freer. J Biol Chem 174:561–573, 1948.
499. P Ostern, T Baranowski, J Terszakowec. Hoppe-Seyler's Z Physiol Chem 151:258–284, 1938.
500. Y Kariya, Y Owada, A Kimura, T Tochikura. Agric Biol Chem 42:1689–1696, 1978.
501. P Biely, S Bauer. Coll Czech Chem Commun 32:1588–1594, 1967.
502. C Neuberg, J Leibowitz. Biochem Z 184:489–495, 1927.
503. A Vercellone, C Neuberg. Biochem Z 280:161–172, 1955.
504. C Neuberg, H Lustig. Arch Biochem 1:511–525, 1945.
505. O Meyerhof, W Kiessling. Biochem Z 276:259–264, 1955.
506. A Crueger, W Crueger. In: K Kieslich, ed. Biotransformations. Weinheim: Verlag Chemie, 1984, vol. 8, pp 421–457.
507. C Blum, N Rehnberg, B Spiess, G Schlewer. Carbohydr Res 302:163–168, 1997.
508. BJ Kallman, AK Andrews. Science 141:1050–1051, 1963.
509. M Utaka, S Konishi, A Takeda. Tetrahedron Lett 27:4737–4740, 1986.
510. M Takeshita, R Yaguchi, Y Unuma. Heterocycles 40:967–974, 1995.
511. A Kamal, MV Rao, AB Rao. Chem Lett 1990; 655–656.
512. RJ Linderman, A Ghannam, I Badejo. J Org Chem 56:5213–5216, 1991.
513. W Baik, TH Park, BH Kim, YM Jun. J Org Chem 60:5683–5685, 1995.
514. W Baik, JL Han, KC Lee, NH Lee, BH Kim, JT Hahn. Tetrahedron Lett 35:3965–3966, 1994.
515. CL Davey, LW Powell, NJ Turner, A Wells. Tetrahedron Lett 35:7867–7870, 1994.

516. M Takeshita, S Yoshida. Heterocycles 30:871–874, 1990.
517. W Baik, DI Kim, S Koo, JU Rhee, SH Shin, BH Kim. Tetrahedron Lett 38:845–848, 1997.
518. W Baik, JU Rhee, SH Lee, NH Lee, BH Kim, KS Kim. Tetrahedron Lett 36:2793–2794, 1995.
519. M Baruah, A Boruah, D Prajapati, JS Sandhu. Synlett 1996; 1193–1194.
520. A Kamal, Y Damayanthi, BSN Reddy, B Lakminarayana, BSP Reddy. Chem Commun 1997; 1015–1016.
521. M Takeshita, S Yoshida, Y Kohno. Heterocycles 37:553–562, 1994.
522. A Rykowski, T Lipinska, E Guzik, M Adamiuk, E Olender. Pol J Chem 71:69–76, 1997.
523. A Kamal, MV Rao, HM Meshram. Tetrahedron Lett 32:2657–2658, 1991.
524. NW Fadnavis, A Deshpande, S Chauhan, UT Bhalerao. J Chem Soc, Chem Commun 1990; 1548–1550.
525. P Ferraboschi, S Casati, E Santaniello. Tetrahedron: Asymmetry 5:19–20, 1994.
526. P Ferraboschi, SR Elahi, E Verza, FM Rivolta, E Santaniello. Synlett 1996; 1176–1178.
527. C Fuganti, G Pedrocchi-Fantoni, A Sarra, S Servi. Tetrahedron: Asymmetry 5:1135–1138, 1994.
528. K Takabe, H Hiyoshi, H Sawada, M Tanaka, A Miyazaki, T Yamada, T Katagiri, H Yoda. Tetrahedron: Asymmetry 3:1399–1400, 1992.
529. G Fronza, C Fuganti, P Grasselli, A Mele, A Sarra, G Allegrone, M Barbeni. Tetrahedron Lett 34:6467–6470, 1993.

20

Biocatalytic Synthesis of Steroids

Miguel Ferrero and Vicente Gotor
Universidad de Oviedo, Oviedo, Spain

I. INTRODUCTION

A number of extensive, general reviews [1], special issues of periodicals [2], and specialized books [3] have been published over the last two decades on the biocatalysis in organic synthesis. This is due to the explosive development and success of this methodology in the area of the synthesis of natural products, pharmaceuticals, and agrochemicals. These reviews and monographs do not usually focus their interest on the target molecule but rather on the structure of the substrate. Moreover, although there are several review articles dealing with natural products such as carbohydrates [4], not much has been reported on biotransformations in the field of steroids, especially from the enzymatic reactions point of view.

It is the aim of this chapter to cover all of the literature from the beginning of the 1980s to the middle of 1998 related to enzyme- and microbial-catalyzed reactions that lead to steroids and their analogs. The subject is treated in a very practical way and the presentation offers abundant tables, schemes, and structures. The work covered in each section is classified by date.

During the last few decades almost no other aspect of organic synthesis has received as much attention as the preparation of enantiomerically pure compounds. The synthesis of optically active materials is an important task and represents a challenge to both academic and industrial chemists alike. This increasing interest in understanding biological processes and the general recognition that chirality plays a crucial role in nature have fostered a tremendous effort in enantioselective synthesis. In addition, there is a strong emphasis on the development of single stereoisomers for the marketing of new drugs, pesticides, fragrances, flavors, and so on.

For organic chemists, enzymatic- or microbial-catalyzed reactions are becoming standard procedures to synthesize enantiomerically pure compounds due to their simple feasibility and high efficiency. Furthermore, in general these catalysts are inexpensive, and in many cases they are able to fit a wide range of substrate structures that are far removed from their natural substrates. Biocatalysts are ecologically beneficial natural catalysts. Due to these advantages, it is to be expected that biocatalyzed reactions will play an increasing role, mainly in the preparation of chiral biologically active compounds in the laboratory, as well as in industrial production.

The commercialization of isolated enzymes in either a crude or partially purified form has greatly developed the use of biocatalysts in synthetic organic preparations, to the extent that they can now be employed like any other shelf chemical catalyst. At present more than 3000 enzymes have been catalogued [5] and several hundred can be obtained commercially. By far the best asymmetrical synthesis is achieved in nature by enzymes, and for almost every type of chemical reaction there exists an enzyme-catalyzed equivalent reaction. A variety of reactions can be mediated, such as oxidations, reductions, hydrolyses, condensations, and isomerizations.

Biotransformations of interest in organic synthesis can be brought about under a variety of conditions. The most appealing to synthetic chemists are those that can be carried out with readily available, isolated enzymes requiring no added cofactors. Among this group, the microbially produced lipases are easily utilized in the hydrolytic direction in the presence of water, or in the synthetic direction in organic solvents.

Those enzymes systems that require cofactors, especially the oxidoreductases, can be handled by coupling enzymatic or nonenzymatic cofactor recycling. Organic chemists might prefer to use naturally coupled enzyme systems and tackle such transformations using whole-cell systems.

The fact that in some cases the extensive understanding of certain enzymatic reactions is still missing does not preclude their utilization, in a pragmatic and somewhat empirical way, once their usefulness has been established. However, in the future more insight into the active sites of biocatalysts, the structural requirements of the substrates, and the physicochemical properties of the reaction media will be known. As a result of this, the current, often high expenditure required for the screening of biocatalysts, suitable substrates, and media in order to find optimal conditions should be significantly shortened.

This chapter represents the desire of the authors to show a range of examples that cover steroid analog syntheses through enzymatic procedures and to summarized and offer an easily accessible, visual, reference review.

II. ENZYME- AND MICROBIAL-CATALYZED STEROID TRANSFORMATIONS

A. Oxidations and Reductions

1. Enzymatic Oxidations and Reductions

For use in synthesis, the oxidoreductases can be roughly divided into two groups, one of which accepts only substrates with limited structural differences. Among these enzymes are the hydroxy steroid dehydrogenases (HSDH), which catalyze the reversible oxidoreduction of the hydroxyl keto groups of steroids with high regio- and stereospecificity.

Carrea and co-workers [6] have systematically investigated the usefulness of HSDH for bile acid synthesis, with cholic acid **1** ($3\alpha,7\alpha,12\alpha$-trihydroxy-5β-cholan-24-oic acid) and dehydrocholic acid **5** (3,7,12-trioxo-5β-cholan-24-oic acid) as model substrates. The results are summarized in Table 1.

The enzymatic approach made it possible on a preparative scale to oxidize or reduce the hydroxyl or keto groups of cholic acid and dehydrocholic acid with very high regio- and stereospecificity, at each of the three possible positions in both compounds. The substrates were quantitatively transformed into the products **2–4** (from cholic acid) and **6–10** (from dehydrocholic acid) in a single step (78–91% isolated yields).

The nicotinamide cofactors [7] were enzymatically regenerated in situ with α-ketoglutarate/glutamate dehydrogenase (GIDH) (to regenerate NAD and NADP), formate/formate dehydrogenase (FDH) (to regenerate NADH), or glucose/glucose dehydrogenase

(GlucDH) (to regenerate NADH and NADPH) systems. The enzymes were employed in the free form or immobilized on Sepharose CL-4B. Full details are given in Table 1.

In the enzymatic process, the coupling of a second enzymatic reaction gave two advantages. First, the costly coenzymes were regenerated at the expense of much cheaper compounds (α-ketoglutarate, formate, glucose). Second, the transformation of the substrates was complete due to the high value of the overall equilibrium constant for the coupled reactions [8]. The enzymes employed for coenzyme regeneration (GlDH, FDH, and GlucDH) and hydroxy steroid dehydrogenases (3α-HSDH, 3β-HSDH, 7α-HSDH, 7β-HSDH, 12α-HSDH) are commercially available.

The potential of enzyme-catalyzed reactions in organic chemistry is now remarkable. This is due partly to the increased number of available enzymes, to their use in the immobilized form, and to the existence of efficient systems of coenzyme regeneration. However, substrates such as steroids, which are poorly soluble in water, present special difficulties in this area since the enzyme-catalyzed transformations must be carried out using large reaction volumes and, consequently, large amounts of biological catalysts and cofactors. Attempts to increase steroid solubility by adding water-miscible organic solvents to the reaction medium have given unsatisfactory results. In contrast, when organic solvents that are practically immiscible with water are used, the situation improves substantially. In this case, a two-phase system consisting of water and an organic solvent is established.

Applying this methodology, Carrea and Cremonesi [9] have shown the specific oxidation–reduction of hydroxyl keto groups of steroids (see Table 2), catalyzed by NAD(P)$^+$-dependent hydroxy steroid dehydrogenase. The reactions are carried out in two-phase systems and are coupled to the enzymatic regeneration of the coenzymes. Thus, entries 1 and 2 show the stereospecific reduction of progesterone **11** and cortisone **13**. The complete reduction of progesterone to the 20β-hydroxy derivative was obtained due to the high value of the overall equilibrium constant (about 10^{15}) for the coupled reactions. The reduction of cortisone **13** to pregn-4-ene-17α,20β,21-trihydroxy-3,11-dione **14** is catalyzed by 20β-HSDH and the regeneration of NADH by alcohol dehydrogenase. Semicarbazide, reacting with acetaldehyde, shifts the equilibrium to the complete conversion of cortisone **13** into the alcohol **14**. In entry 3, the complete oxidation of testosterone **15** to androstenedione **16** was obtained in about 25 h. Entry 4 shows that an almost complete oxidation of cholic acid methyl ester **17** to 12-ketochenodeoxycholic acid methyl ester **18** was achieved in 3 days.

The preparative scale regio- and stereospecific oxidoreduction of the hydroxyl keto groups of a variety of steroids (**11**, **17**, **19**, **21**, **23**, **25**, **26**, and **28**), catalyzed by several hydroxy steroid dehydrogenases (3α-HSDH, 3β,17β-HSDH, 12α-HSDH, 20β-HSDH), has been carried out in a water–organic solvent, two-phase system (phosphate buffer and ethyl acetate or butyl acetate) by Carrea and co-workers [10]. The single-step steroid transformations (oxidizing to **2**, **18**, **20**, **22**, and **24** or reducing **12**, **27**, and **29**) were specific and complete. The results are shown in Table 3. In the systems containing ethyl acetate (entries 1–5 and 7–9, Table 3), the transformation rates were different from those obtained in butyl acetate (entry 6) because of the different effects of the two solvents on such parameters as K_m, V_{max}, K_i, and the partition coefficient of steroids [11]. Various reactors (free enzymes in shaken vessels, immobilized enzymes in shaken vessels, and fixed-bed reactors) have been made and their performance compared. The suitability of two-phase systems of several enzymatic NAD(P)(H)-regenerating systems has also been studied [lactic dehydrogenase (LDH), alcohol dehydrogenase, and formate dehydrogenase (FDH)].

Table 1 Enzymatic Oxidation of Cholic Acid **1** and Reduction of Dehydrocholic Acid **5** by Hydroxy Steroid Dehydrogenases (HSDH)

Entry	Substrate	Product	Enzyme	Cofactors/regeneration system	Yield (%)
1	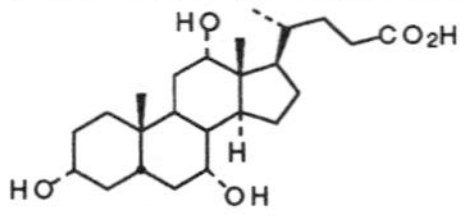 Cholic acid **1**	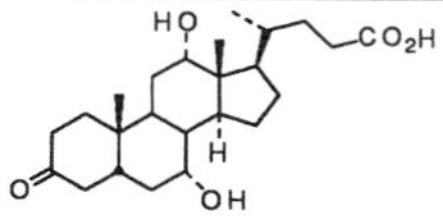 $7\alpha,12\alpha$-Dihydroxy-3-oxo-5β-cholan-24-oic acid **2**	3α-HSDH[a] 3α-HSDH[b]	NADH/NAD$^+$ glutamate dehydrogenase (α-ketoglutarate/ glutamate)	78
2		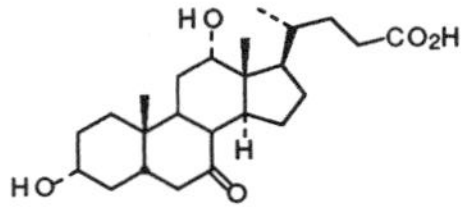 $3\alpha,12\alpha$-Dihydroxy-7-oxo-5β-cholan-24-oic acid **3**	7α-HSDH[a] 7α-HSD[b]	NADH/NAD$^+$ glutamate dehydrogenase (α-ketoglutarate/ glutamate)	82
3		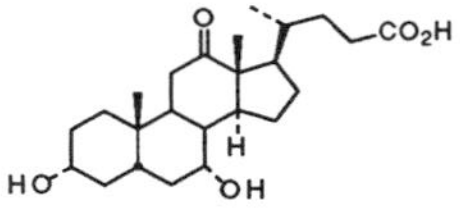 $3\alpha,7\alpha$-Dihydroxy-12-oxo-5β-cholan-24-oic acid **4**	12α-HSDH[a] 12α-HSDH[b]	NADPH/NADP$^+$ glutamate dehydrogenase (α-ketoglutarate/ glutamate)	85
4	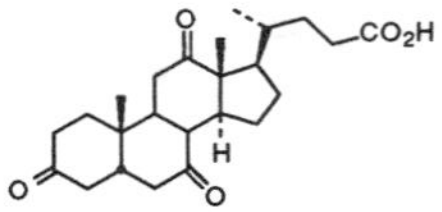 Dehydrocholic acid **5**	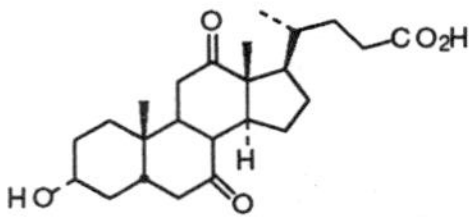 3α-Hydroxy-7,12-dioxo-5β-cholan-24-oic acid **6**	3α-HSDH	NAD$^+$/NADH formate dehydrogenase (HCO$_2$H/CO$_2$)	89

5	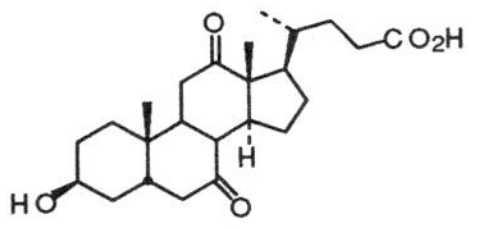 3β-Hydroxy-7,12-dioxo-5β-cholan-24-oic acid **7**	3β-HSDH	NAD$^+$/NADH formate dehydrogenase (HCO$_2$H/CO$_2$)	86
6	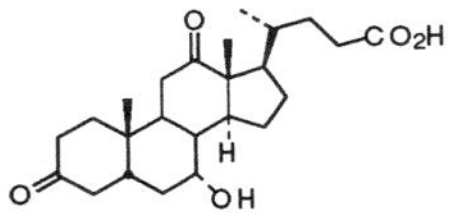 7α-Hydroxy-3,12-dioxo-5β-cholan-24-oic acid **8**	7α-HSDH	NAD$^+$/NADH glucose dehydrogenase (glucose/glucono-δ-lactone)	91
7	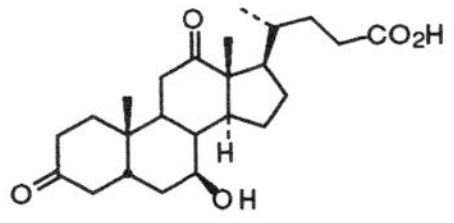 7β-Hydroxy-3,12-dioxo-5β-cholan-24-oic acid **9**	7β-HSDH	NADP$^+$/NADPH glucose dehydrogenase (glucose/glucono-δ-lactone)	86
8	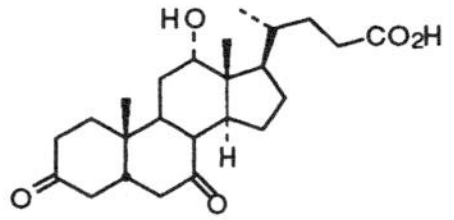 12α-Hydroxy-3,7-dioxo-5β-cholan-24-oic acid **10**	12α-HSDH	NADP$^+$/NADPH glucose dehydrogenase (glucose/glucono-δ-lactone)	87

[a]Free form.
[b]Immobilized on Sepharose CL-4B.

Table 2 Enzyme-Catalyzed Steroid Transformations in Water-Organic Solvent Two-Phase Systems by Hydroxy Steroid Dehydrogenases (HSDH)

Entry	Substrate	Product	Enzyme	Cofactors/regeneration system/turnover #[a]	Solvent	Conv. (%)/t (h)
1	Progesterone **11**	Pregn-4-en-20β-hydroxy-3-one **12**	20β-HSDH	NAD$^+$/NADH formate dehydrogenase (HCO$_2$H/CO$_2$) 400	0.1 M potassium phosphate buffer, pH 7/butyl acetate	100/30
2	Cortisone **13**	Pregn-4-ene-17α,20β,21-trihydroxy-3,11-dior **14**	20β-HSDH	NAD$^+$/NADH alcohol dehydrogenase (ethanol/acetaldehyde) (acetaldehyde + semicarbazide) 20	0.05 M potassium phosphate buffer, pH 7/ethyl acetate	95/3
3	Testosterone **15**	Androstenedione **16**	β-HSDH	NAD$^+$/NADH lactate dehydrogenase (pyruvate/lactate) 350	0.1 M potassium phosphate buffer, pH 8.5/butyl acetate	100/25
4	Cholic acid methyl ester **17**	12-ketochenodeoxycholic acid methyl ester **18**	12α-HSDH	NADP$^+$/NADPH glutamate dehydrogenase (α-keto-glutarate/glutamate) 470	0.1 M potassium phosphate buffer, pH 8/butyl acetate	100/72

[a]Turnover # = turnover number, definition in Ref. 7.

The conclusion was that fixed-bed reactors employing covalently immobilized enzymes are the most advantageous for repeated transformations because of increased enzyme stability, high coenzyme turnover number [12], and satisfactory mass transfer rates. These two-phase reactors demonstrated that cofactor-dependent enzymes can be used in flow systems with no need for macromolecular coenzymes and appear to be the most suitable for scaling up. Free enzymes, which are easier to handle, are preferable for occasional use and small-scale preparations.

The same authors reports [13] the α/β inversion of the C-3 hydroxyl of numerous bile acids with different lengths of the side chain (**1, 31, 34, 37, 40, 43, 46**, and **49**) [6,9,10]. Inversion was obtained in two steps through the sequential use of the commercial enzymes 3α- and 3β-HSDH, employed in the free form, or immobilized on Eupergit C. The effects of product inhibition on reaction rates, as well as the favorable effects produced by organic solvents dissolved in the aqueous buffer, were examined. The results are shown in Table 4.

The approach involved the preparation of the intermediate 3-keto derivatives (**2, 32, 35, 38, 41, 44, 47**, and **50**) through the regiospecific oxidation of the 3α-hydroxyls, catalyzed by 3α-HSDH. The pyruvate/lactic dehydrogenase system was used to regenerate NAD. The keto bile acids obtained were then stereospecifically reduced to the 3β-hydroxyl derivatives (**30, 33, 36, 39, 42, 45, 48**, and **51**) with 3β-HSDH. NADH was regenerated with formate/formate dehydrogenase system. The oxidation of 3α-hydroxyl compounds **1, 31, 34, 37, 40, 43, 46**, and **49** was practically quantitative, and the crude products were 98–99% pure. This confirmed the very high specificity of 3α-HSDH, which ignored the presence of others hydroxyls in the bile acids. Table 4 shows that although the shortening of the bile acid side chain did not affect activity (see **46** and the parent compound **40**), the lengthening of the aliphatic part of the side chain markedly slowed enzymatic activity (see **49** and the parent compound **1**). It is interesting that in two cases (**40** and **46**) the initial reaction rates were higher in the presence of ethyl acetate than in buffer alone.

The 3β-HSDH-catalyzed reductions of the 3-keto bile acids **2, 32, 35, 38, 41, 44, 47**, and **50** to the corresponding 3β-hydroxy derivatives were also obtained with quantitative yields and practically absolute stereospecificity. In this case, all of the various substrates except **32** (about 50% slower) were transformed by the enzyme with similar rates.

More recently, Carrea's group [14] showed a similar transformation (α/β inversion) at the C-7 hydroxyl position of different steroid derivatives. Steroids hydroxylated at C-7 are not so commonly found in nature and are mainly related to the bile acid family. Nevertheless, they do have important pharmaceutical applications due to their ability to dissolve cholesterol gallstones, thus avoiding surgery [15]. This property seems to be displayed to a greater extent by ursodeoxycholic acid **52**.

As Table 5 shows, the four free bile acids **1, 34, 54**, and **31** were quantitatively oxidized to the corresponding 7-keto derivatives **3, 53, 55**, and **56** by action of commercially available, NAD-dependent 7α-HSDH. The cofactor was regenerated in situ by a coupled enzymatic process (lactate/lactate dehydrogenase). The reductions catalyzed by the NADP-dependent 7β-HSDH (cofactor regenerated by the glucose/glucose dehydrogenase system) were also quantitative, affording the 7β-hydroxy derivatives **52, 37**, and **57** with more than 99% purities. However, compound **55** was completely unreactive.

All of the above-described transformations were performed in buffers, as the carboxyl moiety of **1, 34, 54**, and **31** allows the solubilization of these steroids under these particular reaction conditions (pH between 6.5 and 8). However, their solubilities were not particularly high. In order to overcome this limitation, their corresponding amino acid conjugates,

(text continues on pg. 593)

Table 3 Effects of Organic Solvents and Different Reactors on Enzymatic Oxidoreduction of Steroids in Two-Phase Systems

Entry	Substrate	Product	Enzyme	Cofactors/regeneration system	Solvent	Conv. (%)/t (h)/turnover #[a]
1	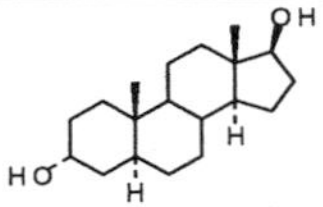 3α,17β-Dihydroxy-5α-androstane **19**	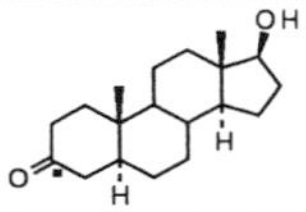 17β-Hydroxy-5α-androstan-3-one **20**	3α-HSDH	NADH/NAD⁺ lactate dehydrogenase (pyruvate/lactate)	0.1 M potassium phosphate buffer, pH 8.5/ethyl acetate	100/12/550[b] 100/22/150[c] 100/16/—[d]
2	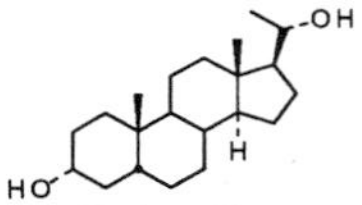 3α,20α-Dihydroxy-5β-pregnane **21**	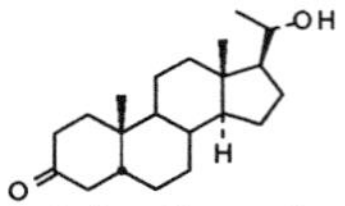 20α-Hydroxy-5β-pregn-3-one **22**	3α-HSDH	NADH/NAD⁺ lactate dehydrogenase (pyruvate/lactate)	0.1 M potassium phosphate buffer, pH 8.5/ethyl acetate	100/48/—[b]
3	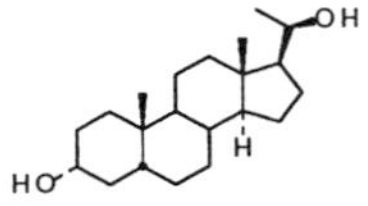 3α,20β-Dihydroxy-5β-pregnane **23**	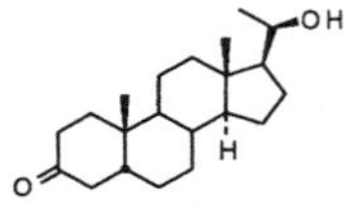 20β-Hydroxy-5β-pregn-3-one **24**	3α-HSDH	NADH/NAD⁺ lactate dehydrogenase (pyruvate/lactate)	0.1 M potassium phosphate buffer, pH 8.5/ethyl acetate	100/96/—[b]
4	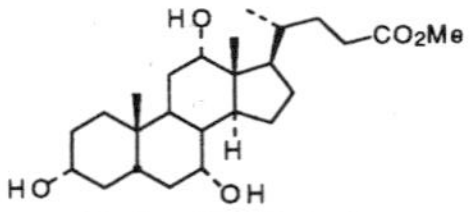 Cholic acid methyl ester **17**	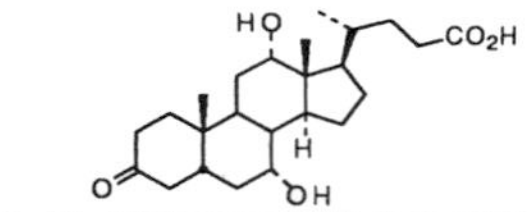 7α,12α-Dihydroxy-3-oxo-5β-cholan-24-oic acid methyl ester **2**	3α-HSDH	NADH/NAD⁺ lactate dehydrogenase (pyruvate/lactate)	0.1 M potassium phosphate buffer, pH 8.5/ethyl acetate	100/48/—[b]
5	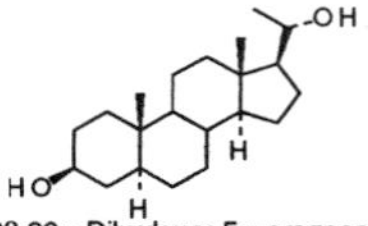 3β,20α-Dihydroxy-5α-pregnane **25**	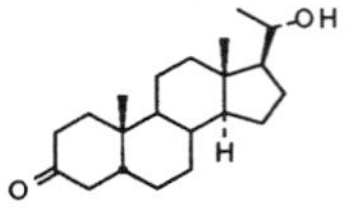 20α-Hydroxy-5β-pregn-3-one **22**	3β,17β-HSDH	NADH/NAD⁺ lactate dehydrogenase (pyruvate/lactate)	0.1 M potassium phosphate buffer, pH 8.5/ethyl acetate	100/48/—[b]

#	Substrate	Product	Enzyme	Cofactor/regeneration system	Conditions	Turnover #
6	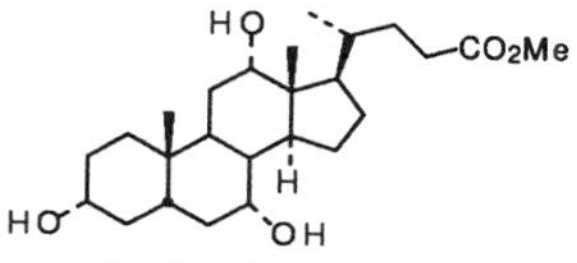 Cholic acid methyl ester 17 (CO$_2$Me)	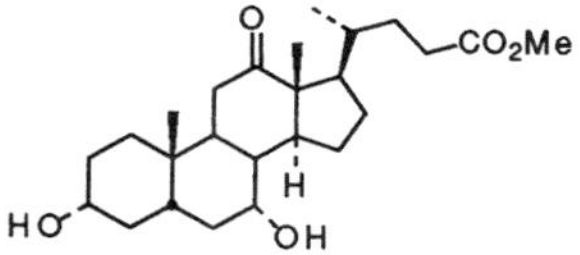 12-Ketochenodeoxycholic acid methyl ester 18 (CO$_2$Me)	12α-HSDH	NADPH/NADP$^+$ alcohol dehydrogenase *T. brockii* system (acetone/isopropanol)	0.05 M potassium phosphate buffer, pH 7.8/butyl acetate	100/72/860[b] 100/72/120[c]
7	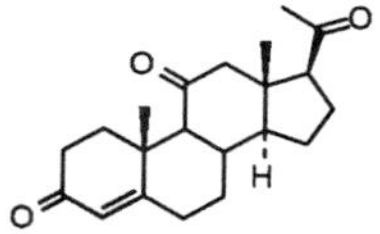 Progesterone 11	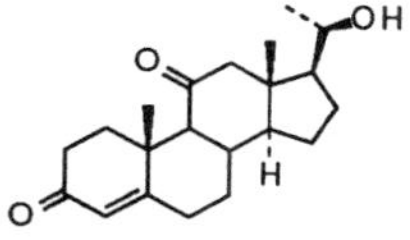 20β-Hydroxypregn-4-en-3-one 12	20β-HSDH	NAD$^+$/NADH formate dehydrogenase (HCO$_2$H/CO$_2$)	0.05 M potassium phosphate buffer, pH 7/ethyl acetate	60/96/—[b] 100/32/210[c]
8	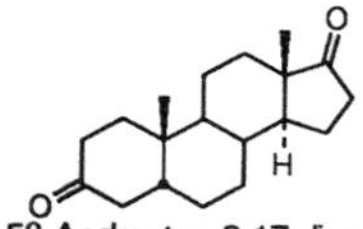 5β-Androstan-3,17-dione 26	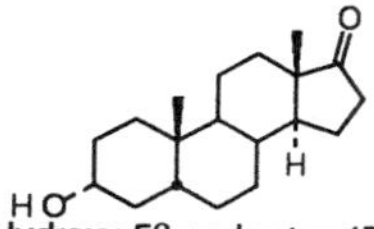 3α-Hydroxy-5β-androstan-17-one 27	3α-HSDH	NAD$^+$/NADH formate dehydrogenase (HCO$_2$H/CO$_2$)	0.05 M potassium phosphate buffer, pH 6.7/ethyl acetate	100/48/44[e]
9	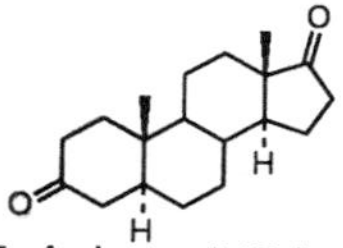 5α-Androstan-3,17-dione 28	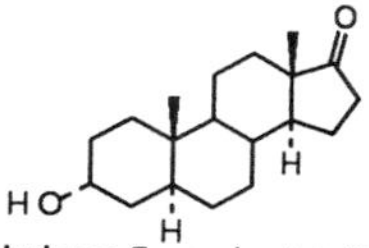 3α-Hydroxy-5α-androstan-17-one 29	3α-HSDH	NAD$^+$/NADH formate dehydrogenase (HCO$_2$H/CO$_2$)	0.05 M potassium phosphate buffer, pH 6.7/ethyl acetate	100/48/44[e]

[a]Turnover # = turnover number, definition in Ref. 7.
[b]Fixed bed reactor; Eupergit C.
[c]Shaken vessel; immobilized onto Eupergit C.
[d]Shaken vessel; immobilized onto Sepharose CL-4B.
[e]Shaken vessel; enzyme-free form.

Table 4 Enzymatic α/β Inversion of C-3 Hydroxyl of Bile Acids and the Effects of Organic Solvents on Reaction Rates[a]

Entry	Substrate	Oxidation product	Relative rates of oxidation		Reduction product
			Buffer	Buffer + 7% EtOAc	
1	Cholic acid **1**	α,12α-Dihydroxy-3-oxo-5β-cholan-24-oic acid **2**	80	70	3β,7α,12α-Trihydroxy-5β-cholan-24-oic acid **30**
2	Hyocholic acid **31**	**32**	100[b]	98	**33**
3	Chenodeoxycholic acid **34**	**35**	83	85	**36**
4	**37**	**38**	100	91	**39**

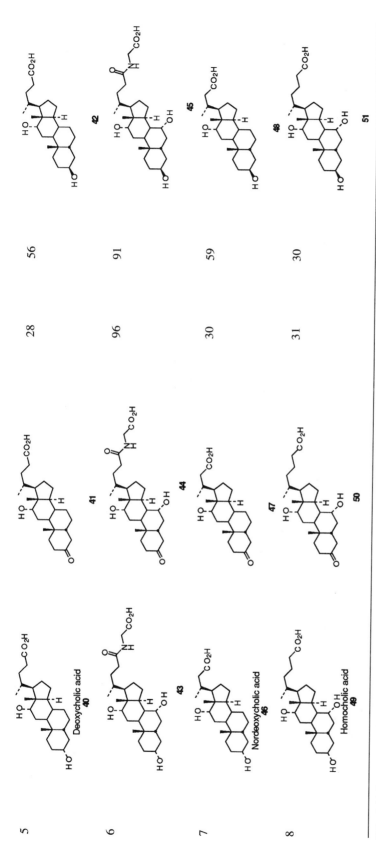

5	**40** Deoxycholic acid	**41**	28	56	**42**
6	**43**	**44**	96	91	**45**
7	**46** Nordeoxycholic acid	**47**	30	59	**48**
8	**49** Homocholic acid	**50**	31	30	**51**

[a]The initial rate determinations were carried out in 0.1 M potassium phosphate buffer, pH 8.5, 25°C, in the presence and absence of 7% (v/v) ethyl acetate.
[b]The oxidation rate of hyocholic acid **31** in buffer was taken as 100.

Table 5 Enzymatic α/β Inversion of C-7 Hydroxyl of Bile Acids

Entry	Substrate	Oxidation product	Oxid. enzymes	Reduction product	Red. enzymes
1	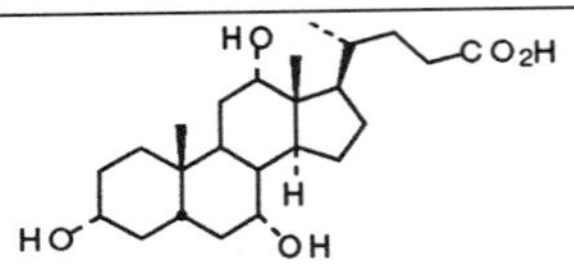 Cholic acid **1**	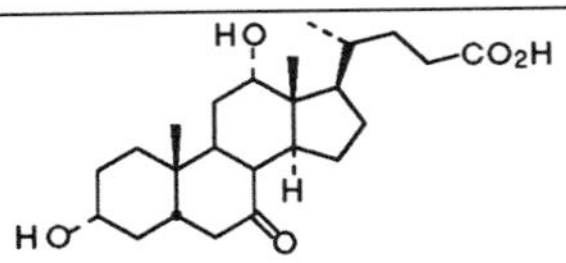 3α,12α-Dihydroxy-7-oxo-5β-cholan-24-oic acid **3**	7α-HSDH NADH/NAD+ lactate dehydrogenase (pyruvate/lactate)	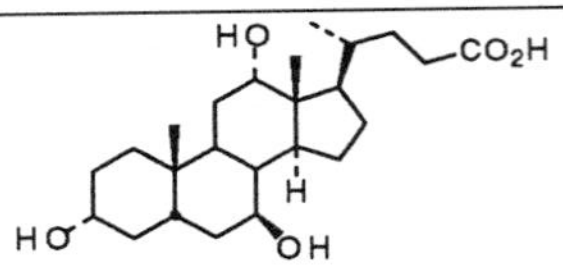 Ursodeoxycholic acid **52**	7β-HSDH NADP+/NADPH glucose dehydrogenase (glucose/glucono-δ-lacone)
2	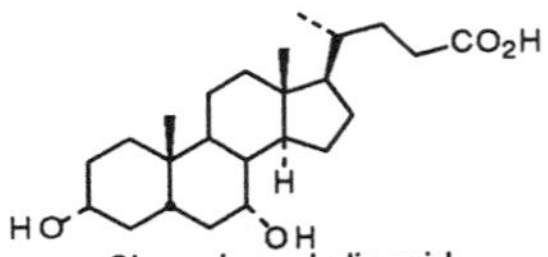 Chenodeoxycholic acid **34**	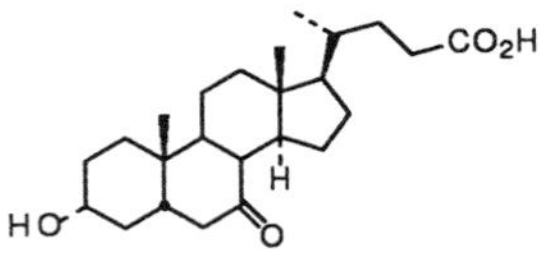 **53**	7α-HSDH NADH/NAD+ lactate dehydrogenase (pyruvate/lactate)	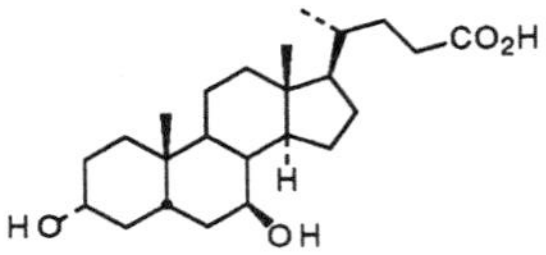 **37**	7β-HSDH NADP+/NADPH glucose dehydrogenase (glucose/glucono-δ-lacone)
3	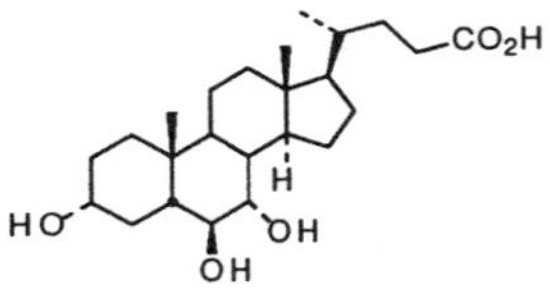 **54**	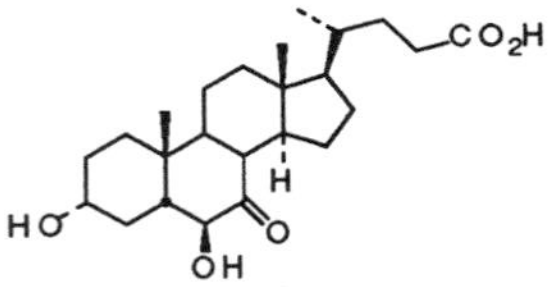 3-Ketoiocholic acid **55**	7α-HSDH NADH/NAD+ lactate dehydrogenase (pyruvate/lactate)		
4	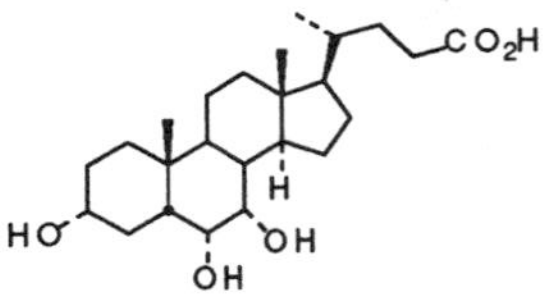 Hyocholic acid **31**	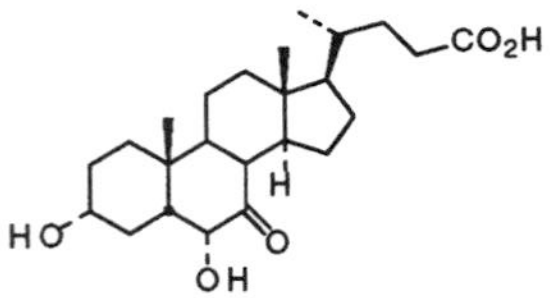 **56**	7α-HSDH NADH/NAD+ lactate dehydrogenase (pyruvate/lactate)	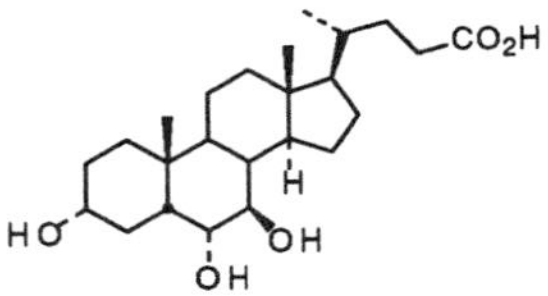 **57**	7β-HSDH NADP+/NADPH glucose dehydrogenase (glucose/glucono-δ-lacone)

590

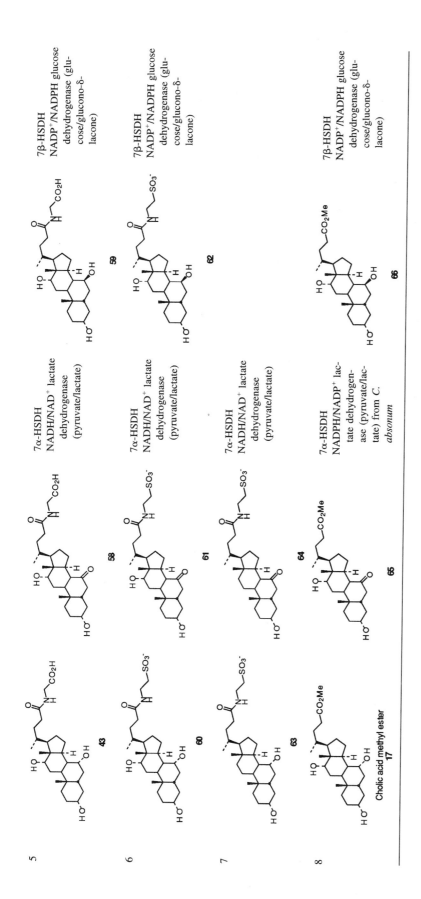

Table 6   Enzymatic Synthesis of Specifically Deuteriated Bile Acids by Hydroxysteroid Dehydrogenases (HSDH)

Entry	Substrate	Product	Enzyme	Cofactors/regeneration system[a]	Yield (%)/ t (h)
1	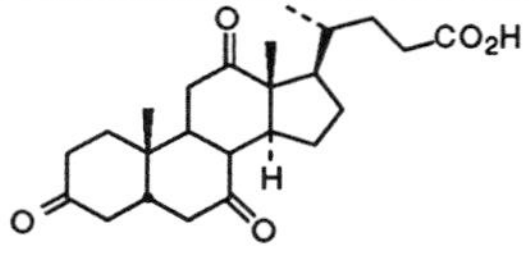 Dehydrocholic acid **5**	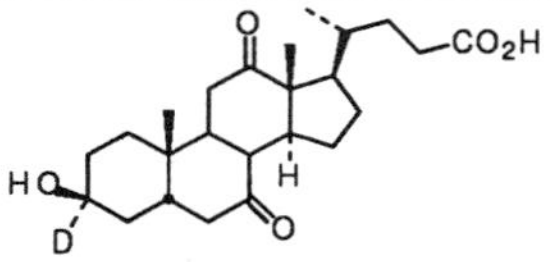 [3α-²H]-3β-Hydroxy-7,12-dioxo-5β-cholan-24-oic acid **67**	3β,17β-HSDH	NAD⁺/NADH glucose dehydrogenase ([1-²H]-glucose/glucono-δ-lactone)	86/8
2		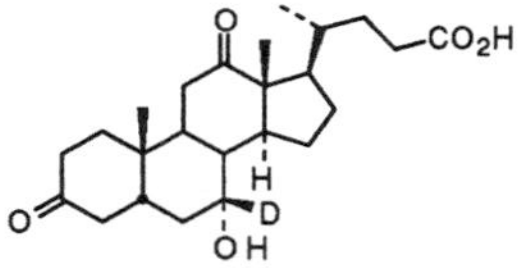 [7β-²H]-7α-Hydroxy-3,12-dioxo-5β-cholan-24-oic acid **68**	7α-HSDH	NAD⁺/NADH glucose dehydrogenase ([1-²H]-glucose/glucono-δ-lactone)	84/9
3		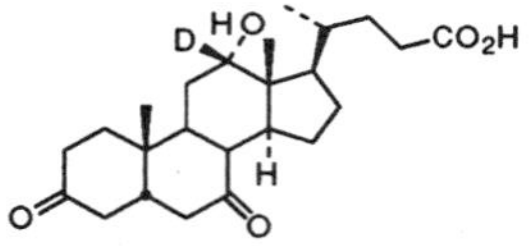 [12β-²H]-12α-Hydroxy-3,7-dioxo-5β-cholan-24-oic acid **69**	12α-HSDH	NADP⁺/NADPH glucose dehydrogenase ([1-²H]-glucose/glucono-δ-lactone)	88/6

[a]Turnover number (see Ref. 7) 150. Reaction conditions: 0.05 M potassium phosphate buffer pH 6.7/10% ethanol at room temperature.

specifically the glyco and tauro conjugates of cholic acid (**43** and **60**) and taurocheno-deoxycholic acid (**63**), were considered. With these compounds, too, the oxidations of **58**, **61**, and **64** catalyzed by 7α-HSDH were very efficient. The 7-keto cholic acid derivatives **58** and **61** were then reduced by 7β-HSDH. Quite unexpectedly, compound **64** was not a substrate for this enzyme under equal experimental conditions. Further experiments showed that both **64** and the expected product 7β-hydroxy become inhibitors of 7β-HSDH.

Finally, attention was turned to neutral steroids, and as a representative model, cholic acid methyl ester **17** was chosen. The initial attempt to oxidize **17** in a biphasic aqueous buffer–ethyl acetate system, by the action of the usual NAD-dependent 7α-HSDH, was unsuccessful. To overcome this problem, a new 7α-HSDH, namely, the NADP-dependent 7α-HSDH from *C. absonum* [16], was considered. Unfortunately, the subsequent reduction of **65** occurred with a low degree of conversion (less than 20%), probably due to enzyme inactivation. The stability of 7β-HSDH was not improved by immobilization on solid supports such as Eupergit C or activated Sepharose [6].

As an application of the previous-described oxidation–reduction procedures, Carrea and co-workers [17] developed an efficient preparative scale method of specifically deuterated bile acids. Deuterated compounds can be synthesized enzymatically on a small scale with procedures involving the stoichiometric consumption of labeled NAD(P)H [18]. These procedures, of course, are not economically convenient, especially for large-scale preparations. To overcome this limitation, a couple–enzymes system was developed in which the coenzyme is regenerated in situ by formate dehydrogenase and deuterated formic acid [19]. However, the system is only usable with NAD-dependent enzymes, since formate dehydrogenase does not accept NADP. Table 6 shows the results of a new coenzyme-regenerating system suitable for both NAD-dependent (3β,17β-HSDH, 7α-HSDH) and NADP-dependent (12α-HSDH) enzymes (glucose dehydrogenase works with both NAD and NADP). Deuterated bile acids **67**, **68**, and **69** were prepared in high yield (≥98%) and isotopic purity (≥94%) from derivative **5**.

2. Microbial Oxidations

Biohydroxylation reactions using microorganisms represent a powerful method for the introduction of hydroxyl groups (an intermediate step for other specific functionalizations) into previously elaborated molecules. Much of the early work has concentrated on the conversion of progesterone **11** into 11α-hydroxyprogesterone **70** by the molds *Rhizopus* sp. [20] or *Rhizopus nigricans* [21]. The importance of these biotransformations was realized and patented when 11α-hydroxyprogesterone was used as intermediate in the synthesis of corticosteroids. Thus, a synthetic substitute of cortisone, prednisone **71**, which is used to relieve rheumatoid arthritis pain, was synthesized using biosynthetically prepared compound **70** (Scheme 1).

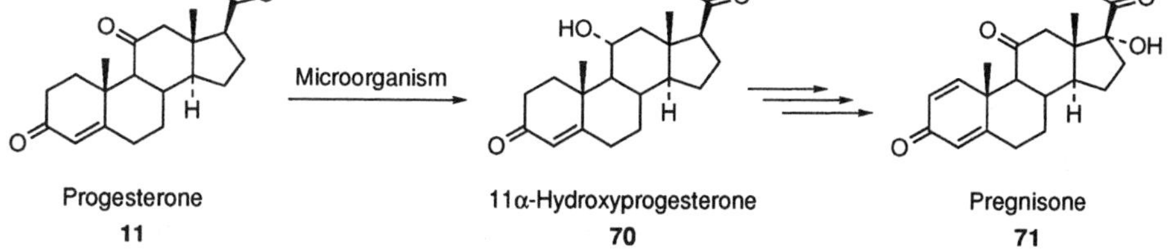

Progesterone **11** → 11α-Hydroxyprogesterone **70** → Prednisone **71**

Scheme 1

Table 7 Regioselective Microbial Oxidation of Bile Acids[a]

Microorganism	Bile acid	Products		
	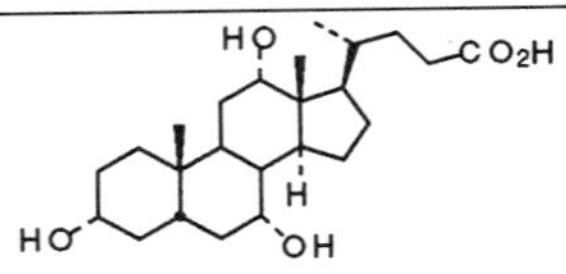 Cholic acid **1**	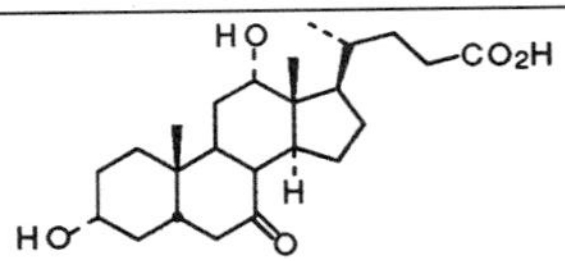 3α,12α-Dihydroxy-7-oxo-5β-cholan-24-oic acid **3**	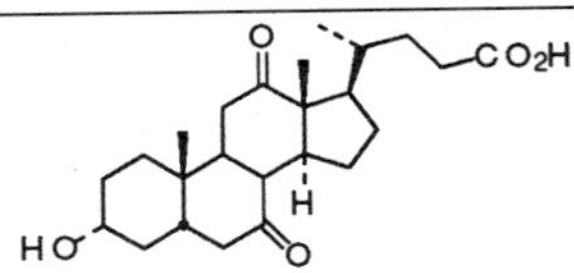 3α-Hydroxy-7,12-dioxo-5β-cholan-24-oic acid **6**	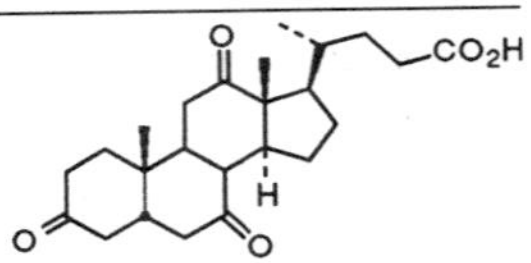 Dehydrocholic acid **5**
ICE BS6		87		
ICE BS7		75		
ICE BS11		52	30	
ICE B37		14	80	
ICE B49		27	30	
ICE B25				25
ICE B26				41
	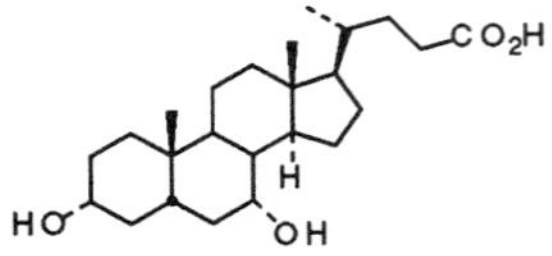 Chenodeoxycholic acid **34**	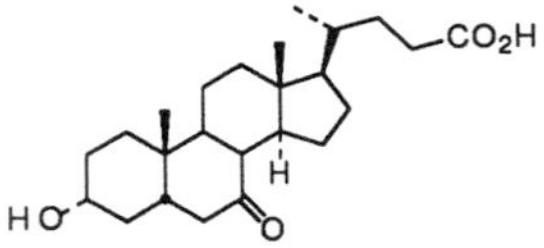 **53**	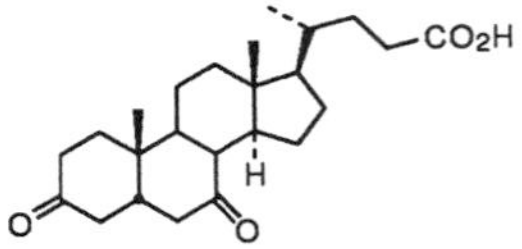 **72**	
ICE BS6		99		
ICE BS11		99		
ICE B37		97	56	
ICE B23		43	93	
ICE B18				

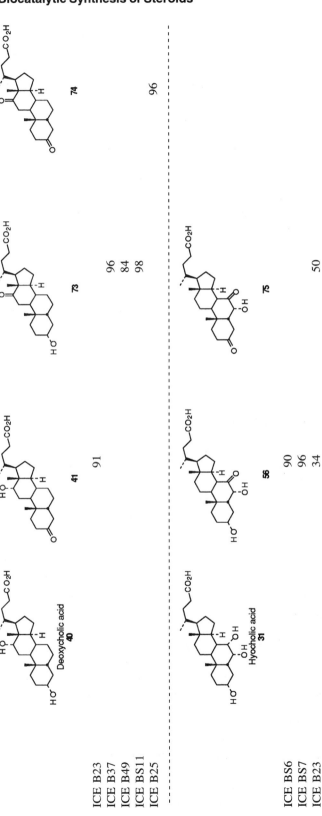

ICE B23
ICE B37
ICE B49
ICE BS11
ICE B25

ICE BS6
ICE BS7
ICE B23

[a]The microorganisms belong to ICE (see Ref. 18) private collection: BS6, *Xantomonas maltophilia*; BS7, B18, B25, and B26, *Pseudomonas fluorescens*; B23, *Bacillus mycoides*; BS11 and B37, *Acinetobacter calcoaceticus lwoffii*; B49, *Acinetobacter calcoaceticus acidovorans*. The values under the structures are isolated yields.

Medici and co-workers [22] isolated and classified several microorganisms from 50 environmental samples withdrawn from ICE industry [23], which extracts and purifies bile acids from raw materials (ox and pig bile). These microorganisms (mostly bacteria) have been screened in oxidation reactions of cholic **1**, chenodeoxycholic **34**, deoxycholic **40**, and hyocholic **31** acids. The most significant results are summarized in Table 7.

The bacteria BS6 and BS7 regioselectively oxidize the C-7 hydroxyl group with almost quantitative yield [formation of **3** (87%), **53** (99%), and **56** (90%) with BS6; and formation of **3** (75%), and **56** (96%) with BS7]. Bacterium B18 probably has the same velocity in the oxidation of the C-7 and C-3 hydroxyl groups, affording only the 3,7-dioxo derivative **72**, whereas B23, with the same regioselectivity, produces the 3-oxo (compound **41**) and 3,7-dioxo (compounds **53/72** and **56/75**) derivatives in about a 1:1 ratio. The strains BS11, B37, and B49 are the microorganisms of choice for the oxidation of the C-7 and C-12 hydroxyl groups, affording, in almost quantitative yields, the 7-oxo derivative **53** from chenodeoxycholic acid **34** (C-12 group absent), and the 12-oxo acid **73** from deoxycholic acid **40** (C-7 hydroxyl group absent). This feature is confirmed by the oxidation of the cholic acid **1**, where the 7-oxo **3** and 7,12-dioxo **6** derivatives are obtained in various ratios. Only the strains B25 and B26 show almost the same selectivity for all C-3, C-7, and C-12 hydroxyl groups, giving the completely oxidized products (compounds **5** and **74**) from cholic **1** and deoxycholic **40** acids.

B. Regioselective Protections and Deprotections

Njar and Caspi [24] reported a mild procedure in neutral reaction conditions, which minimize the detrimental side reactions that often plague the traditional methodology, for the conversion of steroid esters into the corresponding alcohols via a lipase-catalyzed transesterification reaction with octanol in organic solvents. Among all of the lipases screened [porcine pancreas lipase (PPL), acetylcholinesterase, pepsin, and *Candida cylindracea* lipase (CCL)], CCL was revealed as best suited for this transesterification process.

The results are summarized in Table 8. The following observations are noteworthy: (1) the 3β-esters **76**, **78**, **81**, **83**, **86**, **88**, and **93** (entries 1, 2, 4, 5, 7, 8, and 12) were converted to the corresponding alcohols **77**, **79**, **82**, **84**, **87**, **89**, and **94** in high yield, whereas the 3α isomers **80** and **85** (entries 3 and 6) were not affected. (2) The 17β-esters **83**, **88**, **90**, **91**, **93**, and **97** (entries 5, 8–10, 12, and 14) resisted cleavage. Usually considerably less than 5% of the 17β-alcohol could be detected. However, in the case of 17β-propionate **91** (entry 10) about 12% of the 17β-alcohol **15** was produced. (3) 3,17α-Diacetoxyestradiol **99** gave two products: 3,17α-dihydroxyestradiol **100** (60%) and 17α-acetoxy-3-hydroxyestradiol **101** (25%). It is apparent that the 17α-acetate is saponified but at a slower rate than the C-3 acetate. (4) 26-Hydroxycholesterol diacetate **78** (entry 2) gave the 3,26-dihydroxycholesterol **79**. (5) 21-Acetoxycorticosteroids **102**, **103**, and **105** (entries 16–18) were cleaved to their parent corticosteroids **13**, **104**, and **106**. (6) The 19-acetoxy moiety in compounds **92** and **93** (entries 11 and 12) resisted cleavage. (7) Allylic esters (3β-acetoxy-4-enes) **88** and **93** (entries 8 and 12) were cleaved with no undesirable side reactions.

Enzymatic acyl transfer reactions are also practical processes for the acylation of hydroxyl groups in steroids [25]. The lipase from *Chromobacterium viscosum* (CVL), for instance, selectively transfers butyric acid from trifluoroethyl butyrate to equatorial 3β-alcoholic functions being present in a variety of sterols, such as **107**, **110**, and **113** (Table 9). Axially oriented alcohols at C-3 and secondary alcohols at C-17 (**107** and **110**) or in the sterol side chains (**113**) are not derivatized. In addition to the equatorial alcohols, the compounds being accepted as substrates by the lipase must have the A/B ring fusion in the trans configuration. In the B ring a double bond is tolerated (as in **110**), but not in the A ring. In contrast, subtilisin does not recognize the hydroxyl group at C-3 of the steroid nucleus, transferring the acyl moiety to alcoholic groups in the 17-position or in the side chains (form compounds **109** and **112**). Changes in the A ring or the B ring do not dramatically influence the selective mode of action of this biocatalyst. Thus, using these two enzymes complete regioselective protection of either alcoholic group in several steroid diols is possible. This feature opened a route to a new chemoenzymatic process for the oxidation of selected positions of the steroid framework.

When bile acids serve as starting materials, e.g., deoxycholic acid methyl ester **115** (entry 1, Table 10) [26] the cis configuration of the A/B ring fusion prevents the application of CVL, and the aliphatic chain hinders the esterification of the 12α-hydroxyl group by subtilisin. The lipase from *Candida cylindracea* (CCL) has proven to be the most suitable biocatalyst for the enzymatic acylation of bile acids. In hydrophobic solvents, i.e., hexane, toluene, butyl ether, and benzene (except acetone), and by employing trichloroethyl butanoate as acyl donor, the 3α-O-butanoyldeoxycholic acid methyl ester **116** is formed in 96% yield without any byproducts, suggesting that the enzyme is ineffective toward the 12α-hydroxyl group. In addition, the 7α- and 7β-hydroxyl groups, which are present in **117** and **119**, are not esterified by the enzyme in either case, and the 3-butanoates **118** and **120** are formed (entries 2 and 3, Table 10). The same behavior is observed when 7α-hydroxyl and 12α-hydroxyl groups were present simultaneously in cholic acid methyl ester **17**, giving rise exclusively to the 3-butanoate derivative **121** (entry 4, Table 10).

Geometrical isomeric mixtures of steroidal oxime esters have been evaluated for the enzymatic transformations. In this respect, the work by Adamczyk and co-workers [27] describes the use of lipase from CCL in the diastereoselective hydrolysis of 3-(O-carboxymethyl)oxime methyl esters, 17α-hydroxyprogesterone **122**, progesterone **124**, testosterone **126**, and cortisol **128**. CCL proved to be effective in carrying out hydrolysis of methyl esters of steroidal 3-carboxymethyl oximes in a mild manner affording the E and Z diastereoisomers **123/122**, **125/124**, **127/126**, and **129/128** in different ratios. The enzyme exhibited a preference for the anti isomer (Table 11). The faster rate and greater selectivity observed for **128** is probably due to the cortisol derivative's better solubility in the reaction media.

Riva and co-workers [28] demonstrated the regioselectivity of *Candida antarctica* lipase B (CAL) toward the acetylation of polyhydroxylated steroids. The enzyme showed a marked preference for the alcoholic moieties on the A ring and on the steroidal side chain, making selective acylation possible at the 3- or 21-positions of polyhydroxy steroids **130a**, **131a**, **133**, **134**, **136**, and **103** (Table 12). Acylation with the synthetically useful esters chloroacetate and levulinate was also accomplished affording derivatives **130b**, **131b**, and **131c**, whereas esterification with benzoate and pivaloate was unsuccessful.

Baldessari's group reported that lipase from CCL and CAL catalyzes the removal of acetyl groups from 3β-acetoxypregn-5-en-20-one **137** and 3β-acetoxy-20(S)-hydroxycholest-5-en-23-one **138**, through a transesterification reaction in organic solvents [29], giving alcohols **87** and **139** (Table 13). It is known that compound **139** could not be obtained by conventional methods. The enzymatic approach allows its preparation in fair yield.

In a recent paper on the partial synthesis of some minor ecdysteroids and their analogs [30], the selective acetylation of 20(R)-hydroxyecdysone **140** was evaluated in detail. However, difficulties have been encountered in obtaining the pure 2-acetate **141a**, either by direct acetylation or by hydrolysis lof the acetonide moiety of the 20,22-

Table 8 Enzymatic Transesterification of Steroids Esters in Organic Solvents by CCL[a]

Entry	Substrate	Product	Solvent	t (d)	Yield (%)
1	5α-Cholestan-3β-acetate **76**	**77**	Dry i-PrOH	6	87
2	**78**	**79**	Dry i-PrOH	6	90
3	**80**	Starting material	Dry i-PrOH	7	—
4	**81**	**82**	Dry i-PrOH	6	92
5	**83**	**84**	Dry i-PrOH	7	68

Entry	Substrate	Product	Solvent	Time	Yield
6	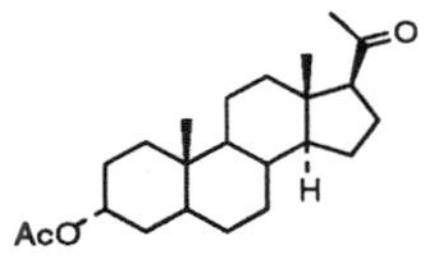 85	Starting material	Dry i-PrOH	6	—
7	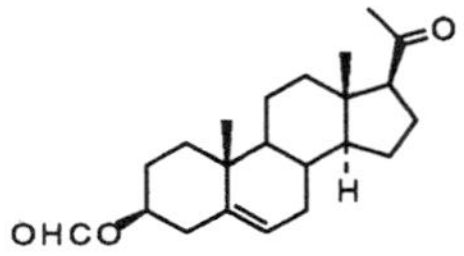 86	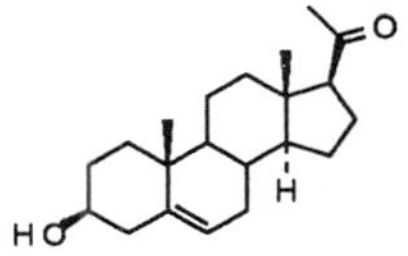 87	Dry i-PrOH	6	90
8	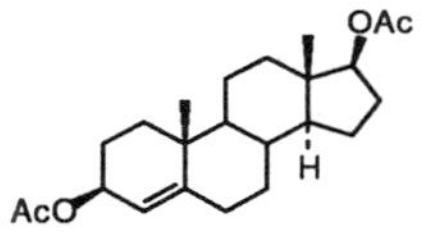 88	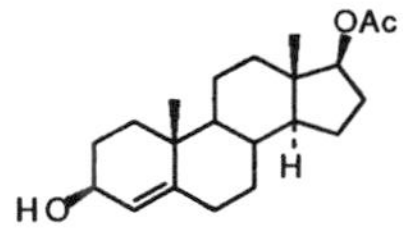 89	Dry i-PrOH	6	80
9	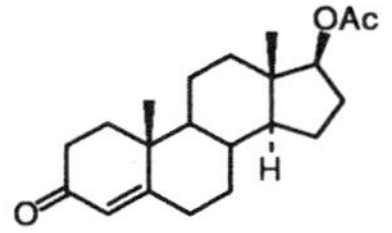 90	Starting material	Dry i-PrOH	10	—
10	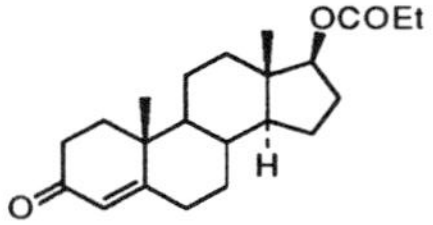 91	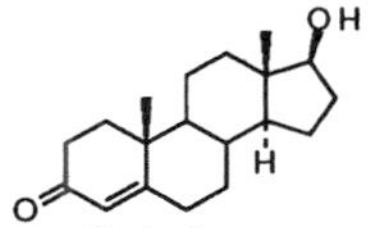 Testosterone 15	Dry i-PrOH	6	(85)[b] 12
11	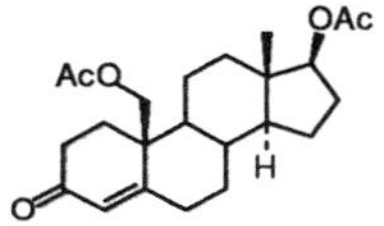 92	Starting material	MeCN/5% distilled water	10	—

Table 8 Continued

Entry	Substrate	Product	Solvent	t (d)	Yield (%)
12	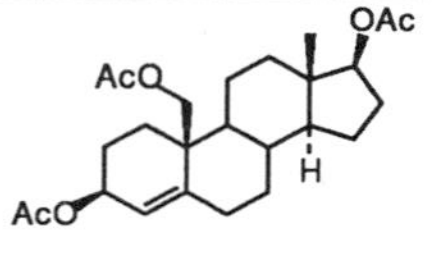 **93**	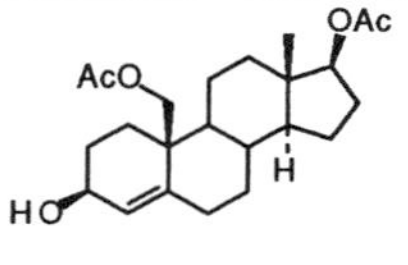 **94**	Dry i-PrOH	6	80
13	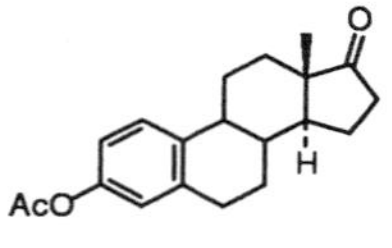 **95**	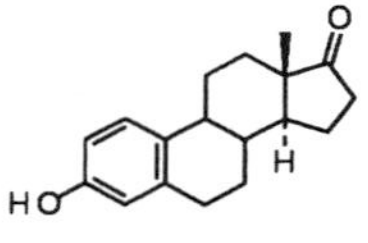 **96**	Dry i-PrOH	6	90
14	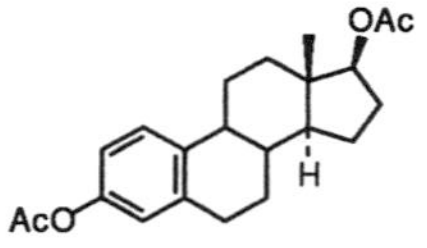 **97**	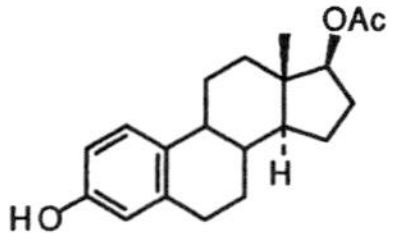 **98**	Dry i-PrOH	6	90
15	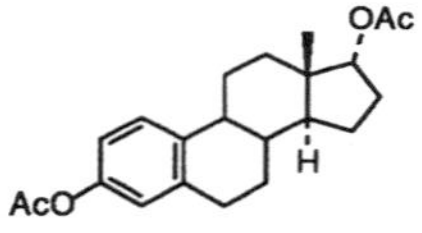 **99**	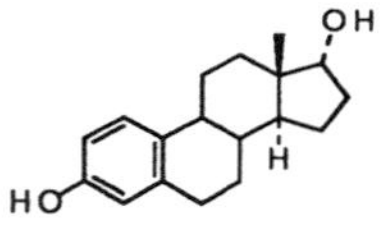 **100** 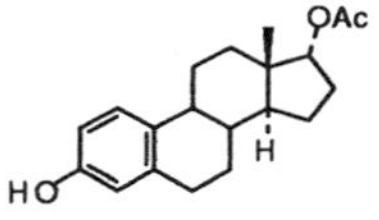 / **101**	Dry i-PrOH	6	**100/101** 60/25

	Substrate	Product	Solvent		
16	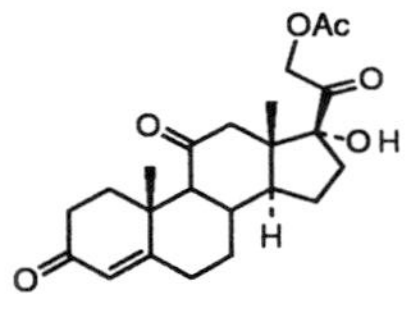 102	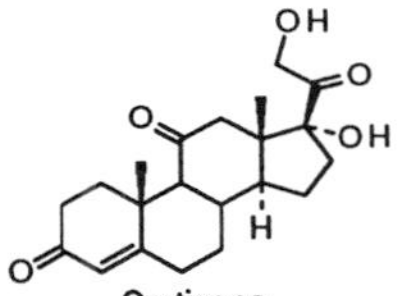 Cortisone 13	MeCN/5% distilled water	6	(20)[b] 62
17	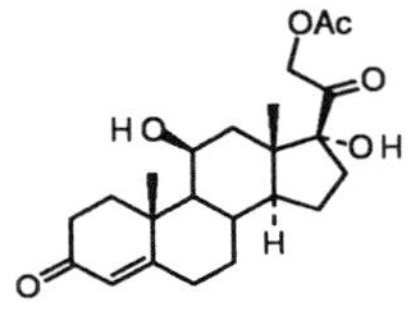 103	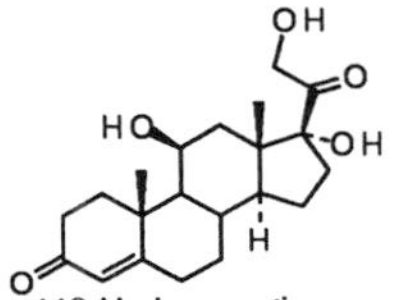 11β-Hydroxycortisone 104	MeCN/5% distilled water	6	(25)[b] 60
18	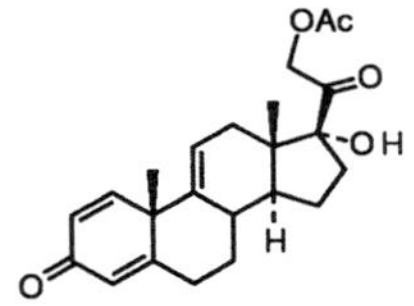 105	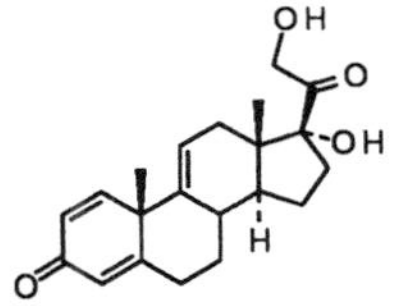 106	MeCN/5% distilled water	6	(25)[b] 60

[a]Conditions not optimized.
[b]Recovered starting material.

Table 9 Enzymatic Regioselective Protection of Steroids with CVL and Subtilisin[a]

Enzyme	Substrate	Products	

| CVL | | 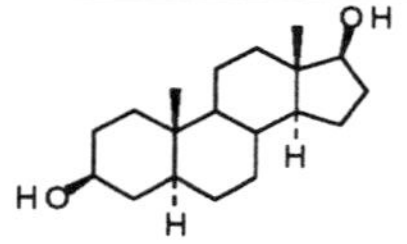 **107** | 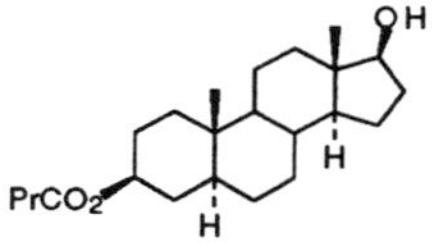 **108** 83 | 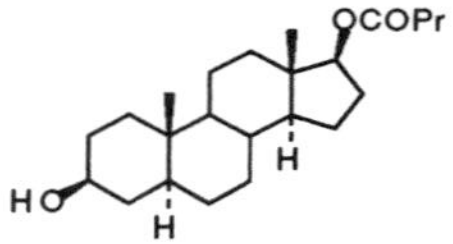 **109** 60 |
| Subtilisin | | | | |

| CVL | | 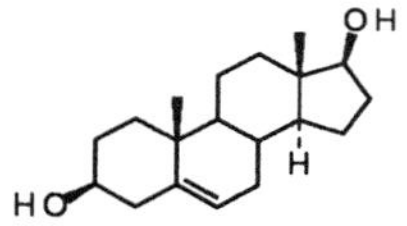 **110** | 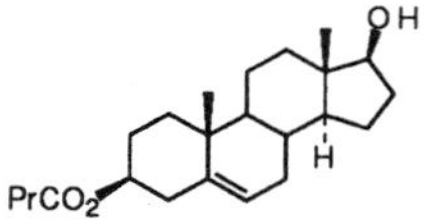 **111** 84 | 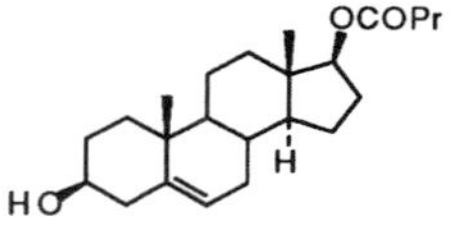 **112** 63 |
| Subtilisin | | | | |

| CVL | | 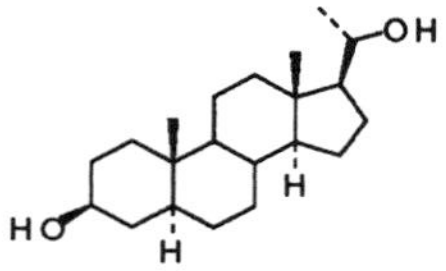 **113** | 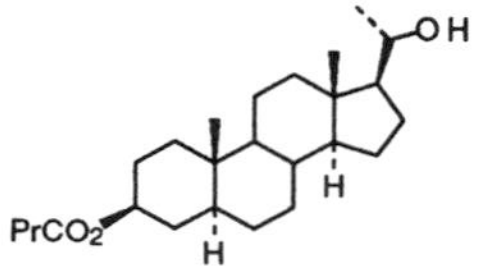 **114** 85 | |

[a]The values under the structures are isolated yields.

Table 10 Regioselective Acylation of Bile Acid Derivatives with CCL in Anhydrous Benzene[a]

Entry	Substrate	Product	Yield (%)[b]
1	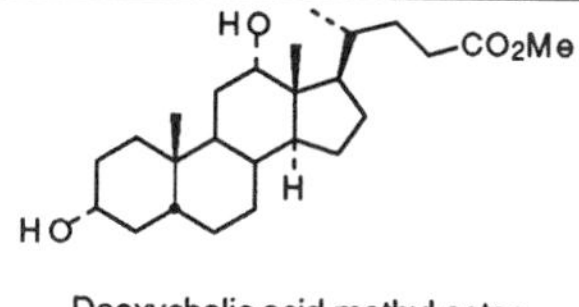 Deoxycholic acid methyl ester **115**	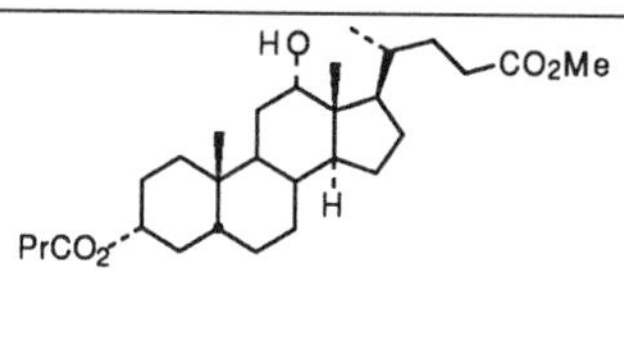 **116**	96
2	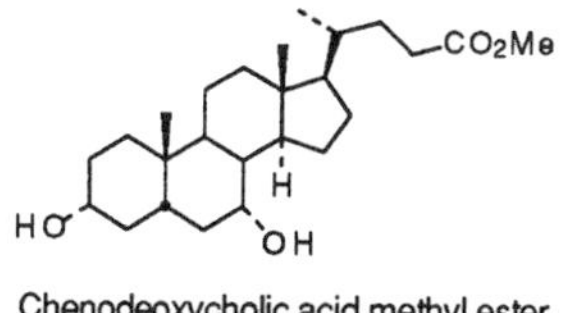 Chenodeoxycholic acid methyl ester **117**	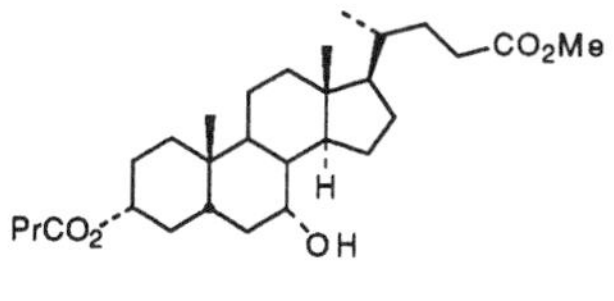 **118**	70–85
3	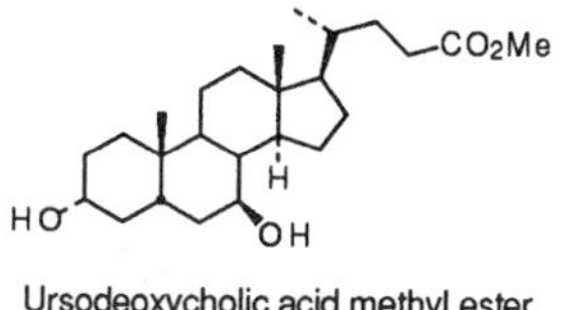 Ursodeoxycholic acid methyl ester **119**	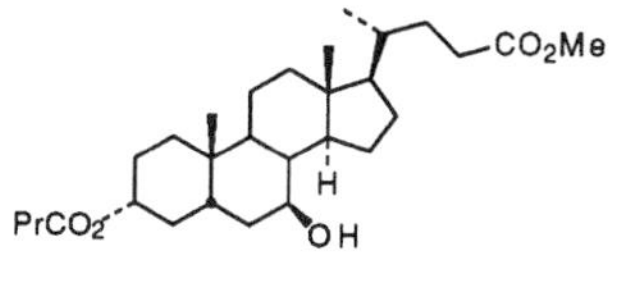 **120**	70–85
4	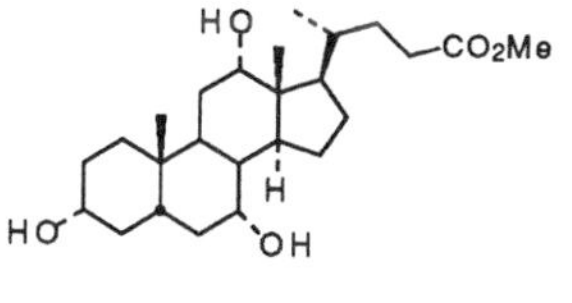 Cholic acid methyl ester **17**	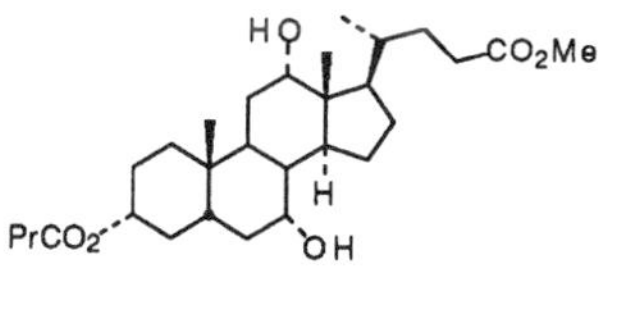 **121**	70–85

[a]Reactions were carried out with trichloroethyl butanoate as an acylating reagent.
[b]Isolated yields.

monoacetonide-2-acetate—a question that is addressed in this report by Danieli, Riva, and co-workers [31]. Thus, immobilized CAL catalyzed the regioselective acylation of 20(R)-hydroxyecdysone **140** and its congeners ecdysone **142**, makisterone **144**, and muristerone **146** at the C-2 hydroxyl group, in high yield and purity (Table 14).

The reaction has been extended to the introduction of long chain acid residues because of the importance of such derivatives. In fact, by reacting **140** with vinyl laureate or trifluoroethyl palmitate in the presence of CAL, the corresponding 2-laureate **141b** and 2-palmitate **141c** were obtained, even though in moderate yield due to lower conversions.

Another important derivative is 2-cinnamate, and for this biotransformation the previously described procedure failed because the bulkiness and the rigidity of the acyl residue encumbered the active site, preventing the approach by a large nucleophile such as **140**. To overcome this failure, a strategy was applied consisting of a combined approach based on the enzymatic introduction of a small malonate unit, which is then chemically converted to the desired cinnamate by reaction with benzaldehyde (Knoevenagel reaction). To this end, dibenzyl malonate was reacted with **140** in the presence of CAL, and the 2-malonate benzyl ester **141d** formed in 62% yield was then transformed to final product by means of chemical methodology. Whereas the enzymatically formed 2-malonate benzyl ester

Table 11 Lipase-Mediated Diastereoselective Hydrolysis of Steroidal Oxime Methyl Esters[a]

Entry	Substrate	Products	
1	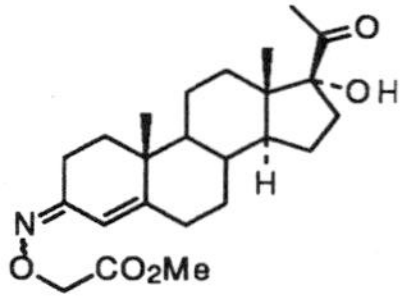 **122**	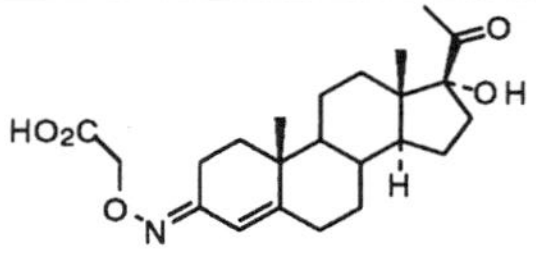 **123**	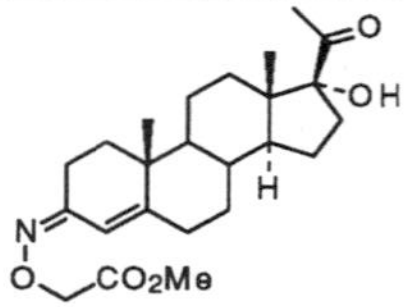 **122**
syn	50	14	89
anti	50	86	11
2	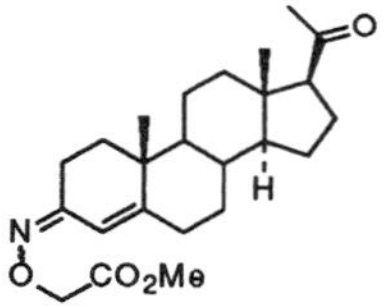 **124**	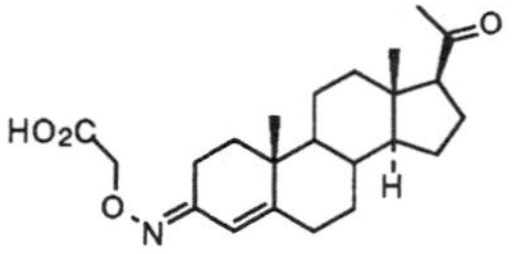 **125**	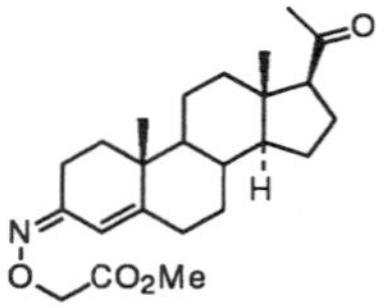 **124**
syn	50	20	80
anti	50	80	20

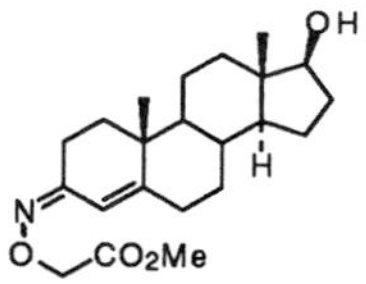

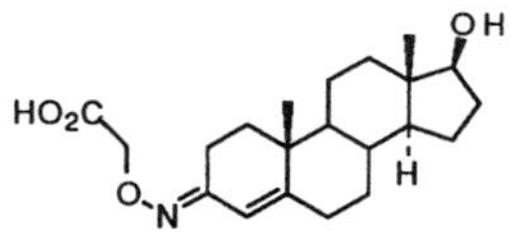

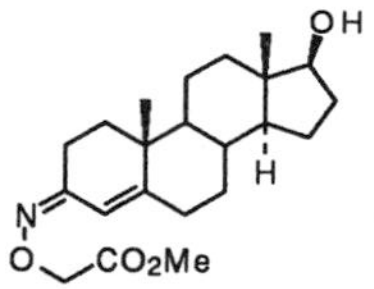

	126	127	126
syn	50	20	81
anti	50	80	19

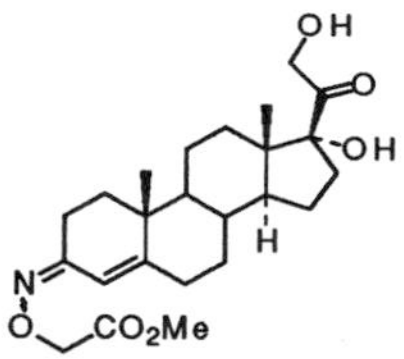

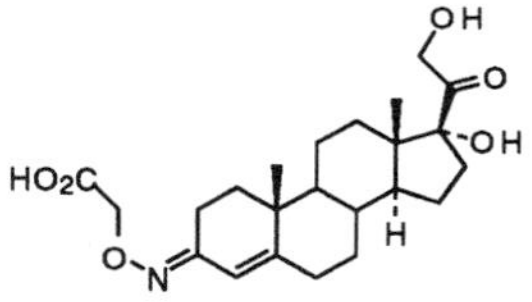

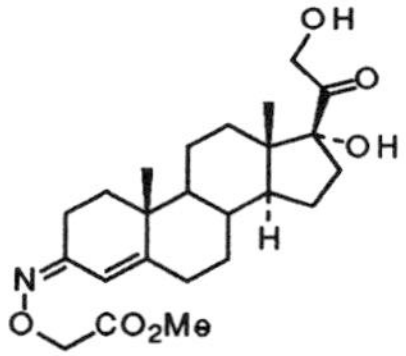

	128	129	128
syn	50	6	98
anti	50	94	2

3

4

[a]Lipase from *Candida cylindracea.*

Table 12 Regioselective Esterification of Polyhydroxylated Steroids by *Candida antarctica* Lipase

Entry	Substrate	Products	T (°C)/t (h)/yield (%)
1	**107**	**130a**, R= Ac **130b**, R= Lev	**130a** 45/40/97 (conv.)
2	**110**	**131a**, R= Ac **131b**, R= Lev **131c**, R= ClCH$_2$CO	**131a** 45/40/97 (conv.) **131b** 45/96/74 **131c** 45/1/100 (conv.)
3	**132**	**133**	45/168/82
4	**17**	**134**	45/1/100 (conv.)
5	**135**	**136**	45/96/100 (conv.)
6	**104**	**103**	45/4/100 (conv.)

Table 13 Enzymatic Deacetylation of Steroids Bearing Labile Functions by CCL and CAL

Entry	Substrate	Product	Enzyme	Solvent	Yield (%)
1	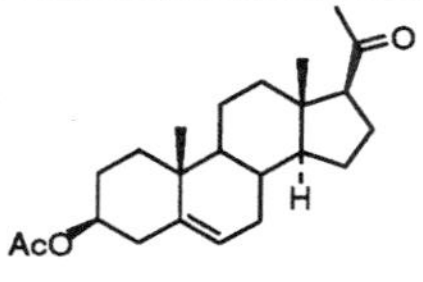 137	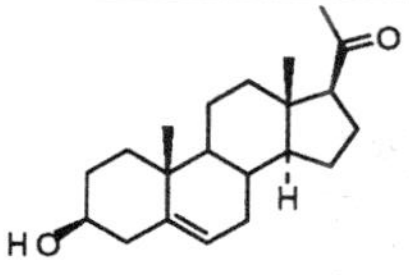 87	CCL	Octanol	84
			CCL	Octanol[a]	87
			CCL	Toluene	91
			CAL	Butanol[a]	88
			CAL	Octanol	82
			CAL	Octanol[a]	93
			CAL	Acetonitrile	82
2	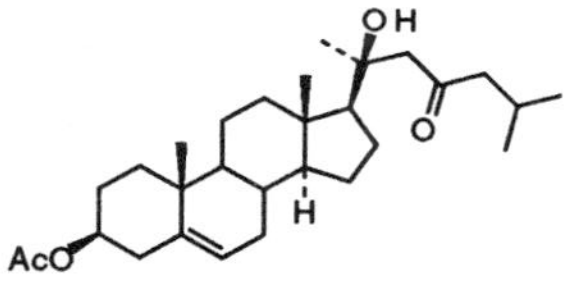 138	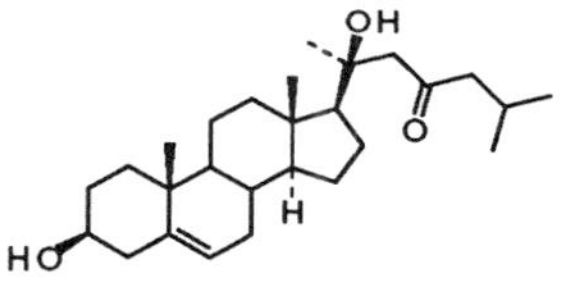 139	CCL	Toluene	68
			CAL	Acetonitrile	56

[a] As nucleophile and solvent.

Table 14   Regioselective Esterification of Ecdysteroids at the C-2 Hydroxyl Group by CAL

Entry	Substrate	Products	t (d)/yield (%)
1	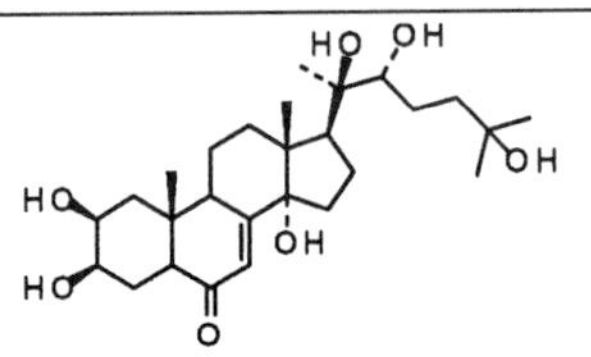 20*R*-hydroxyecdysone **140**	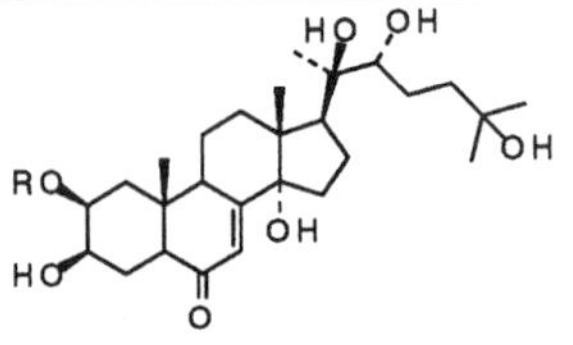 **141a**, R= Ac **141b**, R= $CH_3(CH_2)_{10}CO$ **141c**, R= $CH_3(CH_2)_{14}CO$ **141d**, R= $PhCH_2O_2CH_2CO$	**141a** 7/90 **141b** 6/69 **141c** 6/50 **141d** 7/62
2	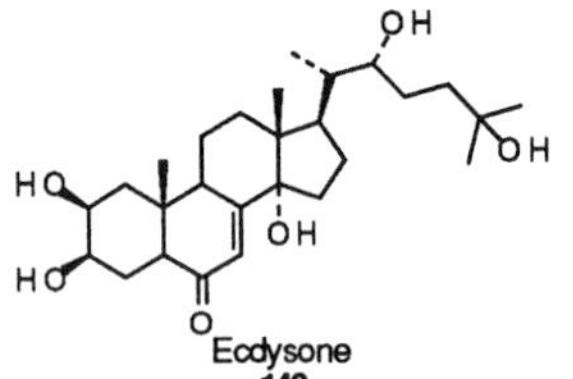 Ecdysone **142**	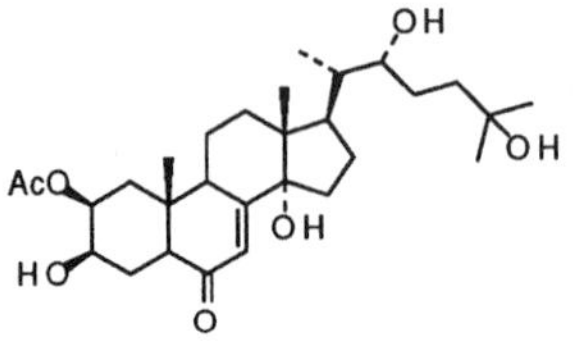 **143**	2/95
3	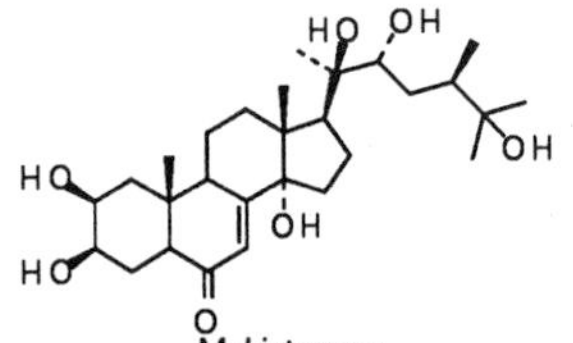 Makisterone **144**	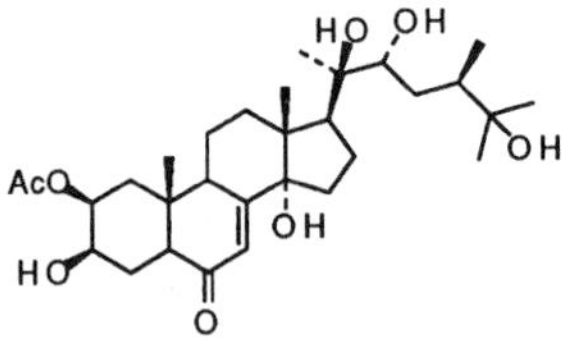 **145**	7/77
4	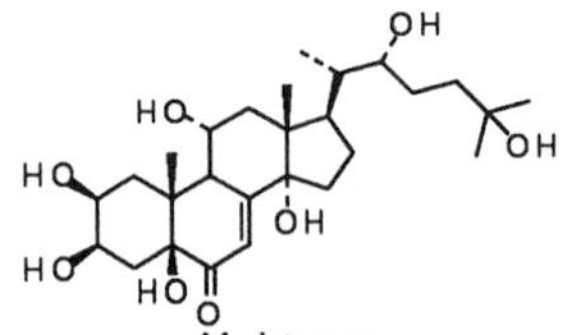 Muristerone **146**	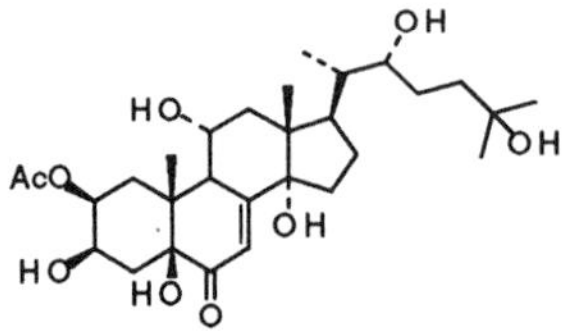 **147**	7/63

141d contained only traces of the isomeric 3-malonate benzyl ester, the final cinnamate was a 7:3 mixture of the 2-cinnamate and the 3-cinnamate, due to acyl migration under the basic conditions employed in the condensation reaction.

C. Modifications in Steroidal Side Chain

The stereospecific synthesis of (25*S*)-26-hydroxycholesterol **156** with a chiral synthon derived from (*S*)-(+)-3-hydroxy-2-methylpropanoic acid **149** is described by Gut and co-workers [32] (Scheme 2). The description of the synthesis of chiral synthons derived from (*S*)-(+)-3-hydroxy-2-methylpropanoic acid by Cohen and co-workers [33] allowed for the synthesis of (25*R*)- and (25*S*)-26-hydroxycholesterol.

      Compound **149** was readily available via the bacterial oxidation of isobutyric acid described by Goodhue and Schaeffer [34]. The synthesis began with the protection of the 3β-hydroxyl group and tosylation of the primary alcohol of 23,24-bisnorchol-5-en-22-ol

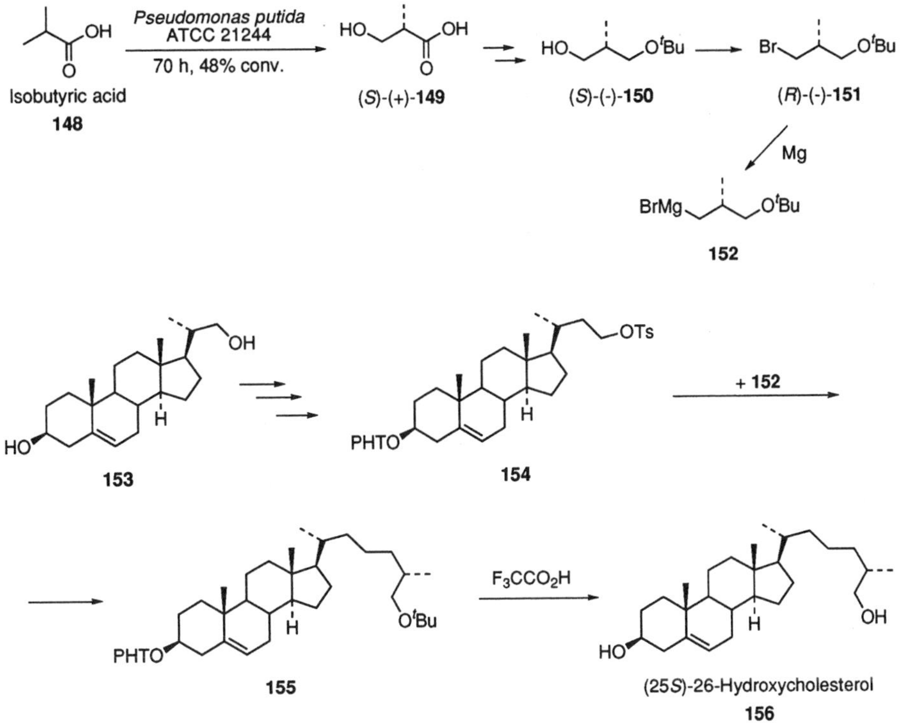

Scheme 2

153. The tosyloxy group was then displaced by iodine to give the corresponding iodo derivative. 1,3-Dithiane was alkylated with this iodine to give the dithiane product, which upon hydrolysis, reprotection with tetrahydropyranyl (THP) ether, and reduction gave 3β-(tetrapyranosyloxy)-24-norchol-5-en-23-ol. The alcohol was then tosylated to afford compound **154**. Ester **154** then reacted with the Grignard solution (intermediate **152**) prepared from (*R*)-(−)-3-*tert*-butoxy-2-methyl-1-bromopropane **151**. The resulting (25*S*)-3β-(tetrapyranosyloxy)-26-hydroxycholesterol 26-*tert*-butyl ether **155** was hydrolyzed with trifluoroacetic acid to give the desired (25*S*)-26-hydroxycholesterol **156**.

An alternative approach to the synthesis of (25*S*)-26-hydroxycholesterol **156** starting from a readily accessible steroid (stigmasterol **160**) and using chiral synthon (*S*)-(−)-ethyl-4-hydroxy-3-methyl butanoate **158** was reported [35] as shown in Scheme 3. The hydroxy ester **158** can easily be prepared by a biohydrogenation process, using baker's yeast as the microbial source [36], and can be subsequently chemically transformed into the sulfone **159**. The steroidal moiety was prepared in a conventional manner from commercially available stigmasterol **160**, which was protected at C-3 as a 3,5-cyclo steroid. The 22-iodo steroid **161** proved to be the best steroidal counterpart for coupling with sulfone **159**, and the reaction was best effected with lithium diisopropylamide (LDA) in tetrahydrofuran (THF). The subsequent removal of the phenyl sulfone moiety with sodium amalgam and careful acidic hydrolysis afforded the C-26 protected steroid **162**. The hydrolysis of the C-26 protecting group required HCl in refluxing THF, after which (25*S*)-**156** was finally obtained in 45% yield, starting from the 22-iodo derivative **161**. Since the optical purity of the final product depends only on the purity or the chiral synthon used, it was established

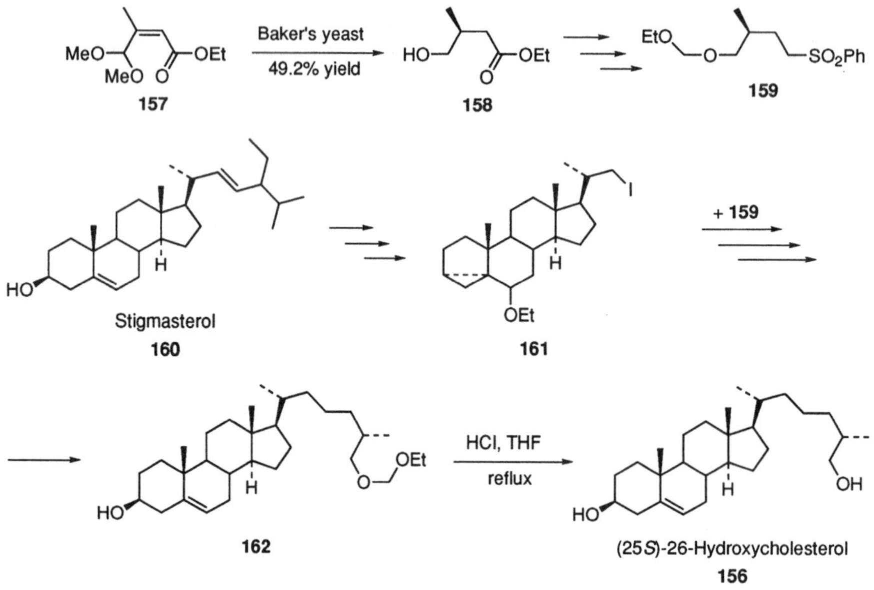

Scheme 3

that the average purity of the hydroxy ester **158** prepared by the fermentative route was generally 90–95%.

Santaniello's group [37] offers an additional example of the regioselective control of the enzymatic reaction and a polyfunctional steroid substrate, opening a new approach to the stereoselective construction of asymmetrical steroid side chains, including access to unnatural (i.e., *20S*) steroids (Scheme 4). The *Pseudomonas cepacia* lipase (PSL) selectively catalyzes the acylation of the (*20S*) isomer of the 22-alcohol group in the C-22 steroid compound (*20R,S*)-**163**, when the transesterification is irreversibly carried out with vinyl acetate in organic solvent. Thus, with vinyl acetate in chloroform at room temperature only (*20S*)-acetate **164** was formed as identified by ¹H-NMR (proton nuclear magnetic resonance) analysis. The result can be explained by considering that, with respect to the 18β-methyl, in the (*20S*) isomer the methyl of the stereogenic center is in a less hindered position than in (*20R*)-**163**.

The lipase-catalyzed transesterification in organic solvents is especially useful when applied to sterols that are highly insoluble in water. Santaniello and co-workers [38] extended the above observation to a primary alcohol in the steroid side chain, bearing the stereogenic center at position C-25, namely, 26-hydroxycholesterol **156** (Scheme 5).

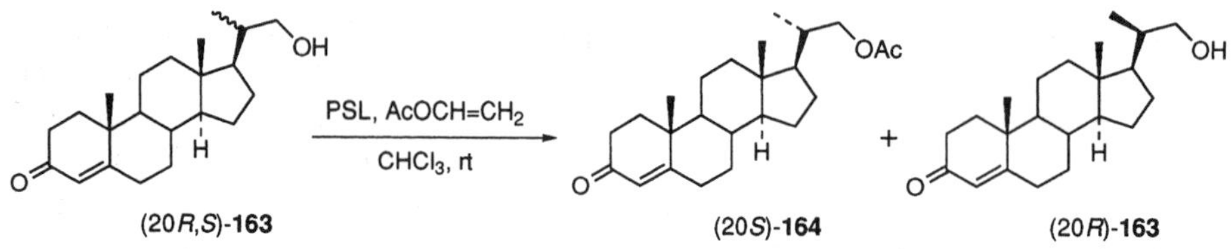

Scheme 4

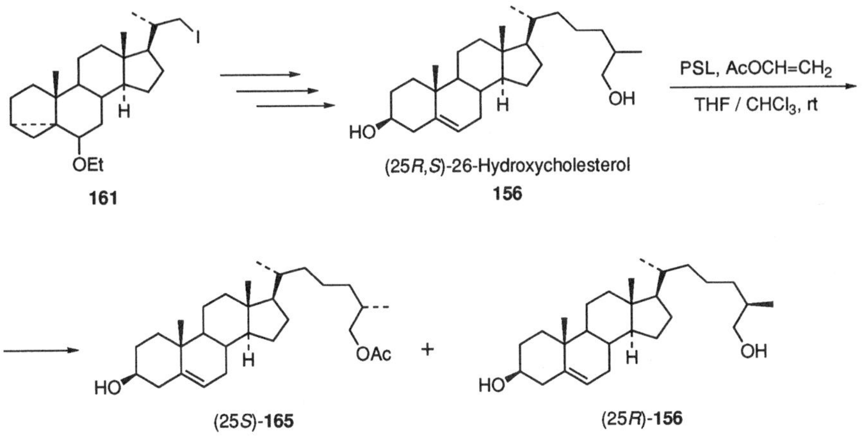

Scheme 5

In order to study the resolution of the stereogenic center present in the side chain, the synthesis of racemic **156** was required and was started from the 22-iodo derivative **161** obtained as previously described (Scheme 3) [35]. Following a straightforward chemical procedure the steroid (±)-**156** was obtained. The (25R,S)-3,26-diol **156** underwent a reaction with PSL and vinyl acetate in chloroform/THF at room temperature. ^{1}H-NMR analysis of the crude detected only (25S)-26-acetate **165**, showing that the enzymatic reaction was highly regio- and enantioselective. This reaction is faster than the formation of the (20S)-acetate **164** (Scheme 4) [37]. However, it should be remembered that, due to different steric hindrance, the C-26 alcohol is more accessible than the C-22 analog. These results offer new approaches to the stereoselective construction of steroid side chains.

D. Syntheses of Steroids from Optically Active Compounds

An efficient construction of the chiral tricyclic dienone **169** in both enantiomeric forms has been reported [39] (Scheme 6), starting from dicyclopentadiene and employing kinetic resolution by lipase in the key reaction. Enzymatic transformation of racemic 1-acetoxy-dicyclopentadiene (±)-**167** (readily accessible in large quantities from dicyclopentadiene (±)-**166** in two steps via the alcohol) with CCL in a phosphate buffer containing acetone at room temperature for 5 days gave a mixture of the alcohol (+)-**168** and the unchanged acetate 16, which was readily separated by silica gel column chromatography. The alcoholic products (+)-**168** and (−)-**168** (after saponification of the unchanged acetate **167**) were further purified by repeated recrystallization. Oxidation of the (+)-alcohol [(+)-**168**] by pyridinium chlorochromate (PCC) furnished the enone (−)-**169**. Owing to its biased structure, (−)-**169** allowed the stereospecific introduction of nucleophiles at the β carbon of the enone system from the convex face of the molecule, a process that has been successfully applied to the enantio-controlled synthesis of the estrogenic steroid (+)-equilenin **174** (as shown in the summary in Scheme 6) [40]. Thus, the reaction of (−)-dienone **169** with Grignard reagent prepared from 6-methoxy-2-bromonaphthalene in the presence of copper(I) iodide gave exclusively the *exo* adduct. Sequential metalloenamine formation and alkylation of this intermediate then afforded an allyl ketone as a mixture of two epimers. The second alkylation with an excess of methyl iodide proceeded stereoselectively to give the *exo*-methyl ketone (+)-**170**. Refluxing (+)-**170** in *o*-dichlorobenzene brought

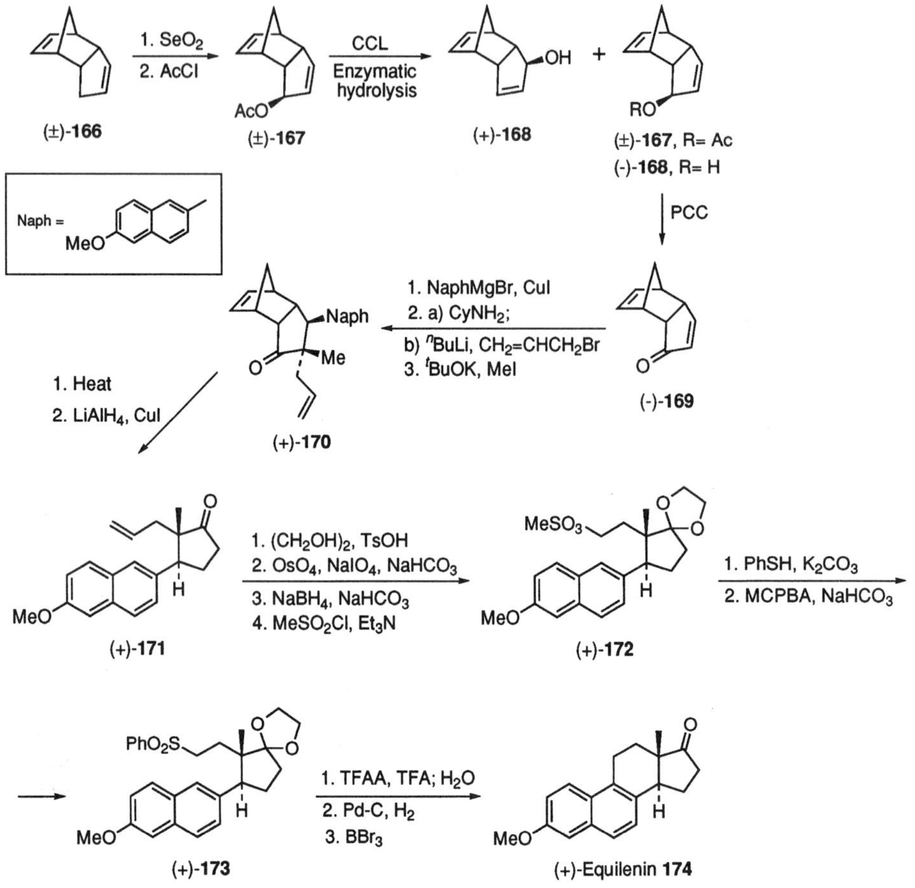

Scheme 6

about a facile retrograde Diels-Alder reaction to give rise to a cyclopentenone, which was next treated with a complex (LiAlH$_4$-CuI) to afford the cyclopentanone (+)-**171** by specific hydrogenation of the enone double bond. In order to continue with this chiral route, (+)-**171** was first converted to a ketal, which was then transformed into a methane sulfonate derivative **172** (through the alcohol via aldehyde formation) by sequential oxidative cleavage and borohydride reduction. Compound **172** was transformed to sulfoxide **173** via the sulfide. Upon exposure to a 1:2 mixture of trifluoroacetic anhydride (TFAA) and TFA, **173** underwent smooth cyclization to furnish Δ^{11}-equilenin methyl ether exclusively. Catalytic hydrogenation yielded (+)-equilenin methyl ether, which was finally treated with boron tribomide to give (+)-equilenin **174**.

An early approach to the asymmetrical synthesis of steroid precursors was undertaken by Kosmol's group [41] using microorganisms such as *Bacillus* spp. and *Saccharomyes* spp. This approach was applied to the synthesis of steroids through reduction of disubstituted cyclopentanediones (Scheme 7). Thus, for example, treatment of cyclopentanedione **175** with actively fermenting baker's yeast resulted in good yields of chiral monoreduced ketol product **176**, as shown in Scheme 7. In all cases the microbial keto products were found to have more than 98% enantiomeric excess (e.e.). It is interesting to recognize the asymmetrical consistency of the microbial reduction of the carbonyl group to provide only

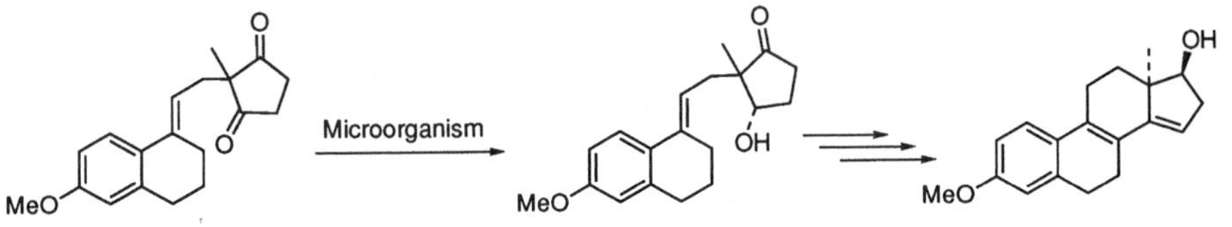

Scheme 7

the (S)-hyroxy configuration. Through further application of this methodology as a valuable element of synthetic design, the chiral steroid derivative **177** has been prepared.

Sugahara and Ogasawara [42] demonstrated the synthesis of (+)-estrone **187** in an enantio-convergent manner from racemic 4-*tert*-butoxy-2-cyclopentenone, with a contrasteric Diels-Alder reaction involving a lipase-mediated kinetic transesterification (Scheme 8). These authors [43] found that the Lewis acid–mediated Diels-Alder reaction between racemic 4-*tert*-butoxy-2-cyclopentenone (±)-**178** and cyclopentadiene occurred in a contrasteric way to give the *endo* adduct (±)-**179**, bearing an *endo*-alkoxy group in 92% yield. Reduction of (±)-**179** with sodium borohydride took place stereoselectively from the convex face of the molecule to give the single *endo*-alcohol (±)-**180**. The reaction of (±)-**180** with vinyl acetate in *tert*-butyl methyl ether (TBME) in the presence of lipase LIP (*Pseudomonas* sp., Toyobo) furnished the (+)-acetate **181** in 48% yield, with recovery (51%) of the unchanged (+)-alcohol **180**. Methanolysis of (+)-**181** gave an excellent yield of the enantiomeric alcohol (−)-**180**. Oxidation of (−)-**180** with sulfur trioxide–pyridine complex in the presence of dimethyl sulfoxide (DMSO) and triethylamine gave the ketone (+)-**182**. In a similar manner, (+)-**180** gave the enantiomeric ketone (−)-**182**. The enantiomer (+)-**182** was treated with iodomethane in the presence of LDA to stereoselectively give the monomethyl product as the single isomer. To carry out β elimination, this compound was first exposed to TiCl$_4$ and then treated directly with aqueous sodium hydroxide to afford the enone (+)-**183**.

On the other hand, (−)-**182** was transformed into the enantiomeric enone (−)-**183** in a similar manner. To invert the stereochemistry, the enantiomer (−)-**183** was treated with alkaline hydrogen peroxide to give an enantiomeric pure epoxide which, through exposure to hydrazine, induced reductive cleavage to give the allyl alcohol. This on oxidation finally gave the key (+)-**183**.

In order to construct (+)-estrone **187**, the enone (+)-**183** was reacted with diene **184** to produce a Lewis acid–mediated Diels-Alder reaction in a regio- and stereoselective manner. Concurrent allylic hydrogen migration occurred to afford the single styryl product **185**. Upon thermolysis, the expected retro-Diels-Alder reaction of **185** gave the tetracyclic compound **186**. Subsequent treatment with lithium hexamethyldisilazide and acetic acid, followed by catalytic hydrogenation and subsequent reduction with trimethylsilane in the presence of TFA, and finally deprotection with boron tribromide, gave (+)-estrone **187** in 8% overall yield from the racemic starting material (±)-**179**.

E. Preparation of Vitamin D Synthons

Vitamin D research has expanded enormously in recent years with the discovery that several metabolites and analogs of 1α,25-dihydroxyvitamin D$_3$ (**188**, Fig. 1), the hormonally active form of vitamin D$_3$, exhibits a broad spectrum of biological activities [44].

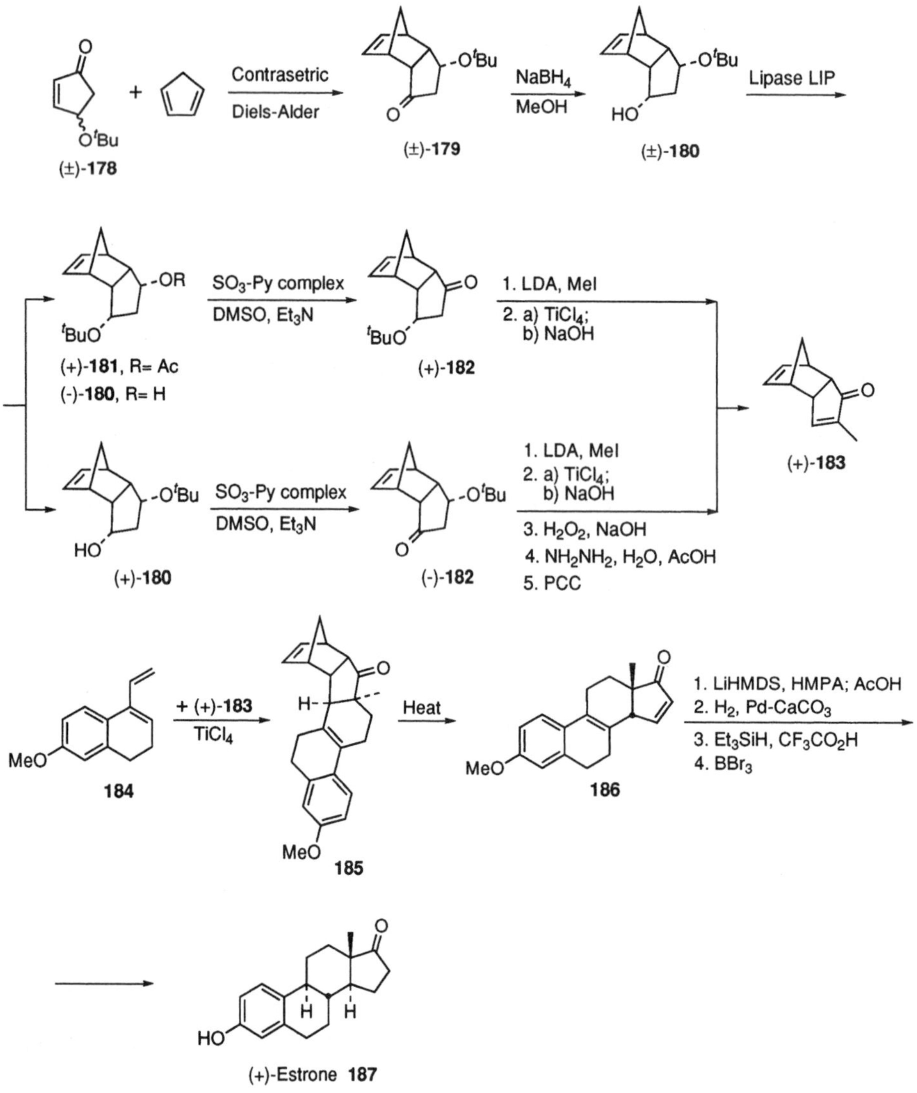

Scheme 8

Parallel to this, new enzymatic techniques (especially in organic solvents) applied to preparation of several synthon precursors of vitamin D analogs have been developed. Here we summarize some of the major synthetic routes utilized in recent years, in which an enzymatic process was developed to synthesize the hormone 1α,25-dihydroxyvitamin D$_3$ and its analogs. There are four main methods, three of which center on the preparation of A-ring synthons (A, B, and C) and one on CD-ring synthons (D).

Method A, phosphine oxide coupling, is probably the most useful approach for producing side chains and other analogs (the A-ring synthon precursors for the purpose of this chapter). In this method, first developed by Lythgoe, the phosphine oxide **189a** is coupled to a Grundmann's ketone derivative **190**, producing the 1α,25-dihyroxyvitamin D$_3$ skeleton directly. The shortcoming of this route, the tedious synthesis of A-ring frag-

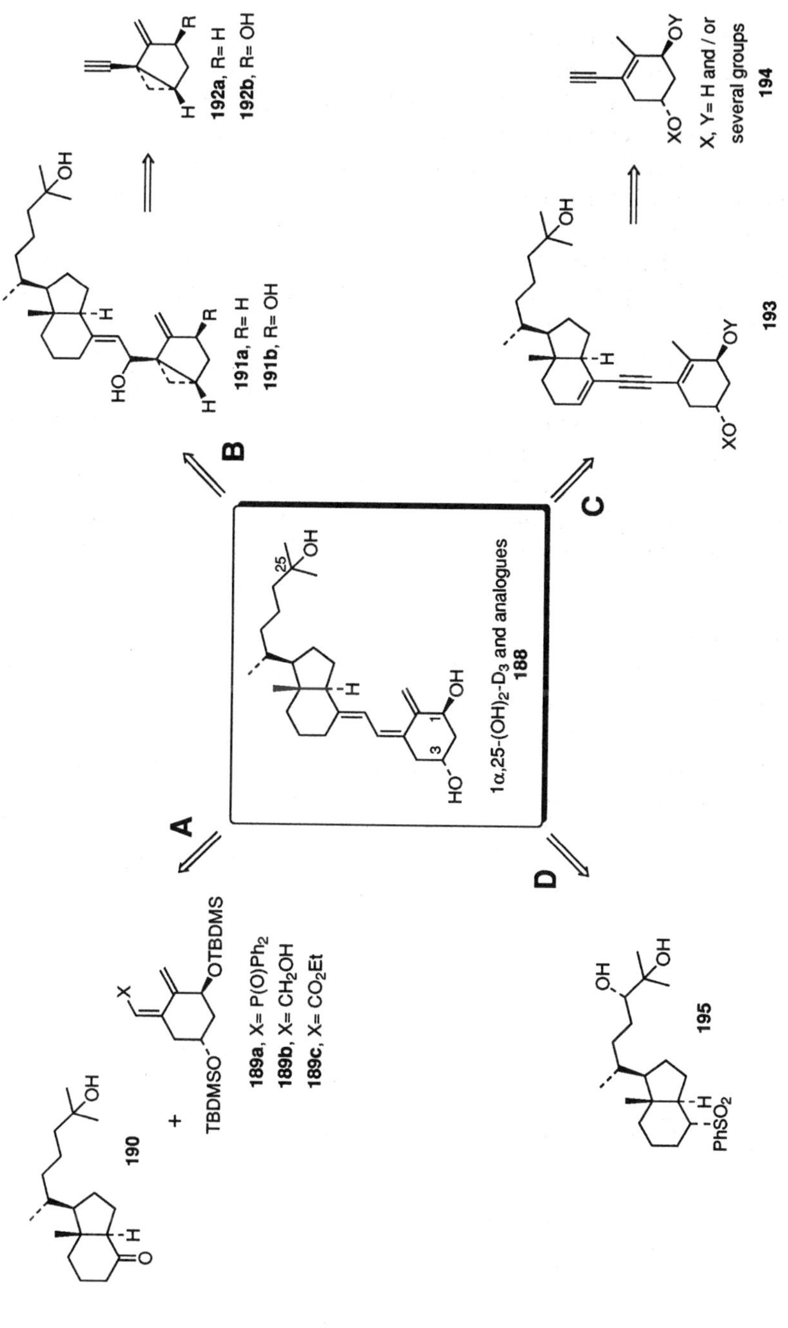
A
B
C
D
1α,25-(OH)$_2$-D$_3$ and analogues
188
190
189a, X= P(O)Ph$_2$
189b, X= CH$_2$OH
189c, X= CO$_2$Et
TBDMSO
OTBDMS
191a, R= H
191b, R= OH
192a, R= H
192b, R= OH
193
X, Y= H and / or several groups
194
195
PhSO$_2$

Figure 1

ment **189a**, has been eliminated, partly due to the development of chemoenzymatic procedures to synthesize the several A rings that will be presented below.

Based on initial studies of Mazur, method B for the production of vitamin D metabolites and analogs has been nicely developed by several groups. In the Mazur approach, the vitamin D isomerized to the *i*-steroid **191**, which can be modified at carbon C-1 and then subsequently back-isomerized under solvolytic conditions to afford 1α,25-dihydroxyvitamin D₃ or analogs. The studies presented here led to the separate production of bicyclo[3.1.0]hexane derivatives **192** in a chemoenzymatic synthetic approach to **191**.

In method C, dienynes **193** are semihydrogenated to a previtamin structure that undergoes rearrangement to the corresponding vitamin D analog. The chemical synthesis developed in the Okamura laboratory for **194** (X = Y = H), starting with (*S*)-cavone, is probably the most efficient approach in this area [45], but the regioselective enzymatic acylations [46] and alkoxycarbonylations [47] of the four stereoisomers of **194** open the doors to a new family of synthon precursors to feed this route.

Method D involves an A-ring precursor that is cross-coupled to 24,25-dihydroxy CD-ring sulfone synthon **195** (prepared through a chemoenzymatic synthesis from easily accessible diketone) leading directly to the 1α,25-dihydroxyvitamin D₃ skeleton.

1. A-Ring Precursors

This first chemoenzymatic synthesis [48] is not included in Fig. 1 but can be considered as a preparation of A-ring synthons in general in the steroid field, and particularly in area of vitamin D (Scheme 9). The stereo control of the quaternary chiral carbon center is one of the important subjects in asymmetrical synthesis. This issue prompted research on the use of chiral monoester **197**, which can be obtained in multihundred gram scale by the pig liver esterase (PLE)–mediated hydrolysis of the corresponding symmetrical diester **196** [49]. Thus, an efficient methodology was developed for the preparation of chiral cyclohexene derivatives from *cis*-diester **196**, using PLE in a biphasic system using phosphate buffer and acetone, to afford the chiral half-ester **197**. Scheme 9 shows the efficient

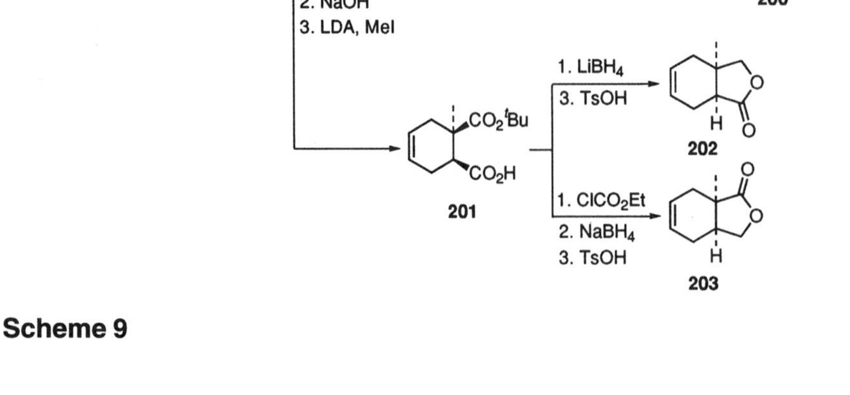

Scheme 9

conversion of monoester **197** into both enantiomers of the methylated bicyclic lactones **199** and **200**, which have the methyl group in the angular position and are considered as valuable starting materials for the synthesis of vitamin D and related compounds.

To introduce the methyl group at C-1, the chiral monoester **197** was treated with LDA and methyl iodide to afford the methylated monoester **198**. In a separate experiment, the chiral monoester **197** was transformed to the *tert*-butyl monoester and was reacted with LDA and MeI in the same manner as for **197** to obtain the methylated product **201**. The conversion of these monoesters **198** and **201** to the bicyclic γ-lactone derivatives was examined. Reaction of the monoester **198** with LiBH₄ and methanol resulted in the reduction of the methoxycarbonyl group, and the subsequent treatment of the crude hydroxy acid with *p*-toluenesulfonic acid (TsOH) afforded the γ-lactone **199**. On the other hand, reduction of the carboxyl group of **198** and acid treatment of the hydroxy ester afforded the isomeric γ-lactone **200**. The chiral monoester **201** was similarly converted to the enantiomeric γ-lactones **202** and **203**.

Among several A-ring synthons so far developed, phosphine oxide **189a** (Fig. 1), or its equivalent dienol **189b**, proves most attractive because of its efficiency in coupling with CD ring synthons such as **190** (Fig. 1, method A). The synthesis of dienol **189b** starting from the chiral monoester **197** is reported [50] (Scheme 10). Monoester **197** (prepared as previously mentioned [49] in Scheme 9) seemed to be an excellent chiral synthon for the synthesis of **189b** considering the carbon skeleton (only one carbon corresponding to C-7 [51] is missing in **197**) as well as a suitable array of functional groups. To introduce a hydroxyl group at C-3, monoester **197** is first converted to lactone. Next the hydrolysis of the lactone is carried out, followed by iodolactonization, silylation of the alcohol, and, finally, heating with 1,8-diazabiclyclo[5.4.0]undec-7-ene (DBU) to give olefin **204**. β-Epoxidation of the latter with *m*-chloroperbenzoic acid (MCPBA), and reduction of the lactone, gave the formation of ketene dithioacetal **205** after reaction with 2-lithio-2-trimethylsilyl-1,3-dithiane, furnishing the carbon skeleton of **189b**. Opening of the epoxide, rearrangement with TsOH, protection of the remaining two hydroxyl groups, and subsequent hydrolysis of dithiane group afforded δ-lactone **206**. With two hydroxyl groups at C-1 and C-3 established, the formation of exocyclic diene with Z stereochemistry was next attempted. For that, LDA, phenylselenide formation and elimination was used, furnishing one of the final exocyclic double bonds. δ-Lactone was clearly opened by alkaline hydrolysis, and the resulting carboxylic acid was methylated. The formation of the second

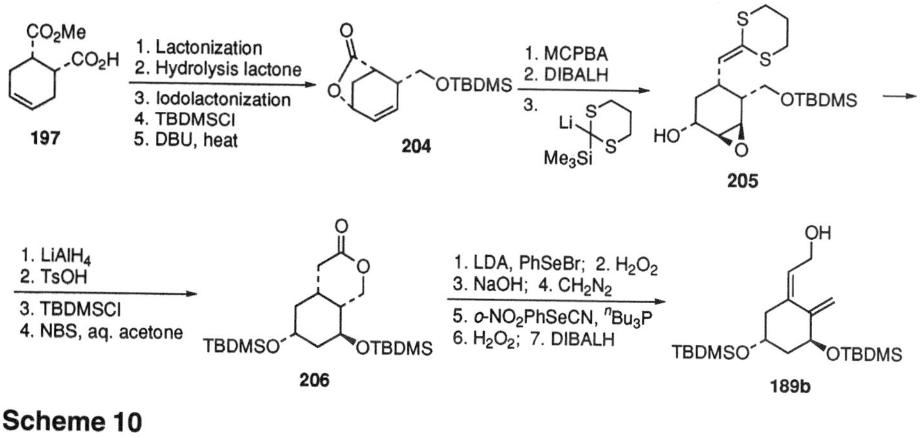

Scheme 10

exocyclic double bond was carried out by treatment with $o\text{-}O_2NPhSeCN$, $^nBu_3P$, and then with hydrogen peroxide, and finally dienal **189b** was obtained by reduction of methyl ester. Thus, the A-ring synthon for vitamin D_3 metabolites with a 1α-hydroxyl group is synthesized from chiral monoester **197** under complete stereo- and regiochemical control.

Using the methodology showed in pathway B (Fig. 1), Takano's group [52] prepared A-ring synthon **213**, precursor of compound **192a**, and immediate precursor of the key intermediate **191a** for the synthesis of $1\alpha,25$-dihydroxyvitamin D_3 or analogs (Scheme 11). They described a practical example that involves a highly efficient kinetic resolution of the readily accessible racemic substrate **209** using lipase in an organic solvent. Thus, racemic 2-ethoxycarbonyl-2-cyclopentenol ($\pm$)-**209**, prepared from aqueous glutaraldehyde **207** and triethyl phosphonoacetate **208** in aqueous potassium carbonate, was treated with vinyl acetate in the presence of PSL in TBME, this being the best combination of enzyme–solvent system out of six lipases (lipase OF, lipase MY, lipase AK, lipase AY, lipase PS, lipase WACO) and three solvents (dichloromethane, benzene, *tert*-butyl methyl ether). Kinetic acetylation of (R)-enantiomer [(R)-**209**] occurred after 96 h with virtually complete stereochemical selection to afford (R)-acetate [(R)-**210**] in 48% yield accompanied by the unreacted (S)-alcohol [(S)-**209**] in 48% yield, after separation by silica gel column chromatography. The (R)-alcohol [(R)-**209**] prepared from the (R)-acetate [(R)-**210**] by alkaline hydrolysis was transformed into the primary alcohol **211** by the sequence: silylation of secondary alcohol in (R)-**209**, reduction with diisobutylaluminum hydride (DIBALH) of the ester moiety, and cyclopropanation with diiodomethane and diethyl zinc. In order to introduce the requisite acetylene side chain, the primary alcohol **211** was oxidized to aldehyde under Swern conditions and then, employing Corey's method, the aldehyde moiety was transformed into acetylene giving rise to compound **212**. Desilylation of the latter followed by Swern oxidation of the resulting secondary alcohol gave the desired ketoacetylene **213**. The overall yield of **213** from (R)-**209** was 30% in eight steps.

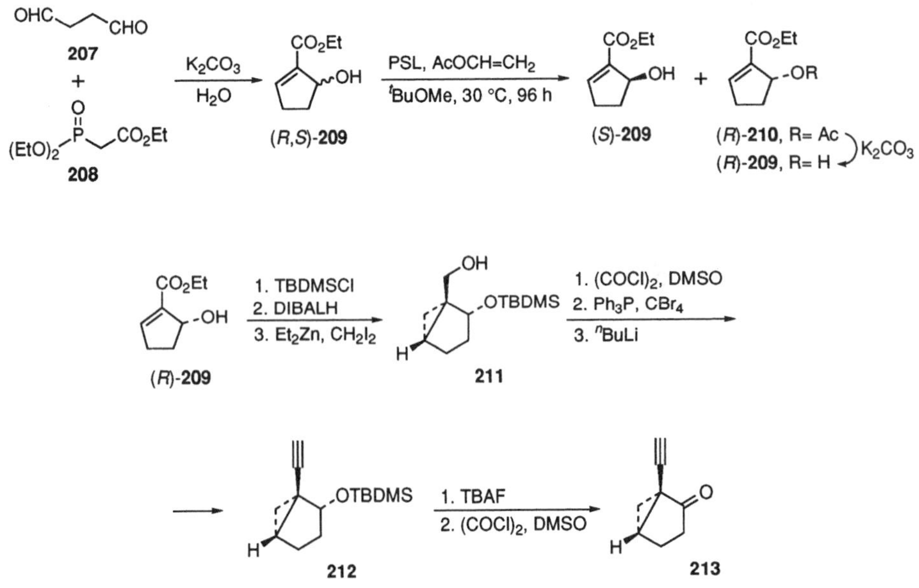

Scheme 11

Also using method B (Fig. 1), Takano's group [53] prepared A-ring synthon **218**, precursor of **192a**, by an oxidation and subsequent double-bond formation. They also performed a detailed study on the enantiocomplementary preparation of optically pure 2-trimethylsilylethynyl-2-cyclopentenol by homochiralization of racemic precursors, alcohol **215** and acetate **216**, respectively (Scheme 12). They applied the basic concept that lipases can both acylate and hydrolyze an appropriate substrate depending on the conditions. The starting racemic alcohol (*R,S*)-**215** was prepared by cross-coupling of the racemic 2-iodo-2-cyclopentenol, obtained from cyclopentenone **214** via treatment with iodide, sodium borohydride, and trimethylsilylacetylene. When (±)-**215** was stirred with two equivalents of vinyl acetate in the presence of lipase PS in toluene at 30°C, 120 h reaction time gave a 1:1 mixture of the unreacted (*S*)-alcohol **215** (47% yield, ≥99% e.e.) and the (*R*)-acetate **216** (48% yield, ≥99% e.e.). On the other hand, when the racemic acetate (*R,S*)-**216** was stirred with the same lipase (PSL) in a buffer solution at 30°C, the reaction terminated after 48 h to give the (*R*)-alcohol **215** (46.2% yield, >99% e.e.) and the unreacted (*S*)-acetate **216** (43% yield, ≥99% e.e.).

Both acylation and deacylation brought about clear-cut enantiotopical discrimination of the specific enantiomers, giving rise to enantiocomplementarity to alcohol **215** and acetate **216** as a 1:1 mixture. These two approaches were made complementary so as to

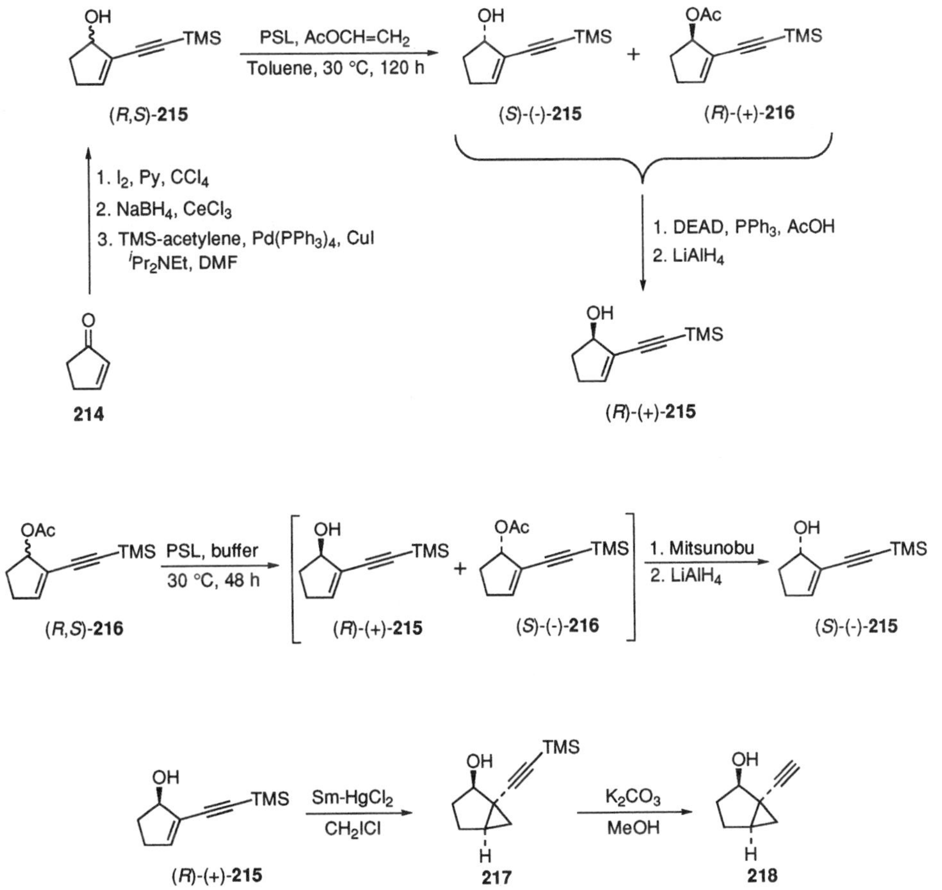

Scheme 12

produce individual products that are enantiomeric to each other. Thus, the racemic alcohol (*R,S*)-**215** was reacted with vinyl acetate as above, in the presence of PSL in toluene. After removal of the insoluble material by filtration, the reaction mixture was subjected to the Mitsunobu reaction to convert the unreacted (*S*)-alcohol **215** to the (*R*)-acetate **216** with inversion of configuration. The (*R*)-acetate was reduced with LiAIH$_4$ to give the (*R*)-alcohol **215**, with an optical purity of 87.5% e.e. and in excellent yield. Purification by a single recrystallization gave an optically pure (*R*)-**215** (≥99% e.e.) in 73% overall yield from the racemic substrates (*R,S*)-**215**. On the other hand, the racemic acetate (*R,S*)-**216** was treated with PSL as above in a phosphate buffer solution to give a mixture of the (*R*)-alcohol **215** and the unreacted (*S*)-acetate **216**. The mixture, without separation, was subjected to Mitsunobu inversion to convert (*R*)-**215** into (*S*)-**216** with inversion of the chirality. The single acetate (*S*)-**216** thus obtained was reductively deacylated with LiAIH$_4$ to give the alcohol (*S*)-**215**, with an 89% e.e. A single recrystallization afforded the optically pure (*S*)-**215** in ≥99% e.e. [75% yield from the racemic acetate (*R,S*)-**216**].

Utilization of the optically pure compounds for the construction of optically active material is exemplified with the preparation of A-ring precursor **218** from alcohol (*R*)-(+)-**215**, which reacts to form cyclopropane compound **217**, which is then desilylated to A-ring synthon **218**.

Recently, Ogasawara and co-workers [54] disclosed an efficient synthesis of the enantiomerically pure tricyclic dienone **222** in both enantiomeric forms by employing lipase-mediated asymmetrization of the *meso*-symmetric precursor, and the novel palladium-mediated elimination reaction of the chiral monoacylated product. With this they show a new synthetic approach to A-ring synthon **189c** (Scheme 13) [55]. Treatment of tricyclic diol **220**, obtained from reduction of diketene **219**, with two equivalents of vinyl acetate in acetonitrile in the presence of PSL furnished the monoacetate **221** in 79% yield after being stirred for 16 days at 28°C. The optically active acetate **221** was treated with ammonium formate in the presence of catalytic amounts of PdCl$_2$(PPh$_3$)$_2$ to furnish the

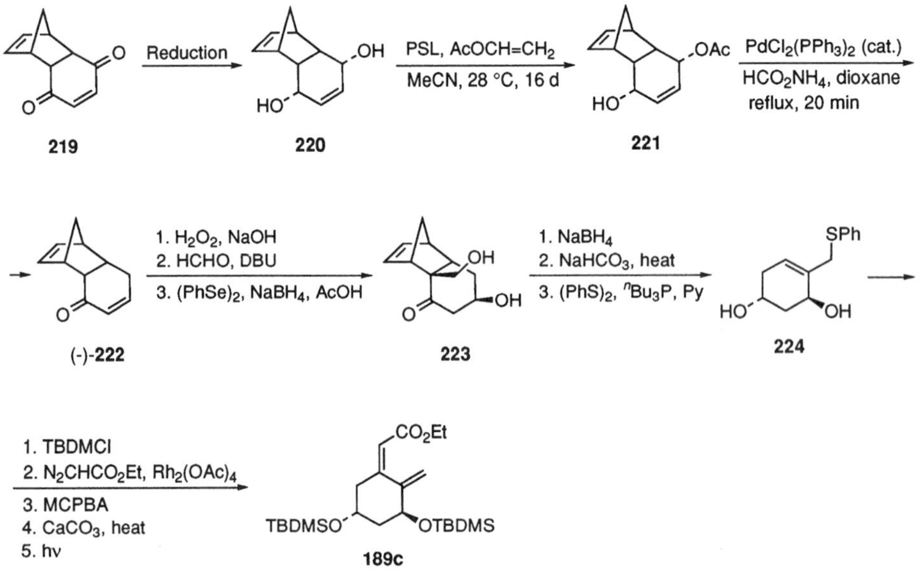

Scheme 13

enone (−)-**222**. β-Epoxidation of the enone (−)-**222** followed by hydroxymethylation and reduction of the *exo*-epoxide generated diol **223**. Reduction of the keto group, retro-Diels-Alder process by heating in NaHCO$_3$, and chemoselective substitution at the primary hydroxy center gave rise the sulfide diol **224**, which was then silylated at the secondary hydroxyl groups and treated with ethyl diazoacetate and a catalytic amount of rhodium(II) acetate to initiate concurrent ylide formation and the 2,3-Wittig rearrangement. The latter compound was oxidized to the sulfoxide, which was heated to give the A-ring precursor **189c** and its double-bond geometrical isomer, the latter of which can be isomerized (hν) into the former in an excellent yield. This new approach to the A-ring precursor **189c** allows us to obtain a 25% overall yield in ten steps from the synthon (−)-**222**.

Okamura, Gotor, and co-workers reported [46] that whereas *Chromobacterium viscosum* lipase selectively catalyzes the acylation of the C-5 hydroxyl of the three stereo-isomeric vitamin D A-ring precursors **225**, **227**, and **229**, only the C-3 hydroxyl of the fourth stereoisomer **231** is acylated under the same conditions in organic solvent (Scheme 14). In a convenient application (Scheme 15), the racemic vitamin D A-ring precursor **233**, possessing only the C-5 hydroxyl, was resolved using suitable conditions identified from the studies of the above-mentioned A-ring precursors of 1α,25-dihydroxyvitamin D$_3$. These studies of enzymatic acylation were first focused on the A-ring fragment **225**, which possesses the natural carbinol stereochemistry (*3S,5R*). Initial experiments concerned screening lipases [*Candida cylindracea* lipase (CCL), *Pseudomonas cepacia* lipase (PSL), *Chromobacterium viscosum* lipase (CVL), porcine pancreas lipase (PPL), and *Candida antarctica* lipase (CAL)] and proteases [subtilisin and papain] as catalysts to determine which enzyme gives the best regioselectivity for acylation. For the initial studies vinyl acetate was used as the acetylating agent and solvent. Of the enzyme-solvent systems studied, CVL in THF gave the highest regioselectivity toward the C-5-(*R*) hydroxyl (**226**), but the use of vinyl acetate as solvent afforded better results.

In order to evaluate other acylating agents and take advantage of the observation that CVL shows excellent acetylation selectivity and yield with vinyl acetate as solvent, the reaction of **225** with other vinyl esters was studied (panel A). The reactions were run at 45°C using an approximately 10:1 ratio of vinyl ester to diol. The vinyl ester was used as solvent whenever possible, but THF was added as solvent in those cases where the high boiling point of the acylating agent rendered cumbersome its removal from the product mixture. Acylation occurred selectively at the C-5-(*R*) hydroxyl except when vinyl benzoate was used as acylating agent. In this case the esterification did not take place even after heating at 45°C for several days (Table 15).

Panel B summarizes the results obtained for the acylation of the enantiomer of (*3S,5R*)-**225**, namely, (*3R,5S*)-**227**, shown in Scheme 14. For direct comparison to the previous data for **225**, CVL was also used, and it was determined that the reaction with **227** also proceeds with quantitative regioselectivity toward the hydroxyl group at the C-5 position. The acylation reaction is, however, faster for **227** with vinyl acetate than its enantiomer **225**. Complete conversion of **227** with vinyl acetate at room temperature occurs in 7 h, whereas that of enantiomer **225** requires heating of the reaction mixture of 45°C for 24 h (see entries **a** in Table 15 vs. Table 16). The regioselectivities obtained with the different vinyl esters are somewhat higher than for **225**. Significantly, **227** reacts with vinyl benzoate at room temperature, in contrasts with the lack of reaction of enantiomer **225** under somewhat more forceful conditions (room temperature vs. 45°C).

Panels C and D in Scheme 14 summarize similar acylation studies of the *cis*-stereoisomers (*3S,5S*)-**229** and (*3R,5R*)-**231**. As shown in panel C, CVL-catalyzed acetylation of the *cis*-diol-**229** occurred with complete regioselectivity (100%) toward the hydroxyl

Panel A

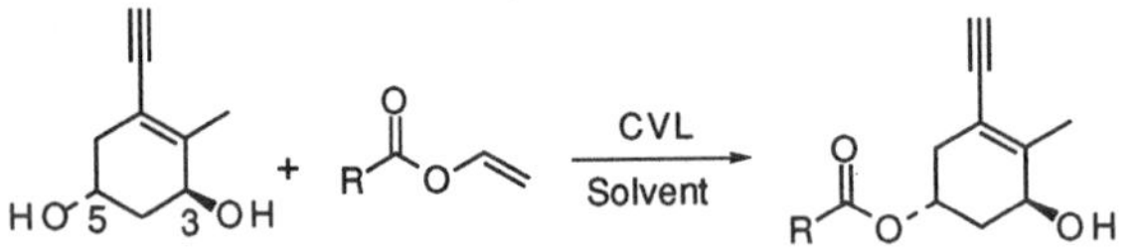

225 + [R= Me, Et, Pr, CH₂Cl, Ph] → 226

Table 15 Reaction of **225** with Vinyl Esters Catalyzed by CVL

Entry	R	Solvent	T (°C)	t (h)	Conv. (%)	C-5 (%)	C-3 (%)
a	Me	none	45	24	100	93 (90)	7
b	Et	none	45	40	100	87 (90)	13
c	Pr	none	45	96	100	84 (85)	16
d	ClCH₂	THF	45	45	100	85 (70)	15
e	Ph	THF	45	113	0	–	–

Panel B

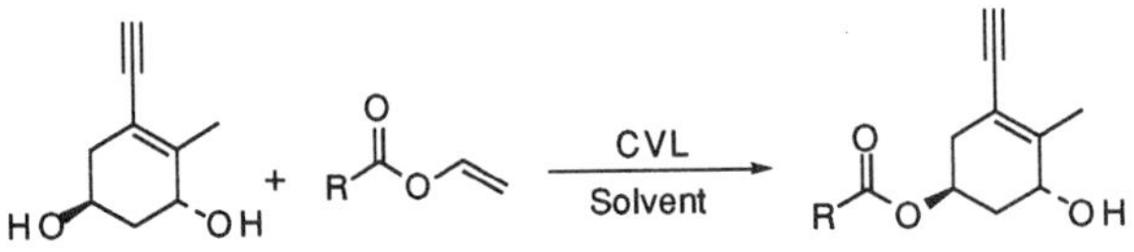

227 + [R= Me, Et, Pr, CH₂Cl, Ph] → 228

Table 16 Reaction of **227** with Vinyl Esters Catalyzed by CVL

Entry	R	Solvent	T (°C)	t (h)	Conv. (%)	C-5 (%)	C-3 (%)
a	Me	none	25	7	100	94 (88)	6
b	Et	none	25	7	100	94 (95)	6
c	Pr	none	25	5	100	95 (70)	5
d	ClCH₂	THF	25	21	100	78 (71)	22
e	Ph	THF	25	119	100	92 (70)	8

Scheme 14

group at the C-5 position. Excellent results were also obtained with vinyl butyrate and vinyl chloroacetate (around 100% and 93% regioselectivity, respectively). As shown in panel D, parallel acylation of the remaining stereoisomer *cis*-diol **231** proved most interesting. Here C-3 hydroxylation dominates, just the opposite of the three C-3,5 stereoisomers **225**, **227**, and **229**. The selectivity and rate were qualitative. Vinyl butyrate, used as both the acylating reagent and solvent, exhibited complete regioselectivity toward the hydroxyl group at the C-3 position.

Given the difference in reaction times observed for enantiomers **225** and **227** as well as **229** and **231**, the possibility that CVL could be useful in the kinetic resolution of very useful racemic A-ring fragment **233** was examined. Using CVL, the resolution (±)-**233** afforded the natural enynol (−) isomer in a single step (Scheme 15). The reaction was carried out at room temperature using CVL with vinyl acetate acting as both acylating

Panel C

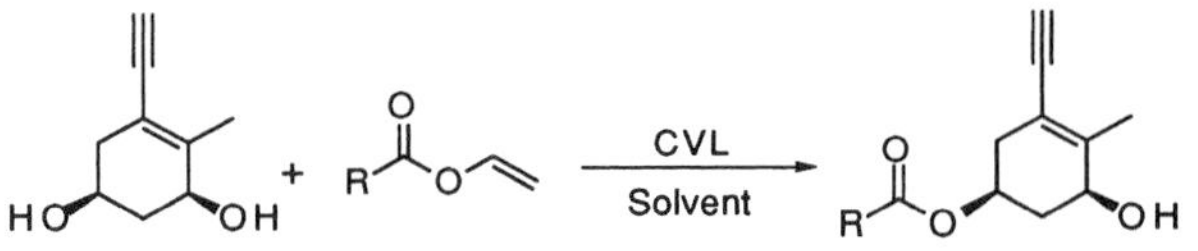

[R= Me, Pr, CH₂Cl]

Table 17 Reaction of **229** with Vinyl Esters Catalyzed by CVL

Entry	R	Solvent	T (°C)	t (h)	Conv. (%)	C-5 (%)	C-3 (%)	C-3,5 (%)
a	Me	none	25	21	100	100 (90)	0	0
c	Pr	none	25	48	100	98 (89)	0	2
d	ClCH₂	THF	25	23	100	93 (86)	7	0

Panel D

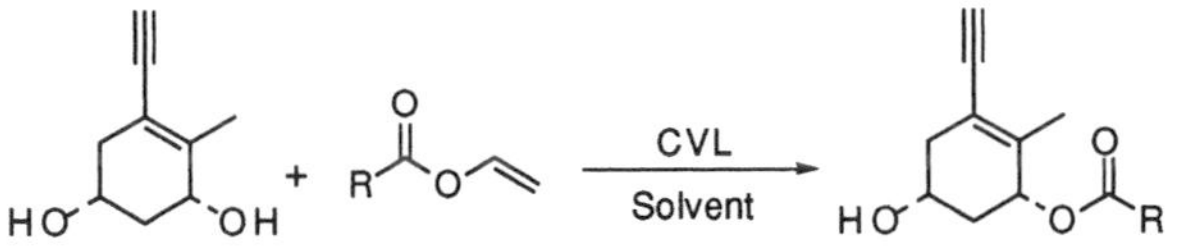

[R= Me, Pr, CH₂Cl]

Table 18 Reaction of **231** with Vinyl Esters Catalyzed by CVL

Entry	R	Solvent	T (°C)	t (h)	Conv. (%)	C-5 (%)	C-3 (%)	C-3,5 (%)
a	Me	none	25	48	100	3.5	90 (80)	6.5
c	Pr	none	25	144	100	<2	>98 (90)	0
d	ClCH₂	THF	25	92	100	2.5	91 (85)	6.5

Scheme 14   Continued

agent and solvent. After 18 h, with gas chromatographic analysis revealing 52% conversion, the reaction was stopped and flash chromatography (silica gel, 20% EtOAc-hexanes) of the product mixture gave the enantiomerically pure (S)-(-)-**233** in high yield (97%, about 100% e.e.). The remaining acetate (99% yield) was treated with sodium methoxide to afford (R)-(+)-**233**, which had 94% e.e. Thus, this route provides a convenient direct route to **233**, which possesses the natural C-5 configuration (corresponding to the steroidal 3β-hydroxyl) of vitamin D₃.

Vandewalle's group [56] have reported the synthesis of 19-*nor*-1α,25-dihydroxyvitamin D₃ **248** from the 25-hydroxy-Grundmann's ketone derivative **246** and an A-ring precursor prepared from *cis*-1,3,5-cyclohexanetriol **235** via a chemoenzymatic approach (Scheme 16). Their interest arises from the fact that among the A-ring modifications, deletion of the 19-exomethylene function has been shown to induce interesting biological activities. They observed that lipase-catalyzed transesterification of using vinyl acetate in

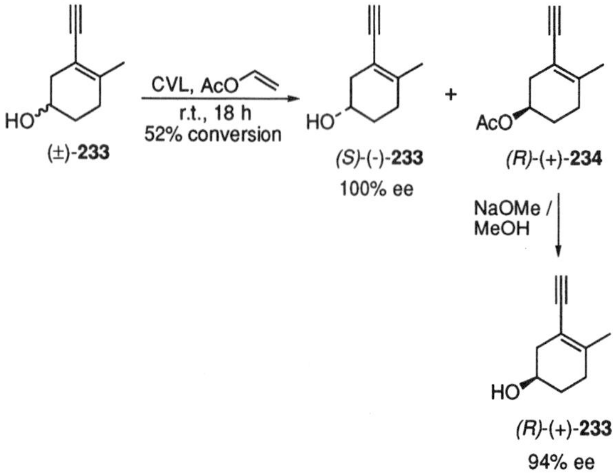

Scheme 15

triol **235** neatly stopped at stage of diacetate **236**. The tosylate **237** was therefore not a substrate for enzyme-catalyzed hydrolysis. The corresponding diol **238**, however, was a substrate for the lipase-catalyzed esterification using vinyl acetate and affording compound **239** with 95% e.e. Substitution of tosylate **239** was uneventful but cyclopropanation of **240** failed because of incompatibility of the acetate function. After protecting-group interconversion, intramolecular alkylation of **241** and deprotection led to **242**. This alcohol is a suitable substrate for Mitsunobu inversion. Methanolysis of the resulting benzoate, TBDMS protection, and DIBALH reduction of **243** afforded aldehyde **244**. Finally, aldehyde **244** was smoothly converted to alkyne **245** using dimethyldiazomethylphosphonate. In order to synthesize 19-*nor*-1α,25-dihydroxyvitamin D₃ **248** the reaction of lithiated alkyne **245** with ketone **246** was carried out giving propargylic alcohol, which was subsequently reduced to the *E*-allylic alcohol **247**. Acid-catalyzed solvolysis of **247**, a vinylog of a cyclopropyl alcohol, gave 19-*nor*-1α,25-dihydroxyvitamin D₃ **248**.

In a new approach to produce 19-*nor*-1α-hydroxyvitamin D₃ **257**, Yong and Vandewalle [57] describe their strategy based on the synthesis of the novel 19-*nor*-A-ring precursor **253** via cyclopropanation of homoallylic alcohol **250** (Scheme 17). The latter was easily obtained from the known 3-cyclopentenol **249** by temporary protection of the hydroxyl group, bromine addition, dehydrobromination, and subsequent desilylation to afford homoallylic alcohol **250**. The cyclopropanation was directed entirely by the homoallylic hydroxy group generating the *endo*-alcohol (±)-**251** stereospecifically. All attempts to carry out the resolution via the formation of diastereoisomers or using enzyme catalysis failed on racemic bicyclo-(±)-**251**. They therefore first carried out the required inversion of the hydroxy function in (±)-**251**. The resulting racemic epimer (±)-**252** was found to be a suitable substrate for lipase-catalyzed transesterification with vinyl acetate. An excellent result was obtained with PSL (*Pseudomonas cepacia* lipase); only the (*R*)-(+)-alcohol was acetylated leaving behind enantiopure alcohol (*S*)-(−)-**252** (more than 99% e.e.), which was then protected as silyl ether affording A-ring precursor **253**. The α,β-unsaturated aldehyde **255**, required for coupling with (*S*)-(−)-**253**, was obtained upon reaction of Grundmann's ketone **254** with lithiated *N*-*tert*-butyl-2-(trimethylsilyl)-acetaldimine followed by acid-catalyzed hydrolysis and the desired *E* isomer of **255** re-

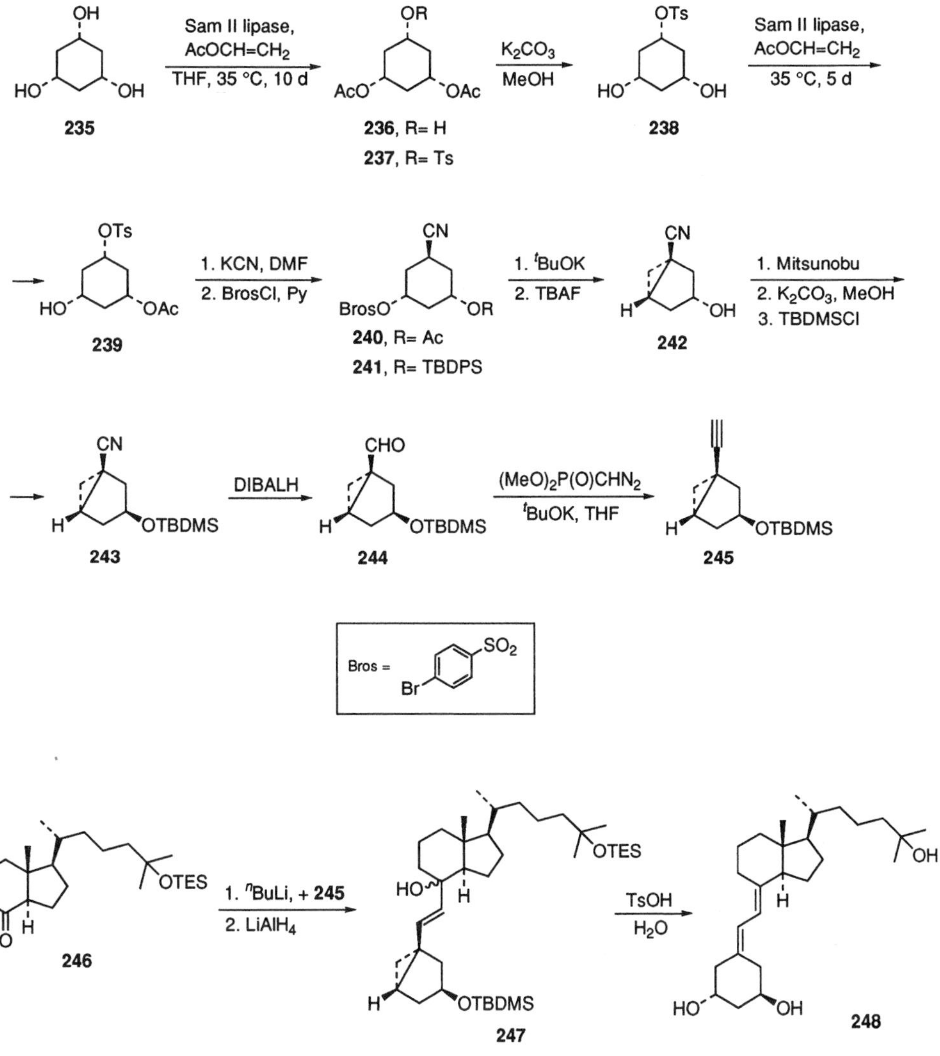

Scheme 16

acted with lithiated **253** to obtain the epimeric mixture of the 19-*nor*-cyclovitamin **256**. Finally, acid-catalyzed solvolysis of **256** and subsequent deprotection of the 1-hydroxy group produced 19-*nor*-1α-hydroxyvitamin D₃ **257**.

Gotor's group [47] have demonstrated the preparation of several A-ring carbamate derivatives of 1α,25-dihydroxyvitamin D₃ through a selective alkoxycarbonylation (Scheme 18) and subsequent chemical transformation. The A-ring modification of 1α,25-dihydroxyvitamin D₃ [**188**, 1α,25-(OH)₂-D₃] is an important area of analog studies in the investigation of the biological activity of vitamin D−related structures. An efficient synthesis of 1α,25-(OH)₂-D₃ C-5 A-ring carbamate derivatives **265** and amino acid derivatives **267** was developed by applying a two-step chemoenzymatic strategy involving the enzymatic synthesis of carbonates followed by reaction with amino derivatives (Scheme 19).

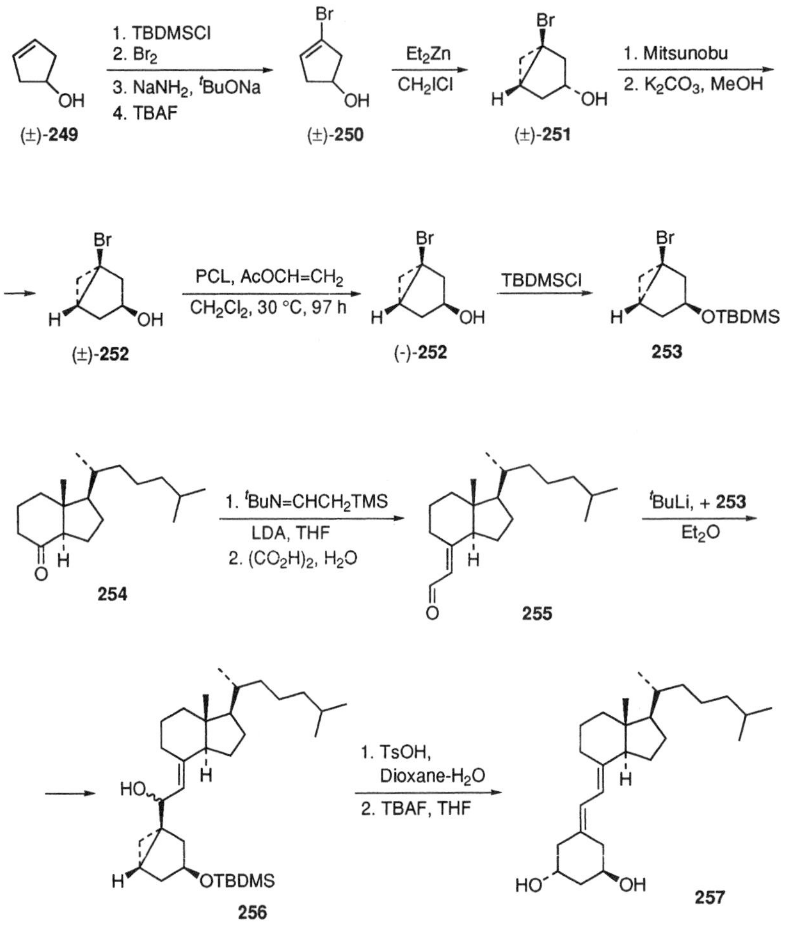

Scheme 17

The studies of enzymatic alkoxycarbonylation were focused on the A-ring fragment **225** (Scheme 18), which possesses the natural enyne stereochemistry (*3S,5R*) used to prepare certain 1α,25-(OH)₂-D₃ analogs. Initial experiments concerned screening *Chromobacterium viscosum* lipase (CVL), *Pseudomonas cepacia* lipase (PSL), and *Candida antarctica* lipase (CAL) as catalysts to determine which enzyme gives the best regioselectivity for alkoxycarbonylation. For the initial studies (vinyloxycarbonyl)oxime **258** was used, and attention was first focused on CVL in THF, using a ratio of carbonate **258** to diol **225** of 10:1, since this gave the best results in previous studies of acylation [46].

In order to find the best enzyme–solvent system for alkoxycarbonylation of the C-5-(*R*) hydroxyl group with total conversion and good yield, other enzymes were tested. Of the enzyme–solvent systems studied, CAL-toluene gave the best result. It catalyzed vinyloxycarbonylation with total conversion in 4 h at 30°C. Only the C-5-(*R*) hydroxyl group was protected and **263a** was obtained in excellent yield (98% of isolated product) (Table 19). It is noteworthy that no C-3 alkoxycarbonylation products were observed in

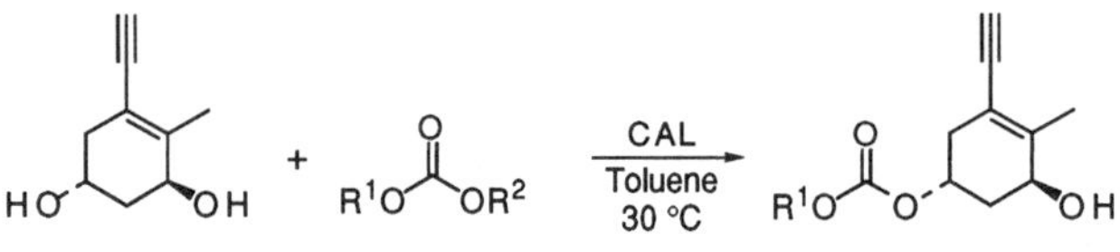

Table 19 Enzymatic Reaction of diol **225** with Carbonate Derivatives **258-262** Catalyzed by CAL

Carbonate	R¹	R²	Ratio 225 : carbonate	t (h)	Product	Yield (%)[a]
258	CH₂=CH	Me₂C=N	1 : 10	4	263a	98
259	Me	Me₂C=N	1 : 10	6	263b	80
259	Me	Me₂C=N	1 : 5	12.5	263b	70
260	PhCH₂	Me₂C=N	1 : 10	72	263c	85
261	Ph	Me₂C=N	1 : 10	36	263d	90
262	Ph	CH₂=CH	1 : 10	24	263a + 263d	b

a. Isolated yield. **b.** As a mixture of carbonates **263a** (33.5% isolated yield) and **263d** (54% isolated yield).

Scheme 18

any of these processes and alkoxycarbonylation did not take place in the absence of enzyme even if stronger conditions were used.

A variety of alkoxycarbonylating agents were subsequently examined to study the utility of CAL in toluene and take advantage of the fact that CAL shows excellent vinyloxycarbonylation selectivity and yield (Table 19). Reactions were run at 30°C using a ratio of carbonates **258–262** to diol **225** of 10:1. Good to excellent yields of carbonates **263b–263d** were achieved in 6–72 h. Alkoxycarbonylation occurred selectively at the C-5-(R) hydroxyl in all cases.

In order to demonstrate the synthetic utility of these A-ring fragment derivatives, the carbamates **265** were synthesized from carbonate **263a** (panel A, Scheme 19). Four kinds of compounds have been tested as a proof of the versatility of the procedure: (1) ammonia (giving place to compound **265a**), (2) amines (form compound **265b**), (3) amino alcohols, and (4) diamines (Scheme 19, Table 20). Interesting products to synthesize are A-ring carbamate derivatives reached from linear amino alcohols or diamines. Carbonate **263a** was allowed to react with 2-amino-1-ethanol (**264c**), 6-amino-1-hexanol (**264d**), and 1,3-propanediamine (**264e**) at 60°C in THF for approximately 2 days to provide carbamates **265c–e** in good yields (entries 3–5; Table 20).

The aforementioned methodology was extended to synthesize some amino acid A-ring precursors. The best results were obtained by submitting the corresponding vinyloxycarbonylated C-5-(R) 1α,25-(OH)₂-D₃ A-ring precursor **263a** to reaction with the amino acid sodium salts **266a–b** in dimethyl formamide at 60°C to provide **267a–b** (panel B, Scheme 19 and Table 21).

2. CD Ring Precursors

Stepanenko and Wicha [58] reported an enantioselective synthesis of (24S),25-dihydroxyvitamin D CD ring synthon from easily accessible optically active hydroxy ketone

Panel A

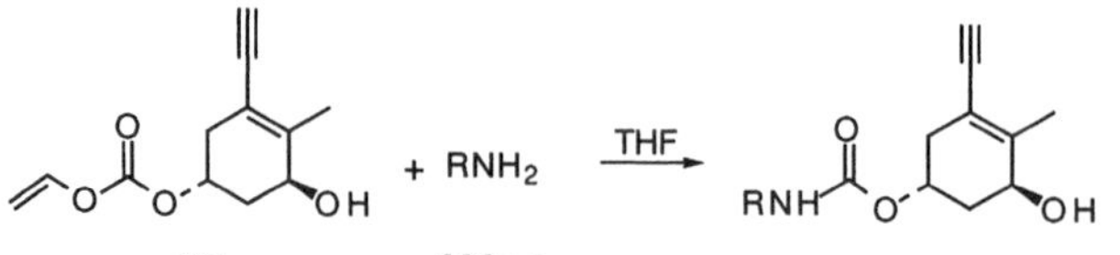

263a   264a-e

265a-e

Table 20   Reaction of Carbonate **263a** with Amino Derivatives **265** in THF

Entry	R	Ratio 263a : 264	T (°C)	t (h)	A-Ring Derivative	Yield (%)[a]
1	H	b	25	18	265a	100
2	Bu	1 : 10	60	24	265b	78
3	HO(CH$_2$)$_2$	1 : 2.5	60	48	265c	80
4	HO(CH$_2$)$_6$	1 : 2.5	60	48	265d	82
5	H$_2$N(CH$_2$)$_3$	1 : 2.5	60	48	265e	76

a. Isolated yield. **b.** Bubbling at 0 °C on the reaction solution.

Panel B

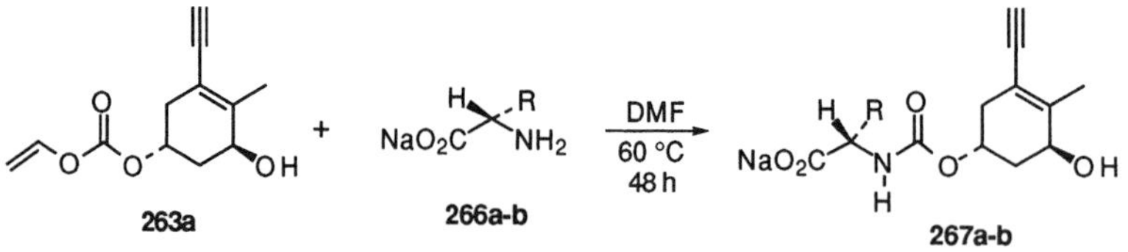

263a   266a-b

267a-b

Table 21   Reaction of Carbonate **263a** with Amino Acid Derivatives **267**

Entry	R	Amino Acid	A-Ring Derivative	Yield (%)[a]
1	H	266a, Glycine	267a	87
2	CH$_2$SMe	266b, S-Methyl-L-Cysteine	267b	91

a. Isolated yield.

Scheme 19

(*S*)-**269** by Baker's yeast reduction of diketone **268** [59]. Thus, oxidation of hydroxy ketone (*S*)-**269** with MCPBA yielded lactone (*S*)-**270** (Scheme 20). The lactone **270** was opened up and then formed the acetonide **271**. Compound **271** was allowed to react with LDA to generate the corresponding enolate, which was quenched with trimethylsilyl chloride. The resulting product **272** was set for two tandem Mukaiyama-Michael conjugate additions. The reaction of **272** with **273** in the presence of TrSbCl$_6$ and then with **274** yielded a mixture whose major component **275** was crystallized and treated first with DIBALH to yield a mixture of the epimeric diols, after which the primary hydroxy group was esterified with tosyl chloride and the sulfide moiety was oxidized to sulfone. This sulfone was then treated with an excess of LiAlH$_4$ and the *trans*-hydrindane derivative was obtained, which yielded the required diol (*24S*)-**277** by removal of the actual protective group.

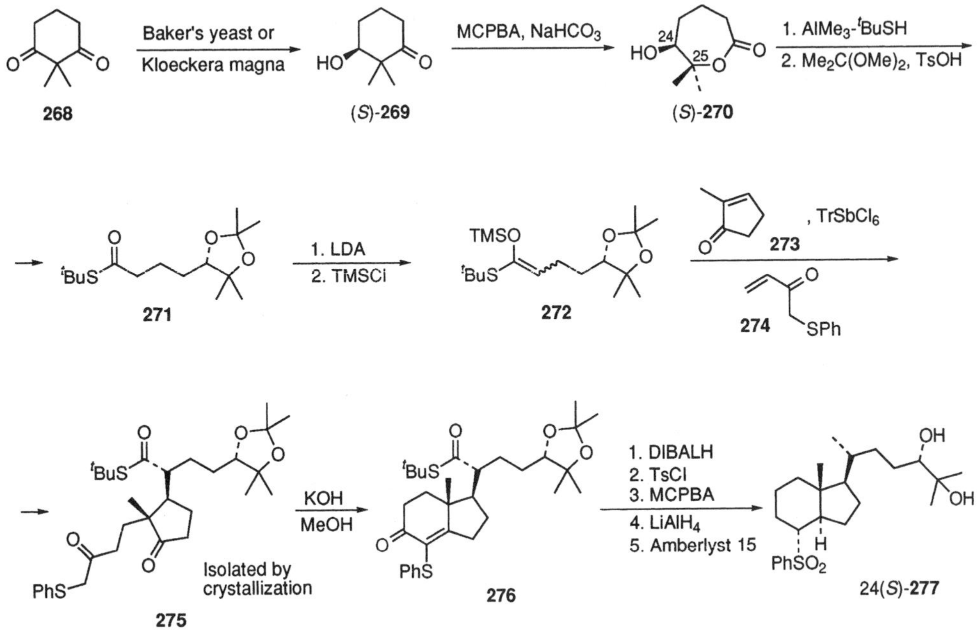

Scheme 20

NOTES AND REFERENCES

1. (a) FS Sariaslani, JP Rosazza. Enzyme Microb Technol 6:242–253, 1984. (b) GM Whitesides, CH Wong. Angew Chem Int Ed Engl 24:617–638, 1985. (c) JB Jones. Tetrahedron 42:3351–3403, 1986. (d) H Yamada, S Shimizu. Angew Chem Int Ed Engl 27:622–642, 1988. (e) R Csuk, BI Glänzer. Chem Rev 91:49–97, 1991. (f) E Santaniello, P Ferraboschi, P Grisenti, A Manzocchi. Chem Rev 92:1071–1140, 1992. (g) NJ Turner. Nat Prod Rep 11:1–15, 1994. (h) R Azerad. Bull Soc Chim Fr 132:17–51, 1995. (i) F Theil. Chem Rev 95:2203–2227, 1995. (j) E Schoffers, A Golebiowski, CR Johnson. Tetrahedron 52:3769–3826, 1996. (k) M Schelhaas, H Waldmann. Agnew Chem Int Ed Engl 35:2056–2083, 1996. (l) SM Roberts. J Chem Soc Perkin Trans 1:157–169, 1998.
2. (a) R Azerad, CJ Sih. Enantioselective Synthesis using Biological Systems. Biocatalysis special issue Vol. 3(1–2), 1990. (b) EM Meijer. Biocatalysis in Organic Chemistry. Recl Trav Chim Pays-Bas special issue Vol. 110(5), 1991. (c) DHG Crout, SM Roberts, JB Jones. Enzymes in Organic Synthesis. Tetrahedron: Asymmetry special issue Vol. 4(5–6), 1993. (d) K Kieslich, HGW Leuenberger, D Seebach. Proceedings of the International Bioorganic Symposium on Biotransformations in Organic Chemistry: Principles and Applications, Interlaken 1993. Chimia special issue vol. 47(4), 1993. (e) DHG Crout, H Griengl, K Faber, SM Roberts. Proceedings of the European Symposium on Biocatalysis, Graz 1993. Biocatalysis special issue Vol. 9(1–4), 1994.
3. (a) J Tramper, HC VanderPlas, P Linko. Biocatalysis in Organic Synthesis, Studies in Organic Synthesis. Amsterdam: Elsevier, 1985, Vol. 22. (b) R Porter, S Clark. Enzymes in Organic Synthesis, CIBA Foundation Symposium Vol. 111. London: Pitman, 1985. (c) MP Schneider. Enzymes as Catalyst in Organic Synthesis, NATO ASI Series C. Dordrecht: Reidel, 1986. (d) HG Davies, RH Green, DR Kelly, SM Roberts. Biotransformations in Preparative Organic Chemistry. The Use of Isolated Enzymes and Whole Cells Systems in Synthesis. London: Academic Press, 1989. (e) K Faber. Biotransformations in Organic Chemistry (3rd ed.). Berlin: Springer-Verlag, 1997. (f) S Servi. Microbial Reagents in Organic Synthesis, NATO ASI Series

C. Dordrecht: Kluwer, 1992. (g) CH Wong, GW Whitesides. Enzymes in Synthetic Organic Chemistry. Oxford: Elsevier, 1994.

4. (a) DG Drueckhammer, WJ Hennen, RL Pederson, CF Barbas III, CM Gautheron, T Krach, CH Wong. Synthesis: 499–525, 1991. (b) HJM Gijsen, L Qiao, W Fitz, CH Wong. Chem Rev 96:443–473, 1996.
5. Enzyme Nomenclature. New York: Academic Press, 1992.
6. S Riva, R Bovara, P Pasta, G Carrea. J Org Chem 51:2902–2906, 1986.
7. Abbreviations: NAD, β-nicotinamide adenine dinucleotide; NADP, β-nicotinamide adenine dinucleotide phosphate.
8. The overall equilibrium constant of the oxidation of cholic acid coupled to the reduction of α-ketoglutarate is higher than 10^6. The overall equilibrium constant of the reduction of dehydrocholic acid coupled to the oxidation of formate is higher than 10^{15}.
9. G Carrea, P Cremonesi. Meth Enzymol 136:150–157, 1987.
10. G Carrea, S Riva, R Bovara, P Pasta. Enzyme Microb Technol 10:333–340, 1988.
11. Abbreviations: K_m, Michaelis constant; V_{max}, maximal enzymatic velocity; K_i, product inhibition constant.
12. Turnover number is the number of moles of product generated per mole of coenzyme.
13. S Riva, R Bovara, L Zetta, P Pasta, G Ottolina, G Carrea. J Org Chem 53:88–92, 1988.
14. R Bovara, E Canzi, G Carrea, A Pilotti, S Riva. J Org Chem 58:499–501, 1993.
15. JH Iser, A Sali. Drugs 21:90–119, 1981.
16. IA Macdonald, BA White, PB Hylemon. J Lipid Res 24:1119–1126, 1983.
17. S Riva, G Ottolina, G Carrea. J Chem Soc Perkin Trans 1:2073–2074, 1989.
18. JB Jones. Tetrahedron 42:3351–3403, 1986.
19. CH Wong, GM Whitesides. J Am Chem Soc 105:5012–5014, 1983.
20. (a) W Charney, HL Herzog. Microbial Transformations of Steroids. New York: Academic Press, 1967. (b) Y Ahronowitz, G Cohen. Sci Am 245:141, September 1981.
21. DH Peterson, HC Murray. J Am Chem Soc 74:1871–1872, 1952.
22. G Fantin, S Ferrarini, A Medici, P Pedrini, S Poli. Tetrahedron 54:1937–1942, 1998.
23. ICE (Industria Chimica Emiliana), Reggio Emilia, Italy.
24. VCO Njar, E Caspi. Tetrahedron Lett 28:6549–6552, 1987.
25. S Riva, AM Kliblanov. J Am Chem Soc 110:3291–3295, 1988.
26. S Riva, R Bovara, G Ottolina, F Secundo, G Carrea. J Org Chem 54:3161–3164, 1989.
27. M Adamczyk, YY Chen, JR Fishpaugh, JC Gebler. Tetrahedron: Asymmetry 4:1467–1468, 1993.
28. A Bertinotti, G Carrea, G Ottolina, S Riva. Tetrahedron 50:13165–13172, 1994.
29. A Baldessari, MS Maier, EG Gros. Tetrahedron Lett 36:4349–4352, 1995.
30. A Suksamrarn, P Pattanaprateep. Tetrahedron 51:10633–10650, 1995.
31. B Danieli, G Lesma, M Luisetti, S Riva. Tetrahedron 53:5855–5862, 1997.
32. CY Byon, M Gut, V Toome. J Org Chem 46:3901–3903, 1981.
33. N Cohen, WF Eichel, RJ Lopresti, C Neukom, C Saucy. J Org Chem 41:3505–3511, 1976.
34. CT Goodhue, JR Schaeffer. Biotechnol Bioeng 13:203–214, 1971.
35. P Ferraboschi, A Fiecchi, P Grisenti, E Santaniello. J Chem Soc Perkin Trans 1:1749–1752, 1987.
36. (a) P Ferraboschi, P Grisenti, R Casati, A Fiecchi, E Santaniello. J Chem Soc Perkin Trans 1: 1743–1748, 1987. (b) HGW Leuenberger, W Bguth, R Barner, M Schmid, R Zell. Helv Chim Acta 62:455–463, 1979.
37. P Ferraboschi, A Molatore, E Verza, E Santaniello. Tetrahedron: Asymmetry 7:1551–1554, 1996.
38. P Ferraboschi, S Rezaelahi, E Verza, E Santaniello. Tetrahedron: Asymmetry 9:2193–2196, 1998.
39. S Takano, K Inomata, K Ogasawara. J Chem Soc Chem Commun: 271–272, 1989.
40. S Takano, K Inomata, K Ogasawara. J Chem Soc Chem Commun: 1544–1546, 1990.
41. (a) H Gibian, K Kieslich, HJ Koch, H Kosmol, C Rufer, E Schröder, R Vössing. Tetrahedron Lett 7:2321–2329, 1966. (b) H Kosmol, K Kieslich, R Vössing, HJ Koch, K Petzoldt, H

Gibian. Liebigs Ann Chem 701:199–205, 1967. (c) C Rufer, H Kosmol, E Schröder, K Kies-lich, H Gibian. Liebigs Ann Chem 702:141–148, 1967. (d) WM Dai, WS Zhou. Tetrahedron 41:4475–4482, 1985. (e) S Schwarz, G Truckenbrodt, M Meyer, R Zepter, G Weber, C Carl, M Wentzke. J Prakt Chem 323:729–736, 1981. (f) DW Brooks, H Mazdiyasni, S Chakrabarti. Tetrahedron Lett 25:1241–1244, 1984.

42. T Sugahara, K Ogasawara. Tetrahedron Lett 37:7403–7406, 1996.

43. T Sugahara, K Ogasawara. Tetrahedron Lett 37:205–208, 1996.

44. (a) GD Zhu, WH Okamura. Chem Rev 95:1877–1952, 1995. (b) R Bouillon, WH Okamura, AW Norman. Endocr Rev 16:200–257, 1995.

45. (a) JM Aurrecoechea, WH Okamura. Tetrahedron Lett 28:4947–4950, 1987. (b) WH Okamura, JM Aurrecoechea, RA Gibbs, AW Norman. J Org Chem 54:4072–4083, 1989.

46. S Fernández, M Ferrero, V Gotor, WH Okamura. J Org Chem 60:6057–6061, 1995.

47. M Ferrero, S Fernández, V Gotor. J Org Chem 62:4358–4363, 1997.

48. M Shimada, S Kobayashi, M Ohno. Tetrahedron Lett 29:6961–6964, 1988.

49. S Kobayashi, K Kamiyama, T Iimori, M Ohno. Tetrahedron Lett 25:2557–2560, 1984.

50. S Kobayashi, J Shibata, M Shimada, M Ohno. Tetrahedron Lett 31:1577–1580, 1990.

51. The carbon numbers are expressed according to the vitamin D numbering in this scheme.

52. S Takano, T Yamane, M Takahashi, K Ogasawara. Synlett: 410–412, 1992.

53. S Takano, M Suzuki, K Ogasawara. Tetrahedron: Asymmetry 4:1043–1046, 1993.

54. S Takano, Y Higashi, T Kamikubo, M Moriya, K Ogasawara. Synthesis: 948–950, 1993.

55. T Kamikubo, K Ogasawara. J Chem Soc Chem Commun: 1951–1952, 1995.

56. P Huang, K Sabbe, M Pottie, M Vanderwalle. Tetrahedron Lett 36:8299–8302, 1995.

57. W Yong, M Vanderwalle. Synlett: 911–912, 1996.

58. W Stepanenko, J Wicha. Tetrahedron Lett 39:885–888, 1998.

59. (a) K Mori, H Mori. Org Synth 68:56–62, 1990. (b) Y Lu, G Barth, K Kieslich, PD Strong, WL Duax, C Djerassi. J Org Chem 48:4549–4554, 1983.

21
Biocatalytic Synthesis of Enantiopure Compounds Using Lipases

Per Berglund and Karl Hult
Royal Institute of Technology, Stockholm, Sweden

I. INTRODUCTION

Our aim with this chapter is to illustrate the diversity and power of lipases as stereoselective biocatalysts. Some representative examples are highlighted from this very broad field. We summarize some examples in kinetic schemes for interpretations of the reaction systems, and for pointing out the difference in kinetic complexity between a system involving a chiral acyl moiety and one involving a chiral nucleophile. We focus our overview on chiral acids, esters, and alcohols as reactants.

A. Lipases

Lipases are carboxyl ester hydrolases (E.C. 3.1.1.3) that have developed the skill to hydrolyze water-insoluble esters such as triglycerides. During the catalysis the active site of a lipase is exposed to the lipid phase, which means that the active site must be designed to work in a hydrophobic environment. This property might explain why many lipases retain high activities when used in organic solvents. Lipases accept a wide range of substrates, which adds to their versatile use in organic stereoselective synthesis. Today many lipases with high stability are commercially available in pure and immobilized form. For a recent extensive review of lipase structure and applications, see Schmid and Verger [1].

B. Lipase Mechanism

Lipases are serine hydrolases and follow a bi-bi ping-pong reaction mechanism [2,3]. During the catalysis (Scheme 1) an acyl enzyme is formed in the acylation step where the serine hydroxyl is acylated by the acyl donor (the first substrate, R_1COOR_2) [4]. The first product (HOR_2) is then released. The acyl acceptor (the second substrate, HOR_3), which in the natural reaction is water but can in principle by any nucleophile, then reacts with the acyl enzyme in the deacylation step to form the second product (R_1COOR_3) and the free enzyme.

Lipase hydrolysis of emulsified esters in water does not obey simple Michaelis-Menten kinetics. The hydrolysis of water-insoluble substrates involves the adsorption of the lipase to the substrate–water interface [5]. This adsorption is in many cases connected

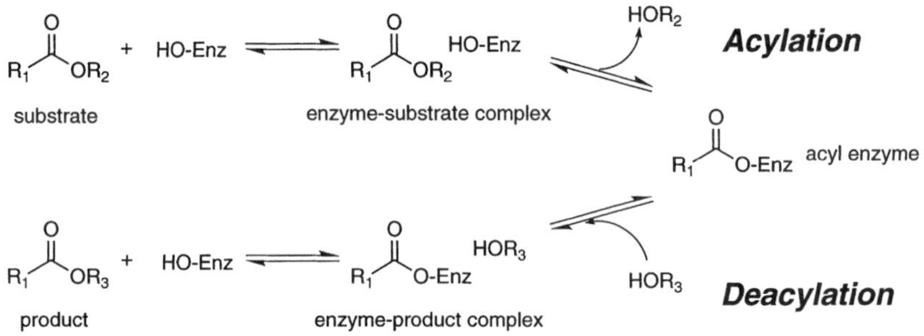

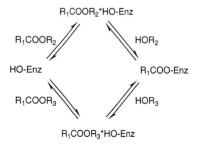

Scheme 1   Reaction mechanism of a serine hydrolase. The serine hydroxyl in the enzyme active site is shown.

to a conformational change of the enzyme. The reaction rate is thus a function of the adsorption equilibrium of the water-soluble lipase to the interface and is dependent on the total interfacial area. Lipases are also active in organic solvents, in which the substrates are soluble. In such systems the kinetics is simplified as the adsorption step is lacking, and normal enzyme kinetic models can therefore be applied [6].

The complications enforced by the partitioning of the lipase will not be treated further here. Instead we will use a reaction mechanism as outlined in Scheme 2. In this representation of the reaction from Scheme 1 the central role of the enzyme (HO-Enz), the competitive substrates for the free enzyme (R_1COOR_2, R_1COOR_3), and the competitive substrates for the acyl enzyme (R_2OH, R_3OH) are emphasized.

C.   Chiral Acyl Acceptors

The resolution of racemic alcohols by esterification can be described with the reaction in Scheme 3. The crucial step for the selectivity is the competition around the acyl enzyme, which involves the two enantiomers of the alcohol, the leaving group of the acyl donor, and water. The intrinsic enantioselectivity of the lipase will be governed by this competition. The use of acyl donors with leaving groups of poor nucleophilicity drives the reaction by abolishing the leaving group as an acyl acceptor. Water will play the role as substrate as long as the water activity is not kept very low and will be noticed as formation

Scheme 2   In this representation of the transesterification reaction shown in Scheme 1, catalyzed by a lipase, the competition between the first substrate (R_1COOR_2) and the product (R_1COOR_3) for the free enzyme (HO-Enz) can be seen. The second substrate (HOR$_3$) and the first product (HOR$_2$) compete for the acyl enzyme (R_1COO-Enz). *Denotes enzyme-bound complexes.

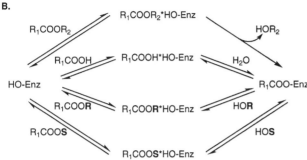

Scheme 3 A: Lipase-catalyzed resolution of a chiral acyl acceptor. B: The scheme shows the competition between two enantiomeric alcohols (**HOR** and **HOS**) and water for the acyl enzyme (R_1COO-Enz). The nucleophile (HOR_2) released from the acyl donor is supposed to tautomerize, evaporate, or otherwise leave the system. The substrates competing for the enzyme (HO-Enz) are the acyl donor (R_1COOR_2), the products formed during the catalysis (R_1COO**R** and R_1COO**S**), and any formed acid (R_1COOH). The enantiomeric alcohol moieties **R** and **S** are shown in boldface.

of acid [7]. The selectivity can depend on what acyl donor is used. Different acyl residues will occupy and fill the active site in the acyl enzyme in different ways, leading to various restrictions for the incoming acyl acceptor.

As can be seen from Scheme 3, all reaction steps are reversible, which has the consequence that the products will start to efficiently compete with the acyl donor for the free enzyme at high conversions of the acyl acceptor. This competition will lead to a decreased enantiomeric excess (e.e.) in the product as it will be used as acyl donor and the acyl acceptor will be transferred back to the substrate pool and has to repeat the competition at the acyl enzyme level. This decrease in enantiomeric excess is not due to an intrinsic property of the enzyme but of the reaction system. The acyl donor can out-compete the products by being used in a high concentration and be chosen to have a high specificity (k_{cat}/K_M) for the enzyme.

D. Chiral Acyl Donors

Reactions involving chiral acyl donors are kinetically more complex as the acyl donor will form diastereomeric acyl enzymes. The selectivity of the enzyme is in this case defined by four different transition states [8], two for each acyl donor one of which leads to the acyl enzyme whereas the other involves its breakdown. As seen in Scheme 4, the formation of the acyl enzymes is a competition between the two enantiomeric acyl donors (**RCOOH** and **SCOOH**) for the free enzyme (HO-Enz). For the breakdown of the acyl enzyme there is a different competitive situation without connection between the two acyl enzymes. The competition involves water and alcohol (in the cases of hydrolysis and esterification) and occurs independently at the two acyl enzymes (**RCOO**-Enz and **SCOO**-Enz).

The structure of the acyl enzymes might differ much more than at the stereocenter of the acyl moieties, as the docking modes of the acyl moieties in the active site can be significantly different [9]. This difference in structure might result in large differences in

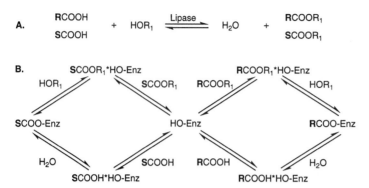

Scheme 4 A: Lipase-catalyzed resolution of a chiral acyl donor (**R**COOH and **S**COOH) by esterification or hydrolysis (the reverse reaction). B: The reaction goes through two diastereomeric acyl enzymes (**R**COO-Enz and **S**COO-Enz). The enantiomeric acyl groups **R** and **S** are shown in boldface.

enantioselectivity when different alcohols are used. This was recently experimentally evidenced for *Candida rugosa* lipase [10] and will be discussed further in Sec. IV.

E. Models for Selectivity Prediction

There are several models for lipase enantioselectivity that can be used to predict which enantiomer of a chiral substrate will react fastest. Most models are substrate-based. This means that the lipase structure was not taken into account when the prediction model was set up. Instead, only common features of the fast-reacting enantiomer were considered. These models have great value when synthetic routes are planned but are less valuable in the understanding of the enantioselective process and design of more selective systems.

In the case of secondary alcohols, the situation is rather simple. Steric effects are predominant and a size preference related to the (*R*) enantiomer of secondary alkanols is followed (Fig. 1). This preference was pointed out by Kazlauskas et al. [11] working with *C. rugosa* and *Pseudomonas cepacia* lipases and has been shown to be valid for most lipases. Two other examples are *Candida antarctica* lipase B [12] and lipase QL from *Alcaligenes* [13]. However, in the case of *C. rugosa* lipase one has to be a little cautious as the rule is valid for cyclic alcohols but not for acyclic ones [11]. It has been suggested that the common steric preference among lipases for secondary alcohols stems from their common tertiary structure [14].

Chiral primary alcohols are only well resolved by porcine pancreatic lipase and *P. cepacia* lipase. No common simple rule based on the size of the substituents at the stereocenter can be found. Instead features more remote from the stereocenter seem to influence the selectivity. For *P. cepacia* lipase, a size-dependent selectivity can be seen (Fig. 1) [15], but if an oxygen instead of a carbon is present at the stereocenter the prediction rule fails.

A model for prediction of enantiopreference in the resolution of chiral acids is only published for *C. rugosa* lipase [16,17] (Fig. 1). This lipase shows high enantioselectivity toward many carboxylic acids, such as the commercial targets 2-arylpropanoic acids and 2-aryloxypropanoic acids, which fit the model. It appears, however, that when the large substituent is extensively branched the substrate no longer fits the model.

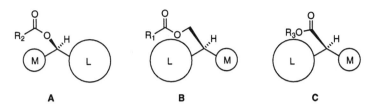

Figure 1 Preferred substrates or preferentially formed products in lipase-mediated acyl transfer reactions of chiral compounds containing a secondary alcohol (A), a chiral primary alcohol (B), or a chiral acyl moiety (C) [11,15,16].

Enantioselectivity is the result of the competition of two enantiomers for one active site. Therefore, models aiming at a molecular understanding of enantioselectivity need to take into account both the enantiomers and the enzyme. Prediction of enantioselectivity has been approached by molecular modeling of tetrahedral intermediates as models of transition states in the active site of lipases [18,19]. The fast-reacting enantiomer can usually be predicted, but a quantitative estimation of the enantioselectivity is much more difficult to achieve [20,21]. The results from molecular modeling [20] and x-ray crystallography [14] show independently that the slow-reacting enantiomer often has a problem in forming all of the hydrogen bonds needed for a functional transition state, when docked in the active site in the same way as the fast-reacting enantiomer.

The molecular modeling methods have limited value as a general tool in prediction of enantioselectivity. However, it is expected that their usefulness will increase as modeling algorithms and computing facilities get better. Probably rational design of enantioselective syntheses will in the future be based on known enzyme structures.

II. REACTION SYSTEM

A. Aqueous Media

The kinetic resolution of racemic esters by hydrolysis catalyzed by lipases in water emulsion systems (Scheme 5) is a simple operational step. To shift the reaction toward hydro-

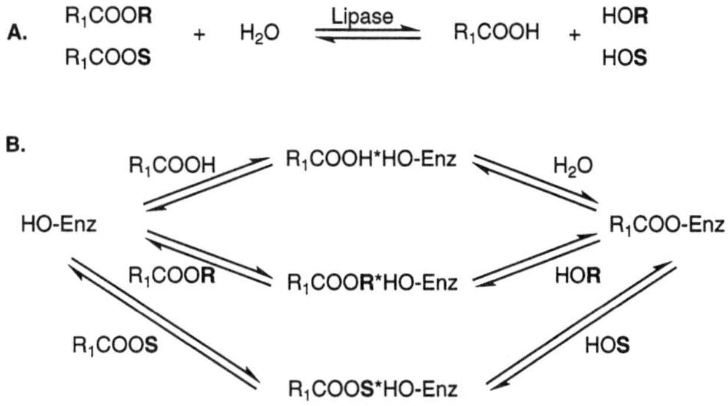

A. R_1COOR R_1COOS $+$ H_2O $\xrightarrow{\text{Lipase}}$ R_1COOH $+$ HOR HOS

B.

HO-Enz $\quad$ R_1COOH $\rightleftharpoons$ $R_1COOH \cdot HO\text{-}Enz$ $\quad$ H_2O

R_1COOR $\rightleftharpoons$ $R_1COOR \cdot HO\text{-}Enz$ $\quad$ HOR $\quad$ $R_1COO\text{-}Enz$

R_1COOS $\quad$ HOS

$R_1COOS \cdot HO\text{-}Enz$

Scheme 5 A: Lipase-catalyzed hydrolysis of an ester with a chiral alcohol moiety (R_1COOR and R_1COOS). B: The scheme shows the competition between the two enantiomeric product alcohols (HOR and HOS) and water for the acyl enzyme ($R_1COO\text{-}Enz$).

lysis the formed acid can be titrated by a buffer or with alkali in a pH-stat instrument. In spite of the seeming simplicity of the system low enantiomeric excess is often achieved due to equilibrium between substrates and products over the acyl enzyme [3]. Produced acids and alcohols with low water solubility will accumulate at the interface of the emulsion. The lipase is also present at the interface and will thus experience a high local concentration of the products. From Scheme 5 it can be seen that a local high concentration of the alcohol around the lipase will decrease the possibility of water acting as nucleophile. The consequence will be that the produced alcohol will react and form the substrate ester. This will cause a decrease in enantiomeric excess in the produced alcohol as the ester formed in the backreaction has to pass through the enantioselective step again.

The high competitive strength of an alcohol was demonstrated by the transesterification of 1-phenyl ethyl butyrate with octanol as the acyl acceptor in an emulsion system [22]. Addition of a water-miscible solvent, which extracts the alcohol from the interface, can increase the enantiomeric excess of the product. This effect was observed in several different systems [23–25].

With a stereocenter in the acyl part of the substrate the accumulation of acid at the interface will in a similar way as above decrease the enantiomeric excess in the product. In this case the produced acid will compete with the substrate ester for the free enzyme (Scheme 4). The concentration of the protonated acid at the interface can be reduced by an increase of pH or complex formation with Ca^{2+}. Holmberg et al. showed that the hydrolysis of octyl 2-methyldecanoate was very poor at pH 7.5, regarding both conversion and enantiomeric excess [26]. The addition of 0.2 M $CaCl_2$ increased both conversion and enantiomeric excess. The best results were obtained at pH 8.0 in the presence of $CaCl_2$. Finally, the authors showed that an emulsion of 2-methyldecanoic acid and 1-octanol in water yielded 8% ester in the presence of *C. rugosa* lipase.

The conclusion is that hydrolysis in water seems to be easy and straightforward, but product accumulation can hamper the enantioselectivity of the enzyme. Low enantiomeric excess of the product may in this case be a consequence of a bad reaction system and not an enzyme with low enantioselectivity.

B. Organic Media

The first reaction catalyzed by a lipase in a nonaqueous solvent was described in 1900 [27]. The purpose was to demonstrate the reversibility of a hydrolytic reaction by switching it to synthesis, nowadays a commonly used lipase-mediated reaction in an organic solvent. More intense studies of enzymes in organic media or in water–nonaqueous solvent mixtures started with a review by Singer in 1962 [28]. The growing need for new, powerful, stereoselective, and environmentally safe synthetic methods then led to an explosive increase in the 1980s, starting in 1981 [29,30] and then followed by an article in *Science* in 1982 [31]. It was not until then the true exploration of the field of nonaqueous enzymology really began. Since then, both introductory [32–35] and recent work [36,37] have been summarized in several review articles. The technique of using a nonaqueous solvent system for enzymatic reactions is now a standard procedure in most synthetic laboratories worldwide.

The hydrophobicity of an organic solvent is characterized by log P [38], the logarithm of the partition coefficient between octanol and water. The most suitable organic solvents for enzyme reactions appear to be nonpolar with a log $P > 2$ [39]. However, deviations from this are not rare and no correlation between log P and enzyme activity has been found, for instance, for the protease chymotrypsin [40]. Physical parameters other than log P have been explored as well in order to find a more satisfactory correlation between solvent polarity and enzyme activity or selectivity. However, it appears that no single parameter can predict enzyme activity in organic media.

Some puzzling results have been obtained regarding lipase enantioselectivity in chlorinated solvents. In the case of the resolution of 3-hydroxybutanoic acid methyl ester in a transesterification with vinyl acetate catalyzed by *P. cepacia* lipase, a reversed enantiopreference was observed in dichloromethane compared to that in other solvents [41]. In the resolution of sulcatol by *P. cepacia* lipase-catalyzed transesterification in various organic solvents, an increasing E value was observed during the reaction in chlorinated solvents. This unexplained phenomenon was not observed in other solvents [42].

The importance of medium engineering (effect of solvent) for altering of enzyme selectivity has been illustrated and rationalized by Carrea et al. in a recent review article [43]. One of the most important issues when comparing solvents might be the effect of water in the solvent. Due to the big difference in water solubility, a comparison of solvent systems should be made at an equal and constant water activity.

C. Water Activity

Virtually all enzymes used in hydrophilic as well as hydrophobic organic solvents are not surrounded by the solvent itself but by water; the proteins are more or less hydrated. Often, less than a molecular monolayer of water is required for an enzyme to be catalytically active. Due to differing solubilities of water in different solvents, different amounts of water are needed to reach this situation. Halling has described the use of water activity (a_w) instead of water content and demonstrates its usefulness in several review articles [44–48]. The water activity describes the mass action of water on an equilibrium and is therefore a better measure of the effect of water than the water content. A polar solvent requires a higher water content to reach a certain water activity than does a nonpolar solvent. Consequently, in order to compare enzyme reactions in different media it is of fundamental importance to do this at an equal and constant water activity.

Several methods of setting and controlling the water activity in enzymatic reactions have been described in the literature. Goderis et al. [49] performed the first reaction under water activity control where the components had been individually pre-equilibrated to a known water activity above saturated salt solutions [50]. However, this commonly used method is not satisfactory when water is consumed or generated in the reaction or when solvation of water is changed during the reaction. Methods for continuous control of the water activity are therefore commonly used. Some examples are circulating a saturated salt solution through the reaction vessel inside a silicone tubing [51], performing the reaction in a headspace above a saturated aqueous salt solution [52,53], continuous controlled drying of the headspace above the reaction mixture [54–57], adding silica particles as an a_w buffer *in situ* [58], and, probably the most convenient, adding an inorganic salt–hydrate pair as a_w buffer *in situ* [59–62].

The rates of enzyme-catalyzed reactions are in many cases strongly dependent on the water activity of the reaction medium. The picture is more complex regarding the effect on enzyme enantioselectivity. Results appearing to be contradictory on how lipase enantioselectivity is influenced by a_w have been published. On lowering the water activity (or water content), both decreased [22,63–67], increased [41,68–73], as well as unaffected [66,68,74–76] E values have been reported (Table 1).

Some of these different observations in Table 1 could be rationalized as a kinetic effect by taking into account if the stereocenter resides in the alcohol part of the substrate

Table 1   Altered E Values at Various Water Activities in Lipase–Catalyzed Resolutions

Presence of stereocenter	E at lowered a_w	Lipase	Solvent	Ref.
Alcohol[a]	Lower[b]	C. rugosa	c-hx/water	22
Alcohol	Higher	C. rugosa	Isooctane	70
Alcohol	Unchanged	C. rugosa	Tributyrin	68
Alcohol	Unchanged	C. rugosa	n-Hexane	76
Alcohol	Lower[b]	P. cyclopium	n-Heptane	64
Alcohol	Higher	PPL[c]	Ether	68
Alcohol	Higher	Lipase PS	Toluene	71
Alcohol	Unchanged	Lipase PS	Toluene	75
Alcohol	Unchanged	Lipase PS	Chloroform	74
Alcohol	Higher	C. antarctica B	n-Hexane	73
Alcohol	Unchanged	C. antarctica B	n-Hexane	76
Alcohol	Unchanged	C. antarctica B	Dichloromethane	73
Alcohol	Unchanged	Pseudomonas	n-Hexane	76
Acyl	Lower[b]	C. rugosa	n-Hexane	63
Acyl	Higher	C. rugosa	n-Hexane	69
Acyl	Higher[b]	C. rugosa	n-Hexane	72
Acyl	Lower	C. rugosa	c-Hexane	65
Acyl	Lower	C. rugosa	c-Hexane	66
Acyl	Lower[b]	C. rugosa	Isopropyl ether	67
Acyl	Higher	P. cepacia	Dodecane	41

[a]Acylation rate limiting.
[b]Water activity not controlled, various amounts of water added.
[c]Porcine pancreatic lipase.

molecule or in the acyl part. This leads to a situation as in either Scheme 3 or Scheme 4. In any lipase-catalyzed resolution water is a competitive nucleophile for the acyl enzyme produced after the first step in the reaction mechanism, acylation (Scheme 1). As previously discussed, only one acyl enzyme is formed in the case of an achiral acyl donor, such as in the resolution of a chiral alcohol (Scheme 3). Water does not participate in an enantioselective step in such a reaction and therefore could not be expected to kinetically influence the enantioselectivity. In line with this, many of the reactions in Table 1 involving chiral alcohols resulted in unchanged E values at different water activities.

In resolution of a chiral acyl donor, however, two different acyl enzymes are formed and, consequently, water participates as a competing nucleophile in an enantioselective step of the reaction. Högberg et al. [65] observed markedly higher E values at higher water activity in the C. rugosa lipase-catalyzed esterification of a chiral 2-methyldecanoic acid in cyclohexane. Similarly, Yasufuku and Ueji [67] added 0.5% water to an esterification of 2-(4-ethylphenoxy)propanoic acid catalyzed by C. rugosa lipase in isopropyl ether and the E value increased from 5 to 45. The latter authors also report the combined effect of water and reaction temperature on the enantioselectivity using the same system and with 2-(4-tert-butylphenoxy)propanoic acid as substrate. They observed a switch in the lipase enantiopreference and the E value increased from 1.53 (S preference) to 7.49 (R preference) when the temperature was increased from 10°C to 57°C and with a simultaneous addition of water up to 1.25%. In Table 1, all reactions with a chiral acyl moiety showed altered E values at different water activity supporting the competitive possibility of water in this situation.

Thus, many of the results in Table 1 can be explained by a kinetic effect involving water. However, not all results in Table 1 can be explained this way, indicating the presence of a more complex situation regarding the role of water in a lipase-catalyzed reaction. This might involve enzyme-bound water, which interacts with one enantiomer only, or water-induced conformational changes in the lipase.

III. APPLIED LIPASE CATALYSIS

For recent extensive reviews on biotransformations with lipases, see Kazlauskas and Bornscheuer [77], Johnson [78], Rubin and Dennis [79], Itoh et al. [80], and Boland et al. [81]. The most widespread and frequently used biocatalytic reaction involving chiral compounds is kinetic resolution of racemates. Other biocatalytic stereoselective methods, although less frequently used, are asymmetrization of prochiral and *meso* compounds. These will be briefly discussed in Secs. C and D, respectively.

A. Kinetic Resolution

In a kinetic resolution one enantiomer of a racemate reacts more rapidly than the other with the resolving agent (Scheme 6). The resolving agent is often used in catalytic amounts and can be a chemical catalyst or a biocatalyst (e.g., an enzyme). The very first kinetic resolution was in fact biocatalyzed and was performed by Louis Pasteur in 1858 [82]. He then obtained a solution of tartaric acid that rotated polarized light from a reaction with the mold *Penicillium glaucum*. Kinetic resolution is the predominant process used with lipases for production of enantiomerically pure alcohols, acids, or esters. For general reviews on kinetic resolution, see Kagan and Fiaud [83] and Sih and Wu [8]. Lipase-catalyzed kinetic resolution has also been extensively reviewed by many authors (see, for instance, Jones [84], Chen and Sih [85], and Santaniello et al. [86]).

Numerous examples exist on the kinetic resolution of chiral acyl acceptors. Among other compounds primary and secondary alcohols, various amines, and peroxides have been resolved. Representative examples are shown in Scheme 7. The secondary alcohol 2-octanol was resolved using *S*-ethyl octanethioate as acyl donor and *C. antarctica* lipase B [12]. The alkyl peroxide was acylated with isopropenyl acetate using *P. cepacia* lipase [87]. The primary amine was resolved by *C. antarctica* lipase B–catalyzed acylation of ethyl octanoate at reduced pressure [88]. The primary alcohol was successfully resolved by acylation of vinyl acetate at $-40°C$ [89].

While amines and alcohols can be used as acyl acceptors in the deacylation of an acyl enzyme, the nucleophilicity of a thiol is low [90]. In order to resolve 1-phenyleth-

$$R \xrightarrow[\text{fast}]{k_R} P$$

$$S \xrightarrow[\text{slow}]{k_S} Q$$

Scheme 6 Classic kinetic resolution where R and S denote the substrate enantiomers and P and Q represent their products, respectively.

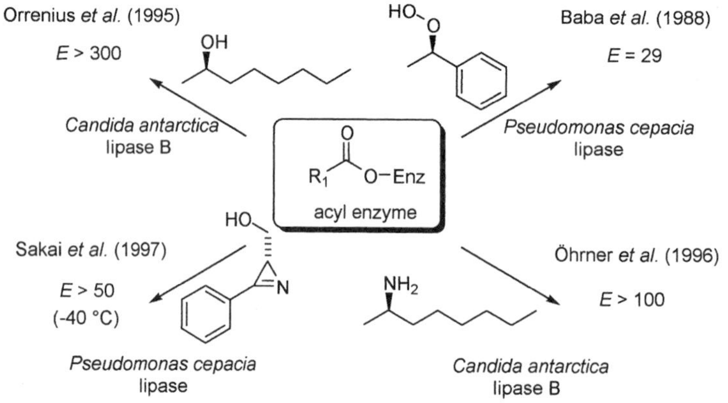

Scheme 7 Examples of lipase-catalyzed kinetic resolution by deacylation of an acyl enzyme (R_1COO-Enz). The chiral acyl acceptors are a secondary alcohol [12], a peroxide [87], a primary amine [88], and a primary alcohol [89]. The fast-reacting enantiomers are shown.

anethiol the corresponding thio ester was instead used as acyl donor in an acyl transfer reaction catalyzed by *C. antarctica* lipase B where racemic 1-phenylethanol was used as acyl acceptor [88]. The thiol was produced with an *E* value of 88 in 95% e.e.

The resolution of chiral acyl donors mainly involves carboxylic acids with the stereocenter at the α position. *Candida rugosa* lipase shows high enantioselectivity to many of these acids in contrast to *C. antarctica* lipase B. To compounds with an electron-withdrawing substituent at the stereocenter, *P. cepacia* lipase shows a high selectivity as well. Two examples are presented in Scheme 8 [91,92].

Pseudomonas, *C. rugosa*, and *C. antarctica* B lipases also resolve acids with a stereocenter at the β position. For instance, 3-hydroxybutanoic acid was esterified with *n*-butanol with an *E* value of more than 50 using *C. rugosa* lipase [93]. *Candida rugosa* lipase also resolves chiral carboxylic acids with remote sulfoxide stereocenters [94].

B. Dynamic Resolutions

Since two enantiomeric substrates or products are physically equal in a symmetrical environment, their different reaction rates in a kinetic resolution process depend only on the free energy difference between their respective diastereomeric transition states during interaction with the chiral resolving agent. Under thermodynamic control, both enantiomers

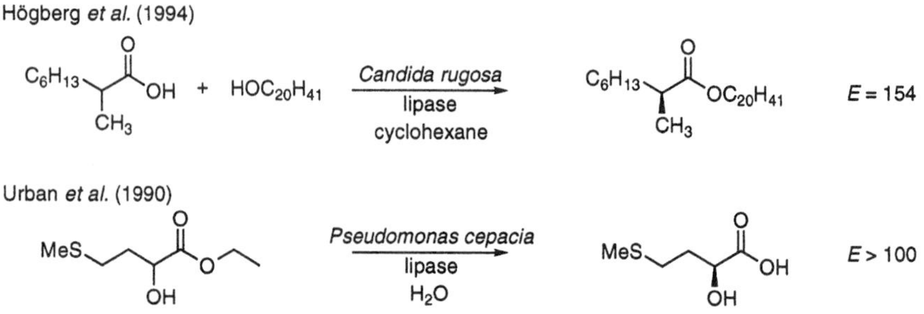

Scheme 8 Kinetic resolution of chiral acyl donors by esterification [91] and hydrolysis [92].

would either reach 100% conversion (irreversible case) or a thermodynamic equilibrium position (reversible case), which in both cases result in equal amounts of the two enantiomeric products. The reaction should therefore be stopped before completion in order to be of any practical use. Consequently, a maximum theoretical yield of 50% of enantiomerically pure product can be obtained from a racemic starting material using this technique. This is not satisfactory in many cases where only one enantiomer is the target compound of interest and therefore the classic resolution technique has recently been further developed and modified to increase its versatility.

Dynamic resolutions are kinetic resolutions modified with an additional feature, i.e., a racemization step (Scheme 9). This can afford 100% yield of the fastest reacting enantiomer if $k_{rac} \geq k_R$, and product racemization does not occur. Furthermore, in contrast to a classic resolution process, the enantiomeric excess of the product in a dynamic resolution process becomes independent of the conversion.

For recent reviews on dynamic resolution, see ref. [95–99]. The crucial step in a biocatalytic dynamic resolution is the in situ racemization [100]. It has to be fully compatible with the biocatalyst, the rate of racemization should be faster than the subsequent biocatalytic transformation of the fastest reacting enantiomer, and the product of the biocatalytic step should not be racemized. These requirements have been fulfilled using transition metal catalysts [101] (Scheme 10). In the case of the dynamic resolution of phenylethylamine [102], the racemic amine was enantioselectively *N*-acetylated by ethyl acetate catalyzed by *C. antarctica* lipase B. The nonacylated *S*-amine was racemized in situ using Pd on charcoal. In this dynamic resolution process (*R*)-*N*-acetylphenylethylamine was produced in 75–77% yield and 99% e.e. The in situ racemization of a secondary alcohol (1-phenylethanol) has been achieved using a rhodium(II) catalyst. This dynamic resolution furnished the acetate in 60% yield and 98% e.e. [103]. Similarly, in situ racemization of nonacetylated 1-phenylethanol has been achieved by a ruthenium-catalyzed hydrogen transfer reaction resulting in the lipase-catalyzed production of (*R*)-1-phenylethyl acetate in 100% yield and more than 99.5% e.e. [104]. The dynamic resolution of an allylic acetate with in situ racemization using a Pd(II) catalyst [PdCl$_2$ (Me$_2$CN)$_2$] gave the allylic alcohol in 96% yield and 96% e.e. [105].

Another approach for the coupled racemization step has been used for compounds having an acidic hydrogen on the stereocenter. Examples of such compounds are chiral acyl donors such as α-substituted esters which are prone to base-catalyzed racemization via an enolate intermediate. This approach has been frequently used and a few examples will be given here to illustrate the utility (Scheme 11). The first examples involve oxazolinones where it was found that porcine pancreatic lipase and lipase from *Aspergillus* sp. exhibited opposite enantiopreferences [106,107]. The remaining oxazolinone was spontaneously racemized via the enolate intermediate and both (L)- and (D)-*N*-benzoyl amino acids could be produced this way in high chemical and optical yields. The pK_a values of thio esters are lower than those of oxo esters [108]. This has been used in the lipase-

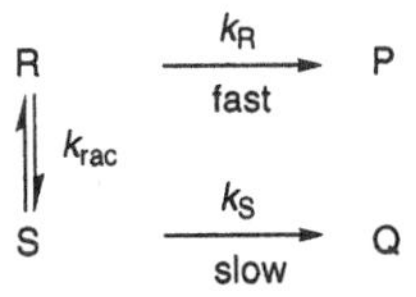

R $\xrightarrow[\text{fast}]{k_R}$ P

k_{rac} ⇅

S $\xrightarrow[\text{slow}]{k_S}$ Q

Scheme 9   Dynamic kinetic resolution where R and S denote the substrate enantiomers and P and Q represent their products, respectively.

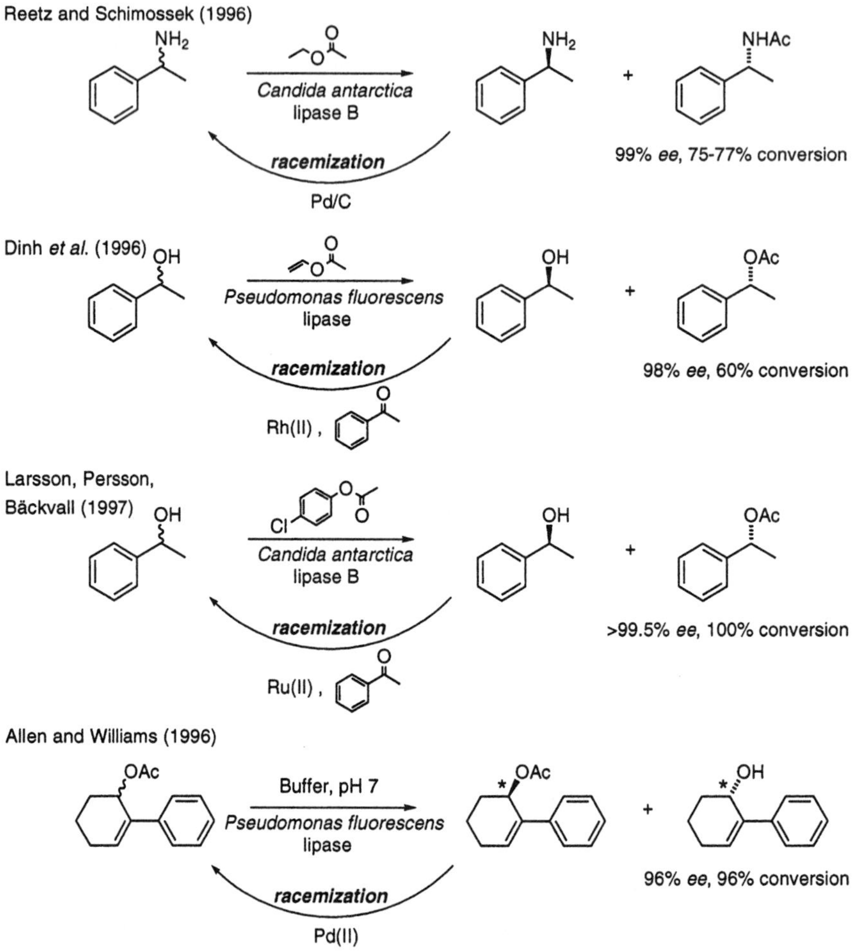

Reetz and Schimossek (1996)

Dinh *et al.* (1996)

Larsson, Persson, Bäckvall (1997)

Allen and Williams (1996)

99% *ee*, 75-77% conversion

98% *ee*, 60% conversion

>99.5% *ee*, 100% conversion

96% *ee*, 96% conversion

Scheme 10 Lipase-catalyzed dynamic resolutions with in situ transition metal-catalyzed racemization [102–105]. *Absolute configuration not determined.

catalyzed dynamic resolution of *S*-ethyl 2-(phenylthio)propanethioate [109]. Due to the water insolubility of the substrate, a two-phase system was used and the racemization of the thio ester was induced in the organic phase using trioctylamine. By this method, the hydrolysis product was produced in quantitative yield and 96.3% e.e. In a transesterification reaction catalyzed by *P. cepacia* lipase, the butyl ester of (*R*)-2,4-dichlorophenoxypropanoic acid could be produced from the trifluoroethyl thio ester in 75% e.e. and 98% yield [110]. The reaction was run in toluene with triethylamine present as the base. The enantiomerically enriched butyl ester was then hydrolyzed using the same enzyme under nonracemizing conditions which gave the (*R*)-acid in 93% e.e. at 81% conversion.

Vörde et al. [111] published a resolution of a 2-methyloctanoate ester using *C. antarctica* lipase B–catalyzed aminolysis by (*R*)-1-phenylethylamine. The resulting (*R,R*)-amide was produced in 30–40% d.e. at 100% conversion, which suggests that an epimerization of the acyl stereocenter was taking place during contact with the enzyme (Scheme 12).

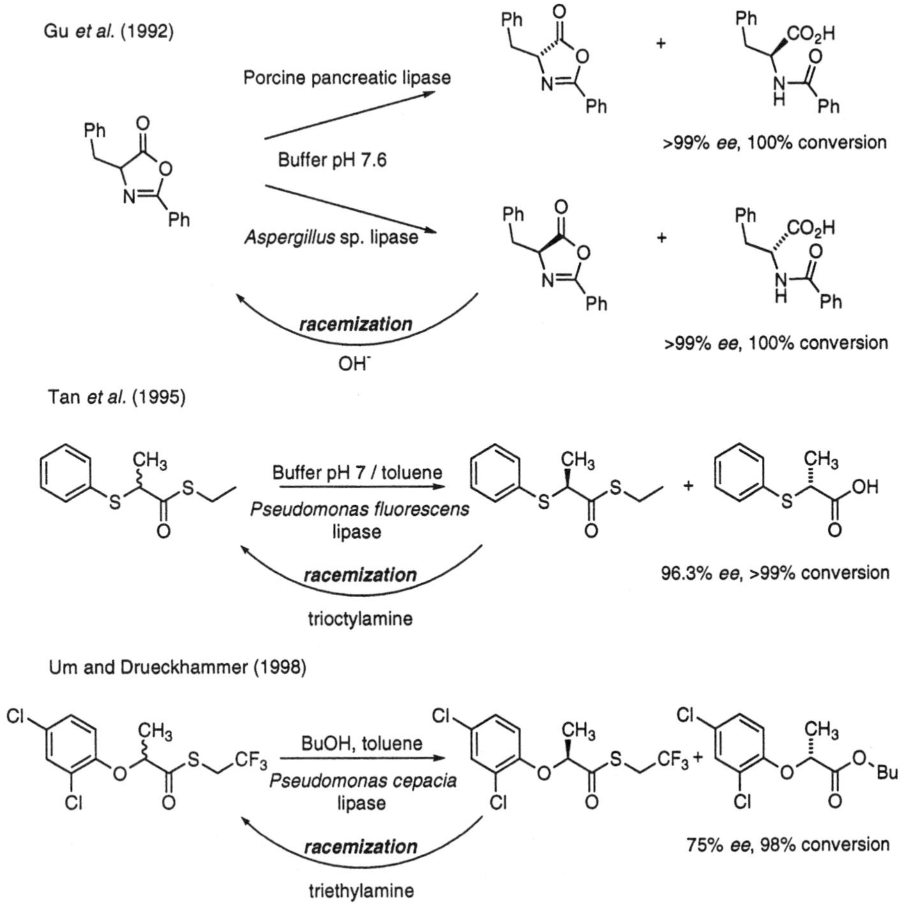

Gu *et al.* (1992)

>99% *ee*, 100% conversion

Porcine pancreatic lipase

Buffer pH 7.6

Aspergillus sp. lipase

>99% *ee*, 100% conversion

racemization

OH⁻

Tan *et al.* (1995)

Buffer pH 7 / toluene

Pseudomonas fluorescens lipase

racemization

trioctylamine

96.3% *ee*, >99% conversion

Um and Drueckhammer (1998)

BuOH, toluene

Pseudomonas cepacia lipase

racemization

triethylamine

75% *ee*, 98% conversion

Scheme 11 Some lipase-catalyzed dynamic resolutions with in situ base-catalyzed racemization [106,107,109,110].

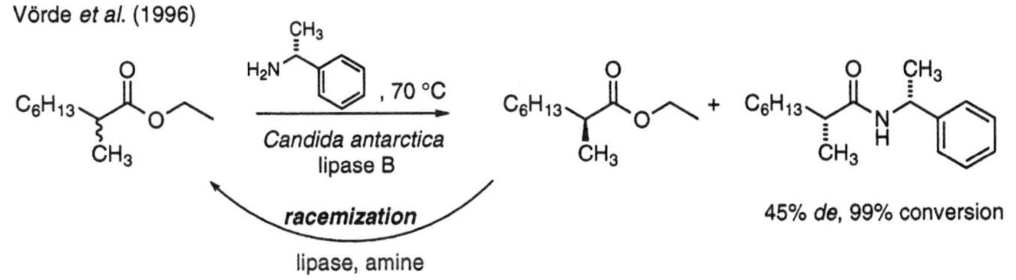

Vörde *et al.* (1996)

, 70 °C

Candida antarctica lipase B

racemization

lipase, amine

45% *de*, 99% conversion

Scheme 12 A lipase-catalyzed dynamic resolution process with in situ enzymatic racemization [111].

No racemization was observed in the presence of phenylethylamine of either the product amide or the substrate ester in the absence of enzyme. This represents a dynamic resolution process with in situ enzymatic racemization. The picture is still unclear as to whether the substrate ester is racemized or if the racemization takes place in the acyl enzyme.

C. Prochiral Substrates

In contrast to a conventional kinetic resolution of a racemate, asymmetrization of prochiral and *meso* compounds can give 100% theoretical yield. Still, the ratio of biocatalyzed asymmetrizations vs. kinetic resolutions has been reported to be only 1:4 [112]. Similar to a dynamic kinetic resolution, the enantiomeric purity of the product of an asymmetrization reaction remains constant and is independent of the extent of conversion. The enantiomeric excess of the product (e.e.$_p$) is given by

$$e.e._p = \frac{E - 1}{E + 1} \qquad (1)$$

For a recent review on enzymatic asymmetrization of prochiral and *meso* compounds, see Schoffers et al. [113]. Enzymatic asymmetrization of prochiral compounds was discussed by Ogston in 1948 [114]. He then rationalized the enzyme's ability to discriminate between enantiotopic groups on a prostereogenic substrate molecule based on the three-point combination model.

One example of a prochiral alcohol is 2-substituted 1,3-propanediol. Guanti et al. investigated the effect of unsaturation adjacent to the prochiral center in lipase-catalyzed hydrolyses of propanediol diacetates (Scheme 13) [115]. The reactions almost stopped at the monoacetate stage and the various products were obtained with (S) configuration in 20–80% isolated yields and in 20 to more than 96% e.e.

Monosubstituted malonic acid monoesters are useful chiral synthons and much work has been done in this area. They cannot be prepared by enzymatic hydrolysis of their corresponding diesters due to the presence of the activated malonic hydrogen, which undergoes fast exchange in aqueous media resulting in racemization. Gutman et al. developed a transesterification reaction in *n*-hexane catalyzed by *C. cylindracea* lipase [116]. Both enantiomers were efficiently produced using the same enzyme but with a switch of the alcohols used as nucleophiles (Scheme 14). The chiral half-esters can then easily be produced by catalytic hydrogenation of the respective enantiomeric product. The yields were low since a considerable slowdown in reaction rate was observed close to 70% conversion, indicating an equilibrium position at approximately 70% conversion.

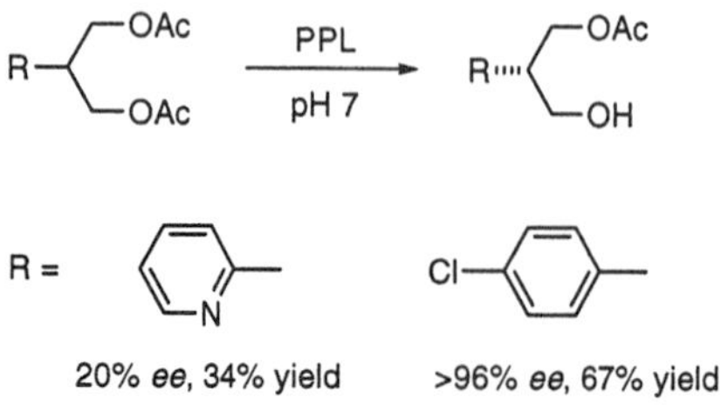

Scheme 13 Hydrolysis of various prochiral propanediol acetates catalyzed by porcine pancreatic lipase [115].

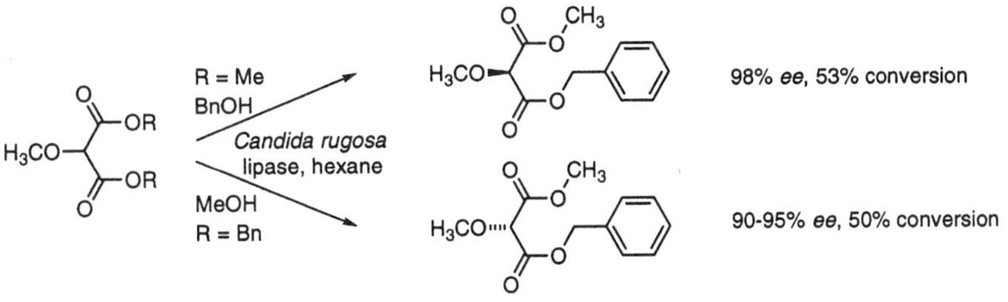

Scheme 14 Lipase-catalyzed transesterification of monosubstituted prochiral malonates [116].

D. Meso Substrates

Similarly as for prochiral substrates, many of the lipase-catalyzed asymmetrizations of *meso* compounds are accompanied by a second reaction step that usually enhances the enantiomeric excess of the product. This second step is a kinetic resolution. For example, in the hydrolysis of a *meso*-diester, the reaction usually does not stop at the monoester stage (Scheme 15). The two enantiomeric monoesters will react further giving the same *meso*-diol. This second step usually favors the minor monoester enantiomer and therefore leads to an increase of the enantiomeric excess of the major monoester, but a decrease in the yield. This has been illustrated and described by Wang et al. for the lipase-catalyzed hydrolysis of *meso*-1,5-diacetoxy-*cis*-2,4-dimethylpentane [117]. The monoacetate was afforded in 89.7% e.e.

A similar e.e. amplification has been described in the lipase-catalyzed asymmetrization of a "double-*meso*" tetraol by Uguen and co-workers [118] (Scheme 16).

The two *pro*-(R)-hydroxyl groups were acetylated in 100% e.e. catalyzed by *Pseudomonas fluorescens* lipase and a mixture of the C$_2$ symmetrical (R,R)-diol and its *meso* isomer was obtained. The mixture could be separated after transformation of the free hydroxyl groups into phenyl sulfides.

IV. TECHNIQUES FOR ENHANCING ENANTIOMERIC EXCESS

Both lipase-catalyzed hydrolyses in water and acyl transfer reactions in an organic medium are more or less reversible reactions that will be detrimental to the enantiomeric excess of the compound of interest. Therefore, many techniques used for enhancing the enantiomeric excess are based on shifting the thermodynamic equilibrium position toward an irreversible situation.

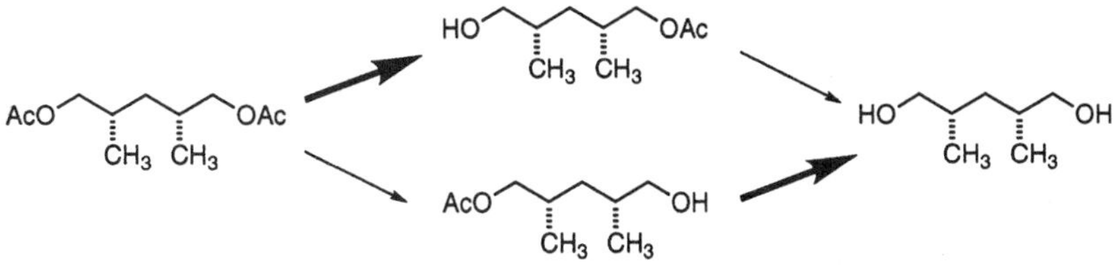

Scheme 15 Enzymatic asymmetrization of a *meso* compound is accompanied by a resolution step that usually enhances the enantiomeric excess of the monoester enantiomer [117]. Boldface arrows indicate the fast reactions.

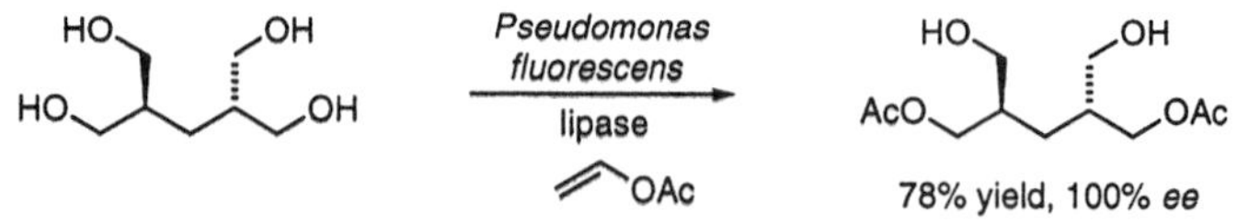

Scheme 16 Asymmetrization of a "double-*meso*" tetraol by *Pseudomonas fluorescens* lipase-catalyzed esterification [118].

A. Chiral Acyl Acceptors

Several techniques of displacing the reaction equilibrium to reach a quasi-irreversible situation have been used previously. For a review of these, see Faber and Riva [119]. The techniques of using activated acyl donors when resolving chiral alcohols afford a more or less irreversible acylation step in the reaction mechanism since the first product is designed to be a poor nucleophile or is supposed to tautomerize or otherwise leave the reaction system (Scheme 3). Some examples of acyl donors frequently used include 2-haloethyl, cyanomethyl, oxime, and enol esters. The rates of the acyl transfer reactions of racemic 2-octanol with various esters catalyzed by porcine pancreatic lipase were one to two orders of magnitude faster when activated esters were used compared with methyl or ethyl alkanoates [120].

Since the first reports appeared on enol esters in lipase reactions [121–123], they have become the most frequently used acyl donors in enantioselective reactions involving lipases [124–126]. The leaving group in vinyl or isopropenyl esters, sometimes referred to as irreversible acyl donors, will tautomerize to acetaldehyde and acetone, respectively. However, it should be pointed out that only the acylation step is irreversible and that the equilibrium involving acyl enzyme formation from the produced ester (Scheme 3) sets a limit to the enantiomeric purity of the product [3,127]. It has been reported that the liberated acetaldehyde from vinyl esters slowly inactivates *C. rugosa* and *Geotrichum candidum* lipases, probably by Schiff base formation with a lysine residue [128–129]. This has recently been overcome by using ethoxy vinyl acetate as the acyl donor, where the leaving group instead will tautomerize to ethyl acetate [130].

In the reaction with 2-octanol catalyzed by *C. antarctica* lipase B, Öhrner et al. compare four acyl donors differing only in their leaving groups [7] (Table 2). In the case of octanoic acid, its low K_M compared with the other acyl donors makes it beneficial at high conversions since it competes well with the product ester for the free enzyme (Scheme 5). However, the initial rate is lower than for other acyl donors, resulting in a slightly longer reaction time than for the thio ester and vinyl ester.

Ethyl octanoate is a good acyl donor provided that the produced ethanol can be evaporated. Ethanol is a good nucleophile that competes for the acyl enzyme resulting in equilibrium conditions. This can be seen in Table 2 as a longer reaction time and lower enantiomeric excess at atmospheric pressure and 39°C. The approach of using reduced pressure to evaporate ethanol has been used successfully for the resolution of many alcohols [7,131].

The thio ester *S*-ethyl octanethioate used as an acyl donor drives the reaction due to the poor nucleophilicity and high volatility of the leaving thiol [131]. However, its high K_M makes it difficult to saturate the enzyme with the acyl donor. This is reflected in Table 2 by the rather slow initial rate. It has been reported that the use of *S*-ethyl octanethioate often leads to products with higher enantiomeric excess than with ethyl octanoate [132].

Table 2 *Candida antarctica* Lipase B–Catalyzed Kinetic Resolution of 2-Octanol Using Various Acyl Donors[a]

Acyl donor	Initial rate (μmol/min mg)	Reaction time (h)	Conversion (%)	e.e. (%)
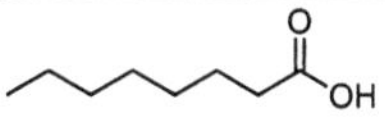	12	11	53	96
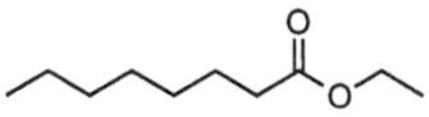	36	60	58	93
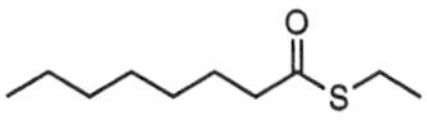	22	5	52	>99
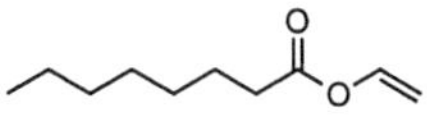	78	5	52	>99

[a]Atmospheric pressure, 39°C.
Source: Ref. 7.

In the case of vinyl octanoate, both a high initial rate and a high enantiomeric excess indicates a good displacement of the equilibrium toward product formation. However, it has been reported that there might be a decreased enantioselectivity if the vinyl ester is used as the solvent [73].

B. Chiral Acyl Donors

The alcohol used as cosubstrate in lipase reactions with chiral acyl donors may act as an enantioselective inhibitor that will be detrimental to the enantiomeric excess. This has been reported for *C. rugosa* lipase-catalyzed kinetic resolution by esterification of 2-methylalkanoic acids (Scheme 17) [134].

A common strategy is to use an excess of the alcohol to prevent equilibrium conditions [66,133]. However, it has been found that by decreasing the heptanol concentration from 900 to 90 mM in esterification of 2-methyldecanoic acid catalyzed by lipase, the E value increased from 37 to 83 [134]. The alcohol was found to be an enantioselective inhibitor influencing $(V_{max})_R$ and $(V_{max})_S$ differently. It has been suggested that an equimolar amount of alcohol is sufficient in these kinds of reactions to avoid a decrease in enantioselectivity [66]. The phenomenon of enantioselective inhibition by the alcohol has been suggested to be connected to the unique presence of the hydrophobic active site tunnel in *C. rugosa* lipase [9]. Molecular modeling studies of the two ester enantiomers bound to the active site of the lipase revealed the presence of two different modes of binding the

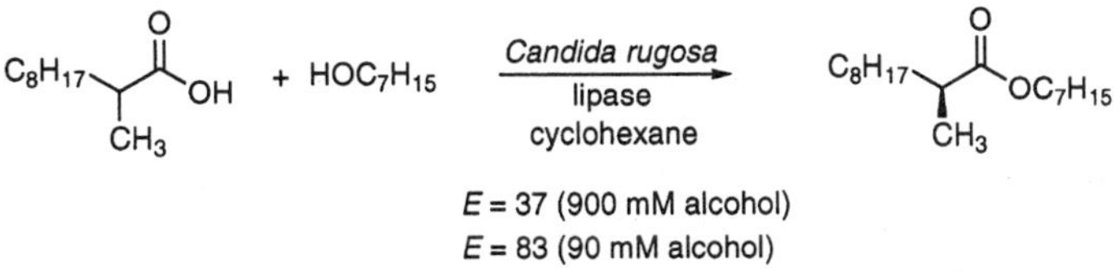

Scheme 17 *Candida rugosa* lipase-catalyzed esterifications of 2-methylalkanoic acids [133, 134]. The E value is higher at a low concentration of alcohol.

enantiomers, where the slow reacting R enantiomer leaves the tunnel empty. The presence of different binding modes has recently been experimentally confirmed by using an engineered substrate with a bulky thiophene group, which is not able to fit the active site tunnel for steric reasons [10]. It is thus possible for the alcohol present in a high concentration to bind to the hydrophobic tunnel and thereby influence the fast-reacting enantiomer binding there.

V. TECHNIQUES FOR ENHANCING THE E VALUE

Changing the E value requires a change of the reaction system such as the solvent, the water activity, and immobilization of the biocatalyst. For instance, it has been reported that lipase immobilized on a hydrophobic carrier makes the E value of *C. rugosa* lipase less sensitive to changes in water activity than the crude enzyme [66]. Other important parameters influencing enantioselectivity are the temperature, the lipase and the lipase formulation, and structural differences in the substrate molecule. Recently, site-directed mutagenesis has attracted increased attention as a tool for altering lipase enantioselectivity [135].

A. Medium Engineering

Common techniques for improving the E value in lipase reactions include medium engineering such as changing the solvent and the water content, as previously discussed. Several changes and modifications can also be made to the biocatalyst, which might induce structural changes in the enzyme leading to a more selective one. For instance, pretreatment of crude *C. rugosa* lipase with 2-propanol has been shown to increase the E value by more than a factor of 10 in hydrolyses of 2-substituted propanoic acid esters [136]. Sih and co-workers increased the E value for the same lipase and in the same order by addition of a chiral base (D or L-3-methoxy-N-methylmorphinane), which acted as an enantioselective noncompetitive inhibitor in hydrolysis of 2-(4-chlorophenoxy)propanoic acid methyl ester [137,138]. More selective lipases have also been obtained by crosslinking and crystallization [139,140], and by imprinting and lipid coating [141].

B. Sequential Resolution

The use of a two-step (tandem) or a sequential resolution procedure would result in a higher apparent (total) E value. A two-step procedure is used to obtain optimal result for both product and substrate in a resolution with a moderate E value. It involves isolation of the product at its optimal e.e. and yield followed by a second identical step for the remaining substrate to reach a similar point where it can be recovered with optimal e.e. and yield [142]. The principle of a sequential resolution is that a substrate is passing the active site of the enzyme twice during the process, resulting in a very high total apparent enantioselectivity. Sequential resolution has been used for resolution of alcohols having two chemically identical groups, such as the C_2-symmetrical compounds binaphthol [143], *threo*-2,3-butanediol [144], and *threo*-2,4-pentanediol (Scheme 18) [145,146].

The sequential resolution approach has also been used in combined hydrolysis-esterification reactions with *Mucor miehei* lipase [147] (Scheme 19). This is possible due to the biocatalyst displaying the same stereochemical preference in both hydrolysis and acyl

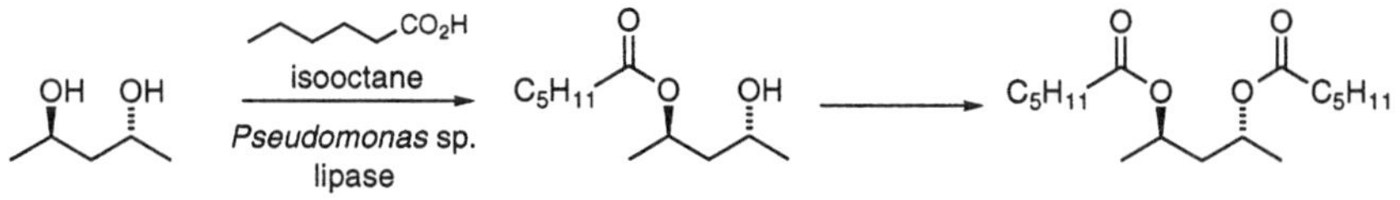

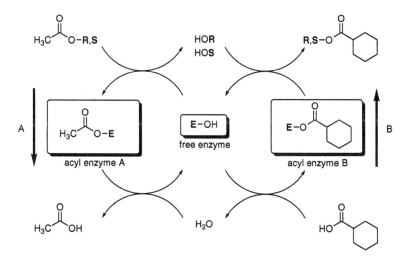

Scheme 18 Sequential lipase-catalyzed resolution of *threo*-2,4-pentanediol by esterification [145]. The fast-reacting enantiomer is shown.

transfer reactions, and therefore a double enantioselection process will take place in situ resulting in a total apparent E value (E_{tot}) [144]

$$E_{tot} \approx \frac{E_A E_B}{2} \qquad (2)$$

where A and B represent the single E values for the hydrolysis and esterification reaction, respectively. By using the sequential approach of Scheme 19, the cyclohexanecarboxylic acid ester of the *endo*-alcohol (**HOR**, **HOS**) was obtained in 27% yield and 99.4% e.e., which corresponds to an E_{tot} value of around 400.

A sequential approach with a combined alcoholysis-esterification reaction has been used for the porcine pancreatic lipase–catalyzed resolution of 2-phenyl-1-propanol [148] (Scheme 20). The alcoholysis reaction (A) was run to 48% conversion followed by the addition of 3 equivalents of vinyl acetate. By this, 2-phenyl-1-propanyl acetate with 92% e.e. and in 20% yield was isolated.

The sequential approach makes the reaction system rather complicated but is an elegant way of improving a low E value by a double-resolution process. However, the complexity makes the system unsuitable for studying the influence of various reaction parameters. The complex mechanistic scheme of the sequential reactions is shown in Scheme 21.

Scheme 19 Mechanism of sequential lipase-catalyzed hydrolysis-esterification in situ. The chiral alcohol (**HOR**, **HOS**) will be involved in two enantioselective steps, i.e., first in the hydrolysis A, and then in the deacylation step of the esterification B [147].

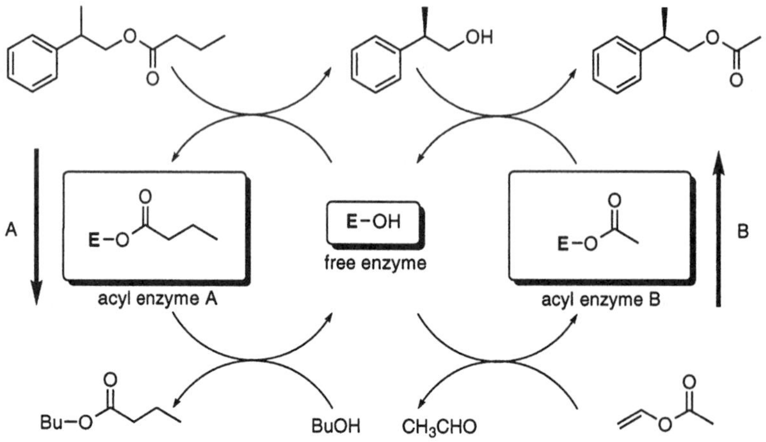

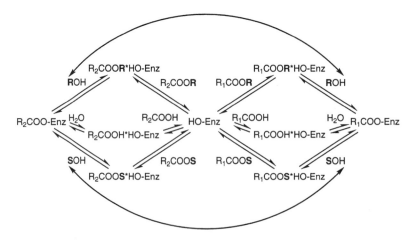

Scheme 20   Mechanism of sequential lipase-catalyzed alcoholysis-esterification. The alcoholysis reaction (A) was followed by a transesterification reaction (B) by the subsequent addition of 3 molar equivalents of vinyl acetate.

VI. CONCLUDING REMARKS

In view of the more complicated reaction systems used today for various reactions involving lipases, it is of fundamental importance to fully understand them in order to correctly interpret experimental results and the influence of various reaction parameters on the enantioselectivity. A very useful tool is therefore to illustrate the different situations of competing species by using the kinetic scheme as exemplified in Scheme 21. This scheme is based on the previously discussed cases of sequential resolutions of an ester bearing a chiral alcohol moiety in Schemes 19 and 20, and shows both the complexity around the free enzyme, where six species compete, and around the two different acyl enzymes, where three nucleophiles compete.

Scheme 21   A kinetic scheme for the sequential resolution of an acyl donor containing a chiral alcohol moiety according to Schemes 19 and 20.

The complexity and the difficulty of correctly interpreting experimental results from this scheme is obvious. Our aim with this chapter has been to illustrate the diversity and utility of lipase catalysis for production of chiral compounds with high enantiomeric purity, to exemplify commonly used reaction systems, and to highlight some representative recent examples.

Lipase catalysis is a very diverse and broad field where, still, little is known about lipase mechanisms on a molecular level. The extensive research currently going on is expected to reveal important information regarding the controlled tailoring of lipase enantioselectivity. Important aspects other than steric effects are the involvement of water, the nature of the solvent, and the entropic influence on substrate binding and transition state stabilization. This knowledge will provide an understanding of the details of lipase catalysis and facilitate the development of quantitative computer models for prediction of enantioselectivity in the very near future.

ACKNOWLEDGMENTS

Financial support from the Swedish Council for Forestry and Agricultural Research, and from the biotechnology program, BIO4-CT95-0231 and BIO4-CT97-2365, of the European Union is gratefully acknowledged.

REFERENCES

1. RD Schmid, R Verger. Angew Chem 110:1694–1720, 1998; Angew Chem Int Ed Engl 37: 1609–1633, 1998.
2. W Chulalaksananukul, JS Condoret, P Delorme, RM Willemot. FEBS Lett 276:181–184, 1990.
3. AJJ Straathof, JLL Rakels, JJ Heijnen. Biocatalysis 7:13–27, 1992.
4. M Kawase, K Sonomoto, A Tanaka. Biocatalysis 6:43–50, 1992.
5. R Verger. Trends Biotechnol 15:32–38, 1997.
6. M Martinelle, K Hult. Biochim Biophys Acta 1251:191–197, 1995.
7. N Öhrner, M Martinelle, A Mattson, T Norin, K Hult. Biocatalysis 9:105–114, 1994.
8. CJ Sih, S-H Wu. In: EL Eliel, SH Wilen, eds. Topics in Stereochemistry, Vol 19. New York: Wiley, 1989, pp 63–125.
9. M Holmquist, F Hæffner, T Norin, K Hult. Protein Sci 5:83–88, 1996.
10. P Berglund, M Holmquist, K Hult. J Mol Catal B: Enzym 5:283–287, 1998.
11. RJ Kazlauskas, ANE Weissfloch, AT Rappaport, LA Cuccia. J Org Chem 56:2656–2665, 1991.
12. C Orrenius, N Öhrner, D Rotticci, A Mattson, K Hult, T Norin. Tetrahedron: Asymmetry 6: 1217–1220, 1995.
13. K Naemura, M Murata, R Tanaka, M Yano, K Hirose, Y Tobe. Tetrahedron: Asymmetry 7: 1581–1584, 1996.
14. M Cygler, P Grochulski, RJ Kazlauskas, JD Schrag, F Bouthillier, B Rubin, AN Serreqi, AK Gupta. J Am Chem Soc 116:3180–3186, 1994.
15. ANE Weissfloch, RJ Kazlauskas. J Org Chem 60:6959–6969, 1995.
16. SN Ahmed, RJ Kazlauskas, AH Morinville, P Grochulski, JD Schrag, M Cygler. Biocatalysis 9:209–225, 1994.
17. MCR Franssen, H Jongejan, H Kooijman, AL Spek, NLFL Camacho Mondril, PMAC Boavida dos Santos, A de Groot. Tetrahedron: Asymmetry 7:497–510, 1996.
18. M Norin, F Hæffner, A Achour, T Norin, K Hult. Protein Sci 3:1493–1503, 1994.

19. J Uppenberg, N Öhrner, M Norin, K Hult, GJ Kleywegt, S Patkar, V Waagen, T Anthonsen, TA Jones. Biochemistry 34:16838–16851, 1995.
20. C Orrenius, F Hæffner, D Rotticci, N Öhrner, T Norin, K Hult. Biocatal Biotransform 16:1–15, 1998.
21. F Hæffner, T Norin, K Hult. Biophys J 74:1251–1262, 1998.
22. E Holmberg, K Hult. Biocatalysis 3:243–251, 1990.
23. A Bosetti, D Bianchi, P Cesti, P Golini. Biocatalysis 9:71–77, 1994.
24. Y Naoshima, M Kamezawa, H Tachibana, Y Munakata, T Fujita, K Kihara, T Raku. J Chem Soc Perkin Trans 1:557–561, 1993.
25. K Lundhaug, A Overbeeke, J Jongejan, T Anthonsen. Tetrahedron: Asymmetry 9:2851–2856, 1998.
26. E Holmberg, M Holmquist, E Hedenström, P Berglund, T Norin, H-E Högberg, K Hult. Appl Microbiol Biotechnol 35:572–578, 1991.
27. JH Kastle, AS Loevenhart. Am Chem Soc 24:491–525, 1900.
28. SJ Singer. Adv Protein Chem 17:1–68, 1962.
29. E Antonini, G Carrea, P Cremonesi. Enzyme Microb Technol 3:291–296, 1981.
30. K Martinek, AN Semenov, IV Berezin. Biochim Biophys Acta 658:76–89, 1981.
31. K Martinek, AV Levashov, YL Khmelnitsky, NL Klyachko, IV Berezin. Science 218:889–890, 1982.
32. A Zaks, AM Klibanov. Proc Natl Acad Sci USA 82:3192–3196, 1985.
33. AM Klibanov. CHEMTECH 354–359, 1986.
34. AM Klibanov. Trends Biochem Sci 14:141–144, 1989.
35. AM Klibanov. Acc Chem Res 23:1249–1251, 1990.
36. JA Jongejan, JBA van Tol, JA Duine. Chimica Oggi/Chemistry Today 12(7/8):15–24, 1994.
37. L Kvittingen. Tetrahedron 50:8253–8274, 1994.
38. A Leo, C Hansch, D Elkins. Chem Rev 71:525–616, 1971.
39. C Laane, S Boeren, K Vos, C Veeger. Biotechnol Bioeng 30:81–87, 1987.
40. E Wehtje, P Adlercreutz, B Mattiasson. Biocatalysis 7:149–161, 1993.
41. U Bornscheuer, A Herar, L Kreye, V Wendel, A Capewell, HH Meyer, T Scheper, FN Kolisis. Tetrahedron: Asymmetry 4:1007–1016, 1993.
42. F Secundo, G Ottolina, S Riva, G Carrea. Tetrahedron: Asymmetry 8:2167–2173, 1997.
43. G Carrea, G Ottolina, S Riva. Trends Biotechnol 13:63–70, 1995. Erratum: Trends Biotechnol 13:122, 1995.
44. PJ Halling. Enzyme Microb Technol 6:513–516, 1984.
45. PJ Halling. Biocatalysis 1:109–115, 1987.
46. PJ Halling. In: C Laane, J Tramper, MD Lilly, eds. Biocatalysis in Organic Media. Amsterdam: Elsevier, 1987, pp 125–132.
47. PJ Halling. Trends Biotechnol 7:50–52, 1989.
48. RH Valivety, PJ Halling, AR Macrae. Indian J Chem 31B:914–916, 1992.
49. HL Goderis, G Ampe, MP Feyten, BL Fouwé, WM Guffens, SM Van Cauwenbergh, PP Tobback. Biotechnol Bioeng 30:258–266, 1987.
50. L Greenspan. J Res Nat Bur St Phys Chem 81A:89–96, 1977.
51. E Wehtje, I Svensson, P Adlercreutz, B Mattiasson. Biotechnol Tech 7:873–878, 1993.
52. S Bloomer, P Adlercreutz, B Mattiasson. Biocatalysis 5:145–162, 1991.
53. A Van der Padt, AEM Janssen, JJW Sewalt, K Van't Riet. Biocatalysis 7:267–277, 1993.
54. JM Cassels, PJ Halling. Enzyme Microb Technol 10:486–491, 1988.
55. SA Khan, PJ Halling, G Bell. Enzyme Microb Technol 12:453–458, 1990.
56. RM Blanco, JLL Rakels, JM Guisán, PJ Halling. Biochim Biophys Acta 1156:67–70, 1992.
57. A Van der Padt, JJW Sewalt, K Van't Riet. In: J Tramper, MH Vermüe, HH Beeftink, U von Stockar, eds. Biocatalysis in Non-Conventional Media. Amsterdam: Elsevier, 1992, pp 557–562.
58. M Goldberg, F Parvaresh, D Thomas, MD Legoy. Biochim Biophys Acta 957:359–362, 1988.
59. P Kuhl, PJ Halling, H-D Jakubke. Tetrahedron Lett 31:5213–5216, 1990.

60. PJ Halling. Biotechnol Tech 6:271–276, 1992.
61. L Kvittingen, B Sjursnes, T Anthonsen, P Halling. Tetrahedron 48:2793–2802, 1992.
62. E Zacharis, IC Omar, J Partridge, DA Robb, PJ Halling. Biotechnol Bioeng 55:367–374, 1997.
63. H Kitaguchi, I Itoh, M Ono. Chem Lett 1203–1206, 1990.
64. JP van der Lugt, J Elfrink, J Evenaar, HJ Doddema. In: S Servi, ed. Microbial Reagents in Organic Synthesis. Dordrecht: Kluwer, 1992, pp 261–272.
65. H-E Högberg, H Edlund, P Berglund, E Hedenström. Tetrahedron: Asymmetry 4:2123–2126, 1993.
66. H Edlund, P Berglund, M Jensen, E Hedenström, H-E Högberg. Acta Chem Scand 50:666–671, 1996.
67. Y Yasufuku, S-I Ueji. Bioorg Chem 25:88–99, 1997.
68. TM Stokes, AC Oehlschlager. Tetrahedron Lett 28:2091–2094, 1987.
69. J Bodnár, L Gubicza, L-P Szabó. J Mol Catal 61:353–361, 1990.
70. M Reslow, P Adlercreutz, B Mattiasson. Biocatalysis 6:307–318, 1992.
71. A Wickli, E Schmidt, JR Bourne. In: J Tramper, MH Vermüe, HH Beftink, U von Stockar, eds. Biocatalysis in Non-Conventional Media. Amsterdam: Elsevier, 1992, pp 577–584.
72. L Gubicza, A Szakács-Schmidt. Biocatalysis 9:131–143, 1994.
73. C Orrenius, T Norin, K Hult, G Carrea. Tetrahedron: Asymmetry 6:3023–3030, 1995.
74. O Nordin, E Hedenström, H-E Högberg. Tetrahedron: Asymmetry 5:785–788, 1994.
75. R Bovara, G Carrea, G Ottolina, S Riva. Biotechnol Lett 15:169–174, 1993.
76. E Wehtje, D Costes, P Adlercreutz. J Mol Catal B: Enzym 3:221–230, 1997.
77. RJ Kazlauskas, UT Bornscheuer. In: DR Kelly, ed. Biotechnology, 2nd ed., Vol. 8a. Biotransformations I. Weinheim: Wiley-VCH, 1998, pp 37–191.
78. CR Johnson. Acc Chem Res 31:333–341, 1998.
79. B Rubin, EA Dennis, eds. Methods in Enzymology, Vol. 286, Lipases, Part B. New York: Academic Press, 1997, pp 1–563.
80. T Itoh, Y Takagi, H Tsukube. J Mol Catal B: Enzym 3:259–270, 1997.
81. W Boland, C Frößl, M Lorenz. Synthesis 1049–1072, 1991.
82. L Pasteur. C R Acad Sci 46:615–618, 1858.
83. HB Kagan, JC Fiaud. In: EL Eliel, SH Wilen, eds. Topics in Stereochemistry, Vol 18. New York: Wiley, 1988, pp 249–330.
84. JB Jones. Tetrahedron 42:3351–3403, 1986.
85. C-S Chen, CJ Sih. Angew Chem 101:711, 1989; Angew Chem Int Ed Engl 28:695–707, 1989.
86. E Santaniello, P Ferraboschi, P Grisenti, A Manzocchi. Chem Rev 92:1071–1140, 1992.
87. N Baba, M Mimura, J Hiratake, K Uchida, J Oda. Agric Biol Chem 52:2685–2687, 1988.
88. N Öhrner, C Orrenius, A Mattson, T Norin, K Hult. Enzyme Microb Technol 19:328–331, 1996.
89. T Sakai, I Kawabata, T Kishimoto, T Ema, M Utaka. J Org Chem 62:4906–4907, 1997.
90. D Cavaille-Lefebvre, D Combes. Biocatal Biotransform 15:265–279, 1997.
91. H-E Högberg, P Berglund, H Edlund, J Fägerhag, E Hedenström, M Lundh, O Nordin, S Servi, C Vörde. Catal Today 22:591–606, 1994.
92. FJ Urban, R Breitenbach, LA Vincent. J Org Chem 55:3670–3672, 1990.
93. S Chattopadhyay, VR Mamdapur. Biotechnol Lett 15:245–250, 1993.
94. SG Allenmark, AC Andersson. Tetrahedron: Asymmetry 4:2371–2376, 1993.
95. R Noyori, M Tokunaga, M Kitamura. Bull Chem Soc Jpn 68:36–56, 1995.
96. RS Ward. Tetrahedron: Asymmetry 6:1475–1490, 1995.
97. S Caddick, K Jenkins. Chem Soc Rev 25:447–456, 1996.
98. H Stecher, K Faber. Synthesis 1–15, 1997.
99. MTEI Gihani, JMJ Williams. Curr Opin Chem Biol 3:11–15, 1999.
100. EJ Ebbers, GJA Ariaans, JPM Houbiers, A Bruggink, B Zwanenburg. Tetrahedron 53:9417–9476, 1997.

101. R Stürmer. Angew Chem 109:1221-1222, 1997, Angew Chem Int Ed Engl 36:1173–1174, 1997.
102. MT Reetz, K Schimossek. Chimia 50:668–669, 1996.
103. PM Dinh, JA Howarth, AR Hudnott, JMJ Williams, W Harris. Tetrahedron Lett 37:7623–7626, 1996.
104. AEL Larsson, BA Persson, J-E Bäckvall. Angew Chem 109:1256–1258, 1997; Angew Chem Int Ed Engl 36:1211–1212, 1997.
105. JV Allen, JMJ Williams. Tetrahedron Lett 37:1859–1862, 1996.
106. R-L Gu, I-S Lee, CJ Sih. Tetrahedron Lett 33:1953–1956, 1992.
107. JZ Crich, R Brieva, P Marquart, R-L Gu, S Flemming, CJ Sih. J Org Chem 58:3252–3258, 1993.
108. TL Amyes, JP Richards. J Am Chem Soc 114:10297–10302, 1992.
109. DS Tan, MM Günter, DG Drueckhammer. J Am Chem Soc 117:9093–9094, 1995.
110. P-J Um, DG Drueckhammer. J Am Chem Soc 120:5605–5610, 1998.
111. C Vörde, H-E Högberg, E Hedenström. Tetrahedron: Asymmetry 7:1507–1513, 1996.
112. W Kroutil, K Faber. Tetrahedron: Asymmetry 9:2901–2913, 1998.
113. E Schoffers, A Golebiowski, CR Johnson. Tetrahedron 52:3769–3826, 1996.
114. AG Ogston. Nature 162:963, 1948.
115. G Guanti, E Narisano, T Podgorski, S Thea, A Willliams. Tetrahedron 46:7081–7092, 1990.
116. AL Gutman, M Shapira, A Boltanski. J Org Chem 57:1063–1065, 1992.
117. Y-F Wang, C-S Chen, G Girdaukas, CJ Sih. J Am Chem Soc 106:3695–3696, 1984.
118. P Breuilles, T Schmittberger, D Uguen. Tetrahedron Lett 34:4205–4208, 1993.
119. K Faber, S Riva. Synthesis 895–910, 1992.
120. G Kirchner, MP Scollar, AM Klibanov. J Am Chem Soc 107:7072–7076, 1985.
121. H Brockerhoff, RJ Hoyle, PC Hwang. Anal Biochem 37:26–31, 1970.
122. HM Sweers, C-H Wong. J Am Chem Soc 108:6421–6422, 1986.
123. M Degueil-Castaing, B De Jeso, S Drouillard, B Maillard. Tetrahedron Lett 28:953–954, 1987.
124. Y-F Wang, C-H Wong. J Org Chem 53:3127–3129, 1988.
125. Y-F Wang, JJ Lalonde, M Momongan, DE Bergbreiter, C-H Wong. J Am Chem Soc 110:7200–7205, 1988.
126. K Laumen, D Breitgoff, MP Schneider. J Chem Soc Chem Commun 1459–1461, 1988.
127. JBA Van Tol, DE Kraayveld, JA Jongejan, JA Duine. Biocatal Biotransform 12:119–136, 1995.
128. B Berger, K Faber. J Chem Soc Chem Commun 1198–1200, 1991.
129. HK Weber, H Stecher, K Faber. Biotechnol Lett 17:803–808, 1995.
130. Y Kita, Y Takebe, K Murata, T Naka, S Akai. Tetrahedron Lett 37:7369–7372, 1996.
131. N Öhrner, M Martinelle, A Mattson, T Norin, K Hult. Biotechnol Lett 14:263–268, 1992.
132. H Frykman, N Öhrner, T Norin, K Hult. Tetrahedron Lett 34:1367–1370, 1993.
133. P Berglund, M Holmquist, E Hedenström, K Hult, H-E Högberg. Tetrahedron: Asymmetry 4:1869–1878, 1993.
134. P Berglund, M Holmquist, K Hult, H-E Högberg. Biotechnol Lett 17:55–60, 1995.
135. M Holmquist, M Martinelle, P Berglund, IG Clausen, S Patkar, A Svendsen, K Hult. J Protein Chem 12:749–757, 1993.
136. IJ Colton, SN Ahmed, RJ Kazlauskas. J Org Chem 60:212–217, 1995.
137. Z-W Guo, CJ Sih. J Am Chem Soc 111:6836–6841, 1989.
138. CJ Sih, Q-M Gu, X Holdgrün, K Harris. Chirality 4:91–97, 1992.
139. JJ Lalonde, C Govardhan, N Khalaf, AG Martinez, K Visuri, AL Margolin. J Am Chem Soc 117:6845–6852, 1995.
140. AL Margolin. Trends Biotechnol 14:219–259, 1996.
141. Y Okahata, A Hatano, K Ijiro. Tetrahedron: Asymmetry 6:1311–1322, 1995.
142. T Oberhauser, M Bodenteich, K Faber, G Penn, H Griengl. Tetrahedron 43:3931–3944, 1987.

143. S-H Wu, L-Q Zhang, C-S Chen, G Girdaukas, CJ Sih. Tetrahedron Lett 26:4323–4326, 1985.

144. G Caron, RJ Kazlauskas. Tetrahedron: Asymmetry 4:1995–2000, 1993.

145. Z-W Guo, S-H Wu, C-S Chen, G Girdaukas, CJ Sih. J Am Chem Soc 112:4942–4945, 1990.

146. K Faber. Biotransformations in Organic Chemistry. A Textbook, 3rd ed. Berlin: Springer-Verlag, 1997, pp 42–46.

147. ELA Macfarlane, SM Roberts, NJ Turner. J Chem Soc Chem Commun 569–571, 1990.

148. C-S Chen, Y-C Liu. J Org Chem 56:1966–1968, 1991.

22

Chemoenzymatic Preparation of Enantiomerically Pure S(+)-2-Arylpropionic Acids with Anti-Inflammatory Activity

Andrés-Rafael Alcántara, José-María Sánchez-Montero, and José-Vicente Sinisterra
Universidad Complutense de Madrid, Madrid, Spain

I. INTRODUCTION

Rheumatoid illness includes many concrete diseases such as arthritis, spinal column pains, lumbago, and so forth, with strong incidence in labor costs. In Spain, near 25% of the population between 46 and 60 years old suffers from some of these pathologies, and the percentage grows to around 30% for the working population older than 60. The high labor costs may be estimated taking into account that 17.5% of those patients have permanent invalidity, and for 25% of the patients some occasional absences from work are due to these diseases [1].

Inflammatory processes are mainly (78% of cases) treated with drugs. Until 1971, glucocorticoids with anti-inflammatory activity were the only drugs used, in spite of their strong secondary effects. The discovery in 1971 of the nonsteroidal anti-inflammatory drugs [2] was the impetus for the development of new anti-inflammatory drugs without secondary effects. Three main chemical structures, shown in Fig. 1, produce inhibition of the synthesis of prostaglandins in a wide variety of cells and tissues [3]:

1. Benzoic acid derivatives, such as aspirin and diflunisal (compounds **1** and **2**)
2. (Hetero)arylacetic acid derivatives, such as indomethacin (**3**) or ibufenac (**4**), and
3. 2-Arylpropionic acids (APAs, the "profen" family) derivatives.

This last type of nonsteroidal anti-inflammatory drugs, such as ibuprofen ([*rac*-2-(4-isobutylphenyl) propionic acid], **5**), flurbiprofen ([*rac*-2-((3-fluoro-4-phenyl)phenyl) propionic acid], **6**), fenoprofen ([*rac*-2-(3-phenoxyphenyl)propionic acid], **7**), suprofen ([*rac*-2-(4(2-thienyloxo)phenyl)propionic acid, **8**], carprofen ([*rac*-2-(6-chloro-9*H*-cabazolyl) propionic acid], **9**), naproxen ([*S*(+)-2-(6-methoxy-2-naphthyl)propionic acid], **10**), Dexketoprofen ([*S*(+)-2-(3-benzoylphenyl)propionic acid], **11**), flunoxaprofen ([*S*(+)-2-(2(4-fluoro-phenyl)benzoxazol-5-yl)propionic acid], **12**), or ketorolac ([*S*(+)-5-benzoyl-2,3-dihidro-1*H*-pyrrolizine-1-carboxylic acid], **13**, thromethamine salt marketed as Toradol), have a more promising future.

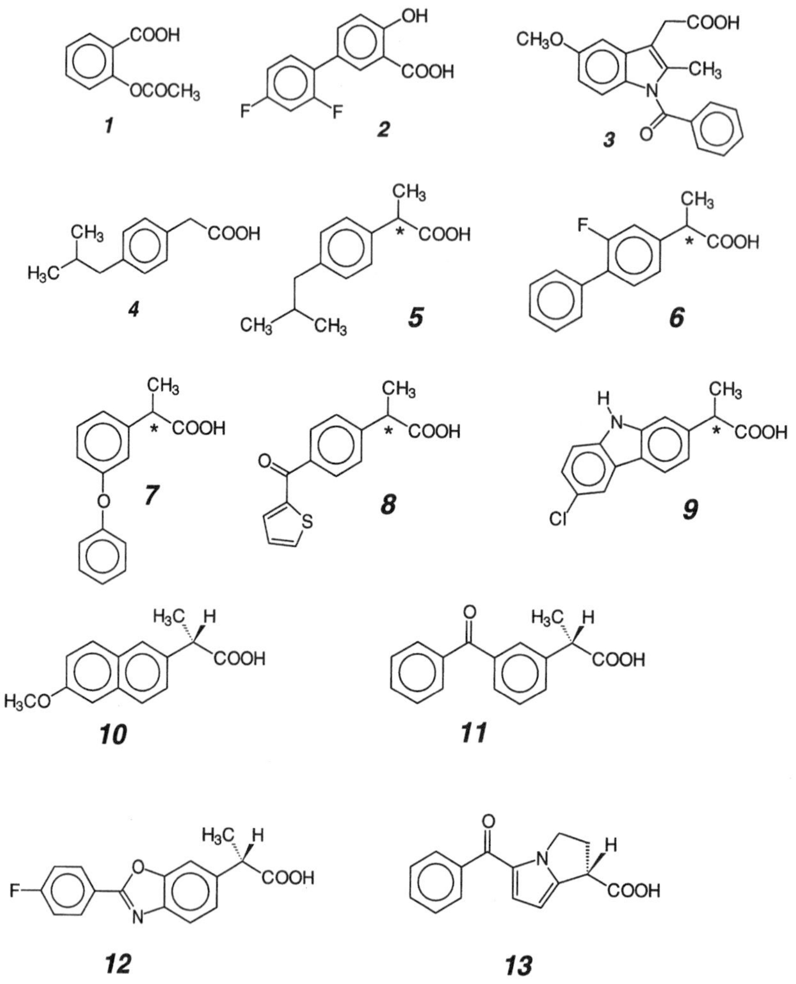

Figure 1 Structures of several anti-inflammatory drugs: aspirin (**1**), diflunisal (**2**), indomethacin (**3**), ibufenac (**4**), ibuprofen (*rac*-2-(4-isobutylphenyl)propionic acid) (**5**), flurbiprofen (*rac*-2-((3-fluoro-4-phenyl)phenyl)propionic acid) (**6**), fenoprofen (*rac*-2-(3-phenoxyphenyl)propionic acid) (**7**), suprofen (*rac*-2-(4(2-thienyloxo)phenyl)propionic acid) (**8**), carprofen (*rac*-2-(6-chloro-9*H*-carbazolyl)propionic acid) (**9**), naproxen (*S*(+)-2-(6-methoxy-2-naphthyl)propionic acid) (**10**), dexketoprofen (*S*(+)-2-(3-benzoylphenyl)propionic acid) (**11**), flunoxaprofen (*S*(+)-2-(2(4-fluorophenyl)benzoxazol-5-yl)propionic acid) (**12**), and ketorolac (*S*(+)-5-benzoyl-2,3-dihidro-1*H*)-pyrrolizine-1-carboxylic acid) (**13**).

The pharmacological activity of these drugs is due to the inhibition of the enzyme cyclooxygenase, by preventing the abstraction of the hydrogen from C-13, thereby avoiding the peroxidation at C-11 and C-15 positions of arachidonic acid and hindering its biotransformation to prostaglandins and thromboxane A_2, which are responsible for the inflammatory mechanism [4–6]. The high activity of ibuprofen is due to the drug displaying a binding affinity for cyclooxygenase similar to that of the natural substrate arachidonic acid [4]. Nevertheless, this process is not well known, although it seems to be caused by the presence of both the aromatic ring and the carboxylic moiety. On the other hand, the time-dependent inactivation of cyclooxygenase is enhanced by the presence in the drug of an arylhalogen group, such as in the case of flurbiprofen [7].

Numerous pharmacological studies above the relative activity of both enantiomers of these compounds have shown that the $S(+)$ isomer not only has a significantly greater therapeutic effect (28-fold in the case of naproxen [8] or 160-fold for $S(+)$ ibuprofen [9] versus the $R(-)$ antipode (which is essentially inactive as an inhibitor of prostaglandin synthesis [10], although it could display some analgesic effect through a different mechanism [11]), but also that the $S(+)$ enantiomer reaches the therapeutic concentration in blood faster than the racemate [11]. Nevertheless, only naproxen (Fig. 1, **10**), dexketoprofen (Fig. 1, **11**), flunoxaprofen (Fig. 1, **12**), and toradol (thromethamine salt of $S(+)$-ketorolac, Fig. 1, **13**) are marketed as the active enantiomer.

It is well documented [12–16] that a certain proportion of $R(-)$-ibuprofen might undergo a metabolic inversion in vivo to the active $S(+)$ enantiomer, as depicted in Fig. 2. The inversion proceeds [17,18] via the stereoselective transformation of the $R(-)$ isomer in the thioester with the coenzyme A, which can be partially transesterified by endogenous triacylglycerols to be stored in the lipidic tissues as "hybrid" triglycerides. R-Ibuprofen-CoA is subsequently isomerized to S-ibuprofen-CoA, which is quickly hydrolyzed to free $S(+)$-ibuprofen. This process is irreversible, so that $S(+)$-ibuprofen is not stored in the lipidic tissue. Recently, the possible formation of an (R)-adenylate intermediate prior to the formation of the (R)-coenzyme thio ester has been described as the stereoselective step

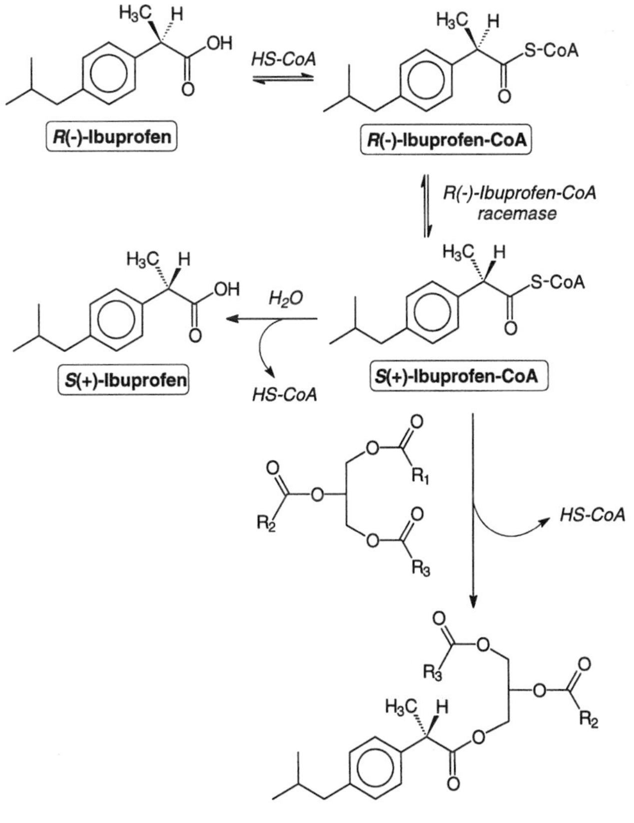

Figure 2 Metabolic inversion of $R(-)$-ibuprofen.

of chiral inversion [19]. In any case, the hybrid triglycerides may display toxic effects due to the disruption produced in the normal lipid metabolism and membrane function. $R(-)$-Ibuprofen undergoes this chiral inversion in vivo in rats [20] and humans [21], and the same process has been described for $R(-)$-ketoprofen in rats [22], while the inversion of $R(-)$-2-(6-methoxy-2-naphthyl)propionic acid and $R(-)$-fenoprofen has been described in rabbits [23]. However, minimal inversion of $R(-)$-ketoprofen has been reported in humans [24], and other drugs, as flurbiprofen, are less prone to enantiomeric conversion in vivo [25,26]. Thus, with the aim of avoiding the problems (dosage, toxicity, and pharmacokinetics) associated with the administration of racemic mixtures, the preparation of homochiral arylpropionic drugs has become a focus of great interest in recent years. The following section will focus on this topic.

II. SYNTHETIC METHODOLOGIES FOR OBTAINING HOMOCHIRAL 2-ARYLPROPIONIC ACIDS

Several synthetic methodologies have recently been described in the literature to obtain the structure of 2-arylpropionic acids (Fig. 3):

1. Preparation of the -COOH group from a precursor X [27]
2. Introduction of a C_1 synthon in the arylacetic chain [28]
3. Introduction of the propionic acid chain in the aromatic ring [29]
4. Insertion of dihalocarbene in C=O bonds in acetophenone using low-valent titanium [30]
5. Propiophenone rearrangement [6]

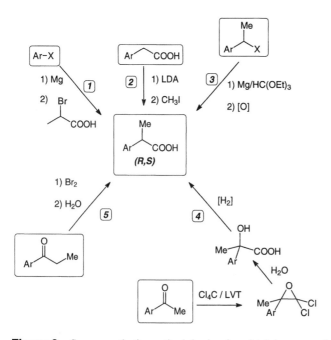

Figure 3 Some synthetic methodologies for obtaining racemic 2-arylpropionic acids.

In all cases, racemic mixtures are produced. Therefore, some other methodologies have been described to achieve the stereoselective synthesis of the active $S(+)$ isomer. Thus, many references can be found in the literature describing different chemical methods for the asymmetrical synthesis of 2-arylpropionic acids [31–47]. Nevertheless, these processes are normally tedious and not very promising from an industrial point of view. Recently, two interesting procedures, presented in Fig. 4, have been published. The first synthetic strategy [46] creates the asymmetry into the molecule by a Sharpless asymmetrical dihydroxylation of the appropriate methyl styrenes, subsequently converting the resulting diols into optically active epoxides, and the required stereogenic center was assembled by means of a catalytic hydrogenolysis of the introduced benzylic epoxide bond, followed by an oxidative cleavage of the optically active primary alcohol. Thus, naproxen and flurbiprofen are obtained with yields ranging from 30% to 35%, and enantiomeric excess of 98% after seven synthetic steps presented in Fig. 4A. In the second methodology (Fig. 4B) [47], a chemoenzymatic synthesis of naproxen and $S(+)$-ibuprofen, after synthesizing the corresponding prochiral 2-aryl-1,3-propanediols, these compounds were

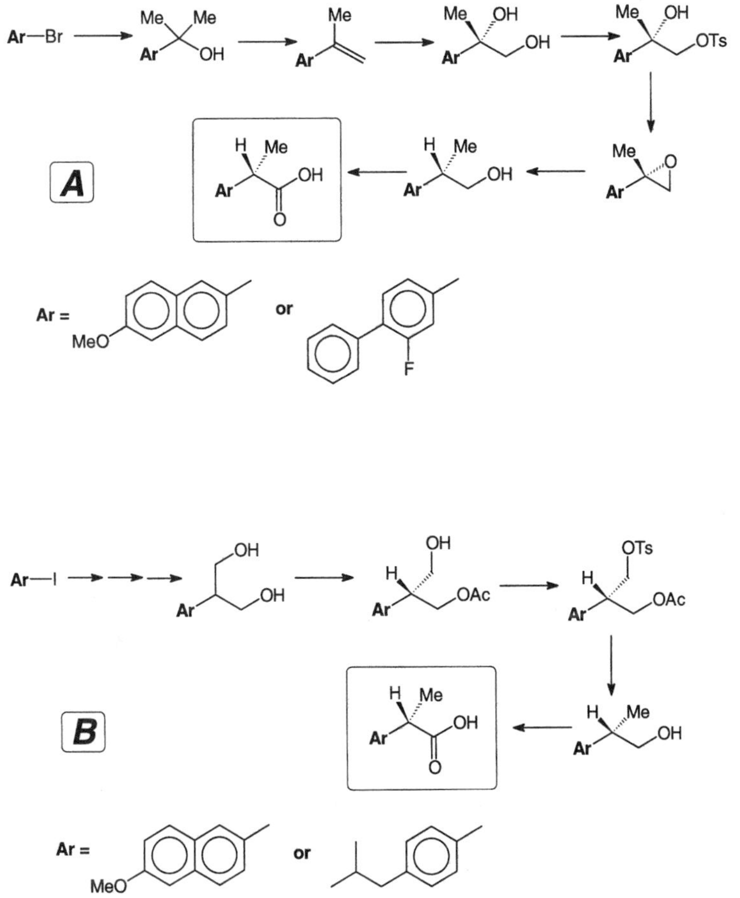

Figure 4 Two recently described strategies for the chemical synthesis of enantiopure 2-arylpropionic acids [46,47].

asymmetrized via a lipase-catalyzed acetylation, leading to optically active hydroxy acetates, which were later deoxygenated on the primary hydroxyl functionality and further oxidized to render the desired pure $S(+)$ enantiomers after nine steps with yields around 25%.

The resolution of the racemates of 2-arylpropionic acids (or their esters) is a very convenient alternative to the stereoselective synthesis, because fairly easy and cheap conventional methodologies can be employed to obtain the racemic mixtures. Thus, several resolution procedures can be found in the literature:

1. Resolution by means of the formation of diastereomeric esters or amides, easy to separate, via reaction of the racemic 2-arylpropionic acid with optically pure alcohol or amines [48]
2. Resolution by means of the formation of diastereomeric salts of the racemic acids with chiral amines such as cinconidine [49], α-methylbenzylamine [50], lysine [51], BOP [(S)-1-(benzyloxymethyl)propylamine, [52]], (−)-ephedryne [52], TPO [(4S, 5R)-(+)-1,2,5,5-tetramethyl-5-phenyl-1,3-oxazolidine, [52]], PTEA [(R)-1-phenyl-2-p-tolylethylamine, [52]] or (S)-phenylglycinol [52].
3. Resolution by esterification with optically pure sugar molecules [53,54].
4. Resolution using high-performance liquid chromatography (HPLC) with chiral stationary phases [55,56]
5. Biotechnological resolution using cells or enzymes. This last methodology will be viewed in this chapter.

III. BIOTECHNOLOGICAL RESOLUTION OF RACEMATES

Two different strategies may be used in order to prepare the pharmacologically active $S(+)$-2-arylpropionic acid, depending on the biocatalyst used for the enantio discrimination:

1. Whole microorganisms
2. Isolated enzymes

A. Kinetic Resolution Using Microorganisms

Several synthetic strategies have been described, depending on the main catalytic activity of the microorganism:

1. Oxidation Reactions Using Microorganisms Capable of Discriminating Between Two Prochiral Methyl Groups and Thus Leading to the S-Enantiomer

This process is interesting, although not very high conversions have been obtained. Thus, a stereospecific oxidative degradation of α-methyl long chain alkyl derivatives of 6-methoxynaphthalene was described using *Rhodococcus* sp. [57] leading to the $R(-)$ acid.

2. Metabolic Resolution of the Racemic Mixture

Metabolic resolution of the racemic mixture has been reported using *Verticillium lecanii* [58]. This microorganism is able to grow on the racemic (*R,S*)-ibuprofen, giving a product mixture enriched in the *S* isomer (*S/R* = 2), after 14 days at pH = 7.0 and 24°C.

3. Metabolic Isomerization of the *R*-Enantiomer

It has been reported [59] that *Candida curvata* IPO1159 can isomerize *R*(−)-2-(6-methoxy-2-naphthyl)propionic acid to a mixture *R/S* = 61:39 after 48 h at pH = 6.8 and 30°C.

4. Biotransformations

The low yield and/or optical purity achieved with the previous methodologies, as well as the possibility to stabilize the cells by immobilization, has opened the possibility of biotransformations using whole cells. In this way, arylmalonic acids were transformed into 2-arylpropionic acids by incubation with *Alcaligenes bronchisepticus* [60].

Nevertheless, the most described methodology is the asymmetrical hydrolysis of nitriles [61] using different microorganisms. The first patents concerning the kinetic resolution of racemic 2-arylpropionic nitriles with resting cells of *Brevibacterium* or *Corynebacterium* were described in the late 1980s [62,63]. The most useful microorganisms are *Rhodococcus butanica* [64], *Rhodococcus equi* A4 [65], and *Rhodococcus rhodochrous* NCIMB 11216, which hydrolyze the racemic substrate and yield naproxen with an enantiomeric excess around 95–98% [66]. Other examples can also be found in the literature [67–72]. The good results obtained have promoted the used of recombinant microorganisms such as the nitrilase gene of *Alcalinegens fecalis* expressed in *Escherichia coli* to improve the productivity [73].

The presence of esterases or amidases in many microorganisms has also been employed to resolve racemic mixtures of the esters or amides of these drugs. In this sense, *Kluyvera oxytoca* SNSM 87 [74] and *Bacillus subtilis* [75] esterases have been successfully used in the resolution of the racemic esters, while *Rhodococcus erythopolis* MP50 has been applied for the resolution of the amides of (*R,S*-2-(6-methoxy-2-naphthyl)propionic acid, yielding 48% yield of naproxen, with an enantiomeric excess more than 99% after 45 h [76].

B. Kinetic Resolution Using Free Enzymes

The use of entrapped whole cells for the resolution of racemic mixtures of 2-arylpropionic acids is only appropriate when working in aqueous media because cells are destroyed by organic solvents. Therefore, these biocatalysts can be employed exclusively for the resolution of racemic nitriles, esters, or amides in a hydrolytic fashion but are useless for esterification or amidation reactions in organic solvents, where the esters and the acids of these lipidic compounds are soluble. Consequently, the experimental conditions seem to be decisive for selecting the optimum biocatalyst, and when using organic solvents, free enzymes are the rational choice. More concretely, lipases are specially suited for these purposes because they exhibit good stability in aqueous media and are able to tolerate high concentrations of substrate (more than 1 M) [77]. Furthermore, due to the fact that these enzymes are naturally designed for catalyzing the hydrolysis of ester bonds at the interface between an insoluble substrate phase (triglycerides) and an aqueous phase in which the enzyme is dissolved, the low water solubility of 2-arylpropionic acids is not a serious problem. On the other hand, as lipases have shown to be equally very resistant to organic solvents [78,79] and able to retain and even increase their stereoselective capability in nonaqueous solvents [80], their use has gained widespread acceptance [81,82].

As can be inferred from the above paragraph, two different strategies can be used for the kinetic resolution of APAs:

1. Enantioselective hydrolysis of their racemic esters or amides, or
2. Enantioselective esterification of the racemic acids

Some examples of the first methodology are presented in Tables 1 to 4. In order to quantify the enantioselective performance in the resolution of enantiomers, the enantiomeric ratio (E) is the parameter used in all of the tables. This value was first defined by Chen et al. [83] and measures the ratio between the specificity constants (V_{max}/K_m) for both enantiomers. (We strongly recommend that the reader consult the recent work of Straathof and Jongejan [84] for an exhaustive review on the meaning and calculation of E.) As it was defined, E values below 15 are insufficient for practical purposes; E values between 15 and 30 are moderate to good, and values above these figures are excellent.

Hydrolytic reactions are completely irreversible because of the high water concentration of the aqueous microenvironment, although this optimum feature is counterbalanced with the poor water solubility of the starting APA's esters, which sometimes makes necessary the employment of some water-miscible organic cosolvents. On the contrary, the esterification and transesterification reactions, carried out in organic media, do not suffer from this problem, and the great stability of lipases in these "nonconventional" media can be exploited for obtaining good yields and enantioselectivities. Some examples are shown in Tables 5–8.

The esterification of 2-APAs, in order to be more efficient, should be carried out with lipases possessing an R-stereo preference; in this way, the active drug, the S-acid, would not suffer any transformation during the reaction course and would be easily recovered from the medium. Nevertheless, this fact forces the reaction yield to be higher than 50% if we want to obtain the drug with high optical purity. Another fact to be considered is that, as can be seen from Tables 5–8, some lipases display different stereobiases depending on the nature of the aromatic moiety of the substrate, or even lipases from the same organism, but from different commercial sources, can show rather different behaviors. Nevertheless, the synthesis of S-esters of profens does not propose any major problem because this compound could be considered as a "prodrug," i.e., a pharmacologically inactive derivative of the parent drug that requires a spontaneous (or enzymatic) transformation within the body to release the active drug. Thus, as the human endogenous organisms are rich in hydrolases capable of transforming the esters [112], with the administration of the prodrug it is possible to obtain bioactivity similar to that one of the acid [113–115]. This reduces the adverse gastrointestinal side effects shown by the profens [114]; the "other" lipase enantiobias does not constitute any serious impediment.

The transesterification methodology shown in Tables 5 and 8 deserves to be taken into account due to the fact that, because of toxicity problems, only ethyl esters should be used, and not all lipases can efficiently use ethanol for esterification. On the other hand, in these processes using leaving groups with reduced nucleophilicity (such as 2-haloethyl; see Table 5), it is possible to force the irreversibility of the reaction, and thus make the process similar to hydrolysis, where the high water concentration pushes the equilibrium toward the desired products.

Table 1A Enantioselective Hydrolysis of *rac*-Ketoprofen Esters: Obtaining the *S*(+)-Acid

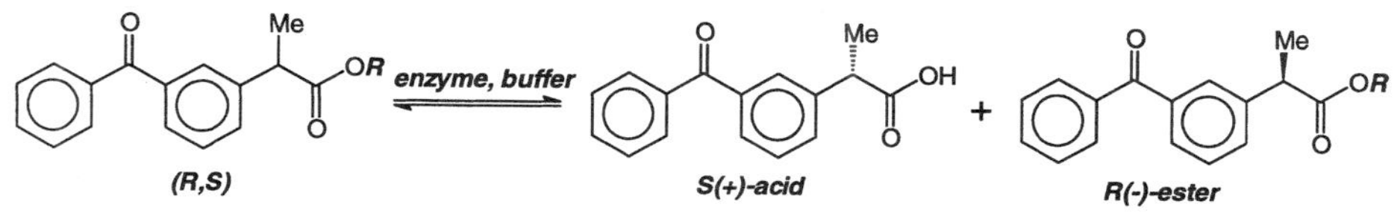

R	Enzyme	t (h)	T (°C)	χ Acid (%)	e.e. Acid (%)	χ Ester (%)	e.e. Ester (%)	E	Ref.
Methyl	*Candida cylindracea* lipase from Meito Sangyo OP-360	168	22	50	60	—	—	7	85
2-Chloroethyl	*Candida cylindracea* lipase from Sigma (L1754, type VII)	120	r.t.	32	52	—	25	4	86
2-Chloroethyl	*Candida cylindracea* lipase from Sigma partially purified (called A form)	—	r.t.	—	—	—	—	>100	85
Ethyl	*Candida cylindracea* lipase from Sigma (type VII)	96	30	20.5	>98	—	—	>100	87
Ethyl	*Candida cylindracea* lipase from Sigma (type VII) dialyzed (20,000 cutoff)	96	30	31.4	>98	—	—	>100	87
2-Chloroethyl	*Candida cylindracea* lipase from Sigma (L1754 type VII) treated with 2-propanol	1.8	—	28	99	—	38	>100	88
Ethyl	*Candida rugosa* lipase B (purified from Sigma type VII starting material)	96	30	40	>99	—	—	>100	89
Ethyl	*Candida rugosa* lipase B from Sigma adding NaCl and CaCl₂	96	30	33	>99	—	—	>100	89
Ethyl	*Candida cylindracea* lipase from Sigma covalently immobilized on Al$_2$O$_3$	168	35	14.5	95	—	—	45	90
Ethyl	*Candida cylindracea* lipase from Sigma covalently immobilized on SiO$_2$	168	35	20.5	97	—	—	84	90
Ethyl	*Candida cylindracea* lipase from Sigma covalently immobilized on agarose	168	35	21.1	94	—	—	41	90
Ethyl	*Candida rugosa* lipase from Sigma	120	30	15	96	—	—	58	91

Table 1B Enantioselective Hydrolysis of *rac*-Ketoprofen Esters: Obtaining the *R*(−)-Acid

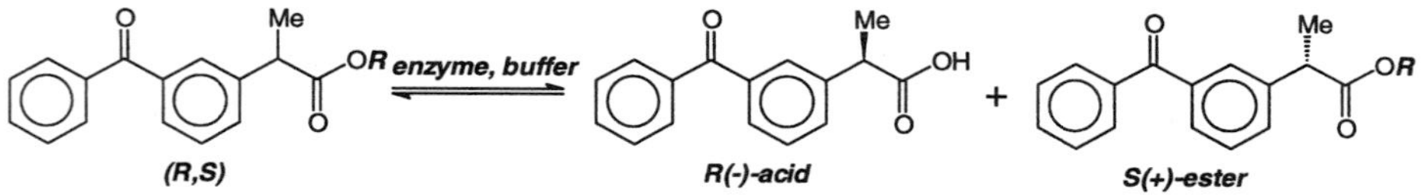

R	Enzyme	t (h)	T (°C)	χ Acid (%)	e.e. Acid (%)	χ Ester (%)	e.e. Ester (%)	E	Ref.
—CH₂—CF₃	*Aspergillus niger* lipase from Amano (AP-6) in different water-saturated organic solvents	48	r.t.	1–47	40→99	—	<1–73	2→100	92
—CH₂—CF₃	*Aspergillus niger* lipase from Amano (AP-6) in mixtures of (*i*-Pr)₂O/ H₂O/water-soluble solvents	24–81	r.t.	24–60	53→99	—	11–79	7→100	92

IV. METHODOLOGIES TO IMPROVE THE YIELD AND/OR THE ENANTIOMERIC PURITY OF THE PRODUCT

Several methodologies have been described in the literature with the aim of improving the enantiomeric purity and/or the yield of the crude product obtained in the enzymatic resolution of racemic 2-APAs, using either the hydrolytic or the synthetic methodologies mentioned in the previous paragraph. These procedures may be classified as follows:

1. Control of the technical variables
2. Control of the purity of the enzymes
3. Selection of the solvent
4. Control of the amount of water
5. Presence of additives
6. Influence of the nature of the support
7. Influence of the chemical structure of the alcohol

A. Control of the Technical Variables

The factorial design of experiments, a multivariant method in which all of the studied parameters are simultaneously changed in a suitable programmed manner [116,117], have been used to determine the main variables that must be optimized in the hydrolysis or in the synthesis of esters using lipases. Thus, in the hydrolysis of (R,S)-ibuprofen ethyl ester using commercial *C. rugosa* lipase immobilized on SiO_2 (1 g of biocatalyst), the following equation was obtained for modeling the obtained yield [90].

$$\text{Yield} = 41.2 + 18.0x_1 - 9.6x_2 - 9.6x_4 \tag{1}$$

where

x_1 = stirring speed (rpm)
x_2 = pH, and
x_4 = substrate concentration [M]

The other experimental variable tested, the ionic strength of the medium ($0.1 < I(M) < 0.7$), did not yield a significant result.

In the case of the enantioselective esterification of 2-APAs in organic solvent, the tested variables were as follows:

x_1 = amount of water (μl/ml solvent)
x_2 = temperature (°C)
x_3 = stirring speed (rpm)
x_4 = amount of catalyst (mg)
x_5 = molar ratio acid/alcohol
x_6 = reaction time (h)
x_7 = amount of solvent (ml)

For this purpose, two different enzymes (adsorbed on resins) were tested: (1) *Candida antarctica* lipase B (SP 4354A) [97] and (2) *Rhizomucor miehei* (lipozyme IM) [98]. The obtained response (yield) in the esterification of (R,S)-ibuprofen were expressed as shown in Eq. (2) and (3), respectively:

(text continues on pg. 677)

Alcántara et al.

Table 2A Enantioselective Hydrolysis of *rac*-2-(6-methoxy-2-naphthyl)Propionic Esters: Obtaining Naproxen

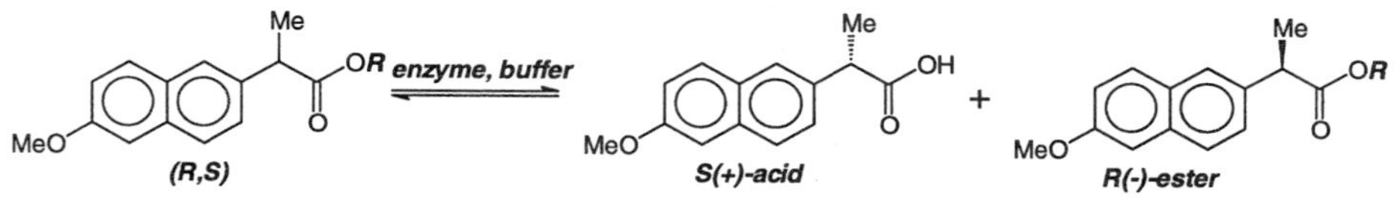

R	Enzyme	t (h)	T (°C)	χ Acid (%)	e.e. Acid (%)	χ Ester (%)	e.e. Ester (%)	E	Ref.
Methyl	*Candida cylindracea* lipase from Sigma (type VII)	216	22	39	>98	—	63	>100	77
2-Chloroethyl	*Candida cylindracea* lipase from Sigma (type VII)	42	22	40.5	>98	47.9	70	>100	77
Methyl	*Candida cylindracea* lipase from Meito Sangyo OP-360	168	22	—	>98	—	42	>100	85
2-Nitropropyl	*Candida cylindracea* lipase from Meito Sangyo OP-360	75	22	—	94	—	63	68	85
Cyanomethyl	*Candida cylindracea* lipase from Meito Sangyo OP-360	37	22	—	>98	—	76	>100	85
2-Chloroethyl	*Candida cylindracea* lipase from Meito Sangyo OP-360	10	22	—	>98	—	39	>100	85
2-Chloroethyl	*Candida cylindracea* lipase from Sigma (type VII, L1754)	48	r.t.	31	99	—	43	>100	86
2-Ethoxyethyl	*Candida cylindracea* lipase from Sigma adsorbed on different supports	1200	35	25	≥95	—	—	≥53	93

Ethyl	Candida cylindracea lipase from Sigma (type VII)	—	30	20.3	—	—	—	—	95
Ethyl	Candida cylindracea lipase from Sigma (type VII)	96	30	19.1	>98	—	—	>100	87
Ethyl	Candida cylindracea lipase from Sigma (type VII) dialyzed (20,000 cutoff)	96	30	23	>98	—	—	>100	87
Ethyl	Candida rugosa lipase B (purified from Sigma type VII starting material)	96	30	38	>99	—	—	>100	89
Ethyl	Candida rugosa lipase B from Sigma adding NaCl and CaCl$_2$	96	30	32	>99	—	—	>100	89
Ethyl	Candida cylindracea lipase from Sigma covalently immobilized on Al$_2$O$_3$	168	35	33.6	98	—	—	>100	90
Ethyl	Candida cylindracea lipase from Sigma covalently immobilized on SiO$_2$	168	35	29	>99	—	—	>100	90
Ethyl	Candida cylindracea lipase from Sigma covalently immobilized on agarose	168	35	20	95	—	—	49	90
Methyl	Candida rugosa lipase from Sigma (type VII, L1754)	47	37	19	>98	—	24	>100	94

Table 2B Enantioselective Hydrolysis of *rac*-2-(6-methoxy-2-naphthyl)Propionic Esters: Obtaining the *R*(−)-Acid

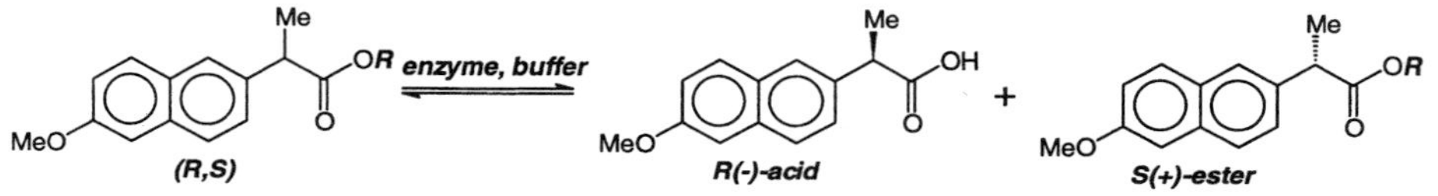

R	Enzyme	t (h)	T (°C)	χ Acid (%)	e.e. Acid (%)	χ Ester (%)	e.e. Ester (%)	E	Ref.
Methyl	*Mucor miehei* lipase from Amano (MAP-10)	120	22	18	95	—	21	51	77,85
Methyl	*Rhizopus arrhizus* lipase from Boehringer Manheim	120	22	11	97	—	13	78	77
Methyl	*Rhizopus* sp. lipase from Serva	120	22	19	92	—	21	27	77,85
Methyl	*Rhizopus oryzae* lipase from Amano (FAP)	120	22	11	76	—	10	8	77
2-Chloroethyl	*Mucor miehei* lipase from Amano (MAP-10)	17	22	—	80	—	65	17	85
2-Chloroethyl	*Rhizopus* sp. lipase from Serva	17	22	—	87	—	81	38	85
Methyl	*Rhizomucor meihei* lipase (Fluka, 62298)	47	37	12	95	—	8	50	94

Table 3 Enantioselective Hydrolysis of *rac*-Ibuprofen Esters: Obtaining the *S*(+)-Acid

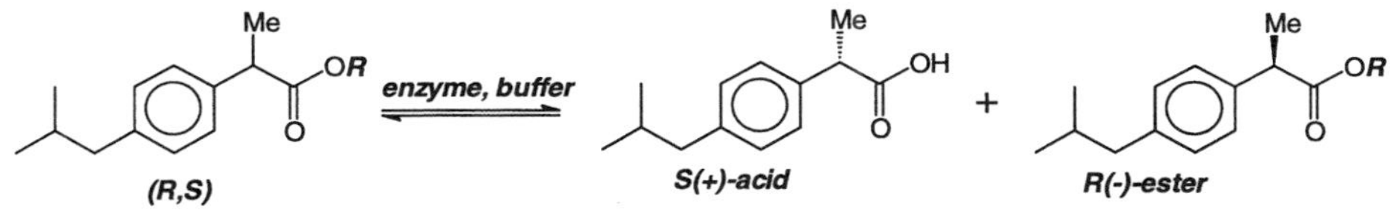

R	Enzyme	t (h)	T (°C)	χ Acid (%)	e.e. Acid (%)	χ Ester (%)	e.e. Ester (%)	E	Ref.
Methyl	*Candida cylindracea* lipase from Meito Sangyo (OP-360)	120	22	42	95	—	70	84	85
Methyl	*Candida cylindracea* lipase from Sigma (type VII, L1754)	48	r.t.	42	95	—	70	84	86
Ethyl	*Candida cylindracea* lipase from Sigma (type VII)	96	30	21.2	>98	—	—	>100	87
Ethyl	*Candida cylindracea* lipase from Sigma (type VII) dialyzed (20,000 cutoff)	96	30	47.2	>98	—	—	>100	87
Ethyl	*Candida rugosa* lipase B (purified from Sigma type VII starting material)	96	30	65	—	—	—	—	89
Ethyl	*Candida rugosa* lipase B from Sigma adding NaCl and CaCl$_2$	96	30	50	—	—	—	—	89
Methyl	*Candida cylindracea* lipase from Sigma with small amounts (1–11%) of water-soluble organic solvents	10	34	26–48	—	—	—	15→100	96
Ethyl	*Candida cylindracea* lipase from Sigma covalently immobilized on Al$_2$O$_3$	168	35	31	>99	—	—	>100	90
Ethyl	*Candida cylindracea* lipase from Sigma covalently immobilized on SiO$_2$	168	35	24	98	—	—	>100	90
Ethyl	*Candida cylindracea* lipase from Sigma covalently immobilized on agarose	168	35	9.5	>99	—	—	>100	90
Ethyl	*Candida rugosa* lipase from Sigma (type VII)	120	30	36	98	—	—	>100	91

Table 4A   Enantioselective Hydrolysis of *rac*-2-(3-fluoro-4-phenyl)Phenyl)propionic Acid Esters: Obtaining Flurbiprofen

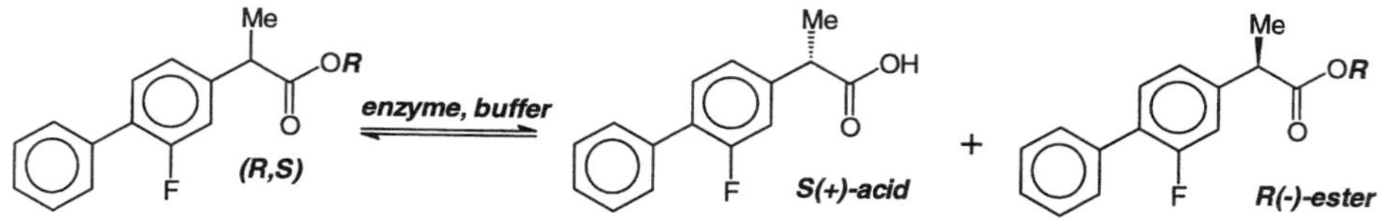

R	Enzyme	t (h)	T (°C)	χ Acid (%)	e.e. Acid (%)	χ Ester (%)	e.e. Ester (%)	E	Ref.
Methyl	*Candida cylindracea* lipase from Meito Sangyo OP-360	168	22	39	65	—	41	7	85
2-Chloroethyl	*Candida cylindracea* lipase from Sigma (L1754, type VII)	72	r.t.	14	80	—	13	10	86
2-Chloroethyl	*Candida cylindracea* lipase from Sigma partially purified (called A form)	—	r.t.	—	—	—	—	>100	86
2-Chloroethyl	*Candida cylindracea* lipase from Sigma (L1754, type VII) treated with 2-propanol	18.6	—	43	97	—	70	>100	88

Table 4B Enantioselective Hydrolysis of *rac*-Suprofen Esters: Obtaining the *S*(+)-Acid

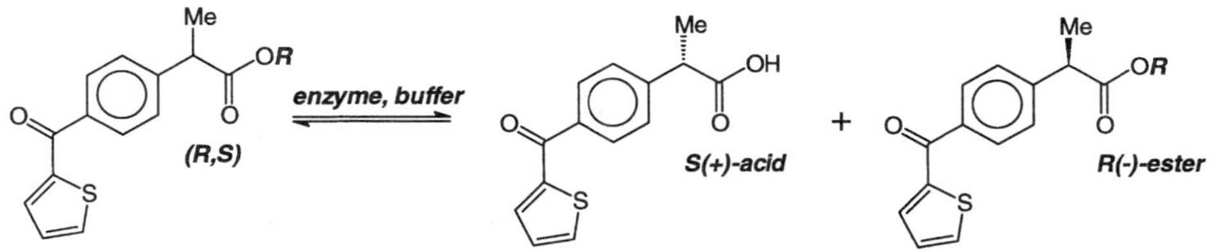

R	Enzyme	t (h)	T (°C)	χ Acid (%)	e.e. Acid (%)	χ Ester (%)	e.e. Ester (%)	E	Ref.
Methyl	*Candida cylindracea* lipase from Meito Sangyo (OP-360)	168	22	49	>95	—	90	>100	85
Methyl	*Candida cylindracea* lipase from Sigma (type VII, L1754)	96	r.t.	48	95	—	90	>100	85

Table 4C Enantioselective Hydrolysis of *rac*-Fenoprofen Esters: Obtaining the *S*(+)-Acid

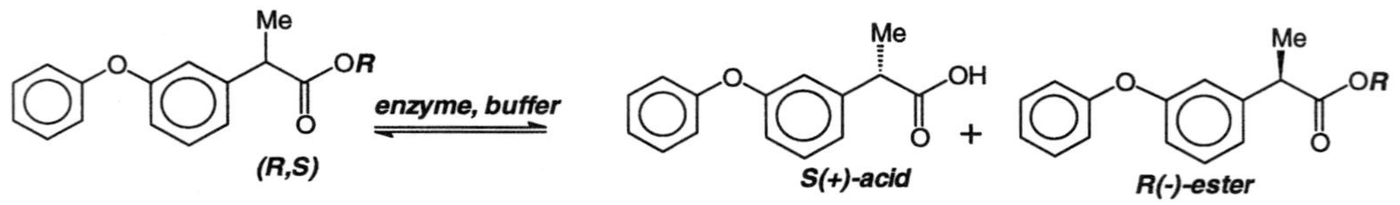

R	Enzyme	t (h)	T (°C)	X Acid (%)	e.e. Acid (%)	X Ester (%)	e.e. Ester (%)	E	Ref.
Methyl	*Candida rugosa* lipase from Sigma (type VII, L1754)	47	37	15	>98	—	11	>100	92

$$\text{Yield} = 38.1 + 39.6x_2 + 27.1x_4 - 25.8x_5 + 13.9x_6 - 8.3x_2x_5 \qquad (2)$$

(solvent = isooctane (log P = 4.5) and alcohol = 1-propanol)

$$\text{Yield} = 18.0 + 17.6x_4 - 25.9x_1 - 10.6x_1x_4 - 10.2x_1x_6 - 9.4x_1x_2 \qquad (3)$$

(solvent = cyclohexane (log P = 3.5) and alcohol = 1-butanol)

From these equations we can deduce that the stirring speed exerts a positive effect in the hydrolysis of esters [x_1 in Eq. (1)], although it is not significant in the enantioselective esterification. This fact can be understood considering that in aqueous environments high stirring speeds would favor the formation and maintenance of the microemulsion oil (ester)/water, thereby increasing the enzymatic activity due to the concomitant increase of the interfacial area. On the other hand, in the esterification both the acid and the alcohol are soluble in the medium, so that we are working with a solid (the biocatalyst) in a homogeneous medium.

In the esterification process, where the enzyme is less active than in the hydrolytic reaction, the amount of biocatalyst (x_4) exerts a positive effect and the amount of water (x_1) introduces a negative [Eq. (2)] or even negligible effect [Eq. (3)], depending on the nature of the support [97,98].

The nature of the solvent plays an important role both in the yield and in the enantioselectivity of the reaction. Thus, in Table 6B we can observe how the presence of Cl$_4$C reduces the yield and the enantiomeric excess in the resolution of esters of trimethylsilyl methyl esters of *rac*-2-(6-methoxy-2-naphthyl)propionic acid with *Candida cylindracea* lipase [100]; another example is shown in Table 7B, where the synthesis of butyl esters of *rac*-ibuprofen is carried out in different organic media [98], obtaining the best results with solvents possessing log P (partition coefficient in the mixture *n*-octanol/water) higher than 2.5. In the same table can be found another, similar example: the esterification with methanol of *rac*-ibuprofen with lipase from *C. cylindracea* [108].

B. Control of Enzyme Purity

The most common limitation for the enantioselective synthetic methods is poor enantioselectivity. Crude *Candida rugosa* (CRL, formerly known as *C. cylindracea*) commercial lipase, one of the most used enzymes for the resolution of racemates, only exhibits poor to moderate enantiodiscrimination for the resolution of some racemic 2-arylpropionic acids and their derivatives (see Table 1A). This might be due to the presence of other proteinaceous impurities (hydrolases and proteases with MW < 62 kDa) in the crude preparation [118], which may display opposite or poor enantioselectivity. A similar situation has also been described for crude lipase from pig pancreas [119].

Therefore, an usual strategy for increasing the enantioselectivity of crude lipases is to purify (up to different homogeneity states) the commercial powder. Generally speaking, pure enzymes, which are of course more expensive, are more active but less stable under operational conditions than commercial powder. Therefore, a partial or complete purification is usually a compromise between the price, the catalytic activity, and the stability of the biocatalyst.

In the case of CRL, we can observe in Table 1A that the higher the purification degree, the higher the yield in hydrolysis of ethylketoprofen. Nevertheless, the commercial lipase (CRCL [87]), the semipurified (dialyzed) lipase (CRSL [87]), and the pure isoenzyme B (CRLB [89]) show the same S enantiopreference. In the same way, Colton et al. [88] and, more recently, Chamorro et al. [120], have shown that the treatment of CRSL

Table 5A Enantioselective Resolution of *rac*-Ketoprofen by the Acyl Transfer Method: Transesterification

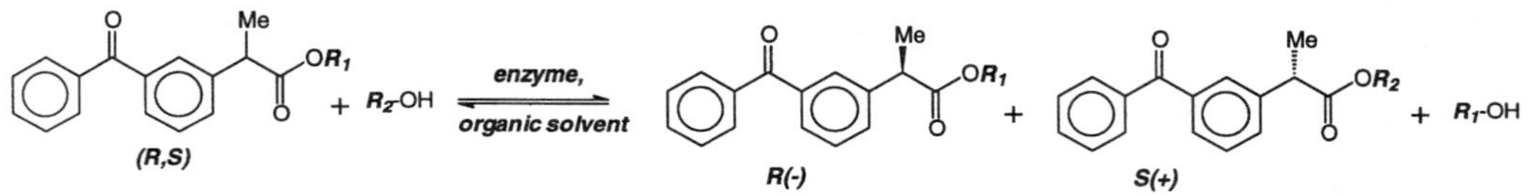

R_1	R_2	Enzyme/solvent	t (h)	T (°C)	χ (%)	e.e. $S(+)$ Product (%)	e.e. $R(-)$ Product (%)	E	Ref.
2-Fluoroethyl	*n*-Butyl	*Pseudomonas* sp. lipase (Amano PS)/*n*-hexane	48	r.t.	37	57	≥99	>100	92
2-Fluoroethyl	*n*-Butyl	*Pseudomonas* sp. lipase (Amano PS)/(i-Pr$_2$)O	48	r.t.	36	46	82	16	92
2-Fluoroethyl	*n*-Butyl	*Aspergillus niger* lipase (Amano AP-6)/*n*-hexane	48	r.t.	27	37	≥99	>100	92
2-Fluoroethyl	*n*-Butyl	*Aspergillus niger* lipase (Amano AP-6)/(i-Pr$_2$)O	48	r.t.	14	16	≥99	>100	92
2-Fluoroethyl	*n*-Octyl	*Aspergillus niger* lipase (Amano AP-6)/*n*-hexane	48	r.t.	7	7	93	30	92
2-Fluoroethyl	*n*-Butyl	*Mucor miehei* lipase (Amano M-AP-10)/*n*-hexane	48	r.t.	33	48	≥99	>100	92
2-Fluoroethyl	2-(2-Pyridyl)ethyl	*Mucor miehei* lipase (Amano M-AP-10)/*n*-hexane	120	r.t.	27	36	≥99	>100	92
2-Fluoroethyl	*n*-Butyl	*Mucor miehei* lipase (Amano M-AP-10)/(i-Pr$_2$)O	48	r.t.	40	65	≥99	>100	92
2-Fluoroethyl	2-(2-Pyridyl)ethyl	*Mucor miehei* lipase (Amano M-AP-10)/(i-Pr$_2$)O	120	r.t.	38	59	≥99	>100	92
2-Fluoroethyl	2-(2-Pyridyl)ethyl	*Mucor miehei* lipase (Amano M-AP-10)/(t-Bu)-O-Me	120	r.t.	37	58	≥99	>100	92
2-Fluoroethyl	2-(2-Pyridyl)ethyl	*Mucor miehei* lipase (Amano M-AP-10)/THF	120	r.t.	31	31	69	7	92
2-Fluoroethyl	2-(2-Pyridyl)ethyl	*Mucor miehei* lipase (Amano M-AP-10)/1,4-dioxane	120	r.t.	3	—	—	—	92

2-Fluoroethyl	n-Butyl	Mucor miehei lipase (Amano M-AP-10)/n-butanol	48	r.t.	2	—	—	—	92
2-Fluoroethyl	n-Butyl	Mucor miehei lipase (Novo Nordisk, Lipozyme 10000L)/n-hexane	48	r.t.	37	51	87	24	92
2-Fluoroethyl	n-Butyl	Mucor miehei lipase (Novo Nordisk, Lipozyme 10000L)/n-hexane	48	r.t.	64	94	53	11	92
2-Fluoroethyl	n-Butyl	Mucor miehei lipase (Novo Nordisk, Lipozyme 10000L)/(i-Pr$_2$)O	48	r.t.	8	9	≥99	>100	92
2-Fluoroethyl	n-Butyl	Mucor miehei lipase (Novo Nordisk, Lipozyme 10000L)/Et$_2$O	48	r.t.	55	77	63	10	92
2-Fluoroethyl	n-Butyl	Mucor miehei lipase (Novo Nordisk, Lipozyme 10000L)/AcOEt	48	r.t.	0	—	—	—	92
2-Fluoroethyl	n-Octyl	Mucor miehei lipase (Novo Nordisk, Lipozyme 10000L)/n-hexane	48	r.t.	28	32	83	14	92
2-Fluoroethyl	n-Octyl	Mucor miehei lipase (Novo Nordisk, Lipozyme 10000L)/Et$_2$O	48	r.t.	2	—	—	—	92
2-Fluoroethyl	2-(2-Pyridyl)ethyl	Mucor miehei lipase (Novo Nordisk, Lipozyme 10000L)/n-hexane or (i-Pr$_2$)O or (t-Bu)-O-Me or THF or 1,4-dioxane	120	r.t.	0 to 6	—	—	—	92
2-Fluoroethyl	n-Butyl	Pig pancreatic lipase (Sigma, type II)/n-hexane	48	r.t.	21	14	52	4	92
2-Fluoroethyl	n-Butyl	Mucor javanicus lipase (Fluka, F62304)/n-hexane	48	r.t.	7	7	≥99	>100	92
2-Fluoroethyl	n-Butyl	Penicillium camembertii lipase (Amano, Lipase G)/n-hexane	48	r.t.	10	11	≥99	>100	92

Table 5B   Enantioselective Resolution of *rac*-Ketoprofen by the Acyl Transfer Method: Esterification

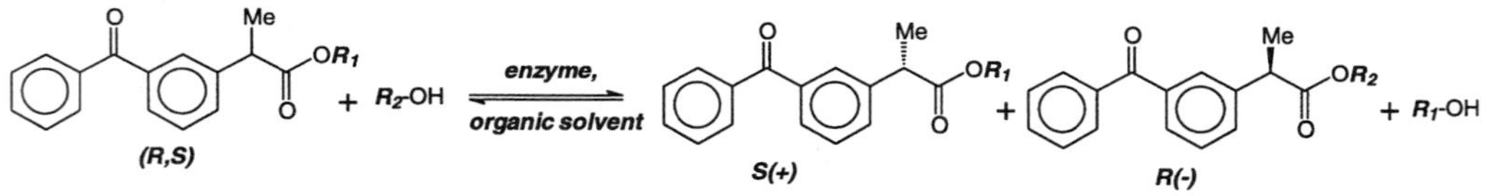

R_1	R_2	Enzyme/solvent	t (h)	T (°C)	X (%)	e.e. $S(+)$ Product (%)	e.e. $R(-)$ Product (%)	E	Ref.
H-	*n*-Propyl	*Candida antarctica* lipase (Novo Nordisk, SP 435A)/isobutyl methyl ketone	3	50	55.6	—	—	3	97
H-	*n*-Butyl	*Rhizomucor miehei* lipase adsorbed on an anion exchange resin (Novo Nordisk, Lipozyme IM)/(*i*-Pr)$_2$O	72	37	20	9.5	—	2.4	98

with short chain polar organic solvents renders a more stereoselective enzyme maintaining the same enantio bias. This positive effect has been explained by the authors due to a conformational change in the protein structure induced by the organic solvent, which would favor the opening of the helical lid that covers the active site of the lipase.

C. Selection of Organic Solvent

We previously mentioned the importance of determining the appropriate solvent for acyl transfer reactions (esterification and transesterification). Nevertheless, it is difficult to select a universal solvent for the esterification of (R,S) 2-arylpropionic acids. In fact, hydrophobic solvents such as cyclohexane [98,102], isooctane [97], or the mixtures isooctane/Cl$_4$C or isooctane/toluene [100] are recommended for the highly hydrophobic substrates naproxen and ibuprofen (see Tables 6 and 7). On the contrary, moderately hydrophilic acids such as ketoprofen (Table 5 [92]) or flurbiprofen (Table 8 [111]) are better esterified in slightly hydrophilic solvents such as *di*isopropyl ether, methyl*iso*butyl ketone, or 1,4-dioxane. Therefore, we can conclude that depending on the hydrophobicity of the substrate we must select the organic solvent in order to obtain the best catalytic performance.

In general terms, the solvent may exert two main effects:

1. Control of the solubility of the substrate in the organic medium, which is strongly dependent of the log P values of both the solvent and the substrate, and
2. Interaction of the molecules of the solvent with each enantiomer of the substrate

The first effect controls mainly the yield, while the second one has to do with the enantioselectivity via formation of two diastereomeric complexes of solvent with R or S enantiomer of the substrate. This last topic was confirmed in the esterification of either R or S-ketoprofen with 1-propanol using R or S-carvone as chiral solvent, catalyzed by lipase B of *C. antarctica* [121].

Figure 5 (pg. 694) shows how the chiral solvent molecules play an important role in the stereoselectivity of the enzyme. Thus, S-($+$)-carvone maintains the S enantiopreference observed for achiral solvents and lipase B of *C. antarctica* [97], whereas R-($-$)-carvone changes the enantioselectivity over long reaction times. A molecular mechanics study of the stability of the diastereomeric complexes formed by both carvones and both enantiomers of the substrate was carried out. The results are shown in Fig. 6 (pg. 694).

The interaction of solvent and ketoprofen takes place through the polar groups (C$=$O and COOH, respectively) through the least sterically hindered face of the carvone. The interactions between S-9($+$)-carvone and R- or S-ketoprofen possess similar energetic values due to the placement of the acid molecule near the flat zone of the enone group. Nevertheless, the interactions of R-($-$) carvone with both acid enantiomers show different energetic levels due to the sterical reasons (higher in **14** than in **15**, Fig. 6), related to the interaction of the α-methyl group of the acid with the hydrogens of the ring of the carvone.

As a clear consequence, R-($-$)- and S-($+$)-ketoprofen would be solvated at the same extent in S-($+$)-carvone, and the substrate–solvent complex would be broken at the same rate. Therefore, the selectivity in S($+$)-carvone would be the same as in achiral solvents, as observed [97], so the stereoselectivity is controlled by the lipase B of *C. antarctica*, which prefers acting on R-($-$)-ketoprofen rather than on S-($+$)-ketoprofen. However, when using R-($-$)-carvone as solvent, the complex with the R-ketoprofen, **14**, would be more stable than the complex with the S-antipode, **15**. Therefore, the molecules of S-($+$)-ketoprofen would be liberated more quickly than the R-($-$) ones, explaining the apparent change in the enantioselectivity observed at long reaction time (Fig. 5).

(text continues on pg. 689)

Table 6A Enantioselective Esterification of *rac*-2-(6-methoxy-2-naphthyl)Propionic Acid: Obtaining Naproxen

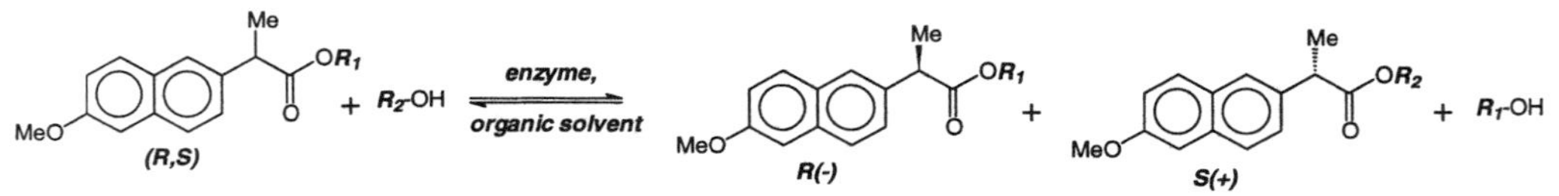

R₁	R₂	Enzyme/solvent	t (h)	T (°C)	X (%)	e.e. S(+) Acid (%)	e.e. R(−) Ester (%)	E	Ref.
H-	*n*-Propyl	*Candida antarctica* lipase (Novo Nordisk, SP 435A)/isobutyl methyl ketone	48	50	31.2	—	—	1.3	97
H-	*n*-Butyl	Porcine pancreas lipase (Sigma)/buffer-saturated isooctane	752	37	8.7	4.1	18.2	1.5	99
H-	*n*-Butyl	*Rhizopus* sp. lipase (Serva)/buffer-saturated isooctane	752	37	15.2	11.2	31.2	2.1	99
H-	*n*-Butyl	*Aspergillus niger* lipase (Fluka)/buffer-saturated isooctane	752	37	23.5	5.2	10.1	1.3	99
H-	*n*-Butyl	*Pseudomonas fluorescens* lipase (Fluka)/buffer-saturated isooctane	752	37	2.7	3.4	24.9	1.7	99
H-	*n*-Butyl	*Mucor javanicus* lipase (Fluka)/buffer-saturated isooctane	752	37	38.8	7.7	8.5	1.3	99
H-	*n*-Methyl	*Mucor javanicus* lipase (Fluka)/buffer-saturated isooctane	—	37	—	—	—	1.2	99

H-	n-Ethyl	—	Mucor javanicus lipase (Fluka)/buffer-saturated isooctane	—	37	—	—	—	2.5	99
H-	n-Propyl	—	Mucor javanicus lipase (Fluka)/buffer-saturated isooctane	—	37	—	—	—	1.7	99
H-	i-Butyl	—	Mucor javanicus lipase (Fluka)/buffer-saturated isooctane	—	37	—	—	—	1.8	99
H-	t-Butyl	—	Mucor javanicus lipase (Fluka)/buffer-saturated isooctane	—	37	—	—	—	1.3	99
H-	n-Hexyl	—	Mucor javanicus lipase (Fluka)/buffer-saturated isooctane	—	37	—	—	—	1.8	99
H-	Cyclohexyl	—	Mucor javanicus lipase (Fluka)/buffer-saturated isooctane	—	37	—	—	—	1.8	99
H-	n-Octyl	—	Mucor javanicus lipase (Fluka)/buffer-saturated isooctane	—	37	—	—	—	1.2	99
H-	3-Chloropropyl	—	Mucor javanicus lipase (Fluka)/buffer-saturated isooctane	—	37	—	—	—	2.5	99
H-	Trimethylsilyl methyl	—	Mucor javanicus lipase (Fluka)/buffer-saturated isooctane	—	37	—	—	—	1.7	99
H-	Trimethylsilyl ethyl	—	Mucor javanicus lipase (Fluka)/buffer-saturated isooctane	—	37	—	—	—	1.4	99

Table 6B Enantioselective Esterification of *rac*-2-(6-methoxy-2-naphthyl)Propionic Acid: Obtaining the Naproxen Esters

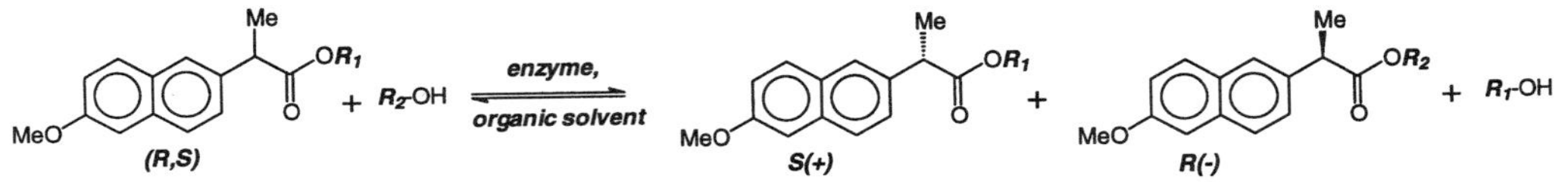

R_1	R_2	Enzyme/solvent	t (h)	T (°C)	X (%)	e.e. $S(+)$ Ester (%)	e.e. $R(-)$ Acid (%)	E	Ref.
H-	Trimethylsilyl methyl	Candida cylindracea lipase (Sigma, type VII)/isooctane	2	37	13.2	>99	12.3	>100	100
H-	Trimethylsilyl methyl	Candida cylindracea lipase (Sigma, type VII)/isooctane	250	37	54.8	78.6	>99	>100	100
H-	Trimethylsilyl methyl	Candida cylindracea lipase (Sigma, type VII)/isooctane/carbon tetrachloride	250	37	50.7	93.8	—	71	100
H-	Trimethylsilyl methyl	Candida cylindracea lipase (Sigma, type VII)/isooctane/toluene	250	37	41.5	92.4	—	50	100
H-	Trimethylsilyl methyl	Candida cylindracea lipase (Sigma, type VII)/isooctane/toluene	250	37	37.1	94.5	—	62	100
H-	Trimethylsilyl methyl	Candida cylindracea lipase (Sigma, type VII)/isooctane/benzene	250	37	21.3	88.4	—	20	100
H-	Trimethylsilyl methyl	Candida cylindracea lipase (Sigma, type VII)/isooctane/1,2-trichloroethane	250	37	19.1	92.0	—	29	100
H-	Trimethylsilyl methyl	Candida cylindracea lipase (Sigma, type VII)/isooctane/dichloromethane	250	37	14.7	92.3	—	29	100
H-	Trimethylsilyl methyl	Candida cylindracea lipase (Sigma, type VII)/isooctane/diethyl ether	5	37	50.7	93.8	—	71	100
H-	n-Butyl	Candida cylindracea lipase (Sigma, type VII)/isooctane/toluene	513	37	61.7	66.5	>99	>20	99
H-	n-Butyl	Rhizopus arrhizus lipase (Fluka)/buffer-saturated isooctane	513	37	3.4	99	2.1	>100	99
H-	Methyl	Candida cylindracea lipase (Sigma, type VII)/buffer-saturated isooctane	—	37	—	—	—	62	99

H-	Ethyl	Candida cylindracea lipase (Sigma, type VII)/buffer-saturated isooctane	—	37	—	6.2	—	—	99
H-	n-Propyl	Candida cylindracea lipase (Sigma, type VII)/buffer-saturated isooctane	—	37	—	15.2	—	—	99
H-	i-Propyl	Candida cylindracea lipase (Sigma, type VII)/buffer-saturated isooctane	—	37	—	10	—	—	99
H-	i-Butyl	Candida cylindracea lipase (Sigma, type VII)/buffer-saturated isooctane	—	37	—	33	—	—	99
H-	t-Butyl	Candida cylindracea lipase (Sigma, type VII)/buffer-saturated isooctane	—	37	—	12.7	—	—	99
H-	n-Hexyl	Candida cylindracea lipase (Sigma, type VII)/buffer-saturated isooctane	—	37	—	38	—	—	99
H-	Cyclohexyl	Candida cylindracea lipase (Sigma, type VII)/buffer-saturated isooctane	—	37	—	58	—	—	99
H-	n-Octyl	Candida cylindracea lipase (Sigma, type VII)/buffer-saturated isooctane	—	37	—	59	—	—	99
H-	3-Chloropropyl	Candida cylindracea lipase (Sigma, type VII)/buffer-saturated isooctane	—	37	—	89	—	—	99
H-	Trimethylsilyl methyl	Candida cylindracea lipase (Sigma, type VII)/buffer-saturated isooctane	—	37	—	>100	—	—	99
H-	Trimethylsilyl methyl	Candida cylindracea lipase (Sigma, type VII)/buffer-saturated isooctane	—	37	—	>100	—	—	99
H-	Trimethylsilyl methyl	Candida cylindracea lipase (Sigma, type VII)/buffer-saturated isooctane	—	25	—	>100	—	—	99
H-	Trimethylsilyl methyl	Candida cylindracea lipase (Sigma, type VII)/buffer-saturated isooctane	—	50	—	60	—	—	99
H-	Trimethylsilyl methyl	Candida cylindracea lipase (Sigma, type VII)/buffer-saturated isooctane	—	65	—	40	—	—	99
H-	Trimethylsilyl methyl	Candida cylindracea lipase (Sigma, type VII)/buffer-saturated isooctane	—	75	—	20	—	—	99
H-	Trimethylsilyl methyl	Candida cylindracea lipase (Sigma, type VII)/mixture (80:20) of buffer-saturated isooctane and toluene	—	50	—	>100	—	—	99
H-	Trimethylsilyl methyl	Candida cylindracea lipase (Sigma, type VII)/reverse micelles AOT/isooctane ($W_0 = 0$)	23	37	5.1	>100	99.9	6.1	101

Table 6B Continued

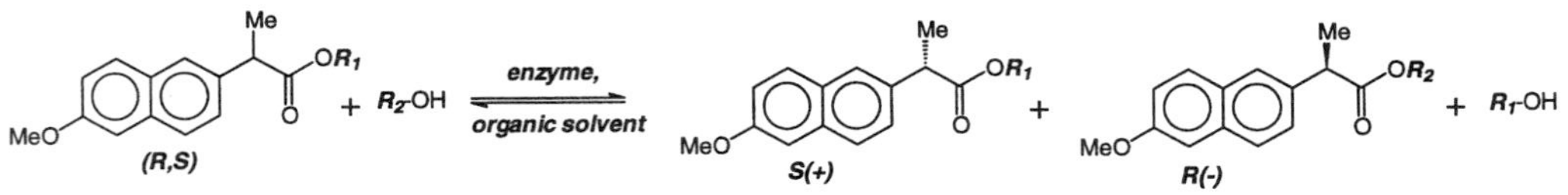

R_1	R_2	Enzyme/solvent	t (h)	T (°C)	χ (%)	e.e. $S(+)$ Ester (%)	e.e. $R(-)$ Acid (%)	E	Ref.
H-	Trimethylsilyl methyl	*Candida cylindracea* lipase (Sigma, type VII)/reverse micelles AOT/isooctane ($W_0 = 1.39$)	7	37	28.7	99.8	18.8	>100	101
H-	Trimethylsilyl methyl	*Candida cylindracea* lipase (Sigma, type VII)/reverse micelles AOT/isooctane ($W_0 = 2.78$)	7	37	67.2	99.8	50.3	>100	101
H-	Trimethylsilyl methyl	*Candida cylindracea* lipase (Sigma, type VII)/reverse micelles AOT/isooctane ($W_0 = 4.17$)	7	37	71.3	99.8	55.2	>100	101
H-	Trimethylsilyl methyl	*Candida cylindracea* lipase (Sigma, type VII)/reverse micelles AOT/isooctane ($W_0 = 5.56$)	7	37	73.0	99.7	56.8	>10	101
H-	Trimethylsilyl methyl	*Candida cylindracea* lipase (Sigma, type VII)/reverse micelles AOT/isooctane ($W_0 = 6.94$)	7	37	56.3	99.8	39.3	>100	101
H-	Trimethylsilyl methyl	*Candida cylindracea* lipase (Sigma, type VII)/reverse micelles AOT/isooctane ($W_0 = 8.33$)	7	37	52.8	99.8	36.4	>100	101

H-	n-Butyl	72	37	27	—	7	1.6	98	*Rhizomucor miehei* lipase adsorbed on an anion exchange resin (Novo Nordisk, Lipozyme IM)/diisopropyl ether
H-	4-Morpholine ethyl	23.9	37	43.0	—	—	>100	102	*Candida rugosa* lipase (Sigma, type VII)/ isooctane
H-	4-Morpholine ethyl	24.4	37	21.3	—	—	32	102	*Candida rugosa* lipase (Sigma, type VII)/ n-heptane
H-	4-Morpholine ethyl	85.8	37	37.7	—	—	29	102	*Candida rugosa* lipase (Sigma, type VII)/ n-hexane
H-	4-Morpholine ethyl	23.3	37	40.8	—	—	102	102	*Candida rugosa* lipase (Sigma, type VII)/ cyclohexane
H-	4-Morpholine ethyl	24.0	37	20	—	—	179	102	*Candida rugosa* lipase (Meito Sangyo Co, Lipase MY)/isooctane
H-	4-Morpholine ethyl	24.3	37	20.4	—	—	52	102	*Candida rugosa* lipase (Meito Sangyo Co, Lipase MY)/n-heptane
H-	4-Morpholine ethyl	85.8	37	35.9	—	—	22	102	*Candida rugosa* lipase (Meito Sangyo Co, Lipase MY)/n-hexane
H-	4-Morpholine ethyl	25.0	37	40.0	—	—	136	102	*Candida rugosa* lipase (Meito Sangyo Co, Lipase MY)/cyclohexane
H-	n-Octyl	193	50	36	>99.9	11.7	>100	103	*Candida cylindracea* lipase (Sigma, type VII)/isooctane
H-	n-Decyl	193	50	39	50.7	17.2	4.1	103	*Candida cylindracea* lipase (Sigma, type VII)/isooctane
H-	n-Dodecyl	193	50	40	—	20.9	2.3	103	*Candida cylindracea* lipase (Sigma, type VII)/isooctane
H-	n-Tetradecyl	193	50	51	44.7	21.2	4	103	*Candida cylindracea* lipase (Sigma, type VII)/isooctane

Table 7A Enantioselective Resolution of *rac*-Ibuprofen Propionic Esters: Obtaining the *S*(+) Pure Enantiomer

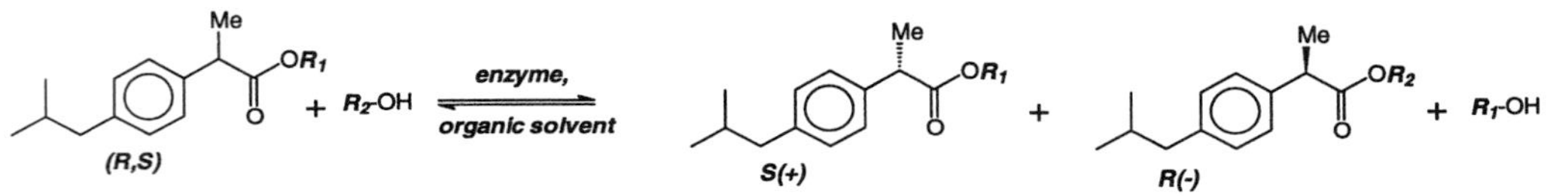

R_1	R_2	Enzyme/solvent	t (h)	T (°C)	χ (%)	e.e. S(+) Ibuprofen product (%)	e.e. R(−) Ester product (%)	E	Ref.
H-	*n*-Propyl	*Candida antarctica* lipase (Novo Nordisk, SP 435A)/isobutyl methyl ketone	3	50	47.6	—	—	2.5	97
H-	*n*-Propyl	*Candida antarctica* lipase (Novo Nordisk, SP 435A)/toluene	30	50	43	28	—	2.8	107
H-	*n*-Propyl	*Candida antarctica* lipase (Novo Nordisk, SP 435A)/Benzo[18]-crown-6	30	50	68	30	—	1.7	107
H-	*n*-Propyl	*Candida antarctica* lipase (Novo Nordisk, SP 435A)/*meso*-teraphenyl porphyrin	30	50	79	53	—	2.0	107
H-	*n*-Propyl	*Candida antarctica* lipase (Novo Nordisk, SP 435A)/*meso*-tetraphenyl porphyrin Cu(II)	30	50	59	27	—	1.8	107
H-	*n*-Propyl	*Candida antarctica* lipase (Novo Nordisk, SP 435A)/*meso*-tetraphenyl porphyrin Zn(II)	30	50	30	24	—	4.5	107
H-	*n*-Propyl	*Candida antarctica* lipase (Novo Nordisk, SP 435A)/*meso*-tetraphenyl porphyrin Fe(II)	30	50	15	17	—	63	107
H-	*n*-Propyl	*Candida antarctica* lipase (Novo Nordisk, SP 435A)/*meso*-tetraphenyl porphyrin V(IV) vanadyl oxide	30	50	29	10	—	1.8	107
H-	*n*-Propyl	*Candida antarctica* lipase (Novo Nordisk, SP 435A)/*meso*-tetra(benzo-[15]-crown-5)porphyrin	30	50	21	20	—	8.6	107

D. Influence of the Nature of the Support

Little work has been done on this topic in the resolution of 2-arylpropionic acids because lipases are generally used either in their native lyophilized form or adsorbed onto commercial resins with similar hydrophobic-hydrophilic characteristics (see Tables 5B and 8). So, when the specific activities of different catalysts are similar, the influence of the support in the reaction must be related to:

1. Diffusional restrictions that reduce the yield, especially for biocatalysts possessing low loading values, or
2. In low-water media, the distribution of the water molecules between the solvent, the support, the enzyme, and the solvation spheres of the substrates. This effect is controlled by the aquaphilicity (Aq) of the support, a parameter that quantifies the water amount retained by a support in water-saturated diisopropyl ether [122].

Besides, the rigidity of the support and the number of enzyme–support surface linkages may play a denaturing effect if the immobilized biocatalysts are lyophilized (Table 9) as described by Arroyo et al. [123]. In this table we can observe how upon lyophilizing the biocatalysts for 48 h the shield of essential water molecules is practically removed from the enzyme microenvironment, leading to inactive enzymatic derivatives for enzymatic reactions in anhydrous solvents. This behavior is not related to the nature of the support as shown in entries 11, 6, and 10. The addition of 500 μl of water to the bulky isooctane phase (10 ml) reactivates the lyophilized derivative on agarose. This result may be explained considering the reversible rehydration of agarose (a highly hydrophilic support whose Aq = 13.8 [122]), which restores the water shield of the enzyme. A similar behavior had previously been described by Martin et al. for α-chymotrypsin covalently bonded to agarose [124]. When the amount of added water is higher than 500 μl, the synthetic activity is not increased because the hydrolytic reaction becomes more prominent. In contrast to agarose, the lyophilization of the immobilized derivatives onto inorganic supports (such as Al_2O_3 or SiO_2, both with Aq < 1) produces a severe dehydration of the immobilized enzyme and a severe distortion of its teritary structure, due to the rigid nature of these solids and the hampered flexibility of the enzyme molecule. Hence, the initial and active conformation is not restored despite the addition of extra water supplied to the medium (Table 9, entries 6 vs. 7 and 10 vs. 11).

Comparing the synthetic activity of *C. rugosa* lipase immobilized on different supports with approximately the same amount of immobilized lipase under the same hydration conditions (entries 4, 8, and 12), we can observe that the enzyme immobilized on silica is more active than on agarose, and the least active is the enzyme immobilized onto alumina.

The effect of nature of the support for lipase immobilization and stereoselectivity is shown in Table 10 [123].

In all cases, the immobilized and the native enzyme stereoselectively esterify the *S*-(+) enantiomer. The enantiomeric excess obtained with the native enzyme increases with the amount of water as described by Hoegberg et al. [125], whereas the immobilized derivatives are not affected by the additional water, except in the case of the derivatives on SiO_2. Finally, we can observe in Table 10 that at the same reaction time (192 h) and at the same amount of water (1 ml buffer), only the immobilized lipase on SiO_2 displayed the same enzymatic activity and enantioselectivity as the native form.

Table 7B Enantioselective Esterification of *rac*-Ibuprofen: Obtaining Ibuprofen Esters

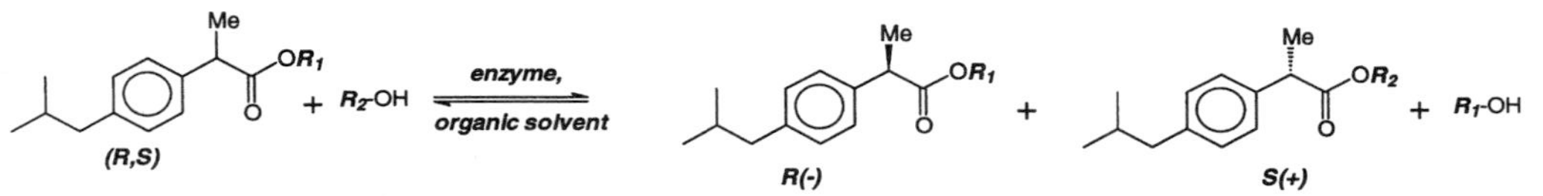

R_1	R_2	Enzyme/solvent	t (h)	T (°C)	χ (%)	e.e. $S(+)$ Ibuprofen product (%)	e.e. $R(-)$ Ester product (%)	E	Ref.
H-	Amyl	*Candida cylindracea* lipase from Biocatalysts/*n*-hexane	24	30	42	99	—	>100	109
H-	Amyl	*Candida cylindracea* lipase from Biocatalysts/*n*-hexane + 0.1% water	24	30	24	98	—	>100	104
H-	Amyl	*Candida cylindracea* lipase from Biocatalysts/*n*-hexane + 0.2% water	24	30	17	90	—	23	104
H-	Amyl	*Candida cylindracea* lipase from Biocatalysts/*n*-hexane + 0.4% water	24	30	13	88	—	18	104
H	*n*-Propyl	Mucor Miehei lipase adsorbed on an anion exchange resin (Novozyme IM2D, Novo Nordisk + 30 mM water/supercritical CO_2	28	50 $P = 150$ bar	15	62	—	4.7	105
H	*n*-Propyl	Mucor Miehei lipase adsorbed on an anion exchange resin (Novozyme IM2D, Novo Nordisk + 30 mM water/supercritical CO_2	28	50 $P = 150$ bar	47	56	—	5.7	105
H	*n*-Propyl	Mucor Miehei lipase adsorbed on an anion exchange resin (Novozyme IM20, Novo Nordisk + 30 mM water/supercritical CO_2	28	50 $P = 150$ bar	60	47	—	5.6	105
H	*n*-Propyl	*Candida cylindracea* lipase from Biocatalyst/isooctane	48	22	41	40	—	3	106

H	n-Propyl	Candida cylindracea lipase from Biocatalyst/AOT/isooctane (3% w/w) $w_0 = 12$	48	22	32	>99	—	>100	106
H	Methyl	Candida cylindracea lipase from Sigma (type VII, L-1754)/ethyl ether	200	30	<0.1	—	—	—	108
H	Methyl	Candida cylindracea lipase from Sigma (type VII, L-1754)/2-heptanone	200	30	1.6	—	—	—	108
H	Methyl	Candida cylindracea from lipase Sigma (type VII, L-1754)/benzene	200	30	33.2	—	—	>100	108
H	Methyl	Candida cylindracea lipase from Sigma (type VII, L-1754)/chloroform	200	30	1.3	—	—	—	108
H	Methyl	Candida cylindracea lipase from Sigma (type VII, L-1754)/anisole	200	30	26.3	—	—	—	108
H	Methyl	Candida cylindracea lipase from Sigma (type VII, L-1754)/toluene	200	30	28.5	—	—	>100	108
H	Methyl	Candida cylindracea lipase from Sigma (type VII, L-1754)/butyl ether	200	30	29.3	—	—	27	108
H	Methyl	Candida cylindracea lipase from Sigma (type VII, L-1754)/carbon tetrachloride	200	30	44.7	—	—	>100	108
H	Methyl	Candida cylindracea lipase from Sigma (type VII, L-1754)/o-xylene	200	30	45.3	—	—	>100	108
H	Methyl	Candida cylindracea lipase from Sigma (type VII, L-1754)/cyclohexane	200	30	44	—	—	>100	108
H	Methyl	Candida cylindracea lipase from Sigma (type VII, L-1754)/n-hexane	200	30	42.2	—	—	>100	108
H	Methyl	Candida cylindracea lipase from Sigma (type VII, L-1754)/n-heptane	200	30	46.5	—	—	>100	108
H	Methyl	Candida cylindracea lipase from Sigma (type VII, L-1754)/n-octane	200	30	45.4	—	—	>100	108
H	n-Butyl	Rhizomucor miehei lipase adsorbed on a macroporous resin (Novo Nordisk, Lypozyme IM)/cyclohexane	72	37	66	—	99	14.5	98
H-	3-Methyl-1-butyl	Lypozyme IM/cyclohexane	72	37	25	—	32	67	98

Table 7B Continued

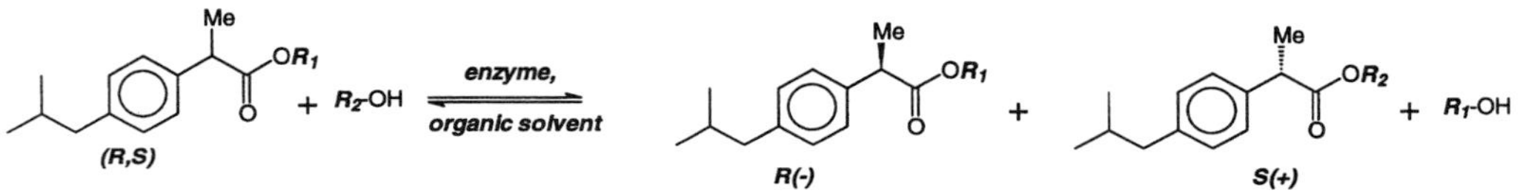

R₁	R₂	Enzyme/solvent	t (h)	T (°C)	χ (%)	e.e. $S(+)$ Ibuprofen product (%)	e.e. $R(-)$ Ester product (%)	E	Ref.
H-	1-Octyl	Lipozyme IM/cyclohexane	72	37	34	—	50	>100	98
H	n-Butyl	Lipozyme IM/(i-Pr)₂O	72	37	42	—	33	3.6	98
H	n-Butyl	Lipozyme IM/1,1,1-tricloroethane	72	37	54	—	67	7.1	98
H	n-Butyl	Lipozyme IM/toluene	72	37	72	—	80	4.2	98
H	n-Butyl	Lipozyme IM/hexane	72	37	72	—	80	4.2	98
H	n-Butyl	Lipozyme IM/heptane	72	37	43	—	66	29.7	98
H	n-Butyl	Lipozyme IM/isooctane	72	37	73	—	77	3.8	98
H	2-N-Morpholine ethyl	Candida rugosa lipase from Meito Sangyo Co. (Lipase MY)/isooctane	42	37	40.7	92.6	63.6	50	109
H	2-N-Morpholine ethyl	Candida rugosa lipase from Meito Sangyo Co. (Lipase MY)/cyclohexane	37	37	34.8	94.3	50.4	56	109
H	2-N-Morpholine ethyl	Candida rugosa lipase from Meito Sangyo Co. (Lipase MY)/n-heptane	45	37	28.9	89.6	36.5	26	109
H	2-N-Morpholine ethyl	Candida rugosa lipase from Meito Sangyo Co. (Lipase MY)/n-hexane	42	37	24.1	87.5	27.6	20	109
H	2-N-Morpholine ethyl	Candida rugosa lipase from Meito Sangyo Co. (Lipase MY)/n-pentane	42	37	19.1	84.3	19.9	14.2	109
H	n-Butyl	Lipozyme IM/cyclohexane	24	4	11.6	—	16.1	>100	98
H	n-Butyl	Lipozyme IM/cyclohexane	24	30	41	—	55	14	98
H	n-Butyl	Lipozyme IM/cyclohexane	24	37	45	—	72	34	98
H	n-Butyl	Lipozyme IM/cyclohexane	24	50	56	—	78	10	98
H	n-Butyl	Lipozyme IM/cyclohexane	24	60	62	—	82	7.2	98
H	n-Butyl	Lipozyme IM/cyclohexane	24	80	49	—	45	4.2	98
H	n-Butyl	Candida cylindracea lipase from Sigma (type VII, L-1754)/isooctane	—	40	45.3	95.3	78.9	100	110

Table 8 Enantioselective Resolution of *rac*-Flurbiprofen Propionic Esters: Obtaining the *R*(−) Pure Enantiomer

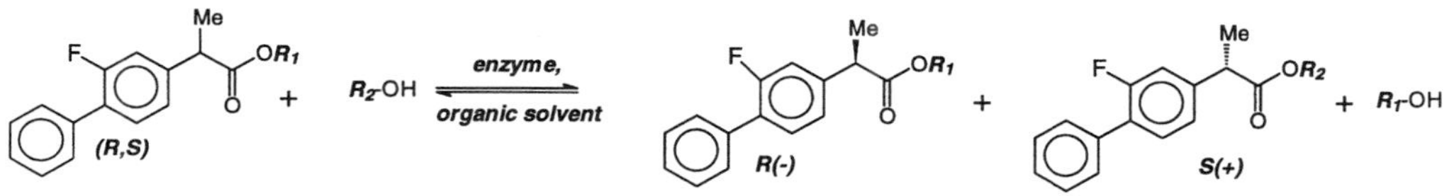

R_1	R_2	Enzyme/solvent	t (h)	T (°C)	X (%)	e.e. $S(+)$ Ibuprofen product (%)	e.e. $R(-)$ Ester product (%)	E	Ref.
H-	*n*-Propyl	*Candida antarctica* lipase (Novo Nordisk, SP 435A)/isobutyl methyl ketone	3	50	40.4	—	—	2.0	97
Me-	*n*-Butyl	*Candida antarctica* lipase Novozyme 435 (Novo/Nordisk)/dioxane	24	45	34	—	82	15.9	111
Me-	*n*-Butyl	*Candida antarctica* lipase Novozyme 435 (Novo Nordisk)/acetonitrile	6	45	25	—	80	11.7	111
Me-	*n*-Butyl	*Candida antarctica* lipase Novozyme 435 (Novo Nordisk)/THF	24	45	13	—	70	6.3	111
Me-	*n*-Butyl	*Candida antarctica* lipase Novozyme 435 (Novo Nordisk)/*terc*-amyl alcohol	12	45	35	—	66	6.7	111
Me-	*n*-Butyl	*Candida antarctica* lipase Novozyme 435 (Novo Nordisk)/*terc*-butyl methyl ether	24	45	31	—	68	7.0	111
H-	*n*-Butyl	*Candida antarctica* lipase Novozyme 435 (Novo Nordisk)/dioxane	24	45	5	—	91	22.2	111
H-	*n*-Butyl	*Candida antarctica* lipase Novozyme 435 (Novo Nordisk)/CH$_3$CN	24	45	19	—	82	12.3	111
H-	*n*-Butyl	*Candida antarctica* lipase Novozyme 435 (Novo Nordisk)/*terc*-amyl alcohol	24	45	4	—	73	6.6	111
H-	*n*-Butyl	*Candida antarctica* lipase Novozyme 435 (Novo Nordisk)/*terc*-butyl methyl ether	24	45	7	—	0	1.0	111
H-	*n*-Butyl	*Candida antarctica* lipase Novozyme 435 (Novo Nordisk)/toluene	24	45	20	—	54	3.8	111
H-	*n*-Butyl	*Candida antarctica* lipase Novozyme 435 (Novo Nordisk)/cyclohexane	4	45	20	—	52	3.6	111

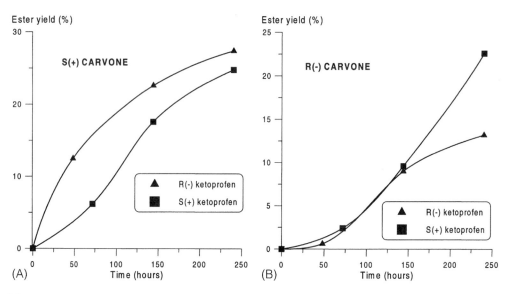

Figure 5 Esterification of $S(+)$- (■) and $R(-)$-ketoprofen (▲) with *n*-propanol in $S(+)$-carvone (A) or $R(-)$-carvone (B). Conditions: 66 mM acid and alcohol in 5 ml of solvent; temperature: 24°C; concentration of enzyme: 7 mg/ml; no addition of water; stirring speed: 300 rpm. (From Ref. 121.)

E. Influence of the Water Amount

Water plays an essential role in enzyme structure and function, especially if enzymes are used in nearly anhydrous organic solvents. In these cases, the small water content in the medium must be shared between all components of the reaction mixture. The strong relation between Aq of the support and the enzymatic water shield was discussed previously.

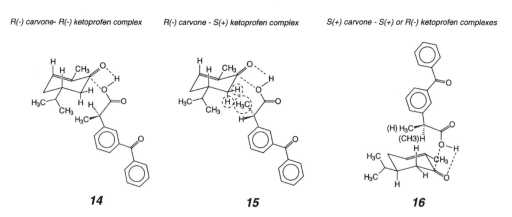

Figure 6 Minimum energy conformers of the interaction of: $R(-)$-carvone with $R(-)$-ketoprofen (**14**); $R(-)$-carvone with $S(+)$-ketoprofen (**15**); and $S(+)$-carvone with either *R*- or *S*-ketoprofen (**16**). (From Ref. 121.)

Table 9 Influence of the Lyophilization and the Presence of Water on the Esterification of Racemic Ibuprofen, Catalyzed by Immobilized Lipase from *C. rugosa* in Isooctane

Entry	Carrier	Mg derivative/10 ml solvent	Lipase loading[a]	Mg active lipase/10 ml isooctane	Tris/HCl buffer added (μl)[b]	Synthetic activity at 96 h[c]
1	Lyophilized agarose	250	296	74	0	0
2	Lyophilized agarose	250	296	74	500	0.015
3	Unlyophilized agarose	250	296	74	1000	0.012
4	Unlyophilized agarose	1000	74	74	1000	0.012
5	Lyophilized silica	1000	74	74	2500	0.016
6	Lyophilized silica	1000	56	56	0	0
7	Unlyophilized silica	2000	56	112	1000	0
8	Unlyophilized silica	2000	40	80	1000	0.031
9	Lyophilized alumina	2000	40	80	1500	0.014
10	Lyophilized alumina	1000	57	57	0	0
11	Unlyophilized alumina	2000	57	114	1000	0
12	Unlyophilized alumina	2000	38	76	1000	0.009
13	Unlyophilized alumina	2000	38	76	1500	0.021

[a]Milligrams of immobilized active lipase per gram of lyophilized or dry support.
[b]The amount of aqueous buffer is added to the organic solution.
[c]The synthetic activity is expressed in μmoles of esterified ibuprofen per mg of active lipase and hour. Conditions: 66 mM ($\pm$) ibuprofen + 66 mM 1-propanol in 10 ml of isooctane; T^a = 30°C; stirring speed: 500 rpm.

Table 10 Influence of Water on the Enantioselectivity of Native and Immobilized Lipase from *C. cylindracea* in the Esterification of Racemic Ibuprofen with *n*-Propanol

Entry	Derivative[a]	Tris/HCl buffer added (μl)/10 ml of isooctane	c (%)[b]	t (h)	e.e. (%)[c]	E[d]
1	Native	200	12	192	15.7	>100
2	Native	1000	40	192	71.7	>100
3	Lyophilized agarose	500	25	144	22.0	6
4	Lyophilized agarose	1000	17	144	27.0	>100
5	Unlyophilized agarose	1000	20	168	35.4	>100
6	Unlyophilized agarose	2500	29	168	37.2	>100
7	Unlyophilized silica	1000	40	192	75.3	>100
8	Unlyophilized silica	1500	21	192	24.6	33
9	Unlyophilized alumina	1000	23	192	16.9	4.2
10	Unlyophilized alumina	1500	32	192	16.0	2.4

[a]The amount of active enzyme (native or immobilized lipase) added to 10 ml of isooctane was 75 mg in all of the experiments. Conditions: 66 mM (±) ibuprofen + 66 mM 1-propanol; T^a = 30°C; 500 rpm.
[b]Ester yield.
[c]Enantiometric excess of remaining $R(-)$-ibuprofen.
[d]Considering the experimental error.

In the case of the native lyophilized enzymes, water plays its crucial role in:

1. Maintaining the active conformation, and
2. Favoring hydrolysis vs. synthesis

The consequences of both effects is the alteration in the yield and/or in the enantio-selectivity as shown in the esterification of (*R,S*)-ibuprofen with 1-pentanol in *n*-hexane (Table 7B [104]). As can be seen, the yield decreases from 42% (no water added) to as low as 13% (0.4% of water present in the reaction medium). Nevertheless, in this case the enantiopreference is not altered by the different amounts of water added. It is possible to find in the literature studies on the water effect, which generally employ water sorption isotherms, measuring values of water activity, a_w [98,126,127]. The use of this thermo-dynamic parameter instead of the water amount is strongly recommended for explaining the water influence [128]. As a general rule, the more hydrophilic the solvent, the higher the initial a_w to be used for obtaining high esterification yields.

F. Presence of Additives

Another methodology used to increase the reaction rate and/or the enantioselectivity is to add some compounds not involved in the reaction, such as crown ethers. In this way we have reported (see Table 7A [107]) the effect of several external compounds on the ester-ification of (*R,S*)-ibuprofen with *n*-propanol, catalyzed by Novozym 435 (*C. antarctica* lipase B immobilized on an anionic resin). As can be observed, the presence of additives does not change the enantiopreference of the lipase B of *C. antarctica*, which again acts preferably on the *R* enantiomer. Therefore, interpretation of data based on a possible interaction of the additives with the active site—which would control the stereoselectiv-ity—should be avoided.

On the other hand, the presence of additives (at 16 mM concentration) does influence the reaction yield and/or the enantiomeric excess of the remaining acid, as shown in Table 7A. Benzo-[18]-crown-6 increases the esterification yield compared to the blank assay, although similar e.e. values are obtained for the remaining acid. The positive effect of the crown ether in the reaction yield may be related to an interaction with the solid (immobilized enzyme), probably via the oxygen of the crown ether, which could interact with the polar zones of the surface on the lipase molecule, especially with the lid (Gly-Pro-Leu-Asp-Ala), increasing the exposure of the lipase active site to the medium. Then this effect would be similar to the role played by the small number of water molecules present in processes in organic solvents. Besides, the apolar moiety of the additive could interact via hydrophobic bonds with the hydrophobic part of the lid. Thus, the overall effect would be an opening of the active site of the immobilized enzymes; this effect could increase the number of active molecules and would increase the reaction yield, but the enantiomeric excess values would remain unaffected because of the absence of interactions between the active site and the additive.

meso-Tetraphenylporphyirin (TPP) increases the esterification yield vs. the blank assay (see Table 7A) and the optical purity of the remnant acid is higher as well. These positive effects are similar to those described by Itoh et al. [129] in the hydrolysis of (*R,S*)-2-acetoxybutyronitrile catalyzed by lipase from *Pseudomonas* sp. using an equimolecular amount of 2-hydroxymethyl-[12]-crown-4 and substrate; in our case, the effect is observed even with a lower amount of additive. The effect could be explained in the same way that for benzo-[18]-crown-6.

meso-Tetraphenylporphyrin-Cu(II) (Table 7A) slightly increases the reaction yield compared to the blank, leading to the same enantiomeric excess. The others metal-porphyrins tested rendered lower reaction yield than the blank, independently of the nature and charge of the cation.

Finally, a complex molecule such as *meso*-tetrabenzo-[15]-crown-5-porphyrin led us to obtain very poor results, both in yield and in optical purity, probably due to the high steric hindrance of the additive, which avoids the interaction with the enzyme.

Several hypotheses could be postulated to explain the positive effect of TPP and the negative effect of the metal-TPP. Probably, the interaction of the anionic resin (the support of Novozym 435) with the metal cation could be the origin of the catalyst deactivation in two ways:

1. By removing the physically adsorbed enzyme from the solid surface
2. By covering the enzymatic derivative with the porphyrin ring

Finally, to confirm the effect of the additive on the stereoselectivity of the lipase B of *C. antarctica*, the esterification rates of pure (*R*)- or (*S*)-ibuprofen were separately measured [105], and the results showed again that these additives act by increasing the esterification rate, not the enantioselectivity of the enzyme. This finding confirmed the interaction of the neutral additive with the lid of the lipase but not with the active site.

G. Influence of the Chemical Structure of the Alcohol

The first point to be mentioned is that usually the structure of the alcohol moiety of the starting profen ester does not have a dramatic influence on the yield and/or the enantio-selectivity in hydrolytic processes (see Table 1A [83,84] or Table 2A [77,85]); on the contrary, the alcohol specificity of the lipase might be the key in the yield achieved in the esterification. This fact is shown in Tables 6B ([100,103], resolution of naproxen esters with *C. cylindracea* lipase) and 7B ([98], resolution of ibuprofen esters catalyzed by *Rhizomucor miehei* lipase). Nevertheless, some exceptions can be observed. In this sense, Tsai and Wei (Table 6A [99]) showed that the change in the alcohol structure (primary or secondary) had almost no effect in the enantiomeric purity obtained by the esterification of (*R,S*)-naproxen catalyzed by *Mucor javanicus* lipase.

This so-called normal behavior, i.e., the much greater importance of the alcohol moiety in the acyl transfer methodology vs. the hydrolytic one, can be easily understood considering that when working in aqueous media the alcohol moiety of the profen ester acts only as a leaving group (following the classical organic chemistry terminology), while the acid moiety must acylate the catalytic serine residue of the active site, according to the lipase and protease catalytic mechanisms [130]. Therefore, the only requirement needed for the alcohol component is to be a good leaving group, i.e., to possess a reduced nucleophilicity in order to avoid the reverse reaction. So, substituents such as 2-haloethyl, cyanomethyl, or 2-ethoxyethyl groups (see Table 1A, 1B, or 2A) would display the best performance. Nevertheless, this effect is masked by the enormous water concentration in these media. In fact, as the number of water molecules is very high, the quick breaking of the acyl-enzyme intermediate almost does not allow one to distinguish between different leaving groups. In the acyl transfer in organic media, the acid moiety of the APA plays exactly the same role as it does in the hydrolytic methodology, but now it is the alcohol that has to break the acyl-enzyme intermediate. It is obvious to consider that the geometry of the acyl-enzyme active site is governed by the structure of the acid part, and now different alcohols would suffer different steric restrictions when approaching the acylated serine. Furthermore, some lipases would be more sensitive to the alcohol structure than others, and this could explain the results obtained by Tsai and Wei [99] with the lipase from *M. javanicus*.

V.   CONCLUSIONS AND PROGNOSIS

The treatment of rheumatoid diseases using homochiral APAs seems to be the best option to control these pathologies, and the obtained profits in pharmacology and pharmacokinetics when using only the active enantiomer are really evident. The reduction of the hepatotoxicity is achieved by reducing the drug concentration. Following these purposes, nowadays four APAs (naproxen, dexketoprofen, flunoxaprofen, and toradol) are marketed as pure enantiomers.

In order to produce this desired stereoisomer, the employment of biocatalytic methods shows more advantages than the classical synthetic methods, due to the excellent yield and optical purity that can be reached using much milder experimental conditions. In this way, the kinetic resolution of racemic mixtures of APAs using lipases constitutes a preferable alternative versus the employment of whole cells because of:

1.   The enhanced resistance of lipases to organic solvent deactivation
2.   The easier control of the enzymatic bioreactors
3.   The easier scaling up of the process, and
4.   The huge number of immobilization methods available for single enzymes

In the future, the rest of the APAs would probably be marketed as the pure enantiomer, following one of the methods mentioned in this chapter, i.e., either enantioselective hydrolysis of the racemic esters in mixtures of water and water-soluble organic solvent, or enantioselective esterification and transesterification (the acyl transfer methodology) in organic solvents. These processes should be respectively catalyzed by covalently immobilized-stabilized lipases, or merely adsorbed on different supports. This affirmation is based on the peculiarities of each strategy, and by simply considering the properties that the immobilization methods confer to the enzyme.

Looking a bit further in time, probably new APAs possessing higher therapeutic effect will be described. In that case, the actually accessible lipases may not be sufficiently effective to resolve the racemic mixtures, so that the screening of new lipase-producing microorganisms appears to be a rational alternative for obtaining biocatalysts with improved efficiency.

REFERENCES

1. Reumatos' 90. Estudio sociosanitario sobre las enfermedades reumáticas en España. Pfizer, SA, Spain, 1992.
2. JR Vane. Nature New Biol 231:232–235, 1971.
3. GA Higgs, EA Higgs, S Moncada. In: C Hansch, JB Taylor, PG Sammes, eds. Comprehensive Medicinal Chemistry, Vol 2. Oxford: Pergamon Press, 1990, pp 147–173.
4. LH Rome, WEM Lands. Proc Natl Acad Sci USA 72:4863–4865, 1975.
5. P Gund, TY Shen. J Med Chem 20:1146–1152, 1977.
6. C Giordano, G Castaldi, F Uggeri. Angew Chem Int Ed Engl 23:413–149, 1984.
7. GJ Roths, PW Majerus. J Clin Invest 56:624–632, 1975.
8. AP Roszkowski, WH Rooks, AJ Tomolonis, LM Miller. J Pharmacol Exp Ther 179:114–123, 1971.
9. SS Adams, P Bresloff, CG Mason. J Pharm Pharmacol 28:156–157, 1976.
10. J Caldwell, AJ Hutt, S Fournel-Gigleux. Biochem Pharmacol 37:105–114, 1988.
11. K Brune, WS Beck, G Geisslinger, S Menzel-Soglowek, BM Peskar, BA Peskar. Experientia 47:257–261, 1991.
12. RFN Mills, SS Adams, EE Liffe, A Dickinson, JS Nicholson. Xenobiotica 3:589–598, 1973.
13. EJD Lee, KM Williams, GG Graham, RO Day, GD Champion. J Pharm Sci 73:1542–1544, 1984.
14. RD Knihinicki, RO Day, KM Williams. Biochem Pharmacol 42:1905–1911, 1991.
15. KM Knights, ME Jones. Biochem Pharmacol 43:1465–1471, 1992.
16. C Reichel, H Bang, K Brune, G Geisslinger, S Menzel. Biochem Pharmacol 50:1803–1806, 1995.
17. KM Knights, UM Talbot, TA Bailie. Biochem Pharmacol 44:2415–2417, 1992.
18. TS Tracy, DP Wirthwein, SD Hall. Drug Metab Dispos 21:114–120, 1993.
19. S Menzel, R Waibel, K Brune, G Geisslinger. Biochem Pharmacol 48:1056–1058, 1994.
20. CS Chen, T Chen, WR Shieh. Biochim Biophys Acta 1033:1–6, 1990.
21. TA Baillie, WJ Adams, DG Kaiser, LS Olanoff, GW Halstead, H Harpootlian, GJ Van Giessen. J Pharmacol Exp Ther 239:517–523, 1989.
22. RT Foster, F Jamali. Drug Metab Dispos 16:623–626, 1998.
23. PJ Hayball, PJ Meffin. J Pharmacol Exp Ther 240:631–636, 1987.
24. F Jamali, AS Russell, RT Foster, C Lemko. J Pharm Sci 79:460–461, 1990.
25. G Geisslinger, SH Ferreira, S Menzel, D Schlott, K Brune. Life Sci 54:173–177, 1994.
26. WJ Wechter. J Clin Pharmacol 34:1036–1042, 1994.
27. F Francalanci, A Gardano, L Abis, T Floriani, M Foa. J Organomet Chem 243:87–94, 1983.
28. N Kasahara, K Suzuki, K Kamiya. (Sanwa Chemical Laboratories) Japan Kokai JP 76 29,466 (Cl. C07C63/52), 24/01/1976, Appl 74/100,021, 31/08/1974); Chem Abstr 85, 77,910j.
29. GJ Matthews, RA Arnold. (Syntex Corporation) Ger Offen 2,805,488 (Cl. C07C63/52), 17/08/1978, US Appl 769,070, 16/02/1977; Chem Abstr, 90, 6,106u.
30. M García, C del Campo, EF Llama, JM Sánchez-Montero, JV Sinisterra. Tetrahedron 49: 8433–8440, 1993.

31. J Crosby. Tetrahedron 47:4789–4846, 1991.
32. JD Brown. Tetrahedron: Asymmetry 3:1551–1552, 1992.
33. HR Sonawane, NS Bellur, JR Ahuja, DG Kulkarni. Tetrahedron: Asymmetry 3:163–192, 1992.
34. WA Nugent, TV RajanBabu, MJ Burk. Science 259:479–483, 1993.
35. KH Ahn, CS Jin, DH Kang, YS Shin, JS Kim, DS Han, DH Kim. J Heterocycl Chem 30: 825–827, 1993.
36. KT Wan, ME Davis. Nature 370:449–150, 1994.
37. TV RajanBabu, AL Casalnuova. Pure Appl Chem 66:1535–1542, 1994.
38. ASC Chan, SA Laneman, RE Miller, JH Wagenknecht, JP Coleman. Chem Ind 53:49–68, 1994.
39. P Camps, S Giménez. Tetradedron: Asymmetry 6:991–1000, 1995.
40. T Uemura, X Zhang, K Matsumura, N Sano, H Kumobayashi, T Ohta, K Nozaki, H Takaya. J Org Chem 61:5510–5516, 1996.
41. K Ishishara, S Nakamura, M Kaneeda, H Yamamoto. J Am Chem Soc 118:12854–12855, 1996.
42. C Garot, T Javed, TJ Mason, JL Turner, JW Cooper. Bull Soc Chim Belg 105:755–757, 1996.
43. DB Pourreau, W Partenheimer. Chem Ind 68:75–85, 1996.
44. W Oppolzer, S Rosset, JD Brabander. Tetrahedron Lett 38:1539–1540, 1997.
45. ME Jung, KL Anderson. Tetrahedron Lett 38:2605–2608, 1997.
46. RC Griesbach, DPG Hamon, RJ Kennedy. Tetrahedron: Asymmetry, 8:507–510, 1997.
47. T Bando, Y Namba, K Shishido. Tetrahedron: Asymmetry, 8:2159–2165, 1997.
48. FJ Lopez, SA Ferriño, MS Reyes, R Román. Tetrahedron: Asymmetry, 8:2497–2500, 1997.
49. IT Harrison, B Lewis, O Nelson, W Rooks, A Roszkowski, A Tomolonis, JH Fried, J Med Chem 13:203–205, 1970.
50. T Manimaran, FJ Impastato. (Ethyl Corp) US 5,015,764 (cl. 562-401;C07B57/00), 14/05/1991, US Appl 539,212, 18/06/1990; Chem Abstr 115, 135,162t.
51. A Bhattacharya, JR Fritch, CD Murphy, LD Zeagler, CA McAdams. (Hoeschst Celanese Corp) US 5,380,867, (Cl. 548-344.1, C07B55/00), 10/01/1995, US Appl 985,083, 02/12/1993. Chem Abstr 122, 222,832z.
52. EJ Ebbers, BJM Plum, GJA Ariaans, B Kaptein, QB Broxterman, A Bruggink, B Zwanenburg. Tetrahedron: Asymmetry 8:4047–4057, 1997.
53. J Svoboda, K Capek, J Palecek. Collect Czech Chem Commun 52:766–774, 1987.
54. G Bellucci, G Berti, R Bianchini, S Vecchiani. Gazz Chim Ital 118:451–456, 1988.
55. SE Brown, JH Coates, CJ Easton, SF Lincoln, Y Luo, AKW Stephens. Aust J Chem 44:855–862, 1991.
56. WH Pirkle, CJ Welch, B Lamm. J Org Chem 57:3854–3860, 1992.
57. T Sugai, K Mori. Agr Biol Chem 48:2501–2504, 1984.
58. GW Hanlon, A Kooloobandi, AJ Hutt. J Appl Bacteriol 76:442–447, 1994.
59. Y Yasohara, Y Tasato, S Takahashi. (Kanegafuchi Chemical Ind) Jpn Kokai Tokkyo Koho JP 05,219,985 [93,219,985] (Cl. C12P41/00), 31/08/1993, Appl 92/25,074, 12/02/1992. Chem Abstr 120, 6,923w.
60. K Miyamoto, SH Tsuchiya, H Ohta. J Fluorine Chem 59:225–232, 1992.
61. M Kobayashi, S Shimizu. FEMS Microbiol Lett 120:217–224, 1994.
62. E Cerbelaud, D Petre. (Rhone-Poulenc Sante) Fr.Demande FR 2,626,288, 28/07/1989, Appl. 88/923, 27/01/1988; Chem Abstr 112, 196,679b, 1990.
63. K Yamamoto, K Otsubo, K Oishi (Asahi Chemical Industry Co., Ltd) Eur Pat Appl EP 348,901, (Cl. C12P7/40) 03/0/.1990, JP JP Appl 88/156,911, 27/06/1988; Chem Abstr 113, 76605y, 1990.
64. F Effenberger, J Bohme. Biorg Med Chem 2:715–721, 1994.
65. L Martinkova, A Stolz, HJ Knackmuss. Biotechnol Lett 18;1073–1076, 1996.
66. ML Gradiely, CJ Knowles. Biotechnol Lett 16:41–46, 1994.

67. K Yamamoto, Y Ueno, K Otsubo, K Kawakami, K Komatsu. Appl Environ Microbiol 56: 3125–3129, 1990.
68. H Kakeya, N Sakai, T Sugai, H Ohta. Tetrahedron Lett 32:1343–1346, 1991.
69. MA Cohen, JS Parrat, NJ Turner. Tetrahedron: Asymmetry 3:1543–1546, 1992.
70. T Beard, MA Cohen, JS Parrat, NJ Turner, J Crosby, J Moilliet. Tetrahedron: Asymmetry, 4: 1085–1104, 1993.
71. T Gilligan, H Yamada, T Nagasawa. Appl Microbiol Biotechnol 39:720–725, 1993.
72. F Effenberger, J Böhme. Bioorg Med Chem 2:715–721, 1994.
73. R Endo, H Yamada, A Shimizu, T Nagasawa, T Kobayashi (Nitto Chemical Industry Co Ltd) Jpn Kokai Tokkyo Koho JP 06,153,968 [94,153,968] (Cl. C12P7/40) 03/06/1994 Appl 92/ 306,663, 17/11/1992. Chem Abstr 121, 177,878k.
74. T Kamyama, Ch Miura, Y Nonoguchi, K Toyoda, K Uzura, T Oonishi. (Nagase & Co. Ltd) Jpn Kokai Tokkyo Koho JP 06,014,797 [94, 14,797], (Cl. C12P41/00) 25/01/1994, Appl 92/ 197,899, 00/06/1992. Chem Abstr 212, 7,440u.
75. WJ Quax, CP Broekhuizen. Appl Microbiol Biotechnol 41:425–431, 1994.
76. F Effenberger, BW Graef, S Oßwald. Tetrahedron: Asymmetry 8:2749–2755, 1997.
77. QM Gu, CS Chen, CJ Sih. Tetrahedron Lett 27:1763–1766, 1986.
78. A Zaks, AM Klibanov. J Biol Chem 263:3194–3201, 1988.
79. DB Volkin, A Staubli, R Langer, AM Klibanov. Biotechnol Bioeng 37:843–853, 1991.
80. CS Chen, CJ Sih. Angew Chem Int Ed Engl 28:695–707, 1989.
81. AM Klibanov. Acc Chem Res 23:114–120, 1990.
82. AL Margolin. Enzyme Microb Technol 15:266–280, 1993.
83. CS Chen, Y Fujimoto, G Girdaukas, CJ Sih. J Am Chem Soc 104:7294–7299, 1982.
84. AJJ Straathof, JA Jongejan. Enzyme Microb Technol 21:559–571, 1997.
85. CJ Sih, QM Gu, DR Reddy. In: E Mutschler, E Winterfeldt, eds. Trends in Medicinal Chemistry. New York: VCH, 1987, pp 181–191.
86. SH Wu, ZW Guo, CJ Sih. J Am Chem Soc 112:1990–1995, 1990.
87. MJ Hernáiz, JM Sánchez-Montero, JV Sinisterra. Tetrahedron 50:10749–10760, 1994.
88. IJ Colton, SN Ahmed, RJ Kazlauskas. J Org Chem 60:212–217, 1995.
89. MJ Hernáiz, JM Sánchez-Montero, JV Sinisterra. J Mol Catal A: Chem 96:317–327, 1995.
90. JM Moreno, JV Sinisterra. J Mol Catal A: Chem 98:171–184, 1995.
91. M García, A Gradillas, C del Campo, EF Llama, JM Sánchez-Montero, JV Sinisterra. Biotechnol Lett 19:999–1004, 1997.
92. A Palomer, M Cabré, J Ginesta, D Mauleón, G Carganico. Chirality 5:320–328, 1993.
93. E Battistel, D Bianchi, P Cesti, C Pina. Biotechnol Bioeng 38:659–664, 1991.
94. M Botta, E Cernia, E Corelli, F Manetti, S Soro. Biochim Biophys Acta 1337:302–310, 1997.
95. C del Campo, M García, A Gradillas, EF Llama, L Salazar, JM Sánchez-Montero, JV Sinisterra. J Mol Catal 84:399–405, 1993.
96. WH Lee, KJ Kim, MG Kim, SB Lee. J Ferment Bioeng 80:141–147, 1995.
97. M Arroyo, JV Sinisterra. J Org Chem 59:4410–4417, 1994.
98. MT López-Belmonte, AR Alcántara, JV Sinisterra. J Org Chem 62:1831–1840, 1997.
99. SW Tsai, HJ Wei. Enzyme Microb Technol 16:328–333, 1994.
100. SW Tsai, HJ Wei. Biotechnol Bioeng 43:64–68, 1994.
101. SW Tsai, CC Lu, CS Chang. Biotechnol Bioeng 51:148–156, 1996.
102. CS Chang, SW Tsai. Enzyme Microb Technol 20:635–639, 1997.
103. YM Cui, DZ Wei, JT Yu. Biotechnol Lett 19:865–868, 1997.
104. A Mustranta. Appl Microbiol Biotechnol 38:61–66, 1992.
105. M Rantakylä, O Aaltonen. Biotechnol Lett 16:825–830, 1994.
106. G Hedström, M Backlund, JP Slotte. Biotechnol Bioeng 42:618–624, 1993.
107. A Gradillas, C del Campo, JV Sinisterra, EF Llama. Biotechnol Lett 18:85–90, 1996.
108. MG Kim, SB Lee. J Ferment Bioeng, 81:269–271, 1996.
109. SW Tsai, JJ Lin, CS Chang, JP Chen. Biotechnol Prog 13:82–88, 1997.
110. YC Xie, HZ Lin, JY Chen. Biotechnol Lett 20:455–458, 1998.

111. R Morrone, G Nicolosi, A Patti, M Piattelli. Tetrahedron: Asymmetry 6:1773–1778, 1995.
112. H Bundgaard. In: H Bundgaard, ed. Design of Prodrugs. Amsterdam: Elsevier, 1985, pp 1–92
113. NM Nielsen, H Bundgaard. J Pharm Sci 77:285–598, 1988.
114. VR Shanbhag, AM Crider, R Gokhale, A Harpalani, RM Dick. J Pharm Sci 81:149–154, 1992.
115. VK Tammara, MM Narurkar, AM Crider, AM Khan. Pharm Res 10:1191–1199, 1993.
116. PD Haaland. Experimental Design in Biotechnology. Vol. 105 of the series Statistics: Textbooks and Monographs. DB Owen, RG Cornell, WJ Kennedy, AM Kshinsagara, EG Schilling, eds. New York: Marcel Dekker, 1989.
117. O Marciset, B Mollet. Biotechnol Bioeng 43:490–496, 1994.
118. ML Rúa, T Díaz-Mauriño, VM Fernández, C Otero, A Ballesteros. Biochim Biophys Acta 1156:181–189, 1993.
119. CH Wong, GM Whitesides. Enzymes in Synthetic Organic Chemistry. Tetrahedron Organic Chemistry Series, Vol. 12, JE Baldwin, PD Magnus, eds. Oxford: Elsevier, 1994, pp 74–82.
120. S Chamorro, JM Sánchez-Montero, AR Alcántara, JV Sinisterra. Biotechnol Lett 20:499–505, 1998.
121. M Arroyo, JV Sinisterra. Biotechnol Lett 17:525–530, 1995.
122. M Reslow, P Adlercreutz, B Mattiasson. Eur J Biochem 172:573–578, 1988.
123. M Arroyo, JM Moreno, JV Sinisterra. J Mol Catal A: Chem 97:195–201, 1995.
124. MT Martin, JV Sinisterra, A Heras. J Mol Catal 80:127–136, 1993.
125. HE Hoegberg, H Edlund, P Berglund, E Hedenstrom. Tetrahedron: Asymmetry 4:2123–2126, 1993.
126. RM de la Casa, JM Sánchez-Montero, JV Sinisterra. Biotechnol Lett 18:13–18, 1995.
127. M Arroyo, JM Sánchez-Montero, JV Sinisterra. Biotechnol Tech 10:263–266, 1996.
128. PJ Halling. Enzyme Microb Technol 16:178–206, 1994.
129. T Itoh, Y Hiyama, A Betchakv, H Tsukube. Tetrahedron Lett 34:2617–2520, 1993.
130. A Fersht. Enzyme Structure and Mechanism, 2nd ed. New York: WH Freeman, 1985, pp 405–426.

23

Biocatalytic Production of Pravastatin, an Anticholesterol Drug

Nobufusa Serizawa
Sankyo Company, Ltd., Tokyo, Japan

I. INTRODUCTION

Coronary heart disease (CHD) is one of the major causes of death in both Western countries and Japan. Among the types of CHD known, ischemic heart disease (IHD) has been singled out as the major cause of death. The number of patients suffering from IHD in Japan has been on the rise in recent years. This trend is partly attributable to a growing elderly population and to increased intake of Western-style food. It is well known that the three major risk factors for IHD are hypercholesterolemia, hypertension, and smoking. Hypercholesterolemia has long been considered to be the most important of these factors [1]. In order to reduce the risk associated with high serum cholesterol levels, the development of several hypolipidemic drugs and therapies has been explored among various research communities.

Cholesterol is present endogenously by both absorption from diet and biosynthesis and is excreted mainly as bile acids into feces [2]. In order to reduce body cholesterol, three major strategies for the inhibition of cholesterol biosynthesis can be considered.

The biosynthesis of cholesterol constitutes an extensive biochemical process exceeding more than 20 steps starting from acetyl-coenzyme A (CoA). The rate-limiting enzyme of this pathway is 3-hydroxy-3-methylglutaryl(HMG)-CoA reductase [mevalonate: $NADP^+$ oxidoreductase (CoA-acylating), EC 1.1.1.34], which catalyzes the reduction of HMG-CoA to mevalonate.

In 1971, Sankyo Co., Ltd. launched a major research initiative to screen inhibitors for the cholesterol synthesis from the culture broth of microorganisms using a cell-free enzyme system from rat liver. After much rigorous screening of microorganisms, ML-236B (mevastatin, Fig. 1) was discovered in the culture broth of *Penicillium citrinum* in 1975 [3].

It is noteworthy that compactin, a compound identical to ML-236B, was independently isolated by Beecham Pharmaceuticals as a weak antifungal antibiotic [4] 2 years after Sankyo's patent was filed.

As shown in Fig. 1, a portion of the ML-236B structure resembles the structure of 3-hydroxy-3-methylglutarate, the part of HMG-CoA that serves as the substrate of HMG-CoA reductase. Accordingly, ML-236B and the related compounds shown in Fig. 1 inhibit the enzyme in a competitive manner with respect to HMG-CoA.

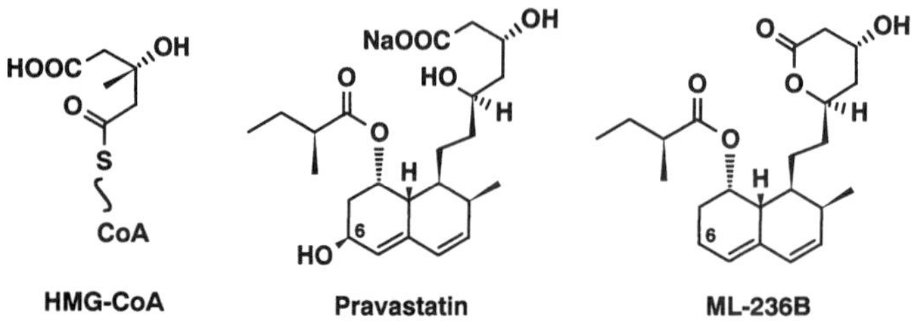

Figure 1   Structures of pravastatin, ML-236B, and HMG-CoA.

It has been well known that the liver and intestine are the major organs involved in de novo cholesterogenesis. Since target organ–directed cholesterol inhibition could be expected to minimally disturb cholesterol metabolism in other organs, including hormone-producing organs, our drug discovery program has focused on finding a compound with enhanced target organ–directed characteristics. After screening microbial products as well as chemically and biologically modified derivatives of ML-236B, pravastatin was finally chosen as the candidate for development. Indeed, pravastatin displays a stronger and more tissue-selective inhibition of cholesterol synthesis than ML-236B [5,6].

Pravastatin contains a hydroxyl group at the 6β position of its decaline structure (Fig. 1). In 1979, this drug was first found to be a minor urinary metabolite of ML-236B in dogs. Chemical synthesis was attempted for the industrial hydroxylation of ML-236B, but this method was assessed as economically infeasible, and microbial hydroxylation was adapted for the production of pravastatin. After screening for microorganisms capable of converting ML-236B for pravastatin, *Streptomyces carbophilus* was selected for the second step of the fermentation process [7,8].

Pravastatin was chosen for development as a hypolipidemic drug in 1981, and clinical trials were started in 1984. Pravastatin was launched in 1989 as ''Mevalotin'' in Japan. The drug was licensed to the Bristol-Myers Squibb Company, and as a result of worldwide development effort, it is currently available on the market in 76 countries including Japan.

As mentioned above, pravastatin was produced by a two-step fermentation process: the first step is production of ML-236B and the second is hydroxylation of ML-236B. This chapter surveys the biocatalytic production of pravastatin with a particular focus on the molecular mechanism of hydroxylation of ML-236B by cytochrome P450 (Cyt P450) from *S. carbophilus*.

II. STEREOSELECTIVE FORMATION OF PRAVASTATIN

Since the hydroxylation reaction that forms pravastatin cannot be easily brought about through conventional synthetic methods, we set out to form pravastatin by isolating microorganisms capable of hydroxylating ML-236B at the 6β. This chapter focuses on the production of pravastatin by microbial hydroxylation.

Various microorganisms were tested for their capability to hydroxylate ML-236B at the 6β position [5,8,10]. In an extensive screening program, *Mucor hiemalis* was found to be one of the most effective microorganisms for this type of hydroxylation. *Mucor hiemalis* converted ML-236B mainly to pravastatin. However, none of these fungi tolerated high concentrations of ML-236B in the culture broth, probably because of the antifungal activity of the substrate.

ML-236B does not exhibit antibiotic activity in many strains of actinomycetes and bacteria. Identified cultures and isolates of these strains from soil samples were therefore tested for their ability to hydroxylate ML-236B. The soil samples were collected from several areas of Australia.

Isolates from soil samples collected in Australia had potent abilities to hydroxylate ML-236B at the 6β position. The strains were identified as *Amcolata autotrophica* and were found to tolerate ML-236B and to have strong transformation activity; however, all of the strains tested produced dihydroxylated byproduct.

From the experience we gained in isolating *A. autotrophica*, microorganisms with potent hydroxylating activity may preferentially live in low-nutrient soil. Generally, microorganisms take energy using a common carbon source such as glucose under a rich nutrient condition. However, microorganisms that live in poor and highly oxygenated soil will not be able to obtain energy using chemically inactive compounds, and these microorganisms cannot survive. These microorganisms have been used as a source of monooxygenases, which catalyze the hydroxylation reaction.

In intensive studies on petroleum fermentation in the 1970s, many microorganisms with strong hydroxylation activity were isolated from oil field areas.

III. *STREPTOMYCES CARBOPHILUS* AS A POTENT CONVERTER OF ML-236B [7–9,11]

In order to produce pravastatin on an industrial scale, we screened for microorganisms that had strong hydroxylating activity and produced few byproducts. *A. autotrophica* grew well on D-trehalose media. After further screening for fresh isolates of *Actinomycetes* that grew on D-trehalose, one strain isolated from Australian soil was discovered as a potent converter that produced only limited byproducts. The morphological and physiological characteristics suggested that this strain belonged to the genus *Streptomyces*. Taxonomically this strain was classified as a new species of the genus *Streptomyces* as *S. carbophilus*.

A. Two-Component-Type Cyt P450 in Prokaryotes That Catalyzes Hydroxylation of ML-236BNa to Pravastatin [7,8]

In our previous work, a most noteworthy finding was the discovery of *S. carbophilus*, a potent converter of ML-236B which permitted the production of pravastatin on an industrial scale. Nevertheless, an understanding of the hydroxylation mechanism for conversion of ML-236B to pravastatin is still requisite for industrial scale production. As described previously, pravastatin was first found as a minor hydroxylated metabolite of ML-236B in dogs, suggesting the participation of hepatic Cyt P450s in hydroxylation reaction. When pravastatin was later obtained from microbes [3], a microbial counterpart to Cyt P450 was anticipated to play a role in this hydroxylation.

In the last two decades, microbial Cyt P450s have been studied extensively. In prokaryotes, especially *Actinomycetes*, Cyt P450–dependent monooxygenase systems have been known to participate in the detoxification of xenobiotics. Here the induction of Cyt P450s was also observed [12,13].

Since the Cyt P450s [8] involved in pravastatin production originate from *S. carbophilus*, it was named as Cyt P450sca-1,-2. These Cyt P450s are closely related to the mitochondrial-type Cyt P450 and other bacterial Cyt P450s based on amino acid analysis.

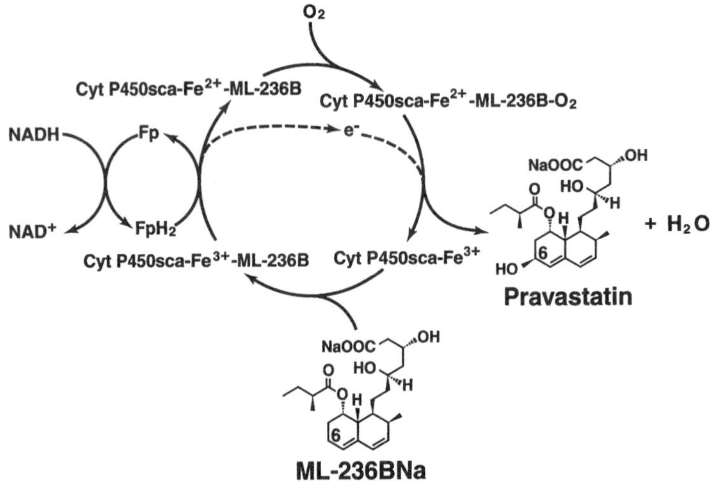

Figure 2 A production mechanism for the synthesis of pravastatin in *S. carbophilus*. Fp, flavoprotein.

Generally, Cyt P450 monooxygenase systems can be classified as either two- or three-component systems [14]. The three-component systems are composed of Cyt P450 reductase, Cyt P450, and iron-sulfur protein, whereas the two-component systems are composed of Cyt P450 reductase and Cyt P450. The results from many previous studies have shown that with the exception of Cyt P450BM-3 from *Bacillus megaterium*, almost all prokaryotes have the three-component (mitochondrial) Cyt P450 systems [15]. The Cyt P450sca monooxygenase system is composed of P450 and flavoprotein. The addition of ferredoxin did not cause any stimulation of the hydroxylation activity. NADH-Cyt P450 reductase from *S. carbophilus* contained both FAD and FMN. All two-component-type Cyt P450 systems are membrane-bound. However, the Cyt P450sca monooxygenase system exists in soluble fraction. Therefore, a novel two-component system from *S. carbophilus* was discovered. Our finding represents the first evidence of a two-component type of Cyt P450 system in prokaryotes (Fig. 2).

Recently, we found that Cyt P450sca is induced not only by its substrate, ML-236BNa [7], but also by phenobarbital (PB), a typical inducer of many Cyt P450 proteins [16]. We are interested in the mechanism of Cyt P450sca induction from *S. carbophilus* because it might serve as a model for mammalian systems.

B. Cloning and Expression of Cytochrome P450sca-2 Gene (*cyt P450sca-2*) from *S. carbophilus* [17]

Cyt P450sca *S. carbophilus* is induced 30-fold by its substrate ML-236BNa [7], as well as by PB [16]. Although PB is one of the major inducers for mammalian Cyt P450s, its induction mechanism has only been fully elucidated in the case of barbiturate-inducible Cyt P450BM-3 from *B. megaterium* [18]. This Cyt P450sca induction may serve as a model for mammalian PB-inducible Cyt P450s [2]. In order to understand the induction mechanism, it is required to clone the *cyt P450sca*. In this chapter, we describe the cloning of *cyt P450sca-2* and its expression in *S. lividans*, a strain generally used as a host in the *Streptomyces* gene manipulation.

1. Cloning and Sequence Analysis of *S. carbophilus cyt P450sca-2*

The entire *cyt P450sca-2* was cloned. An open reading frame (ORF) of 1233 bp encoding a 410-amino-acid residue protein is preceeded by a potential ribosome binding site. We also compared the HR2 region of Cyt P450sca-2, which contains the heme-binding cysteine, with that of several prokaryotic and eukaryotic Cyt P450s. Cyt P450sca-2 had the greatest homology (79%) with Cyt P450SU1 derived from *Streptomyces griseolus* [19].

2. Expression of *cyt P450sca-2* in *S. lividans* TK21/pSCA205

Plasmid pSCA205, containing the ORF of the *cyt P450sca-2* and its 1.0-kb 5'-noncoding region, was constructed from pIJ702 [20] and *S. lividans* TK21 was transformed using this plasmid [21]. The *S. lividans* TK21/pSCA205 converted ML-236BNa to pravastatin.

3. Substrate Induction of Cyt P450sca-2 [7,16]

ML-236BNa is known to induce the production of Cyt P450sca-2, but the mechanism of induction remains unclear. Therefore in attempts to clarify this situation, we assayed transcripts of *cyt P450sca-2* both in the presence and absence of ML-236BNa by northern hybridization. No transcripts of *cyt P450sca-2* mRNA were detected in the absence of ML-236BNa, but transcripts of *cyt P450* mRNA were observed in its presence. The above findings indicated that substrate induction is possibly regulated at the transcriptional level.

C. Molecular Mechanism of *cyt P450sca* Expression [16,22]

Although the mechanism whereby PB and "PB-like" inducers such as ML-236B activate transcription of the Cyt P450 remains unclear, possible mechanisms can be broadly categorized into two classes: (1) receptor-dependent induction mechanisms and (2) Cyt P450–dependent induction mechanisms [23].

The most actively studied barbiturate-inducible Cyt P450 is Cyt P450BM-3 from bacterial *B. megaterium* [24]. In *B. megaterium*, PB and other lipophilic barbiturates induce the expression of two Cyt P450s P450BM-1 (gene *CYP106*) [25] and P450BM-3 (gene *CYP102*), a naturally occurring and catalytically self-sufficient Cyt P450-NADPH (a P450 reductase fusion protein of unusually high activity). Thus, the responsiveness of P450s to PB is not confined to eukaryotes and may have certain features that are broadly conserved in nature. A barbiturate-responsive regulatory region of the bacterial *CYP102* spanning the region between 0.8 and 1.1 kb upstream from the transcription start site has been localized through transformation studies of a DNA segment [26]. Although the functional importance of this DNA segment has yet to be demonstrated, it seems likely that it contributes to PB-dependent P450 regulation-negative regulation in the case of the *B. megaterium* genes and positive regulation in the case of the rat genes.

Interestingly, in vitro treatment of bacterial or rodent liver extracts with PB mimicked the effects of in vivo PB treatment on the retardation patterns [27]. This implies that, in the case of the rat liver extracts, (1) the effects of the PB on the binding protein involved in these interactions do not require a new protein synthesis, and (2) putative PB recepter protein may be present in the nucleus. Further study is needed to confirm the validity and generality of these conclusions, and also may reveal the extent to which other mechanistic aspects of the PB induction process are conserved in prokaryotic and eukaryotic systems. As described earlier, we have found that Cyt P450sca is induced not only by its substrate, ML-236BNa, but also by PB, a typical inducer of many Cyt P450s. We are interested in the mechanism of Cyt P450sca induction in *S. carbophilus* since it may serve as a model for mammalian systems. In this section, we have described the characteristics of induction of PB and ML-236BNa-inducible Cyt P450sca-2 from *S. carbophilus*, an industrial pravastatin-producing strain.

1. Transcriptional Activation by ML-236BNa and PB in *S. carbophilus*

To determine the induction mechanism of Cyt P450sca-2 by PB and ML-236BNa, we determined the nucleotide sequence belonging to the 5′ noncoding region of *cyt P450sca-2*.

PB and ML-236BNa, a substrate of Cyt P450sca, are typical inducers of many Cyt P450s. The effect of ML-236BNa and PB on the transcription of Cyt P450sca-2 of *S. carbophilus* and *S. lividans* was analyzed by northern hybridization. The mRNA amount of *cyt P450sca-2* was gradually elevated by inducers such as ML-236B or PB. This finding suggests that induction of Cyt P450sca-2 in *S. carbophilus* was regulated at the transcriptional level.

These results are indicative of the possibility that the same induction mechanism of Cyt P450sca-2 exists in both *S. carbophilus* and *S. lividans* TK21/pSCA205. The result from Cyt P450sca-2 induction in *S. lividans* TK21/pSCA205 enabled us to construct chimeric genes that contain various lengths of the 5′ upstream region of the *cyt P450sca-2*. Moreover, induction of Cyt P450sca-2 by ML-236BNa could thus be analyzed further.

2. Regulation Mechanism of *cyt P450sca-2* Expression

In order to examine the region involved in the regulation of the transcriptional induction by ML-236BNa, a series of plasmids containing the deleted 5′ noncoding region of *cyt P450sca-2* was constructed and *S. lividans* TK21 was transformed using these plasmids.

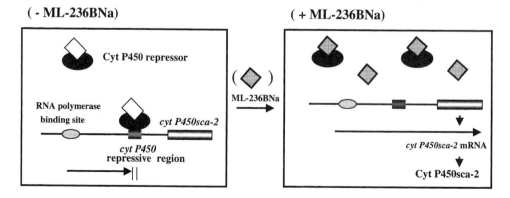

Figure 3   A proposed mechanism of Cyt P450sca induction by ML-236BNa.

These transformants were cultivated in the presence or absence of ML-236BNa. Strong transcriptional enhancement was observed in *S. lividans* TK21/pSCA205 harboring 1013 bp of the 5′ noncoding region in the presence of ML-236BNa.

On the other hand, *S. lividans* TK21/pSCA1013 (Δ−1013-428) and *S. lividans* TK21/pSCA1013 (Δ−1013-320) showed strong mRNA production both in the presence and in the absence of ML-236BNa. Moreover, the amount of transcript was similar to that in *S. lividans* TK21/pSCA205 in the presence of ML-236BNa. Based on this observation, we can infer that the transcription is repressed by the 5′ noncoding region between -1013 bp and -428 bp.

The induction mechanism of Cyt P450sca-2 was suggested to be a negative regulation at the transcriptional level, and the palindromic sequence possibly functions as an operator to regulate the transcription. In light of the above observations, we proposed a mechanism of Cyt P450sca induction by ML-236BNa (Fig. 3). A DNA-binding protein possibly exists in the cell and interacts with the palindromic sequence (in the domain of -46 bp and -24 bp) of the 5′ noncoding region. In the absence of ML-236BNa, the protein possibly represses the transcription. In the presence of ML-236BNa, the protein separates from the palindromic sequence by changing its affinity to the sequence. As a result, RNA polymerase can transcribe the *cyt P450sca-2*.

As described above, PB is one of the well-known inducers of Cyt P450s. It is interesting that in spite of the absence of any obvious structural relationship between PB and ML-236BNa both induce Cyt P450sca-2. However, PB induction was not apparent in *S. lividans* TK21/pSCA205. We can offer no explanation for this disparity at this time.

Fulco et al. [28] described the presence of a repressor (Bm3R1) which interacts with an operator (Barbie box: -ATCAAAAGCTGGAGG-) in the 5′ noncoding region of *CYP102* in *B. megaterium*. Two similar sequences resembling that of the Barbie box in *S. carbophilus* were detected. We searched for a DNA-binding protein to locate these sequences by means of a gel shift mobility assay, but no gel shift band was detected. It is likely, therefore, that a different protein regulate the Cyt P450sca induction by PB in *S. carbophilus*.

Cyt P450sca-2 plays a critical role in the production of pravastatin. We have shown that the Cyt P450sca-2 is functional in *S. lividans*. Our future studies will concentrate on improving the efficiency of pravastatin production by genetic engineering of *Streptomyces*.

IV. DISCOVERY OF MICROORGANISMS, 6α-HYDROXYLATION ACTIVITY

In the course of studies on the hydroxylation of ML-236B to pravastatin, we found that 6α-hydroxylation of ML-236B was performed by *Syncephalastrum nigricans* or *S. racemosum* (6α/6β = 10:1) [29]. Based on these results, we have succeeded in obtaining a stereospecific 6α- or 6β-hydroxylation enzyme (Fig. 4). This result indicated the high degree of selectivity and stereospecificity of the microbial enzymes.

Cyt P450sca has already been purified, cloned, and crystallized [30], and the mechanism of this hydroxylation has been clarified.

If we are able to characterize the 6α-hydroxylation enzyme from *Syncephalastrum* in the future, a complete understanding of the stereospecific mechanism of 6α- and 6β-hydroxylation of ML-236B will be available by comparing the enzymes.

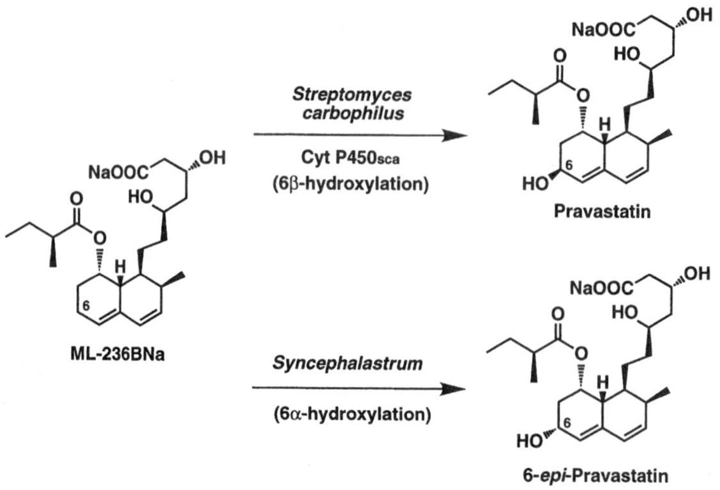

Figure 4   Stereospecific 6α- or 6β-hydroxylation of ML-236B.

REFERENCES

1. HI Page, JN Berrettoni, A Butkus, FM Sones Jr. Prediction of coronary heart disease based on clinical suspicion, age, total cholesterol, and triglyceride. Circulation 42:625–645, 1970.
2. JM Dietschy, JD Wilson. Regulation of cholesterol metabolism. N Engl J Med 282:1179–1183, 1970.
3. A Endo, M Kuroda, Y Tsujita, A Terahara, C Tamura. Jap Pat Kokai 50–155690, 1975.
4. AG Brown, TC Smale, TJ King, R Hasenkamp, RH Thompson. Crystal and molecular structure of compactin, a new antifungal metabolite from *Penicillium brevicompactnm*. J Chem Soc Perkin Trans 1:1165–1170, 1976.
5. N Serizawa, K Nakagawa, K Hamano, Y Tsujita, A Terahara, H Kuwano. Microbial hydroxylation of ML-236B (compactin) and monacolin K. J Antibiot 36:604–607, 1983.
6. Y Tsujita, M Kuroda, Y Shimada, K Tanzawa, M Arai, I Kaneko, M Tanaka, H Masuda, C Tarumi, Y Watanabe, S Fujii. CS-514, a competitive inhibitor of 3-hydroxy-3-methylglutaryl coenzyme A reductase: tissue-selective inhibition of sterol synthesis and hypolipidemic effect on various animal species. Biochim Biophys Acta 877:50–60, 1986.
7. T Matsuoka, S Miyakoshi, K Tanzawa, M Hosobuchi, K Nakahara, N Serizawa. Purification and characterization of cytochrome P-450sca from *Streptomyces carbophilus*. Eur J Biochem 184:707–713, 1989.
8. N Serizawa, T Matsuoka. A two-component-type cytochrome P-450 monooxygenase system in prokaryotes that catalyzes hydroxylation of ML-236B to pravastatin, a tissue-selective inhibitor 3-hydroxy-3-methylglutaryl coenzyme A reductase. Biochim Biophys Acta 1084:35–40, 1991.
9. N Serizawa, S Serizawa, K Nakagawa, K Furuya, T Okazaki, A Terahara. Microbial hydroxylation of ML-236B (compactin): Studies on microorganisms capable of 3β-hydroxylation of ML-236B. J Antibiot 36:887–891, 1983.
10. T Okazaki, N Serizawa, R Enokita, A Torikata, A Terahara. Taxonomy of actinomycetes capable of hydroxylation of ML-236B (compactin). J Antibiot 36:1176–1183, 1983.
11. T Okazaki, R Enokita, H Miyaoka, H Otani, A Torikata. Streptomyces darwinensis sp. nov., S. cactaceus sp. nov., from Australian soils. Annu Rep Sankyo Res Lab 41:123–133, 1989.
12. J Romesser, DP O'Keefe. Induction of cytochrome P-450-dependent sulfonylurea metabolism in *Streptomyces griseolus*. Biochem Biophys Res Commun 140:650–659, 1986.

13. FS Sariaslani, DA Kunt. Induction of cytochrome P-450 in *Streptomyces griseus*. Biochem Biophys Res Commun 141:405–410, 1986.

14. N Serizawa. Cytochrome P-450 of actinomycetes (in Japanese). Nippon Nogeikagaku Kaishi 66:149–153, 1992.

15. FS Sariaslani. Microbial cytochrome P-450 and xenobiotics metabolism. Ad Appl Microbiol 36:133–178, 1991.

16. I Watanabe, N Serizawa. Molecular approaches for production of pravastatin, HMG-CoA reductase inhibitor: transcriptional regulation of the cytochrome P450$_{sca}$ gene from *Streptomyces carbophilus* by ML-236B sodium salt and phenobarbital. Gene 210:109–116, 1998.

17. I Watanabe, F Nara, N Serizawa. Cloning, characterization and expression of the gene encoding cytochrome P-450sca-2 from *Streptomyces carbophilus* involved in production of pravastatin, a specific HMG-CoA reductase inhibitor. Gene 163:81–85, 1995.

18. GC Shaw, AJ Fulco. Inhibition by barbiturates of the binding of Bm3R1 repressor to its operator site on the barbiturate-inducible cytochrome P-450BM-3 gene of *Bacillus megaterium*. J Biol Chem 268:2297–3004, 1993.

19. CA Omer, R Lenstra, PJ Litle, C Dean, JM Tepperman, KJ Leto, JA Romesser, DP O'Keefe. Genes for two herbicide-inducible cytochromes P-450 from *Streptomyces griseolus*. J Bacteriol 172:3335–3345, 1990.

20. E Katz, CJ Thompson, DA Hopwood. Cloning and expression of the tyrosinase gene from *Streptomyces* antibioticus in *Streptomyces lividans*. J Gen Microbiol 129:2703–2714, 1983.

21. DA Hopwood, MJ Bibb, KJ Chater, T Kieser, CJ Bruton, HM Kieser, DJ Lydiate, CP Smith, JM Ward, H Schrempf. Genetic Manipulation of *Streptomyces*: A Laboratory Manual. John Innes Foundation, Norwich, England, 1985.

22. I Watanabe, N Serizawa. Molecular approaches for production of pravastatin, a HMG-CoA reductase inhibitor: transcriptional regulation of the cytochrome P450sca gene from Streptomyces carbophilus by ML-236B sodium salt and phenobarbital. Gene 210:109–116, 1998.

23. DJ Waxman, L Azaroff. Phenobarbital induction of cytochrome P-450 gene expression. Biochem J 281:577–592, 1992.

24. GC Shaw, AJ Fulco. Barbiturate-medicated regulation of expression of the cytochrome P450BM-3 gene of *Bacillus megaterium* by Bm3R1 protein. J Biol Chem 267:5515–5526, 1992.

25. AJ Fulco. P450BM-3 and other inducible bacterial P450 cytochromes: biochemistry and regulation. Annu Rev Pharmacol Toxicol 31:177–203, 1991.

26. L-P Wen, RT Ruettinger, AJ Fulco. Requirement for a 1-kilobase 5'-flanking sequence for barbiturate-inducible expression of the cytochrome P-450BM-3 gene in *Bacillus megaterium*. J Biol Chem 264:10996–11003, 1989.

27. J-S He, AJ Fulco. A barbiturate-regulated protein binding to a common sequence in the cytochrome P450 genes of rodents and bacteria. J Biol Chem 266:7864–7869, 1991.

28. AJ Fulco, J-S He, Q Liang. Induction by barbiturates of P450 cytochromes and other drug-metabolizing enzymes in bacteria and eukaryotes. Abstracts of 10th International Symposium on Microsomes and Drug Oxidations, Toronto, Canada, 1994; July 18–21 (University of Toronto).

29. N Serizawa, K Nakagawa, Y Tsujita, A Terahara, H Kuwano. 3α-Hydroxy-ML-236B (3α-hydroxycompactin), microbial transformation product of ML-236B (compactin). J Antibiot 36: 608–610, 1983.

30. S Ito, T Hata, I Watanabe, T Matsuoka, N Serizawa. Crystal structure of cytochrome P-450sca from *Streptomyces carbophilus* involved in production of pravastatin, a cholesterol-lowering drug. FASEB J 11:A776, 1997.

24

Application of PEG-Modified Enzymes in Biotechnological Processes

Yoh Kodera, Misao Hiroto, Ayako Matsushima, Hiroyuki Nishimura, and Yuji Inada
Toin University of Yokohama, Yokohama, Japan

I. INTRODUCTION

Chemical modification of proteins with a synthetic macromolecule has been extensively studied for the purpose of applying proteins in biomedical and biotechnological processes [1–8]. Among the synthetic macromolecules, polyethylene glycol (PEG), with nontoxic, amphipathic, and nonimmunogenic properties, is reported to be the most suitable for protein modification. It is a linear synthetic polymer with the general formula HO-$(CH_2CH_2O)_n$-CH_2CH_2OH. PEG with a low molecular weight is a viscous liquid, whereas that with a high molecular weight is a wax-like solid that has been used in lubricants, vehicles, solvents, binders, and intermediates in the rubber, food, pharmaceutical, cosmetic, agriculture, textile, paper, petroleum, and other industries [9]. The purpose of the chemical modification of proteins with PEG includes the alteration of immunoreactivity or immunogenicity, as well as suppression of immunoglobulin E production in biomedicine [5,10–13]. Another purpose is preparation of the novel catalyst PEG–enzyme, which is soluble and active in organic solvents, and exploration of its application in biotechnological processes [14,15].

Since the 1970s, many papers on the chemical modification of proteins and enzymes by PEG ("pegylation") have been published in medical and technological fields. As is shown in Fig. 1, the number of reports is increasing greatly year by year, and since 1994 more than 100 reports have been published yearly in the fields of biomedicine and biotechnology. Figure 2 shows the further classification of these reports in each field. In the medical field, antitumor substances, metabolic enzymes, anti-inflammation enzymes, allergens, antithrombotic enzymes, and blood substitutes have been modified with PEG. Among them, PEG-modified adenosine deaminase (PEG–ADA) [16] was the first PEG–protein conjugate approved by the U.S. Food and Drug Administration (FDA) in 1991, and has been used for the treatment of children with ADA-deficient severe combined immunodeficiency. More than 40 proteins, including asparaginase [17], interleukins [18], superoxide dismutase [19], hemoglobin [20] and thrombopoietin [21], have been modified with PEGs for clinical evaluation.

In biotechnological fields, PEG–enzymes are endowed with the amphipathic property of PEG, a synthetic polymer, so as to become soluble in a hydrophobic environment.

713

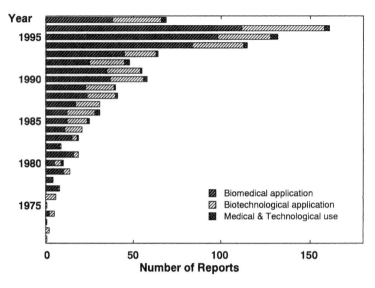

Figure 1 Number of reports on PEG-hybrids.

Pollak and Whitesides [22] applied PEG–enzyme to aqueous two-phase system composed of aqueous solutions of PEG and dextran: glucose-6-phosphate dehydrogenase modified with PEG was preferably distributed into PEG phase and active in the phase with the aim of the enzyme recovery from the reaction system.

Inada and co-workers [23] demonstrated in 1984 that enzymes modified with PEG become soluble and remain active in organic solvents. Since PEG is an amphipathic macromolecule, its hydrophilic nature makes it possible to modify enzymes in aqueous solution and its hydrophobic nature makes it possible to make PEG-enzymes soluble in a hydrophobic environment. In fact, PEG–enzymes such as PEG–catalase [24] and PEG–peroxidase [25] exhibited markedly high activity in organic solvents. Furthermore, PEG-modified hydrolytic enzymes such as PEG–lipase [23] and PEG–protease [26] catalyzed the reverse reaction of hydrolysis effectively in hydrophobic media, namely, ester synthesis or ester exchange reaction and acid-amide formation in transparent organic solvents or in hydrophobic substrates [14]. Table 1 shows the list of PEG–enzymes with their references for biotechnological processes.

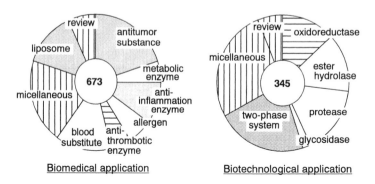

Figure 2 Classification of reports on PEG–hybrid materials in medical and technological fields.

Table 1 List of PEG-Modified Enzymes for Biotechnological Use

Oxidoreductase: catalase [24,97], dehydrogenases [22,98,99], glucose oxidase [100], peroxidase [25,101], superoxide dismutase [102].

Hydrolase: lipase [58,103], phospholipase [104], alkaline proteinase [105], chymotrypsin [26,106,107], papain [49,108], pepsin [56], subtilisin [51,109], thermolysin [52], trypsin [51,110], cellulase [76,111,112], β-galactosidase [113].

II. PREPARATION OF PEG–ENZYME

As stated in Sec. I, PEG is a linear synthetic polymer with a general formula HO-$(CH_2CH_2O)_n$-CH_2CH_2OH and a molecular weight of 500–20,000. Although PEG has two reactive hydroxyl groups at the end of the molecule, a monofunctional derivative, mono-methoxypolyethylene glycol (mPEG), is often used for protein modification to avoid cross-linking between target protein molecules by PEG modifiers. PEG is amphipathic and is soluble in an aqueous solution as well as organic solvents except for ethyl ether or aliphatic hydrocarbons due to its flexible chain with polar (C-O) and nonpolar (C-C) parts [9]. In addition, the PEG polymer is nontoxic and harmless to active proteins or cells. These properties of PEG allowed us to prepare PEG–enzymes that are soluble and active in hydrophobic media in biotechnological fields [8].

A. PEG Derivative for Protein Modification

Almost all proteins are soluble in aqueous solution but are insoluble in organic solvents. Therefore, chemical modification of proteins with PEG should be performed in aqueous solutions. Pollak and Whitesides [22] coupled a PEG derivative with amino groups, α,ω-di-*p*-aminobenzoxypolyethylene glycol, into carboxyl groups on the protein surface by using water-soluble carbodiimide (Scheme 1). Lee and Sehon [27] prepared PEG–protein

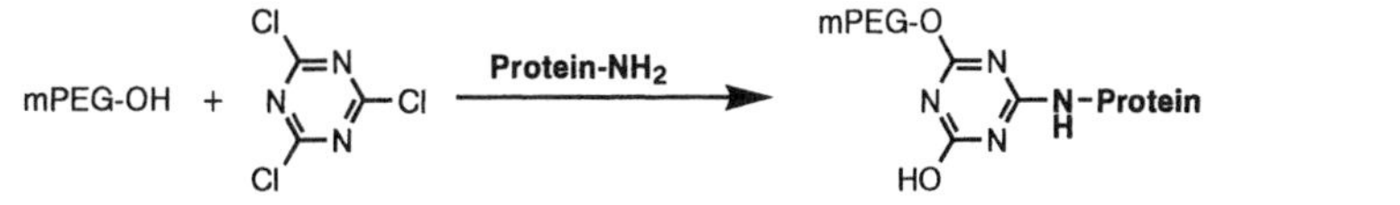

$$\tag{1}$$

by adding cyanuric chloride, a coupling agent, to an alkaline solution containing a protein and polyethylene glycol (Scheme 2). Although these papers clearly altered the properties

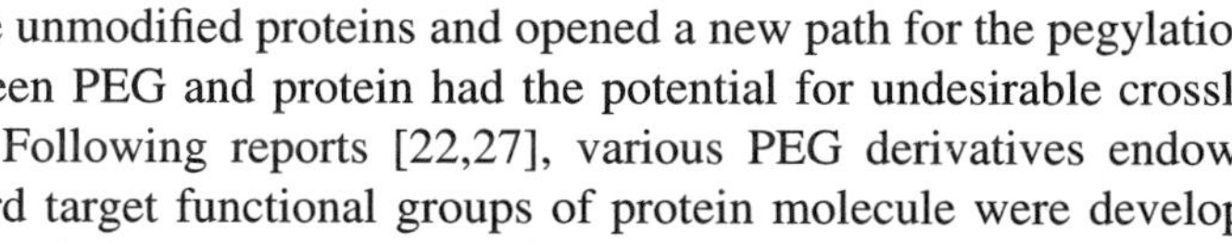

$$\tag{2}$$

of the unmodified proteins and opened a new path for the pegylation, the coupling reactions between PEG and protein had the potential for undesirable crosslinking.

Following reports [22,27], various PEG derivatives endowed with high reactivity toward target functional groups of protein molecule were developed, as shown in Fig. 3. Some of them are commercially available from Shearwater Polymers, Inc. (Huntsville, AL), Sigma Chemical Co. (St. Louis, MO), and others.

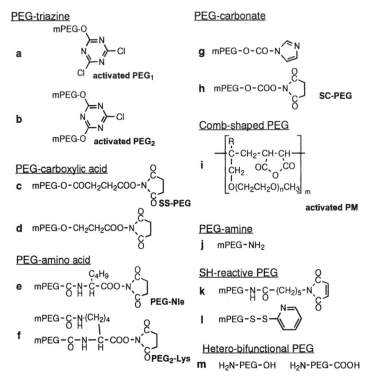

Figure 3 Polyethylene glycol derivatives (PEGs) frequently used for modifying proteins and bioactive substances.

B. Activated PEG₁

Abuchowski et al. [28] prepared a reactive intermediate, activated PEG₁ (Fig. 3a). To cyanuric chloride (0.03 mol) dissolved in anhydrous benzene containing sodium carbonate was added monomethoxypolyethylene glycol (mPEG-OH with MW of 5000, 0.01 mol) and the mixture was stirred overnight at room temperature. The modifier thus prepared, activated PEG₁, directly reacts with amino groups on the surface of protein molecule in an aqueous buffered solution at pH 9.2 and 4°C for 1 h (Scheme 3).

$$\text{(3)}$$

C. Activated PEG₂

Matsushima et al. [29] prepared activated PEG₂ (Fig. 3b) which was prepared from mPEG–OH (MW 5000) and cyanuric chloride at a molar ratio of 2:1 by refluxing for 44 h in anhydrous benzene using sodium carbonate as a catalyst. Two of three chlorine atoms in a triazine ring were replaced by PEG chains and the remaining third chlorine atom is the reactive site to amino groups of proteins. The reaction between activated PEG₂ and amino groups in the protein molecule is carried out in mild alkaline conditions (pH 9.5–

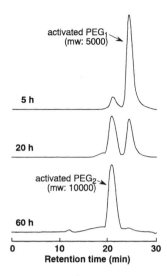

Figure 4 Gel filtration profile during the course of activated PEG_2 synthesis using zinc oxide. (From Ref. 30.)

10.0 at 37°C). Activated PEG_2 seems to have an advantage over activated PEG_1; two PEG chains attach to one amino group in the protein molecule. In 1991, Ono et al. [30] synthesized activated PEG_2 in a homogeneous state by using zinc oxide as a catalyst (Fig. 4; Scheme 4). Activated PEG_2 is now available from Seikagaku Corp. (Tokyo, Japan).

$$\text{(4)}$$

D. PEG–Carboxylic Acid and PEG–Amino Acid

Monomethoxypolyethylene glycol having a terminal carboxyl group (mPEG-COOH) is a versatile derivative to make PEG–protein conjugates. One of the most practical methods for introducing carboxyl groups at a terminus of PEG may be the reaction with succinic anhydride, which is then activated by N-hydroxysuccinimide (NHS) using dicyclohexyl-carbodiimide to form mPEG succinimidyl succinate (SS-PEG; Fig. 3c) [31]. Alternatively, carboxyethyl (Fig. 3d) or carboxymethyl derivative of PEG is also available. The NHS esters of PEG react directly with amino groups in the protein molecule at a pH ranging from 7.0 to 9.0 [31] (Scheme 5).

$$\text{(5)}$$

Yamasaki et al. [32] prepared some amino acid derivatives of PEG. mPEG–carboxylic acid activated with NHS was introduced to the α-amino group of norleucine or to the α- and ε-amino groups of lysine, and the carboxylic group of the amino acid was activated with NHS again. The former (Fig. 3e) is convenient to direct determination of the number of amino groups coupled with it by means of acid hydrolysis followed by amino acid analysis [33]. The latter (Fig. 3f) is a branched PEG modifier just like activated PEG$_2$.

A hydroxyl group of mPEG was activated with 1,1'-carbonylimidazole to yield a reactive intermediate (Fig. 3g), which reacts with amino groups of the protein molecule [34]. Veronese et al. [35] also prepared phenylchloroformate-activated PEG. Recently, Zalipsky et al. [36] activated a hydroxyl group of mPEG by treatment with phosgene followed by NHS. The resulting modifier, PEG succinimidyl carbonate (SC-PEG, Fig. 3h), reacts with an amino group of protein at pH 7–10 to form a stable urethane linkage (Scheme 6).

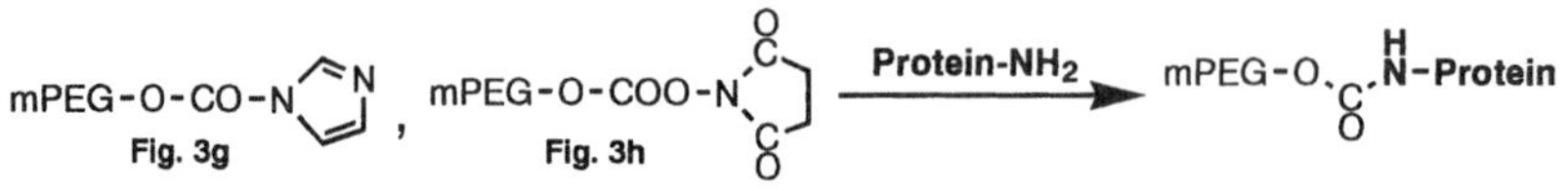

$$\text{mPEG-O-CO-N} \quad , \quad \text{mPEG-O-COO-N} \quad \xrightarrow{\text{Protein-NH}_2} \quad \text{mPEG-O} \cdot \overset{\text{H}}{\underset{\text{O}}{\text{C}}} \cdot \text{N-Protein} \qquad (6)$$

Fig. 3g   Fig. 3h

E. Activated PM

A copolymer of polyethylene glycol and maleic anhydride with a comb-shaped form, activated PM, was developed by NOF Corp. (Nippon Oil and Fats Co., Tokyo, Japan) [37]. Two activated PM copolymers have been prepared (Fig. 3i); activated PM$_{13}$ (MW 13,000, $m \approx 8$, $n \approx 33$, R = H) and activated PM$_{100}$ (MW 100,000, $m \approx 50$, $n \approx 40$, R = CH$_3$). Amino groups in a protein molecule are coupled with maleic anhydride residues in the PM modifier to form acid–amide bonds (Scheme 7). These comb-shaped modifiers possess unique properties: covering a protein molecule with the reaction between amino groups in a protein and multivalent acid anhydrides in the modifier.

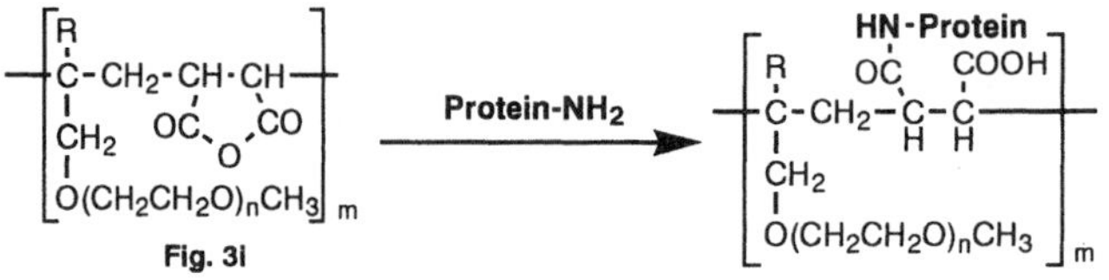

$$\left[\begin{array}{c}\text{R}\\ \text{C-CH}_2\text{-CH·CH}\\ |\\ \text{CH}_2 \quad \text{OC} \diagdown_\text{O} \diagup \text{CO}\\ |\\ \text{O(CH}_2\text{CH}_2\text{O)}_n\text{CH}_3\end{array}\right]_m \xrightarrow{\text{Protein-NH}_2} \left[\begin{array}{c}\text{HN·Protein}\\ \text{R} \quad \text{OC} \quad \text{COOH}\\ \text{C-CH}_2\text{-C-C}\\ |\qquad\quad \text{H} \ \text{H}\\ \text{CH}_2\\ |\\ \text{O(CH}_2\text{CH}_2\text{O)}_n\text{CH}_3\end{array}\right]_m \qquad (7)$$

Fig. 3i

F. Other PEG Derivatives

Apart from the amino group modification, various strategies for protein modification have been published. PEG–amine (Fig. 3j) is also utilized for the modification of bioactive compounds with carboxyl groups [38,39] and for a highly reactive intermediate of various modifier syntheses [40] such as PEG–maleimide (Fig. 3k). Urrutigoity et al. [41] coupled PEG–amine to the periodate-oxidized carbohydrate moiety of a glycoprotein.

Sulfhydryl group is another target of modification. A maleimidyl derivative of PEG (Fig. 3k) is one of the modifiers, to which a sulfhydryl group causes the addition reaction to form a stable thio ether linkage [42] (Scheme 8). Goodson et al. [43] coupled a protein

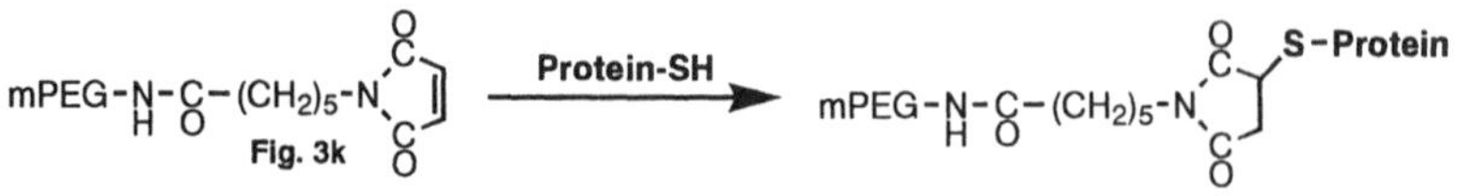

$$\text{mPEG-}\overset{\text{}}{\underset{\text{H}}{\text{N}}}\text{-}\overset{\text{O}}{\underset{\text{O}}{\text{C}}}\text{-(CH}_2)_5\text{-N} \quad \xrightarrow{\text{Protein-SH}} \quad \text{mPEG-}\overset{\text{}}{\underset{\text{H}}{\text{N}}}\text{-}\overset{\text{O}}{\underset{\text{O}}{\text{C}}}\text{-(CH}_2)_5\text{-N} \diagup^{\text{S-Protein}} \qquad (8)$$

Fig. 3k

molecule with only one PEG chain at a sulfhydryl group of the only cysteine residue introduced by genetic engineering technique. In the case of site-specific modification, a free sulfhydryl group of cysteine residues, the number of which is usually much less than that of amino groups of lysine or N-terminal amino acid residues, may have advantage as the target functional group over an amino group if the sulfhydryl group is not essential to the protein function. Furthermore, the sulfhydryl group can be attached with a PEG chain through disulfide linkage by using the 2-pyridyldithio derivative of PEG (Fig. 3l) [44], which is a unique reversible modification (Scheme 9).

$$mPEG-S-S-\underset{\text{Fig. 3l}}{\diagdown} \xrightarrow{\text{Protein-SH}} mPEG-S-S-Protein \qquad (9)$$

Heterofunctional PEG derivatives have also been developed as a new type of modifier (Fig. 3m) [45–47], which will be convenient for well-designed crosslinking agents.

III. ENZYME REACTION IN A HYDROPHOBIC ENVIRONMENT

PEG-modified enzymes prepared by coupling proteins with PEG in aqueous buffer solution are soluble not only in aqueous solution but in hydrophobic environments. Table 2 summarizes the list of PEG–enzymes prepared by the authors for the application to biotechnological processes.

A. PEG–Oxidoreductase

In 1984, the authors reported that catalase from bovine liver modified with activated PEG_2 became soluble in benzene and catalyzed the hydrolysis of hydrogen peroxide [24]. Table 3 represents the degree of modification of the amino groups of various PEG–catalase preparations and their solubilities in benzene, together with their enzymic activities both in aqueous solution and in benzene. Although unmodified enzyme hardly dissolved in benzene, PEG-modified catalase became more soluble by increasing the degree of modification. PEG–catalase with a degree of modification of 46% could dissolve at 0.64 mg protein in 1 ml of benzene. The enzymic activity in aqueous solution was decreased by increasing the degree of modification, whereas in benzene the activity was considerably enhanced, and then decreased by too much modification. The activity of PEG–catalase with the degree of modification of 42% in benzene was, surprisingly, 1.4 times as high as that of unmodified enzyme in aqueous solution.

Horseradish peroxidase was also modified with activated PEG_2 [25]. PEG–peroxidase, in which 60% of the amino groups were coupled with PEG, had 70% of the original enzymic activity in aqueous solution and was found to be active in benzene. Using PEG–peroxidase and PEG–cholesterol oxidase, cholesterol could be photometrically determined in transparent benzene solution by measuring the absorbance increase at 490 nm [37] as shown by the following equations:

Cholesterol $+ O_2 \rightarrow$ 4-cholestene-3-one $+ H_2O_2$

$H_2O_2 + o$-phenylene diamine $\rightarrow$ oxidized o-phenylene diamine

B. PEG–Protease

A similar line of work was done using PEG–protease and PEG–lipase. They are soluble and active in organic solvents. In particular, these PEG-modified hydrolytic enzymes ef-

Table 2 Application of PEG–Enzymes to Biotechnological Processes

Enzyme	Origin	Modifier	Purpose	Solvent	Ref.
Lipase	*P. fluorescens*	Activated PEG$_2$	Ester synthesis, ester exchange reaction	Benzene	23, 58, 61, 65, 66, 72, 79, 87
Lipase	*P. fragi*	Activated PEG$_2$	Synthesis of terpene alcohol ester and gefarnate, enantioselective esterification	Benzene	60, 80, 86, 114
Lipase	*P. cepacia*	Activated PEG$_2$	Lactonization, enantioselectivity, regioselectivity	Benzene, trichloroethane, decanol	63, 88, 89
Lipase	*C. cylindracea*	SS-PEG	Ester synthesis and exchange reaction, deacetylation	Benzene, trichloroethane	62, 81, 84
Chymotrypsin	Bovine pancreas	Activated PEG$_2$	Formation of oligopeptide	Benzene	26
Papain	Papaya latex	Activated PEG$_2$	Formation of oligopeptide	Benzene, trichloroethane	49, 54, 108
Proteases		Activated PEG$_2$	Solid phase peptide synthesis	Methanol	56
Lipase	*P. fluorescens*	Activated PM	Stabilization	Benzene	64
Trypsin	Bovine pancreas	Activated PM	Stabilization	Aqueous	75
Catalase	Bovine liver	Activated PEG$_2$	Decomposition of H$_2$O$_2$	Benzene	24
Peroxidase	Horseradish	Activated PEG$_2$	Couple with cholesterol oxidase	Benzene	25
Cholesterol oxidase	*Nocardia* sp.	Activated PM	Oxidation of cholesterol	Benzene	37
Hemin, porphyrin, chlorophyllin		PEG amine	Peroxidation, photo-oxidation	Aqueous solution, benzene	39, 115–117
Avidin	Chicken egg-white	Activated PEG$_2$	Affinity partitioning	Aqueous two-phase system	118
Lipase and others		Activated PEG$_2$	Effect of water and solvent	Benzene, trichloroethane, dimethylsulfoxide	59, 67, 68
Lipase and others		PEG + magnetite	Recovery of enzyme with magnetic force	Aqueous solution, benzene	73, 74, 119, 120
Lipase and others		PEG derivatives	Reviews on PEG–enzymes in biotechnological field	Organic solvents	1–5, 14, 15, 57, 96, 121–123

Table 3 Solubility and Activity of PEG–Catalases in Benzene with Differing Degrees of Modification Together with Those of Unmodified Catalase

Degree of modification[a] (%)	Solubility in benzene (mg/ml)	Activity (μmol/min/mg protein)	
		In aqueous solution	In benzene
0	0	4.7×10^4	0×10^4
21	0.14	3.0	2.3
34	0.25	2.2	2.7
42	—	1.8	7.3
46	0.64	1.4	6.2

[a]The total number of amino groups in the catalase molecule is 112.

fectively catalyzed the reverse reaction of hydrolysis in hydrophobic media, i.e., ester synthesis, ester exchange, and acid–amide bond formation reactions in organic solvents.

Various proteases such as chymotrypsin [26,48], papain [49], subtilisin [50,51], thermolysin [52], and trypsin [51,53] have been coupled with PEG derivatives for the purpose of peptide synthesis in organic solvents.

Matsushima et al. [26] succeeded in the formation of acid–amide bonds by the catalytic action of PEG–chymotrypsin in benzene, such as benzoyl tyrosine butylamide and benzoyl tyrosine–oligophenylalanine ethyl ester.

$$Bz–Tyr–OEt + H_2N-C_4H_9 \rightarrow Bz–Tyr–NH-C_4H_9 + EtOH$$

$$Bz–Tyr–OEt + nPhe–OEt \rightarrow Bz–Tyr–(Phe)_n–OEt + nEtOH$$

Ohwada et al. [54] synthesized a derivative of the salty peptide N^α,N^δ-dicarbobenzoxy-L-ornithyl–β-alanine benzyl ester (Zz–Orn(Z)–β-Ala–OBzl) in 1,1,1-trichloroethane using PEG–papain (Fig. 5).

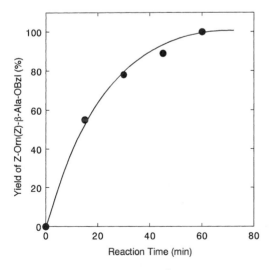

Figure 5 Synthesis of N^α,N^δ-dicarbobenzoxy-L-ornithyl-β-alanine benzyl ester, a derivative of salty peptide, with PEG–papain in 1,1,1-trichloroethane. (From Ref. 54.)

Z–Orn(Z)–OEt + β-Ala–OBzl → Z–Orn(Z)–β-Ala–OBzl + EtOH

Nakajima et al. [55] synthesized a bioactive tetrapeptide, L-arginyl–glycyl–L-aspartyl–L-serine (Arg–Gly–Asp–Ser), from amino acids carrying suitable protecting groups by applying PEG–papain, PEG–chymotrypsin, and PEG–trypsin in organic solvents. Sakurai et al. [56] also prepared a series of PEG–proteases, such as PEG–trypsin, PEG–chymotrypsin, PEG–papain, PEG–thermolysin, and PEG–pepsin, and synthesized various oligopeptides by solid phase synthesis (Table 4).

C. PEG–Lipase

Lipase has an important role in hydrolyzing various kinds of esters in an emulsified aqueous system in vivo. As the substrates are generally soluble in organic solvents but not in aqueous solutions, enzymatic reactions catalyzed by lipase are expected to proceed in hydrophobic environments such as benzene, toluene, and chlorinated hydrocarbons. Therefore, PEG–lipase may be a potent catalyst of the following reactions in hydrophobic media.

Ester synthesis:

$$R_1COOH + R_2OH \rightarrow R_1COOR_2 + H_2O$$

Ester exchange:

$$R_1COOR_2 + R_3OH \rightarrow R_1COOR_3 + R_2OH$$

$$R_1COOR_2 + R_3COOH \rightarrow R_3COOR_2 + R_1COOH$$

$$R_1COOR_2 + R_3COOR_4 \rightarrow R_1COOR_4 + R_3COOR_2$$

In fact, various esters were synthesized in transparent benzene solution using PEG–lipase [57]. For example, dodecyl dodecanoate was synthesized from dodecanol and dooic acid in organic solvents [58]. The reaction rate in 1,1,1-trichloroethane was 3.6 times as

Table 4 Solid Phase Synthesis of Tripeptides by a Series of PEG–Proteases

Boc-based tripeptide synthesis	Yield (%)	Fmoc-based tripeptide synthesis	Yield (%)
H–Arg–Gly–Phe–OH ↑ ↑ TRY THR	53.6	Z–Gly–Phe–Gly–OH ↑ ↑ PAP CHY	68.9
H–Leu–Phe–Gly–OH ↑ ↑ THR CHY	54.3	Z–Phe–His–Leu–OH ↑ ↑ CHY PAP	58.7
H–Lys–Phe–Phe–OH ↑ ↑ TRY THR	37.5	Z–Val–Phe–Gly–OH ↑ ↑ PAP CHY	56.2
H–Ala–Ala–Ala–OH ↑ ↑ PAP PAP	76.1	Z–Phe–Leu–Gly–OH ↑ ↑ PEP PEP	57.5

PEG–proteases used were abbreviated as follows: CHY, PEG–chymotrypsin; PAP, PEG–papain; PEP, PEG–pepsin; THR, PEG–thermolysin; TRY, PEG–trypsin.

Table 5 Peroxide Value (POV) of Retinyl Esters Sythesized with PEG–Lipase in Benzene at 25°C

Method		POV (mEq/kg)	
		Retinyl palmitate	Retinyl oleate
Enzymic synthesis	under N_2 gas	2.5 ± 0.3	9.0 ± 0.4
	under air	4.1 ± 0.3	11 ± 0.1
Organic synthesis[a]	under N_2 gas	43 ± 4.3	200 ± 16

[a]Organic synthesis was carried out by refluxing with *p*-toluenesulfonic acid for 3 h.
Source: Ref. 61.

high as that in benzene [59]. Furthermore, various esters, such as terpene alcohol esters [60], have been synthesized by PEG–lipase in organic solvents.

Enzymic reaction catalyzed by PEG–lipase proceeded under mild conditions at room temperature in a homogeneous reaction system. Therefore, esterification of unstable substrates proceeded successfully without extensive oxidation of unsaturated substrates [61]: PEG–lipase catalyzed an ester exchange reaction from retinyl acetate and saturated or unsaturated fatty acid in benzene solution.

Retinyl acetate + palmitic acid → retinyl palmitate + acetic acid

Retinyl acetate + oleic acid → retinyl oleate + acetic acid

Enzymically synthesized retinyl palmitate had a lower peroxide value (POV = 2.5 mEq/kg under N_2 gas) than that obtained by the conventional organic synthesis (43 mEq/kg under N_2) as is shown in Table 5. In the case of retinyl oleate synthesis from retinyl acetate and oleic acid, a similar difference was observed between enzymic and organic syntheses: the POV of retinyl oleate was 9.0 mEq/kg for enzymic synthesis under N_2, which was one-twentieth as much as that for organic synthesis (200 mEq/kg under N_2). Therefore, the synthesis of biologically interesting substances, such as eicosapentaenoic acid esters [62], will become possible under quite mild conditions by the use of enzymes such as PEG–lipase.

The rate of the reaction catalyzed by PEG–lipase was quite high because the enzyme is soluble in the reaction mixture [63,64], in comparison with that catalyzed by unmodified enzyme suspended in an organic solvent. The high rates of synthesis observed with PEG–enzymes may account for the solubilization of the enzymes and substrates in the organic solvents, together with enhancing internal diffusion and ensuring high conformational flexibility [48].

D. Reaction in Substrates Without Organic Solvent

PEG–lipase catalyzed ester synthesis and ester exchange reactions not only in organic solvents but also in straight hydrophobic substrates. These reactions between an ester and an alcohol, between an ester and an acid, and between two esters were catalyzed by PEG–lipase in the mixture of straight substrates without using any other organic solvents [65]. Matsushima et al. [66] reported that PEG–lipase catalyzed ester exchange reactions between trilaurin and triolein in the mixture of these substrates. As is shown in Table 6, the two substrates, trilaurin (LLL) and triolein (OOO), were decreased by increasing the in-

Table 6 Triglyceride Composition and Melting Temperature of the Mixture of Trilaurin and Triolein During the Incubation with PEG–Lipase at 58°C

Incubation time (h)	Melting temp. (°C)	Triglyceride composition (μmol)			
		LLL	LLO	LOO	OOO
0	33–36	180	0	0	154
17	23–26	94	33	28	94
40	12–16	89	42	37	87
65	11–13	79	58	44	71

LLL, trilaurin; LLO, dilauroyl-monooleoyl-glycerol; LOO, monolauroly-dioleoyl-glycerol; OOO, triolein.
Source: Ref. 66.

cubation time at 58°C in the presence of PEG–lipase, and new triglycerides, dilauroyl-monooleoyl-glycerol (LLO) and monolauroyl-dioleoyl-glycerol (LOO), were synthesized. The melting temperature of the reaction mixture was lowered with increasing the incubation time from 33–36°C to 11–13°C, which may be applicable to alter the melting temperature of oils and fats in food technology. This type of reaction may have several advantages over the conventional methods for environmental safety by not using any harmful organic solvents.

E. Effect of Water on Enzymic Activity

In ester synthesis and exchange reactions, as well as in hydrolysis reactions induced by PEG–lipase in hydrophobic media, the existence of a trace amount of water in the reaction system was most important in terms of the reactions proceeding. Matsushima et al. [67] carried out a kinetics study of PEG–lipase in transparent benzene solution to estimate the K_m value of water, one of the substrates of lipase in the ester hydrolytic reaction. Indoxyl acetate was hydrolyzed by PEG–lipase to form acetic acid and 3-hydroxyindole, which was photometrically determined. A double-reciprocal plot of the velocity of the indoxyl acetate hydrolysis against water concentration at a given concentration of indoxyl acetate indicated that the hydrolysis took place as a double-displacement reaction (ping-pong reaction). The apparent Michaelis-Menten constant of water and the maximum velocity were calculated to be $K_m^{H_2O} = 7 \times 10^{-2}$ M and $V_{max} = 4700$ μmol/min/mg of protein, respectively.

In the case of ester synthesis reaction by PEG–lipase, a nonaqueous environment might be suitable from a viewpoint of chemical equilibrium. However, an ester synthetic reaction from dodecanol and dodecanoic acid proceeded effectively in water-saturated benzene solution rather than dry solvent without containing water. Takahashi et al. [68] demonstrated that PEG–lipase trapped water molecules on the surface of the enzyme molecule and exerted its enzymic activity in organic solvents. PEG–lipase showed the highest activity in water-saturated benzene, which contains approximately 30 mM water. The addition of dimethylsulfoxide to the enzyme solution in benzene took the trapped water away from the enzyme surface and reduced its activity. A trace amount of water would be necessary to exhibit the enzymic activity in water-immiscible organic solvents. Similar phenomena were also observed, e.g., PEG–chymotrypsin activity was enhanced by the addition of a trace amount of water in a water-immiscible solvent [69–71].

F. Recovery and Recycling of PEG–Enzyme

Although the high solubility of PEG–lipase in various organic solvents is beneficial to the reaction rate, it will be necessary to develop a protocol for enzyme recovery and reuse. Yoshimoto et al. [72] demonstrated that PEG–lipase can be precipitated by adding hexane or petroleum ether into a benzene solution containing the reaction mixture. PEG is soluble in aqueous or various organic solvents but not in hydrocarbons; therefore, PEG–enzymes are also insoluble in these hydrocarbons. PEG–lipase could be recovered from 1 ml of benzene solution containing pentanol and dodecanoic acid (substrates) together with pentyl dodecanoate (product) by adding 2 ml of hexane to the mixture. The PEG–enzyme thus recovered retained enzymic activity even after three cycles of recovery and reuse, as is shown in Table 7.

PEG–enzymes can also be conjugated to magnetite (Fe_3O_4) particles, which can be an alternative way to recover biocatalysts. The magnetic lipase dispersed in both organic solvents and aqueous solutions, and it catalyzed ester synthesis reactions in organic solvents [73,74], which could be readily recovered by magnetic force without loss of enzymic activity (Fig. 6).

G. Stabilization of PEG–Enzyme

Almost all of the PEG derivatives have a chain-shaped form with one or a few reactive site(s) in the modifier. Activated PM (Fig. 3i) possesses a comb-shaped form with multiple PEG chains and multivalent reactive sites in one modifier molecule. Therefore, modification of enzymes with activated PM leads to the stabilization of enzymes due to the retention of the protein conformation [14].

In the case of lipase [64], acid anhydride residues of activated PM with a molecular weight of 13,000 were coupled to the amino groups on the surface of the enzyme molecule. PM–lipase thus prepared was soluble and active in hydrophobic media, and catalyzed not only an ester hydrolytic reaction in aqueous solution but also ester synthesis or ester exchange reactions in organic solvents. These characteristics of PM–lipase were the same

Table 7 Recovery of PEG–Lipase from Benzene Solution Containing the Reaction Mixture of Pentyl Dodecanoate Synthesis

	Recovery (%)	
Run	Protein conc.	Activity[a]
Initial	100	100
1st	100	97
2nd	100	89
3rd	100	71

[a]The initial activity for pentyl dodecanoate formation was 3.0 μmol/min/mg protein.
Source: Ref. 72.

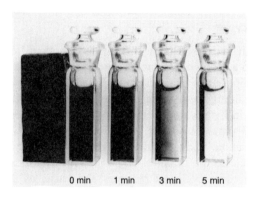

Figure 6 Magnetic separation of magnetic lipase from benzene after dodecyl dodecanoate synthesis. The magnetic separation was carried out by magnetic force of 6000 Oe for 5 min in benzene. (From Ref. 96.)

as those of PEG–lipase mentioned above. Furthermore, PM–lipase was much more stable than unmodified enzyme (Fig. 7); although unmodified lipase lost its activity completely by 150 min incubation at 55°C, PM–lipase retained more than 60% of its initial activity after the same incubation time.

In the same way, trypsin from bovine pancreas was also stabilized by chemical modification with activated PM with a molecular weight of about 100,000 [75]. PM–trypsin was much more stable than unmodified enzyme not only toward autolytic digestion but toward heat or urea denaturation as well. Table 8 represents the stability of the PM-modified and unmodified enzymes together with succinylated trypsin. In the case of trypsin preparations incubated at 37°C and pH 7.6, unmodified trypsin markedly reduced its enzymic activity; its residual activities were only 28% and 15% after 60 min and 120 min

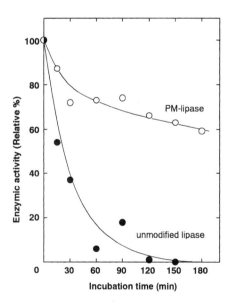

Figure 7 Heat stability of PM–lipase. Ester hydrolytic activities of PM–lipase and unmodified lipase in olive oil emulsion after incubation at 55°C.

Table 8 Stabilization of Trypsin Modified with Activated PM Toward Autolysis, Urea, and Heat Treatments (%)[a]

Treatment	Autolysis (37°C, pH 7.6)		4 M urea	60°C
	60 min	120 min	60 min	60 min
Unmodified trypsin	28	15	6	2
Succinylated trypsin	100	107	31	14
PM–trypsin	104	104	85	69

[a]Activities of each trypsin preparation are relatively displayed after the treatments for 60 min or 120 min in comparison with that of corresponding enzyme without treatment.

incubation. Succinylated and PM–trypsin preparations, of which the amino groups were coupled with the modifiers, completely retain their activities. The modification of amino groups, which are the cleavage sites during trypsin digestion, clearly protected the enzymes from the inactivation due to autolysis. In the same way, unmodified trypsin was almost inactivated by the treatment of urea or heat for 60 min, whereas PM–trypsin retained high enzymic activity even by these treatments, as is shown in Table 8. The enzymic activity of PM–trypsin was much higher not only than that of unmodified enzyme but also than that of succinylated enzyme. The circular dichroism (CD) analysis indicated that the conformation of PM–trypsin was not disrupted by 4 M urea treatment.

Kajiuchi et al. [76,77] succeeded in the stabilization of cellulase by the chemical modification with activated PM. These phenomena would suggest that the comb-shaped modifier, activated PM, plays an important role in the maintenance of the conformation of polypeptide chain(s) of the enzyme molecule to avoid its denaturation.

The chemical modification of enzymes as well as immobilization on the surface of solid supports or hydrogel polymer networks has been widely investigated to stabilize enzymes. In some natural glycoproteins, it has been reported that the oligosaccharide moiety of glycoprotein is closely associated with protein stability. In fact, saccharide chains on the protein surface were depleted from the invertase molecule, and its stability was markedly reduced [78]. As is shown in Fig. 8, PM–enzymes may be covered through hydrogen bonds or hydrophobic interactions between multiple PEG chains and functional groups on the enzyme surface [4]. In addition, the carboxylic groups of activated PM may interact with the functional groups of protein surface through covalent and ionic bonds. These hypotheses contribute to the protection of the protein conformation from disturbance due to heat, proteases, or chemical denaturants.

IV. ENZYME REACTION WITH THE SUBSTRATE SPECIFICITIES

Generally, native enzymes never dissolve in organic solvents. We have reported that PEG-modified enzymes dissolved not only in various organic solvents but also in hydrophobic substrates and can be used as catalysts in chemical reactions in these hydrophobic media. These reactions proceed under very mild conditions, which enable us to apply PEG–enzymes to various organic syntheses of unstable or chiral compounds. Moreover, the enzymic reactions with high substrate specificity are suitable for the synthesis of various useful compounds.

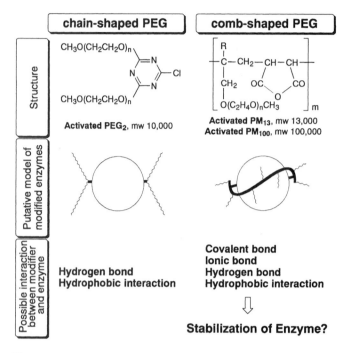

Figure 8   Putative models of structural relationships between the modifiers and the protein.

A.   Substrate Specificities of PEG–Lipase

Since the reactions catalyzed by PEG–lipase from *Pseudomonas fluorescens* were conducted in a homogeneous solvent system rather than in a suspended system, it was possible to study the kinetics for the ester synthesis reaction in benzene to determine its substrate specificity [23]. By changing the concentration of a substrate in the reaction mixture, K_m and V_{max} values were determined according to the Michaelis-Menten equation. As shown in Table 9, V_{max} values for fatty acids and normal alcohols increased by increasing the carbon chain length of either fatty acid or alcohol, whereas K_m values (0.1–0.3 M) were hardly affected. In the case of esterification of 1-pentanol with a branched carboxylic acid, 2- or 3-methylpentanoic acid, both of which have a branched methyl group close to the carboxylic group, did not serve as substrate but rather as a competitive inhibitor for the esterification reactions catalyzed by PEG–lipase. Esterification of 1-pentanol with benzoic acid could not take place. On the other hand, branched alcohols such as 2-, 3-, and 4-methylpentanols as well as 1-methylpentanol served as substrate for the esterification of pentanoic acid, except for 1,1-dimethylpropanol, a tertiary alcohol. In the case of alcohols with short carbon chains, the optimum temperature of the esterification reaction with decanoic acid was extremely low at $-3°C$ [79]. The substrate specificity of the modified lipase from *P. fragi* 22.39B were similar to that from *P. fluorescens* [80].

PEG–lipase from *C. cylindracea* catalyzes esterification in water-saturated benzene and its substrate specificity was compared with that of PEG–lipase from *P. fluorescens* [81] and *P. fragi* [80]. First, esterification from a normal alcohol and a fatty acid with different carbon chain length was tested using three kinds of PEG–lipase, as shown in Fig. 9. Although PEG–lipases from *Pseudomonas* preferentially catalyzed the esterification reactions from *n*-alcohol and fatty acid with longer carbon chains, PEG–lipase from *C.*

Table 9 k_m and V_{max} Values for Ester Synthesis by PEG–Lipase from *P*. fluorescens in Benzene

Substrates	K_m (M)	V_{max} (μmol/min/mg)
 pentanol-OH + (C4) COOH	0.17	2.8
 (C5) COOH	0.22	8.0
 (C6) COOH	0.14	8.1
 (C7) COOH	0.31	10.3
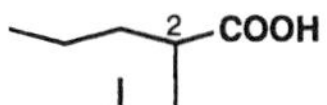 2-methyl COOH	—[a]	—[a]
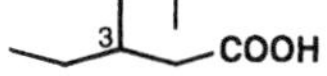 3-methyl COOH	—[a]	—[a]
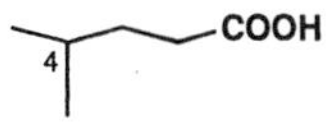 4-methyl COOH	0.19	0.2
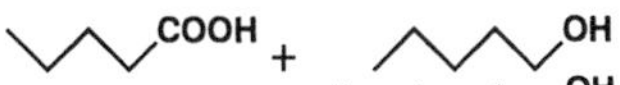 (C5) COOH + n-propanol OH	0.23	2.6
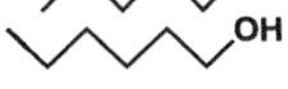 butanol OH	0.25	2.2
 pentanol OH	0.24	5.4
 hexanol OH	0.33	8.0
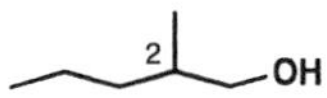 2-methyl OH	0.21	2.5
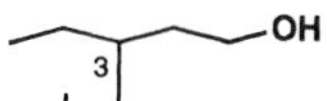 3-methyl OH	0.62	2.0
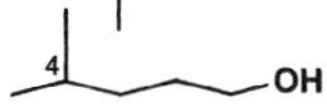 4-methyl OH	0.19	1.0
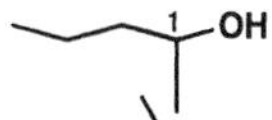 1-methyl OH	0.31	0.1
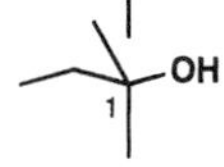 1,1-dimethyl OH	—[b]	—[b]

[a]These compounds acted not as substrates but as competitive inhibitors.
[b]1,1-Dimethylpropanol did not act as a substrate or an inhibitor.

cylindracea showed a quite different profile. In the case of an alcohol as substrate (left-side panels in Fig. 9), the ester synthetic activity of PEG–lipase from *Candida* was relatively high when using pentanoic acid and an *n*-alcohol with a lower carbon number (C_1–C_5) such as methanol and pentanol. In the case of fatty acids (right-side panels in Fig. 9), PEG–lipase from *Candida* catalyzed the esterification reactions from 1-pentanol and a fatty acid with four carbon atoms (butyric acid) and fatty acids with longer chain length (C_8–C_{12}) as suitable substrates, while the PEG–lipase poorly catalyzed the esterification from pentanoic acid (C_5) and hexanoic acid (C_6).

Ester synthesis from 1-pentanol and 2-, 3-, or 4-methylpentanoic acid was also investigated with these PEG–lipases [81]. Although PEG–lipase from *P*. *fluorescens* could only catalyze the esterification of 4-methylpentanoic acid with 1-pentanol as shown in Table 10, PEG–lipase from *C*. *cylindracea* catalyzed the esterification of these substituted carboxylic acids with 1-pentanol. In particular, esterification of 2-methylpentanoic acid

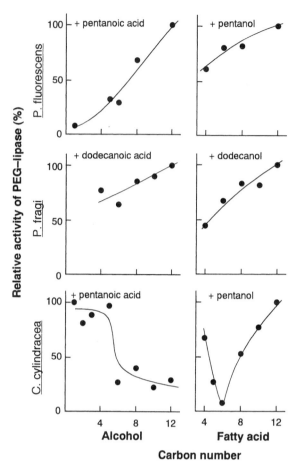

Figure 9 Ester synthesis from alcohols and fatty acids with various carbon numbers in benzene by PEG–lipases from *P. fluorescens*, *P. fragi*, and *C. cylindracea*. Left-side and right-side panels: ester synthesis with varying carbon number of alcohol and fatty acid, respectively.

was catalyzed by the PEG–lipase half as fast as that of 4-methylpentanoic acid. Therefore, methyl benzoate synthesis from methanol and benzoic acid in benzene was catalyzed only by PEG–lipase from *C. cylindracea* but not by PEG–lipase from *P. fluorescens* (Fig. 10). Similarly, methyl retinoate was synthesized in benzene from methanol and retinoic acid, an α-substituted carboxylic acid, only with PEG–lipase from *C. cylindracea*.

B. Regioselective Deacetylation of Acetylated Monosaccharide

Enzyme-catalyzed regioselective or enantioselective reactions are useful techniques in the organic synthesis of various bioactive substances. Numerous investigators have reported on such reactions in aqueous [82] or microaqueous organic environments [83]. Recently, the authors reported the regioselective deacetylation of peracetylated monosaccharide derivative by PEG–lipase from *C. cylindracea* [84]. Using a series of peracetylated methyl hexopyranosides as the substrates, it was demonstrated that PEG–lipase catalyzed the ester hydrolytic reaction only at C-4 and C-6 and never hydrolyzed the acetyl groups at C-2

Table 10 Effect of Branching Carboxylic Acid on Esterification Reaction with PEG–Lipases from *Pseudomonas* and *Candida*

Substrate (methylpentanoic acid)	Relative activity (%)	
	P. lipase	*C. lipase*
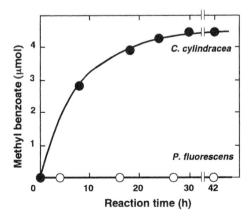	100[a]	100[a]
	0	4
	0	43

[a]Using a benzene solution containing 4-methylpentanoic acid (0.6 M) and pentanol (0.6 M), the activities were 0.20 μmol/min/mg for PEG–lipase from *P. fluorescens* and 0.053 μmol/min/mg for PEG–lipase from *C. cylindracea*. These values were taken as 100%.

and C-3. The resulting products were identified to be 6-mono-deacetylated and 4,6-di-deacetylated products with a small amount of 4-mono-deacetylated product. Furthermore, peracetylated methyl xyloside, a derivative of pentose, was hydrolyzed only at the position of C-4 by PEG–lipase. From the results obtained above, regioselective disaccharide synthesis was conducted by using PEG–lipase (Fig. 11). First, (*N*-carbobenzoxy-L-serine benzyl ester)-2,3,4-tri-*O*-acetyl-β-D-xylopyranoside (XylAc$_3$-(Z)Ser-OBzl, compound **1**) was hydrolyzed by PEG–lipase in water-saturated 1,1,1-trichloroethane solution. PEG–lipase catalyzed the hydrolysis only at the C-4 position of the xylopyranoside ring, and (*N*-carbobenzoxy-L-serine benzyl ester)-2,3-di-*O*-acetyl-β-D-xylopyranoside (XylAc$_2$-(Z)Ser-OBzl, compound **2**) was obtained. The hydroxyl group of compound **2** was then coupled with another monosaccharide derivative (2,3,4,6-tetra-*O*-acetyl-β-D-galactofluoride, GalAc$_4$F) by using silver trifluoromethanesulfonate and hafnecene dichloride to yield (*N*-

Figure 10 Ester synthesis from benzoic acid and methanol catalyzed by PEG–lipase from *C. cylindracea* or *P. fluorescens*. (From Ref. 81.)

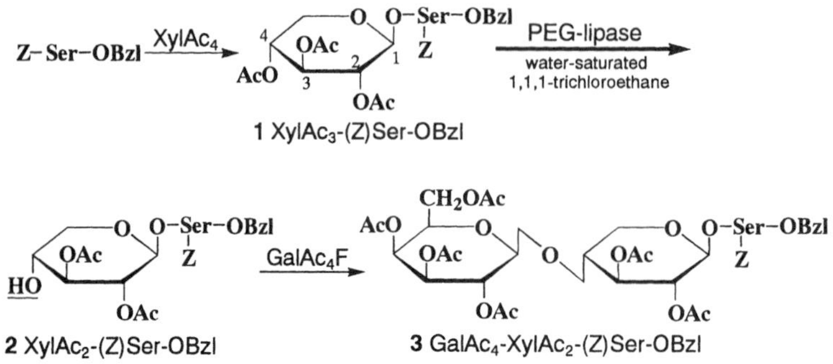

Figure 11 Scheme of regioselective deacetylation of acetylated monosaccharide and disaccharide synthesis using PEG–lipase.

carbobenzoxy-L-serine benzyl ester)-2,3-di-*O*-acetyl-4-*O*-(2,3,4,6-tetra-*O*-acetyl-β-D-gal-actopyranosyl)-β-D-xylopyranoside (GalAc$_4$-XylAc$_2$-(Z)Ser-OBzl, compound **3**). The resulting product constitutes the carbohydrate–protein linkage region of proteoglycans. These reactions lead to the synthesis of valuable oligosaccharide derivatives.

C. Substrate Specificity of PEG–Chymotrypsin

Gaertner et al. [48] investigated peptide bond formation catalyzed by PEG–proteases in benzene. The substrate specificity of PEG–chymotrypsin for ester hydrolytic activity in aqueous solution was almost the same as that of unmodified enzyme [51]: both PEG-modified and unmodified chymotrypsins catalyzed the hydrolysis of aromatic amino acid esters but these enzymes did not hydrolyze the basic amino acid ester. In the case of dipeptide synthesis in benzene, however, not only Bz-Tyr-Phe-NH$_2$ but also Bz-Lys-Phe-NH$_2$ was synthesized from corresponding amino acid derivatives by PEG–chymotrypsin. The change in the substrate specificity, which has been reported in a few cases, might be due to a surrounding environment on the enzyme molecule.

Sinisterra et al. [85] reported the stereoselective hydrolysis of racemic *N*-benzoyl-phenylalanine methyl ester catalyzed by PEG–chymotrypsin in aqueous methanolic solution. The hydrolyzed products, *N*-benzoylphenylalanine, obtained by PEG–chymotrypsin as well as unmodified enzyme were extracted from the reaction mixture and analyzed by both polarimetry and proton nuclear magnetic resonance spectrophotometry. The enantiomeric ratios of the products were 98:2 and 50:50 (*S:R*) for PEG–enzyme and unmodified enzyme, respectively. In this case, chymotrypsin acquired enantioselectivity by the chemical modification with PEG.

D. Esterification of Chiral Secondary Alcohols by PEG–Lipase

Using PEG–lipase from *P. fragi* 22.39B, a kinetic study of esterification of chiral secondary alcohols with fatty acids in transparent organic solvent was carried out [86]. PEG–lipase, of which 49% of the total amino groups in the enzyme molecule were coupled with activated PEG$_2$, was soluble and active in hydrophobic organic solvents. The ester synthetic activity of PEG–lipase, which retained 43% of the hydrolytic activity of the unmodified enzyme, was 13.6 μmol/min/mg protein in benzene and 30.5 μmol/min/mg protein in 1,1,1-trichloroethane, when dodecanol and dodecanoic acid were used as substrates.

Esterification of chiral secondary alcohols with dodecanoic acid by PEG–lipase was tested in benzene [86]. Figure 12 shows the esterification activity of PEG–lipase in 1,1,1-trichloroethane with varied concentrations of (*R*)- or (*S*)-2-octanol at a fixed concentration (0.5 M) of dodecanoic acid. In the case of each isomer, the enzymic activity was enhanced by increasing the alcohol concentration and tended to approach a constant level (panel a). However, a remarkable difference in the activities was observed between (*R*)- and (*S*)-2-octanol, indicating that PEG–lipase preferentially catalyzed the esterification of (*R*) isomer. Reciprocal plot of the reaction rate (1/*v*) against the alcohol concentration (1/*S*) gave a straight line (panel b), from which K_m and V_{max} values could be obtained. Table 11 summarizes the K_m and V_{max} values obtained for the esterification of various chiral secondary alcohols, including aliphatic and aromatic alcohols, with dodecanoic acid. As is clear from the table, the K_m and V_{max} values of (*R*) isomers were relatively constant even when the number of carbon atoms (C_4–C_9) in the secondary alcohol was changed. On the other hand, the K_m value of (*S*) isomers was increased and the V_{max} value was lowered by increasing the number of carbon atoms in the alcohol. Furthermore, each (*R*) isomer showed a smaller K_m value and a larger V_{max} value than those of the corresponding (*S*) isomer. However, there was an exception of 2-butanol, in which no apparent differences in K_m and V_{max} values between (*R*) and (*S*) isomers were observed.

A secondary aromatic alcohol in (*S*) form, (*S*)-α-phenylethanol, did not serve as a substrate of the esterification by PEG–lipase. From the result obtained above, it can be concluded that PEG–lipase exhibited higher stereoselectivity for chiral secondary alcohols with a longer carbon chain or a phenyl group. In order to test the applicability of PEG–lipase to optical resolution of racemic alcohols, we conducted the esterification with PEG–lipase in 1,1,1-trichloroethane using racemic α-phenylethanol and dodecanoic acid [86]. As is shown in Fig. 13, the amount of the substrate, (*R,S*)-α-phenylethanol, decreased

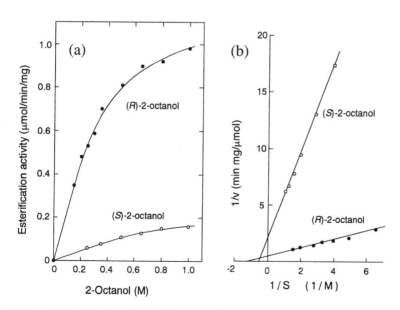

Figure 12 Kinetics of esterification of (*R*)- or (*S*)-2-octanol with decanoic acid by PEG–lipase from *P. fragi* in benzene at 25°C. (a) Plot of an esterification activity against various concentrations of (*R*)- or (*S*)-2-octanol; (b) reciprocal plot of reaction rate (1/*v*) against substrate concentration (1/*S*). (From Ref. 86.)

Table 11　K_m and V_{max} Values for Esterification of Secondary Alcohols with Dodecanoic Acid Catalyzed by PEG–Lipase from *P. fluorescens* in Benzene

Substrates (secondary alcohols)		K_m (M)		V_{max} (μmol/min/mg)	
		(R)-	(S)-	(R)-	(S)-
2-propanol		0.43	0.43	1.18	1.23
2-pentanol		0.43	1.34	1.25	0.45
2-octanol		0.54	1.50	1.68	0.42
2-nonanol		0.50	1.66	1.37	0.30
α-phenylethanol		0.47	—[a]	0.90	—[a]

[a](S)-α-Phenylethanol did not serve as a substrate.

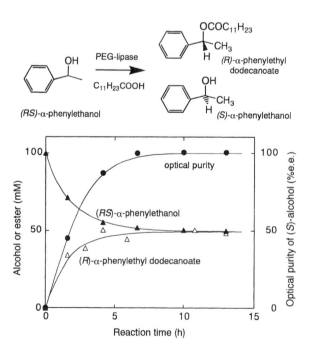

Figure 13　Time course of optical resolution of α-phenylethanol. Esterification of (R,S)-α-phenylethanol with dodecanoic acid by PEG–lipase in 1,1,1-trichloroethane. (From Ref. 86.)

with time and tended to approach a constant level of 50%. The yield of the product, which was identified as (R)-α-phenylethyl dodecanoate, kept a constant level of 50% over more than 10 h of incubation. The optical purity of the nonreacted alcohol, (S)-α-phenylethanol, reached 99% e.e. in 7 h. This indicates that (R) and (S) forms of α-phenylethanol can be discriminated by PEG–lipase in 1,1,1-trichloroethane.

E. Optical Resolution of Chiral Lactone

A linear polymer of 10-hydroxydecanoic acid was synthesized with PEG–lipase in benzene [87] and a lactone with a 17-membered ring was also synthesized from 16-hydroxyhexadecanoic acid ethyl ester [88]. These are inter- and intramolecular esterifications of each substrate, respectively. Among them, bioactive lactones, which are chiral and widely utilized in the fields of foods, perfumes, and pharmaceuticals, are synthesized by the enzymic method rather than by organic reaction.

Alcoholysis of ε-decalactone, a racemic lactone with a seven-membered ring, catalyzed by PEG–lipase from *P. cepacia* was carried out in 1,1,1-trichloroethane with the hope of the optical resolution [63], as shown in Fig. 14. The degree of modification of amino groups in the lipase molecule was 70% and the hydrolytic activity in emulsified olive oil was 600 μmol/min/mg protein. PEG–lipase catalyzed the alcoholysis of ε-decalactone in 1,1,1-trichloroethane containing ethanol as a substrate to form 6-hydroxydecanoic acid ethyl ester which was determined by HPLC. Panel a of the figure represents the time-course of the alcoholysis of (R,S)-ε-decalactone at 65°C using PEG–lipase. The amount of (S)-ε-decalactone was not changed during the reaction time at all, whereas that

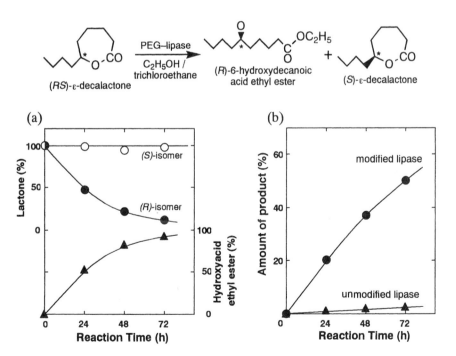

Figure 14 Alcoholysis of ε-decalactone with PEG–lipase in 1,1,1-trichloroethane at 65°C. (a) Time course of the reaction; (b) rate of the reaction with PEG–lipase and unmodified lipase. (From Ref. 63.)

of (*R*)-ε-decalactone was sharply decreased with reaction time and tended to approach a constant level of zero. Also, (*R*)-6-hydroxydecanoic acid ethyl ester appeared with time and approached a constant level of 100%. These results indicate that (*R*)-ε-decalactone is preferentially changed to (*R*)-6-hydroxydecanoic acid ethyl ester by alcoholysis using PEG –lipase. The reaction rate of the alcoholysis catalyzed by PEG–lipase was 20 times higher than that with unmodified lipase (Fig. 14b). The reaction rate with PEG–lipase was enhanced by increasing the reaction temperature ranging from 15°C to 65°C and was sharply decreased by heat denaturation above 80°C. The effect of carbon number in alcohols on the formation of hydroxydecanoic acid alkyl esters at 25°C was also tested. Increasing the carbon number of alcohols from ethanol (C_2) to hexadecanol (C_{18}) gives rise to a higher yield of the products for 72-h incubation, suggesting that PEG–lipase recognizes the carbon chains of alcohol substrates in 1,1,1-trichloroethane.

In the same way, optical resolution of δ-decalactone was also demonstrated [89]. Although the stability of the substrate with a six-membered ring was much higher than that of ε-decalactone with a seven-membered ring, similar enantioselectivity in the ring opening reaction was also observed when a mixture of substrates without any organic solvents was added by PEG–lipase.

V.   FUTURE PROSPECTS

Since it was found in 1984 that oxidoreductase [24,25] and hydrolase [26,58] modified with amphipathic macromolecule of PEG become soluble and active in organic solvents, the following advantages have been demonstrated:

1. Progress of ester synthesis, ester exchange, and peptide bond formation reactions in hydrophobic media at room temperature.
2. Characteristics of PEG–enzymes having substrate specificities such as regioselectivity and stereoselectivity.
3. Stabilization of PEG–enzymes by covering with PEG modifier, especially in the case of activated PM modifier.
4. Materials soluble in organic solvents and insoluble in aqueous solution being available as substrates.

On the contrary, it will be desirable that the enzymic activity is not reduced by the modification. To avoid the inactivation, it would be necessary to determine the optimum degree of modification in the PEG–enzyme molecule with respect to the enzymic activity.

Apart from PEG–enzymes as "stereoselective biocatalysts," this chemical modification technique will be expanded to other fields as shown in Fig. 15. Not only enzymes or proteins but also low molecular weight biosubstances could be conjugated with synthetic macromolecules such as PEG. These bioactive molecules are also conjugated with inorganic compounds, such as magnetite, glass, and insoluble supports. In fact, chlorophyll isolated from green leaves, which is quite unstable toward light, became photostable and catalyzed light-induced reactions by the conjugation with a synthetic macromolecule, polyvinylpyrrolidone, and a clay mineral, smectite [90]. Similar conjugates have also been prepared with a new function or an advantageous property as biohybrid materials: PEG– NAD [91,92] for cofactor regeneration and reusability in bioreactor with various dehydrogenases, PEG–porphyrin [93] and PEG–chlorophyllin [39] for photosensitizers, PEG –melanin for solubilization of insoluble pigment [94], and PEG–hemoglobin [95] for an

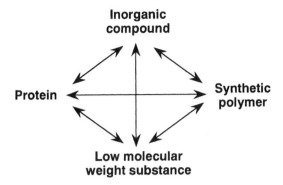

Figure 15 Biohybrid materials.

electron carrier in electrochemical sensors. This approach will open a new avenue for the development of biomedical and biotechnological processes.

REFERENCES

1. Y Inada, T Yoshimoto, A Matsushima, Y Saito. Trends Biotechnol 4:68–73, 1986.
2. Y Inada, A Matsushima, Y Kodera, H Nishimura. J Bioact Compat Polym 5:343–364, 1990.
3. Y Inada, A Matsushima, M Hiroto, H Nishimura, Y Kodera. In: YC Lee, RT Lee, eds. Methods in Enzymology. San Diego: Academic Press, 1994, pp 65–90.
4. Y Inada, M Furukawa, H Sasaki, Y Kodera, M Hiroto, H Nishimura, A Matsushima. Trends Biotechnol 13:86–91, 1995.
5. Y Inada, A Matsushima, M Hiroto, H Nishimura, Y Kodera. In: A Fiechter, ed. Advances in Biochemical Engineering and Biotechnology. Berlin: Springer-Verlag, 1995, pp 129–150.
6. IN Topchieva. Pharm Chem J 23:511–517, 1989.
7. C Delgado, GE Francis, D Fisher. Crit Rev Ther Drug Carrier Syst 9:249–304, 1992.
8. JM Harris, ed. Poly(ethylene glycol) Chemistry. New York: Plenum Press, 1992.
9. GO Curme Jr, F Johnston, eds. Glycols. New York: Reinhold, 1952.
10. S Zalipsky. Adv Drug Deliv Rev 16:157–182, 1995.
11. RAG Smith, JM Dewdney, R Fears, G Poste. Trends Biotechnol 11:397–403, 1993.
12. ML Nucci, R Shorr, A Abuchowski. Adv Drug Deliv Rev 6:133–151, 1991.
13. G Fortier. Biotechnol Genet Eng Rev 12:329–356, 1994.
14. A Matsushima, Y Kodera, M Hiroto, H Nishimura, Y Inada. J Mol Catal B: Enzymatic 2:1–17, 1996.
15. Y Inada, K Takahashi, T Yoshimoto, A Ajima, A Matsushima, Y Saito. Trends Biotechnol 4: 190–194, 1986.
16. MS Hershfield. Clin Immunol Immunopathol 76:S228–S232, 1995.
17. Y Kamisaki, H Wada, T Yagura, A Matsushima, Y Inada. J Pharm Exp Ther 216:410–414, 1981.
18. JC Yang, SL Topalian, DJ Schwartzentruber, DR Parkinson, FM Marincola, JS Weber, CA Seipp, DW White, SA Rosenberg. Cancer 76:687–694, 1995.
19. RA Greenwald. Free Rad Res Commun 12-13:531–538, 1991.
20. R Bradley, S Sloshberg, K Nho, B Czuba, D Szesko, R Shorr. Art Cells Blood Subs Immob Biotech 22:657–667, 1994.
21. TR Ulich, JD Castillo, G Senaldi, O Kinstler, S Yin, S Kaufman, J Tarpley, E Choi, T Kirley, P Hunt, WP Sheridan. Blood 87:5006–5015, 1996.
22. A Pollak, GM Whitesides. J Am Chem Soc 98:289–291,1976.
23. K Takahashi, T Yoshimoto, A Ajima, Y Tamaura, Y Inada. Enzyme 32:235–240, 1984.

24. K. Takahashi, A Ajima, T Yoshimoto, Y Inada. Biochem Biophys Res Commun 125:761–766, 1984.

25. K Takahashi, H Nishimura, T Yoshimoto, Y Saito, Y Inada. Biochem Biophys Res Commun 121:261–265, 1984.

26. A Matsushima, M Okada, Y Inada. FEBS Lett 178:275–277, 1984.

27. WY Lee, AH Sehon. Nature 267:618–619, 1977.

28. A Abuchowski, T van Es, NC Palczuk, FF Davis. J Biol Chem 252:3578–3581, 1977.

29. A Matsushima, H Nishimura, Y Ashihara, Y Yokota, Y Inada. Chem Lett 773–776, 1980.

30. K Ono, Y Kai, H Maeda, F Samizo, K Sakurai, H Nishimura, Y Inada. J Biomater Sci Polym Ed 2:61–65, 1991.

31. A Abuchowski, GM Kazo, CR Verhoest Jr, T van Es, D Kafkewitz, ML Nucci, AT Viau, FF Davis. Cancer Biochem Biophys 7:175–186, 1984.

32. N Yamasaki, A Matsuo, H Isobe. Agric Biol Chem 52:2125–2127, 1988.

33. L Sartore, P Caliceti, O Schiavon, C Monfardini, FM Veronese. Appl Biochem Biotechnol 31:213–222, 1991.

34. CO Beauchamp, SL Gonias, DP Menapace, SV Pizzo. Anal Biochem 131:25–33, 1983.

35. FM Veronese, R Largajolli, E Boccu, CA Benassi, O Schiavon. Appl Biochem Biotechnol 11:141–152, 1985.

36. S Zalipsky, R Seltzer, S Menon-Rudolph. Biotechnol Appl Biochem 15:100–114, 1992.

37. T Yoshimoto, A Ritani, K Ohwada, K Takahashi, Y Kodera, A Matsushima, Y Saito, Y Inada. Biochem Biophys Res Commun 148:876–882, 1987.

38. S Furukawa, N Katayama, T Iizuka, I Urabe, H Okada. FEBS Lett 121:239–242, 1980.

39. T Itoh, A Ishii, Y Kodera, M Hiroto, A Matsushima, H Nishimura, Y Inada. Res Chem Intermed 22:129–136, 1996.

40. JM Harris. Macromol Chem Phys C25:325–373, 1985.

41. M Urrutigoity, J Souppe. Biocatalysis 2:145–149, 1989.

42. TP Kogan. Synth Commun 22:2417–2424, 1992.

43. RJ Goodson, NV Katre. Biotechnology 8:343–346, 1990.

44. C Paul, J Vincentelli, J Brygier, T Musu, D Baeyens-Volant, C Guermant, Y Looze. Phytochemistry 35:1413–1417, 1994.

45. S Zalipsky, G Barany. Polym Prepr 27(1):1–2, 1986.

46. M Yokoyama, T Okano, Y Sakurai, A Kikuchi, N Ohsako, Y Nagasaki, K Kataoka. Bioconj Chem 3:275–276, 1992.

47. Y Nagasaki, M Iijima, M Kato, K Kataoka. Bioconj Chem 6:702–704, 1995.

48. HF Gaertner, AJ Puigserver. Proteins: Struct Funct Genet 3:130–137, 1988.

49. H Lee, K Takahashi, Y Kodera, K Ohwada, T Tsuzuki, A Matsushima, Y Inada. Biotechnol Lett 10:403–407, 1988.

50. A Ferjancic, A Puigserver, H Gaertner. Appl Microbiol Biotechnol 32:651–657, 1990.

51. H Gaertner, A Puigserver. Eur J Biochem 181:207–213, 1989.

52. A Ferjancic, A Puigserver, H Gaertner. Biotechnol Lett 10:101–106, 1988.

53. O Munch, D Tritsch, J-F Biellmann. Biocatalysis 5:35–47, 1991.

54. K Ohwada, T Aoki, A Toyota, K Takahashi, Y Inada. Biotechnol Lett 11:499–502, 1989.

55. A Nakajima, Y Hirano, T Terai, K Goto, T Hayashi, Y Ikada. J Biomater Sci Polym Ed 1: 183–190, 1990.

56. K Sakurai, K Kashimoto, Y Kodera, Y Inada. Biotechnol Lett 12:685–688, 1990.

57. Y Inada, A Matsushima, K Takahashi, Y Saito. Biocatalysis 3:317–328, 1990.

58. Y Inada, H Nishimura, K Takahashi, T Yoshimoto, AR Saha, Y Saito. Biochem Biophys Res Commun 122:845–850, 1984.

59. K Takahashi, A Ajima, T Yoshimoto, M Okada, A Matsushima, Y Tamaura, Y Inada. J Org Chem 50:3414–3415, 1985.

60. T Nishio, K Takahashi, T Yoshimoto, Y Kodera, Y Saito, Y Inada. Biotechnol Lett 9:187–190, 1987.

61. A Ajima, K Takahashi, A Matsushima, Y Saito, Y Inada. Biotechnol Lett 8:547–552, 1986.

62. T Yoshimoto, M Nakata, S Yamaguchi, T Funada, Y Saito, Y Inada. Biotechnol Lett 8:771–776, 1986.
63. M Furukawa, Y Kodera, T Uemura, M Hiroto, A Matsushima, H Kuno, H Matsushita, Y Inada. Biochem Biohys Res Commun 199:41–45, 1994.
64. M Hiroto, A Matsushima, Y Kodera, Y Shibata, Y Inada. Biotechnol Lett 14:559–564, 1992.
65. K Takahashi, Y Kodera, T Yoshimoto, A Ajima, A Matsushima, Y Inada. Biochem Biophys Res Commun 131:532–536, 1985.
66. A Matsushima, Y Kodera, K Takahashi, Y Saito, Y Inada. Biotechnol Lett 8:73–78, 1986.
67. A Matsushima, M Okada, K Takahashi, T Yoshimoto, Y Inada. Biochem Int 11:551–555, 1985.
68. K Takahashi, H Nishimura, T Yoshimoto, M Okada, A Ajima, A Matsushima, Y Tamaura, Y Saito, Y Inada. Biotechnol Lett 6:765–770, 1984.
69. MJ Cabezas, C del Campo, E Llama, JV Sinisterra, H Gaertner. J Mol Catal 71:261–278, 1992.
70. H Gaertner, T Watanabe, JV Sinisterra, A Puigserver. J Org Chem 56:3149–3153, 1991.
71. G Ljunger, P Adlercreutz, B Mattiasson. Biocatalysis 7:279–288, 1993.
72. T Yoshimoto, K Takahashi, H Nishimura, A Ajima, Y Tamaura, Y Inada. Biotechnol Lett 6:337–340, 1984.
73. T Mihama, T Yoshimoto, K Ohwada, K Takahashi, S Akimoto, Y Saito, Y Inada. J Biotechnol 7:141–146, 1988.
74. K Takahashi, Y Tamaura, Y Kodera, T Mihama, Y Saito, Y Inada. Biochem Biophys Res Commun 142:291–296, 1987.
75. M Hiroto, M Yamada, T Ueno, T Yasukohchi, A Matsushima, Y Kodera, Y Inada. Biotechnol Tech 9:105–110, 1995.
76. T Kajiuchi, JW Park. J Chem Eng Jpn 25:202–206, 1992.
77. T Kajiuchi, JW Park. Enzyme Eng News 31:14–18, 1994.
78. FK Chu, RB Trimble, F Maley. J Biol Chem 253:8691–8693, 1978.
79. K Takahashi, T Yoshimoto, Y Tamaura, Y Saito, Y Inada. Biochem Int 10:627–631, 1985.
80. T Nishio, K Takahashi, T Tsuzuki, T Yoshimoto, Y Kodera, A Matsushima, Y Saito, Y Inada. J Biotechnol 8:39–44, 1988.
81. Y Kodera, K Takahashi, H Nishimura, A Matsushima, Y Saito, Y Inada. Biotechnol Lett 8:881–884, 1986.
82. GM Whitesides, C-H Wong. Angew Chem, Int Ed Engl 24:617–638, 1985.
83. AM Klibanov. Acc Chem Res 23:114–120, 1990.
84. Y Kodera, K Sakurai, Y Satoh, T Uemura, Y Kaneda, H Nishimura, M Hiroto, A Matsushima, Y Inada. Biotechnol Lett 20:177–180, 1998.
85. JV Sinisterra, EF Llama, C del Campo, MJ Cabezas, JM Moreno, M Arroyo. J Chem Soc Perkin Trans 2:1333–1336, 1994.
86. S Kikkawa, K Takahashi, T Katada, Y Inada. Biochem Int 19:1125–1131, 1989.
87. A Ajima, T Yoshimoto, K Takahashi, Y Tamaura, Y Saito, Y Inada. Biotechnol Lett 7:303–306, 1985.
88. Y Kodera, M Furukawa, M Yokoi, H Kuno, H Matsushita, Y Inada. J Biotechnol 31:219–224, 1993.
89. T Uemura, M Furukawa, Y Kodera, M Hiroto, A Matsushima, H Kuno, H Matsushita, K Sakurai, Y Inada. Biotechnol Lett 17:61–66, 1995.
90. T Itoh, A Ishii, Y Kodera, A Matsushima, M Hiroto, H Nishimura, T Tsuzuki, T Kamachi, I Okura, Y Inada. Bioconj Chem 9:409–412, 1998.
91. T Yomo, I Urabe, H Okada. Eur J Biochem 203:533–542, 1992.
92. AF Bückmann, M Morr, M-R Kula. Biotechnol Appl Biochem 9:258–268, 1987.
93. T Sagawa, H Ishida, K Urabe, K Yoshinaga, K Ohkubo. J Mol Catal 81:L13–L17, 1993.
94. A Ishii, M Furukawa, A Matsushima, Y Kodera, A Yamada, H Kanai, Y Inada. Dyes Pigments 27:211–217, 1995.
95. NY Kawahara, H Ohno. Bioconj Chem 8:643–648, 1997.

96. Y Inada, K Takahashi, T Yoshimoto, Y Kodera, A Matsushima, Y Saito. Trends Biotechnol 6:131–134, 1988.
97. H Joo, YJ Yoo, DDY Ryu. Enzyme Microb Technol 19:50–56, 1996.
98. A Nakamura, I Urabe, H Okada. J Biol Chem 261:16792–16794, 1986.
99. T Yomo, I Urabe, H Okada. Ann NY Acad Sci 613:313–318, 1991.
100. S Yabuki, F Mizutani. Biosens Bioelectron 10:353–358, 1995.
101. L Yang, RW Murray. Anal Chem 66:2710–2718, 1994.
102. K Taniguchi, N Hara, T Shibuya. Meiji Daigaku Nogakubu Kenkyu Hokoku 97:41–62, 1993.
103. M Basri, AB Salleh, K Ampon, WMZ Yunus, CNA Razak. Biocatalysis 4:313–317, 1991.
104. H Matsuyama, R Taguchi, H Ikezawa. Chem Pharm Bull 39:743–746, 1991.
105. Y Yamagata, E Ichishima. Curr Microbiol 19:307–311, 1989.
106. M-T Babonneau, R Jacquier, R Lazaro, P Viallefont. Tetrahedron Lett 30:2787–2790, 1989.
107. IN Topchieva, NV Efremavo, NV Khvorov, NN Magretova. Bioconj Chem 6:380–383, 1995.
108. H-H Lee, H Fukushi, M Uchino, K Sato, K Takahashi, Y Inada, K Aso. Chem Express 4: 253–256, 1989.
109. Z Yang, M Domach, R Auger, FX Yang, AJ Russell. Enzyme Microb Technol 18:82–89, 1996.
110. P Caliceti, O Schiavon, L Sartore, C Monfardini, FM Veronese. J Bioact Compat Polym 8: 41–50, 1993.
111. A Garcia III, S Oh, CR Engler. Biotechnol Bioeng 33:321–326, 1989.
112. JW Park, T Kajiuchi. Kagaku Kogaku Ronbunshu 19:702–704, 1993.
113. JE Beecher, AT Andrews, EN Vulfson. Enzyme Microb Technol 12:955–959, 1990.
114. A Mizutani, K Takahashi, T Aoki, K Ohwada, K Kondo, Y Inada. J Biotechnol 10:121–125, 1989.
115. K Takahashi, A Matsushima, Y Saito, Y Inada. Biochem Biophys Res Commun 138:283–288, 1986.
116. A Ajima, SG Cao, K Takahashi, A Matsushima, Y Saito, Y Inada. Biotechnol Appl Biochem 9:53–57, 1987.
117. Y Kodera, A Ajima, K Takahashi, A Matsushima, Y Saito, Y Inada. Photochem Photobiol 47:221–223, 1988.
118. H Nishimura, N Munakata, K Hayashi, M Hayakawa, H Iwamoto, Y Takahata, Y Kodera, H Tsurui, T Shirai, Y Inada. J Biomater Sci Polym Ed 7:289–296, 1995.
119. Y Tamaura, K Takahashi, Y Kodera, Y Saito, Y Inada. Biotechnol Lett 8:877–880, 1986.
120. T Yoshimoto, T Mihama, K Takahashi, Y Saito, Y Tamaura, Y Inada. Biochem Biophys Res Commun 145:908–914, 1987.
121. K Takahashi, Y Saito, Y Inada. J Am Oil Chemists Soc 65:911–916, 1988.
122. Y Inada, Y Kodera, A Matsushima, H Nishimura. In: Y Imanishi, ed. Synthesis of Biocomposite Materials. London: CRC Press, 1992, pp 85–108.
123. Y Kodera, H Nishimura, A Matsushima, M Hiroto, Y Inada. J Am Oil Chemists Soc 71:335–338, 1994.

25

New Avenues to High-Performance Immobilized Biosystems: From Biosensors to Biocatalysts

Iqbal Singh Gill
Roche Vitamins Inc., Nutley, New Jersey

Francisco J. Plou and Antonio Ballesteros
Consejo Superior de Investigaciones Científicas, Madrid, Spain

I. INTRODUCTION

The utilization of proteins and whole cells as versatile and highly selective catalysts and recognition elements is now firmly established in analytical and synthetic organic chemistry [1–11]. With the burgeoning of chemical and biotechnological applications of biocatalysis has come the swift realization that the consolidation of biocatalysis/biorecognition in the synthetic and analytical arenas depends critically on the provision of efficient, immobilized biological constructs [11–21]. In particular, synthetic applications now require affordable, stable, and adaptable "off-the-shelf" heterogeneous biocatalysts, and the new generation of biosensors and biodiagnostics demands highly sensitive and reliable bioarrays amenable to microelectronics interfacing.

The importance and success of bioimmobilization is evident from the existence of commercial processes for synthesizing glycerides, alkyl glycoside esters, peptides, amino acids, high-fructose syrups, antibiotics, and various other bioactives and chiral synthons [1–10,12–21]. Similarly, one can also point to the increasing availability of clinical, industrial, and home-use biosensors for monitoring analytes as diverse as sugars, lactate, ascorbate, amino acids, sterols, antigens, ammonia, oxygen, nitrogen monoxide, food toxins, organophosphate pesticides, and pathogens [22–35], and the recent expansion of biochip and bioelectronic technologies [11,35].

Traditional techniques such as physical adsorption and covalent linkage onto solid supports, entrapment in polymer matrices, and microencapsulation have long been used for immobilizing such enzymes as lipases, proteases, hydantoinases, acylases, amidases, oxidases, isomerases, lyases, and transferases [12–18]. Encapsulation and adsorption have also proved their utility in the immobilization of bacterial, fungal, animal, and plant cells [12–21]. However, as biocatalysis applications have grown, so the drawbacks and limitations of traditional approaches have become increasingly evident. The forefront issues now facing bioimmobilization are indicated in Table 1.

741

Table 1 Immobilizate Requirements for Biocatalyst and Biosensor Applications

Requirement	Issues and hurdles for bioimmobilization protocols
Generic technologies	Efficient protocols are essential for addressing the large diversity of biological catalysts and their manifold applications. Thus far, usages have developed much faster than immobilization technologies, considerably hindering laboratory and industrial applications.
Robust, operationally flexible heterogeneous biocatalysts	Biocatalysts now typically face aqueous-organic, low-water, anhydrous solvent, supercritical, and gas phase operation, in some cases at elevated temperature, pressure and shear conditions. Immobilization should ensure efficient biocatalytic function under these conditions, and the supporting immobilization chemistry must also be robust.
Tailored protocols for biocatalysts with special requirements	Lipases, phospholipases, cutinases, membrane-associated proteins, and coupled enzymes have specific activation and/or stabilization needs, which must be met for efficient functioning. Whole-cell systems require biocompatible protocols, which can support long-term viability.
Protocols for fabricating highly defined catalytic elements	Advanced micro-/nanoreactor, and array-based catalyst, sensor and diagnostic applications demand the fabrication of highly ordered protein films with high packing densities that can be produced en masse to stringent specifications.
Multicomponent biocatalysts	Lucrative and synthetically demanding targets such as oligosaccharides and glycoconjugates often require multicatalytic schemes employing expensive biocatalysts, thus necessitating highly efficient coimmobilization protocols.
High catalytic density materials	Many usages need high catalytic densities, especially biosensors, combinatorial biocatalysis, and high productivity and "Green" continuous processes operated with low-volume reactors.
Fabrication flexibility	Bioimmobilizates need to be fabricated in different forms to satisfy the diverse requirements of the catalyst and sensor markets—thin/thick films, printed/molded coatings, foams, glasses, and polymeric composites are now required in addition to monoliths and particulates.

In this chapter, we overview some of the more advanced and promising protein immobilization technologies that have become available in the last decade, especially the fabrication of organized thin film architectures, sol-gel encapsulation, crosslinked crystals, and protein-polymer composites. The current status and applicability of these methods in the catalyst, sensor, and diagnostics arenas, as well as prospective applications and future developments are discussed.

II. THIN-FILM TECHNOLOGIES

The attachment of proteins onto solid surfaces in a highly organized manner is critical to the development of efficient and operationally flexible sensor and diagnostic arrays and micro-/nanoreactors [11,22–35]. Two features of proteins make them particularly useful for thin-film (TF) technologies—their capacity to form highly packed homo- and heteroassemblies in vivo and in vitro to generate a wide variety of 2D and 3D structures, and their ability to recognize and bind other molecules with great specificity [36–38].

The accelerating drive toward device minaturization and integration of biologicals with microelectronics has focused attention on protein TF architectures [36–38]. Mono- and multilayer protein constructs are proving to be eminently suited to advanced microfabrication techniques, and the efficient integration of biocatalytic and biorecognition elements with electronic interfaces is now being realized. The main TF technologies under development at present are self-assembled monolayer and bilayer films, layer-by-layer adsorbates, electropolymerized assemblies, and crystalline-based conjugates.

A. Self-Assembled Films

Protein self-assembly processes give rise to an array of 2D and 3D nanostructures, which are often characterized by an extraordinary degree of extended spatial order, combined with high packing densities and elevated stabilities [36–38]. In order to support such extended molecular layers on solid substrates such as silicon, quartz, glass, gallium arsenide, gold, platinum, and mica, two technological approaches have been utilized, namely, the Langmuir-Blodgett (LB) method, and self-assembly (SA) techniques.

1. Langmuir-Blodgett and Langmuir-Schaeffer Films

The Langmuir-Blodgett (LB) and Langmuir-Schaeffer (LS) techniques and their various modifications have proved highly effective for the production of layered films based on amphiphilic species exemplified by saturated and unsaturated fatty acids, long chain amines, glycerides, phospholipids, glycolipids, polyether lipids, and membrane fragments [36–40]. Recently, it has proved possible to generate inorganic and composite LB/LS films consisting of metal phosphonates, metalloporphyrins, metallophthalocyanins, and amphiphilic metal complexes [41,42]. Protein LB/LS structures can be produced via two distinct avenues [36–38,43–45], either the direct assembly of proteins or protein-lipid conjugates into LB films, or the immobilization of proteins onto precast reactive LB layers. Examples of such assemblies are given in Table 2.

Membrane-associated proteins such as bacteriorhodopsins, phycobiliproteins, photosynthetic reaction centers (PRCs), and ferritins are unique in that they can self-assemble into 2D crystals in the presence of membrane fragments or surfactants [36–38,43–51]. However, the aqueous solubility and nonamphiphilic nature of most proteins does not lend them to LB/LS film formation per se, due to their unfavorable partitioning behavior, and their unfolding and denaturation at the air–water interface [36–38,43–45]. However, soluble proteins can be assembled into bidimensional arrays using several approaches:

1. Protein denaturation during LB/LS film generation can be avoided by the use of a modified casting technique, which avoids protein contact with the air–water interface [43]. This has been successfully applied to protein-surfactant LB composites of cytochrome c, alcohol dehydrogenase, and glutathione *S*-transferase.
2. Mutagenesis can be used to modify protein–protein, protein–lipid, and protein–substrate interactions so as to induce monolayer formation, increase stability at the interface, and enhance packing. This method has provided highly packed and structured films for cytochrome P450 and superoxide dismutase [52–54].

Table 2 Examples of Supported LB/LS Thin-Film Biosystems

Protein	Analyte/application	LB/LS system	Sensing	Ref.
Allophycocyanin	Optotransduction	Membrane fragments; surfactants	Optical	47, 48
Photosynthetic reaction centers	Phototransduction	Membrane fragments; surfactants	Optical	48, 49, 51
Alkaline phosphatase	Organic phosphates	—	—	55
GOx	Glucose	Behenic acid; N-lauroyl-PEI; N-lauroyl-PVP; phospholipids; CTAB; stearylamine; cetylamine	Amperometric	58, 59, 61, 62
Monoamine oxidase	Tyramine	N-Lauroyl-PEI; N-lauroyl-PVP	Amperometric	59
Urease	Urea	Stearylamine; cetylamine	ISFET	62
Alcohol dehydrogenase	Alcohols	Stearic acid	Amperometric	63
Choline oxidase	Choline	Behenic acid; phospholipids	Amperometric	64, 65
Acetylcholinesterase	Acetylcholine	Viologen-lipids	Optical	66
L-Glutamate dehydrogenase	L-Glutamate	Behenic acid	—	67
Phycoerythrin	Optotransduction	Avidin/streptavidin-biotinylated lipids	Optical	70
Anti-IgG-fluorescein	IgG	Protein A; protein G; protein A/G	Optical	72–74

3. Specific, exposed arginine, lysine, methionine, cysteine, and tryptophan residues on the surface of proteins can be modified with lipids to introduce amphiphilicity, followed by LB assembly of the lipoconjugate [36–38,43–45,52]. This route has been used for derivatizing unique surface arginine on cytochrome c, cytochrome c551, and azurin. When a unique, modifiable site is not present or the residue is critical, single or multiple point mutations can be used to engineer a site for lipid attachment, as has been done for cytochrome P450 [36–38,43–45].

4. Alternatively, native proteins such as cytochromes can be enclosed in reverse micelles of glycerides, glycolipids, phospholipids, or other surfactants capable of organizing biomolecules at the air–water interface [36–38,43–45]. Collapse of the micelles and LB formation with multiple compression–decompression cycles furnishes orientated lipid-supported protein monolayers displaying high packing densities [55].

5. Soluble proteins can be linked to amphiphilic proteins capable of LB formation. Thus, a cytochrome c–PRC chimera has been shown to form a stable and highly organized LB in the presence of phospholipids [49].

6. Manipulation of the subphase composition during fabrication. For instance, the use of high ionic strength subphases furnishes highly thermostable alkaline phosphatase LB films [56].

Apart from autoassembly, proteins can also be immobilized onto reactive LB/LS films [36–38,43–45] as follows:

1. *Adsorption.* The protein is immobilized on an LB film by electrostatic, H bonding, and/or hydrophobic interactions. This has been used to produce biosensors based around glucose oxidase (GOx), monoamine oxidase, urease, alcohol dehydrogenase, choline oxidase, acetylcholinesterase, glutamate dehydrogenase, and luciferase adsorbed onto fatty acid, fatty amine, phospholipid, polyelectrolyte, and polymerizable surfactant LB/LS films [57–67].

2. *Covalent coupling.* Reactive LB/LS membranes with dense assemblies of amphiphiles displaying activated ester, anhydride, imide, and disulfide functionalities have been developed for direct or linker-mediated coupling to proteins, especially for biosensors [68,69]. Thus, α-chymotrypsin can be bound to LB films of succinimidyl behenoate to give supported planar monolayers [70]. Alternatively, functionalized polymers such as polyethyleneimine and polyallylamine can be adsorbed to fatty acid LB/LS membranes, followed by covalent attachment of proteins to the polymers [68]. However, covalent immobilization suffers from the drawback that it may lead to undesirable conformational disturbances in the protein and partial denaturation.

3. *Avidin-biotin systems.* Stable and highly active protein-LB/LS film composites can be accessed using avidin-biotin conjugation techniques. Thus, sensor elements displaying biotinylated GOx and phycoerythrin, and streptavidin and avidin conjugates of phycoerythrin, have been bound to avidin-lipid and biotinylated lipid-streptavidin LB films, or biotinylated lipid layers have been produced [71–73].

4. *Antibody–antigen systems.* Highly specific antigenic interactions can also be used to direct the assembly of proteins. For example, IgGs have been assembled onto lipid LB films containing protein A, protein G, or protein A/G chimeras for immunosensing applications [74–76]. Similarly, an amperometric glucose

sensor has been developed around lipid-IgG fragment LB membranes directing the assembly of GOx labeled with an antigenic determinant [77].

The majority of proteins, whether soluble or membrane-bound, appear to retain their structures and activities when deposited onto solid substrates as LB/LS films [36–38,43–45]. Indeed, a common feature of protein LS films is their extraordinary thermal stability in terms of both protein structure and catalytic/recognition function. Most proteins examined to date, including alkaline phosphatase, glutathione S-transferase, alcohol dehydrogenase, urease, thioredoxins, phycoerythrin, bacteriorhodopsins, azurins, cytochrome P450, photosynthetic reaction centers, and immunoglobulins, show large elevations in structural and functional melting temperatures, with preservation of native function up to 150–200°C in LS films [36–38].

Enhanced stability is of obvious interest for biotransformations, where biocatalysts are increasingly required to function efficiently in high-temperature, low-water environments. Indeed, heterogeneous biocatalysts that remain functional up to 150°C can be produced by depositing enzyme-lipid LS films onto polysiloxane-activated glass beads, as shown for urease and glutathione S-transferase [78].

Although protein–LB/LS technology is still relatively new, it is notable for the extraordinary degree of structural organization that can be realized for biomolecule assemblies, the exceptional stabilities displayed by proteins in such films, and the availability of tools for manipulating TF architecture and biomolecular interactions. Indeed, it offers an avenue to true nanoscale biomolecular engineering, an area that is pivotal to the successful development of bioelectronic devices for biochemical sensing and diagnostics usages.

2.   Self-Assembled Monolayers

Self-assembled monolayers (SAMs) are formed spontaneously via the specific chemoadsorption of functionalized molecules such as alkanethiols, dialkylsulfides, thioamino acids, chlorosilanes, substituted pyridines, and even pyridine ligand–based transition metal complexes onto complementary surfaces exemplified by gold, platinum, silver, silicon, glass, and graphite [79–90]. The use of species terminated with reactive carboxyl, amino, disulfide, and other functionalities enables the direct or mediated anchoring of biomolecules. Several methodologies exist for SAM construction, differing in the type of anchor site, adsorbate chemistry, and supporting substrate. SAM technology is proving particularly useful for accomplishing direct electron transfer between electrode materials and redox proteins, a critical requirement for producing effective mediated and mediator-free oxidase and reductase biosensors.

For instance, glucose bisensors have been built from GOx adsorbed onto alkylthiol, ferrocenylalkylthiols, and cellulose acetate propionate-alkanethiol SAMs, or covalently linked to ω-carboxyalkanethiol, aminoalkanethiol, and cysteamine SAMs [79,91–96]. High sensitivities and linearities have been observed for these systems, deriving from efficient electrical transduction to the underlying metal substrate. Application of GOx together with a polyelectrolyte such as ferrocenyl-poly(4-vinylpyridine) provides highly responsive wired, mixed films with good resistance to interference from urate and ascorbate [94]. The use of avidin- or biotin-modified GOx in combination with biotinylated dialkyldisulfide SAMs makes it possible to build stratified films with higher catalytic densities, and thus better responses [97]. It is interesting to note that studies with cytochrome c displays on single component and mixed SAMs of alkanethiols, ω-carboxyalkanethiols, α,ω-dithioalkanes, and thioalkylsiloxanes have shown the importance of monolayer composition and immobilization chemistry in determining packing density, biomolecule orientation, and film organization, and thereby bioactivity and electron transfer efficiency [98–103].

A variety of other oxidase and reductase biosensors have been described [104–113]. Thus, highly active and sensitive 2D and 3D sensors for hydrogen peroxide have been created from horseradish peroxidase, microperoxidase B11, cytochrome c oxidase and hemin, adsorbed or covalently tethered to alkanethiol, cysteine, viologen-functionalized alkanethiol, and ω-carboxyalkanethiol SAMs on gold and silver [104–108]. Similarly, fumarate reductase, glutathione reductase, malate dehydrogenase, and lactate dehydrogenase have been bound to alkanethiol, PQQ-cysteamine, dye-alkanethiol, and succinimidyl–cysteic acid SAMs, to provide electrosensors for fumarate/oxaloacetate, malate, lactate, and oxidized glutathione, respectively [109–113]. Intriguingly, the glutathione reductase electrosensor can be converted to a photoactivated sensor by incorporating a Ru(II)bipyridyl photosensitizer and using an enzyme- or polymer-linked bipyridinium electron relay system [113]. Representative examples of SAM-based biosensors are indicated in Table 3.

Nonredox proteins can also be utilized in SAM-based sensors by employing substrates giving rise to electroactive products or by use of optical transduction [114–117]. The attachment of a N-(acetylmuramoyl)-L-alanine oxidase-β-galactosidase chimera onto choline-functionalized SAMs enables electrochemical determination of β-D-galactopyranosides [114]. Likewise, conjugation of acetylcholinesterase onto cysteamine SAMs provides a sensitive biosensor for the determination of organophosphate and carbamate pesticides, based on the inhibition of enzyme activity toward 4-aminophenyl acetate [115]. Alternatively, the hemoproteins, hemin, myoglobin, and synthetic iron(III) protoporphyrin IX–binding peptides can be adsorbed onto silylated silicas to furnish optosensors for oxygen and carbon monoxide [116,117].

Antibodies can also be bound to SAMs. For instance, linkage of anti-human chorionic gonadotropin (anti-hCG) monoclonal antibodies to thioctate or biotinylated alkanethiol-streptavidin-biotinylated Fab-hCG constructs on gold-coated nylon allows the highly sensitive colorimetric (ELISA) determination of hCG [118,119]. A novel immunoassay with electrochemical detection has also been designed for hCG; an anti-hCG-alkaline phosphatase chimera is tethered to thioctate SAMs on gold and, when exposed to a mixture of free conjugate, hCG, and 4-aminophenyl phosphate, generates an anodic response proportional to the hCG concentration [120].

The facility of SAM construction—the capacity to generate highly organized monolayers, engineer the functional displays, fabricate complex 2D and 3D patterns, and immobilize biomolecules in a selective manner on the resulting surfaces—is a notable attribute [79–90]. Interestingly, the thermal properties of SAM-immobilized proteins are intermediate to those of the native species in aqueous environments and the LB/LS-supported biomolecules. These features lend the technique to macro- and microarray bioelectrosensor construction, and clinical units for the analysis of glucose, lactate, amino acids, urea, cholesterol oxidase, and so forth are currently being realized.

3. Bilayer Lipid Membranes

Natural phospholipids, sterols, glycopepetides, lipopeptides, polyether lipids, and sulfolipids, as well as synthetic amphiphiles and even certain silanes, can assemble to form bilayer lipid membranes (BLMs), which can be free or supported on mercury, gold, silver, graphite, clay, etc. [121–123]. Amphiphilic membrane proteins and soluble proteins mod-

Table 3 Examples of SAM-Based Biocatalytic Architectures

Protein	Analyte/application	SAM construct	Support	Sensing	Ref.
Catalase	Hydrogen peroxide	ω-Carboxyalkanethiols	Ag	Amperometric	86
GOx; biotinylated GOx	Glucose	Alkanethiols; ω-carboxyalkanethiols; cysteamine; ferrocenylalkanethiols; aminoalkanethiols; avidin-biotinylated dialkyl disulfide	Au	Amperometric	91–97
Cytochrome c	Hydrogen peroxide; electrotransduction	Alkanethiols; ω-carboxyalkanethiols	Au	Amperometric Potentiometric	98, 99, 101, 103, 107, 108
Cytochrome c + photosynthetic reaction center	Optotransduction	Siloxyalkanethiol	Au	Optical	102
Cytochrome c oxidase	Electrotransduction	ω-Carboxyalkanethiols	Au	Potentiometric	100, 107
Horseradish peroxidase	Hydrogen peroxide	L-Cysteine; viologen-alkanethiol; ω-carboxyalkanethiols	Au	Amperometric	104, 105, 107, 108
Microperoxidase-MP	Hydrogen peroxide; optotransduction	Alkanethiols	Ag; Au	Optical; amperometric	106, 108
Fumarate reductase	Fumarate/oxaloacetate	Alkanethiols	Au	Amperometric	109
Malate dehydrogenase	Malate	PQQ-cysteamine	Au	Amperometric	110
Lactate dehydrogenase	L-Lactate	Cibacron blue F3G-A-alkanethiols	Au	Amperometric	111, 112
β-Galactosidase	β-Galactosides	Choline-alkanethiols	Au	Amperometric	114
Acetylcholinesterase	Acetylcholine; organophosphates; carbamates	Cysteamine	Au	Amperometric	115
Anti-hCG IgG; biotinylated IgG	hCG	Thioctic acid; biotinylated alkanethiol-Streptavidin	Au	Optical; amperometric	118, 119

ified with suitable lipidic anchors are capable of spontaneous partial or total insertion into BLMs, whereas nonmodified soluble proteins can be attached via covalent chemistries or by way of polyelectrolyte complexes.

BLMs have proved to be quite useful for constructing bioelectrochemical devices, as indicated in Table 4. Thus, electrorestrictive coupling of avidin-GOx chimeras to metal-supported biotinylated phospholipid BLMs together with electrochemical detection of hydrogen peroxide provides glucose microsensors displaying good linearities [124,125]. Cytochrome c oxidase can be effectively incorporated into deoxycholate-alkanethiol BLMs on gold [126]. In an interesting example of redox enzyme–electrode coupling, the membrane-associated hydrogenase from *Desulfovibrio gigas* has been shown to insert spontaneously into mixed siloxane-lipid BLMs, where it efficiently catalyzes the oxidation of hydrogen over a wide pH range [127]. Maleimide-functionalized lipid BLMs and sandwich-type biotinylated phospholipid BLM-streptavidin constructs are very effective for tethering cytochrome b5 and biotinylated cytochrome c into ordered protein assemblies suitable for redox reporter usages [128,129].

As with SAMs, nonredox enzymes can also be utilized for BLM sensors. For instance, highly sensitive electrochemical and ISFET sensors for urea, penicillin, acetylcholine, thyroxine, and triazine herbicides have been produced from phospholipid BLMs with urease, penicillinase, acetylcholinesterase, and antibodies attached electrostatically, covalently, or via avidin-biotin conjugation [130–132].

Of course, the most promising application of BLM arrays is toward membrane protein-based devices [133–138]. In this respect, a rather intriguing use is sensor systems based on membrane transport proteins [133–135]. Lactose permease, sodium/glucose cotransporter protein, sodium/potassium ATPase, glutamate receptor ion channel proteins can be embedded into lipid BLMs supported on standard electrodes and pH ISFET elements, and used as highly selective transmembrane sensors for lactose, glucose, ATP, and glutamate. PRCs and bacteriorhodopsins have also been successfully integrated with planar phospholipid BLMs for use as photoelectronic devices and photocatalysts [136–138].

Despite the generally lower stability of protein-BLM systems as compared to LB/LS and SAM assemblies, this technology offers some unique advantages. Thus, BLMs are particularly effective for immobilizing membrane-associated proteins and biomolecules that rely on cross-membrane transduction processes. Also, a wide variety of lipids can be used for BLM construction, and the architectures can be readily engineered by varying the lipid components and via doping with functional molecules/macromolecules. In addition, as with LB/LS and SAM films, BLMs can be easily microfabricated and integrated with microelectronic devices.

B. Multicomponent Films by Layer-by-Layer Adsorption

The fabrication of nanostructured, multilayer protein films is a prime target for the development of highly efficient sensors or bioreporters, especially where "wiring" of the biological to a transducer or coupled catalytic sequences are needed [11,22–35]. This has become possible with the advent of layer-by-layer adsorption (LLA) [78,139–146], a procedure based on the deposition of alternate layers of functional macromolecules and proteins, or proteins and proteins, held together by electrostatic, hydrophobic, and/or molecular recognition interactions.

Silicon, glass, quartz, gold, and alkylthiol-modified gold have been used as substrates, and the intervening noncatalytic layers have been assembled from organic polyelectrolytes such as polyethyleneimine and polystyrene sulfonate, expanded inorganic sil-

Table 4 Examples of Supported Protein-BLM Assemblies

Protein	Analyte/application	BLM construct	Support	Sensing	Ref.
GOx (avidin-GOx)	Glucose	Biotinylated phospholipid	SS	Amperometric	124, 125
Cytochrome c oxidase	Electrotransduction	Deoxycholate-alkanethiol	Ag/Au	Potentiometric	126
Hydrogenase	Hydrogenation	Alkylviologen-alkyltrichlorosilane	Pt	Potentiometric	127
Cytochrome b5	Electrotransduction	Biotinylated phospholipidsipids-streptavidin; maleimide-phospholipids	Au	Amperometric	128
Urease; avidin-urease	Urea	Phospholipids; biotinylated phospholipids	SS; PPY; PP; ISFET; glass	Amperometric; pH ISFET	130–132
Penicillinase	Penicillin	Mixed phospholipids	ISFET; glass; PP	pH ISFET; amperometric	131, 132
Acetylcholinesterase	Acetylcholine; triazine herbicides	Mixed phospholipids	Glass; PP	Amperometric	132
Lactose permease	Lactose	Mixed phospholipids	ISFET	pH ISFET	133
Na/glucose permease	Glucose	Mixed lipids	ISFET	ISFET	134
Na/K ATPase	5'-ATP	Mixed lipids	ISFET	ISFET	134
Glutamate ion channel	Glutamate	Mixed lipids	ISFET	ISFET	134, 135

icates such as montmorillonites, and streptavidin/avidin [139–146]. By this method, stratified multicomponent films consisting of up to a dozen consecutive intercalator-protein layers can be assembled. Proteins such as GOx, horseradish peroxidase, choline oxidase, lactate oxidase, myoglobin and lysozyme, and antibodies, antigens, and DNA have been incorporated into mono- or bicomponent optical, amperometric, and microgravimetric sensors for glucose, hydrogen peroxide, choline, acetylcholine, lactate, immunoglobulins, and DNA [139–146].

Although LLA does not offer the level of organization of LB or BLM assemblies, its technical simplicity and high biocompatibility are notable. Indeed, the accessibility of spatially patterned nanolattices incorporating functionalized intercalators, the low diffusional cross-sections, and the ability to "wire" proximate transducer–protein or protein–protein layers should provide fast response constructs for TF sensors and diagnostics.

C. Protein Immobilization on 2D Bacterial Crystallines

Previous sections have dealt with the production of supported bioassemblies by transfer of precast constructs to solid supports, by structured assembly on functionalized surfaces, or by nonstructured incorporation into polymeric films. An intriguing avenue to the realization of structured nanoscale biocatalytic films is the utilization of crystalline bacterial cell surface layers (S layers or crystallines) as preorganized and functionalized substrates for biomolecule attachment [147–151].

These hexagonally ordered protein/glycoprotein lattices are obtained from *Bacillus*, *Caulobacter*, and *Clostridium* spp. as nanoporous 2D crystals with precisely defined spatial distributions of amino, hydroxyl, and carboxyl functionality, which are amenable to chemical modification. Importantly, S layers can be dispersed and then recrystallized in a highly organized manner on planar and nonplanar silicon, glass, quartz, gold, and platinum surfaces. These features make crystallines ideal preorganized/prefunctionalized substrates for the geometrically defined immobilization of monolayers of proteins and nucleic acids [147–151].

S layers have found particular utility in the preparation of optical and amperometric single and multiple component biosensors based around such enzymes as GOx, β-fructosidase, and invertase, as well as immunoglobulins and DNA [147–157], but have also been used for producing immobilized β-glucosidase, α-rhamnosidase, and peroxidase biocatalysts [147–151,158,159]. Interestingly, supported crystallines can be patterned by ultraviolet and plasma etching, and protein-crystalline conjugates can be coated with platinum or gold using sputtering and laser deposition techniques, thereby allowing the fabrication of nanoengineered bioarrays [147–151,160,161].

Although commercialization has as yet to be seen, the close structural definition of crystalline conjugates, their isoporosity, together with the precise, surface organization of immobilized proteins offers a very promising avenue to high-speed catalytic elements with high signal-to-noise ratios for demanding sensor usages.

D. Electropolymerized Films

Electrochemical polymerization, specifically the oxidative polymerization of suitable monomers, provides a facile approach for the immobilization of proteins in conducting and nonconducting polymer films at electrode surfaces [162–166]. A unique aspect of this approach is that the enzyme can be efficiently wired to the underlying electrode material via the encapsulating polymer matrix, a feature that is very valuable when using redox proteins.

The rate and extent of polymerization, degree of oxidation, and film deposition can be controlled by varying the electrode material and its surface structure, the applied potential, and the solution mix, thereby enabling regulation of film microstructure and the amount of enzyme immobilized. In particular, it is possible to generate composite films and multilayers using copolymerization and sequential deposition techniques, and thus fabricate doped and multicatalytic stratified films [162–165].

In the simplest, one-step approach, electropolymerization is carried out in a solution of the monomer and enzyme, at a pH sufficiently removed from the pI such that the protein is incorporated in the charged polymer matrix via electrostatic interactions [162–167]. By this method, multilayer architectures with distinct electropolymer and/or encapsulated proteins can be fabricated. Alternatively, or in combination with this, biologicals can simply be adsorbed onto or covalently attached to predeposited electropolymer surfaces.

Enzymes, especially oxidases, notably GOx, have been entrapped in single- and multicomponent conducting and nonconducting electropolymerized films, as well as carbon paste, clay, and metal powder composites, for applications in biosensing, molecular electronics, and molecular recognition [162–167].

Among the conducting polymers, the most intensively studied are homo- and copolymers based on pyrrole, substituted pyrroles such as N-alkylpyrroles, N-(2-carboxyethyl)pyrrole, pyrrole-2-carboxylic acid, and 1-pyrrolylpropanol, as well as nitrogen- and sulfur-containing systems such as polyaniline (PAN), poly(1,2-diaminobenzene), polythianaphthene, and poly(metallophthalocyanins) [168–203]. Depending on the monomer type, electrostatic encapsulation, surface adsorption, covalent coupling, and biotin-avidin/streptavidin conjugation can be employed, and mixed or stratified multienzyme sensors can be readily constructed on substrates such as platinum, gold, graphite, electrodeposited metals, etc.

Most studies have focused on amperometric glucose sensors based on GOx or GOx/peroxidase [168–182]. However, a wide variety of other enzymes, including peroxidase, tyrosinase, choline oxidase, sarcosine oxidase, cholesterol oxidase, xanthine oxidase, glutamate oxidase, glutamate dehydrogenase, nitrate reductase, cholinesterase, cholesterol esterase, uricase, urease, creatinase, and creatininase, have also been employed for bioelectrosensor construction [183–203]. Representative examples are outlined in Table 2.

Nonconducting polymers are exemplified by overoxidized polypyrrole (oxPPy), poly(o-phenylenediamine) (o-PPD), poly(m-phenylenediamine) (m-PPD), poly(1,2-diaminobenzene) (1,2-PDAB), poly(1,3-diaminobenzene) (1,3-PDAB), poly(3,4-dihydroxybenzaldehyde), poly(3,3′-diaminobenzidine), poly(o-aminophenol), and other polyphenols [162–166]. The electropolymers, especially PPD and PDAB, can readily be obtained via one-step procedures as highly adherent and mechanically robust films that show high working stabilities.

Mann enzymes of relevance to clinical analyses, e.g., GOx, L-glutamate oxidase, L-lactate oxidase, L-glutamate dehydrogenase, aldehyde dehyrogenase, urease, and creatininase, have been encapsulated within, adsorbed onto, or covalently attached to nonconducting polymers, especially PPD and PDAB films, deposited onto conducting organic salt, graphite, platinum, or polyimide electrodes [204–217]. Results obtained to date demonstrate that many nonconducting film-based biosensors display good permselectivities, thereby conferring resistance to fouling by proteins, and providing protection from interference by contaminating electroactive species such as ascorbate, urate, and acetaminophen

[162–167,204–217]. These are very significant advantages for clinical sensors, which have to maintain high sensitivities, linearities, and long-term stabilities when operated with complex biological samples. In addition, flexible operation in aqueous and aqueous-organic media, as well as polar organic solvents, has also been reported [162–167]. Examples are cited in Table 5.

Despite the low degree of structural organization of electropolymerized assemblies, the low cost, technical simplicity, and versatility of the technique are highly attractive for commercial applications. Indeed, the high stability, efficient wiring, and rugged operation of sandwich-type conducting/nonconducting electropolymer constructs makes them eminently suitable to the mass fabrication of clinical use bioelectrosensors.

E. Microfabrication: Bioarrays and Biochips

The aforementioned thin film fabrication techniques, especially LB/LS films, BLMs, and protein-electropolymer films, have found lucrative applications in the biochip and micro-biosensor arenas [37,38,218–227]. In particular, advances in genomics have created a need for high-sensitivity, high-throughput methods based on protein and nucleic acid microarray technology, for such tasks as DNA sequencing and comparative nucleic acid hybridization analysis [218–225,228–235]. Likewise, advanced diagnostics now require reliable, fast-response, nano- and microsensor systems for in vivo and in vitro clinical use [37,38,226, 227,236–242].

These applications rely critically on micropatterning for creating spatially defined arrays using SAM patterning, photolithography, photopolymerization, etching, electrodeposition, microwriting, mechanical microspotting, ink jetting, etc. [37,38,218–245]. The introduction of soft lithographic techniques, especially microcontact printing and microstamping, is expected to considerably expand the avenues for bioarray construction, as this provides an economic and technically facile method for the mass production of micrometer scale, patterned arrays based on SAM, LB/LS, BLM, and LLA technologies [246–256].

Most commercial focus has concentrated on the functional genomics market, as evident from the availability of over 20 DNA and RNA array technologies with optical or electrical transduction for genome sequencing, gene expression studies, gene discovery, polymorphism mapping, mutation detection, transcription analysis, and pathogen detection and typing [37,38,218–225,228–235].

The sensor arena has also attracted much interest, especially with regard to implantable clinical devices, arrays for the detection and quantification of analytes in biological fluids, in vivo biomonitoring and drug testing, probes for pathogen typing, and reporter elements for biorecognition studies [37,38,226,227,236–245,257–271]. Although serious commercialization has as yet to be seen, the construction of micro-/nanoarrays of proteins such as laccase, peroxidase, GOx, cytochrome oxidase, urease, lactate oxidase, choline oxidase, ascorbate oxidase and luciferase, antibodies, cytochromes, bacteriorhodopsin, and even living cells [236–245,257–271] holds great promise for prospective sensor applications.

In addition, microreactor technologies based on etching and lithographic techniques are now sufficiently developed to allow the construction of microcapillary and microgroove reactor systems, and allow for their integration with microelectronics [247,249,272–274]. Thus, GOx, urease, lactate oxidase, trypsin, carboxypeptidase Y, and DNA polymerase have been used in microreactors for the continuous determination of glucose, urease, and lactate, the microscale digestion/sequencing of proteins, and micro-PCR [272–283].

Table 5 Examples of Amperometric Biosensors Based on Electropolymerized Thin Films

Protein	Analyte/application	Electropolymer film type	Ref.
GOx	Glucose	Polypyrrole	169, 170, 172, 174
GOx	Glucose	Overoxidized polypyrrole	206
GOx	Glucose	Poly(N-(carboxyalkyl)pyrrole)	168, 175
GOx	Glucose	Poly(pyrrolealkylcarboxylate)	173
GOx	Glucose	Polyaniline	174, 177, 179
GOx	Glucose	Poly(o-phenylenediamine)	174, 204, 206
GOx	Glucose	Poly(metallophthalocyanin)	176, 211
GOx	Glucose	Polythianaphthene	178
GOx	Glucose	Poly(o-aminophenol)	208
GOx	Glucose	Poly(3,3'-diaminobenzidine)	210
Tyrosinase	Phenols	Substituted polypyrrole	180
Tyrosinase	Phenols	Amphiphilic polypyrrole	187, 188
Tyrosinase	Phenols	Amphiphilic polypyrrole-laponite	186
Choline oxidase	Choline	N-substituted polypyrrole	180
GOx + peroxidase	Glucose; hydrogen peroxide	Amphiphilic polypyrrole	181
Urease	Urea	Polypyrrole	182, 198
Horseradish peroxidase	Hydrogen peroxide	Polyaniline–polyvinyl sulfonate	183, 184
Uricase	Uric acid	Polyaniline–polypyrrole	185
Creatininase + creatinase + sarcosine oxidase	Creatinine	Poly(1,3-diaminobenzene)	189
L-Glutamate oxidase + horseradish peroxidase	L-Glutamate	Polypyrrole	190
Xanthine oxidase	Xanthine	Poly(p-benzoquinone)	192
Choline oxidase	Choline	Poly(1,2-diaminobenzene)	195
Acetylcholinesterase	Acetylcholine	Poly(1,2-diaminobenzene)polyresorcinol	195
Cholesterol oxidase	Cholesterol	Polypyrrole	196
Lactate oxidase	Lactate	Poly(o-phenylenediamine)	205
Nitrate reductase	Nitrate	Viologen-polypyrrole	199–202
Nitrate reductase	Nitrate	Poly(thiophene-bipyridinium)	203
Aldehyde dehydrogenase	Aliphatic aldehydes	Poly(3,4-dihydroxybenzaldehyde)	214

Relying as it does on the pool of available TF and ordered bioassembly techniques, the microfabrication arena has seen a surge of commercialization in recent years. Applications will undoubtedly expand in the lucrative molecular biotechnology arena as bioarrays and microelectronics are more closely integrated, but rapid developments are also expected in the micro- and nanoanalysis and biomonitoring sectors as high-performance on-chip array technologies are realized.

III. TECHNOLOGIES FOR PRODUCING BULK BIOCATALYST ELEMENTS

The previous sections have dealt with technologies aimed at structurally defined TF catalytic elements, which, at least until now, have been applied almost exclusively in the sensor and diagnostic domains. Although these technologies will undoubtedly carry over to some extent into the biocatalyst arena, their technological costs are likely to restrict them to specialized applications, such as microarrays for catalyst screening and combinatorial biocatalysis, flow-through reactors for the modification of analytes, and multicatalytic micro- and nanoreactors for the multistep production of lucrative bioactive targets and biosynthetic intermediates.

For bulk catalyst usages, cheaper and less refined protocols are needed, and to this end several emerging approaches are expected to play a very prominent role in complementing and perhaps even ultimately replacing existing immobilized catalyst technologies.

A. Sol-Gel Bioencapsulates

One of the most notable developments in bioimmobilization has undoubtedly been the demonstration in the early 1990s that proteins such as cytochrome c, hemoglobin, and myoglobin could be entrapped within silica glasses to produce bioactive ceramics retaining the biological properties of the native proteins [284,285]. This remarkable extension of traditional sol-gel processing, which has long been used to fabricate optics, sensors, catalyst supports, coatings, and specialty polymers [281–289], has opened up the possibility of producing robust hybrid materials encapsulating the characteristic highly selective bioactivities of labile biologicals within mechanically and chemically robust inorganic glasses [290–305].

A large amount of work has since been done on integrating sol-gel chemistry and fabrication with biological catalysis and recognition, and to date enzymes, catalytic antibodies, noncatalytic proteins, polynucleic acids, and even live microbial, plant, and animal cells have been encapsulated, for applications as biocatalysts [291–323] and biosensors [284,285,291–304,324–368]. Most of the work to date has focused on silica glasses, but more recently encapsulation has been extended to Ti, Zr, and Al metallosilicates, metal oxides, alkylsiloxanes, functional siloxanes with hydroxyalkyl, aminoalkyl, amino acid, polyethylene glycol (PEG), ferrocene, and other moieties, and interpenetrating composites with organic polymers such as polyvinyl alcohol (PVA), polyglyceryl methacrylate, polyvinylpyrrolidone, and alginate [291–304]. Representative examples of sol-gel-encapsulated biologicals and their applications are given in Tables 6 and 7.

In its simplest form, sol-gel bioencapsulation proceeds as follows [291–304]:

1. An alkoxysiloxane precursor or mix (usually the methyl or ethyl ester) is partially or completely hydrolyzed to produce a sol composed of linear, branched and/or cyclic polysiloxane and/or polysilicate species. Acidic prehydrolysis, fol-

Table 6 Examples of Sol-Gel Encapsulated Biocatalysts

Biological catalyst	Matrix	Synthetic application	Ref.
Organophosphate hydrolase	Fn-Siloxane-SiO$_2$	Paraoxon detoxification	305
Proteinase K	FnS-Metallosilicate	Peptide synthesis	305
C. rugosa lipase-1	AlkS-SiO$_2$	Resolution of acids	305
Oxynitrilase	AlkS-IFnS-SiO$_2$	Hydrocyanation of aldehydes	305
Sialic acid aldolase + myokinase + pyruvate kinase + pyrophosphatase + CMP-sialate synthetase + sialyl transferase	FnS-Metallosilicate	Sialyl-trisaccharide synthesis	305
Subtilisin	SiO$_2$, metallosilicate	Peptide synthesis	307
Rabbit muscle aldolase	SiO$_2$, metallosilicate	Aldol condensations	307
Lipoxygenase	SiO$_2$	Hydroperoxidation of PUFAs	309
Cytochrome c, myoglobin, hemoglobin, peroxidase	SiO$_2$	Dibenzothiophene oxidation	311
Bacterial, fungal, and animal lipases	AlkS-SiO$_2$	Ester synthesis, resolutions	313–319
β-Glucosidase	Alginate-SiO$_2$	Glucoside hydrolysis	320
Catalase, trypsin, acid phosphatase, thermostable β-glucosidase	SiO$_2$	—	306, 309, 312
α-Chymotrypsin, α-galactosidase	SiO$_2$	Glycoside and peptide synthesis	305
Glycerol 3-phosphate oxidase, pyruvate decarboxylase	Metallosilicate	—	305
Thermolysin, alcohol dehydrogenase, pig liver esterase, phospholipase D	AlkS-SiO$_2$, Alk-siloxane-metallosilicate	—	305
Aspartate aminotransferase, tyrosinase, carboxypeptidase Y	FnS-SiO$_2$	—	305
β-Glucuronidase, penicillin acylase	IPN SiO$_2$ composites	—	305
Pseudomonas spp. cell extracts	SiO$_2$	Atrazine degradation	321
Pseudomonas spp. cells	SiO$_2$	Atrazine degradation	191
S. cerevisiae cells	FnS-metallosilicate	Ethyl 3-oxobutanoate reduction	305
C. vaginalis cells	SiO$_2$	Metabolite production	323
R. miehei cells	Metallosilicate	Hydroxylation	305
P. oleovorans cells	AlkS-SiO$_2$	Epoxide and sulfoxide synthesis	175
A. niger spores, S. salmonicolor cells	IPN FnS/SiO$_2$ composites	β-Oxidation and hydroxylation of PUFAs	305

FnS, functional siloxane; AlkS, alkylsiloxane.

Table 7 Examples of Sol-Gel Biosensor Systems

Biological catalyst	Matrix	Analyte	Sensor	Ref.
Cytochrome c	SiO_2	O_2; NO	OBS	285, 344–347
Myoglobin	SiO_2	O_2; CO; NO	OBS	285, 346–349
Hemoglobin	SiO_2	O_2; CO; NO	OBS	285, 346–349
Cu-Zn SOD	SiO_2	Cyanide	OBS	285, 348, 349
Phycobiliproteins	SiO_2	—	OBS	366, 367
Bacteriorhodopsin	SiO_2	—	OBS	362–365
Aequorin	SiO_2	Calcium	OBS	361
Nitrate reductase	SiO_2	Nitrate	OBS	360
Bacteriorhodopsin + urease/acetylcholine esterase/penicillinase	SiO_2	Urea/acetylcholine/penicillin	OBS	365
Alcohol dehydrogenase	SiO_2	Alcohols; aldehydes	OBS	353
Lactate dehydrogenase	SiO_2	Pyruvate	OBS	354
Glucose 6-phosphate, dehydrogenase	SiO_2, IPN PVA-*FnS*-SiO_2	Glucose 6-phosphate	OBS	305
Oxalate oxidase	SiO_2	Oxalate	OBS	350
GOx	SiO_2, C-*FnS*-/*AlkS*-SiO_2, C-Ru-SiO_2, Au-*AlkS*, V_2O_5	Glucose	OBS, EBS	324, 326–339
Laccase	SiO_2	Phenols	OBS	352
Lactate oxidase	SiO_2, C-SiO_2	Lactate	OBS; EBS	352
Peroxidase (HRP)	SiO_2, C-SiO_2	H_2O_2; cyanide	OBS; EBS	342, 343, 355
Glucose oxidase + peroxidase	SiO_2, C-SiO_2	Glucose	OBS; EBS	325, 340, 341
Glucose oxidase + galactose oxidase + lactate oxidase + peroxidase	IPN PVA-*FnS*-SiO_2 composite	Glucose, galactose, lactose, lactate	OBS	305
Catalase	SiO_2	H_2O_2	OBS	345
Acetylcholine esterase + choline oxidase	IPN PVA-Pd-C-*FnS*-SiO_2	Acetylcholine, choline	EBS	305
Urease	SiO_2	Urea	OBS	359
Parathion hydrolase	SiO_2	Organophosphate	OBS	356, 357
Cholinesterase	SiO_2	Organophosphate	OBS	126, 127

FnS, functional siloxane; *AlkS*, alkylsiloxane; OBS, optical biosensor; EBS, electrobiosensor.

lowed by partial or complete removal of the formed alcohol, generally gives the best results.

2. Addition of a buffered solution or suspension of the biological to this sol, together with additives (PVA, PEG, graphite, etc.), followed by storage until gelation occurs to give the "hydrogel," consisting of a crosslinked polysiloxane/polysilicate matrix with the entrapped biological.

3. Aging of the hydrogel in a closed system, to allow further matrix crosslinking, and pore network development and maturation.

4. Where dense and/or dry sol-gels are required, the hydrogel is dried under controlled conditions (pinhole drying, freeze-drying, etc.) so as to dehydrate the matrix and allow for completion of crosslinking and structural consolidation, to give the final "xerogel" product. When employing live cells, the aged hydrogel is used as such, as drying typically leads to partial or complete loss of viability.

The evidence gathered to date suggests that depending on the precursor, biological, and sol-gel synthesis conditions, encapsulation can occur via adsorption onto macromonomeric polysilicate/polysiloxane species during particle formation, by adsorption/entrapment in the course of network formation, or through surface capture and partial embedding into the consolidating matrix [291–304]. However, it is equally clear that little is known for certain about the exact nature of protein–matrix interactions during sol-gel production or, indeed, how or even if these can be usefully manipulated to influence the activity and stability profiles of the resulting biogel.

Standard sol-gel bioencapsulation protocols are hampered by several problems [288–305]:

1. The low water solubility and reactivity of traditional sol-gel precursors typically necessitates the use of cosolvents and catalysts respectively, both of which can adversely affect the biological and lead to inactivation [286–305].

2. Hydrolysis liberates alcohols, which are deleterious to bioactivity, and their resulting evaporation generates shrinkages (30–80% v/v) and pore collapse during xerogel formation, resulting in steric compression of the entrapped biological and reduced mass transfer, thereby compromising biocatalytic efficiency by 20–90% [291–304].

3. Such complications together with difficulties encountered in controlling aging effects have meant that the reproducible production of stable, high-activity bioencapsulates have until recently proved somewhat elusive.

Recent advances in prehydrolysis protocols, preparation of alcohol-free sols, and optimization of gelation, aging, and drying conditions have substantially improved the performance of bioencapsulates, to the stage where one can expect to gain immobilization efficiencies of 40–80% for proteins and 20–50% for cells [290–304]. Also, we have described biocompatible sol-gel precursors, based on polyol esters of silicates and siloxanes, that appear to address several of the above issues, and permit the fabrication of highly active and stable sol-gel polymers doped with proteins and whole cells, with activity recoveries of 74–99% [305].

However, structure development, templating phenomena, pore formation and consolidation, and aging processes are still poorly understood, and the need for their effective manipulation and control is critical to the production of efficient encapsulates. In spite of the aforementioned difficulties, sol-gel encapsulation has many unique advantages when compared with other immobilization protocols, as indicated in Table 8.

Table 8 Critical Features of Sol-Gel Bioencapsulation

Aspect	Sol-gel feature	Advantages	Drawbacks/hurdles
Protocols	Facile methodology	No modification of biological required. Enables entrapment under very mild conditions that minimize toxicity effects and denaturation.	—
Compatibility	Biocompatible precursors/protocols		More efficient methods required for live cells
Flexibility	Wide variety of organosilicons available for sol-gel use	(i) Wide range of chemistries possible—silica, metallosilicates, functional siloxanes, and composites; (ii) polymer can be tailored to the biological and the intended application.	Rather few applications reported for the advanced sol-gel matrices
Loading	Loadings can be varied	Low or high (up to 20% w/w) catalytic density elements can be fabricated—important for sensors/arrays, combinatorial biocatalysis, and high-throughput bioreactors.	Higher loadings can reduce bioactivity, efficiency, and matrix integrity
Operation	Compatible with a range of media	Operable in aqueous, mixed, low-water, and anhydrous liquid media, and with gaseous substrates.	Few applications in nonconventional media
Encapsulation	Entrapment of biological in rigid polymer cage	(i) Leaching of biological almost eliminated; (ii) operational and storage stability enhanced; (iii) degradative/denaturation processes inhibited; (iv) sensitive membrane proteins, and even complete biosynthetic pathways can be immobilized.	(i) Steric restriction can reduce activity; (ii) issue of internal diffusional resistance
Multicatalytics	Multicatalytic elements possible	Coupled reactions for synthetic applications, and multiuse elements for sensors can be realized.	Problem of divergent encapsulation conditions
Advanced fabrication	Fabrication in a variety of forms	(i) Monoliths, films, and micromolded/microprinted structures accessible; (ii) glasses, micro-/macroporous foams, flexible rubbers, and rigid plastics available; (iii) microstructure and bulk properties can be engineered via precursor mix, hydrolysis and aging conditions, and templating methods.	Full demonstration of advanced sol-gel biofabrication required

Developments in the 1990s have firmly demonstrated the utility and applicability of sol-gel bioencapsulation, and a range of commercial applications are expected in the near future in bioorganic synthesis, biosensors, diagnostics, and environmental technology. Indeed, the commercial introduction of sol-gel lipase biocatalysts [313–316] and the demonstration of the robustness and flexibility of sol-gel biosensors are clear indications of the viability of the technology. Although the relative infancy of the field has meant that issues such as porosity, mechanical robustness, and the effective long-term stabilization of encapsulated biologicals are far from fully addressed, the prospect for engineering highly efficient, stable, and robust biodoped sol-gels is fully realizable.

B. Crosslinked Enzyme Crystals

Following the demonstration of the exceptional stability and activity of crosslinked crystals (CLECs) of thermolysin in aqueous and organic media, and its application in peptide synthesis [369], this novel technology has been rapidly commercialized and shown to be applicable to a variety of enzymes [370–383].

Currently, *Candida rugosa* and *Pseudomonas cepacia* lipases, thermolysin, subtilisin, and penicillin acylase are commercially available from Altus Biologics Inc. and have proved to be very effective catalysts for the synthesis of peptides and esters, and the resolution of acids, alcohols, amines, and amino acids in a variety of media [370–372,374–377,379–381]. In addition, proteinase K, lactate dehydrogenase, horse liver alcohol dehydrogenase, and rabbit muscle aldolase have also been crosslinked and shown to operate efficiently in synthetic reactions [305,374,378,382,383]. The unique features, and the advantages and drawbacks of using CLECs, are outlined in Table 9.

Although CLECs are of recent advent, their commercial introduction has generated enormous industrial interest. Indeed, despite earlier concerns about the high costs of CLEC catalysts, in-use cost estimates suggest that the high activities and stabilities of CLECs enable very high specific productivities such that economic feasibility can be achieved not only for low-volume, high-value products, but also for high-volume, low-cost targets with critical economic margins. In fact, it is clear that most companies in the biocatalysis and biosensor arenas are conducting investigatory or pilot trials with CLECs, and it is expected that major commercial applications will soon be forthcoming.

C. Protein-Polymer Composites

Several groups have reported on the production of biocomposite materials based on covalent protein-polymer conjugates or enzymes entrapped in bulk polymers [390–407]. For example, the proteases subtilisin, thermolysin, proteinase K, papain, and α-chymotrypsin, as well as *Eco*RI and an MAB, can be modified via reductive amination-mediated coupling to poly(2-*N*-methacrylamido-2-deoxyglycose) polymers of glucose, galactose, and ribose, to give the corresponding carbohydrate-protein conjugates (CPCs) as dispersed polymerics [390–395]. These composites are highly active in aqueous and mixed media, as well as in low-water hydrophilic solvents such as alcohols and acetonitrile, and demonstrate elevated thermostabilities as compared to the native enzymes. The protease CPCs have proved to be efficient catalysts for performing amino acid and peptide fragment condensations in low-water media, where the native enzymes and conventional immobilizates often show reduced activities and stabilities [390–395]. Although their dispersed nature complicates preparative applications of CPCs, suitable solid catalysts could in principle be obtained by copolymerization or by compounding into a second polymer phase.

Table 9 Advantages and Limitations of Crosslinked Enzyme Crystals

Aspect	CLEC feature	Advantage	Hurdles	Ref.
Purity	CLECs are pure	Critical factor for resolutins where the absence of contaminating enzymes/isozymes greatly increases enantioselectivities, e.g., lipase-mediated reactions.	Purification can be the most expensive and least productive step in CLEC manufacture.	370–375, 377, 379, 381
Density	Highest attainable catalytic densities	Asset for (i) low-volume, fast-flow bioreactors; (ii) biosensors and microcatalytic arrays.	Many applications require a low catalytic density.	370–375
Stability	Broad operational range	(i) Efficient operation in aqueous, aqueous-organic, low-water organic, and anhydrous organic media; (ii) often notable operational and storage stabilities; (iii) catalyst/medium engineering techniques used for free/immobilized enzymes are valid for CLECs.	(i) Performance can depend on crystallization, crosslinking, storage, and conditioning regimens; (ii) lack of engineerable matrix limits catalyst modifications.	370–388
Generality	Based on standard 3D crystallization methods	A vast number of proteins have been crystallized. This knowledge base should provide routes to CLECs for most enzymes of interest.	To date, only nine enzymes have been produced as CLECs.	370–385
Fabrication	Standard 3D crystallization protocols	(i) Many enzymes are polymorphic—prospect of tailoring CLEC morphology to application; (ii) crystallization with activators/cofactors possible for lipases, oxidoreductases, etc.	Little is currently known about the influence of crystal macro-/microstructure on CLEC activity and stability.	373, 377, 379, 381
Advanced fabrication	Fabrication flexibility	(i) Prospect of 2D crystallization for sensors/arrays, and cocrystallization for multicatalytics; (ii) CLECs can be encased in polymers to give more robust and easily handled catalytic composites.	(i) 2D- and cocrystallization remain to be applied to CLECs; (ii) small crystal size required for composites—also issue of multiple diffusional limitations.	383, 389

Parallel approaches have been described for the preparation of polyacrylate-protease conjugates [396–400]. Acryloylation of subtilisin and α-chymotrypsin, followed by mixed polymerization with methyl methacrylate, vinyl acetate, styrene, or ethylvinyl ether, provides insoluble, doped polymethyl methacrylate, polyvinyl acetate, polystyrene, and poly-ethylvinyl ether polymers [396]. These biocatalytic plastics perform especially well in hydrophilic and hydrophobic solvents, and have been used for peptide synthesis and the regioselective acylation of sugars and nucleosides. Similarly, modification of subtilisin and thermolysin with PEG monomethacrylate, then copolymerization with methyl methacrylate and trimethylolpropane trimethacrylate furnishes protease-polymethyl methacrylate plastics, which show good activities and stabilities in aqueous, mixed, and low-water and anhydrous organic media [397–400]. The protein-acrylate composites are unique in that they enable catalytic densities as high as 50% w/w.

We have recently found that lipases from *C. rugosa, P. cepacia, Rhisopus arrhizus, Mucor javanicus, Aspergillus niger, Pseudomonas fluorescens, P. roquefortii, Rhizomucor miehei,* and *Candida antartica,* along with other enzymes such as β-glucosidase, thermolysin, papain, and organophosphate hydrolase, can be efficiently converted to catalytic silicone biocomposites [401]. The technique relies on initial adsorption or covalent binding onto polyhydroxymethylsiloxane, followed by incorporation into a condensation-cure silicone polymer. This route provides efficient and rugged heterogeneous catalysts with catalytic densities of up to 10% w/w, which exhibit excellent activity and stability profiles in a variety of media. The composites can be obtained as rigid monoliths; tough, elastic sheets and films; hard particulates; and even as macroporous foams if a blowing agent such as polymethylhydrogensiloxane is included in the composition.

Foamed polyurethane-enzyme conjugates can be produced via the aqueous polymerization of isocyanate-polyol polyurethane compositions in the presence of proteins [402–405]. This novel approach has been used to generate macroporous sponges bearing acetylcholinesterase, butyrylcholinesterase, and organophosphate hydrolases. These materials are very promising biocatalysts for use in the continuous detoxification of organophosphate pesticide and nerve agent feeds, air filters for personal protection, and sponges for personal decontamination and environmental cleanup applications [402–405].

The use of conventional polymerization protocols, the accessibility of high catalytic density elements in a variety of forms, and the excellent performances of the biocomposites in nonaqueous media are particular attractions of these technologies. Also, the nonaqueous polymerization chemistries of the acrylate and silicone composites lends them to specialty, high-performance biocatalytic coatings and filters, and, more intriguingly, for micromolding catalyst and sensor arrays. Indeed, a variety of applications can be envisaged—from organic phase biosensors and microarrays for combinatorial biocatalysis, to high-performance bulk catalysts for synthetic and environmental applications.

IV. CONCLUSION

The bioimmobilization domain has seen some remarkable advances over the last decade, principally in the sensor and diagnostic arenas. The introduction and refinement of thin film fabrication methods based on LB/LS, SAM, BLM, LLA, crystallines, and EPs, together with advances in ultrastructural characterization techniques, have enabled the realization of exquisitely ordered protein assemblies, which are requisite for advanced sensing applications. Furthermore, the increasing drive toward device minaturization has resulted in the development of efficient commercial LB/LS, SAM, and EP microarray systems for diagnostic and analytical applications. Commercial products will undoubtedly flourish in molecular biotechnology and clinical analysis. Furthermore, and in addition, a new generation of biochips based on nanoscale assemblies should appear, as TF technologies are further scaled down and fully integrated with electronics.

Although, by comparison, the bulk catalyst has seen rather modest technical advances, the advent of sol-gel encapsulation, CLECs, and protein–polymer composites has nonetheless made a dramatic impact on the immobilized biocatalyst arena. This sector increasingly demands more versatile and rugged biocatalysts that can function efficiently under a variety of aggressive operational conditions yet demonstrate high long-term stabilities, especially for high-volume, low-value targets, as well as low-volume, high-value specialties requiring exotic catalytic schemes. Results obtained to date with the new technologies clearly demonstrate their commercial potential, not only for bulk catalysts as evident from the rapid industrial uptake of sol-gels and CLECs, but also, to a lesser extent, for micro- and macroscale biosensors.

Finally, one can also expect some carryover from the high-technology, TF arena into the industrial biocatalysis sector. Most immediately, one can envisage efficient syntheses of lucrative pharmaceuticals based on modular, microfabricated bioreactors; EP-protein assemblies for bioelectrosyntheses; "off-the-shelf" thermostable, multicomponent heterogenous catalysts derived from supported LB/LS and SAM architectures; and perhaps even solvent- and high-temperature-resistant 2D enzyme crystal assemblies for harsh operating environments.

REFERENCES

1. K Drauz, H Waldmann. Enzyme Catalysis in Organic Synthesis. Weinheim: VCH, 1995.
2. C-H Wong, GM Whitesides. Enzymes in Synthetic Organic Chemistry, Tetrahedron Organic Chemistry Series, Vol. 12. Oxford: Elsevier, 1994.
3. L Poppe, L Novák. Selective Biocatalysis. Weinheim: VCH, 1992.
4. HW Blanch, DS Clark. Applied Biocatalysis, Vol. 1. New York: Marcel Dekker, 1991.
5. HG Davies, RH Green, DR Kelly, SM Roberts. Biotransformations in Preparative Organic Chemistry. London: Academic Press, 1989.
6. HL Holland. Organic Synthesis with Oxidative Enzymes. New York: VCH, 1992.
7. A Gabelman. Bioprocess Production of Flavour, Fragrance, and Colour Ingredients. New York: John Wiley, 1994.
8. T Scheper. Biotechnology of Aroma Compounds, Advances in Biochemical Engineering Biotechnology, Vol. 55. Heidelberg: Springer-Verlag, 1997.
9. MP Schneider. Enzymes as Catalysts in Organic Synthesis, NATO ASI Series C, Vol. 178. New York: NATO ASI, 1986.
10. S. Servi. Microbial Reagents in Organic Synthesis. Berlin: Kluwer Academic, 1992.
11. E Kress-Rogers. Handbook of Biosensors and Electronic Noses. Boca Raton, FL: CRC Press, 1996.
12. GF Bickerstaff. Immobilization of Enzymes and Cells. Totowa, NJ: Humana Press, 1997.
13. RF Taylor. Protein Immobilization. New York: Marcel Dekker, 1991.
14. K Mosbach. Methods in Enzymology, Vol. 135. New York: Academic Press, 1987.
15. J Woodward. Immobilized Cells and Enzymes: A Practical Approach. Oxford: IRL Press, 1985.
16. AI Laskin. Enzymes and Immobilized Cells in Biotechnology. London: Benjamin Cummings, 1985.
17. K Mosbach. Methods in Enzymology, Vol. 44. New York: Academic Press, 1976.
18. CA White, JF Kennedy. Enzyme Microb Technol 2:82–90, 1980.

19. C Webb. Studies in Viable Cell Immobilization. New York: Academic Press, 1996.
20. A Groboillot, DK Boadi, D Poncelot, RJ Neufeld. Crit Rev Biotechnol 14:75–107, 1994.
21. K Nilsson. Trends Biotechnol 5:73–78, 1987.
22. APF Turner. Ann Chim 87:255–260, 1997.
23. RE Kunz. Sens Actuators B B38:13–28, 1997.
24. Y Liu, T Yu. J Macromol Sci Rev Macromol Chem Phys C37:459–500, 1997.
25. P Skladal. Electroanalysis 9:737–745, 1997.
26. A Sharma, KR Rogers. Food Technol Biotechnol 34:113–123, 1996.
27. L Campanella, M Tomassetti. Food Technol Biotechnol 34:131–141, 1996.
28. JR Woodward, RB Spokane. Chem Anal 148:227–264, 1998.
29. AF Collings, F Caruso. Rep Prog Phys 60:1397–1445, 1997.
30. DR Purvis, D Pollard-Knight, PA Lowe. Chem Anal 148:165–224, 1998.
31. G Palleschi, G Lubrano, GG Guilbault. Chem Anal 148:77–98, 1998.
32. KR Rogers, EN Koglin. NATO ASI Series 2 38:335–349, 1997.
33. S Ramanathan, M Ensor, S Daunert. Trends Biotechnol 15:500–516, 1997.
34. SS Pathak, HFJ Savelkoul. Immunol Today 18:464–467, 1997.
35. V Tvarozek. NATO ASI Series 2 38:1431–1471, 1997.
36. C Nicolini. Trends Biotechnol 15:395–401, 1997.
37. C Nicolini. Biosens Bioelectron 10:105–127, 1995.
38. C Nicolini. Molecular Bioelectronics. Singapore: World Scientific Publishing Co, 1996.
39. K Kurihara. Colloids Surf A 123–124:425–432, 1997.
40. T Richardson. In: W Jones, ed. Org Mol Solids. Boca Raton, FL: CRC, 1997, pp 63–90.
41. CT Seip, H Byrd, JK Pike, S Whipps, DR Talham. NATO ASI Series C 485:27–37, 1996.
42. MK DeArmond, GA Fried. Prog Inorg Chem 44:97–142, 1997.
43. TS Berzina, VI Troitsky, A Petrigliano, D Alliata, AY Tronin, C Nicolini. Thin Solid Films 284–285:757–761, 1996.
44. C Nicolini. Ann NY Acad Sci 799:297–310, 1996.
45. V Erokhin, LA Feigin. Prog Colloid Polym Sci 85:47–51, 1991.
46. G Decher. Science 277:1232–1237, 1997.
47. JA He, LJ Jiang, L Jiang, ZC Bi, JR Li. Langmuir 12:1840–1850, 1996.
48. IM Pepe, C Nicolini. J Photochem Photobiol B 33:191–200, 1996.
49. J Miyake, M Hara. Mater Sci Eng C C4:213–219, 1997.
50. J Miyake, M Hara. Adv Biophys 34:109–126, 1997.
51. Y Yasuda, H Sugino, H Toyotama, Y Hirata, M Hara, J Miyake. Bioelectrochem Bioenerg 34:135–139, 1994.
52. A Riccio, M Lanzi, C Antolini, C De Nitti, C Tavani, C Nicolini. Langmuir 12:1545–1549, 1996.
53. C Nicolini, V Erokhin, F Antolini, P Catasti, P Facci. Biochim Biophys Acta 1158:273–278, 1993.
54. C Nicolini. Biosens Bioelectron 10:105–127, 1995.
55. V Erokhin, S Vacula, C Nicolini. Thin Solid Films 238:88–94, 1994.
56. A Petrigliano, A Tronin, C Nicolini. Thin Solid Films 284–285:752–756, 1996.
57. L Marron-Brignone, RM Morelis, PR Coulet. Langmuir 12:5674–5680, 1996.
58. N Dubreuil, S Alexandre, C Fiol, JM Valleton. J Colloid Interface Sci 181:393–398, 1996.
59. A Eremenko, I Kurochkin, S Chernov, A Barmin, A Yaroslavov, T Moskvitina. Thin Solid Films 260:212–216, 1995.
60. SY Zaitsev. Sens Actuators B B24:177–179, 1995.
61. SY Zaitsev. Colloids Surf A 75:211–216, 1993.
62. S Arisawa, T Arise, R Yamamoto. Thin Solid Films 209:259–263, 1992.
63. P Pal, D Nandi, TN Misra. Thin Solid Films 239:138–143, 1994.
64. AP Girard-Egrot, RM Morelis, PR Coulet. Langmuir 13:6540–6546, 1997.
65. AP Girard-Egrot, RM Morelis, PR Coulet. Langmuir 14:476–482, 1998.
66. JW Choi, J Min, JW Jung, HW Rhee, WH Lee. Mol Cryst Liq Cryst Sci Technol A 295: 451–454, 1997.

67. AP Girard-Egrot, RM Morelis, PR Coulet. Thin Solid Films 292:282–289, 1997.
68. T Osa. Appl Biochem Biotechnol 41:41–49, 1993.
69. JJ Ramsden, S Karrasch. Sens Mater 8:469–476, 1996.
70. J Anzai, S Lee, T Osa. Bull Chem Soc Jpn 62:3018–3020, 1989.
71. LA Samuelson, DL Kaplan, KA Marx, P Miller, DM Galotti, J Kumar, SK Tripathy. Mater Res Soc Symp Proc 218:159–163, 1991.
72. S Lee, J Anzai, T Osa. Sens Actuators B 12:153–158, 1993.
73. KA Marx, LA Samuelson, M Kamath, JO Lim, S Sengupta, D Kaplan, J Kumar, SK Tripathy. Adv Chem Ser 240:395–412, 1994.
74. K Owaku, M Goto, Y Ikariyama, M Aizawa. Anal Chem 67:1613–1616, 1995.
75. A Tronin, T Dubrovsky, G Radicchi, C Nicolini. Sens Actuators B B34:276–282, 1996.
76. T Dubrovsky, A Tronin, S Dubrovskaya, A Guryev, C Nicolini. Thin Solid Films 284–285: 698–702, 1996.
77. W Schuhmann, SP Heyn, HE Gaub. Adv Mater 3:388–391, 1991.
78. S Paddeu, V Erokhin, C Nicolini. Thin Solid Films 284–285:854–858, 1996.
79. SE Creager, KG Olsen. Anal Chim Acta 307:277–289, 1995.
80. E Delamarche, B Michel, HA Biebuyck, C Gerber. Adv Mater 8:719–729, 1996.
81. D Mandler, I Turyan. Electroanalysis 8:207–213, 1996.
82. DR Jung, AW Czanderna. Crit Rev Solid State Mater Sci 19:1–54, 1994.
83. DL Allara. Biosens Bioelectron 10:771–783, 1995.
84. A Ulman. Chem Rev 96:1533–1554, 1996.
85. RM Crooks, AJ Ricco. Acc Chem Res 31:219–227, 1998.
86. N Patel, MC Davies, M Hartshorne, RJ Heaton, CJ Roberts, SJB Tendler, PM Williams. Langmuir 13:6485–6490, 1997.
87. J Spinke, M Liley, FJ Schmitt, HJ Guder, L Angermaier, W Knoll. J Chem Phys 99:7012–7019, 1993.
88. YW Lee, J Reed-Mundell, JE Zull, CM Sukenik. Langmuir 9:3009–3014, 1993.
89. P Wagner, M Hegner, P Kernen, F Zaugg, G Semenza. Biophys J 70:2052–2066, 1996.
90. M Maskus, J Tirado, J Hudson, R Bretz, HD Abruna. NATO ASI Series C 485:337–353, 1996.
91. AJ Guiomar, SD Evans, JT Guthrie. Supramol Sci 4:279–291, 1997.
92. S Sampath, O Lev. Adv Mater 9:410–413, 1997.
93. XD Dong, J Lu, C Cha. Bioelectrochem Bioenerg 42:63–69, 1997.
94. SF Hou, HQ Fang, HY Chen. Anal Lett 30:1631–1641, 1997.
95. I Willner, A Riklin, B Shoham, E Katz. Adv Mater 5:912–915, 1993.
96. S Rubin, JT Chow, JP Ferraris, TA Zawodzinski Jr. Langmuir 12:363–370, 1996.
97. P He, J Ye, Y Fang, J Anzai, T Osa. Talanta 44:885–890, 1997.
98. S Arnold, ZQ Feng, T Kakiuchi, W Knoll, K Niki. J Electroanal Chem 438:91–97, 1997.
99. ZQ Feng, S Imabayashi, T Kakiuchi, K Niki. J Chem Soc Faraday Trans 93:1367–1370, 1997.
100. J Li, G Cheng, S Dong. J Electroanal Chem 416:97–104, 1996.
101. LL Wood, SS Cheng, PL Edmiston, SS Saavedra. J Am Chem Soc 119:571–576, 1997.
102. SM Amador, JM Pachence, R Fischetti, JP McCauley Jr, AB Smith III, JK Blasie. Langmuir 9:812–817, 1993.
103. M Collinson, EF Bowden, MJ Tarlov. Langmuir 8:1247–1250, 1992.
104. C Ruan, F Yang, C Lei, J Deng. Anal Chem 70:1721–1725, 1998.
105. J Li, J Yan, Q Deng, G Cheng, S Dong. Electrochim Acta 42:961–967 1997.
106. TM Cotton, J Zheng. Spectrosc Biol Mol Mod Trends 7:171–172, 1997.
107. S Dong, J Li. Bioelectrochem Bioenerg 42:7–13, 1997.
108. T Loetzbeyer, W Schuhmann, HL Schmidt. Bioelectrochem Bioenerg 42:1–6, 1997.
109. KT Kinnear, HG Monbouquette. Langmuir 9:2255–2257, 1993.
110. I Willner, A Riklin. Anal Chem 66:1535–1539, 1994.
111. L Bertilsson, HJ Butt, G Nelles, DD Schlereth. Biosens Bioelectron 12:839–852, 1997.

112. DD Schlereth. Sens Actuators B B43:78–86, 1997.
113. I Willner, N Lapidot, A Riklin, R Kasher, E Zahavy, E Katz. J Am Chem Soc 116:1428–1441, 1994.
114. J Madoz, BA Kuznetzov, FJ Medrano, JL Garcia, VM Fernandez. J Am Chem Soc 119:1043–1051, 1997.
115. F Pariente, C La Rosa, F Galan, L Hernandez, E Lorenzo. Biosens Bioelectron 11:1115–1128, 1996.
116. DL Pilloud, F Rabanal, BR Gibney, RS Farid, PL Dutton, CC Moser. J Phys Chem B 102:1926–1937, 1998.
117. MA Firestone, ML Shank, SG Sligar, PW Bohn. J Am Chem Soc 118:9033–9041, 1996.
118. C Duan, ME Meyerhoff. Mikrochim Acta 117:195–206, 1995.
119. J Spinke, M Liley, HJ Guder, L Angermaier, W Knoll. Langmuir 9:1821–1825, 1993.
120. C Duan, ME Meyerhoff. Anal Chem 66:1369–1377, 1994.
121. JM Kaufmann. NATO ASI Ser 2 38:107–114, 1997.
122. YC Yu, T Pakalns, Y Dori, JB McCarthy, M Tirrell, GB Fields. Meth Enzymol 289:571–587, 1997.
123. JJ Ramsden. Biosens Bioelectron 11:523–528, 1996.
124. M Snejdarkova, M Rehak, M Babincova, DF Sargent, T Hianik. Bioelectrochem Bioenerg 42:35–42, 1997.
125. T Hianik, M Snejdarkova, VI Passechnik, M Rehak, M Babincova. Bioelectrochem Bioenerg 41:221–225, 1996.
126. JD Burgess, MC Rhoten, FM Hawkridge. Langmuir 14:2467–2475, 1998.
127. T Parpaleix, JM Laval, M Majda, C Bourdillon. Anal Chem 64:641–646, 1992.
128. C Yeung, D Leckband. Langmuir 13:6746–6754, 1997.
129. PL Edmiston, SS Saavedra. J Am Chem Soc 120:1665–1671, 1998.
130. T Hianik, Z Cervenanska, T Krawczynski vel Krawczyk, M Snejdarkova. Mater Sci Eng C C5:301–305, 1998.
131. DP Nikolelis, MG Tzanelis, UJ Krull. Biosens Bioelectron 9:179–188, 1994.
132. DP Nikolelis, CG Siontorou. J Autom Chem 19:1–8, 1997.
133. D Ottenbacher, R Kindervater, P Gimmel, B Klee, F Jaehnig, W Goepel. Sens Actuators B B6:192–196, 1992.
134. M Sugawara, N Sugao, Y Umezawa, Y Adachi, K Taniguchi, H Minami, M Uto, K Odashima, EK Michaelis, T Kuwana. Proc Electrochem Soc 11:268–279, 1993.
135. M Sugawara, A Hirano, R Ayumi, J Nakanishi, K Kawai, H Sato, Y Umezawa. Biosens Bioelectron 12:425–439, 1997.
136. J Miyake, M Hara. Adv Biophys 34:109–126, 1997.
137. G Puu, I Gustafson, E Artursson, PÅ Ohlsson. Biosens Bioelectron 10:463–476, 1995.
138. C Ganea, J Tittor, E Bamberg, D Oesterhelt. Biochim Biophys Acta 1368:84–96, 1998.
139. A Diederich, M Loesche. Adv Biophys 34:205–230, 1997.
140. G Decher, B Lehr, K Lowack, Y Lvov, J Schmitt. Biosens Bioelectron 9:677–684, 1994.
141. Y Lvov, K Ariga, T Kunitake. Chem Lett 12:2323–2326, 1994.
142. M Onda, Y Lvov, K Ariga, T Kunitake. Biotechnol Bioeng 51:163–167, 1996.
143. A Riklin, I Willner. Anal Chem 67:4118–4126, 1995.
144. J Hodak, R Etchenique, EJ Calvo, K Singhal, PN Bartlett. Langmuir 13:2708–2716, 1997.
145. F Mizutani, S Yabuki, Y Hirata. Anal Chim Acta 314:233–239, 1995.
146. F Caruso, K Niikura, DN Furlong, Y Okahata. Langmuir 13:3427–3433, 1997.
147. UB Sleytr, M Sara, P Messner, D Pum. Ann NY Acad Sci 745:261–269, 1994.
148. D Pum, M Sara, UB Sleytr. In: UB Sleytr, ed. Immobilised Macromolecules. London: Springer-Verlag, 1993, pp 141–160.
149. UB Sleytr, D Pum, M Sara. Ady Biophys 34:71–79, 1997.
150. UB Sleytr, D Pum, M Sara, P Messner. Electr Magn Biol Med Rev Res Pap World Congr 195–196, 1993.
151. S Kupcu, A Neubauer, C Hoedl, D Pum, M Sara, UB Sleytr. NATO ASI Ser E 252:57–66, 1993.

152. K Taga, R Kellner, U Kainz, UB Sleytr. Anal Chem 66:35–39, 1994.
153. A Neubeuer, D Pum, UB Sleytr. Anal Lett 26:1347–1360, 1993.
154. A Neubauer, C Hoedl, D Pum, UB Sleytr. Anal Lett 27:849–865, 1994.
155. D Pum, A Neubauer, UB Sleytr, S Pentzien, S Reetz, W Kautek. Ber Bunsen-Ges 101:1686–1689, 1997.
156. A Neubauer, S Pentzien, S Reetz, W Kautek, D Pum, UB Sleytr. Sens Actuat B B40:231–236, 1997.
157. A Breitweiser, S Kuepcue, S Howorka, S Weigert, C Langer, K Hoffmann-Sommergruber, O Scheiner, UB Sleytr, M Sara. Biotechniques 21:918–925, 1996.
158. M Sara, UB Sleytr. Appl Microbiol Biotechnol 38:147–151, 1992.
159. S Kuepcue, C Mader, M Sara. Biotechnol Appl Biochem 21:275–286, 1995.
160. D Pum, G Stangl, C Sponer, K Riedling, P Hudek, W Fallmann, UB Sleytr. Microelectron Eng 35:297–300, 1997.
161. D Pum, G Stangl, C Sponer, W Fallmann, UB Sleytr. Colloids Surf B 8:157–162, 1997.
162. PH Treloar, IM Christie, PM Vadgama. Biosens Bioelectron 10:195–201, 1995.
163. M Trojanowicz, T Krawczynski, T Krawczynski vel Krawczyk. Mikrokhim Acta 121:167–181, 1995.
164. ER Reynolds, RJ Geise, AM Yacynch. ACS Symp Ser 487:186–200, 1992.
165. PN Bartlett, JM Cooper. J Electroanal Chem 362:1–12, 1993.
166. SA Wring, JP Hart. Analyst 117:1215–1229, 1992.
167. R Teasdale, GG Wallace. Analyst 118:329–334, 1993.
168. SE Wolowacz, BFY Yon-Hin, CR Lowe. Anal Chem 64:1541–1545, 1992.
169. BFY Yon-Hin, M Smolander, T Crompton, CR Lowe. Anal Chem 65:2067–2071, 1993.
170. MC Shin, HS Kim. Anal Lett 28:1017–1031, 1995.
171. A Cirulli, G Palleschi. Electronanalysis 9:1107–1112, 1997.
172. NF Almeida, EJ Beckman, MM Ataai. Biotechnol Bioeng 42:1037–1045, 1993.
173. M Yasuzawa, N Matsushita, H Satake, A Kunugi. Sens Actuat B 14:665–666, 1993.
174. U Ruedel, O Gaschke, K Cammann. Electroanalysis 8:1135–1139, 1996.
175. BFY Yon-Hin, CR Lowe. J Electroanal Chem 374:167–172, 1994.
176. Z Sun, H Tachikawa. Anal Chem 64:1112–1117, 1992.
177. PN Bartlett, PR Birkin. Anal Chem 66:1552–1559, 1994.
178. D Compagnone, G Federici, JV Bannister. Electroanalysis 7:1151–1155, 1995.
179. SY Lu, CF Li, DD Zhang, Y Zhang, ZH Mo, Q Cai, AR Zhu. J Electroanal Chem 364:31–36, 1994.
180. L Coche-Guerente, S Cosnier, C Innocent, P Mailley, JC Moutet, RM Morelis, B Leca, PR Coulet. Electroanalysis 5:647–652, 1993.
181. IC Popescu, S Cosnier, P Labbe. Electroanalysis 9:998–1004, 1997.
182. T Hianik, M Snejdarkova, Z Cervenanska, A Miernik, T Krawczylnski vel Krawczyk, M Trojanowicz. Chem Anal 42:901–906, 1997.
183. EI Iwuoha, DS De Villaverde, NP Garcia, MR Smyth, JM Pingarron. Biosens Bioelectron 12:749–761, 1997.
184. Y Yang, SJ Mu. Electroanal Chem 432:71–78, 1997.
185. S Uchiyama, H Sakamoto. Talanta 44:1435–1439, 1997.
186. JL Besombes, S Cosnier, P Labbe. Talanta 44:2209–2215, 1997.
187. S Cosnier, C Innocent. J Electroanal Chem 328:361–366, 1992.
188. S Cosnier, C Innocent. Bioelectrochem Bioenerg 31:147–160, 1993.
189. MB Madaras, RP Buck. Anal Chem 68:3832–3839, 1996.
190. S Yoshida, H Kanno, T Watanabe. Anal Sci 11:251–256, 1995.
191. T Tatsuma, M Gondaira, T Watanabe. Anal Chem 64:1183–1187, 1992.
192. G Arai, S Takahashi, I Yasumori. Denki Kagaku Oyobi Kogyo Butsuri Kagaku 61:893–894, 1993.
193. Q Li, S Zhang, J Yu. Anal Lett 28:2161–2174, 1995.
194. B Leca, RM Morelis, PR Coulet. Ann NY Acad Sci 750:109–111, 1995.

195. Z Huang, R Villarta-Snow, GJ Lubrano, GG Guilbault. Anal Biochem 215:31–37, 1993.
196. W Trettnak, I Lionti, M Mascini. Electroanalysis 5:753–763, 1993.
197. JL Besombes, S Cosnier, P Labbe, G Reverdy. Anal Chim Acta 317:275–280, 1995.
198. SB Adeloju, SJ Shaw, GG Wallace. Anal Chim Acta 281:611–620, 1993.
199. S Cosnier, C Innocent, Y Jouanneau. Anal Chem 66:3198–3201, 1994.
200. LM Moretto, P Ugo, M Zanata, P Guerriero, CR Martin. Anal Chem 70:2163–2166, 1998.
201. S Cosnier, B Galland, C Innocent. J Electroanal Chem 433:113–119, 1997.
202. Q Wu, GD Storrier, F Pariente, Y Wang, JP Shapleigh, HD Abruna. Anal Chem 69:4856–4863, 1997.
203. I Willner, E Katz, N Lapidot, P Baeuerle. Bioelectrochem Bioenerg 29:29–45, 1992.
204. D Centonze, I Losito, C Malitesta, F Palmisano, PG Zambonin. J Electroanal Chem 435:103–111, 1997.
205. E Dempsey, J Wang, MR Smyth. Talanta 40:445–451, 1993.
206. D Centonze, A Guerrieri, F Palmisano, L Torsi, PG Zambonin. J Anal Chem 349:497–501, 1994.
207. S Eddy, K Warriner, I Christie, D Ashworth, C Purkiss, P Vadgama. Biosens Bioelectron 10:831–839, 1995.
208. Z Zhang, H Liu, J Deng. Anal Chem 68:1632–1638, 1996.
209. RJ Geise, SY Rao, AM Yacynych. Anal Chim Acta 281:467–473, 1993.
210. Z Zhang, C Lei, J Deng. Analyst 121:971–975, 1996.
211. TF Kang, GL Shen, RQ Yu. Anal Lett 30:647–662, 1997.
212. MB Madaras, RP Buck. Anal Chem 68:3832–3839, 1996.
213. A Curulli, I Carelli, O Trischitta, G Palleschi. Biosens Bioelectron 12:1043–1055, 1997.
214. F Pariente, E Lorenzo, F Tobalina, HD Abruna. Anal Chem 67:3936–3944, 1995.
215. SH Si, YJ Xu, LH Nie, ZZ Yao. Talanta 42:469–474, 1995.
216. E Dempsey, J Wang, MR Smyth. Talanta 40:445–451, 1993.
217. MO Berners, MG Boutelle, M Fillenz. Anal Chem 66:2017–2021, 1994.
218. ML Simpson, GS Sayler, BM Applegate, S Ripp, DE Nivens, MJ Paulus, GE Jellison Jr. Trends Biotechnol 16:332–338, 1998.
219. M Schena, RA Heller, TP Theriault, K Konrad, E Lachenmeier, RW Davis. Trends Biotechnol 16:301–306, 1998.
220. H Thorp. Trends Biotechnol 16:117–121, 1998.
221. KA Giuliano, DL Taylor. Trends Biotechnol 16:135–140, 1998.
222. RW Wallace. Mol Med Today 3:384–389, 1997.
223. A Marshall, J Hodgson. Nat Biotechnol 16:27–31, 1998.
224. EM Southern. Trends Genet 12:110–115, 1996.
225. G Ramsay. Nat Biotechnol 16:40–44, 1998.
226. C Nicolini. Biosens Bioelectron 10:105–127, 1995.
227. P Connolly. Trends Biotechnol 12:123–127, 1994.
228. T Pastinen, A Kurg, A Metspalu, L Peltonen, AC Syvanen. Genome Res 7:606–614, 1997.
229. CF Edman, DE Raymond, DJ Wu, E Tu, RG Sosnowski, WF Butler, M Nerenberg, M Heller. Nucleic Acids Res 25:4907–4914, 1997.
230. JG Hacia, W Makalowski, K Edgemon, MR Erdos, CM Robbins, SPA Fodor, LC Brody, FS Collins. Nat Genet 18:155–158, 1998.
231. J Wang, X Cai, G Rivas, H Shiraishi, N Dontha. Biosens Bioelectron 12:587–599, 1997.
232. A de Saizeiu, U Certa, J Warrington, C Gray, W Keck, J Mous. Nat Biotechnol 16:45–48, 1998.
233. S Solinas-Toldo, S Lampel, S Stilgenbauer, J Nickolenko, A Benner, H Dohner, T Cremer, P Lichter. Genes Chromosomes Cancer 20:399–407, 1997.
234. T Livache, B Fouque, A Roget, J Marchand, G Bidan, R Teoule, G Mathis. Anal Biochem 255:188–194, 1998.
235. M Schena, D Shalon, R Heller, A Chai, PO Brown, RW Davis. Proc Natl Acad Sci USA 93:10614–10619, 1996.

236. E Di Fabrizio, M Gentili, P Morales, R Pilloton, J Mela, S Santucci, A Sese. Appl Phys Lett 69:3280–3282, 1996.
237. P Morales, M Sparendei. Appl Phys Lett 64:1042–1044, 1994.
238. VS Bannikov. Biosens Bioelectron 11:933–945, 1996.
239. S Hiroyuki. Electron Biotechnol Adv Forum Ser 2:157–174, 1996.
240. I Willner, E Katz, B Willner, R Blonder, V Heleg-Shabtai, AF Buckmann. Biosens Bioelectron 12:337–356, 1997.
241. FH Hong, M Chang, NRB Baofu, FT Hong. Mater Res Soc Symp Proc 330:257–262, 1994.
242. VB Pizziconi, DL Page. Biosens Bioelectron 12:287–299, 1997.
243. DW Conrad, AV Davis, SK Golightly, JC Bart, FS Ligler. Proc SPIE-Int Soc Opt Eng 2978: 12–21, 1997.
244. M Koudelka-Hep, NF De Rooij, DJ Strike. Methods Biotechnol 1:87–92, 1997.
245. I Moser, T Schalkhammer, E Mann-Buxbaum, G Hawa, M Rakohl, G Urban, F Pittner. Sens Actuators B B7:356–362, 1992.
246. JM Calvert. J Vac Technol B 11:2155–2163, 1993.
247. GP Lopez, HA Biebuyck, R Harter, A Kumar, GM Whitesides. J Am Chem Soc 115:10774–10781, 1993.
248. HA Biebuyck, NB Larsen, E Delamarche, B Michel. IBM J Res Dev 41:159–170, 1997.
249. A Kumar, NL Abbott, HA Biebuyck, E Kim, GM Whitesides. Acc Chem Res 28:219–226, 1995.
250. GM Whitesides, AJ Black, PF Nealey, JL Wilbur. Proc Robert A Welch Found Conf Chem Res 39:109–121, 1995.
251. JL Wilbur, A Kumar, E Kim, GM Whitesides. Adv Mater 6:600–604, 1994.
252. M Mrksich, GM Whitesides. Trends Biotechnol 13:228–235, 1995.
253. D Qin, Y Xia, JA Rogers, RJ Jackman, XM Zhao, GM Whitesides. Top Curr Chem 194:1–20, 1998.
254. PF Nealey, AJ Black, JL Wilbur, GM Whitesides. Mol Electron 343–367, 1997.
255. Y Xia, GM Whitesides. Poly Mater Sci Eng 77:596–598, 1997.
256. JL Wilbur, A Kumar, HA Biebuyck, E Kim, GM Whitesides. Nanotechnology 7:452–457, 1996.
257. T Vopel, A Ladde, H Muller. Anal Chim Acta 251:117–120, 1991.
258. D Wilke, H Mueller. Fresenius J Anal Chem 349:661–665, 1994.
259. C Jiminez, J Bartrol, NF de Rooij, M Koudelka-Hep. Anal Chim Acta 351:169–176, 1997.
260. Y Liusong, G Urban, I Moser, G Jobst, H Gruber. Polym Bull 35:759–765, 1995.
261. DJ Pritchard, H Morgan, JM Cooper. Anal Chem 67:3605–3607, 1995.
262. N Dontha, WB Nowall, WG Kuhr. Anal Chem 69:2619–2625, 1997.
263. M Hengsakul, AEG Cass. Bioconjugate Chem 7:249–254, 1996.
264. MC Pirrung, CY Huang. Bioconjugate Chem 7:317–321, 1996.
265. AW Flounders, DL Brandon, AH Bates. Biosens Bioelectron 12:447–456, 1997.
266. C Bourdillon, C Demaille, J Moiroux, JM Saveant. Acc Chem Res 29:529–535, 1996.
267. SK Bhatia, JL Teixeira. Anal Biochem 208:197–205, 1993.
268. MN Wybourne, M Yan, JF Keana, JC Wu. Nanotechnology 7:302–305, 1996.
269. RS Chittock, J Cooper, CW Wharton, N Berovic, NS Parkinson, JB Jackson, TD Beynon. Thin Solid Films 284–285:776–779, 1996.
270. JK Chang, CY Park, JH Choi, YK Kim, DC Han. Proc SPIE-Int Soc Opt Eng 2978:31–40, 1997.
271. Y Ito, G Chen, Y Imanishi. Bioconjugate Chem 9:277–282, 1998.
272. MU Kopp, HJ Crabtree, A Manz. Curr Opin Chem Biol 1:410–419, 1997.
273. I Karube, Y Murakami, K Yokoyama. Trans Mater Res Soc Jpn 18A:45–50, 1994.
274. MA Roberts, JS Rossier, P Bercier, H Girault. Anal Chem 69:2035–2042, 1997.
275. DJ Strike, P Thiebaud, P Arquint, M Koudelka-Hep, NF De Rooij, MGH Meijerink. DECHEMA Monogr 132:125–137, 1996.
276. M Son, F Peddie, D Mulcahy, D Davey, MR Haskard. Sens Actuators B B34:422–428, 1996.

277. Y Murakami, T Takeuchi, K Yokoyama, E Tamiya, I Karube, M Suda. Anal Chem 65:2731–2735, 1993.
278. M Suda, T Sakuhara, Y Murakami, I Karube. Appl Biochem Biotechnol 41:11–15, 1993.
279. LN Amankwa, WG Kuhr. Anal Chem 64:1610–1613, 1992.
280. M Licklider, WG Kuhr. Anal Chem 70:1902–1908, 1998.
281. M Licklider, WG Kuhr. Anal Chem 67:4170–4177, 1995.
282. B Danielsson, B Xie. EXS 81:71–85, 1997.
283. J Drott, K Lindstroem, L Rosengren, T Laurell. J Micromech Microeng 7:14–23, 1997.
284. S Braun, S Rappoport, R Zusman, D Avnir, M Ottolenghi. Mater Lett 10:1–8, 1990.
285. LM Ellerby, CR Nishida, F Nishida, FA Yamanaka, B Dunn, JB Valentine, JI Zink. Science 255:1113–1115, 1992.
286. LL Hench, JK West. Chem Rev 90:33–79, 1990.
287. JE Mark. Heterog Chem Rev 3:307–326, 1996.
288. D Avnir. Acc Chem Res 28:328–337, 1995.
289. N Hüsing, U Schubert. Angew Chem Int Ed 37:22–37, 1998.
290. J Livage. Mater Sci Forum 152–153:43–54, 1994.
291. D Avnir, S Braun. Biochemical Aspects of Sol-Gel Science and Technology. Hingham, MA: Kluwer, 1996.
292. J Livage. C R Acad Sci Ser II 322:417–427, 1996.
293. D Avnir, S Braun, O Lev, D Levy, M Ottolenghi. Sol-Gel Opt 539–582, 1994.
294. D Avnir, S Braun, O Lev, M Ottolenghi. Chem Mater 6:1605–1614, 1994.
295. BC Dave, B Dunn, JS Valentine, JI Zink. Anal Chem 66:1120A–1127A, 1994.
296. BC Dave, B Dunn, JS Valentine, JI Zink. ACS Symp Ser 622:351–365, 1996.
297. J Lin, CW Brown. Trends Anal Chem 16:200–211, 1997.
298. S Barreau, JN Miller. Anal Commun 33:5H–6H, 1996.
299. BD MacCraith, CM McDonagh, G O'Keefe, AK McEvoy, T Butler, FR Sheridan. Sens Actuators B B29:51–57, 1995.
300. JI Zink, SA Yamanaka, LM Ellerby, JS Valentine, F Nishida, B Dunn. J Sol-Gel Sci Technol 2:791–795, 1994.
301. R Armon, C Dosoretz, J Starosvetsky, F Orshansky, I Saadi. J Biotechnol 51:279–285, 1996.
302. JI Zink, B Dunn, S Yamanaka, E Lan, JS Valentine, KE Chung. Mater Res Soc Symp Proc 346:1017–1026, 1994.
303. D Avnir, S Braun, M Ottolenghi. ACS Symp Ser 499:384–404, 1992.
304. BC Dave, B Dunn, JS Valentine, JI Zink. Proc Electrochem Soc 96:296–304, 1996.
305. I Gill, A Ballesteros. J Am Chem Soc 120:8587–8598, 1998.
306. S Sheltzer, S Rappoport, D Avnir, M Ottolenghi, S Braun. Biotechnol Appl Biochem 15:227–235, 1992.
307. I Gill, A Ballesteros. Ann NY Acad Sci 799:697–700, 1996.
308. CG Kaufmann, RT Mandelbaum. J Biotechnol 51:219–225, 1996.
309. AF Hsu, TA Foglia, GJ Piazza. Biotechnol Lett 19:71–74, 1997.
310. JM Miller, B Dunn, JS Valentine, JI Zink. J Non-Cryst Solids 202:279–289, 1996.
311. S Wu, J Lin, SI Chan. Appl Biochem Biotechnol 47:11–20, 1994.
312. O Ariga, T Suzuki, Y Sano, Y Murakami. J Ferment Bioeng 82:341–345, 1996.
313. MT Reetz. Adv Mater 9:943–954, 1997.
314. MT Reetz, A Zonta, J Simpelkamp. Angew Chem Int Ed Engl 34:301–303, 1995.
315. MT Reetz, A Zonta, J Simpelkamp. Biotechnol Bioeng 49:527–534, 1996.
316. MT Reetz, A Zonta, J Simpelkamp, W Koenen. Chem Commun 11:1397–1398, 1996.
317. G Kuncova, M Sivel. J Sol-Gel Sci Technol 8:667–671, 1997.
318. K Kawakami, S Yoshida. Biotechnol Tech 8:441–446, 1994.
319. K Kawakami, S Yoshida. J Ferment Bioeng 82:239–245, 1996.
320. O Heichal-Segal, S Rappoport, S Braun. Bio/Technology 13:798–800, 1995.
321. M Rietti-Shati, D Ronen, RT Mandelbaum. J Sol-Gel Sci Technol 7:77–79, 1996.
322. L Inama, S Dire, G Carturan, A Cavazza. J Biotechnol 30:197–210, 1993.

323. R Campostrini, G Carturan, R Caniato, A Piovan, R Filippini, G Innocenti, EM Cappelletti. J Sol-Gel Sci Technol 7:87–97, 1996.
324. S Braun, S Sheltzer, S Rappoport, D Avnir, M Ottolenghi. J Non-Cryst Solids 147:739–743, 1992.
325. SA Yamanaka, F Nishida, LM Ellerby, CR Nishida, B Dunn, JS Valentine, JI Zink. Chem Mater 4:495–497, 1992.
326. S Sheltzer, SA Braun. Biotechnol Applied Biochem 19:293–305, 1994.
327. U Narang, PN Prasad, FV Bright, K Ramanathan, ND Kumar, BD Malhotra, MN Kalmasanan, S Chandra. Anal Chem 66:3139–3144, 1994.
328. U Kuenzelmann, H Boettcher. Sens Actuators B B39:222–228, 1997.
329. T-M Park, EI Iwuoha, MR Smyth, BD MacCraith. Anal Commun 33:271–273, 1996.
330. J Wang, PVA Pamidi, DS Park. Anal Chem 68:2705–2708, 1996.
331. I Pankratov, O Lev. J Electroanal Chem 393:35–41, 1995.
332. J Wang, DS Park, PVA Pamidi. J Electroanal Chem 434:185–189, 1997.
333. S Sampath, O Lev. Electroanalysis 8:1112–1116, 1996.
334. PVA Pamidi, DS Park, J Wang. Polym Mater Sci Eng 76:513–514, 1997.
335. S Sampath, I Pankratov, J Gun, O Lev. J Sol-Gel Sci Technol 7:123–128, 1996.
336. J Wang, PVA Pamidi, DS Park. Electroanalysis 9:52–55, 1997.
337. J Wang, PVA Pamidi. Anal Chem 69:4490–4494, 1997.
338. S Bharathi. Anal Commun 35:29–31, 1998.
339. V Glezer, O Lev. J Am Chem Soc 115:2533–2534, 1993.
340. SA Yamanaka, F Nishida, LM Ellerby, CR Nishida, B Dunn, JS Valentine, JI Zink. Chem Mater 4:495–497, 1992.
341. L Coche-Guerente, S Cosnier, P Labbe. Chem Mater 9:1348–1352, 1997.
342. SL Chut, J Li, SN Tan. Analyst 122:1431–1434, 1997.
343. J Li, SN Tan, H Ge. Anal Chim Acta 335:137–145, 1996.
344. JW Aylott, DJ Richardson, DA Russell. Chem Mater 9:2261–2263, 1997.
345. JM Miller, B Dunn, JS Valentine, JI Zink. J Non-Cryst Solids 202:279–289, 1996.
346. DJ Blyth, JW Aylott, DJ Richardson, DA Russell. Analyst 120:2725–2730, 1995.
347. EH Lan, MS Davidson, LM Ellerby, B Dunn, JS Valentine, JI Zink. Mater Res Soc Symp Proc 330:289–294, 1994.
348. JI Zink, B Dunn, S Yamanaka, EH Lan, JS Valentine, KE Chung. Mater Res Soc Symp Proc 346:1017–1026, 1994.
349. SA Yamanaka, LM Ellerby, EH Lan, CR Nishida, F Nishida, B Dunn, JS Valentine, JI Zink. Mater Res Soc Symp Proc 277:99–104, 1994.
350. SA Yamanaka, NP Nguyen, B Dunn, JS Valentine, JI Zink. J Sol-Gel Sci Technol 7:117–121, 1996.
351. RA Simkus, V Laurinavicius, L Boguslavsky, T Skotheim, SW Tanenbaum, JP Nakas, DJ Slomczynski. Anal Lett 29:1907–1919, 1996.
352. T-M Park, EI Iwuoha, MR Smyth, R Freaney, AJ McShane. Talanta 44:973–978, 1997.
353. AK Williams, JT Hupp. J Am Chem Soc 120:4366–4371, 1998.
354. K Ramanathan, MN Kamalasanan, BD Malotra, DR Pradhan, S Chandra. J Sol-Gel Sci Technol 10:309–316, 1997.
355. TM Park, EI Iwuoha, MR Smyth. Electroanalysis 9:1120–1123, 1997.
356. DA Navas, PMC Ramos. Sens Actuators B B39:426–431, 1997.
357. F Akbarian, A Lin, B Dunn, JS Valentine, JI Zink. J Sol-Gel Sci Technol 8:1067–1070, 1997.
358. U Narang, PN Prasad, FV Bright, A Kumar, ND Kumar, BD Malhotra, MN Kamalasanan, S Chandra. Chem Mater 6:1596–1598, 1994.
359. C Dosoretz, R Armon, J Starosvetsky, N Rothschild. J Sol-Gel Sci Technol 7:7–11, 1996.
360. JW Aylott, DJ Richardson, DA Russell. Analyst 122:77–80, 1997.
361. DJ Blyth, SJ Oynter, DA Russell. Analyst 121:1975–1978, 1996.
362. S Wu, LM Ellerby, JS Cohan, B Dunn, MA El-Sayed, JS Valentine, JI Zink. Chem Mater 5:115–120, 1993.

363. HH Weetall. Biosens Bioelectron 11:327–333, 1996.
364. HH Weetall, B Robertson, D Cullin, J Brown, M Walch. Biochim Biophys Acta 1142:211–213, 1993.
365. PC Pandey, S Singh, B Upadhyay, HH Weetall, PK Chen. Sens Actuators B B36:470–474, 1996.
366. Z Chen, LA Samuelson, J Akkara, DL Kaplan, H Gao, J Kumar, KA Marx, SK Tripathy. Chem Mater 7:1779–1783, 1995.
367. Z Chen, DL Kaplan, K Yang, J Kumar, KA Marx, SK Tripathy. J Sol-Gel Sci Technol 7:99–108, 1996.
368. E Bressler, S Braun. J Sol-Gel Sci Technol 7:129–133, 1996.
369. NL St Clair, MA Navia, J Am Chem Soc 114:7314–7316, 1992.
370. J Lalonde. CHEMTECH 27:28–35, 1997.
371. J Lalonde. Chem Eng 108–112, 1997.
372. MD Grim. Pharm Manuf Int 49, 1997.
373. T Zelinski, H Waldmann. Angew Chem Int Ed Engl 36:722–724, 1997.
374. AL Margolin. Trends Biotechnol 14:223–230, 1996.
375. Y-F Wang, K Yakovlevsky, N Khalaf, B Zhang, AL Margolin. Ann NY Acad Sci 799:777–783, 1996.
376. Y-F Wang, K Yakovlevsky, B Zhang, AL Margolin. J Org Chem 62:3488–3495, 1997.
377. N Khalaf, CP Govardhan, JJ Lalonde, RA Persichetti, Y-F Wang, AL Margolin. J Am Chem Soc 118:5494–5495, 1996.
378. SB Sobolov, M Draganoiu, A Bartoszko-Malik, KI Voivodov, F McKinney, J Kim, AJ Fry. J Org Chem 61:2125–2128, 1996.
379. Y-F Wang, K Yakovlevsky, AL Margolin. Tetrahedron Lett 37:5317–5320, 1996.
380. RA Persichetti, NL St Clair, JP Griffith, MA Navia, AL Margolin. J Am Chem Soc 117:2732–2737, 1995.
381. JJ Lalonde, C Govardhan, N Khalaf, AG Martinez, K Visuri, AL Margolin. J Am Chem Soc 117:6845–6852, 1995.
382. SB Sobolov, A Baroszki-Malik, TR Oeschger, MM Montelbano. Tetrahedron Lett 35:7751–7754, 1994.
383. AJ Fry, SB Sobolov, MD Leonida, KI Voivodov, J Fenton. Proc Electrochem Soc 97-6:115–122, 1997.
384. PA Fitzpatrick, D Ringe, AM Klibanov. Biochem Biophys Res Commun 198:675–681, 1994.
385. J Partridge, GA Hutcheon, BD Moore, PJ Halling. J Am Chem Soc 118:12873–12877, 1996.
386. JL Schmitke, CR Wescott, AM Klibanov. J Am Chem Soc 118:3360–3365, 1996.
387. CR Wescott, H Noritomi, AM Klibanov. J Am Chem Soc 118:10365–10370, 1996.
388. K Xu, AM Klibanov. J Am Chem Soc 118:9815–9819, 1996.
389. I Gill. Unpublished results.
390. CA Wartchow, P Wang, MD Bednarski. J Org Chem 60:2216–2226, 1995.
391. P Wang, TG Hill, CA Wartchow, ME Huston, LM Oehler, MB Smith, MD Bednarski. J Am Chem Soc 114:378–380, 1992.
392. TG Hill, P Wang, ME Huston, CA Wartchow, LM Oehler, MB SMith, MD Bednarski. Tetrahedron Lett 32:6823–6826, 1991.
393. P Wang, TG Hill, MD Bednarski, MR Callstrom. Tetrahedron Lett 32:6827–6830, 1991.
394. P Wang, TG Hill, MD Bednarski, MR Callstrom. Mater Res Soc Symp Proc 218:23–30, 1991.
395. TG Hill, P Wang, ME Huston, CA Wartchow, LM Oehler, MB Smith, MD Bednarski. Mater Res Soc Symp Proc 218:7–15, 1991.
396. P Wang, MV Sergeeva, L Lim, JS Dordick. Nat Biotechnol 15:789–791, 1997.
397. JL Panza, KE LeJaune, S Venkatasubramanian, AJ Russell. ACS Symp Ser 680:134–139, 1998.
398. KE LeJeune, JL Panza, S Venkatasubramanian, AJ Russell. Polymer Prepr 38:563–564, 1997.
399. Z Yang, AJ Mesiano, S Venkatasubramanian, SH Gross, JM Harris, AJ Russell. J Am Chem Soc 117:4843–4850, 1995.

400. Z Yang, D Williams, AJ Russell. Biotechnol Bioeng 45:10–17, 1995.
401. I Gill, A Ballesteros. J Am Chem Soc, submitted.
402. KE LeJeune, AJ Mesiano, SB Bower, JK Grimsley, JR Wild, AJ Russell. Biotechnol Bioeng 54:105–112, 1997.
403. KE LeJeune, DS Frazier, GR Caranto, DM Maxwell, G Amitai, AJ Russell, BP Doctor. Med Def Biosci Rev Proc 1:223–230, 1996.
404. KE LeJeune, AJ Russell. Biotechnol Bioeng 51:450–457, 1996.
405. PL Havens, HF Rase. Ind Eng Chem Res 32:2254–2258, 1993.

26

Enzymatic Protecting Group Techniques in Organic Synthesis

Tanmaya Pathak
National Chemical Laboratory, Pune, Maharasthra, India

Herbert Waldmann
Max-Planck-Institute of Molecular Physiology and University of Dortmund, Dortmund, Germany

I. INTRODUCTION

The proper introduction and removal of protecting groups is one of the most important and widely carried out transformations in synthetic organic chemistry. For the manipulation of protecting groups under the mildest possible conditions numerous classical chemical methods are available [1–3]. Nevertheless, there still remain severe problems during the synthesis of complex, polyfunctional molecules that cannot or can only with great difficulty be solved by chemical methods alone. In addition to their stereodiscriminating properties, enzymes offer the opportunity to carry out highly chemo- and regioselective transformations. They often operate at neutral, weakly acidic, or weakly basic pH values and in many cases combine a high selectivity for the reactions they catalyze and the structures they recognize with a broad substrate tolerance. Therefore, the use of enzyme labile protecting groups has offered viable alternatives to classical methods. On the other hand, the application of biocatalysts in combination with the classical chemical methods has substantially enriched the arsenal of protecting group chemistry [3–7].

II. ENZYMES AND PROTECTING GROUPS

Although the development of enzymatic protecting group techniques has been intensively studied only in the last few years, a number of interesting enzyme-labile blocking functions have been developed for organic synthesis (Fig. 1). For example, the phenylacetamide (PhAc) and the 4-acetoxybenzylgxycarbonyl (AcOZ) urethane have been used as enzyme-labile protecting groups for amino groups [3–5]. Carboxylic acid functionalities can be selectively demasked through enzyme-mediated cleavage of heptyl, choline, or (methoxyethoxy)ethyl (MEE) esters and hydroxyl groups protected as the acetate, benzoate, butyrate, and even pivaloate can be liberated enzymatically [3–6]. In this way, protection and deprotection can be carried out for polyfunctional molecules such as carbohydrates, peptides [7], nucleosides [8], steroids, alkaloids, and phenolic natural products.

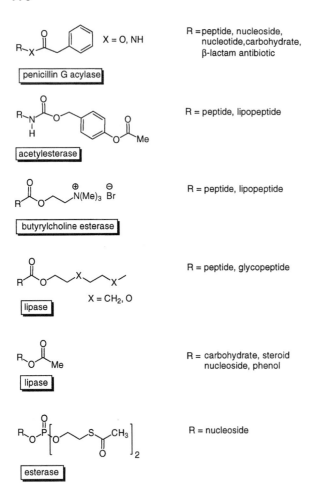

R = peptide, nucleoside,
nucleotide, carbohydrate,
β-lactam antibiotic

R = peptide, lipopeptide

R = peptide, lipopeptide

R = peptide, glycopeptide

R = carbohydrate, steroid
nucleoside, phenol

R = nucleoside

Figure 1   Selected enzyme-labile protecting groups.

Owing to the distinct chemo-, regio-, and stereoselective nature of the enzyme-mediated transformations, enzyme-labile protecting groups can be combined with other blocking groups removable only by nonenzymatic methods. An application of this concept is illustrated [9] by the chemoenzymatic synthesis of the *S*-palmitoylated and *S*-farnesylated C-terminal lipohexapeptide of the human N-Ras protein (Fig. 2). This type of modified lipopeptide cannot be deblocked under basic conditions because the labile palmitic acid thio ester group would be preferentially hydrolysed. The C terminus of the peptide chain was successfully deprotected selectively via removal of the choline ester (Cho) by the use of choline esterase without affecting the palmitic acid thio ester bond. The observed chemoselectivity here is exactly opposite to that found in nonenzymatic conversions.

Detailed discussions on the use of enzymatic protecting group techniques for various functional groups have already been published [4,5]. This chapter will, therefore, highlight the increasing applications of the enzymatic protecting group techniques that have taken place in recent years encompassing a wide variety of organic molecules.

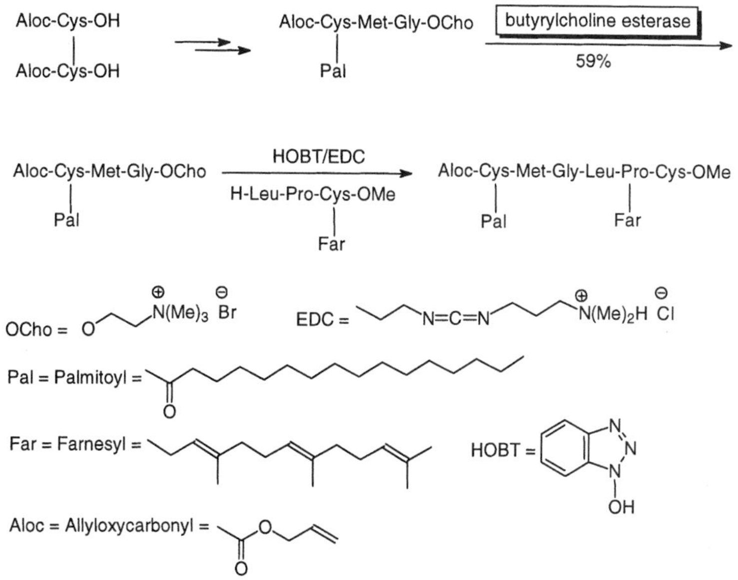

Aloc-Cys-OH   ⟶   Aloc-Cys-Met-Gly-OCho | butyrylcholine esterase |
| | ⟶
Aloc-Cys-OH   Pal   59%

Aloc-Cys-Met-Gly-OCho   —— HOBT/EDC ——→   Aloc-Cys-Met-Gly-Leu-Pro-Cys-OMe
| H-Leu-Pro-Cys-OMe | |
Pal   Far   Pal   Far

OCho = O—CH$_2$CH$_2$—$\overset{\oplus}{N}$(Me)$_3$ $\overset{\ominus}{Br}$ EDC = CH$_2$CH$_2$CH$_3$—N=C=N—CH$_2$CH$_2$CH$_2$—$\overset{\oplus}{N}$(Me)$_2$H $\overset{\ominus}{Cl}$

Pal = Palmitoyl =

Far = Farnesyl =   HOBT =

Aloc = Allyloxycarbonyl =

Figure 2 Chemoenzymatic synthesis of *S*-palmitoylated and *S*-farnesylated C-terminal lipohexa-peptide of the human N-Ras protein.

III. FORMATION AND HYDROLYSIS OF AMIDES

Enzymatic protecting group techniques have recently been extended [10,11] to the synthesis of phosphopeptides, in which the phenylacetamide group is removed from the respective PhAc amides by means of penicillin G acylase (PGA). In the course of the ensuing enzymatic transformations only the phenylacetamide blocking group was attacked; the peptide bonds, the C-terminal ester, and the phosphates remained intact (Fig. 3). More importantly, the reaction conditions were so mild that no trace of β elimination was observed [10,11].

The phenylacetamide group has been used for the temporary masking of nucleobases in the synthesis of oligonucleotides, both in solution and on a solid support. PhAc-protected oligonucleotides were synthesized on controlled-pore glass beads. After release of

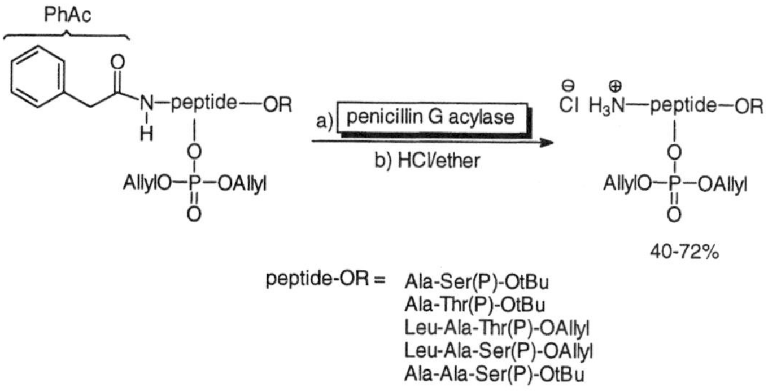

peptide-OR = Ala-Ser(P)-OtBu
Ala-Thr(P)-OtBu
Leu-Ala-Thr(P)-OAllyl
Leu-Ala-Ser(P)-OAllyl
Ala-Ala-Ser(P)-OtBu

40-72%

Figure 3 The PGA-mediated selective N-terminal deprotection of phosphopeptides.

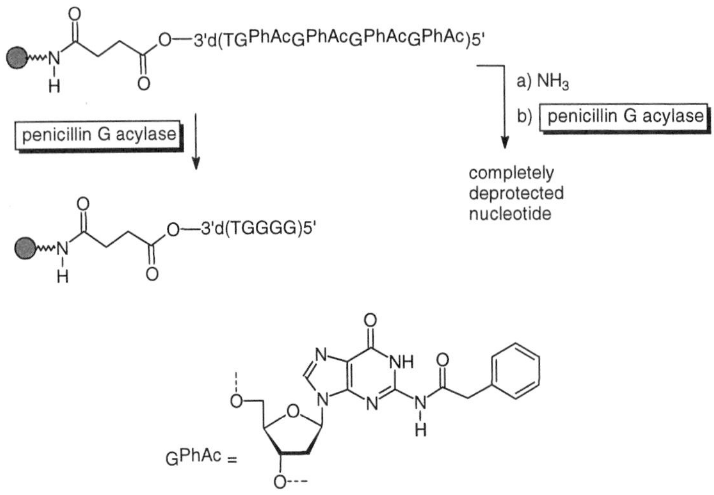

Figure 4 The use of PGA-labile protecting group in the synthesis of oligonucleotides, both in solution and in solid phases.

the oligonucleotides from the solid support and deprotection of internucleotide phosphates, they were incubated with PGA to remove all PhAc groups from the nucleobases. On the other hand, treatment of the solid phase–anchored oligonucleotide with PGA resulted in the selective unmasking of the base amino groups (Fig. 4). This result demonstrates that biocatalysts like PGA may also be applied on solid phases [12].

PGA-mediated deprotections of 1-[N-(phenylacetyl)amino]-4,5-dioxohexane and 1-[N-(phenylacetyl)amino]-4,5-dioxoheptane were used as the key steps in the synthesis (Fig. 5) of the strong natural roast odorants 2-acetyl-1-pyrroline and 2-propionyl-1-pyrroline [13].

The immobilized PGA-mediated selective removal of the PhAc group to unmask a primary amino group in the presence of an acetate at the final step was the method of choice [14] in the synthesis of a tyramine spacer–containing immunodeterminant tetra-saccharide (Fig. 6). Deblocking of N-phenylacetyl-protected ethanolamine phosphate

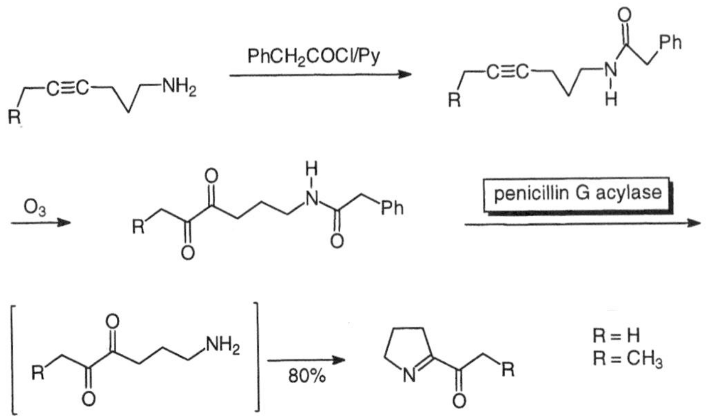

Figure 5 The PGA-mediated deprotection as the key step in the synthesis of natural roast odorants.

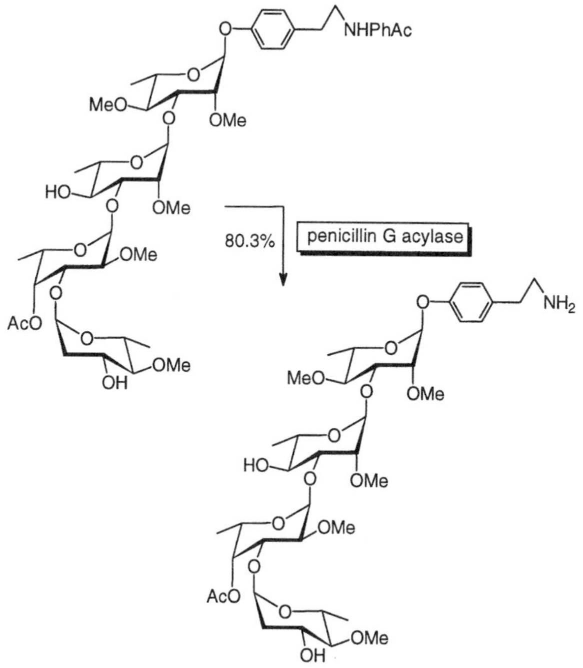

Figure 6 Use of PGA-assisted selective deprotection in the last step of the synthesis of an immunodeterminant tetrasaccharide.

heptosyl disaccharide, a protected spacer–containing fragment of the inner core of the lipopolysaccharide region of *Neisseria meningitidis*, immunotype L3, was readily accomplished [15] with immobilized PGA yielding the target dimer (Fig. 7).

A one-pot chemoenzymatic synthesis of 3'-functionalized cephalosporin C or cephazolin involving three consecutive biotransformations has been reported [16]. The enzymatic deacylation of cephalosporin C in one batch was performed by the use of D-amino acid oxidase (DAO) from *Trigonopsis variabilis*, glutaryl acylase (GA) from *Acetobacter* sp., and a continuous flow of oxygen. DAO catalyzed the oxidative deamination of the α-

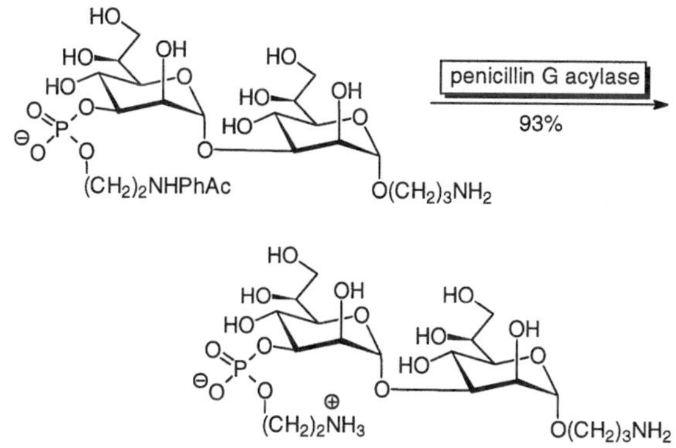

Figure 7 Deblocking of *N*-phenylacetyl-protected ethanolamine phosphate heptosyl disaccharide with immobilized PGA.

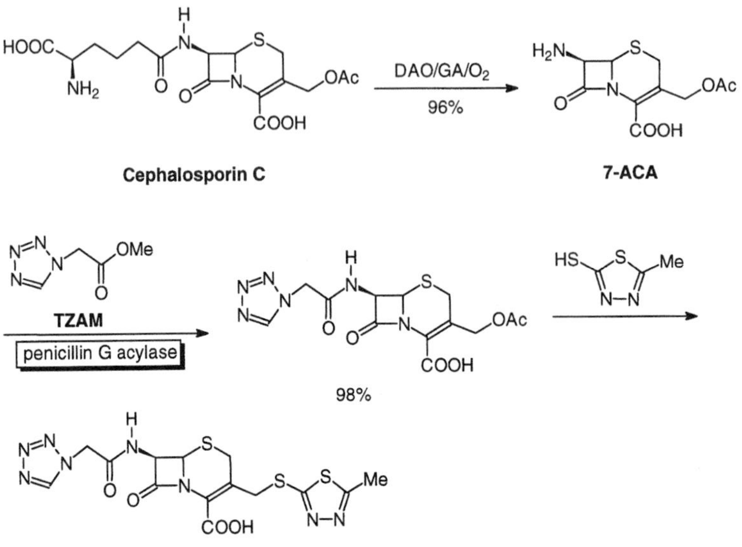

Figure 8 Chemoenzymatic synthesis of cefazolin.

aminoadipic acid side chain of cephalosporin C to give the α-ketoadipic derivative. Its decarboxylation and further oxidation gave the glutaryl analog, which was then deacylated by GA to 7-ACA. The subsequent acylation was performed directly on the solution of the crude 7-ACA, after filtration, using the immobilized PGA isolated from *E. Coli* A and a 3:1 molar excess of the ester TZAM. The product was obtained in 98% yield at pH 6.5 and 4°C (Fig. 8).

Lipase from *Rhizomucor miehei* has recently been used for the selective N-acylation of 1-deoxy-1-methylamino-D-glucitol by fatty acids in hexane [17]. The methodology has been extended further to the enzymatic synthesis of biosurfactants (glycamides) by trans-acylation reaction (Fig. 9) catalyzed by immobilized lipase from *Candida antarctica* in organic media [18].

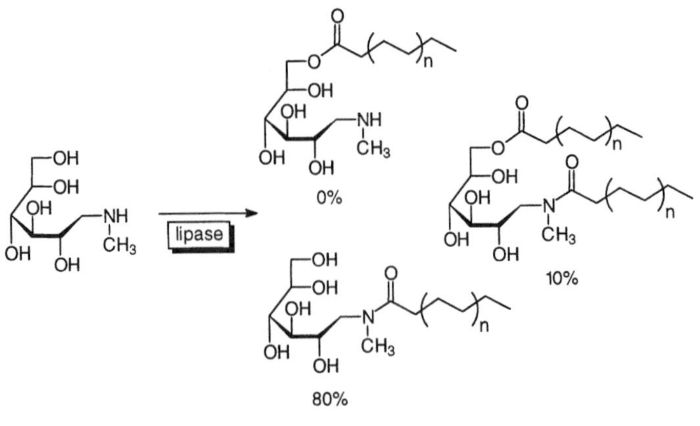

Figure 9 Lipase-mediated synthesis of biosurfactants.

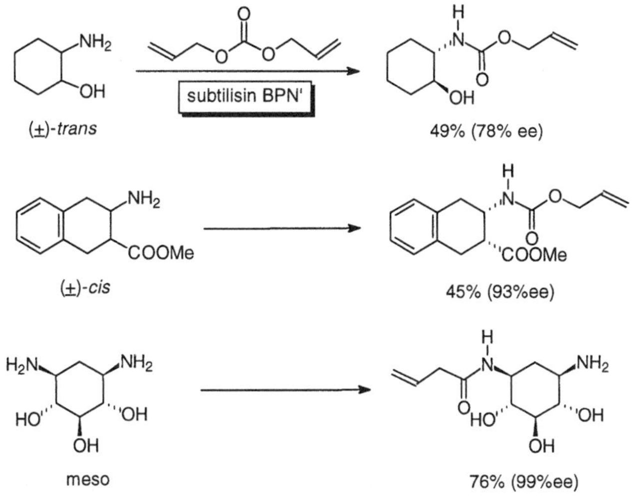

Figure 10 Enantioselective enzymatic protection of amines.

The phthalimido group has always been an attractive alternative to amine protection in chemical synthesis. Native phthalyl amidase was isolated from *Xanthiobacter agilis* and subsequently cloned and overexpressed in *Streptomyces lividans*. Phthalyl amidase selectively deprotects phthalimido groups under very mild conditions (30°C in 200 mM potassium phosphate buffer, pH 8.0) in a one-pot reaction to produce phthalic acid and the free amine. The enzyme has been shown to deprotect several primary amines of distinctly different structures [19].

The use of unsymmetrical vinyl carbonates to protect amines using *Candida antarctica* lipase (CAL) (SP 435) was reported earlier [20–22]. Recently, commercially available, low-cost homocarbonates have been used (Fig. 10) as substrates for the enantioselective enzymatic protection of amines [23]. Bis(phenylacet)amides of *meso*-diamino-di(tri, tetr)ols were deprotected enantioselectively (Fig. 11) with penicillin amidase from *Escherichia coli* to produce several enantiomerically pure cyclic amines [24]. (±)-Trans-2-aminocyclohexanol has been acylated (Fig. 12) with dimethylglutarate in the presence of CAL in 1,4-dioxane in 60% yield [25]. Acylase I from *Aspergillus oryzae* hydrolyzes natural N-acetyl-DL-amino acids at useful reaction rates and in very high enantiomeric purity. A number of sulfur- and selenium-containing amino acids can be obtained in a similar fashion [26].

Although lipases are well-established catalysts for transesterification and ester hydrolysis, they may also be used for the transformation of benzyl esters to amides. Amano lipase PS-30 isolated from *Pseudomonas capacia* catalyzes the amidation (Fig. 13) by

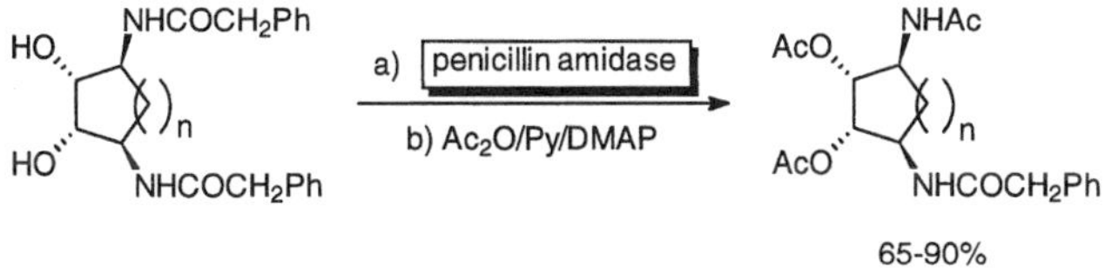

Figure 11 Enantioselective enzymatic deprotection of amines.

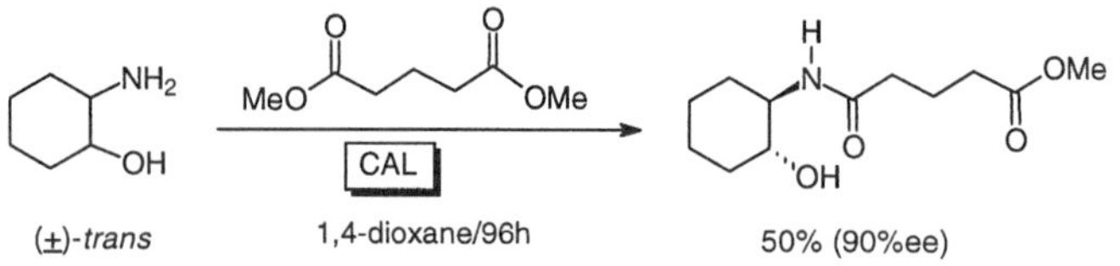

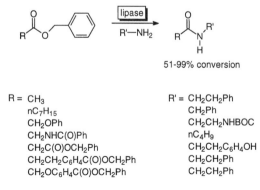

Figure 12 Enzyme-mediated selective N-acylation.

accommodating a variety of amines, including those with protective groups and other functionalities [27]. Another regioselective CAL-catalyzed amidation of several diethyl L-glutamate derivatives that only differ in the N-blocking group has been reported [28]. Pentylamine and diisopropyl ether were used as unique nucleophile and medium, respectively. α-Monoamides were obtained in all cases (Fig. 14). The CAL active site accepted derivatives bearing an N group as big as diphenylacetyl but not trityl. However, the existence of some electronic interactions that produced a positive effect on the reaction rate has also been postulated. These atypical uses of lipase might have important implications in the regioselective transformation of complex organic molecules containing different functionalities.

IV. HYDROLYSIS OF URETHANES

A new PGA-removable protecting group, namely, *p*-phenylacetoxybenzyloxycarbonyl (PhAcOZ), has been developed. The group is a urethane that embodies a functional group (here a phenylacetate) that is recognized by the biocatalyst (here PGA) and that is bound by an enzyme-labile linkage (here an ester) to a functional group that undergoes a spontaneous fragmentation upon cleavage of the enzyme-sensitive bond (here a *p*-hydroxybenzylurethane) resulting in the liberation of a carbamic acid derivative, which decarboxylates to give the desired peptide or peptide conjugate (Fig. 15). The usefulness of this protecting group was demonstrated by its application in the synthesis of a phosphoglycohexapeptide (Fig. 16) [29,30].

51-99% conversion

R = CH_3
nC_7H_{15}
CH_2OPh
$CH_2NHC(O)Ph$
$CH_2C(O)OCH_2Ph$
$CH_2CH_2C_6H_4C(O)OCH_2Ph$
$CH_2OC_6H_4C(O)OCH_2Ph$

R' = CH_2CH_2Ph
CH_2Ph
CH_2CH_2NHBOC
nC_4H_9
$CH_2CH_2C_6H_4OH$
CH_2CH_2Ph
CH_2CH_2Ph

Figure 13 Lipase-catalyzed amidation of benzyl esters.

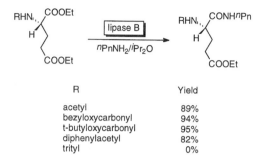

R	Yield
acetyl	89%
bezyloxycarbonyl	94%
t-butyloxycarbonyl	95%
diphenylacetyl	82%
trityl	0%

Figure 14   Lipase-catalyzed regioselective amidation.

V.   FORMATION AND HYDROLYSIS OF ESTERS

A series of acetoxy derivatives of androstane were deacylated in organic solvents by several lipases. The most satisfactory results were obtained with lipase from *Candida cylindracea* (CCL) and *Candida antarctica* (CAL). In some derivatives, CCL and CAL showed an overwhelming regioselectivity toward the removal of the 3β- or the 17β-acetyl group and three new steroid derivatives were obtained [31] through this approach (Fig. 17).

Lipase AL (*Achromobacter* sp., Meito) preferentially hydrolyzes the diacetate of (*E*)-2-substituted-2-butene-1,4-diol to generate 4-hydroxy-2-butenyl acetate with 88% regioselectivity. The regioselectivity was enhanced up to 94% when the reaction was carried out [32] at 0°C (Fig. 18). Lipase from pig pancreas (PPL) has been shown to selectively catalyze (Fig. 19) the hydrolysis of bisacetates of polymethylene glycols and also of unsaturated diols to monoacetates in 48–95% yields [33]. An interesting observation on the considerable improvement of the rate of partial hydrolysis of 4-acetoxy-2-methylbut-2-enyl acetate by *Candida rugosa* lipase (CRL) or *Pseudomonas cepacia* lipase (PCL) in the presence of catalytic amounts of a thia crown ether has been reported [34]. The presence of crown ether also gave rise to a highly regioselective deacetylation to produce 4-hydroxy-2-methylbut-2-enyl acetate in a buffer-free medium (Fig. 20). The amount of thia crown ether was only 5 mol% of the substrate. A porcine pancreatic lipase (PPL)–catalyzed reaction, in contrast, was not influenced by thia crown ether even in the presence of 33 mol% of the crown ether.

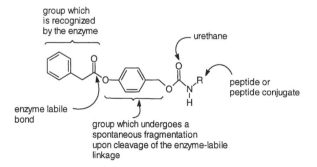

Figure 15   Principle behind the designing of the PhAcOZ group.

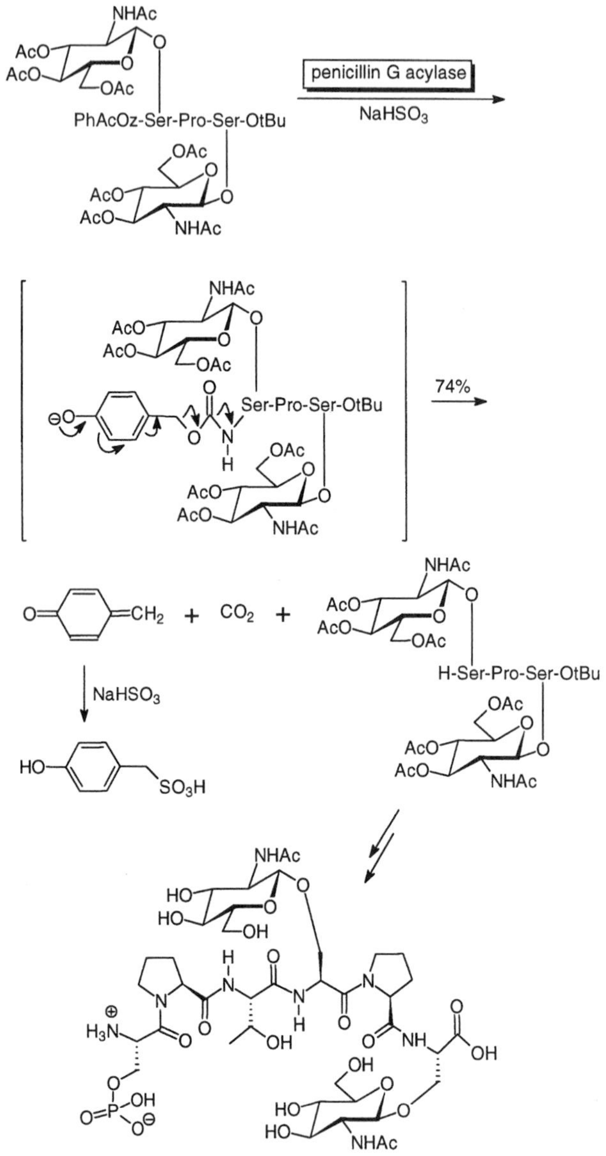

Figure 16 Actual application of the PhAcOZ group in the synthesis of a phosphoglycohexapeptide.

A regioselective saponification of dicarboxylic acid esters has also been reported [35]. When dimethyl (R,S)-2-methylglutarate was treated with the esterase from *Pseudomonas putida* MR 2068, only the regioselectively hydrolyzed product methyl (R)-4-carboxyvalerate was obtained (Fig. 21). 4-Methoxycarbonylvaleric acid or 2-methylglutaric acid could hardly be detected. The esterase could hydrolyze neither dimethyl 3-methylglutarate nor dimethyl 2,4-dimethylglutarate.

Dipyrazolic diester crown ethers have been synthesized via lipase-mediated transesterification [36]. The key and the first step of this synthesis was the highly regioselective

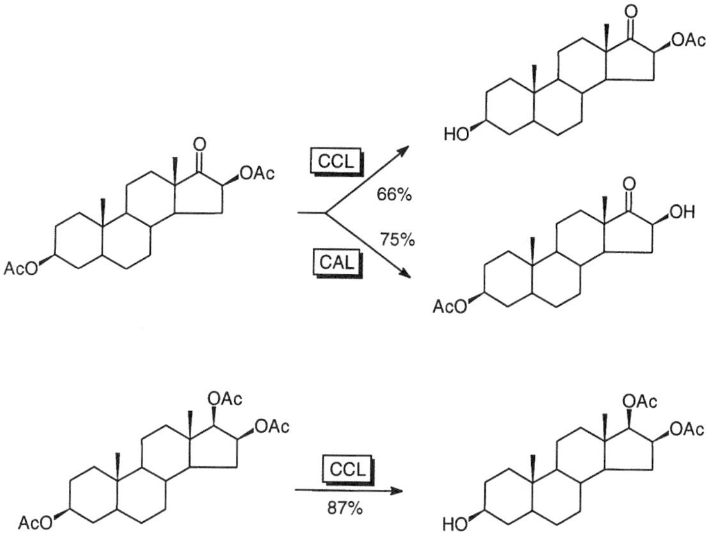

Figure 17 Lipase-mediated regioselective deprotection of acetoxy derivatives of androstane.

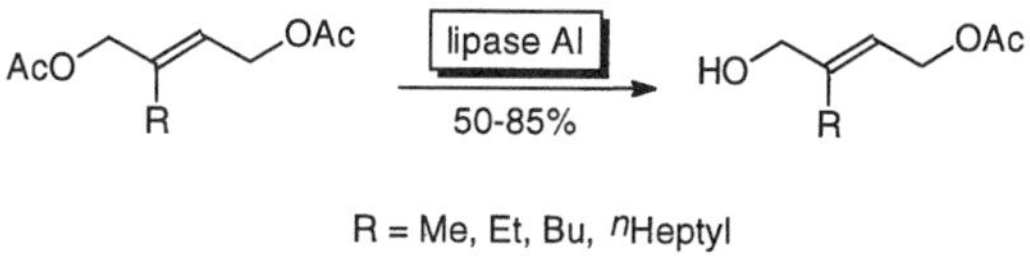

R = Me, Et, Bu, nHeptyl

Figure 18 The regioselective hydrolysis of the diesters of (*E*)-2-substituted-2-butene-1,4-diols.

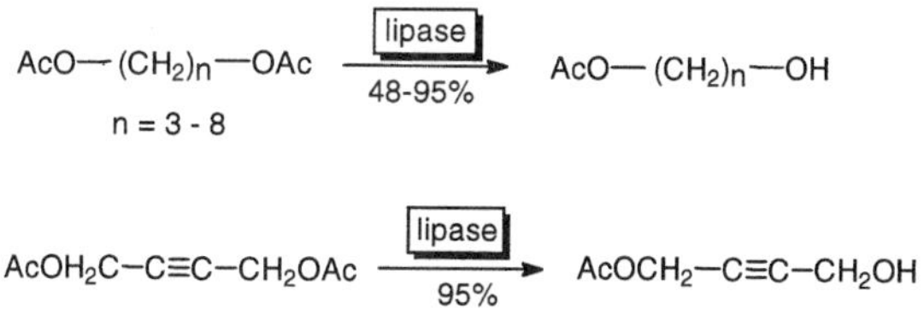

Figure 19 Lipase-assisted regioselective saponification of the esters of polymethylene glycols and an unsaturated diol.

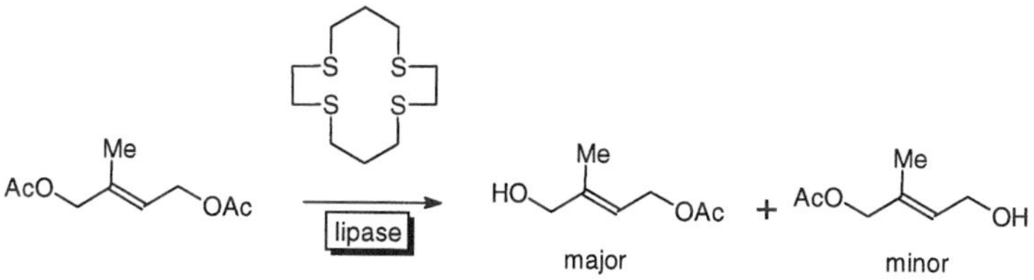

Figure 20 Thia crown ether–assisted enzymatic regioselective deacetylation.

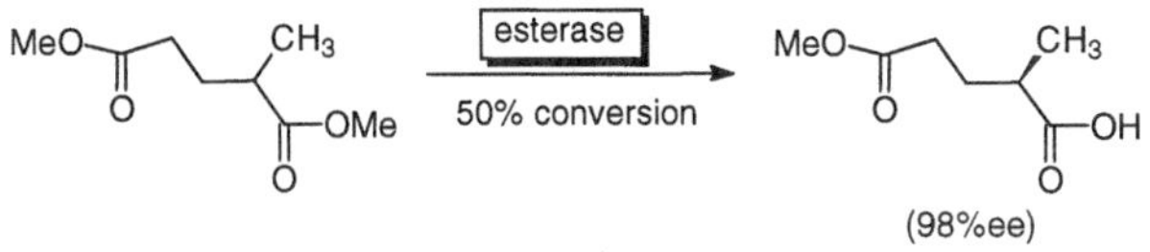

Figure 21   Esterase-mediated regioselective saponification of a dicarboxylic acid ester.

transesterification, which produced only the 3-substituted analog (Fig. 22). The second step, i.e., intramolecular cyclization, produced mainly the 3,3′ cycles.

The 6′-OH group of benzyl lactoside was acetylated selectively by CAL in high yields. Only the 2′-OH group of 6′-O-acetyl benzyl lactoside was levulinylated by the same lipase in the presence of trifluoroethyl levulinate as the acylating agent [37]. Lipozyme or immobilized CAL was used for selective esterification of the 6-OH group of 1,2-cyclohexylidene-α-D-glucofuranoside by fatty acids of varying chain lengths [38]. The acids acted as both solvent and reactant. The reaction involving short chain (C_8) fatty acids gave higher yield of 75% whereas the yield dropped to 50% with C_{18} (Fig. 23). Acylated methyl β-D-xylopyranosides (acyl being pivaloyl, acetyl, or a combination of both) were subjected to hydrolysis catalyzed by rabbit serum and esterase isolated from rabbit serum [39]. Compounds acylated by a combination of acetyl and pivaloyl groups with an acetyl group at O-2 hydrolyzed in good yields and in a chemoselective manner. In all three cases a pivaloyl group was removed preferentially. When chemoselectivity was not possible as in peracylated compounds, regioselective hydrolysis occurred and the acetyl group at O-3 was removed preferentially (Fig. 24).

A highly efficient *Pseudomonas* lipase–catalyzed regioselective acetylation method has been used [40] as the key step in the synthesis of *N*-acetyl-D-allosamine and its derivatives from tri-*O*-acetyl-D-glucal (Fig. 25). The same selective enzymatic acetylation also worked successfully for the protection of the primary hydroxyl groups of a dimeric glucal (Fig. 26). A fluorescence-labeled trisaccharide, Gal (β1–3) Gal (β1–4) Xyl (β)-MU (MU = 4-methyl-2-oxo-2*H*-chromen-7-yl or 4-methylumbelliferyl), corresponding to the linkage region between glycosaminoglycans and core proteins in proteoglycans, was

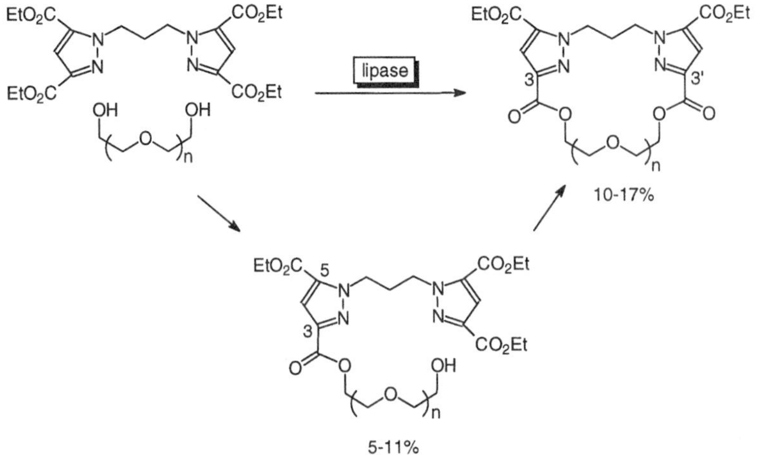

Figure 22   Synthesis of dipyrazolic diester crown ethers via lipase-mediated transesterification.

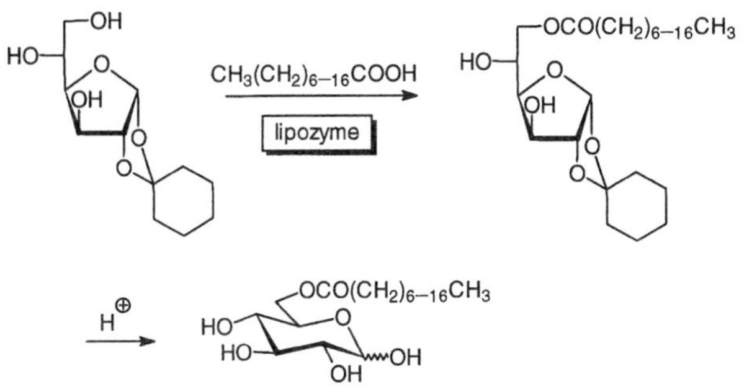

Figure 23 Selective esterification of a partially protected glucofuranose derivative by lipozyme.

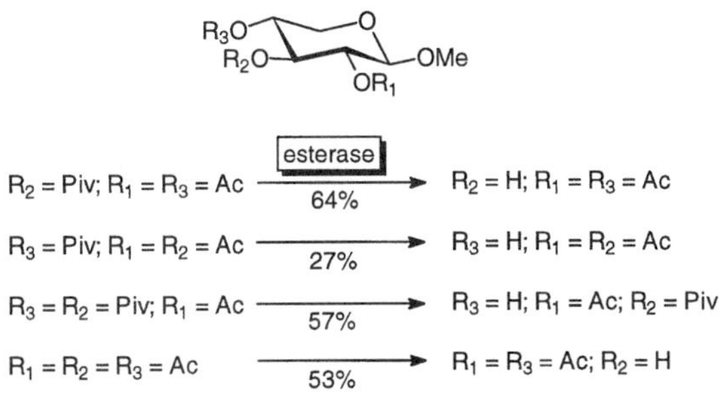

Figure 24 Regioselective saponification of acylated methyl xylopyranosides, catalyzed by esterase.

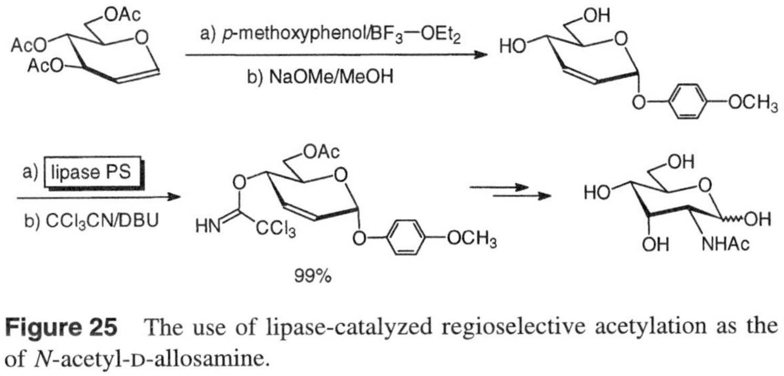

Figure 25 The use of lipase-catalyzed regioselective acetylation as the key step in the synthesis of *N*-acetyl-D-allosamine.

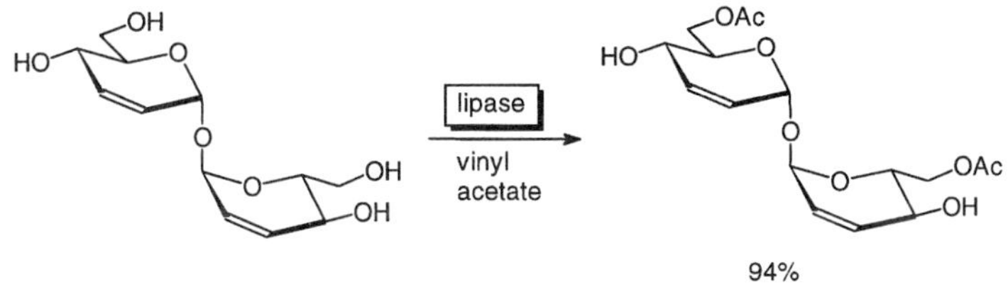

Figure 26 Lipase-catalyzed selective protection of primary hydroxyl groups of a dimeric glucal.

synthesized using enzymatic transglycosidation methodology [41]. Introduction of the second galactosyl residue at the 3' position of Gal-Xyl-MU was achieved by minimal protection of the disaccharide intermediate whose reactive primary hydroxy function was selectively protected with an acetyl group by using lipase (Amano PB)–catalyzed trans-acetylation (Fig. 27).

Porcine pancreatic lipase immobilized in microemulsion-based gels was used for the regioselective hydrolysis of perpropanoates of di-/trihydroxylic phenols in benzene. For example, 2,6-dipropanoyloxy toluene was converted to 2-hydroxy-6-propanoyloxy toluene in 40% yield [42]. The lipase from CCL deacylates the acetoxy functionality of alkyl acetoxybenzoates faster than the benzoyloxy one. The difference in reactivity between the two ester functionalities is less in aryl than in alkyl acetoxybenzoates and a methoxy group in the meta position of the aryl group reverses the reactivity [43].

Lipase (*Aspergillus niger*)–mediated saponification of heptyl esters was employed successfully for the C-terminal deprotection of several phosphopeptides. The methodology has been implemented in the synthesis (Fig. 28) of a consensus sequence of the Raf-1 kinase [11,44]. A report has appeared [45] on the selective C-terminal deprotection of *O*-glycopeptide MEE esters. This is achieved under mild conditions (pH 6.6. 37°C) by enzymatic hydrolysis using papain or lipase M from *Mucor javanicus* to prepare building blocks useful for extending the chain of gylcopeptides. On the other hand, the selective removal of acetyl protecting groups from the saccharide portion of glycopeptides is accomplished by alternative enzymatic hydrolysis with lipase WG from wheat germ (Fig. 29). Several model substrates for enzymatic glycosyl transfer reactions were accessible that may be employed to extend the carbohydrate side chain of these conjugates [45]. Biocatalysts have been used recently for the synthesis of sensitive biomolecules such as nucleopeptides [46,47,47a]. The central theme of this chemoenzymatic synthesis was to use various enzymes to unmask the different protected functionalities present in the mol-

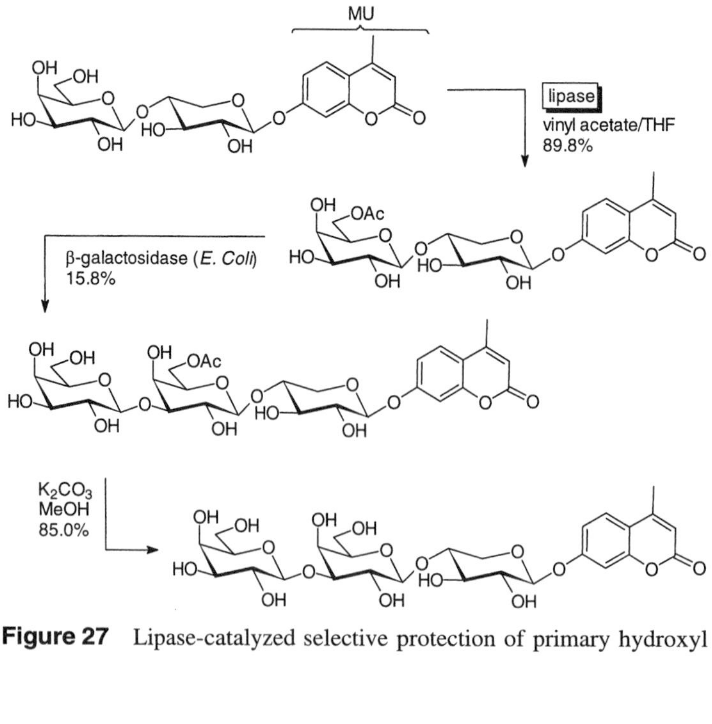

Figure 27 Lipase-catalyzed selective protection of primary hydroxyl groups of a trisaccharide.

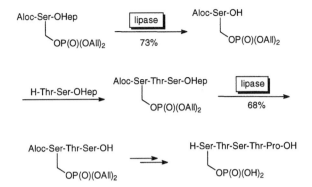

Figure 28 Use of lipase-labile heptyl protecting group in the synthesis of a phosphopeptide.

ecule (Fig. 30). An alternative approach to the synthesis of nucleopeptides has been reported recently [47]. The 3'-hydroxyl function of the fully protected nucleoserine ester was unmasked selectively by means of lipase WG–catalyzed removal of the acetyl group. On the other hand, the C terminus of the corresponding nucleopeptide choline ester was deprotected smoothly by butyrylcholine esterase (Fig. 31). Neither the acetate, the N-terminal urethane, the allyl phosphate, the phenylacetamide and the peptide bonds, nor the acid- and base-labile purine nucleosides and serine phosphates were attacked [47,47a].

The enzymatic deprotection principle has also been applied in solid phase synthesis to create an anchoring structure that contains two cleavable positions: an allylic ester on the one hand and a hydrophilic ester on the other hand. The release of peptide using a palladium(0) catalyst delivers the peptide that is still bound to the hydrophilic structure. The carboxylic function can be liberated [48] using lipase catalysts (Fig. 32). This strategy was successfully demonstrated in the solid phase synthesis of peptides as well as *O*- and *N*-glycopeptides [48].

In an attempt to construct enzyme-labile linkers for combinatorial synthesis, the 4-acetoxy-3-carboxybenzyloxy moiety has recently been employed [49]. For the design of

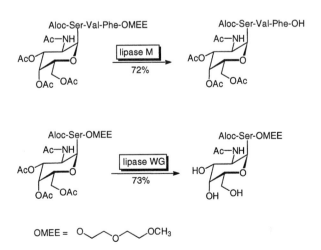

Figure 29 Lipase M–catalyzed unmasking of the C terminal and lipase WG–catalyzed saponification of hydroxyl groups of glycopeptides.

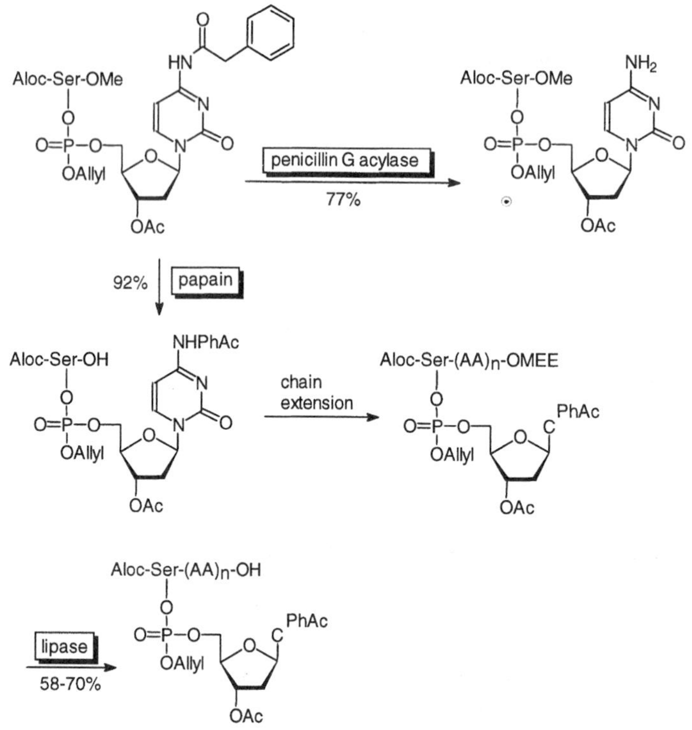

Figure 30 Selective deprotection of different functionalities of nucleopeptides by various enzymes.

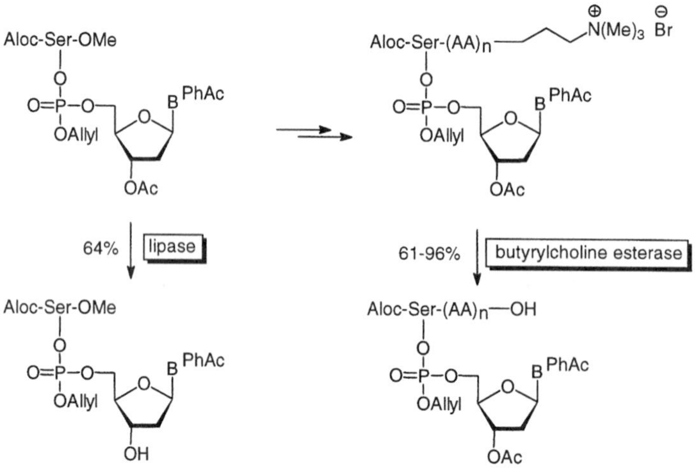

Figure 31 An alternative chemoenzymatic approach to the synthesis of nucleopeptides.

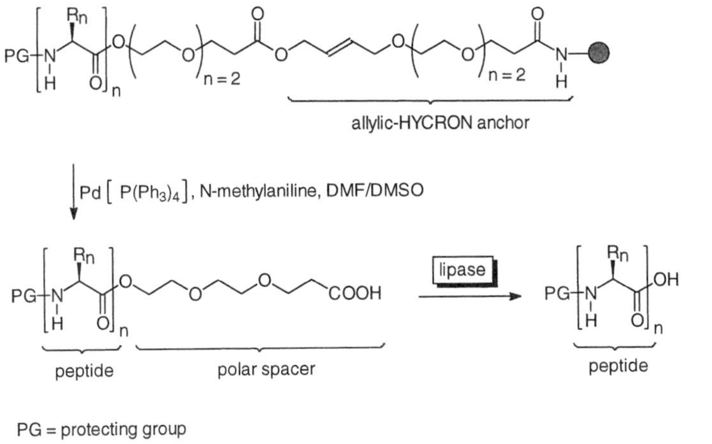

PG = protecting group

Figure 32 Enzymatic hydrolysis of a polar spacer in solid phase synthesis.

such a linker, experience gained in the development of enzymatically removable protecting groups served as the lead. The basic principle is similar to what has been described for the design of the PhAcOZ group (Fig. 15). The only additional feature required for the enzyme-labile linker was another functional group that was needed to link the phenolic moiety to the solid support. The linker was attached to TentaGelS-NH$_2$. The usability of such a system has been demonstrated by carrying out the synthesis of tetrahydro-β-carbolines by means of the Pictet-Spengler reaction and their subsequent enzyme-mediated release from the solid support (Fig. 33).

For the deprotection of an antibiotic p-nitrobenzyl ester in aqueous-organic solvent, the evolution of an esterase has been directed through sequential generations of random mutagenesis and screening. One p-nitrobenzylesterase variant performs as well in 30% dimethylformamide as the wild-type enzyme in water, reflecting a 16-fold increase in esterase activity. The positions of the effective amino acid substitutions have been identified in a p-nitrobenzylesterase structural model [50].

VI. HYDROLYSIS OF THIO ESTERS

Thiols are usually prepared from thio esters by hydrolysis under basic conditions or by reduction with metal hydrides. However, in the case of keto thio esters, such methods cannot be applied. In a recently reported [51] methodology, this problem was circumvented by subjecting these esters to lipase-catalyzed hydrolysis (Fig. 34). Thus, S-2-benzoylbutylacetothioate was treated separately with four lipases, e.g., PL (*Alcaligenes* sp.), AH (*Pseudomans* sp.), PS (*Pseudomonas cepacia*), and OF (*Candida cylindracea*) in isopropyl ether saturated with water. Lipase AH was well suited for this hydrolysis, giving 43% yield of enantiomerically pure (R)-2-benzoylbutane thiol. Only lipase OF showed the opposite enantioselectivity. However, in general, this lipase-catalyzed hydrolysis is a fairly useful method for the synthesis of thiols from the thio esters with chemically sensitive groups even if they do not have any stereogenic center.

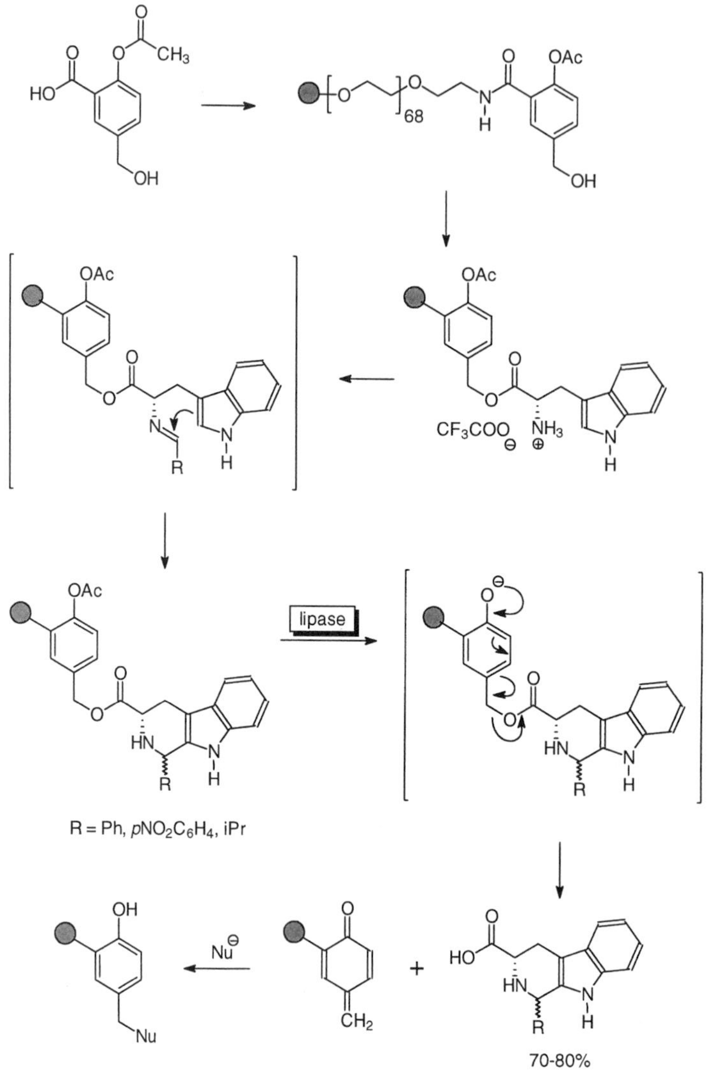

Figure 33   Use of enzyme-labile solid phase linker in the synthesis of tetrahydro-β-carbolines.

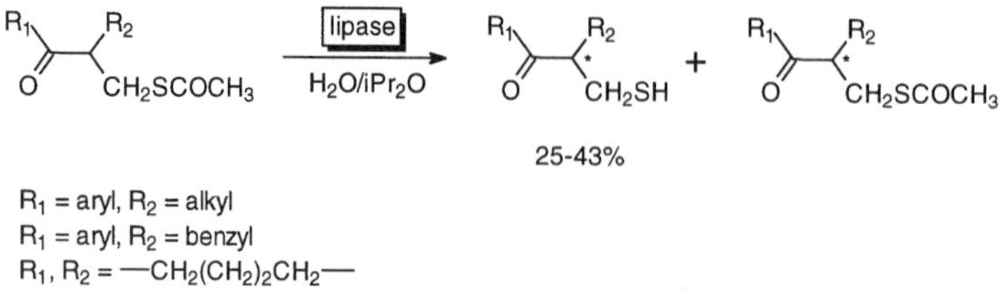

R₁ = aryl, R₂ = alkyl
R₁ = aryl, R₂ = benzyl
R₁, R₂ = —CH₂(CH₂)₂CH₂—

Figure 34   Lipase-catalyzed hydrolysis of keto thio esters.

VII. HYDROLYSIS OF PHOSPHATE ESTERS IN PRODRUGS

In spite of the increasing popularity of the enzymatic protecting group strategies, only a few attempts have been made to use biolabile phosphate protecting groups in the synthesis of phosphate containing biomolecules. The application of enzyme-labile phosphate protecting groups has gained momentum in recent years in the area of nucleotide prodrugs. In order to exert their biological effects, the $2',3'$-dideoxynucleoside analogs need to be converted to their $5'$-phosphates by enzymes present in the cells. In many cases, among the three successive phosphorylation steps, the monophosphorylation is rate-limiting and further conversions to the di- and triphosphates are catalysed by less specific kinases. The dependence on phosphorylation for activation of a nucleoside analog may, therefore, be a problem in cells where the activity of phosphorylating enzymes is known to be low or even lacking. This problem of phosphorylation may be circumvented by using suitably protected lipophilic nucleotides directly where the negative charges of the nucleoside phosphates are masked. This structural modification will make the nucleotides capable of penetrating the cell surface and will also make them stable to extracellular enzymes. The most important part of this whole process, however, is the ability of the phosphate protecting group to undergo enzymatic deprotection in order for the nucleotide triester to revert back to the charged nucleoside monophosphate under the influence of intracellular biocatalysts. To demonstrate the practical utility of this concept, $3'$-azido-$3'$-deoxythymidine (AZT) was converted to the corresponding bis(S-acetyl-2-thioethyl)phosphotriester derivative. The lipophilic AZT-phosphotriester was shown [52,53] to liberate the AZT-monophosphate inside the cell through a carboxylate esterase–mediated activation process (Fig. 35). This concept has been extended in the area of nucleotides derived from isodideoxynucleoside [54] and 9-[2(phosphonomethoxy)ethyl]adenine diester derivatives [55] incorporating carboxyesterase-labile S-$tert$-butyl-2-thioethyl moieties as transient protecting group. An interesting variation of this methodology utilized the neighboring group catalysis of a salicyl-based nucleotide prodrug where the carboxylate function unmasked by pig liver esterase attacked electron-deficient phosphorus intramolecularly to liberate [56] the nucleotide monophosphate (Fig. 36).

Enzyme-labile phosphate groups have also been utilized in the area of antisense oligonucleotides. Thus, enzymatically reversible dinucleotide phosphorothioate triesters were synthesized and the porcine liver esterase mediated cleavage of the acyloxyalkyl groups to generate the charged dinucleotide phosphorothioate was demonstrated [57]. In a more recent report, on the basis of the "pseudorotation principles," the same group of authors have selected acyloxyaryl functions as the phosphate protecting groups of dinu-

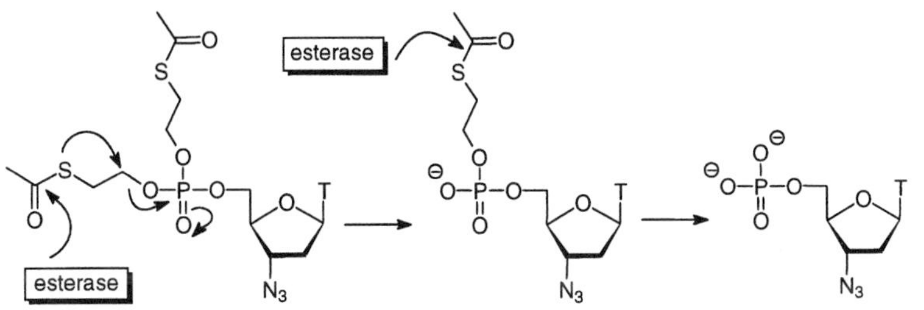

Figure 35 The conversion of AZT-phosphotriester to AZT-phosphomonoester by pig liver esterase.

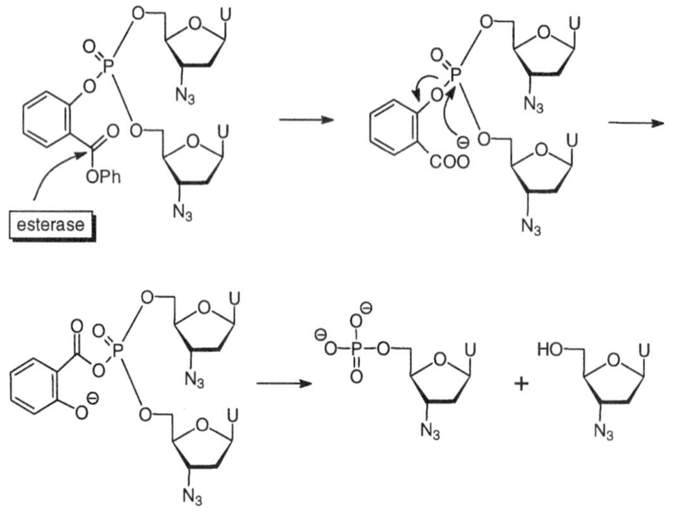

Figure 36 Neighboring group–catalyzed formation of a nucleoside-phosphomonoester initiated by esterase.

cleotide phosphorothioates. The biocatalytic removal of this class of protecting groups (Fig. 37) was more selective and reduced the S-P bond cleavage to a minimum [58]. It remains to be seen as to whether the knowledge generated in this area can be applied in the synthesis of phosphate-containing organic molecules in the future.

VIII. ENZYMES AND NO PROTECTING GROUPS

The fundamental reason for using protecting groups is to attach a particular functional group or a relatively large molecule to a defined region of a polyfunctionalized substrate. An unavoidable consequence of the introduction and removal of protecting groups, however, is that the synthesis is lengthened, leading to lower overall yield of the target molecules. Because enzymes are capable of combining biomolecules regioselectively in nature, it may be argued that unprotected or minimally protected polyfunctional molecules could be coupled in the presence of properly selected biocatalysts directly to generate the required product. This strategy will not only reduce the length but also lower the cost and increase the efficiency of a synthetic methodology. It has, for example, been established that the protection of the α-amino group of the acyl donor ester and of the carboxyl group

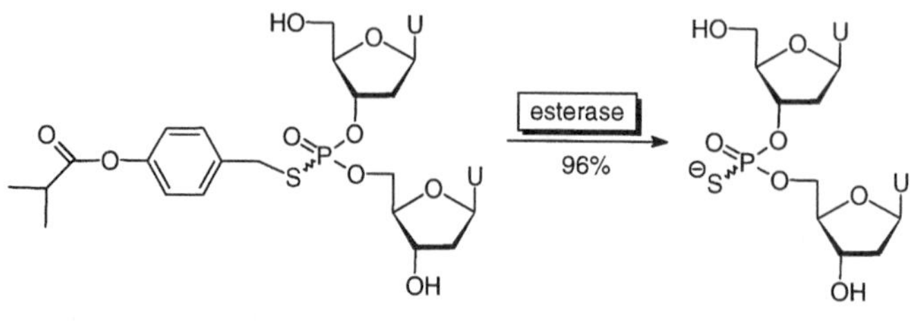

Figure 37 Esterase-mediated deprotection of a dinucleotide phosphorothioate triester.

of the amino component is not necessary for chymotrypsin-catalyzed peptide synthesis reactions. This strategy has been verified for penta- and hexapeptides by single-step segment condensation [59]. The concept of using biocatalysts for combining two unprotected molecules has gained popularity in recent years in the area of carbohydrates [60,61]. For example, aldolase-catalyzed synthesis of novel carbohydrate mimetics has been utilized [62,63] to generate a wide variety of polyfunctionalized molecules. Progress has also been made in the area of oligosaccharide synthesis using glycosyltransferases [60,61,64] and glycosidases [65–68] or multienzyme systems [60,61]. A combination of enzymatic protecting group techniques and regioselective enzymatic coupling has been utilized [41] in the synthesis of a fluorescence-labeled trisaccharide (Fig. 27). Enzyme-catalyzed synthesis of oligosaccharides that contain functionalized sialic acid has recently been reported [69]. An interesting report has appeared in which the properties of a bifunctionalized bacterial antibiotic resistance enzyme that catalyzes regioselective ATP-dependent 2″-phosphorylation and acetyl-CoA-dependent 6′-acetylation of aminoglycosides [70] is discussed. However it should be pointed out that the application of such methodologies is currently limited to a specific group of organic molecules.

IX. CONCLUSIONS

The use of protecting groups is going to remain an active area of research in the construction of polyfunctionalized molecules for the foreseeable future. In this regard, not only do the biocatalyzed transformations complement the arsenal of nonenzymatically removable protecting groups, but in many cases they additionally offer the opportunity to carry out useful functional group interconversions with selectivities that cannot or can hardly be matched by classical chemical techniques. However, the use of biocatalysts in protecting group chemistry in the sense of a general method deserves further intensive development. Thorough and methodical investigations on the applications of the known enzymes need to be carried out to expand the frontiers of protecting group chemistry. On the other hand, owing to the advances in recombinant DNA technology and structural biology, it is now possible to produce virtually any enzyme and its engineered variants in large quantities. Continued exploration of new techniques such as microwave-promoted biocatalysis [71], enzyme reactions under high pressure [72], mechanically stable heterogeneous biocatalysts [73], and crosslinked enzyme crystals (CLECs) [74] are increasing the usability of biocatalysts under unnatural conditions. Sequential cycles of random mutagenesis and screening have been used to direct the evolution of a subtilisin which is hundreds of times more active than its wild-type ancestor in high concentrations of dimethylformamide [75]; directed evolution of a *para*-nitrobenzyl esterase has already been discussed [50]. Application of nonconventional enzyme catalysis [76] or discovery of enzymes from microorganisms growing in extreme environments [77,78] could change the face of biocatalysis in the near future. All of these new techniques and discoveries in combination with existing knowledge will continue to expand the scope and use of enzymatic protecting group chemistry in complex multistep synthesis of polyfunctionalized organic molecules.

ACKNOWLEDGMENTS

This research was supported by the Deutsche Forschungsgemeinschaft and the Fonds der Chemischen Industrie. Tanmaya Pathak thanks the Alexander von Humboldt Foundation for a fellowship.

REFERENCES

1. WT Greene, PGM Wuts. Protective Groups in Organic Synthesis, 2nd ed. New York: John Wiley and Sons, 1991.
2. PJ Kocienski. Protecting Groups. Stuttgart: Thieme, 1994.
3. M Schelhaas, H Waldmann. Angew Chem 35:2192–2219, 1996; Angew Chem Int Ed Engl 35:2057–2083, 1996.
4. H Waldmann, D Sebastian. Chem Rev 94:911–937, 1994.
5. K Drauz, H Waldmann. Enzyme Catalysis in Organic Synthesis. Weinheim: VCH, 1995, pp 851–890.
6. T Pathak, H Waldmann. Curr Opin Chem Biol 2:112–120, 1998.
7. T Kappes, H Waldmann. Liebigs Ann/Recueil 803–813, 1997.
8. AK Prasad, J Wengel. Nucleosides Nucleotides 15:1347–1359, 1996.
9. M Schelhaas, S Glomsda, M Hänsler, H-D Jakubke, H Waldmann. Angew Chem 108:82–85, 1996; Angew Chem Int Ed Engl 35:106–109, 1996.
10. H Waldmann, A Heuser, S Schulze. Tetrahedron Lett 48:8725–8728, 1996.
11. D Sebastian, A Heuser, S Schulze, H Waldmann. Synthesis 1098–1108, 1997.
12. H Waldmann, A Reidel. Angew Chem 109:642–644, 1997; Angew Chem Int Ed Engl 36:647–649, 1997.
13. TF Favino, G Fronza, C Fuganti, D Fuganti, P Grasselli, A Mele. J Org Chem 61:8975–8979, 1996.
14. K Zegelaar-Jaarsveld, SAW Smits, NCR Van Straten, GA Van Der Marel, JH Van Boom. Tetrahedron 52:3593–3608, 1996.
15. NCR Van Straten, HI Duynstee, E De Vroom, GA Van Der Marel, JH Van Boom. Liebigs Ann/Recueil 1215–1220, 1997.
16. OH Justiz, R Fernandez-Lafuente, JM Guisan, P Negri, G Pagani, M Pregnolato, M Terreni. J Org Chem 62:9099–9106, 1997.
17. T Maugard, M Remaud-Simeon, D Petre, P Monsan. Tetrahedron 53:7587–7594, 1997.
18. T Maugard, M Remaud-Simeon, D Petre, P Monsan. Tetrahedron 53:7629–7634, 1997.
19. CA Costello, AJ Kreuzmann, MJ Zmijewski. Tetrahedron Lett 37:7469–7472, 1996.
20. M Pozo, R Pulido, V Gotor. Tetrahedron 48:6477–6484, 1992.
21. M Pozo, V Gotor. Tetrahedron 49:4321–4326, 1993.
22. M Pozo, V Gotor. Tetrahedron 49:10725–10732, 1993.
23. B Orsat, PB Alper, W Moree, C-P Mak, C-H Wong. J Am Chem Soc 118:712–713, 1996.
24. S Grabowski, J Armbruster, H Prinzbach. Tetrahedron Lett 38:5485–5488, 1997.
25. A Maestro, C Astorga, V Gotor. Tetrahedron: Asymmetry 8:3153–3159, 1997.
26. AS Bommarius, K Drauz, K Gunther, G Knaup, M Schwarm. Tetrahedron: Asymmetry 8:3197–3200, 1997.
27. M Adamczyk, J Grote. Tetrahedron Lett 37:7913–7916, 1996.
28. S Conde, P Lopez-Serrano, M Fierros, M Isabel Biezma, A Martinez, MI Rodriguez-Franco. Tetrahedron 53:11745–11752, 1997.
29. T Pohl, H Waldmann. Angew Chem 108:1829–1832, 1996; Angew Chem Int Ed Engl 35:1720–1723, 1996.
30. T Pohl, H Waldmann, J Am Chem Soc 119:6702–6710, 1997.
31. A Baldessari, AC Bruttomesso, EG Gros. Helv Chim Acta 79:999–1004, 1996.
32. T Itoh, A Uzu, N Kanda, Y Takagi. Tetrahedron Lett 37:91–92, 1996.
33. O Houille, T Schmittberger, D Uguen, Tetrahedron Lett 37:625–628, 1996.
34. T Itoh, K Mitsukura, K Kaihatsu, H Hamada, Y Takagi, H Tsukube. J Chem Soc Perkin Trans 1:2275–2278, 1997.
35. E Ozaki, K Sakashita. Chem Lett 741–742, 1997.
36. S Conde, M Fierros, I Dorronsoro, ML Jimeno, MI Rodriguez-Franco. Tetrahedron 53:11481–11488, 1997.
37. L Lay, L Panza, S Riva, M Khitri, S Tirendi. Carbohydr Res 291:197–204, 1996.
38. I Redmann, M Pina, B Guyot, P Blaise, M Farines, J Graille. Carbohydr Res 300:103–108, 1997.

39. V Petrovic, S Tomic, D Ljevakovic, J Tomasic. Carbohydr Res 302:13–18, 1997.
40. T Sugai, H Okazaki, A Kuboki, H Ohta. Bull Chem Soc Jpn 70:2535–2540, 1997.
41. T Yasukochi, K Fukase, Y Suda, K Takagaki, M Endo, S Kusumoto. Bull Chem Soc Jpn 70: 2719–2725, 1997.
42. VS Parmar, HN Pati, SK Sharma, A Singh, S Malhotra, A Kumar, KS Bisht. Bioorg Med Chem Lett 6:2269–2274, 1996.
43. A Cipiciani, F Fringuelli, AM Scappini. Tetrahedron 52:9869–9876, 1996.
44. D Sebastian, H Waldmann. Tetrahedron Lett 38:2927–2930, 1997.
45. J Eberling, P Braun, D Kowalczyk, M Schultz, H Kunz. J Org Chem 61:2638–2646, 1996.
46. H Waldmann, S Gabold. Chem Commun 1861–1862, 1997.
47. V Jungmann, H Waldmann. Tetrahedron Lett 39:1139–1142, 1998.
47a. S Flohr, V Jungmann, H Waldmann. Chem Eur J 5:669–681, 1999.
48. M Gewehr, P Braun, O Seitz, H Kunz. Poster presented at the 24th Symposium of the European Peptide Society, Edinburgh, September 8–13, 1996.
49. B Sauerbrei, V Jungmann, H Waldmann. Angew Chem 110:1187–1190, 1998; Angew Chem Int Ed Engl 37:1143–1146, 1998.
50. JC Moore, FH Arnold. Nat Biotech 14:458–467, 1996.
51. T Izawa, Y Terao, K Suzui. Tetrahedron: Asymmetry 8:2645–2648, 1997.
52. C Perigaud, J-L Girardet, I Lefebvre, M-Y Xie, A-M Aubertin, A Kirn, G Gosselin, J-L Imbach, J-P Sommadossi. Antiviral Chem Chemother 7:338–345, 1996.
53. C Thumann-Schweitzer, G Gosselin, C Perigaud, S Benzaria, J-L Girardet, I Lefebvre, J-L Imbach, A Kirn, A-M Aubertin. Res Virol 147:155–163, 1996.
54. G Valette, A Pompon, J-L Girardet, L Cappellacci, P Franchetti, M Grifantini, P La Colla, AG Loi, C Perigaud, G Gosselin, J-L Imbach. J Med Chem 39:1981–1990, 1996.
55. S Benzaria, H Pelicano, R Johnson, G Maury, J-L Imbach, A-M Aubertin, G Obert, G Gosselin. J Med Chem 39:4958–4965, 1996.
56. S Khamnei, PF Torrence. J Med Chem 39:4109–4115, 1996.
57. RP Iyer, D Yu, S Agrawal. Bioorg Chem 23:1–21, 1995.
58. RP Iyer, D Yu, T Devlin, N-H Ho, S Agrawal. Bioorg Med Chem Lett 6:1917–1922, 1996.
59. S Gerisch, H-D Jakubke. J Peptide Sci 3:93–98, 1997.
60. C-H Wong. Acta Chem Scand 50:211–218, 1996.
61. GJ McGarvey, C-H Wong, Liebigs Ann/Recueil 1059–1074, 1997.
62. F Moris-Varas, XH Qian, C-H Wong. J Am Chem Soc 118:7647–7652, 1996.
63. L Qiao, BW Murray, M Shimazaki, J Schultz, C-H Wong. J Am Chem Soc 118:7653–7662, 1996.
64. L Panza, PL Chiappini, G Russo, D Monti, S Riva. J Chem Soc Perkin Trans 1:1255–1256, 1997.
65. S Singh, M Scigelova, DHG Crout. Chem Commun 993–994, 1996.
66. G Vic, M Scigelova, JJ Hastings, OW Howarth, DHG Crout. Chem Commun 1473–1474, 1996.
67. I Matsuo, H Fujimoto, M Isomura, K Ajisaka. Bioorg Med Chem Lett 7:255–258, 1997.
68. U Gambert, J Thiem. Carbohyd Res 299:85–89, 1997.
69. MD Chappell, RL Halcomb. J Am Chem Soc 119:3393–3394, 1997.
70. E Azucena, I Grapsas, S Mobashery. J Am Chem Soc 119:2317–2318, 1997.
71. J-R Carrillo-Munoz, D Bouvet, E Guibe-Jampel, A Loupy, A Petit. J Org Chem 61:7746–7749, 1996.
72. S Kunugi. Ann NY Acad Sci 672 (Enzyme Eng XI):293–304, 1992.
73. MT Reetz, A Zonta, J Simpelkamp, W Könen. Chem Commun 1397–1398, 1996.
74. T Zelinski, H Waldmann. Angew Chem 109:746–748, 1997. Angew Chem Int Ed Engl 36: 722–724, 1997.
75. L You, FH Arnold. Protein Eng 9:77–83, 1995.
76. H-D Jakubke, U Eichhorn, M Hänsler, D Ullmann. Biol Chem 377:455–464, 1996.
77. MWW Adams, RM Kelly. Chem Eng News 18:32–42, 1995 (Dec.).
78. MB Brennan. Chem Eng News 14:31–33, 1996 (Oct.).

27

Enzymatic Reactions in Supercritical Carbon Dioxide

Thorsten Hartmann, Eckhard Schwabe, and Thomas Scheper
University of Hannover, Hannover, Germany

Didier Combes
Centre de BioIngénierie Gilbert Durand, INSA, Toulouse, France

I. INTRODUCTION

Over the last several years supercritical carbon dioxide ($SCCO_2$) has attracted attention as an alternative solvent to replace conventional solvents. For applications in the food industry and pharmacy, CO_2 has considerable advantages over commonly used solvents like hexane or dichloromethane. It is cheap, neither combustible nor explosive, and is a GRAS (Generally Regarded As Safe) substance acceptable for all uses in foodstuff production (EC directive 88/344/EEC). It is environmentally harmless because it is easily separated from the products and recycled. The low critical temperature of 31°C makes $SCCO_2$ an ideal solvent for heat-sensitive substances and biocatalytic conversions. The potential of using $SCCO_2$ as a reaction medium for enzymatic reactions has been investigated since 1985 [1]. By using CO_2 as a solvent in both the reaction and the subsequent separation by SFE, the design of integrated processes is possible with higher efficiency in terms of engineering and economy. The drastic changes in the thermodynamic state of the supercritical phase caused by minor variations of temperature and/or pressure can be exploited to control reaction selectivity. $SCCO_2$ also offers the same advantage as organic solvents, such as solubilization of hydrophobic compounds, ease of recovery of the enzyme, and the possibility of performing reactions which are thermodynamically unfavorable in water (e.g., synthesis of esters and amides). The use of a supercritical solvent that returns to its gaseous state at atmospheric pressure avoids the need for additional treatment to completely remove the residual solvent.

II. UNUSUAL PROPERTIES OF SUPERCRITICAL FLUIDS

The phase diagram of CO_2, showing the conditions of solid, liquid, and gaseous state as well as the critical point, can be seen in Fig. 1. At conditions below the critical point the liquid phase can be in coexistence with its gaseous phase. Moving along the curve of coexistence in the direction of increasing pressure, the gaseous phase is compressed until its density reaches that of the liquid phase. Beyond the critical point, both phases are

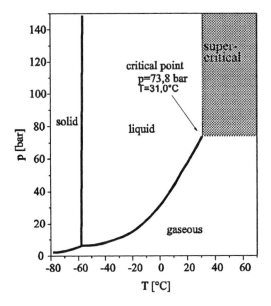

Figure 1 Phase diagram of CO_2.

indistinguishable. Some of the properties of supercritical fluids (SCFs) are typical for liquids; others are typical for gases. This interesting combination make SCFs particularly favorable solvents for extraction and separation processes. The most important properties are as follows.

A. Controllable Solvating Power

The solvating power of an SCF resembles that of a liquid. It varies dramatically with variations of pressure and temperature near the critical point. This property is utilized for various separation processes [2,3]. Dielectric constant, solubility parameter, diffusivity, and viscosity can be adjusted simply by changing the reactor pressure or temperature [4]. This will also change enzyme stereoselectivity [5–7]. Without changing solvent and using pressure as the sole adjustable parameter, both enzyme activity and stereoselectivity can be predictably tailored in supercritical environments [8].

The main advantage of $SCCO_2$ lies in the easiness of the possible downstream processing. Separating products from reactants can be greatly facilitated by the ease with which the solvent power of SCF solvents can be adjusted. By coupling a series of depressurization steps at the outflow of the reaction vessel, it is possible to selectively fractionate products and residual substrates of a reaction [9]. The fluid can be recycled after repressurizing. Different methods are described for the calculation of solubility in SCFs [10–13].

Pure $SCCO_2$ is quite unpolar (like hexane), limiting its application as a solvent for bioprocesses mainly to lipids and other unpolar compounds. By introduction of modifiers the polarity of $SCCO_2$ can be increased, thus making it possible to solute polar compounds which are merely soluble in the pure fluid [14].

Table 1 Physical Properties of Carbon Dioxide [15]

Parameter	Gas	Liquid	Supercritical phase
Density [g/cm^3]	10^{-3}	0.6 to 1.6	0.4 to 1.0
Diffusion coefficient [cm^2/s^1]	10^{-1}	2×10^{-6}	7×10^{-4}
Viscosity [g/cm^{-1}/s^1]	10^{-4}	3×10^{-2}	3×10^{-4}

B. Low Viscosity

Supercritical fluids exhibit significantly lower viscosity than liquids (typically one order of magnitude lower), which provides favorable flow properties. This permits the fluids to penetrate matrices with low permeability more readily than conventional solvents.

In Table 1 the properties of supercritical CO_2 are compared to those of the liquid CO_2. $SCCO_2$ has unlimited miscibility with gases. Oxygen has already been used for oxidation reactions.

The major disadvantage of $SCCO_2$ is the high pressure and the resulting need for expensive and sophisticated equipment, but higher investment costs are compensated by the many advantages of the supercritical state. $SCCO_2$ is already used in industrial scale for the extraction of caffeine and flavor compounds [16]. It is also used for wastewater treatment ($SCCO_2$ dissolves ketones, chlorinated solvents, and oils), polymer manufacturing, soil remediation, and as a solvent for pharmaceuticals. Furthermore, it is the mobile phase in "supercritical fluid chromatography" [17].

C. High Rates of Mass Transfer

High rates of diffusion (typically an order of magnitude higher than liquids) permit faster extraction than with liquid solvents. This is an important property for extraction processes, as the rate of extraction is limited by the transport of molecules from the matrix to the bulk fluid. The excellent mass transfer conditions during a reaction also lead to high production rates.

III. ENZYMATIC REACTION IN SUPERCRITICAL CO_2

A. New Strategies for Enzymatic Reactions: The Use of Biocatalysts in Supercritical Media

In the last few years the use of enzymes in nonaqueous media, especially in SFCs has received much attention. As described in Sec. II, SCFs offer some advantages of organic solvents, such as dissolving properties for hydrophobic compounds, ease of enzyme recovery, separation of products and unreacted substrates, and the possibility of performing reactions that are thermodynamically unfavorable in water. One of the best advantages of SCFs is their excellent transport properties and adjustable solvent power, due to their low viscosity and high diffusivity.

Many studies followed the first attempt of Randolph et al. in 1985 [1] to operate enzymatic reactions in SCFs. In the same year, Hammond et al. [18] reported the oxidation of p-cresol and p-chlorophenol to the corresponding benzoquinones with oxygen catalyzed by a polyphenol oxidase in fluoroform and in carbon dioxide as supercritical solvents. In the last few years the development of the enzymatic resolution of chiral compounds in

$SCCO_2$ was pushed to the fore. The separation of pure enantiomers with the help of enzymes in $SCCO_2$ is interesting for the pharmaceutical industry, where the control of organic solvent residues is crucial. Table 2 shows a list of works about enzymatic reactions in SFC. Besides nontoxicity, CO_2 offers numerous advantages. It is nonexplosive and inflammable. Its critical pressure is relatively low (73.8 bar), and its critical temperature (31.0°C) is consistent with the use of biochemical compounds, especially with the activity of enzymes. Its natural character is of interest, in many cases, as a replacement for organic solvents.

In most cases all of these reactions take place in batch processes (Fig. 2). The substrates are dissolved, the enzyme is added into the reactor, and the vessel is closed. After the reaction the reactor is cooled in dry ice or liquid nitrogen. The solid CO_2 sublimates and the products remain in the reactor.

The difficulty of this procedure is the unknown start time of the reaction. Marty et al. [54] proposed a novel experiment device that allows the gathering of numerous data in a single run. They use a recirculating reactor with a sample loop. This vessel is made of saphire, so a visual control is possible.

In Fig. 2B a semicontinuous process is shown. It was performed by Hammond et al. [18], Miller et al. [74], and Randolph et al. [64]. $SCCO_2$ flows through a "saturation vessel" in which substrates are stored. The substrates can be impregnated on glasswool. The CO_2 dissolves the reactants and transports them into the reactor. It is possible to produce a monophasic and saturated mixture. At the outlet of the reactor $SCCO_2$ is expanded and all dissolved components condense. The CO_2 can be recycled.

A continuous process was performed by van Eijs et al. [69] and Marty et al. [54]. Substrates can be continuously injected into the $SCCO_2$ flow by a high-performance liquid chromatography (HPLC) pump. The substrate concentration is known, an adjustment to a fixed value is possible. In this case the CO_2 was not recycled. They also developed a continuous process with recycling of CO_2 and post reactional fractionation (Fig. 2) [2]. Table 3 shows studies of important parameters for reactivity and productivity of enzymatic reactions in supercritical media. Furthermore, some authors compare supercritical media and conventional organic solvents:

Nakamura et al., 1986 [61]
Chi, Nakamura, Yano, 1988 [75]
Randolph, Blanch, Prausnitz, 1988 [64]
Pasta et al., 1989 [63]
Kamat et al., 1992 [46]
Marty et al., 1990, 1992, [54,976]
Dumont et al., 1992 [33]

Martins et al. [52] investigated the enantioselective esterification of glycidol with butyric acid at 35°C and 140 bar. They received a enantiomeric excess of (S)-glycidyl butyrate with lipase from porcine pancreas of 72–82%. The conversions were between 11% and 23%. With immobilized lipase they optimize the enantiomeric excess to 88–90% e.e. and a conversion of 20%. The e.e. of the substrate was 20%. In his dissertation, Bornscheuer [83] described the lipase-catalyzed transesterification of 3-hydroxymethyloctanoate with cyclohexyl acetate in organic solvents and supercritical CO_2 (40°C/100 bar). After 400 h reaction time the author yielded 20% substrate and 96% product e.e. The conversion was 17% and no remaining lipase activity was detected. With vinyl acetate Bornscheuer found an e.e. of 37% for the substrate, 3% for the product, and a turnover of 92%. Three years later Bornscheuer et al. [24] described the lipase-catalyzed kinetic resolution of the 3-hydroxymethyloctanoate with vinyl acetate in $SCCO_2$. The process was monitored on-line measuring the absorbance of the byproduct acetic aldehyde by an ultraviolet (UV) photometer. The $SCCO_2$, the reactants, and the products flow through a high-pressure flow cell. They achieved optical purities and conversions (75% conversion, 32% e.e. product, 130 h) similar to the values in *n*-hexane or *n*-dodecane (90–99% e.e., 8 h). The reaction time in CO_2 was much higher compared to the resolution in organic solvents.

The use of supercritical carbon dioxide for enzymatic reactions is limited by the difficulty of dissolving relatively polar substrates. To increase the solubility of polar compounds, Castillo et al. [84] complexed the substrate with phenylboronic acid or immobilized polar reactants on silica gel. As model reaction the esterification of oleic acid with glycerol or D-fructose was used.

B. Results of the Investigation of Different Parameters Influencing Reactions in $SCCO_2$

1. Pressure

The increase in pressure leads to an increase in the density of the fluid, improving the solubilizing power. In addition, the dielectricity constant and its polarity will be modified. Randolph et al. studied these effects [64]. They investigated reactant solubility effects on the oxidation rate of cholesterol, as the solubility of cholesterol in $SCCO_2$ increases with pressure.

Some other studies were made by Nakamura et al. [85], Chi et al. [75], and Miller et al. [74]. In the same year, Erikson et al. [35] studied the interesterification between palmitic acid and trilaurine, catalyzed by immobilized lipase of *Rhizopus arrhizus* in $SCCO_2$. When increasing the pressure from 8 to 10 MPa they observed a decrease of initial velocity. They proposed an influence of acidification by dissolving CO_2 in the aqueous phase. The authors concluded that the partition of reactants between supercritical phase and the enzyme vicinity was affected by pressure. With increasing pressure the reactants dissolve better in the supercritical phase, so their local concentration decreases in the enzyme vicinity leading to a reduced activity.

Van Eijs et al. [69] investigated the continuous transesterification of isoamyl acetate in $SCCO_2$ in a fixed-bed reactor. A productivity drop with increasing pressure was explained by decreasing of the diffusion coefficient of the solutes, resulting in an internal mass transfer limitation. Another hypothesis dealt with a modification of the protein structure at higher pressures.

Ikushima et al. [77] observed a strong change in enantioselectivity for the kinetic resolution of citronellol depending on pressure (76–190 bar) and temperature yielding racemic or highly optical pure product (98.9% e.e.).

Martins et al. [78,80] described the enzymatic resolution of racemic glycidol in $SCCO_2$. The chosen reaction route was the esterification with butyric acid catalyzed by either free or immobilized PPL (porcine pancreas lipase). The solubility of glycidol was measured in CO_2 at 35°C and pressures in the range of 70–180 bar for butyric acid concentrations up to 500 mM. The partitioning of water between CO_2 and the enzyme

Table 2 List of Some Previous Published Works on Enzymatic Reactions and Enzyme Behavior in Supercritical Fluids, Especially CO_2

Ref.	Enzyme	Reaction, reactants	Solvent	Reactor
Adshiri et al. [19]	Lipase	Interesterification	CO_2	Extractive reaction
Balaban et al. [20]	Pectinesterase	Activity	CO_2	
Berg et al. [21]	Lipase, immobilized	Interesterification of edible fat	CO_2	Solid phase extraction cartridge
Bernard and Barth [22]	Lipase from *M. miehei*	Esterification of myristic acid	CO_2; hexane	
Bolz et al. [23]	Lipase	Different reactions	CO_2, hexane	
Bornscheuer et al. [24]	Lipase	Transesterification	CO_2	Batch
Cai [25]	Lipase	Esterification of butyric acid	CO_2, hexane	
Capewell et al. [26]	Lipase	Transesterification	CO_2, *n*-hexane	Batch, batch with recirculating fluid
Cernia and Palocci [27]	Review, lipase	Esterification	CO_2 and others	
Chrisochoou and Schaber [28]	Different enzymes	Transesterification of ethylacetate and isoamyl alcohol	CO_2	Cascade
Chrisochoou, Schaber, Bolz [29]	Different enzymes	Transesterification	CO_2	
Chrisochoou, Schaber, Stephan [30]	Lipase	Transesterification	CO_2	
Combes [31]	Lipozyme	Esterification of oleic acid	CO_2, hexane	
Diaz; Cobos; de la Hoz; Ordonez [32]	Review, different enzymes	Enzyme activity	CO_2	
Dumont et al. [33]	Lipase	Esterification	CO_2	Batch
Endo, Fujimoto, Arai [34]	Immobilized lipase	Interesterification between tricaprylin and methyl oleate	CO_2	
Erickson et al. [35]	Lipase	Interesterification	CO_2	Batch
Glowacz et al. [36]	Lipase	Hydrolysis of triolein and its partial glycerides	CO_2	
Goddard, Bosley, Al-Duri [37]	Lipozyme	Esterification between oleic acid and ethanol	CO_2	Continuous packed bed

Reference	Enzyme	Reaction/Topic	Fluid	Reactor
Gunnlaugsdottir and Sivik [38]	Lipase	Alcoholysis of cod liver oil	CO_2	Stirred-tank reactor
Habulin, Krmelj, Knez [39]	Different enzymes	Synthesis of oleyl oleate	CO_2	Semicontinuous
Hammond et al. [18]	Polyphenol oxidase	Oxidation	CO_2, fluoroform	
Ikariya, Jessop, Noyori [40]	Review, different enzymes	Different reactions	CO_2, water	
Ikushima [41]	Review, lipase	Ester synthesis	CO_2	
Ikushima [42]	Lipase	Ester synthesis	CO_2	
Ishikawa and Osajima [43]	Review	Enzyme inactivation	CO_2	
Ishikawa et al. [44]	Glucoamylase, acid and alkaline protease, lipase	Activity of the enzymes	Microbubbled system of $SCCO_2$	
Jackson et al. [45]	Lipase, immobilized	Triglycerides, palm olein, high-stearate soybean oil	CO_2	
Kamat, Beckman, Russell [8]	Review	Activity of enzymes	Supercritical fluids	
Kamat et al. [46]	Lipase	Transesterification	Ethane, ethylene, CO_2, fluoroform, propane, sulfur, hexafluoride	Batch
Kamat, S. [47]	Different enzymes	Enzyme activity and selectivity	Supercritical fluids	
Lee et al. [48]	α-Amylase, glucoamylase	Hydrolysis	CO_2	
Lin, C. [49]	Review, different enzymes	Reactions, activity, stability of enzymes, reaction rate	Supercritical fluids	
Lin, Qiu, Wang [50]	Review, different enzymes	Interesterifications	CO_2	
Lozano et al. [51]	Immobilized-α-chymotrypsin	Stability of the enzyme	CO_2	
Martins et al. [52]	Lipase	Esterification	CO_2	Batch
Marty et al. [53]	Different enzymes	Reaction-fractionation process	CO_2	
Marty et al. [54,55]	Lipase	Esterification	CO_2	
Michor et al. [56,57]	Different lipases and one esterase	Transesterification of racemic citronellol and menthol	CO_2	Batch, continuous

Table 2 Continued

Ref.	Enzyme	Reaction, reactants	Solvent	Reactor
Michor, Gamse, Marr [58]	Lipases, esterase	Racemate separation of D,L-Menthol	CO_2	
Miller et al. [59]	Lipase	Interesterification	CO_2	Continuous
Nakamura [60]	Review, lipase	Trans- and esterifications	CO_2	
Nakamura, Chi [61]	Lipase	Interesterification	CO_2	Batch
Parve et al. [62]	Lipase (lipolase)	Hydrolysis of bicyclo[3.2.0]heptanol esters	CO_2	
Pasta et al. [63]	Subtilisin	Transesterification	CO_2	Batch
Randolph et al. [64]	Cholesterol oxidase	Oxidation	CO_2	Semicontinuous
Rantakylä et al. [65]	Lipase	Hydrolysis	CO_2	Batch
Russell, Beckman, Chaudhary [66]	Review, different enzymes	Activity, specificity and stability of enzymes	Supercritical fluids	
Sereti; Stamatis, Kolisis [67]	*Fusarium solani* cutinase	Stability, reactivity, esterification of hexanoic acid with hexanol	CO_2	
Snyder, King, Jackson [68]		Methylation of lipids	Supercritical fluids	
Van Eijs et al. [69]	Lipase	Transesterification	CO_2	Continuous and extractive reaction
Vija, Telling, Tougu [70]	Lipase	Esterification	CO_2, *n*-hexane	
Xu, Zhu, Yang [71]	Review, different enzymes	Stability, reactions	Supercritical fluids, organic solvents	
Yoon et al. [72]	Lipase (lipozyme)	Transesterification between triolein and ethylbehenate	CO_2	
Zheng and Tsao [73]	Cellulase	Hydrolysis	CO_2	

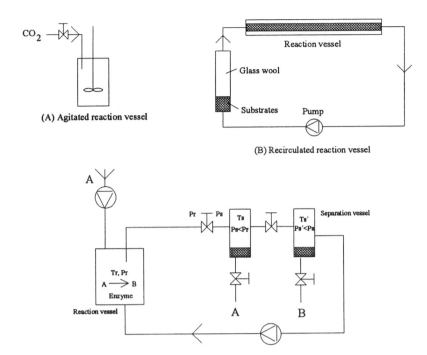

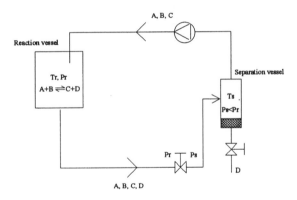

Figure 2   Some schematic set ups for SCCO$_2$ reactions.

preparations was quantified for various amounts of added water. A series of supports covering a wide range of hydrophilicities were screened for enzyme activity.

Rantakylä and Aaltonen [79] studied the enantioselective esterification of racemic ibuprofen with *n*-propanol by immobilized lipase in SCCO$_2$. The enantiomeric excess of the product was 70% at 15–20% conversion. The enantioselectivity was affected by temperature and concentration of ibuprofen and lipase. The initial reaction rate increased with pressure, but enantioselectivity was not affected by pressure changes. The reaction rates in SCCO$_2$ were similar in *n*-hexane.

Table 3 Parameters for Reactivity and Productivity of Enzymatic Reactions in Supercritical Media

Parameter	Ref.
Pressure	Nakamura et al. [61]
	Chi, Nakamura, Yano [75]
	Randolph, Blanch, Prausnitz [64]
	Erikson, Schyns, Cooney [35]
	Marty et al. [9]
	Ikushima et al. [77]
	Capewell et al. [26]
	Martins et al. [78]
	Rantakylä and Aaltonen [79]
Mass transfer limitations	van Eijs et al. [69]
	Randolph, Blanch, Prausnitz [64]
	Dumont et al. [33]
Water content	Nakamura et al. [61]
	Chi, Nakamura, Yano [75]
	van Eijs et al. [69]
	Marty et al. [9,54]
	Dumont et al. [33]
	Glowacz et al. [36]
	Martins et al. [80]
	Michor, Marr, Gamse [56]
	Bernard and Barth [81]
Cosolvent's effects	Randolph, Blanch, Prausnitz [64]
	Capewell et al. [26]
Temperature	Kamat et al. [82]
	Michor et al. [58]
	Rantakylä and Aaltonen [79]

2. Water Content

The first indication about the influence of water content was found by Randolph et al. [64]. They investigated a continuous oxidation of cholesterol in a packed-bed reactor, fed with anhydrous $SCCO_2$. They observed a rapid decrease in conversion. After addition of 0.1% (v/v) water, the authors obtained the initial activity.

Van Eijs et al. [69] proved a direct influence of water on the activity of an enzyme. They examined the productivity as a function of the water added to $SCCO_2$. The continuous transesterification between ethyl acetate and isoamyl acetate was used as a model reaction. The results proved that the enzyme needs water to maintain the enzymatically active form, but to avoid hydrolysis as a side reaction it would be appropriate to work in anhydrous solvents.

An optimal water concentration in the reaction phase must be found. The authors determined an isotherm for water adsorption between $SCCO_2$ (10 MPa, 60°C) and Lipozyme (a lipase from *Mucor miehei*, immobilized on a macroporous hydrophilic support (Duolite) and commercialized by Novo Nordisk).

Marty et al. [54] studied esterification activity vs. water content of the enzyme. They used Lipozyme/$SCCO_2$ and Lipozyme/*n*-hexane. They obtained sharp, bell-shaped curves, an evidence that enzymes need a small quantity of water to maintain their active conformation.

Glowacz et al. [36] examined the stereoselectivity of PPL during enzymatic hydrolysis of triolein and its partial glycerides in the presence of $SCCO_2$. The water content of the immobilized lipase was varied. The stereoselectivity depends on the reaction time, the substrates, and the enzyme water content. The authors suggested that the effect of the enzyme water content on the activity and selectivity of PPL is based on a modification of the microenvironment of the enzyme by the solution of CO_2 in water, causing a decrease of the pH value. The PPL with 1.5% water content showed a preference for the *sn*-3 position. With 5% water content only the racemate of these glycerides could be detected. The lipase with 15% water content showed an *sn*-3 enantioselectivity only in the analysis of the dioleins and *sn*-1,2-dioleins were performed preferentially.

3. Mass Transfer

One great advantage of $SCCO_2$ is the very good mass transfer property due to its low viscosity and high diffusivity (sec. 2). Table 4 lists some articles about the mass transfer properties.

4. Cosolvent Addition

The restricted solubility of polar compounds is a disadvantage of $SCCO_2$. A solution for using hydrophobic substrates is the addition of small amounts of polar cosolvents (also known as "modifiers") like methanol or ethanol. This increases the polarity of the whole solvent. Table 5 gives some references for cosolvents in supercritical fluids.

5. Comparison SFC/Organic Solvents

Some investigations were made to compare supercritical fluids with organic solvents like *n*-hexane. In 1988, Chi [75] examined the hydrolysis and esterification of triolein. They found a fourfold increase of the initial reaction rate when using $SCCO_2$ instead of *n*-hexane. These results were explained with the better mass transfer properties of $SCCO_2$ and the higher dielectric constant of water in the microenvironment of the enzyme.

Kamat et al. [46] observed a 10-fold higher reaction rate in *n*-hexane for the transesterification between methyl methacrylate and ethyl hexane catalyzed by a free lipase from *Candida rugosa*. The direct effect of CO_2 was studied by performing the reaction in *n*-hexane with bubbled CO_2. The carbon dioxide can cause a drop of pH in the aqueous phase surrounding the enzyme, leading to a decrease of enzyme activity.

Marty et al. [54] and Dumont et al. [33] compared the kinetic parameters for two different systems. Dumont et al. examined the esterification of myristic acid with ethanol, Marty et al. of oleic acid with ethanol in $SCCO_2$ and *n*-hexane. The reaction was catalyzed by an immobilized lipase. The enzyme is inhibited by the alcohol and the kinetic follows a ping-pong bi-bi mechanism. The kinetic parameters in both media were in the same range.

The affinity for oleic acid is better in $SCCO_2$ than in *n*-hexane because $SCCO_2$ shows better transfer properties. The solubility of ethanol is much better in $SCCO_2$, but initial velocity was smaller in $SCCO_2$ than in *n*-hexane.

Chulalaksananukul et al. [88] compared the transesterification between propyl acetate and geraniol in $SCCO_2$ and in *n*-hexane. The optimum adsorbed water content is 10% g/ g of support. They found an inhibition by geraniol and a ping-pong bi-bi mechanism. The

Table 4 Mass Transfer Properties of SCCO$_2$

Ref.	Reaction	Reactor	Investigation	Result
Randolph, Blanch, Prausnitz [64]	—	Packed bed	SCCO$_2$ flow against conversion	No effects—absence of external diffusional resistance
Miller, Blanch, Prausnitz [74]	Continuous interesterification of trilaurin and myristic acid	Packed bed	SCCO$_2$ flow against conversion	Productivity increases with increasing flow—diffusion is not limiting step of process
Van Eijs et al. [69]	Transesterification of isoamyl acetate			External limitation
Marty, Condoret, Combes [54]	—	Packed bed	Modeling with plug flow model	No influence of external diffusion, but internal limitations
Chulalaksananukul et al. [86]	Esterification by immobilized lipase in n-hexane	Stirred vessel	Comparison with SCCO$_2$	External mass transfer limitations
Marty et al. [9]	Fatty acid esterification in supercritical carbon dioxide	Stirred vessel	Comparison with n-hexane	No external mass transfer limitations

Table 5 Cosolvents in Supercritical Media

Ref.	Cosolvent	Reaction	Results
Wong, Johnston [87] Randolph, Blanch, Prausnitz [64]	Various alcohols, acetone	Cholesterol oxidation	Ethanol increases reaction rate, methanol and acetone show no effect; butanol, less polar than ethanol, increases reaction rate significantly
Marty et al. [55]	Different concentrations of ethanol	—	Influence of the partition of water between enzymatic support and the solvent

maximum initial velocity was 5 times higher in *n*-hexane than in $SCCO_2$. The apparent affinity constant of both substrates was 5 times better in supercritical CO_2, demonstrating better mass transfer properties.

Comparison of the transesterification and esterification shows a difference in the conversion of the substrate. For esterification reaction, the conversions are similar in both solvents (99% in hexane and 95% in $SCCO_2$). For transesterification reaction, the conversion is different in both media: 85% in hexane and only 30% in $SCCO_2$. The authors suggested that an intrinsic effect of the solvent was involved in the differences in reaction yield.

Production of polar compounds should gain energetic advantage in a less unpolar solvent [89]. Consequently, *n*-hexane, being less polar than $SCCO_2$, should favor transesterification.

Cernia et al. [90] found an increase in enantioselectivity, conversion, and remaining activity of the lipase for acylation of different secondary alcohols in $SCCO_2$ compared to reactions in organic solvents.

Capewell et al. [26] compared some transesterifications in *n*-hexane and $SCCO_2$. Furthermore, they investigated the influences of cosolvents, water content, immobilization, and different reaction procedures on the enzyme activity and the enantiomeric excess.

In 1994, Kamat et al. [82] published an article about transesterification between methyl methacrylate and 2-ethylhexanol with lipase from *Candida cylindracea*. They demonstrated that the activity of the enzyme is low compared to the similar reaction in *n*-hexane. The reaction rates were compared at different temperatures in both media. In CO_2 temperature showed no effects in the range of 40–55°C. Above 55°C the reaction rates increased. It appeared that carbon dioxide formed reversible complexes with the free amino groups on the surface of the enzyme (carbamate formation). The authors suggested that below 55°C the enzyme is inhibited by CO_2.

6. Coupled Processes in $SCCO_2$: Enzymatic Reactions with Integrated Separation

The density of SCF is a function of pressure and temperature. The solvent power can be varied and this can be used for coupling of a reaction with an integrated separation. The first scientists who tried this method were Doddema et al. [91]. They investigated the transesterification of nonanol with ethyl acetate. At the outlet of the reactor they reduced pressure and the solutes separated. Ethyl acetate and ethanol were extracted from the fluid in a column with $SCCO_2$ (7.5 MPa/60°C). In a following step fresh ethyl acetate was added to the remaining mixture; then it was pumped into a second enzyme reactor with $SCCO_2$ (20 MPa/60°C). The authors achieved a conversion of over 90% for nonanol, since removal of ethanol avoided a product inhibition.

Another example for removal of a side product, which leads to the reverse reaction, was described by Adshiri et al. [19]. They examined the lipozyme-catalyzed interesterification of tricaprylin with methyl oleate. In a stirred reactor without solvent, 60% of fatty acid was produced. To optimize the degree of conversion they bubbled $SCCO_2$ (10 MPa/ 40°C) through the reactor, followed by a reflux column with temperature distribution (40°C bottom → 100°C top) at the outlet of the reactor (Fig. 2D). The aim was to extract the methyl caprylate selectively from the reaction mixture. The authors reached a shifting of the equilibrium and the degree of conversion raised to 87.3%.

Marty et al. [55] studied a continuous reaction and separation process for enzymatic esterification in $SCCO_2$. They wanted to use the variable solvent power for a multistage depressurization in which a selective recovery is expected leading to a separation of products and nonreacted substrates.

The transformation was carried out at 40°C and 15 MPa in a tubular fixed-bed reactor. To separate products and reactants a cascade of four depressurizations in separators was used. The recycling of CO_2 in the process was possible. The conversion exceeded 95% after 15 h. They obtained 92.1% of ester recovery and the purity of oleic acid was about 93 wt%. For the same attempt in n-hexane the authors achieved a purity of 49%. The ratio of ester concentration vs. oleic acid concentration was 5 times higher in $SCCO_2$ than in n-hexane.

Chrisochoou and Schaber [28] described a supercritical fluid extraction method to separate multicomponent mixtures incurred in enzyme-catalyzed reactions. As a model system the transesterification of ethyl acetate and isoamyl alcohol yielding ethanol and isoamyl acetate in $SCCO_2$ is chosen. The separating strategy is based on separating two fractions of components with significantly different solubility in carbon dioxide. The separation was realized by a reactor cascade at different pressures and temperatures. An efficient scheme is proposed combining the reaction and the separation in an integrated process using CO_2 as the unique solvent.

Marty et al. [2] published an extended study of the enzymatic esterification of oleic acid with ethanol based on previous works. The reaction was conducted in a tubular fixed bed, followed by a separation process in a series of depressurizations to recover the products and the nonreacted substrates with the greatest selectivity and yield.

In 1996, Michor et al. [57] examined the resolution of racemic citronellol and menthol by enzymatically catalyzed transesterification in $SCCO_2$. Different lipases and one esterase with various acylating reagents were employed. While the transesterification of (±)-menthol is reasonably fast and gave high enantiomeric excess, resolution of (±)-citronellol was not feasible.

Michor et al. [58] published another study of the enzymatic racemate separation of D,L-menthol. First the enzyme (four lipases and one esterase) catalyzed the selective formation of L-menthyl acetate. Different acid esters were used (isopropenyl acetate, triacetin, n-butyl acetate). Furthermore the solubility of L-menthol and L-menthyl acetate in $SCCO_2$ and the effects of pressure and temperature on the initial reaction rate were investigated. The solubility increased with higher pressures, but the reaction rate decreased. These studies show that an integrated reaction–separation process is possible.

C. Stability of Enzymes in $SCCO_2$

For the use of enzymes in different media the stability of the biocatalyst is important. It is important to know many parameters about the enzyme, e.g., pH and temperature optimum. For an industrial application the biocatalyst must be more stable to withstand harsh operating conditions.

Different models were developed to describe the influence of nonaqueous media on the stability of enzymes:

log P concept, Laane et al. [92]
Denaturation capacity, Khmelnitzky et al. [107]
Multidimensional solubility index, Schneider [93]
Water-attracting capacity, Reslow, Adlercreutz, Mattiasson [94]

To stabilize enzymes in organic phases, many methods can be used:

Immobilization on a support
Addition of compounds
Chemical or biochemical modification
Protein engineering

Table 6 gives an overview of methods for enzyme stabilization in organic solvents.

Both the purity of the enzyme and the hydrophobicity or polarity of the solvent are important parameters for the stability of enzymes. Wehtje et al. [108] found an interesting process for stabilization of immobilized enzymes. They added stabilizing agents (polyethylene glycol, gelatin, casein, tryptone, peptone, or albumin) when immobilizing the enzyme on the support.

The same methods as described for enzyme stabilization in organic solvents were investigated on their application for stabilization processes in $SCCO_2$.

1. Enzyme Stability and Substrate Solubility in $SCCO_2$

The stability and activity of enzymes in supercritical media depends on following parameters:

Pressure
Temperature
Cosolvents
Water content

A summary is given in Table 7. The first surveys were made from Taniguchi et al. [109]. They reported a stability time of only 1 h for 10 different enzymes. The enzymes tolerate up to 6% ethanol or 0.1% water as cosolvent. Marty et al. [9,55] showed that an immobilized lipase of *Rhizomucor miehei* lost 75% of its activity within a day at a water concentration between 0.4 and 1% (v/v).

Cholesterol oxidase was stable for 3 days [64]. This stability is insufficient for use in technical processes. Similar to experiments in organic solvent an immobilization could be useful for stabilizing the enzymes.

In the enzymatic hydrolysis of racemic 3-(4-methoxyphenyl)glycidic acid methyl ester in $SCCO_2$ the initial rate of the (*2S,3R*) isomer was faster than the rate of the (*2R,3S*) isomer [65]. An immobilized lipase from *Mucor miehei* was used. The stereoisomeric excess of the (*2R,3S*) form reached 87% at 53% total conversion level. No effects of water content of the reaction mixture and initial concentration of the racemate on isomeric purity was found. The reaction rate was faster in $SCCO_2$ than in toluene/water.

Another point of interest is the water content in the microenvironment of the enzyme. The used supports adsorbed water on their surface. This water concentration must be added to the water content of the enzyme. The investigation of the "real concentration" is complicated by the fact that the solubility of water in $SCCO_2$ depends on pressure and temperature. So the exact concentration is unknown.

Systematic analyses using supports of different hydrophilicities are necessary to find an optimum in enzyme stabilization and make the activity of enzymes independent from small changes in water content. Besides the "unknown water concentrations," the solubility of substrates is important for a technical use of $SCCO_2$. Chrastil reported in 1982 [111] that the solubility of substrates decreases significantly at higher temperature and pressure below 200 bar, e.g., palmitic oil is soluble from 0.1% to 1.5% (w/w). Such low concentrations are not feasible for technical processes. By adding of, say, 10% ethanol, the solubility of palmitic oil increases up to 5% [112].

Van Tol et al. [113] described the relationship between enzyme activity and the physical properties of organic media. The hydrolysis of propyl butyrate and tributyrin, the hydrolysis of decyl acetate, the transesterification of vinyl acetate with glycidol, and the esterification of propanol and butyric acid were chosen as model reactions. Although $SCCO_2$ was not investigated, the article describes strategies to determine the dependence of enzymatic reactions on different parameters like transfer properties.

Ishikawa et al. [44] described the investigation of the stability of glucoamylase, acid and alkaline protease, and lipase in systems with microbubbles of $SCCO_2$. The bubbles could increase the carbon dioxide concentration in the sample solution from 0.4 to 0.92 mol/L (25 MPa/35°C). Alkaline protease and lipase were completely inactivated by treatment at 35°C and 15 MPa for 30 min. Glucoamylase and acid protease showed a strong decrease in activity with increasing density of CO_2.

D. New Research Areas

New methods of conducting and analyzing enzymatic reactions in $SCCO_2$ are presently under investigation in our research groups. The introduction of new analytical procedures enables the on-line control of the reaction and coupled separation processes. Thus a characterization of the systems and a planning of the upscale for industrial applications is enabled. Applied analytical methods are gas chromatography (GC), 2D fluorescence, and UV spectroscopy. Further targets of interest are reactions in two-phase systems of $SCCO_2$ and water and enhancement of enantioselectivity by variation of enzymes in different types of reactions. The appropriate combination of reaction and separation as well as higher enzyme stability may lead to new innovative technical processes. The enantioselective transesterification and hydrolysis of isopropylidene glycerol, 3-hydroxy esters, and *p*-nitrophenol acetate are carried out in pure $SCCO_2$ or in a biphase system of $SCCO_2$ and aqueous buffer solution containing lipases in the phase boundary. Combinations of transport, extraction, and enzymatic reaction are investigated for an innovative application of supercritical fluids in chemical engineering. Esterification of polyhydroxy compounds with oleic acid was achieved by complexation of the unpolar substrates.

1. Analysis

Analysis of the reaction mixture may be done on-line (coupling of reaction system with analytical instruments) or off-line (by withdrawing samples out of the system). While off-line analysis can be regarded as state of the art, different on-line analysis systems are presently under investigation.

(*a*) *Off-line Analysis.* The first experiments with enzymatic reactions in $SCCO_2$ were stopped by cooling the vessel with liquid nitrogen. The reaction mixture condenses on the bottom of the reactor and the pressure decreases. After sublimating the CO_2 the residue can be analyzed [1]. This method has the great disadvantage of allowing a frequency of only one sample per reaction.

To achieve more information in a shorter time, analysis during the reaction has to be established. By pumping the reaction mixture through a port valve and sample loop, small amounts of the solution can be decompressed for further analysis without disturbing the reaction [114].

Table 6 Examples for the Stabilization of Enzymes Used in Organic Solvents

Enzyme	Conditions	Studied factor	Effect	Ref.
IMMOBILIZATION				
Lipase porcine pancreas	Benzene, 40°C	Adsorption to glass, kieselguhr, alumina, and agar beads	With agar beads 50–100% higher transesterification but only 11% higher esterification	Cao et al. [95]
Lipase *Candida rugosa*	≤7.5% butanol, 50°C	Time and temperature during immobilization on aldehyde-activated Sepharose CL-6B	5 h, 25°C, 75-fold stabilization	Otero, Ballesteros, Guisàn [96]
	Hexane, immobilization on alginate beads	By immobilization 4-fold slower inactivation		Hertzberg et al. [97]
	Isooctane, immobilization on Celite	Addition of 2 mM DTT	Loss of activity after 5 reactions reduced from 78% to 17%	Kawase, Tanaka [98]
	Crosslinking with dimethyl suberimidate		Partial increase in the thermostability for 15 min	Kawase, Sonomoto, Tanaka [99]
Trypsin	Dioxane/butanediol 60/30%, 25°C	Attachment on agarose • one-point • multipoint	Loss in activity: • in 10 d 50% • in 60 d 10%	Blanco, Alvaro, Guisàn [100]
Penicillin G acylase	Water and 50% DMF, pH 5.8, 28°C	One and multipoint attachment	Thermostabilization maximum 20,000 fold	Fernandez-Lafuente, Rosell, Guisàn [101]
Lipase *Geotrichum candidum*		Immobilization on polyallyl amine beads	"Increase" in thermostability	Sugihara, Shimada, Tominaga [102]

IMMOBILIZATION AND ADDITION OF SOLVENTS AND OTHER EFFECTORS

Enzyme	Conditions	Addition/Treatment	Effect	Reference
Lipase *Candida rugosa*	Immobilized on DEAE-Sephadex A50, 30°C	5 and 20% olive oil	Operational half-life 20 and 220 h	Yang and Rhee [103]
		Addition of 15% glycerol	Half life doubled from 220 to 450 h	
Lipase *Pseudomonas fluorescence*	Immobilized on Celite, 28°C	Ethanolysis and isopropanolysis	Residual activity after 10 d 19% and 83%	Shaw, Wang, Wang [104]
Lipase porcine pancreas, *Candida rugosa* (dried powder)	Buffer or 2 M heptanol in tributyrin	Heat stabilization at 200°C by a reduced water content from 100 to 0.8 and 0.015%	Loss of activity: 100% in 1 min, 85% in 36 min, 7% in 60 min	Zaks and Klibanov [105]
Alcohol dehydrogenase (horse liver)	Free and immobilized on CNBr-Sepharose or decylamine agarose at 4 and 30°C, pH 7 and 9	Addition of 5 mM AMP	Relative increase in stability, in maximum 38 fold, at 30°C and pH 9	Görisch and Schneider [106]
α-Chymotrypsin	Water/methanol 80:20%/%	Hydrophilization by pyromellitic dianhydride	$V_{max}/2$ observed at 52% methanol instead of 33%	Khmelnitzki et al. [107]

Table 7 Enzyme Stability and Productivity in $SCCO_2$

Enzyme	Experimental parameters	Effect on stability or productivity	Cosolvent influence	Ref.
Alcohol dehydrogenase, α-amylase, catalase	Commercial enzymes with 5–7 w%[b], water at 35°C, 200 bar in $SCCO_2$ and	Residual activity after 1 h:		Taniguchi, Kamihara, Kobayashi [109]
Glucoamylase, β-galactosidase, glucose oxidase, glucose isomerase, lipase, thermolysin	1) no cosolvent 2) 3 wt%[c] EtOH 3) 6 wt%[c] 4) 0.1 wt%[c] H_2O	94–102% 87–98% 82–101% 92–106%	S≈ S≈ S≈ S≈	
Lipase, α-amylase, glucose oxidase	see 1) enzyme with 50 wt%[b] water	33% 100%	W– W≈	
α-Chymotrypsin/trypsin/penicillin amidase	37°C, 100 bar in presence of	Residual activity after 24 h and 3 cycles of increasing and decreasing the pressure:		Kasche, Schlothauer, Brunner [110]
	3% (w/v) H_2O <1% (w/v) H_2O	32/11/0% 94/109/0%	W–	
Lipase *Rhizopus delemar*	1/150/300 bar, free lipase 35°C	50% loss of activity after: >>12/8.6/7 h (in two steps)	P– t–	Nakamura [85]
	50°C	1.6/1/0.6 h (linear with time)	t–	
Lipase *Rhizomucor miehei* (lipozyme)	130 or 180 bar	Residual activity after 6 d, that is independent of the pressure:	T– P≈	Marty et al. [54,55]
	30–40°C 60°C	90% 80%	T–	
	130 bar, 40°C and addition of 0.4–1% (v/v) water	The residual activity after 1 d decreased from 80 to 20%	W–	

Enzyme / Support	Conditions	Observation		Reference
Lipase of *Rhizopus arrhizus*				
• immobilized on hydrated Hyflo Supercel	100 bar, 60°C and 0.08 to 0.3% (v/v) water	The relative initial velocity ($\geq 80\%_{max}$) has an optimum at water content between 2 and 11%[a]	W –	van Eijs et al. [69]
• immobilized on glass beads	40°C, dry SCCO$_2$, increasing pressure from 86 to 300 bar	50% reduction of esterification	P –	Erikson, Schyns, Cooney [35]
	1) 35°C, 96.5 bar	70% decrease in the transesterification rate of trilaurin	t≈	Miller, Blanch, Prausnitz [59]
	2) increasing water content up to 1.7 g/kg CO$_2$	Residual activity after 80 h 100%	W	
	3) increasing pressure (86 to 114 bar)	Enzyme activity unchanged but increased hydrolysis		
	• in water-saturated SCCO$_2$	Increased hydrolysis and interesterification	P,W	
	• in dry SCCO$_2$	Reduced hydrolysis but constant interesterification		
Lipase of	50°C, 294 bar	Extent of interesterification with a maximum at a water content between		Chi, Nakamura, Yano [75]
• *Rhizopus japonicus*	Immobilized on Celite	20 and 45%[a]	W	
• *Rhizopus delemar*	Immobilized on Celite	7 and 40%[a]	W	
• *Alcaligenes* sp.	Free	3 and 9%[a]	W	
• *Rhizomucor miehei*	Immobilized on Duolite	Rate similar high from 0.1 to 50%[a], but the initial rate increases linearly from 0–20%[a] water	W	

Table 7 Continued

Enzyme	Experimental parameters	Effect on stability or productivity	Cosolvent influence	Ref.
Cholesterol oxidase (immobilized on glass beads)	35°C, 101 bar, 10% O_2 and addition of	Turnover number (which is independent on substrate solubility):		Randolph, Blanch, Prausnitz [64]
• *Gloecysticum chrysocreas*	1) no cosolvent	75 s^{-1} (stable for 3 d)	S≈	
	2) 2% (v/v) methanol	62 s^{-1}	S++	
	3) 2% (v/v) ethanol	165 s^{-1}	S+	
	4) 2% *n*-butanol	100 s^{-1}	S	
	5) 2% *sec-, tert*-butanol	238, 274 s^{-1}	+++	
	6) no cosolvent, removal of water	Reversible inactivation within 60 min	W	
• *Streptomyces* sp. (powder)	see 1)	80% loss in activity within 1.5 h	t−	
Subtilisin Carlsberg	45°C, 150 bar	Transesterification with ethanol		Pasta et al. [63]
	2/5% (v/v) ethanol	• decreased at higher concentration	S−	
	45/80°C	• increased at higher temperature	T+	
Lipase, glucoamylase, acid protease, alkaline protease	35°C, 25 MPa	Treatment with microbubbles of SCCO$_2$	t−	Ishikawa et al. [44]

^{a} % (w/w) solvent/carrier.
^{b} % water in the enzyme preparation.
^{c} Cosolvent in SCCO$_2$.
The resulting effects after increase of cosolvent (S), water (W), temperature (T), and time (t) are shown in the fourth column. +, −, or ≈ indicates the influence.

An advantage of off-line measurement is that in general different analytical methods can be applied to the same sample, matching the properties of different compounds in the system (GC, HPLC, SFC, mass spectrometry (MS), nuclear magnetic resonance (NMR), infrared spectroscopy (IR), and capillary electrophoresis). The analytes become diluted in a solvent, so that higher concentrations of compounds can be examined. However, dilution may also be a disadvantage because trace analysis becomes problematic and the used solvent may disturb analysis. Broad solvent peaks in GC or HPLC may mask underlying substances.

Analysis of the hydrolysis of 3-hydroxy-4-pentenoicacid ethyl ester (HPAE) and isopropylidene glycerol ester (IPG) was performed utilizing Shimadzu GC 14A and the β-Hydrodex-3P columns, manufactured by Macherey-Nagel, Düren.

In GC assays the racemic mixture is separated into two peaks representing each enantiomer. The ratio of the peak area of both peaks can be used to calculate the enantiomeric excess given by the following equations:

$$eeS = \left| \frac{(F_R - F_S)}{(F_R + F_S)} \times 100\% \right| \tag{1}$$

$$eeR = \left| \frac{(F_R - F_S)}{(F_R + F_S)} \times 100\% \right| \tag{2}$$

where

eeS = enantiomeric purity of the substrate
eeR = enantiomeric purity of the product
F_R = peak area of the R enantiomer (product, substrate)
F_S = peak area of the S enantiomer (product, substrate)

For analysis of the conversion U in irreversible reactions Equation (3) can be used:

$$U = \frac{eeS}{(eeS + eeP)} \times 100\% \tag{3}$$

For the enantioselectivity E follows with Eqs. (1)–(3):

$$E = \frac{\ln[(1 - U)(1 - eeS)]}{\ln[(1 - U)(1 + eeS)]} \tag{4}$$

$$E = \frac{\ln[1 - U(1 + eeS)]}{\ln[1 - U(1 - eeS)]} \tag{5}$$

In order to achieve high yields, the enantioselectivity and activity of the enzyme under given conditions should be as high as possible.

(*b*) *On-line Analysis.* For on-line analysis the analytical instrument is directly coupled with the reaction system and either has to withstand the high pressure above 73 bar or needs a decompression by valves or restrictors. Direct coupling of the reaction system with analytical systems is difficult for some type of instruments (e.g., HPLC columns can be destroyed by gas bubbles).

By avoiding the dilution step in off-line analysis, trace analysis is accessible. At the same time, higher concentrations of analytes in $SCCO_2$ are harder to handle in on-line analysis.

Supercritical fluid chromatography (SFC, with packed or capillary columns) can be utilized for product fractionation and analysis. By coupling with SFC an easy way of combining reaction with separation and analysis of compounds is feasible, allowing easy automation of sampling runs. For higher analyte concentration cryogenic traps and heated valve interfaces can collect higher sample amounts. Higher dead volumes of pressure-resistant couplings can lead to "memory effects," especially in trace analysis. Direct coupling may limit the sample throughput of the analytical instruments.

UV-adsorbing compounds can be measured by circulating reaction media through a high-pressure-capable UV-Vis detector cell. Capewell [115] observed the lipase-catalyzed reaction of 3-hydroxymethyloctanoate with vinyl acetate. The byproduct, acetaldehyde, was detected at 320 nm to measure the conversion. This noninvasive method proved to give an easy access to reaction kinetics.

The easy removal of solvent leads to a good compatibility to mass spectrometry, which is already used in coupling of SFC-MS [116]. SFE-SFC-GC coupling is described in [117].

(*c*) *In Situ Analysis by 2D Fluorescence Spectroscopy.* Due to the high pressure (100 bat at 45°C) on-line monitoring of reaction compounds (e.g., HPLC, UV spectroscopy) is quite difficult. Thus, the use of 2D fluorescence spectroscopy with open-end detection as a new analytical tool for this application was tested by Lindemann [118]. In fluorescence spectroscopy the emitted light of molecules under excitation with light of a fixed wavelength is measured. In 2D fluorescence spectroscopy the excitation wavelength is varied and the emitted light intensity is measured at different wavelengths to obtain a 2D spectrum of the compound. The complete spectrum is obtained within 1 min.

High-pressure-resistant quartz glass windows make it possible to survey the reaction directly in the vessel ("in situ"), using light guides for the excitation and emission light (Fig. 3). The pH and oxygen concentration in the reaction mixture can be measured by the use of fluorescent indicators.

In Fig. 4 the result of a solubility test of a coumarin derivative in $SCCO_2$ is shown. As long as the CO_2 pressure is below the critical point, no solubility of the substrate is found. When reaching the critical point, the coumarin dissolves in the $SCCO_2$ phase and the fluorescence of the dissolved substrate increases.

In addition, enzymatic reactions with nonfluorescent substrates and fluorescent products as well as systems with hydrophobic but $SCCO_2$-soluble substrate and hydrophilic

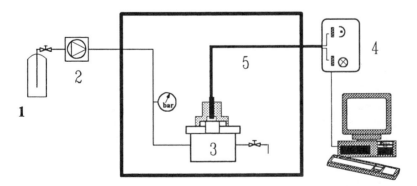

Figure 3 In situ 2D fluorescence. 1 liquid CO_2 tank, 2 cooled high pressure pump, 3 quartz-window equipped reaction vessel, 4 2d fluorescence spectrometer, 5 light guide.

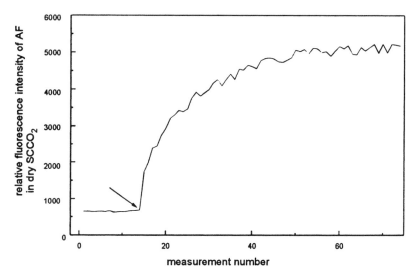

Figure 4 7-Amino-4-trifluoromethylcoumarin fluorescence at Ex./Em.: 370/510 nm in $SCCO_2$. Pressure of 100 bar is reached after the 14th measurement.

products have been tested. As a result, the reactions can be measured by the fluorescence spectrometer and from the fluorescence measurement kinetic data can be derived.

An important question for enzymatic reactions with $SCCO_2$–water systems is the pH value of the water/buffer phase because CO_2 dissolves in the water and the pH becomes acidic. This might cause problems because the enzyme activity depends on the pH value. The fluorescence sensor can be used in combination with pH-dependent fluorophores like fluorescein, which must be dissolved in the aqueous phase inside the reactor to measure the pH. Then the fluorescence intensity is measured through the quartz window. In experiments it was possible to survey the enzymatic hydrolysis of a fluorescing ester and to detect the decreasing pH of aqueous solutions under a carbon dioxide atmosphere.

2. Reactions of Polar Substrates in Supercritical CO_2

New successful enzymatic reactions were carried out with different polyhydroxyl compounds such as fructose and sucrose in supercritical CO_2 catalyzed by an immobilized lipase (Lipozyme) with the help of two original methods. The high polarity of both substrates makes them nearly insoluble in the unpolar $SCCO_2$. By complexation of polar substrates with PBAC or adsorption onto silica gel the enzymatic esterification of oleic acid was conducted.

Firstly, by experiments that involve HPLC analysis and visual observations, we have established the optimal substrate/PBAC ratio for complexing fructose and glycerol with PBAC to obtain the maximal solubilization in $SCCO_2$. We have found an optimal ratio of 1 for the glycerol and of 0.5 for the fructose. Table 8 presents the results of this solubilization test.

The solubilizations were tested at 14 MPa and 40°C. Only one product has been identified after enzymatic reactions with oleic acid. The products formed were monoolein in the case of reaction with glycerol and monooleyl-1-fructose in the case of reaction with fructose. The recovery of reaction products was easily made by liquid–liquid extraction with a slightly acidified solution of water. In the case of the reactions with glycerol re-

Table 8   Maximal Solubilization of
Substrates in *n*-Hexane and SCCO₂

	SCCO₂	
	Substrate/PBAC[a]	mmol/L[b]
Glycerol	1	30
Fructose	0.5	20

[a]Molar ratio substrate PBAC.
[b]Maximal concentration of solubilized substrate.

covery of pure monoolein was possible with this method (90% of total produced). In the case of fructose reactions, 80% of pure fructose monoester was recovered.

By adsorption of sucrose and glucose on silica gel these compounds are able to interact with the supercritical phase. First in this work, we have found the ratio glycerol/silica gel equal to 1, which allows high yields of products. Thus, we have tested the esterification reactions of adsorbed sucrose or fructose with oleic acid in SCCO₂ as solvent and using Lipozyme as catalyst. After 10 h of reaction, kinetics equilibria were achieved for both reactions and conversions of substrate around 50% for fructose and 45% for sucrose were obtained. In controlled reactions carried out without silica gel no conversion of substrate was observed. The products were identified as a mixture of hydrophobic compounds as monoesters, diesters, and triesters of sucrose and fructose. Traces of monoesters were only found in reaction with sucrose. Lipozyme in these conditions did not show any substrate regiospecificity although reactivity remains important.

It is clear that the optimization of the reaction medium and the reaction conditions will be necessary to improve yield and the regioselectivity of these reactions. Nevertheless, the feasibility of enzymatic esterification reactions with polar substrates in near-anhydrous solvents opens a broad spectrum of possibilities of new enzymatic synthesis reactions.

3.   Two-Phase Systems

Enzymatic Hydrolysis of p-Nitrophenol Acetate.   Product recovery from SCCO₂ involves the precipitation of the solute by reducing the pressure. This method yields the product in pure form but makes it necessary to recompress the CO₂ for recycling, which is energy consuming when done on a large scale. An alternative is to extract the product from the SCCO₂ into an aqueous solution. A setup with two phases was established by filling the room over a buffer solution with SCCO₂. Lipases will catalyze reactions in the phase boundary.

The enzymatic hydrolysis of *p*-nitrophenyl acetate (PNPA) to *p*-nitrophenol (PNP) was used to investigate a process whereby the extraction is combined with an enzymatic reaction. The idea is that the unpolar PNPA is easily extracted by SCCO₂, while the more polar PNP is better soluble in the aqueous solution. This system is a model for a process whereby a nonpolar substrate is extracted by SCCO₂ from a raw material and converted to a polar product, which is reextracted into an aqueous solution. Pure PNPA was placed in the reactor and dissolved as carbon dioxide reached the supercritical state. The pressure was kept at 100 bar; the temperature was 40°C.

First a controlled experiment without enzyme was carried out using only 0.1 M phosphate buffer to maintain a pH of 5 in the aqueous phase. Figure 5 shows that PNPA

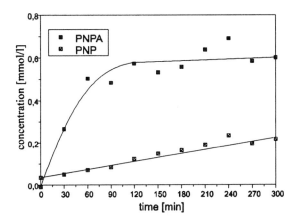

Figure 5   Concentration of PNPA and PNP in the trap solution.

is transported into the aqueous solution until an equilibrium concentration is reached at 0.6 mmol/L. The slight increase of PNP indicates that hydrolysis also occurs without enzymatic catalysis. In the second experiment Lipozyme (*Mucor miehei*) was added to the buffer solution to catalyze the reaction.

The results show that with enzymatic catalysis the PNP concentration increases faster, whereas the PNPA concentration remains constant at a low level. This means that the rate of PNPA transported to the trap vial and the rate consumed in the enzymatic reaction in the aqueous phase are equal after 60 min. The higher concentration of PNP (1.6 mmol/L compared to 0.6 mmol/L after 210 min) in the second experiment (Fig. 6) proves the possibility of combining reaction and separation by exploiting the different distribution of products and substrate in the two-phase system water/SCCO₂.

4.   Enzymatic Hydrolysis of 3-Hydroxy Esters

Basic studies about enzymatic reactions in SCCO₂ are dealing with esterification, trans- and interesterification. In the following experiments the enantioselective enzymatic hydro-

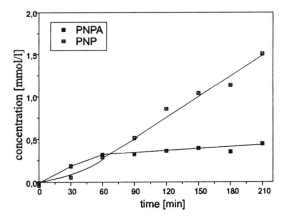

Figure 6   Concentration of PNPA and PNP in the trap solution with enzymatic catalysis of the hydrolytic reaction.

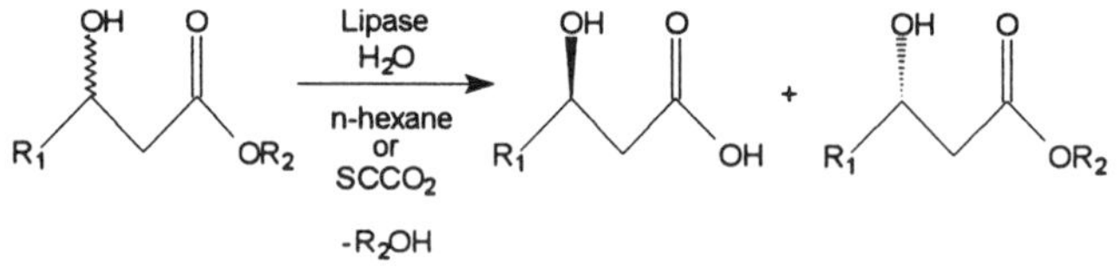

Figure 7 Hydrolysis of 3-hydroxy esters with lipase.

lysis of HPAE in a two-phase system of water/supercritical phase was studied. The goal was to separate one ester-enantiomer from the product-acid, by introducing a separation step into the process exploiting the different solubilities in the two phases.

Some lipases and one esterase were used as enzymes for the reaction. The catalysis was performed in both *n*-hexane and $SCCO_2$. The ester-racemate as substrate is soluble in hexane or $SCCO_2$. The deprotonated acid should be soluble in the aqueous phase. Figure 7 shows the general reaction scheme for the enzymatic reaction. The class of 3-hydroxy esters represents useful chiral precursors for the synthesis of a wide variety of natural compounds as well as pharmaceutically active substances such as beta blockers. The solubility of HPAE (Fig. 8) in *n*-hexane was investigated since its solvating properties are similar to those of $SCCO_2$ and allow easy estimation of the accessible substrate concentration in the reaction mixture under supercritical conditions.

In an enzyme screening of 15 commercially available lipases and one esterase the enzymes with the highest enantioselectivity for the given substrates were determined (Table 9). A screening was done to find enzymes with high substrate enantiomeric excess and short catalyzing times to achieve short reaction times. The substrate (50 μl) was added to a solution of 950 μl hexane and 750 μl 0.5 M KPP-buffer (pH 8). The enzyme ($\approx$15 mg) was added afterwards. Samples were taken in constant intervals from the hexane phase and analyzed by GC. Table 8 shows the enzyme used, the preferred enantiomer, the area ratio, and the reaction time. The best results were achieved with lipase (porcine pancreas) from Sigma, the Amano PS (*Pseudomonas* sp.), Lipozyme (*M. miehei*, immobilized), the Pl (*Alcaligenes* sp.) from Meito, the *Chromobacterium viscosum* and the esterase (*M. miehei*). Figure 9 shows the enzyme conversions, the reaction time, and the preferred enantiomer. These enzymes were used for all further investigations.

Solubility Experiments.   For an optimal separation effect, the solubility of the ester in $SCCO_2$ should be high, whereas the solubility of the free acid in $SCCO_2$ should be much lower than in water. The distribution coefficient of the product was investigated under different conditions.

The distribution coefficient of HPAE between hexane and $SCCO_2$ was determined as follows: The concentration of HPAE in the buffer solution was measured under normal pressure. A batch reactor was then filled with KPP-buffer and $SCCO_2$ at similar volumes

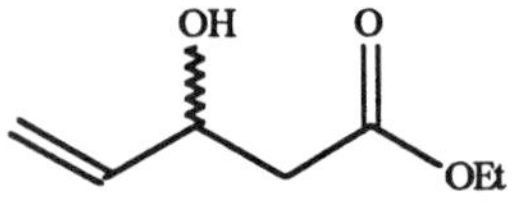

Figure 8 3-Hydroxy-4-pentenoic acid ester.

Table 9 Enzyme Screening

Enzyme[a]	Preferred enantiomer	Reaction time (min)	Ratio F_R/F_S
Lipase PPL (porcine pancreas, Sigma)	S	1740	3.32
Lipase PS (*Pseudomonas* sp., Amano)	R	345	0.76
Lipase (*Candida cylindracea*)	S	2810	1.24
Lipase A (*Aspergillus niger*, Amano)	—	2820	0.9
Lipozyme (*Mucor miehei*, immobilized, Novo Nordisk)	R	2770	0.11
Lipase AY (*Candida cylindracea*, Amano)	S	2780	1.24
Lipase CE 5 (Amano)	S	1335	1.51
Lipase PL (*Alcaligenes* sp., Meito)	R	1335	0.45
Lipase 2212E (*Mucor miehei*, Röhm)	—	1335	1.05
Lipase B (H&R)	R	2790	1.52
Lipolase (Novo Nordisk)	—	2865	0.94
Lipase (*Chromobacterium viscosum*)	S	1390	2.65
Lipase G (*Penicillium camembertii*, Amano)	—	2860	0.92
Lipase 2212F (Röhm)	—	2860	0.94
Lipase 7023 (Röhm)	S	2860	1.71
Esterase (*Mucor* miehei)	R	1350	0.18

[a]Not all information (species, firm) is available for all enzymes.

and samples were taken from the buffer phase after 24 h. The concentration was determined afterward and the distribution coefficient was derived from the difference of both analyses. The same procedure was used for hexane:

$$\alpha_{hexane/KPP} = 1948\ (20°C) \qquad \alpha_{SCCO_2/KPP} = 1547\ (45°C/100\ bar)$$

In principle, the solubility of HPAE can be determined at different pressures. The

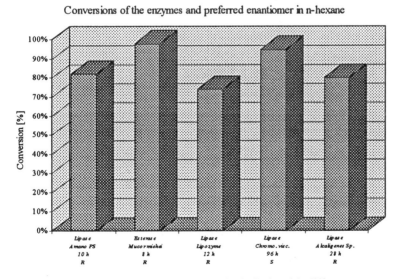

Figure 9 Conversions in the enzymatic hydrolysis with different enzymes in *n*-hexane.

process can be monitored by on-line detection of the absorbance of the substrate using a UV-photometer with a high-pressure cell. The CO_2 and dissolved substances flow through this cell and the absorbance can be measured vs. different pressures.

The dependence of the solubility on temperature and pressure was determined by the following method: 0.5 ml of HPAE was dissolved in 31 ml 0.5 M KPP-buffer and the reactor was filled with $SCCO_2$. Samples were taken out of the KPP phase at different temperatures. With increasing temperatures the pressure increased too, so only the mean values of pressure are plotted in Fig. 10. Moreover, with increasing pressure the solubility of HPAE in the KPP-buffer decreased. In conclusion of this result the solubility increases in $SCCO_2$. Deviations in the single graphs are due to the complicated analysis procedure.

Additional experiments were done in *n*-hexane to compare the enantioselective hydrolysis of HPAE in $SCCO_2$ with *n*-hexane as solvent. In batch reactions the enzyme (0.3 g) was used in a two-phase system of hexane (15 ml) and KPP-buffer (19 ml, 0.5 M, pH 7.5). The reaction was started by adding 1 ml of HPAE. The mixture was continuously stirred with 1000 rpm. At different times samples were withdrawn from the aqueous phase. After acidification with HCl the neutral species of HPAE is soluble in organic solvents. For GC analysis the HPAE and the acid were extracted with ether. Figures 11–13 show the results. In all cases conversions over 80% were achieved. The enzyme started to catalyze the unpreferred enantiomer after several hours. For the lipase from *Chromobacterium viscosum* and PL the *eeS* values obtained over 75%.

To compare these reactions in hexane with those in $SCCO_2$ the same concentration ratio of solvent, buffer, and substrate was chosen for $SCCO_2$. Unlike hexane, the reactions in $SCCO_2$ were performed in two reaction vessels to achieve a combined reaction–separation process. The first flask (source) contained the substrate HPAE. It is soluble in $SCCO_2$ and was transported to the second vessel (trap) filled with enzyme dissolved in KPP-buffer at pH 7.5. The solubility of CO_2 in water under high pressure leads to a decrease of pH. Tservistas [119] showed that the pH under this condition decreases to 5.0–5.5. Nonreactants or contaminating substances remain in the source vessel if they are not soluble in $SCCO_2$. Figure 14 shows the scheme of the reactor, including both reaction vessels.

The substrate HPAE is soluble in $SCCO_2$ and is transported to the trap vessel. Insoluble compounds remain in the source vessel, so the substrate can be purified. For

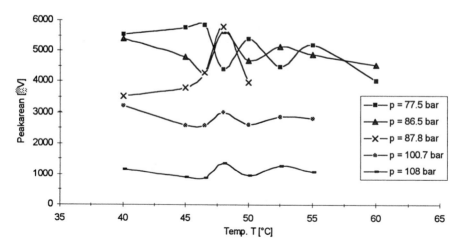

Figure 10   Solubility of HPAE independent of temperature and pressure.

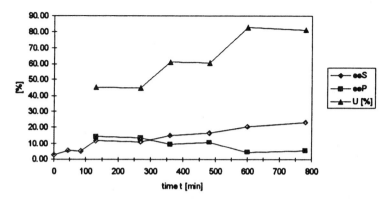

Figure 11 The e.e. values and conversion data of the enzymatic hydrolysis of HPAE with lipase PS (Amano).

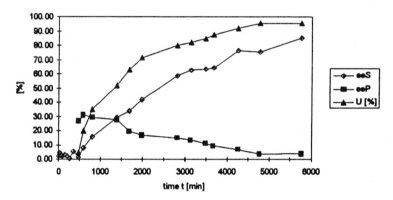

Figure 12 The e.e. values and conversion data of the enzymatic hydrolysis of HPAE with lipase from *Chromobacterium viscosum*.

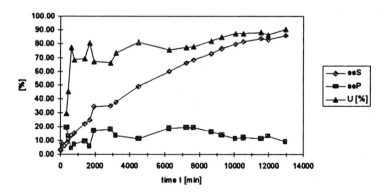

Figure 13 The e.e. values and conversion of the enzymatic hydrolysis of HPAE with lipase PL from Meito.

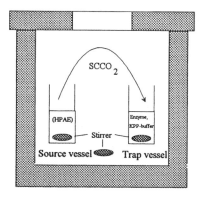

Figure 14 Reaction–separation process.

example, it is possible to extract the substrate from a mixture so it will be transported by the $SCCO_2$ to the aqueous phase and the reaction will take place. The aim of this experiment is to show that an integrated separation–reaction process is possible. The results could be compared with the experiments in *n*-hexane regarding enantiomeric excess, conversion, and reaction time. The reactor was placed in a thermostat oven to keep the temperature constant. A sample could be withdrawn from the KPP phase, which is circulated by an HPLC pump via a sample loop. The enzyme was held back by a microfiltration membrane (Fig. 15).

The enzymatic catalysis of HPAE in $SCCO_2$ was investigated with the esterase from *M. miehei*. The results are shown in Figs. 16 and 17.

In the $SCCO_2$ experiment the product was formed after 24 h with an e.e. of 27%. After 117 h no more substrate peaks were detectable with the GC, so the experiment was stopped at a conversion of 71% with an *eeS* of 26% and an *eeP* of 10%. In *n*-hexane/ KPP-buffer (pH 5,3) conversion reached 92% and an *eeS* of 60% and an *eeP* of 4% were achieved after 138 h. These results show that a transport process can be coupled with an

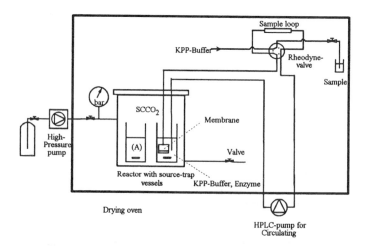

Figure 15 Experimental setup.

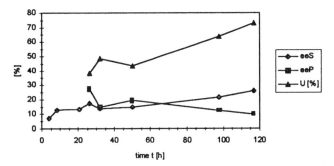

Figure 16 The e.e. values and conversion of the enzymatic hydrolysis of HPAE with the esterase (*Mucor miehei*) in SCCO$_2$.

enzymatic reaction. The substrate is transported from a source through the supercritical phase to a vessel filled with the enzyme solution where the reaction takes place.

5. Transesterification of Isopropylidene Glycerol and Acetic Acid Vinyl Ester

Catalysis of transesterification in the supercritical phase as well as in a two-phase system was examined with the same setup as in Fig. 15 [119]. The reaction between isopropylidene glycerol (IPG) and acetic acid vinyl ester was used as a model reaction. The product is a precursor to monoglycerols, which are widely used as emulsifiers for medicine, pharmacy, cosmetics, and food processing. The reaction is catalyzed by lipase from *Pseudomonas cepacia*. The byproduct of the transesterification is a vinyl alcohol, which is immediately tautomerized to give the corresponding aldehyde, so the equilibrium of the enzymatic reaction is completely shifted to the product side. The result of these investigations is a slower reaction in SCCO$_2$ than in hexane, but the enantiomeric excess is higher in SCCO$_2$ with low enantioselectivity in both solvents. Lipase from *P. cepacia* is more stable in SCCO$_2$ than in hexane (Fig. 18).

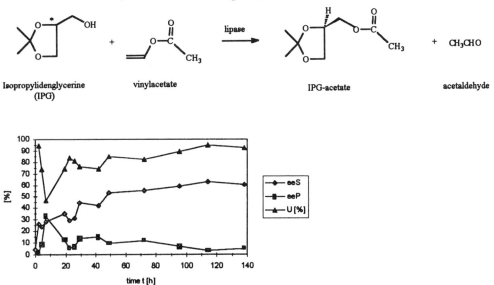

| Isopropylidenglycerine (IPG) | vinylacetate | IPG-acetate | acetaldehyde |

Figure 17 The e.e. values and conversion of the enzymatic hydrolysis of HPAE with the esterase (*Mucor miehei*) in hexane and KPP-buffer (pH 5.3).

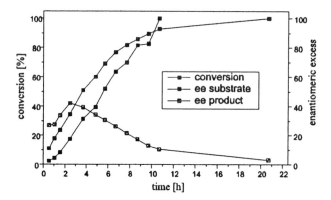

Figure 18 Synthesis of isopropylidene glycerol ester in supercritical carbon dioxide (T = 40°C, p = 100 bar).

6. Sequential Resolution of Racemates

To increase the enantioselectivity the different distribution coefficients of substrate and product between two phases with different polarity can be used. The enantioselective hydrolysis of the ester was investigated in a combination of hydrolysis and following transesterification in a two-phase system.

In the aqueous phase enzymatic hydrolysis of racemic IPGA leads to IGP (mainly one enantiomer), which can be extracted into the supercritical fluid. Buffering of the pH in the aqueous solution is necessary to provide optimal conditions for the enzyme. In the following lipase-catalyzed esterification IPG reacts with vinyl butyrate to IPG-butyrate (lipase was from *P. cepacia*). In this sequential resolution of racemates no processing of intermediates is necessary; both reactions take place in different phases (Fig. 19).

The degree of conversion was only 15%, mainly because unpolar IPG-acetate was extracted into the SCCO$_2$ and therefore not hydrolyzed. Transport limitations due to a phase transition of the IPG leads to longer reaction times.

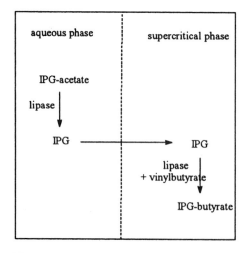

Figure 19 Reaction scheme of sequential resolution.

The enantioselectivity was measured in preceding experiments to $E = 1.8$ for the hydrolysis and $E = 3.6$ for the esterification with vinyl butyrate in hexane, and reached $E = 10.3$ in the overall sequential reaction. By this new method of conducting a resolution of racemates a high enantioselectivity can be reached.

IV. CONCLUSIONS

The increasing number of publications dealing with enzymatic reactions in supercritical carbon dioxide displays a new confidence in economical applications of this technology. The problem of inactivation of enzymes is a great challenge for the economic application of biocatalyzed reactions, and therefore a number of recent publications have investigated the nature and dependencies of enzyme activity in $SCCO_2$.

Supercritical carbon dioxide shows interesting properties for organic and biotechnical chemistry. A large number of investigations demonstrated that supercritical fluids might be suitable to perform enzymatic reactions. Good enzyme stability, increased solubility of hydrophobic compounds, and easy enzyme recovery in supercritical fluids (especially in carbon dioxide) were observed. $SCCO_2$ is the most suitable supercritical fluid because of its relatively low critical pressure, its nonhazardous character, and its low cost. The low critical temperature prevents thermal denaturation when processing biological compounds and its nontoxic nature is greatly appreciated in the food and pharmaceutical industries.

In most studies lipases were used. These enzymes are active at the phase boundary of oil/water mixtures. Some authors investigated reactions in "water-free" media and showed that most enzymes are inactivated. Adding water led to an activation of the enzymes. One problem is to find the optimal water concentration for reactions like transesterifications and all water-free catalysis. With increasing water concentration hydrolysis is preferred. An important factor for applying $SCCO_2$ as reaction phase for enzymatic processes is the good mass transfer property of $SCCO_2$ due to its low viscosity and high diffusitivity. The solvent power can be controlled by changing pressure and temperature of the fluid, allowing integrated reaction–separation for technical processes. It offers another advantage that, upon depressurization of the reaction mixture, products and non-reacted substrates can be easily recovered.

All parameters influencing enzyme activity and stability, solvent power, or reaction rates are described in the literature. An increase in pressure leads to an increase of the density of $SCCO_2$ and in most cases to an increase of the reaction rate. Enantioselectivity did not change by increasing pressure. The properties of this unconventional reaction phase can be adjusted to the needs of the biotransformation system. Modern on-line analysis will help us to understand, describe, and control the complex reaction systems in a better way.

REFERENCES

1. TW Randolph, HW Blanch, JM Prausnitz, CR Wilke. Enzymatic catalysis in a supercritical fluid. Biotechn Lett 7(5):325–328, 1985.
2. A Marty, D Combes, JS Condoret. Continuous reaction–separation process for enzymatic esterification in supercritical carbon dioxide. Biotech Bioeng 43:497–504, 1994.
3. E Reverchon, F Senatore. J Agric Food Chem 42:154–458, 1994.
4. B Wu, M Klein, S Sandler. Ind Eng Chem Res 30:822–828, 1991.
5. S Kamat, E Beckmann, A Russel. J Am Chem Soc 105:8845, 1993.

6. Y Ikushima, N Saito, O Sato, K Hatakeda, S Ito. Proceedings of the 3rd International Symposium on Supercritical Fluids, Strasbourg, 1994, Vol. 3, 131–136.
7. A Russel, E Beckmann, A Chaudhary. Chemtech March, 33–37, 1994.
8. S Kamat, EJ Beckman, AJ Russell. Enzyme activity in supercritical fluids. Crit Rev Biotechnol 15(1):41–71, 1995.
9. A Marty, JS Condoret, D Combes. Fatty acid esterification in supercritical carbon dioxide. In: Tramper et al., eds. Biocatalysis in Nonconventional Media, Vol. 8. New York, Elsevier, 1992, pp 425–432.
10. JC Giddings, MN Meyers, L McLaren, RA Keller. Science 162:67, 1968.
11. Jm Valle, JM Aguilera. Ind Eng Chem Res 27:1551–1553, 1988.
12. JW King, JP Friedrich. J Chromatogr Sci 517:449–458, 1990.
13. DY Peng, DB Robinson. Ind Eng Chem Fund 15:59, 1976.
14. JM Levy, E Storozynsky, M Ashraf-Khorassani. In: FV Bright, MEP McNally, eds. ACS Symposium Series 488, 336–362, 1992.
15. MD Luque de Castro, M Valcarcel, MT Tena. Analytical Supercritical Fluid Extraction. Berlin: Springer-Verlag, 1994.
16. K Kerrola. Food Re Int 11(4):547–573, 1995.
17. H Black. Supercritical carbon dioxide: The "greener" solvent. Environ Sci Technol News 30(3):124A–127A, 1996.
18. DA Hammond, M Karel, AM Klibanov, VJ Krukonis. Enzymatic reactions supercritical gases. Appl Biochem Biotechn 11:393–400, 1985.
19. T Adshiri, H Akiya, LC Chin, K Arai, K Fujimoto. Lipase-catalyzed interesterification of triglyceride with supercritical carbon dioxide extraction. J Chem Eng of Japan 25:104–105, 1992,
20. MO Balaban, AG Arreola, M Marshall, A Peplow, CI Wei, J Cornell. Inactivation of pectin esterase in orange juice by supercritical carbon dioxide. J Food Sci 56(3):743–746, 750, 1991.
21. BE Berg, EM Hansen, S Gjorven, T Greibrokk. On line enzymatic reaction, extraction, and chromatography of fatty acids and triglycerides with supercritical carbon dioxide. J High Resolut Chromatogr 16(6)358–363, 1993.
22. P Bernard, D Barth. Internal mass transfer limitation during enzymic esterification in supercritical carbon dioxide and hexane. Biocatal Biotransform 12(4):299–308, 1995.
23. U Bolz, K Stephan, P Stylos, A Riek, M Rizzi, M Reuss. Comparison of enzymic catalyzed reactions in organic solvents and in supercritical fluids. Biochem Eng --Stuttgart, [Proc. Int. Symp.], 2nd Meeting Date 1990, 82–5. Edited by: Reuss, Matthias. Fischer: Stuttgart, Fed. Rep. Ger., 1991.
24. U Bornscheuer, A Capewell, Wendel, T Scheper. On line determination of the conversion in a lipase-catalyzed kinetic resolution in supercritical carbon dioxide. J Biotechnol 46:139–143, 1996.
25. W Cai. Experimental research on enzyme-catalyzed esterification in n-hexane and supercritical carbon dioxide. Shiyou Daxue Xuebao, Ziran Kexueban 16(3):122–125, 1992.
26. A Capewell, V Wendel, U Bornscheuer, HH Meyer, T Scheper. Lipase-catalyzed kinetic resolution of 3-hydroxyesters in organic solvents and supercritical carbon dioxide. Enzyme Microb Technol 19:181–186, 1996.
27. E Cernia, C Palocci. Lipases in supercritical fluids. Meth Enzymol, 286 (Lipases, Part B), 495–508, 1997.
28. A Chrisochoou, K Schaber. Design of a supercritical fluid extraction process for separating mixtures incurred in enzyme-catalyzed reactions. Chem Eng Proc 35:271–282, 1996.
29. A Chrisochoou, K Schaber, U Bolz. Phase equilibria for enzyme-catalyzed reactions in supercritical carbon dioxide. Fluid Phase Equilib 108(1–2):1–14, 1995.
30. A Chrisochoou, K Schaber, K Stephan. Phase equilibria with supercritical carbon dioxide for the enzymic production of an enantiopure pyrethroid component. J Chem Eng Data 42(3): 551–557, 1997.

31. D Combes. Reaction/separation process in supercritical CO2 using lipases, NATO ASI Ser., Ser. E, 317(Engineering of/with Lipases), 613–618, 1996.

32. O Diaz, A Cobos, L de la Hoz, JA Ordonez. Supercritical carbon dioxide in the production of food from plants. Other application; Aliment Equipos Tecnol 16(8):55–63, 1997.

33. T Dumont, D Barth, C Corbier, G Branlant, M Perrut. Enzymatic reaction kinetic: Comparison in an organic solvent and in supercritical carbon dioxide. Biotechnol Bioeng 40(2):329–333, 1992.

34. Y Endo, K Fujimoto, K Arai. Enzyme reaction in supercritical fluid; Dev. Food Eng., Proc. Int. Congr. Eng. Food, 6th, Meeting Date 1993, Volume Pt. 2, 849–51. Edited by: Yano, Toshimasa, Matsuno, Ryuichi, Nakamura, Kozo. Blackie: Glasgow, UK, 1994.

35. JC Erickson, P Schyns, CL Cooney. Effect of pressure on an enzymatic reaction in a supercritical fluid. AIChE J 36(2):299–301, 1990.

36. G Glowacz, M Bariszlovich, M Linke, P Richter, C Fuchs, J-T Mörsel. Stereoselectivity of lipases in supercritical carbon dioxide. I. Dependence of the regio- and enantioselectivity of porcine pancreas lipase on the water content during hydrolysis of triolein and partial glycerides. Chem and Phys Of lipids 79:101–106, 1996.

37. RD Goddard, JA Bosley, B Al-Duri. The application of supercritical carbon dioxide as a solvent in lipase catalyzed reactions using a continuous packed bed reactor. Jubilee Res. Event, Two-Day Symp., Volume 2, pp 993–996, 1997.

38. H Gunnlaugsdottir, B Sivik. Integration of lipase and catalysis and product separation in supercritical carbon dioxide. Proc Technol Proc 12 (High Pressure Chem Eng), 79–84, 1996.

39. M Habulin, V Krmelj, Z Knez. Supercritical carbon dioxide as a medium for enzymically catalyzed reaction. Proc Technol Proc 12 (High Pressure Chem Eng), 85–90, 1996.

40. T Ikariya, PG Jessop, R Noyori. Chemical reactions in supercritical fluids. Yuki Gosei Kagaku Kyokaishi 53(5):358–369, 1995.

41. Y Ikushima. Supercritical fluids: an interesting medium for chemical and biochemical processes. Adv Colloid Interface Sci 71–72:259–280, 1997.

42. Y Ikushima. Supercritical fluids as novel media for chemical reactions. Koatsuryoku no Kagaku to Gijutsu 6(2):86–93, 1997.

43. H Ishikawa, Y Osajima. Application of supercritical carbon dioxide method to aqueous system. Novel sterilization and enzyme inactivation technique. Kagaku to Seibutsu 35(9):632–637, 1997.

44. H Ishikawa, M Shimoda, T Kawano, Y Osajima. Inactivation of enzymes in an aqueous solution by micro-bubbles of supercritical carbon dioxide. Biosci Biotech Biochem 59(4): 628–631, 1995.

45. MA Jackson, JW King, GR List, WE Neff. Lipase-catalyzed randomization of fats and oils in flowing supercritical carbon dioxide. J Am Oil Chem Soc 74(6):635–639, 1997.

46. S Kamat, J Burrera, EJ Beckman, AJ Russell. Biocatalysis synthesis of acrylates in organic solvents and supercritical fluids: I: Optimization of enzyme environment. Biotechnol Bioeng 40(1):158–166, 1992.

47. S Kamat. Biocatalysis in supercritical fluids: control of enzyme activity and selectivity by solvent engineering. Diss Abstr Int B 58(3):1405, 1996.

48. HS Lee, WGI Lee, SW Park, H Lee, H Chang. Starch hydrolysis using enzyme in supercritical carbon dioxide. Biotechnol Tech 7(4):267–270, 1993.

49. C Lin. Enzymic reactions in supercritical fluid. Zhejiang Gongye Daxue Xuebao 24(4):335–342, 1996.

50. Z Lin, A Qiu, X Wang. Advances of enzymic interesterification in supercritical carbon dioxide. Zhongguo Youzhi 22(2):26–28, 1997.

51. P Lozano, A Avellaneda, R Pascual, JL Iborra. Stability of immobilized-α-chymotrypsin in supercritical carbon dioxide. Biotechnol Lett 18(11):1345–1350, 1996.

52. JF Martins, TC Sampaio, IB Carvalho, MN da Ponte, Barreiros (Eds: C Balny, R Hayashi, K Heremans, P Masson). Lipase catalyzed esterification of glycidol in chloroform and in supercritical carbon dioxide. High Press Biotechnol 224:411–415, 1992.

53. A Marty, S Manon, DP Ju, D Combes, J-S Condoret. The enzymic reaction-fractionation process in supercritical carbon dioxide. Ann NY Acad Sci 750 (Enzyme Eng XII), 408–411, 1995.
54. A Marty, W Chulalaksananukul, JS Condoret, RM Willemot, G Durand. Comparison of lipase-catalyzed esterification in supercritical carbon dioxide and in n-hexane. Biotechn Lett 12(1):11–16, 1990.
55. A Marty, JS Chulalaksananukul, RM Willemot, W Condoret. Kinetics of lipase-catalyzed esterification in supercritical CO_2. Biotechnol Bioeng 39(3):273–280, 1992.
56. H Michor, R Marr, T Gamse, I Schilling, E Klingsbichel, H Schwab. Enzymatic catalysis in supercritical carbon dioxide: comparison of different lipases and a novel esterase. Biotechnol Lett 18(1):79–84, 1996.
57. H Michor, R Marr, T Gamse. Enzymic catalysis in supercritical carbon dioxide. Effect of water activity. Proc Technol Proc 12 (High Pressure Chem Eng), 115–120, 1996.
58. H Michor, T Gamse, R Marr. Enzymkatalyse in überkritischem Kohlendioxid: Racematspaltung von D,L-Menthol, Chemie Ingenieurtechnik (69)5197:690–694, 1997.
59. DA Miller, HW Blanch, JM Prausnitz. Enzyme-catalyzed interesterification of triglycerides in superficial carbon dioxide. Ind Eng Chem Res 30(5):939–946, 1991.
60. K Nakamura. Enzymic Synthesis in Supercritical Fluids; Supercrit Fluid Technol Oil Lipid Chem., 306–320. Edited by: King, Jerry W.; List, Gary R. AOCS Press: Champaign, Ill.
61. N Nakamura, YM Chi, Y Yamada, T Yano. Lipase activity and stability in supercritical carbon dioxide. Chem Eng Commun 45:207–212, 1986.
62. O Parve, I Vallikivi, L Lahe, A Metsala, U Liile, V Tougu, H Vija, T Pehk. Lipase-catalyzed enantioselective hydrolysis of bicyclo[3.2.0]heptanol esters in supercritical carbon dioxide. Bioorg Med Chem Lett 7(7):811–816, 1997.
63. P Pasta, G Mazzola, G Carrea, S Riva. Subtilisin-catalyzed transesterification in supercritical carbon dioxide. Biotechn Lett 2(9):643–648, 1989.
64. TW Randolph, HW Blanch, JM Prausnitz. Enzyme-catalyzed oxidation of cholesterol in supercritical carbon dioxide. AIChE J 34(8):1354–1360, 1988.
65. M Rantakylä, M Alkio, O Aaltonen. Stereospecific hydrolysis 3-(4-methoxyphenyl)glycidic ester in supercritical carbon dioxide by immobilized lipase. Biotechnol Lett 18(9):1089–1094, 1996.
66. AJ Russell, EJ Beckman, AK Chaudhary. Studying enzyme activity in supercritical fluids. CHEMTECH 24(3):33–37, 1994.
67. V Sereti, H Stamatis, FN Kolisis. Improved stability and reactivity of Fusarium solani cutinase in supercritical CO_2. Biotechnol Tech 11(9):661–665, 1997.
68. JM Snyder, JW King, MA Jackson. Analytical supercritical fluid extraction with lipase catalysis: conversion of different lipids to methyl esters and effect of moisture. J Am Oil Chem Soc 74(5):585–588, 1997.
69. AMM van Eijs, JPJ de Jong. HJ Doddema, DR Lindeboom. Enzymatic transesterification in supercritical carbon dioxide. In: M Perrut, ed. Proceedings of the International Symposium on Supercritical Fluids (Nice, France) (M Perrut, Ed.) 933–942, 1988. AMM van Eijs, JPJ de Jong, HHM Oostrom, HJ Doddema, MA Visser and R Stoop. Enzymatic synthesis of nonylacetate and isoamylacetate in supercritical carbon dioxide and organic solvents. In: Breeler H, et al., eds., Proceedings 2nd Netherlands Biotechnology Congress 1988, Amsterdam. Netherlands Biotechnology Society, 1988.
70. H Vija, A Telling, V Tougu. Lipase-catalyzed esterification in supercritical carbon dioxide and in hexane. Bioorg Med Chem Lett 7(3):259–262, 1997.
71. H Xu, Z Zhu, L Yang. Enzymic catalysis in supercritical fluids. Huaxue Fanying Gongcheng Yu Gongyi. 12(3):232–237, 1996.
72. S-H Yoon, O Miyawaki, K-H Park, K Nakamura. Transesterification between triolein and ethylbehenate by immobilized lipase in supercritical carbon dioxide. J Ferment Bioeng 82(4):334–340, 1996.
73. Y Zheng, GT Tsao. Avicel hydrolysis by cellulase enzyme in supercritical CO_2. Biotechnol Lett 18(4):451–454, 1996.

74. DA Miller, HW Blanch, JM Prausnitz. Enzymatic interesterification of triglycerides in supercritical carbon dioxide. Ann NY Acad Sci 613:534–537, 1990.

75. YM Chi, K Nakamura, T Yano. Enzymatic interesterification in supercritical carbon dioxide. Agric Biol Chem 52:1541–1550, 1988.

76. A Marty, JS Condoret, D Combes. Fatty acid esterification in supercritical carbon dioxide; Biocatalysis in Non Conventional Media (Ed. Tramper et al., Elsevier Science Publishers), Vol 8, 425–432, 1992.

77. Y Ikushima, N Saito, T Yokohama, K Hatakeda, S Ito, M Arai, HW Blanch. Solvent effects on an enzymatic ester synthesis in supercritical carbon dioxide. Chem Lett 109–112, 1993.

78. J Martins, IB de Carualho, TC de Sampaio, S Barreiros. Lipase-catalyzed enantioselective esterification of glycidol in supercritical carbon dioxide. Enzyme Microb Techn 16:785–790, 1994.

79. M Rantakylä, O Aaltonen. Enantioselective esterification of ibuprofen in supercritical carbon dioxide by immobilized lipase. Biotechnol Lett 16(8):825–830, 1994.

80. J Martins, IB de Carualho, TC de Sampaio, S Barreiros. Lipase-catalyzed enantioselective esterification of glycidol in supercritical carbon dioxide. Enzyme Microb Techn 16:785–790, 1994.

81. P Bernard, D Barth. Enzymic reaction in supercritical carbon dioxide internal mass transfer limitation. Proc Technol Proc 12 (High Pressure Chem Eng), 103–108, 1996.

82. S Kamat, G Critchley, EJ Beckman, AJ Russel. Biocatalytic Synthesis of acrylates in organic solvents and supercritical fluids: III. Does carbon dioxide covalently modify enzymes? Biotechn Bioeng 46:610–620, 1994.

83. U Bornscheuer. Reaktionstechnische Untersuchungen zur enzymatischen Racematspaltung verschiedener 3-Hydroxysäureester in unkonventionellen Lösemitteln; Dissertation: Fortschr.-Ber. VDI Reihe 17, Nr. 87, VDI-Verlag.

84. E Castillo, A Marty, D Combes, JS Condoret. Polar substrates for enzymatic reactions in supercritical CO_2. How to overcome the solubility limitation. Biotechn Lett 16(2):169–174, 1994.

85. K Nakamura. Supercritical fluid bioreactor. In: Fiechter, Okada, Tanner, eds., Bioproducts and Bioprocesses. Berlin: Springer-Verlag, 1989, pp 257–265.

86. W Chulalaksananukul, JS Condoret, P Delorme, RM Willemot. Kinetic study of esterification by immobilized lipase in n-hexan. FEBS Lett 276(1,2):181–184, 1990.

87. JM Wong, KP Johnston. Solubilization of biomolecules in carbon dioxide based supercritical fluids. Biotechnol Proc 2:29–38, 1986.

88. W Chulalaksananukul, JS Condoret, D Combes. Kinetics of geranyl acetate synthesis by lipase-catalyzed transesterification in n-hexane. Enz Microb Techn 14:293–297, 1992.

89. PJ Halling. Solvent selection for biocatalysis in mainly organic solvent systems: predictions of effects on equilibrium position. Biotechn Bioeng 35:691–701, 1990.

90. E Cernia, C Palocci, F Gasparrin, D Misiti, N Fagano. Enantioselectivity and reactivity of immobilized lipase in supercritical carbon dioxide. J Mol Catal 89:L11–L18, 1994.

91. HJ Doddema, RJJ Janssens, JPJ de Jong, JP van der Lugt, HHM Oostrom. Enzymatic reactions in supercritical carbon dioxide and integrated product-recovery; 5th European Congress on Biotechnology, Copenhagen (Ed. Christiansen et al.), 239–242, 1990.

92. C Laane, S Boeren, K Vos, C Veeger. Rules for optimization of biocatalysis in organic solvents. Biotechnol Bioeng 30:81–87, 1987.

93. LV Schneider. A three dimensional solubility parameter approach to nonaqueous enzymology. Biotechn Bioeng 37:627–638, 1991.

94. M Reslow, P Adlerkreutz, B Mattiasson. On the importance of the support material for bioorganic synthesis. Eur J Biochem 172:573–578, 1988.

95. SG Cao, ZB Liu, Y Feng, L Ma, ZT Ding, YH Cheng. Esterification and transesterification with immobilized lipase in organic solvents. Appl Biochem Biotechnol 32:1–6, 1992.

96. C Otero, A Ballesteros, JM Guisàn. Immobilization/stabilization of lipase from Candida rugosa. Appl Biochem Biotechnol 19:163–175, 1988.

97. S Hertzberg, L Kvittingen, T Anthonson, G Skjak-Braek. Alginate as immobilization matrix and stabilization agent in a two-phase liquid system: Application in lipase-catalyzed reactions. Enzyme Microb Technol 14:42–47, 1992.
98. M Kawase, A Tanaka. Improvement of Operational Stability of yeast lipase by chemical treatment. Biotechnol Lett 10:393–396, 1988.
99. M Kawase, K Sonomoto, A Tanaka. Effects of non-covalent interaction of dimethyl suberimidate on lipase stability. Agric Biol Chem 54:2605–2609, 1990.
100. RM Blanco, G Alvaro, JM Guisàn. Enzyme reaction engineering: design of peptide synthesis by stabilized trypsin. Enzyme Microb Technol 13:573–583, 1991.
101. R Fernandez-Lafuente, CM Rosell, JM Guisàn. Enzyme reaction engineering: synthesis of antibiotics catalysed by stabilized penicillin G acylase in the presence of organic solvents. Enzyme Microb Technol 13:898–905, 1991.
102. A Sugihara, Y Shimada, Y Tominaga. Enhanced stability of a microbial lipase immobilized to a novel synthetic amine polymer. Agric Biol Chem 52:1589–1590, 1988.
103. D Yang, JS Rhee. Stability of the lipase immobilized on DEAE-sephadex for continuous lipid hydrolysis in organic solvents. Biotechnol Lett 13:553–558, 1991.
104. J-F Shaw, D-L Wang, YJ Wang. Lipase-catalyzed ethanolysis and isopropanolysis of triglycerides with long-chain fatty acids. Enzyme Microb Technol 13:544–546, 1991.
105. A Zaks, A Klibanov. Enzymatic catalysis in organic media at 100°C. Science 224:1249–1251, 1984.
106. H Görisch, M Schneider. Stabilization of soluble and immobilized horse liver alcohol dehydrogenase by adenosine 5-monophosphate. Biotechnol Bioeng 26:998–1002, 1984.
107. YL Khmelnitzki, AB Belova, AV Levashov, VV Mozhaev. Relationship between surface hydrophilicity of a protein and its stability against denaturation by organic solvents. FEBS Lett 284:267–269, 1991.
108. E Wehtje, P Adlercreutz, B Mattiasson. Stabilization of adsorbed enzymes used as biocatalysis in organic solvents. In: Progress in Biotechnology 8, Biocatalysis in Non-Conventional Media, edited by J. Tramper et al., 1992, pp 373–382. Elvesier, Amsterdam.
109. M Taniguchi, M Kamihara, T Kobayashi. Effect of treatment with supercritical carbon dioxide on enzymatic activity. Agric Biol Chem 51:593–594, 1987.
110. V Kasche, R Schlothauer, G Brunner. Enzyme denaturation in supercritical carbon dioxide: stabilizing effect of disulfide bonds during the depressurization step. Biotechnol Lett 10:569–574, 1988.
111. J Chrastil. Solubility of solids and liquids in supercritical gases. J Phys Chem 86:3016–3021, 1982.
112. G Brunner, S Peter. On the solubility of glycerides and fatty acids in compressed gases in the presence of an entrainer. Sep Sci Technol 17:199–214, 1982.
113. JBA van Tol, RMM Stevens, WJ Veldhizen, JA Jongejan, JA Duine. Do organic solvents affect the catalytic properties of lipase? Intrinsic kinetic parameters of lipases in ester hydrolysis and formation in various organic solvents. Biotechnol Bioeng 47:71–81, 1995.
114. A Marty, W Chulalaksanukul, JS Condoret, RM Willemot, G Durand. Biotechnol Lett 12(1):11–16, 1990.
115. A Capewell. Fortschrittberichte 17, Nr. 118, VDI-Verlag Düsseldorf 1994.
116. R Smith, JC Fjeldsted, M Lee. Direct fluid injection interface for capillary supercritical fluid chromatography-mass spectrometry. J Chromatogr 247:231–243, 1987.
117. JM Levy, WM Ritchey. J High Res Chromatogr Chromatogr Commun 10:493–496, 1987.
118. C Lindemann. Anwendungen der 2D-Fluoreszenzspektroskopie zur Bioprozessbeobachtung, PhD dissertation, Institut für technische Chemie, Universität Hannover, 1998.
119. M Tservistas. Dissertation, Untersuchungen zum Einsatz von überkritischem Kohlendioxid als Medium für biokatalysierte Reaktionen, 1987.

28

Dehydrogenases in the Synthesis of Chiral Compounds

M.-R. Kula

Institut für Enzymtechnologie der Heinrich-Heine-Universität Düsseldorf im Forschungszentrum Jülich, Jülich, Germany

U. Kragl*

Institut für Biotechnologie 2, Forschungszentrum Jülich, Jülich, Germany

I. INTRODUCTION

Oxidoreductases play a central role in the metabolism and energy conversion of living cells. As early as 1897 Buchner demonstrated in his famous experiment that such biocatalysts can also function in cell-free extracts. About 25% of the presently known enzymes are oxidoreductases [1]. Among those the most interesting catalysts for preparative and possible industrial applications are dehydrogenases or reductases classified in the groups EC 1.1.1 (acting on the CH-OH group of donors), EC 1.2.1 (acting on the aldehyde or oxo group of donors), EC 1.4.1 (acting on the CH-NH$_2$ group of donors), and using NAD or NADP as an acceptor (Scheme 1). The enzyme-catalyzed reduction is of prime interest yielding chiral compounds, e.g., alcohols, hydroxy acids and their esters or amino acids, respectively, which may be the desired product or more often serve as key intermediate in the synthesis of pharmaceuticals, agrochemicals, or flavors. Dehydrogenases as all other enzymes catalyze a reversible reaction, where hydrogen and two electrons are abstracted from the donor and transferred together to the nicotinamide moiety of the coenzyme in the oxidative direction regarding the donor. The structure of NAD/NADP is shown in Fig. 1. In the reverse direction, hydrogen and two electrons are transferred from the reduced nicotinamide to the carbonyl group to effect a reduction of the substrate. The reaction is in general highly stereospecific, the dehydrogenase delivers the hydride either from the *si* or *re* side to the carbonyl yielding in the reduction of an asymmetrical ketone the corresponding (*R*)- or (*S*)-alcohol, respectively (Fig. 2). Depending on the binding site of the NADH/NADPH on the dehydrogenase either the pro-*R* or the pro-*S* hydrogen at C-4 is abstracted. The stereochemistry of the hydrogen transfer is quite well understood [2–4] and results from the specific binding of substrates as well as coenzymes on the chiral enzyme surface in close proximity and in a defined geometry.

Current affiliation: Lehrstuhl für Technische Chemie, Universität Rostock, Rostock, Germany.

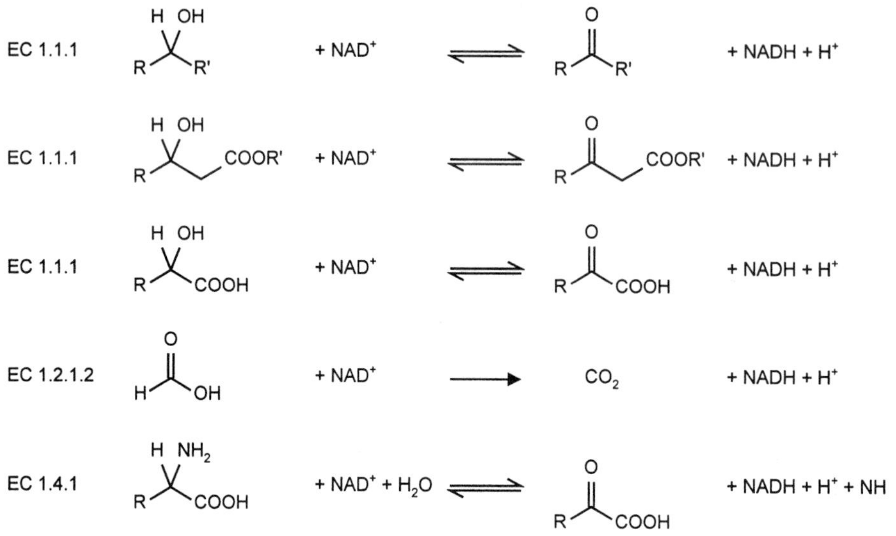

EC 1.1.1   R–CH(OH)–R' + NAD⁺ ⇌ R–CO–R' + NADH + H⁺

EC 1.1.1   R–CH(OH)–CH₂–COOR' + NAD⁺ ⇌ R–CO–CH₂–COOR' + NADH + H⁺

EC 1.1.1   R–CH(OH)–COOH + NAD⁺ ⇌ R–CO–COOH + NADH + H⁺

EC 1.2.1.2   H–CO–OH + NAD⁺ → CO₂ + NADH + H⁺

EC 1.4.1   R–CH(NH₂)–COOH + NAD⁺ + H₂O ⇌ R–CO–COOH + NADH + H⁺ + NH

Scheme 1   Enzyme-catalyzed reductions of potential interest in the production of fine chemicals.

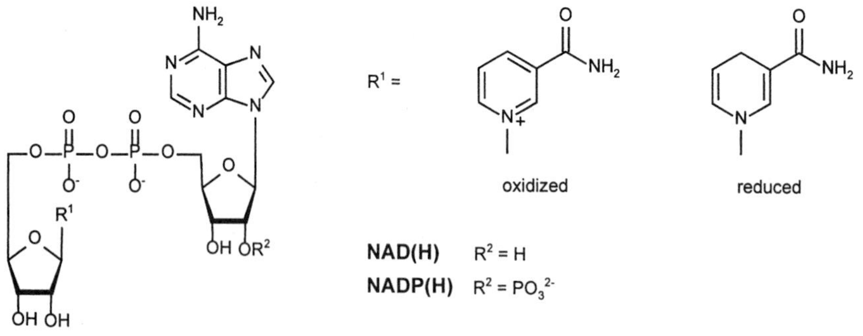

R¹ = (oxidized / reduced nicotinamide moiety)

NAD(H)   R² = H
NADP(H)   R² = PO₃²⁻

Figure 1   Structure of NAD(H) and NADP(H).

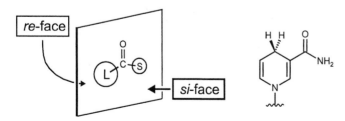

Figure 2   Delivery of the "hydride" from NADH to the *re* or *si* side of a prochiral ketone yielding either (*S*)- or (*R*)-alcohols. Depending on the dehydrogenase involved either the pro (*R*)- or the pro (*S*)-hydrogen at C-4 of the nicotinamide moiety of NAD(P)H is selectively transferred in the course of the reduction. *L* denotes the larger and *S* the smaller substituent at the central carbon.

It should be noted that the enzyme classification system is based on the general reaction catalyzed and not on the producing organism. The same enzyme, e.g., an alcohol dehydrogenase, derived from different organisms might have different properties: binding sites, stereochemistry, substrate recognition, kinetic constants, stability, pH optimum, etc. This adds to the versatility of enzymatic reactions but increases the complexity of the approach for the organic chemist [4].

The vast reservoir of wild-type microorganism can be tapped to find a suitable enzyme by screening procedures and constantly new enzymes including dehydrogenases are described in the literature. A number of dehydrogenases are commercially available (see Tables 1 and 2) and numerous substrates have been tested with those as well as other not-yet-commercialized enzymes. A useful compilation of such data is found in books [5,6] or on a CD ROM produced by the Warwick Biotransformation Club [7]. Besides, two recent reviews [4,8] provide additional information.

Modern tools of molecular genetics are applied to make enzymes available in larger quantities and at lower cost in recombinant hosts. The same techniques are utilized in protein engineering to alter the amino acid sequence of enzymes. By directed evolution and rational design, mutant enzymes are derived with altered properties such as coenzyme specificity, thermostability, and substrate recognition. This way tailor-made catalysts can be generated, although this requires special skills and will take time. The difficulties encountered in a single case are not easily predictable, but the scientific knowledge in protein engineering is increasing rapidly and in the future more tailor-made enzymes will be utilized in industrial processes.

II. COENZYME REGENERATION

As pointed out above, dehydrogenases are involved in numerous metabolic reactions in living cells balancing carbon flux and free energy in a complicated network mediated by the coenzymes involved. NAD(P)/NAD(P)H are usually not very tightly bound; K_M values are found in the range $10^{-6}-10^{-4}$ M. Therefore, coenzymes dissociate freely and are regenerated by another reaction within the network. If isolated dehydrogenases are to be utilized for the reduction of a single compound, the network is disrupted, and NADH or NADPH is required in stoichiometric amounts to be converted to NAD/NADP in the course of the reaction (Scheme 1). Since NAD(P)H is expensive, an efficient and cost-effective in situ regeneration from the oxidized form is prerequisite for large-scale applications in order to meet economic constraints [9].

Application of whole cells might possibly avoid these difficulties. Many examples have been described using "fermenting yeast" as a reducing agent [10]. The yeast cells utilize the sugar metabolism to generate the reducing equivalents. Also growth on ethanol has been observed to generate reducing equivalents by metabolism [11]. "Baker's yeast" (*Saccharomyces cerevisiae*) is readily available locally; it is cheap and might be the catalyst of choice to synthesize moderate amounts of a desired enantiomer. Scale-up of many protocols to the multikilogram range appear difficult or impossible. Reproducible performance by baker's yeast is not guaranteed because many competing dehydrogenases are present in the cell and their relative abundance and activity depends on genetic variability between strains, growth conditions, storage and reaction conditions, etc. [4,10]. In addition, substrates and products must be transported efficiently over the cell membrane to get reasonable reaction rates and productivity. The conversion of precursors into the wanted product may be impaired if educt and/or product are transformed by other enzymes present in the yeast cell in unwanted directions. Some selected examples of microbial transformations will be discussed below

Many of the difficulties associated with the application of whole cells as catalyst can be avoided using isolated enzymes. The necessary in situ regeneration and high conversion of NADH/NADPH can be accomplished in different ways [12]. In every case the stoichiometric link between coenzyme consumption and product formation is shifted toward the consumption of a cosubstrate and the concomitant production of a second product. The basic concept is derived from the metabolism but is simplified from the network to a single reaction step, which is more amenable to optimization by reaction engineering techniques [9,13].

In substrate-coupled NADH/NADPH regeneration (Fig. 3) a single enzyme is involved, usually an alcohol dehydrogenase (ADH), which accepts small aliphatic alcohols as secondary substrates. In the oxidation of, say, isopropanol to acetone, the coenzyme is regenerated in the same active site that carries out the reductive conversion of the substrate of interest. The substrate-coupled reaction system is limited by the thermodynamically controlled equilibrium between the two substrates and products and the necessity of using an enzyme with alcohol dehydrogenase activity. To achieve high conversions the isopropanol concentration usually has to be raised to rather high levels, which may improve the solubility of educts but more often is detrimental with regard to enzyme stability or impairs kinetic constants. Cosubstrate and coproduct are achiral and in general can be easily separated from the compound of interest by distillation. The substrate-coupled approach was of special interest in the utilization of (R)-ADH for stereospecific reductions yielding (R)-alcohols and related compounds [14]. All presently know (R)-ADHs need NADPH as coenzyme, which could not be regenerated otherwise in a more economic way. However, a new and efficient NADPH regeneration system based on a mutant formate dehydrogenase has been described recently, offering a choice [15].

In an enzyme-coupled NADH (NADPH) regeneration two enzymes are employed as shown in Fig. 3. This allows for creation of novel cycling systems and avoidance of constraints from thermodynamic equilibrium encountered in the substrate-coupled regeneration discussed above. Amino acid dehydrogenases or 2-hydroxy acid dehydrogenases do not possess ADH activity; therefore, only an enzyme-coupled approach is possible, unless highly complicated reaction mixtures are separated. For NADH regeneration the enzyme of choice is formate dehydrogenase (FDH; EC 1.2.1.2), which is obtained from methanol utilizing microorganisms [16,17] and catalyzes the oxidation of formate to CO_2 with the concomitant reduction of NAD to NADH. The equilibrium of the reaction is very favorable and lies far on the side of CO_2 and NADH [16]. Formate is a cheap and safe

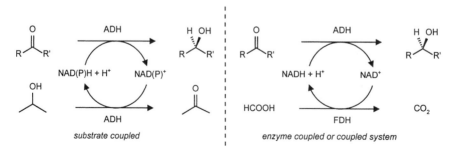

Figure 3   Methods for in situ coenzyme regeneration.

hydrogen source; the coproduct CO_2 can be easily removed and is not obnoxious to proteins. Numerous batch and continuous dehydrogenase-catalyzed reduction or reductive aminations have been described in the literature using formate dehydrogenase (FDH) for regeneration [18–21]. By an excess of formate high conversions of the educt can be achieved in the coupled system. Total turnover numbers (ttn) (mole product/mole coenzyme) for NADH up to 600,000 have been realized [22], demonstrating that in principle the coenzyme costs are no longer prohibiting large-scale application.

The production of FDH with the methylothropic yeast *Candida boidinii* has been studied in 200-L continuous culture producing 2300 U/(L × d) [23,24]. An efficient isolation procedure is also available [25,26]. The FDH gene of *C. boidinii* has been cloned recently and expressed in *Escherichia coli*. By a rational design the stability of the enzyme against oxidative stress has been considerably improved [27]. Besides the FDH from *C. boidinii* the bacterial enzyme from *Pseudomonas spec* has been intensively studied [28,29]. The three-dimensional structure has been solved to high resolution [30]. The catalytic mechanism is investigated by protein engineering [31,32]. There are no detailed reports on the application of the bacterial FDH for in situ NADH regeneration so far, but the optimization of expression in a recombinant *E. coli* and pilot scale production have been carried out recently [17]. In general, the FDH from methylothrophs does not accept NADP as a coenzyme. Therefore, an efficient in situ NADPH regeneration was lacking. The use of ADH from *Thermoanaerobicum brockii* in an enzyme-coupled process has been investigated [33]. Tishkov et al. generated an NADP-dependent FDH by a rational design strategy based on the bacterial FDH gene [17,28]. The wild-type bacterial enzyme already has a low activity with NADP (about 5% reaction rate compared to NAD and a high K_M value of 35 mM). The mutant enzyme exhibits about 25% of the reaction rate compared to NAD. The K_M value for NADP has been lowered by more than 100-fold by site-directed mutagenesis. The new NADP-dependent FDH has been produced in pilot scale using a recombinant *E. coli* strain [17]. Successful applications for in situ NADPH regeneration have been reported together with an (*R*)-ADH yielding (*R*)-2-phenylethanol [15] and in a coupled synthesis of chiral lactones by a monooxygenase [34].

In the long run, protein engineering can be used to alter the coenzyme specificity of other NADP-dependent dehydrogenases. Also, in terms of cost, NAD is about sevenfold cheaper than NADP. The new FDH accepting NADP as substrate can be universally applied and buys time in process development. It can be safely stated that for water-soluble substrates the in situ coenzyme regeneration has been solved, and at least in the case of NADH it is economically feasible for chiral products or intermediates. L-*tert*-leucine is produced commercially in ton scale using leucine dehydrogenase for reductive amination of 2-oxo-3,3-dimethylbutanoic acid with NADH regeneration by FDH (Fig. 4) [35–37]. NAD is introduced in catalytic amounts and is efficiently recycled during the production of the L-amino acid. Some aspects of reaction engineering to achieve high total turnover numbers are discussed in Sec. V.

By comparison with microbial transformations, in situ coenzyme regeneration in general yields much fewer if any byproducts from the precursor. Therefore, product recovery is easier and continuous processing is possible, making this an attractive option.

III. MICROBIAL TRANSFORMATIONS

Baker's yeast is produced in large amounts and is readily available locally throughout the world. It has long been considered a cheap catalyst by organic chemists. The first reduction

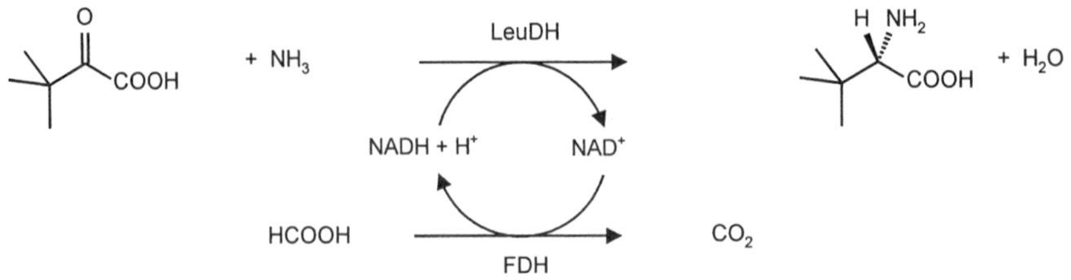

Figure 4 Reductive amination of 2-oxo-3,3-dimethyl butanoic acid with NADH regeneration (LeuDH, leucine dehydrogenase; FDH, formate dehydrogenase).

of a ketone by "fermenting yeast" was reported in 1918 by Neuberg and Lewite [38]. In large-scale applications the initial cost of the catalyst becomes less important than added value, product yield, and separation cost. Yeast cells contain many oxidoreductases [10], and substrates and/or products might be converted in an undesired direction diminishing yield and often creating a difficult separation problem. Therefore, screening for a suitable microbial strain and operating conditions is mandatory in process development to achieve high selectivity. Three examples are discussed below whereby microbial reductions were selected for scale-up and different solutions could be found for various problems arising in recovery and process operation.

A. Flavor Production

The first example relates to the production of natural "green note" flavor compounds developed by the flavor company Firmenich SA, Geneva, CH [39]. Natural flavors have high consumer appeal and are desired in the food industry. Products obtained by fermentation or enzymatic catalysis usually qualify for the label "natural" and can be sold at higher prices than nature identical flavors derived by chemical synthesis. The green notes are a mixture of hexenal, hexan-1-ol, E-2-hexenal, E-2-hexen-1-ol, and Z-3-hexen-1-ol, with the latter dominating the impact of freshness on the sensory perception. Natural green notes are in high demand in the food industry, exceeding the supply traditionally extracted from plants such as mint (*Mentha arvensis*). The company developed a process starting with polyunsaturated plant fatty acids, which are converted to six-carbon hexenals and hexanal by the combined action of lipoxygenase and hydroperoxide lyase. These enzymes were screened from a large number of plant tissues. Soybean flour was selected possessing the key enzyme for the first conversion, the hydroperoxide lyase, in abundance. In the second conversion the carbonyl group had to be reduced to yield the main components in the flavor, while the double bond in the hexenals should be preserved. This was accomplished by microbial reduction employing a 20% baker's yeast suspension as catalyst.

Usually such a high yeast concentration would present a severe problem in product recovery. But flavor compounds are sufficiently volatile and can be isolated by steam distillation. The process is operated in ton scale [39]. Flavors and fragrances are high-priced specialties of high potency, often containing chiral synthons. Although in the process discussed the reduction leads to a primary alcohol, the biocatalyst provides the selectivity for the carbonyl group vs. the double bond. The process is intensified using high catalyst concentrations. The yeast is discarded after the biotransformation. The essential feature of the development is the product recovery by steam distillation, stripping a purified concentrate in amounts of 1–5 g only per kilogram of highly complex reaction mixture. It appears that steam-volatile compounds are especially suited for reductions or other biotransformations using baker's yeast, provided the selectivity of the reaction is satisfactory.

B.   Reduction of a Ketosulfone

The asymmetrical reduction of the ketosulfone shown in Fig. 5 provides another example for a biotransformation developed to plant scale at Zeneca [40]. The resulting hydroxy sulfone is a key intermediate in the synthesis of a new glaucoma drug. It contains two chiral centers, both required in (S) configuration. The stereochemistry at the carbon in the ring structure carrying the methyl group is obtained by chemical synthesis from the chiral pool starting with methyl-3-hydroxybutyrate. Chemical reduction of the ketosulfone leads predominantly to the undesired (4R,6S)-hydroxy sulfone. A screening was carried out using a limited number of strains obtained from culture collections comprising bacteria, yeast, and fungi. Several strains were identified as reducing the ketosulfone with varying yields and stereoselectivities. In the course of the investigation it was noted that the ketosulfone is easily racemized in water even at neutral pH values, whereas the hydroxy sulfone is configurationally stable. Racemization could be suppressed at low pH. Therefore the biotransformation had to be carried out at pH values of 4.0–4.5. A fungus, *Neurospora crassa*, was finally selected as catalyst for the biotransformation, which grew well at low pH and yielded a product of high optical purity. In the optimized biotransformation *N. crassa* is grown in submersed culture to a specific stage in the growth cycle. Then the ketosulfone is fed at a controlled rate to match the catalytic activity of the biomass. This way the ketosulfone concentration is kept low in the reactor and the chemical side reaction is minimized. After completion of the reaction the fungal biomass is separated by filtration and the product is extracted from the broth with an appropriate organic solvent. After crystallization the (4S,6S)-hydroxy sulfone is obtained in more than 80% yield and high optical purity [40]. The successful development illustrates that the desired enzymatic activity can often be found with relatively little effort in culture collection strains, provided the expertise of a microbiologist is available. Besides, optimal reaction conditions and stereoselectivity do not depend solely on the biocatalyst or microbial cells involved; rather, chemical side reaction, racemization of educt and product, and so forth have to be taken into consideration. This in turn leads to the necessity of a careful biochemical engineering design to achieve a viable process at plant scale.

C.   Synthesis of a Chiral Alcohol

A team effort is also evident in the development of a biocatalytic reduction of 3,4-methylene-dioxyphenyl-acetone to the corresponding (S)-alcohol at Lilly Research Laboratories (Fig. 6) [41–43]. The chiral alcohol is a key intermediate in the synthesis of an anticon-

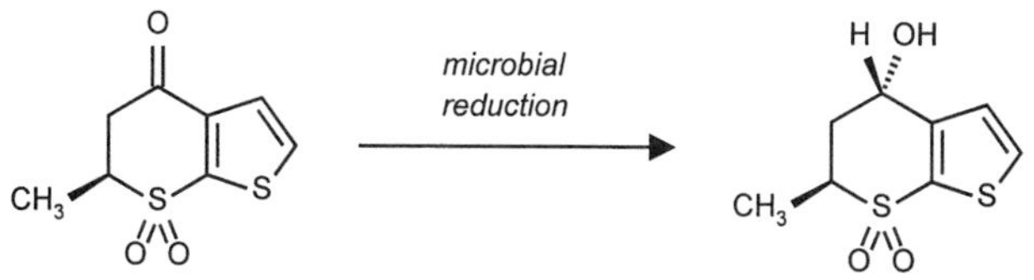

Figure 5   Asymmetrical reduction of a ketosulfone to the desired (4S,6S)-hydroxysulfone, a key intermediate in the synthesis of a drug against glaucoma [40].

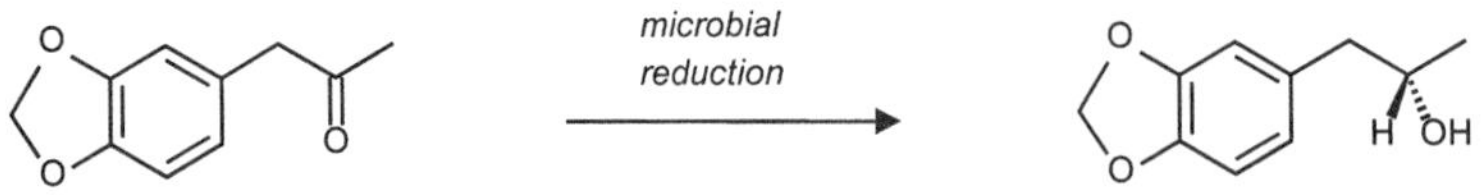

Figure 6 Asymmetrical reduction of 3,4-methylene-dioxyphenylacetone to the (S)-alcohol by *Zygosaccharomyces rouxii* [42,43].

vulsant drug. The yeast *Zygosaccharomyces rouxii* was finally selected after extensive screening because it exhibited no side reactions and produced the desired (S) enantiomer in high optical purity [43]. The lipophilic substances, educt as well as product, diffused over the cell membrane without significant transport resistance. These properties are favorable for a microbial transformation. It turned out, however, that educt and product are toxic to the cells. *Zygosaccharomyces rouxii* tolerates only a maximal concentration of 6 g/L. While educt concentrations can be maintained at low levels by a feeding strategy as discussed above, product will accumulate in the reactor unless an in situ product removal can be accomplished. In situ solvent extraction was discounted for reasons of toxicity and/ or unfavorable partition coefficients and emulsion problems. Macroporous polymeric resins are used extensively for the isolation and purification of natural products. A screening was performed to find a nonpolar resin combining high capacity for both educt and product in water with a useful phase equilibrium. Amberlite XAD-7 was selected for in situ product removal. In practice, the ketone is preadsorbed to XAD-7 in concentrations of 80 g/L resin; then glucose and wet cells of *Z. rouxii* are added and the mixture is gently stirred for 8–12 h until the reaction has been completed. The initial educt concentration in equilibrium with the adsorber in the aqueous phase is approximately 2 g/L and is replenished from the resin reservoir as the reaction proceeds, while at the same time the product generated is adsorbed to the resin. At the end of the process 75–80 g of the alcohol is found on the resin and about 2 g/L remains in the aqueous phase. The resin is separated from the yeast by a simple procedure using a steel screen in an agitated filter. After washing with water the product can be desorbed by acetone and further purified by crystallization. The in situ yield of 3,4-methylene-dioxyphenyl-(S)-isopropanol was 99–100%. The isolated yield of product was 85–90% with an enantiomeric excess (e.e.) of ≥99.9%. In large-scale experiments the ratio educt/resin/catalyst (on a dry weight basis) was approximately 1:10:1. A productivity of 75 g/(L × d) was reached [42].

The problem of product toxicity could be solved using in situ adsorption. Apparently, coenzyme regeneration with glucose is impaired in damaged cells of *Z. rouxii*. Later work demonstrated that NADPH serves as the coenzyme for the ADH involved in the reduction. The molecular event(s) leading to the toxic effect remain unclear.

IV. REDUCTIONS WITH ISOLATED ENZYMES

A. Alcohol Dehydrogenases

For the production of chiral compounds secondary alcohol dehydrogenases are of special interest. To identify the different enzymes in abbreviated form, it is customary to add a one- or two-letter combination identifying the original source of the enzyme in front of ADH. For example, YADH denotes the alcohol dehydrogenase from yeast (*Saccharomyces cerevisiae*). HLADH is the alcohol dehydrogenase from horse liver and TBADH the alcohol dehydrogenase from *Thermoanaerobium brockii*. All of these and others are classified in the enzyme nomenclature as EC 1.1.1.1. Depending on the initial screening and to some extent on the initial characterization, alcohol dehydrogenases may be called carbonyl- or ketoreductases in the literature.

In the author's laboratories a screening for ADHs reducing aromatic asymmetrical ketones was carried out, resulting in the identification of a suitable enzyme in a strain of *Rhodococcus erythropolis* (DSM 43297) [44]. Later in an independent screen another *R. erythropolis* (DSM 743) was found to possess an enzyme capable of reducing β-keto esters as well as γ- and δ-keto esters stereoselectively to the corresponding (*S*)-hydroxy esters, which was characterized as a carbonyl reductase [45–47]. When the substrate range was investigated more extensively it turned out that both enzymes are quite similar although with characteristic differences. The close relationship has been confirmed by partial sequence analysis (B. Riebel, W. Hummel, unpublished data). The example is given to illustrate the point that you only find what you look for. Besides, it should emphasize that not only enzymes called ADH reduce ketones but that, once again, enzymes from different, even closely related, sources may have different properties, which cannot necessarily be deduced from available information.

The organic chemist prefers commercially available enzymes. A number of ADHs as well as other dehydrogenases can be obtained from specialized suppliers; most were originally developed for analytical purposes exploiting differences in the ultraviolet/visible (UV-VIS) spectrum between NAD(P) and NAD(P)H to measure metabolites. Biochemical data for commercial dehydrogenase preparations are compiled in Table 1. In the analysis of complex mixtures in clinical diagnostics or the food industry the enzyme has to ensure the specificity of the signal produced. This requirement often necessitates an extensive purification of analytical enzymes and is, besides a small total market volume, the reason for the rather high prices of these catalysts. For preparative and industrial application, the educt is added in more or less pure form and the enzyme can be used as a rather crude preparation. However, the activity should be stable enough and the educt and/or product not converted in an undesired direction. The latter can be tested quite easily via mass balance. The current catalogue price of enzymes should not deter scientists from experiments. If an enzyme is needed in larger amounts it can generally be produced much more cheaply by available technology. In this context it is worth remembering that high fructose corn syrup (HFCS) is produced from starch annually in amounts exceeding 5×10^6 tons. The process requires three enzymes and competes successfully with saccharose, which is sold in the supermarket for about 1 \$/kg.

Commercial ADHs are usually checked first for the conversion of a ketone of interest. Based on published data, Table 2 generalizes the substrate range of commercially available ADHs and may be used to guide experiments. Often a better catalyst can be obtained by screening. Examples are discussed above. Besides, there are many enzymes sufficiently well described in the literature that may be suitable. The producing organisms are often accessible from culture collections or from the specific laboratories involved in the original work. Some basic skills in microbiology are required and some special equipment to handle microbial cultures at least in small scale. There are also small biotechnology companies offering services in screening or enzyme production.

1. Reductions of Poorly Soluble Substrates

Ketones of potential interest for stereoselective reduction by ADH are often not sufficiently water-soluble. This in turn will influence the kinetics and productivity of the reaction. In recent years new approaches have been tried to overcome these limitations.

To improve the solubility of educts addition of water-miscible solvents is possible. It has to be tested whether the enzyme is stable under these conditions and to what extent

Table 1 Biochemical Data of Some Commercially Available Dehydrogenases and Carbonyl Reductases

Enzyme	Coenzyme Utilized	Configuration of products	Temperature stability [°C]	PH optimum for reduction	Specific activity for reduction [U/mg]
YADH	NADH	S	<25	7.2	250–500
HLADH	NADH	S	25	7.0	1–3
TBADH	NADPH	S	≤65	7.5–8.0	5–90
LKADH	NADPH	R	<25	6.5–7.5	180
CPCR	NADH	S	≤35	7.5–8.5	40/>1000
RECR	NADH	S	≤40	5.5–6.5	20/>1000
L-LDH	NADH	S	25	7.2	150–500
D-LDH	NADH	R	25	7.2	250–300
L-HicDH	NADH	S	≤40	6.5–8.0	30
D-HicDH	NADH	R	≤40	6.0–7.0	25
AlaDH	NADH	S	≤75	7.0–9.0	30–50
GluDH	NAD(P)H	S	20	7.8–9.0	10–40
PheDH	NADH	S	≤30	8.5–9.5	6–30
LeuDH	NADH	S	≤50	8.5–9.5	60–200

kinetic parameters, notably K_M values, are affected by the kind and concentration of solvent under consideration. It is well known that YADH is readily destabilized by water-miscible solvents and rapidly loses activity at ambient temperatures.

Operation in biphasic mixtures using water-immiscible solvents introduces a linked equilibrium in the partition of educt and product and possible transport limitations at the interface, which have to be considered. Besides, enzyme deactivation at the interface and possible effects of the residual solvent solubility in aqueous buffers on enzyme stability have to be checked. Table 3 summarizes some data on stability of ADHs dissolved in aqueous buffers in a biphasic mixture with organic solvents [48]. Two different reactor concepts for continuous operation and enzyme catalysis in homogeneous phase have been studied—a bimembrane reactor [13,14] and an emulsion reactor [49]—which are discussed below with regard to reaction engineering. Using water-immiscible solvents one can make use of the fact that NAD(P)/NAD(P)H are charged molecules and practically insoluble in apolar solvents. The coenzyme introduced in the reaction is therefore confined and physically "immobilized" with the enzymes in the aqueous phase. This facilitates efficient use of the coenzyme, especially if the volume fraction of the aqueous phase is kept low [13].

The activity and stability of HLADH in reverse micelles have also been studied by Larsson et al. [51]. In prolonged operations pH control in the micelles is problematic. A totally different approach has been investigated by Zelinski and Kula to avoid undesirable influences from solvents [52]. The apparent solubility of hydrophobic compounds in aqueous solutions can be improved by the addition of cyclodextrins (CDs). In this case the equilibrium and kinetics of the host–guest complex formation increases the complexity. In the case of 2-acetylnaphthalene it has been shown that the presence of the CD derivative heptakis-(2,6-di-O-methyl)-β-cyclodextrin (DIMEB) had no detectable effect on enzyme kinetics and stability; the k_{dis} of the complex between the ketone and DIMEB was found to be about two orders of magnitude higher than the K_M of the enzyme used. The kinetics

Table 2 Substrate Range of Commercially Available Alcohol Dehydrogenases and Carbonyl Reductases

Enzymes	Sources	Coenzyme needed	Aldehydes	Acyclic ketones	Simple cyclic ketones	Bulky cyclic ketones	Aromatic ketones	Diketones	Unsaturated ketones	3-Oxo-acid esters	4-Oxo-acid esters	5-Oxo-acid esters
YADH	S. cerevisiae	NADH	x						x			
HLADH	Horse liver	NADH	x		x	x	x	x		x	x	
TBADH	T. brockii	NADPH	x	x	x	x		x	x	x	x	x
CPCR	C. parapsilosis	NADH	x	x	x		x	x		x	x	x
RECR	R. erythropolis	NADH					x	x	x	x	x	x
LKADH	L. kefir	NADPH		x	x	x	x				x	

Table 3 Activity and Stability of Alcohol
Dehydrogenases in Biphasic Mixtures with
Organic Solvents

Enzyme sources	Limiting log p values for	
	Activity	Stability
S. cerevisiae	complete deactivation observed up to log P of 6.6	
Horse liver	≥ 1.2	≥ 1.2
T. brockii	≥ 3	≥ 1.5
R. erythropolis DSM 43297	≥ 3.5	≥ 4.0
L. kefir	≥ 2.5	≥ 2.5

Log P is the logarithm of the partition coefficient of a solvent
in an octanol/water system and a measure for the relative hydrophobicity of the solvent. In general apolar solvents (log P
> 4) are compatible with enzyme activity [50]. Solvents with
log P values <2 usually deactivate because they interfere with
the hydration layer on the enzyme surface, compatibility with
solvents of log p 2–4 are unpredictable and depend on the
single case.

of complex formation and dissociation was fast enough to support high reaction rates. The
solubility of the ketone was enhanced about 150-fold at 100 mM DIMEB concentrations.
High conversion (97%) and excellent enantioselectivity (e.e. more than 99.5) were obtained in coupled reactions using FDH for NADH regeneration [52].

2. Heterogenization of Alcohol Dehydrogenases

Enzymes are often immobilized on solid supports in order to reutilize the catalyst and/or
to improve the stability. Reports on the immobilization of ADHs are few and scattered.
Earlier investigations indicated that YADH and HLDH lose activity upon covalent immobilization on a support. Grunwald et al. deposited HLADH together with the coenzyme
on the surface of glass beads by drying. The beads were used under nearly anhydrous
conditions in apolar organic solvents for enzymatic reductions. Very high total turnover
numbers of 10^5–10^6 for the coenzyme were achieved in substrate-coupled processes [53].
Regeneration with FDH in an enzyme-coupled process is more complicated because the
cosubstrate formate cannot be delivered in the organic solvent; titration with formic acid
might be possible. Due to the insolubility of the coenzyme in the organic solvents used,
NAD(H) was not leached from the support and continuous processing possible. As seen
with all heterogeneous catalysts, mass transport limitations are unavoidable if a dehydrogenase is immobilized on a solid support. This has further consequences on the kinetics
and productivity of the reaction. The consequences are particularly severe in NAD(P)H-
dependent enzymatic synthesis if a second enzyme is involved for coenzyme regeneration.
The possible gain in stability by covalent immobilization of an enzyme has to be judged
against the loss of activity in the procedure as well as the loss in effectivity in operation.

Covalent coupling of TBADH to cyanogen bromide–activated agarose and on Eupergit C was studied by Keinan et al., who reported 30% or 15% residual activity, re-

spectively [54,55]. The immobilized enzyme preparation was used in column reactors in aqueous buffer solutions containing 10–20% 2-propanol to achieve NADPH regeneration. During 30 days of continuous reduction of 2-pentanone no decrease in activity was observed with the TBADH Eupergit column, indicating high operational stability. The feed solution had to be supplemented with 5×10^{-6} M NADP to maintain maximal conversion rates. Pulsing of the coenzyme into the feed solution improved the total turnover numbers to values as high as 10^4–10^5 [55]. The immobilized preparation was considered to be especially useful for the conversion of reactive compounds such as ω-chloroketones. Keeping the reservoir of the feed solution and the collected product solutions at low temperatures and minimizing storage time in aqueous buffers, decomposition of the starting material as well as the product could be avoided. The stereoselectivity of TBADH was not affected by the immobilization [54,55].

B. Carbonyl Reductases

Chiral β-hydroxy esters are versatile building blocks that may be obtained by enzymatic or microbial reduction of the corresponding keto esters. The yeast *S. cerevisiae* has been used extensively with variable results regarding the stereoselectivity of the reaction. Different enzymes have been isolated and characterized from *S. cerevisiae*, all requiring NADPH as coenzyme and yielding either (*R*)- or (*S*)-configurated 3-hydroxy esters [56,57]. The substrate range of these enzymes has not been explored extensively and application studies are lacking. NADH-dependent keto ester reductases with a broad substrate range have been described recently from *Candida parapsilosis* (CPCR) [58] and *R. erythropolis* (DSM 743) (RECR) [46]. In addition, it has been found that the NADH-dependent (*S*)-ADH from *R. erythropolis* (DSM 43297) [44] and the NADPH-dependent (*R*)-ADHs from *Lactobacillus kefir* and *L. brevis* also reduce keto esters [48] (see Table 2). Since the carbonyl reductases accept 2-propanol as substrate, a substrate-coupled regeneration of the coenzyme is possible as well as enzyme-coupled regeneration with FDH. Using isopropanol as cosubstrate, the e.e. value of the product is often slightly lower (94–97%) than that obtained with formate as cosubstrate (more than 99%) [59]. Nevertheless, the optical purity is high enough to warrant consideration of this option as well. Employing FDH for NADH regeneration, CPCR is the enzyme of choice to produce (*S*)-configurated compounds, since the pH optimum of FDH overlaps better with the pH activity profile for reduction of CPCR than RECR. The (*R*)-ADHs from *Lactobacillus* have a pH optimum at 7.0 and can be coupled with the NADPH-dependent variant of FDH. Using these enzymes continuous operation has been achieved over prolonged periods of time with high conversions and space time yields up to 200 g/(L × d).

An extensive screening for microbial strains reducing ketopantoyl lactone has been carried out in Japan. From *C. parapsilosis* an enzyme was isolated and further characterized as a polyketone reductase (EC 1.1.1.184) [60,61]. The polyketone reductase requires NADPH as coenzyme and is different from the CPCR mentioned above. The enzymatic reduction of ethyl 4-chloro-3-oxobutanoate yielding the corresponding (*S*)-β-hydroxy ester has been described from *Candida magnolia* and *Candida macedoniensis* [62]. The (*R*)-configured products were obtained using a reductase from *Sporobolomyces salmonicolor*. These enzymes require NADPH as coenzyme and reduce a variety of carbonyl compounds besides ethyl 4-chloro-3-oxobutanoate with an overlapping but distinctly different substrate range [63]. More structural and kinetic data are needed to categorize the ketoreductases and to predict the conversion of untested substrates with higher accuracy.

Since chiral 2-hydroxy acids are available by fermentation or enzymic processes (see below), the reduction of 2-oxo acid esters has rarely been studied. Matzinger et al. described the microbial reduction of N-protected 4-amino-2-oxobutyric acid esters in 200-L scale to the corresponding (*R*) enantiomer by *C. parapsilosis*. The (*S*) enantiomer could be obtained using *Toropsilosis magnoliae*. The enzymes involved have not been identified [64].

C. Reduction of 2-Oxo Acids

The reduction of pyruvate to lactate is an important metabolic reaction and very well studied. L- and D-lactic acid are produced by fermentation. Besides mammalian tissues, lactobacilli or other bacteria are the preferred sources of lactate dehydrogenase (LDH). The stereoselectivity of LDHs is very high. Apparently the carboxylate group is involved for the correct positioning of the substrate in the active site; 2-oxo esters are usually not accepted as substrates. D-LDH as well as L-LDH are commercially available (Table 1) and are used extensively in food industry and clinical analysis. Unfortunately, these enzymes have a very limited useful substrate range; the relative activity drops drastically for side chains larger than butyrate [65,66]. For the reduction of 2-oxo acids with bulky hydrophobic side chains different enzymes, i.e., 2-hydroxyisocaproate dehydrogenases (HicDH), have been found in *Lactobacillus* sp., which are more useful than LDH for preparative applications. Either enantiomer can in principle be produced from the corresponding 2-oxo acid using the NADH-dependent D-HicDH [67,68] or L-HicDH [69], respectively. Continuous processes in an enzyme membrane reactor have been studied and space–time yields of 410 g/(L × d) have been obtained with L-HicDH in the production of L-2-hydroxyisocaproate [70] and 700 g/(L × d) with D-HicDH in the production of (*R*)-mandelic acid [71]. The ttn for PEG-NADH were 69,000 and 24,000, respectively, for L-HicDH and D-HicDH under nonoptimized conditions. (*R*)-Mandelate dehydrogenase is found as a synonym for D-HicDH in the literature. It appears that other enzymes described in the patent literature for the reduction of 2-oxo acids with bulky side chains resemble HicDH rather than LDH [72]. Because of their potential for the production of chiral 2-hydroxy acids the L-HicDH as well as the D-HicDH from *L. casei* have been cloned and expressed in *E. coli* [73,74] and can be produced in large amounts if needed. The substrate range for the D-HicDH from *L. casei* has been studied by Kallwass [75] and Krix et al. [76]. Some data on the substrate range of L-HicDH and studies to change substrate specificity by site directed mutagenesis have been reported [76–78]. From the primary product, chiral hydroxy acids, chiral diols, and other derivatives can be obtained by chemical procedures preserving the chirality.

D. Reductive Amination of 2-Oxo Acids by Amino Acid Dehydrogenases

2-Oxo acids can be converted to amino acids by reductive amination using an appropriate amino acid dehydrogenase (EC 1.4.1). Only L-amino acids are accessible this way, since no D-specific amino acid dehydrogenases are known (see Scheme 1). Glutamate dehydrogenase (GluDH) and alanine dehydrogenase (AlaDH) are involved in the nitrogen metabolism of cells and appear to have a limited substrate range, whereas leucine dehydrogenase (LeuDH) and phenylalanine dehydrogenase (PheDH) have mainly a catabolic function and accept a wide variety of 2-oxo acids as substrates [76,79]. Reductive amination is one of several options to synthesize L-amino acids. It is of special interest to produce nonproteinogenic amino acids, which cannot be obtained by fermentation, or analogs of proteinogenic amino acids, or to introduce nitrogen isotopes such as [11]N, ^{13}N, or ^{15}N into amino acids for analytical purposes. For example, 35 g of ^{15}N-Glu has been prepared by repetitive batch operations with a coupled enzyme system, recovering the enzymes between batches by ultrafiltration [80]. For the introduction of radioactive ^{13}N and especially [11]N the speed of the reaction and subsequent product recovery are the most important considerations, whereas for the synthesis of stable ^{15}N-amino acids the cost of the ^{15}N salt and yield are the key parameters. Using equimolar concentrations of 2-oxoglutarate and [15]NH$_4$Cl, L-glutamate was isolated in 86% yield at high optical purity (e.e. more than 99.5%) and a ^{15}N abundance of 98% equal to the starting material [15]NH$_4$Cl [80]. The equilibrium of the enzyme-catalyzed reaction strongly favors the amino acid synthesis. Therefore, high conversion close to 100% can be obtained, raising the ammonia concentration. These features are important when considering reductive amination of a prochiral precursor vs. resolution of a racemate of amino acids by enzymatic or other methods giving 50% yield. Reductive amination has been studied extensively [19,22,81,82] and is now applied in the commercial production of L-*tert*-leucine [36]. An economically feasible NADH regeneration has been developed in parallel. During continuous synthesis of L-phenylalanine 600,000 ttn for NADH have been achieved [22]. In situ regeneration of the coenzyme is possible only by using a second enzyme, e.g., FDH as shown in Fig. 4 and discussed above. AlaDH, LeuDH, and PheDH are NADH-dependent enzymes. Besides NADPH, GluDH also accepts NADH depending on the source of the enzyme. The kinetics of the coupled reaction with FDH have been determined in detail and different reactor designs experimentally and theoretically evaluated on the basis of a detailed kinetic model [9]. Determining the kinetic parameters of a variety of 2-oxo acids using LeuDH isolated from different strains of *Bacillus* showed that the enzyme from *B. cereus* tolerated best branches in the C-3 position and was the best catalyst for L-*tert*-leucine production as well as other L-amino acids with bulky hydrophobic side chains such as neopentylglycine [76,83]. Wild-type *B. cereus* might produce toxins and therefore is not considered a safe production organism for an enzyme. To establish a secure supply of BcLeuDH the structural gene has been cloned into *E. coli* [84]. In optimized inducible expression systems 40–60% of the soluble proteins in recombinant *E. coli* represents BcLeuDH [85]. The enzyme is obtained in an active form and can be easily purified if desired. The BcLeuDH production has been scaled to 200-L culture volume yielding 1.2×10^8 U in a single cultivation. The amount of catalyst obtained is sufficient to produce approximately 100 tons of L-*tert*-leucine, taking into account an experimental value of 930 U/kg for the product-specific enzyme consumption rate [82]. The substrate range for PheDH and LeuDH overlap to some extent. The phenylalanine analogs with substitutions in the aromatic ring, homophenylalanine [(S)-2-amino-4-phenyl-butanoic acid] [86] and some amino acids carrying very bulky side chains are preferably synthesized using PheDH [76]. The PheDH gene from *Rhodococcus* as well as other microbial strains [79,87,88] has been cloned and can be expressed in active form in *E. coli*. The expression systems have not been fully optimized yet. Due to the very high intrinsic specific activity of PheDH from *Rhodococcus* M4 (more than 2000 U/mg), even poor substrates can be converted with useful rates [22,76].

V. REACTION ENGINEERING FOR OXIDOREDUCTASES

Discussing specific examples, it has already been pointed out that (bio)chemical reaction engineering can be used with advantage for process optimization. Besides the space–time yield and product-specific enzyme consumption, the cost for the coenzyme has to be taken into account, employing oxidoreductases for the production of fine chemicals. To lower the product-specific coenzyme costs a high ttn has to be reached. Depending on the substrate solubility, different cases have to be considered as are shown schematically in Fig. 7. Examples are summarized in Table 4.

- A high substrate concentration together with a low coenzyme concentration will yield high ttn values; for example, with 500 mmol/L substrate and 0.1 mmol/L coenzyme a ttn of 5000 can be reached, after complete conversion.
- With low or intermediate substrate solubility additional measures for the repeated use of the coenzyme have to be taken into consideration. For example, recirculation of the water phase in a multiphase reactor or the use of nanofiltration or charged membranes are possible strategies.

Besides an efficient coenzyme regeneration, scale-up of an enzymatic process requires general investigations as summarized in Fig. 8. For a more extensive introduction into this field, please see Ref. 89. For the final optimization, a complete model based on kinetic and thermodynamic properties of the whole reaction system is of great importance. Nevertheless, it should be the simplest model that sufficiently describes the observed kinetic behavior. When the reaction mechanism is known, an appropriate kinetic model may be developed [90,91]. Very often the concentration range where kinetic data have been estimated is much lower than the desired concentration for the practical application. In such a case it must be ascertained as to whether an extrapolation of the model to those high concentrations is still giving good agreement between prediction and experimental results.

Even in the early stages some basic information about the kinetic properties of the enzymes involved is of great interest, as are data about enzyme and coenzyme stability under reaction conditions. In addition, the chemical stability of educts and product in aqueous buffers must be considered [4,40]. If a strong substrate surplus inhibition occurs, the reaction rate is very low when high substrate concentrations are applied at the start of

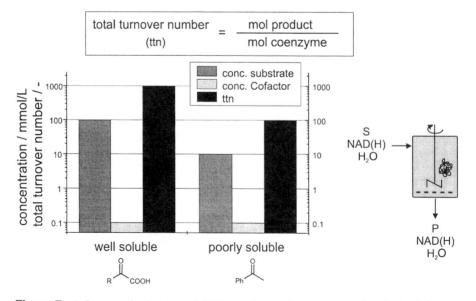

Figure 7   Influence of substrate solubility on the total turnover number (ttn) of the coenzyme.

Table 4 Selected Examples for Different Strategies to Increase the Total Turnover Number (ttn) of the Coenzyme Involved

Enzyme	Substrate	Solubility	Method	ttn	Reference
Leucine dehydrogenase	2-Oxo acid	High	High substrate concentration	4250	13
Leucine dehydrogenase	2-Oxo acid	High	Nanofiltration membrane	7920	98
Phenylalanine dehydrogenase	2-Oxo acid	High	PEG-enlarged NADH	600000	22
Alcohol dehydrogenase	1-Phenyl-propan-2-one	Low	Bi-membrane reactor	1350	14
Carbonyl reductase	2-Octanone	Low	Emulsion reactor	124	49
Carbonyl reductase	2-Octanone	Low	Homogeneous phase	14	49
Menthone reductase	1-Menthone	Low	Emulsion reactor	2500	109
Mannitol dehydrogenase	Fructose	High	Charged membrane	150000	99, 100, 101
Glucose dehydrogenase	Glucose	High	Addition of polyethyleneimine to an UF membrane reactor	Not reported	102

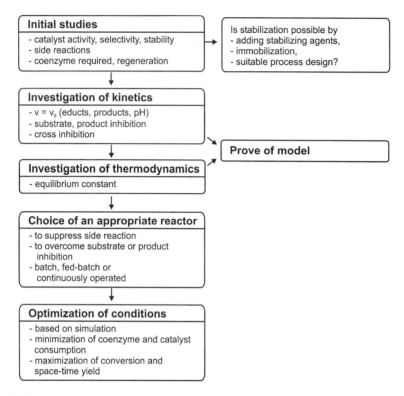

Figure 8 Flow scheme of process development.

the reaction and increases gradually with the conversion. This problem can be easily overcome by a repeated dosing of the substrate to ensure low concentration in a batch process or by performance of the reaction in a continuously operated stirred-tank reactor (Fig. 9) [89,92]. Also, a strong product inhibition is often found requiring special considerations. In an initial estimation the ratio of the Michaelis-Menten constant K_M and the parameter describing the product inhibition K_i can be used. If the ratio K_M/K_i is less than 1, the reaction can be easily exploited for synthesis, whereas a ratio of more than 1 indicates a strong product inhibition [93,94]. In such a case the most appropriate reactor is a batch reactor, a plug flow reactor, or a cascade of continuously operated stirred-tank reactors, which allow the most efficient use of the catalyst employed. However, one has to keep in mind that there might be other tools available to allow the effective operation of an enzymatic reaction even with high product inhibition, e.g., by selective removal of the inhibiting product [94,95]. To collect these basic data for dehydrogenases, only a few experiments are necessary. These are easily carried out using a spectrophotometer or a microplate reader, following the NAD(P)H consumption rate at 340 nm under various reaction conditions. In addition, any problems of cross-inhibition of the substrates and products of the desired reaction and the regeneration system have to be checked [71].

The repeated use of the biocatalyst is often necessary to reduce the catalyst costs. Soluble enzymes can be easily recovered from the reaction mixture using an ultrafiltration step [18,96,97], thereby combining homogeneous catalysis with selective enzyme recovery and reuse. Ultrafiltration can be performed from a microliter to a multi cubic meter scale with commercially available ultrafiltration membranes and equipment. For first experi-

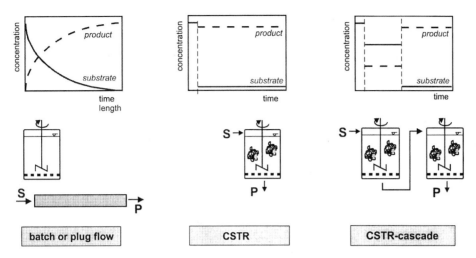

Figure 9 Concentration profiles of different reactors.

ments a working volume of 5–50 ml is appropriate employing single flat sheet membrane in a stirred ultrafiltration cell (Fig. 10a). For larger volumes a loop reactor consisting of a reaction vessel and an external filtration unit (flat membrane stack or hollow fiber module) can be used (Fig. 10b). The educts might be pumped through the cell or the feed reservoir can be pressurized with an inert gas. In these types of reactors the ultrafiltration membrane may be replaced by a nanofiltration membrane or a charged membrane allowing partial retention of the cofactor as well [98–101]. Charged polymers, e.g., polyethylene imine, can be added initially to the reactor to create a charged barrier in the gel polarization

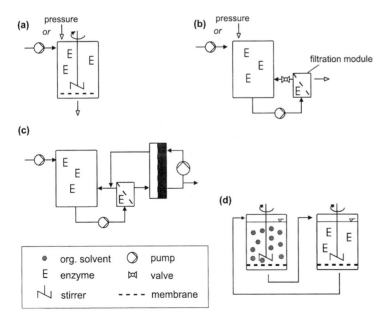

Figure 10 Different types of membrane reactors for well (a,b) or poorly soluble substrates (c,d).

layer of an ultrafiltration membrane if charged membranes are not readily available [102]. Reactors based on membrane filtration are especially suited for substrates showing a high solubility in the reaction medium. For a high retention of the coenzyme, covalent binding to a water-soluble polymer [22,103,104] has been described, allowing separation of the coenzyme together with the enzymes from the reaction mixture. If a substrate with a low solubility is to be converted a bimembrane reactor as shown in Fig. 10c can be employed [13,14]. The pure substrate is pumped into the reactor keeping the concentration just below saturation in the aqueous buffer. The enzyme concentration and residence time in the first loop has to be adjusted in such a way (by trial and error or preferably by a simulation based on the kinetic data) that high conversion is reached. The product formed is extracted in a membrane contactor into an immiscible organic solvent, a process called *perstraction*. The water phase containing the coenzyme and the highly soluble regeneration substrate is fed back into the reactor. In principle, due to the complete retention of the coenzyme in the aqueous phase an infinite ttn would be possible. Performance is limited, however, by the thermal deactivation of the coenzyme, which has been found to be around 1% per hour at 25°C [13], depending on the reaction conditions. It is therefore advisable to operate a dehydrogenase-catalyzed reaction at ≤30°C. The reduction in total ttn due to the thermal deactivation is shown in Fig. 11. For retention values up to 0.95 the improvement in ttn is dominated by the retention alone. The higher the retention the more pronounced is the influence of coenzyme stability because it stays and is recycled in the reactor for a long time. Besides temperature, the pH also strongly influences the stability of the nicotinamide coenzymes summarized in Fig. 12. Generally, the oxidized form is more stable at lower pH values, whereas the reduced form is more stable under slightly alkaline conditions [105]. Thus, it is important to find the most suitable pH for the process with respect not only to enzyme activity and stability, but to the coenzyme as well.

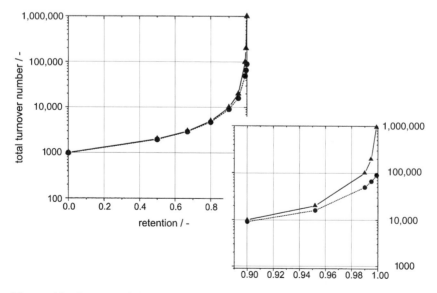

Figure 11 Increase of the total turnover number (ttn) of the coenzyme by retention and influence of the thermal deactivation. Concentrations: substrate 100 mmol/L, coenzyme 0.1 mmol/L.

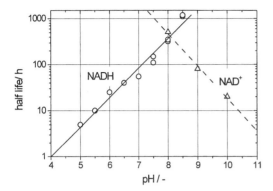

Figure 12 Half-life of NAD$^+$ and NADH as function of pH at 25°C.

The water phase can also be recharged with fresh substrate in an emulsion reactor, where a hydrophilic membrane is used to cleave the emulsion (Fig. 10d) [49]. In this type of reactor as well as in the bimembrane reactor the enzyme is well separated from the organic phase. This is important as interphases, e.g., between two immiscible solvents or between a liquid and a gas differing widely in the dielectric constants, may lead to protein denaturation. A more detailed description of these reactors can be found in the references given or for membrane reactors in general [35,106–108].

For L-*tert*-leucine [12,13] and some other amino and hydroxy acids [20,22,71], as well as for chiral alcohols [49], detailed reaction engineering studies have been performed and published in recent years [82]. The main targets were to increase the ttn of the coenzyme involved, mainly NADH, and to decrease the product-specific catalyst consumption. Typical results are compiled in Table 5 for L-*tert*-leucine as a highly soluble substrate, or product, and (*S*)-1-phenyl-2-propanol and (*S*)-2-octanol as poorly soluble products obtained in different types of reactors. The reactions analyzed are shown in Figs. 4 and 13, respectively.

If a readily soluble substrate such as a 2-oxo acid is converted to a highly soluble product as in the case of L-*tert*-leucine, it is possible to apply a continuous feed of the native coenzyme in its cheaper oxidized form into a membrane reactor. Normally only very low coenzyme concentrations are needed to saturate the enzymes involved. The kinetic investigation of this reaction system revealed a substrate surplus and product inhibition. On one hand, a high initial substrate concentration is favorable with respect to a high ttn of the coenzyme. On the other hand, a high substrate concentration increases the inhibition problems. A possible solution is the use of a membrane reactor cascade where the first membrane reactor operates under reduced substrate surplus inhibition and moderate product inhibition, whereas the second reactor in the cascade is not limited by substrate surplus inhibition thus compensating for the increasing product inhibition. Due to the high substrate concentration even without coenzyme retention a ttn of 4230 was reached. Inserting a nanofiltration membrane (with a retention for the coenzyme of $R = 0.67$) in the reactor, even in a single stage with a lower initial substrate concentration a total turnover number of 7920 was reached (Fig. 14).

For substrates and products with low solubility in aqueous buffer the ttn that can be reached is somewhat lower. Two examples for sparingly soluble ketones are shown in Fig.

Table 5 Conditions and Results for Oxido-Reductase Catalyzed Reductions with Coenzyme Regeneration; the Reaction Schemes Are Shown in Figs. 4 and 13, Respectively; the Enzyme Consumption Is Given Only for the Production Enzyme

	L-tert-leucine	L-tert-leucine	(S)-1-phenyl-2-propanol	(S)-2-octanol
Reactor	UF-membrane, two stage cascade	NF-membrane	Bi-membrane	Emulsion
Substrate/mmol/L	900	500	9	68
2nd Substrate/mmol/L	—[a]	—[a]	75	225
NAD^+/mmol/L	0.2	0.06	0.12	0.5
Enzyme/U/mL	—[a]	—[a]	ADH 0.9 FDH 2.6	CPCR 0.48 FDH 1.44
Conversion/%	94	95	72	91
Space-time yield/g/(L × d)	290	373	63	11
Total turnover number/–	4230	7920	1350	124
Enzyme consumption/U/kg$_{product}$	—[a]	—[a]	3540	22380
Reference	12	20	14	49

[a]Data are not available due to industrial process.

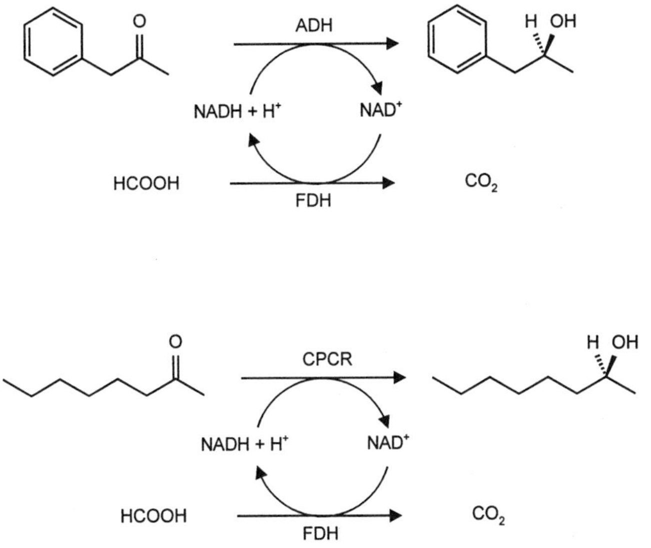

861

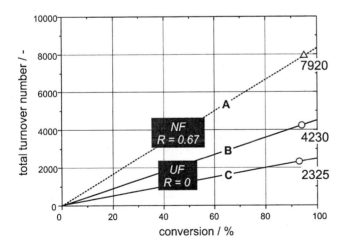

Figure 13 Reaction schemes for the synthesis of (*S*)-1-phenyl-2-propanol and (*S*)-2-octanol (ADH, alcohol dehydrogenase; CPCR, *Candida parapsilosis* carbonyl reductase; FDH, formate dehydrogenase).

13. The results are summarized in Fig. 15, which shows the improvement of the ttn over a simple enzyme membrane reactor, as in Fig. 10a operated only with a saturated feed solution.

In general, the coenzyme concentration strongly influences the ttn, which can be achieved and therefore reduces the catalyst costs. This is shown in Fig. 16 for the systems

Figure 14 Total turnover number (ttn) as a function of the ratio of substrate/coenzyme concentration and coenzyme retention for the enzymatic synthesis of L-*tert*-leucine. The data points are experimental values (compare Table 5). Feed concentrations: 2-oxo-3,3-dimethylbutanoic acid (A) 500 mmol/L, (B) 900 mmol/L, (C) 500 mmol/L; NAD^+ (A) 0.06 mmol/L, (B) 0.2 mmol/L, (C) 0.2 mmol/L. UF, ultrafiltration; NF, nanofiltration; R, retention.

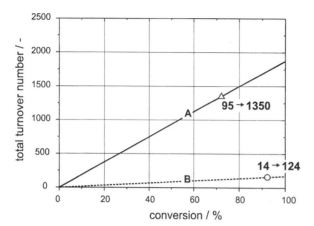

Figure 15   Total turnover number (ttn) as a function of the ratio substrate/coenzyme concentration and coenzyme retention of the enzymatic reduction of poorly soluble ketones. The data points are experimental values (compare Table 5) and indicate the increase of the ttn compared to the operation without additional means for coenzyme retention. (A) Synthesis of (*S*)-1-phenyl-2-propanol; (B) synthesis of (*S*)-2-octanol.

discussed previously. The coenzyme costs per mole of product were calculated using the price for bulk quantities of the cheaper NAD. For the enzymatic synthesis presented the coenzyme costs are well below 1 U.S. $ per mole product, for the highly soluble substrates even below 25 cents! It also becomes obvious that for reactions with educts of a lower solubility there is a bigger reward on the side of coenzyme costs when an appropriate reactor is selected.

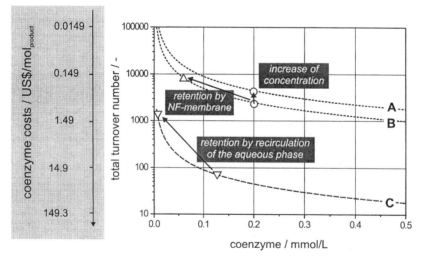

Figure 16   Total turnover number and coenzyme costs as a function of coenzyme concentration, initial substrate concentration, and coenzyme retention. Coenzyme costs were calculated with 2.25 US $/g$_{NAD}$ (Oriental yeast, 1997) and a molecular weight of 663 g/mol. (A,B) Synthesis of L-*tert*-leucine; (C) synthesis of (*S*)-2-phenyl-1-propanol; substrate concentrations: (A) 900 mmol/L, (B) 500 mmol/L, (C) 9 mmol/L.

VI. CONCLUSION

Numerous stereoselective reductions using various enzymes have been described in the literature. To utilize such reactions for the commercial synthesis of fine chemicals an efficient regeneration of the coenzyme(s) involved is needed. This can be achieved in the metabolic network of intact cells, feeding glucose or another appropriate cosubstrate in a biotransformation process. Otherwise using isolated enzymes, formate, or alcohols might serve as hydrogen donors to accomplish in situ NAD(P)H regeneration. Because of the complex coupled reactions a thorough biochemical reaction engineering approach is advantageous to design an economic and robust process. For compounds of high solubility economic constraints can be met today. For poorly soluble educts/products new reactor concepts have to be implemented in pilot scale and improved to increase ttn and decrease cost. Modern genetic techniques are increasingly applied to make dehydrogenases cheaply available and to adapt enzyme properties to processing needs. With an increasing demand for chiral compounds in the pharmaceutical and agro industry, more coenzyme-dependent enzymatic reactions will be utilized in the future.

ACKNOWLEDGMENTS

We thank our many collaborators and students for their contributions to the field as well as the Bundesministerium für Bildung, Wissenschaft, Forschung und Technologie, Deutsche Forschungsgemeinschaft, Sonderforschungsbereich 380, "Asymmetrische Synthesen mit chemischen und biologischen Methoden," and Degussa AG for financial support.

REFERENCES

1. Enzyme Nomenclature. San Diego: Academic Press, 1992.
2. V Prelog. Pure Appl Chem 9:119–130, 1964.
3. J Retey, JA Robinson. Sterospecificity in Organic Chmistry and Enzymology. Weinheim: VCH, 1982.
4. J Peters. In: HJ Rehm, G Reed, eds. Biotechnology, 2nd ed., Vol. 8a. Weinheim: Wiley-VCH, 1998, pp 391–474.
5. K Drauz, H Waldmann, eds. Enzyme Catalysis in Organic Chemistry. Weinheim: VCH, 1995.
6. K Faber. Biotransformations in Organic Chemistry, 3rd ed. Berlin: Springer-Verlag, 1997.
7. Warwick Biotransformation Club. Biotransformations CD-ROM. London: Chapman and Hall.
8. R Devaux-Basseguy, A Bergel, M Comtat. Enz Microb Technol 20:248–258, 1997.
9. U Kragl, D Vasic-Racki, C Wandrey. Bioproc Eng 14:291–297, 1996.
10. S Servi. In: HJ Rehm, G Reed, eds. Biotechnology, 2nd ed., Vol. 8a. Weinheim: Wiley-VCH, 1998, pp 363–389.
11. T Kometani, H Yoshii, R Matsuno. J Mol Catal B Enzymatic 1:45–52, 1996.
12. W Hummel, M-R Kula. Eur J Biochem 184:1–13, 1989.
13. U Kragl, W Kruse, W Hummel, C Wandrey. Biotechnol Bioeng 52:309–319, 1996.
14. W Kruse, W Hummel, U Kragl. Recl Trav Chim Pays-Bas 115:239–243, 1996.
15. K Seelbach, B Riebel, W Hummel, M-R Kula, V Tishkov, A Egorov, C Wandrey, U Kragl. Tetrahedron Lett 37:1377–1380, 1996.
16. H Schütte, J Flossdorf, H Sahm, M-R Kula. Eur J Biochem 62:151–160, 1976.
17. VI Tishkov, AG Galkin, VV Fedorchuk, PA Savitsky, AM Rojkova, H Gieren, M-R Kula. Biotechnol Bioeng 64:187–193, 1999.
18. M-R Kula, C Wandrey. Meth Enzymol 136:9–21, 1987.

19. R Wichmann, C Wandrey, AF Bückmann, M-R Kula. Biotechnol Bioeng 23:2789–2802, 1981.
20. C Wandrey, B Bossow. Biotechnol Bioind 3:8–13, 1986.
21. M-R Kula. In: Proceedings of the Chiral Europe 1994 Symposium. Stockport: Spring Innovations Ltd, 1994, pp 27–33.
22. W Hummel, H Schütte, E Schmidt, C Wandrey, M-R Kula. Appl Microbiol Biotechnol 26: 409–416, 1987.
23. D Weuster-Botz, H Paschold, B Striegel, H Gieren, M-R Kula, C Wandrey. CET 17:131–137, 1994.
24. D Weuster-Botz, C Wandrey. Proc Biochem 30:563–571, 1995.
25. KH Kroner, H Schütte, W Stach, M-R Kula. J Chem Tech Biotechnol 32:130–137, 1982.
26. A Cordes, M-R Kula. J Chromatogr 376:375–384, 1986.
27. H Slusarczyk, M Pohl, M-R Kula. In: A Ballesteros, FJ Plou, JL Iborra, P Halling, eds. Stability and Stabilization of Biocatalysts. Amsterdam: Elsevier, 1998, pp 331–336.
28. VI Tishkov, AG Galkin, GN Marchenko, YD Tsygankov, AM Egorov. Biotechnol Appl Biochem 18:201–207, 1993.
29. AV Mesentsev, VS Lamzin, VI Tishkov, TB Ustinnikova, VO Popov. Biochem J 321:475–480, 1997.
30. VS Lamzin, Z Dauter, VO Popov, EH Harutyunyan, KS Wilson. J Mol Biol 236:759–785, 1994.
31. VI Tishkov, AD Matorin, AM Rojkova, W Fedorchuk, AP Savitzky, LA Dementieva, VS Lamzin, AV Metzentzer, VO Popov. FEBS Lett 390:104–108, 1996.
32. AG Galkin, VS Lamzin, AV Mezentzev, VO Popov, DV Shelukho, TB Ustinnikova, VI Tishkov. Biochem J (in press).
33. J Peters, M-R Kula. Biotechnol Appl Biochem 13:363–370, 1991.
34. S Rissom, U Schwarz-Linetz, M Vogel, VI Tishkov, U Kragl. Tetrahedron: Asymmetry 8: 2523–2526, 1997.
35. AS Bommarius, K Drauz, U Groeger. In: AN Collins, GN Sheldrake, J Crosby, eds. Chirality in Industry. London: John Wiley and Sons, 1992, pp 371–397.
36. AS Bommarius, M Schwarm, K Stingl, M Kottenhahn, K Huthmacher, K Drauz. Tetrahedron: Asymmetry 6:2851–2888, 1995.
37. AS Bommarius, M Estler, KH Drauz. INBIO 98 Conference Papers, Spring Innovations Ltd., Stockport, UK, 1998.
38. C Neuberg, A Lewite. Biochem Zschr 91:257–266, 1918.
39. BL Muller, C Dean, IM Whitehead. Symposium: Plant Enzymes in the Food Industry, 26. Lausanne, 1994.
40. RA Holt. Chimica Oggl (Sept), 17–20, 1996.
41. BA Anderson, MM Hansen, AR Harkness, CL Henry, JT Vicenzi, MJ Zmijewski. J Am Chem Soc 117:12358–12359, 1995.
42. JT Vicenzi, MJ Zmijewski, MR Reinhard, BE Landen, WL Muth, PG Marler. Enzyme Microbial Technol 20:494–499, 1997.
43. MJ Zmijewski, J Vicenzi, BE Landen, W Muth, P Marler, B Anderson. Appl Microbiol Biotechnol 47:162–166, 1997.
44. W Hummel, C Gottwald. Ger Pat Appl DEP 42.09.022.9, 1991.
45. J Peters, T Zelinski, MR Kula. Appl Microb Biotech 38:334–340, 1992.
46. T Zelinski, J Peters, M-R Kula. J Biotechnology 33:283–292, 1994.
47. T Zelinski, M-R Kula. Bioorg Med Chem 6:421–428, 1994.
48. W Hummel. Biochem Eng Biotechnol 58:145–184, 1997.
49. A Liese, T Zelinski, M-R Kula, H Kierkels, M Karutz, U Kragl, C Wandrey. Mol Catal B: Enzymatic 4:91–99, 1998.
50. C Laane, S Boeren, K Vos, C Veeger. Biotechnol Bioeng 30:81, 1987.
51. KM Larsson, P Adlercreutz, B Mattiasson. Eur J Biochem 166:157–161, 1987.
52. T Zelinski, M-R Kula. Biocatal Biotransf 15:57–74, 1997.
53. J Grunwalk, B Wirz, MP Scollar, AM Klibanov. J Am Chem Soc 108:6732–6734, 1986.
54. E Keinan, EK Hafeli, KK Seth, R Lamed. J Am Chem Soc 108:162–169, 1986.

55. E Keinan, KK Seth, RJ Lamed. Ann NY Acad Sci 501:130–150, 1987.
56. WR Shieh, AS Gopalan, CJ Sih. J Am Chem Soc 107:2993–2994, 1985.
57. J Heidlas, K-H Engel, R Tressl. Eur J Biochem 172:633–639, 1988.
58. J Peters, T Minuth, M-R Kula. Enzyme Microb Technol 15:950–958, 1993.
59. J Peters, T Zelinski, T Minuth, M-R Kula. Tetrahedron: Asymmetry 4:1173–1182, 1993.
60. H Hata, S Shimizu, S Hattori, H Yamada. Biochem Biophys Acta 990:175–181, 1989.
61. H Hata, S Shimizu, S Hattori, H Yamada. J Org Chem 55:4377, 1990.
62. M Wada, M Kataoka, H Kawalsata, Y Yasohard, N Kizaki, J Hasegawa, S Shimizu. Biosci Biotechnol Biochem 62:280–285, 1998.
63. M Kataoka, H Sakai, T Morikawa, M Katoh, T Miyoshi, S Shimizu, H Yamada. Biochim Biophys Acta 1122:57–62, 1992.
64. PK Matzinger, B Wirz, HGW Leuenberger. Appl Microbiol Biotechnol 32:533–537, 1990.
65. MJ Kim, GM Whitesides. J Am Chem Soc 110:2959, 1988.
66. ES Simon, R Plante, GM Whitesides. Appl Biochem Biotechnol 22:169, 1989.
67. W Hummel, H Schütte, M-R Kula. Appl Microbiol Biotechnol 21:7–15, 1985.
68. W Hummel, H Schütte, M-R Kula. Appl Microbiol Biotechnol 28:433–439, 1988.
69. H Schütte, W Hummel, M-R Kula. Appl Microbiol Biotechnol 19:167–176, 1984.
70. R Wichmann, C Wandrey, W Hummel, H Schütte, AF Bückmann, M-R Kula. Enzyme Eng 7; Ann NY Acad Sci 434:87–90, 1984.
71. D Vasic-Racki, M Jonas, C Wandrey, W Hummel, M-R Kula. Appl Microbiol Biotechnol 31: 215–222, 1988.
72. O Ghisalba, D Gygax, R Lattmann, H Schär, P Schmidt, G Sedelmeier. Swiss Pat 0,347,374, 1989.
73. H-P Lerch, H Blöcker, H Kallwass, J Hoppe, H Tsai, J Collins. Gene 78:47–57, 1989.
74. H-P Lerch, R Frank, J Collins. Gene 83:263–270, 1989.
75. H Kallwass. Enzyme Microb Technol 14:28, 1992.
76. G Krix, AS Bommarius, K Drauz, M Kottenhahn, M Schwarm, M-R Kula. J Biotechnol 53: 29–39, 1997.
77. IK Feil, HP Lerch, D Schomburg. Eur J Biochem 223:857–863, 1994.
78. IK Feil, J Hendle, D Schomburg. Protein Eng 10:255–262, 1997.
79. Y Asano, A Yamada, Y Kato, K Yamaguchi, Y Hibino, K Hirai, K Kondo. J Org Chem 55: 5567–5571, 1990.
80. U Kragl, A Gödde, C Wandrey, W Kinzy, JJ Cappon, J Lugtenburg. Tetrahedron: Asymmetry 4:1193–1202, 1993.
81. W Hummel, E Schmidt, C Wandrey, M-R Kula. Appl Microbiol Biotechnol 25:175–185, 1986.
82. U Kragl, D Vasic-Racki, C Wandrey. Chem Ing Tech 64:499–509, 1993.
83. H Schütte, W Hummel, H Tsai, M-R Kula. Appl Microb Biotechnol 22:306–317, 1985.
84. T Stoyan, A Recktenwald, M-R Kula. J Biotechnol 54:77–80, 1997.
85. M Ansorge. PhD dissertation, Heinrich-Heine-Universität Düsseldorf, 1998.
86. CW Bradshaw, C-H Wong, W Hummel, M-R Kula. Bioorg Chem 19:29–39, 1991.
87. NMW Brunhuber, A Banerjee, WR Jacobs, JS Blanchard. J Biol Chem 269:16203–16211, 1994.
88. H Takada, T Yoshimura, T Ohshima, N Esaki, K Soda. J Biochem (Tokyo) 109:371, 1991.
89. Chapter A4.4: Reaction engineering for enzyme-catalyzed biotransformations. In: K Drauz, H Waldmann, eds. Enzyme Catalysis in Organic Synthesis: A Comprehensive Handbook. Weinheim: VCH, 1995, pp 89–155.
90. A Cornish-Bowden. Fundamentals of enzyme kinetics 1995, 2nd ed. Portland Press.
91. IH Segel. Enzyme Kinetics. New York: John Wiley and Sons, 1975.
92. E Schmidt, E Fiolitikakis, C Wandrey. Ann NY Acad Sci 501:434–437, 1987.
93. LG Lee, GM Whitesides. J Am Chem Soc 107:6999–7008, 1985.
94. A Liese, M Karutz, J Kamphuis, C Wandrey, U Kragl. Biotechnol Bioeng 51:544–550, 1996.
95. A Liese. PhD dissertation, Bonn University, 1998.

96. E Flaschel, C Wandrey, M-R Kula. Biochem Eng Biotechnol 26:73–142, 1983.
97. G Belfort. Biotechnol Bioeng 33:1047–1066, 1989.
98. K Seelbach, U Kragl. Enzyme Microb Technol 20:389–392, 1997.
99. B Nidetzky, D Haltrich, KD Kulbe. Chemtech 31–36, 1996.
100. B Nidetzky, W Neuhauser, D Haltrich, KD Kulbe. Biotechnol Bioeng 52:387–396, 1996.
101. M Ikemi, Y Ishimatsu. J Biotechnol 14:211–220, 1990.
102. JM Obón, MJ Almagro, A Manjón, JL Iborra. J Biotechnol 50:27–36, 1996.
103. AF Bückmann, M-R Kula, R Wichmann, C Wandrey. J Appl Biochem 3:301–315, 1981.
104. AF Bückmann, G Carrea. Adv Biochem Eng Biotechnol 39:98–152, 1989.
105. HK Chenault, GM Whitesides. Appl Biochem Biotechnol 14:147–197, 1987.
106. U Kragl. In: T Godfrey, S West, eds. Industrial Enzymology: Application of Enzymes in Industry, 2nd ed. London: Macmillan, 1996, pp 271–283.
107. DMF Prazeres, JMS Cabral. Enzyme Microb Technol 16:738–750, 1994.
108. C Salagnad, A Gödde, B Ernst, U Kragl. Biotechnol Progr 13:810–813, 1997.
109. S Kise, M Hayashida. J Biotechnol 14:221–228, 1990.

29

Stereoselective Microbial Baeyer-Villiger Oxidations

Amit Banerjee
Searle, St. Louis, Missouri

I. BAEYER-VILLIGER REACTION

About a century ago, Baeyer and Villiger reported [1] the reaction of cyclic ketones with peroxymonosulfuric acid to produce lactones. The oxidation of cyclopentanone to ε-caprolactone is carried out by the nucleophilic attack of a peroxide to the ketone to generate the "Criegee" intermediate (Fig. 1). The tetrahedral intermediate then rearranges via migration of the alkyl group to the proximal oxygen of the peroxide concomitant with the cleavage of the weak peroxy linkage. A cyclic ketones do not readily undergo Baeyer-Villiger (BV) reactions with carboxylic peracids.

II. MICROBIAL BAEYER-VILLIGER REACTIONS

One of the early reports of microbial BV reaction was reported [2] by Fried et al., where *Penicillium chyrysomogenum* and other microorganisms were used to prepare testololactone from progesterone. Other microbial BV reactions are implicated in the microbial degradation of eburicocic acid by *Glomerella fusarioides* [3], fenchone to the 1,2-fencholide by *Corynebacterium* sp. [4] (Fig. 2), and the degradation of 6-oxocineole by *Rhodococcus* sp. [5]. The oxidation of 2-heptylcyclopentanone or 2-pentylcyclopentanone by *Pseudomonas oleovorans* to the corresponding lactones was also reported [6]. These microbial BV reactions were carried out by monooxygenases that uses 1 mol of reductant, NADH or NADPH, and 1 mol of O_2. Only one atom of oxygen is incorporated in the product, the lactone in these reactions, and 1 mol of H_2O is produced (Fig. 3).

Trudgill and co-workers [7] purified the Baeyer-Villiger monooxygenase (BVase) from *Pseudomonas putida* ATCC 17453, grown on (+)-camphor that catalyzes 2,5-diketocamphane to the corresponding lactone in the catabolism of camphor. The first enzyme activity (MO1) consists of two enzymes of two nonidentical subunits (M_r 78,000): 2,5-diketocamphane 1,2-monooxygenase (2,5-DKCMO) and 3,6-diketocamphane monooxygenase (3,6-DKMCO). The enzymes 2,5-DKMO and 3,6-DKMO are NADH-dependent. The enzymes bind one molecule of FMN. These enzymes are involved in the degradation of bicyclic intermediate (Fig. 4). However, when the enzyme activity MO1 is reacted with (+)-camphor, the corresponding lactone is produced, which is stable (Fig. 5). Subse-

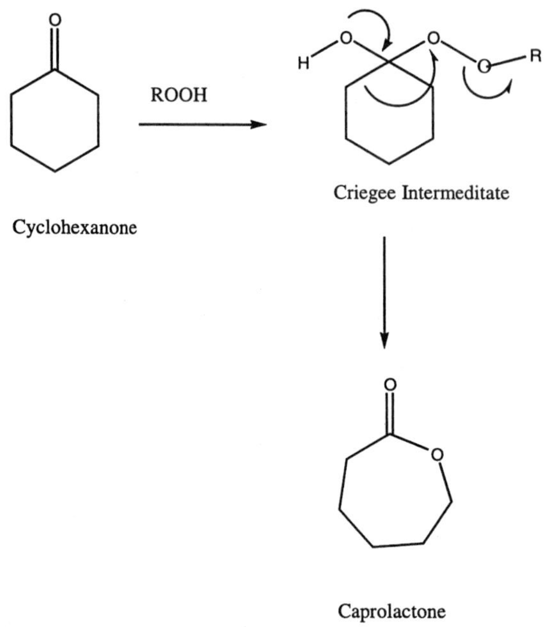

Figure 1   Baeyer-Villiger reaction.

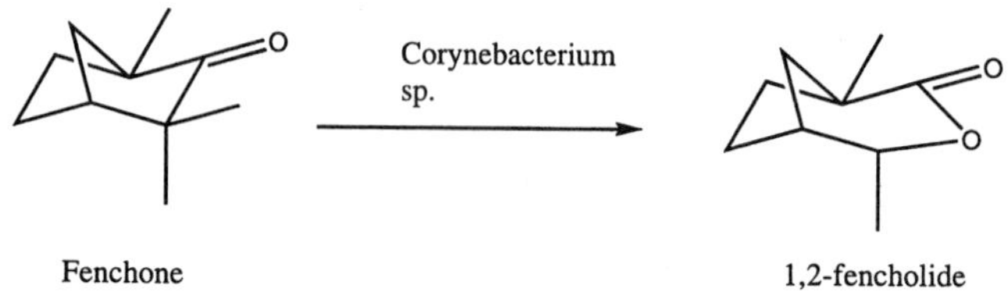

Figure 2   Transformation of fenchone to 1,2-fencholide. (From Ref. 4.)

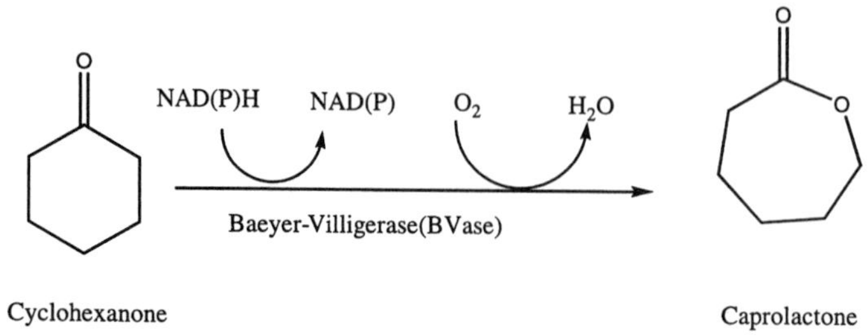

Figure 3   Bioconversion of cyclohexanone to ε-caprolactone. (From Ref. 11.)

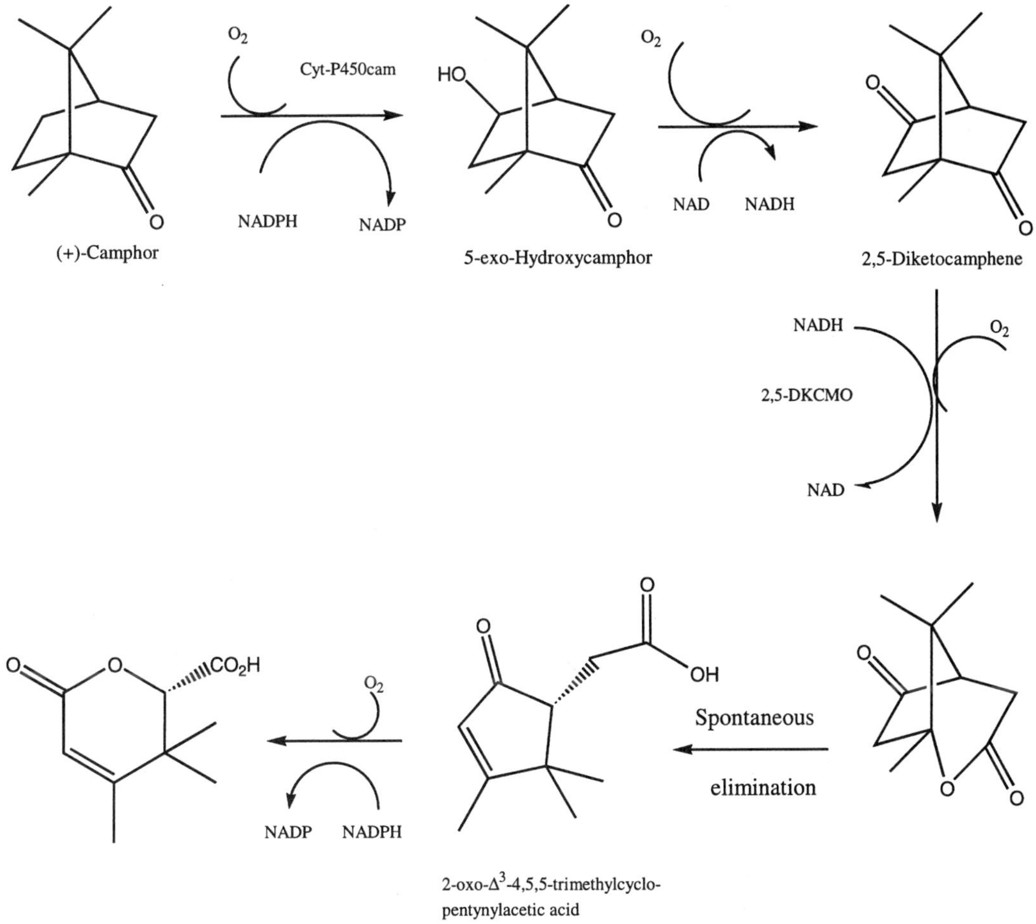

Figure 4 Catabolism of (+)-camphor by *Pseudomonas putida* ATCC 17453. (From Ref. 8.)

Figure 5 Baeyer-Villiger transformation of (+)-camphor by 2,5-DKCMO from *Pseudomonas putida*. (From Ref. 6.)

quently, a second BVase (MO2) was purified from the (+)-camphor catabolic pathway by *P. putida* [8] involved in the oxygenation of monocyclic intermediate 2-oxo-Δ^3-4,5,5-trimethylcyclopentenylacetic acid in the later stages of camphor metabolism (Fig. 4). This enzyme (M_r 106,000) is a homodimer that binds a single molecule of flavin adenine dinucleotide (FAD), and uses NADPH and dioxygen as cosubstrates.

The culture *Pseudomonas* sp. NCIB 9872 was isolated from soil using cyclopentanol as the sole carbon source [9]. Cell-free extracts of the organism displayed alcohol dehydrogenase activity and BVase activity. To obtain energy from this carbon source, the pivotal step is ring fragmentation, made possible by oxidation of the alcohol to the cyclopentanone, followed by insertion of oxygen atom to δ-valerolactone, and subsequent enzymatic hydrolysis to aliphatic hydroxy acids which are then converted to metabolites. The purified enzyme consists of four subunits (M_r 200,000), binds two to four molecules of FAD, and requires NADH as reductant.

Two bacterial species were also isolated from soil using cyclohexanol as the sole carbon source, i.e., *Nocardia globerula* CL1 [10] and *Acinetobacter* sp. NCIMB 9871 [11]. These two microorganisms were capable of converting cylohexanone to ϵ-caprolactone. The enzyme activity cyclohexanone monooxygenase (CHMO) of *N. globurela* CL1 was purified [12]. It is a single polypeptide chain (M_r 53,000) with one molecule of FAD bound as the prosthetic group and requires NADPH as electron donor. The cyclopentanol monooxygenase (CPMO) from *Pseudomonas* sp. NCIB 9872 and the two bacterial CHMO can use a variety of ketones as substrate. *Acinetobacter* TD63 was isolated from soil using 1,2-*trans*-cyclohexanediol [13] as the sole carbon energy source. The cell-free extracts showed monooxygenase activity. It has been reported [14] that the organism when reacted with racemic menthone only metabolizes (+)-menthone. It has been speculated that this organism lacks lactone hydrolase activity.

A *Xanthobacter autotrophica* DSM 431 species capable of growth on cyclohexane as sole carbon energy source was isolated [15]. A soluble CHMO was purified, which is a single polypeptide (M_r 50,000) associated with one molecule of flavin mononucleotide (FMN) and requires NADPH as electron donor [16]. This CHMO also exhibits a broad substrate specificity.

BVase activity in whole cells of the filamentous fungi *Curvularia lunata* NRRL 2380 was reported [17] using substituted cyclohexanone as substrate. However, no enzyme was isolated and characterized from any filamentous fungi to date.

III. CYCLOHEXANONE MONOOXYGENASE

The CHMO from *Acinetobacter* sp. NCIMB 9871 has been studied quite extensively. The enzyme is a monomer (M_r 66,000) contains a noncovalently bound FAD molecule and uses NADPH as the electron donor. This enzyme was cloned and expressed in *E. coli* [18]. The mechanism of action of CHMO was proposed by Walsh et al. [19] (Fig. 6). It proposes that the enzyme-FAD-4α-OOH is the nucleophilic peroxide equivalent that produces the tetrahedral adduct with bound substrate. The adduct then decomposes by the migration of carbon–carbon bond to the proximal peroxide oxygen as the O-O bond breaks. This produces the ring expansion to lactone and the enzyme-FAD-4α-OH spontaneously eliminates H_2O to regenerate the oxidized FAD.

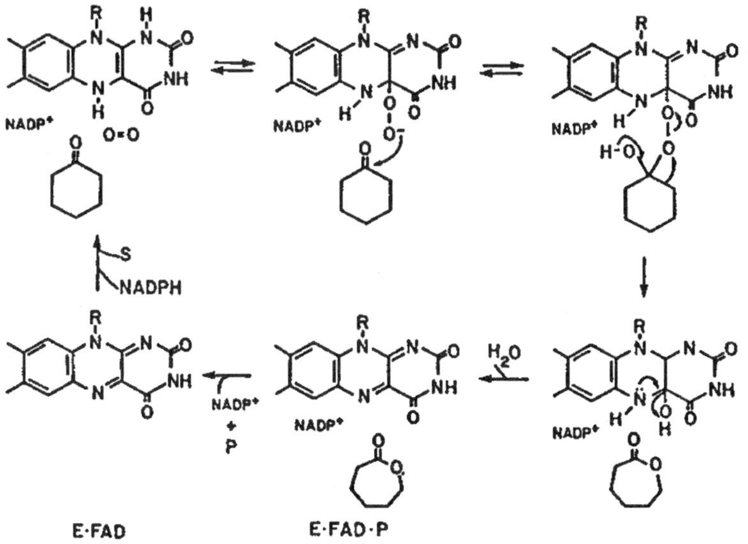

[Figure 6 reaction mechanism scheme]

Figure 6 Mechanism proposed for the CHMO-catalyzed Baeyer-Villiger reaction. (From Ref. 19.)

IV. STEREO- AND REGIOSELECTIVE BIOCONVERSIONS USING BAEYER-VILLIGER MONOOXYGENASES

The stereoselectivity and regioselectivity of microbial BVases has drawn a lot of attention from a wide number of research laboratories. An active site model for BVase was proposed by Kelly et al. [20] based on the configuration of the Criegee intermediate. The priority assignment is based on peroxide > hydroxyl (or ether) > migrating group > nonmigrating group. Based on this unconventional system, Kelly et al. assigns *si* or *re* faces to the enzyme-bound Criegee intermediate (Fig. 7). The model was tested with success by the proponents using bicyclo[3.2.0]hept-2-en-6-one and tricyclo[4.2.1.0]nonan-2-one using purified CHMO from *Acinetobacter* sp. NCIMB 9871, *Pseudomonas* sp. NCIMB 9872, *P. putida* NCIMB 10007, as well as the two 2,5-DKCMO and 3,6-DKCMO (MO1 activity) from *P. putida* NCIMB 10007 grown on camphor. An important outcome of this postulate is that the *S*-Criegee intermediates can only be formed on the *si* face of the flavin ring and the *R*-Criegee intermediates can be formed on the *re* face.

S or si face   (a)   **R or re face**   (b)

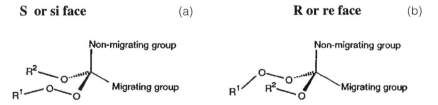

Figure 7 Active side model of BVase proposed by Kelly. R^1, flavin or acyl residue; R^2, flavin or proton. MO1 is NADH-dependent, MO2 is NADPH-dependent. (a) *Acinetobacter* sp. NCIMB 9871, *P. putida* NCIMB 9872, *P. putida* NCIMB 10007 (MO2). (b) 2,5-diketocamphane 1,2-monooxygenase and 3,6-diketocamphane 1,2-monooxygenase from *P. putida* NCIMB 10007 (MO1). (From Ref. 21.)

Kelly et al. [21] also proposes that there are two classes of BVases: those that use FAD as the prosthetic group and NADPH as an electron donor and utilize S-Criegee intermediate; and those that use FMN and NADH and the R-Criegee intermediate. They observed a high degree of homology in the N-terminal sequences among the FAD/NADPH enzymes. There was no homology between the FAD/NADPH- and the FMN/NADH-linked enzyme.

Ottolina [22] and subsequently Holland [23] proposed an active site model of BVase based on cubic space description. The model was initially proposed for CHMO-mediated asymmetrical sulfoxidation of sulfides. The Ottolina model (Fig. 8) places the enzyme substrate interaction in a cubic space due to the lack of x-ray data on these enzymes. The stereochemical predictions using the active site model requires some guidelines:

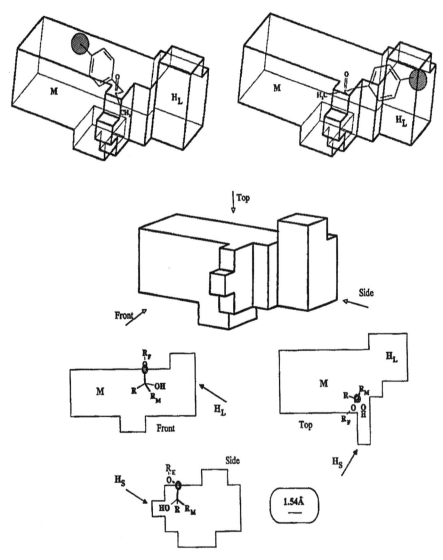

Figure 8 Active site model for CHMO as proposed by Ottolina et al. (From Ref. 22.) R_F and R_M are flavin and migrating groups, respectively. H_L and H_S are large and small hydrophobic pockets.

1. The oxidized carbon atom must be place in the catalytic site aligned along the C-O-O- axis.
2. The migrating group (R_m) must be placed *anti* periplanar to the O-O bond.
3. The hydroxyl group (C-OH) must be oriented toward the small hydrophobic pocket (Hs).
4. For cyclic substrate, the peroxide group will be *exo* or equatorial.

The author's claim that this model can be used to predict the stereochemical outcome of BVase-catalyzed reactions. The model was tested with 40 different substrates.

The enantioselective sulfoxidation by CHMO from *Acinetobacter* NCIB 98 with benzyl alkyl sulfides as substrate was primarily influenced by the alkyl moiety [24]. The configuration of the sulfoxide changed from (R) to (S) as the alkyl group changed from methyl to ethyl and other bulkier group (Fig. 9). The introduction of a *para*-alkyl substituent decreased the (R) selectivity with the exception of n-butyl. However, introduction of polar groups at the para position on the benzyl ring interfered with enantioselectivity of CHMO [25].

The first exploitation of microbial BVase to produce chiral lactones from cyclic racemic bicyclo[2.2.1]hept-2-en-7-ones (Fig. 10) was reported by Ouazzani et al. [17]. A report of preparative scale bioconversion of cyclic ketone was made [26]. The preparative scale bioconversion of using *Acinetobacter* TD 63 provides lactones in high optical purity [27]. Subsequently [28], whole-cell bioconversions of bicyclic lactones was made, using *Acinetobacter* sp. NCIB 9871 and *Pseudomonas* sp. NCIB 9872 to produce the same chiral synthons. In all cases, the racemic ketone was oxidized, and equal amounts of the regioisomeric lactones (−)-(*1S,5R*)-2-oxabicyclo[3.3.0]oct-6-en-3-one and (−)-(*1R,5S*)-3-oxabicyclo[3.3.0]oct-6-en-3-one were produced in high enantiomeric purity.

Purified CHMOs from *Acinetobacter* sp. NCIB 9871 and *Pseudomonas* NCIMB 10007 coupled with dehydrogenases to generate NADPH/NADH in situ was used to prepare (*1R,5S*)- and (*1S,5R*)-lactones [29,30]. These chiral lactones are important synthons used in the synthesis of prostaglandins.

Monocylic alkyl cyclopentanones were converted to (S)-lactones with high enantiomeric excess (e.e.), using partially purified 3,6-DKCMO from *P. putida* NCIMB 10007 grown on camphor [31]. The same enzyme produced (S)-lactones from α-substituted alkylcyclohexanone and (R)-lactones (Fig. 11) for aryl substituents [32].

CHMO from *Acinetobacter* sp. NCIB 9871 and in situ−generated NADPH was used to oxidize a series of bicylo[2.2.1]hept-2-en-7-ones to produce highly enantiopure lactones

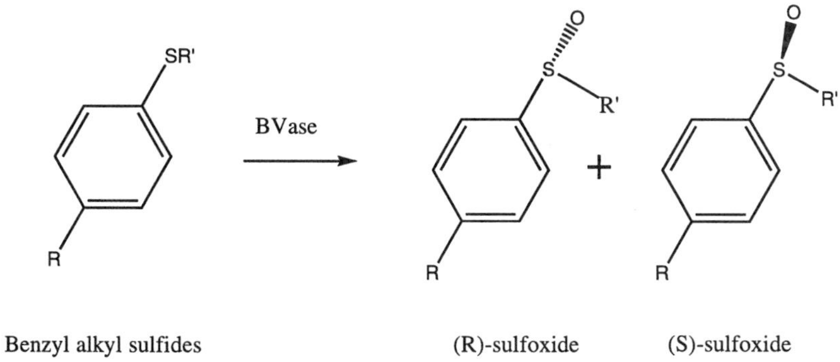

Benzyl alkyl sulfides   (R)-sulfoxide   (S)-sulfoxide

Figure 9   Enantioselective oxidation of sulfides to sulfoxides. (From Ref. 20.)

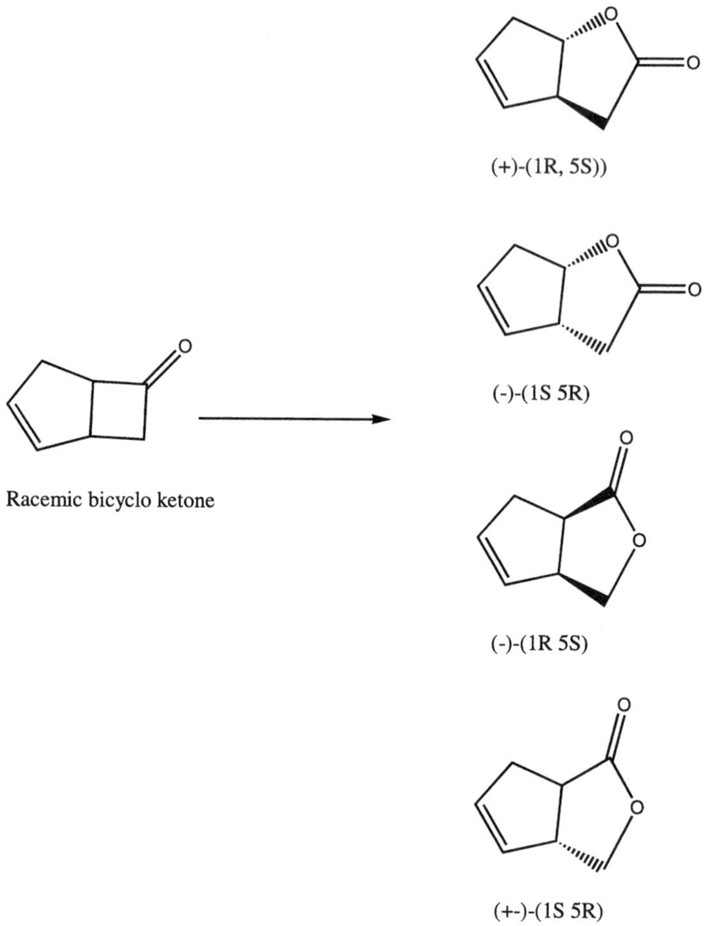

(+)-(1R, 5S))

(-)-(1S 5R)

(-)-(1R 5S)

(+-)-(1S 5R)

Racemic bicyclo ketone

Figure 10 Possible lactone products from BVase oxidation of racemic bicyclo[3.2.0]hept-2-en-6-one. (From Ref. 17.)

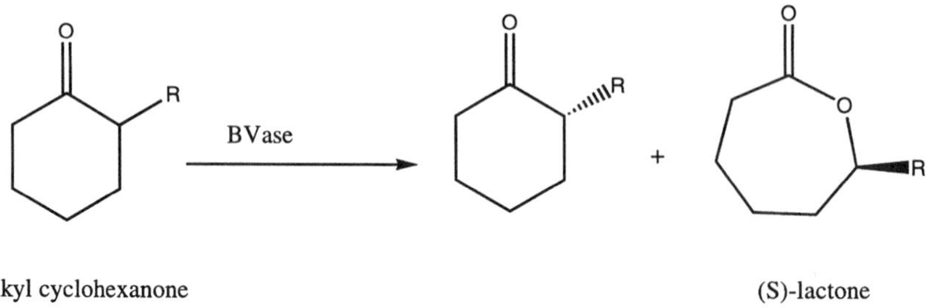

α-Alkyl cyclohexanone

(S)-lactone

Figure 11 Biotransformations of α-substituted cyclohexanones by *Acinetobacter* TD63 and *Pseudomonas putida* NCIMB 10007. (From Ref. 31.)

[33]. CHMOs have been used to oxidize cycloalkanones to produce chiral lactones which were used in the total synthesis of the natural product (R)-$(+)$-lipoic acid [34].

Stewart et al. [35] reported the cloning and expression of CHMO from *Acinetobacter* sp. NCIB 9871 in baker's yeast (*Saccharomyces cerevisiae*) for the use of organic chemists attempting BV reactions. The "designer yeast" was used to convert prochiral 2- and 3-substituted cylohexanones to lactones with high enantiomeric purity.

The BVase-mediated oxidation of prochiral and racemic ketones to chiral lactones has been a useful tool for the preparation of chiral synthons used in the synthesis of a variety of interesting compounds.

ACKNOWLEDGMENTS

The author thanks Melissa Howe for her help in the preparation of this manuscript.

REFERENCES

1. A Baeyer, V Villiger. Chem Ber 32:3625–3633, 1899.
2. J Fried, RW Thomas, A Klingberg. J Am Chem Soc 75:5764–5767, 1953.
3. AI Laskin, P Grabowich, C DeLisle Meyers. J Med Chem 7:406–409, 1964.
4. PJ Chapman, G Meerman, IC Gunsalus. Biochem Biophys Res Commun 20:106–108, 1965.
5. DR Williams, PW Trudgill. Biochem Soc Trans 17:912–913, 1989.
6. R Shaw. Nature 209:1369, 1966.
7. PW Trudgill, R Dubus, IC Gunsalus. J Biol Chem 241:1194–1205, 1966.
8. HJ Ougham, DG Taylor, PW Trudgill. J Bacteriol 153:140–152, 1983.
9. M Griffin, PW Trudgill. Biochem J 129:595–603, 1972.
10. DB Norris, PW Trudgill. Biochem J 121:363–370, 1971.
11. NA Donoghue, PW Trudgill. Eur J Biochem 60:1–7, 1975.
12. NA Donoghue, DB Norris, PW Trudgill. Eur J Biochem 63:175–192, 1976.
13. V Alphand, R Furstoss, S Pedragosa-Moreau, SM Roberts, AJ Willets. J Chem Soc Perkin Trans 1:1867–272, 1996.
14. V Alphand, R Furstoss. Tetrahedron: Asymmetry 3:379–382, 1992.
15. KM Tower, MR Buckland, M Griffin. Biochem Soc Trans 463–464, 1985.
16. KM Tower, MR Buckland, M Griffin. Eur J Biochem 181:199–206, 1989.
17. J Ouazzani-Chahdi, D Buisson, R Azerad. Tetrahedron Lett 28:1109–1112, 1987.
18. Y-CJ Chen, OP Peoples, CT Walsh. J Bacteriol 170:781–789, 1988.
19. CT Walsh, Y-CJ Chen. Acc Chem Res 13:148–155, 1988.
20. DR Kelly, CJ Knowles, JG Mahdi, IN Taylor, MA Wright. J Chem Soc Chem Commun 729–730, 1995.
21. DR Kelly, CJ Knowles, JG Mahdi, MA Wright, IN Taylor. J Chem Soc Chem Commun 2333–2334, 1996.
22. G Ottolina, G Carrea, S Colanna, A Ruckelmann. Tetrahedron: Asymmetry 7:1123–1136, 1996.
23. G Ottolina, P Piero, D Varley, HL Holland. Tetrahedron: Asymmetry 7:3427–3630, 1996.
24. P Pasta, C Giacomo, HL Holland, S Dallavalle. Tetrahedron: Asymmetry 6:933–936, 1995.
25. G Ottollina, P Pasta, G Carrea, S Colonna, S Dallavalle, HL Holland. Tetrahedron: Asymmetry 6:1375–1386, 1995.
26. O Abril, CC Ryerson, C Walsh, GM Whitesides. Biorg Chem 17:41–52, 1989.
27. V Alphand, A Archelas, R Furstoss. Tetrahedron Lett 30:3663–3664, 1989.
28. AJ Carnell, SM Roberts, V Sik, AJ Willets. J Chem Soc Chem Commun 20:1438–1439, 1990.
29. AJ Willets, CJ Knowles, MS Levitt, SM Roberts, H Sandey, NF Shipston. J Chem Soc Perkin Trans 1:1608, 1991.

30. G Grogan, S Roberts, A Willets. Biotechnol Lett 14:1125–1130, 1993.
31. R Gagnon, G Grogan, MS Levitt, SM Roberts, PWH Wan, AJ Willets. J Chem Soc Perkin Trans 1:2537–2543, 1994.
32. MT Bes, R Villa, SM Roberts, PWH Wan, A Willets. J Mol Cat B1:127–134, 1996.
33. MJ Tashner, L Peddada, P Cyr, Q-Z Chem, DJ Black. In: S Servi, ed. Microbial Reagents in Organic Synthesis. Dordrecht: Kluwer Academic, 1992, pp 347–360.
34. B Adger, MT Best, G Grogan, R McCague, SP Moreau, SM Roberts, R Villa, PWH Wan, AJ Willets. Biorg Med Chem 5:253–261, 1997.
35. JD Stewart, KW Reed, CA Martinez, J Zhu, G Chem, MM Kayser. J Am Chem Soc 120: 3541–3548, 1998.

30

Stereoselective Biocatalysis: Amino Acid Dehydrogenases and Their Applications

Toshihisa Ohshima
University of Tokushima, Tokushima, Japan

Kenji Soda
Kansai University, Osaka-fu, Japan

I. INTRODUCTION

Amino acid dehydrogenases (EC 1.4.1.-) catalyze the reversible deamination of L-amino acids to their corresponding oxo acids in the presence of the pyridine nucleotide coenzymes NAD and NADP.

$$\text{L-Amino acid} + \text{NAD(P)} + H_2O \rightleftharpoons \text{oxo acid NH}_3 + \text{NAD(P)H} + H^+$$

An amino group and ammonia participate in the reactions, and accordingly the enzyme is situated at a turning point between organic and inorganic metabolisms. The enzymes are considerably different from alcohol dehydrogenases in their structure and properties. As shown in Table 1, more than 10 kinds of amino acid dehydrogenases have been found in various kinds of organisms. Threonine-3-dehydrogenase, serine dehydrogenase, and phenylserine dehydrogenase catalyze the oxidation of a hydroxyl group to a carbonyl group in their side chains; they are similar to alcohol dehydrogenases in the reaction and usually do not fall into the category of amino acid dehydrogenase. The metabolic role of amino acid dehydrogenase consists of regulation of the synthesis of amino acids and oxo acids. In spite of their metabolic roles, the equilibrium of amino acid dehydrogenase reactions lies far to the amination of oxo acid: the K_{eq} values are 10^{-14}–10^{-18}. Thus, the reactions are favorable for asymmetric synthesis of amino acids from their prochiral oxo analogs and ammonia. Since it was shown that leucine dehydrogenase (LeuDH) is useful for the continuous production of L-leucine, L-*tert*-leucine, and their analogs with the membrane reactor, much attention has been paid to enzymatic synthesis of various L-amino acids with amino acid dehydrogenases. In addition, amino acid dehydrogenase have been used for analysis of amino acids, oxo acids, and ammonia, and assay of some enzymes acting on the substrate L-amino acids.

On the other hand, studies of amino acid dehydrogenases with techniques of genetic engineering, protein engineering, and x-ray crystallography have given us detailed information about the relationship between the structure and function of the enzymes. In this chapter, we describe the functional and structural characteristics of several amino acid dehydrogenases such as glutamate dehydrogenase, leucine dehydrogenase, and phenylal-

Table 1   NAD(P)-Dependent Amino Acid Dehydrogenases

EC number	Enzyme	Coenzymes	Major source
1.4.1.1	AlaDH	NAD	Bacteria (*Bacillus, Streptomyces, Anabena, Pseudomonas, Rhodobacter, Arthrobacter, Thermus, Enterobacter, Phormidium*) chrorella
1.4.1.2	GluDH	NAD	Plants, fungi, yeasts, bacteria
1.4.1.3	GluDH	NAD(P)	Animals (bovine liver, chicken liver), tetrahymena, bacteria (*Clostridium, Thiobacillus*)
1.4.1.4	GluDH	NADP	Plants, *Euglena gracilis, Chrorella sarokiniana*, fungi, yeasts, bacteria
1.4.1.5	L-Amino acidDH	NADP	Bacteria (*Clostridium sporogenes*)
1.4.1.7	SerDH	NAD	Plants (parsley)
1.4.1.8	ValDH	NAD, NADP	Bacteria (*Streptomyces, Alcaligenes faecalis, Planococcus*), plants (pea, wheat)
1.4.1.9	LeuDH	NAD	Bacteria (*Bacillus, Clostridium, Thermoactinomyces*)
1.4.1.10	GlyDH	NAD	Bacteria (*Mycobacterium tuberculosis*)
1.4.1.11	DAHDH	NAD, NADP	Bacteria (*Clostridium, Brevibacterium*)
1.4.1.12	DAPDH	NAD(P)	Bacteria (*Clostridium*)
1.4.1.15	LysDH (cylizing)	NAD	Human liver
1.4.1.16	DAPMDH	NADP	Bacteria (*Corynebacterium glutamicum, Brevibacterium* sp., *Bacillus sphaericus*
1.4.1.17	MethylAlaDH	NADP	Bacteria (*Pseudomonas* sp.)
1.4.1.18	LysDH (Lys-6-DH)	NAD	Bacteria (*Agrobacterium tumefaciens, Klebsiella pneumoniae*)
1.4.1.19	TryDH	NAD(P)	Plants (*Nicotiana tabacum, Pisum sativum, Spinacia oleracea*)
1.4.1.20	PheDH	NAD	Bacteria (*Sporosarcina ureae, Bacillus sphaericus, Rhodococcus marinas, Thermoactinomyces intermedius*)
1.4.1.-	AspDH	NADP	Bacteria (*Klebsiella pneumoniae*)

DH, dehydrogenase; NAD(P), NAD and NADP-nonspecific; DAHDH: L-*erythro*-3,5-diaminohexanoate dehydrogenase; DAPDH, 2,4-diaminopentanoate dehydrogenase; DAPMDH, *meso*-2,6-diaminopimelate dehydrogenase; MethylAlaDH, *N*-methyl-L-alanine dehydrogenase.

anine dehydrogenase, and enantiospecific synthesis of amino acids and the applications of the enzymes to the synthesis and determination of L-amino acids.

II. PURIFICATION AND CHARACTERISTICS OF AMINO ACID DEHYDROGENASES

A. Purification of Enzymes

In general, highly purified preparations of enzyme are preferably used for the enzymatic syntheses and analyses of various compounds to avoid undesirable side reactions with impure enzymes. How pure enzymes should be used is dependent on the purpose of the enzyme application, as well as the properties and purity of the substrates and products. Usually screening is carried out for the purpose of finding microorganisms that produce the enzyme effectively, and the medium and growth conditions that lead to abundant enzyme production are searched. The addition of enzyme inducers is usually very effective. For example, the addition of alanine and branched chain amino acids to the medium promotes the production of alanine dehydrogenase (AlaDH) and LeuDH, respectively, by *Bacillus* sp. [1]. Cloning of the gene of amino acid dehydrogenases into *Escherichia coli* often enhances the enzyme productivity. In particular, gene cloning is advantageous for the large-scale production and rapid purification of thermostable amino acid dehydrogenases from thermophiles. The *E. coli* clone cells produce a large quantity of a themostable amino acid dehydrogenase from a thermophile and various inherent thermolabile proteins. Accordingly, the dehydrogenase is effectively purified from the cell extract by heat treatment [2].

In addition to the conventional purification methods like gel filtration, ion exchange, and adsorption chromatographies, affinity chromatography using triazine dye ligands such as Red Sepharose CL-4B (e.g., dye is Reactive Red 120) is very useful for the purification of amino acid dehydrogenases. For example, glutamate dehydrogenases (GluDHs) of *Pyrococcus woesei* and *P. furiosus* can be easily purified to homogeneity by only two sequential Red affinity chromatographies from the crude extracts. The dye ligand affinity resin is very stable and cheap as an affinity ligand. The stereo structures of triazine dyes are very similar to that of the adenosine moiety of NAD(P), and accordingly the affinity is widely applicable to the purification of various amino acid dehydrogenases. Thus, a specific affinity elution with the mixture of coenzyme and substrate as an eluent is used for the effective purification of enzyme as shown in Fig. 1 [3].

The preparative polyacrylamide slab gel electrophoresis is also effective for the purification of amino acid dehydrogenases due to the high separation ability and the easy

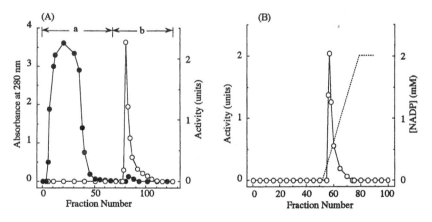

Figure 1 Purification of *Pyrococcus woesei* NADP-dependent GluDH by two sequential Red Sepharose CL-4B column chromatographies. (A) Elution profile of the first chromatography. Crude enzyme (235 mg, 63.5 U, specific activity 0.27, in the 10 mM phosphate buffer, pH 7.2) was applied on the first column and the enzyme (4.41 mg, 44.7 U, specific activity 10.1) was eluted with 20 mM phosphate buffer (pH 8.0) containing 0.5 M NaCl. (B) Elution profile of the second chromatography. Enzyme solution (pH 7.2) dialyzed was applied on the affinity column and GluDH (0.88 mg, 21.4 U, specific activity 24.3, yield 33.7%) was eluted with a linear gradient of NADP concentration (0–2 mM) in the presence of 10 mM L-glutamate [36].

detection of the enzyme activity on the gel by activity staining after the electrophoretic run [4]. Two typical purification procedures of LeuDH from *Bacillus stearothermophilus* [5] and phenylalanine dehydrogenase (PheDH) from the recombinant *E. coli* [6] cells that produce *Thermoactinomyces intermedius* PheDH are described in Table 2. In the latter case, the enzyme content in the crude extract from the clone cells is about 30-fold higher than that from *T. intermedius* cells [7] and heat treatment of the crude extract of the transformant cells at 70°C for 30 min is useful to the effective purification because impurity proteins in the crude extract are mostly denatured during the treatment. The enzyme can be purified efficiently from the cell extract: it takes only a day to obtain about 3 mg of the pure enzyme from a 2-L cell culture. In addition, this procedure can easily be scaled up, and the enzyme is commercially available (Yunitika Ltd., Osaka, Japan).

B. Basic Molecular and Catalytic Properties

1.   Alanine Dehydrogenase (EC 1.4.1.1, AlaDH)

AlaDH catalyzes the reversible deamination of L-alanine to pyruvate in the presence of NAD. The enzyme occurs in various microorganisms including bacteria, archaea, and eukaryotes, but not in higher plants and animals. The enzyme of *Bacillus* species is inducibly produced by alanine and plays a central role in L-alanine catabolism [8,9]. In contrast, in the cells of nitrogen-fixing microorganisms such as *Bradyrhizobium japonicum* in soybean nodules, a photosynthetic bacterium, *Rhodobacter capsulatus*, and a blue-green alga, *Anabena clyndrica*, the enzyme functions for the formation of L-alanine to store ammonia after converting N_2 gas to ammonia [10]. In these microorganisms, the enzyme formation is not induced by alanine. The thermostable AlaDH is found in various thermophilic microorganisms such as *Bacillus* sp. and *Thermus thermophilus* [8,11]. The enzyme has been purified from several microorganisms and characterized (Table 3). AlaDH shows high substrate and coenzyme specificities in the oxidative deamination. L-2-Aminobutyrate, L-serine, L-norvaline, and L-valine, besides L-alanine, are oxidatively deaminated, though very slowly: the relative activities of the *B. sphaericus* enzyme for these lower than 2% of that for L-alanine, and D-alanine is inert. The substrate specificity for the reductive amination is lower than that for the deamination. In addition to pyruvate,

Table 2 Purification of LeuDH from *T. intermedius* (A) and the *T. intermedius* PheDH from the recombinant *E. coli* JM109/pKPDH2 (B)

Steps	Total protein (mg)	Total units	Specific activity (units/mg)	Yield (%)
(A)				
Crude extract	3610	701	0.194	100
Heat treatment	1290	553	0.429	78
Red Sepharose CL-4B	81.9	559	7.31	85
Preparative electrophoresis	25.0	516	20.5	74
HiLoad Superdex 200	5.10	501	102	72
(B)				
Crude extract	247	1061	4.3	100
Heat treatment	132.8	965	29.5	86
DEAE-Toyopearl	11.1	710	64.3	67
Ammonium sulfate fractionation	3.1	612	200	58

Table 3 Properties of Some AlaDHs

Properties	B. sphaericus IFO 3525	B. sphaericus DSM 462	H. cultirubrum	Bra. Japonicum	P. lapidum
Native M_r (subunit M_r), kDa	230 (38 × 6)	230 (38 × 6)	72.5 (monomar)	190	240 (41 × 6)
Optimum pH:					
Deamination	10.5	10.5	9.0	10.0	9.2
Amination	9.0	8.2	9.0	8.4	8.4
Thermostability (°C)	55	75			50
K_m (mM)					
NAD	0.23	0.26	0.5	0.2	0.04
L-Ala	18.9	10.5	7.0	1.0	5.0
NADH	0.01	0.10	0.2	0.086	0.02
Pyruvate	1.7	0.5	0.8	0.49	0.33
Ammonia	28.2	38	0.82	8.9	60.6

B, Bacillus; Bra., Bradyrhizolium; M, Mycobacterium.

the *B. sphaericus* enzyme catalyzes the amination of 2-oxobutyrate, 2-oxovalerate, 3-hydroxypyruvate, glyoxylate, 3-fluoropyruvate, and 3-chloropyruvate, although their catalytic efficiencies (V_{max}/K_m) are lower than 16% of pyruvate. This low amino accepter specificity is advantageous to the enzymatic synthesis of the corresponding L-amino acid as described later. The equilibrium constant for the oxidative deamination is 10^{-14}–10^{-17} M [12,14]. AlaDHs from some *Bacillus* strains including *B. stearothermophilus* are commercially available.

2. Leucine Dehydrogenase (EC 1.4.1.9, LeuDH)

This enzyme catalyzes the reversible deamination of branched chain L-amino acids such as L-leucine and L-valine, and the straight chain amino acids such as L-norvaline and L-norleucine to their oxo analogs in the presence of NAD. The enzyme is found in limited types of bacteria. It occurs mainly in endospore-forming bacteria such as Bacilli and Clostridia, and the nonsporing bacterium *Corynebacterium pseudodiphtheriticum* [15]. The enzyme in the *Bacillus* cells probably functions for the formation of branched chain oxo acids, ammonia, and NADH from L-branched chain amino acids and NAD during spore germination. The enzyme has been purified from *B. sphaericus* [15], *B. cereus* [16], *B. stearothermophilus* [5], *Clostridium thermoaceticum* [18], *B. subtilis* [17], *Corynebacterium pseudodiphtheriticum* [19], *Thermoactinomyces intermedius* [20] and the halophilic thermophile *B. licheniformis* [21], and characterized (Table 4). Enzyme production is highly enhanced by gene cloning of the enzymes from several microorganisms and their expression in *E. coli*; the specific activities in the extracts of *E. coli* cells producing *B. stearothermophilus* [22], *C. thermoaceticum* [18], *T. intermedius* [20], and *B. licheniformis* [21] enzymes were 50, 900, 9.8, and 3.9 times higher, respectively, than those of the parent cells. In particular, *E. coli* MV1184 with a vector plasmid, pUC119, can produce a large amount of the *B. stearothermophilus* LeuDH, which corresponds to about 60% of the total soluble protein. The thermostable enzyme can be purified to more than 95% homogeneity

Table 4 Properties of Several Microbial LeuDHs

	Sources			
Properties	B. sphaericus	B. stearothermophilus	B. celeus	T. intermedis
Native M_r (subunit M_r), kDa	245 (41 × 8)	300 (41 × 6)	320 (39 × 8)	340 (42 × 8)
Specific activity of final preparation (U/mg protein):	25[a]	120[b]	60[c]	102[b]
Optimum pH:				
Deamination	10.7	11.0	11.5	10–10.5
Amination	9.0–9.5	9.0–9.5	8.5–9.0	9.0–9.6
Thermostability (°C)	<50	<70	<50	<70
Coenzyme:				
NAD: K_m (mM)	0.39	0.49	0.34	0.36
NADH: K_m (mM)	0.035	—	0.034	0.042
Substrate specificity (Rel. act.) (K_m, mM) (deamination)				
L-Leucine	100 (1.0)	100 (4.4)	100 (1.5)	100 (2.0)
L-Valine	74 (1.7)	98 (3.9)	61 (2.5)	87 (2.4)
L-Isoleucine	58 (1.8)	73 (1.4)	61 (1.0)	89 (0.4)
L-Norvaline	41 (3.5)	—	28 (2.9)	27 (—)
L-2-Aminobutyrate	14 (10)	—	24 (22)	7.8 (—)
L-Norleucine	10 (6.3)	—	6 (1.5)	3.6 (—)
DL-*tert*-leucine	1.6 (—)	—	—	0.5 (—)
D-Leucine (amination)	0	0	0	0
2-Oxoisocaproate	100 (0.31)	100	100 (0.45)	100 (0.63)
2-Oxoisovalerate	126 (1.4)	167	154 (2.1)	125 (4.4)
2-Oxo-3-methylvalerate	—	—	104 (0.4)	82 (2.2)
2-Oxovalerate	76 (1.7)	86	51 (0.40)	58 (0.91)
2-Oxocaproate	46 (7.0)	—	51 (1.2)	46 (1.4)
2-Oxobutyrate	57 (7.7)	45	51 (1.5)	28 (7.2)
4-Methyl-2-oxo-5,5,5,-trifluoropentnoate	—	—	—	26 (2.7)
2-Oxo-4-mercaptobutyrate	—	—	37 (2.1)	—
Pyruvate	0	0	—	1.5
2-Oxoglutarate	0	0	0	0
Phenylpyruvate	0	0	0	—

[a]Activity was measured at 25°C, [b]at 50°C, and [c]at 30°C.

by only one step, heat treatment of the cell extract, with an average yield of 75 mg/g of wet cells (obtained from 100 mL of the culture [23].

The substrate specificity of LeuDH in the oxidative deamination is relatively low. Branched chain L-amino acids such as L-leucine and L-valine are the preferred substrates, and the straight chain L-amino acids such as norvaline and 2-aminobutyrate are also deaminated to some extent. In the reverse reaction, the enzyme exhibits high reactivity for all of the oxo analogs of the amino acid substrates, which served as good substrates for the amination. In general, the reactivity of LeuDH for 2-oxoisovalerate, the oxo analog of valine, is higher than that for 2-oxoisocaproate, the oxo analog of leucine, although the catalytic efficiency (V_{max}/K_m) for 2-oxoisovalerate is somewhat low. The amination rate of

2-oxoisocaproate is more than 10 times higher than the L-leucine deamination rate. The equilibrium of the LeuDH reaction lies so far to the reductive amination like those of other amino acid dehydrogenases (K_{eq}: in the region of 10^{-15} M). Ammonia is an exclusive amino donor; hydroxyamine, methylamine, glutamine, and asparagine are inert as substrate [15]. NAD(H) is almost exclusive as the coenzyme, though some of the unnatural NAD analogs serve as a coenzyme as found in other amino acid dehydrogenases; 3-acetylpyridine-NAD exhibits a higher coenzyme reactivity than NAD [15], and NADP(H) is inert.

The enzyme shows the maximum pH at the fairly alkaline side for the deamination and amination. The enzymes from the thermophilic *Bacillus* species such as *B. stearothermophilus* and *T. intermedius* are stable under the conditions at high temperature such as 70°C and a wide range of pH, and can be stored for a longer time than that from the mesophilic bacilli [2]. The enzymes from *T intermedius* and the halophilic thermophile *B. licheniformis* are more stable at high concentrations of salts such as NaCl and KCl. The enzyme is inhibited with sulfhydryl reagents such as p-CMB and HgCl$_2$ [24]. Therefore, 2-mercaptoethanol is added to the enzyme solution to protect the oxidation of the sulfhydryl group of the enzyme. Pyridoxal 5'-phosphate inactivated enzyme activity by the formation of Schiff base at the amino group of the active site lysine [25]. LeuDH from *B. stearothermophilus* is commercially available (Unitika Ltd.).

3. L-Phenylalanine Dehydrogenase (EC 1.4.1.20, PheDH)

At the beginning of the 1980s, the wide screening of aromatic amino acid dehydrogenases led to the discovery of PheDH in *Brevibacterium* species [26]. The enzyme was isolated from several mesophiles—*Bacillus sphaericus* [27], *Sporosarcina ureae* [27], *B. badius* [28], *Rhodococcus* sp. [29], *Nocardia* sp. [30] and *Microbacterium* sp. [31], and also from the thermophile *T. intermedius* [32]—and characterized (Table 5). The enzyme acts on L-norleucine, L-methionine, L-norvaline, and L-tyrosine besides L-phenylalanine in the presence of NAD, although slowly. L-Tryptophan, L-alanine, and D-phenylalanine are inert as the substrate. The enzyme shows lower substrate specificity for 2-oxo acids than that for amino acids like AlaDH and LeuDH. The K_m values for ammonia are more than 70 mM. The *T. intermedius* PheDH is the most thermostable and a useful catalyst for industrial and clinical applications. The enzyme is easily and effectively purified from the recombinant *E. coli* [6] and commercially available (Unitika Ltd.).

4. Glutamate Dehydrogenase (EC 1.4.1.2–4, GluDH)

GluDHs reversibly catalyze the oxidative deamination of L-glutamate to 2-oxoglutarate with concomitant reduction of NAD(P). GluDHs are categorized into three groups based on coenzyme specificity: NAD-specific (EC 1.4.1.2), NAD- and NADP-nonspecific (EC 1.4.1.3), and NADP-specific (EC 1.4.1.4). GluDHs occur ubiquitously in animals, plants, and microorganisms except for certain bacteria, such as *Bacillus*, as reflecting their metabolic importance [33,34].

Many *Bacillus* strains lack GluDH and possess AlaDH and LeuDH. GluDHs have been purified from various organisms and characterized. Table 6 shows properties of several GluDHs. GluDHs from bovine liver and *Proteus* sp. [35] are commercially available (Toyobo, Osaka, Japan; and Sigma Chemicals). Almost all GluDHs except for the *Neurospora crassa* NAD-GluDH (a tetramer of 116 kDa) are hexameric with a molecular mass in the range of 220–330 kDa. Extremely thermostable GluDHs were recently isolated

Table 5 Comparison of Properties of Microbial PheDHs

Properties	Sources			
	B. sphaericus	*S. ureae*	*B. badius*	*R. maris*
Native M_r (subunit M_r), kDa	340 (41 × 8)	310 (41 × 8)	335 (41 × 8)	70 (36 × 2)
Specific activity of final preparation (U/mg protein):	111	84	68	65
Optimum pH:				
Deamination	11.3	10.5	10.4	10.8
Amination	10.3	9.0	9.4	9.8
Thermostability (°C)	55	<40	<55	35
K_m (mM)				
NAD	0.17	0.14	0.15	0.25
L-Phe	0.22	0.096	0.088	3.80
NADH	0.025	0.072	0.21	0.043
Phenylpyruvate	0.40	0.16	0.106	0.50
Ammonia	78	85	127	70

	Nocardia sp.	*Thermoactinomyces intermedius*	*Microbacterium* sp.
Native M_r (subunit M_r), kDa	42 (42 × 1)	270 (41 × 6)	330 (41 × 8)
Specific activity of final preparation (U/mg protein):	30	86	37
Optimum pH:			
Deamination	10	11	12
Amination	—	9.2	12
Thermostability (°C)	<53	70	55
K_m (mM)			
NAD	0.23	0.07	0.20
L-Phe	0.75	0.22	0.10
NADH	—	0.025	0.07
Phenylpyruvate	0.06	0.045	0.30
Ammonia	96	106	85

from several hyperthermophiles including archaea and bacteria. The NADP-GluDH from *Pyrococcus furiosus* [36] and NAD-GluDH from *Pyrobaculum islandicum* [37] are highly thermostable and are not inactivated even by incubation at 100°C for 2 h. In addition, the *Pyrobaculum islandicum* enzyme exhibits high stability in some denaturants and water-miscible organic solvents such as methanol and dimethylsulfoxide (DMSO) [37]. The hyperthermophilic enzymes may have great potential in applications to bioprocesses. For the majority of GluDHs, the optimum pH values for the deamination and amination are on the alkaline side like those of other amino acid dehydrogenases. The pH optima for amination lie between 7.4 and 8.7, and the corresponding pH optima for deamination are between 8.2 and 9.8. Although amino acid dehydrogenases generally exhibit high K_m values for ammonia, the value of *Proteus* sp. GluDH is relatively low (1.1 mM) [35].

Table 6 Properties of Some GluDHs

Properties	Sources			
	Bovine liver	*Neurospora crassa*		*Proteus* sp.
Coenzyme	NAD(P)	NAD	NADP	NADP
Native M_r (subunit M_r), kDa	330 (55 × 6)	480 (116 × 4)	288 (48 × 6)	300 (50 × 6)
Specific activity of final preparation (U/mg protein):	40	—	—	300
Optimum pH:				
Deamination	8.5–9.0	—	8.6–9.0	9.8
Amination	7.8.3	—	7.6	8.5
Thermostability (°C)				<50
K_m (mM)				
NAD, NADP	0.7, 0.047	0.33	0.050	0.015
L-Glu	1.8	5.5	4.5	1.2
NADH, NADPH	0.024	0.55	0.125	0.043
2-Oxoglutarate	0.7	4.6	5.3	0.34
Ammonia	3.2	17	10	1.1
	Clostridium symbiosum	*Thermotoga maritima*	*Pyrococcus furiosus*	*Pyrobaculum islandicum*
Coenzyme	NAD	NAD, NADP	NADP	NAD
Native M_r (subunit M_r), kDa	295 (42 × 6)	265 (47 × 6)	300 (47 × 6)	220 (36 × 6)
Specific activity of final preparation (U/mg protein):		4[b]	10.3[a]	3.5[a]
Optimum pH:				
Deamination		8.8[b]	8.2[a]	9.7[a]
Amination	—	8.3[b]	7.4[a]	8.7[a]
Thermostability (°C)	<53	80	105	100
K_m (mM)				
NAD, NADP		—	0.035	0.025
L-Glu		—	0.95	0.17
NADH, NADPH		0.022, 0.058	0.0071	0.0050
2-Oxoglutarate		0.7	0.11	0.066
Ammonia		40	6.3	9.7

[a,b]The activity is measured at 50°C and 60°C, respectively.

5. Other Amino Acid Dehydrogenases

Nearly 10 other amino acid dehydrogenases in addition to the above-mentioned four enzymes have been characterized as summarized in Table 1. The presence of NAD-specific glycine dehydrogenase (EC 1.4.1.10), serine dehydrogenase (EC 1.4.1.7), and NADP-specific L-amino acid dehydrogenase (EC 1.4.1.5) have been reported, but they were not substantially characterized. Tryptophan dehydrogenase (EC 1.4.1.19, TryDH) was shown to be present in spinach leaves and to catalyze the NAD- or NADP-dependent conversion of tryptophan into indolepyruvate [38]. The enzyme is activated by the addition of calcium ions. It has not been purified and is only poorly characterized. N-Methyl-L-alanine dehydrogenase has been isolated from *Pseudomonas* MS [39]. The enzyme catalyzes the formation of N-methyl-L-alanine from pyruvate and methylamine. The enzyme was separated from AlaDH and exhibited no activity for amination of pyruvate in the presence of ammonia and NADPH. The enzyme catalyzes the NADP-specific oxidation of N-methyl-L-alanine, and the optimum pH is around 8.4. Besides pyruvate, the enzyme acts on oxaloacetate and 2-oxobutyrate. The enzyme is stable only in the presence of dithiothreitol. DAHDH (EC 1.4.1.11) and 2,4-DAPDH (EC 1.4.1.12) were purified from anaerobic *Clostridium* species and partially characterized [40]. Besides these poorly characterized amino acid dehydrogenases, ValDH (EC 1.4.1.8), *meso*-2,5-diaminopimelate DH (EC 1.4.1.8), LysDH (EC 1.4.1.18, lysine-6-dehydrogenase), and AspDH were purified and their basic properties were reported. AspDH was recently discovered and characterized [41]. We describe below the properties of the four amino acid dehydrogenases.

(*a*) *ValDH (EC 1.4.1.8).* ValDH acts mainly on branched chain amino acids and is similar to LeuDH. ValDH was first found in *Streptomyces* species [42] and plays an important role in the synthesis of the oligoketide antibiotic tylosin [43]. A marked difference between ValDH and LeuDH is their substrate specificity. Besides substrate specificity, the amino acid composition and immunochemical properties of the *Alcaligenes faecalis* ValDH are clearly different from those of *B. sphaericus* and *Clostridium thermoaceticum* LeuDHs [44]. The properties of ValDH from several microorganisms are summarized in Table 7. The subunit structure of each of ValDHs from various microorganisms is diverse, with most being dimers, but tetramers and dodecamers have been demonstrated. The pH optimum of ValDH is 10.5–10.8 for the oxidative deamination and 8.8–9.0 for the reductive amination.

(*b*) *meso-2,6-Diaminopimelate D-dehydrogenase (EC 1.4.1.16, DAPDH).* This enzyme catalyzes the reversible NADP-dependent oxidative deamination of the D-dehydrogenation of *meso*-2,6-diaminopimelate to generate L-2-amino-6-oxopimerate (L-tetrahydrodipicolinate) [45]. The enzyme is specific for the *meso* stereoisomer of 2,6-diaminopimelate and thus distinguishes between the two chemically identical, but stereochemically opposite, centers in an achiral molecule. The enzyme was first isolated from *B. sphaericus* [45], and then found in the cells of *Brevibacterium* sp. [46] and *Corynebacterium glutamicum* [47]. The enzyme functions in the synthesis of *meso*-2,6-diaminopimelate, which is a key intermediate in the bacterial biosynthesis of L-lysine from L-2-amino-6-oxopimerate [48]. The enzyme was purified from three bacteria—*Bacillus sphaericus*, *Brevibacterium* sp., and *C. glutamicum*—and enzymologically characterized [46,47]. The *ddb* gene encoding the *C. glutamicum* enzyme was cloned and expressed at a high level in *E. coli* [49]. The enzyme is a homodimer with a molecular mass of about 70 kDa. The enzyme is specific for NADP(H) as the coenzyme. All three enzymes have pH optima in the alkaline pH region for both the oxidative deamination (9.8–10.5) and reductive amination reactions (7.5–8.5).

(*c*) *LysDH (EC 1.4.1.18, Lysine-6-dehydrogenase).* Two different kinds of LysDH were demonstrated. One is lysine 2-deaminating dehydrogenase (EC 1.4.1.15), which catalyzes the oxidative α-deamination of L-lysine to 1,2-didehydropiperidine-2-carboxylate. There is only one report on the enzyme from human liver [50], but the properties are unknown. Another LysDH (EC 1.4.1.18) catalyzes the oxidative ε-deamination of L-lysine to 2-aminoadipate-5-semialdehyde (allysine), which is spontaneously converted to 1,6-didehydropiperidine-2-carboxylate. The reaction is substantially irreversible. The enzyme is found in several microorganisms [51] and was purified from a plant-pathogenic microor-

Table 7 Number of Amino Acids and Subunit Molecular Weights of Main Amino Acid Dehydrogenases Whose Primary Structures Have Been Determined

Enzymes and their sources	Number of amino acids	Molecular weight of subunit
AlaDH		
B. sphaericus	372	39460
B. stearothermophilus	372	39694
B. subtilis	378	39683
Enterobacter aerogenes	377	39803
Mycobacterium tuberculosis	371	38713
Vibrio proteolyticus	374	39802
Shewanella sp. Ac10	371	39257
Carnobacterium sp.	371	39142
Phormidium lapideum	361	38557
Synechocystis sp.	360	38267
LeuDH		
B. stearothermophilus	429	46903
B. celeus	366	39866
B. licheniformis	364	40040
B. subtilis	364	39991
Thermoactinomyces intermedius	366	40586
PheDH		
B. sphaericus	381	41578
B. badius	380	41352
Thermoactinomyces intermedius	366	40488
Sporosarcina ureae	379	41328
GluDH		
Human	558	61432
Bovine liver	500	55393
Chicken liver	503	55658
Escherichia coli	447	48581
Salmonella typhimurium	447	48560
Trypanosoma brucei	992	111981
Haemophilus influenzae Rd.	449	48660
B. subtilis	426	47211
Neurospora crassa (NAD)	1047	118265
N. crassa (NADP)	454	48915
Clostridium symbiosum	449	49164
Sulforobus shibatae	391	42206
Sulforobus solfataricus	421	46982
Pyrococcus furiosus	420	46739
Pyrococcus KOD1	421	47002
Thermococcus litoralis	419	46739
Pyrobaculum islandicum	421	46905
Thermotoga maritima	416	45802
ValDH		
Streptomyces cinnamonensis	358	37750
Streptomyces coelicolor	284	31866
Streptomyces fradiae	371	38766
Streptomyces albus	364	38398
DAPDH		
Corynebacterrium glutamicum	320	35199

Data were from the Entrez at NCBI Databases.

ganism, *Agrobacterium tumefaciens* [52], and a yeast, *Candida albicans* [53]. The enzyme plays an important role in lysine catabolism of *A. tumefaciens* and is produced inducibly by adding lysine to the growth medium. The molecular weight of the enzyme is about 70,000 and the enzyme consists of two identical subunits. The active form (tetramer) is reversibly formed by the exogenous addition of lysine to the enzyme [54]. The enzyme requires NAD, and NADP is inert. L-Lysine is the preferred substrate and only the lysine analog *S*-(β-aminoethyl)-L-cysteine is oxidized with a low reaction rate (2.9% of L-lysine). The optimum pH for the oxidation is 9.7 and K_m values for NAD and L-lysine are 0.059 and 1.5 mM, respectively. The *C. albicans* enzyme is specific in oxidative deamination: the relative activities for *S*-(β-aminoethyl)-L-cysteine, L-leucine, δ-hydroxylysine, and ornithine are 42, 21, 14, and 5%, respectively.

(*d*) *AspDH.* AspDH was recently discovered in vitamin B_{12}–producing *Klebsiella pneumoniae* [41]. The enzyme catalyzes the oxidative deamination of L-aspartate to form oxaloacetate. The enzyme is specific to L-aspartate and NADP as substrates, and other amino acids such as L-glutamate, D-aspartate, L-serine, and NAD are inert. The enzyme has a molecular mass of 124 kDa and consists of two identical subunits. The optimum pH for the oxidative deamination is 7–8.

III. CATALYTIC MECHANISM AND STRUCTURE

A. Catalytic Mechanism

Amino acid dehydrogenases, like other NAD(P)H-dependent dehydrogenases and reductases, show either pro-R (A-) or pro-S (B-) stereospecificity for hydrogen transfer from the C-4 position of the nicotinamide moiety of NAD(P)H to the substrate amino acids. The stereospecificity for the hydrogen transfer of amino acid dehydrogenases has been investigated by means of NAD(P)H labeled stereospecifically with ^{2}H or ^{3}H. ^{2}H at the C-4 position of the nicotinamide moiety of NAD(P)H is analyzed by proton-NMR and the transfer of ^{3}H from NADH to substrate is monitored with radioactivity. Pro-S-specific enzymes are GluDH, LeuDH, PheDH, DAPDH, and ValDH, and AlaDH and LysDH are pro-R-specific enzymes. The stereospecificity is an inherent characteristic of individual NAD(P) dehydrogenase, and independent of the catalytic reaction and enzyme source.

A series of steady-state kinetic analyses provides information about the reaction mechanism. The oxidative deamination catalyzed by an amino acid dehydrogenase proceeds via the formation of a ternary complex with sequential and random substrate-binding mechanisms. Diversity is found in the manner of substrate binding and product release. The AlaDH reactions of *Bacillus* proceed through different kinetic mechanisms. The oxidative deamination of alanine catalyzed by a mesophilic *B. sphaericus* proceeds through a sequentially ordered binary-ternary mechanism in which NAD and L-alanine bind to the enzyme in that order and three products—pyruvate, ammonia, and NADH—are released from the enzyme in that order after dehydrogenation [12]. AlaDHs from the *B. subtilis* and the thermophilic *B. sphaericus* are different from the mesophilic *B. sphaericus* enzyme in terms of product release. In the reaction by the *B. subtilis* enzyme [55], the order in the release of pyruvate and ammonia is the reverse of that of the mesophilic *B. sphaericus* enzyme [14]. The random fashion of the release of pyruvate and ammonia from the enzyme is attributable to the thermophilic *B. sphaericus* enzyme (Fig. 2). In addition, an abortive ternary dead-end inhibition by the formation of an enzyme–NAD–pyruvate complex was shown to be present in the reaction of thermophilic *B. sphaericus* enzyme. This inhibition may reflect an important physiological role in preventing the enzyme functioning in the

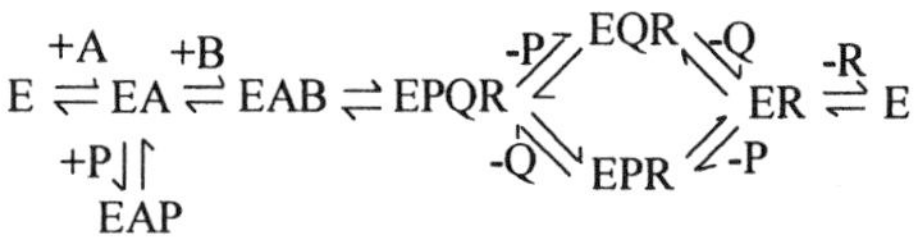

$$+A \quad +B \qquad\qquad -P \diagup EQR \diagdown Q \quad -R$$
$$E \rightleftharpoons EA \rightleftharpoons EAB \rightleftharpoons EPQR \diagup \diagdown \quad ER \rightleftharpoons E$$
$$+P \Big\downarrow\Big\uparrow \qquad -Q \diagdown EPR \diagup -P$$
$$EAP$$

Figure 2   Kinetic mechanism for the oxidative deamination of AlaDH from *B. sphaericus* DSM 462 [14]. A, NAD; B, L-alanine; P, pyruvate; Q, ammonia; R, NADH.

L-alanine synthesis. Similar difference in terms of substrate binding and product release was observed in reactions by other amino acid dehydrogenases such as GluDH and LeuDH [2,33,56].

Pyridoxal 5′-phosphate (PLP) inhibits the LeuDH reaction by the formation of Schiff base with a reactive lysine residue [25]. The *B. stearothermophilus* enzyme is inactivated by the modification at K80 that is conserved in the "GGGK" region in various amino acid dehydrogenases. The optimum pH for inactivation is 8.5, and the enzyme is protected from inactivation by either NAD and leucine, or NADH alone [57]. NAD alone cannot afford significant protection and modification of enzyme with PLP does not inhibit the NADH binding. Sekimoto et al. [58] examined the pH dependence of kinetic constants, V and V/K, of the wild-type enzyme and K80A, K80N, and K80R to understand the chemical mechanism of the enzyme reaction. The wild-type enzyme has $V/K_{leu}pK$ values of 8.9 and 10.7, while the V profile depends on one group having pK value of 8.6. The group of at 8.9/8.6 was assigned to K80 and it was concluded that the K80 was acting as a general base to raise the nucleophilicity of an attacking water. In addition, another lysine residue exhibiting the $V/K_{leu}pK$ values of 10.7 was assigned as the K68 by disappearance of the high pK of the K68A and K68R mutants [59]. As a result, a model for the chemical mechanism of LeuDH was proposed (Fig. 2). In the deamination, after the binding of NAD, leucine is bound to K68 with the carboxyl group of the substrate by a hydrogen bond. Hydride transfer from the α carbon of L-leucine to NAD forms the amino acid intermediate and NADH. An active site water molecule is hydrogen-bonded to the K80, activates it, and enhances its attack on the α carbon to form a carbinolamine. The carbinolamine intermediate decomposes to the oxo analog and ammonia. This model is similar to the mechanisms proposed for some other amino acid dehydrogenases such as GluDH and AlaDH.

B.   Structure

Extensive developments of the techniques in gene cloning and related fields have enabled rapid determination of the primary structures of amino acid dehydrogenases. In addition, x-ray crystallographic analyses of several amino acid dehydrogenases have been undertaken and revealed their ternary and quaternary structures in detail.

Among amino acid dehydrogenases, primary structures of GluDHs, AlaDH, LeuDH, PheDH, ValDH, and DAPDH have so far been determined by peptide and DNA sequencing methods (Table 8). The molecular weight of subunits of amino acid dehydrogenases is in the range of 30,000–55,000 except for GluDHs from a few organisms such as *Neurospora crassa* and *Trypanosoma brucei*. Although a computer-aided search of a protein sequence database revealed somewhat low similarities of overall sequences among amino acid dehydrogenases, a common partial sequence of about 30 residues in NAD(P) binding domain was observed in all of the enzymes [6]. In addition to the determination of primary struc-

Table 8 Enzymatic Determination with Amino Acid Dehydrogenases

Amino acid dehydrogenase	Amino acids, keto acids, ammonia, enzyme activity
AlaDH	L-Alanine, pyruvate, D-alanine, 3-fluoropyruvate, alanine aminotransferase, γ-glutamylcyclotransferase, amino acylase, aminopeptidase, stereospecificity of hydrogen transfer of NADH (alanine racemase)
LeuDH	L-Leucine, L-valine, L-branched chain amino acids, 2-oxoisovalerate, 2-oxo branched chain acids, 2-oxoisocaproate, L-methionine (methioine γ-lyase), aminopeputidase, D-amino acid aminotransferase, tripeptide aminopeptidase, stereospecificity of hydrogen transfer of NADH (low substrate amino acid racemase), maple syrup urine disease, hyperinsulinemic euglycemic clamps
PheDH	L-Phenylalanine, phenylpyruvate, phenylketonuria, cell destruction by shock wave
GluDH	L-Glutamate, 2-oxoglutarate, ammonia, urea (urease), D-glutamate (glutamate racemace), L-glutamine (glutaminase), L-citrulline (citrulline hydrolase), creatinine (creatinine diaminase), L-proline, aspartate aminotransferase
DAPADH	*meso*-2,6-diaminopimerate
LysDH	L-Lysine

tures, recently the numerous three-dimensional x-ray crystallographic analyses were done. The first molecular model reported for an amino acid dehydrogenase is that of the hexameric GluDH from *Clostridium symbiosum* [60]. The three-dimensional fold of each monomer of other NAD(P)-dependent dehydrogenases is shown to be clearly bilobal with a deep cleft separating the N- and C-terminal domains (Fig. 3). The x-ray analysis of crystals soaked in a solution of NAD showed that the C-terminal domain is responsible for the binding and contained a modified Rossmann fold with seven rather than six strands of β sheet and one of the strands running the reverse direction [61]. The structure of the GluDH–glutamate binary complex revealed the residues involved in glutamate positioning as well as the large conformational change associated with substrate binding [62,63]. In a subsequent study, the three-dimensional structure of hyperthermostable *Pyrococcus furiosus* GluDH was determined. The denatured temperature of *P. furiosus* enzyme is 113°C and much higher than that of the mesophilic bacterium *C. symbiosum* (denatured temperature 55°C) (Fig. 4).

The structural comparison of the hyperthermostable enzyme with the *C. symbiosum* showed a key role for intersubunit and intrasubunit ion pair networks in maintaining the enzyme stability at extremely high temperatures (100°C) [64]. In addition, the structure of GluDH from the hyperthermophilic bacterium *Thermotoga maritima* was determined [65]. The denaturation temperature (93°C) of *T. maritima* enzyme is slightly lower than that of the *P. furiosus* enzyme and much higher than that of the *C. symbiosum* enzyme. The structural study of the *T. maritima* enzyme showed that hydrophobic interactions play a more important role than ion pair networks in maintaining high thermostability. The three-dimensional structure of the octameric LeuDH from *B. sphaericus* also was determined. Comparison of the structure of LeuDH with *C. symbiosum* GluDH showed that these two enzymes share a related fold and possess a similar catalytic chemistry, and a mechanism for the basis of different substrate specificities for amino acids involves point mutations in the amino acid side chain specificity pocket and subtle changes in the shape of this

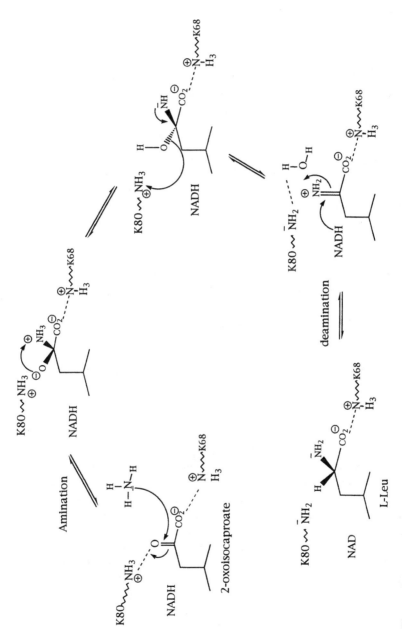

Figure 3 Reaction mechanism of LeuDH [58].

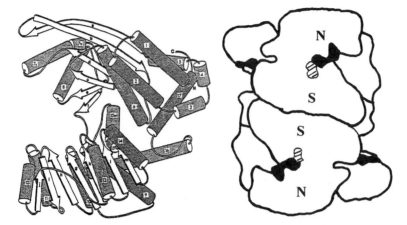

Figure 4 Schematic representation of the *Clostridium symbiosum* GluDH structure. Left: Three-dimensional structure of the subunit. Right: Subunit arrangement within the hexamer. N, Nucleotide binding domain; S, substrate binding domain. The positions of NAD and the substrate glutamate are shown by the black filled and hatched segments, respectively [60,62].

pocket caused by the differences in quaternary structure [66]. For the third structural analysis of amino acid dehydrogenase, the three-dimensional structure of the complex of dimetric *meso*-DAPADH from *Corynebacterium glutamicum* including NADP was shown [67]. This is a unique amino acid dehydrogenase acting on the chiral center with D configuration of *meso*-diamino acid. The enzyme is a homodimer of structurally nonidentical subunits, with each subunit composed of three domains in contrast with those of GluDH and LeuDH having two domains. The N-terminal domain contains a modified dinucleotide binding domain, or Rossmann fold. The second is the dimerization domain, which contains two α helices and three β strands forming the monomer−monomer interface of the dimer. The C-terminal domain is composed of five α helices and six β strands forming the substrate binding site. In both GluDH and LeuDH, the substrate binding domain and NAD(P) binding domain are located inside of the N and C termini of their subunits, respectively. Thus, this is a marked difference between *meso*-DAOADH and GluDH and LeuDH in terms of location of the coenzyme and substrate binding domains. The next enzyme of which the structure is determined is the hexameric AlaDH from a cyanobacterium, *Phormidium lapideum*, in the form of a binary complex with either NAD or pyruvate [68]. The AlaDH shares little primary structural homology with other amino acid dehydrogenases, and is unique in carrying out stereospecific transfer of the 4*R* hydrogens of NADH. The structures of the dimeric PheDH from *Rhodococcus* sp. in the two ternary complexes with enzyme−NAD−phenylpyruvate and enzyme−NAD−β-phenylpropionate were recently reported [69]. This is the first example of structures of the amino acid dehydrogenase with a ternary complex. Studies of the ternary complexes probably give us useful information for understanding the catalytic mechanism of enzymes. PheDH is a homodimeric and each monomer is composed of distinct globular N- and C-terminal domains, which form a deep cleft containing the active site (Fig. 5). The N-terminal domain binds the substrate and plays an important role in the reaction on the subunit−subunit interface. The C-terminal domain forms a typical Rossmann fold responsible for NAD binding as found for GluDH and LeuDH. This shows that amino acid dehydrogenases are composed of structurally independent coenzyme and substrate binding domains. On the

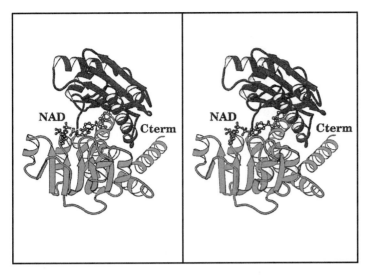

Figure 5 Ribbon representation of the structure of PheDH–NAD–phenylpyruvate ternary complex [69].

basis of the comparison of sequence homology between the *T. intermedius* PheDH and the *B. stearothermophilus* LeuDH, a high similarity (47%) is found between the two enzymes. Thus, a chimeric enzyme consisting of an N-terminal domain of PheDH containing the substrate binding region and a C-terminal domain of LeuDH containing the NAD binding region has been constructed by genetic engineering and characterized [70]. Although the catalytic efficiency of the chimeric enzyme on L-phenylalanine is 6% of that of the parental PheDH, the chimeric enzyme shows a similar K_m value for L-phenylalanine, pH optimum, and the same stereospecificity for hydrogen transfer at the C-4 position of the NADH. In contrast, the substrate specificity of the chimeric enzyme differs from PheDH: the chimeric enzyme showed a lower substrate specificity than the parental PheDH (Fig. 6). In addition to phenylalanine and its derivatives, it acts on poor substrates of both parent enzymes such as L-methionine, L-tryptophan, and L-phenylglycine in the oxidative deamination. Furthermore, the chimeric enzyme acts on L-branched chain amino acids such as L-valine and L-isoleucine. The specificity of the chimeric enzyme in the reductive amination is an admixture of the specificities of the two parent enzymes. This suggests that we created a amino acid dehydrogenase that exhibits new substrate specificity.

IV. APPLICATIONS

Extensive research on characteristics and structure of amino acid dehydrogenases reflects their usefulness for application in industry and other fields. In particular, L-amino acids, which are substrates in oxidative deamination and products in reductive amination, are a very important nutrient, and are also starting materials in pharmaceutical compounds. Amino acid dehydrogenases have been used for the stereospecific synthesis of amino acids from achiral substrates, 2-oxo acids, and ammonia, and for analysis of L-amino acids, oxo acids, ammonia, and assay of enzymes of which amino acids and oxo acids are their substrates or products. Some applications of amino acid dehydrogenases are described in this section.

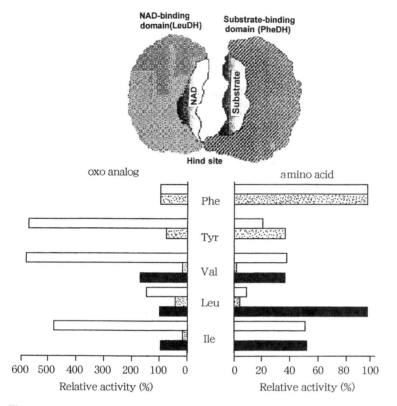

Figure 6 Scheme of the chimeric enzyme consisting of an amino terminal domain of PheDH and a carboxy terminal domain of LeuDH. Comparison of substrate specificity of PheDH (□), chimeric enzyme (▨), and LeuDH (■) on both amination and deamination [70].

A.   Synthesis of L-Amino Acids and Their Analogs

Wandrey and co-workers first studied the continuous production of L-leucine and other aliphatic L-amino acids from 2-oxoisocaproate and other corresponding oxo analogs, and ammonium formate by means of the enzyme membrane reactor system containing the *B. sphaericus* LeuDH, the *Candida boidinii* formate dehydrogenase (FDH), and polyethylene glycol–bound NADH (PEG-NADH, M_r about 20,000) [71]. In this system, the two enzymes and PEG-NADH are retained in free state in the reactor with an ultrafiltration membrane (molecular cutoff about 5000) as shown in Fig. 7 [72]. Yeast FDH is used for the effective in situ regeneration of NADH with formate; the enzyme is relatively stable and cheaply available from yeast cells, and its catalytic reaction is irreversible. PEG-NADH, the coenzyme chemically bound to PEG, which is a hydrophilic and lipophilic polymer, is retained in the membrane reactor. The coenzyme activity of PEG-NADH is comparable with NADH; the values of V/K_m for NADH and PEG-NADH are 400 and 397 U/mg mM, respectively [72]. The substrates 2-oxoisocaproate and ammonium formate are continuously pumped into the reactor, and the products containing L-leucine and CO_2 are released from the membrane reactor because of low molecular weight of the products. CO_2 that is coproduced can be easily removed under acidic conditions. In contrast to batch operation, the continuous process is characterized by higher space–time yields, as well as

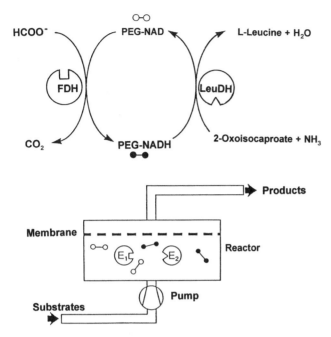

Figure 7 Enzyme membrane reactor for the continuous L-leucine production from 2-oxoisoca-proate and ammonium formate as the substrates with LeuDH (E1), FDH (E2), and PEG-NAD(H) [72].

easy and quick control of reaction conditions. In addition, this membrane reactor can be operated under sterile conditions [73].

The enzyme reactor is industrially used for the asymmetric production of an unnatural amino acid, L-*tert*-leucine, which is useful for syntheses of a variety of anti-AIDS and anticancer compounds, and L-[^{15}N]leucine, which is used in the investigation of protein balance in medicine as a nonradioactive tracer [74,75]. L-*tert*-Leucine is produced from 3,3-dimethylpyruvate and ammonia, and L-[^{15}N]leucine from 2-oxoisocaproate and [^{15}N]ammonium chloride. The thermostable *B. stearothermophilus* enzyme is more useful than the *B. sphaericus* enzyme for the industrial production due to the high stability [76].

L-Leucine, L-valine, and L-isoleucine are produced from ammonia and the corresponding 2-oxo acids with semipermeable nylon-polyethyleneimine artificial cells containing LeuDH, alcohol dehydrogenase, and dextran-NAD [77]. In this system, the recycling of dextran-NAD is achieved with alcohol dehydrogenase. Another coimmobilization method of LeuDH and FDH, a droplet gel-entrapping method, was developed for L-leucine production from its oxo analog and ammonia [78]. In this method, the two enzymes are freeze-dried with bovine serum albumin, dextrin, and stabilizers. The powder obtained is dispersed in methylcellosolve containing PEG-4000 diacrylate, *N,N'*-methylenebisacrylamide, and 2-hydroxyethyl acrylate, and then the suspension is gelled with initiators. After the gel has been cut up, the pieces are washed with buffer to remove the methylcellosolve and the dextrin. As the result, many droplets retaining the two enzymes and bovine serum albumin are enclosed in the gel, and the diffused substrates are converted to the products in the presence of NAD(H) in the droplets. In addition, this method was applied for the enantioselective synthesis of L-selenomethionine. L-Selenomethionine was produced in 95% yield from 2-oxo-4-methylselenobutyrate and ammonium formate in the presence of

B. stearothermophilus LeuDH, yeast FDH, and NAD(H) [79]. Hanson et al. utilized a similar system consisting of *B. sphaericus* LeuDH, FDH, and NADH for the production of L-3-hydroxyvaline, which is a key intermediate needed for the synthesis of tigemonam, an orally active monobactam antibiotic [80]. The LeuDH catalyzes the reductive amination of 2-oxo-4-hydroxyisovalerate with a 41% of relative activity compared with the best substrate, i.e., 2-oxoisovalerate.

AlaDH is used for the syntheses of L-alanine and its analogs. L-Alanine is produced continuously from pyruvate and ammonium formate with a membrane reactor containing AlaDH, FDH, and PEG-NADH [81]. The same method is applicable to the continuous production of 3-fluoro-L-alanine from 3-fluoropyruvate and ammonium formate in a similar membrane reactor. 3-Chloro-L-alanine was produced from 3-chloropyruvate by the same method [82]. AlaDH can be used for L-[^{15}N]alanine synthesis from pyruvate and [^{15}N]ammonia [83]. Another enzyme reaction system for L-alanine production using AlaDH and malic enzyme was reported [84]. In this system, the regeneration of NADH from NAD with malic enzyme is accompanied by the production of pyruvate from L-malate, and the product pyruvate is converted to L-alanine in the presence of ammonia with AlaDH.

Similar methods have been used for the syntheses of L-phenylalanine and its analogs. An enzyme membrane reactor system containing *Brevibacterium* sp. PheDH, yeast FDH, and PEG-NADH was developed for the syntheses of L-phenylalanine from phenylpyruvate and ammonium formate [85]. Asano and Nakazawa synthesized L-phenylalanine, tyrosine, and some other L-amino acids using a dialysis tube containing the *Sporosarcina ureae* PheDH and yeast FDH [86]. In addition, optically pure three-substituted pyruvates with bulky substituents, such as *S*-2-amino-4-phenylbutyrate and *S*-2-amino-5-phenylvalerate, were synthesized from their oxo analogs in a similar way [87].

The synthesis of various D-amino acids by a multienzyme system has been developed. In this system, D-amino acids are produced from the corresponding 2-oxo acids and ammonia by coupling four enzyme reactions catalyzed by D-amino acid aminotransferase, glutamate racemase, GluDH, and FDH as shown in Fig. 8 [88]. This is based on the high substrate specificity of glutamate racemase and the strict enantioselectivity and low structural specificity for the substrates of D-amino acid aminotransferase. Various D-amino acids are produced from their corresponding 2-oxo acids by catalysis of D-amino acid aminotransferase with the consumption of D-glutamate. D-Glutamate is regenerated with GluDH and glutamate racemase from 2-oxoglutarate, ammonia, and NADH. NADH is continuously formed from formate and NAD with FDH. The reactions of D-amino acid aminotransferase, GluDH, and glutamate racemase are reversible, but the FDH reaction is substantially irreversible. Thus, the whole system is expected to produce D-amino acids and CO_2 from the oxo analogs, ammonia, and formate. The D-amino acids produced, except

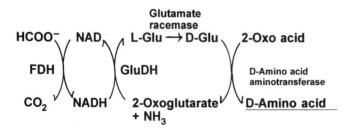

Figure 8   Enzymatic synthesis of D-amino acids by a multienzyme system consisting of the coupling reaction of four enzymes.

for D-glutamate, are not racemized by glutamate racemase. Another multienzyme system for the synthesis of D-amino acids from the corresponding 2-oxo acids was also developed [89]. In this system, four thermostable enzymes—D-amino acid aminotransferase, alanine racemase, AlaDH from *Bacillus* species, and FDH from *Mycobacterium vaccae*—are used. The *M. vaccae* FDH is more stable than the yeast enzyme [90]. The genes of these thermostable enzymes are cloned into *E. coli* cells and the four enzymes are easily isolated from the *E. coli* cells, which are also used as a catalyst for D-amino acid production.

Galkin et al. synthesized optically active amino acids from 2-oxo acids with the recombinant *E. coli* cells that express the heterogeneous genes [91]. L-Amino acids were produced with thermostable L-amino acid dehydrogenase and FDH from 2-oxo acid and ammonium formate with only an intracellular pool of NAD for the regeneration of NADH. The plasmids containing the gene of *M. vaccae* FDH and amino acid dehydrogenases including LeuDH, AlaDH, and PheDH were used for the production of L-leucine, L-valine, L-norvaline, L-methionine, L-phenylalanine, and L-tyrosine, with a yield (more than 80%) and a high optical purity (up to 100% enantiomeric excess). Stereospecific conversion of various 2-oxo acids to D-amino acids was also carried out using the recombinant *E. coli* cells containing the four heterogeneous genes of the thermostable D-amino acid amino-transferase, alanine racemase, AlaDH, and FDH. Optically pure D-enantiomers of gluta-mate and leucine are produced.

B. Preparation of Stereoselectively Deuterated NADH and NADPH

NAD(P)H-dependent dehydrogenases and reductases show either pro-R (A-) or pro-S (B-) stereospecificity for hydrogen removal from the C-4 position of the nicotinmide moi-ety of NAD(P)H. As mentioned above, LeuDH and AlaDH show pro-R and pro-S ster-eospecificity, respectively. The stereospecificity of hydrogen transfer is examined by means of stereospecifically C-4 deuterium-labeled NAD(P)H, which is prepared enzymatically from NAD and deuterium-labeled substrate. The coupling of amino acid dehydrogenase and amino acid racemase reactions results in preparation of [4R-^{2}H]- and [4S-^{2}H]NAD(P)H [92]. Amino acid racemases catalyze the exchange of α-H of the substrate amino acids with deuterium in ^{2}H$_2$O. The reaction system for preparation of [4R-^{2}H]NADH contains AlaDH (pro-R stereospecific), alanine racemase, and ω-amino acid aminotransferase in ^{2}H$_2$O (Fig. 9). D-Alanine is racemized to yield α-deuterio-D- and L-alanines. The deuterium of α-deuterio-L-alanine is stereoselectively transferred to the C-4 of the nicotinamide ring of NAD by catalysis of AlaDH, and [4R-^{2}H]NADH, pyruvate, and ammonia are produced.

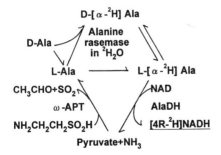

Figure 9 Enzymatic preparation of [4R-^{2}H]NADH by coupling reaction of AlaDH, alanine race-mace, and ω-amino acid aminotransferase (ω-APT) [92].

Pyruvate is transaminated with hypotaurine by ω-amino acid aminotransferase to form alanine, and acetaldehyde and sulinate, which are formed irreversibly from sulfinoace-toaldehyde and primarily produced [4R-^{2}H]NADH in a high yield. In contrast, [4S-^{2}H]NADH is produced with LeuDH (pro-S stereospecific) and amino acid racemase with low substrate specificity in a similar manner. [4S-^{2}H]NADPH can also be synthesized by means of NADP-dependent GluDH (pro-S stereospecific) and glutamate racemase in a similar method [93].

We developed a simple method for determination of the stereospecificity for hydrogen transfer of NAD(P)H by an NAD(P)-dependent oxidoreductase. After incubation of the reaction mixture containing AlaDH, alanine racemase, D-alanine, NAD, and NAD-dependent oxidoreductase such as dehydrogenase and reductase whose stereospecificity is to be determined, and the oxidized form of substrate in ^{2}H$_2$O, the C-4 hydrogen of NAD is determined by ^{1}H-NMR. If the stereospecificity for hydrogen transfer of a dehydrogenase is the same as that of AlaDH, the C-4 hydrogen of NAD is fully retained and a doublet that is specific for it appears in the ^{1}H-NMR spectra, and the final product is deuterated (Fig. 10). In the combination of oxidoreductases having different stereospecificities, the C-4 deuterium in NAD remains; consequently, no peak for the C-4 hydrogen of NAD appears in the ^{1}H-NMR spectra. The stereospecificity of hydrogen transfer by oxidoreductase is determined in the same manner with the alternative enzyme system. LeuDH and amino acid racemase with low substrate specificity are substituted for AlaDH and alanine racemase, respectively, in the reaction system. The stereospecificity of NADPH hydrogen transfer can be determined in a similar method using pro-S stereospecific NADP-dependent

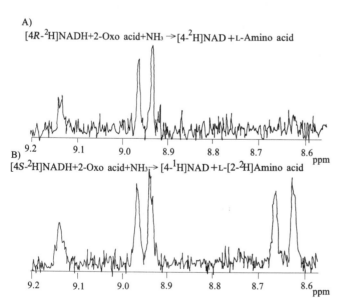

Figure 10 Determination of stereospecificity in the hydrogen transfer of NADH by ^{1}H NMR. After incubation of [4R-^{2}H]NADH (A) or [4S-^{2}H]NADH (B), 2-oxo acid as the substrate, ammonia and amino acid dehydrogenase, aromatic region of the ^{1}H NMR spectra of NAD was detected to identify [4-^{2}H]NAD or [4-^{1}H]NAD. In the spectrum of [4-^{2}H]NAD (A), no resonance doublet around δ-8.8 for the C-4 position of NAD appears. In contrast, the resonance doublet is observed in the case of [4-^{1}H]NAD. The amino acid dehydrogenase described in this figure shows Pro-S stereospecificity from both spectra [7].

GluDH. Thus, we can determine the stereospecificity of various NAD(P)-dependent oxidoreductases by ^{1}H-NMR measurement without isolation of coenzymes and products.

C. Application of Amino Acid Dehydrogenases to Enzymatic Analysis

The reversible deaminations of amino acids catalyzed by amino acid dehydrogenases are accompanied with the formation of NADH and NADPH from NAD and NADP, respectively. Spectra of NADH and NADPH have an absorption maximum at 340 nm (molar absorption coefficient 6220). Thus, we can readily determine spectrophotometrically the amounts of amino acids, oxo acids, and ammonia from the formation and disappearance of NAD(P)H. Enzymatic analysis of amino acids, oxo acids, and ammonia is an important tool in clinical chemistry, bioprocess control, and nutrition studies. For example, the concentration of the product L-leucine in the membrane reactor for the continuous production of L-leucine from 2-oxoisocaproate and ammonia was enzymatically determined by measuring an increase in absorbance at 340 nm of NADH with LeuDH [71]. For the control of continuous 3-fluoro-L-alanine production in the membrane reactor, the substrate 3-fluoropyruvate remaining in the reaction mixture was determined by measuring a decrease in absorbance at 340 nm of NADH [81]. It is simpler and less expensive than ion exchange high-performance liquid chromatography. The methods are also applicable to the assay of enzymes that are related to human disease and used as a bioprocess control.

LeuDH is also used for the assay of serum leucine aminopeptidases, which is related to liver function disorders [94]. AlaDH is successfully applied to the measurements of urinary dipeptidase [95], erythrocytic γ-glutamylcyclotransferase [96], serum γ-glutamyltransferase [97], and serum aminopeptidase activities [98]. PheDH is used for the specific determination of L-phenylalanine and phenylpyruvate [99,100], and is therefore applicable to the process control of phenylalanine fermentation and diagnosis of neonatal hyperphenylalaninemia and phenylketonuria [101]. The coupling of PheDH and phenylpyruvate:glutamate transaminase reactions is applicable to the microdetermination of L-phenylalanine in human blood [102]. Assay of PheDH is also useful to the monitoring of the level of cells disrupted by shock wave; destruction of the spheroplast of recombinant cells is monitored sensitively by measuring PheDH activity leaked from the cells [103].

The K_m values of amino acid dehydrogenases for ammonia are in general very high (e.g., *Bacillus sphaericus* LeuDH 200 mM and AlaDH 28.2 mM, and *Thermoactimyces* PheDH 106 mM). In contrast, NADP-dependent GluDH from *Proteus* sp. exceptionally exhibits a relatively low K_m value (1.1 mM) for ammonia. Therefore, the GluDH is used for the determination of ammonia, as well as urea, by coupling of the urease reaction [104].

V. SUMMARY

We have described the recent advances in the biochemical and biotechnological aspects of NAD- and NADP-dependent amino acid dehydrogenases, with emphasis on their application to the synthesis of optically active amino acids. In particular, based on recent information of three-dimensional structures of amino acid dehydrogenases (e.g., GluDH, LeuDH, and PheDH), we have shown the relationship between the structure and function of the enzymes. In addition, gene engineering, site-directed mutagenesis [105], and directed evolution of amino acid dehydrogenase will provide us more exciting new developments in application of the enzymes.

REFERENCES

1. T Ohshima, C Wandrey, M Sugiura, K Soda. Biotechnol Lett 7:87–876, 1985.
2. T Ohshima, K Soda. Trends Biotechnol 7:210–214, 1989.
3. T Ohshima, N Nishida. Biosci Biotech Biochem 57:945–951, 1993.
4. T Ohshima, M Ishida. Protein Express Purif 3:121–125, 1992.
5. T Ohshima, S Nagata, K Soda. Arch Microbiol 141:407–411, 1985.
6. H Takada, T Yoshimura, T Ohshima, N Esaki, K Soda. J Biochem 109:371–376, 1991.
7. T Ohshima, H Takada, T Yoshimura, N Esaki, K Soda. J Bacteriol 173:3943–3948, 1991.
8. T Ohshima, C Wandrey, M Sugiura, K Soda. Biotechnol Lett 7:871–876, 1985.
9. KJ Siranosian, K Ireton, AD Grossman. J Bacteriol 175:6789–6796, 1993.
10. JC Murell, H Daltpon. J Gen Microbiol 129:1197–1206, 1983.
11. Z Vali, F Kilar, S Lalatos, SA Venyaminov, P Zavodoszky. Biochim Biophys Acta 615:34–47, 1980.
12. T Ohshima, K Soda. Eur J Biochem 100:29–39, 1979.
13. A Yoshida, E Freese, Biochim Biophys Acta 96:248–262, 1965.
14. T Ohshima, M Sakane, T Yamazaki, K Soda. Eur J Biochem 191:715–720, 1990.
15. T Ohshima, H Misono, K Soda. J Biol Chem 253:5719–5725, 1978.
16. H Shuette, W Hummel, H Tsai, M-R Kula. Appl Microbial Biotechnol 22:306–317, 1985.
17. H Shimoi, S Nagata, N Esaki, H Tanaka, K Soda. Agric Biol Chem 51: 3375–3381, 1987.
18. G Livesey, P Lund. Meth Enzymol 166:282–288 (1988).
19. H Misono, K Sugihara, Y Kuwamoto, S Nagata, S Nagasaki. Agric Biol Chem 54:1491–1498.
20. T Ohshima, N Nishida, S Bakthavatsalam, K Kataoka, H Takada, T Yoshimura, N Esaki, K Soda. Eur J Biochem 222:305–312, 1994.
21. S Nagata, S Bakthavatsalam, AG Galkin, H Asada, S Sakai, N Esaki, K Soda, T Ohshima, S Nagasaki, H Misono. Appl Microbiol Biotechnol 44:432–438, 1995.
22. S Nagata, K Tanizawa, N Esaki, Y Sakamoto, T Ohshima, H Tanaka, K Soda. Biochemistry 27:9056–9062, 1988.
23. M Oka, YS Yang, S Nagata, N Esaki, H Tanaka, K Soda. Biotechnol Appl Biochem 11:307–311, 1989.
24. T Ohshima, T Yamamoto, H Misono, K Soda. Agric Biol Chem 42:1734–1743, 1978.
25. T Ohshima, K Soda. Agric Biol Chem 48:349–354, 1984.
26. W Hummel, N Weiss, M-R Kula. Arch Microbiol 137:47–52, 1984.
27. Y Asano, A Nakazawa, K Endo. J Biol Chem 262:10346–10354, 1987.
28. Y Asano, A Nakazawa, K Endo, Y Hibino, M Ohmori, N Numao, K Kondo. Eur J Biochem 168:153–159, 1987.
29. H Misono, J Yonezawa, S Nagata, S Nagasaki. J Bacteriol 171:30–36, 1989.
30. L De Boer, M Van Rijissel, GJ Euverink, L Dijkhuizen. Arch Microbiol 153:12–18, 1989.
31. Y Asano, M Tanetani. Arch Microbiol 169:220–224, 1998.
32. T Ohshima, H Takada, T Yoshimura, N Esaki, K Soda. J Bacteriol 173:3943–3948, 1991.
33. T Ohshima, K Soda. Adv Biochem Eng/Biotechnol 42:187–209, 1990.
34. EL Smith, BM Austen, KM Blumenthal, JF Nye. The Enzymes 11:292–367, 1968.
35. H Shimizu, T Kuratsu, F Hirata. J Ferment Technol 57:428–434, 1979.
36. T Ohshima, N Nishida. Biosci Biotech Biochem 57:945–951, 1993.
37. C Kujo, T Ohshima. Appl Environ Microbiol 64:2152–2157, 1998.
38. K Vackova, A Mehta, M Kutacek. Biol Plant 27:154–158, 1985.
39. MCM Lin, C Wagner. J Biol Chem 250:3746–3751, 1975.
40. TC Stadtman. Adv Enzymol 38:413–448, 1973.
41. T Okamura, H Noda, S Fukuda, M Ohsugi. J Nutr Sci Vitaminol 44:483–490, 1998.
42. ND Priestley, JA Robinson, Biochem J 261:853–861, 1989.
43. S Omura, Y Tanaka, H Hamada, R Masuma, J Antibiot 36:1792–1794, 1983.
44. T Ohshima, K Soda. Biochim Biophys Acta 1162:221–226, 1993.

45. H Misono, H Tagawa, T Yamamoto, K Soda. J Bacteriol 137:22–27, 1979.
46. H Misono, M Ogasawara, S Nagasaki. Agric Biol Chem 50:1329–1330, 1986.
47. H Misono, M Ogasawara, S Nagasaki. Agric Biol Chem 50:2729–2734, 1986.
48. B Shurumpf, A Schwarzer, J Karinowiski, A Puhler, L Eggeling, H Sahm. J Bacteriol 173: 4510–4516, 1991.
49. SG Reddy, G Scappin, JS Blanchard. Proteins: Struct Funct Genet 25:514–516, 1996.
50. W Buerigi, R Richiterich, JP Colombo. Nature 211:854–855, 1966.
51. H Misono, S Nagasaki. Agric Biol Chem 47:631–633, 1983.
52. H Misono, S Nagasaki. J Bacteriol 150:398–401, 1982.
53. T Hammer, R Bode, D Birnbaum. J Gen Microbiol 137:711–715, 1991.
54. H Misono, H Hashimoto, H Uehigashi, S Nagata, S Nagasaki. J Biochem 105:1002–1008, 1989.
55. CE Grimshaw, WW Cleland. Biochemistry 20:5650–5655, 1981.
56. T Ohshima, N Nishida, S Bakthavatsalam, K Kataoka, H Takada, T Yoshimura, N Esaki, K Soda. Eur J Biochem. 222:305–312, 1994.
57. T Matsuyama, K Soda, T Fukui, K Tanizawa. J Biochem 112:258–265, 1992.
58. T Sekimoto, T Matsuyama, T Fukui, K Tanizawa. J Biol Chem 268:27039–27045, 1993.
59. T Sekimoto, T Fukui, K Tanizawa. J Biol Chem 269:7262–7266, 1994.
60. PJ Baker, KL Britton, PC Enge, GW Farrants, KS Lilley, DW Rice, J Stillman. Proteins: Struct Funct Genet 12:75–86, 1992.
61. MG Rossmann, A Liljas, CI Branden, LJ Banaszak. The Enzymes, 3rd ed., Vol. 11, Academic Press, New York, 1975, pp. 62–102.
62. DW Rice, KSP Yip, TJ Stillman, KL Britton, A Fuentes, I Coonerton, A Pasquo, R Scandurra, PC Engel. FEMS Microbiol Rev 18:105–117, 1996.
63. TJ Stillman, PJ Baker, KL Britton, DW Rice. J Mol Biol 234:1131–1139, 1993.
64. KSP Yip, TJ Stillman, KL Britton, PJ Artymiuk, PJ Baker, SE Sedelnikova, PC Engel, A Pasquo, R Chiaraluce, V Consalvi, R Scandurra, DW Rice. Structure 3:1147–1158, 1995.
65. S Knapp, WM deVos, DW Rice, R Laden. J Mol Biol 267:916–932, 1997.
66. PJ Baker, AP Turnbull, SE Sedelnikova, TJ Stillman, DW Rice. Structure 3:693–705, 1995.
67. G Scapin, SG Reddy, JS Blanchard. Biochemistry 35:13540–13551, 1996.
68. PJ Baker, Y Sawa, H Shibata, SE Sedelnikov, DW Rice. Nature Struct Biol 5:561–567, 1998.
69. JL Vanhooke, JB Thodon, NMW Brynhuber, JS Blanchard, HM Holden. Biochemistry 38: 2326–2339, 1999.
70. K Kataoka, H Takada, K Tanizawa, T Yoshimura, N Esaki, T Ohshima, K Soda. J Biochem 116:931–936, 1994.
71. R Wichmann, C Wandrey, AF Wuchmann, M-R Kula. Biotechnol Bioeng 23:2789–2802, 1981.
72. M-R Kula, C Wandrey. Meth Enzymol 136:9–21, 1987.
73. U Kragl, DV Racki, C Wandrey. Ind J Chem 32B:103–117, 1993.
74. U Kragl, DV Racki, C Wandrey. Bioprocess Eng 14:291–297, 1996.
75. C Wandrey. Forum Mikrobiologie 7, Sonderheft Biotechnol Sept, 33–39, 1984.
76. T Ohshima, C Wandrey, M-R Kula, K Soda. Biotechnol Bioeng 27:1616–1618, 1985.
77. KF Gu, TMS Chang. Biotechnol Appl Biochem 12:227–236, 1990.
78. S Kajiwara, H Maeda. Agric Biol Chem 51:2873–2879, 1987.
79. N Esaki, H Shimoi, Y-S Yang, H Tanaka, K Soda. Biotechnol Appl Biochem 11:312–317, 1989.
80. KR Hanson, J Singh, TP Kissick, RN Patel, LJ Szarka, RH Mueller. Bioorg Chem 18:116–130, 1990.
81. T Ohshima, C Wandrey, D Conrad. Biotechnol Bioeng 34:394–397, 1989.
82. Y Kato, K Fukumoto, Y Asano. Appl Microbiol Biotechnol 39:301–304, 1993.
83. A Mocanu, G Niac, A Ivanof, V Gorun, N Palibroda, E Vargha, M Bologa, O Barzu. FEBS Lett 143:153–156, 1982.
84. S Suye, M Kawagoe, S Inuta. Can J Chem Eng 70:306–312, 1992.

85. W Hummel, E Schmidt, C Wandrey, M-R Kula. Appl Microbiol Biotechnol 25:175–185, 1986.
86. Y Asano, A Nakazawa. Agric Biol Chem 51:2035–2036, 1987.
87. Y Asano, A Yamada, Y Kato, K Yamaguchi, Y Hibino, K Hirai, K Kondo. J Org Chem 55: 5567–5571, 1990.
88. N Nakajima, K Tanizawa, T Tanaka, K Soda. J Biotechnol 8:243–248, 1988.
89. A Galkin, L Kulakova, H Yamamoto, K Tanizawa, T Tanaka, N Esaki, K Soda. J Ferment Bioeng 83:299–300, 1997.
90. A Galkin, L Kulakova, V Tishkov, N Esaki, K Soda. Appl Microbiol Biotechnol 44:479–483, 1995.
91. A Galkin, L Kulakova, T Yoshimura, K Soda, N Esaki. Appl Environ Microbiol 63:4651–4656, 1997.
92. N Esaki, H Shimoi, N Nakajima, T Ohshima, K Yonaha, H Tanaka, K Soda. J Biol Chem 264:9750–9752, 1989.
93. N Nakajima, K Tanizawa, H Tanaka, K Soda. Agric Biol Chem 50:2823–2830, 1986.
94. S Takamiya, T Ohshima, K Tanizawa, K Soda. Anal Biochem 130:266–270, 1983.
95. Y Ito, Y Watanabe, K Hirano, M Sugiura, S Sawaki, T Ogino. J Biochem (Tokyo) 96:1–8, 1984.
96. T Takahashi, T Kondo, H Ohno, S Minato, T Ohshima, S Mikuni, K Soda, N Taniguchi. Biochem Med Metab Biol 38:311–316, 1987.
97. H Kondo, M Hashimoto, K Nagata, K Tomita, H Tsubota. Clin Chim Acta 207:1–9, 1992.
98. Y Sakamoto, H Kondo, K Soda. Biosci Biotech Biochem 58:1675–1678, 1994.
99. W Hummel, H Shutte, M-R Kula. Anal Biochem 170:397–401, 1988.
100. T Ohshima, H Sugimoto, K Soda. Anal Lett 21:2205–2215, 1988.
101. U Wendel, M Koppelkamm, W Hummel. Clin Chim Acta 201:95–98, 1991.
102. K Nakamura, T Fujii, Y Kato, Y Asano, AJ Looper. Anal Biochem 234:19–22, 1996.
103. K Teshima, T Ohshima, S Tanaka, T Nagai. Shock Waves 4:293–297, 1995.
104. T Murachi, K Tabata. Meth Enzymol 137D:260–271, 1986.
105. A Galkin, L Kulakova, T Ohshima, K Soda. Protein Eng 10:687–690, 1997.

Index